2014 IEEE International Conference on Semiconductor Electronics

(ICSE 2014)

Kuala Lumpur, Malaysia
27-29 August 2014

IEEE Catalog Number: CFP14421-POD
ISBN: 978-1-4799-5761-3

Copyright © 2014 by the Institute of Electrical and Electronic Engineers, Inc
All Rights Reserved

Copyright and Reprint Permissions: Abstracting is permitted with credit to the source. Libraries are permitted to photocopy beyond the limit of U.S. copyright law for private use of patrons those articles in this volume that carry a code at the bottom of the first page, provided the per-copy fee indicated in the code is paid through Copyright Clearance Center, 222 Rosewood Drive, Danvers, MA 01923.

For other copying, reprint or republication permission, write to IEEE Copyrights Manager, IEEE Service Center, 445 Hoes Lane, Piscataway, NJ 08854. All rights reserved.

***This publication is a representation of what appears in the IEEE Digital Libraries. Some format issues inherent in the e-media version may also appear in this print version.**

IEEE Catalog Number: CFP14421-POD
ISBN 13:   978-1-4799-5761-3

Additional Copies of This Publication Are Available From:

Curran Associates, Inc
57 Morehouse Lane
Red Hook, NY 12571 USA
Phone: (845) 758-0400
Fax: (845) 758-2633
E-mail: curran@proceedings.com
Web: www.proceedings.com

PROCEEDINGS

ICSE 2014

2014 IEEE INTERNATIONAL CONFERENCE ON SEMICONDUCTOR ELECTRONICS (ICSE)

27th – 29th August 2014
Berjaya Times Square Hotel,
Kuala Lumpur, Malaysia

Organized by:

Co-organized by:

2014 IEEE INTERNATIONAL CONFERENCE ON SEMICONDUCTOR ELECTRONICS (ICSE)

Proceedings

27 – 29 August 2014
Berjaya Times Square, Kuala Lumpur, Malaysia

Organized by
Electron Devices Chapter of IEEE Malaysia Section

Co-organized by
Institute of Microengineering & Nanoelectronics (IMEN),
Universiti Kebangsaan Malaysia

**Message from IEEE ED Malaysia Chapter Chair**

I would like to extend my warmest welcome to the participants of the 11th IEEE International Conference on Semiconductor Electronics. The conference is a good platform for fellow colleagues and students to share and discuss knowledge and findings while expanding networks.

I would like to express my gratitude to the conference chair, Prof. Dato' Dr. Burhanuddin Yeop Majlis for his effort in making this conference a reality. The staff and students from IMEN, Universiti Kebangsaan Malaysia are also acknowledged for their co-operation. I would also like to thank the editorial staff, members of the organizing committee, secretarial staff, and everybody who have worked hard to make this conference a success. Last but not least, I would also like to thank main sponsor, IMEN UKM, for their financial contribution. I wish all of you a pleasant stay in this country and have fruitful discussions during ICSE2014.

Terima kasih.

Assoc. Prof. Dr. Mohd Nizar Hamidon, MIEEE
IEEE Electron Devices Malaysia Chapter Chair

**Message from ICSE2014 Conference Chair**

Selamat Datang to Kuala Lumpur and ICSE2014.

On behalf of the organizing committee, it is with great pleasure to welcome you to the 11th IEEE International Conference on Semiconductor Electronics 2014 (ICSE2014). Over the last twenty two years, ICSE has become the preeminent international forum on semiconductor electronics embracing all aspects of the semiconductor technology from circuit device, modelling and simulation, photonics and sensor technology, MEMS technology, process and fabrication, packaging technology and manufacturing, failure analysis and reliability, material and devices and nanoelectronics.

This time, the conference offers six keynote lectures by distinguished persons in their own fields, 129 oral presentations and over 14 posters contributed by participants. We are proud to have Distinguished Prof. Dr. Asad Madni from University of California, Los Angeles, USA; Prof. Dr. Park Young June from Seoul National University; Prof. Dr. Edward Yi Chang from National Chiao Tung University, Taiwan R.O.C; Prof. Dr. Arokia Nathan from Cambridge University, United Kingdom; Prof. Dr. Abdelkrim Khelif from FEMTO-ST Institute, France and Prof. Dr. Hiroshi Inokawa from Shizuoka University, Japan as our keynote speakers. We would also like to thank Dato' Hjh Ida Suraya, the Managing Director of Opals Group of Companies for her generous sponsorship towards our conference.

The conference proceedings provide views into the current advances in semiconductor electronics in the region and we look forward to the presentations of our participants. We hope participants would appreciate knowledge imparted by the lectures.

To all participants, I hope we can gain knowledge and benefits from the conference while making new contacts with other participants. To participants from overseas, I wish you a pleasant stay in this country and we will endeavor to make your stay here enjoyable.

Terima kasih.

Prof. Dato' Dr. Burhanuddin Yeop Majlis, SMIEEE
Chair, 11th IEEE International Conference on Semiconductor Electronics

IEEE-ICSE 2014 Committee

Chair:
Prof. Dato' Dr. Burhanuddin Yeop Majlis   *Universiti Kebangsaan Malaysia*

Co-Chair:
Assoc. Prof. Dr. Mohd Nizar Hamidon   *Universiti Putra Malaysia*

Hon. Secretary:
Assoc. Prof. Dr. P. Susthitha Menon   *Universiti Kebangsaan Malaysia*

Assistant Hon. Secretary:
Pn. Maizatul Zolkapli   *Universiti Teknologi MARA*

Treasurer:
Assoc. Prof. Dr. Roslina Mohd. Sidek   *Universiti Putra Malaysia*

Technical Chairs:
Assoc. Prof. Dr. Badariah Bais   *Universiti Kebangsaan Malaysia*
Assoc. Prof. Dr. Norhayati Soin   *Universiti Malaya*

Publicity Chair:
Dr. Zubaida Yusoff   *Multimedia University*

Logistic Chair:
Dr. Wan Zuha Wan Hasan   *Universiti Putra Malaysia*

Committee Member

Prof. Dr. Muhamad Mat Salleh	*Universiti Kebangsaan Malaysia*
Prof. Dr. Ibrahim Ahmad	*Universiti Tenaga Nasional*
Prof. Dr. Norani Muti Mohamed	*Universiti Teknologi Petronas*
Prof. Dr. Uda Hashim	*Universiti Malaysia Perlis*
Prof. Dr. Razali Ismail	*Universiti Teknologi Malaysia*
Prof. Dr. A.H.M Zahirul Alam	*International Islamic University Malaysia*
Assoc. Prof. Dr. Jumril Yunas	*Universiti Kebangsaan Malaysia*
Assoc. Prof. Dr. Azlan Abdul Aziz	*Universiti Sains Malaysia*
Assoc. Prof. Dr. Zaliman Sauli	*Universiti Malaysia Perlis*
Assoc. Prof. Dr. Dee Chang Fu	*Universiti Kebangsaan Malaysia*
Dr. Ismail Saad	*Universiti Malaysia Sabah*
Dr. Zainal Arif Burhanudin	*Universiti Teknologi Petronas*
Dr. Mohd Ambri Mohamed	*Universiti Kebangsaan Malaysia*
Dr. Asrulnizam Abdul Manaf	*Universiti Sains Malaysia*
Dr. Norhana Arsad	*Universiti Kebangsaan Malaysia*
Mr. Hazian Mamat	*MIMOS Berhad*
Dr. Mohd Zainizan Sahdan	*Universiti Tun Hussein Onn*
Dr. Hashimah Hashim	*Universiti Teknologi MARA*
Dr. Sharifah Fatmadiyana Wan Mohd Hatta	*Universiti Malaya*
Dr. Kim Shyong Siow	*Universiti Kebangsaan Malaysia*
Dr. Suhaidi Shafie	*Universiti Putra Malaysia*
Dr. Rosminazuin Ab Rahim	*International Islamic University Malaysia*
Ir. Nor Azhadi Hj. Ngah	*TMR&D Sdn. Bhd.*

Secretariat

Mr. Mohd Farhanulhakim Mohd Razip Wee
Mr. Muhamad Ramdzan Buyong
Miss Hayati Hussin
Mr. Mohd Faizal Aziz
Mrs. Marzieana Ab Rahman
Mr. Shafii Abdul Wahab
Mrs. Aishah Fauthan
Mrs. Mimiwaty Mohd Noor
Mrs. Dilla Duryha Berhanuddin

Secretariat Office

Secretariat of IEEE-ICSE2014
Electron Devices Malaysia Chapter
Institute of Microengineering & Nanoelectronics (IMEN)
Universiti Kebangsaan Malaysia, 43600 UKM Bangi, Selangor, MALAYSIA
Telephone +603 8921 6987
Fax +603 8925 0439
Email: edsmalaysia@gmail.com

Objective

The IEEE-ICSE2014 is aiming at bringing together researchers from industry and academia to gather and explore various issues and trends in the field of micro and nano electronics. ICSE has become the prominent international forum on semiconductor electronics embracing all aspects of the semiconductor technology from circuits, device modeling and simulation, photonics and sensor technology.

Scopes of Conference

- MEMS / NEMS
- Device Modeling & Simulation
- Nanoelectronics
- Device physics and characterization
- Opto-electronics and photonics technology
- Microwave Device and MMIC
- IC Packaging and Testing
- Reliability and Failure Analysis
- Semiconductor Manufacturing & Process
- Microelectronics Application in Product Development
- Process Technology (CMOS, Bipolar, BiCMOS, GaAs, etc)
- Electronics Materials and Device Fabrication

TABLE OF CONTENTS

Message from IEEE ED Malaysia Chapter Chair	iii
Message from ICSE2014 Conference Chair	iv
IEEE-ICSE 2014 Committee	v
Scope of Conference	vii
Table of Contents	viii

No.   Papers   **Page**

K1   **Convergence of Emerging Technologies to Address the Challenges of the**   A1
      21st Century
      Distinguished Prof. Dr. Asad M. Madni (*Distinguished Adjunct
      Professor/Distinguished Scientist, Electrical Engineering Department,
      University of California Los Angeles, USA*)

K2   **There is Plenty of Room at the Silicon**   A2
      Prof. Dr. Young June Park (*Professor, Physical Electronics Laboratory, Seoul
      National University*)

K3   **Realization of GaN-Based Technology for High Power and High**   A3
      Frequency Applications
      Prof. Dr. Edward Yi Chang (*Vice Pesident of Research & Development,
      National Chiao Tung University, Taiwan, R.O.C.*)

K4   **Amorphous Oxide Electronics**   A4
      Prof. Dr. Arokia Nathan (*Professor and Chair, Photonic Systems and Displays,
      Electrical Engineering Division, Cambridge University, United Kingdom*)

K5   **Acoustic Metamaterials and Phononic Crystals: Towards the Total**   A5
      Control of the Wave Propagation
      Prof. Dr. Abdelkrim Khelif (*Senior researcher, CNRS French National Centre
      for Scientific Research (FEMTO-ST), Besancon, France*)

K6   **SOI Photodiode with Surface Plasmon Antenna: From Sensitivity**   A6
      Enhancement to Refractive Index Measurement for Biosensing
      Prof. Dr. Hiroshi Inokawa
      (*Professor, Research Institute of Electronics, Shizuoka University, Japan*)

Cluster 1

1   **Numerical estimation of self-sputtering effect in ionized physical vapor**   1
     deposition system
     Nafarizal Nayan, Jais Lias, Mohd Zainizan Sahdan, Mohd Khairul Ahmad, Lim Huey
     Sia, Low Jia Wei, Ahmad Shuhaimi Abu Bakar, Mohamad Rusop
     Universiti Tun Hussein Onn Malaysia, MALAYSIA

2	**Synthesis and Characterization of Carbon Nano Structures on Gallium Phosphate** Aishah Fauthan, Zainab Yunusa, Mohd Nizar Hamidon, Burhanuddin Yeop Majlis *Universiti Putra Malaysia, MALAYSIA*	5
3	**Contactor Characterization Methodology on Pin Inductance** Ling Li Ong, Chan Yee Kit, You Ah Heng *Multimedia University Melaka, MALAYSIA*	9
4	**Two-Stage Small-Signal Amplifier with Darlington and Sziklai pairs** SachchidaNand Shukla, Beena Pandey *Dr. Ram Manohar Lohia Avadh University, Faizabad, INDIA*	13
5	**A 20 GHz Power Amplifier Design On Novel Nonlinear Model** Zonghua Zheng, Lingling Sun, Jun Liu *Zhejiang University, Hangzhou, CHINA*	17
6	**Investigation of TMAl preflow to the properties of AlN and GaN film grown on Si(111) by MOCVD** Franky Lumbantoruan, Yuan-Yee Wong, Yue-Han Wu, Wei-Ching Huang, Niraj Man Shrestra, Tung Tien Luong, Tran Binh Tinh, Edward Yi Chang *National Chiao Tung University, TAIWAN*	20
7	**The Impact of Minority Carrier Lifetime and Carrier Concentration on the Efficiency of CIGS Solar Cell** M.F.M. Fathil, M.K. Md Arshad, U. Hashim, A.R Ruslinda, R.M. Ayub, A.H. Azman, M. Nurfaiz, M.Z.M. Kamarudin, M. Aminuddin, A.R Munir *Universiti Malaysia Perlis, MALAYSIA*	24
8	**Modelling of Hybrid Energy Harvester with DC-DC Boost Converter using Arbitary Input Sources for Ultra-Low-Power Micro-devices** Michelle S.M.Lim, Sawal H.M.Ali, S. Jahariah and MD.Shabiul Islam *Universiti Kebangsaan Malaysia, MALAYSIA*	28
9	**Correlation between the microstructure of copper oxide thin film and its gas sensing response** Jia Wei Low, Nafarizal Nayan, Mohd Zainizan Sahdan, Mohd Khairul Ahmad, Ali Yeon Md Shakaff, Ammar Zakaria, Ahmad Faizal Mohd Zain *Universiti Tun Hussein Onn Malaysia, MALAYSIA*	32
10	**Quantum Ballistic Simulation Study of InGaAs/InAs/InGaAs Quantum Well MOSFET: Effects of Doping and Physical Device Parameters** Sudipta Romen Biswas, Kanak Datta, Ehsanur Rahman, Abir Shadman, Quazi D. M. Khosru *Bangladesh University of Engineering and Technology, BANGLADESH*	36
11	**Low Dimensional Simulator for Carbon-based Devices** Chin Lin Ng, Michael Loong Peng Tan *University Teknologi Malaysia, MALAYSIA*	40

12   **Design and Implementation of a 1-bit FinFET Full Adder Cell for ALU in Subthreshold Region**   44
'Aqilah binti Abdul Tahrim, Michael Loong Peng Tan
University Teknologi Malaysia, MALAYSIA

13   **The RF Power Effect on the Surface Morphology of Titanium Dioxide (TiO$_2$) Film**   48
S. Norhafiezah, RM Ayub, M. K. Md Arshad, A.H. Azman, M. F. Fatin, M.A. Farehanim and U. Hashim
Universiti Malaysia Perlis, MALAYSIA

14   **Development of Silicon Carbide MEMS Capacitive Pressure Sensor Operating at 500°C**   52
Noraini Marsi, Burhanuddin Yeop Majlis, Azrul Azlan Hamzah, Ummikalsom Abidin, Faisal Mohd-Yasin
Universiti Kebangsaan Malaysia, MALAYSIA

15   **Effects of high-K Dielectric with Metal gate for Electrical Characteristics of 18nm NMOS Device**   56
Norani Bte Atan, Ibrahim Bin Ahmad, Burhanuddin Bin Yeop Majlis
Universiti Tenaga Nasional, MALAYSIA

16   **Design of a 4-Bit Adder using Reversible Logic in Quantum-Dot Cellular Automata (QCA)**   60
Darushini Kunalan, Chee Lee Cheong, Chien Fat Chau, Azrul Bin Ghazali
University Tenaga Nasional, MALAYSIA

17   **Top-down Fabrication of Silicon Nanowire Sensor Using Electron Beam and Optical Mixed Lithography**   64
S.F.A. Rahman, N.A. Yusof, M.N. Hamidon, R.M. Zawawi, U. Hashim
Universiti Malaysia Perlis, MALAYSIA

18   **I-V Characteristic Effects of Fluidic-based Memristor for Glucose Concentration Detection**   68
Nor Shahanim Mohamad Hadis, Asrulnizam Abd Manaf, Sukreen Hana Herman
Universiti Teknologi MARA, MALAYSIA

19   **Aluminum Nitride Thin Film Deposition Using DC Sputtering**   72
Mohd H.S Alrashdan, Azrul Azlan Hamzah, Burhanuddin Yeop Majlis, Mohd Faizal Aziz
Universiti Kebangsaan Malaysia, MALAYSIA

20   **Lower DIBL in Inverted Substrate of UTBB SOI n-MOSFETs**   76
Noraini Othman, M.K. Md Arshad, U. Hashim
Universiti Malaysia Perlis, MALAYSIA

21   **In$_x$Ga$_{1-x}$As Surface Channel Quantum Well MOSFET: Quantum Ballistic Simulation Using Mode Space Approach**   80
Ehsanur Rahman, Abir Shadman, Sudipta Romen Biswas, Kanak datta, Quazi

D. M. Khosru
Bangladesh University of Engineering and Technology, BANGLADESH

22 **Condition Based Engine Oil Degradation Monitoring System, Synthesis and Realization on ASIC**
M.F.M Idros, Sawal Ali, Md.Shabiul Islam
Universiti Teknologi MARA, MALAYSIA
84

23 **Impact of different ground planes of UTBB SOI MOSFETs under the single-gate (SG) and double-gate (DG) operation mode**
Noraini Othman, M.K. Md Arshad, S.N. Sabki, and U. Hashim
Universiti Malaysia Perlis, MALAYSIA
88

24 **Physical and Optical Studies on ZnO Films by Sol-gel**
A. R. Nurulfadzilah, S. N. M. Tawil, C. A. Norhidayah, N. Nafarizal, M. Z. Sahdan
Universiti Pertahanan Nasional Malaysia, MALAYSIA
92

25 **Process Development of 40 nm Silicon Nanogap for Sensor Application**
M.S. Nur Humaira, U. Hashim, , T. Nazwa, S.T. Ten, S. Ahmad, N. A. Yusof
Universiti Malaysia Perlis, MALAYSIA
96

26 **Synthesis and Characterization of Cupric Oxide Thin Films by Sol-Gel Method**
M.F. Nurfazliana, S.A. Kamaruddin, N. Nafarizal, H. Saim, and M.Z. Sahdan
Universiti Tun Hussein Onn Malaysia, MALAYSIA
100

27 **Shear Horizontal Surface Acoustic Wave COMSOL Modeling on Lithium Niobate Piezoelectric Substrate**
S. T. Ten, U. Hashim, W.W. Liu, K.L.Foo, C.H. Voon, H.Hisham, T. Nazwa, A. Sudin and M.S. Nur Humaira
Universiti Malaysia Perlis, MALAYSIA
104

28 **A Study on the Role of Solvent on Properties of $Gd_xZn_{1-x}O$ films synthesized by Sol-gel Method**
C.A Norhidayah, S.A Kamaruddin, N.Nafarizal, M.Z Sahdan, A.R Nurulfadzilah, S.N.M Tawil
Universiti Tun Hussein Onn, MALAYSIA
108

29 **The Impact of Channel Doping in Junctionless Field Effect Transistor**
B.S Lim, M.K. Md Arshad, Noraini Othman, M.F.M. Fathil, M.F. Fatin and U. Hashim
Universiti Malaysia Perlis, MALAYSIA
112

30 **Experimental Analysis on SU-8 Micromolding Structure of PDMS (Poly-Dimethylsiloxane) Based Microfluidic Channel**
Marianah Masrie, Burhanuddin Yeop Majlis and Jumril Yunas
Universiti Kebangsaan Malaysia, MALAYSIA
115

31 **Mechanical Characteristics of Porous Silicon Membrane for Filtration in Artificial Kidney** 119
Norhafizah Burham, Azrul Azlan Hamzah, Burhanudin Yeop Majlis
Universiti Kebangsaan Malaysia, MALAYSIA

32 **Bonding Stress Analysis on Palladium Bond Pad Between Cu and Au FAB** 123
Sauli. Z., Retnasamy, V. and Mamat, H.
Universiti Malaysia Perlis, MALAYSIA

33 **PCB Depanelling Stress Distribution Simulation Analysis Using Designated Thru Holes** 126
Retnasamy, V., Sauli. Z.,Vairavan, R. and Mamat, H.
Universiti Malaysia Perlis, MALAYSIA

34 **High Power LED Heat Dissipation Simulation Analysis via Heat Sink Fin Variation** 130
Retnasamy, V., Sauli. Z., Vairavan, R., Taniselass, S., Mamat, H.
Universiti Malaysia Perlis, MALAYSIA

35 **The Impact of Scaled Channel Length in Tunneling Field Effect Transistors (TFETs)** 134
Nurul Huda Abdul Rahman, M. K. Md Arshad, Noraini Othman, M.F.M. Fathil, M.S. Nur Humaira and U. Hashim
Universiti Malaysia Perlis, MALAYSIA

36 **Performance Evaluation of Dual-Channel Armchair Graphene Nanoribbon Field-Effect Transistor** 138
Adila Syaidatul Azman, Zaharah Johari and Razali Ismail
Universiti Teknologi Malaysia, MALAYSIA

37 **Novel Low-Voltage RF-MEMS Switch: Design and Simulation** 142
Ma Li Ya, Norhayati Soin, Anis Nurashikin Nordin
University of Malaya, MALAYSIA

38 **Three-Input Four-Output Voltage-Mode Multifunction Filter Using Two DDCCs** 146
Montree Kumngern and Usa Torteanchai
King Mongkut's Institute of Technology Ladkrabang, THAILAND

39 **Comparative study of subthreshold characteristics of different antimonide-based and nitride-based Dual Material Gate (DMG) HEMTs** 150
Arman-Ur-Rashid, Md. Aynal Hossain, Tanvir Rahman and Farseem M. Mohammedy
Bangladesh University of Engineering and Technology, BANGLADESH

40 **Characterization of Vertical Strained SiGe Impact Ionization MOSFET for Ultra-Sensitive Biosensor Application** 154
Ismail Saad, Mohd. Zuhir H., Bun Seng C., Khairul A.M, Bablu Ghosh, N. Bolong, Razali Ismail

Universiti Malaysia Sabah, MALAYSIA

41 **Investigation on Iodine Flow Rate in MEH-PPV: I-MWCNT** 158
Nanocomposite Thin Film
M.S.P. Sarah, M. Mudasir, S.S. Shariffudin, H. Hashim, M.Rusop
Universiti Teknologi MARA, MALAYSIA

42 **Impact of Implantation Methods on Speed and Accuracy Trade-off in** 162
Calibrated TCAD Tool
Muhamad Amri Ismail
MIMOS Berhad, MALAYSIA

43 **FDCCTA: The Active Building Block for Analog Signal Processing** 166
Applications
Montree Kumngern
King Mongkut's Institute of Technology Ladkrabang, THAILAND

44 **Dependency of electrical characteristics on nano gap variation of pinch off** 170
lateral gate transistors
Farhad Larki, Arash Dehzangi, Sawal Hamid Md Ali, Azman Jalar, Md. Shabiul
Islam, Burhanuddin Y. Majlis, Mohd Nizar Hamidon, Elias B Saion, Sabar D.
Hutagalung
Universiti Kebangsaan Malaysia, MALAYSIA

45 **The Annealing Temperature Effect on the Structure and Electrical** 174
Properties of Titanium Dioxide (TiO$_2$) Film Deposited by Reactive RF
Sputtering
S. Norhafiezah, RM Ayub, M. K. Md Arshad, A.H. Azman, M. F. Fatin, M.A.
Farehanim and U. Hashim
Universiti Malaysia Perlis, MALAYSIA

46 **Effect of Process Parameter Variability on the Threshold Voltage of** 178
Downscaled 22nm PMOS using Taguchi Method
Afifah Maheran A.H., Menon, P.S., S. Shaari, I. Ahmad, Noor Faizah Z.A.,
Kalaivani T., P.R. Apte
Universiti Kebangsaan Malaysia, MALAYSIA

47 **An Efficient ROM Compression Technique for Linear- Interpolated Direct** 182
Digital Frequency Synthesizer
Qahtan Khalaf Omran, Mohammad Tariqul Islam, Norbahiah Misran,
Mohammad Rashed Iqbal Faruque
Universiti Kebangsaan Malaysia, MALAYSIA

48 **Modeling the Velocity Saturation Region of Graphene Nanoribbon** 186
Transistor
M. Hosseinghadiry, Razali Ismail, F. Fotovvatikhah, M. Khaledian, M.
Saeidmanesh
Universiti Teknologi Malaysia, MALAYSIA

xiii

49 **Transistor Sizing Methodology for Low Noise Charge Sensitive Amplifier With Input Transistor Working in Moderate Inversion**   189
N. Aimaier, R. M. Sidek, M. N. Hamidon, N. Sulaiman
University Putra Malaysia, MALAYSIA

50 **Fabrication of Interdigitated Microelectrodes for CuO Nanowires I-V Measurement**   193
Tiong Teck Yaw, Dee Chang Fu, Azrul Azlan Hamzah, Burhanuddin Yeop Majlis, Muhamad Mat Salleh, Mohd Faizal Aziz, Saadah Abdul Rahman
Universiti Kebangsaan Malaysia, MALAYSIA

51 **Spice Model Design for Carbon Nanotube Field Effect Transistor (CNTFET)**   197
Soheli Farhana, AHM Zahirul Alam and Sheroz Khan
International Islamic University Malaysia, MALAYSIA

52 **Study of Interdigitated Electrode Sensor for Lab-on-Chip Applications**   201
Mohammadmahdi Vakilian, Burhanuddin Yeop Majlis
Universiti Kebangsaan Malaysia, MALAYSIA

53 **Simulation of Temperature Dependent Characterization of Charge Transport in Organic Transistor**   205
Umar Faruk Shuib, Khairul Anuar Mohamad, Ismail Saad, Bablu Kumar Gosh, Tamer A. Tabet
Universiti Malaysia Sabah, MALAYSIA

54 **Process-induced NBTI-imbalance of high- /metal-gate deep-submicron CMOS**   209
Y. Abdul Wahab, N. Soin, S. Shahabuddin, H. Hussin
University of Malaya, MALAYSIA

55 **Synthesization of Nickel Nanoparticles Embedded in SU8 Polymer for Electromagnetic Actuator Membrane**   213
Muzalifah Mohd Said, Jumril Yunas, Roer Ekapawinto, Burhanuddin Yeop Majlis, Badariah Bais
Universiti Kebangsaan Malaysia, MALAYSIA

56 **Characterization of MEMS Structure on silicon wafer using KrF Excimer Laser Micromachining**   217
M. Mazalan, S. Johari, B. P. Ng and Y. Wahab
Universiti Malaysia Perlis, MALAYSIA

57 **A Study on the gas sensing effect on current-voltage characteristics of ZnO Nanostructures**   221
W.H.Khoo and S.M.Sultan
Universiti Teknologi Malaysia, MALAYSIA

xiv

58	**Memristive Behavior of HF-Etched Sputtered Titania Thin Films** Z. Aznilinda, M. M. Ramly, NS Kamarozaman and S.H.Herman *Universiti Teknologi MARA, MALAYSIA*	225
59	**Investigation of the Inserted LT-AlGaN Interlayer in AlGaN/GaN/AlGaN DH-FET Structure on Si Substrates** Yu-Lin Hsiao, Chia-Ao Chang and Edward Yi Chang *National Chiao-Tung University, TAIWAN*	229
60	**Statistical Process Modelling For 32nm High-K/Metal Gate PMOS Device** Afifah Maheran A.H., Noor Faizah Z. A., P. S. Menon, I. Ahmad, P.R. Apte, Kalaivani T., Salehuddin, F. *Universiti Kebangsaan Malaysia, MALAYSIA*	232
61	**Finite Element Modeling of Dielectrophoretic Microelectrodes Based on a Array and Ratchet Type** Muhamad Ramdzan Buyong, Norazreen Abd Aziz, Azrul Azlan Hamzah, M.F. Mohd Razip Wee, Burhanuddin Yeop Majlis *Universiti Kebangsaan Malaysia, MALAYSIA*	236
62	**Dielectrophoretic Characterization of Array Type Microelectrodes** Muhamad Ramdzan Buyong, Norazreen Abd Aziz, Azrul Azlan Hamzah, Burhanuddin Yeop Majlis *Universiti Kebangsaan Malaysia, MALAYSIA*	240
63	**Structural damage of Si-implanted in the $In_{0.53}Ga_{0.47}As$ thin film** Muhammad Zulkhairi Roslan, , Dilla D. Berhanuddin, Mohd Ambri Mohamed, M.F. Mohd Razip Wee, Farhad Larki, Burhanuddin Yeop Majlis *Universiti Kebangsaan Malaysia, MALAYSIA*	244
64	**Corner compensation mask design on (MEMS) accelerometer structure** Norliana Yusof , Abdullah C.W. Noorakma, Norhayati Soin *Universiti Sultan Zainal Abidin, Terengganu, MALAYSIA*	248
65	**Electronic state transition in cooperatively interacting point-defects in semiconductor crystals** Mohd Ambri Mohamed, B. Y. Majlis, M. H. Ani *Universiti Kebangsaan Malaysia, MALAYSIA*	252
66	**Effects of Material and Membrane Structure on Maximum Temperature of Microheater for Gas Sensor Applications** Mimiwaty Mohd Noor, Gandi Sugandi, Mohd Faizal Aziz and Burhanuddin Yeop Majlis *Universiti Kebangsaan Malaysia, MALAYSIA*	255
67	**FEM analysis of wavelength effects in piezoelectric substrate** Norazreen Abd Aziz, Badariah Bais, Muhamad Ramdzan Buyong, Burhanuddin Yeop Majlis, Anis Nurashikin Nordin *Universiti Kebangsaan Malaysia, MALAYSIA*	259

68   **Performance Analysis of Zinc Oxide Piezoelectric MEMS Energy Harvester**   263
Umi Milhana Jamain, Nur Hidayah Ibrahim, Rosminazuin Ab Rahim
International Islamic University, MALAYSIA

69   **Channel length effect on the saturation current and the threshold voltages of CNTFET**   267
Abu Hanifah Muhamad Ali and M. H. Ani, Mohd Ambri Mohamed
International Islamic University Malaysia, MALAYSIA

70   **Polydimethlsiloxane (PDMS) Microchannel with Trapping Chamber for BioMEMS Applications**   270
Ummikalsom Abidin, Burhanuddin Yeop Majlis and Jumril Yunas
Universiti Kebangsaan Malaysia, MALAYSIA

71   **Influence of Gadolinium Doping on The Crystalline Structure and Optical Properties of Zinc Oxide Thin Films**   274
N. Sarip, F. Mahmud, M.Z. Sahdan, S.N.M. Tawil
Universiti Tun Hussein Onn Malaysia, MALAYSIA

72   **Structural and Electronic Properties of Li doped Decahedral $Ag_{12}Li_1$ Bimetallic Nanoclusters**   278
Shaikat Debnath, Suhana Mohd Said
University of Malaya, MALAYSIA

Cluster 2

73   **Dispersion Characteristics of Twisted Clad Chiral Nihility Fibers**   282
N. Iqbal, M.A. Baqir and P.K. Choudhury
Universiti Kebangsaan Malaysia, MALAYSIA

74   **A 48GHz-78GHz MMIC Sub-Harmonic Pumped Image Rejection Mixer**   286
Shengzhou Zhang, Lingling Sun, Jincai Wen, Jun Liu
Zhejiang University, CHINA

75   **Gold Nanoplates as Sensing Material for Plasmonic Sensor of Formic Acid**   290
Marlia binti Morsin, Muhamad Mat Salleh and Akrajas Ali Umar
Universiti Kebangsaan Malaysia, MALAYSIA

76   **Design of 40GHz Multilayer End Coupled Band Pass Filter using LTCC Technology**   294
Zulkifli Ambak, Azmi Ibrahim, Hizamel Mohd Hizan, Ahmad Ismat Abdul Rahim, Mohd Zulfadli Mohamed Yusoff, Razali Ngah and Young Chul Lee
TM Research and Development, MALAYSIA

77   **Effect of Different Tunnel diodes on The Efficiency of Multi-junction III-V Solar cells**   298
Hung-Wei Yu, Hong-Quan Nguyen, Chen-Chen Chung, Ching-Hsiang Hsu, Chih-Jen Hsiao and Edward Yi Chang
National Chiao Tung University, TAIWAN

| 78 | Design and optimization of a Mach-Zehnder Interferometer (MZI) for optical modulators
Hanim Abdul Razak, Hazura Haroon, P Susthitha Menon, Sahbudin Shaari, Norhana Arsad
Universiti Teknikal Malaysia Melaka, MALAYSIA | 301 |

Cluster 3

| 79 | An Ultra-low Power and Area Efficient 10 bit Digital to Analog Converter Architecture
Iffa Binti Sharuddin, L. Lee
Multimedia University, MALAYSIA | 305 |

| 80 | Implementation of Low Power Compressed ROM for Direct Digital Frequency Synthesizer
Salah Hasan Alkurwy, Sawal Hamid Md Ali, Md. Shabiul Islam
Universiti Kebangsaan Malaysia, MALAYSIA | 309 |

| 81 | Adhesion Enhancement for Electroless Plating on Mold Compound for EMI Shielding with Industrial Test Compliance
Tai, Min Fee, Lee, Swee Kah, Goh, Soon Lock, Kok, Swee Leong, Kenichiroh.Mukai, Tafadwa Magaya
Universiti Teknikal Malaysia Melaka, MALAYSIA | 313 |

| 82 | A Low Power 0.18μm CMOS Technology Integrating Dual-Slope Analog-to Digital Converter
Ili Shairah Abdul Halim, Nor Syazwana Mohd Yusof, Siti Lailatul Mohd Hassan
Universiti Teknologi MARA, MALAYSIA | 317 |

| 83 | Comparative Study on 8T SRAM with Different Type of Sense Amplifier
Siti Lailatul Mohd Hassan, Idalailah Dayah, Ili Shairah Abdul Halim
Universiti Teknologi MARA, MALAYSIA | 321 |

| 84 | Discrete Event Simulation Modeling for Semiconductor Fabrication Operation
MA Chik, SR AB Rahim, AZ MD Rejab, K Ibrahim, U. Hashim
Silterra Malaysia Sdn Bhd, MALAYSIA | 325 |

| 85 | Horrendous Capacity Cost of Semiconductor Wafer Manufacturing
K Ibrahim, MA Chik, U. Hashim
Silterra Malaysia Sdn Bhd, MALAYSIA | 329 |

| 86 | Low power and low voltage SRAM design for LDPC codes hardware applications
Rosalind Deena Kumari Selvam, C. Senthilpari, Lee Lini
Multimedia University, MALAYSIA | 332 |

| 87 | Gold-Free Cu-Metallized III-V Solar Cell
Ching-Hsiang Hsu, Hsun-Jui Chang, Hung-Wei Yu, Hong-Quan Nguyen, Jer- | 336 |

Shen Ma and Edward Yi Chang
National Chiao Tung University, MALAYSIA

88 **On-Wafer Scattering Parameter Characterization of Differential Four-Port Networks LNA using Two-Port Vector Network Analyzer**   339
Maizan Muhamad, Norhayati Soin, Harikrishnan Ramiah, Norlaili Mohd Noh, Chong Wee Keat
University Teknologi MARA, MALAYSIA

89 **Strain Rate Effect on Micromechanical Properties of SnAgCu Solder Wire**   343
I. Abdullah, R. Ismail and A. Jalar
Universiti Kebangsaan Malaysia, MALAYSIA

Cluster 4

90 **Epitaxial Lift-Off of Large-Area GaAs Multi-Junction Solar Cells For High Efficiency Clean and Portable Energy Power Generation**   347
N. Pan
MicroLink Devices, Inc., USA

91 **Memristor Applied In Delay Locked Loop For High Lock Speed And Wide Frequency Range**   350
Siti Musliha Ajmal Binti Mokhtar, Wan Fazlida Hanim Abdullah
Universiti Teknologi MARA, MALAYSIA

92 **Photoluminescence from Localized States in GaAsBi Epilayers**   354
A. R. Mohmad, B. Y. Majlis, F. Bastiman, C. J. Hunter, R. D. Richards, J. S. Ng, J. P. R. David
Universiti Kebangsaan Malaysia, MALAYSIA

93 **Growth Parameters Optimization of GaN High Electron Mobility Transistor Structure on Silicon Carbide Substrate**   358
Y. Y. Wong, S. C. Huang, W. C. Huang, F. Lumbantoruan, Y. S. Chiu, H. C. Wang, H. W. Yu, E. Y. Chang
National Chiao Tung University, TAIWAN

94 **Fabrication of anodic aluminum oxide template by using a single step anodization process**   N/A
Asma Fatehi, Dee Chang Fu, Burhanuddin Yeop Majlis, Azrul Azlan Hamzah
Universiti Kebangsaan Malaysia, MALAYSIA

95 **Characterisation of Nickel Germanide formed on Amorphous and Crystalline Germanium**   366
Fahid Algahtani, Elena Pirogova, Anthony Holland, Mark Blackford, Jeffrey C McCallum, Brett C Johnson
RMIT University, Melbourne, AUSTRALIA

96 **A Field-Effect Device Based on an Exfoliated Thin Film of Few-Layer Graphene** 370
S. R. Kasjoo, M. M. Ramli, M. R. Zakaria, M. K. Md Arshad, R. Mat Ayub, R. A. Rahim and U. Hashim
Universiti Malaysia Perlis, MALAYSIA

97 **Effects of Annealing Temperature on Dielectric Property of Nano-Films Lead Titanate Prepared on ITO Glass by Spin Coating Method** 373
Z. Nurbaya, I.H.H. Affendi, N.A. Azhar, M.H Wahid and M. Rusop
Universiti Teknologi MARA, MALAYSIA

98 **Surface Functionalization of Multiwalled Carbon Nanotube for Biosensor Device Application** 377
M.F. Fatin, A. Rahim Ruslinda, M. K. Md Arshad, U. Hashim, S. Norhafizah, M. A. Farehanim
Universiti Malaysia Perlis, MALAYSIA

99 **Fabrication of SAW Device By Using Zno Thin Film as a Sensing Area** 380
S.L.Lai, M.R. Zakaria, U. Hashim, K.L Foo, R. Haarindra Prasad
Universiti Malaysia Perlis, MALAYSIA

100 **Fabrication of a CMOS-Compatible Surface Acoustic Wave Device for Application in Pathogen Sensing** 384
S. T. Ten, U. Hashim, F.Malek, W.W. Liu, K.L.Foo, C.H. Voon, F. H. Wee, Y.S. Lee, H.Hisham, A. Sudin and M.S. Nur Humaira
Universiti Malaysia Perlis, MALAYSIA

101 **Preparation Of Zinc Oxide Piezoelectric for Surface Acoustic Wave Biosensor Device** 388
M. R. Zakaria, F. Hamzah, M. F. Omar, U. Hashim, R. Mat Ayub, M. A. Farehanim, A Wesam Al-Mufti
Universiti Malaysia Perlis, MALAYSIA

102 **Controlling growth rate of ultra-thin Silicon Dioxide layer by incorporating Nitrogen gas during dry thermal oxidation** 392
A.H. Azman, RM Ayub, M. K. Md Arshad, S. Norhafiezah, M.F.M. Fathil, M.Z. Kamarudin, M. Nurfaiz,U. Hashim
Universiti Malaysia Perlis, MALAYSIA

103 **Fabrication and Characterization of Undoped Polysilicon Nanowire for pH sensor** 396
C.C. Yee, M. K. Md Arshad, M. Nuzaihan M. N, M.F.M. Fathil and U. Hashim
Universiti Malaysia Perlis, MALAYSIA

104 **Annealing Temperature Effect on Nanostructured TiO2 Films** 400
I.H.H. Affendi, M.S.P. Sarah, M. Rusop
Universiti Teknologi MARA, MALAYSIA

105 **Fabrication and Characterization of Polysilicon for DNA Detection**   404
Y. M. Ang, M. K. Md Arshad, K. L. Foo, M. Nuzaihan Md. N., A.H. Azman,
U. Hashim
Universiti Malaysia Perlis, MALAYSIA

106 **Atomic Force Microscope base nanolithography for reproducible micro**   408
and nanofabrication
Arash Dehzangi, Farhad Larki, Burhanuddin Y. Majlis, Zainab Kazemi,
MohammadMahdi Ariannejad, A Makarimi Abdullah, Mahmood Goodarz
Nasery, Manizheh Navasery, Elias B Saion, Halimah Mohamed. K, Nasrin
Khalilzadeh, Sabar D. Hutagalung
Universiti Kebangsaan Malaysia, MALAYSIA

107 **Effects of Ethanol in Oxalic Acid on the Synthesis of Porous Anodic**   412
Alumina
N.H.M. Saleh, B.Y.Lim, C.H.Voon, S.T.Ten, M.N.Derman, K.L. Foo, M. K.
Md Arshad, U.Hashim
Universiti Malaysia Perlis, MALAYSIA

108 **Performance of Inverted Organic Solar Cell using Different Metal**   416
Electrodes
M. S. Alias, S.A. Kamaruddin, N. Nafarizal, M.Z. Sahdan
Universiti Tun Hussein Onn, MALAYSIA

109 **Synthesis of Zinc Oxide Thin Film by Anodizing**   420
C.H.Voon, N.Tukemon @ Tukiman, B.Y.Lim, U.Hashim, S.T.Ten,
M.N.Derman, K.L. Foo, M. K. Md Arshad
Universiti Malaysia Perlis, MALAYSIA

110 **Properties of Boron Doped Carbon Films by Carbon Palm Oil for Carbon**   424
Based Solar Cell Applications
A. Ishak, M. F. Malek, and M. Rusop
Universiti Teknologi MARA, MALAYSIA

111 **Optical performance of MEH-PPV/ZnO Nanocomposite at Different**   428
Weight Percent for OLED Applications
N. E. A. Azhar, S. S. Shariffudin, Z. Nurbaya, I. H.H. Affendi, A. K. Shafura,
M.Rusop
Universiti Teknologi MARA, MALAYSIA

112 **Glass Etching for Cost-effective Microchannels Fabrication**   432
N.M Salih, N. Nafarizal, C.F. Soon, M.Z. Sahdan, A. Tijjani, U. Hashim
Universiti Tun Hussein Onn, MALAYSIA

113 **Fabrication of SONOS Flash Memory Device By Using Engineered Tunnel**   436
Barrier Technique
M. R. Zakaria, Shahrir R. Kasjoo, A. F. Mahyidin, A Wesam Al-Mufti, R. Mat
Ayub, U. Hashim
Universiti Malaysia Perlis, MALAYSIA

114 **Integrated Constant Voltage Constant Current Readout Interfacing Circuit for EGFET Electrochemical Sensing**   440
H. Guliga, W. F. H. Abdullah, S. H. Herman
Universiti Teknologi MARA, MALAYSIA

115 **Effect of Indium Concentration on Optical and Electrical Properties of In Doped ZnO Thin Films for Gas Sensing Application**   444
M.Hannas, Azrif Manut, S.H.Herman, M.Rusop
Universiti Teknologi MARA, MALAYSIA

116 **Switching Behavior of Lateral-Structured Zinc Oxide-Based Memristive Device**   448
Raudah Abu Bakar, Mohd Nur Nazmi, 'Awatif Harun, Nur Syahirah Kamarozaman, Nor Azira Akmar Shaari, Shafaq Mardhiyana Mohamat Kasim, Sukreen Hana Herman
Universiti Teknologi MARA, MALAYSIA

117 **Annealing Temperature Dependence of Resistive Switching Behavior for Sol-gel Spin Coated Zinc Oxide Thin Films**   452
Raudah Abu Bakar, Ahmad Faiz Mohamad Zohaimi, Nur Syahirah Kamarozaman, Nor Azira Akmar Shaari, Shafaq Mardhiyana Mohamat Kasim, Sukreen Hana Herman
Universiti Teknologi MARA, MALAYSIA

118 **Effect of V/III Ratios on Surface Morphology in a GaSb Thin Film Grown on GaAs Substrate by MOCVD**   456
Chih-Jen Hsiao, Chun-Kuan Liu, Sa-Hoang Huynh, Thien-Huu Ha Minh, Hung-Wei Yu, Hong-Quan Nguyen, Jer-Shen Maa, Shoou-Jinn Chang and Edward Yi Chang
National Cheng Kung University, TAIWAN

119 **Electrical and Optical Properties Characterization of MEH-PPV Thin Film using Sol-Gel Method**   459
H. Hashim, S. S. Shariffudin, A. M. Khairuddin, M. S. P. Sarah and M. Rusop
Universiti Teknologi MARA, MALAYSIA

120 **Dye-Sensitized Solar Cell Using Aligned ZnO Nanorod Grown on SZO Films at Different Solution Molarities**   463
I. Saurdi, A.K. Shafura, M.H. Mamat and M. Rusop
Universiti Teknologi MARA, MALAYSIA

121 **The Effect of Softbaking Temperature on SU-8 Photoresist Performance**   467
Shazlina Johari, Nithiyah Tamilchelvan, Mohammad Nuzaihan Md Nor, Muhammad Mahyiddin Ramli, Bibi Nadia Taib, Mazlee Mazalan and Yufridin Wahab
Universiti Malaysia Perlis, MALAYSIA

xxi

122 **2.4GHz WLAN RF Energy Harvester for Passive Indoor Sensor Nodes** 471
Fatima Alneyadi, Maitha Alkaabi, Salama Alketbi, Shamsa Hajraf, and Rashad Ramzan
UAE University, United Arab Emirates

123 **Theoretical Study of On-Chip Meander Line Resistor to Improve Q-factor** 475
Wong Goon Weng, Norhayati Binti Soin
University of Malaya, MALAYSIA

124 **Temperature Dependence of Ga:Zno Film Deposited By RF Magnetron Sputtering** 479
Farah Lyana Shain, Azmizam Manie @ Mani, Lam Mui Li, Saafie Salleh and Afishah Alias
University Malaysia Sabah, MALAYSIA

125 **Memristor to Control Delay of Delay Element** 483
Siti Musliha Ajmal Binti Mokhtar, Wan Fazlida Hanim Abdullah
Universiti Teknologi MARA, MALAYSIA

126 **Design and Fabrication of PCB Based Planar Micro-coil For Magnetic MEMS Actuator** 487
Roer Eka Pawinanto, Jumril Yunas, Muzalifah Mohd. Said, Mimiwaty Mohd. Noor, BurhanuddinYeop Majlis
Universiti Kebangsaan, MALAYSIA

127 **TiO_2-based Extended Gate FET pH-sensor: Effect of Annealing Temperature on its Sensitivity, Hysteresis and Stability** 491
K. A. Yusof, S. H. Herman, W. F. H. Abdullah
Universiti Teknologi MARA, MALAYSIA

128 **Modeling and FEM Simulation Using Fluid- Structures Interaction of Flexible Micro-Bridge Bending Within PDMS Micro-Channel** 495
Nadir Belgroune, Abdelkader Hassein-Bey, Burhanuddin Yeop Majlis, Mohamed El-Amine Benamar
University Saad Dahlab Blida, ALGERIA

129 **The effects of Growth Parameters on the Electrical Properties in InAlN/AlN/GaN High-Electron-Mobility Transistors (HEMTs)** 499
Wei-Ching Huang, Yuen-Yee Wong, Kuan-Shin Liu, Chi-Feng Hsieh, Edward Yi Chang
National Chiao Tung University, TAIWAN

130 **High Quality Ge Epitaxial Films Grown on In0.51Ga0.49P/GaAs and GaAs Substrates by Ultra High Vacuum Chemical Deposition** 502
Yung-Hsuan Su, Shih-Hsuan Tang, Chi Lang Nguyen, Ching-Wen Kuan, Hung-Wei Yu and Edward Yi Chang
National Chiao Tung University, TAIWAN

131 **Preparation and Characterization of MWCNT Dispersed in Various Solutions**   505
M. A. Farehanim, U. Hashim, S. Norhafiezah, M.F. Fatin, RM Ayub, Norhayati Soin, Fatimah Ibrahim
Universiti Malaysia Perlis, MALAYSIA

132 **Design Optimization and Finite Element Analysis of AlN/3C-SiC Piezoelectric Bio-Sensors**   509
Abid Iqbal, Faisal Mohd-Yasin, Sima Dimitrijev
Griffith University, AUSTRALIA

133 **Synthesizing SnAgCu Nanoparticles by Electrodeposition of Reverse Microemulsion Electrolyte**   513
S.P. Foo, K.S. Siow, A. Jalar
Universiti Kebangsaan Malaysia, MALAYSIA

134 **Hydrothermal growth of ZnO nanotube on InGaP/GaAs/Ge Solar Cells**   517
Chen-Chen Chung, Kung-Liang Lin, Hung-Wei Yu, Nguyen-Hong Quan, Chang-Fu Dee and Edward Yi Chang
National Chiao Tung University, TAIWAN

135 **Effects of Thermal Cycling on the Mechanical Properties of Gold Wire Bonding**   521
Wan Yusmawati Wan Yusoff, Azman Jalar, Norinsan Kamil Othman, Irman Abdul Rahman
Universiti Kebangsaan, MALAYSIA

136 **Parametric Analysis of Boost Converter for Energy Harvesting using Piezoelectric for Micro Devices**   525
Shafii A. Wahab, M. S. Bhuyan, Jahariah Sampe, Sawal H. Md. Ali
Universiti Kebangsaan, MALAYSIA

137 **Nanostructured Al-doped ZnO-Based Gas Sensor Prepared Using Sol-Gel Spin-Coating Method**   529
A.K. Shafura, I. Saurdi, N.E.A. Azhar, M.H. Mamat, Uzer M, M. Rusop and A. Shuhaimi
Universiti Teknologi MARA, MALAYSIA

138 **Effects of the Fin Width Variation on the Performance of 16nm FinFETs with Round Fin Corners and Tapered Fin Shape**   533
S.Wan Muhamad Hatta, N.Soin, S.H.Abdul Rahman, Y.Abdul Wahab, H.Hussin
University of Malaya, MALAYSIA

xxiii

Convergence of Emerging Technologies to Address the Challenges of the 21st Century

Distinguished Prof. Dr. Asad M. Madni
Electrical Engineering Department,
University of California, Los Angeles,
56-125B Engineering IV Building
Box 951594
Los Angeles, CA 90095-1594, USA
Email: asad.madni@sloan.mit.edu

Abstract

There are numerous "Grand Challenges" facing humanity that will have to be addressed by us as a global society in order to maintain our well-being from the standpoint of quality of life, healthcare, environment, energy needs, manufacturing efficiencies, etc., if we are to continue humanity's trajectory of progress. Traditional technologies based on classical disciplines and thought processes of the past several decades are no longer viable in addressing these challenges, and a new approach based on interdisciplinary thinking is necessary. Fortunately, numerous emerging technologies are advancing at an unimaginable rate and it is the convergence of these technologies that demonstrate the potential to have a major impact on our lives, businesses, government, society and our planet. These emerging technologies are establishing the basis for a new paradigm in the development and commercialization of next generation intelligent, miniaturized, highly robust complex systems. This lecture will address some of these major technologies and their applications including, intelligent sensors and wireless sensor networks, intelligent cars and smart highways, tele-health (wireless healthcare), micro-electromechanical systems (MEMS), nanotechnology, clean technology, robotics and automation, smart grid, and ultra high throughput and wide bandwidth instrumentation.

There is Plenty of Room at the Silicon

Prof. Dr. Young June Park
Physical Electronics Laboratory, School of Electrical Engineering
Seoul National University
Shinlim Dong, Kwanak-Gu, Seoul 151-744, Korea
Email: ypark@snu.ac.kr

Abstract

As the silicon technology is scaled down to sub. 10nm and most advanced chips are fabricated mostly in the industry, silicon related researches draw less attention from young researchers. Rather, academia researches on electronics are shifted to exploring new materials such as graphene, 2D metal oxide and carbon nanotube.

However, there is plenty of room at the Silicon. The 'plenty of room' implies more than the 'more than Moore' paradigm, where the silicon chip technology is expanded to applications such as photonics, bio sensing, MEMS, etc. Rather, the 'plenty of room' paradigm implies that the operational principles of silicon devices are applied (rather than simple integration) to expanding the capability of the 'more than Moore' and even 'beyond the Moore' paradigm.

As examples of the new paradigm, we present three examples developed at Seoul National University. Firstly, the electron tunneling phenomenon from silicon to gate through gate insulator is used as means of injecting electrons to water solution(water solution is gate in this case). In this way, well controlled electrons in terms of numbers and energies can be supplied to water to understand and control the surface chemistry at the water-solid interface [1].

Secondly, well known electronic feedback taken place in the bipolar transistor under the BVceo condition is applied to the UVLED based spectroscopy. With an electron detouring from the PD(Photo detector diode) to the photo generation at the PD junction by way of LED, the 'bipolar like snap back' in the current voltage characteristics of the PD can be obtained by the impact ionization feedback. As the BVceo is critically sensitive to the transistor α, the PD snap back is sensitive to the detouring path, where the light from LED is absorbed by the molecules before arriving at the PD junction. Using the system, orders of magnitude enhancement in the sensitivity has been obtained for the detection of the (contaminating) molecules in the water.

In the third example, the pulse technique is applied between the electrical channel and water solution to alleviate the signal degradation from the bio molecules due to the charge screening effect, which is similar to the technique used in the CCD(Charge Coupled Device) to transfer the electrons from the optical excitation[2]. We are applying the pulse technique to sensing the DNA as the biomarker of the Denge virus in collaboration with the IMEN(Institute of Microengineering and Nanoelectronics of Malaysia.

With the experiences, we propose that the 'plenty of room at the silicon' paradigm to maximally utilize the benefit of the silicon technology.

Realization of GaN-Based Technology for High Power and High Frequency Applications

Prof. Dr. Edward Yi Chang
408, Microelectron and Information Research Center,
Department of Materials Science and Engineering,
National Chiao Tung University
1001 Ta-Hsueh Rd., Hsinchu,
Taiwan, R.O.C.
E-mail: edc@mail.nctu.edu.tw

Abstract

Wide band gap semiconductor of GaN and its related materials are promising for future power and high frequency applications. In particular, the GaN high electron mobility transistor (HEMT) grown on large-size Si substrate is suitable for low-lost and high power switching applications. The GaN HEMT could be fabricated into convertors and invertors for electrified vehicle (EV). In order to achieve GaN HEMT device with high efficiency, various issues have to be considered. These include the careful design of material structure and device layout. Furthermore, surface passivation techniques are critical for reducing dynamic on-resistance (R_{on}) and improving reliability. For safety purpose, a normally-off device is required. Thus, the pros and cons of normally-off device fabrication approaches such as gate-recessed, p-GaN cap and F-plasma treatment will be discussed. The possibility of using fully-copper-based metallization will also be addressed.The copper metallization can reduce the fabrication cost effectively by replacing the conventional gold metallization. Finally, power module is demonstrated by employing the GaN HEMTs and Schottky barrier diodes.

For future RF power application, GaN HEMTs on SiC substrate are fabricated. The GaN material grown on SiC can achieve better crystal quality andthe HEMT devices arealso beneficial from better thermal dissipation due to high thermal conductivity SiC substrate. GaN HEMT on SiC could be used in future high frequency applications such as formilitary phased array radar and civilian 4th-generation base station. Besides the AlGaN/GaN HEMT structure, new material structures such as InAlN/GaN and AlN/GaN are also demonstrated. These structures have great potential for very high frequency (>300 GHz) and high power applications.

Amorphous Oxide Electronics

Prof. Dr. Arokia Nathan
Electrical Engineering Division, Cambridge University
9 JJ Thomson Avenue, Cambridge CB3 0FA, United Kingdom
Email: an299@cam.ac.uk

Abstract

Oxide semiconductors are known for their optical transparency and high electron mobility even when processed at room temperature, making them a promising candidate for the next-generation thin film transistor (TFT) technology. Compared to existing well-established TFT technologies, the oxide transistor shows superiority in terms of process simplicity and cost, and stable device behaviour in the dark. While its non-uniformity over large areas is comparable to that of thin film silicon transistors, its photo-instability at low wavelengths can be an issue due to persistence in photoconductivity. This talk will discuss progress and issues related to oxide transistors for large area applications, and in particular, show how the material can be tuned for displays and imaging applications.

Acoustic Metamaterials and Phononic Crystals: Towards the Total Control of the Wave Propagation

Prof. Dr. Abdelkrim Khelif

CNRS French National Centre for Scientific Research (FEMTO-ST),
UMR CNRS 6174, 32 Avenue de l'Observatoire,
Besancon, France
Email : abdelkrim.khelif@femto-st.fr

Abstract

Classical waves, including elastic waves (acoustic waves) and electromagnetic waves (optical waves and microwaves), are described by conventional wave-propagation functions. Elastic waves were the first waveforms to be understood in condensed matter and have a wide range of applications from industry to defense, from healthcare to entertainment. In 1987, the photonic crystal was proposed to describe the propagation of optical waves in refraction index-modulated periodic structures analogous to the propagation of electrons in real crystals. This situation recalls the classical work by Brillouin. Brillouin considered elastic waves in periodic strings, electromagnetic waves in electrical circuits, and electrons in crystals as a system, resulting in some important common concepts, such as the Brillouin zone, band gap, etc., which are generally shared by the various forms of waves: electrons as scalar waves, optical waves as vector waves, and elastic waves as tensor waves. Following on from photonic crystals, the concept of phononic crystals was conceived with elastic waves propagating in periodic structures modulated with periodic elastic moduli and mass densities. These artificially structured materials possess a number of important properties, such as band gaps, band edge states[7], and the ability to slow the velocity of sound (slow wave effect). Furthermore, by creating artificially designed structures on a deep subwavelength scale, artificial acoustic 'atoms' can be purposely engineered into acoustic metamaterials to dramatically change the excitation and propagation of acoustic waves, and thus give rise to subdiffraction-limited resolution and its related myriad novel effects, such as negative, negative elastic modulus, and negative mass density.

Finally, Acoustic metamaterials and phononicc crystal are a newly emerging field, which have inherently abnormal and interesting physical effects that are important to basic research, and offer potential for applications in everyday life that might revolutionize acoustic materials.

SOI Photodiode with Surface Plasmon Antenna: From Sensitivity Enhancement to Refractive Index Measurement for Biosensing

Prof. Dr. Hiroshi Inokawa
Research Institute of Electronics
Shizuoka University,
3-5-1 Johoku, Naka-ku, Hamamatsu, 432-8011 Japan

Abstract

In order to enhance the performance and functionality of the pn-junction photodiode in silicon on insulator (SOI), metallic line-and-space (L/S) surface plasmon (SP) antenna is introduced. In case of the 100-nm-thick silicon, external quantum efficiency (QE) of the photodiode can be increased by nearly an order of magnitude. The antenna shows wavelength and polarization selectivities, and incident angle dependence, and the peak wavelength and the polarization angle can be tailored only by changing the layout design. This is especially beneficial for integrating a variety of photodiodes with different characteristics in a single chip.

Wavelengths of the QE peaks for various L/S pitches and light incident angles were examined in detail, and it was suggested that the coupling between the diffracted light from the L/S and the propagation modes in the SOI slab waveguide caused the QE enhancement.

Related to this operation mechanism, it was found that the photodiode could detect the refractive index of the medium around the SP antenna as the shift of the peak wavelength when the incident light was tilted. This may lead to the fluorescence-label-free biosensing, featuring simple optics and high throughput attained by the parallelism with a large number of integrated photodiodes.

Consequently, a wider application of this photodiode can be expected along with the evolution of the SOI-based large-scale integrated circuits (LSIs).

Numerical estimation of self-sputtering effect in ionized physical vapor deposition system

Nafarizal Nayan, Jais Lias, Mohd Zainizan Sahdan,
Mohd Khairul Ahmad, Lim Huey Sia, Low Jia Wei,
Microelectronic and Nanotechnology - Shamsuddin Researh
Centre (MiNT-SRC)
Universiti Tun Hussein Onn Malaysia
Batu Pahat, Johor, Malaysia
nafa@uthm.edu.my

Ahmad Shuhaimi Abu Bakar
Low Dimensional Materials Research Centre
Department of Physics, Faculty of Science,
Universiti Malaya, Kuala Lumpur, Malaysia
shuhaimi@um.edu.my

Mohamad Rusop
Centre for Nano-Electronic Technology
Faculty of Electrical Engineering
Universiti Teknologi MARA, Shah Alam, Malaysia
nanouitm@gmail.com

Abstract—**During the fabrication of ultra large scale integrated (ULSI) circuits, Ti and TiN thin films are used as diffusion seed and barrier layers in Cu metal contacts. They are often deposited using magnetron sputtering technique where energetic ions bombard the target surface to release the target material. In ionized physical vapor deposition (IPVD) system, the sputtered atoms are ionized in the plasma and thus accelerated into narrow trenches for Ti and TiN thin film fabrication. In IPVD, the density of ionized sputter Ti atom and Ar discharge gas may be at the same range. Therefore, the self-sputtering effect from ionized sputter Ti atoms is important. In the present work, the sputtering yields of Ti and TiN target materials with 100-800 eV at normal incident Ar and Ti ions are calculated. In addition, the influence of N ions in TiN sputtering is also considered. The simulation results are calculated from TRIM, which is a vectorized Monte Carlo code simulation of ion-surface interaction using a binary collision mode. The depth phenomenon of sputtered target incident is also discussed.**

Keywords—sputtering; thin films; IPVD; titanium; TRIM simulation

I. INTRODUCTION

Magnetron sputtering deposition has been widely used in the formation of thin film[1]. Magnetron sputtering process is a process where ions bombard the solid surface to remove the target material for deposition. A permanent magnet is located at the back of sputter target to concentrate the discharge ion in certain region near the target surface. It will thus increase the sputtering rate of target material. This physical collision process involved has been studied for long time. One of the most important data in sputtering process is sputtering yield. The sputtering yield of particular materials are of interest for

practical applications in plasma etching and plasma sputtering devices[2], [3].

In recent ultra-large-scale-integrated (ULSI) circuits, thin film diffusion and barrier layers are deposited to prevent the contact and intermixing between two reactive materials of interconnections. Especially in Cu interconnect material, where Cu is known to have high diffusivity in Si and SiO_2. In order to prevent this phenomenon, Ti and TiN are used as seed and barrier layer for Cu interconnect. Typically, Ti and TiN are deposited into trenches of interconnect using magnetron sputtering deposition system. However, due to the decreasing of device feature size, conventional magnetron sputtering deposition was not capable to deposit Ti and TiN into narrow trenches. An ionized physical vapor deposition (IPVD) was developed to overcome this problem[4], [5]. In IPVD system, sputtered atoms are ionized in the gas phase of plasma and are accelerated toward the substrate by the sheath potential between the plasma and the substrate. This results in a directional deposition and conformal formation of seed and barrier layers[6], [7]. In our previous work, we have evaluated the spatial distribution of Ti^+ and Ar^+ ion densities in magnetron sputtering plasma system. We have shown that the Ti^+ ions density was almost the same with Ar^+ ions density in IPVD system[8]. Therefore, the Ti^+ and Ar^+ ions are exist in the gas phase of plasma at the same time and the Ti^+ ions may also bombard Ti target at the same energy of Ar^+ ions. This condition is known as self-sputtering effect. This paper will discuss the numerical simulation of the sputtering yield of Ti from Ti and TiN sputter target by Ar^+, Ti^+ and N^+ ions at normal incidence and low energies. It will help in understanding the physical processes involved in IPVD system and may be applied for etching and focus ion beam system applications.

II. SIMULATION METHOD

TRIM (transport of ions in matter) is a free software developed by J. F. Zeigler[9]. It is a Monte Carlo computer program, which simulates the quantum mechanical treatment of ion-target collision using a binary collision mode and calculates the physical sputtering, reflection, energy deposition, target damage, and three-dimensional trajectory of energetic particles[3], [10], [11]. This program is vectorized, which means that instead calculating the sputtering yield of one material in mixed compound, it could calculate many sputtering yield of mixed compound in parallel. As mentioned, it also support simulation of compound targets which one can alter its stoichiometry. TRIM could accept a compound target up to eight layers of different materials. For current simulation, 75% of Ti was set for TiN target material. This is to compare with the previous work done by Ranjan *et. al.* In the simulations, the number of ion hit were fixed at 10 000 times and was repeated at various energies. The density of Ti and TiN were set at 4.51 gcm^{-3} and 5.22 gcm^{-3}, respectively. Ar$^+$, Ti$^+$ and N$^+$ ions at normal incidence was used to evaluate the sputtering yield of Ti from Ti and TiN target. In addition, TRIM simulation also plot the two-dimensional collision trajectory in target materials due to ion bombardment. The projected range of target depth was set at 10 nm and collision trajectory for Ar$^+$, Ti$^+$ and N$^+$ ions incidence at various energies were observed. Note that the thermal effect that may cause the changes to the final damage of target material is ignored.

III. RESULTS AND DISCUSSIONS

Figure 1 shows the computational result for Ar$^+$ and Ti$^+$ ions bombardment on Ti target at normal incidence angle. The sputtering yield is dependent on incident ion-beam energy. Ar$^+$ ion has slightly higher sputtering yield values compared to that of Ti$^+$ ion. The calculation is confirmed since the sputtering yield of Ti from Ar$^+$ ion was similar with the work reported by Ranjan *et. al.*[3]. However, they did not report the contribution of Ti$^+$ ion sputtering effect. In a conventional laboratory magnetron sputtering system, the target size is at the range of 2~4 inches and their supplied voltage is at the range of 100~400 V. At these sputtering conditions, figure 1 shows that the sputtering yield of Ti from Ti$^+$ ion is only 0.6~0.8 times lower than that of Ar$^+$ ion. Therefore, significant contribution of Ti$^+$ ion sputtering are expected in IPVD system when the density of Ti$^+$ ion in plasma may be at the same range with Ar$^+$ ion. On the other hand, one may sustain the plasma by self-sputtering effect and reduce or switch off the Ar gas flow. The self-sputtering technique will allow an uncontaminated Ti thin film fabrication using Ti$^+$ ions only[12]–[14].

In general, during the deposition of TiN thin film, there are two sputtering techniques available. One is by using a reactive sputtering of Ti solid metal target with the combination of Ar and N$_2$ discharge. The other is by using the TiN compound target in Ar only or Ar+N$_2$ discharge. In both techniques, the production of Ni$^+$ ions are expected[15], [16]. Figure 2 shows the computational result for Ar$^+$, Ti$^+$ and Ni$^+$ ions bombardment on TiN target at normal incidence angle. The

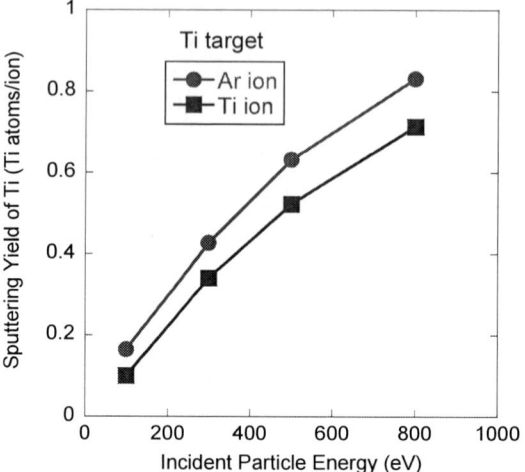

Fig. 1. Sputtering yield of Ti from Ar$^+$ and Ti$^+$ bombardment on Ti target in the range of 100-800 eV at normal incidence angle.

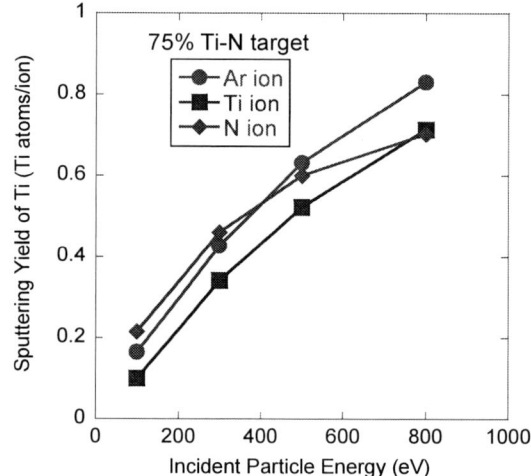

Fig. 2. Sputtering yield of Ti from Ar$^+$, Ti$^+$ and Ni$^+$ bombardment on TiN target in the range of 100-800 eV at normal incidence angle.

sputtering yield of Ti from Ar$^+$ and Ti$^+$ ions bombardment in figure 2 are almost the same with that in figure 1. This may be due to the dominant composition of Ti in TiN target, as we have set it at 75% of Ti in TiN compound target. However, it is very interesting in figure 2 that at bombardment energy below 400 eV, the sputtering yield of Ti from Ni$^+$ ion are higher than that from Ar$^+$ and Ti$^+$ ions. Although in the real sputtering plasma system Ni$^+$ ion density may not as much as Ar$^+$ and Ti$^+$ ions, figure 2 shows an important phenomenon to be taken into account especially in reactive Ar+N$_2$ sputtering plasma.

In order to discuss in more details on the self-sputtering effect in IPVD system, the collision cascade of Ar$^+$, Ti$^+$ and N$^+$ ions into TiN target are shown in figure 3 and 4. Figure 3

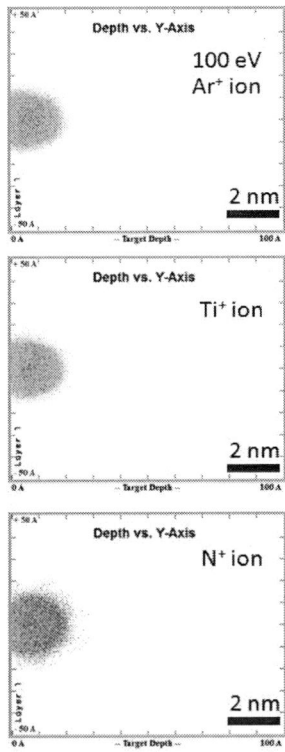

Fig. 3. Collision cascade of Ar⁺, Ti⁺ and N⁺ ions on 75%Ti-N target at 100 eV with normal incidence angle. (in color)

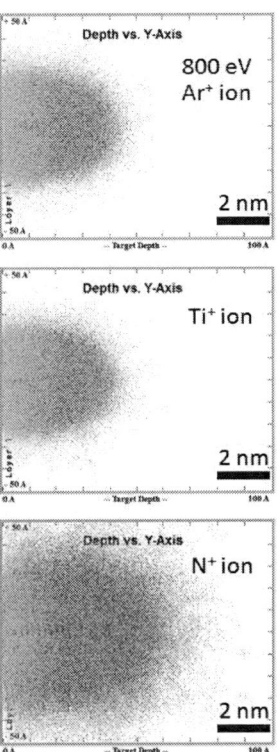

Fig. 4. Collision cascade of Ar⁺, Ti⁺ and N⁺ ions on 75%Ti-N target at 800 eV with normal incidence angle. (in color)

shows the collision cascade of Ar^+, Ti^+ and N^+ ions on TiN target at 100 eV of bombardment energy. Collision cascade in target is useful to understand the target damage. The incident angle ion is at 90° from target surface and its start from the center of left y-axis. The green color refer to target atom, while the red color refer to incident ion (in color). As shown in figure 3, the peak depth of penetration for Ar^+, Ti^+ and N^+ ions on TiN target are approximately the same at 2 nm from the target surface. The penetration of N^+ ion on TiN target is much significant as one can see larger volume of N^+ ion (red color) in figure 3. This is the reason why the sputtering yield of Ti from N^+ ion at low energy was higher than that from Ar^+ and Ti^+ ions in figure 2.

Figure 4 shows the collision cascade of Ar^+, Ti^+ and N^+ ions on TiN target at 800 eV of bombardment energy. As shown in figure 4, the peak depth of penetration for Ar^+ and Ti^+ ions on TiN target are approximately the same at 4 nm from the target surface. On the other hand, the penetration of N^+ ion on TiN target is much deeper at peak 6 nm from the target surface. The radial penetration of N^+ ion is also wider than that of Ar^+ and Ti^+ ions on TiN target. Figure 4 shows that at high energy N^+ ion penetrate the TiN target and implanted inside it without sputtering the Ti atom.

As shown in figure 2, the sputtering yield of Ti by N^+ ion at 800 eV is lower than that of Ar^+ ion. These incorporation of background atoms into the films is often detrimental to the mechanical, electrical and optical properties of the functional thin film coating. Although the simulation results have shown a significant effect of Ti^+ and N^+ ions in sputtering processes, further investigation on the influence of TiN target composition is essential. Higher value of N in TiN target will increased the hardness of TiN target and thus reduce the sputtering yield[17]. In addition, an experimental surface analysis of damaged target is also essential. This studies will be in our further investigations.

IV. CONCLUSIONS

Self-sputtering effect from Ti^+ ions have been successfully investigated using TRIM Monte Carlo simulation. At low incident energy, simulation results showed that the sputtering yield from Ti^+ was comparable to that of Ar^+. Therefore, sputtering plasma could sustain its condition if the density of Ti^+ ion and Ar^+ ion are comparable in plasmas. The so-called self-sputtering technique is an advanced technique that will allow an uncontaminated Ti thin film fabrication using Ti^+ ions only. From the simulation results, this technique is applicable for single metal target and multi compound target.

ACKNOWLEDGMENT

This study was supported by Ministry of Education Malaysia and Universiti Tun Hussein Onn Malaysia under the Fundamental Research Grant Scheme and Malaysian Technical Universities Research Grant Scheme.

REFERENCES

[1] P. Kelly and R. Arnell, "Magnetron sputtering: a review of recent developments and applications," *Vacuum*, vol. 56, no. 3, pp. 159–172, Mar. 2000.

[2] P.M. Bérubé, J.-S. Poirier, J. Margot, L. Stafford, P. F. Ndione, M. Chaker, and R. Morandotti, "Correlation between surface chemistry and ion energy dependence of the etch yield in multicomponent oxides etching," *J. Appl. Phys.*, vol. 106, no. 6, p. 063302, 2009.

[3] R. Ranjan, J. P. Allain, M. R. Hendricks, and D. N. Ruzic, "Absolute sputtering yield of Ti/TiN by Ar$^+$/N$^+$ at 400–700 eV," *J. Vac. Sci. Technol. A Vacuum, Surfaces, Film.*, vol. 19, no. 3, p. 1004, 2001.

[4] S. Rossnagel and J. Hopwooda, "Magnetron sputter deposition with high levels of metal ionization," *Appl. Phys. Lett.*, vol. 63, no. 24, pp. 3285–3287, 1993.

[5] J. T. Gudmundsson, "The high power impulse magnetron sputtering discharge as an ionized physical vapor deposition tool," *Vacuum*, vol. 84, no. 12, pp. 1360–1364, Jun. 2010.

[6] N. Nafarizal, N. Takada, K. Shibagaki, K. Nakamura, Y. Sago, and K. Sasaki, "Two-Dimensional Distributions of Ti and Ti + Densities in High-Pressure Magnetron Sputtering Discharges," *Jpn. J. Appl. Phys.*, vol. 44, no. No. 23, pp. L737–L739, May 2005.

[7] N. Nafarizal, N. Takada, K. Nakamura, Y. Sago, and K. Sasaki, "Deposition profile of Ti film inside a trench and its correlation with gas-phase ionization in high-pressure magnetron sputtering," *J. Vac. Sci. Technol. A Vacuum, Surfaces, Film.*, vol. 24, no. 6, p. 2206, 2006.

[8] N. Nafarizal, N. Takada, and K. Sasaki, "Investigations of Production Processes of Ti + in High-Pressure Magnetron Sputtering Plasmas," *Jpn. J. Appl. Phys.*, vol. 48, no. 12, p. 126003, Dec. 2009.

[9] J. F. Ziegler, M. D. Ziegler, and J. P. Biersack, "SRIM – The stopping and range of ions in matter (2010)," *Nucl. Instruments Methods Phys. Res. Sect. B Beam Interact. with Mater. Atoms*, vol. 268, no. 11–12, pp. 1818–1823, Jun. 2010.

[10] A. Rahmati, "Reactive magnetron sputter deposition of (Ti,Cu)N nano-crystalline thin films: modeling of particle and energy flux toward the substrate," *Phys. Scr.*, vol. 86, no. 2, Aug. 2012.

[11] U. Vogel, C. Klaus, C. Nobis, and J. W. Bartha, "Analysis of the energy input during wire coating from a cylindrical magnetron source," *Thin Solid Films*, vol. 520, no. 20, pp. 6404–6408, Aug. 2012.

[12] O. V Vozniy, D. Duday, A. Lejars, and T. Wirtz, "Double magnetron self-sputtering in HiPIMS discharges," *PLASMA SOURCES Sci. Technol.*, vol. 20, no. 6, Dec. 2011.

[13] A. Wiatrowski, W. M. Posadowski, G. Jóźwiak, J. Serafińczuk, R. Szeloch, and T. Gotszalk, "Standard and self-sustained magnetron sputtering deposited Cu films investigated by means of AFM and XRD," *Microelectron. Reliab.*, vol. 51, no. 7, pp. 1203–1206, Jul. 2011.

[14] H. Wu and A. Anders, "Energetic deposition of metal ions: observation of self-sputtering and limited sticking for off-normal angles of incidence," *J. Phys. D. Appl. Phys.*, vol. 43, no. 6, p. 065206, Feb. 2010.

[15] Y. Liu, L. Li, M. Xu, X. Cai, Q. Chen, Y. Hu, and P. K. Chu, "Effects of nitrogen ion implantation and implantation energy on surface properties and adhesion strength of TiN films deposited on aluminum by magnetron sputtering," *Mater. Sci. Eng. A*, vol. 415, no. 1–2, pp. 140–144, Jan. 2006.

[16] A. P. Ehiasarian, A. Vetushka, Y. A. Gonzalvo, G. Sáfrán, L. Székely, and P. B. Barna, "Influence of high power impulse magnetron sputtering plasma ionization on the microstructure of TiN thin films," *J. Appl. Phys.*, vol. 109, no. 10, p. 104314, 2011.

[17] B. Subramanian, R. Ananthakumar, A. Kobayashi, and M. Jayachandran, "Effect of nitrogen ion implantation on structural and microstructural properties of reactive magnetron sputtered TiN thin films," *Trans. Inst. Met. Finish.*, vol. 89, no. 1, pp. 28–32, Jan. 2011.

Synthesis and Characterization of Carbon Nano Structures on Gallium Phosphate

Aishah Fauthan, Zainab Yunusa, Mohd Nizar
Hamidon,
Institute of Advanced Technology,
Universiti Putra Malaysia, 43400 Serdang, Selangor,
Malaysia

Aishah Fauthan, Burhanuddin Yeop Majlis
Institute of Microengineering and Nanoelectronics
Universiti Kebangsaan Malaysia, 43600, UKM Bangi,
Malaysia
Email: aishahfauthan@gmail.com

Abstract— Carbon nano structures were grown on Gallium Phosphate substrate by using Alcohol Catalytic Chemical Vapor Deposition (ACCVD) method. The aim of this paper is to study the structure and the morphology of the carbon nano structures growth on Gallium Phosphate. Gallium Phosphate is known as piezoelectric materials which are more stable and similar to quartz in its crystal structure. The ACCVD is chosen because of its simplicity and economical method for the growth of carbon nano structure. Mixture of ethanol and Iron Nitrate in a ratio of 1:25 was used as the catalyst to impregnate the carbon nano structures. The carbon nano structures were grown at 800°C. The ethanol liquid which was used as a carbon source was injected into the furnace tube with flow rate of 2.0 ml/min. The furnace was flowed by Argon gasses throughout the experiment. FE-SEM and EDX are used to investigate the morphology of the carbon structure. Finally Raman measurements have been performed and equipped with laser diode emitting at 632nm.

Keywords— Gallium Phosphate; Carbon Nanostructures; Raman Spectrum

I. INTRODUCTION

Gallium Phosphate ($GaPO_4$) is known as piezoelectric materials which has a lot of interesting properties. It is similar to quartz in crystal structure but tends to be more stable even at temperature up to 930°C. It has high electrochemical coupling, existence of temperature-compensated orientation for bulk acoustics waves (BAWs) and SAWs, lower SAW velocity and also high electric resistivity [1, 2]. This makes it a suitable piezoelectric material for Surface Acoustic Waves devices which are suitable for applications in chemical sensors, pressure sensors, biosensors, temperature sensors and mass sensors [2].

The beginning of new area in carbon material science was discovered by Harry Kroto when he found C60 molecules in 1985 [3]. While in 1991, Sumio Iijima continued this research and discovered its special properties and become famous among researcher after he observed the thin needle-like material under TEM. It was proved had a graphite structure [4]. Carbon nanotubes (CNTs) are currently known as one of the strongest and stiffest fiber due to its C-C bond. Therefore, CNTs can be used in many applications such as sensors, energy storage, biotechnology, thin-film electronics, field emission devices and electromagnetic shield [5]. Many researches have been made that the CNTs can be characterized by its strong covalent bonding, a unique structure and nanometer size. It also have special properties including exceptionally high tensile strength, high resilience, electronic properties ranging from metallic to semiconducting, high current carrying capacity, and high thermal conductivity [6].

There are many fabrication processes that have been developed for carbon nano structures. The carbon nano structures can be synthesized by several conventional methods such as: laser ablation, chemical vapor deposition (CVD) and electric arc discharge [7]. Alcohol Chemical Catalyst Vapor Deposition (ACCVD) is the most popular technique because of its high yield, high purity, low cost and flexibility in temperature and hydrocarbon gasses adjustment to control the structure of carbon nano structures produced. Different parameters are considered for the synthesis of CNTs such as temperature, pressure, reaction time and gas flow rate.

In this project, $GaPO_4$ is used as a substrate and ACCVD process for the synthesis of carbon nano structures. The substrate and the process are suitable for high temperature conditions. Iron Nitrate is used as catalyst to impregnate the carbon nanostructures. The surface morphologies, element composition and properties of carbon nanostructures using ACCVD synthesis on the $GaPO_4$ will be discussed thoroughly.

II. MATERIAL AND METHOD

A. Sample Preparation

The carbon nano structures have been growth by ACCVD method. The catalyst was obtained from 0.5 g of Iron (III) Nitrate Nonahydrate, 98+% mixed with 12.5ml ethanol from GmbH chemicals. To achieve the thick mixture, the solution was sonicated for 10 minutes. Then the mixture of the solution kept under room temperature for 1 day. Before using the solution, the catalyst has to be sonicated back for another 5 minutes. Lastly, the solution was drop casted on the Gallium Phosphate and was dried under ultraviolet light for another 10 minutes. During the ACCVD process, 50ml ethanol was injected into the furnace tube. The flow rate was 2.0 ml/min and the furnace was flowed by Argon gasses throughout the experiment.

978-1-4799-5761-3/14 $31.00 © 2014 IEEE

B. Characterization

The characterization of the carbon nano structures have been done using JEOL Field Emission Scanning Electron Microscopy (FE-SEM), MERLIN Energy Dispersive X-ray Analysis (EDX) and Raman Spectroscopy. The morphology of the carbon nanostructures were provided by FE-SEM. It used to investigate the types of carbon nano structures. Followed by EDX technique has been done for identifying the elemental composition of the specimen. Finally Raman measurements have been performed and equipped with laser diode emitting at 632nm.

III. RESULT AND DISCUSSION

A. Surface Morphologies

Fig. 1 shows FESEM images with different magnifications for carbon nano structures on the Gallium Phosphate substrate. FESEM images show coil and straight carbon nano structures of carbon nano structures have been grown. From the previous research, metal catalyst is one of the important factors for the formation of carbon coils [8-12]. In this study, Iron Nitrate is used as the catalyst to impregnate carbon nano structures. This shown that metal-support interactions and concentrations are determined the morphological and structural characteristics of carbon nano structures. It also affects the catalyst dispersion, crystallographic orientation and shape of the carbon nanostructures form.

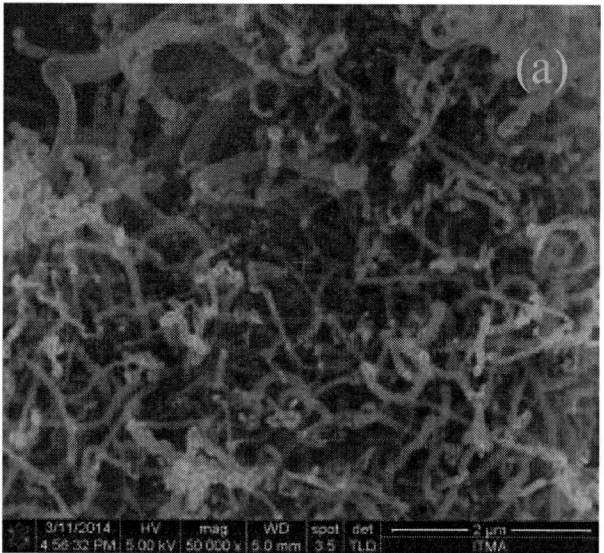

Fig. 1: FESEM image of carbon nano coil and straight like structures with different magnifications (a) 50 000 magnification (b) 150 000 magnification

It is noteworthy that the catalyst coated has to be uniform on the Gallium Phosphate. The thickness growth of carbon nanostructures are based on this uniformity based on the Fig. 2. It can be seen, some of the carbon nano structure with diameter around 60nm – 200nm. The screws pitch some of these nano structures are high compared to the structures with the smaller diameter.

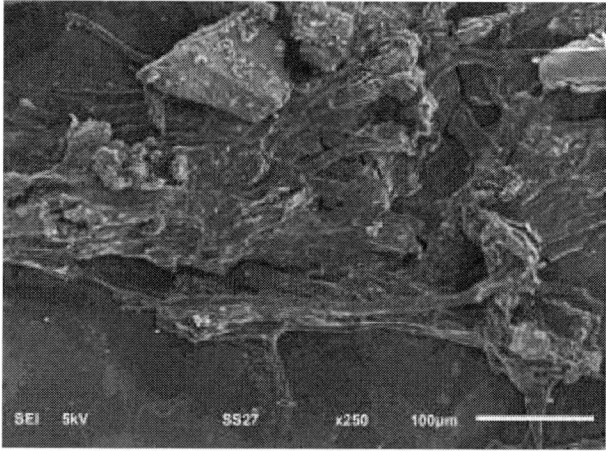

Fig. 2: FESEM image of carbon nano structures scattered based on the catalyst uniformity.

B. Elemental Composition

EDX spectra were done from a bundle of carbon nano structures in order to confirm the features were element of carbon nano structures and Gallium Phosphate. The image for EDX analysis is shown in Fig. 3.

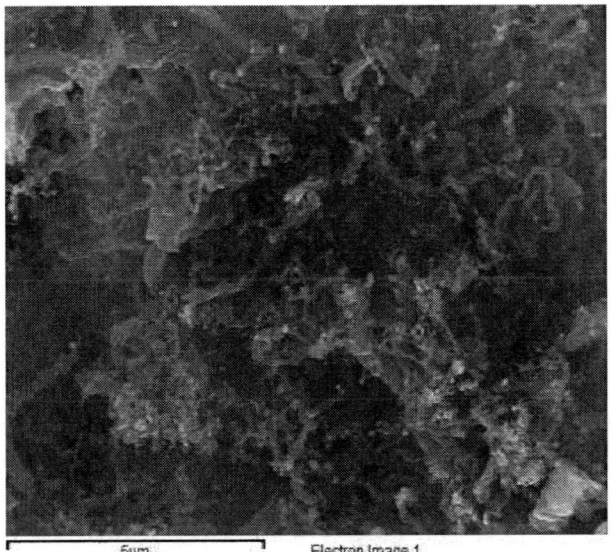

Fig. 3: Carbon nanostructures image for EDX analysis

The spectrum from the data set is shown in Fig. 4. In this figure, a distinct peak due to the spectrum rays of Fe can be seen at about 0.5 keV. In addition, the small spectrum peak of Fe at about 6.4 keV also can be seen. The existing of Fe in this spectrum shows that there are only small amounts of catalyst left throughout this process. The large peak due to the spectrum rays of carbon at about 0.2 keV. This carbon spectrum weight contains 84.77% from the overall element on the substrate. Therefore, these EDX results confirm that carbon nano structures can be growth on the Gallium Phosphate using this synthesis procedure.

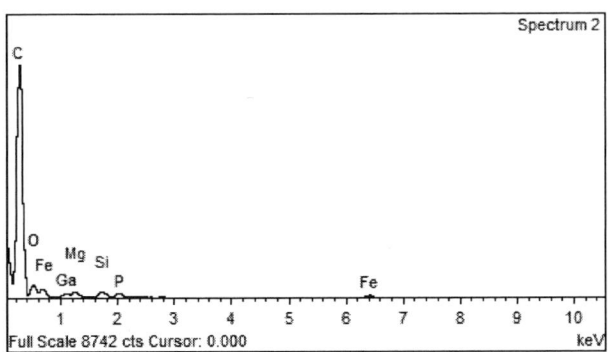

Fig. 4: An energy dispersive x-ray fluorescence spectrum of a carbon nano structures bundle showing a clear signature of Carbon in the sample.

C. Carbon Nanostructures Properties

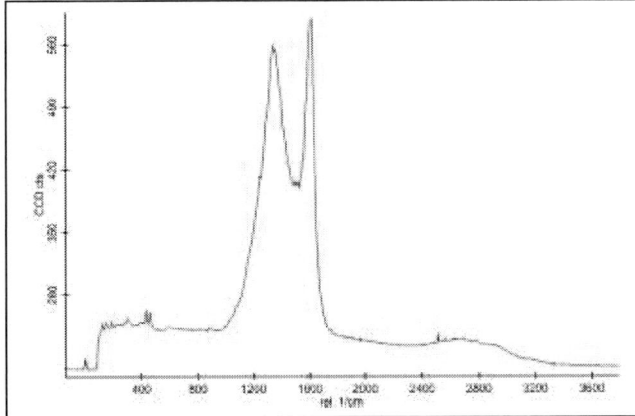

Fig. 5: Raman spectra of a sample of Carbon nano tubes on a Gallium Phosphate substrate

Raman spectroscopy is a powerful tool used for the characterization of carbon nanostructure in order to detect the prominent peaks for the classification of different carbon nanostructures. An area of 150 micron by 150 micron was selected for the Raman analysis. There is a prominent band around 1336cm-1 and 1598cm-1 as shown in Fig. 5. This band is known as the D band and G band. The graphite is composed of sp2 bonded carbon in planar sheet [13]. The D band can be referred as the defect band and its intensity relative to the G band is used as a measure of the quality with the nanotubes [13]. The I_d to I_g ratio was found to be 1.1; this high value could be explained as a result of the densely concentric layers of the carbon nanostructures with very large diameters. The I_d to I_g ratio of 1.1 is very close to the value of 1.02 as obtained by [14]. Usually, the I_d to I_g ratio is used for determination of degree of purity of the Carbon Nanostructures, a higher value greater than 1 indicates a that lots of defective walls are present in the structures [15].

The strongest feature in Carbon Nanotubes Raman spectra is the lower frequency radial breathing mode (RBM). The RBM is appearing only in carbon nanotubes and is a unique phonon mode [15]. Therefore, from the observation with the RBM in the Raman spectrum in Fig. 5 proved direct evidence that a sample contains carbon nanotubes.

IV. CONCLUSION

In this study, the Carbon nanostructures are growth on the Gallium Phosphate. As mention in the introduction, Gallium Phosphate is known as the piezoelectric material and suitable for the BAWs and SAWs fabrication. The method of the synthesis used can be done to growth the carbon nanostructure based on the characterization shown.

For the conclusion, the synthesized carbon nano structures using alcohol catalytic chemical deposition of ethanol and Iron Nitrate on the Gallium Phosphate is successful. The diameter of the carbon nano structures is in a range of 60nm – 200nm. The catalyst has to be coated uniformly on the substrate to get the better thickness of carbon nanostructures. The growth of carbon nano structures is successful in the present of ethanol and Iron Nitrate throughout the process. FESEM, EDX and Raman characterization indicates that carbon nanotube also growth in this process.

ACKNOWLEDGMENT

Authors are thankful for the grant provided by Ministry of Science, Technology and Innovation (MOSTI) Malaysia. My thanks also for supervisors, lecturers, technicians and all that involve in this paper works for helpful advice and discussions, and also Institute of Microengineering and Nanoelectronics (IMEN), UKM for providing the facilities.

REFERENCES

[1] P. Krempl, G. Schleinzer, W. Wallnöfer. Gallium phosphate, GaPO$_4$: A New Piezoelectric Crystal Material for High-Temperature Sensories. Sensors and Actuators A61 (1997), 361-363.

[2] M.N. Hamidon, V. Skarda, N.M. White, F. Krispel, P. Krempl, M. Binhack, W. Buff. Fabrication of High Temperature Surface Acoustic Wave Devices for Sensor Applications, Sensors and Actuators A 123–124 (2005) 403–407.

[3] Kroto, H.W., J.R. Heath, S.C. O'Brien, R.F. Curl, R.E. Smalley. C60: Buckminsterfullerene. Nature (1985), 318:162-163.

[4] Iijima, S., Helical Microtubules of Graphitic Carbon. Nature (1991), 354:56-58.

[5] Michael F. L. De Volder, Sameh H. Tawfick, Ray H. Baughman, A. John Hart. Carbon Nanotubes: Present and Future Commercial Applications. Science Magazines, Vol. 339 No. 6119, pp. 535-539.

[6] Ali Nasir Imtani and V. K. Jindal. Bond Lengths of Single-Walled Carbon Nanotubes. arxiv.org/pdf/cond-mat/0611484 (10 April 2014).

[7] Motoo Yumura. Synthesis and Purification of Multi-Walled Carbon Nanotubes. The Science and Technology of Carbon Nanotubes (1999), 2-13.

[8] N. M. Rodriguez, "A review of catalytically grown carbon nanofibers," Journal of Materials Research, vol. 8, no. 12, pp. 3233–3250, 1993.

[9] M. Kawaguchi, K. Nozaki, S. Motojima, and H. Iwanaga, "A growth mechanism of regularly coiled carbon fibers through acetylene pyrolysis," Journal of Crystal Growth, vol. 118, no. 3-4, pp. 309–313, 1992. View at Scopus

[10] Q. Zhang, L. Yu, and Z. Cui, "Effects of the size of nano-copper catalysts and reaction temperature on the morphology of carbon fibers," Materials Research Bulletin, vol. 43, no. 3, pp. 735–742, 2008. View at Publisher · View at Google Scholar · View at Scopus

[11] S. Hokushin, L. Pan, and Y. Nakayama, "Diameter control of carbon nanocoils by the catalyst of organic metals," Japanese Journal of Applied Physics, Part 1, vol. 46, no. 8, pp. 5383–5385, 2007. View at Publisher · View at Google Scholar · View at Scopus

[12] N. Tang, J. Wen, Y. Zhang, F. Liu, K. Lin, and Y. Du, "Helical carbon nanotubes: catalytic particle size-dependent growth and magnetic properties," ACS Nano, vol. 4, no. 1, pp. 241–250, 2010. View at Publisher · View at Google Scholar · View at Scopus

[13] Joe Hodkiewicz. Characterizing Carbon Materials with Raman Spectroscopy. Thermo Fisher Scientific, (Application Note: 51901).

[14] Zainab Yunusa, Mohd Nizar Hamidon, Suraya Abdul Rashid "Growth of Multiwalled Carbon Nanotubes on Platinum IEEE RSM 2013 Conference

[15] L. Bokobza "Raman spectroscopic characterization of multiwall carbon nanotubes and its composites" Express Polymer Letters, vol 6, no 7, pp 601-608, May 2012.

Contactor Characterization Methodology on Pin Inductance

Ling Li Ong
Faculty of Engineering and Technology
Multimedia University
Melaka, Malaysia
onglingli@gmail.com

Chan Yee Kit, You Ah Heng
Faculty of Engineering and Technology
Multimedia University
Melaka, Malaysia

Abstract—**Contactor pin is commonly used as the test tooling to enable million times of repeatability functional testing in manufacturing. Simulation should be performed before contactor sent for prototyping. The objective is to optimize the design according to device performance and avoid design faulty which may lead to higher testing cost. This paper introduces the methodology on contactor pin characterization specifically the pin inductance that are helpful for high speed tooling engineer to access their contactor's performances. The impact of the contactor pin design, topology and contactor housing design will be explained in this paper. Indirectly, this is to emphasize the importance of assessing the contactor by considering many design factors instead of assessing in ideal way. It is widely applicable to any electronic devices testing that need contactor.**

Keywords—contactor; characterization; pogo pin; inductance;

I. Introduction

Modern technology has getting more robust. The design specification on the devices becomes more stringent on the frequency and package size. Manufacturing test tooling must be designed to support the new device specification so the device performance can be correctly tested [1]. This paper discusses the test contactor, the important part of test tooling.

Test contactor provides signal connection between package and the printed circuit board. It consists of three major components that are pin, housing body and clamp. Contactor pin is the main conductor to enable the signal transmission. There are many contactor characterization methodologies implemented by different suppliers. This has created the variety of data on the same contactor [2]. To ensure the contactor characterization methodology is reliable, it involves the understanding of the device specification and the design. For instance, the frequency of the device, package type, pin topology and distance between pins. This paper explains the impact of some parameters that may influence the inductance of the contactor pin through the 3D modeling. The customized contactor design will also be discussed.

II. Methodology of Contactor Pin Characterization

A. Basic understanding of contactor pin modeling

There are many types of contactor pins such as pogo pin and stamp pin available in the industrial to meet the different application. Pogo pin is the most commonly used in the manufacturing testing due to the simplicity of design and higher reliability. This type of contactor pin provides perfect conductivity that is typically coated with material Beryllium Copper (BeCu) [3]. The structure of pogo pin can be divided into three major parts that are top plunger, barrel and bottom plunger. Spring is built inside the barrel for the compression purposes. However, spring can be eliminated during modeling due to the complexity of spiral shape and it does not contribute a significant impact to the overall inductance. This phenomenon is known as Skin Effect and can be illustrated with Figure 1(a). Figure 1(b) is the 3-Dimension model of pogo pin used in this analysis.

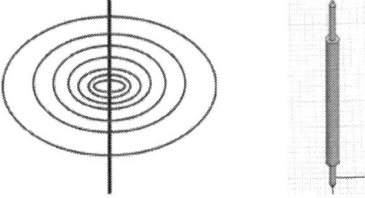

Fig. 1(a).[Left] The magnectic field line ring generated when current flows through conductor. Fig. 1(b). [Right] 3D modeling of pogo pin using ANSYS Q3D tools

Figure 1(a) shows the magnetic field line ring generated when current flows through the conductor. The farther from the surface of the current, the lesser the magnetic field line ring that also indicates the lower resistance. This explains the signal propagation in the outer surface of the conductor in high frequency. Similarly, skin effect explains the contactor pin modeling only concern on barrel dimension during high speed transmission. Most of the signal flows through the barrel from top tip to the bottom tip of a pogo pin instead of flowing internally through spring.

978-1-4799-5761-3/14 $31.00 © 2014 IEEE

B. Contactor Pin Modeling

This section discusses the construction of a basic contactor models and the extrapolation into more complex contactor model in next section. This intends to analyze the external factor that may impact the mutual inductance and lead to the change of total inductance. The factors include the distance between pin, number of ground pin populated and the pin topology.

Basically, contactor pin can be divided into 3 major parts: Top plunger, barrel and bottom plunger. Pin tip has very minimal impact to the overall inductance and therefore, it can be simplified to a flat surface. The pin length, pin dimension and the air gap between pin and housing body are usually displayed in the mechanical drawing. When model the contactor pin, it should be modeled on the compressed length to reflect the actual electrical path.

The simplest model must consume at least one signal pin and one ground pin to form a complete signal path from source to ground. Figure 2(a) is the sample 3D contactor pin model. Ground plane serves as the return path between signal and ground. Otherwise, channel will be open-ended. To better correlate with the actual contactor model, build the housing to enclose the contactor pin, something in the Figure 2(b). As the contactor housing is the solid block and may overlap with the pin, relief to "drill" the 3D hole is necessary to isolate the contactor pin from housing. After all, the 3D construction is completed.

ANSYS Q3D [4] is the common software tools used in 2D and 3D electromagnetic field simulation for parasitic extraction. It is based on the method of moment (integral equation) and FEMs to compute the capacitive, conductance, inductance and resistance matrices. The accuracy of this tool has been verified in many technical papers.

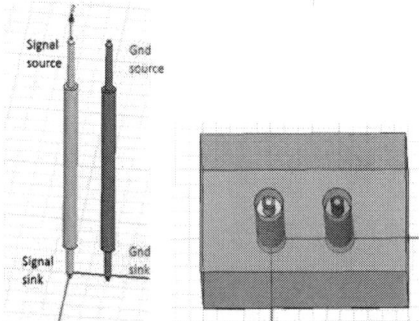

Fig. 2.[Left] Basic 2-pins model. [Right]Top view of basic contactor model with housing turned on the visibility

III. Result and Discussion

This section demonstrates the factors which can influence the total loop inductance of contactor. It is beneficial to understand the mutual coupling between pins and how it changes by pin population, pin pitch and pin count.

A. Number of Ground pin versus Loop Inductance

The intention of this study is to simulate the response of signal integrity by adding the ground pin. For the industrial design, adding more ground pins lead to higher design cost. Somehow, certain high speed transmission design required more ground pins to optimize the channel stability. To achieve optimum number of ground without over designing and impose additional cost, pin modeling provides basic guidance to study the required ground pin.

Generally, ground pin in the contactor design provides the return path from the signal to ground. When the number of ground pin increases, it means signal provides more channels returning to ground. To relate with the loop inductance, the increase of ground pin divides the signal to return at different ground pin. One signal pin is being designed to reduce the complexity and number of ground pin is to increase the iteration until the value of inductance becomes saturated.

Figure 3 illustrates the modeling topology where the ground pin is extrapolates in a row from one pin to nine pins in a consistent distance from signal pin horizontally. Figure 4 summarizes the loop inductance process out by Q3D and the graph is plotted to explain the relationship between loop inductance responded to the ground pins. It illustrates when the number of ground pin increases, loop inductance reduced exponentially and became saturated after seven ground pins being added. Nevertheless, the loop inductance has shown less significant improvement on the loop inductance after the 4th pins. This hint the engineer by adding 4 ground pin in this design, loop inductance improved approximately 27% compared to 1 ground pin. This simple analysis is useful for engineer to design the high speed test board and intend to reduce the transmission loss with optimum ground pins. This helps in cost reduction at the same time.

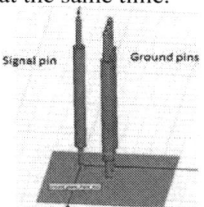

Fig. 3.One signal pin to 3 ground pins topology

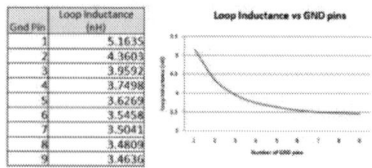

Gnd Pin	Loop Inductance (nH)
1	5.1635
2	4.3603
3	3.9592
4	3.7498
5	3.6269
6	3.5458
7	3.5041
8	3.4809
9	3.4636

Fig. 4.Summary table and graph to show the relationship of signal to multiple ground pins

978-1-4799-5761-3/14 $31.00 © 2014 IEEE

B. Pitch versus Loop Inductance

This section demonstrates the response of inductance to the pitch sweeping. Modeling setup is as simple as only two pins in a contactor block, that are one ground pin and one signal pin. Similar to the earlier model, shorting plane designed at the bottom to provide a complete signal path between signal and ground pin. Pitch can be explained as the distance between two pins counting from the center of the pin. It can be illustrated from the Figure 5.

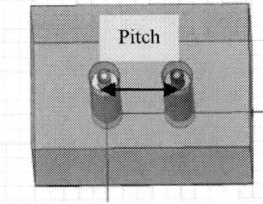

Fig. 5.Contactor block with pitch illustration

Theoretically, loop inductance is the proper term used to describe the total inductance of the contactor model since the complete path of signal being routed. Loop inductance is formed by the matrix of self-inductance and mutual inductance. For this model setup, loop inductance can be explained as the total number of field line rings around the signal pin minus the mutual-field line rings from the ground pin or vice versa because signal flows in opposite directions. Equation (1) express:

$$Lloop = \begin{bmatrix} Lself11 & Lmutual12 \\ Lmutual21 & Lself22 \end{bmatrix}, \text{thus}$$
$$= [L11 - (L12 + L21) + L22] \quad (1)$$

Where
Lself11 = self-inductance of the Pin1
Lself22 = self-inductance of the Pin2
Lmutual12/Lmutual21 = Mutual inductance between Pin1 and Pin2

Note that the Pin1 can be the signal pin and Pin2 can be the ground pin or vice versa. This can be explained as the identical pin usually applied on signal and ground pins unless special application which required unique ground pin design to accommodate the impedance control perspective.

Figure 6 is the result extracted from Q3D modeling. The printed result is somehow matches to the loop inductance matrix. That are Lself11 = 4.4328nH, Lself22 = 4.4398nH, Lmutual12 = Lmutual21 = 1.9903nH. Note that the Lself11 is close to Lself22 when same pin design being referred. Lmutual12 is same with the Lmutual21 because of identical number of magnetic-field line rings surroundings between Pin1 and Pin2. Self-inductance is not impacted by pitch. The magnetic field line influences the mutual inductance when pitch reduces because the more overlapping of magnetic line. Therefore, when the distance reduces, self inductance remains unchanged but mutual inductance increases, this explains the reason of pitch is inversely proportional to the loop inductance.

	GND_pin:Source2	Signal_pin:Source1
Freq: 1 (GHz)		
GND_pin:Source2	4.4328	1.9903
Signal_pin:Source1	1.9903	4.4398

Fig. 6.Extraction of self inductance and mutual inductance in Q3D tools

The intention of sweeping pitch is to demonstrate the trend of graph by pitch change. The analysis swept from 1mm pitch down to 0.4mm pitch. These numbers of pitch are chosen based on the popularity of the high speed devices especially the SOC (System on Chip) with higher pin count. Contactor designed for pitch smaller than 0.3mm is still not commonly used in industrial

From the trend of graph in Figure 7, the loop inductance reduced non-linearly with the pitch reduced. The closer the distance between 2 pins means the higher of mutual inductance between them while self inductance of signal and ground pins are always consistent. This is correlated with equation (1), the smaller the pitch, the lower of the loop inductance. The loop inductance will continuously reduce until no gap in between signal and ground pins. However, it will not be realistic when pitch is smaller than 0.2mm. It is beyond the latest capability of current technology to manufacture such a small pogo pin for the design lesser than pitch 0.2mm. Similar analysis is applicable on any new contactor pin design to guesstimate the loop inductance at the design pitch.

TABLE I.Result summary of loop inductance versus pitch

Pitch (mm)	Loop Inductance (nH)
1.0	4.36
0.9	4.12
0.8	3.87
0.7	3.57
0.6	3.26
0.5	2.90
0.4	2.50

Loop Inductance vs Pitch

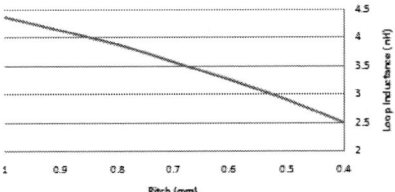

Fig. 7.Graph of Loop Inductance Versus Pitch with Data from TABLE.I

C. Pin Topology

Pin topology or pin map on the device impacts the system performance. It is advisable to characterize the contactor pin with actual pin map from the device that going to interconnect with this contactor. The intension is to reflect the real performance of this contactor used and avoid too optimistic characterization result [5].

Basically, pin topology determines the mutual coupling between pins. When many signal pins populated at the same location, mutual inductance will be relatively higher and increase the signal interference. Likewise, when populate some ground pins surrounding the signal pins, the mutual inductance will be significantly dropping.

This section demonstrates the different scenario of mutual inductance changed by pin topology. It has some similarity to the Section A by discussing the interaction between signals to ground pins. However, model constructed in this section are in block 3x3 matrixes to make closer to the industrial devices especially the BGA packages. Three models simulated for the case study in Figure 8.

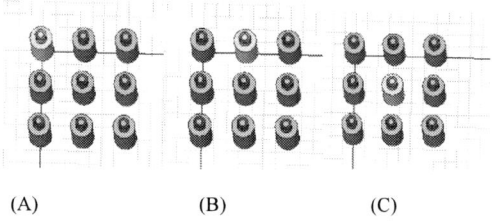

(A)   (B)   (C)

Fig. 8. The diagram of 3x3 matrix of contactor pin models with signal pin designed located at different pin location . Yellow pin represents is the signal pin and brown pin represents the ground pin.

TABLE II. Result summary of simulation in Figure10

Topology	Loop Inductance (nH)
Model A (corner)	2.26
Model B (edge)	1.89
Model C (center)	1.71

The TABLE II captures the loop inductance from three different models. Signal pin of the Model A located at the top right corner, this is the worst case scenario among the three of them. The electromagnetic lines of the signal pin in Model A interfere directly with only three ground pins. This also indicates the majority current from signal return to ground through these three ground pins. By looking at Model B, signal pin located at the edge and the loop inductance is slightly lower. Current from signal pin in Model B has better return path as it interferes directly with 5 ground pins. Best case scenario demonstrated in Model C has the lowest loop inductance. The signal pin located in the middle and surrounded by an island of ground pins. Current in Model C returns to ground through multiple ground pins and hence,

magnetic lines interference between signal pin to ground pin reduces total loop inductance.

Loop inductance impacts the system performance by the ground bounce effect. Ground bounce is the voltage drops between signals to ground due to changing of current in loop. It is always the goal to lower the loop inductance and decrease of risk of ground bounce in system that may lead to the switching noise and EMI. From Equation (1), two main items derive the loop inductance are mutual inductance and self-inductance. Mutual inductance depends on the pitch between 2 pins as explained in Section B, but when the pitch becomes the constant based on design of device, self-inductance is another entity that can be tweaked. Decrease the self-inductance for this contactor study can be increasing the number of return path, just like populating more ground pin surrounding the signal pin.

Because of this reason, it is advisable to study the pin map of the actual device, highlight the worst case corner where lesser ground pins are populated to be the characterization topology. This can be best representing the performance of the contactor when applying in system testing.

IV. Summary

This paper explains the methodology of contactor pin characterization particularly the pogo pin that is commonly used in the manufacturing testing. Mutual coupling of the contactor relies on the few factors, such as number of ground pin populated, pitch between pin and pin topology. Understanding the importance of these factors can accurately predict the performance of pin when applied in the real manufacturing. To avoid of getting the surprise when using the contactor in the final manufacturing testing, it is strongly encouraged to characterize the contactor according to actual device. Over optimistic methodology used to characterize the contactor may contribute to the unwanted channel distortion later.

References

[1] Bahadir Tunaboylu, "Electrical Characterization of Test Socket with Novel Contactors", IEEE Trans. on Device and Materials Reliability, vol. 14, no.1, pp.580-582, 2012

[2] Eric Bogatic, Signal and Power Integrity – Simplified, 2nd Edition, 2010

[3] David G. Figueroa, Chee Yee Chung, "High Performance Socket Characterization Technique for Microproccessors," IEEE Electrical Performance of Electronic Packaging, pp. 125 – 128, 1999

[4] ANSYS Q3D, http://www.ansys.com/Products/Simulation+Technology/Electromagnetics/Signal+Integrity/ANSYS+Q3D+Extractor

[5] J.Zhou and J.Diller, "Are spring contact probes valid for fine pitch", in Proc. BiTS Workshop, pp. 1-8, March 2012, [Online]. Available: http://www.bitsworkshop.org/archive/archive2012/2012s5.pdf

[6] Ruey-Bo Sun, Ruey-Beei Wu, "Compromise Impedance Match Design for Pogo Pins with Different Single-Ended and Differential Signal-Ground Patterns," IEEE Trans. on Advanced Packaging, vol. 33, no.4, pp 953-960, Nov 2010

Two-Stage Small-Signal Amplifier with Darlington and Sziklai pairs

SachchidaNand Shukla, Beena Pandey

Department of Physics and Electronics, Dr. Ram Manohar Lohia Avadh University
Faizabad - 224001, U.P., India
sachida_shukla@yahoo.co.in

Abstract— A New circuit model of two-stage small-signal amplifier using Darlington and Sziklai pairs is proposed for the first time. Proposed circuit is obtained by cascading a small-signal Darlington pair amplifier with that of Sziklai pair based small-signal amplifier with minor modifications. Proposed amplifier essentially uses two additional biasing resistances in its design. It crops considerably high voltage and current gain (345.523 and 464.357 respectively) with audible range bandwidth of 43.363 KHz for AC input signals swinging in 10-30mV range at 1KHz or lower frequency. Proposed circuit-architecture successfully removes the poor response problem of conventional Darlington pair amplifiers at higher frequencies and narrow bandwidth problem of recently announced (by authors) circuits of small-signal Sziklai pair amplifier. Variations in voltage gain with frequency and various biasing components like biasing resistances, DC supply voltage and coupling capacitors are discussed in length. Temperature sensitivity of performance parameters, THD and small-signal AC equivalent circuit analysis of the proposed circuit are elaborately studied. Proposed design can be implemented in cascadable gain blocks for receivers, 715Hz-44KHz frequency range power sources and in the circuits where reproduction of signal with simultaneously high voltage and current gain is the prime requirement.

Keywords— Two-stage small-signal amplifiers, Sziklai pair Amplifiers, Darlington Pair, Circuit design, Simulation

I. Introduction

Small-signal amplifiers are key contributors to a number of important electronics and communication circuits [1]-[2]. Sometimes voltage/current gain of single-stage amplifiers are found insufficient for practical applications, and therefore, two or more amplifier stages are used to get the desired level of gain [1]-[3].

In the last few decades, electronics industry has witnessed Darlington pairs as useful audible-range device for single/multi-stage small-signal and power amplifiers [3]-[5]. This paired unit of identical BJTs realizes higher β values but suffers from poor-response-problem at higher frequencies [3]-[5]. In recent years, Darlington pairs are being gradually replaced by Sziklai pairs in single or multi-stage power amplifiers due to having almost similar range of current gain, input resistance, output resistance and voltage gain with extra advantages of better linearity and only half of the base turn-on voltage than Darlington's unit [5]-[9]. However, use of Sziklai pairs in single or multi-stage small-signal amplifiers is still under the developmental phase [3],[6]-[9].

Unlike Darlington pairs, Sziklai pairs hold one NPN and other PNP transistor in CE-CE connection [6]-[7]. Polarity of this unit is determined by the *driver* transistor. Thus, Sziklai pair with PNP driver and NPN output transistor acts as a PNP transistor and vice versa [6]-[7]. Principally, both Darlington and Sziklai pairs enjoy high input resistance, low output resistance and voltage gain approximately equal to unity [4]-[6]. However, the current gain factor β of Sziklai pair $(\beta_{Q1}\beta_{Q2}+\beta_{Q1})$ is slightly less than Darlington pair $(\beta_{Q1}\beta_{Q2}+\beta_{Q1}+\beta_{Q2})$ topology but at higher β values both are approximated to $\beta_{Q1}\beta_{Q2}$ [7].

Recently, authors have developed three different circuits of Sziklai pair based small-signal amplifiers [3],[8]-[9]. Present exploration carries an intention to combine desirable features of small-signal Darlington and Sziklai pair amplifiers, and therefore, focuses around the development and analysis of a two-stage small-signal amplifier using Darlington pair in stage-1 and Sziklai pair in stage-2.

II. Description of Circuits

The present investigation includes a comparative study of two-stage small-signal proposed amplifier (Fig.3) with conventional Darlington pair amplifier (Fig.1) and a modified Sziklai pair amplifier (Fig.2) [5],[8].

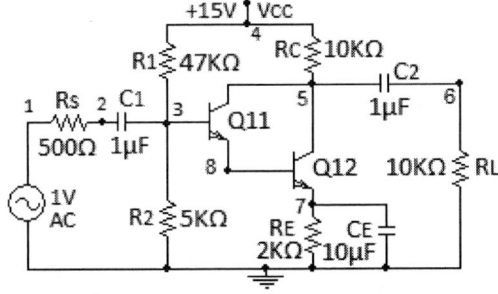

Fig.1. Small-signal Darlington pair amplifier (Reference circuit)

The amplifiers of Fig.1 [5] and Fig.2 [8] are treated herein as reference circuits. However, proposed two stage amplifier (Fig.3) is obtained by cascading reference amplifier circuits of Fig.1 and Fig.2 with minor modifications [1]-[3]. Devices Q11, Q12 and Q22 in respective circuit designs of Fig.1, Fig.2 and Fig.3 amplifiers are NPN BJTs (Q2N2222 with β=255.9) whereas Q21 is a PNP BJT (Q2N2907A with β=231.7) [10].

PSpice simulation [10] is used to receive respective observations. Every amplifier is fed by 1V AC input signal source from which, 1mV for Fig.1 whereas 10mV for Fig.2 and Fig.3 amplifiers at 1KHz frequency are drawn as input for present studies. Fig.1 amplifier provides significant response for 1-10mV of AC input whereas Fig.2 and Fig.3 amplifiers provide this for 10-30mV AC input signals.

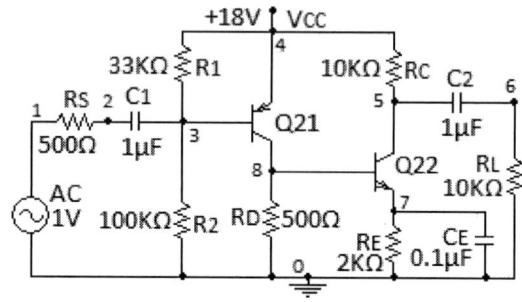

Fig.2. Modified small-signal Sziklai pair amplifier (Reference circuit)

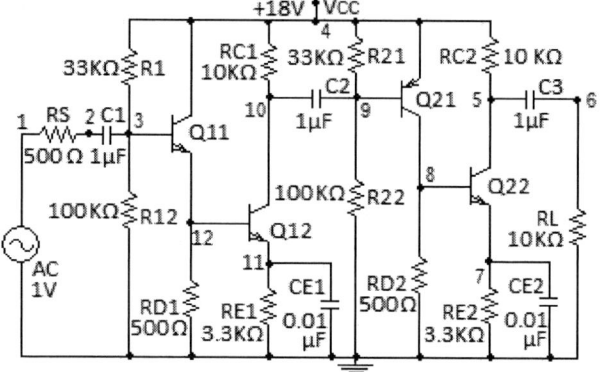

Fig.3. Proposed two-stage amplifier with Darlington and Sziklai pairs

III. **Results and Discussions**

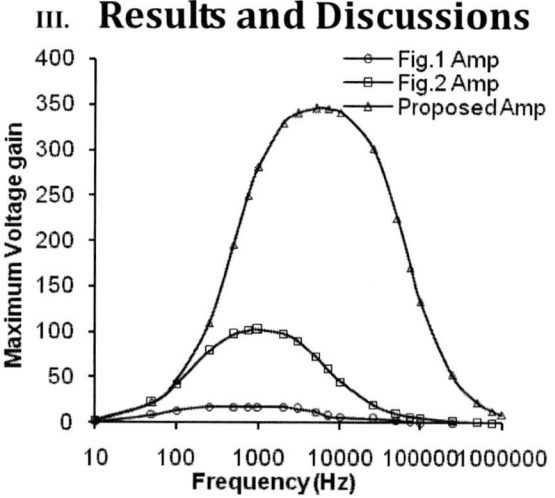

Fig.4. Variation of voltage gain with frequency

Variation of maximum voltage gain (A_{VG}) as a function of frequency for Fig.1, Fig.2 [3],[8]-[9] and Fig.3 amplifiers is depicted in Fig.4. Based on Fig.4; the observed values of various performance parameters of respective circuits are listed in TABLE-I. Values corresponding to Maximum voltage gain A_{VG}, peak output voltage V_{OP}, peak input voltage

V_{IP}, maximum current gain A_{IG}, peak output load current I_{OP}, peak input source current I_{IP}, bandwidth B_W, upper cut off frequency f_H, lower cut off frequency f_L, total harmonic distortion THD and output waveform phase change θ in TABLE-I present a qualitative comparison of performance of proposed amplifier with the other two.

TABLE-I: PERFORMANCE PARAMETERS OF AMPLIFIERS

Parameters	Fig.1 Amplifier	Fig.2 Amplifier	Fig.3 Amplifier
A_{VG}	16.98	102.309	345.523
A_{IG}	8.51	7.345	464.357
B_W	102.058KHz	4.80KHz	43.364KHz
F_L	79.913Hz	224.453Hz	714.189Hz
F_H	102.138 KHz	5.0556 KHz	44.077KHz
V_{OP}	17.939mV	1.106V	2.906V
V_{IP}	1mV	10mV	10mV
I_{OP}	1.7939 μA	109.434 μA	290.457 μA
I_{IP}	190.975 nA	13.574 μA	666.005 nA
Θ	180°	180°	180°
THD	0.626%	1.72%	2.938%

Refer Fig.4 and TABLE-I. The proposed amplifier is found to produce considerably large A_{VG} and A_{IG} than reference amplifiers (Fig.1 and Fig.2) with a fair response in 715Hz to 44KHz frequency range. The proposed circuit obtained by cascading Fig.1 and Fig.2 amplifiers bears an enhanced THD. However, this circuit (Fig.3) with desirable features of Fig.1 and Fig.2 amplifiers becomes free from poor-response-problem of small-signal Darlington's amplifier (Fig.1) at higher frequencies and narrow bandwidth restrictions of Sziklai pair amplifier (Fig.2). The proposed circuit can be used in designing cascadable gain blocks for receivers and 715Hz-44KHz frequency range power sources.

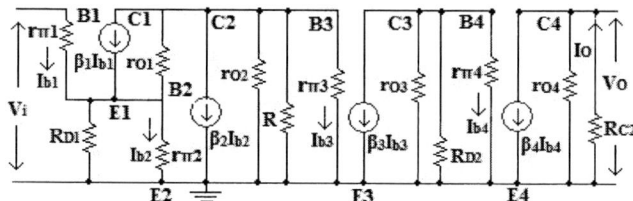

Fig.5. Small-signal AC equivalent circuit of proposed Fig.3 amplifier

Fig.5 shows small-signal AC equivalent circuit of the proposed amplifier. Based on simulation results, Fig.5 consists base-emitter resistance r_π with $r_{\pi1}=245\Omega$, $r_{\pi2}=358\Omega$, $r_{\pi3}=177\Omega$ and $r_{\pi4}=195\Omega$, AC current gain factor β with $\beta_1=194$, $\beta_2=36.2$, $\beta_3=222$ and $\beta_4=28.6$, collector-emitter resistance r_o with $r_{O1}=3.72K\Omega$, $r_{O2}=2.41\Omega$, $r_{O3}=3.54K\Omega$ and $r_{O4}=1.26\Omega$ for respective transistors in circuit design. Additionally, I_b is base current and R=$R_{21}\| R_{22}$ in Fig.5.

Fig.5 can further be reduced with exclusion of r_{o1} and r_{o3} due to high values. Because of an extremely low value, r_{o2} would realize $\beta_2I_{b2,}$ and therefore, is to be ignored. Additionally, the combination of R and r_{o2} provides impedance matching to stage-2 in the proposed design. Moreover, parallel combination of r_{o4}-R_{C2} brings effective output resistance of the circuit to very low value ($\approx1.2598\Omega$). Due to these features the proposed design can be successfully used in sound amplifier circuits. Analysis of Fig.5 with mentioned approximation reveals following expression for small-signal AC voltage gain of the proposed amplifier [1]-

$$A_{VG} \approx \frac{-\beta_3\beta_4\left\{\beta_1 + \frac{\beta_2(1+\beta_1)}{\left(1+\frac{r_{\pi 2}}{R_{D1}}\right)}\right\}(R_{C2}\|r_{o4})}{\left(1+\frac{r_{\pi 2}}{R}\right)\left(1+\frac{r_{\pi 4}}{R_{D2}}\right)\left\{r_{\pi 1} + \frac{(1+\beta_1)r_{\pi 2}}{\left(1+\frac{r_{\pi 2}}{R_{D2}}\right)}\right\}} \quad (1)$$

The negative sign in (1) shows phase reversal in output voltage waveform. With mentioned values of β, r_π, r_o, R, R_D and R_C for respective transistors and resistances of Fig.3, A_{VG} is computed to be about 388.50. This is sufficiently close to the simulated value 345.523, and therefore, justifies the authenticity of design and Equation (1).

TABLE-II: VARIATION OF A_{VG} AND A_{IG} WITH TEMPERATURE

Temp. (°C)	Fig.1 Amplifier		Fig.2 Amplifier		Fig.3 Amplifier	
	A_{VG}	A_{IG}	A_{VG}	A_{IG}	A_{VG}	A_{IG}
-30	8.95	4.48	83.433	5.7417	314.275	345.053
0	13.51	6.77	93.704	6.5964	332.573	410.117
10	14.86	7.45	96.965	6.8762	337.725	430.748
27	16.98	8.51	102.309	7.3448	345.523	464.357
50	19.53	9.79	109.125	7.9622	354.212	506.459
80	22.38	11.22	117.221	8.7337	97.960	83.167

Variation of A_{VG} and A_{IG} with temperature is listed in Table-II. Both A_{VG} and A_{IG} increase with temperature for Fig.1 and Fig.2 amplifiers. However, for Fig.3 amplifier A_{VG} and A_{IG} increase with temperature up to a critical limit of 50°C, thereafter, both type of gains fall rapidly. In addition, the bandwidth of respective amplifiers does not alter significantly with temperature (hence not shown in Table-II). Observations in Table-II verify the usual behaviour of transistor parameter h_{FE} with temperature [11] for respective amplifiers. However, beyond critical temperature, the proposed amplifier deviates from usual behaviour perhaps either due to additional biasing resistances R_{D1} and R_{D2} in the circuit or due to frequent collision of excessively generated carriers [12].

TABLE-III: VARIATION OF A_{VG} AND A_{IG} WITH LOAD RESISTANCE R_L

R_L (KΩ)	Fig.1 Amplifier		Fig.2 Amplifier		Fig.3 Amplifier	
	A_{VG}	A_{IG}	A_{VG}	A_{IG}	A_{VG}	A_{IG}
1	3.13	15.69	78.112	54.80	260.116	3.49
10	16.98	8.52	102.309	7.35	345.523	464.36
50	28	2.81	105.96	1.52	355.928	95.67
100	30.47	1.53	105.453	0.76	357.948	48.02
300	32.88	0.54	105.692	0.25	358.287	16.05
500	32.79	0.33	105.74	0.15	358.4	9.63
600	-	-	105.752	0.13	358.432	8.03

Variations in A_{VG} and A_{IG} with load resistance R_L for respective amplifier circuits are listed in Table-III. A_{IG} attains maxim for Fig.1 and Fig.2 amplifiers at 1KΩ, thereafter falls rapidly with elevation of R_L and reaches below unity at 300KΩ for Fig.1 amplifier and at 100KΩ for Fig.2 amplifier. However proposed amplifier shows linear increment in A_{IG} up to 10KΩ, thereafter A_{IG} decreases rapidly at higher R_L values but remains greater than unity till R_L=600 KΩ. Similarly, A_{VG} rises almost linearly up to certain values of R_L for respective amplifiers, and thereafter, gradually tends towards saturation. These critical values for A_{VG} are 100 KΩ for Fig.1, 50 KΩ for Fig.2 and 300 KΩ for Fig.3 amplifiers. Rising and saturation of A_{VG} and declination in A_{IG} at elevated R_L is found in

accordance with the usual behaviour of small-signal BJT amplifiers [1].

Variations in A_{VG} and A_{IG} with supply voltage V_{CC} are recorded in Table-IV. Both A_{VG} and A_{IG} for Fig.1 amplifier rise almost linearly with V_{CC}. For Fig.2 amplifier, A_{IG} gradually decreases at higher V_{CC} whereas A_{VG} rises with V_{CC} up to a critical limit of 30V and thereafter dips to a lower value. In addition, both A_{VG} and A_{IG} show nonlinear gradual enhancement at higher V_{CC} for Fig.3 amplifier. Fig.1 amplifier switches ON at 10V of V_{CC} whereas the similar is observed for Fig.2 and Fig.3 amplifiers at lesser V_{CC} (i.e. 6V). As a unique feature, considerably higher values of A_{VG} and A_{IG} for proposed Fig.3 amplifier than other two ensure the simultaneous availability of high voltage, current and power gains in permissible range of V_{CC} (6V-40V). Refer Table-IV. The separate DC biasing to each stage due to coupling capacitor C_2 reveals I_C of respective stages to enhance with rising V_{CC} in accordance with I_C=(V_{CC}−V_{CE})/(R_C+R_E). Correspondingly an elevation in V_{CC} provokes r_π of respective transistors to reduce with simultaneous reduction in β values except for Q_{11}. These deliberations are in accordance with (1) and cause load current (across R_L) and therefore A_{VG} to enhance with V_{CC}.

TABLE-IV: VARIATION OF A_{VG} AND A_{IG} WITH DC SUPPLY V_{CC}

V_{CC} (Volts)	Fig.1 Amplifier		Fig.2 Amplifier		Fig.3 Amplifier	
	A_{VG}	A_{IG}	A_{VG}	A_{IG}	A_{VG}	A_{IG}
6	-	-	55.15	7.84	78.78	138.87
10	1.11	0.56	84.77	7.82	182.31	289.63
15	16.98	8.52	98.74	7.53	290.40	413.22
20	36.28	18.18	103.65	7.22	378.61	491.68
30	72.32	37.19	104.35	6.66	511.91	572.82
40	110.54	55.25	101.23	6.17	602.84	604.04

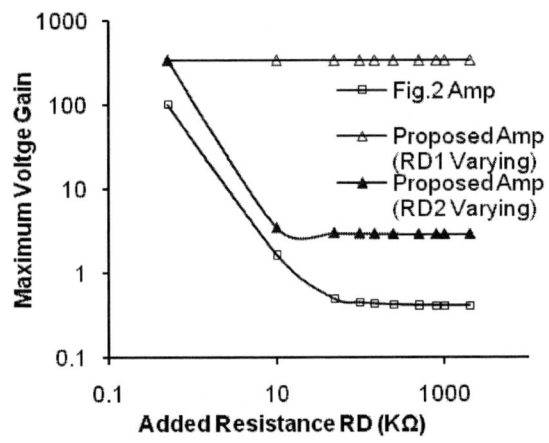

Fig.6. Variation of Voltage gain with Additional biasing resistance R_D

Variations in A_{VG} with additional biasing resistances R_D, R_{D1} and R_{D2} (Fig.2 and Fig.3) are shown in Fig.6. The maxim of A_{VG} corresponding to added resistance R_D for Fig.2 amplifier is observed at R_D=0.5kΩ. However A_{VG} linearly decreases up to R_D=50kΩ thereafter tend towards saturation. Thus, Fig.2 amplifier produces considerable response for $R_D \leq$ 1kΩ. However in the proposed amplifier, A_{VG} shows 'no change' with the increasing values of R_{D1} (R_{D2} constant) whereas A_{VG} decreases linearly with increasing values of R_{D2} (R_{D1} constant) up to 10KΩ, thereafter tends towards saturation. Observations suggest that increase in R_{D1} (R_{D2} constant) causes marginal reduction in β values of Q11-Q12, r_π of Q11, I_C and I_{RL} of the stage-1 amplifier (Fig.3) with a considerable enhancement in r_π of Q12. This in view of (1) causes minor reduction in A_{VG} with elevation in R_{D1}. However, increase in R_{D2} causes marginal reduction in β value but considerable enhancement in r_π of Q21 transistor. This in agreement with (1) causes significant reduction in R_C and load current across R_L and therefore in A_{VG}.

Additionally, if R_{D1} is removed from the circuit, A_{VG} trims down to 2.95, A_{IG} to 2.22 whereas THD enhances to 14.56% and bandwidth considerably widens to 263.90KHz. However, if R_{D2} is removed, A_{VG} goes downs to 342.76, bandwidth nominally shrinks to 42.13KHz and THD to 2.90% whereas A_{IG} enhances to 531.30. Instead, if both R_{D1} and R_{D2} are simultaneously removed, A_{VG} reduces to 2.81, A_{IG} to 2.30, THD to 1.46% whereas bandwidth widens to 264.05KHz. Thus, the simultaneous presence of R_{D1} and R_{D2} is essential for the proposed design to maintain amplification property.

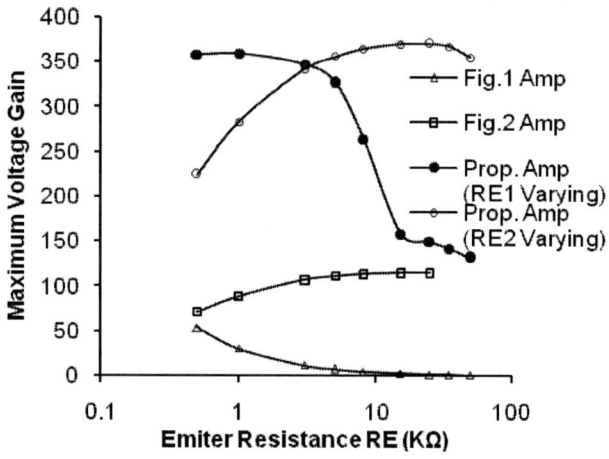

Fig.7. Variation of Voltage gain with Emitter resistance R_E

Fig.7 depicts variations in A_{VG} with R_E for Fig.1, Fig.2 and Fig.3 amplifiers. A_{VG} for Fig.1 amplifier has decreasing tendency (almost exponentially) at elevated R_E. However, A_{VG} for Fig.2 amplifier increases with R_E and the respective curve is found more-or-less inverted replica of the curve for Fig.1 amplifier. Simultaneously, A_{VG} for Fig.3 amplifier falls gradually at increasing values of R_{E1} (R_{E2} constant) whereas it rises nonlinearly with increasing values of R_{E2} (R_{E1} constant) up to 25 KΩ, and thereafter begins to fall with slow pace. Since for BJTs $I_C \approx I_E$, the increase in R_{E1} causes simultaneous reduction in I_C of Q_{11}-Q_{12} with significant enhancement in r_π of Q_{11}-Q_{12} and β of Q_{12}. These observations, in agreement to (1), responsibly decrease current across R_L and therefore A_{VG} for an elevated R_{E1}.

Coupling capacitors play important role in determining the performance of multistage amplifiers [1]. These capacitors preferably provide separate DC biasing to each stage. Table-V carries variations in A_{VG}, A_{IG}, f_L and f_H of Fig.3 amplifier with coupling capacitors C_1, C_2 and C_3.

Refer Tables V. Variations in C1 considerably affect A_{VG}, A_{IG} and f_L of proposed amplifier but f_H is maintained at about

43KHz. However, variations in C2 affect f_L and f_H and hence the bandwidth with almost constant values of A_{VG} (≈ 345) and A_{IG} (≈ 464) whereas similar order variations in C3 do not produce any considerable changes in A_{VG} (≈ 345), A_{IG} (≈ 464) but affects f_L.

TABLE-V: VARIATIONS IN PERFORMANCE PARAMETERS FOR FIG.3 AMPLIFIER WITH COUPLING CAPACITORS

Capacitance	C1			C2		C3
	A_{VG}	A_{IG}	f_L (Hz)	f_L (Hz)	f_H (KHz)	f_L (Hz)
22nf	180.98	270.23	1013.9	14551	92.84	1056
50nf	249.44	356.37	818.46	9150	66.23	806.79
100nf	291.50	405.47	767.28	5517	55.71	759.63
1µf	345.52	464.36	714.18	714.2	44.08	714.18
10µf	352.06	471.14	707.96	80.53	41.57	

IV. Conclusion

The proposed design is free from poor response problem of small-signal Darlington pair amplifier at higher frequencies and narrow bandwidth restriction of small-signal Sziklai pair amplifier. Moreover, multifold enhancement in A_{VG} and A_{IG} is received with additive nature of THDs. Design can be used to fabricate cascadable gain blocks for receivers, sound amplifiers and low frequency power sources. Additionally, use of R_{D1} and R_{D2} is essential to maintain amplification property for the proposed design.

References

[1] M.M.E. Fahmi, A.M.H. Sayed El Ahl and S.N. Mohammad, "Implication of non-ideality of Coupling capacitors in the performance of multistage amplifiers", Solid State Electronics, Vol. 45, p-1843, 2001

[2] R.R. Johnston, "Designing Spice-predictable, D.C. coupled, multistage, bipolar-junction-transistor amplifiers", Proceeding of 40th IEEE Midwest Symposium on Circuits and Systems, Vol.1, p-217, 1997

[3] S.N. Shukla and S. Srivastava, "Development and Qualitative Analysis of a New Circuit Model of Two-stage Small-signal Sziklai pair Amplifier", International Journal of Computer Theory and Engineering, Vol.5, No. 4, p-668-672, 2013

[4] Hodges David A., "Darlington's Contribution to Transistor Circuit Design", IEEE Transactions on Circuits and Systems-I, Vol.46, No.-1, p-102, 1999

[5] S.N. Shukla and S. Srivastava, "A New Circuit Model of Small-signal Sziklai pair amplifier", International Journal of Applied Physics and Mathematics, Vol.3, No.4, p-231, 2013

[6] G. C. Sziklai, "Push-pull complementary type transistor amplifier", U.S. Patent 2,762,870, September 11, 1956

[7] Elliott Rod, "Compound pair Vs Darlington pair", http://sound.westhost.com/articles/cmpd-vs-darl.htm, 2011

[8] B. Pandey, S. Srivastava, S. N. Tiwari, J. Singh and S. N. Shukla, "Qualitative Analysis of Small Signal Modified Sziklai Pair Amplifier", Indian Journal of Pure and Applied Physics, Vol. 50, p-272, 2012

[9] S. N. Shukla, B. Pandey and S. Srivastava, "Qualitative Study of a New Circuit Model of Small Signal Amplifier using Sziklai Pair in Compound Configuration", Proc. 10th IEEE International Conference on Semiconductor Electronics, September 19-21, 2012, Kuala Lumpur, p-600, 2012

[10] M. H. Rashid, "Introduction to PSpice Using OrCAD for Circuits and Electronics", Pearson Education, 3rd Ed., p-255, 2004

[11] A. G. Barua and B. Tiru, "Variation of width of the hysteresis loop with temperature in an emitter-coupled Schmitt trigger", Indian Journal of Pure & Applied Physics, 44, p-482, 2006

[12] I. Zherebtsov, "Basic Electronics", Mir Publishers, Moscow, 1988, p-8

A 20 GHz Power Amplifier Design On Novel Nonlinear Model

Zonghua Zheng
Institute of VLSI Design
Zhejiang University
Hangzhou, China
MOE Key Laboratory of RF Circuits and Systems
Hangzhou Dianzi University
Hangzhou, China

Lingling Sun*, Jun Liu
MOE Key Laboratory of RF Circuits and Systems
Hangzhou Dianzi University
Hangzhou, China
sunll@hdu.edu.cn

Abstract—This paper designs a 20 GHz power amplifier based on a novel model which is the quadratic polynomial expansion of nonlinear scattering function. The novel model is capable of accurately computing the voltage of device under any fundamental and harmonic frequency load impedance changing implemented with Frequency Domain Defined Device (FDD) component in the ADS. Through the new model, it is easy to find out the optimal fundamental and harmonic load impedance. The designed PA achieves a peak power-added efficiency (PAE) of 35.1% and the saturated output power of 21 dBm in PP1010MS EEHEMT operating at 20 GHz.

Keywords—Power amplifier; Nonlinear scattering function; EEHEMT; Quadratic polynomial

I. Introduction

Recently, high frequency nonlinear measurement systems have gained increasing market with the release of instruments such as NVNA and ZVxPlus, used for extracting the X Parameters or S function, which are the state-of-the-art technologies for commercial RF nonlinear modeling [1-5]. X Parameters is a registered trademark of Agilent Technologies, and was considered as Taylor first order expansion of nonlinear scattering functions which is a black box modeling technique on frequency domain. Strictly speaking, the first order expansion model is found to be a weak function of load impedance, so it will become much complex when dealing with measurement data covering a large area of the smith chart. The challenges increasingly incurred when accurately modeling high power transistors under variable complex impedance matching conditions. Although the load-pull + NVNA measurement approach provided an initial validation of the non-linear mismatching characterization of the DUT [6], the experiments to be performed are too expensive and complex. It can be foreseen that the collection of suitable datasets and their conversion into functional parameterized data blocks for use in simulation packaged suitable for the

design of large signal power amplifiers will gain popularity within the industry and become the default design flow.

In this paper, the new model frameworks is the quadratic polynomial expansion of nonlinear scattering functions that considering the harmonic load highly mismatched environments. The model can provide an analytical solution for the transistor matching conditions to fulfill a set of design criteria such as power gain, input and output match, efficiency in designing microwave amplifiers. One can find the optimal load reflection coefficient and source reflection coefficient based on the novel model, which is facilitates to the design of power amplifier.

II. Model Description

The basic concept of nonlinear scattering functions was first introduced by professor Jan Verspecht in 1998 as (1) in [7].

$$B_{pm} = F_{pm}(A_{11}, A_{12},..., A_{21}, A_{22},...) \qquad (1)$$

Where the multivariate complex function $F_{pm}(.)$ (also called description function) correlates all of the relevant input spectral components A_{qn} with the output spectral components B_{pm}, whereby p and q ranges from one to the number of signal ports, and m and n ranges from zero to the highest harmonic index.

In designing of RF power amplifier, the influences of higher order harmonic components can be ignored for simplicity while some dominant harmonic spectral components should be taken into account. We can get the reduced nonlinear scattering functions considering only some important first three order harmonic components as (2).

$$B_{pm} = F_{pm}(A_{11}, A_{21}, A_{22}, A_{23}) \qquad (2)$$

Then using second order polynomial to substitute it as (3) shows.

$$B_{pm} = k_{pm11}A_{11} + \sum_{n=1}^{3} K_{pm2n}A_{2n} + \sum_{n=1}^{3} L_{pm2n}A_{2n}^{*}$$

$$+ \sum_{n=1}^{3} KS_{pm2n}A_{2n}^{2} + \sum_{n=1}^{3} LS_{pm2n}(A_{2n}^{*})^{2} \qquad (3)$$

Where the A_{qn}, B_{pm} are mentioned above representing the n harmonic incident wave of port q and the m harmonic reflection wave of port p, and superscript * denotes the conjugation operator. The K_{pm2n}, L_{pm2n}, KS_{pm2n}, LS_{pm2n} are the coefficients of polynomial relative to the m harmonic frequency of p port induced by n harmonic frequency of the 2 port.

Assuming the transistor is a nonlinear time-invariant system, which implied that if the incident wave goes some phase shifted, the other waves will undergo the same phase shifted. Let us define $P = e^{j\varphi(A_{11})}$ as the phase of the incident wave A_{11}. we get

$$B_{pm} \cdot P^{-m} = k_{pm11}A_{11} \cdot P^{-1} + \sum_{n=1}^{3} K_{pm2n}A_{2n} \cdot P^{-n}$$

$$+ \sum_{n=1}^{3} L_{pm2n}A_{2n}^{*} \cdot P^{n} + \sum_{n=1}^{3} KS_{pm2n}A_{2n}^{2} \cdot P^{-2n} \quad (4)$$

$$+ \sum_{n=1}^{3} LS_{pm2n}(A_{2n}^{*})^{2} \cdot P^{2n}$$

$$B_{pm} = k_{pm11}A_{11} \cdot P^{m-1} + \sum_{n=1}^{3} K_{pm2n}A_{2n} \cdot P^{m-n}$$

$$+ \sum_{n=1}^{3} L_{pm2n}A_{2n}^{*} \cdot P^{m+n} + \sum_{n=1}^{3} KS_{pm2n}A_{2n}^{2} \cdot P^{m-2n} \quad (5)$$

$$+ \sum_{n=1}^{3} LS_{pm2n}(A_{2n}^{*})^{2} \cdot P^{m+2n}$$

We can use the (5) to establish a black box circuit model based on the measured magnitude and phase of each port harmonic spectral components. It is more accurate for considering the second order relationship of each incident and reflect wave than X parameters.

The output power is analytically described as:

$$P_{out} = \frac{1}{2}\left(|B_{21}|^{2} - |A_{21}|^{2}\right) \qquad (6)$$

Theoretically, from (5) and (6) we can obtain the optimal harmonic reflection coefficient by taking derivation, but it is a little complicated for find the peak value of multiple variables complex function. So we implemented it in ADS by means of a FDD built-in nonlinear block. This element allows ADS users to easily add custom behavioral, non-linear system models, by describing the output spectral components in terms

of arbitrary functions of the input spectral components. The identified parameters are stored in a muti-dimensional file format (.mdf), correspond to the independent VAR. And a data-access component (DAC) links the tabulated data to the model and performs multidimensional interpolation during the simulation. The implemented model in ADS is shown in Fig.1.

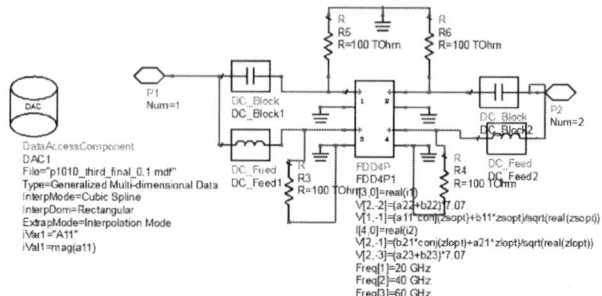

Fig.1. The proposed model implemented with a FDD built-in component

III. Circuit Design

A two stage cascade common source EEHEMT power amplifier is designed based on the new transistor model as shown in Fig.2. The transistors is 04x075 of EEHEMT which provided by WIN Semiconductors and was replaced by the proposed model in design amplifier, and its optimal fundamental load impedance is 17.7+j*11.8, the optimal second load impedance is 22-7*j, the optimal fundamental source impedance is 10+j*15, the optimal second source impedance is 5.3-10*j. The input/output impedance matching and interstage impedance matching network was implemented with the micro-strip transmission line and capacitor. All the transmission lines were done with 50 Ω characteristic impedance, except for the shorted stubs to the drain power supply, corresponding to an characteristic impedance of 25 Ω. which are helpful to short down the length and avoid electromigration and Joule heating. A shunt resistor and capacitor on the transistor gate are employed to ensure stability.

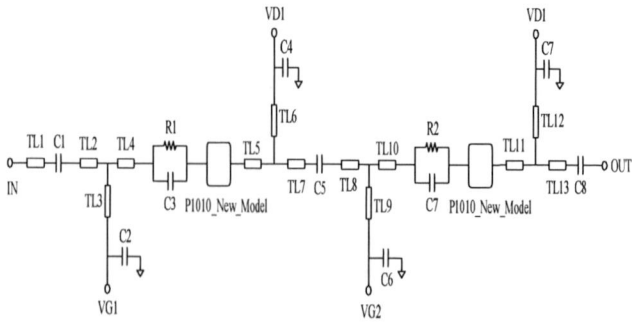

Fig.2. Schematic of the designed PAs

978-1-4799-5761-3/14 $31.00 © 2014 IEEE

IV. Amplifier Performance

S parameter simulation shows the amplifier's linear gain and power matching characteristics. Fig.3 shows a linear gain of 17 dB is achieved over 16 GHz to 21 GHz. The S11 and S22 are both below -10 dBm demonstrate good matching. The simulated stability factor K for the whole circuit is bigger than 1 across DC-40 GHz as shown in Fig.4.

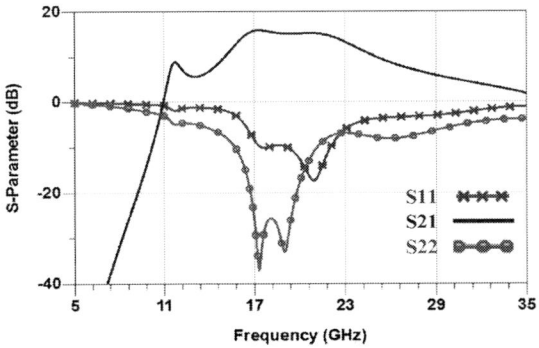

Fig.3. S parameter of the designed power amplifier

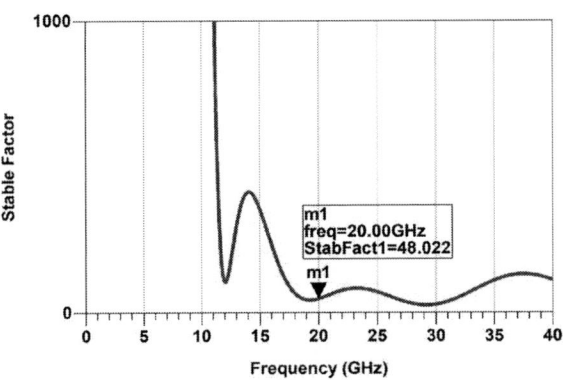

Fig.4. K factor of the whole circuit

Fig.5 shows the results of output power and power added efficiency as a function of the input power at 20 GHz. The amplifier reaches a saturated output power of 21 dBm when driven with 8 dBm input power with a gain of 13 dB. And the PAE can reach the peak value of 35.1% at input power of 8 dBm.

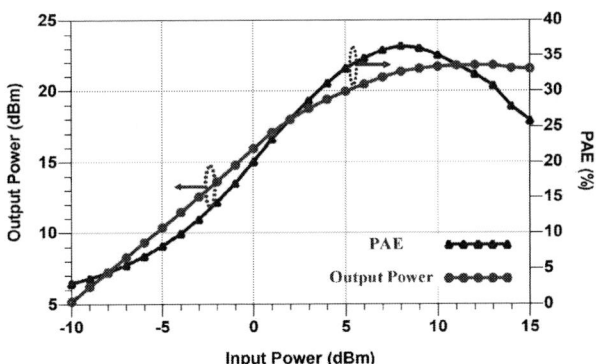

Fig.5. Output power and PAE versus input power at 20 GHz

V. Conclusion

A new nonlinear transistor model was proposed and used to design a two stage cascade EEHEMT power amplifier .The quadratic polynomial expansion of nonlinear scattering function can present the analytical description of the output power of each harmonic. Simulation results show the small signal gain is 17 dB over 16 GHz to 21 GHz. And the power-added efficiency (PAE) of 35.1%, the saturated output power of 21 dBm in PP1010MS EEHEMT operating at 20 GHz.

Acknowledgment

The authors especially wish to thank the WIN Semiconductors for giving access to the EEHEMT technology.

References

[1] D. E. Root, J. Xu, J. Horn, M. Iwamoto, and G. Simpson, "Device Modeling with NVNAs and X-parameters," 2010IEEE MTT-S INMMiC Conference, Gotenborg, Sweden, April 26, 2010.

[2] D. E. Root, J. Verspecht et al., "X-Parameters: Characterization, Modeling, and Design of Nonlinear RF and Microwave Components," Cambridge, U.K.: Cambridge University Press, 2013.

[3] Jang H, Zai A, Reveyrand T, et al. "X-parameter applications for characterizing and modeling power amplifiers for envelope tracking applications," International Microwave Symposium. 2013.

[4] Wang, Yelin, et al. "X-parameter-based modelling of polar modulated power amplifiers." IET Microwaves, Antennas & Propagation 7.14 (2013): 1161-1167.

[5] NMDG newsletter - IMS Special Edition, 2009 [On-line available:http://www.nmdg.be/newsletters/attach/S Functions.pdf].

[6] G. Simpson,J. Horn,D. Gunyan , and D. E. Root, "Load-Pull + NVNA=Enhanced X-Parameters for PA Designs with High Mismatch and Technology-Independent Large-Signal Device Models, "ARFTG Microwave Measurement Symposium, pp. 88-91, 2008.

[7] J. Verspecht and P. Van Esch, "Accurately characterizing hard nonlinear behavior of microwave components with the nonlinear network measurement system: Introducing 'nonlinear scattering functions' ," in Proc. 5th Int. Workshop Integrated Nonlinear Microwave Millimeterwave Circuits, Germany, Oct. 1998, pp. 17−26.R. Nicole, "Title of paper with only first word capitalized," J. Name Stand. Abbrev., in press.

Investigation of TMAl preflow to the properties of AlN and GaN film grown on Si(111) by MOCVD

Franky Lumbantoruan, Yuan-Yee Wong, Yue-Han Wu, Wei-Ching Huang, Niraj Man Shrestra, Tung Tien Luong, Tran Binh Tinh, Edward Yi Chang[a], *Fellow Member*, IEEE

Department Of Materials Science and Engineering
National Chiao Tung University
1001 University Rd., Hsinchu 30010 Taiwan

Abstract—The influence of TMAl preflow to the AlN buffer layer and GaN thin film was studied by Optical Microscope, Atomic Force Microscope, X-ray diffraction and Transmission Electron Microscope. Different duration of TMAl preflow lead to substantially differences of the AlN buffer layer and GaN film properties in terms of surface morphology and crystal quality. It was found without TMAl preflow the crystal quality of AlN buffer layer and GaN deteriorated due to the formation of amorphous interlayer between Si and AlN. Meltback etching and cracks was observed on the surface of GaN grown on AlN without TMAl preflow. However, overlong duration of TMAl preflow degraded the properties of AlN buffer layer and the subsequent GaN layer. GaN grown with longer TMAl preflow suffer of poor crystal quality, high density cracks and rough surface morphology. With the optimum duration of TMAl preflow, crystal quality and surface roughness of GaN can be improved.

Keywords—GaN on Si(111); AlN buffer layer; TMAl preflow; MOCVD

I. INTRODUCTION

III-nitride materials has attracted much interest due to its properties for the future high power electronic devices and shortwave optical devices[1]–[3]. GaN as one of the promising semiconductor device is usually grown heteroepitaxially on sapphire and SiC which are expensive and difficult for device fabrication. Silicon is promising alternative substrate for the growth of GaN related materials due to its high thermal conductivity and availability at large size with lower cost. Yet, there still many challenging problems need to be solved i.e large mismatch in lattice parameter, thermal stress due to different thermal expansion coefficient, plastic deformation of silicon and melt back etching. Various thin films such as AlAs[4], SiC[5], AlN[6] and SCN[7] has been introduced as Gan buffer layer to reduce the large tensile stress and meltback etching for GaN epitaxy on Si. AlN is commonly adopted buffer layer due to its small lattice constant and its large band gap. An important challenge of AlN growth on Si is to prevent the formation of amorphous interlayer at the interface while having a good crystal quality and smooth surface morphology as buffer layer. Thus, within this study we carry out an experiment to find out the effect of TMAl preflow to the properties AlN layer and GaN film on Si (111) by MOCVD.

[a]electronic mail: edc@mail.nctu.edu.tw

II. EXPERIMENTAL PROCEDURE

All the samples were grown on 2-inch Si (111) substrates using a VEECO D180 MOCVD rotating disk reactor. The substrate thickness, resistivity and warp were 380 ± 20 µm, 1-20 Ω-cm and < 40 µm, respectively. The trimethyl gallium (TMGa), trimethyl aluminum (TMAl) and ammonia (NH$_3$) were used as precursor gas for Ga, Al and N, respectively. H$_2$ was used as carrier gas. The 2-inch Si (111) were cleaned with BOE to remove contaminant and native oxide for 1 min[8] and annealed under H$_2$ ambient at 1015 °C for 10 min to remove surface oxide thermally. Silicon was introduced to trimethyl aluminum (TMAl) with a flow rate of 1.09 µmol/sec for 0 min; 0.5 min; 1 min and 1.5 min at 1015 °C prior to AlN growth. The V-III ratio of AlN was fixed at 3400. To investigate the influence of TMAl preflow to the GaN layer, 600 nm thick GaN was grown on the AlN buffer layer with 0 sec, 30 sec and 60 sec TMAl preflow. Structural properties of AlN buffer layers and GaN layers were investigated using X-ray Diffraction (XRD). The interface of AlN and Si was examined by Transmission Electron Microscope (TEM), while surface morphology and RMS roughness were measured by optical microscope and atomic force microscopy (AFM).

III. RESULT AND DISCUSSION

A. Surface Morphology and XRD analysis of AlN

The AlN layer were measured by AFM with scan size of 5µm x 5 µm^2 to investigate the effect of the TMAl preflow. Figure 1 shows the surface morphology of the AlN buffer layer at different TMAl preflow through AFM images. The root mean square roughness (RMS) value for 0 sec, 30 sec, 60 sec and 90 sec TMAl preflow were 3.42 nm, 3.09 nm, 4.48 and 6.28 nm. The non-linear relation of TMAl preflow and surface roughness was in agreement to those reported by Jianxing Cao et al[6]. It was observed, the surface of AlN with 0 sec TMAl preflow covered by small grain and small pits. AlN with 30 sec TMAl preflow has a smooth surface with uniformly smaller grain across the surface and no pits can be observed in the surface of AlN. Prolonging the TMAl preflow would result in discontinue and rough surface due to accumulation of aluminum on silicon surface that too large to facilitate the 2D growth. These effect can be seen clearly for AlN with 90 sec TMAl preflow where pits with diameter of 500 nm can be observed on the surface. AFM results indicate that the TMAl preflow for 30 sec was beneficial for the smooth surface of AlN layer.

In order to gain further insight the influence of TMAl-preflow times to the AlN buffer layers, the four sample were measured by XRD in figure 2. The full width at half maximum (FWHM) of the XRD rocking curve was used to quantify the relative crystalline imperfection. The FWHM (002) plane values for 0, 30, 60 and 90 sec were 2.38, 1.90, 1.48 and 1.04 degree respectively. The linear relation between the TMAl-preflow time and crystal quality signifying the better crystal quality with prolong TMAl-preflow times. Degradation of crystal quality without TMAl-preflow was caused by the formation of amorphous interlayer 2 nm thick at the interface of Si and AlN (figure 3). With 30 sec TMAl pre-flow, the silicon substrates was likely to be protected from the growth of amorphous interlayer. This amorphous interlayer was likely to be a SixNy layer as a result of nitridation of silicon.

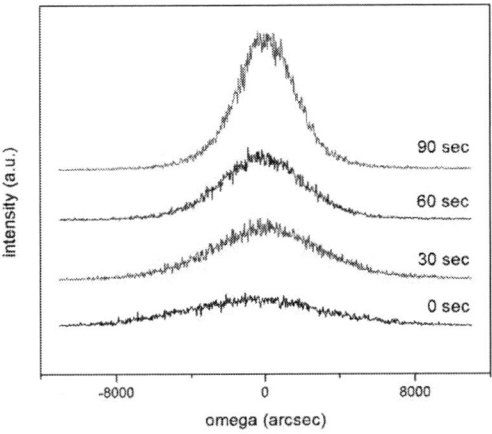

Fig. 2. FWHM XRD RC (002) AlN buffer layer on Si(111) substrates with different TMAl preflow

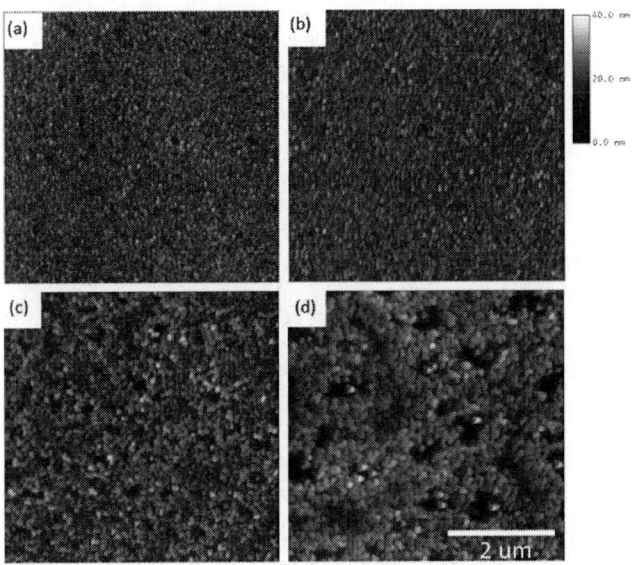

Fig. 1. AFM images of the AlN buffer layers at different TMAl pre-flow times (a) 0 sec, (b) 30 sec, (c) 60 sec and (d) 90 sec

The growth of AlN epitaxial layers on Si is initiated with 3D growth due to the large lattice mismatch between AlN and Si substrate (18.9%). Variation in TMAl preflow may induced the two dimensional growth. TMAl form a nucleation site on the silicon to enhance the AlN wetting on Si [9]. As the number of nuclei density increased, the lateral and vertical growth of AlN will undergone with a higher rate. As a result, the overall growth rate of AlN will be higher with the number of nucleation density[10]. The reflectance curve confirm slightly higher growth rate of AlN with smaller oscillation amplitude for higher duration of TMAl preflow (not shown here). These results indicating that TMAl-preflow influences the growth of AlN and effect surface morphology and crystal quality of AlN.

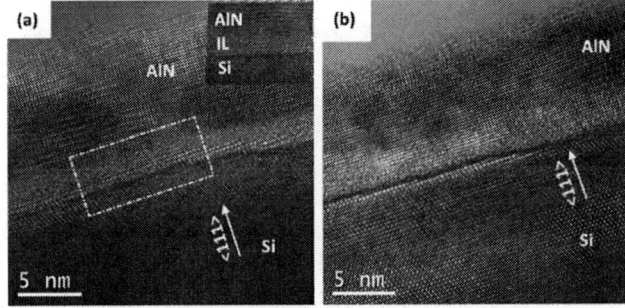

Fig. 3. Figure 1Cross-sectional TEM images of AlN and Si with (a) 0 sec and (b) 30 sec TMAl preflow

B. Surface Morphology and XRD analysis of GaN

To investigate the influence of TMAl-preflow to the subsequent layer, GaN was grown on AlN buffer layers with different duration of TMAl preflow. Figure 4 shows the optical images and AFM images of GaN growth on silicon with different TMAl-preflow. All samples shows a mirror-like characteristic with cracks at the edge of sample. Higher tensile stress in GaN film caused by large differences of lattice constant and large differences in thermal expansion coefficient between GaN and Si. This will lead to cracks generation on GaN film during cooling down process. High crack density can be observed for samples with 0 sec and 60 sec TMAl preflow possibly resulting from the degraded quality of AlN layer, which could not fully deplete the stress in GaN epilayer. Meltback etching pattern was dominant for sample without TMAl preflow and less for sample with TMAl preflow. Meltback etching causes a deterioration of GaN as a result of reaction between Ga and Si due to direct exposed of Ga to silicon or diffusion through crack[11]. Rough surface can be observed for sample with 60 sec TMAl preflow due to 3D growth AlN buffer layer. TMAl preflow for 30 sec gives less melt-back etching pattern and almost crack free. The value of RMS carried out by the samples with 0 sec, 30 sec and 60 sec

TMAl preflow were 2.74 nm, 1.91 nm and 3.84 nm respectively. These results indicate that the duration of TMAl preflow affect the stress generation and surface roughness of GaN layers. As shown by images on figure 4, almost crack free GaN with smooth surface can be obtained by 30 sec TMAl preflow.

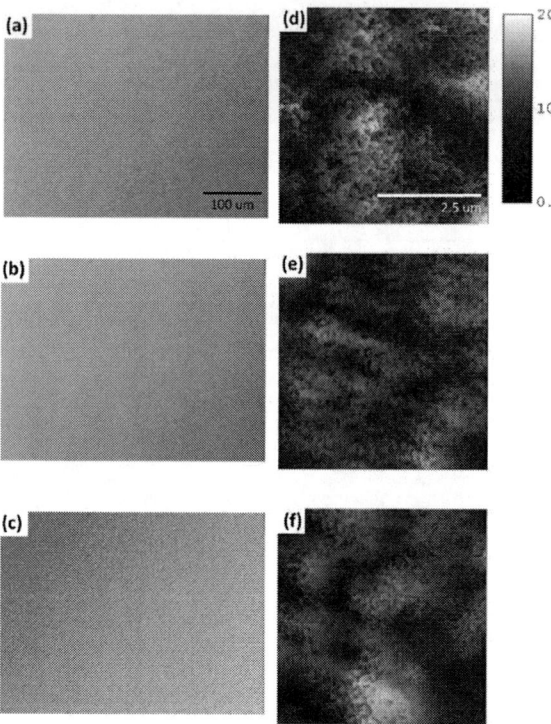

Fig. 4. 5x5 μm² surface morphologies of GaN observerd by Optical Microscope (a,b and c) and AFM (d,e and f) with different TMAl pre-flow duration (a) and (d) 0 sec; (b) and (e) 30 sec; (c) and (f) 60 sec

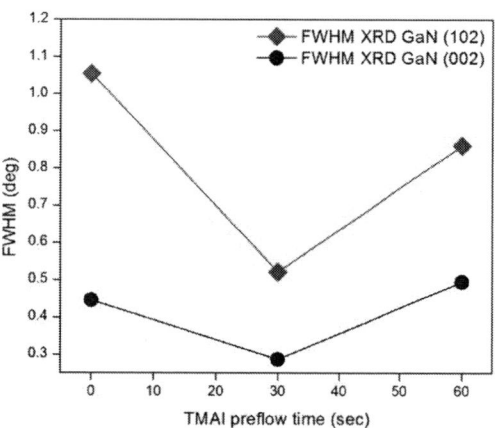

Fig. 5. FWHM GaN (002) and (102) plane with different duration of TMAl preflow

Figure 5 shows the FWHM of XRD Rocking Curve for the symmetrical (002) and asymmetrical (102) planes results of

GaN growth with different TMAl preflow. The rocking curves of the (002) symmetrical planes are normally responsive only to screw and mixed type threading dislocation. However, the rocking curves of the (102) asymmetric plane indicating crystalline distortions caused by all threading dislocations, including pure dislocations. The FWHM for (002) and (102) plane were 0.45 and 1.06 degree respectively for GaN without TMAl preflow. Increasing the duration of TMAl preflow to 30 sec improve the crystal quality of GaN. However, prolonging the TMAl preflow degraded the crystal quality of GaN. The results of FWHM XRD RC for (002) plane and (102) plane indicate the high crystal quality can be obtained by using the proper TMAl-preflow time which was 30 sec for this experiment.

IV. CONCLUSIONS

High quality of GaN on Si (111) substrate was obtained by optimizing TMAl-preflow prior to AlN growth. Without TMAl-preflow, AlN crystal quality will be degraded by the formation of amorphous interlayer at the interface of Si and AlN layer, while prolong TMAl-preflow lead to discontinue and rough surface due to accumulation of Al on silicon surface. Meltback etching can be observed for GaN sample without TMAl-preflow and less obtained with TMAl-preflow. GaN tensile stress can be reduced by implementing proper TMAl preflow leading to almost crack free GaN. Improvement on GaN crystal quality and surface morphology can be optimized with suitable duration of TMAl-preflow.

ACKNOWLEDGMENT

This work was sponsored by the NCTU-UCB I-RiCE program, Ministry of Science and Technology, Taiwan, and TSMC, under Grant No. NSC 103-2911-I-009-302.

REFERENCES

[1] K. Cheng, M. Leys, J. Derluyn, S. Degroote, D. P. Xiao, a. Lorenz, S. Boeykens, M. Germain, and G. Borghs, "AlGaN/GaN HEMT grown on large size silicon substrates by MOVPE capped with in-situ deposited Si3N4," *J. Cryst. Growth*, vol. 298, pp. 822–825, Jan. 2007.

[2] F. Schulze, A. Dadgar, J. Bläsing, A. Diez, and A. Krost, "Metalorganic vapor phase epitaxy grown InGaN/GaN light-emitting diodes on Si(001) substrate," *Appl. Phys. Lett.*, vol. 88, no. 12, p. 121114, 2006.

[3] A. Dadgar, M. Poschenrieder, J. Bläsing, O. Contreras, F. Bertram, T. Riemann, a. Reiher, M. Kunze, I. Daumiller, a. Krtschil, a. Diez, a. Kaluza, a. Modlich, M. Kamp, J. Christen, F. a. Ponce, E. Kohn, and a. Krost, "MOVPE growth of GaN on Si(111) substrates," *J. Cryst. Growth*, vol. 248, pp. 556–562, Feb. 2003.

[4] A. Strittmatter, a. Krost, M. Straßburg, V. Türck, D. Bimberg, J. Bläsing, and J. Christen, "Low-pressure metal organic chemical vapor deposition of GaN on silicon(111) substrates using an AlAs nucleation layer," *Appl. Phys. Lett.*, vol. 74, no. 9, p. 1242, 1999.

[5] J. Komiyama, Y. Abe, S. Suzuki, H. Nakanishi, and A. Koukitu, "MOVPE of AlN-free hexagonal GaN/cubic SiC/Si heterostructures for vertical devices," *J. Cryst. Growth*, vol. 311, no. 10, pp. 2840–2843, May 2009.

[6] J. Cao, S. Li, G. Fan, Y. Zhang, S. Zheng, Y. Yin, J. Huang, and J. Su, "The influence of the Al pre-deposition on the properties of AlN buffer layer and GaN layer grown on Si (111) substrate," *J. Cryst. Growth*, vol. 312, no. 14, pp. 2044–2048, Jul. 2010.

[7] M. a. Moram, M. J. Kappers, T. B. Joyce, P. R. Chalker, Z. H. Barber, and C. J. Humphreys, "Growth of dislocation-free GaN islands on

Si(111) using a scandium nitride buffer layer," *J. Cryst. Growth*, vol. 308, no. 2, pp. 302–308, Oct. 2007.

[8] M. Grundmann, A. Krost, and D. Bimberg, "Low-temperature metalorganic chemical vapor deposition of InP on Si(001)," *Appl. Phys. Lett.*, vol. 58, no. 3, p. 284, 1991.

[9] K. . Zang, L. . Wang, S. . Chua, and C. . Thompson, "Structural analysis of metalorganic chemical vapor deposited AlN nucleation layers on Si (111)," *J. Cryst. Growth*, vol. 268, no. 3–4, pp. 515–520, Aug. 2004.

[10] H. Jia, Y. Chen, X. Sun, D. Li, H. Song, H. Jiang, G. Miao, and Z. Li, "Effect of Trimethyl-aluminum Preflow on the structure and strain properties of AlN Films," *Chinese J. Lumin.*, vol. 33, no. 1, pp. 82–87, 2012.

[11] a. Dadgar, a. Strittmatter, J. Bläsing, M. Poschenrieder, O. Contreras, P. Veit, T. Riemann, F. Bertram, a. Reiher, a. Krtschil, a. Diez, T. Hempel, T. Finger, a. Kasic, M. Schubert, D. Bimberg, F. a. Ponce, J. Christen, and a. Krost, "Metalorganic chemical vapor phase epitaxy of gallium-nitride on silicon," *Phys. Status Solidi*, vol. 1606, no. 6, pp. 1583–1606, Sep. 2003.

The Impact of Minority Carrier Lifetime and Carrier Concentration on the Efficiency of CIGS Solar Cell

M.F.M. Fathil, M.K. Md Arshad, U. Hashim, A.R
Ruslinda, R.M. Ayub, A.H. Azman, M. Nurfaiz,
M.Z.M. Kamarudin
Institute of Nano Electronic Engineering, Universiti
Malaysia Perlis (UniMAP), Kangar, Perlis, Malaysia
faris.fathil@yahoo.com

M. Aminuddin, A.R Munir
IC Microsystems Sdn. Bhd. Cyberjaya, Selangor, Malaysia

Abstract— This paper deals with minority carrier lifetime and carrier concentration of $Cu(In,Ga)Se_2$ (CIGS)-based thin film solar cells with a ZnS(n)/CIGS(p) heterojunction structure. The structure is simulated in commercial numerical simulation and the impact of minority carrier lifetime in the CIGS absorber layer on the open circuit voltage, short circuit current density, fill factor and efficiency of the CIGS solar cell are investigated. The increase of minority carrier lifetime has also increased the CIGS solar cell performance. Similar effects are also observed at different carrier concentrations of CIGS layer. All these simulated results give a helpful indication for a practical fabrication process.

Keywords— CIGS; solar cell; minority carrier lifetime; carrier concentration; open circuit voltage; short circuit current density; fill factor; efficiency

I. INTRODUCTION

The Cooper-Indium-Gallium-Selenide (CIGS) solar cell has the best conversion efficiency of thin film solar cells. CIGS solar cell has surpassed 20 % under standard test conditions (STC) and it holds the current world record efficiency for a single junction, thin-film, and polycrystalline solar cell [1]. Nonetheless, for a single junction cell, this efficiency could be expanded in the future since the present efficiency record represents about two thirds of thermodynamic limit [2, 3]. Currently, by depositing localized back contacts, the efficiency of CIGS cells can be increased [4]. The method has been supported by a research on crystalline silicon solar cell; the efficiency can be increased by using this concept [5]. The efficiency of a solar cell is depended on both open circuit voltage and the short circuit current density which depends on the minority carrier lifetime. The importance of effective minority carrier lifetime to a silicon solar cell's efficiency is reflected in its crucial impact on both open circuit voltage and short circuit current density [6]. Based on this fact, this paper investigates on the effect of the minority carrier lifetime on the performance of solar cells. This includes the observation on the open circuit voltage (V_{oc}) i.e. when the current is zero, short circuit current density (J_{sc}) i.e. when the voltage is zero, fill factor (FF), and efficiency (eff) of the CIGS solar cell after the minority carrier lifetime is altered. At the same time, the carrier concentration of CIGS is altered to see any significant impact on the solar cell performances.

II. METHODOLOGY

A. Simulated CIGS solar Cell structure

The CIGS solar cell structure is simulated by using software 2D-Silvaco Athena. The structure consists of silicon dioxide that is used as a substrate layer for this solar cell. Silicon dioxide has the ability to act as an insulator layer which avoids current leakage from the solar cell. A molybdenum layer with thickness of 300 nm coats on the substrate, which served as a back contact and to reflect most unabsorbed light back into absorber layer. Molybdenum also has good conductivity. Next, CIGS layer with thickness of 2 μm is deposited on the back contact. The buffer layer zinc sulphide with thickness 50 nm is next deposited on top of the CIGS layer. The transparent conductive layer consists of a zinc oxide layer, ZnO and aluminum-doped ZnO, ZnoAl with thickness of 50 nm and 300 nm, respectively, is formed on the solar cell. Fig. 1 illustrates the schematic diagram of the simulated structure.

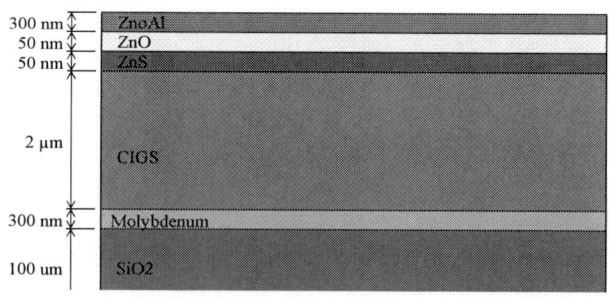

Fig. 1. Structure of CIGS Solar Cell.

B. Model Parameters

The parameter of the each material in the CIGS solar cell structure is then needs to be assigned in order to proceed with simulations using SILVACO Atlas. Some of the material properties, i.e. ZnoAl, ZnO, ZnS, and CIGS are not in available in the standard library. These material properties need to be assigned manually in the software coding and the properties was obtained from [7–9]. This is important aspects since every material carries its own characteristic and the chosen parameters will affect the characteristics of the solar cell (V_{oc}, J_{sc}, FF, and eff) later on. In this simulation, the minority carrier lifetime and carrier concentration of CIGS absorber layer is determined to observe its impact on the CIGS solar cell. Thus, we consider Schockley-Read-Hall as the recombination model. The solar cell is simulated under two conditions, i.e. without illumination (no light source), and under the illumination (a light beam using solar spectrum air mass (AM) of 1.5) in order to make sure the structure work accordingly as a theoretical solar cell. The light beam located on the top of the CIGS solar cell, at the angle of 90°, with a wavelength from 350 nm to 890 nm, and with number of samples equal to 50. The current-voltage (I-V) characteristic curve is used for extraction of V_{oc}, J_{sc}, FF and eff.

III. RESULT AND DISCUSSION

A. Effect of Minority Carrier Lifetime on the I-V Characteristic of CIGS Solar Cell

The minority carrier lifetime is the average time, which a minority carrier can spend in an excited state after electron-hole generation but before recombination to occur. Fig. 2 shows the effect of minority carrier lifetime on the I-V curve of a solar cell at a standard CIGS carrier concentration, 1e+16 cm^{-3}.

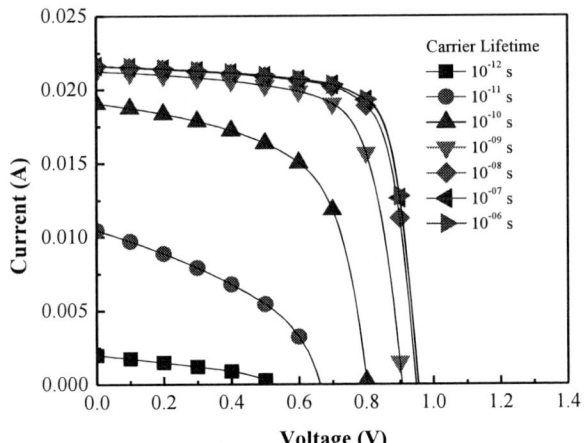

Fig. 2. Effect of the minority carrier lifetime on the I-V characteristic of CIGS solar cell at standard CIGS carrier concentration, 1.00E+16 cm^{-3}.

One can see, the minority carrier lifetime affected the I-V curve of the solar cell. As the minority carrier lifetime becomes longer, the J_{sc} and V_{oc} is increased, hence produce much ideal I-V curve of a solar cell. This straightforwardly based on the fact that the longer a minority carrier exists, the greater are its chances of being collected by the electric field at the junction and of contributing to the photocurrent [10]. Thus, this is the reason why the 1 μs minority carrier lifetime, which has the longest minority carrier lifetime and produced higher J_{sc} compared to the other solar cell with a smaller minority carrier lifetime. However, the I-V becomes saturated

when the minority carrier lifetime reaches a certain period i.e. 1e-8 s to 1e-6 s in this case. This is due to the solar cell has reached its limit in device performance, thus made the I-V saturated even with the increase of minority carrier lifetime as shown in Fig. 2.

B. Effect of Minority Carrier Lifetime and Carrier Concentration on Current Density, J_{sc}

As previously mentioned, the J_{sc} is the current density through the solar cell when the voltage across the solar cell is zero (i.e., when the solar cell is short circuited). It is due to the generation and collection of light-generated carrier [11]. Fig. 3 shows the effect of minority carrier lifetime on the current density, J_{sc} for the CIGS solar cell with different CIGS carrier concentration.

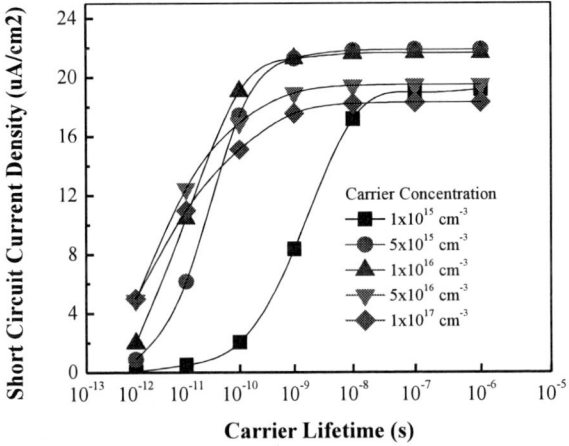

Fig. 3. Effect of CIGS Absorber Layer Carrier Lifetime on Short Circuit Current Density of CIGS Absorber Layer Solar Cell with Different CIGS Carrier Concentration.

When the carrier lifetime increased from 1 ps to 1 μs, the current density also increased with the maximum J_{sc} is 21.8561 uA/cm^2. The lifetime of the minority carriers in the absorber layer of the cell has greatly influenced the J_{sc}. This straightforwardly based on the fact that the longer a photo-generated carrier exists, the greater are its chances of being collected by the electric field at the junction and of contributing to the photocurrent [10]. Even with different concentration, the current density increases with the rise of carrier lifetime. However, as the CIGS carrier concentration increased to 5e+16 cm^{-3} and 1e+17 cm^{-3}, which is equivalent to the carrier concentration of window layer, ZnS (1.808e+17 cm^{-3}), the smallest minority carrier lifetime, 1ps, the J_{sc} is the highest, while at the 1 μs minority carrier lifetime, the J_{sc} is the lowest compared to the other CIGS concentration. This is due to the Auger recombination, usually dominates in high concentration solar cell [12]. Auger recombination occurs when an electron from the conduction band recombines with a hole from the valence band giving its energy to another electron. The Auger recombination rate increases as the cube of the carrier density, and is generally the dominant recombination mechanism when the carrier concentrations exceed 10^{17} cm^{-3} [12].

C. Effect of Carrier Lifetime and Carrier Concentration on the Open Circuit Voltage, V_{oc}

The open-circuit voltage, V_{oc}, is the maximum voltage available from a solar cell, and this occurs at zero current. The open-circuit voltage corresponds to the amount of forward bias on the solar cell due to the bias of the solar cell junction with the light-generated current. Fig. 4 shows the effect of CIGS absorber layer carrier lifetime on the V_{oc} for the CIGS solar cell with different CIGS carrier concentration.

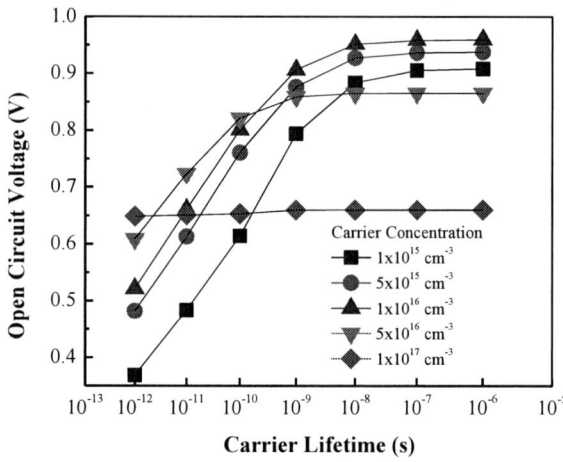

Fig. 4. Effect of CIGS Absorber Layer Carrier Lifetime on Open Circuit Voltage of CIGS Solar Cell with Different CIGS Carrier concentration.

The V_{oc} of the solar cell increased as CIGS carrier lifetime increased with the maximum V_{oc} is 0.95984 V for CIGS carrier concentration of 1e+16 cm^{-3}. Furthermore, as the CIGS carrier concentration of the solar cell increased by 1e+16 cm^{-3}, the Voc also increased. However, for CIGS carrier concentrations of 5e+16 cm^{-3} and 1e+17 cm^{-3}, at 1 ps, both carrier concentrations produce a very high V_{oc} but at 1 µs, both carrier concentrations produce the lowest V_{oc} when compared to the other CIGS carrier concentration. According to Werner et al [13], the increase in carrier concentration causes the grain boundaries in microcrystalline layer to become weakly n-type, resulting in higher recombination velocity. This has the effect of causing the V_{oc} to decrease. It is also essential to control the grain boundary in order to avoid the reduction of V_{oc} with the increase of carrier concentration.

D. Effect of Carrier Lifetime and Carrier Concentration on the Fill Factor

Fill factor, FF is the ratio of the maximum power from the solar cell to the product of V_{oc} and J_{sc}. It determines the "squareness" of the I-V curve. A solar cell with a higher voltage has a larger possible FF since the "rounded" portion of the IV curve takes up less area [14]. Equation (1) shows the formula for fill factor.

$$FF= (V_m.J_m) / (V_{oc}. J_{sc}) \qquad (1)$$

Based on Fig. 1, as the minority carrier lifetime increased, the V_{oc} and J_{sc} also increased. This included the maximum

power from the solar cell (product of V_m and J_m) on the I-V curve. The I-V curve has become more "squareness" as the carrier lifetime increased. Thus, when the value of V_m, J_m, V_{oc}, and J_{sc} inserted into the equation (1), the fill factor result is shown as in Fig. 5. Significant increase can be seen on the fill factor, as the minority carrier lifetime is increased. The minority carrier lifetime of 1 µs and CIGS carrier concentration of 1e+16 cm^{-3} produced the best fill factor on the I-V curve.

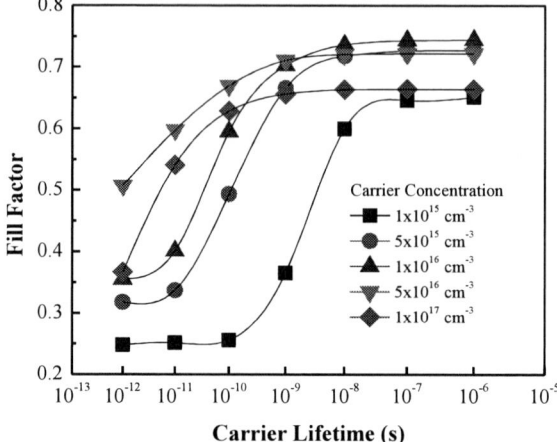

Fig. 5. Effect of CIGS Absorber Layer Carrier Lifetime on Fill Factor for CIGS Absorber Layer Solar Cell with Different CIGS Carrier Concentration.

E. Effect of Carrier Lifetime and Carrier Concentration on the Efficiency

The parameter most commonly used to compare the performance of one solar cell to another solar cell is called efficiency. Efficiency (Eff) is the ratio of energy output from the solar cell to input energy from the sun. So, the efficiency of a solar cell is determined as the fraction of incident power which is needed in order to convert light energy into electricity. Equation (2) shows the formula for efficiency, η for solar cell.

$$Eff, \eta=(V_{oc}.J_{sc}.FF)/P_{in} \qquad (2)$$

Based on the equation (2), the Eff of the solar cell is directly proportional to the V_{oc}, J_{sc} and FF. So the efficiency of the CIGS solar cell depends on the result from Fig. 3, Fig 4 and Fig. 5. By inserting the values of the V_{oc}, J_{sc} and FF, the efficiency is calculated and plotted as in Fig 6. The longest minority carrier lifetime, 1 µs produced the best efficiency among the other minority carrier lifetime with an efficiency of 11.3265 %, 14.8937 %, 15.4502 %, 12.1792 %, and 8.01815 % for CIGS carrier concentration of 1e+15 cm^{-3}, 5e+15 cm^{-3}, 1e+16 cm^{-3}, 5e+16 cm^{-3}, and 1e+17 cm^{-3}, respectively. Although from Fig. 6, the graphs show that the minority carrier lifetime is already saturated at 1.00E-7 s, or less, but there is a small increase of efficiency as the minority carrier lifetime is increased. Table 1 summarizes the solar cell efficiency performance which clearly shows 1 µs produces small improvements in the efficiency of the CIGS solar cell which should not be neglected.

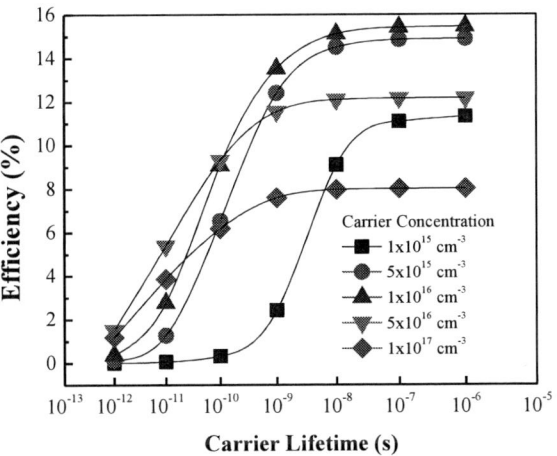

Fig. 6. Effect of CIGS Absorber Layer Carrier Lifetime on Efficiency for CIGS absorber Layer Solar Cell with Different CIGS Carrier Concentration.

TABLE I.   EFFICIENCIES OF THE CIGS SOLAR CELL AT DIFFERENTS MINORITY CARRIER LIFETIME AND CARRIER CONCENTRATION

Minority Carrier Lifetime (s)	Efficiency, Eff (%) at different Carrier Concentration				
	$1E+15$ cm^{-3}	$5E+15$ cm^{-3}	$1E+16$ cm^{-3}	$5E+16$ cm^{-3}	$1E+17$ cm^{-3}
10^{-12}	0.00615	0.13272	0.36726	1.5142	1.18346
10^{-11}	0.06175	1.27424	2.77055	5.38802	3.85644
10^{-10}	0.32166	6.53176	9.08419	9.30038	6.19065
10^{-09}	2.42365	12.4139	13.5252	11.5546	7.5971
10^{-08}	9.10156	14.5047	15.1194	12.1045	7.96705
10^{-07}	11.0910	14.8540	15.4146	12.1723	8.01338
10^{-06}	11.3265	14.8937	15.4502	12.1792	8.01815

IV. CONCLUSION

The minority carrier lifetime of the CIGS absorber layer has a significant impact on the performance of the solar cell. The performance of the solar cell in terms of V_{oc} and J_{sc} has improved with the increased of the minority carrier lifetime. This information has opened up new possibilities in fabrication process steps for fabricating CIGS solar cell at the industry in order to maintain or increase the minority carrier lifetime for improving the efficiency of the CIGS solar cell.

Although the increase of carrier concentration of the absorber layer of CIGS solar cells also helped in the optimization of the CIGS solar cell, there is a limitation. As the carrier concentration increased further than $1e+16$ cm^{-3}, the efficiency of the solar cell also reduced due to the effect of increasing the carrier concentration causes the grain

boundaries in microcrystalline layer to become weakly n-type, resulting in higher recombination velocity. Thus, it is also essential to control the grain boundary. Based on this, further simulation is needed in order to overcome this problem and optimize the efficiency of the solar cell.

From the simulation, the minority carrier lifetime of 1 μs with carrier concentration of $1e+16$ cm^{-3} produced the best performance compared to the other parameter, with the efficiency of 15.4402 %.

ACKNOWLEDGMENT

The authors would like to acknowledge the contribution from other members of the Institute of Nano Electronic Engineering, Universiti Malaysia Perlis (UniMAP) and IC Microsystems Sdn. Bhd., Cyberjaya for providing training and usage of Silvaco's software for the duration of the device simulation and characterization, hence, making the simulation process possible.

REFERENCES

[1] M. A. Green, K. Emery, Y. Hishikawa, and W. Warta, "Solar cell efficiency tables (version 36)," *Prog. Photovoltaics Res. Appl.*, vol. 18, no. 5, pp. 346–352, Jun. 2010.

[2] H. J. Queisser, "Detailed balance limit for solar cell efficiency," *Mater. Sci. Eng. B*, vol. 159–160, pp. 322–328, Mar. 2009.

[3] J. Mattheis, U. Rau, and J. Werner, "Finite Mobility Effects on the Radiative Efficiency Limit of Pn-Junction Solar Cells," *2006 IEEE 4th World Conf. Photovolt. Energy Conf.*, vol. 0, no. 3, pp. 95–98, 2006.

[4] M. Nerat, "Copper–indium–gallium–selenide (CIGS) solar cells with localized back contacts for achieving high performance," *Sol. Energy Mater. Sol. Cells*, vol. 104, pp. 152–158, Sep. 2012.

[5] J. Zhao, A. Wang, and M. A. Green, "High-e $ ciency PERL and PERT silicon solar cells on FZ and MCZ substrates," vol. 65, pp. 429–435, 2001.

[6] G. Kumaravelu, M. M. Alkaisi, D. Macdonald, J. Zhao, B. Rong, and a. Bittar, "Minority carrier lifetime in plasma-textured silicon wafers for solar cells," *Sol. Energy Mater. Sol. Cells*, vol. 87, no. 1–4, pp. 99–106, May 2005.

[7] H.-H. Yang and G.-C. Park, "A Study of the Properties of CuInS 2 Thin Film by Sulfurization," *Trans. Electr. Electron. Mater.*, vol. 11, no. 2, pp. 73–76, Apr. 2010.

[8] M. I. Alonso, K. Wakita, J. Pascual, M. Garriga, and N. Yamamoto, "Optical functions and electronic structure of CuInSe 2 , CuGaSe 2 , CuInS 2 , and CuGaS 2," *Phys. Rev. B*, vol. 63, 2001.

[9] D. A. Miller, *Optical Properties of Solid Thin Films by Spectroscopic Reflectometry and Spectroscopic Ellipsometry*. New York, NY: ProQuest, 2008, p. 220.

[10] D. L. Meier, J. Hwang, and R. B. Campbell, "The Effect of Doping Density and Injection Level on Minority -Carrier Lifetime as Applied to Bifacial Dendritic Web Silicon Solar Cells," no. I, 1988.

[11] A. Kosarian and P. Jelodarian, "Modeling and Optimization of Advanced Single- and Multijunction Solar Cells Based on Thin-Film a-Si : H / SiGe Heterostructure," vol. 2011, 2011.

[12] P. J. Verlinden, "High Efficiency Concentrator Silicon Solar Cells," in *Practical Handbook of Photovoltaics : Fundamental and Applications*, T. Markvart and C. Luis, Eds. Oxford: Elsevier Ltd, 2003, pp. 436–455.

[13] J. H. Werner and R. Bergmann, "Perspectives of Crystalline Silicon Thin Film Solar Cells," *Tech. Dig. 11th Intern. Photov. Sci. Engin. Conf.*, p. 923, 1999.

[14] M. A. Green, "Solar Cell Fill Factors: General Graph and Empirical," *Solid State Electron.*, vol. 24, no. 8, pp. 788–789, 1981.

Modelling of Hybrid Energy Harvester with DC-DC Boost Converter using Arbitary Input Sources for Ultra-Low-Power Micro-devices

Michelle S.M.Lim[1,3], Sawal H.M.Ali[2], S. Jahariah[1] and MD.Shabiul Islam[1]

[1]Institute of Microengineering and Nanoelectonics (IMEN)
[2]Department of Electrical, Electronics and Systems, Faculty of Engineering and Built Environment
Universiti Kebagnsan Malaysia (UKM),
43600 Bangi, Selangor, Malaysia
Email: mlsm_2002@yahoo.com/shabiul@ukm.edu.my

Michelle S.M.Lim[1,3]

[3]Faculty of Applied Sciences and Computing (FASC),
Tunku Abdul Rahman University College (TAR UC)
Jalan Genting Kelang, Setapak,
53300 KL, Malaysia
Email: mlsm_2002@yahoo.com

Abstract— This work involves the modeling of three arbitrary input sources representing Hybrid Energy Harvesters (HEH) using a DC-DC Boost converter. These sources are combined in parallel and targeted at scavenging passive human power, therefore the three suitable ambient sources are motion, thermal and indoor light. Multiple sources mitigate limitations caused by single source harvesters but suffer impedance mismatches which greatly limit the total combined power that could have been harvested. A Boost Converter with suitable parameters has been designed and integrated to the HEH, and PSPICE software has been used for both the modeling of arbitrary sources as well as the integration with the Boost Converter. An input source as low as 18 mV to 907 mV was able to be boosted into a 310 mV-27.9 V output when suitable parametric values were selected for the Ultra Low Power (ULP) HEH. A duty ratio of 0.5, with 10 kΩ load, 22 μH inductor as well as a switching frequency of 25 kHz was selected to be slightly above the audio range as well as being high enough to reduce passive component sizes. While V_O/V_S of the boost converter is linear, P_O/P_{IN} is a function of third order polynomial. Therefore, at the HEH's lowest combined configuration of 1 K temperature difference, 0.25 g of vibration and 100 lux of indoor lighting, a combined 14 μW can be harvested. At its maximum of 10 K heat difference, 1 g vibration and 1000 lux of indoor lighting a combined 187 μW can be harvested. At its minimum, this enables possibility of battery-less applications in powering a quartz watch at 5 μW while at its maximum capacity powering a pace maker of ~50 μW as well as micro devices of ~100 μW solely from passive human activity. Once a 33 mF input capacitor is placed between the sources and converter, an output power of between 9.61 μW- 78 mW can be obtained.

Keywords—Modelling Arbitary Sources; Hybrid Energy Harvester(HEH); Ultra Low Power (ULP)

I. INTRODUCTION

Autonomous systems such as wearable devices, biomedical implants and wireless sensor nodes are essentially the gist of Ultra Low Power (ULP) energy harvesting. With considerable motivations in ULP consumer products and the need for lifetime lasting power supply, energy harvesting is

the solution. As such, for reliability purposes, multi-input ambient sources have been proposed by several literatures [1]-[4] to ensure continuous flow of energy when either one or two sources are absent. Past researchers for multiple inputs use duty cycling, pulse-width, switching frequency and time multiplexing [1], [4], selecting the input with maximum power [5] or parallel combinations of harvesters [2]. This paper focuses on the parallel approach alike [2] to increase the combined current, albeit for three ubiquitous ambient sources from piezoelectric (PZT), Photovoltaic (PV) and Thermoelectric (TEG). These three commercially available sources will be arbitrarily modeled both individually and combined, in which Table I distinguishes between the three different harvesters in terms of material, impedance for Optimum Power Transfer and Maximum Output Power. These sources, once combined, will then be fed into a DC-DC Boost converter with considerations of suitable parametric analysis such as parasitic inductive resistance, capacitive resistances, rise and fall times, load resistance, duty cycle and switching frequencies in accordance to the three combined ambient sources.

TABLE I. COMPARISONS OF ENERGY HARVESTER CHARACTERISTICS

Parameters	Energy Harvester Types[a]		
	PV [AM1417]	*PZT [V21BL]*	*TEG [CP60133]*
Material	Amorphous	Cantilever	Peltier Module
Size	4.865 cm^2	13.09 cm^2	2.25 cm^2
Operating Range	50 - 1000 lux	0.25 – 1 g	$\Delta T = 1 - 10$ K
Open Circuit Voltage, V_{OC}	1.177 – 2.8 V	1.98 - 6.05 V	53 - 530 mV
Optimum Impedance, Z_{OP}	31 k – 293 kΩ	64.941 kΩ	0.5 Ω
Operating Voltage, V_{MPP}	0.709 - 2 V	1 – 3 V	26.5 – 265 mV
Maximum Power, P_{MPP}	1.71 - 126.8 μW	15.01 - 140.876 μW	1.368 - 139.8 mW
Maximum Power Extraction	Requires Tracking	P_{MPP} at Voc/ 2	P_{MPP} at Voc/ 2

[a.] Note that lower values are obtained during circuit operation due to PMOS diode drop and impedance mismatches.

II. Modelling Of Hybrid Energy Harvesters

Firstly, the power that can be obtained from the PV cell is $P_{PV} = V_{PV} I_{PV}$. A plot of P-V, I-V and P-R characteristic between 50 lux to 1000 lux of fluorescence illumination based on modeling of (1) gives a maximum power, P_{MPP} between 2.4-128.7 µW at an operating voltage of 1.31 - 2.24 V as summarized in Table I. Secondly, a temperature difference of between 1-10 K is simulated leading to 26 - 200 mV of operating voltage and a possibility of 1.368 mW -139.8 mW if impedance were matched at a constant 0.5 Ω as shown in Fig. 1 and summarized in Table I. For the cantilever type MIDE PZT of V21BL, f_P = 110 Hz at 0g tip mass and only 1 of the 2 PZT connected, the input amplitude range of 0.25 g to 1 g provides a V_{MPP} of 1-3 V and P_{MPP} of between 15.01 µW - 140.876 µW at an optimum impedance of 64.941 kΩ as summarized in Table I and verified by the PSPICE simulation in Fig.1.

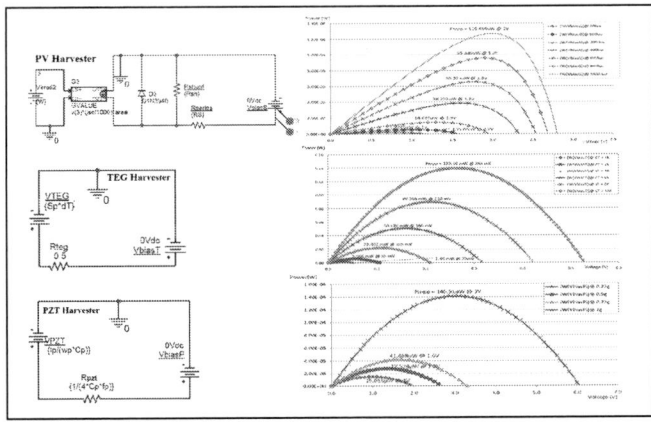

Fig. 1. Equivalent electrical model of energy harvesting sources with corresponding P-V curve using PSPICE.

All three transducers in this literature has been modeled in PSPICE according to the datasheet parameters [9]-[11] as shown in Fig.1 where the single diode model has been used for the modeling of an amorphous PV cell, similar to [1],[2],[6],[9]. The equation representing the solar model is as given by (1) referring to the PV circuit model in Fig. 1 with a Voltage Controlled Current Source (VCCS) indicated as G-value and derivation using the Kirchhoff Current Law and Shockley Equation of ideal diode [2], [6], we have

$$I_{PV} = I_L - I_O[\exp(\frac{V_{PV} + I_{PV}R_S}{nV_T}) - 1] - (\frac{V_{PV} + I_{PV}R_S}{R_{SH}})$$

(2)

where $V_T = kT_C / q$ is the thermal voltage with k = 1.381 × 10-23, q = 1.602 × 10^{-19} C and T_C = operating Temperature in Kelvin. The other five parameters in (1) are $I_L \approx I_{SC}$ = light current/ short circuit current, I_O = diode reverse saturation current, R_S = series resistance, R_{SH} = shunt resistance and n = diode ideality factor. The TEG module however is based on the equation as shown in (2) [2],

$$V_{TEG(OC)} = S * \Delta T = n_T * \alpha(T_H - T_C)$$

(2)

which is made up of n_T thermocouples connected thermally in parallel but electrically in series [1], [2],[7],[10], where α and S are the Seebeck's Coefficient of a single thermocouple and TEG respectively [2] in which S is modeled as 52.8 mV/ K based on [10].

Finally, the PZT harvester together with Mide datasheet values is modeled based on the Thevenin Equivalent circuit of a sinusoid with full bridge rectification as suggested in [1], [8]. A PZT represented normally by a sinusoid current source, given as

$$I_{PZT} = I_P \sin(2\pi f_P)$$

(3)

at a resonant frequency of f_P shunt with internal Capacitance, C_P. and full bridge rectification with ideal diodes has been modeled here as a voltage source of $V_{PZT(OC)} = I_{PZT}/ 2\pi f_P C_P$ with an effective series resistor, $R_{PZT} = 1/4*C_P *f_P$ just as indicated in [1],[8] and as shown in Fig.1.

The corresponding P-V characteristics that has been simulated is shown in Fig. 1. This generally shows that PV harvester requires tracking to obtain the Maximum Power Point (MPP) due to its constantly varying internal impedance in regard to varying Illumination levels. Previous Maximum Power Point Tracking (MPPT) approaches include the fractional open circuit and fractional short circuit [6], [14] or more precise techniques that require extra computation such as perturb and observe (P&O) [1], Incremental Conductance Method [14] as well as particle swarm method [17] and fuzzy algorithm [18] to track its MPP. The TEG and PZT harvesters however have its MPP at half of its open circuit voltage with a fixed internal impedance, thus normally only requires one time setting of impedance [1] and do not require any complex tracking algorithm.

Previously, power conditioning architectures were built to cater for specific harvesters but today the challenge lies in the interactions between these sources which causes impedance mismatches as the TEG harvester has generally very low impedances in the range of Ω while PV and PZT harvesters has impedances in the range of kΩ as can be observed from datasheet values [9]-[11] and the P-R curves of all three harvesters where its optimum impedance for different input source levels, R_{MPP} has been tabulated in Table I. Therefore, it has been stated that no common MPPT technique exists between these three harvesters [12], however, past literatures has already proposed time multiplexing [1] or pulse counting method [4] to adaptively adjust harvester impedance individually before entering any DC-DC converter topology.

Here, these three harvesters will be connected in parallel, as has been proposed previously to combine two sources by [2], [12]. Each of these sources will have a resized PMOS transistor configured as diode to avoid reverse current flow thus contributing to certain amount of losses. The combination of these three sources has been tested for all ranges both individually and combined to obtain optimum impedances in preparation for Boost Converter's parametric analysis in PSPICE. As in [2], the combined HEH voltage, $V_{HEH} = V_{PV} + V_{DPV} = V_{TEG} + V_{DTEG} = V_{PZT} + V_{DPZT}$, while the accumulated

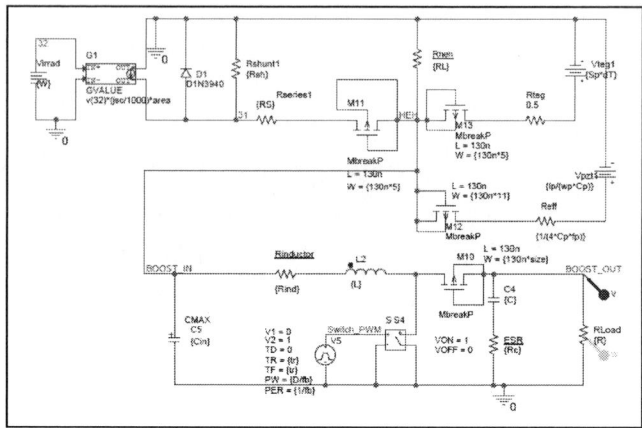

Fig. 2. Equivalent circuit model of boost converter with HEH sources configured in parallel

current, $I_{HEH} = I_{PV} + I_{PZT} + I_{TEG}$. The power, P_{HEH} is given by (3),

$$P_{HEH} = P_{PV} + P_{TEG} + P_{PZT}$$
$$= V_{HEH} * [I_{SC} - I_O \exp(\frac{V_{PV} + I_{PV} R_S}{n V_T}) - 1] \qquad (3)$$
$$+ [V_{HEH} * (\frac{V_{TEG(OC)} - V_{HEH}}{R_{TEG}})] + [V_{HEH} * (\frac{V_{PZT(OC)} - V_{HEH}}{R_{PZT}})])$$

with all its parameters defined previously. Fig. 2, on the other hand, shows the combined electrical model of the proposed HEH in PSPICE with its corresponding P-V, P-R and transient analysis values in Table II when individual or all three sources are present at 100 lux of illumination, 0.25 g amplitude and 1K thermal difference. It should be noted that the low voltage levels and impedance of TEG limits its domination over the other two sources. The R_{MPP} values are obtained during a DC sweep of parametric R values so that optimum impedances are used during transient analysis for Maximum Power Transfer (MPT). It can be observed that due to the additional PMOS diode and parallel combinations for accumulated current values from all sources, the R_{MPP} values will no longer correspond to the originally simulated values as shown in Table I previously. Finally, these combined harvesters will be fed into the input of the Boost converter to be explained in the next section.

TABLE II.   OPTIMUM POWER, VOLTAGE AND IMPEDANCE VALUES FOR VARIOUS HARVESTERS COMBINATIONS

Active Energy Harvesters (Levels)	Optimum Operation Values @ 27°C		
	P_{MPP}	R_{MPP}	V_{MPP}
PV (100 lux)	3.707 µW	185 kΩ	829 mV
TEG (dT = 1K)	1.0899 nW	284 kΩ	18 mV
PZT (0.25g)	10.407 µW	79 kΩ	907 mV
PV + TEG	3.707 µW	185 kΩ	829 mV
PZT + TEG	10.407 µW	79 kΩ	907 mV
PV + PZT	14.085 µW	54 kΩ	876 mV
PV + PZT + TEG	14.085 µW	54 kΩ	876 mV

TABLE III.   EFFECTS OF SOURCE VOLTAGE VARIATIONS

Simulation Condition: D= 0.5, t_r=1 ns, R_{IND}=0.1 Ω, R_C = 0.01 Ω, f_{SW} = 25 kHz, R_{LOAD}=10 kΩ, W/L = 1:1000, T=27 °C			
Active Energy Harvesters Sources (Levels)	Vs	Optimum Boost Output Values	
		Vo,mpp	Po,mpp
PV (100 lux)	829 mV	25.530 V	65.180 mW
TEG (dT = 1 K)	18 mV	310.003 mV	9.610 µW
PZT (0.25 g)	907 mV	27.945 V	78.092 mW
PV + TEG	829 mV	25.547 V	65.266 mW
PZT + TEG	907 mV	27.945 V	78.092 mW
PV + PZT	876 mV	26.926 V	72.501 mW
PV + PZT + TEG	876 mV	26.921 V	72.474 mW

III.   HEH INTEGRATION WITH DC-DC BOOST CONVERTER

Energy harvesters from ambient environment need to be conditioned to an appropriate voltage level either for a fixed load requirements or variable load situation such as battery charging [15] or transmission in a WSN. Therefore, most literature will use either one [4] or multiple DC-DC converters [1], [2], [5] units for these input sources to obtain desired output levels. Here, a single conventional DC-DC Boost converter is considered for the three parallel sources in this literature. Several parameters are analyzed to obtain an optimum output value from a variety of input voltages scavenged from the ambient environment. The parameters to be considered for this Boost converter are inductive resistance, capacitive resistance, rise and fall times, switching frequency, W/L ratios, switch voltage drops and load changes. Based on Table II, the worst case scenario is when only TEG is present with a mere 18 mV with a maximum power transfer when an optimum impedance of 284 kΩ is present. We will proceed to use this voltage condition for the following analysis and finally integrating the sources to the optimized Boost converter is shown in Fig. 2. Several parametric effects contributing to the converter's energy losses will be investigated in future.

Table III summarizes the optimum values of the Boost converter when different configurations of harvesting sources are available at $T = 27$ °C. Source Voltages between 18 mV to 907 mV can be boosted to 310 mV to 27.9 V with power output ranging from 9.6 µW to 78 mW when an input capacitor of 33 mF and a 22 µH inductor is used with 25 kHz of switching frequency while Fig. 3 shows the boosted voltage and output power curve when all input sources are active at various room temperature conditions, with 26.921 V and 72.474 mW at 27 °C as tabulated in Table III.

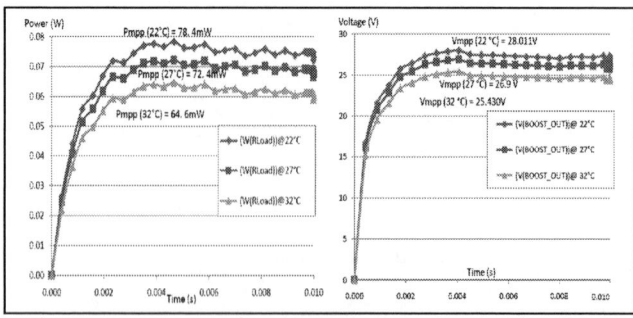

Fig. 3. Ouput power (Left) and boosted voltage curves (Right) of HEH during dT=1 K + 100 lux + 0.25 g vibration at 22 °C, 27 °C and 32 °C

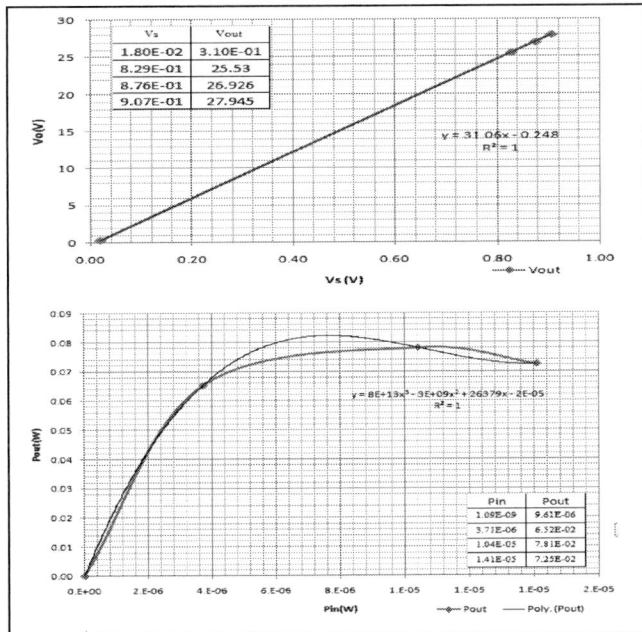

Fig. 4. Plot of V_O/V_S (left) and P_O/P_{IN} (Right)

In its entirety, in order to integrate these three sources optimally, the combined and chosen parameters for an optimum boost converter design for the combined harvesters with an input of 18 mV to 907 mV requires a sampling frequency of 25 kHz, D = 0.5, ESR = 0.01 Ω, t_R = 1 ns, R_L = 0.1 Ω, PMOS W/L = 1000:1 when C_{IN} = 33 mF, C_{LOAD} = 0.2 μF and L = 22 μH in order to obtain an output voltage of ~20 times the source voltages and an output power range of 9.61 μW to 78 mW. All of which are simulated at a room temperature of T = 27 °C. A plot of V_O/V_S is represented by a linear regression trend line while a plot of the integrated P_O/P_{IN} for this boost converter is represented by a third order trend line with corresponding equations shown in Fig. 4.

IV. CONCLUSION

Arbitrary source modeling of PV, PZT and TEG harvesters which is commercially available has been investigated and its generic characteristics achieved from as low as ΔT = 1 K from the TEG to a combined ΔT = 1 K, 100 lux of illumination and 0.25 g of vibration amplitude. It has been verified via PSPICE simulation, mathematical model and datasheet parameters that all harvesters bear wide range of impedance from low Ω values of TEG to kΩ values of the PV and PZT harvesters. The mismatching of these impedances causes loss in overall maximum harvested power, therefore it is pivotal to match these harvesters impedance to its load, in this case a Boost converter. These harvesters are fed into a boost converter to investigate optimum boost converter design for these three harvesters which includes a switching frequency of 25 kHz, duty cycle of, D = 0.5, load of 10 kΩ, inductive resistance of 0.1 Ω, rise times of 1ns, load capacitance of 0.2 μF, L=22 μH, C_{IN} = 33 mF and a ripple below 2.09 % . This boost topology with three harvesting sources achieves almost 20 times input

voltage values from 18 mV-907 mV to output voltage of 0.3 – 27.9 V. A total output power of between 9.61 μW-78 mW can be achieved when a 33 mF input capacitor is inserted before the converter. Also, a plot of V_O/V_S shows output voltage as a linear function of input source voltages while output power is a function of third order polynomial. With this amount power, a possibility if powering Quartz watches, cardiac pace makers, WSN and hearing aids are possible.

REFERENCES

[1] S. Bandyopadhyay, and A. P. Chandrakasan, "Platform architecture for solar, thermal, and vibration energy combining with MPPT and single inductor," *Solid-State Circuits, IEEE Journal of*, vol. 47, no. 9, pp. 2199-2215, 2012.

[2] Y. K. Tan, and S. K. Panda, "Energy harvesting from hybrid indoor ambient light and thermal energy sources for enhanced performance of wireless sensor nodes," *Industrial Electronics, IEEE Transactions on*, vol. 58, no. 9, pp. 4424-4435, 2011.

[3] S. Mi, M. Lim, M. Ali, S. Hamid, and M. Shabiul Islam, "A Novel Architecture of maximum power point tracking for ultra-low-power based hybrid energy harvester in ubiquitous devices: a review," *American Journal of Applied Sciences*, vol. 10, no. 10, 2013.

[4] C. Shi, B. Miller, K. Mayaram, and T. Fiez, "A multiple-input boost converter for low-power energy harvesting," *Circuits and Systems II: Express Briefs, IEEE Transactions on*, vol. 58, no.12, pp.827-831, 2011.

[5] H. Lhermet, C. Condemine, M. Plissonnier, R. Salot, P. Audebert, and M. Rosset, "Efficient power management circuit: From thermal energy harvesting to above-IC microbattery energy storage," *IEEE Journal of Solid-State Circuits*, vol. 43, no. 1, pp. 246-255, 2008.

[6] L. Castaner, and S. Silvestre, *Modelling photovoltaic systems using PSpice*: John Wiley and Sons, 2002.

[7] Y. K. Ramadass, and A. P. Chandrakasan, "A batteryless thermoelectric energy-harvesting interface circuit with 35mV startup voltage." *IEEE Journal of Solid-State Circuits*, vol. 46, no. 1, pp. 333-341, 2011.

[8] Y. K. Ramadass, and A. P. Chandrakasan, "An efficient piezoelectric energy harvesting interface circuit using a bias-flip rectifier and shared inductor," *Solid-State Circuits, IEEE Journal of*, vol.45, no.1, pp.189-204, 2010.

[9] "SANYO Amorphous Solar Cell datasheet: Amorton AM-1417". 2008.

[10] "CUI Inc. CP60 series Peltier Module". 2012.

[11] "MIDE piezoelectric Energy Harvesters". 2013.

[12] Y. K. Tan, Energy Harvesting Autonomous Sensor Systems: Design, Analysis, and Practical Implementation: CRC Press, 2013.

[13] N. A. C. Mustapha, A. Alam, S. Khan, and A. W. Azman, "Parametric analysis for designing low voltage and low frequency energy harvester booster." *2013 IEEE Regional Symposium on Micro and Nanoelectronics (RSM)*. pp. 122-125. Sept, 2013.

[14] N.Femia, G. Petrone, G.Spagnuolo, M.Vitelli, *Power Electronics and Control Techniques for Maximum Energy Harvesting in Photovoltaic Systems*. CRC Press, 2013.

[15] F. Ongaro, S. Saggini, and P. Mattavelli, "Li-ion battery-supercapacitor hybrid storage system for a long lifetime, photovoltaic-based wireless sensor network," *Power Electronics, IEEE Transactions on*, vol. 27, no. 9, pp. 3944-3952, 2012.

[16] A. Richelli, S. Comensoli, and Z. M. Kovacs-Vajna, "A DC/DC boosting technique and power management for ultralow-voltage energy harvesting applications," *Industrial Electronics, IEEE Transactions on*, vol. 59, no. 6, pp. 2701-2708, 2012.

[17] K.Lian, J.Jhang, and I.Tian, "A Maximum Power Point Tracking Method Based on Perturb-and-Observe Combined With Particle Swarm Optimization," *IEEE Journal of Photovoltaics*,vol.4,no.2, pp.626-633, 2014.

[18] B. N. Alajmi, K. H. Ahmed, S. J. Finney, and B. W. Williams, "Fuzzy-logic-control approach of a modified hill-climbing method for maximum power point in microgrid standalone photovoltaic system," *Power Electronics, IEEE Transactions on*, vol. 26, no. 4, pp. 1022-1030, 2011.

Correlation between the microstructure of copper oxide thin film and its gas sensing response

Jia Wei Low, Nafarizal Nayan, *Member, IEEE,*
Mohd Zainizan Sahdan, Mohd Khairul Ahmad
Microelectronic and Nanotechnology – Shamsuddin
Research Centre (MiNT-SRC),
Universiti Tun Hussein Onn Malaysia
86400, Batu Pahat, Johor, Malaysia
Email: jwlow88@hotmail.com

Ali Yeon Md Shakaff, Ammar Zakaria
Centre of Excellence for Advanced Sensor
Technology
Universiti Malaysia Perlis
02600 Arau Perlis, Malaysia
Email: aliyeon@unimap.edu.my

Ahmad Faizal Mohd Zain
Faculty of Manufacturing Engineering
Universiti Malaysia Pahang
26300 Gambang Kuantan, Pahang Darul Makmur, Malaysia
Email: afaizal@ump.edu.my

Abstract—Copper oxide gas sensor was prepared on silicon wafer by sputtering of copper target at different oxygen flow rate of 0, 4, 8 and 16 sccm using RF magnetron sputtering technique. Argon flow rate, RF power, working pressure and substrate bias voltage were fixed at 50 sccm, 400 W, 22.5 mTorr and -40 V, respectively. The effect of varying the oxygen flow rate towards the time response of the copper oxide gas sensor was investigated. In addition, the influence of the copper oxide thin films microstructure and I-V characteristic was also considered. Based on the result, copper oxide gas sensor fabricated at 8 sccm of oxygen flow rate provide a better response of 0.024V/s compare to those fabricated at 0, 4 and 16 sccm.

Keywords—*sputtering; thin films; copper oxide; gas sensor*

I. Introduction

As a p-type semiconductor, copper oxide (Cu_2O and CuO) is one of the promising materials for applications such as dye sensitized solar cells, photo catalysis, photochromic devices and gas sensing devices [1]. Copper oxide has excellent advantages compared to ZnO, SnO_2 and In_2O_3 for its non-toxic and inexpensive material [2]. In addition, it has strong sensing properties especially for gas sensor which has been proven by A. S. Zoolfakar *et. al.* and P. Samarasekara *et. al.* [3-4].

To date, copper oxide thin films have been deposited and grown using several methods, such as chemical vapor deposition [5], sol-gel [6], pulsed laser deposition [7] and radio frequency (RF) magnetron sputtering [8]. Among these available techniques, RF magnetron sputtering deposition has

been widely used in the formation of copper oxide thin film due to its low deposition rates and uniformity of the thin films [8]. Besides, RF magnetron sputtering techniques is more simple and repeatability performance. In general, the characteristics of the deposited copper oxide thin films are influenced by the deposition parameter such as oxygen flow ratio, sputtering power, working pressure and deposition time and substrate bias voltage. Hence, in this work the influence of oxygen flow rate was taken into consideration.

They have been known for a long time that electrical resistance of a copper oxide is very sensitive to the changes on the surface [9]. Other than that, copper oxide attracted much attention in the field of gas sensing under atmospheric conditions due to their low cost, flexibility in production, simplicity of their use and most important is that it detect large number of gases [10]. Throughout this year, numerous researchers have shown interest in the characteristic of copper oxide to be able to produce a reaction with gases on the surface [11]. However, the reaction in between the gas and solid-state copper oxide often influenced by various causes, such as the base properties of the material used, surface areas and microstructure of the sensing layers. Based on M. Parmar and K. Rajanna [12], when certain concentration of methanol/ethanol is introduced on the CuO thin film, a chemisorbed process will happen on the surface of the CuO thin film. After that, the CuO thin film will undergo dehydrogenation in order to breakdown CO_2 and H_2 using pre-absorbed oxygen atom. Electron that is releases from the dehydrogenation process causes an increase in the number of minority charge carriers and reducing majority charge carriers in the p-type semiconducting film. This in turn increases the resistance of the thin film. Therefore, in this paper, we will discuss the influence of microstructure of the copper oxide thin films towards the time response performance of the copper oxide gas sensor and the I-V characteristic of the copper oxide thin films.

II. Experimental Setup

Fig. 1 shows the experimental setup for copper oxide thin film deposition using RF magnetron sputtering plasma. First of all, the magnetron sputtering source is made of a conventional system that is made of cylindrical permanent magnets attached to an indirect water cooling system [13]. The sputtering plasma was produced by a 13.6 MHz RF magnetron discharges with an automated matching network and rf discharges power was fixed at 400 W during the whole deposition process. In order to produce a high quality thin film, the sputter chamber was evacuated to base pressure of 5×10^{-6} Torr using vacuum turbo pump and backed by rotary pump. As for the substrate, the Cu_2O and CuO thin film was deposited on 4-inch p-type silicon wafer at a distance of 13 cm from the target. The silicon wafer was oxidized to form 1 μm thick SiO_2 layer as the insulator. The sputter target is a 3-inch in diameter pure (99.99% purity) copper target. The required sputtering gas, argon, and reactive gas, oxygen was admitted into the chamber by using the mass flow controller which is attached to the top of the vacuum chamber. The Ar flow rate was fixed at 50 sccm. The total working pressure was also fixed at 22.5 mTorr during the sputtering process. As for the investigated parameter, the oxygen flow rate was varied at 0, 4, 8 and 16 sccm. Lastly, the substrate bias voltage was fixed at -40 V and the whole deposition time was 4 minutes. A summary of the parameter used during the deposition of Cu_2O and CuO thin film are tabulated in table 1.

As for the microstructure of the copper oxide thin films, it was evaluated using high-resolution field emission scanning electron microscope from Joel JSM-7600F series. In order to perform an investigation on the response time of the gas sensor, a simple circuit is used to conduct the measurement. Fig. 2 is the schematic diagram for the measurement setup.

III. Results And Discussion

Fig. 3 shows the microstructure image of copper oxide thin films surface observed using FE-SEM at 80,000 magnification and 5 kV acceleration voltages. The microstructure of the copper oxide thin films deposited at 0 sccm oxygen flow rate is noticed that the thin film produced was not homogeneous and it forms a lot of cracks within the thin films. As the oxygen flow rate was increased to 4 sccm, the thin

TABLE 1. SUMMARY OF THE PARAMETER USED DURING SPUTTERING PROCESS

Working Distance	13 cm
Substrate Temperature	Room Temperature
Argon Flow Rate	50 sccm
Oxygen Flow Rate	0, 4 ,8, 16 sccm
Working pressure	22.5 mTorr
RF dissipation power	400 W
Target	3-inch Copper target
Substrate bias voltage	-40 V
Deposition Time	4 mins

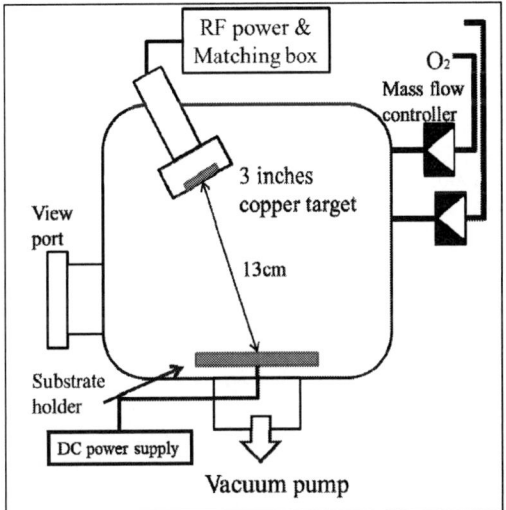

Fig. 1. Experimental setup for copper oxide thin film deposition by reactive magnetron sputtering system.

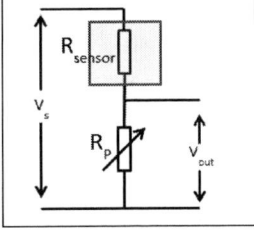

Fig. 2. The schematic diagram for measurement setup.

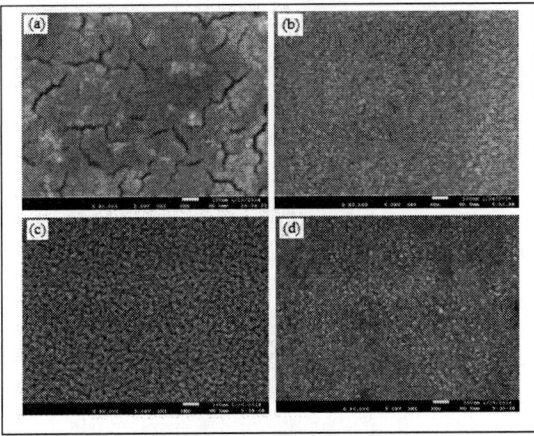

Fig. 3. FE-SEM image of copper oxide thin films at various oxygen flow rate (a) 0sccm, (b) 4sccm, (c) 8sccm and (d) 16sccm

film appears to form smaller particles size due to the existence of $Cu_2O + CuO$. On the other hand, as it forms particles on the thin films, the surface area of the thin films was also greatly enhanced. When the oxygen flow rate was increased further to 8sccm, the particle size is getting larger.

Finally, when the oxygen flow rate was increased to 16 sccm, it retains the same crystallite shape of pure CuO characteristic but smaller in size compared to the copper oxide thin film formed at oxygen flow rate of 8 sccm.

Fig. 4 displays the *I-V* characteristic of the copper oxide at various oxygen flow rate. Since that the thin films deposited using 0 sccm oxygen flow rate exhibit Cu characteristic, the thin films shows a linear graph as shown in fig. 5. On the other hand, the copper oxide thin films deposited at 4, 8 and 16 sccm demonstrated a curve which is similar to those of semiconductor characteristic. Basically, *I-V* characteristic helps to describe the material to be a metal or a non-metal. Based on the *I-V* characteristic, the material can be distinguished whether it is a metal-metal contact or metal-semiconductor contact by applying a layer of metal above the sputtered copper oxide thin films. Due to its characteristic, the metal-semiconductor contact will form a Schottky barrier. The barrier between a metal and semiconductor is foreseen by the Schottky-Mott rule to be proportional to the difference of the metal-vacuum work function and the semiconductor-vacuum electron affinity. Work function is the thermodynamic work obtained by moving the electron from the vacuum to the Fermi level, as in (1).

$$W = E_{VAC} - E_F \qquad (1)$$

Electron affinity, typically denoted by E_{EA}, is defined as the energy required by moving an electron from the vacuum just outside the semiconductor to the bottom of the conduction band just inside the semiconductor, as in (2).

$$E_{EA} = E_{VAC} - E_C \qquad (2)$$

Since that the work function for copper oxide is similar it does not play a vital point in this different in *I-V* characteristic. The main point that brings to the different in *I-V* characteristic is the electron affinity which depends on the surface termination, for example the crystal facet, surface chemistry and crystallinity of the sample. Based on the *I-V* characteristic curve, the barrier height of the semiconductor is higher when it is nearer to the x-axis. Figure 4 shows that the thin films deposited at 8 sccm oxygen flow rate have the highest barrier height followed by the one deposited at 4 sccm oxygen flow rate and 16 sccm oxygen flow rate. This is acceptable as the crystallinity of the thin films deposited at 8 sccm oxygen flow rate is more noticeable.

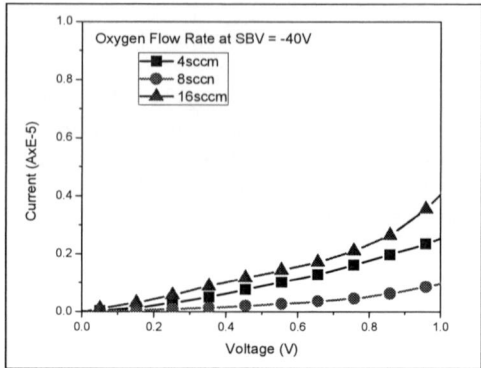

Fig. 4. I-V characteristic of the copper oxide thin films at various oxygen flow rate.

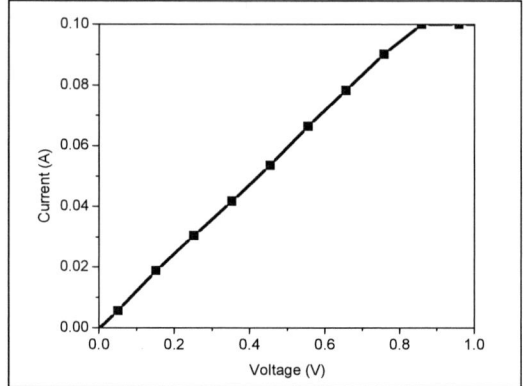

Fig. 5. I-V characteristic of the copper oxide thin films at 0sccm oxygen flow rate.

Fig. 6 is the time response of the copper oxide gas sensor deposited at various oxygen flow rate. The circuit for measurement is design using the theory of voltage divider rule as follows (3):

$$V_{OUT} = \frac{R_P}{R_P + R_{sensor}} V_S \qquad (3)$$

First of all, V_{OUT} is the output voltage measured using oscilloscope. V_S is the voltage sources that will be supplied to the circuit. In this case, the voltage source was fixed at 5 V. R_S is the sensor resistance and R_P is the potentiometer that is used to alter the resistance of the circuit. It is understandable that higher resistance value for R_P will produce a better output which is in good agreement with the Ohm's Law (4).

$$V = IR \qquad (4)$$

Based on M. Parmar and K. Rajanna [12], when ethanol gas is reacted with CuO gas sensor, the resistance of the CuO thin film increases.

As indicated earlier, the thin films deposited at 0 sccm oxygen flow rate were a conductive material due to the existence on Cu characteristic. Therefore, it does not show any response when it is reacted with ethanol vapour. As for the rest, the copper oxide thin films deposited at 8 sccm oxygen flow rate show a

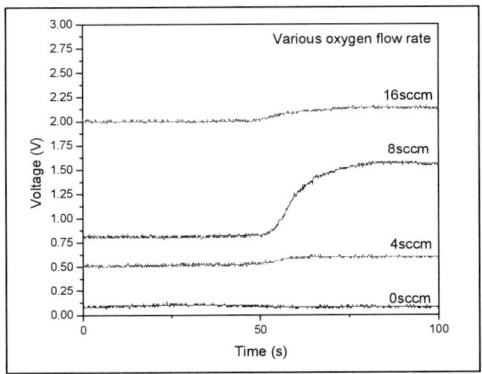

Fig. 6. Time response of the copper oxide gas sensor deposited at various oxygen flow rate.

remarkably time response of 0.024 V/s which is the highest among the gas sensor tested. The copper oxide thin films deposited at 4 sccm oxygen flow rate have a time response of 0.0055 V/s and the copper oxide thin films deposited at 16 sccm oxygen flow rate have a time response of 0.0058 V/s. Therefore, it is best to accept that the copper oxide thin film deposited at 8 sccm oxygen flow rate was the suitable parameter to produce a copper oxide gas sensor.

IV. Conclusion

The response time for various oxygen flow rate for the deposition of copper oxide thin film was successfully analyzed using simple circuit design. The relationship between amount of oxygen flow rate during the deposition of copper oxide thin film and the response time for individual copper oxide gas sensor was studied. On the other hand, the effect of microstructure and I-V characteristic towards the time response was also studied. Based on the result, it is clearly shown that Cu_2O/CuO gas sensor that were deposited at 8 sccm oxygen flow rate provides the best response time compared to the 0, 4 and 16 sccm deposited copper oxide gas sensor. This is due to the structure of the sample have the highest surface area.

Acknowledgement

The present work was supported by Malaysian Technical Universities Network: Centre of Excellence Grant and Short Term Grant of Universiti Tun Hussein Onn Malaysia. The authors would also like to thank Universiti Tun Hussien Onn Malaysia

for the financial support through the MTUN-COE C023.

References

[1] Y. Zhang, X. He, J. Li, H. Zhang, and X. Gao, "Gas-sensing properties of hollow and hierarchical copper oxide microspheres," *Sensors Actuators B Chem.*, vol. 128, no. 1, pp. 293–298, Dec. 2007.

[2] D. Jundale, S. Pawar, M. Chougule, P. Godse, and S. Patil, "Nanocrystalline CuO Thin Films for H_2S Monitoring☐: Microstructural and Optoelectronic Characterization," *J. Sens. Technol.*, vol. 1, pp. 36–46.

[3] A. S. Zoolfakar, M. Z. Ahmad, R. A. Rani, J. Z. Ou, S. Balendhran, S. Zhuiykov, K. Latham, W. Wlodarski, and K. Kalantar-zadeh, "Nanostructured copper oxides as ethanol vapour sensors," *Sensors Actuators B. Chem.*, 2013.

[4] P. Samarasekara, N. T. R. N. Kumara, and N. U. S. Yapa, "Sputtered copper oxide (CuO) thin films for gas sensor devices," *J. Phys. Condens. Matter*, vol. 18, no. 8, pp. 2417–2420, Mar. 2006.

[5] D. Barreca, E. Comini, A. Gasparotto, C. Maccato, C. Sada, G. Sberveglieri, and E. Tondello, "Chemical vapor deposition of copper oxide films and entangled quasi-1D nanoarchitectures as innovative gas sensors," *Sensors Actuators B Chem.*, vol. 141, no. 1, pp. 270–275, Aug. 2009.

[6] A. Y. Oral, E. Menşur, M. H. Aslan, and E. Başaran, "The preparation of copper(II) oxide thin films and the study of their microstructures and optical properties," *Mater. Chem. Phys.*, vol. 83, no. 1, pp. 140–144, Jan. 2004.

[7] W. Seiler, E. Millon, J. Perrière, R. Benzerga, and C. Boulmer-Leborgne, "Epitaxial growth of copper oxide films by reactive cross-beam pulsed-laser deposition," *J. Cryst. Growth*, vol. 311, no. 12, pp. 3352–3358, Jun. 2009.

[8] M. Hari Prasad Reddy, A. Sreedhar, and S. Uthanna, "Structural, surface morphological and optical properties of nanocrystalline Cu_2O films prepared by RF magnetron sputtering: substrate bias effect," *Indian J. Phys.*, vol. 86, no. 4, pp. 291–295, May 2012.

[9] V. E. Bochenkov and G. B. Sergeev, "Sensitivity , Selectivity , and Stability of Gas-Sensitive Metal-Oxide Nanostructures," vol. 3, pp. 31–52, 2010.

[10] C. Wang, L. Yin, L. Zhang, D. Xiang, and R. Gao, "Metal oxide gas sensors: sensitivity and influencing factors.," *Sensors (Basel).*, vol. 10, no. 3, pp. 2088–106, Jan. 2010.

[11] G. Korotcenkov, "Metal oxides for solid-state gas sensors: What determines our choice?," *Mater. Sci. Eng. B*, vol. 139, no. 1, pp. 1–23, Apr. 2007.

[12] M. Parmar and K. Rajanna, "Copper (II) oxide thin film for methanol and ethanol sensing," vol. 4, no. 4, pp. 710–725, 2011.

[13] N. Bin Nayan, "Studies on high – pressure magnetron sputtering plasmas using laser – aided diagnostic techniques," 2008.

IEEE-ICSE2014 Proc. 2014, Kuala Lumpur, Malaysia

Quantum Ballistic Simulation Study of InGaAs/InAs/InGaAs Quantum Well MOSFET: Effects of Doping and Physical Device Parameters

*Sudipta Romen Biswas, Kanak Datta, Ehsanur Rahman, Abir Shadman, Quazi D. M. Khosru

Department of Electrical and Electronic Engineering
Bangladesh University of Engineering and Technology
Dhaka–1000, Bangladesh
*Email: sudipta_r_biswas@yahoo.com

Abstract—In this work, simulation study of device parameter variation on quantum ballistic Current-Voltage (I-V) characteristics of a $In0.7Ga0.3As/InAs/In_{0.7}Ga_{0.3}As$ Quantum Well (QW) MOSFET is presented. Doping density and various physical device parameters like channel thickness, gate dielectric thickness affect ballistic performance of nanoscale transistors. To simulate Current-Voltage (I-V) characteristics in quantum ballistic regime, nonequilibrium Green's function formalism (NEGF) has been used. In this work, the effect of device parameters on subthreshold and short channel performance is also demonstrated. It is observed that scaling of gate dielectric material and channel thickness could provide better electrostatic control at the expense of ballistic device current. However ballistic current can be improved by increasing doping density.

Keywords—*Quantum Well MOSFET, 2D Electrostatics, III-V Semiconductors, Delta Doping, Ballistic Transport*

I. INTRODUCTION

With silicon CMOS technology reaching its bottleneck due to practical and fundamental limitations [1, 2], research on new and alternate device structures and device materials has become very important. III-V materials and their associated device structures [3] have emerged as potential candidates for future low power, high speed device applications. A 30nm QW device structure that incorporates a thin pure InAs sub-channel and extremely scaled HfO2 as gate dielectric has been reported in recent literature [4]. In this paper, we observe the effect of various device parameters like doping density, channel thickness, gate dielectric thickness on ballistic performance of the device structure. The simulation approach is evolved from nanoMOS 2.5 [5]. Similar approach has already been applied for the simulation of III-V HEMT device structures [6]. We follow the same approach for the quantum ballistic simulation in this study.

II. DEVICE STRUCTURE AND SIMULATION APPROACH

In this work, we simulate a slightly simplified version of the device structure reported in [4]. This structure consists of $In_{0.7}Ga_{0.3}As/InAs/In_{0.7}Ga_{0.3}As$ channel. High-k material HfO2 has been used as the gate dielectric. To improve interface properties, a 1 nm thin InP layer is grown on top of the channel. 1 nm Si δ-doped layer is placed 5 nm below the $In_{0.7}Ga_{0.3}As$ channel region.

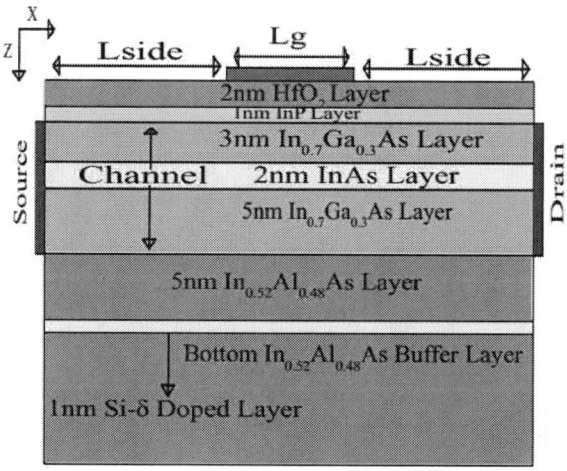

Fig. 1. Simplified simulation region for the 30 nm QW device used in this work. Simulation region is similar to the region used in[6].

5 nm $In_{0.52}Al_{0.48}As$ layer is placed between the channel and δ-doped region to reduce ionized dopant scattering. The bottom $In_{0.52}Al_{0.48}As$ buffer layer is above 300 nm thick as reported in [4]. In actual devices current flow takes place from raised drain contact through the heterostructure stack and to the channel. However, to make the simulation process simpler we place ideal contacts at two ends of the channel. Although this approach does neglect some practical aspects of device operation like source access issues, source starvation phenomenon mentioned in [9, 10], it could provide us with an upper level projection of ballistic performance [6].

Fig. 1 shows the simplified simulation region used in this study. Here Lg refers to the gate length. Lside refers to the distance from gate to n^+ layer edge of the device. To maintain numerical stability we choose Lside to be 44 nm and doping in the doped δ-doped Si region to be 1.5×10^{16} /m². Simulation procedure can be found in details in [6].

The simulation starts by assuming a flat band potential profile. We determine spatial charge density in the device from this initial assumption. Using this charge profile, 2D Poisson equation is solved using a nonlinear algorithm [7].

978-1-4799-5761-3/14 $31.00 © 2014 IEEE

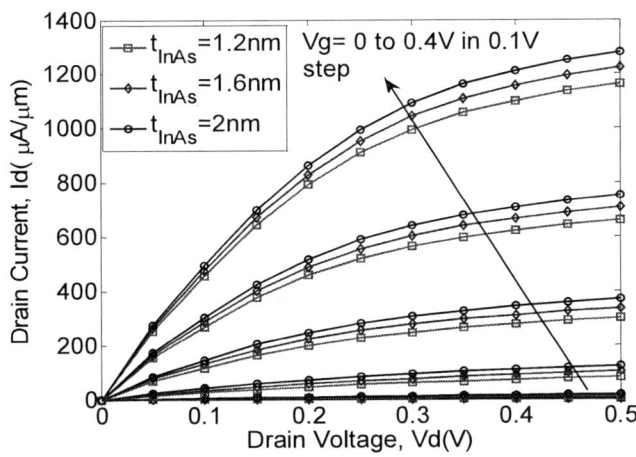

Fig. 2. Id-Vd characteristics for the device at three different channel thickness conditions. Device with higher channel thickness leads to higher carrier accumulation in the channel and therefore higher drain current.

Using the solution of Poisson equation, Schrodinger equation is solved along gate confinement direction at different channel positions to extract wave function and subband profiles in the channel. Using extracted subband profiles, transport equations are solved along the channel using NEGF formalism, spatial charge profile is determined in the structure and self-consistent simulation continues. Once self consistency is achieved, quantum ballistic current is calculated using NEGF formalism. For current calculation, we consider 2 subbands in the QW channel.

III. RESULTS AND DISCUSSION

A. Effect of Channel Thickness

Fig. 2 shows the effect of channel thickness variation on intrinsic Id-Vd characteristics for the 30 nm gate length device structure. Here the gate voltage is varied from 0 volt to 0.4 volt. In this study, we varied the thickness of middle InAs region in steps.

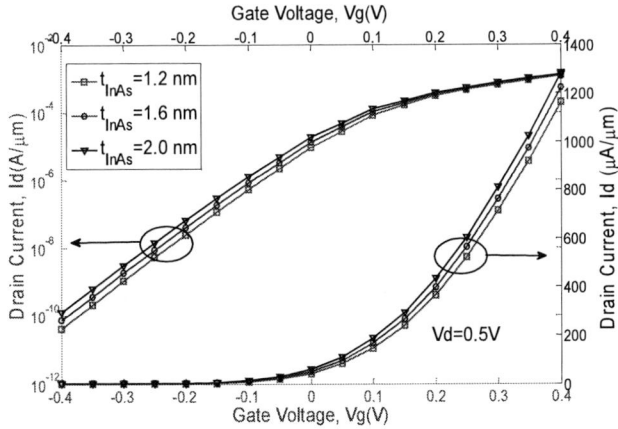

Fig. 3. Effect of InAs layer thickness on Id-Vg characteristics of the 30 nm device at drain bias, Vd= 0.5 volt. Device with higher channel thickness gives higher current in subthreshold region and also at high gate bias as a result of higher carrier accumulation in the channel.

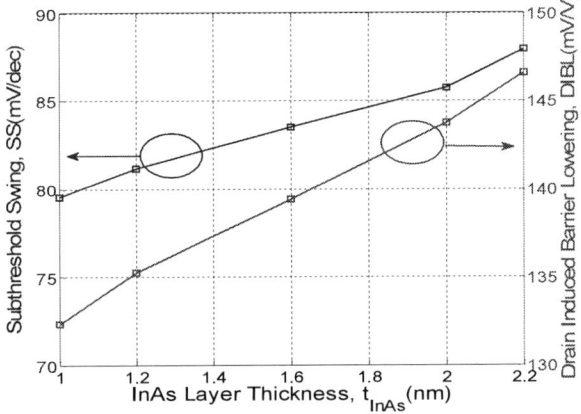

Fig. 4. Effect of InAs layer thickness on SS and DIBL parameters. Increased channel thickness degrades subthreshold performance of the device and makes the device more prone to short channel effects.

From Fig. 2 it is observed that, as we increase channel thickness, the drain current increases due to higher carrier accumulation in the channel. Fig. 3 shows Id-Vg characteristics for various channel thickness conditions at a drain bias of 0.5 volt in both normal and log scale. The logId-Vg characteristics give us an idea of the subthreshold performance of the device. Fig. 4 shows the effect of channel thickness on subthreshold swing (SS) and drain induced barrier lowering (DIBL) calculated from ballistic simulation results. Fig. 4 reveals that DIBL increases with increasing channel thickness. This could be attributed to the lowering of gate control over the spreading of carriers with increasing channel thickness and 2D electrostatics.

B. Effect of Gate Oxide

We have observed the effect of gate oxide on ballistic characteristics of the device. We vary the HfO_2 dielectric thickness from 1.6 nm to 2.4 nm and observe the effect on device characteristics.

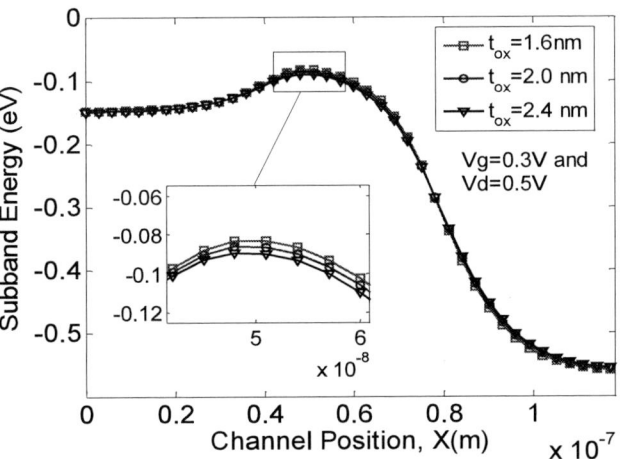

Fig. 5. First subband profile in the channel for intrinsic gate bias, Vg= 0.3 volt and drain bias, Vd= 0.5 volt. According to 2D electrostatics, increased oxide thickness results in weaker gate control over the source to channel potential barrier. Higher oxide thickness results in lower potential barrier.

IEEE-ICSE2014 Proc. 2014, Kuala Lumpur, Malaysia

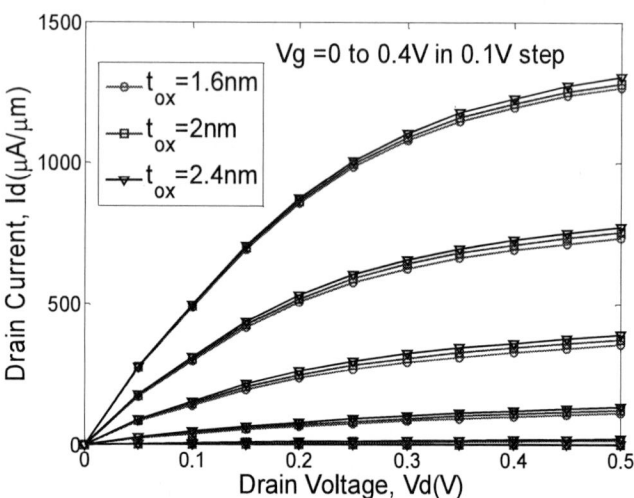

Fig. 6. Id-Vd characteristics in ballistic regime for different oxide thickness conditions for the 30 nm device. Increased oxide thickness lowers potential barrier for the source injected carriers and results in higher drain current.

Fig. 5 shows the effect of gate oxide thickness on the first subband profile for the 30 nm device. As oxide thickness is increased, for the same applied gate bias, the voltage drop across oxide increases and therefore control of gate voltage over the potential barrier in the channel becomes weaker. Due to 2D electrostatics, this could lead to a lowered potential barrier for the source injected carriers. Lowered potential barrier could lead to a higher current for higher gate oxide thickness condition.

Fig. 6 shows Id-Vd characteristics for the device at different gate oxide thickness conditions. Here, the gate voltage is being varied from 0 volt to 0.4 volt in 0.1 volt step. Although the device with higher oxide thickness have a higher drain current, drain current increases more rapidly with channel thickness than oxide thickness.

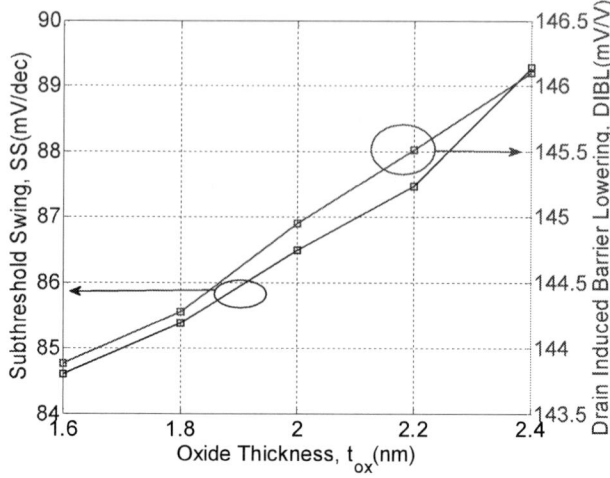

Fig. 8. SS and DIBL values extracted from simulation at different oxide thickness conditions. Increased oxide thickness degrades electrostatic control and therefore degraded subthreshold performance is observed.

Fig. 7 shows the Id-Vg characteristics for the device at different oxide thickness conditions. Device with higher oxide thickness shows higher drain current in subthreshold region. Fig. 8 shows the effect of oxide thickness on device parameters like subthreshold swing (SS) and drain induced barrier lowering (DIBL) calculated from ballistic simulation.

C. Effect of Doping Density

We have also observed the effect of doping density on the device performance. In this study, we vary the doping density in the Si δ-doped region from 1.5×10^{16} /m^2 to 1.8×10^{16} /m^2 and observe the results. Fig. 9 shows the first subband profile in the channel for various doping concentrations at a gate voltage 0.3 volt and drain voltage 0.5 volt. It can be seen that, device with higher doping density shows a lower subband profile.

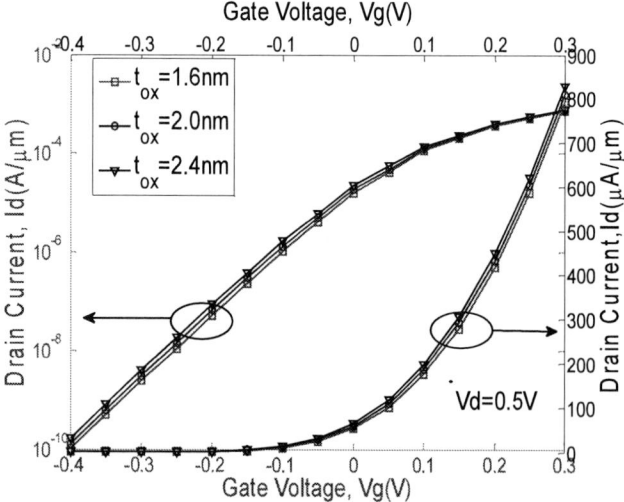

Fig. 7. Id-Vg characteristics for the simulated device at intrinsic drain bias, Vd= 0.5 volt. Device with higher oxide thickness leads to higher current in the subthreshold regime.

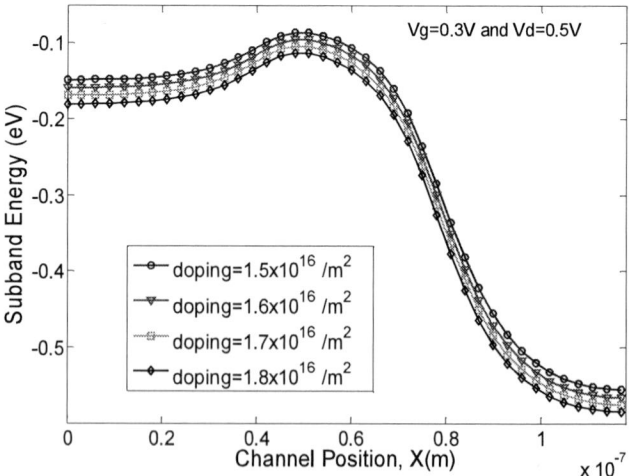

Fig. 9. First subband profile in the channel for various doping concentrations at gate bias, Vg=0.3 volt and drain bias, Vd= 0.5 volt. Device with higher doping concentration has a slightly lower subband profile to account for higher carrier accumulation in the channel.

978-1-4799-5761-3/14 $31.00 © 2014 IEEE   38

IEEE-ICSE2014 Proc. 2014, Kuala Lumpur, Malaysia

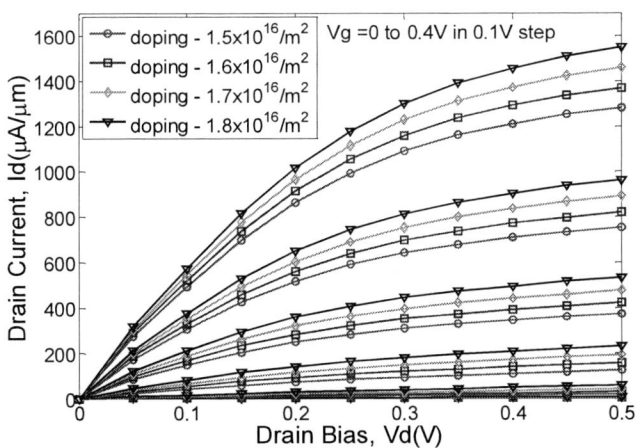

Fig. 10. Id-Vd characteristics for the 30 nm device under various doping concentrations. Device with higher doping has more carriers in the channel and therefore higher current for same bias conditions.

MOSFETs usually operate by modulating the potential barrier between source and channel [8]. As gate voltage increases, the barrier gets lowered and more carriers are injected from source. This energy barrier also depends on doping density of the device. For same gate and drain bias condition, device with a higher doping density shows a slightly higher potential barrier for the source injected carriers.

Fig. 10 shows the effect of doping on Id-Vd characteristics of the device. The device with higher doping density shows higher current for same bias conditions. Fig. 11 shows the effect of doping density on Id-Vg characteristics of the device. At subthreshold region, device with higher doping has higher current. Fig. 12 shows the effect of doping density on subthreshold swing and drain induced barrier lowering which reveals that both subthreshold swing and drain induced barrier lowering increase with increased doping. At high doping density, the energy barrier at the virtual source position is lowered more with increasing drain bias voltage. This could lead to an increase in drain induced barrier lowering (DIBL) of the device according to 2D electrostatics.

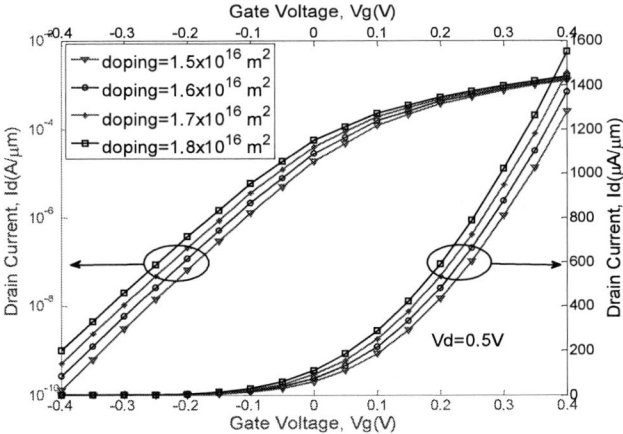

Fig. 11. Id-Vg and logId-Vg characteristics for the 30 nm device under various doping concentrations at drain bias, Vd=0.5 volt. Higher doping density increases subthreshold current and degrades subthreshold performance.

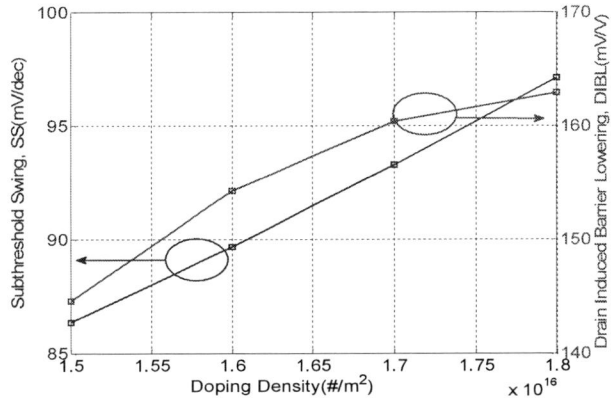

Fig. 12. SS and DIBL values extracted from simulation for various doping concentrations. Increased doping density leads to degraded subthreshold performance and makes the device more prone to drain induced barrier lowering.

IV. CONCLUSION

In this work, a simulation study of device parameter variation on quantum ballistic performance of a novel 30nm $In_{0.7}Ga_{0.3}As/InAs/In_{0.7}Ga_{0.3}As$ QW MOSFET has been reported. Doping density exerts more pronounced effect on Current-Voltage characteristics among observed three parameters. Scaling thin InAs sub-channel could improve short channel performance of the device, but at the same time ballistic performance could get compromised. Increasing gate dielectric material thickness could improve ballistic device performance but could also lead to degraded short channel performance. However, ballistic current can be improved by increasing doping density.

REFERENCES

[1] International Technology Roadmap for Semiconductors [Online]. Available:http://public.itrs.net

[2] Oktyabrsky,Ye, et. al. Fundamentals of III-V Semiconductor MOSFETs. Springer US, 2010.

[3] J. A. del Alamo, "Nanometre-scale electronics with III-V compound semiconductors," Nature, vol. 479, pp. 317-323, November 2011.

[4] Lin, J., D. A. Antoniadis, and J. A. Del Alamo. "Sub-30 nm InAs Quantum-Well MOSFETs with self-aligned metal contacts and Sub-1 nm EOT HfO 2 insulator." In Electron Devices Meeting (IEDM), 2012 IEEE International, pp. 32-1. IEEE, 2012.

[5] Z. Ren, R. Venugopal, S. Goasguen, S. Datta, M. Lundstorm, "nanoMOS 2.5: A two-dimensional simulator for quantum transport in double-gate MOSFETs," Electron Devices, IEEE Transactions on, vol 50, pp. 1914-1925, Sept. 2003.

[6] Y. Liu, and M. Lundstrom, "Simulation of III-V HEMTs for high-speed low-power logic applications," ECS Transactions 19.5 (2009): 331-342.

[7] Z. Ren, "Nanoscale MOSFETs: Physics, Simulation, and Design," Ph.D. Thesis,Purdue University, West Lafayette, IN (2001).

[8] M. Lundstrom, and Z. Ren, "Essential physics of carrier transport in nanoscale MOSFETs," Electron Devices , IEEE Transactions on, vol. 49, pp. 133-141, Jan. 2002.

[9] Y. Naveh, and K. K. Likharev. "Modeling of 10-nm-scale ballistic MOSFET's," Electron Device Letters, IEEE, vol. 21, pp. 242-244, May 2000.

[10] M. V. Fischetti, et al. "Simulation of electron transport in high-mobility MOSFETs: Density of states bottleneck and source starvation." In Electron Devices Meeting, 2007. IEDM 2007. IEEE International, pp. 109-112. IEEE, 2007.

Low Dimensional Simulator for Carbon-based Devices

Chin Lin Ng
Faculty of Electrical Engineering
Universiti Teknologi Malaysia
81310 Skudai, Johor, Malaysia
clng_0927@hotmail.com

Michael Loong Peng Tan
Faculty of Electrical Engineering
Universiti Teknologi Malaysia
81310 Skudai, Johor Malaysia
michael@fke.utm.my

Abstract— Carbon-based devices such as carbon nanotubes (CNT) and graphene nanoribbon (GNR) have been explored rigorously as the potential successor to conventional metal–oxide–semiconductor field-effect transistors (MOSFET). The limitations of silicon-based devices have catalysed much breakthrough research on carbon-based devices. In this paper, a comprehensive quantum simulation tool based on carbon devices is developed as a graphical user interface (GUI) using MATLAB. It is known as Low Dimensional Simulator (LODISI). This simulation tool allows the user to reach a trade-off between precision and time, as it provides impromptu analysis either by graph or direct calculation values. In addition, the auto generation of the voltage transfer curve from the complementary nanotransistor drain characteristics is one of the significant feature of LODISI.

Keywords— graphene; nanoribbon; MATLAB; CMOS; quantum simulator

I.

In 2004, the groundbreaking work carried out by the two physicists Andre Geim and Kostya Novoselow on two-dimensional graphene won the 2010 Nobel Prize for Physics [1]. As a result, scientists around the world became increasingly interested in this material and its potential to break through the limitations of silicon technology. Despite these encouraging prospects, carbon-based devices are still in their relative infancy, and much research is still required. A simulation tool that predicts electronic characteristics and provides impromptu analysis is essential in the innovation of this technology. Undoubtedly, a few simulation tools for carbon-based device are available in the market, such as NEMO-VN1 [2], Quantum Wise [3] and FETToy [4, 5]. However, these simulations tools do not include the both p-type and n-type simulation of carbon nanotube (CNT) and graphene nanoribbon (GNR) field-effect transistors (FETs). In addition, our simulation tool allow the assessment of a digital inverter using the voltage transfer curve (VTC) characteristic curve to obtain the gain, and operating logic-levels. Hence, the new simulation tool, known as Low Dimensional Simulator (LODISI), is proposed and designed to fulfil these demands.

II.

A. Device Modelling

This simulation tool is based on the top-of-the-barrier modelling approach along the nanostructure channel and the source Fermi energy. The utilized top-of-the-barrier method is one of the alternative device modeling technique which uses a simpler ballistic model that is able to capture and solve the device physics effectively and efficiently. The framework originated from the analytical MATLAB script codenamed FETToy, as proposed by Rahman [4, 6]. FETToy can be assessed as a open source online simulator through nanoHUB.org [7]. Similar to FETToy, our simulator, Low Dimensional Simulator (LODISI) is based on the top-of-the-barrier modelling. The modeling framework has been benchmarked and verified against simulation and experimental data and simulation data in previous work [8, 9].

$$I_{ds} = G_{ON} \frac{K_B T}{q} \left[\Im_0 \left(\frac{U_{SF}}{K_B T} \right) - \Im_0 \left(\frac{U_{DF}}{K_B T} \right) \right] \quad (1)$$

$$I_{ds}(V_G, V_d, V_s) = G_{ON} \frac{K_B T}{q} \left[\log \left(1 + \exp \left(\frac{q(E_F - V_{SC}(V_G, V_d, V_s))}{K_B T} \right) \right) \right]$$
$$- G_{ON} \frac{K_B T}{q} \left[\log \left(1 + esp \left(\frac{q(E_F - V_{SC}(V_G, V_d, V_s) - V_d - V_s)}{K_B T} \right) \right) \right] \quad (2)$$

Equation (1) is a quasi one-dimensional structure for net drain current due to the difference between positive and negative currents, based on the Landauer-Buttiker formalism [10]. It can be rewritten as (2) in the function of the drain voltage, source voltage and gate voltage coefficient [11] where G_{ON} is the ON-conductance and $\Im_0$ is the Fermi-Dirac integral function [8, 9]. The quantum conductance limit of a ballistic SWCNT and GNR is $G_{ON} = 4\ q^2/h$ and $G_{ON} = 2q^2/h$ respectively [12, 13].

Though the carbon-based device can be modeled using the atomistic Green's function (AGF) method, the computing cost of a system with massive grid points, can be enormous. Nevertheless, the mesosopic device modeling using a non-equilibrium Green's function (NEGF) will be explored in future work.

B. Graphical User Interface (GUI)

The quantum device simulator codenamed LODISI is designed by using MATLAB interactive layout editor in GUI Development Environment (GUIDE). The main interface features of LODISI is illustrated in Fig. 1. From the main LODISI interface, user allow to choose the type of simulation accordingly. The interface for LODISI CNT is shown in Fig 2. LODISI is a user friendly simulator that provides control over every aspect of the simulation parameter shown in Table I. The tool allows users to insert their own n-type and p-type parameters for both CNT and GNR. Flexibility and ease of use are hard to compromise, but for a quantum device simulator it is an essential factor for the researcher. Consequently, a few inevitable buttons are added to this simulator, including a reset button, run button, default button and close button.

The reset button allows users to set all the parameters back to zero. When the default button is pressed, all the parameters are set to the present value. The default value for each parameter for n-type and p-type is show in Table I. The close button is different to the exit button of the window because it will not exit immediately after being pressed- it will instead ask your permission before closing the simulator in order to prevent any accidental data loss. The run button will automatically run the program with the parameters that the user has inserted.

TABLE I. PARAMETER REQUIRED BY USERS IN LODISI AND THE DEFAULT VALUE FOR EACH PARAMETERS.

Parameter	Symbol	Default Value	
		N-type	P-type
Fermi Energy (Source)	E_F	-0.32	0.32
Gate control	alphag	0.88	
Drain control	alphad	0.032	
Gate insulator thickness	Gate thickness	1.0×10^{-8}	
Gate insulator dielectric constant	epsr	25	
Diameter (CNT only)	diameter	1.5437×10^{-9}	
Chirality (GNR only)	chirality	19	
Temperature	temperature	300	
No of bias point	NV	20	
Initial voltage	VI	0	
Final voltage	VF	1	-1

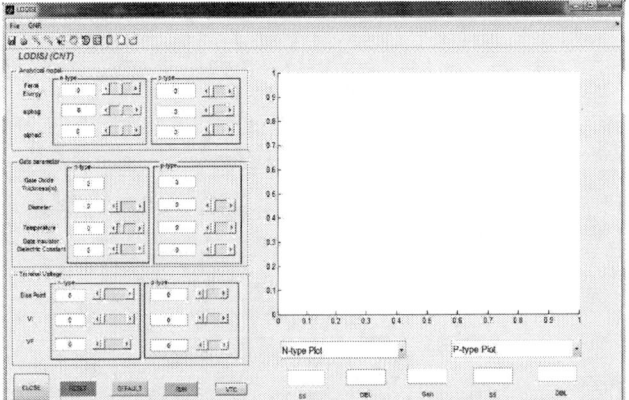

Fig. 2. Graphical User Interface of Low Dimensional Simulator (LODISI)

Another unique characteristic of the LODISI is that it will display the voltage transfer characteristic graph spontaneously after the voltage transfer characteristic (VTC) is pressed based on the rain characteristic graph of PMOS and NMOS. In addition, LODISI also allow users to determine the drain-induced-barrier-lowering (DIBL), subthreshold swing (SS) and gain from the drain current versus gate voltage characteristic. The DIBL for p-type and n-type is an added feature of LODISI. DIBL is calculated using (3) and expressed by

$$DIBL = \frac{\partial V_T}{\partial V_{DS}} \qquad (3)$$

where V_T is the threshold voltage and V_{DS} is the drain voltage. Subthreshold swing for P-type and N-type will automatically show in its text box after the run button is clicked. SS is given as

$$SS = \frac{\partial V_{GS}}{\partial (log_{10} I_{DS})} \qquad (4)$$

where V_T is the threshold voltage, I_{DS} is the drain current.

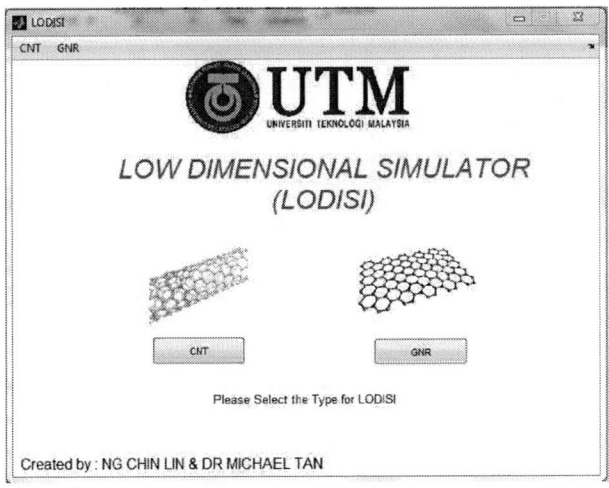

Fig. 1. Main interface of LODISI

GAIN for the CMOS circuit will automatically show in the text box after the VTC button is press.

$$GAIN = \frac{V_{out}}{V_{in}} \quad (5)$$

where V_{out} is the drain voltage and V_{in} is gate voltage in the inverter circuit. The gain obtain is a negative value as the inverter is the inverting amplifier, obtaining a negative gradient in VTC. LODISI provides various types of plots for n-type and n-type FET. This feature is mainly to increase the efficiency and ease of usage for users to do on-spot analysis and design verification. The types of graphs available in this LODISI are show in Table II.

III.

LODISI provides a variety of graphs by which users may benchmark their simulation results. The graph options supported by the LODISI are show in Table II. With the various type of graphs, users can reduce the time constraints and errors by changing their parameters on the simulator in real time.

TABLE II.   LIST OF PLOT OPTIONS AND METRIC PERFORMANCE OF LODISI

Type of Plots
Drain Current vs Drain Voltage
Drain Current vs Gate Voltage
Drain Current vs Gate Voltage (semilog)
NP-FET (n-type and p-type plot)
Voltage Transfer Characteristic
Average Velocity vs Gate voltage
Quantum Capacitance vs Gate voltage
List of Metric Performance
DIBL, SS and GAIN

A. Drain Current Versus Drain Voltage Characteristic

The relationship of the drain current of a ballistic carbon nanostructure is vital to measure the drive strength of a transistor for a higher or reduced performance. The drain current as a function of drain voltage (I-V) for a wide range of gate voltage is shown in Fig. 3.

B. Drain Current Versus Gate Voltage Characteristic

Drain current versus gate voltage depicted in Fig. 4 and Fig. 5 are another important plot for further analysis of a nanotransistor. It provides the value of threshold voltage point for the transistor to turn on at different drain voltages. From the drain current against gate voltage graph, the current on/off (I_{on}/I_{off} ratio), DIBL and SS can be calculated.

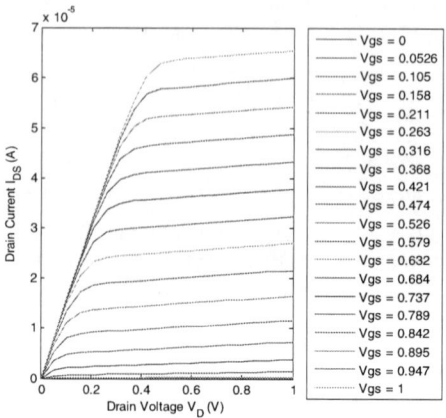

Fig. 3. Drain current versus drain voltage of a n-type CNTFET using the default parameter in tabulated in Table I.

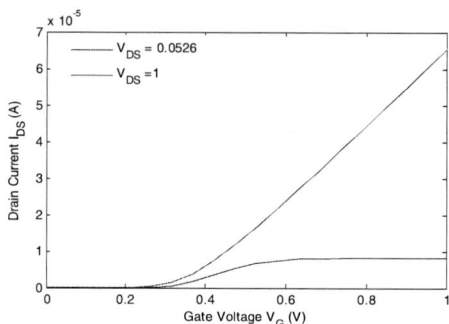

Fig. 4. Drain current versus gate voltage for d n-type CNTFET with default parameters.

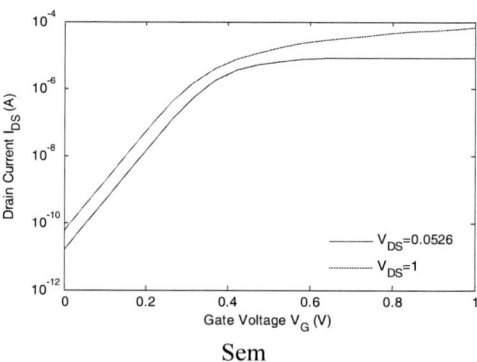

Sem

Fig. 5. Semilog plot of drain current, I_{DS} against gate voltage, V_G for the default parameters of n-type CNT. The I_{ON}/I_{OFF} ratio for $V_{DS}=1$ is 10^6

The NP-FET graph is one of special features of LODISI. This graph shows drain characteristics by two different types of CNT plot simultaneously. The graph is plotted by inverting the PMOS graph result in a mirror image of the drain characteristics. The number of bias points must be equal for both transistors. In Fig. 6, the red dashed line represents the CMOS voltage transfer characteristic (VTC) for different gate voltages. This curve is then used to re-plot the voltage transfer characteristics curve.

IEEE-ICSE2014 Proc. 2014, Kuala Lumpur, Malaysia

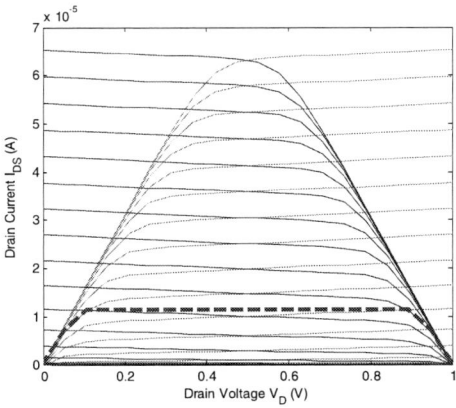

Fig. 6. I-V characteristics for n-type (green line) and p-type (blue line) CNTFET for $V_G = 0V$ to $1V$ (bottom up) in 0.0526V spacing. The red dashed line represents the drain current in the CNT-based *CMOS inverter*. NMOS-PMOS

C. Voltage Transfer Characteristic (VTC)

The voltage transfer characteristic graph will be automatically generated when the VTC button is pressed. The data is taken from the red dashed line in the NMOS-PMOS graph. A new algorithm is used to re-plot the red dashed line by changing the x-axis to its gate voltage values and the y-axis to its drain voltage values. From the VTC curve, gain is calculated by the gradient of the steepest curve. LODISI also accept the simulation of CMOS-like CNT with different source Fermi energy for n-type and p-type CNTFET. The VTC curve of CNTFET for default value as shown in Fig. 7.

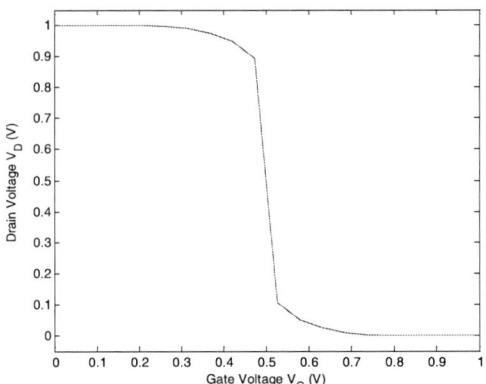

Fig. 7. Voltage transfer characteristic for an almost symmetrical I-V for n-type and p-type CNTFET with source fermi energy, $E_F = -0.32$ (n-type) or 0.32 eV(p-type)

IV.

Based on the modeling framework of top-of-the-barrier approach, a comprehensive simulation tool for carbon based devices is presented. The differences between LODISI and other simulations tools are that LODISI includes unique features such as a VTC graph, auto-calculating DIBL, SS and

Gain. LODISI provides a platform for users to benchmark their project and design verification. For future work, the tool will be improved further to include non-idealities such as phonon scattering and benchmarking against Predictive Technology Model (PTM) of short channel MOSFET of various technology process.

Acknowledgment

The authors acknowledge the financial support from Research University Grants (vote no: Q.J130000.2523.05H64) and Fundamental Research Grant Scheme (vote no: R.J130000.7823.4F247) of the Ministry of Higher Education (MOHE), Malaysia. They acknowledges the support of the UTM Research Management Centre (RMC) for providing an excellent environment to advance the thought processes that allowed result-oriented team to deliver an output of high standards worthy of consideration by the industry.

References

[1] A. K. Geim and K. S. Novoselov, "The rise of graphene," *Nat Mater,* vol. 6, pp. 183-191, 2007.

[2] H. Dinh Sy, L. Nguyen Thi, M. Le Hoang, P. Tran Tien, T. Pham Thanh, D. Bui An, T. Huynh Lam Thu, T. Nguyen Van Le, T. Thi Tran Anh, T. Huynh Hoang, N. Nguyen Thi Thanh, and N. Dinh Viet, "Development of quantum device simulator NEMO-VN1," *Journal of Physics: Conference Series,* vol. 187, p. 012088, 2009.

[3] A. Zienert, J. Schuster, and T. Gessner, "Comparison of quantum mechanical methods for the simulation of electronic transport through carbon nanotubes," *Microelectronic Engineering,* vol. 106, pp. 100-105, 6// 2013.

[4] J. W. A. Rahman, J. Guo, S. Hasan, Y. Liu, A. Matsudaira, S.S. Ahmed, S. Datta, and M. Lundstrom. (2014). *FETToy.* Available: http://nanohub.org/resources/fettoy

[5] A. Rahman, J. Guo, S. Datta, and M. S. Lundstrom, "Theory of ballistic nanotransistors," *IEEE Transactions on Electron Devices,* vol. 50, pp. 1853-1864, Sep 2003.

[6] A. Rahman, J. Guo, S. Datta, and M. S. Lundstrom, "Theory of ballistic nanotransistors," *Electron Devices, IEEE Transactions on,* vol. 50, pp. 1853-1864, 2003.

[7] G. Klimeck, Michael, S. Brophy, George, and M. Lundstrom, "nanoHUB.org: Advancing Education and Research in Nanotechnology," *Computing in Science & Engineering,* vol. 10, pp. 17-23, 2008.

[8] M. L. P. Tan, G. Lentaris, and G. A. Amaratunga, "Device and circuit-level performance of carbon nanotube field-effect transistor with benchmarking against a nano-MOSFET," *Nanoscale Research Letters,* vol. 7, p. 467, 2012.

[9] M. L. P. Tan, "Long Channel Carbon Nanotube as an Alternative to Nanoscale Silicon Channels in Scaled MOSFETs," *Journal of Nanomaterials,* vol. 2013, p. 5, 2013.

[10] Y. Imry and R. Landauer, "Conductance Viewed as Transmission," in *More Things in Heaven and Earth,* B. Bederson, Ed., ed: Springer New York, 1999, pp. 515-525.

[11] M. L. Tan, "Device and Circuit-level Models For Carbon Nanotube and Graphene Nanoribbon Transistors," *University of Cambridge,* 2011.

[12] M. L. P. Tan, L. L. Lim, W. S. Wong, H. C. Chin, E. L. M. Su, and C. F. Yeong, "Nanoscale Device Modeling and Circuit-level Performance Projection of Top-gated Graphene Nanoribbon Field-Effect Transistor for Digital Logic Gates," *Science of Advanced Materials,* vol. 6, p. In Press, 2014.

[13] H. C. Chin, C. S. Lim, W. S. Wong, K. A. Danapalasingam, V. K. Arora, and M. L. P. Tan, "Enhanced Device and Circuit-Level Performance Benchmarking of Graphene Nanoribbon Field-Effect Transistor against a Nano-MOSFET with Interconnects," *Journal of Nanomaterials,* vol. 2014, p. 14, 2014.

978-1-4799-5761-3/14 $31.00 © 2014 IEEE

Design and Implementation of a 1-bit FinFET Full Adder Cell for ALU in Subthreshold Region

'Aqilah binti Abdul Tahrim
Faculty of Electrical Engineering
Universiti Teknologi Malaysia
81310 Skudai, Johor, Malaysia
aqilah.tahrim@yahoo.com

Michael Loong Peng Tan
Faculty of Electrical Engineering
Universiti Teknologi Malaysia
81310 Skudai, Johor, Malaysia
michael@fke.utm.my

Abstract— The FinFET based Full Adder in various cell designs is investigated in terms of performance and energy efficiency. Additionally, the performance of the FinFET Full Adder in the subthreshold region reveals significant results in low power technology. The 1-bit FinFET based Full Adder is designed in four different cell designs, and while the average power dissipated, delays, power-delay-product (PDP) and energy-delay-product (EDP) of all four topologies were analyzed based on the types of transistors used i.e. conventional Field Oxide Transistor (MOSFET) and FinFET. Based on this study, FinFET based Full Adder shows an average of 94 % reduction in delay, 97 % reduction in power dissipation and 99 % reduction for both PDP and EDP over the conventional FET, giving FinFET advantages in energy efficiency.

Keywords— *1-bit Full Adder; conventional FET; FinFET; subthreshold region; delays; power dissipation; PDP.*

I.

Addition is one of the basic arithmetic operations used in the arithmetic logic unit (ALU), which is the most important unit in a data path unit of RTL design. In digital systems, a full adder with a low power consumption, fast speed, high energy efficient and good reliability is desirable. A full adder can be implemented with different types of transistor, such as FinFET and Metal Oxide Semiconductor Field-Effect Transistor (MOSFET), which may result in different performances across variations in voltage. In this paper, we propose that FinFET will reveal significant improvements in performance over the conventional MOSFET in full adder cell design. Both FinFET and MOSFET full adders will incorporate multiple cell designs, that will be further discussed in Section II, where all cell designs will also be demonstrated in subthreshold region.

The main objective of this project is to investigate the circuit performance of FinFET as a full adder in a subthreshold region that is 0.2V. Besides that, the metric performance of FinFET as full adder cells will be computed based on multiple cell designs such as Complementary MOS (CMOS), Transmission Gate (TG), Complementary Pass-Transistor Logic (CPL) and Hybrid CMOS (HCMOS). The purpose of this research is also to benchmark FinFET low-power full adder cells with 16nm MOS Predictive Technology Model (PTM). The FinFET structure as a new alternative for conventional MOSFET, as stated by The International Technology Roadmap for Semiconductor (ITRS), would give

great advantages over the conventional MOSFET. This project would contribute to the optimized values of parameters such as width and length of the FinFET model, which are the most important parameters in designing a high performance and low power 1-bit full adder cell. In addition, this project functions as a benchmark to determine which type of full adder designs with the structure of FinFET will give the best performance in terms of delay, average power dissipation and power delay product (PDP).

II.

This scaling process of MOSFET is now approaching the technology limit, as a length below 20nm will cause the electrical parameters to start to degrade, and the silicon process variations will impact heavily on performance. FinFET is a new alternative structure for MOSFET which allows transistors to be scaled down to smaller sizes, promising advantages over conventional MOSFET such as higher drain current, lower switching voltage and significantly less static leakage current.

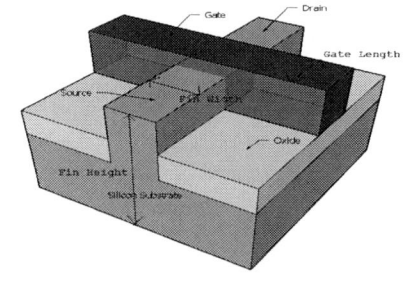

Fig. 1. Basic Structure of FinFET [1].

FinFET is categorized as a multi-gate device where the mode of operation of FinFET is almost the same as the traditional MOSFET transistors. Typically, it has a source, a drain and a gate to control the current flow. A channel between the source and drain of FinFET is created as a three-dimensional bar on top of a silicon substrate known as the 'fin'. The gate is covered around the channel as shown in Fig 5, with the purpose of forming several gate electrodes on each side, aiming to both reduce leakage effects and enhance drive current. Previous research on FinFET observed that FinFET has good parameter-variation immunity due to self-alignment of two gates[2].

Based on Fig. 1, the effective channel width 'W' of a single-fin FinFET must be twice the silicon fin height, H_{fin}. For a larger W, the design can be practically implemented by using multiple fins aligned in parallel, but is limited to increments of double amounts of H_{fin}. The design flexibility of FinFET may be a trade-off for a taller fin which allows the layout area to be used efficiently [1].

III.

A. Boolean Algebra Expression

A full adder is commonly used in many digital systems, such as the ALU, to add any values bit by bit; given the inputs of A, B and C_{in} to be added together to produce the output of Sum and C_{out}. The expression of Sum and C_{out} is given by:

$$Sum = A \oplus B \oplus C_{in} \qquad (1)$$

$$C_{out} = A \cdot B + C_{in} \cdot (A \oplus B) \qquad (2)$$

The output of addition can be expected based on the binary values as tabulated in the truth table in Table 1. In this research, both MOSFET and FinFET were implemented in four different cell designs i.e. CMOS, TG, CPL and HCMOS.

TABLE I.   TRUTH TABLE OF 1-BIT FULL ADDER

A	B	C_{in}	C_{out}	Sum
0	0	0	0	0
0	0	1	0	1
0	1	0	0	1
0	1	1	1	0
1	0	0	0	1
1	0	1	1	0
1	1	0	1	0
1	1	1	1	1

B. 1-bit Complementary MOS (CMOS) Full Adder

The CMOS full adder with 28 transistors is designed based on (1) and (2). The advantage of this cell design is high noise margin, which makes the design very reliable with low voltage [3]. However, a high number of transistors may result in a large power consumption and high input loads, as well as requiring a larger Silicon area in a wafer. Researchers in [3] also stated that this design may introduce more delay, because Sum is generated from C_{out} as input, as can be observed from Fig 2.

C. 1-bit Complementary Pass-Transistor Logic (CPL) Full Adder

This cell design uses 32 transistors and is made up of NMOS pass-transistors and cross-coupled PMOS transistors. The advantages of this design are high speed performance and full swing operation [4]. Additionally, this design has a good driving capability due to the output of static inverters, as shown in Fig. 3 [4]. On the other hand, this design may generate many intermediate nodes and complement results to generate the final outputs. It is also mentioned in [5] that

overloading input from this design produces high capacitance. The high number of transistors also may increase power dissipation.

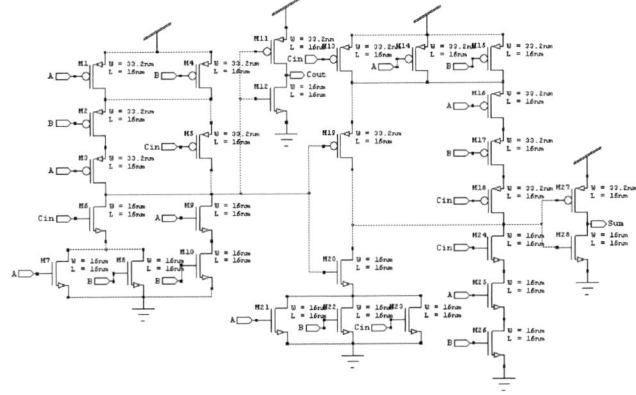

Fig. 2.   1-bit Complementary MOS (CMOS) Full Adder [3].

D. 1-bit Transmission Gate (TG) Full Adder

This system consists of 20 transistors, made up of transmission gate, PMOS, and NMOS transistors as illustrated in Fig. 4. Transmission gates are used in the design because of their high speed operation and low power dissipation [4]. The TG full adder circuit is much simpler than CMOS and CPL with balanced generation of Sum and C_{out} output signals, lower transistor count, fewer intermediate nodes and lower input loading. In contrast, if compared to CMOS full adder, the TG full adder has higher power dissipation [3]. However, if the TG full adder is cascaded in series, the propagation delay may also increase [2].

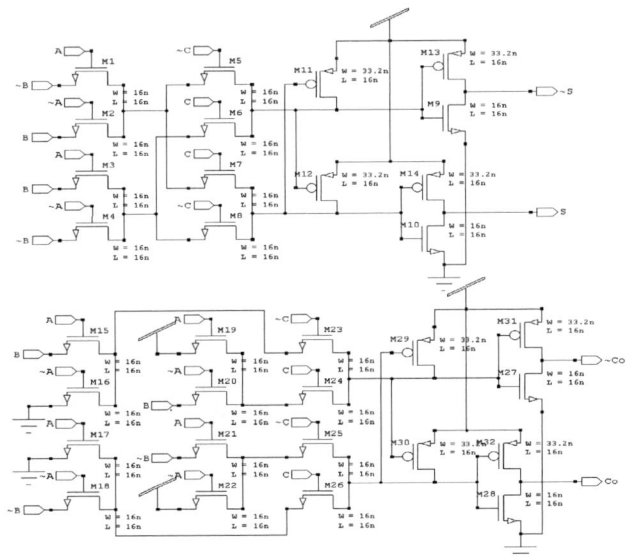

Fig. 3.   1-bit Complementary Pass-Transistor Logic (CPL) Full Adder [4].

E. 1-bit Hybrid CMOS (HCMOS) Full Adder

This design of full adder was proposed in [4]. It combines transmission gates, pass-transistors, PMOS and NMOS with a total of 24 transistors, as shown in Fig. 5. This design has a lower transistor count with higher noise immunity [4]. The work reveals that this HCMOS design produces less power dissipation compared to a CPL full adder. However, the design has a relatively high delay, almost equal to CMOS and CPL full adders [2].

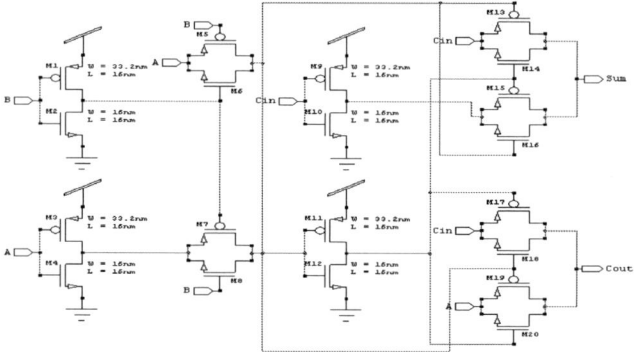

Fig. 4.   1-bit Transmission Gate (TG) Full Adder [6].

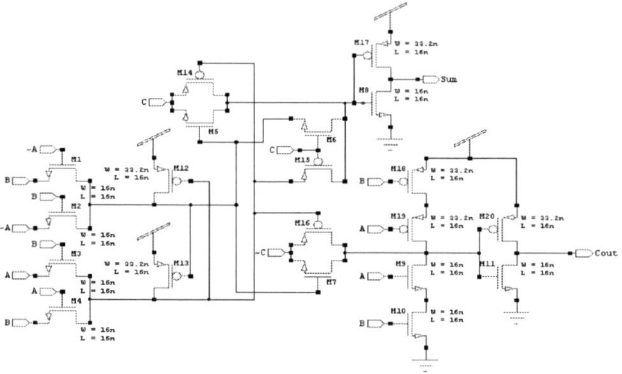

Fig. 5.   1-bit Hybrid CMOS (HCMOS) Full Adder [4].

IV.

A. The 1-bit Full Adder

Fig. 6 shows the inputs and outputs of CMOS 1-bit Full Adder cell design, with the SPICE model simulated in HSPICE. It is shown that different cell designs have a wide range of delays, average power dissipations, Power Delay Product (PDP) and Energy Delay Product (EDP) as tabulated in Table I for MOSFET based full adder and Table II for FinFET based full adder respectively.

B. Average Power Dissipation

The average power dissipated in all four full adder cell designs was measured from 0 to 40μs. As shown in Fig 7, TG FinFET full adder has a higher average power dissipation,

followed by the HCMOS and CMOS FinFET adder, while the CPL FinFET full adders have the least power dissipation.

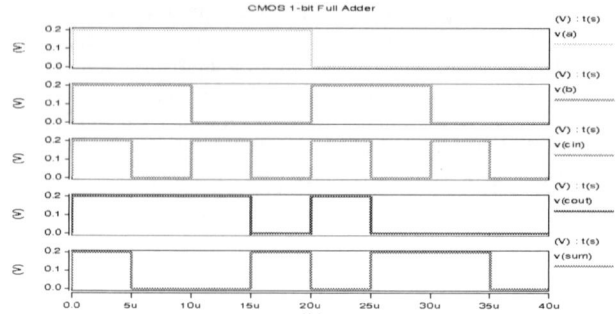

Fig. 6.   The Inputs and Outputs of 1-bit Full Adder.

TABLE II.   METRIC PERFORMANCE PARAMETERS FOR 1-BIT MOSFET FULL ADDER CELLS AT $V_{DS} = 0.2V$.

Full Adder Cell	Average Power (W)	Delay (s)	Power Delay Product (J)	Energy Delay Product (Js)
CMOS	4.58×10^{-10}	2.41×10^{-08}	1.10×10^{-17}	2.66×10^{-25}
CPL	4.44×10^{-10}	2.02×10^{-08}	8.98×10^{-18}	1.82×10^{-25}
TG	4.81×10^{-10}	2.71×10^{-08}	1.30×10^{-17}	3.52×10^{-25}
HCMOS	4.62×10^{-10}	2.49×10^{-08}	1.15×10^{-17}	2.87×10^{-25}

TABLE III.   METRIC PERFORMANCE PARAMETERS FOR 1-BIT FINFET FULL ADDER CELLS AT $V_{DS} = 0.2V$.

Full Adder Cell	Average Power (W)	Delay (s)	Power Delay Product (J)	Energy Delay Product (Js)
CMOS	1.31×10^{-11}	1.25×10^{-09}	1.64×10^{-20}	2.05×10^{-29}
CPL	1.21×10^{-11}	1.20×10^{-09}	1.45×10^{-20}	1.75×10^{-29}
TG	1.53×10^{-11}	1.83×10^{-09}	2.81×10^{-20}	5.14×10^{-29}
HCMOS	1.39×10^{-11}	1.47×10^{-09}	2.05×10^{-20}	3.01×10^{-29}

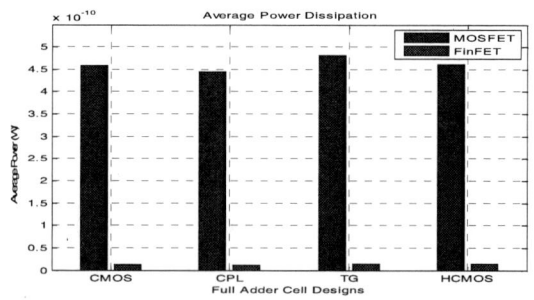

Fig. 7.   Average Power Dissipation of CMOS, CPL, TG and HCMOS Full Adder.

C. Propagation Delay

The propagation delay is the time taken for the input to change its output at a certain time. Fig. 8 shows that CPL has the highest speed of operation with least delay among other full adders, while TG full adder was the slowest. Unlike TG FinFET adder, the CPL adder cell design allows high speed performance with full-swing operation.

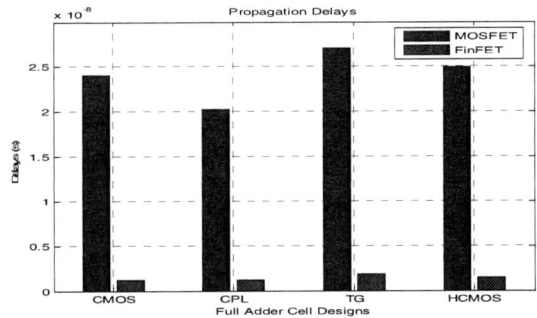

Fig. 8. Propagation Delays of CMOS, CPL, TG and HCMOS Full Adder.

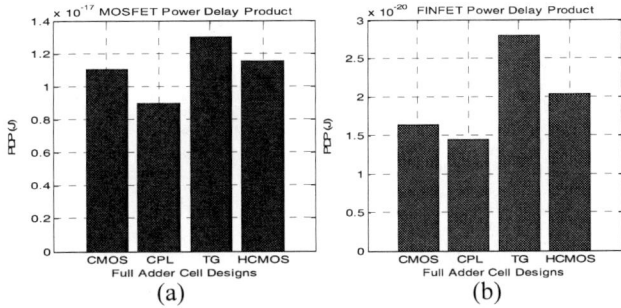

(a)   (b)

Fig. 9. The power-delay-product (PDP) of (a) MOSFET and (b) FinFET based full adder.

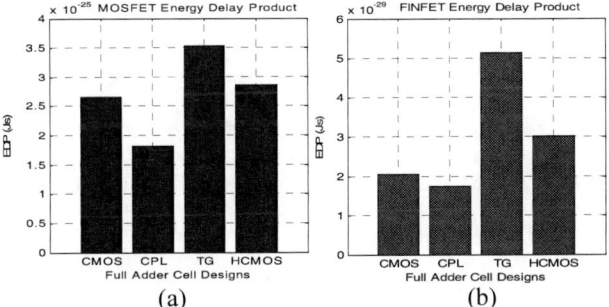

(a)   (b)

Fig. 10. The energy-delay-product (EDP) of (a) MOSFET and (b) FinFET based full adder.

D. Power Delay Product (PDP)

The PDP is the product of average power dissipation and propagation delays. As shown in Fig. 9, TG full adder has the highest PDP due to its high delay, while CPL has the lowest PDP owing to its high speed operation.

E. Energy Delay Product (EDP)

The Energy Delay Product (EDP) for the full adder was measured as shown in (3)

$$EDP = \text{Average Power Dissipation} \times (\text{Delay})^2 \qquad (3)$$

Fig. 10 shows the EDP produced by all four full adder cell designs, where TG has three times greater EDP than CPL. In (3), the EDP was measured by taking the square of delays. Since the delay of TG was the highest among the other cell designs, EDP doubled the delay, giving TG the highest EDP. In contrast, CPL has the lowest delay compared to other cell designs, which contributes to the lowest EDP.

Based on the four parameters measured; average power dissipation, delays, PDP and EDP, the 1-bit CPL full adder cell shows the best performance compared to other cell designs. The fast speed performance and full-swing both contribute to the least PDP and EDP.

V.

In conclusion, FinFET technology has more advantages over the conventional MOSFET in terms of performance and energy efficiency. Using FinFET technology in 1-bit Full Adder circuitry will provide higher performance compared to MOSFET technology. However, different types of circuit design or topology may produce varying performance, based on their advantages and disadvantages as previously discussed. Based on this study, FinFET based Full Adder shows an average of 94 % drop in delays, 97 % decrease in power dissipation and 99 % reduction for both PDP and EDP over the conventional FET, thus giving FinFET advantages in energy efficiency. Moreover, 1-bit FinFET based Full Adder of Complementary Pass-Transistor Logic (CPL) shows less delay and power delay product (PDP) compared to others circuit designs due to its high speed and full swing operation.

Acknowledgment

The authors acknowledge the financial support from Research University Grants (vote no: Q.J130000.2523.05H64) and Fundamental Research Grant Scheme (vote no: R.J130000.7823.4F247) of the Ministry of Higher Education (MOHE), Malaysia. Michael acknowledges the support of the UTM Research Management Centre (RMC) for providing an excellent environment to advance the thought processes that allowed result-oriented team to deliver an output of high standards worthy of consideration by the industry.

References

[1] T.-J. King, "FinFETs for nanoscale CMOS digital integrated circuits," in *Computer-Aided Design, 2005. ICCAD-2005. IEEE/ACM International Conference on*, 2005, pp. 207-210.

[2] A. Islam, M. Akram, and M. Hasan, "Variability Immune FinFET-Based Full Adder Design in Subthreshold Region," in *Devices and Communications (ICDeCom), 2011 International Conference on*, 2011, pp. 1-5.

[3] S. Panda, A. B. B. Maji, and A. Mukhopadhyay, "Power and Delay Comparison in between Different types of Full Adder Circuits," *International Journal of Advanced Research in Electrical, Electronics and Instrumentation Engineering*, vol. 1, pp. 168-172, September 2012.

[4] S. Goel, A. Kumar, and M. A. Bayoumi, "Design of robust, energy-efficient full adders for deep-submicrometer design using hybrid-CMOS logic style," *Very Large Scale Integration (VLSI) Systems, IEEE Transactions on*, vol. 14, pp. 1309-1321, 2006.

[5] A. M. Shams, T. K. Darwish, and M. A. Bayoumi, "Performance analysis of low-power 1-bit CMOS full adder cells," *Very Large Scale Integration (VLSI) Systems, IEEE Transactions on*, vol. 10, pp. 20-29, 2002.

[6] M. Damle, S. Limaye, and M. Sonwani, "Comparative Analysis of Different Types of Full Adder Circuits," *IOSR Journal of Computer Engineering (IOSR-JCE)*, vol. Vol. 11, pp. pp. 01-09, 2013.

The RF Power Effect on the Surface Morphology of Titanium Dioxide (TiO$_2$) Film

S. Norhafiezah, RM Ayub, M. K. Md Arshad, A.H. Azman, M. F. Fatin, M.A. Farehanim and U. Hashim

Institute of Nano Electronic Engineering (INEE),
Universiti Malaysia Perlis (UniMAP)
01000 Kangar, Perlis, Malaysia.
mizfyza@gmail.com

Abstract—In this paper, we present the influence of deposition process parameter on the morphological properties of titanium dioxide (TiO$_2$). Thin film of TiO$_2$ was deposited on Si (100) substrate using the reactive Radio Frequency (RF) sputtering technique with different RF power. The XRD analysis showed that only Anatase structure was obtained during low RF power deposition, while both; Anatase and Rutile structure were obtained at high RF power. It was also observed that when the RF power is increased from 100W to 300W, the surface roughness and the particle size of the TiO$_2$ film measured by using AFM were shown to be decreased from 0.39 to 0.25 nm and 68.3 to 59.6 nm respectively. Consequently, the diffuse transmittance measured using UV-vis spectroscopy shown degradation of transmittance percentage from 85% to 60% and the E$_g$ also reduce from 3.24 eV to 2.55 eV. Moreover, the small particle size with the acceptable surface roughness, the less percentage of transmittance and the reduction of the band gap were successfully achieved.

Keywords— Reactive RF sputtering; TiO$_2$ thin film; Deposition parameter effects; RF power

I. INTRODUCTION

Titanium Dioxide or also known as Titania, is an inexpensive material. It had been widely used in cosmetic, food and medical industry since it has an excellent characteristic such as chemically stable, non-toxicity, high refractive index, etc. [1, 2]. TiO$_2$ is basically a photocatalytic material compared to other semiconductor metal oxide material since it has contradictory properties of the conduction and valence bands. Therefore, it has emerged a behavior that can be applied in bio-sensor application due to its activation of electron either by an applied voltage or UV light [3].

TiO$_2$ can existed in three crystalline phases namely; anatase, rutile and brookite. Both anatase and rutile were in a tetragonal form whereas brookite can be normally found in an orthorhombic form. Rutile structure has been proven as the most stable structure for bio-sensor application [4] while anatase and brookite are metastable and transform to rutile at high annealing temperature. The large band gap of energy between 1.8 and 4.1 eV makes TiO$_2$ more sensitive and suitable for bio-sensor application.

In order to obtain the rutile structure, the deposition technique of TiO$_2$ thin film has played a major role since it has

dramatic influence on optical, physical and electrical behavior [5]. In addition to technological applications, researchers have strived to deposit a thin film layer uniformly on top of the substrate. There were various methods introduced by researchers to successfully deposit a thin layer of TiO$_2$, which is by using sputtering technique, sol-gel method, atomic layer deposition, etc. [6]. Different deposition processes dealt with some drawbacks and certain deposition has its own advantages.

By sol-gel approach, thin-film can be coated on various substrates by using several techniques such as dip coating, spin coating, roll coating, etc. Therefore, utilizing the sol-gel method could produce a homogeneous, inexpensive, non-complexity of fabrication process and enable the synthesis at lower temperature [6, 7]. However, this technique still has some drawbacks which deal with higher raw material costs, the existence of microspores after synthesis and longer reaction time. Another deposition method that captures the researcher attention is atomic layer deposition (ALD). ALD can indicate to produce excellent step coverage and conformal deposition of high aspect ratio structures [9]. Thus, by using the ALD technique, the thickness of the deposit layer can be precisely controlled, but due to extremely high equipment cost, makes it rarely used method. Owing to its simplicity [10] and uniformly [11] deposition method, RF sputtering was chosen as deposition technique for this work.

The TiO$_2$ thin film was deposited on silicon (100) substrates by using the reactive RF sputtering technique with a 10:1 ratio of argon to oxygen gas. Argon acts as inert gas involved in the process of target sputtering while oxygen acts as a reactive gas throughout the whole deposition process. Surface morphological was characterized in term of surface roughness, crystal structure and particle size of the TiO$_2$ film for the bio-sensor application.

II. EXPERIMENTAL PROCEDURE

The titanium dioxide thin film was deposited by reactive RF sputtering technique with a titanium target (purity 99.9%, thickness 3 inches) using the Penta Vacuum E303 sputter CVD system. The system was operated in pure argon plasma with the RF generator operating at 13.56 MHz. In this experiment, the deposition time, the substrate temperature and target-to-substrate distance was fixed to 10 minutes, 150°C and 14 cm, respectively. Throughout the sputtering process, the RF powers were varied to two different values of 100W and 300W. The RF power variation was chosen based on the sputtering machine constraints, which is the system, cannot operate higher than 300W during deposition. The machine chamber was evacuated to 5.5×10^{-5} Torr before the mixture of 50 sccm Argon (purity 99.99%) and 5 sccm Oxygen (99.99%) gases was introduced prior deposition process. The p-type silicon wafer with <100> orientation was first cleaned before the deposition process takes part. The wafer was immersed in piranha solution to remove the organic contamination. Then, the mixture of Ammonium Hydroxide, Hydrogen Peroxide and deionized water (RCA-1) with Hydrochloric Acid, Hydrogen Peroxide and deionized water (RCA-2) cleaning was performed to eliminate the residue and metallic contamination on the wafer surface. Lastly, the wafer was being dipped into diluted Hydrofluoric Acid (HF) to etch the native oxide that is form naturally on top of the silicon surface. The effect of the deposition parameter was studied in term of changes in surface characteristic and the thickness by using X-Ray Diffractometer (XRD), UV-visible (UV-vis) spectroscopy and Atomic Force Microscopy (AFM), respectively. The crystalline phases of the deposited material were examined using an X-Ray Diffractometer (Bruker D2 Phaser) with the 2θ scan range of 20° to 80° at 0.02 seconds time per step value. The crystallite size was calculated from the Debye-Scherer equation (Equation 1):

$$D = 0.9 \, \lambda \, / \, \beta \cos \theta \qquad (1)$$

Where D is the crystallite size, k is the constant normally the value is close to the unity, λ is the wavelength of the X-Ray (0.154 nm), β is the full-wave-half-maximum (FWHM) and θ is the Bragg angle. Then, the UV-Vis spectrometer (Perkin Elmer Lambda 35) was used to measure the optical transmittance of the TiO_2 film deposited on silicon substrate with the wavelength range of 300 to 800 nm. The absorbance value, A was calculated from transmittance percentage using Equation 2:

$$A = - \log (\% \text{ transmittance} / 100) \qquad (2)$$

The energy band gap, E_g of as-deposited TiO_2 film with different RF power was determined by using Equation 3:

$$E_g = hc \, / \, \lambda \qquad (3)$$

The Planck's constant is given by 6.626×10^{-34} m^2 kg/s, c is the speed of light and λ is cut off wavelength. The Atomic Force Microscopy SPA-400 was utilized to determine the surface roughness and the grain size of TiO_2 film. The 2 μm scanning area and 2 Hz scanning speed was governed to perform the AFM analysis of the TiO_2 structure.

III. RESULT AND DISCUSSION

Fig. 1 shows the XRD pattern of as-deposited TiO_2 film on silicon substrate with different RF power (100W and 300W) variation. As it can be seen in Fig. 1, by using 100W of RF

power, the TiO_2 layer was getting crystallized and the crystal structure of Anatase (004) was obtained (PDF 49-1433) at 32.96°. As the power increased up to 300W, the formation of both Anatase (004) and Rutile (001) (PDF 29-1360) structure at 32.96° and 22.42° was obtained as shown in Fig. 1. The significant changes of the crystallographic directions were dependent on the change of RF power which is proportional to the deposition rate. At lower deposition rate, the substrate has more time to reach the equilibrium state which exhibits a high intensity of 004 structures [12]. Furthermore, higher deposition rate will improve the growth of the crystal structure because of the concentration of nucleated increase, so that more peaks will be obtained at lower intensity which is 004 and 001 [13].

Apart from that, the changes of the crystal structure from anatase to rutile were highly stimulated by the variation of sputtering parameter i.e. RF power [14]. Normally the stress that is caused by the non-equilibrium microstructure formation was the main factor of the transformation crystal growth [15]. The tensile or compressive stress was associated with the energetic titanium atom bombardment during the sputtering process. Hence, the low RF power will only indicates 004 structure compared to high RF power that contribute both 004 and 101 structures due to the increasing of the deposition rate which is turn contributed to increment of the TiO_2 thickness. It was also proved that the TiO_2 film was homogenously deposit on the silicon substrate since there was no peak shifted shown in Fig. 1.

The crystallite size of the peak broadening was calculated using Scherer equation using the strongest peak which is 004. Furthermore, the crystallite size was determined by the sharp or broad reflections of the peaks. The broad-ranging of the peak will result in smaller crystallite size of the structure. From Table 1, it can be observed that the crystallite size decrease from 179.34 to 111.84 nm as the RF power increase from 100W to 300W. The XRD result reveals that the deposited TiO_2 thin film at 300W exhibit a good crystalline with mixed anatase and rutile phases compared to low RF power which only consist of tetragonal anatase structure.

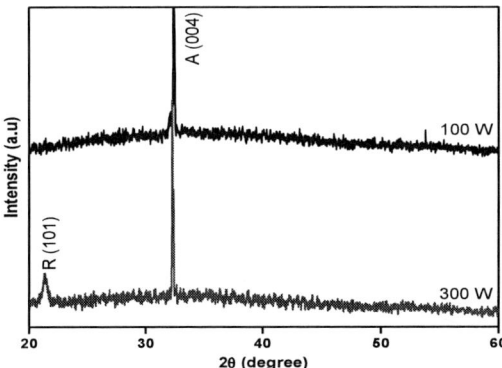

Fig. 1. XRD pattern of as-deposited TiO_2 film with different RF power

TABLE I. XRD PARAMETER TO OBTAINED CRYSTALLITE SIZE

RF Power (Watt)	FWHM (rad)	2θ (degree)	Crystallite Size (nm)
100	805.97×10^{-6}	16.49	179.34
300	1.29×10^{-3}	16.14	111.84

Fig. 2 shows the diffuse transmittance spectra of as-grown TiO_2 film on silicon substrate with different RF power and fixed substrate temperature, which is 150°C. From the diagram, the spectrum shows a dissimilar trend of the TiO_2 film due to the deposition parameter effect. Fig. 2 (a) indicates at 100W, the high transmittance was exhibited in the range of 80 to 90% through UV-visible regions. However, the transmittance was significantly decreased in the range of 60 to 70% as the RF power rise to 300W. The changes of the transmittance is due to the different deposition rate at 100W and 300W, in which higher RF power will produce high energy sputtering and the TiO_2 film will become thick. As the film surfaces become rougher, the light losses in the films grow up resulting in low transmittance range. For lower RF power, the obtained spectrum does not really match the TiO_2 spectra due to its very thin layer and non-uniformly deposited material. It can be verified by measuring the roughness average of the two samples using AFM.

Fig. 3 depicts the absorbance spectra of TiO_2 film deposited on 100 silicon wafer substrate. From Fig. 3, it can be observed that the calculated absorbance coefficient of the film is very low in the wavelength range. This is due to the very thin layer of deposited film and the absorbance range almost reached the silicon value. The calculated energy band gap using equation 3, shows that low RF power (100W) indicates the large value of E_g which is 3.24 eV compared to high RF power (300W) which only 2.55 eV. This result is in agreement in [16] which show that the sample with pure anatase structure indicates larger band gap compared to mixed structure (anatase and rutile).

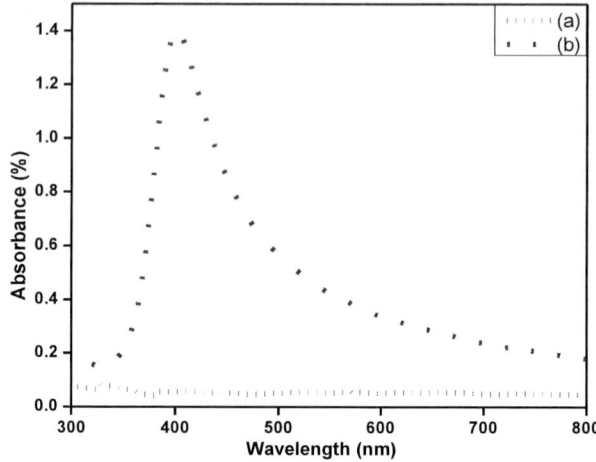

Fig. 3. Diffuse absorbance spectra of as-deposited TiO2 (a) 100W (b) 300W

Fig. 4 (a) and (b) shows the three dimensional images of TiO_2 film as-deposited on silicon substrate by using different RF powers of 100W and 300W, respectively. It was shown that the thickness of the film increases from 2 to 3 nm as the power rise from 100 to 300W. Fig. 4 (a) depicts the relatively smooth surface compared to Fig. 4 (b). This is due to the high energy of adatoms on the substrate surface that was resulting from high RF power supply. Therefore, there was sufficient mobility of the titanium atom to form a smooth surface onto the substrate. On the other hand, Table 2 shows the result obtained from AFM. The measured grain size shows a slight decrement from 68.6 to 59.3 nm as the RF power increase from 100W to 300W. The RMS value also reduced from 0.51 to 0.34 nm. It has a good agreement with the calculated crystallite size of XRD peaks which shows that the higher RF power will produce high crystalline structure which give good precursor for obtaining smaller particle sizes. Thus, the AFM result confirmed that by using the reactive RF sputtering technique at high RF power, a good crystalline and homogenous structure is achievable. Furthermore, in order to obtain the nanoparticle size, further post-annealing will be required for the TiO_2 thin film in the future [17].

TABLE II. THE COMPARISON OF ROUGHNESS AVERAGE, ROOT-MEAN-SQUARE AND GRAIN SIZE WITH DIFFERENT RF POWER

RF Power (Watt)	Roughness Average, Ra (nm)	Root Mean Square, RMS (nm)	Grain Size (nm)
100	0.3989	0.5156	68.63
300	0.2552	0.3443	59.33

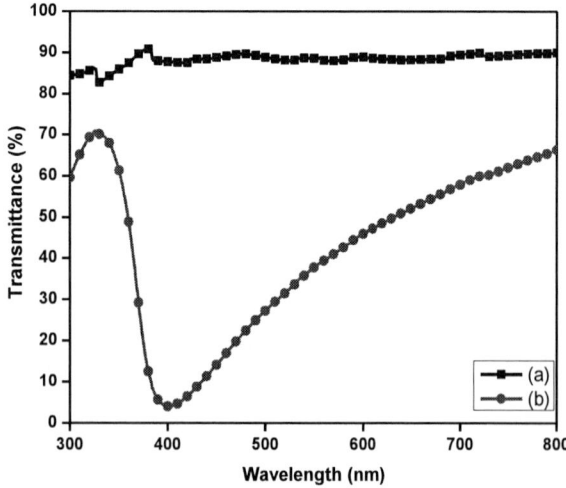

Fig. 2. Diffuse transmittance spectra of as-deposited TiO_2 (a) 100W (b) 300W

IEEE-ICSE2014 Proc. 2014, Kuala Lumpur, Malaysia

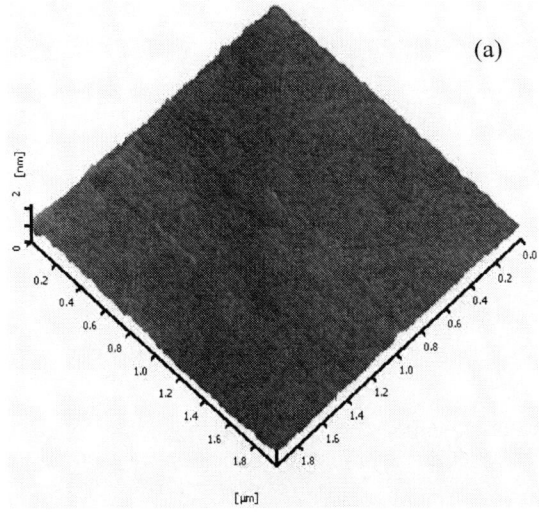

(a)

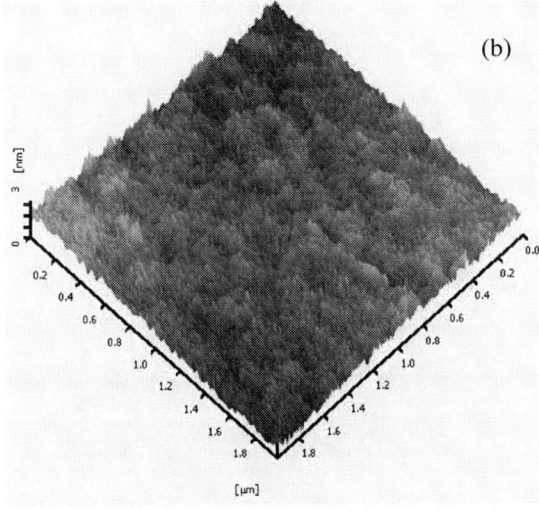

(b)

Fig. 4. AFM surface morphology of TiO$_2$ film deposited with different RF power (a) 100W (b) 300W

IV. CONCLUSION

In this study, we have found that the surface morphology of deposited TiO$_2$ were highly affected by the variation of RF power during sputtering process. It has been proven that high RF power of 300 W, improves the cystallinity of anatase and rutile structure with the smaller crystallite size i.e. 111.84 nm. In addition to that, smooth surface roughness average and smaller grain size i.e. 0.25 nm and 59.33 nm were achieved respectively. Moreover, the percentage value of transmittance were highly caused by the film thickness and surface roughness in the TiO$_2$ films.

ACKNOWLEDGMENT

The authors are grateful to the Department of Higher Education, Ministry of Higher Education, (KPT) for funding this research through the Fundamental Research Grant Scheme (FRGS) with the grant number 9003-00360. Special thanks to Advanced Materials Research Centre (AMREC) lab for the equipment facilities during the deposition process. The author also would like to acknowledge all the team members in Institute of Nano Electronic Engineering (INEE) for their guidance and help.

REFERENCES

[1] "Titanium Dioxide Coatings in Medical Device Applications," in *Surface Engineered Surgical Tools and Medical Devices*, pp. 49–63.

[2] L. Bedikyan, "Titanium Dioxide Thin Films: Preparation and Optical Properties," *J. Chem. Technol. Metall.*, vol. 6, no. 48, pp. 555–558, 2013.

[3] J. Torres, J. Sweeney, and A. Barreto, "An Investigation of the Electrical Properties of Titanium Oxide Coatings for Potential Use in Biosensors," *South. Biomed. Eng. Conf.*, pp. 143–144, 2009.

[4] M. Ahmadi, "Study of Different Parameters in TiO2 Nanoparticles Formation," *J. Mater. Sci. Eng.*, vol. 5, pp. 87–93, 2011.

[5] C. H. Wei and C. M. Chang, "Polycrystalline TiO2 Thin Films with Different Thicknesses Deposited on Unheated Substrates Using RF Magnetron Sputtering," *Mater. Trans.*, vol. 52, no. 3, pp. 554–559, 2011.

[6] E. Buta, P. Pascariu, F. Prihor, and L. Vlad, "Characterization of Sputtered TiO2 Thin Films," no. 11, pp. 1–6, 2008.

[7] S. Nadzirah and U. Hashim, "Annealing effects on titanium dioxide films by Sol-Gel spin coating method," *RSM 2013 IEEE Reg. Symp. Micro Nanoelectron.*, pp. 159–162, Sep. 2013.

[8] M. Alam and D. Cameron, "Preparation and characterization of TiO2 thin films by sol-gel method," *J. sol-gel Sci. Technol.*, pp. 137–145, 2002.

[9] E. Cartier, T. Ando, and M. Hopstaken, "Characterization and Optimization of Charge Trapping in High-k Dielectrics," *IEEE*, pp. 1–7, 2013.

[10] S. Jeong, B. Kim, B. Lee, and J. Kim, "Structural and Optical Properties of TiO2 Films Prepared Using Reactive RF Magnetron Sputtering," *J. Korean Phys. Soc.*, vol. 41, no. 1, pp. 67–72, 2002.

[11] H. Yao, W. Wu, M. Chiua, and F. Shieua, "Characterization and Microstructure of Titanium Dioxide Prepared by Radio Frequency Magnetron Sputtering," pp. 1–18.

[12] D. Mardare, M. Tasca, M. Delibas, and G. I. Rusu, "On the Structural Properties and Optical Transmittance of TiO2 RF Sputtered Thin Films," *Appl. Surf. Sci.*, pp. 200–206, 2000.

[13] D. Kaczmarek, J. Domaradzki, D. Wojcieszak, and B. Gornicka, "XRD and AFM Studies of Sanocrystalline TiO2 Thin Films Prepared by Modified Magnetron Sputtering," *Int. Spring Semin. Electron. Technol.*, pp. 159–162, May 2008.

[14] A. R. Bushroa, R. G. Rahbari, H. H. Masjuki, and M. R. Muhamad, "Approximation of crystallite size and microstrain via XRD line broadening analysis in TiSiN thin films," *Vacuum*, vol. 86, no. 8, pp. 1107–1112, Feb. 2012.

[15] V. Chawla, R. Jayaganthan, a. K. Chawla, and R. Chandra, "Microstructural characterizations of magnetron sputtered Ti films on glass substrate," *J. Mater. Process. Technol.*, vol. 209, no. 7, pp. 3444–3451, Apr. 2009.

[16] M. Yamagishi, S. Kuriki, P. K. Song, and Y. Shigesato, "Thin film TiO2 photocatalyst deposited by reactive magnetron sputtering," *Thin Solid Films*, vol. 442, no. 1–2, pp. 227–231, Oct. 2003.

[17] D. Yoo, I. Kim, S. Kim, C. H. Hahn, C. Lee, and S. Cho, "Effects of annealing temperature and method on structural and optical properties of TiO2 films prepared by RF magnetron sputtering at room temperature," *Appl. Surf. Sci.*, vol. 253, no. 8, pp. 3888–3892, Feb. 2007.

Development of a Silicon Carbide MEMS Capacitive Pressure Sensor Operating at 500 °C

Noraini Marsi, Burhanuddin Yeop Majlis, Azrul Azlan
Hamzah, Ummikalsom Abidin
Institute Microengineering and Nanoelectronics (IMEN)
Universiti Kebangsaan Malaysia
43600 UKM Bangi, Selangor, Malaysia
Email: burhan@vlsi.ukm.my

Faisal Mohd-Yasin
Queensland Micro- and Nanotechnology Centre (QMNC)
Griffith University, 4111, Brisbane,
Queensland, Australia
Email: f.mohd-yasin@griffith.edu.au

Abstract—In this paper, we present development of MEMS capacitive pressure sensor based silicon carbide (3C-SiC) materials. The sensor is made up of four elements: a 3C-SiC diaphragm, silicon substrate, a reliable stainless steel (SS) o-ring and (SS) vacuum clamper as the package. The designed are inherent simplicity and ruggedness of this physical configuration that acceptably performed for extreme environment applications such as in gas turbine engine. This study reported a reliability testing of a prototype package MEMS capacitive pressure sensor verified up to 500 °C through high temperature lab testing. At 500 °C, the reliability test results show that the sensitivity of 0.826 pF/MPa is achieved. Experimentally, sensor nonlinearity of 0.61 % is found with hysteresis of 3.13 %. The maximum temperature coefficient of output change is 0.073 %/°C measured at 5 MPa.

Keywords— MEMS; silicon carbide; pressure sensor; sensitivity; hysteresis

I. Introduction

The silicon-based pressure sensor is the most prevalent and commercially successful pressure sensor. Dry air and benign media are typical environments for these pressure sensors [1]. Typically, the media being sensed is in direct contact with the silicon sensing diaphragm especially in harsh environment applications [2]. In fact, corrosive media etch the diaphragm and corrode metallic components on the sensors. This phenomenon tends to affect the performances of the silicon-based sensor diaphragm [3]. The silicon carbide has been use to replace silicon due to its attractive mechanical, electrical and tribological properties that can outstanding wear and heat resistance even at extreme temperature [4]. The main advantage a silicon carbide diaphragm affords capable of operating at high temperature up to 500 °C with high accurately and reliably [5].

Assembly-type pressure sensors device involve many engineering and design tools and supporting infrastructure for Microsystems. MEMS capacitive pressure sensor based silicon carbide (3C-SiC) is expertly designed adaptations of a simple, durable and fundamentally stable with the electrical capacitor. The designed consists a compact housing contains two closely spaced (air gap) parallel, electrically-isolated metallic surfaces which is one of essentially a diaphragm capable of slight flexing under applied pressure [6]. The diaphragm is constructed of a thin layer of 3C-SiC is deposited using

LPCVD techniques on silicon substrate. This diaphragm are mounted so that a slight mechanical flexing of the assembly caused by a change in applied pressure, alters the gap between them creating in effects with a variable deflection and capacitance. The resulting change in capacitance is detected by accuracy, linearity, sensitivity and hysteresis outputs [7].

II. Fabrication And Packaging

A. Fabrication

The sensor consists of three parts fabrication process as shown in Fig.1, Fig.2 and Fig.3.

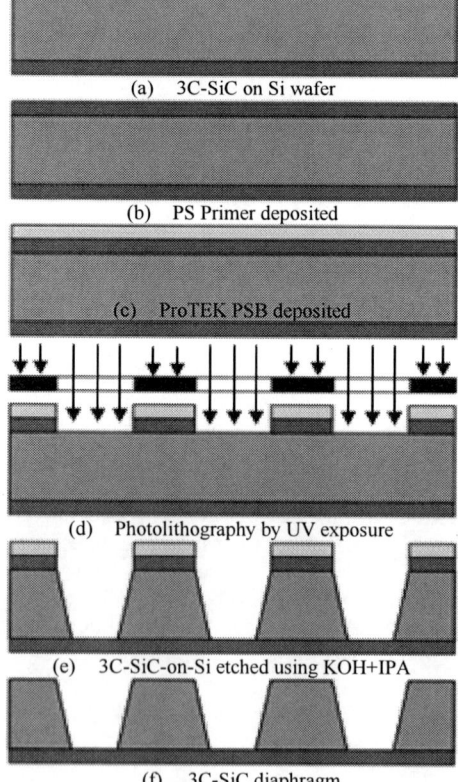

Fig. 1. Process flow for the fabrication of 3C-SiC diaphragm for MEMS capacitive pressure sensor

(a) 3C-SiC on Si wafer
(b) PS Primer deposited
(c) ProTEK PSB deposited
(d) Photolithography by UV exposure
(e) 3C-SiC-on-Si etched using KOH+IPA
(f) 3C-SiC diaphragm

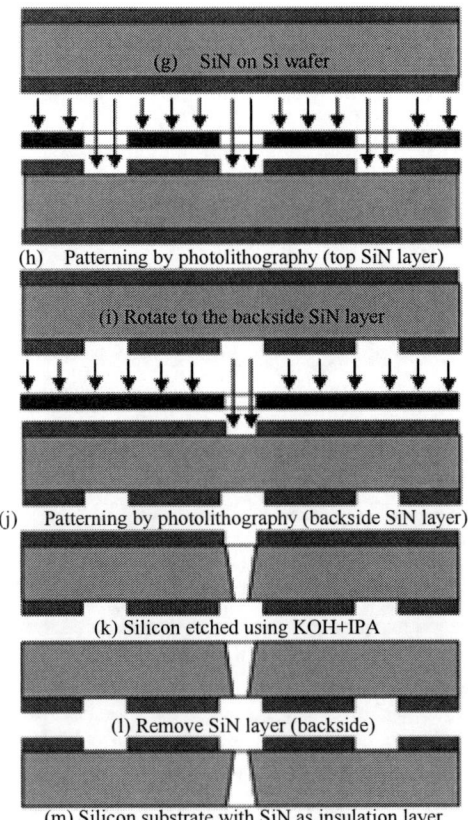

(g) SiN on Si wafer

(h) Patterning by photolithography (top SiN layer)

(i) Rotate to the backside SiN layer

(j) Patterning by photolithography (backside SiN layer)

(k) Silicon etched using KOH+IPA

(l) Remove SiN layer (backside)

(m) Silicon substrate with SiN as insulation layer

Fig. 2. Process flow for the fabrication of substrate for MEMS capacitive pressure sensor.

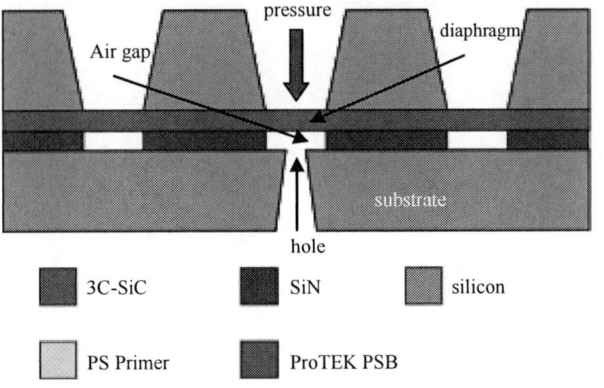

Fig. 3. MEMS capacitive pressure sensor of diaphragm and substrate by bonding process

The fabrication process steps involves (i) fabrication of 3C-SiC diaphragm; (ii) fabrication of silicon substrate; (iii) bonding process of MEMS capacitive pressure sensor between diaphragm and substrate as shown in Fig. 3 is realized as follows; a 1.0 μm-thick 3C-SiC thin film is deposited using LPCVD on 680 μm silicon substrate on a 6-inch. The 3C-SiC on silicon wafer is patterned to define geometry into 2 mm x 2 mm square shape to define the 3C-SiC diaphragm. The first layer of ProTEK PS Primer coating is deposited on the silicon substrate with the spin-coat speed of 1000 rpm for 60 seconds

in Fig. 1(b). For the second layer of ProTEK PSB coating is applied on the wafer with the spin-coat speed at 3000 rpm for 60 seconds in Fig.1(c). Then, the ProTEK PSB is baked at temperature of 120 °C in 60 seconds, as shown in Fig.1 (c). Afterwards, the photosensitive coating layer is exposed using a photolithography process to pattern the diaphragm for approximately 300 seconds through a photomask, as shown in Fig. 1(d). The ethyl lactate (EL) is used to develop the ProTEK PSB mask by immersing it into the EL bath for 30 seconds until the geometric pattern from the photomask is copied onto the Si substrate. Fig.1(e) shows the Si as been back-etched in Potassium Hydroxide (KOH) to leave the 3C-SiC thin film. Finally, PS Primer and ProTEK PSB are removed using io-4000 polishing pad and 180 nm slurry as a lubricant [8].

Fabrication process of substrate MEMS capacitive pressure sensor involves silicon nitride (SiN) with the thicknesses of 200 nm in both the top and bottom layer on silicon wafer as shown in Fig.2 (g). The top of SiN layer is patterned by etching process using Buffered Oxide Etch (BOE) to get the air gap as shown in Fig. 2(h). Then, Fig. 2(j) explained the bottom of SiN layer is patterned and etched with KOH + IPA to determine a little hole for sensing pressure that allow pressure though the air passage. In the end, the bottom layer is polished to remove SiN layer. In Fig.3 shows the fabrication of bonding process of diaphragm and substrate using direct bonding techniques. Both wafers were cleaned with fuming nitric acid (100%) and heated in the hot nitric acid with 70% concentration at 90 °C to the cleaning temperature. Wafers are then rinsed in a quick dump rinse (QDR) which provides a fast method rinsing with DI water and spin dried before direct bonding process. It has been aligned that the two pairs of wafers with the flat. The second wafer (Fig.1) is placed and bonded in top of the first wafers (Fig.2) carefully. A small pressure is applied to the bonded wafers at both side and finally, the two pairs of wafers are completely bonded in Fig.3. The chemical treatments were performed of 5% HF solution for 5 minutes to remove the native oxide layer deposition which is left after cleaning the wafers [9].

B. Packaging

Fig. 4 presents the exploded view of the MEMS capacitive pressure sensor prototype packaged with stainless steel vacuum chamber. The MEMS capacitive pressure sensor is realized from a thin 3C-SiC diaphragm and a substrate as backing plate with electrically insulating silicon carbide thin film deposited on another silicon wafer. The capacitive pressure sensor is packaged in a set of stainless steel chamber which also provides structural support in Fig. 5. The MEMS capacitive pressure sensor is realized with simple assembly approach with a reliable stainless steel o-ring packaging concept. The resulting pressure possesses several appealing traits, including robustness to extreme temperature, alkaline and saline environments and reduces cost.

The characterization of the packaged sensor, LCR meter and a 4-wire measurement technique is used to measure pressure sensor response. The packaged sensor with a standard ½-inch thread is connected through a SS tube to the media pressure source. For testing with nitrogen gas, a high pressure nitrogen cylinder is connected to the packaged sensor through a pressure regulator. The test setup includes a bypass to release the pressure following the test. Fig.5 shows the in-house designed experimental instrumentation used for pressure testing. The commercial pressure gauge is used as a reference for tests with gas. The sensor was packages in a high temperature and characterized under static pressures up to 5 MPa and a temperatures of 500 °C is measured directly from identical sensor elements using LCR meter as shown in Fig.6.

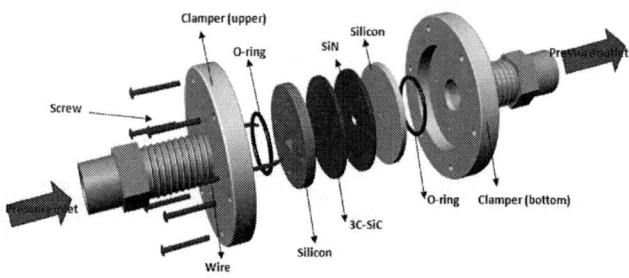

Fig. 4. Exploded view of MEMS capacitve pressure sensor

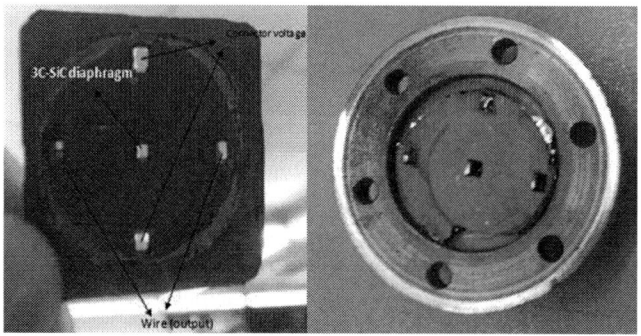

Fig. 5. MEMS capacitive pressure sensor components: 3C-SiC sensor with a 2.0 x 2.0 mm^2 sensing diaphragm

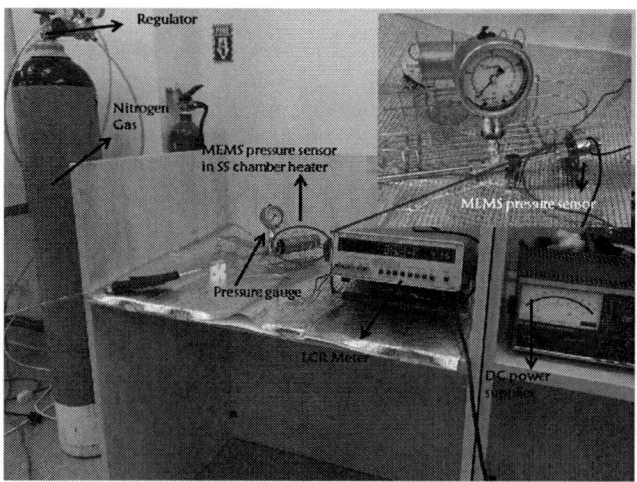

Fig. 6. Experimental set-up for prototype MEMS capacitive pressure sensor testing

III. RESULTS AND DISCUSSIONS

The packaging of MEMS capacitive pressure sensor has been demonstrated for operation at temperatures up to 500 °C. The MEMS packaged has been characterized in a room temperature to 500 °C. In Fig. 7 and tabulated in Table 1, the preliminary measurements indicate that experimental results, at room temperature (27 °C), the sensitivity of the sensor was 0.774 pF/MPa in the range of (1.0 − 5.0 MPa), with nonlinearity of 0.78% and hysteresis of 0.67%. At 300 °C, the sensitivity is 1.289 pF/MPa, the nonlinearity and hysteresis of 0.99 % and 1.73 %, respectively. The sensitivity decreased by 2.1 pF/MPa (0.515 % decreased from room temperature), corresponding temperature coefficient of sensitivity is 0.058 %/°C as shown in Fig. 8. In Fig 9, at 500 °C, the maximum temperature coefficient of output change is 0.073 %/°C measured at 5 MPa.

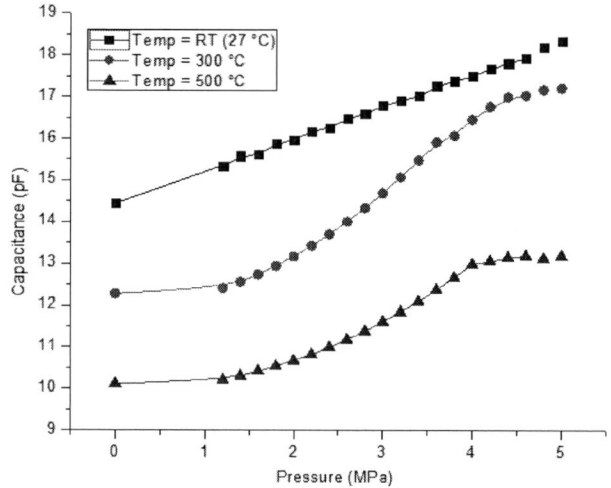

Fig. 7. Capacitance vs. pressure at different temperatures

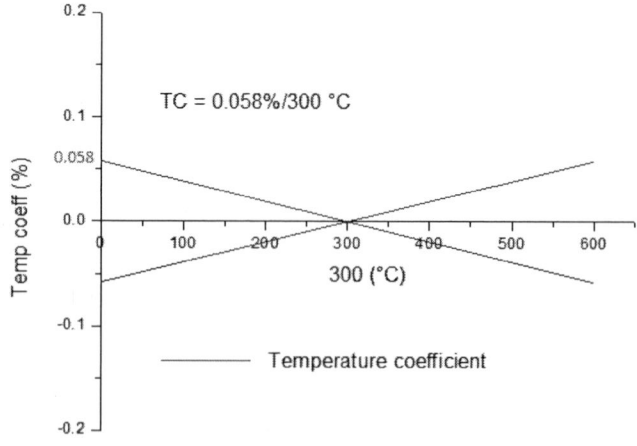

Fig. 8. Temperature coefficient at temperature of 300 °C

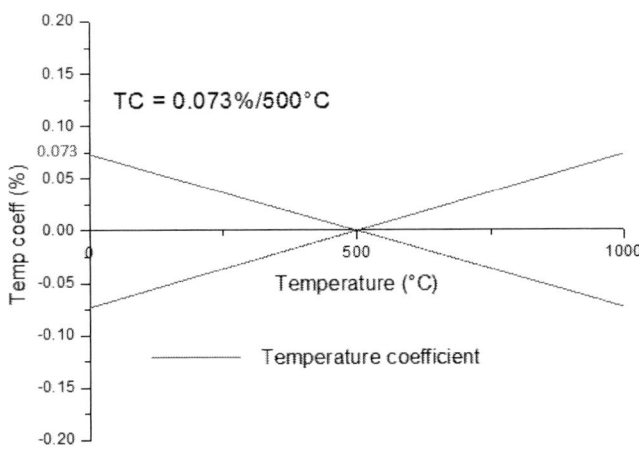

Fig. 9. Temperature coefficient at temperatures of 500 °C

TABLE I.   EXPERIMENTAL RESULTS OF PROTOTYPES MEMS CAPACITIVE PRESSURE SENSOR

Temperatures	*RT (27 °C)*	*300 °C*	*500 °C*
Range Pressure (MPa)	0 – 5	0 – 5	0 – 5
Sensitivity (pF/MPa)	0.774	1.289	0.826
Nonlineairty (%)	0.78	0.98	0.61
Hysteresis (%)	0.67	1.73	3.13
Temp. Coeff. (%)	N/A	0.058	0.073

The results indicates that the performance of 3C-SiC at 300 °C shows high nonlinearity with better sensitivity compared to temperature of 500 °C. This is consistent with the observations in other sensors. The difference nonlinearity is related to the higher thermal stability of the electrical properties of 3C-SiC diaphragm [10]. At temperatures over 500 °C, the readout is extremely noisy. This is likely caused by packaging issues as shorts between bonding wires and metal traces. The hysteresis degrades beyond 500 °C, as the leakage current increased [11]. In order to test long-term performance of these sensors in a high temperature environment and eventually commercialize the MEMS sensors, a high temperature durable and long-term reliable packaging is essential. The testing results indicate that this packaging is suitable for applications with temperatures up to 500 °C. This packaging is primarily designed for high temperature SiC capacitive pressure sensors and applies to other high temperature sensors, especially in different pressure environments.

IV.   CONCLUSION

It is evident that the fundamentals of emerging fabrication of MEMS capacitive pressure sensor based 3C-SiC using ProTEK PSB is capabilities in bulk micromachining a viable technology in MEMS devices. A reliable O-ring packaging concept designed in prototype package devices is designed to provide a simple assembly approach and reduces the manufacturing cost. By comparing the existing data, which works was achieved its proprietary performance through the data that have been identified. The importance of 3C-SiC materials are reveals its durability and potential operating at high temperature (~500 °C) compare previous research was tested only 300 °C, power levels and in harsh environments.

ACKNOWLEDGMENTS

The authors would like to thanks the Institute of Microengineering and Nanoelectronics (IMEN) of Universiti Kebangsaan Malaysia (UKM), Sciencefund MOSTI for supporting this project under grant 03-01-02-SF0849 and Queensland Micro- and Nanotechnology Centre (QMNC) of Griffith University for providing the resources and facilities in part of the Queensland node of the Australian National Fabrication Facility, a company established under the National Collaborative Research Infrastructure Strategy to provide nano and micro-fabrication facilities for Australia's researchers.

REFERENCES

[1] H. Shih-Shian, S. Rajgopal, and M. Mehregany, "Stainless steel capacitive pressure sensor for hostile environments: sample-to-sample variability and reliability characterization," IEEE Sensors, 2011, pp. 452-455.

[2] S. Fricke, A. Friedberger, H. Seidel, and U. Schmid, "High temperature measurement set-up for the electro-mechanical characterization of robust thin systems," Measurement Sceince and Technology, IOP Science, vol.25(1), 2014.

[3] M. Soeda, T. Kataoka, Y. Ishikura, S. Kimura, T. Masuda, Y. Yoshikawa, and M. Nagata, "Sapphire-based capacitive pressure sensor for high temperature and harsh environment applciation," Proceedings of IEEE Sensors, 2002,vol.2, pp. 950-953.

[4] B. S. Virgil, "Application of silicon carbide for high temperature electronics and sensors," Jet Propulsion Laboratory, 1995, pp. 1-9.

[5] D. J. Young, D.U. Jiangang, C. A. Zorman, and W. H. Ko, "High-temperature single-crystal 3C-SiC capacitive pressure sensor,"IEEE Sensors Journal, vol.4 (4), 2004, pp. 464-470.

[6] R. H. Krodorfer, and Y. K. Kim,"A Packaging effect on MEMS pressure sensor performance," IEEE Transactions on Components and Packaging Technologies, vol. 30(2),2007, pp. 285-293.

[7] F. Deeba, S. K. Mohammed, and M. S. Islam." A complete analytical model for square diaphragm capactive sensor with clamped edge," 8th IEEE International Conference on Nano/Micro Engineered and Molecular Sysems, 2013, pp.1311-1314.

[8] N. Marsi, B. Yeop Majlis, Hamzah, A. A., and Mohd-Yasin, F, "Characterization of proTEX PSB thin film as photosensitive layer for MEMS capacitive pressure sensor diaphragm based on SiC-on-Si wafer," Key Engineering Materials, vols. 594-595, 2014, pp. 1083-1086.

[9] N. Marsi, B. Yeop Majlis, Hamzah, A. A., and Mohd-Yasin, F, "Characterization of direct bonding of SiC/SiN layer on Si wafer for MEMS capacitive pressure sensor," IEEE Regional Symposium on Micro and Nanoelectronics, 2013, pp. 50-53.

[10] Z. Min-Xin, H. Qing-An, G. Ming, and Z. Wei, "A novel capacitive pressure sensor based on sandwich strcutures," Journal of Microelectromechanical Systems, vol 14(6), 2005, pp. 1272-1282.

[11] B. Wang, J. de Coster, A. Witvrouw, S. Simone, M. Wevers, and I. de Wolf, "A new method to determine the internal pressure and leakage rate of MEMS packages,"The 17th International Conference on Solid-State Sensors, Actuators and Microsystems, 2013, pp. 558-561.

EFFECTS OF HIGH-K DIELECTRICS WITH METAL GATE FOR ELECTRICAL CHARACTERISTICS OF 18NM NMOS DEVICE

Norani Bte Atan,Ibrahim Bin Ahmad
Centre of Micro and Nano Engineering (CeMNE)
Universiti Tenaga Nasional (UNITEN)
43000 Kajang, Selangor, Malaysia
norani@uniten.edu.my

Burhanuddin Bin Yeop Majlis
Institute of Microengineering and Nanoelectronics (IMEN)
Universiti Kebangsaan Malaysia (UKM)
43600 Bangi, Selangor, Malaysia
burhamems@gmail.com

Abstract— This paper presents a systematic study of various high-K materials on metal gate MOSFET for 18nm NMOS. From the study, we find a suitable combination materials between the high-K and metal gate, which has beneficial effects on the electrical characteristics of 18nm NMOS. The device shows a good improvement on its result of sub-threshold leakage current, I_{OFF}, and drive current, I_{ON} for different dielectric constants (k). The virtual design and fabrication of the device were performed by using Athena module. While electrical characteristic performance was simulated by using Atlas module of SILVACO software. Physical models of the 18nm NMOS were used for simulation from Al_2O_3, HfO_2, and TiO_2 as the material gate dielectric, with $TiSi_2$ as the metal gate, which provide higher physical thickness that able to reduce the sub-threshold leakage current I_{OFF}. Thus, excellent dielectric properties such as high-K constant, low I_{OFF}, higher I_{ON}, threshold voltage V_{TH}, and electrical characteristics were demonstrated. From the simulation results of I_{ON} and I_{OFF}, it was proven that HfO_2 is the best dielectric material with combination of metal gate, $TiSi_2$.

Keywords—18 nm NMOS, High-K dielectric, Metal gate, Silvaco.

I. INTRODUCTION

New technologies have forced many industries to rely on a progression of smaller, denser, faster, cheaper and good quality of MOSFET. By having this criteria, the main target in industry is to reduce the production cost and at the same time able to produce big quantities of MOSFET at a time. With increasing in global competition, modern industries have to adapt their production process to be more efficient and competitive. To do this, more advanced technologies have to be employed to scale down the MOSFET into nanometer regime. As outlined in the International Technology Roadmap for Semiconductor (ITRS) year 2011, the size of gate length (Lg) is targeted to be able to reduce from 18nm (in year 2014) to 12.8nm by year 2018. While MOSFET can be scaled down to smaller dimension with higher performance, the gate length (Lg) and oxide thickness would also be reduced. However, as the thickness scales of SiO_2 is below 2nm, the leakage currents due to tunneling will increase drastically, leading to unfavourable high power consumption and reduce device

reliability. To offset this therefore, replacing the SiO_2 with a high-K material will allow the gate capacitance increase [2]. The electrical characteristics of the device performance are analyzed with several of high-K materials, and the gate oxide thickness is scaled to get the same Equivalent Oxide Thickness (EOT). Recently, many researchers are focusing on metal oxide materials with high-K values that have the ability to be integrated in MOSFET process flow. There are many high-K materials that are being studied nowadays such as Al_2O_3, HfO_2, and TiO_2 [3]. TiO_2 has high dielectric constant property, k ~85, but the band alignment is not favorable at 0.4eV. Although Al_2O_3 has high band at 4.3eV of band alignment, its dielectric constant is low at k~9 which offsets the advantage of high band-gap. In the other hand HfO_2 has the moderate dielectric constant property k~25 with high band gap at 5.68 eV. Whereby HfSiON has high dielectric constant, k~14 and high band gap at 4.3eV. By looking the property values, it is easy to compare the three materials Al_2O_3, HfO_2 and TiO_2 rather than taking HfSiON. That is the reason why the three materials Al_2O_3, HfO_2 and TiO_2 have been chosen in the study. The best characteristics of gate dielectric should have high dielectric constant, large band gap with a favorable band alignment, low interface state density, and good thermal stability. Among the high-K materials some are compatible with silicon, whereby other high-K materials have too low or too high dielectric constant that may not be suitable to use for alternative gate dielectric [4]. Other research paper has concentrated on 18nm gate length with a TiN/HfO2 gate stack by V. Barral (and etc) IEEE, 2007 [5]. However, in this particular paper, we will compare the electrical characteristics results of high-K materials for Titanium Silicide ($TiSi_2$) fabricated on Al_2O_3 (k~9), HfO_2 (k~25) and TiO_2 (k~85) as gate dielectric. Thus, the performance for all high-K materials with $TiSi_2$ are explored and presented in the following sections.

II. MATERIALS AND METHODS

The specification of the sample used in this experiment was p-type (boron doped) silicon substrate with doping concentration of 7×10^{14} atoms cm^{-3} and <100> orientation.

978-1-4799-5761-3/14 $31.00 © 2014 IEEE

The next procedure is to develop the P-well with growing a 200 Å oxide screen on the top of bulk silicon. This method using dry oxygen at high temperature of 970°C followed by Boron as dopant with a dose of 3.75×10^{13} atoms/cm^3. The oxide layer as a mask was etched after the complete process of P-well doping. The next step was annealing process to ensure all boron atoms being spread uniformly in the wafer at 900°C with nitrogen and followed by 950°C with dry oxigen. The next step was preparing the isolated neighboring transistor or Shallow Trench Isolator (STR) of 130 Å thicknesses [6]. Then, the wafer was oxidized with dry oxygen for 25 minutes at 900°C. There were two important processes involved in developing STI such as Low Pressure Chemical Deposition (LPCVD), and Reactive Ion Etching (RIE). The LPCVD process was applied with 1000 Å nitride layer which was deposited on top of the STI oxide layer, followed by a photo resistor that was deposited on the wafer. Then, RIE process etching the unnecessary part on the top of STI area. Chemical Mechanical Polishing (CMP) was implemented to etch extra oxide on the wafer. STI was annealed for 15 min at 900°C. A sacrificial oxide layer was developed and etched to eliminate defects on the surface [7]. The process of STI then was considered completed.

Between the channel regions, the deposition of high-K materials (Al$_2$O$_3$, HfO$_2$, and TiO$_2$) process with gate oxide thickness is scaled-adjust so that they have the same EOT with SiO$_2$ by analyzing electrical characteristics of the device. The length of the high-K material was scaled and adjusted to get 18nm same value as the gate length of transistor. Then, the implantation dose of boron on the N-well active area for the threshold voltage (V$_{TH}$) adjustment process was done. Next, the process done for Titanium Silicide (TiSi$_2$) deposited on the top of high-K materials (Al$_2$O$_3$, HfO$_2$, and TiO$_2$), and followed by halo implantation with adjusted of indium dose to obtain the optimum value of the NMOS device [8, 9]. The next process was the formation of the sidewall spacer. It was used as a mask for source and drain implantation. In this case, the implantation was with arsenic dose and followed by phosphor dose. It was to ensure the smooth current flow in NMOS device [10].

The next step was the development 0.5 μm layer of Boron Phosphor Silicate Glass (BPSG). This layer acted as the Pre Metal Dielectric (PMD) [11]. Again, annealing process was done on the wafer to strengthen the structure under temperature 950°C. The last process of wafer was the compensation implantation which using the phosphor dose. From the above experiment, the dosage quantities for boron, indium, arsenic, and phosphor were different which based on the high-K materials (HfO$_2$, TiO$_2$ and Al$_2$O$_3$). The last step was deposition of aluminum layer as the metal contact for source and drain. Therefore, under ATHENA module the design of 18nm NMOS structure already completed. There are many factors influenced the input process parameters on the threshold voltage 18nm NMOS such as Gate oxide thickness, Substrate implant dose, Pocket-halo implant tilt angle, Gate oxide diffusion temperature, V$_{TH}$ implant dose, V$_{TH}$ implant energy, Pocket-halo implant energy, Pocket-halo implant dose,

S/D implant dose, S/D implant energy, Compensation implant dose and Compensation implant energy. Now, from here we will proceed the simulation process under ATLAS module to measure the electrical characteristic such as I$_D$ versus V$_{DS}$, I$_D$ versus V$_{GS}$, I$_{ON}$, I$_{OFF}$, and V$_{TH}$.

III. RESULTS AND DISCUSSIONS

The complete fabrication of 18nm NMOS has been modeled and simulated successfully in Silvaco Simulink. Figure 1 shows the complete 18nm gate length NMOS. Figure 2 shows clearly on the doping profile of one of the design structure with gate length 18nm NMOS.

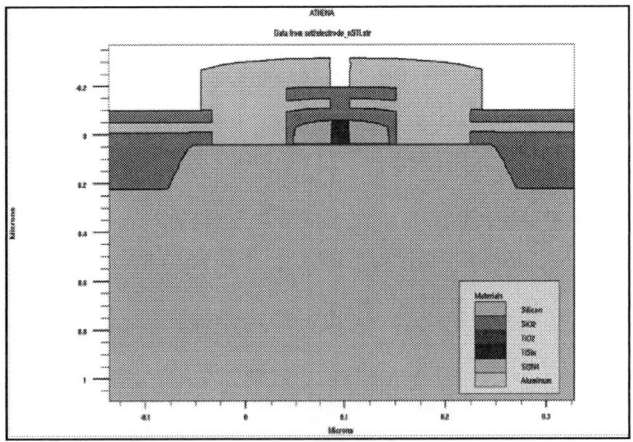

Figure 1. A 18nm NMOS cross section after Fabrication Simulation

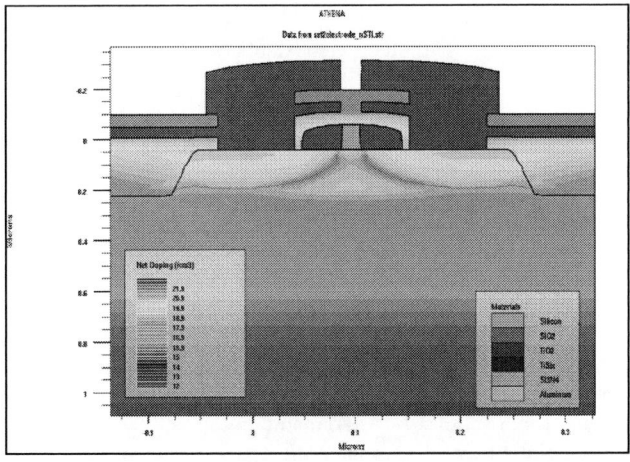

Figure 2. Doping profile of 18nm gate length NMOS with TiO$_2$ dielectric and TiSi$_2$ metal gate

Results of electrical characteristic simulation are obtained in Figure 3 for I$_D$ –V$_{DS}$ and Figure 4 for I$_D$ – V$_{GS}$ with different materials of high-K parameter such as Al$_2$O$_3$ (k~9), HfO$_2$ (k~25) and TiO$_2$ (k~85). Voltage, V$_{GS}$ = 2.6 volts is applied for I$_D$ –V$_{DS}$ graph with different voltage of V$_{DS}$. While, V$_{DS}$ =

1.4 volts is supplied for $I_D - V_{GS}$ graph with different V_{GS} voltage. The threshold voltage (V_{TH}), state on current (I_{ON}) and state off current (I_{OFF}) can be extracted from $I_D - V_{GS}$ curve.

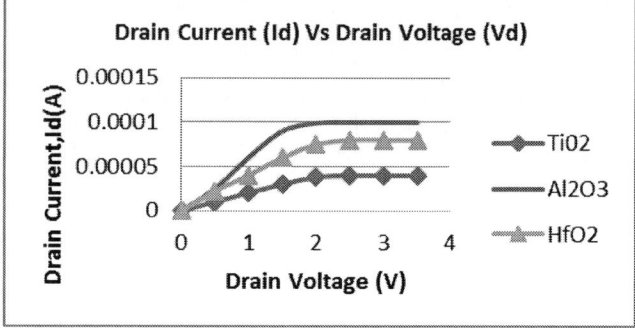

Figure 3. I_D Versus V_{DS} characteristic for different of high-K dielectric constants (k)

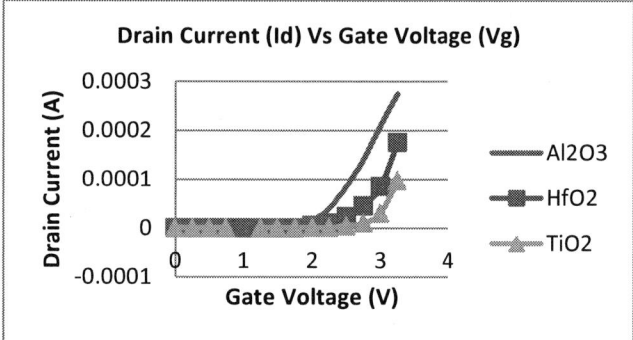

Figure 4. I_D Versus V_{GS} characteristic for different of high-K dielectric constants (k)

A good doping concentration is a main factor to ensure the transistor works well with fewer leakage current and enhance gate control [12]. There are four factors that influence in the threshold voltage counter measure such as Threshold voltage ajustment implant, Halo implant, Channel implant and Compensation implant. But for this research, by changing the various of dielectric materials (Al_2O_3, HfO_2 and TiO_2) on the $TiSi_2$ of transistor, the Threshold voltage ajustment implant which is best doping concentration will be obtained to get a threshold voltage (V_{TH}) 0.302651 with 6.03036 x 10^{13} atom/cm^3 boron for Al_2O_3. To maintain the same value of V_{TH}, due to the physically thicker dielectric layer, therefore the boron doping for HfO_2 and TiO_2 were adjusted and increased to 8.53256 x 10^{13} atom/cm^3 and 9.73654 x 10^{13} atom/cm^3 respectively. The increase of V_{TH} adjustment implant doping concentration was proportional with increasing the value of high-K dielectric, and at the same time the values of drive current (I_{ON}) were decreased as shown in Figure 5.

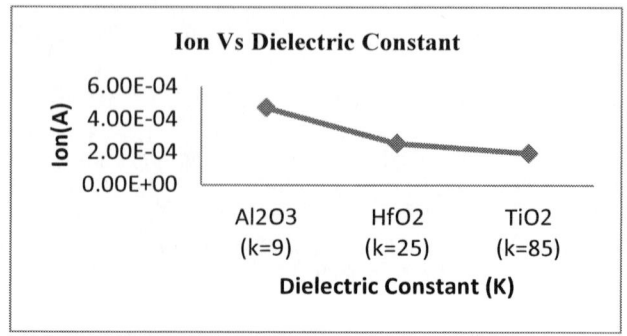

Figure 5. I_{ON} current for Al_2O_3, HfO_2 and TiO_2 dielectrics

Drain leakage current (I_{OFF}) or sub-threshold leakage current occurs when the gate voltage (V_{GS}) is lower than the threshold voltage (V_{TH}). In ideal case, when the transistor is turned off, $V_{GS} = 0$ volt and $V_{DS} = V_{DD}$ (voltage supply), there is no current can go through into the channel ($I_{OFF} = 0$). Refer to Figure 6, the sub-threshold leakage current for HfO_2 dielectric is lowest compared with Al_2O_3 and TiO_2 dielectrics. It indicates, HfO_2 dielectric material is the most compatible with silicon and the most stable oxide with the highest heat of formation [13, 14].

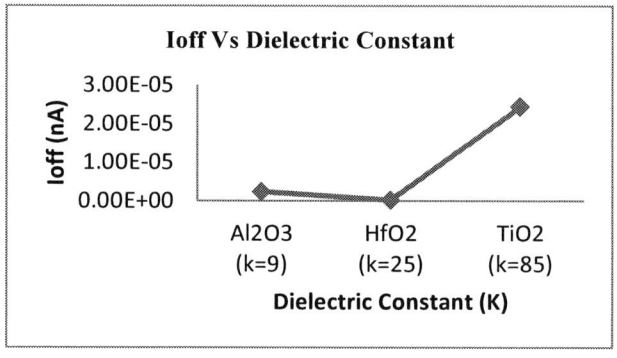

Figure 6. I_{OFF} current for Al_2O_3, HfO_2 and TiO_2 dielectrics

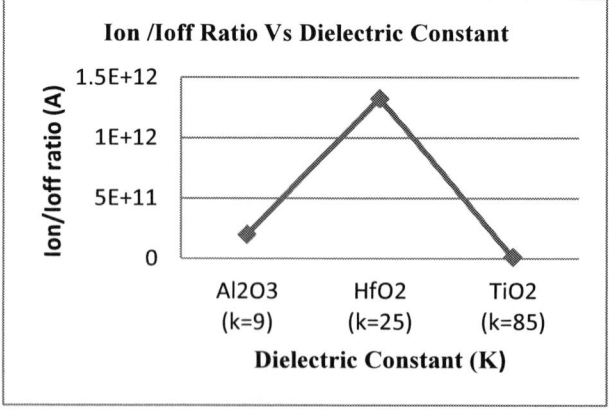

Figure 7. I_{ON}/I_{OFF} current for Al_2O_3, HfO_2 and TiO_2 dielectric

Figure 7 shows the plotted graph of I_{ON}/I_{OFF} current ratio for different materials of high-K dielectric. The result of HfO_2 dielectric gives the highest I_{ON}/I_{OFF} ratio compared with Al_2O_3 and TiO_2 dielectrics. Hence, better performance of device can be obtained by using HfO_2 dielectric as gate. This device shows the higher I_{ON} and I_{OFF} ratio at sub-threshold region of operation. So, the device is suitable for low power application [15].

Table 1. Simulated Results of Various Dielectric materials With ITRS 2011 Prediction

Parameter	Simulation			ITRS 2011 Prediction
	Al_2O_3	HfO_2	TiO_2	
V_{TH}(V)	0.302651	0.302651	0.302651	0.302
I_{ON} (A/μm)	4.721×10^{-4}	2.536×10^{-4}	1.934×10^{-4}	1.0×10^{-7}
I_{OFF} (A/μm)	2.365×10^{-15}	1.912×10^{-16}	2.432×10^{-14}	1.496×10^{-6}
I_{ON}/I_{OFF} ratio	1.996×10^{11}	1.326×10^{12}	7.952×10^{9}	6.685×10^{-2}

Table 1 shows the simulated results for Al_2O_3, HfO_2 and TiO_2 of 3 dielectric materials with $TiSi_2$ as the metal gate for 18nm NMOS. The I_{ON} values resulted from the simulation are bigger values compared to prediction value. While the simulation results for I_{OFF} are lower than prediction value. The good performance of design MOSFET is when I_{OFF}=0 and higher of I_{ON} value. Therefore, it is concluded that all the 3 dielectric materials of high-K materials are suitable with metal gate and compatible with silicon of transistor. However, it is found the best choice material is HfO_2 as the dielectric of transistor. Why HfO_2 gives the better result? It is because HfO_2 has high enough dielectric constant, high enough band-gap and band offsets with silicon, good scalability and low leakage current. Heat of formation for HfO_2 is 271 Kcal/mol and its band gap is 5.68eV and high enough to obtain sufficient barrier height. HfO_2 is thermodynamically stable with silicon substrate, high dielectric constant (~25), resistive to impurity diffusion because of its high density (9.68g/cm3), with lattice parameter similar to that of Si with a small lattice misfit (<5%).

IV. CONCLUSION

The NMOS structure with 18nm were successfully designed and simulated to study the various of dielectric materials on metal gate of device performance. The performance of the three dielectric materials, Al_2O_3, HfO_2 and TiO_2 with $TiSi_2$ as metal gate were compared and it was found HfO_2 is the best dielectric material for the future nano scale MOS devices technology. It based on the highest value of I_{ON}/I_{OFF} ratio, and lowest value of sub-threshold leakage current (I_{OFF}). Hence, this device is suitable for low power application.

ACKNOWLEDGEMENT

The authors would like to thank to Ministry of Higher Education (MOE), Institute of Microengineering and Nanoelectronics (IMEN) Universiti Kebangsaan Malaysia (UKM), and Centre of Micro and Nano Engineering (CeMNE) Universiti Tenaga Nasional (UNITEN) for financial, facilities and moral throughout the project.

REFERENCES

[1] ITRS 2011.www.ITRS2011.net.
[2] M.K.Bera and C.K. Maiti, "Electrical Properties of SiO_2/TiO_2 High-K Gate Dielectric Stack", Materials Science in Semiconductor Processing, pp.909-917, 2006.
[3] Hei Wong and Hiroshi Iwai, "On the Scaling Issues and High-k Replacement of Ultrathin Gate Dielectric for Nanoscale MOS Transistor," Microelectronic Engineering 83, pp.1867-1904, 2006.
[4] Byoung Hun Lee, Laegu Kang, We-Jie Qi, Renee Nieh, Yongjoo Jeon, Katsunori Onishi and Jack c.Lee, "Ultrathin Hafnium Oxide Low Leakage and Excellent Reliability for Alternative Gate Dielectric Application", IEEE Electron Device, pp. 133-136, Dec 1999.
[5] V.Barral,T.Poiroux, F.Andrieu et al.,"Strained FDSOI CMOS technology scalability down to 2.5nm film thickness and 18nm gate length with a TiN/HfO2 stack", IEEE 2007.
[6] J.W. Sleight, I.Lauer, O.Dokumaci, D.M. Fried, D.Guo, B Harun, S Narasimha, C Sheraw, D Singh, M Steigerwalt, X Wang, P Olaiges, D Sadana, C Y Sung, W Haensch, M Khare, " Challenges and Opurtunities for High Performance 32nm CMOS Technology," IEDM Tech Digital, 2006.
[7] F.salahuddin, I.Ahmad, F.A Hamid and A.Zaharim, "Analyze and Optimize the Silicide Thickness in 45nm CMOS Technology Using Taguchi Metho," ICSE Proc.2010.
[8] H.A.Elgomati, B.Y.Majlis, I.Ahmad, F.Salahuddin ,F.A.Hamid,A.Zaharim, T.Ziad Mohamad and P.R.Apte, "Investigation of the Effect for 32nm PMOS Transistor and Optimizing Using Taguchi Method," Asian Jounal of Applied Science, 2011
[9] Shuan-Yuan Chen et al., "Using simulation to Characterize High performance 65nm Node Planar," International Symposium on Nano Science and Technology, Taiwan, 20-21 Nov 2004.
[10] F.salahuddin, I.Ahmad, and A.Zaharim, "Impact of Different Dose and Angle in Halo structure for NMOS Device," ICMST 2010, pp. 81-85, 26-28 Nov 2010.
[11] Sarcona G.T., M. Stewart, and M.K. Hatalis, " Polysilicon Thin Film Transistor Using Self-Aligned Cobalt and Nickel Silicide Source and Drain Contacts," IEEE Electron Device let, vol. 20, 1999
[12] Ashok K. Goel et. al., (1995), "Optimization of Device Performance Using Semiconductor TCAD Tools", Silvaco International Product Description, Silvaco International http://www.silvaco .com/products/descriptions.
[13] Hubbard, K.J., Schlom, D.G, "Thermodynamic stability of binary oxides in contact with silicon", J.Mater.Res, 11, 2757-2776. 1996
[14] Chattererjee, S., Kuo, Y., Lu, J., Tewg, J.Y., Majhi, P., "Electrical reliability aspects of HfO2 high-K gate dielectric with TaN metal gate electrodes under constant voltage stress", Microelectronics Reliability, 46, 69-76.
[15] Nirmal,Vijaya Kumar,Patrick Chella Samuel, Siruthi, Divya Mary Thomas Mohan Kumar, "Analysis of Dual Gate Mosfets Using Higk-K Dielectrics", IEEE Electron Device Proc., pp 22-25 2011.

Design of a 4-bit Adder using Reversible Logic in Quantum-Dot Cellular Automata (QCA)

Darushini Kunalan, Chee Lee Cheong, Chien Fat Chau, Azrul Bin Ghazali

Center for Micro & Nano Engineering (CeMNE), College of Engineering
University Tenaga Nasional
Selangor, Malaysia

Abstract— **Both quantum-dot cellular automata (QCA) and reversible logic are emerging technologies that are promising alternatives to overcoming the scaling and heat dissipation issues, respectively, in the current CMOS designs. Here, the fundamentals of QCA and reversible logic are studied; the feasibility of incorporating reversible logic in QCA designs is also demonstrated. Based on two existing designs, an improved version of the reversible gates, namely the Feynman Gate and the Toffoli Gate, were implemented in QCA technology using QCADesigner. The proposed design of the QCA-based Feynman Gate is faster by ½ cycle as compared to the existing design; while the proposed Toffoli Gate has the same latency as the existing design but it is readily to be cascaded into a more complex design. A 4-bit ripple carry adder in QCA is then designed using the proposed Feynman and Toffoli gates to realize a reversible QCA full adder. This 4-bit QCA adder with reversible logic consists of 2030 QCA cells, has a latency of 7 clock cycles and 8 garbage outputs.**

Keywords—Adder, Ripple Carry, quantum-dot cellular automata, reversible logic, Feynman, Toffoli

I. INTRODUCTION

For the past few decades, the number of transistors in the integrated circuit (IC) have increased tremendously consistent with the Moore's law [1], [2]. However, increase in the number of transistors in IC leads to size reduction of the transistor due to scaling. Unfortunately, device scaling is reaching its limits which affects the device's performance due to constraints like heat dissipation, interconnects and power consumption [3]. To solve these problems, many alternative technologies were explored and quantum-dot cellular automata (QCA) is considered to be one of the promising alternatives that have potentials to replace the CMOS technology. QCA was first introduced by C.S. Lent *et al.* in 1993 [3]. Unlike standard transistors that channel information through current or voltage, QCA transfers information based on cell-to-cell or Coulombic interactions [4]. Due to its features, QCA is expected to have high switching or processing speed, is capable of achieving high device density and have low power consumption [5], which overcomes the performance degradation and scaling problems faced by the conventional CMOS transistors.

Apart from that, reversible logic technology [6] is also considered as an alternative technology to help improve the heat dissipation problem in CMOS devices. This is because most of current circuits are irreversible whereby input cannot be retrieved from the output since it is erased after every logical operation. Erasing the input information dissipates heat [6]. This process does not occur in reversible circuits which ideally reduces the heat dissipation problem in device. This paper introduces the fundamental concepts of QCA, QCA devices and reversible logic. Since the adder is an essential component in microprocessors, a 4-bit adder with reversible logic was developed using QCA to illustrate the feasibility of implementing of reversible logic concepts in QCA adder circuits.

II. QCA BACKGROUND

A. QCA Cell

A basic QCA cell is in the form of a square, consisting of four quantum dots each placed at the corner of the cell. Each quantum dots represents a potential well in which electron resides in. Each cell consist of two electrons which are free to move to neighbouring dots via tunneling. The two electrons will localize at the opposite corners of the cell since it is the largest possible separation distance within the cell itself, due to the Columbic repulsive force of the electrons. This leads to only two possible configurations with equivalent polarizations designated as +1 (logic 1) and -1 (logic 0) [4], [7], as shown in Fig. 1. Using these logics, binary information can be encoded. As the state of one cell can be strongly influenced by the state of its neighbouring cells, the transfer of information along an array of QCA cells is possible [4]. A QCA device is implemented via sensible arrangements of these cells which develops the behaviour of logic gates.

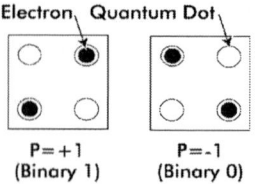

Fig. 1. Basic QCA cell showing two possible cell polarization (after [4]).

The authors gratefully acknowledge the Ministry of Education Malaysia for providing the financial assistance through ERGS grant (ERGS/1/2013/TK02/UNITEN/02/05).

B. QCA Wire

An array of QCA cells placed side by side forms a QCA wire which transfers binary information from one end (the input cell) to the other (the output cell). There are two types of QCA cell orientation to form a wire, which are 90° QCA wire and 45° QCA wire [7], as depicted in Fig. 2. In a 90° QCA wire, polarization is fixed at the leftmost cell which is regarded as the input cell and the state of the neighbouring cells will align similarly as the state of the newly polarised input cell and continues till the entire length of the wire [7]. In a 45° QCA wire also known as an inverter wire, the cells are rotated 45° and demonstrate anti-aligning behaviour. This mean that the state of its neighbouring cells will be the invert the state of the input cell [7].

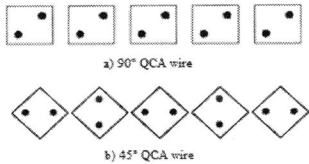

Fig. 2. QCA wires (after [7])

C. QCA Crossover

There are two types of wire crossing in QCA, which are coplanar wire crossing and multilayer wire crossing [8]. Coplanar wire crossing is a single layer wire crossing design making use of both 90° and 45° QCA wire. Those wires are arranged perpendicular to each other. Typically, horizontal wire uses 90° QCA wire whereas the vertical wire consists of 45° QCA wire [8]. Multilayer crossing involves building a bridge-like wire in layers that will form a crossing over another wire. Fig. 3, shows a multilayer wire crossing whereby it consists of three layers which are the main layer (layer 1), the interconnect layer (layer 2) and the bridge layer (layer 3) which crosses over the vertical wire. Both wires are able to transmit signal '0' and signal '1' respectively without any interference [8].

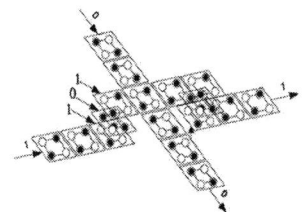

Fig. 3. Multilayer wire crossing (after [8])

D. QCA Majority Gate

The QCA three-input majority gate is used as the basic building block for complicated circuit. It has three inputs A, B and C which are connected to a cell in the center known as the device cell [4], as depicted in Fig. 4. The purpose of this cell is to compute the majority polarization of the majority of the input cells. The logic function of the majority gate can be expressed in the Boolean form as $M(A, B, C) = AB + BC + CA$ [4]. From this majority gate, both the QCA AND and OR gates can be designed by fixing the polarization of one of the input cells as logic 0 and logic 1, respectively.

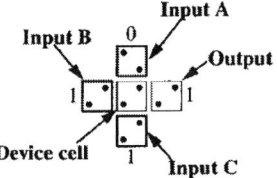

Fig. 4. A QCA three-input majority gate (after [4])

E. QCA Clocking

In QCA devices, clocking system is used to provide synchronization in implementing pipelining as well as to monitor the flow of information [2]. The process whereby the QCA array is gradually and smoothly switched in order to maintain it in its instantaneous ground state is called adiabatic switching which consists of four phases namely switch, hold, release and relax phases [8]. Together, this four phases is equal to one clock cycle [8]. The entire cycle is shown in Fig. 5.

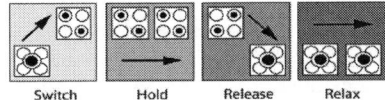

Fig. 5. The four phases of QCA clocking zone (after [8])

III. REVERSIBLE LOGIC

According to C.H. Bennett in 1973, reversible circuits is the solution to overcome the heat dissipation problem in irreversible circuits which typically erase input information each time after a logical operation is carried out [6]. The Landauer's Principle (refer to [9]) elaborates that erasing the values involves the compression of the states which will release heat. Reversible circuits do not erase information, thus prevent heat dissipation problems. In addition, logically reversible circuits have a one-to-one mapping which means the number of input must be equal to the number of output [11]. This unique mapping between inputs and outputs enables the input information to be retrieved from the output information which prevents information loss problem. In this project, the two common reversible gates used are Feynman and Toffoli gates [6].

A. Feynman Gate

Feynman gate [12, 13] is a (2, 2) reversible gate that was proposed by Richard Feynman in 1985. Feynman gate has input vector (A, B) and output vector (X, Y) and realises the equations whereby when $A = 0$, Y returns the value of B and when $A = 1$, Y returns the inverted value of B:

$$X = A \qquad (1)$$

$$Y = A \oplus B \qquad (2)$$

B. Toffoli Gate

Toffoli gate [14, 15] is a (3, 3) reversible gate that was proposed by Tommaso Toffoli in 1982. Toffoli gate has input vector *(A, B, C)* and output vector *(X, Y, Z)* and realises the equations whereby when *A = 1* and *B = 1*, output *Z* returns the inverted value of *C*. Otherwise, output *Z* returns the value of *C*:

$$X=A \qquad (1)$$

$$Y=B \qquad (2)$$

$$Z=C \oplus BA \qquad (3)$$

IV. 4-BIT ADDER IN QCA WITH REVERSIBLE LOGIC

To demonstrate the feasibility of reversible logic in QCA designs, a 4-bit ripple carry adder (RCA) with reversible logic gates was developed and implemented in QCADesigner [10].

A. Comparison of the Existing and Proposed Design of the Reversible Gates

Modifications are done on the previous designs of Feynman and Toffoli gates based on [13] and [15], respectively, as depicted in Fig. 6. The white circles indicate some of the changes done on the designs. For the Feynman gate, simple diagonally placed cells are used as inverters instead of a full inverter design which reduces the number of cells used. In contrast, a different inverter layout is used for Toffoli gate where the cells are displaced by 10nm. This configuration is used to obtain a more stable inverting signal. Parameters such as number of cell used, area, latency and advantage of the proposed design are compared and summarised in Table I. The modified Feynman gate is faster by ½ cycle as compared to [13]. The proposed Toffoli gate has the same latency but it is readily to be cascaded into a more complex design as compared to [15].

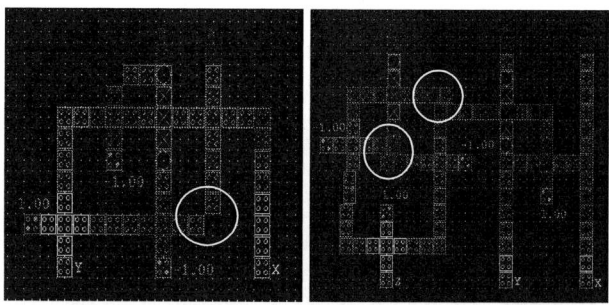

Fig. 6. Proposed Feynman Gate (Left) and Toffoli Gate (Right)

TABLE I. COMPARISON OF THE REVERSIBLE GATES

	Feynman Gate Design		Toffoli Gate Design	
	Proposed	[13]	Proposed	[15]
Cell Count	62	65	108	101
Area	$0.11\mu m^2$	$0.09\mu m^2$	$0.14\mu m^2$	$0.12\mu m^2$
Latency	1 cycle	1 cycle	1 cycle	$1\,^1/_2$ cycle
Design Advantage	Readily cascaded in circuits	-	Faster by ½ cycle	-

B. 1-Bit Adder in QCA with Reversible Logic

The proposed 1-bit adder is implemented based on the design in [11]. It uses two each of the proposed Feynman gates *(FG1 and FG2)* and Toffoli gates *(TG1 and TG2)*. The inputs are denoted as *A*, *B* and *Cin* whereas the outputs are *Sum* and *Cout*. As depicted in Fig. 7, there are two garbage outputs which are denoted as *G1* and *G2*. These outputs are unused outputs which normally do not affect the actual outputs of the adder. Garbage outputs are useful only when retrieving the input information from the outputs generated. Changes were done on the design in [11], thus enables the less significant bit adder to be easily connected as *Cin* of the more significant bit adder without the use of crossover wires. The QCA design of the proposed 1-bit adder consists of 396 QCA cells with the size $0.65\mu m^2$ as depicted in Fig. 8. Each of the reversible gates are assigned one clock cycle, hence the total is four clock cycles. The simulation result of the 1-bit adder using reversible gates is shown in Fig. 9. The correct results of the 1-bit adder were obtained after 4 cycles which agrees with the design.

C. 4-Bit Adder in QCA with Reversible Logic

The 4-bit RCA is developed by cascading four 1-bit adder as depicted in Fig. 10, whereby the *Cout* of the less significant bit adder is the *Cin* of the more significant bit adder (for example *Cout0* is connected to *Cin1*) The outputs of this 4-bit adder are *Sum3* to *Sum0*, arranged from left to right which is from the most significant to the least significant bit, respectively. The finalized QCA-based 4-bit adder with reversible gates consists of 2030 cells with the size of $4.62\mu m^2$ as depicted in Fig. 11. The adder requires a total of 7 clock cycles because extra clock zones were added to the inputs (*B3, A3, B2, A2, B1* and *A1*) and the outputs (*Sum0, G1, Sum1, G3, Sum2* and *G5*) to synchronise the final output. Therefore, the first correct output of the 4-bit adder is obtained after 7 clock cycles as depicted in Fig. 12. This design generate 8 garbage outputs (denoted as *G1* to *G8*) since there are four 1-bit adder cascaded. These outputs are used to retrieve information of inputs (*Ai* and *Bi* where *i* represent 0 to 3) since the outputs (*Sumi* and *Couti* where *i* represent 0 to 3) are known. For verification purposes, the 4-bit adder was tested by setting *Cin* as alternating '0'and '1', inputs *A* and *B* as a range of values from 0 to 15. The output obtained, as depicted in Fig. 12. agrees with the testbench used. The first input values of testbench are *Cin, A, B = 0*, thus the output is *Sum = Cin + A + B = 0* with *Cout = 0* as well. The last observable output in this testbench is *Sum = 1 + 9 + 9 = 19 (1 0011_2)* with *Cout = 1*.

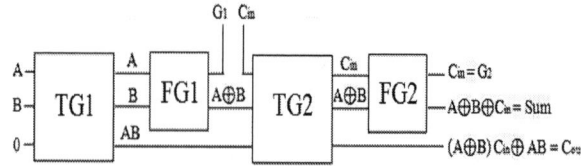

Fig. 7. Proposed 1-bit adder in QCA with reversible logic

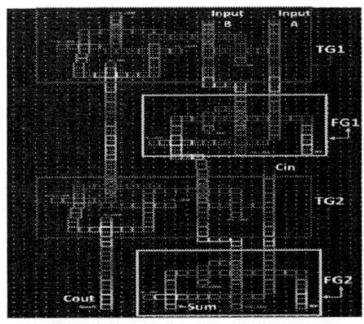

Fig. 8. QCA design of the proposed 1-bit adder

Fig. 9. Simulation result of the 1-bit adder

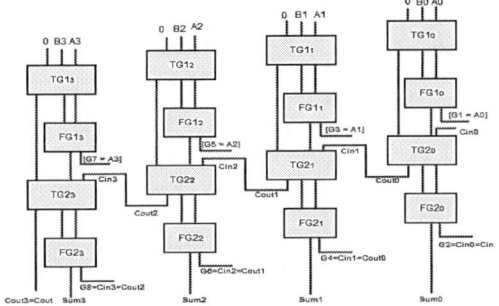

Fig. 10. Proposed 4-bit adder using reversible logic

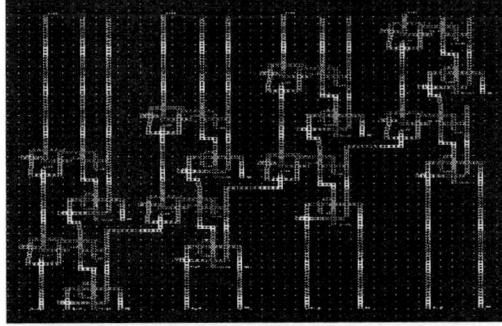

Fig. 11. QCA design of the proposed 4-bit adder

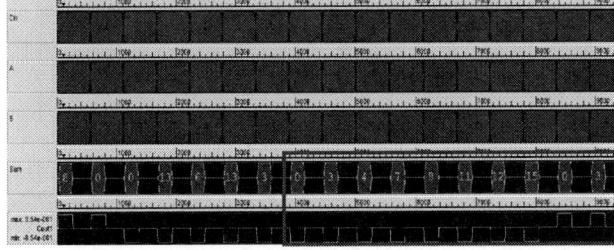

Fig. 12. Simulation results of 4-bit adder

CONCLUSION

This paper has explored the basic principles and devices in QCA, as well as the fundamentals of reversible logic. QCADesigner software was used for the design and simulation of all QCA-based gates and circuits. The Ripple Carry architecture was chosen for the design of the 4-bit adder with reversible logic as it is the simplest adder architecture to be implemented. The theoretical values of the 1-bit and 4-bit adder using reversible logic matches the simulation results obtained in QCADesigner. The 4-bit adder using reversible logic uses 16 reversible gates and has a total of eight garbage outputs, which is relatively small and has a latency of 7 clock cycles. No performance comparison has been carried for the proposed design of the 1-bit and 4-bit adder since in [11], the existing adder designs are not implementations of QCA technology using QCADesigner. This project proposes improvements on the designs of Feynman and Toffoli gates in QCA and proves that these designs can be incorporated into the design of QCA 1-bit and 4-bit adders.

REFERENCES

[1] N.H.E. Weste and D.M. Harris, Integrated Circuit Design, 4th ed. Boston: Pearson, 2011.

[2] G. Snider, A. Orlov, C.S. Lent, G. Bernstein, M. Lieberman and T. Fehlner, "Implementations of Quantum-dot Cellular Automata," in *IEEE Int. Conf. Nanoscience and Nanotechnology (ICONN'06)*, Brisbane, Australia, 3-7 July 2006.

[3] C.S. Lent, P.D. Tougaw, W. Porod, and G.H. Bernstein, "Quantum-dot Cellular Automata," Nanotechnology, vol. 4, no. 1, pp. 49-57,1993.

[4] W. Porod, "Quantum-dot devices and Quantum-dot Cellular Automata," J. Franklin Institute, vol. 334, no. 5-6, pp. 1147-1175, 1997.

[5] S.T.Y. Chan, C.F. Chau, and A. bin Ghazali, "Design of a 4-bit ripple adder using Quantum-dot Cellular Automata (QCA)," In *IEEE Int. Conf. Circuits and Systems (ICCAS)*, Kuala Lumpur, Malaysia, pp. 33-38, Sept 2013.

[6] W.D. Pan and M. Nalasani, "Reversible Logic," IEEE Potentials, vol. 24, no. 1, pp. 38-41, 2005.

[7] T. Cole and J.C Lusth, "Quantum-dot Cellular Automata," Progress in Quantum Electronics, vol. 25, no. 4, pp. 165-189, 2001.

[8] D. Rajeswari, K. Paul and M. Balakrishnan, "Clocking-Based Coplanar Wire Crossing Scheme for QCA," in *23rd Int. Conf. VLSI Design (VLSID '10)*, Bangalore, India, pp 339-344, 3-7 January 2010.

[9] R. Landauer, "Irreversibility and heat generation in the computing process," IBM J. Research and Development, vol. 5, no.3, pp. 183-191, 1961.

[10] K.Walus, T.J. Dysart, G.A. Jullien and R.A. Budiman, "QCADesigner: a rapid design and simulation tool for Quantum-dot Cellular Automata," IEEE. Trans. Nanotechnol. Canada. vol 3, no.1, pp 26-31, March 2004.

[11] H.M.H. Babu, M.R. Islam, A.R. Chowdhury and S.M.A. Chowdhury, "Reversible Logic Synthesis for Minimization of Full-Adder Circuit," *in Proc. Euromicro Symposium on Digital System Design Conf.*, Belek-Antalya, Turkey, pp 50 – 54, 1-6 Sept 2003.

[12] R. Feynman, "Quantum Mechanical Computers," Optics News, Vol.11, pp. 11–20, 1985.

[13] M.A. Rahman, F. Khatun, A. Sarkar and M.F. Huq, "Design and Implementation of Feynman Gate in Quantum-dot Cellular Automata (QCA)," Int. J. Computer Science Issues (IJCSI), vol. 10, no. 1, pp 167-170, July 2013.

[14] E. Fredkin and T. Toffoli, "Conservative Logic," in Collision-based Computing, London, UK: Springer London, 2002, pp. 47-81.

[15] B. Cvetkovska, I. Kostadinovska, and J. Danek, "Implementing the Toffoli gate in Quantum-dot Cellular Automata," Final Report for Seminar Work in Unconventional Information Processing Methods and Platforms Course, University of Ljubljana, Slovenia, November 2013.

Top-down Fabrication of Silicon Nanowire Sensor Using Electron Beam and Optical Mixed Lithography

S.F.A. Rahman, N.A. Yusof, M.N. Hamidon, R.M.
Zawawi
Universiti Putra Malaysia,
43400 Serdang, Selangor, Malaysia.
siti_fatimah0410@yahoo.com

U. Hashim
Institute of Nanoelectronics Engineering,
Universiti Malaysia Perlis,
01000 Kangar, Perlis, Malaysia.
uda@unimap.edu.my

Abstract—**The realization of reliable nanobiosensor devices requires the improvement of fabrication techniques to form the nanometer-sized structures and patterns, which were used to attach nano materials such as DNA for the device elements. This study demonstrates the sensitivity of silicon nanowires (SiNWs)as a sensing element in sensor application. Starting with silicon on insulator (SOI) material, the SiNWswith <100nm in width were fabricated using electron beam lithography combined with conventional CMOS process. Different numbers of SiNWs which are single, 10 arrays of nanowires and 20 arrays of nanowires were developed. Subsequently, the two metal electrodes which are designated as source (S) and drain (D) were fabricated on top of individual SiNWs using optical lithography process. Optical and electrical characteristic have been proposed to verify the outcome of the fabricated structures. One major part is to observe the SiNWs optically in order to meet the nano-scale variation by using High Power Microscope (HPM) inspection and Field Emission Scanning Electron Microscope (FESEM) imaging.Finally, the samples will be tested electrically using I-V measurement system. The results show thatdevice with single SiNW with 60nm in width give the highest resistivity value due to surface to volume ratio.**

Keywords—silicon nanowires; electron beam lithography; electrical characterization

I. Introduction

Semiconductor nanowires are attractive components for future nanoelectronics since they can exhibit a range of device function and at the same time serve as bridging wires that connect to larger scale metallization. SiNWs are increasing being used as buildings blocks for future highly sensitive, label-free biosensor using electrical detection. Additionally, nanowires are referred to as one-dimensional nanostructures since their lengths are orders of magnitude larger than their diameter[1]. The properties of this one-dimensional nanowire provide a sensing modality for label-free and real-time electrical readout when the nanowire is configured as a field effect transistor (FET), which exhibits conductivity or resistivity change in response to variations in the electric field or surface potential at the surface[2]. In such sensing devices, the two metal electrodes which are designated as source (S) and drain (D) of a device is bridged by a nano-object (nanowire) [1-2]. Binding to the surface of these nanowires

alters their ability to conduct, which serves as the detection mechanism [1].The schematic in Fig. 1 shows the principal of the SiNWs sensor where the current is charged, $I \rightarrow I + \Delta I$, due to the attached charges molecules.

However, existing SiNW sensor have had some technical problems such as poor controllability of electrical properties and difficulty of integration to other existing microelectronic components [3].This is due to bottom-up synthesized SiNW which have large statistical variation in the electrical properties and difficult to manipulate for the reliable integration to other microelectronic component [4]. In addition, good ohmic contact between the synthesized SiNW with metal interconnection is difficult to achieve. Besides, there is also a weak point of the synthesized SiNW which is the incompatibility with CMOS process. SiNW grown via metal-catalyzed chemical vapor deposition (CVD) process contain of metal elements such as gold or copper. Therefore integration of these SiNW is not compatible to the CMOS process[4-5].

For this reason, this research work demonstrates the top-down approach for device fabrication to create SiNW sensor which shows significant advantages in small size and low cost. The top-down nanofabrication method by electron beam lithography (EBL) and integrated with CMOS process such as reactive ion etching (RIE) is implemented for device structure formation. In addition, a simple surface modification of SiNW is introduced in order to create the device responses to the analyte detection (DNA) and enable to generate a reasonable electrical readout.

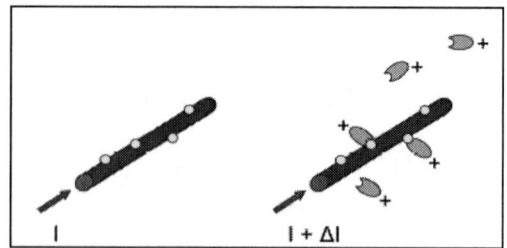

Fig. 1.Schematic showing the principle of the NW sensor. The surface is coated with receptor molecules (yellow). As charged molecules (green) attaches to the receptor molecules the current is changed.

978-1-4799-5761-3/14 $31.00 © 2014 IEEE

II. Experimental

A. Device Fabrication

The 4 inch (100 mm in diameter) p-type silicon on insulator (SOI) wafer which initially had a 60 nm silicon layer on a 160nm buffered-oxide (BOX) insulating layer with resistivity 1-20 Ωcm was used as the starting material in this work. Standard cleaning procedure using RCA1, BOE, and RCA 2 was introduced to remove organic and inorganic contaminant on the samples surface. A set of three nano-sensor devices with different numbers of SiNWs (singleSiNWs, 10 SiNWs and 20 SiNWs) are fabricatedusing electron beam lithography (EBL) on each die to determine device sensitivity with respect to numbers of SiNWs. The patterned nanowire will act as the bridge with 400nm length to connect between the source and drain metal pad. These micro-sized electrode pads are fabricated via conventional photolithography process. Fig. 2 illustrates the general steps of EBL and optical lithography which are implemented in this work. The process started with EBL process to fabricate nanowire as shown in Fig. 2 part A and followed by optical lithography to employ the source and drain pad as illustrates in Fig. 2 part B. Next, the electrical measurements were performed using a DC analyzer system. Electrical performances of the fabricated nanowires were carried out by measuring the current-voltage (I-V) properties of nanowires with varying width and numbers.

B. Device Characterization

The High power microscope was used to observed the fabricated silicon nanowires structures. Field emission scanning electron microscope was carried out to measure the structures diameterand identification of elements.Electrical characterization of SiNWs devices were performed using a DC analyzerprobe station. The swept voltage was applied from 0V to 2V to the SiNWs device and the current through the SiNWs was measured.

III. Result and Discussion

The p-type SiNWs used in this work were fabricated based on top-down approach. The SiNWs were developed in single and array through electron beam lithography combined with CMOS process. SiNW is etched anisotropically producing nice and straight walls as shown in Fig. 3 (a). The height seen in the measurement corresponds to the top Si layer thickness of the SOI substrate which is approximately 50nm.Fig. 3 (b) shows an optical image of fabricated single NW and two arrays of NWs with different numbers (10 NWs and 20NWs). The individual SiNW has a dimension of 60nm in diameter and a length of 400μm.In such sensing device, the two metal electrodes which are designated as source (S) and drain (D) of a device is bridge by nanowire as shown in Fig. 3 (c) and (d). These micro-sized contact pads allowed electrical transport measurements of the silicon nanowires.

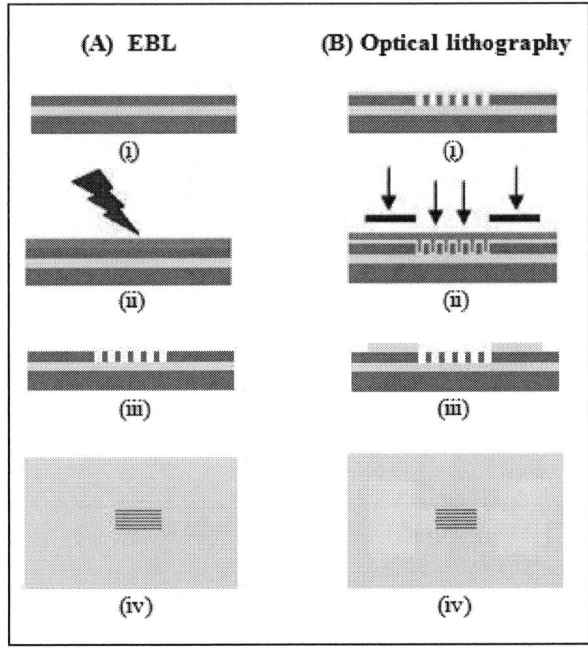

Fig. 2. Experimental scheme of electron beam lithography (EBL) coupled with optical lithography. Part A (i) cleaned SOI sample (ii) coated sample is exposed using EBL (iii) after Si etching process (iv) top view of SiNWs. Part B (i) Deposit Au onto the sample (ii) photolithography process using electrode pad mask (iii) after Au etching process (iv) top view of aligned SiNWs electrode pad.

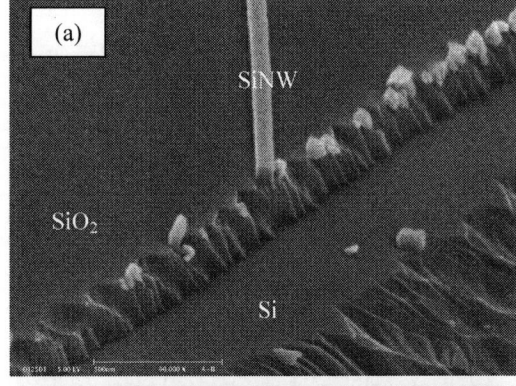

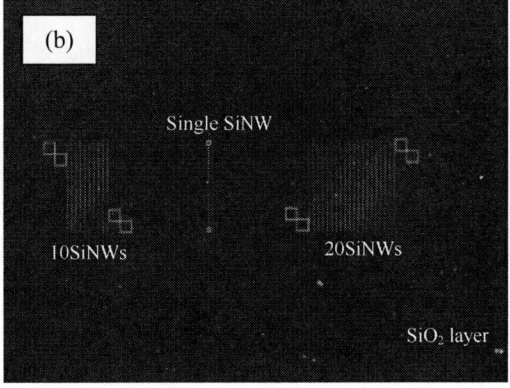

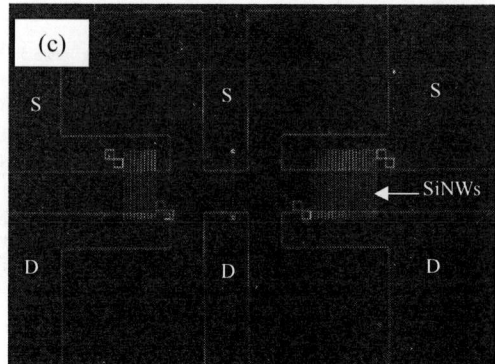

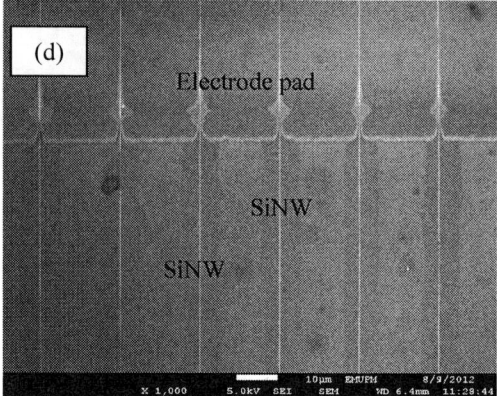

Fig. 3. (a) cross section view of the silicon nanowire fabricated on the SiO_2 layer, (b)HPM image of the fabricated single and multiple nanowire, (c) a dark field (DF) image of aligned SiNWs structures with gold electrode pads (source and drain) observed using HPM and (d) SEM image of the fabricated nanowire under Au pad layer.

Once the SiNWs has been formed as a two-terminal sensing device, its current-voltage (I-V) characteristic is then measured by using DC analyzer system. During the measurement set-up, a typical resistor setting is employed on to the two terminal NW devices by supplying voltages to the source (S) region and obtains the output current at the drain (D). For this experimental, devices with different width (W) of wires approximately (W = 60nm, 70nm, 80nm, 90nm and 100nm) and different numbers (N) of wires (N = 1SiNW,10SiNWs and 20SiNWs) were characterized. The purpose was to analyze the size and number dependence of wires to the device's electrical properties. The results of these electrical characterizations are presented in Fig. 4.

As shown by the data points in Fig. 4 (a), indicating that the device's resistivity values are related to the width of the Si wire. The resistivity (ρ) of the SiNW sensor can be expressed as $\rho=RA/L$, where R is the resistance value, A is the area size value (diameter x height) and L is the length of nanowire between electrode channel. By using this equation, the ρ values of the fabricate devices with the wire width approximately of 60 nm, 70 nm, 80 nm, 90 nm and 100 nm were 13.06×10^{-4} Ωm, 12.88×10^{-4} Ωm, 9.07×10^{-4} Ωm, 5.58×10^{-4} Ω m and 2.06×10^{-4} Ω m, respectively. It is shown that, the resistance of the wire is inversely proportional to its width. As the wire gets smaller in dimension, the electrical resistance becomes higher. This is due to the increased surface-area-to-volume ratio for smaller dimensions of NW which has resulted

greater surface effects on the electrical conduction [6]. The surface contamination effects of SiNWs such as surface damage, defects and trapped charges developed through the plasma etching process give significant roles for smaller NW due to the larger surface-to-volume ratio [7]. Thus, it has been shown that the electrical characterization of the devices depends strongly on the width of the wires; resistance value increases with the decreasing NW width due to the surface-to-volume ratio effect.

Previous studies suggested that the number of bridging nanowires incorporated into a device may be an important parameter in determining sensitivity [8-9]. Thus, for this research, three different quantities of nanowires which are single nanowire, an array of 10 nanowires and 20 nanowires were fabricated in order to investigate the dependence of nanowire number on the device's electrical properties. All the devices were characterized in response to the APTES functionalize of SiNWs surface. As shown in Fig. 4 (b), it was found that single nanowire give the maximal value of the resistance, followed by an array of 10 nanowires and 20 nanowires.A device that consists of parallel multinanowires resulting in a larger current flow compared to the single nanowire. Therefore, the resistivity value of device also depends on the number of nanowire, resistance value decreases with the increasing of number of nanowire.

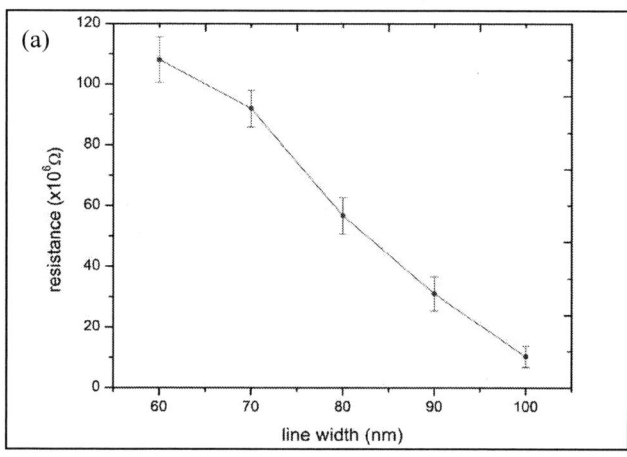

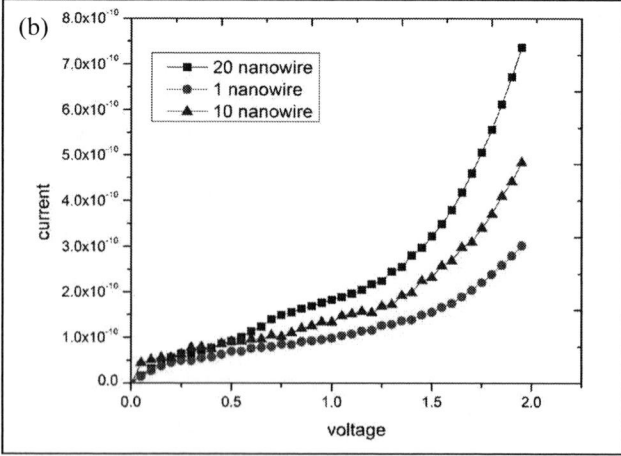

Fig. 4 Electrical response of the device for (a) different width of SiNW and (b) single and multiple SiNW

IV. Conclusion

This research demonstrated the SiNWs sensor device have been fabricated by aligning an array of nanowires across paired of Ti/Au electrodes using advanced nanolithography process coupled with conventional lithography process. The comparison among the devices with different SiNW width demonstrates the resistance of the NW is inversely proportional to its width. Smaller NW exhibit high resistance due to the increase of the surface-area-to-volume ratio.In addition, it is found that the resistance of nanowires also depends on their quantity. As the number of nanowires increases, the resistance value decrease. This study may serve as a design reference for the optimization of SiNWs sensor performance.

Acknowledgment

The appreciation goes to all the team members in the Institute of Nanoelectronics Engineering, Universiti Malaysia Perlis (UniMAP) and Chemistry Lab, Universiti Putra Malaysia (UPM) for technically and theoretically support.

References

[1] D. Grieshaber, *et al.*, "Electrochemical Biosensors - Sensor Principles and Architectures," *Sensors,* vol. 8, pp. 1400 - 1458, 2008.

[2] M. Curreli, *et al.*, "Real-Time, Label-Free Detection of Biological Entities Using Nanowire-Based FETs," *Nanotechnology, IEEE Transactions on,* vol. 7, pp. 651-667, 2008.

[3] X. Y. Zhang, *et al.*, "Fabrication and characterization of highly ordered Au nanowire arrays," *Journal of Materials Chemistry,* vol. 11, pp. 1732-1734, 2001.

[4] J. B. Grunes, "Reaction selectivity studies on nanolithographically-fabricated platinum model catalyst arrays," Other Information: TH: Thesis (Ph.D.); Submitted to the University of California, Berkeley, CA (US); PBD: 15 May 2004, 2004.

[5] Y. Civale, *et al.*, "Aspects of Silicon Nanowire Synthesis by Aluminum-Catalyzed Vapor-Liquid-Solid Mechanism," *Paper presented at the Seventh Annual Workshop on Semiconductor Advances for Future Electronics (SAFE), Netherland.,* 2004.

[6] P. Inkyu, *et al.*, "Top-down fabricated silicon nanowire sensors for real-time chemical detection," *Nanotechnology,* vol. 21, p. 015501, 2010.

[7] I. Park, *et al.*, "Towards the silicon nanowire-based sensor for intracellular biochemical detection," *Biosensors and Bioelectronics,* vol. 22, pp. 2065-2070, 2007.

[8] D. Zhang, *et al.*, "Detection of NO2 down to ppb Levels Using Individual and Multiple In2O3 Nanowire Devices," *Nano Letters,* vol. 4, pp. 1919-1924, 2004/10/01 2004.

[9] J. Li, *et al.*, "Effect of Nanowire Number, Diameter, and Doping Density on Nano-FET Biosensor Sensitivity," *ACS Nano,* vol. 5, pp. 6661-6668, 2011/08/23 2011.

I–V Characteristic Effects of Fluidic-based Memristor for Glucose Concentration Detection

Nor Shahanim Mohamad Hadis[1], Asrulnizam Abd Manaf[2]

Advanced Integrated System Device (AISDe)
School of Electrical and Electronic Engineering,
Universiti Sains Malaysia, Engineering Campus,
14300 Nibong Tebal, Penang, Malaysia
[1]norsh713@ppinang.uitm.edu.my,
[2]eeasrulnizam@eng.usm.my

Nor Shahanim Mohamad Hadis[1], Sukreen Hana Herman[3]

NANO-ElecTronic Centre (NET),
Faculty of Electrical Engineering,
Universiti Teknologi MARA Malaysia,
40450 Shah Alam, Selangor, Malaysia
[1]norsh713@ppinang.uitm.edu.my,
[3]hana1617@salam.uitm.edu.my

Abstract—The I–V characteristic effect of thin film TiO_2 fluidic-based memristor sensor utilized in sensing various glucose concentrations is described in this paper. Four different glucose concentrations, namely, 5, 10, 20, and 30 mM, are prepared and applied to the sensor. The device is then characterized with Keithley 4200-SCS semiconductor characterization system. Results show that different concentration levels of glucose affect the I–V characteristic of the sensor device. The difference is observed at the first voltage sweep of 0 V to 3 V after glucose was applied. A uniform change in current was recorded for small voltages below 0.9 V. The current decreases as the glucose concentration increases. Analysis shows that the resistance of the memristor sensor increases with the increase in glucose concentration through a quadratic relation.

Keywords—memristor sensor; I–V characteristic; memristive behavior; glucose concentration

I. Introduction

The memristor is a new circuit device or element commercialized in various fields, such as memory, neural networks, computing, and bio-sensing. The memristor was first introduced in 1971 by Leon O. Chua [1], who discovered this new element by theory. He classified the memristor as another new passive element (the fourth) after resistor, inductor, and capacitor. This element behaves like a normal resistor at a high frequency. This new element is different because of its capability to remember the past at a low frequency level. Thus, it is suitable to use in memory devices or as a switching element. The element generates pinched hysteresis at a low frequency. In 2008, Hewlett-Packard (HP) researchers, Strukov et al., published the first physical model of their memristor device in the journal Nature; they described the construction of the memristor element [2].

The physical HP model consists of a thin film of TiO_2 sandwiched between two metal contacts (platinum contact) [2, 3]. The TiO_2 film is fabricated in two layers. The first layer is perfect TiO_2, and the second is a layer of TiO_2 with oxygen vacancies [3]. Several researchers, including Tedesco et al.

and Gale et al., employed aluminum as a contact material [4, 5]. Prodromakis et al. utilized gold as a contact material [6]. Previous research shows that different contact materials contribute minimal change in memristive behavior [2, 5, 6].

The capability of the memristor device to detect various glucose concentrations in bio-sensing applications is investigated in this study. The study is based on the memristive behavior effect (I–V characteristic) of the memristor device. Currently, most bio-sensing devices detect a virus through other parameter observations, such as surface effect observation [7], cyclic voltammetry measurement [8, 9], and amperometric response observation [9].

II. Memristor Sensor Structure

A. Fabrication

An indium tin oxide (ITO)-coated glass substrate was cleaned with methanol and distilled water. The substrate was then dried with nitrogen gas. TiO_2 was sputtered on of the surface of the ITO-coated glass substrate with a radio frequency (RF) sputtering machine at 200 $^\circ$C and 300 W, with additional 50 sscm argon gas. High-purity (99.99%) TiO_2 was employed as a target. Pt contact was then deposited through an RF sputtering machine, and the target utilized was high purity (99.99%) Pt. A metal mask was employed to pattern Pt and TiO_2 during deposition. Fig. 1 shows the top and cross-sectional views of the memristor sensor. For characterization, the exposed area of TiO_2 was covered with epoxy for glucose injection. The cover was attached to the memristor sensor device after fabrication was completed.

978-1-4799-5761-3/14 $31.00 © 2014 IEEE

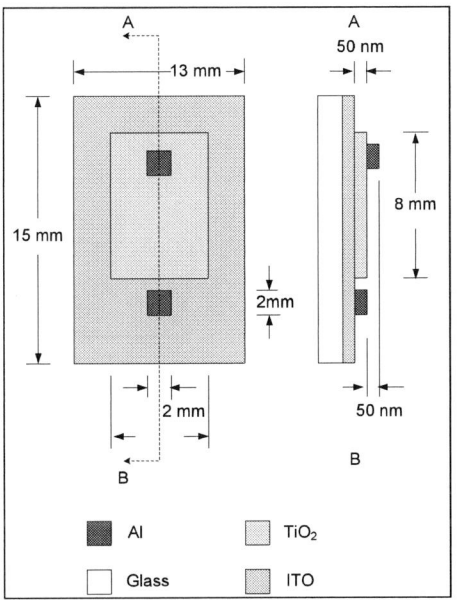

Fig. 1.   Structure of the memristor sensor

B.   Characterization

The characterization for the memristor sensor was conducted with a Keithley 4200-SCS semiconductor characterization system. The measurement setup, which comprise the characterization system and a workstation, is shown in Fig. 2. The workstation consists of a chuck, micro-probe, cable, microscope, and others. The memristor sensor was placed on the chuck, and the micro probe was positioned on the top and bottom contacts of the device. An SMU1 probe was placed on the top contact, and a Gnd probe was placed on the bottom contact.

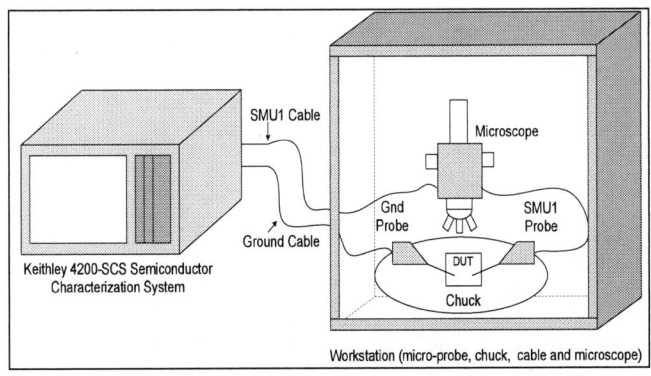

Fig. 2.   Measurement setup for memristor sensor characterization

Three voltage ranges were applied in sweeps to observe the hysteresis behavior. The first sweep was from 0 V to 3 V, followed by 3 V to □3V, and then □3 V to 0 V. This process was performed with the I–V characterization software of the Keithley 4200-SCS setup. For the three voltage sweeps, the

currents were measured and labeled as shown in Table 1. Fig. 3 shows the detailed expected characterization output, which pinched at zero and thus produced pinched hysteresis. I1 is the first current measured in the first voltage sweep (0 V to 3 V). I2 is the second current measured in the voltage sweep from 3 V to □3V, and I3 is the third current measured in the last sweep voltage (□3 V to 0 V).

TABLE 1. MEASURED CURRENT FOR THE THREE RANGES OF VOLTAGE SWEEP

Current	Sweep Voltage Range
I1	0 V to 3 V
I2	3 V to □3 V
I3	□3 V to 0 V

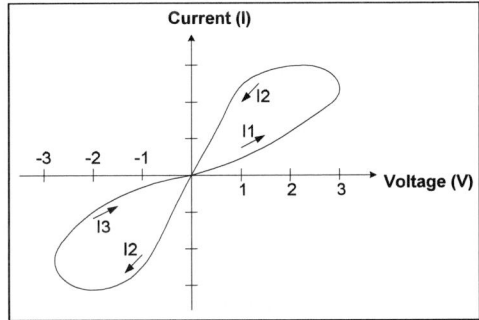

Fig. 3.   Basic operation of the memristor device

Six glucose concentrations were prepared for characterization. The selected concentrations are 5, 10, 15, 20, 25, and 30 mM. However, the analyses in this study highlights only four concentrations to avoid congestion. The selected concentrations are 5, 10, 20, and 30 mM. Initially, characterization was conducted for the memristor sensor itself without glucose to be used as reference. Then, the memristor sensor was characterized with 5 mM glucose on the TiO_2 surface and followed by the three other concentrations.

III.   Results and Discussion

Fig. 4 shows the I–V characteristics of the memristor sensor for no glucose and for the four glucose concentrations, namely, 5, 10, 20, and 30 mM.

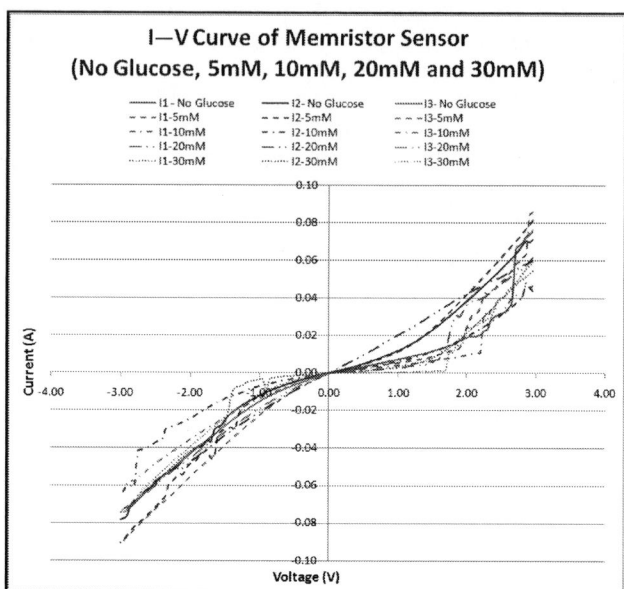

Fig. 4. Comparison of I–V characteristics for no glucose and the four glucose concentrations (5, 10, 20, and 30 mM)

The memristor sensor with glucose behaved like a normal memristor. However a slight change in current was observed in each voltage sweep when different glucose concentrations were applied. The hysteresis pattern of the I–V curve pinched at zero reflects normal memristive behavior. The observations clearly indicate that current I1 for the four glucose concentrations has uniform change. Currents I2 and I3 did not exhibit any uniformity in terms of current change. The graph of I1 versus voltage is illustrated in Fig. 5.

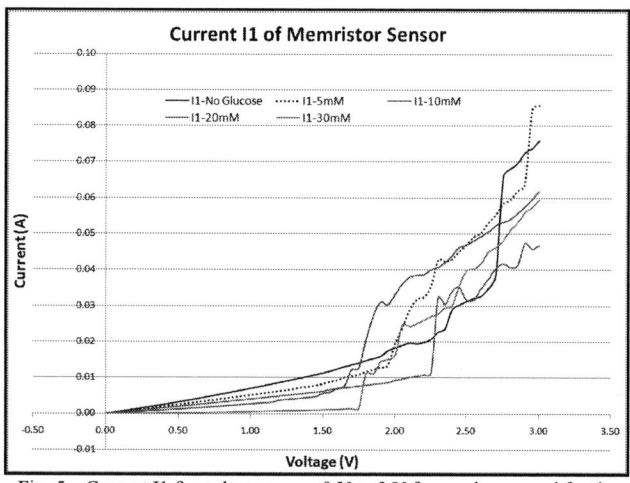

Fig. 5. Current I1 for voltage sweep 0 V to 3 V for no glucose and for the four glucose concentrations (5, 10, 20, and 30 mM)

The I–V curves in Fig. 5 exhibited a linear relationship between current I1 and voltage at a small voltage range; however, the relationship became unstable after 1.7 V. These findings show that the device can be used in glucose detection for the first range of small voltage sweep.

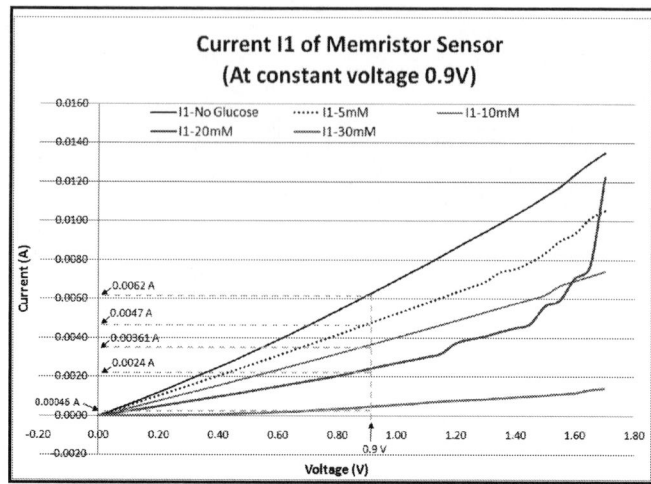

Fig. 6. Magnification of Figure 5 for the range of 0 V to 1.7 V for no glucose and the four glucose concentrations (5, 10, 20, and 30 mM)

Fig. 6 shows a large scale graph, which is a magnification of the voltage sweep from 0 V to 1.7 V. The graph shows that the curves form a straight line for sweep voltage from 0 V to 0.9 V; thus, the resistance of the memristor can be determined. Ohm's Law V = IR was utilized to calculate the resistance values of the four glucose concentrations and no glucose. The values are presented in Table 2. The formula used is shown in Equation 1.

$$Resistance\ (\Omega) = Voltage\ (V) / Current\ (A) \qquad (1)$$

TABLE 2. RESISTANCE OF THE MEMRISTOR SENSOR WITH GLUCOSE AT CONSTANT VOLTAGE OF 0.9 V (OBSERVED FROM I1)

Glucose Concentration	No Glucose	5 mM	10 mM	20 mM	30 mM
Current (A)	0.0062	0.0047	0.00361	0.0024	0.00046
Resistance (Ω)	145.16	191.49	249.31	375	1956.52

Table 2 shows that the resistance of the memristor sensor increases as the glucose concentration increases. The resistance of the memristor device without glucose is only 145.16 Ω. The resistance increases to 191.49 Ω when 5 mM glucose is applied on the exposed area of TiO_2. As the glucose concentration further increases to 30 mM, the resistance likewise increases to 1.96 kΩ.

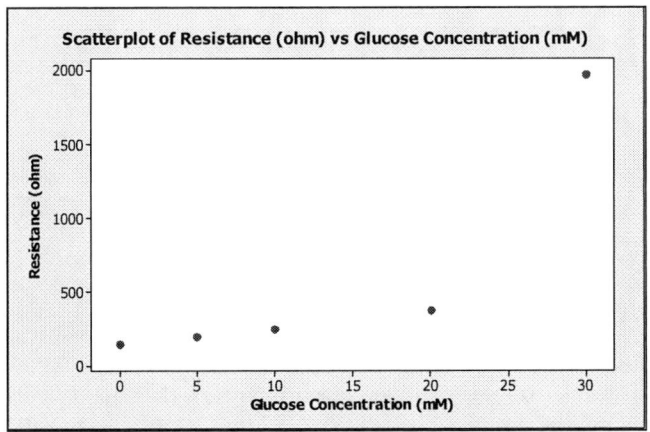

Fig. 7. Relationship between memristor sensor resistance and glucose concentrations

The data in Table 2 are plotted in Fig. 7 through a scatter plot. Memristor sensor resistance was set to Y-axis, and glucose concentration was set to X-axis. The scatter plot reveals the quadratic relationship between memristor sensor resistances and glucose concentrations. To confirm this condition, curve fitting regression analysis was conducted with Minitab software. Fig. 8 proves that memristor sensor resistance has a quadratic relationship with glucose concentration, as presented in Equation 2.

$$Resistance\ (\Omega) =$$
$$270.7 - 55.89\ Glucose\ Concentration\ (mM) +$$
$$3.662\ Glucose\ Concentration\ (mM)^2 \qquad (2)$$

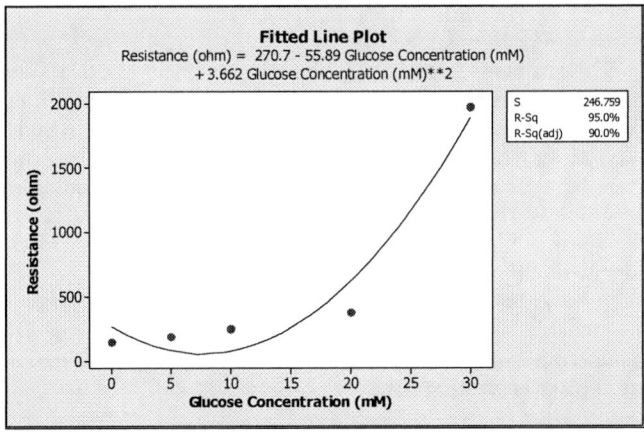

Fig. 8. Relationship between memristor sensor resistance and glucose concentration

The fitted line plot in Fig. 8 shows that the standard deviation of the model is 246.759, which is high because of the high resistance value. Regression-Sq (adjusted) indicates that 90% of the variation in resistance depends on glucose concentration, whereas the other 10% depends on an unexplained factor.

IV. Conclusion

The memristor sensor can detect various glucose concentrations with different I–V behaviors at a voltage range of 0 V to 3 V. From 0 V to 0.9 V, the I–V curve produces a uniform difference among the four glucose concentrations. The current level decreased as glucose concentration increased. The resistance of the memristor sensor at this point was calculated. The resistance of the memristor sensor increased as the glucose concentration increased. Minitab regression analysis indicates that memristor resistance has a quadratic relationship with glucose concentration. However, the quadratic relation is not strong.

Acknowledgments

The authors would like to thank Universiti Sains Malaysia Research University (RU) grant no. 1001/PELECT/814168 and Universiti Teknologi MARA grant no. ERGS/1/2013/TK04/UITM/03/05.

References

[1] L. O. Chua, "Memristor - The Missing Circuit Element," *IEEE Transactions on Circuit Theory*, vol. 18, p. 13, 1971.
[2] D. B. Strukov, G. S. Snider, D. R. Stewart, and R. S. Williams, "The missing memristor found," *Nature*, vol. 453, p. 5, 2008.
[3] R. S. Williams. (2008) How We Found The Missing Memristor. *IEEE Spectrum*. 7.
[4] J. L. Tedesco, L. Stephey, M. Hernandez-Mora, C. A. Ritcher, and N. Gergel-Hackett, "Switching mechanisms in flexible solution-processed TiO$_2$ memristors," *Nanotechnology*, vol. 23, p. 6, 2012.
[5] E. Gale, B. d. L. Costello, and A. Adamatzky, "The Effect of Electrode Size on Memristor Properties: An Experimental and Theoretical Study," presented at the 2012 International Conference on Electronics Design, Systems and Applications (ICEDSA), Kuala Lumpur, 2012.
[6] T. Prodromakis, K. Michelakis, and C. Toumozou, "Fabrication and Electrical Characteristics od Memristors with TiO$_2$/TiO$_{2+x}$ active layers," in *Proceedings of 2010 IEEE International Symposium on Circuits and Systems (ISCAS)*, 2010, p. 3.
[7] P. Xiao, Y. Zhang, and G. Cao, "Effect on surface defect on biosensing properties of TiO$_2$ nanotube arrays," *Elsevier: Sensor and Actuators B: Chemical*, vol. 155, p. 6, 2011.
[8] R. R. Pandey, K. K. Saini, and M. Dhayal, "Using Nano-Arrayed Structures in Sol-Gel Derived Mn^{2+} Doped TiO$_2$ for High Sensitivity Urea Biosensor," *Journal of Biosensors and Bioelectronics*, vol. 1, p. 4, 2010.
[9] H. D. Jang, S. K. Kim, H. Chang, K.-M. Roh, J.-W. Choi, and J. Huang, "A glucose biosensor based on TiO$_2$–Graphene composite," *Elsevier: Biosensors and Bioelectronics*, vol. 38, p. 5, 2012.

Aluminum Nitride Thin Film Deposition Using DC Sputtering

Mohd H.S Alrashdan, Azrul Azlan Hamzah, Burhanuddin Yeop Majlis , Mohd Faizal Aziz .
Institute of Microengineering and Nanoelectronics (IMEN), Universiti Kebangsaan Malaysia
Selangor, 43600, Malaysia

E-mail: azlanhamzah@ukm.my

Abstract—**Aluminum nitride thin film depositions at a low temperature become one of the most promising fields in micro-electro mechanical systems and in the semiconductor industry; because of its good compatibility with designs on silicon substrates, its mechanically strong, chemically stable, wide band-gap energy ($\approx$6.2 eV), and has a large electro-mechanical coupling constant. An AlN thin film deposition using DC Magnetron sputtering have the advantage over other deposition methods due to its simplicity, better parameter control, cheapness, and requires a low deposition temperature . The NTI nano film DC sputtering system was used to deposit the AlN thin film with 99.99% pure aluminum target material and 100 silicon substrates ,the working temperature is at 20C°,there is a 10Cm separation distance between the target and the substrate , 335~351 V cathode voltage, the foreline and base pressures are 2×10^{-2} T, 4×10^{-5} T respectively, and uses 200W DC power. We vary the time and nitrogen/argon gas flow ratio. Deposited film was characterized by X-ray diffraction and (002) of wurtzite hexagonal phase of AlN thin film was found with beak intensity of 800 count per second for 50% nitrogen content. Field Emission Scanning Electron Microscopy was used to study thin film cross section, film thicknesses and deposition flow rate at different times and gas flow ratio ,there is inverse relationship between nitrogen gas percentage deposition and flow rate . Deposition flow rate are 4.12 nm/ min for 50% nitrogen and 2.217 nm/min for 75% of nitrogen content.**

Keywords—Aluminum Nitride; DC Sputtering; Thin Film Deposition.

I.

Aluminum nitride (AlN) thin films deposition has been the topic of research in micro-electro mechanical systems(MEMS) and in the semiconductor industry because of its good compatibility with designs on silicon (Si) substrate, also its mechanically strong, and has a high thermal conductivity. It is thermally and chemically stable, especially in an inert environment and oxidation starts at its surface in air at a temperature above 800 C°[56]. AlN has a high electrical resistivity (1016 Ω.cm) [1] which prevent AlN thin film from electrical failure and charge leakage, high Curie temperature (1150 C°), high melting point is at 2200 C° ,it has a good electro-mechanical coupling constant [2], a wide band-gap energy ($\approx$6.2 eV) [3], therefore AlN thin film become one of the most promising piezoelectric material in MEMS power harvesting devices especially at a low frequency range, and the piezoelectric signal produced can overcome the DC leakage during its operating time in power harvester devices . AlN is covalently bonded, therefore it will not introduce any ionic conduction and due to the large band gap, the electronic conduction at high temperature will remain minimal.

UKM-GUPNBT-08-25-084.

AlN thin films deposition can be done by numerous techniques which include chemical vapor deposition (CVD) [4,5] , physical vapor deposition (PVD)[6], reactive molecular beam epitaxy [7], metal-organic chemical vapor deposition (MOCVD)[8], and DC/RF reactive sputtering [9].

DC sputtering have the advantage over other methods of AlN thin films deposition for electro mechanical system and VLSI applications due to its simplicity, better parameter control, cheapness, and requires a low deposition temperature compared to MBE and CVD techniques, which require a normally high substrate temperature of more than 500C° [3]. AlN thin films deposition at a low temperature is recommended due to a difference in the thermal expansion coefficients of the AlN thin films and silicon substrate thus can cause a film detachment during the cooling process to room temperature. This problem can be solved easily by using the sputtering method, since high-energy plasma overcomes the temperature limitations of other deposition methods so a variety of substrate materials with low melting point can be used. DC sputtering is cheaper than RF magnetron sputtering and also has a higher deposition rate, so it's preferred because low cost and its thin film properties [9].

Mechanical, electrical ,thermal ,chemical, optical and electro mechanical properties of deposited AlN thin films using DC sputtering varies based on thin film microstructure, crystal structure, and chemical composition. These properties can be defined during deposition by controlling sputtering conditions such as distance between the source and target, sputtering power, gas flow ratio and its pressure [10], deposition time and temperature.

In this study we deposit (AlN) thin films using DC sputtering at different sputtering powers: Ar, N_2 gas ratio, time for power harvesting devices.

II.

A. DC sputtering

DC Sputtering is a physical phenomenon based on ion acceleration using a high DC voltage source and the bombardment of a 'target' or cathode. Through momentum transfer, atoms close to the surface of the target metal become

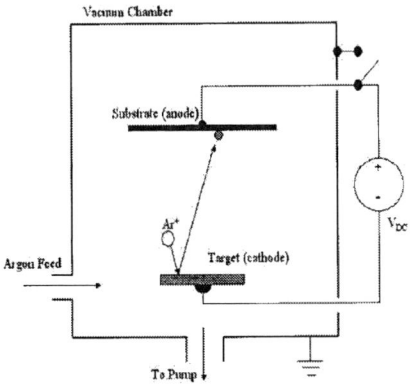

Fig 1: DC sputtering system.

TABLE 1. THE PROCESS PARAMETER

Process parameter	Numerical value
Tempreture	20C°
Target substrate distance	10Cm
Cathode voltage	335~351 V
Foreline pressure	2×10^{-2} T
Base pressure	4×10^{-5} T

volatile and are transported as a vapor to a (Anode) substrate. A thin film grows at the surface of the substrate via deposition.

Fig. 1 shows a classic DC sputtering system that consists of a high DC voltage as a power supply , a vacuum chamber, a sample holder, and a sputtering target of the desired film. After evacuating the chamber down to a pressure of 10^{-6} to 10^{-8} Torr, an inert gas such as argon is interred into the chamber at pressure of a few mTorr of plasma of the inert gas is then ignited. The accelerated ions with enough energy of the plasma bombard the surface of the conductive target such as Aluminum. In our case, Aluminum makes some of the conductive target atoms to be released from their position and react with other reactive gases in the chamber such as Nitrogen, producing new atoms or compounds. Some of these atoms land on the sample's surface and makes a new thin film such as Aluminum Nitride.

III.

Silicon wafer with (100) orientation is cleaned for 5 minutes with acetone and ultrasonic, rinse it for 2 minutes with Deionized (DI) water, and another 5 minutes with methanol and ultrasonic and rinse it again for 2 minutes with deionized (DI) water, and finally 5 minutes with Hydrofluoric acid (HF) 10% to remove any native oxide on the silicon substrate.

NTI nano film DC sputtering system was used to deposit the AlN thin film with 99.99% pure Aluminum was used as a target material; the process parameter we use is illustrated in table 1. We vary DC power, deposition time nitrogen and argon gas flow ratio.

For the first group, which is composed of four samples; I used DC power of 200 W, and the same gas flow ratio between Ar: N_2 equals to 20:20 sccm. The times vary from 20 min to 60 min.

The Second group of two samples using 200W of DC power and a time of 30 min and 60 min, the gas flow ratio between Ar:N2 equal 10:30 sccm.

The deposited film was characterized by X-ray diffraction (XRD) to study thin film orientation and film structure. Field Emission Scanning Electron Microscopy (FESEM) was used to study thin film cross section, film thicknesses, and construction details such as structure uniformity, determination small contamination feature geometry. Energy-dispersive X-ray spectroscopy (EDX) was used to study chemical characterization and elemental composition measurements.

IV.

The XRD pattern is shown in *Fig. 2*, The results show AlN(002) thin film is deposited for all samples with a peak intensity for all samples about 800 count/ second. The Bragg angle equals 38, this indicates good crystal orientation of AlN (002) thin film with c axis normal to silicon substrate. It has been found that, for nitrogen/argon gas flow ratio of 20:20 sccm and 200 W, only the (002) reflection of Wurtzite hexagonal phase of AlN thin film was found with a very small amount of shift. This confirms the good crystalline of thin films with small amounts of residual stress, one of the samples have AlN(100) with a peak intensity of about 200 count/ second . This peak at a smaller diffraction angel indicates that there is a large spacing between the lattice plane and shows not good properties same other sample maybe due to its place inside the sputtering machine or some defect in the wafer. .*Fig.3* show the XRD pattern for AlN, this film sputtered at 200 W and 30:10 sccm flow rate ratio between nitrogen and argon gas, we could not obtain (002) reflection at any time. Because of high nitrogen content and then decrease the momentum at silicon substrate due to smaller nitrogen mass compared to argon. And also low power and target temperature (20C°).

Fig. 4 shows the relationship between the deposition rate of the ALN thin film and nitrogen gas concentration for different ratios of nitrogen and argon. The deposition rate in tables 2, 3 is calculated as thickness measured by FESEM divided by the deposition time.

It is clear that there is an inverse relationship between the deposition rate and Nitrogen: Argon flow . The deposition rate decreased with an increasing N2:Ar + N2 ratio, when the gas ratio of N+ or N2 + ions increased compare to argon gas, the total mass decreased since the mass of Argon is higher than the mass of nitrogen, this leads to a lower momentum transferred to the target and causes a lower energy and efficiency of the target atoms. This causes a decrease in the ALN deposition rate [11], also high Nitrogen content in the chamber makes the chemical reaction faster at the surface of target and develop some AlN compounds at the surface of the Aluminum target. The surface binding energy of AlN is 9.1 ev while for Al it is 3.35 ev which lowers the atom released from the target and then lowers the deposition rate at the silicon wafer [12] .

AlN thin films deposited using DC sputtering at different gases flow ratios had different colors because of the variation in the crystal structures and stereochemistry [13]. Thin Films deposited at a low flow ratio between Nitrogen and Argon are yellow , while the violet color is produced from films deposited at a higher flow ratio, and the blue color is produced from moderate flow ratios.

In general, the secondary electron emission coefficient for Aluminum's target is lower than AlN, *Fig.5* shows the relationship between gas flow ratio and the cathode voltage during the process. It was clear that when the Nitrogen gas content increases, it leads to an increase in Nitrogen's partial pressure that causes a drop in the cathode voltage and an increase of the secondary electron emission coefficient of the original Al target as a result of AlNx compounds that was formed at the surface of the Al target. Figures 5 show the cathode voltages during AlN deposition as functions of the deposition time for different gas ratios.

A Strong relationship between the cathode voltage and the percentage of AlNx products covered the surface of the Al target during sputtering [14]. Initially the cathode voltage for 50% gas flow ratio was higher than that at 75% gas flow ratio because of more Al surface target purity, on the other hand, the initial cathode voltage decreased as a result of increasing gas flow ratio. Then normally for both gas flow ratio, the cathode voltage exhibit same manner and reaches an intermediate state and then increases slowly again with time to reach a steady state again. There is a small variation of cathode voltage during this process about 15 V.

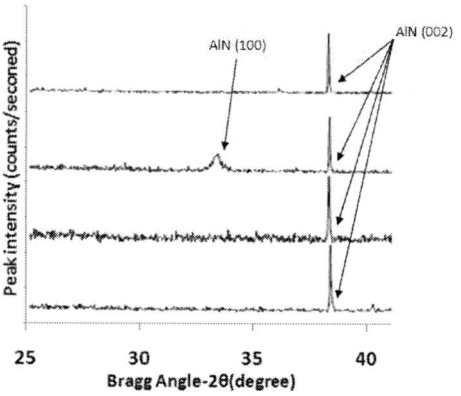

Fig 2: XRD result at power of 200 w and gas flow rate 20:20 sccm.

Table 2: measured results for 200 w and gas flow rate 20:20 sccm

Wafer #	Measured Results		
	Time (min)	Thickness (nm)	Deposition rate (nm/min)
1	20	80	4
2	30	130	4.33
3	50	190	3.8
4	60	207	4.35
Average			4.12

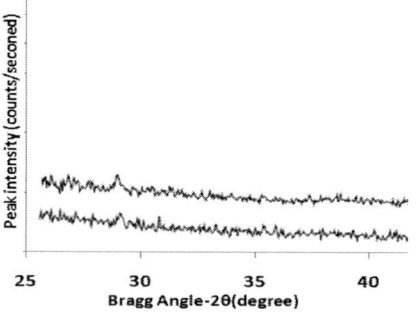

Fig 3: XRD result at power of 200 w and gas flow rate N_2: Ar 30:10 sccm.

Table 3: measured results for 200 w and gas flow rate 30:10 sccm

Wafer #	Measured Results		
	Time (min)	Thickness (nm)	Deposition rate (nm/min)
1	30	75	2.5
2	60	116	1.93
Average			2.217

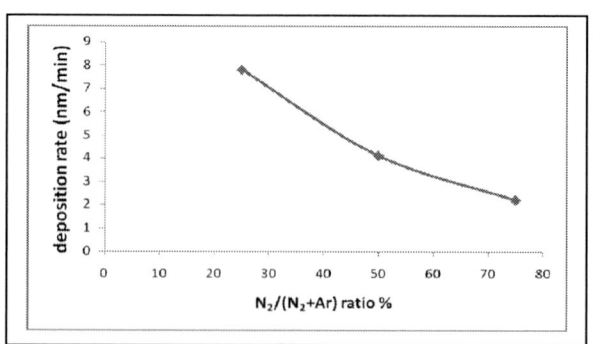

Fig 4: Deposition rates for the AlN thin films grown as functions of the N_2/Ar + N_2 ratio.

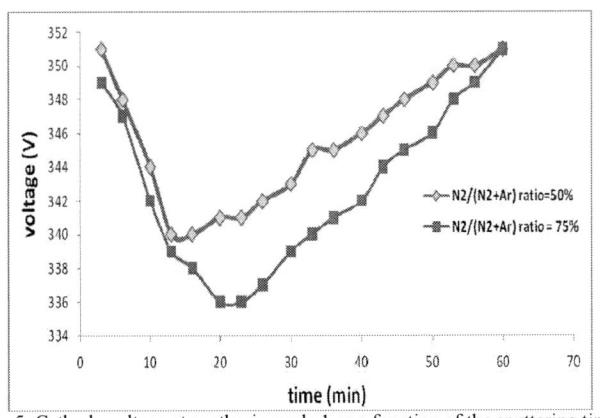

Fig 5: Cathode voltage at synthesis mode 1 as a function of the sputtering time for various $N_2/Ar + N_2$ ratios.

AlN thin films were deposited on (100) silicon substrates using DC sputtering. The deposited film has shown Wurtzite crystal structure with (002) reflection with a peak intensity of 800 counts per seconed when DC power is 200 W and a Nitrogen gas flow ratio of 50%. When flow ratio of nitrogen has been increased to 75% the thin film not deposited. There is an inverse relationship between nitrogen gas percentage deposition flow rates. Deposition flow rate are 4.12 nm/ min for 50% nitrogen and 2.217 nm/min for 75% of nitrogen content. The color of the thin film depends on the Nitrogen flow ratio. The cathode voltage decreased as a result of the Nitrogen flow rate increasing.

Acknowledgment

The authors would like to thank University Kebangsaan Malaysia for supporting this project under grant UKM-GUPNBT-08-25-084.

References

[1] K. Tsubouchi and N. & Mikoshiba, "Zero temperature coefficient saw delay line on AlN epitaxial films," in *Ultrasonics Symposium, Proc. IEEE*, 1980, p. 299.

[2] Xu, H. Wu, F. Zhang, J. Duan, andZ. Li, "The research progress of AlN piezoelectric thin films, " Rare Metal Materials and Engineering, vol. 31, pp. 456-9, 2002.

[3] V. Dimitrova, D. Manova, and E. Valcheva, "Optical and dielectric properties of dc magnetron sputtered AlN thin films correlated with deposition conditions," *Materials Science and Engineering B*, vol. 68, no. 1, pp. 1–4, 1999.

[4] Hoffman D., Prakash S., Athavale S. Economov D., Liu J. Zheng Z. and Chu W., "Chemical Vapor Deposition of Aluminum and Gallium Nitride Thin Films from Metalorganic Precursors", J. Vac. Sci Technol. A., Vol.14, No.2, 306, (1996).

[5] Gordon R., Hoffman D., and Riaz U., "Atmospheric Pressure Chemical Vapor Deposition of Aluminium Nitride Thin Films at 200-250 °C", J, Mater. Res, Vol.6, No.1, 5,(1991).

[6] L. La Spina, H. Schellevis, N. Nenadović, and L. K. Nanver, "PVD aluminium nitride as heat spreader in silicon-on-glass technology," in Proc. IEEE MIEL, 2006, pp. 365-368.

[7] H. Takikawa, K. Kimura, R. Miyano et al., "Effect of substrate bias on AlN thin film preparation in shielded reactive vacuum arc deposition," *Thin Solid Films*, vol. 386, no. 2, pp. 276–280,2001.

[8] J. P. Kar, G. Bose, and S. Tuli, "Influence of nitrogen concentration on g rain growth, structural and electrical properties of sputtered aluminum ni tride films," Scripta Materialia, vol. 54, pp. 1755, 2006.

[9] P. J. Kelly and R. D. Arnell, "Magnetron sputtering: a review of recent developments and applications," *Vacuum*, vol. 56, no. 3, pp. 159–172, 2000.

[10] S. Cho, "Effect of nitrogen flow ratio on the structural and optical properties of aluminum nitride thin films," *Journal of Crystal Growth*, vol. 326, no. 1, pp. 179–182, 2011.

[11] M. A. Moreira, I. Doi, J. F. Souza, and J. A. Diniz, "Electrical characterization and morphological properties of AlN films prepared by dc reactive magnetron sputtering," *Microelectronic Engineering*, vol. 88, no. 5, pp. 802–806, 2011.

[12] J. B. Malherbe, Critical Reviews in Solid State and Material Science **19(2)**, 55 (1994).

[13] D.-Y. Wang, Y. Nagahata, M. Masuda, and Y. Hayashi, "Effect of nonstoichiometry upon optical properties of radio frequency sputtered Al-N thin films formed at various sputtering pressures," Journal of Vacuum Science and Technology A, vol. 14, no. 6, pp. 3092–3099, 1996.

[14] R. McMahon, J. A_nito and R. R. Parsons, J. Vac. Sci. Technol. **20**, 376 (1982).

Lower DIBL in Inverted Substrate of UTBB SOI n-MOSFETs

Noraini Othman, M.K. Md Arshad, U. Hashim

Institute of Nano Electronic Engineering
School of Microelectronic Engineering
Universiti Malaysia Perlis (UniMAP)
01000 Kangar, Perlis, Malaysia.

Abstract— **In this paper, based on simulations, we present the impact of the different substrate space-charge conditions, on the Drain Induced Barrier Lowering (DIBL) of Ultra-Thin Body (UTB) SOI MOSFETs with two buried oxide (BOX) thicknesses as a function not only of substrate bias but also of drain lateral field penetration into the BOX and underlying substrate. We notice that the lowest DIBL is achieved when the substrate is in inversion regime. In the case of Ultra-Thin BOX (UTBB), the devices exhibit a significant reduction of DIBL which, with substrate inversion, becomes comparable to the use of ground plane, for gate length down to 25 nm. The simulation results are well supported and explained by the simulated potential variations at the substrate / BOX interface.**

Keywords— *UTBB, Substrate biasing, Inverted substrate, space-charge conditions.*

I. INTRODUCTION

Fully Depleted Silicon-on-Insulator (FD-SOI) appears as an attractive solution for transistor scaling deficiency as well as enhanced performance, especially in low operating power and low standby power applications [1]–[3]. The introduction of thin body with undoped channel in planar SOI MOSFETs technology provides an efficient reduction of the short-channel effects without random doping variability [4]. Thin BOX adds additional scaling benefit by reducing lateral electrostatic coupling between source and drain through the BOX [5]. This results in improved Drain-Induced Barrier Lowering (DIBL), better subthreshold slope (SS), higher I_{On}/I_{Off} ratios, and excellent V_{Th} roll-off [6]–[11] for short channel devices. However, with thin BOX, the role of the underlying substrate on fringing fields cannot be neglected anymore. In [12]–[14], they discussed and modelled the drain potential that modifies the properties of both back and front channels, via interface coupling, through the thick BOX and substrate.

In this paper we investigate, by the crossed influence of the substrate and drain biases and resulting substrate space charge conditions on the lateral field penetration through the BOX and low-doped underlying substrate, which in turn affects the DIBL, of 25 nm-gate-long, ultra-thin-body with UTB and UTBB SOI MOSFETs with underlapped and undoped channel.

II. SIMULATION PROCEDURE

ATLAS 2D-Device simulation is used to simulate the devices. The simplified structure consists of ultra-thin Si film with a thickness of 11.5 nm on ultra-thin BOX of 10 nm (UTBB) or standard BOX of 150 nm (UTB). The underneath Si substrate is p-type doped at 6.5×10^{14} cm^{-3}. The gate length is 25 nm and channel doping is close to as in experimental devices, $N_A = 6.5 \times 10^{14}$ cm^{-3}. The source and drain are doped n-type, with abrupt 0.9×10^{21} cm^{-3} and gaussian doping profiles, reaching 10^{19} cm^{-3} near the channel. The structure is designed to have two 15 nm-long spacers extending from the gate edges to source and drain junction creating an underlap structure.

III. RESULTS AND DISCUSSIONS

Fig. 1 shows the simulated current-voltage (I_d-V_g) characteristics which will be used to extract the DIBL. The simulated result is in correlation with experimental data as in [7], [15], [16]. The DIBL is defined as gate voltage shift due to the drain variation $\Delta V_g / \Delta V_d$ extracted at a constant normalized drain current, $I_{dnorm} = I_d / (W/L)$ of 10^{-7} A. In this work, $W = 1$ μm and $L_g = 25$ nm are used with drain voltages chosen as $V_d = 50$ mV and 1 V.

The simulated DIBL curves as a function of substrate bias (V_{Sub}) i.e. translates into different space-charge conditions at the BOX / substrate interface, as explained in [7] are shown in Fig. 2. Several important DIBL characteristics and differences between UTBB and UTB are sufficiently fairly reproduced by the simulations so that to use them for qualitative discussion. Fig. 2 firstly indicates that the UTBB device globally performs better than UTB as expected. However, the UTBB curve shows very significant variations as a function of substrate bias in correspondence with different BOX / substrate space charge conditions. Rather unexpected, the best DIBL is achieved when the substrate is in inversion ($V_{Sub} = -2.0$ V) and the highest is when the substrate is in depletion ($V_{Sub} = 0.5 / 1.0$V), while the accumulation regime ($V_{Sub} = 2.0$ V) lies in between.

We then further investigate these behaviours by performing horizontal ($\psi_{Horizontal}$) and vertical potential

($\psi_{Vertical}$) cuts in the structure. The $\psi_{Vertical}$ cut is performed near the source through the gate, channel, BOX and substrate. The $\psi_{Horizontal}$ cut is done exactly at the substrate / BOX interface. The gate voltages are chosen to correspond to $I_{dnorm} = 10^{-7}$ A, as for DIBL extraction.

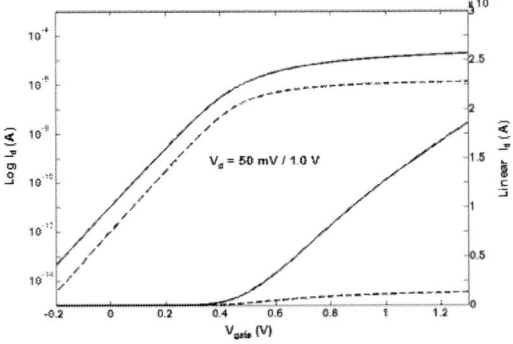

Fig. 1 Current-voltage characteristics of simulated UTBB devices for 25 nm gate length and width of 1 μm, showing in log-linear (left) and linear-linear (left) scale.

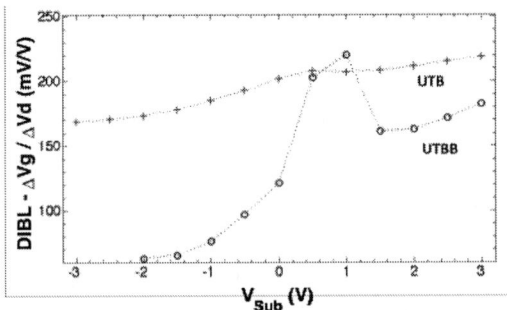

Fig. 2 Simulated results of DIBL extracted at normalized constant current of $I_{dnorm} = 10^{-5}$ A for standard BOX (UTB) and ultra-thin-BOX (UTBB) SOI MOSFETs.

a) $V_{sub} = 0.5\ V$

At the substrate / BOX interface, the surface potential referred to quasi-neutral substrate, i.e. $\psi_s = \psi_{Horizontal} - (V_{Sub} - \Phi_F)$, where Φ_F is the Fermi level here equal to 267 mV, allows for examining the potential rise due to drain bias in terms of the obtained operating regime, i.e. $\psi_s < 0$ indicates accumulation, $0 < \psi_s < 0.53$ V depletion and $\psi_s > 0.53$ V strong inversion.

In Fig. 3, for UTBB devices, we observe that:
- for low V_d, the substrate / BOX interface is in strong depletion below source and drain but in weaker depletion below channel;
- for high V_d, this transforms into strong depletion below the channel and source, and eventually strong inversion below the drain. With the V_d increase from 0 mV to 1 V, ψ_S under drain region increases from 0.3 V to 0.75 V and by about 200 mV in the mid-channel area.

For UTB, with similar increase of drain bias, ψ_S also increases from 0.12 V to 0.25 V below the source, and to

almost 0.38 V and 0.54 V for mid-channel and below the drain areas, respectively. This suggests that the substrate below mid-channel and drain areas goes from weak depletion for low V_d to strong depletion and strong inversion respectively for high V_d. In this case, high DIBL is obtained as shown in Fig. 2.

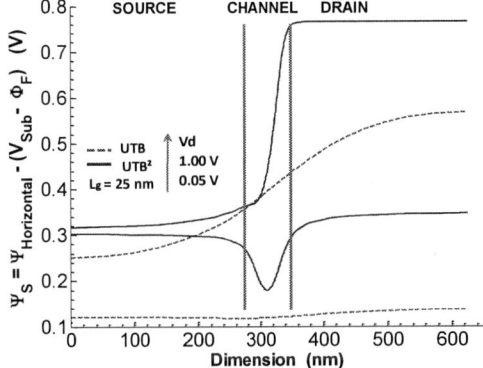

Fig. 3   Horizontal potential cut ($\psi_s = \psi_{Horizontal} - (V_{Sub} - \Phi_F)$) at the substrate / BOX interface for standard BOX (UTB) and ultra-thin-BOX (UTBB) SOI MOSFETs at $V_{Sub} = 0.5$ V.

In both cases, the lateral electrostatic coupling induced by the drain penetrates through the BOX and the substrate strongly affects the properties in the channel. From the vertical cut, the $\psi_{Vertical}$ (Fig. 4) extends in the substrate, up to 1 μm maximum depletion width in inversion, from the front gate surface. This contributes to an increase of BOX equivalent oxide thickness and worsens the DIBL due to the maximum depletion effects. The increase of $\psi_{Vertical}$ reaches a peak (~ 50 nm distance from the front gate as shown in Fig. 4) in the BOX region or in the substrate for UTB or UTBB respectively, suggesting that the $\psi_{Vertical}$ peaks occur at similar location.

The similarity of potential (ψ_S and $\psi_{Vertical}$) distribution as shown in Fig. 3 and Fig. 4 for UTB and UTBB translates into closer DIBL trends in substrate depletion as shown in Fig. 2.

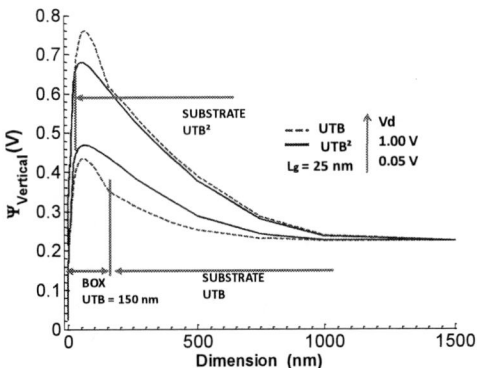

Fig. 4   Vertical potential ($\psi_{Vertical}$) cut at 3 nm from the gate edge, near the source, through the gate, channel, BOX and substrate for standard BOX (UTB) and ultra-thin-BOX (UTBB) SOI MOSFETs at $V_{Sub} = 0.5$ V.

b) $V_{Sub} = -2.0\ V$

In Fig. 5, the ψ_S for UTBB is a bit higher than for UTB but in both cases, at any location below the source, channel and drain, the substrate / BOX interface is in strong inversion.

For high V_d, the resulting ψ_S increases below the drain and below mid-channel areas, for both devices, are not significant because in strong inversion regime, the overall ψ_S is always pinned at $2\Phi_F$ plus a few kT/q. However for UTB, we notice that the potential below the source is also increased with V_d, revealing drain lateral field penetration up to the source area in relation with the thick BOX.

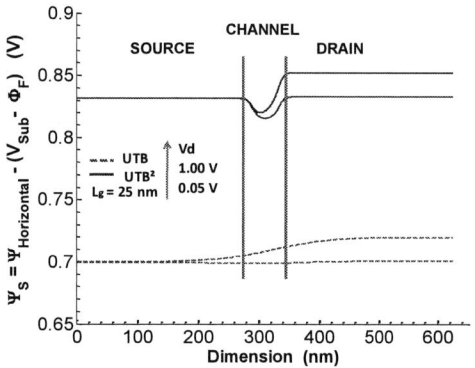

Fig. 5   Horizontal potential cut ($\psi_s = \psi_{Horizontol} - (V_{Sub} - \Phi_F)$) at the substrate / BOX interface for standard BOX (UTB) and ultra-thin BOX (UTBB) SOI MOSFETs at $V_{Sub} = -2.0$ V.

On the contrary to Fig. 4, in Fig. 6 there is no significant $\psi_{vertical}$ increase in substrate when the two devices are subjected to variation from low to high V_d. But there is a slight increase in the BOX potential for UTB, while it is not noticeable for UTBB. This indicates that the lateral field penetration is confined within the BOX thickness and the substrate behaves like quasi-ground plane in preventing the lateral field penetration through the substrate. In both devices, the overall DIBL results are the best.

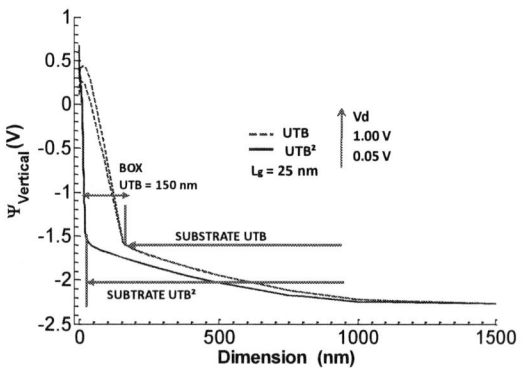

Fig. 6   Vertical potential ($\psi_{vertical}$) cut at 3 nm from the gate edge, near the source, through the gate, channel, BOX and substrate for standard BOX (UTB) and ultra-thin-BOX (UTBB) SOI MOSFETs at $V_{Sub} = -2.0$ V.

c) $V_{Sub} = 2.0$ V

In Fig. 7, the trends of the ψ_S value are opposite when compared to $V_{Sub} = -2.0$ V. At low V_d, the ψ_S below the source, mid-channel and drain are always negative ($\psi_S < 0$ V), indicating that all the locations are in strong accumulation. The UTBB device is in stronger accumulation than UTB device as ψ_S is lower by approximately 125 mV.

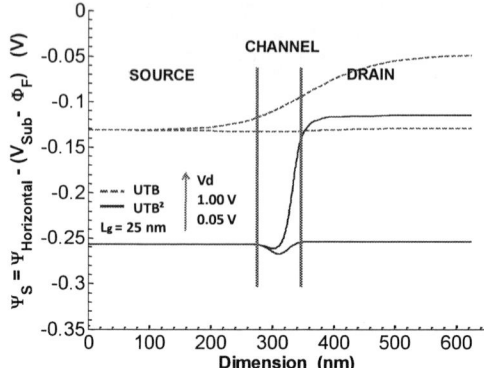

Fig. 7   Horizontal potential cut ($\psi_s = \psi_{Horizontol} - (V_{Sub} - \Phi_F)$) at the substrate / BOX interface for standard BOX (UTB) and ultra-thin-BOX (UTBB) SOI MOSFETs at $V_{Sub} = 2.0$ V.

For high V_d, the ψ_S below the drain area increases significantly when compared with the increase of ψ_S for the same locations when the substrate / BOX is in inversion as shown in Fig. 5, especially for UTBB devices. This is due to the fact that there is no pinning of ψ_S below the drain when the p-type substrate / BOX interface starts in accumulation and goes towards depletion with V_d increase. Again as in Fig. 5 for substrate inversion, the drain lateral field penetrates towards the source area for UTB, clearly influencing the channel and hence provoking DIBL, while this is strongly attenuated in UTBB.

For the $\psi_{Vertical}$, high V_d only slightly increases the $\psi_{Vertical}$ in the BOX thickness as shown in Fig. 8, and similar remarks as for substrate / BOX in inversion (Fig. 6) can be made, i.e. that lateral field penetration is confined within the BOX thickness and the substrate behaves like quasi-ground plane in preventing the lateral field penetration through the substrate. However as discussed above, this is less efficient than with substrate inversion.

Fig. 9 finally recapitulates the ψ_S for UTBB for various substrate biases and regimes. Whatever V_d, the substrate / BOX interface is in accumulation when $V_{Sub} = 2.0$ V ($\psi_S < 0$ V), and strong inversion when $V_{Sub} = -2.0$ V and $V_{Sub} = 0$ V ($\psi_S > 0.53$ V) for any location below source, channel and drain. However for $V_{Sub} = 0.5$ V, the substrate is kept in depletion below source and mid-channel ($0 < \psi_S < 0.53$) while close and below the drain it goes in strong inversion for high V_d. This translates into very different DIBL, with $V_{Sub} = 0.5$ V and $V_{Sub} = -2.0$ V resulting in highest and lowest DIBL respectively according to Fig. 2 and explained by higher potential variations induced by V_d in depleted substrate versus strong potential pinning in strong inversion.

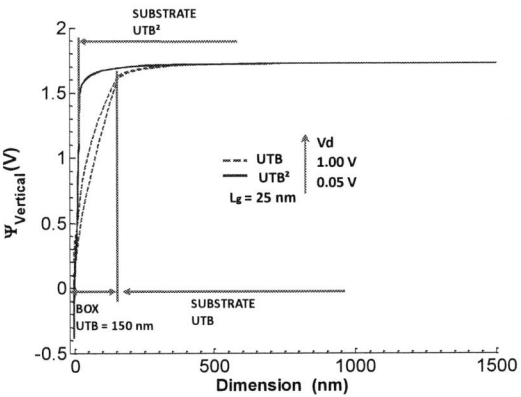

Fig. 8   Vertical potential ($\psi_{Vertical}$) cut at 3 nm from the gate edge, near the source, through the gate, channel, BOX and substrate for thin-BOX (UTBB) and standard BOX (UTB) SOI MOSFETs at $V_{Sub} = 2.0$ V.

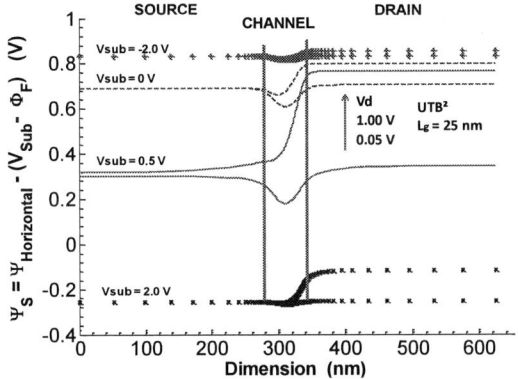

Fig. 9   Horizontal potential cut ($\psi_s = \psi_{Horizontol} - (V_{Sub} - \Phi_F)$) at substrate / BOX interface for ultra-thin-BOX (UTBB) SOI MOSFETs at $V_{Sub} = 0.0, 0.5, 2.0$ and -2.0 V.

IV.   CONCLUSION

In this work, we present the cross influence of the substrate and drain biases on the DIBL of ultra-thin body with standard buried oxide (UTB) SOI MOSFET devices. We extended the investigation to 25 nm-long devices with ultra-thin-BOX (UTBB) and low-doped p-type substrates and explained it by horizontal potential variations at the substrate / BOX interface. The DIBL value is always better in UTBB than in UTB, especially in accumulation and inversion regimes, not only thanks to thin BOX but also due to a quasi-ground plane action that reduces both the lateral field penetration within the BOX thickness and through the substrate. When the substrate is depleted, a change from low to high drain voltage can induce substrate surface potential variations, from weak depletion below mid-channel and drain to strong depletion and strong inversion, respectively. The lowest DIBL appears when the substrate keeps in strong inversion all the way from source to channel and drain, thanks to potential pinning. In substrate accumulation, surface potential pinning is not as efficient for a positive V_d increase with a p-type substrate, resulting in DIBL values lying between the substrate inversion and depletion cases.

ACKNOWLEDGEMENT

The authors also would like to acknowledge all the team members in Institute of Nano Electronic Engineering (INEE) for their guidance and help.

REFERENCES

[1]   K. Cheng, A. Khakifirooz, P. Kulkarni et al. "Extremely thin SOI (ETSOI) technology: Past, present, and future," in *Proc. IEEE International SOI Conference*, 2010, pp. 1–4.

[2]   O. Faynot, F. Andrieu, C. Fenouillet-Beranger et al. "Planar FDSOI technology for sub 22nm nodes," in *IEEE Symposium on VLSI Technology (VLSIT)*, 2010, pp. 26–27.

[3]   L. Chang, Y.-K. Choi, D. Ha et al. "Extremely scaled silicon nano-CMOS devices," *Proc. IEEE*, vol. 91, no. 11, pp. 1860–1873, 2003.

[4]   T. Ohtou, N. Sugii, and T. Hiramoto, "Impact of parameter variations and random dopant fluctuations on short channel fully depleted SOI MOSFETs with extremely thin BOX," *IEEE Electron Device Lett.*, vol. 28, no. 8, pp. 740–742, 2007.

[5]   M.K Md Arshad, J. Raskin, V. Kilchytska et al. et al, "Extended MASTAR modeling of DIBL in UTB and UTBB SOI MOSFETs," *IEEE Trans. Electron Devices*, vol. 59, no. 1, pp. 247–251, 2012.

[6]   J.-P. Colinge, "Fully-depleted SOI CMOS for analog applications," *IEEE Trans. Electron Devices*, vol. 45, no. 5, pp. 1010–1016, May 1998.

[7]   S. Burignat, D. Flandre, M. K. Md Arshad et al. "Substrate impact on threshold voltage and subthreshold slope of sub-32nm ultra thin SOI MOSFETs with thin buried oxide and undoped channel," *Solid. State. Electron.*, vol. 54, no. 2, pp. 213–219, 2010.

[8]   G. Tsutsui, M.saitoh, T. Nagumo, and T. Hiramoto, "Impact of SOI Thickness Fluctuation on Threshold Voltage Variation in Ultra-Thin Body SOI MOSFETs," *IEEE Trans. Electron Dev.*, vol. 4, no. 3, pp:369–373, 2005.

[9]   C. Fenouillet-Beranger, P. Perreau, T. Benoist et al. "Impact of local back biasing on performance in hybrid FDSOI/bulk high-k/metal gate low power (LP) technology," *Solid. State. Electron.*, vol. 88, pp. 15–20, Oct. 2013.

[10] A. S. Mohd Zain, S. Markov, B. Cheng, and A. Asenov, "Comprehensive study of the statistical variability in a 22nm fully depleted ultra-thin-body SOI MOSFET," *Solid. State. Electron.*, vol. 90, pp. 51–55, Dec. 2013.

[11] N. Sugii, R. Tsuchiya, and T. Ishigaki, "Local Vth Variability and Scalability in Silicon-on-Thin-BOX (SOTB) CMOS," *IEEE Trans. Electron Dev.*, vol. 57, no. 4, pp. 835–845, 2010.

[12] R. Ritzenthaler, S. Cristoloveanu, O. Faynot et al. "Lateral coupling and immunity to substrate effect in XFET devices," *Solid. State. Electron.*, vol. 50, pp. 558–565, 2006.

[13] T. Ernst, R. Ritzenthaler, O. Faynot, and S. Cristoloveanu, "A Model of Fringing Fields in Short-Channel Planar," vol. 54, no. 6, pp. 1366–1375, 2007.

[14] T. Ernst, C. Tinella, C. Raynaud, and S. Cristoloveanu, "Fringing fields in sub-0.1 µm fully depleted SOI MOSFETs: Optimization of the device architecture," *Solid State Electron.*, vol. 46, no. 3, pp. 373–378, 2002.

[15] M.K. Md Arshad, S. Makovejev, and S. Olsen et al, "UTBB SOI MOSFETs analog figures of merit: Effects of ground plane and asymmetric double-gate regime," *Solid-State Electron*, vol. 90, pp. 56 - 64. 2013.

[16] M. K. Md Arshad, V. Kilchytska, M. Emam, et al. "Effect of parasitic elements on UTBB FD SOI MOSFETs RF figures of merit," *Solid. State. Electron.*, vol. 97, pp. 38–44, Jul. 2014.

IEEE-ICSE2014 Proc. 2014, Kuala Lumpur, Malaysia

$In_xGa_{1-x}As$ Surface Channel Quantum Well MOSFET: Quantum Ballistic Simulation Using Mode Space Approach

*Ehsanur Rahman, Abir Shadman, Sudipta Romen Biswas, Kanak datta, Quazi D. M. Khosru

Department of Electrical and Electronic Engineering
Bangladesh University of Engineering and Technology,
Dhaka-1000, Bangladesh
*E-mail: EhsanEEEBuet@gmail.com

Abstract—In this paper, I-V characteristics of High k stack InGaAs surface channel Quantum Well MOSFET were simulated using Matlab. To simulate the device in quantum ballistic regime we used non equilibrium Green's function formalism. Self-consistent solution of Schrodinger-Poisson equation was performed using FDM method taking 2D electrostatics into account. The charge is well confined in the quantum well. So, mode space approach is applicable & efficient. Simulation results for the Quantum Well device with 55nm gate length have been reported. Variation of I-V characteristics with the change in channel length has also been reported.

Keywords—*Quantum well, Surface channel, High-k stack, IV Characteristics, 2D Electrostatics.*

I. Introduction

Silicon based MOSFETs have reached their fundamental limit due to the requirements of power constrained scaling. So, new alternative structures and materials are required for future technologies. III-V materials is a potential candidate due to their high electron mobility providing high drive current at low supply voltage[1]. $In_xGa_{1-x}As$ ($x \geq 0.53$) is a leading candidate as a channel material because of its low electron effective mass (m*) and high saturation velocities (v). Large inter-valley separation in $In_{0.53}Ga_{0.47}As$ (InGaAs) also reduces inter-valley scattering. So electron velocities remain high even at high electric fields [2]. Although new p and n channel InGaAs MOSFET is a matter of high research experimentally, a thorough self-consistent simulation based I-V characterization is of major requirement for proper understanding of device physics. We described in this paper a highly efficient, two-dimensional quantum ballistic NEGF simulator based on effective masses in mode space for III-V Quantum Well MOSFET structure. The simulation approach used in this work is evolved from Nanomos 2.5, a two dimensional quantum transport simulator used for Si MOSFETs [3]. The simulation approach is introduced in next section in detail, followed by some key simulation results.

II. Device Structure & Simulation Approach

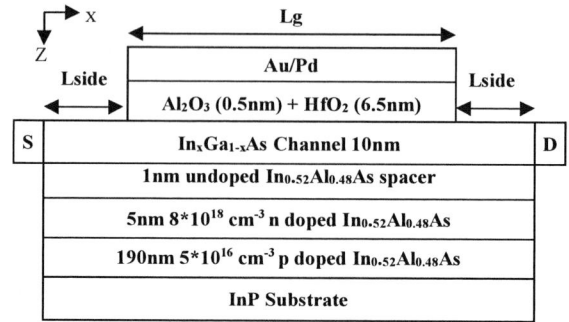

Fig. 1. Rectangular intrinsic simulation region where quantum ballistic transport dominates, used in this study.

Schematic view of the device studied in this paper is shown in Fig. 1. The device has 5nm delta-doped n type $In_{0.52}Al_{0.48}As$ layer separated from channel by 1nm $In_{0.52}Al_{0.48}As$ spacer. High-k oxide stack consists of 0.5nm Al_2O_3 & 6.5nm HfO_2. We placed ideal contact on 2 sides to ignore real contact & its associated resistance. Although this approach does neglect some practical aspects of device operation like source access issue, source starvation etc. it provides us with an upper level projection of device performance.

Fig. 1 shows the rectangular 'intrinsic' simulation region used in this study. Here, Lg refers to the gate length; Lside refers to the distance from gate to n+ layer (source/drain) edge of the device. To maintain numerical stability we have considered length of Lside region to be equal of Lg. The total simulation region is divided into Nx*Nz number of nodes, where Nx and Nz refers to the total number of nodes in X and Z direction respectively. The 2D simulation starts by solving the Poisson equation with the charge from an initial flat band as the guess potential. The updated potential leads to a new set of spatial charge distribution. This process is repeated until desired convergence is achieved. The ballistic current is then readily calculated using the NEGF formalism in mode space for each subband. This approach of quantum ballistic simulation has already been applied for III-V HEMT structures [4][5]. We have applied the same approach to simulate the reported QW device [6]. For current calculation; we consider 2 subbands in the QW channel. The Poisson equation is solved using proper boundary conditions applying a non-linear algorithm [7]. In

978-1-4799-5761-3/14 $31.00 © 2014 IEEE   80

this approach, the carrier concentration $n_{m,n}$ at node [m, n] is expressed in terms of the to-be-solved potential $V_{m,n}$ and the quasi-Fermi potential energy, Fn[8]:

$$n_{m,n} = N_C \mathfrak{I}_{1/2}\left[\frac{(F_n)_{m,n} + qV_{m,n}}{k_B T}\right] \quad (1)$$

Where $\mathfrak{I}_{1/2}$ stands for the Fermi-Dirac integral of order ½ and N_C is the effective density of states. The quasi-Fermi potential is pre-calculated from the input potential and charge density:

$$(F_n)_{m,n} = -q(V_{old})_{m,n} + k_B T \mathfrak{I}_{1/2}^{-1}\left(\frac{n_{m,n}}{N_c}\right) \quad (2)$$

We can obtain the charge distribution by solving the 1D effective mass Schrödinger equation along the Z directed slice in the 2D discretization mesh

$$-\frac{h^2}{2m_z^*}\frac{d^2}{dz^2}\Psi_i(x,z) - qV(x,z)\Psi_i(x,z) = E_i(x)\Psi_i(x,z) \quad (3)$$

Where, m_z^* ıe electron effective mass in the z direction, $\Psi_i(x,z)$ and $E_i(x)$ are the wavefunction and eigen energy for subband i at slice x. Using an uncoupled mode space approach, we divide this 2D problem into 2 1D problems. We find out the total number of subbands in the quantum well channel and calculate charge density in each of these subbands using NEGF formalism. If a is the grid spacing in X direction, using the Green's function formalism [9], the 2D charge density matrix of mode i is obtained as

$$n_{2D} = \int_0^\infty \frac{1}{a\hbar}\sqrt{\frac{m_y^* k_B T}{2\pi^3}}\left[\mathfrak{I}_{-1/2}(\mu_S - E_l)A_S \right. \quad (4)$$
$$\left. + \mathfrak{I}_{-1/2}(\mu_D - E_l)A_D\right]dE_l$$

Where

$$A_S = G\Gamma_S G^+, \qquad A_D = G\Gamma_D G^+$$

are the spectral density functions due to source/drain contacts[10].Γ_S and Γ_D are self energy terms for source & drain. Physically the electron exchange rates between the source/drain reservoirs and the active device region are determined by this term. But in general it can be viewed as the measure of interaction strength due to any perturbation source. Although the device itself may be in non-equilibrium state, electrons are injected from the equilibrium source/drain reservoirs. G is the retarded green function [7]. The 3D charge density at slice x can be obtained by multiplying $n_{2D}(x)$ by the wavefunction probability in z direction and summing for all the subbands:

$$n(x,z) = \sum_i n_{2D}(x,i)|\Psi_i(x,z)|^2 \quad (5)$$

Once self-consistency is achieved, the intrinsic ballistic current is calculated as

$$I_D = \int_0^\infty \frac{q}{\hbar^2}\sqrt{\frac{m_y^* k_B T}{2\pi^3}}\left[\mathfrak{I}_{-1/2}(\mu_S - E_l)\right. \quad (6)$$
$$\left. - \mathfrak{I}_{-1/2}(\mu_D - E_l)\right]T_{SD}(E_l)dE_l$$

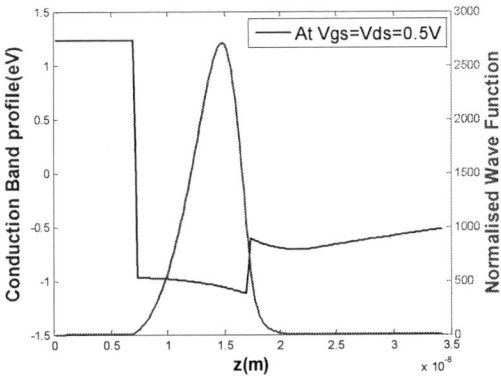

Fig. 2. Conduction band profile along with normalized wavefunction

Where, T_{SD} (E_l) is the transmission coefficient as a function of the longitudinal energy calculated from the Green's function. For the device in Fig. 1, the conduction band profile and normalized wavefunction are shown in Fig. 2. From the figure, we observe that the carriers are well confined within the quantum well channel even in transport condition.

III. Results & Discussions

Fig. 3(both a & b) shows the self-consistent conduction subband profiles versus position under various V_{gs} & V_{ds} bias points keeping one constant at a time. Actually we plot the bottom of the first subband versus position, not the conduction band edge. As the gate bias increases, energy at the top of the barrier is expected to decrease. So more and more carriers get injected into the channel that leads to an increase in drain current.

Subband energy at the drain end decreases with increasing drain bias for a fixed gate bias voltage. As the drain bias is increased, the energy barrier at the source end gets modulated by the increasing drain bias voltage. This DIBL effect becomes important for short channel MOSFETs that reduces the threshold voltage for nanoscale devices.

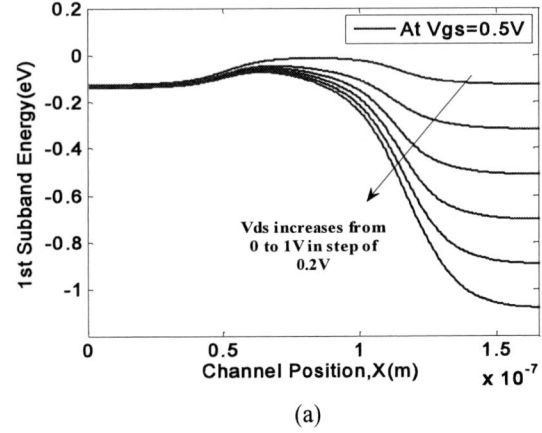

(a)

IEEE-ICSE2014 Proc. 2014, Kuala Lumpur, Malaysia

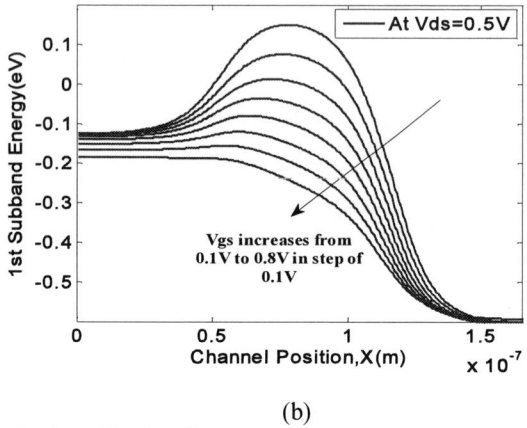

(b)

Fig. 3. First subband profile versus channel position for (a) Vgs = 0.5 V and Vds varying from 0.0 V to 1.0 V (b) Vds = 0.5 V and Vgs varying from 0.1 V to 0.8 V.

Fig.4 (a) shows the 1st subband energy & Fig. 4(b) shows transmission coefficient for the 55nm gate length device at intrinsic gate bias of 0.5V. As we are simulating the device at quantum ballistic regime, the transmission coefficient should go to a value of 1 around the top of the energy barrier in the channel. This energy at the barrier top is being modulated by the gate bias voltage. At the same time due to DIBL effects, increasing drain bias also modulates barrier height in the QW channel. Accordingly, transmission coefficient also gets modulated by the increasing drain bias voltage. The inset figure shows the overall transmission coefficient for the2 subbands together. This profile shows two steps that correspond to two subbands. From this figure, it can be observed that 2nd subband energy is higher than first subband energy.

Fig. 5 shows the average velocity versus position at different drain to source voltages. As expected in this ballistic

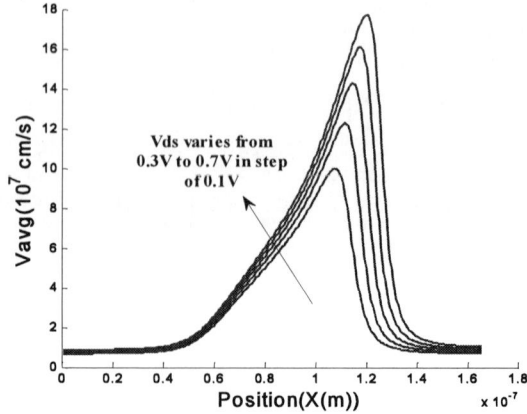

Fig. 5. Average velocity versus channel position for the device of Fig. 1 under ballistic conditions..

transistor, the velocity near the drain increases without limit. Also we did not take into account of band structure limit. Velocity saturation occurs in a ballistic MOSFET [11], but here we have a virtual source approximation which will result in some discrepancy between original & simulated outputs. Also this velocity saturates at thermal limit at the top of the barrier which is different from velocity saturation in bulk MOSFET.

In our simulation, we consider two subbands to calculate ballistic current.Fig.6 shows intrinsic I_d-V_d characteristics for the device structure used in this work. The figure shows simulation results for intrinsic gate bias voltage ranging from 0.3V to 0.9V with 0.1V step and drain bias voltage ranging from 0V to 1.0 V. The ballistic current at intrinsic gate bias of 0.6V is approximately 2.5 times of the reported experimental value for the same device [6] due to absence of any inelastic scattering in our simulation.

Fig.7 shows the I_d-V_{gs} characteristics for the simulated device with 55 nm gate length. For particularly this figure, no metal-semiconductor work function has been taken into account. Here we present I_d-V_{gs} characteristics for two drain bias voltages.

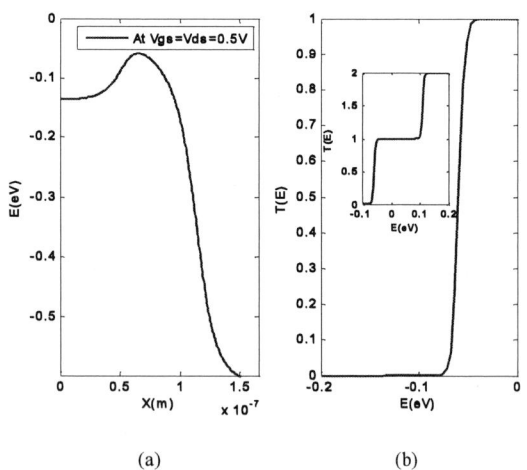

(a)   (b)

Fig. 4. (a) 1st subband energy profile (b) transmission co-efficient for intrinsic gate bias of 0.5V and drain bias 0.5V. The inset figure shows transmission coefficient for combined 1st and 2nd subband.

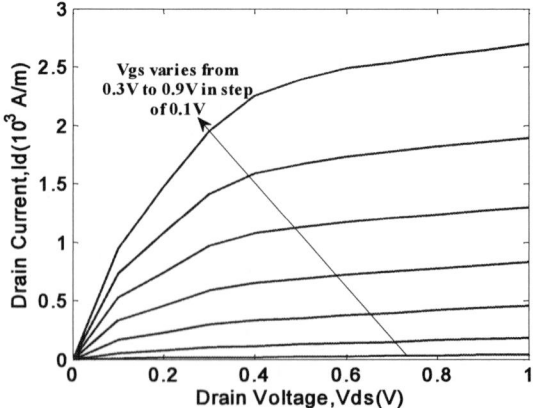

Fig. 6. The intrinsic I_d-V_{ds} characteristics for the simulated device structure.

978-1-4799-5761-3/14 $31.00 © 2014 IEEE   82

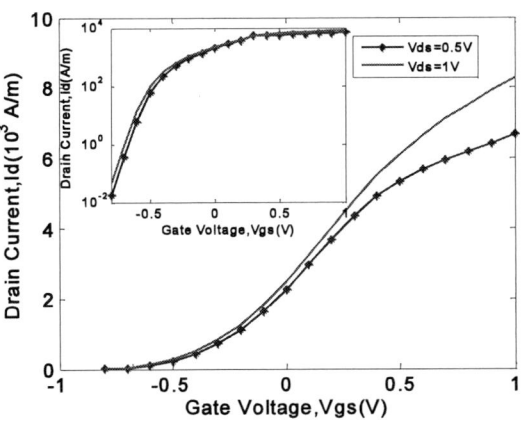

Fig. 7. The intrinsic I_d-V_{gs} characteristics for the simulated device. No metal-semiconductor work function has been taken into account for this figure.

The Fig. shows intrinsic I_d-V_{gs} characteristics in both normal scale and log scale. Sub threshold characteristics of the device can be better understood from the log scale plot. Several important device parameters like Sub threshold Swing (SS), DIBL can be measured from the sub threshold characteristics of the simulated device. For the 55 nm device our simulation shows a DIBL value of approximately 262mV/V and SS value of 83mV/dec. Discrepancy between simulated and reported value may be attributed to the avoidance of non-ideal conditions in our quantum ballistic simulation.

Fig. 8 shows impact of gate length on the intrinsic I_d-V_d characteristics of the device. The increment of current with lower gate length can be attributed to the lowering of energy barrier in the channel with increasing drain bias. With decreasing gate length, short channel effect like DIBL becomes more prominent. As a result, for lower gate length devices transmission co-efficient goes to a value of one at lower subband energy. This phenomenon leads to a higher current for lower gate lengths.

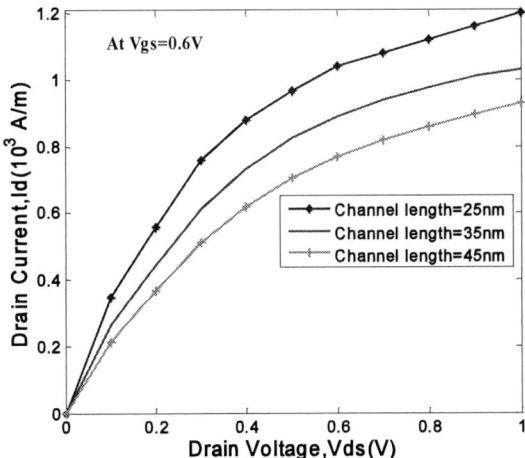

Fig. 8. The intrinsic I_d-V_{ds} characteristics of the simulated device structure for channel length varying from 25 nm to 45nm at Vgs= 0.6V

IV. Conclusion

We report quantum ballistic simulation results of 55 nm Quantum Well device reported in [6]. By keeping the length of side spacer region equal to gate length of the structure, the simulation approach used in [5] has been applied in this work. 2D Poisson equation coupled with Schrodinger equation is solved in uncoupled mode space approach. Using NEGF formalism, we solve transport equations in quantum ballistic regime. We determine subband profile in the $In_{0.53}Ga_{0.47}As$ channel and transmission coefficient in ballistic regime. Intrinsic I_d-V_{ds}, I_d-V_{gs} characteristics are presented for the Quantum Well device reported in [6]. Also variation of I-V characteristics with the change in channel length has been reported.The simulated ballistic current is approximately 2 times the values obtained from experimental measurements. This discrepancy can be attributed to the avoidance of scattering phenomenon in the QW channel. Also, in this approach, the effect of doped heterostructure stack along with contact resistance has been neglected. To get a more realistic picture of transport characteristics, the effect of heterostructure contact stack should be taken into account.

References

[1] B. P. Tinkham, B. R. Bennett, R. Magno, B. V. Shanabrook, and J. B. Boos,"Growth of InAsSb-channel high electron mobility transistor structures" *J. Vac. Sci. Technol.* B 23, 1441 (2005).

[2] Uttam Singisetti, et al. " InGaAs channel MOSFET with self-aligned source/drain MBE regrowth technology "P hys. Status Solidi C 6, No. 6, 1394–1398 (2009).

[3] Z. Ren, R. Venugopal, S. Goasguen, S. Datta, M. Lundstorm, "nanoMOS 2.5: A two-dimensional simulator for quantum transport in double-gate MOSFETs," Electron Devices, IEEE Transactions on, vol 50, pp. 1914-1925, Sept. 2003.

[4] Liu, Yang, and Mark Lundstrom. "Simulation of III-V HEMTs for highspeedlow-power logic applications." ECS Transactions 19.5 (2009): 331-342.

[5] Neophytou, Neophytos, Titash Rakshit, and Mark S. Lundstrom. "Performanceanalysis of 60-nm gate-length III–V InGaAs HEMTs: Simulationsversus experiments." Electron Devices, IEEE Transactions on 56.7 (2009):1377-1387.

[6] M. Egard et. al., "High Transconductance Self-Aligned Gate-Last Surface Channel Ino.53Gao.47AsMOSFET",IEEE IEDM, p.303(2011).

[7] Z. Ren, Nanoscale MOSFETs: Physics, Simulation, and Design, Ph.D.Thesis, Purdue University, West Lafayette, IN (2001).

[8] Liu, Yang, and Mark Lundstrom. "Simulation of III-V HEMTs for highspeed low-power logic applications." ECS Transactions 19.5 (2009): 331-342

[9] S. Datta,"Nanoscale device modeling: the Green's functionmethod." Superlattices and Microstructures 28.4 (2000): 253-278.

[10] S. Datta, Quantum Transport: Atom to Transistor, Cambridge UniversityPress(2005)

[11] M. S. Lundstrom, "Elementary scattering theory of the MOSFET," IEEE Electron Device Lett., vol. 18, pp. 361–363, Nov. 1997.

Condition Based Engine Oil Degradation Monitoring System, Synthesis and Realization on ASIC

M.F.M Idros[1,2]

[1]Faculty of Electrical Engineering
Universiti Teknologi MARA (UiTM)
40450 Shah Alam, Malaysia Country
faizul_san287@salam.uitm.edu.my

Sawal Ali[2], Md.Shabiul Islam[3]

[2]Department of Electrical, Electronics & System Engineering
[3]Institute of Microengineering and Nanoelectronics (IMEN)
Universiti Kebangsaan Malaysia (UKM)
43600 UKM, Bangi, Malaysia

Abstract—**This paper presents the realization and synthesis process of condition based monitoring system of lubricant degradation in *Application Specific Integrated Circuit* (ASIC) for condition based monitoring system of lubricant degradation. With the awareness of the environmental effect of wasted lubricant, a condition based monitoring has been explored recently. For this work, an enhancement of a new algorithm for ASIC implementation is introduced. The Least Square Method (LSM) is used as an algorithm for lifetime prediction of lubricant where it written in verilog language including the FLC. The functional simulation had been applied for functional checking in Quartus II software. Then, a Design Compiler from Synopsys is used as a floor plan simulator and RTL implementation of the FLC and LSM. By implementing the prediction algorithm and decision making, the usage of lubricant can be prolonged and it capable to be implemented in a real monitoring system. By using ASIC, the design is predicted to use $26954.386357\mu m^2$ of cell area and a total dynamic power of 2.8590mW.**

Keywords—*ASIC; Fuzzy Logic; lubricant; degradation, verilog*

I. INTRODUCTION

This In a recent century, a system for lubricant condition monitoring is highly demanded and it purposed for a reduced amount of wasted lubricant. The unnecessary change of lubricant contributes in increasing the amount of wasted engine oil [1, 2]. A lot of recent research has done in a lubricant monitoring technique, including chemical contents, mechanical system and statistical analysis.

Total Acid number (TAN), Total Base Number (TBN), water contents, viscosity and soot contamination are the parameters that have been identified as a main contributor for the degradation factor. The effect of using whether the monograde or multigrade engine oil is the viscosity effect that will influence the fuel consumption of the vehicle [3]. A different types of lubricants were tested in a single-cylinder, four-stroke, direct injection diesel engine to monitor the effect of viscosity to the fuel consumption. It concluded that the fuel consumption was reduced by reducing High Temperature High Shear (HTHS) viscosity, limit friction coefficient and pressure viscosity coefficient. The cause of viscosity variation has been explored by [4]. They focused on studying the effect of soot contamination to the viscosity of engine oil. The result

shows that the higher percentage of soot in engine oil, the viscosity also increased.

An optical sensing technique is widely used in degradation monitoring of engine oil. An equipment, namely Fourier Transform Infrared (FTiR) had been used in monitoring the effect of infrared light to the chemical properties of used lubricant. M. A. Al-Ghouti et al. [5] Another technique is to differentiate the condition between virgin and recycled lubricant. From their approaches of evaluating the carbonyl containing degradation products and polymethacrylate stretching vibrations at $1716cm^{-1}$ and wave band from $1169cm^{-1}$ to $1732cm^{-1}$, the contents of lactones, esters, aldehydes, ketones, carboxylic acids, salts, viscosity modifiers and pour point depressant additives have been identified. Another optical approach has been discussed in [6] from lower wave band and higher waveband. The low waveband region from $918cm^{-1}$ to $1052cm^{-1}$ were selected for ethylene glycol determination while the region from $3598cm^{-1}$ to $3732cm^{-1}$ were used for water determination. A variation value of absorbance or transmittance can be monitored by measuring the voltage difference between the interval of measurement. The lower value of transmittance gives the low value of voltage [7].

As for degradation prediction takes part in the decision making of changing the engine oil, some literature have been done for the analysis and prediction purposes. The quantitative analysis of engine oil degradation due to the running mileage and temperature effect has been studied in [8]. The results show that the reflectance percentage (%R) decreased as the mileage increased and the time of engine oil exposed to the high temperature is inversely proportional to the %R. The user of a Vector Support Machine (VSM) for classification and regression of data has been introduced in [9]. The VSM is applied to differentiate the multi-grade oil from the single grade according to the classification from the manufacturer. As for validation purposes, the full cross validation has been performed by algorithm implementations in Matlab. In the prediction of the suitable time to change the engine oil, the used of the mission profile in analyzing the degraded status has been introduced by in [10]. The prediction maintenance is based on the TAN, TBN and viscosity where the data were collected from sensors and sent to the ground station for analysis. An algorithm involving the some statistical analysis such as discriminant analysis, factor analysis and multiple regressions has been implemented as a decision maker of the engine oil changing time. Another statistical approached for lubricant oil monitoring has been proposed by [11]. Their approached were focused on qualitative and quantitative techniques in monitoring the diesel engines at near and middle infrared spectroscopy.

Even though there were many techniques have been introduced, none of the previous technique can be worked as standalone devices. They were focused on the offline analysis technique. Thus, an implementation of optical color sensor with blue LED has been implemented in [12] for engine oil degradation monitoring with critical limit determination. The sensor has been assembled on a car engine for online monitoring purposes and data collection. The analysis shows that, there was a substantial decreased of transmittance and the critical limit had been introduced [13]. However, the approached is focused on single degradation parameter, whereas the blue light on optical has been used for the sensing purposes.

Hence, the most effective analysis technique for online monitoring is proposed. The optical data based on the waveband discussed in [14] are used for validation purposes. The variation of slope is the main purpose of this work to see the degradation trend as suggested in [13]. Then, ASIC flow is implemented to identify the performance of the circuit in term of validity and power consumption.

II. METHODOLOGY

This project uses a combination of statistical analysis and design technique as shown in Fig.1 below. The statistical data are taken from the actual aging of engine oil on a 4 stroke motorcycle engine. The samples are taken up to 2000km in a constant interval of 50km. An optical sensor with white emitted light is used as a light source to be transmitted through the sample. The amount of light passed through the sample determines the level of degraded lubricant. The collected data will be used for validation purposes. The data will then convert into a digital number to form the input signal to the LSM algorithm that had been coded in Verilog language and generated RTL blocks. The RTL of LSM consists of a combination of arithmetic and logic component such as multiplier, divider, adder, subtraction and flip flop as a register circuit. Quartus II software is used for verilog coding, RTL conversion and data validation while Design Compiler from Synopsis is used for circuit level simulation.

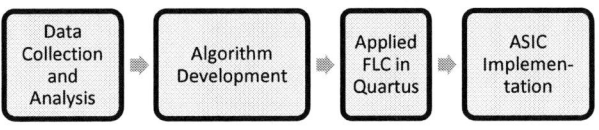

Fig. 1. Flow of works

A. Data Collection and Analysis

Data are taken in a voltage forms from the receiver of the optical sensor circuit. The data collections are purposely to ensure the validity of the prediction algorithm. Therefore, this works need more data for better analysis. Even though the validity will be proven in offline condition, the algorithm is expected to be used in the online monitoring system.

Figure 2 shows the variation voltage of used engine oil after running for a certain mileage. Here, we can see that the slope dropped drastically at the beginning 250km due to the drastic changes of colour darkness and the became stable after 750km. Here, the real task is started to monitor the variation of slope at a certain interval to prove the suggested warning limit of lubricant at a drastic increased of slope by Saurabh Kumar et.al [12].

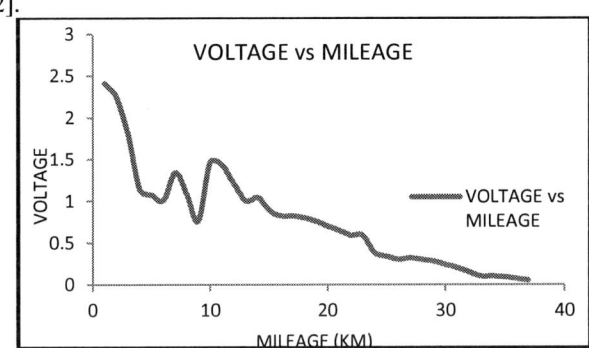

Fig. 2. Voltage versus Mileage from optical sensor

B. Algorithm and Register Transfer Logic Development of LSM

The algorithm in this work is based on the linear regression model where the objective of this algorithm is to monitor the variation of slope over the time. This model is requiring the unknown parameters of slope (β_o) and interception of (β_1). The best fitting are required to get the best result of slope from a set of time series input data. Then, the straight line model for the response y in terms of x can be represented as

$$y = \beta_o + \beta_1 x + \varepsilon \qquad (i)$$

The values of $\hat{\beta}_o$ and $\hat{\beta}_1$ that minimize equation (iv) are obtained by assuming the two partial derivatives $\partial SSE/\partial\hat{\beta}_o$ and $\partial SSE/\partial\hat{\beta}_1$, equal to 0 and solving the resulting simultaneous linear system of least-squares equations.

$$SS_{xy} = \sum_{i=1}^{n}(x_i - \bar{x})(y_i - \bar{y}) = \qquad (ii)$$

$$SS_{xx} = (x_i - \bar{x})^2 = \sum_{i=1}^{n} x_i^2 - \frac{\left(\sum_{i=1}^{n} x_i\right)^2}{n} \qquad (iii)$$

Where n=sample size

Therefore, the slope of the least-squares line is

$$\hat{\beta}_1 = \frac{SS_{xy}}{SS_{xx}} \qquad \text{(iv)}$$

Then, the equation (i) – (iv) are written in Verilog code using Quartus II software where the equation is divided into sub-block and recombined it as a main project as shown in Figure 3. There were 2 main input data, namely transmittance as **Data in** and running hours as **Time in**. The main objective of this result is to get the time series variation values of slope.

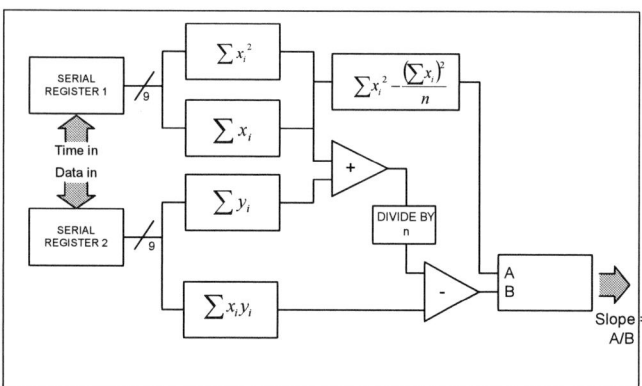

Fig. 2. Combination of arithmatic and logic sub-block.

C. Fuzzy Logic Controller in Verilog Language as a Decision Maker

The hardware development of FLC requires 9-bit data input to get the total of 512 data. The system is designed to have a 512 interval which requires 1 hour for each interval. The previous research shows that the warning limit and critical limit occurred between 180 hours and 300 hours [14], [12], [15]. The technique of using hours to monitor the degradation rather than mileage is because of the driving conditions in Malaysia that frequently facing traffic jammed. During the traffic jammed, the engine still running and the heat from outside contributes to the degradation rate of the engine oil. Moreover, the lifetime of engine oil for the highway driving condition is longer than the urban driving conditions [13].

The Fuzzy Inference System (FIS) has been chosen from Mamdani type because it is well suited to the human input. All the input had been coded into digital formed in three different conditions either 'normal', 'warning' and 'critical'. Each condition was divided into three different intervals and it is depending on the first reading of the data. The maximum value of the membership function occurred at a range 0-5.5 for 'normal', 5.6-8.5 for 'warning' and '8.6-10' for critical conditions.

The Verilog language for the Membership Functions (MF) is shown below. The crisp inputs to the FLC are programmed for 26 bits. Therefore, it will have 226 types of input numbers and all numbers are written in decimal formats. The output of MF shows '00' as an error if the crisp input is in

range between d'200 and d'500. This is because it is impossible to have the worst condition of engine oil at the beginning running hours. The normal condition was programmed in a range between the sign decimal numbers from d'67108863 to d'67108856. The warning condition was programmed to be occurred between d'67100000 and d'67108855 while the critical limit was set between d'15 and d'20. The partial verilog codes of thr MF block is given in below as an example.

```
// program
mf_out  , // Data output
read_en , // Read Enable
ce      // Chip Enable
);
        begin
        if ((crisp_in < 26'd500) && (crisp_in > 26'd200))
                mf_out = 00;
                else if ((crisp_in < 26'd67108863) && (crisp_in >
26'd67108856))
        mf_out = 01;
    else if ((crisp_in < 26'd67108855) && (crisp_in > 26'd67100000))
    mf_out = 10;
  else if ((crisp_in < 26'd20) && (crisp_in > 26'd15))
    mf_out = 11;
        end
endmodule
```

[Verilog language to determine the level of MF]

D. ASIC Implementation

The ASIC implementation is purposely to simulate the actual performance of Integrated Circuit (IC). Speed, power and area are the main parameters to be monitored in these works. Synopsis tools are used to synthesise the designed circuit as required from the designer. This software is capable in any design of IC from the front end up to backend process. There were three main stages in developing IC uses ASIC flow, including login synthesis, gate level netlist and physical layout.

This work is focuses for logic synthesis and the floor plan area in the RTL - description. Here, the Verilog codes are transferred from the Quartus II and run on the ASIC for the synthesis process in Design Compiler tools. The tools will check the compatibility of the Verilog language with the available standard cell in the library. This is to ensure that the hardware is capable to be developed. Once the logic synthesis is done, the power analysis can be applied using Power Compiler tools.

III. RESULT AND DISCUSSION

This section shows the results from the work done. The first result is the functional simulation test from Quartus II tools. The second result is the comparison between the manual calculation of slope and digital functional simulation. The ASIC implementation shows two results, cell area and total dynamic power.

The functional simulation test for the system is shown in Fig. 4. Here, the slope and hours are used as the main parameters of degradation to be monitored by the system. *SLOPE*, *SSxx* and *SSxy* are the parameters of the LSM where have been programmed in 'sign-decimal' numbers. It is because the value of *SLOPE* varies from negative to positive, according to the equation of slope. in_data is the number of hours that counted automatically and shown in decimal number while in_data5 is the optical data in transmittance percentage value. The FLC is written as *mf_hour*, *mf_percentage* for the degree of membership function for hours and slope respectively.

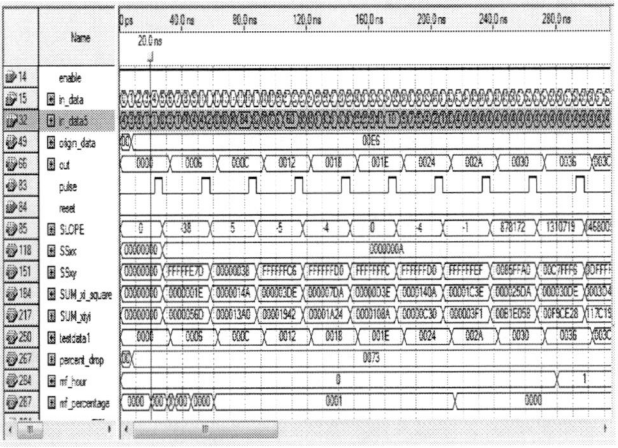

Fig. 3. Functional simulation test

The logic synthesis in ASIC determined the floor plan size and power consumption. The total cell area of the design is using 26954.386357μm² from the combinational and non-combinational area as shown in Table 2. This design consumed 2.8590mW of total dynamic power.

Table 1. Area usage of the overall design.

	μm²
Combinational Area	248442.167084
Non-combinational Area	48512.219273
Cell Area	296954.386357

Table 2. Total of dynamic power

	W
Total dynamic power	2.8590 x10⁻³

IV   CONCLUSION

The small difference between the normal calculation of slope work and functional simulation prove that the coded LSM using verilog in Quartus II tools in successfully implemented. Then, the ASIC design that produced the 6954.386357μm² of cell area and a total dynamic of 2.8590mW are concluded that the implementation of FLC for lubricant monitoring is successfully achieved. The design is ready for gate level netlist analysis after the overall circuits are combined together. This is the common flow of the ASIC in digital IC development. The target of this work is to get the actual physical layout of the overall design before the Graphic Data System II (GDS II) are configured for layout fabrication. using

REFERENCES

1. M.F.M.Idros, S.H. Ali, and S. Islam, *Design of Intelligent SoC Controller for Engine Oil Sensing and Monitoring System.* Asian Journal of Scientific Research, 2012. 5(3): p. 70-77.

2. M.F.M.Idros, S.H. Ali, and S. Islam. *Optical analysis for condition based monitoring of oxidation degradation in lubricant oil.* in *Intelligent and Advanced Systems (ICIAS), 2012 4th International Conference on.* 2012.

3. De Carvalho, M.J.S.S., Peter Rudolf; Belchior, Carlos Rodrigues Pereira, *Lubricant viscosity and viscosity improver additive effects on diesel fuel economy.* Tribology International, 2010. 43(12): p. 2298-2302.

4. George, S., et al., *Effect of diesel soot on lubricant oil viscosity.* Tribology International, 2007. 40(5): p. 809-818.

5. Al-Ghouti, M.A. and L. Al-Atoum, *Virgin and recycled engine oil differentiation: A spectroscopic study.* Journal of Environmental Management, 2009. 90(1): p. 187-195.

6. Borin, A. and R.J. Poppi, *Application of mid infrared spectroscopy and iPLS for the quantification of contaminants in lubricating oil.* Vibrational Spectroscopy, 2005. 37(1): p. 27-32.

7. Idros, M.F.M., et al., *Capability of optical approach in condition based monitoring of lubricant oil.* Sensors and Transducers, 2012. 17(SPL 12): p. 125-134.

8. M.F.M.Idros, S.H. Ali, and S. Islam. *Quantitative Analysis of Spectroscopy's Study for Engine Oil Degradation Monitoring Due to Temperature Effect.* in *2012 Third International Conference on Intelligent Systems Modelling and Simulation.* 2012.

9. Guan, L., X.L. Feng, and G. Xiong, *Engine lubricating oil classification by SAE grade and source based on dielectric spectroscopy data.* Analytica Chimica Acta, 2008. 628(1): p. 117-120.

10. Jun, H.-B., et al., *Predictive algorithm to determine the suitable time to change automotive engine oil.* Computers & Industrial Engineering, 2006. 51(4): p. 671-683.

11. Caneca, A.R., et al., *Assessment of infrared spectroscopy and multivariate techniques for monitoring the service condition of diesel-engine lubricating oils.* Talanta, 2006. 70(2): p. 344-352.

12. Saurabh Kumar, P.S.M., *Online condition monitoring of engine oil.* Industrial Lubrication and Tribology, 2005. 57: p. 260-267.

13. Shirley, S. and S. Donald, *Automotive Engine Oil Condition Monitoring,* in *Tribology Data Handbook.* 1997, CRC Press. p. 931.

14. M.F.M.Idros, et al., *Capability of Optical Approach in Condition Based Monitoring of Lubricant Oil.* Sensors & Transducers Journal, 2012. 17(Special Issue): p. 125-134.

15. Guan, L., et al., *Application of dielectric spectroscopy for engine lubricating oil degradation monitoring.* Sensors and Actuators A: Physical, 2011. 168(1): p. 22-29.

Impact of different ground planes of UTBB SOI MOSFETs under the single-gate (SG) and double-gate (DG) operation mode

Noraini Othman, M.K. Md Arshad, S.N. Sabki U. Hashim

Institute of Nano Electronic Engineering
School of Microelectronic Engineering
Universiti Malaysia Perlis (UniMAP)
01000 Kangar, Perlis, Malaysia.

Abstract—— In this work, we investigate the impact of different ground planes of UTBB SOI MOSFETs under the single-gate (SG) and double-gate (DG) operation modes by numerical simulations. Simulations were performed for 10 nm gate length UTBB SOI MOSFET of 7 nm thin body (T_{si}) and 10 nm thin buried oxide (T_{BOX}) for V_d = 20 mV and 1.0 V. For DG operation mode, the back-gate (BG) and front-gate (FG) were swept simultaneously from 0 to 1.5 V with 10 mV incremental steps. Results reported are on key device parameters such as the threshold voltage (V_{th}), drain induced barrier lowering (DIBL), drive current (I_{on}), subthreshold swing (SS) and transconductance (g_m). Ground Plane (GP) – B structure which employed a p+ doping under the channel shows the best results under the DG operation mode in terms of the lowest DIBL and SS. However, it recorded a slightly higher V_{th} while the results of I_{on} and g_m are comparable with its other GP counterparts.

Keywords— UTBB SOI MOSFETs, ground planes, single-gate (SG) and double-gate (DG)

I. INTRODUCTION

In Ultra-Thin Body and BOX (UTBB), better control of short channel effects (SCEs) is achieved through the reduction of BOX thickness which allows suppression of fringing electric fields through the BOX, thus leads to a better electrostatic coupling [1]-[2]. However, a very thin BOX can cause a depleted zone to extend into the underlying substrate which causes the thin BOX to behave as a thick-BOX due to an increase in the BOX equivalent thickness. In order to counter this effect, ground plane (GP) doping has been introduced to suppress the field penetration under the BOX from propagating to the source [3]-[4]. These were done by heavy doping of the substrate [5]–[7]. The incorporation of GP also allows for V_{th} modulation if the GP contacts are back-biased [8]-[9]. For improved performance, different operation mode which includes double-gate (DG) configuration has been investigated for both dc and ac figures-of-merit [10]-[11].

In this work, we investigate the impact of different ground planes of UTBB SOI MOSFETs under the single-gate (SG) and double-gate (DG) operation modes. Better control of SCEs has been reported with DG configuration since the gate shields the channel from both sides, thus suppressing the penetration of the field from both sides [12]–[14].

II. SIMULATION PROCEDURE

The simulations were carried out using Sentaurus Device simulator [15]. Models used include the hydrodynamic (energy balance) transport model, Shockley-Read Hall recombination model and mobility models consisting of PhuMob, HighFieldSaturation and Enormal models. As the silicon body thickness, T_{si} of the SOI structure in this work is of only 7 nm thick, the thin-layer mobility model is also included in the simulations of both SG and DG operation-mode.

Graphs of drain current versus gate voltage, I_d - V_g were plotted at V_d = 20 mV and 1.0 V. Under SG operation-mode, the back-gate voltage (V_{bg}) is kept at 0 V while for DG operation-mode, front-gate voltage (V_g) and back-gate voltage (V_{bg}) were swept simultaneously from 0 V to 1.5 V in 10 mV incremental steps. Quantum confinement effect was not considered in these simulations, which may lead to some underestimation of threshold voltage values if the devices are of T_{si} < 5 nm [6], [16] which is not investigated here.

III. SIMULATED DEVICE STRUCTURE

Three different structures consisting of SOI with different GP were simulated in this work. The standard-GP UTBB SOI MOSFET is made up of buried oxide, T_{BOX} = 10 nm and Si-body thickness, T_{si} = 7 nm. The channel is undoped with acceptor concentration of 6.5×10^{14} cm^{-3} while S/D donor concentrations doping are of 1×10^{20} cm^{-3}. A thin layer of P+ doping of 1×10^{18} cm^{-3} is employed under the BOX region. The standard-GP structure is then compared with 2 other GP structures as illustrated in Fig. 1. In GP - A structure, P+ doping of 1×10^{20} cm^3 were created under the S/D area which are similar to [17], whereas GP - B structure formed P+ doping of 200 nm deep under the channel with concentration of 1×10^{18} cm^{-3}. For all the GP structures, metal gate with work

978-1-4799-5761-3/14 $31.00 © 2014 IEEE

function of 4.65 eV and an effective oxide thickness (EOT) of 1.2 nm were used.

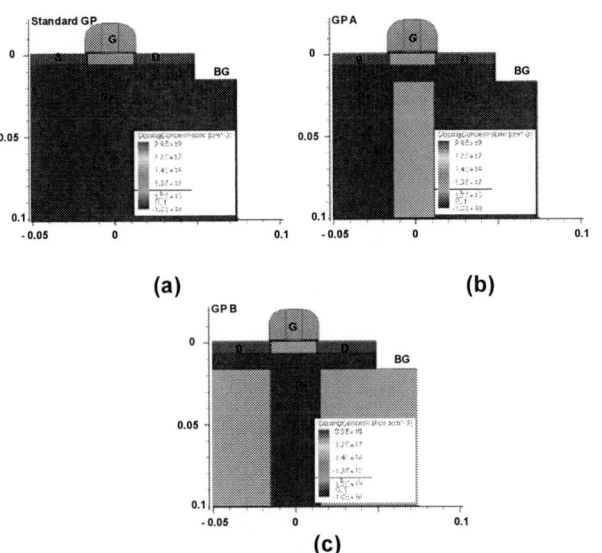

(a)

(b)

(c)

Fig. 1 Different ground plane structures of UTBB SOI MOSFETs (a) standard GP (b) GP – A and (c) GP –B. All shows the back-gate (BG) electrode near the drain.

IV.   SIMULATION RESULTS

A. *Current-voltage (I-V) characteristics*

Fig. 2 shows the drain current, I_d as a function of gate voltage, V_g. It can be seen that different operation mode of SG and DG does not have any impact on the off-current, I_{off}. However, higher I_{on} is produced with DG configurations as the gate shield the channel from both sides, thus improving the charge control by the top and bottom gates. Meanwhile, variations in the results of I_{on} and I_{off} with the implementations of different GP structures were found to be negligible.

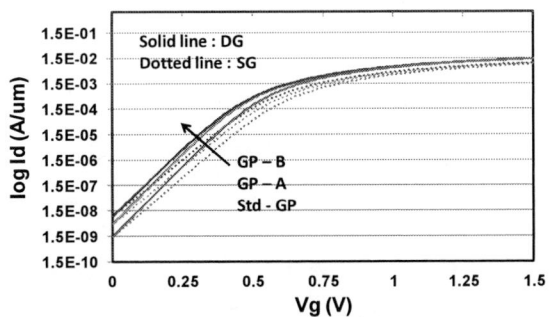

Fig. 2 log I_d as a function of V_g at V_d= 1 V for different ground plane (GP) structures under SG and DG operation mode.

B. *Threshold voltage, V_{th}*

Fig. 3 shows the results of V_{th} extracted by using the constant-current method corresponding to drain current, of $1x10^{-7}$ A. Lower V_{th} were recorded under DG operation mode with V_{th} variations between the different GP structures of only ~ 0.03 mV and bears no significant impact.

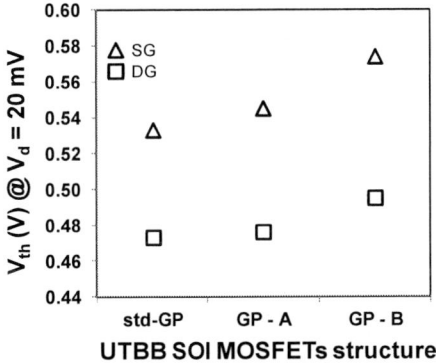

Fig. 3 V_{th} of different GP structures of UTBB SOI MOSFETs under SG and DG operation-mode.

C. *Drain-Induced Barrier Lowering (DIBL)*

Fig. 4 shows the results of DIBL extracted as (V_{thlin} - V_{thsat}) / ΔV_d [18] at drain current of $1x10^{-7}$ A for two drain voltages i.e. V_d = 20 mV and 1.0 V. GP – B shows the lowest DIBL recorded both under SG and DG operation-mode.

Fig. 5 shows the horizontal potential distribution from the source through the channel and ends at the drain for both SG and DG operation-mode for low and high drain voltage i.e. V_d = 20 mV and 1.0 V. By referring to the SG operation mode, it can be seen that GP – B shows no increase of potential under the channel or under the drain region when the drain is biased from 20 mV to 1 V. This reflects the superiority of GP – B in shielding the channel from the influence of drain electric field when a high drain bias is applied. This in turn is translated into the lowest DIBL values by GP – B. Similar observations are recorded for the DG operation-mode, only which the horizontal potential is higher due to the presence of the additional BG. In addition to that, the potential between low and high drain voltage is negligible which translates into lower DIBL in DG compared with SG operation mode.

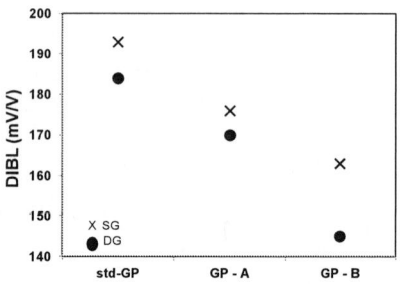

Fig. 4 Results of DIBL as a function of single-gate (SG) and double-gate (DG) for standard-GP, GP –A and GP – B.

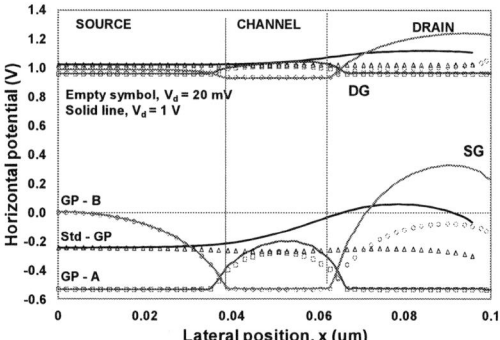

Fig. 5 Horizontal potential along the BOX/substrate interface at linear (V_d = 20 mV - empty symbol) and saturation (V_d = 1 V – solid line) for SG and DG for standard-GP, GP-A and GP-B.

D. I_{on} and I_{off}

Results for on-current, I_{on} and off-current, I_{off} were extracted from the plot of log I_d-V_g. I_{on} is defined as the maximum current, whereas I_{off} is defined as I_d at V_g= 0 V. It can be seen from Fig. 6 that the configuration of SG and DG had no effects on the results of I_{off} in contrast with I_{on}. Higher I_{on} is obtained for DG operation mode which is in agreement with other previous works [19]. However, implementing different GP structures were found to have insignificant impact on the results of individual I_{on}.

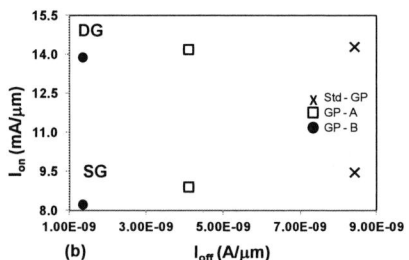

Fig. 6. On-current, I_{on} and off-current, I_{off} as a function of SG and DG

E. Subthreshold Slope, SS

Fig. 7 shows the results of subthreshold slope, SS at V_d = 20 mV as a function of SG and DG configurations. Lower SS were recorded under DG operation which can be explained by referring to Fig. 8 that shows the electric field lines under SG and DG operation-mode for a standard-GP structure. It can be seen that under DG operations, the extra shielding plate i.e BG protects the channel region from the electric field lines from the drain. This in turns improves the SS as the gates acquire better control of the channel region which is in agreement with [14],[19]. Similar results that showed superiority of DG configuration in shielding the channel regions were also observed for GP – A and GP – B structure (thus not shown here).

Further investigation of GP – A and GP - B structures under the DG operation mode showed negligible differences in the electric field lines distributions as compared to the std-GP as shown in Fig. 8b. These were reflected in the results whereby eventhough GP – B shows the best SS under the DG configuration, its improvement was very minor (only between 3-5 mV/dec improvements) as compared to its other GP counterparts.

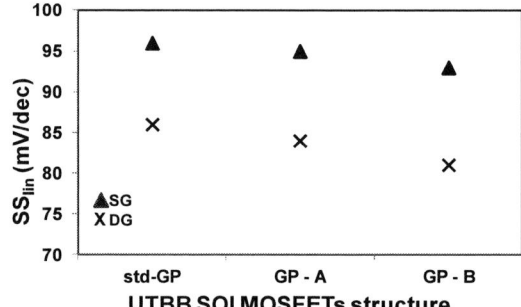

Fig. 7. Linear subthreshold slope, SS for SG and DG operation-mode

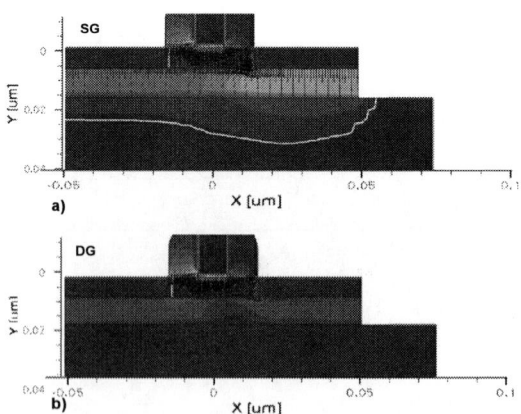

Fig. 8. Contour of electric field lines at V_d = 1 V under SG and DG operation-mode for a std-GP structure.

F. Transconductance, g_m

Transconductance, g_m refers to the measures of the drain current (I_{ds}) variation with a gate-source voltage (V_{gs}) variation while keeping the drain-source voltage (V_{ds}) constant. In this work, it is extracted as:

$$g_m = \Delta I_{ds}/\Delta V_{gs} \text{ at } V_{ds} = 1 \text{ V} \qquad (1)$$

Results in Fig. 9 show that the use of DG increases the g_m in agreement with [20] as both gates are strongly coupled to the channel. However, the use of different GP structures was found to have only marginal impact on the results of the corresponding g_m.

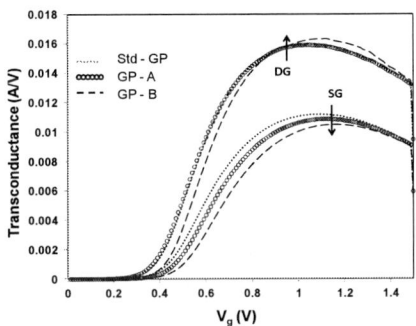

Fig. 9. Transconductance, g_m at $V_d = 1$ V for SG and DG configuration.

V. CONCLUSIONS

In this work, implications of different ground plane (GP) of UTBB SOI MOSFET under the single-gate (SG) and double-gate (SG) operation-mode were evaluated. Different GPs in UTBB SOI MOSFETs were found to have only a significant impact on the DIBL which depends on the position of dopants being introduced in the substrate. GP – B which produces the lowest DIBL is a simple method but is found to be effective in suppressing substrate depletion. Other key device parameters such as V_{th}, I_{on}, SS and g_m showed that the introduction of different GPs only have minor impacts to their values. In all cases, DG operation-mode showed better results than the SG operation-mode.

REFERENCES

[1] S. Burignat, D. Flandre, M. K. Md Arshad, V. Kilchytska, F. Andrieu, and et al, "Substrate impact on threshold voltage and subthreshold slope of sub-32nm ultra thin SOI MOSFETs with thin buried oxide and undoped channel," *Solid. State. Electron.*, vol. 54, no. 2, pp. 213–219, 2010.

[2] N. Xu, F. Andrieu, B. Ho, B.-Y. Nguyen, O. Weber, C. Mazuré, O. Faynot, T. Poiroux, and T.-J. K. Liu, "Impact of back biasing on carrier transport in ultra-thin-body and BOX (UTBB) Fully Depleted SOI MOSFETs," in *2012 Symposium on VLSI Technology (VLSIT)*, 2012, pp. 113–114.

[3] H. P. Wong, S. Member, D. J. Frank, P. M. Solomon, C. H. J. Wann, and J. J. Welser, "Nanoscale CMOS," in *Proc. of the IEEE*, 1999, vol. 87, no. 4, pp. 537–570.

[4] T. Ernst, C. Tinella, C. Raynaud, and S. Cristoloveanu, "Fringing fields in sub-0 . 1m fully depleted SOI MOSFETs : optimization of the device architecture," *Solid. State. Electron.*, vol. 46, pp. 373–378, 2002.

[5] H. P. Wong, D. J. Frank, P. M. Solomon, I. B. M. T. J. Watson, P. O. Box, Y. Heights, and N. York, "Device Design Considerations for Double-Gate, Ground-Plane, and Single-Gated Ultra-Thin SO1 MOSFET's at the 25 nm Channel Length Generation," in *IEDM Technical Digest*, 1998, pp. 407–410.

[6] R. Yan, R. Duane, P. Razavi, A. Afzalian, I. Ferain, C. Lee, N. D. Akhavan, B. Nguyen, K. K. Bourdelle, and J. Colinge, "LDD and Back-Gate Engineering for Fully Depleted Planar SOI Transistors with Thin Buried Oxide," in *IEEE Transactions on Electron Devices*, 2010, vol. 57, no. 6, pp. 1319–1326.

[7] P. Perreau, L. Tosti, O. Thomas, J. Noel, T. Benoist, O. Weber, F. Andrieu, A. Bajolet, S. Haendler, M. Cassé, X. Garros, K. K. Bourdelle, F. Boedt, O. Faynot, and F. Boeuf, "Low power UTBOX and Back Plane (BP) FDSOI technology for 32nm node and below," in *IEEE Intl. Conference on IC Design & Technology (ICICDT)*, 2011, pp. 1–4.

[8] R. Tsuchiya, M. Horiuchi, S. Kimura, M. Yamaoka, T. Kawahara, S. Maegawa, T. Ipposhi, Y. Ohji, and H. Matsuoka, "Silicon on Thin BOX: A New Paradigm of the CMOSFET for low power and High-Performance Application Featuring Wide-Range Back-Bias Control," in *IEDM Technical Digest*, 2004, pp. 631–634.

[9] L. Grenouillet, M. Vinet, J. Gimbert, B. Giraud, J. P. Noël, Q. Liu, P. Khare, M. A. Jaud, Y. Le Tiec, R. Wacquez, T. Levin, P. Rivallin, S. Holmes, S. Liu, K. J. Chen, O. Rozeau, P. Scheiblin, E. Mclellan, M. Malley, J. Guilford, A. Upham, R. Johnson, M. Hargrove, T. Hook, S. Schmitz, S. Mehta, J. Kuss, N. Loubet, S. Teehan, M. Terrizzi, S. Ponoth, K. Cheng, T. Nagumo, A. Khakifirooz, F. Monsieur, P. Kulkarni, R. Conte, J. Demarest, O. Faynot, W. Kleemeier, S. Luning, B. Doris, and V. D. D. S. M. A. X. Rbb, "UTBB FDSOI transistors with dual STI for a multi-V t strategy at 20nm node and below," in *IEDM*, 2012, pp. 64–67.

[10] M. K. MdArshad and U. Hashim, "Emulation of double gate transistor in ultra-thin body with thin buried oxide SOI MOSFETs," *RSM 2013 IEEE Reg. Symp. Micro Nanoelectron.*, pp. 162–165, Sep. 2013.

[11] M. K. Md Arshad, V. Kilchytska, M. Emam, F. Andrieu, D. Flandre, and J.-P. Raskin, "Effect of parasitic elements on UTBB FD SOI MOSFETs RF figures of merit," *Solid. State. Electron.*, vol. 97, pp. 38–44, Jul. 2014.

[12] A. Majumdar, Z. Ren, S. J. Koester, and S. Member, "Undoped-Body Extremely Thin SOI MOSFETs With Back Gates," *IEEE Trans. Electron Devices*, vol. 56, no. 10, pp. 2270–2276, 2009.

[13] V. Kilchytska, D. Bol, J. De Vos, F. Andrieu, and D. Flandre, "Quasi-double gate regime to boost UTBB SOI MOSFET performance in analog and sleep transistor applications," *Solid. State. Electron.*, vol. 84, pp. 28–37, Jun. 2013.

[14] M. K. Md Arshad, S. Makovejev, S. Olsen, F. Andrieu, J.-P. Raskin, D. Flandre, and V. Kilchytska, "UTBB SOI MOSFETs analog figures of merit: Effects of ground plane and asymmetric double-gate regime," *Solid. State. Electron.*, vol. 90, pp. 56–64, Dec. 2013.

[15] V. E-, "Sentaurus Device User," *System*, no. December, 2010.

[16] M. Arshad, J. Raskin, V. Kilchytska, F. Andrieu, P. Scheiblin, and et al, "Extended MASTAR modeling of DIBL in UTB and UTBB SOI MOSFETs," *IEEE Trans. Electron Devices*, vol. 59, no. 1, pp. 247–251, 2012.

[17] H. Makiyama, Y. Yamamoto, T. Tsunomura, T. Iwamatsu, K. Sonoda, H. Oda, N. Sugii, and Y. Yamaguchi, "Novel Local Ground-Plane Silicon on Thin BOX (SOTB) for Improving Short-Channel-Effect Immunity," in *Euro SOI*, 2012, vol. 2, no. January, pp. 27–28.

[18] J.-P. Colinge and C. A. Colinge, *Physics of Semiconductor Devices*. Kluwer Academic Publishers, 2002.

[19] J. Colinge, "Multiple-gate SOI MOSFETs," *Solid. State. Electron.*, vol. 48, no. 6, pp. 897–905, Jun. 2004.

[20] D. J. Frank, S. E. Laux, and M. V Fischetti, "Monte Carlo Simulation of a 30 nm Dual-Gate MOSFET: How Short Can Si Go?," *IEDM*, pp. 553–556, 1992.

Physical and Optical Studies on ZnO Films by Sol-gel

A. R. Nurulfadzilah, S. N. M. Tawil
Department of Electrical and Electronics Engineering
Universiti Pertahanan Nasional Malaysia
57000 Sungai Besi, Kuala Lumpur, Malaysia
nooraya@upnm.edu.my

C. A. Norhidayah, N. Nafarizal, M. Z. Sahdan
Microelectronics and Nanotechnology Shamsuddin Research Center (MiNT-SRC)
Universiti Tun Hussein Onn Malaysia
86400 Parit Raja, Batu Pahat, Johor, Malaysia
zainizan@uthm.edu.my

Abstract—**Zinc oxide (ZnO) is a new material used in electronics applications such as spintronics, diluted magnetic semiconductor (DMS) and thin films. In this paper, the findings in fabrication and characterization of ZnO nanoparticles are reported. ZnO thin films were deposited onto glass substrates using a spin coating techniques at different pre-annealing temperatures. The optical transmittance measurement of the ZnO nanostructures was examined using an ultraviolet-visible spectrophotometer (UV-Vis) with wavelength ranging from 300-800 nm. On the other hand, the topography and film roughness were, measured by an atomic force microscope (AFM). The surface morphologies was obtained using a field emission scanning electron microscope (FESEM) and showed that the ZnO nanoparticles are distributed uniformly.**

Keywords ZnO, sol-gel, spin coating, optical measurement, topography

I. INTRODUCTION

Nowadays, the world is facing one of the biggest problems which is global warming. As a consequence, the earth's temperature will increase. There are various causes which can lead it to happen. Therefore, the need for green energy sources and renewable energy to replace fossil fuels is very important. Fossil fuels are hazardous to the earth's atmosphere as it releases carbon dioxide (CO_2) into the environment and attenuate the ozone layer. Various disasters have occurred which prompt the community to use energy sources that are safe and more environmentally friendly. Thus, the use of solar cells employing photovoltaic effect was introduced, in which satisfy the desired purposes. Among the commercially available solar cell technologies, thin film solar cells such as cadmium telluride (CdTe) and copper indium diselenide ($CuInSe_2$) are most famous [1-3]. However, CdTe is not suitable for the alternative green energy because cadmium is highly toxic. On the other hand, $CuInSe_2$ has complex manufacturing for solar device and maintaining at low cost is difficult [4]. As a result, a non-toxic and abundance materials in the earth are required for green energy devices. Zinc oxide (ZnO) has been considered as one of the best candidate because it has the above properties. Besides, ZnO is much cheaper than most of the non-toxic elements. Due to that reasons, ZnO has been the most investigated material for its optical, acoustic and electric properties which could extend into applications in the field of electronics, optoelectronics and sensors [5]. The ZnO thin films are grown by various methods such as spray pyrolysis (PLD), chemical vapor deposition (CVD), magnetron sputtering, metal organic chemical vapour deposition (MOCVD) etc. [6-8]. Sol-gel method is well known for being simple as there is no need of costly vacuum system, the easy control of chemical components and fabrication of thin films at low cost. In order to investigate its structural and optical properties, large area coating capabilities is feasible by this technique [9-10]. In this work, various characterizations have been employed such as atomic force microscope (AFM, XE-100 Park System), field-emission scanning electron microscope (FESEM, JSM-7600 JEOL) and ultraviolet-visible spectrophotometer (UV-Vis, UV-1800 Shimadzu). Furthermore, we have investigated the effects of different pre-annealing temperatures on ZnO thin films. AFM is utilized to observe the topography and roughness of the thin films.

II. EXPERIMENTAL SETUP

ZnO thin films were deposited by a sol-gel spin coating technique onto glass substrates. Aqueous solution containing same molarity of 0.4 mol zinc acetate dehydrate (ZAD) and monoethanolamine (MEA) was diluted in 30 ml of 2-methoxyethanol. The solution was prepared at 60 °C and stirred for 24 h to maximize reaction process. After 24 h, the ZnO solution was coated onto the glass substrates. The films were pre-annealed at different temperature of 200 °C, 225 °C, 250 °C, and 275 °C for 3 minutes using a hot plate to evaporate the solvent and remove organic residuals. The procedures from coating to pre-annealing were repeated 10 times. Then, the ZnO films were annealed at 500 °C for 1 h. The films were underwent slow cooling until achieving room temperature before characterization.

The procedure for preparing ZnO thin films is shown in Fig. 1. The topography and the roughness of the films were scanned by the AFM. The optical transmittance measurement of the thin film was observe at room temperature using UV-

Vis spectrophotometer with wavelength range from 300 to 800 nm [2]. The surface morphology was observed by the FESEM.

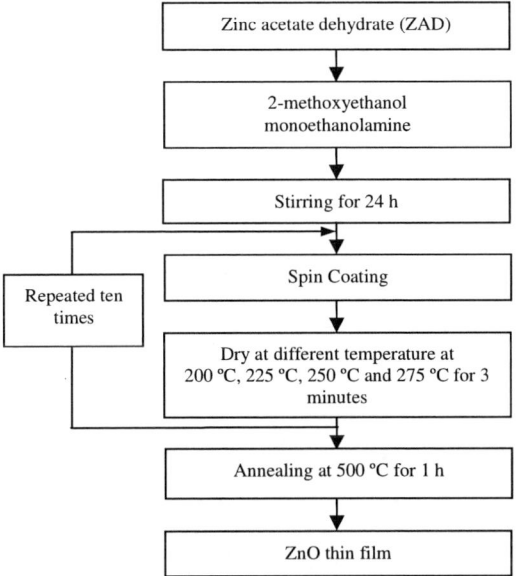

Fig. 1. The flowchart showing the procedure for preparing the ZnO thin films.

III. RESULT AND DISCUSSION

A. Topological Analysis

Fig. 2 shows the three dimensional view of sample surface over a 1 μm x 1 μm scan area of the thin film. As can be seen in Fig. 2 (a-d), generally the films changed its topology when temperature is varied from 250 °C to 275 °C. The film's topology was changed from hillock-like structure at 200 °C into wavy spheres at 225 °C. Then, further increase on the temperature to 250 °C has resulted to the formation of uniform particles with the optimum roughness. Finally, increasing the temperature to 275 °C results the formation of hillocks between the spheres. The average roughness for this film is summarize in Table 1. From the experimental data, it was found that the temperature affects the surface morphology of ZnO thin films [11].

Table 1. The average roughness of ZnO thin films

Sample	Temperatures (°C)	Roughness (nm)
a.	200 °C	4.278
b.	225 °C	8.911
c.	250 °C	1.931
d.	275 °C	2.274

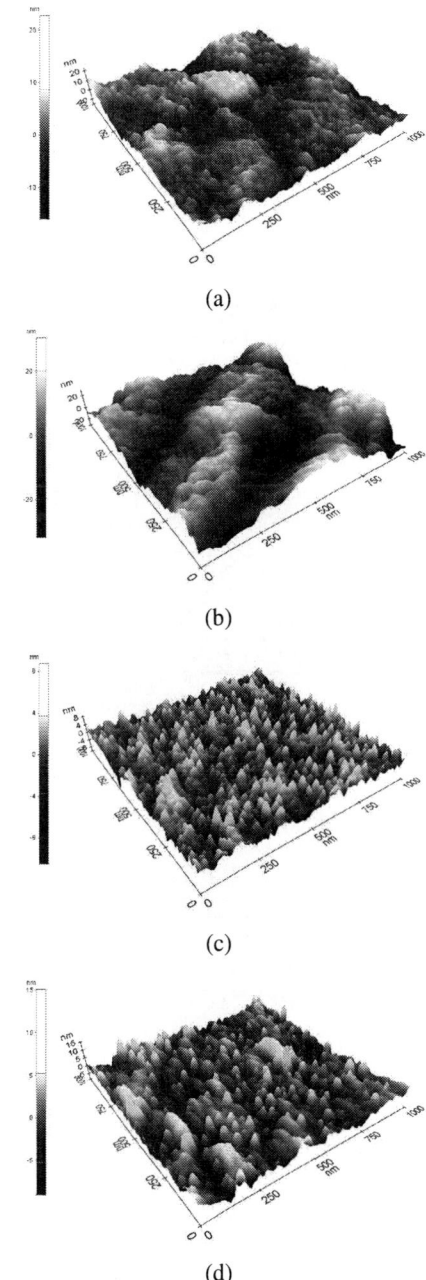

Fig. 2. AFM topography of ZnO thin film by spin coating techniques at different temperature for pre-annealing. (a) 200°C (b) 225°C (c) 250°C (d) 275°C

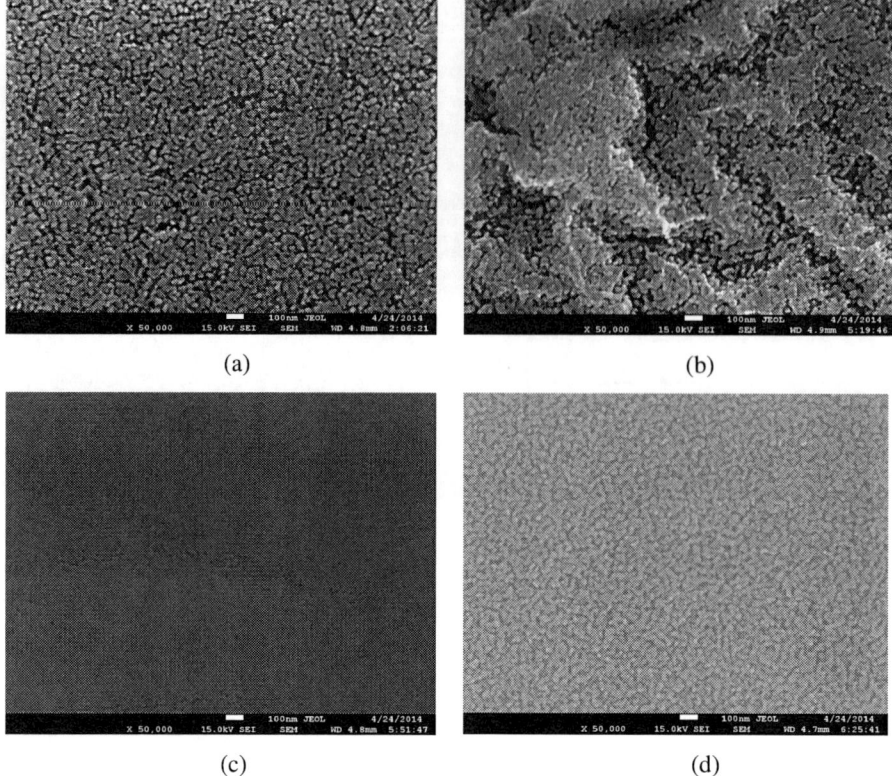

Fig. 3. FESEM images of ZnO nanostructures at different pre-annealing temperatures. (a) 200°C (b) 225°C (c) 250°C (d) 275°C

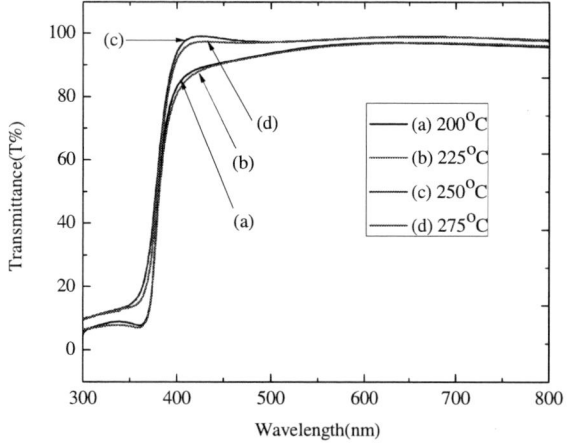

Fig. 4. Optical transmittance measurement of ZnO thin films

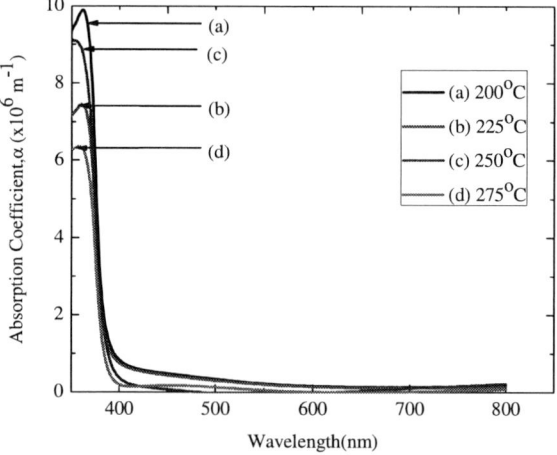

Fig. 5. Absorption coefficient, α, of ZnO thin films prepared at different temperature of pre-annealing. (a) 200 °C (b) 225 °C (c) 250 °C (d) 275 °C

B. Morphological Analysis

The surface morphology to support the AFM was indicated by Fig. 3 show that nanoparticles were formed on the films [4]. The size of the nanoparticles is homogeneous and there are porous granular surface caused by different temperature during the pre-annealing process [8]. When the pre-annealing temperature was increased, the surface morphology was changed; the size of the nanoparticles and granular surface become smaller. From Fig. 3(b), it was obviously shows that there are large cluster grains on the film. But then, when the temperature was changed from 250 °C to 275 °C, the size of the grains become smaller. Fig. 3(c) shows the nanoparticles on the film were extremely small which is in good agreement with the surface roughness.

C. Optical Analysis

Fig. 4 indicates the transmittance and absorption coefficient spectra of the thin films using the UV-Vis spectrophotometer. The transmission decreased sharply near the ultraviolet region at approximately 380 nm [12]. It is clearly seen that the transmittance spectra was divided into two forms which are A (a and b) and B (c and d). For A the signal is bouncing which indicates the film is highly crystalline whereas for B, the films have a very nice slope which indicates the crystallinity is lower compared to A. The lower average transmittance properties of ZnO thin films may be due to the presence of voids around the grains, as being discussed with reference to FESEM results [12].

Fig. 5 shows the absorption coefficient of ZnO thin films. The absorption coefficient, α, can be calculated using Lambert's Law based on relation (1):

$$\alpha = \frac{1}{t}\ln(\frac{1}{T}) \qquad (1)$$

Where t is the thickness of the thin film and T is the transmittance of the thin film. As we can see, a steep decrease occurred in the absorption coefficient around the absorption edge. Due to that reason, it might be due to the increase of the surface roughness caused by higher temperature. The highest absorption coefficient for the thin film was observed at 200 °C and the lowest was at 275 °C.

IV. CONCLUSION

ZnO nanoparticles film was synthesized using a spin coating technique at different pre-annealing temperatures. The structural, topography and optical properties were investigated. It was found from the AFM results that, when the temperature was increased, the surface roughness has decreased and improving the hillock structures on the films. FESEM confirmed that the sample preheated at 250 °C has smooth and uniform surface. Based on UV-Vis results, the film has good crystallinity when the temperature is above 250 °C which conclude that higher temperature has higher crystallinity. The properties of the ZnO thin film can be tailored to suit variety of application such as optoelectronic devices.

ACKNOWLEDGMENT

The authors would like to acknowledge Ministry of Education for the financial support through FRGS grants vote 0823 from UTHM and FRGS/1/2013/SG06/UPNM/02/1 from UPNM.

REFERENCES

[1]   V. A. Chaudhari and C. S. Solanki, "A novel two step metallization of Ni/Cu for low concentrator c-Si solar cells," *Solar Energy Materials and Solar Cells,* vol. 94, pp. 2094-2101, 2010.

[2]   B. Li, *et al.*, "Review of recent progress in solid-state dye-sensitized solar cells," Solar Energy Materials and Solar Cells, vol. 90, pp. 549-573, 2006.

[3]   A. Shah, et al., "Microcrystalline silicon and 'micromorph'tandem solar cells," Thin Solid Films, vol. 403, pp. 179-187, 2002.

[4]   M. A. Green, "Solar cells: operating principles, technology, and system applications," Englewood Cliffs, NJ, Prentice-Hall, Inc., 1982. 288 p., vol. 1, 1982.

[5]   S. Ilican, et al., "Preparation and characterization of ZnO thin films deposited by sol-gel spin coating method," Journal of optoelectronics and advanced materials, vol. 10, pp. 2578-2583, 2008.

[6]   Z. R. Khan, et al., "Optical and structural properties of ZnO thin films fabricated by sol-gel method," Materials Sciences and Applications, vol. 2, p. 340, 2011.

[7]   H. Li, et al., "Sol–gel preparation of transparent zinc oxide films with highly preferential crystal orientation," Vacuum, vol. 77, pp. 57-62, 2004.

[8]   M. Habibi and M. K. Sardashti, "Structure and morphology of nanostructured zinc oxide thin films Prepared by dip-vs. spin-coating methods," Journal of the Iranian Chemical Society, vol. 5, pp. 603-609, 2008.

[9]   M. Shahzad, et al., "A study on the Al doping behavior with sol aging time and its effect on structural and optical properties of sol–gel prepared ZnO thin films," Thin Solid Films, vol. 534, pp. 242-248, 2013.

[10]   J. Neamtu and M. Volmer, "The influence of doping with transition metal ions on the structure and magnetic properties of zinc oxide thin films."

[11]   S. Nagarajan and K. A. Kuppusamy, "Extracellular synthesis of zinc oxide nanoparticle using seaweeds of gulf of Mannar, India," Journal of nanobiotechnology, vol. 11, p. 39, 2013.

[12]   M. Malek, et al., "Sonicated sol–gel preparation of nanoparticulate ZnO thin films with various deposition speeds: The highly preferred< i> c</i>-axis (002) orientation enhances the final properties," Journal of Alloys and Compounds, vol. 582, pp. 12-21, 2014.

Process Development of 40 nm Silicon Nanogap for Sensor Application

M.S. Nur Humaira, U. Hashim, , T. Nazwa, S.T. Ten
Institute of Nano Electronic Engineering
UniMAP
Kangar, Malaysia
humaira@mardi.gov.my

S. Ahmad
Mechanization and Automation Research Center
MARDI
Serdang, Malaysia

N. A. Yusof
Faculty of Science
UPM
Serdang, Malaysia

Abstract—A recent breakthrough in nanotechnology provides a great extent in sensor fabrication and application. The technology has emerged as a powerful technique to minimize the size of devices; amount of materials, energy and time consumption. Nanogap based sensor is one of the sensor that capable of characterizing and quantifying molecules selectively and sensitively with good electrical behavior. In this manuscript, we present a collaboration work between UniMAP, MARDI and UPM in the process development of 40 nm silicon nanogap for sensor application. The process consists of a combination of electron beam lithography (EBL) method and conventional photolithography method. Both methods were for nanogap and electrodes pattern respectively. Silicon on insulator (SOI) substrate was used to fabricate the nanogap structure and gold was used for the electrode. The ability of EBL system to fabricate a gap in nanometer scale with direct lithography technique on SOI substrate gives advantages in this development work. The developed silicon nanogap device was physically characterized with scanning electron microscope (SEM). The sensor application was accomplished by testing the device with different level of pH solutions using a dielectric analyzer.

Keywords—*nanogap;sensor;electron beam lithography;pH sensing*

I. INTRODUCTION

Nanotechnology is the application of science, engineering and technology to produce materials and devices with the size within 1 -100 nanometers scale. Nanotechnology has the ability to minimize the size of devices; amount of materials; energy, and time and cost consuming. In these last two decades, nanotechnology has brought significant progress in the development process of nanogap device for sensor application. Nanogap is a device having a pair of electrodes facing each other with a nanometer size range of gap. Molecule trapped between the two electrodes will be identified by observing the electrical characterization such as capacitance, permittivity, conductance, resistance and impedance [1].

Many methods have been introduced by researchers in order to fabricate nanogap device for sensor application with excellent sensitivity and selectivity. Early years, photolithography process has been applied in nanogap development. Photolithography is a process used in microfabrication to transfer a pattern from mask on a substrate using light. However, to get a desired gap in a nanometer scales are beyond the capability of microfabrication technologies. Several techniques such as size expansion or size reduction are a compulsory additional method to overcome this obstacle [1-3]. Dhahi reported a 50 nm size of nanogap fabrication using conventional photolithography combined with patterned-size reduction technique. The nanogap pattern was transfer through a chrome mask using UV-lithography mask aligner and the nanogap size was achieved by performed a dry-etching RIE process of the silicon layer [3]. However to achieved the desired nanogap size, the process must be in a control condition of time.

Next-generation lithography has been introduced in the lithography technology as early as in 1970s. However, the next-generation lithography such as electron-beam lithography, focused ion beam, shallow evaporation, electromigration, electrochemical deposition, mechanical break junction and so on only been used in these recent decades to develop nanogap devices. Each method described its own advantages and characteristics with successfully achieved a gap as small as 10 nm. Several methods has been integrated and adopted into the development process to obtain a good size of gap for improving molecules detection function in nanogap based sensor [4-8].

In this manuscript, we present our work on silicon on insulator (SOI) for development of nanogap device using electron beam lithography (EBL) system. EBL is a direct lithography technique where the design of nanogap is directly patterned to the SOI substrate without a used of a mask. The developed device was then tested with electrolyte pH solutions for sensor application purpose.

II. METHODOLOGY

A. Design

The silicon nanogap was fabricated using a p-type SOI wafer substrate with thickness of Si, SiO2 and Si were 50 nm, 200 nm and 725 μm respectively. The nanogap and electrodes were designed using AutoCAD software. Fig. 1 (Right) shows a schematic of the nanogap device with size of 20.00 mm × 20.00 mm square and (Left) the nanogap pattern, which was used as a pattern design for EBL. The nanogap pattern design with size 2.54 mm length and 0.58 mm width had a 40 nm size of gaps between it.

Fig. 2a is a schematic design for gold electrode mask. The distance between the electrodes is 0.70 mm; it is fitted with nanogap design in sensing area. The mask are printed onto chrome glass surface and purchased from Photonic Pte. Limited Singapore. The actual mask printed on chrome glass shown in Fig. 2b.

B. Development of Nanogap Device

In this study, we had combined two lithography system methods in the device development process; EBL system and conventional photolithography system for nanogap develop.

The EBL system, ELS – 7000 combined with field-emission scanning electron microscope, ESM – 9000 was used for the nanogap development. It was equipped with ZrO/W thermal field emission electron gun. The parameter of the EBL system is provided in table 1.

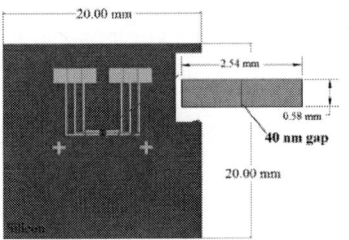

Fig. 1. Schematic of (Left) nanogap device and (Right) the nanogap pattern.

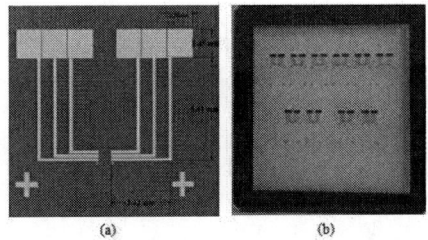

Fig. 2. (a) Design specification and (b) chrome mask for gold electrode.

The process of defining a nanogap pattern using EBL consists of a few steps which are very similar with conventional photolithography. The process started by coated the pre-cleaned SOI substrate with e-beam resist which is sensitive to electrons (Fig. 3b). The e-beam resist type ZEP 520 was dispensed uniformly onto the SOI surface and spin coated at 500 rpm for 5 seconds followed by 3000 rpm for 10 seconds to form a 100 nm thickness. ZEP 520 is a high performance positive e-beam resist with high resolution; high sensitivity and dry etch resistance. Then the SOI substrate was baked at 180 °C for 3 minutes on a hot plate to remove solvent from the resist. The substrate then was loaded into the e-beam chamber and exposed to electrons using a 50 pA beam current (Fig. 3c). EBL uses a focused beam of electrons to directly draw a pattern on the resist surface in series according to the pattern designed as shows in Fig. 1(Right). After exposure process, the substrate undergoes developing proses using a resist developer MEK and MIBK for 30 seconds (Fig. 3d). This process removed the corresponding exposure region where the resist become soluble. Then dry etching process is performed to the substrate using Reactive Ion Etching Oxford Plasma for 15 minutes (Fig. 3e). Finally the resist mask was removed to obtain the final nanogap pattern (Fig. 3f).

To fabricate the gold electrode on top of the nanogap using the conventional lithography system, a mask from Fig. 2(b) was used for pattern transfer. The substrate was deposited with 50 nm titanium layer and 150 nm gold layer using Auto 306 thermal evaporator (Fig. 3g). Titanium was used as a glue material to attach the SOI substrate and gold. Then positive photoresist PA-4000 was dispensed onto the gold surface and spin coated at 3000 rpm for 25 seconds to form a 1200 nm layer (Fig. 3h). The substrate subsequently baked with hot plate at 100 °C for 90 seconds. After the substrate had cooled down to room temperature, the lithography process was initiated by exposing the substrate to mask (Fig. 2(b)) using UV-lithography mask aligner (Fig. 3i). The substrate was then developed using RD 6 developer to remove the unexposed substrate (Fig. 3j). Finally the gold electrode with 40 nm nanogap was obtained by wet-etching process of the titanium and gold layer (Fig. 3k) and removal process of photoresist (Fig. 3l).

TABLE I. PARAMETER OF EBL SYSTEM

Parameter	Details
Acceleration voltage	100 keV
Beam current	50 pA
Beam spot size	1.8 nm for 10 pA at 100 kV
Scanning speed	10 MHz
Beam position	18bit
Laser interferometer	Stage positioning with 0.61 nm resolution
Area step size	600 μm at 6000 dots map
Area dose	320
Exposure speciality	Normal

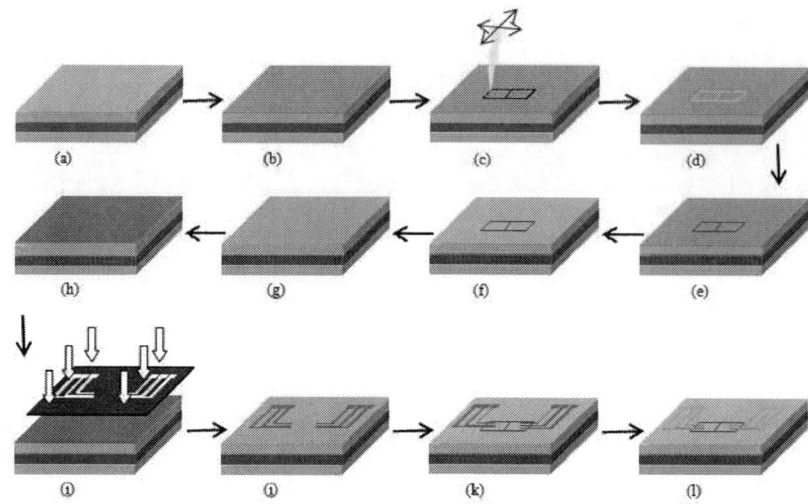

Fig. 3. Schematic of the process development of nanogap device.

C. Nanogap Device Characterization and Measurement

The fabricated nanogap device was morphologically characterized with scanning electron microscope (SEM). In this study, we used Dielectric Analyzer (Alpha-A High Performance Frequency Analyzer, Novocontrol Technologies, Germany), where we connected the device with Dielectric Analyzer at the probing area. The measurement was held in clean room condition with voltage and frequency range was set up to 30 mV and 100 Hz to 1 MHz range respectively to measure the capacitance.

D. Nanogap Device for Sensor Application

The nanogap device sensitivity was tested using 4 different levels of electrolyte solutions with pH 3, 5, 7, and 9. All the solutions were certified standard solutions (certified by the National Institute of Standards and Technology) and were purchased from QREC, Malaysia. 0.5 µL of each solution was dropped at the sensing area and the capacitance was measured.

III. RESULT AND DISCUSSION

A. Developed Nanogap Device

The morphological characterization results are obtained from the EBL technique is shown in a series of Scanning Electron Microscope (SEM) images in Fig. 4. In Fig. 4a, the overall pad of developed nanogap is shown with nanogap located in the middle of the pad. The nanogap size was confirmed by SEM images shown in Fig. 4b is 40 nm. The gold electrode fabricated using method conventional photolithography shows in Fig. 5.

B. Electrical Properties of the Developed Nanogap Device

The function of nanogap device was measured using dielectric analyzer with bare gap and DI water. Fig. 6 shows

the capacitance-frequency profile in current experiment. It is observed there is small or no changes in capacitance when measurement was performed with air. Meanwhile, when DI water was dropped onto the nanogap sensing area, the result was increased significantly to 28 n F. This trend was explained by Nevill, where both air and water is a neutral molecule. However, the polar character of water has made it a better contributor of charge compare to air [9].

Fig. 7 shows the result of the nanogap device sensitivity when different level of electrolyte solutions was dropped to the device sensing area. The capacitance measurement shows that capacitance increased when the device tested with higher pH solution. This is due to a large electrostatics forces which originate from the dissociation of polar groups at higher pH level. The presence of hydroxide ion in the solutions contributed to the capacitance changing. The ions reduced the double layer capacitance at the interfaces of the electrodes and solution, increasing overall capacitance of the device. This can be explained by capacitor equation, $C = \varepsilon A / d$, where C represents the capacitance value, ε represent dielectric constant, A represents the surface area of the nanogap and d is

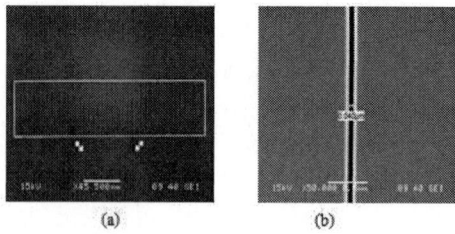

Fig. 4. SEM image of (a) developed nanogap area, (b) 40 nm size of nanogap

Gold
Electrode

Nanogap

Fig. 5. Developed nanogap device.

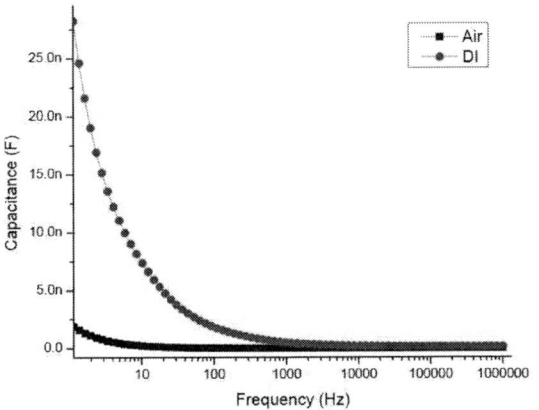

Fig. 6. Capacitance – frequency profile for nanogap device functionalize inspection.

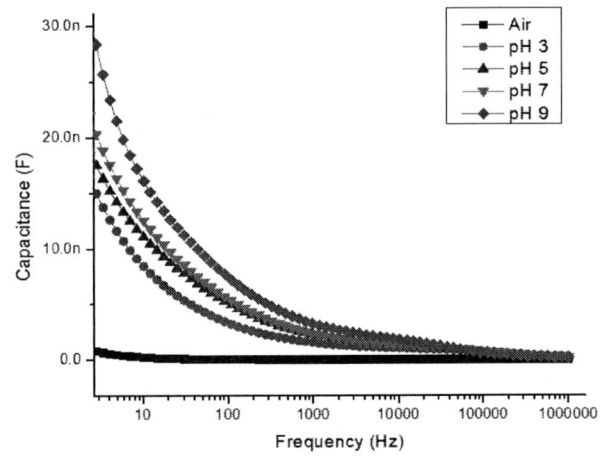

Fig. 7. Capacitance – frequency profile of a different level of pH solutions.

the size of gap or the distance between two electrodes [10]. Nanogap based sensor using a dielectric mechanism where the dielectric properties of molecule depend in electron transfer, atomic bonds and the large-scale molecular structure.

Therefore, the molecular structures will respond differently depending on the frequency [11].

IV. CONCLUSION

A silicon nanogap device using a combination of EBL system and conventional photolithography is fabricated. The fabrication of the nanogap is direct and can be achieved with access to a standard EBL system. The electrical properties characterized from the nanogap device shows that nanogap could be used for sensor application. Future experiments will focus on the Escherichia coli (*E.coli*) detection by coupling the DNA of *E.coli* within the nanogap with dimension close to its persistence length.

ACKNOWLEDGMENT

Author M.S. Nur Humaira thankfully acknowledges collaborators at INEE, MARDI and UPM. This work was supported by MARDI, through the Development of Nano-photonic Sensors for Detection of Toxins and Microbes in Foods project.

REFERENCES

[1] T. S. Dhahi, U. Hashim, N.M. Ahmed, Md. Eaqub Ali, and T. Nazwa, "Electrical characterization of in-house fabricated polysilicon micro-gap for yeast concentration measurement," Journal of Engineering and Technology Research, vol. 3, pp. 246 - 254, 2011.

[2] B. S. Rao, M. N. Asri, U. Hashim, and T. Adam, "Conventional photolithography and process optimization of pattern-size expansion technique for nanogap biosensor fabrication," Advanced Materials Research, 2014. 832: p. 89 - 94.

[3] T. S. Dhahi, U. Hashim, and N.M Ahmed, "Fabrication and charactherization of 50 nm silicon nano-gap structures," Science of Advance Materials, vol. 3, pp. 233 - 238, 2011.

[4] L. Yan, W. Xiaofeng, amd Z. Jiayomg, "Lithography-indeendent and large scale fabrication of metal electrode nanogap," Journal of semiconductors, vol. 30, p. 4, 2009.

[5] B. Liu, J. Xiang, and J. H. Tian, "Controllable nanogap fabrication on microchip by chronopotentiometry," Electrochimica Acta, vol. 50, pp. 3041 - 3047, 2005.

[6] T. Nagase, T. Kubota and S. Mashiko, "Fabrication of nano-gap electrodes for measuring electrical properties of organic molecules using a focused ion beam," Thin solid films, vol. 438 - 439, pp. 374 - 377, 2003.

[7] T. Li, W. Hu, and D. Zhu, Nanogap Electrodes. Advance Materials, 2010. 22: p. 286 - 300.

[8] Humayun Q. and U. Hashim, Fabrication techniques of electrical nanogap biosensor. Australian Journal of Basic and Applied Sciences, 2013. 7(1): p. 278 - 285.

[9] J. T. Nevill, K.H. Jeong and L.P. Lee, "Ultrasensitive nanogap biosensor to detect changes in structure of water and ice," in Proc. Transducers, Seoul, Korea, 2005, pp. 1577 - 1580.

[10] B. S. Rao, U. Hashim, T.S. Dhahi, and T. Adam, "pH sensing using in house fabricated polysilicon nanoelectrode based transducer," in 2012 IEEE EMBS International Conference on Biomedical Engineering and Sciences, Langkawi, Malaysia, 2012, pp. 122 - 125.

[11] Mingqiang Yi, Ki-Hun Jeong, and L.P. Lee, "Theoretical and experimental study towards a nanogap dielectric biosensor," Biosensors and bioelectronics, 2004.

Synthesis and Characterization of Cupric Oxide Thin Films by Sol-Gel Method

M.F. Nurfazliana, S.A. Kamaruddin, N. Nafarizal, H. Saim, and M.Z. Sahdan

Microelectronic and Nanotechnology - Shamsuddin Research Center
Universiti Tun Hussein Onn Malaysia
86400 Parit Raja, Batu Pahat, Johor, Malaysia
zainizan@uthm.edu.my

Abstract—**Cupric oxide (CuO) thin films were successfully grown on glass substrates by low cost sol-gel method. In this work, non-contact atomic force microscope (AFM) and current mode atomic force microscope (C-AFM) was used to physically characterize the CuO thin films. Two different samples using molar concentration of 1 mM and 5 mM were used. The surface morphology of CuO thin films were observed by a field emission scanning electron microscopy (FESEM). Meanwhile, the structural property of the films was measured by an X-ray diffractometer (XRD). From the results, XRD showed the formation of single-phase CuO films with monoclinic structures while FESEM morphological image found that the films with different molar concentration produced different shapes of morphological surface structure. In addition, it was noticed that when the molar concentration increased, the surface of the films become smoother and denser. The microstructures characterization revealed that the films were substantially affected by precursor concentration.**

Keywords☐ cupric oxide; sol-gel; microstructures; precursor concentration

I. INTRODUCTION

In the past decade, a large amount of research is carried out on synthesis of inorganic materials because of their outstanding properties and potential for various device applications [1-2]. Cupric oxide (CuO) is one of the promising materials which has unique features such as a p-type semiconductor with monoclinic structure and has a narrow band gap in the range of 1.2 - 2.1 eV [3]. In addition, CuO is a non toxicity in nature, low cost, the abundant availability of copper, relatively simple formation of the oxide layer, high optical and electrical properties and the theoretical solar cell efficiency is 18% with ZnO [2, 4]. Mostly, CuO thin films have received much attention for its various applications such as semiconductors [5], gas sensors [6], photo catalysis [7], batteries [8], magnetic storage [9], biosensors [10], solar cells [11] and capacitors [12]. To date, micro- and nano-scale of CuO thin films have been prepared and grown by a number of techniques including radio frequency magnetron sputtering, spray pyrolisis, electrodeposition, pulsed laser deposition, chemical vapor deposition and hydrothermal [3]. Among these techniques, chemical bath deposition (CBD) is a promising technique because it is economical, safe and simple. Furthermore, this method offers the opportunity of doping the host ions with impurities on different shapes and sizes on glass substrates. In this study, we report our work on the characterization of CuO thin films synthesized by CBD technique from an aqueous solution mixture containing copper(II) nitrate ($Cu(NO_3)_2$) and hexamethylenetetramine (HMT) at diferent precursor concentrations (1 mM and 5 mM). The surface morphological, structural and optical properties are measured using field emission scanning electron microscopy (FESEM), X-ray diffractometer (XRD) and UV-Vis spectrophotometer, respectively. The surface topology, roughness, grain boundaries and current measurement were specially characterized by a non-contact atomic force microscope (AFM) and a current mode atomic force microscope (C-AFM).

II. EXPERIMENTAL DETAILS

CuO thin fillms were grown by chemical bath deposition (CBD) technique on glass substrates from an aqueous solution that containing copper(II) nitrate and hexamethylenetetramine (HMT). CBD is an aqueous analogue of chemical vapor deposition (CVD) technique. This technique is based on slow controlled precipitation of the desired compound from its ions in a reaction bath solution [13]. The synthesis of CuO films onto glass substrate were describe as follows:

A. Substrate preparation

At first, the glass substrates (2.5 cm × 2.5 cm) were cleaned in three steps which are cleaning in absolute acetone, absolute ethanol and de-ionized (DI) water for 10 min each in an ultrasonic bath to remove the impurities. Then, the glass substrates were dried in ambient. A catalyst (gold) was deposited at 20 mA for 30 sec on the glass substrates in order to adherent the deposition of CuO films.

B. Solution preparation

Copper(II) nitrate, hexamethylenetetramine (HMT) and de-ionized (DI) water were used as a precursor, stabilizer and solvent, respectively. An aqueous solution with different molar concentration (1 mM and 5 mM) was prepared by dissolving copper(II) nitrate in 100 ml DI water under stirring in a magnetic stirrer at room temperature. After 5 minutes, 5 mM HMT was added into both aqueous solution and stirred for 1 hour to ensure that the solutions was completely mixed precisely and a blue solution produced.

C. CuO thin films deposition

Cleaned glass substrate with a gold catalyst was attached horizontally by a rubber tape in a reagent bottle. Then, an aqueous solution was poured into the reagent bottle and heating process was introduced at 80 °C for 3 h using annealing furnace. The glass substrates were taken out from the bath solution after 3 h and rinsed with DI water for 1 min and preheated at 60 °C for 5 min to remove the residual impurities. Fig. 1 shows a flow diagram of the experimental setup.

D. Characterization

The crystal structures of the CuO fillms were analyzed by X-ray diffractometer system (Bruker Advance D8). Surface morphological study was carried out using FESEM (JEOL JSM-7600) and AFM (XE-100 Park System) was used to investigate the surface topology and current properties of the films. Optical properties of the samples were analyzed employing an UV-Vis spectrophotometer (UV-1800, Shimadzu).

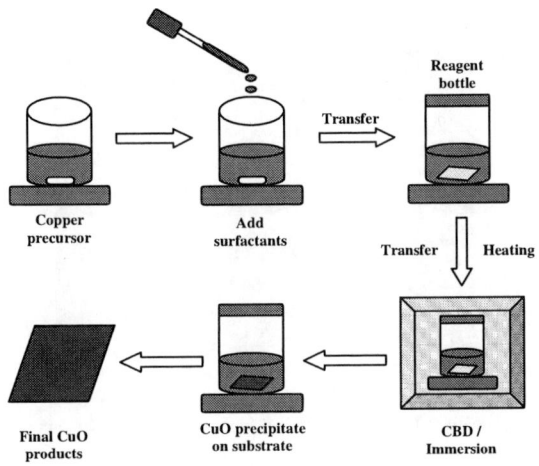

Fig. 1. Experimental setup for CuO thin films deposition by chemical bath deposition (CBD) technique

III. RESULTS AND DISCUSSION

Fig. 2 shows the surface topography of the monoclinic-phase CuO films realized at different precursor concentration, analyzed using atomic force microscope (AFM) at 3 μm × 3 μm scan size. The two conditions of different precursor concentration were 1 mM and 5 mM. The AFM results indicate that the surface topography of the CuO films is distributed within nanometer-sized grain around 95 nm to 105 nm. In addition, the surface morphologies of the films changed with an increasing precursor concentration. It can be seen that the surface of CuO films grown with densely packed and regular shaped grains using 5 mM concentration of precursor. Furthermore, the grain size decreases slightly when the precursor concentration was increased from 1 mM to 5 mM. The roughness of the deposited film at 5 mM precursor concentration was also observed to be smaller which is about

24.3 nm compared to films deposited at 1 mM precursor concentration (37.8 nm).

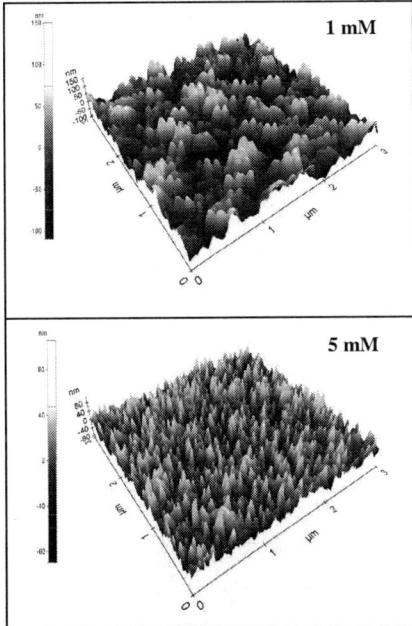

Fig. 2. AFM micrographs of the CuO films fabricated at different precursor concentration

Fig. 3 shows the current mode atomic force microscope (C-AFM) measurement which indicates different behavior for sample deposited using 1 mM and 5 mM of precursor concentration. It is obviously seen in Fig. 3(a) that the current response is in negative plane even though positive voltage was applied and it indicates a significant and more effectively current response for CuO thin films. The resistance of the films can be calculated using Ohm's Law as shown in the following equation (1):

$$V = IR \qquad (1)$$

where;
V = voltage (V),
I = current (A),
R = resistance (R)

The resistance of CuO films in Fig. 3(a) is around 62.72 nΩ while the resistance of CuO films in Fig. 3(b) cannot be determined due to the non-linearity curve obtained. The slope (m) of CuO films deposited at 1 mM precursor concentration was calculated at 10.08 as shown in Fig. 3(a). However, it can be seen that the I-V characteristics was unstable and contains noise which is caused by the porosity surface and the current rectification on the films. Fig. 3(b) shows a stable and well pattern I-V curve films. In the future, the electrical properties of nanostructures CuO films will be further examined. Thus, C-AFM is expected to be very effectively useful for highly efficient solar cells developments. The theoretical of C-AFM state that this method is very useful to characterize the conductive thin films which it can examined the material functionality, especially for the films electrical properties [14].

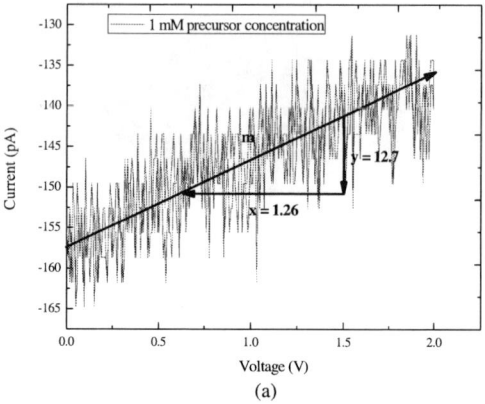

(a)

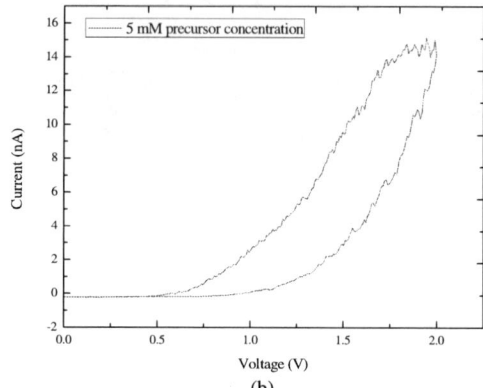

(b)

Fig. 3. C-AFM *I-V* characteristics curve of CuO films at different precursor concentration: (a) 1 mM; and (b) 5 mM

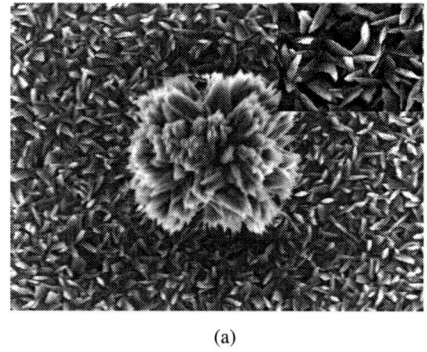

(a)

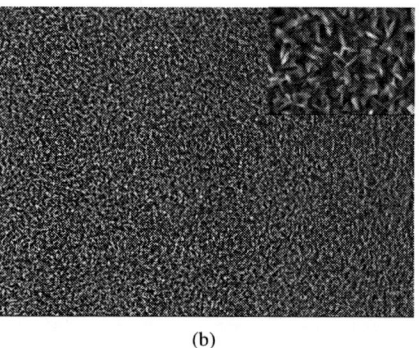

(b)

Fig. 4. FESEM micrographs of CuO films: (a) 1 mM precursor concentration at 10 K magnification; (b) 5 mM precursor concentration at 10 K magnification; (inset is the magnification at 100 K)

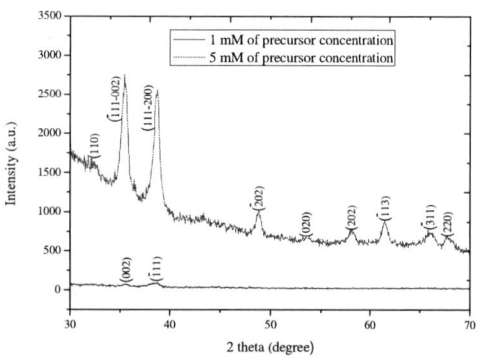

Fig. 5. The XRD pattern C-AFM *I-V* characteristics curve of CuO films at different precursor concentration: (a) 1 mM; and (b) 5 mM

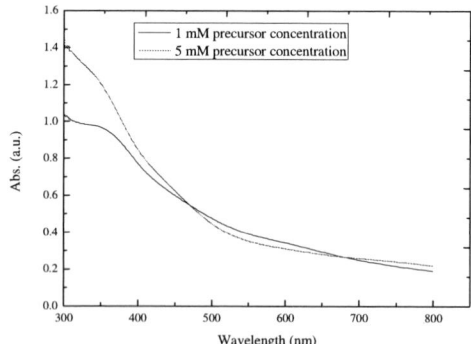

Fig. 6. UV-visible spectra of CuO thin films deposited at different precursor concentrations

Fig. 4(a-b) presents typical FESEM of CuO nanostructures thin films deposited at 1 mM and 5 mM copper(II) nitrate concentrations. The entire image indicates that the films structures were defined as a triangular shaped petals (nanosheet) structures and the surface morphology of the films was almost uniform and homogenous especially for CuO films deposited at 5 mM precursor concentration. However, there were few flower-like shaped nanostructures exhibits from the FESEM image which is shown in Fig. 4(a). The clear view of flower-shaped structures reveals that the flowers are consists of many triangular shaped petals. In addition, this structures can be considered as a monoclinic crystal structure as was reported by others [13]. The width of the nanosheet structures for both thin films (1 mM and 5 mM) are in the range of 150 nm - 240 nm and 95 nm - 100 nm, respectively. This happens because the CuO films growth rate in Fig. 4(a) is faster than the films in Fig. 4(b). When the growth rate is faster, it can produce a highly packed and dense films compared to the slower growth rate. The slower growth rate will yielded a huge, thick and porosity surface structures. It can be concluded that the increasing of copper(II) nitrate affects the grain size of the CuO films.

Fig. 5 shows the typical XRD patterns of the CuO thin films that were prepared at different concentrations of precursor which is 1 mM and 5 mM. Each XRD pattern (Fig. 5) of the products obtained are identical to the single-phase of CuO films with a monoclinic structure and the diffraction data were in good agreement with JCPDS card of CuO (JCPDS 05-0661). The major peak for XRD pattern of CuO films (1 mM) is located at $2\theta = 35.45°$ and $35.55°$ while at $2\theta = 35.45°$ indicates the major peaks for 5 mM films. As we can see, when the precursor concentration increased, the intensity also was increases and the films become highly conductive. Additionally, there is no side products such as $Cu(OH)_2$ and Cu_2O are detected from both patterns.

The optical absorption experiment of CuO films obtained by 1 mM and 5 mM concentration of precursor was further carried out by a UV-Vis spectrophotometer. As shown in Fig. 6, the highest optical absorption peak is at 5 mM precursor concentration of CuO films. However, there is a broad absorption band from 300 nm to 340 nm were seen for 1 mM precursor concentration of CuO films while the absorption band of 5 mM CuO deposited films was obviously not appeared. We observed that the broad peak centred for both absorption spectrums of CuO films was in the range of 300 nm to 550 nm.

IV. CONCLUSION

The CuO thin films have been successfully synthesized by a simple solution method via an aqueous solution mixture of $Cu(NO_3)_2$ and HMT. The effects of $Cu(NO_3)_2$ concentration on topographical, electrical, morphological, structural and optical properties of CuO thin films were investigated. The width of the nanosheet grains was found to be ~100 nm at 5 mM prepared film. The densely packed and regular shaped grains are observed in topographical image of 5 mM prepared film. The formation of porosity and bigger structure from high growth rate will obtained poor I-V characteristics as shown in Fig. 3(a). Furthermore, CuO is sensitive to light energy and naturally absorbs light energy higher than its band gap which is around 1.4 eV. Due to the composition of the sample 1 mM is much smaller than 5 mM, it is expected that in this results the absorption for film 1 mM is lower than films at 5 mM. The

structural of both CuO deposited films was showed that the films were formed as the single-phase CuO films with monoclinic structures. In addition, the surface of the films become smoother and denser when the molar concentration increased. Overall results indicate that the CuO films with increasing precursor concentration exhibits a thin and smooth films.

Acknowledgment

A special thanks to Universiti Tun Hussein Onn Malaysia and Ministry of Education for the financial support through Fundamental Research Grant Scheme Vote 1059 and 1210.

References

[1] G. Cheng, "Synthesis and characterisation of CuO nanorods via a hydrothermal method," *Micro & Nano Letters*, vol. 6, pp. 774-776, 2011.

[2] V. Dhanasekaran, *et al.*, "Electrochemical deposition and characterization of cupric oxide thin films," *Thin Solid Films*, vol. 520, pp. 6608-6613, 2012.

[3] V. Dhanasekaran, *et al.*, "SEM and AFM studies of dip-coated CuO nanofilms," *Microscopy research and technique*, vol. 76, pp. 58-65, 2013.

[4] S. Ghosh, *et al.*, "Deposition of thin films of different oxides of copper by RF reactive sputtering and their characterization," *Vacuum*, vol. 57, pp. 377-385, 2000.

[5] M. Singhai, *et al.*, "Synthesis of ZnO nanoparticles for varistor application using Zn-substituted aerosol OT microemulsion," *Materials Research Bulletin*, vol. 32, pp. 239-247, 1997.

[6] K. J. Choi and H. W. Jang, "One-dimensional oxide nanostructures as gas-sensing materials: review and issues," *Sensors*, vol. 10, pp. 4083-4099, 2010.

[7] J. Liu, *et al.*, "Tailoring CuO nanostructures for enhanced photocatalytic property," *Journal of colloid and interface science*, vol. 384, pp. 1-9, 2012.

[8] L. Chen, *et al.*, "Electrochemical performance of polycrystalline CuO nanowires as anode material for Li ion batteries," *Electrochimica Acta*, vol. 54, pp. 4198-4201, 2009.

[9] R. V. Kumar, *et al.*, "Sonochemical synthesis and characterization of nanometer-size transition metal oxides from metal acetates," *Chemistry of Materials*, vol. 12, pp. 2301-2305, 2000.

[10] G. L. Luque, *et al.*, "Glucose biosensors based on the immobilization of copper oxide and glucose oxidase within a carbon paste matrix," *Talanta*, vol. 66, pp. 467-471, 2005.

[11] S. Anandan, *et al.*, "Room temperature growth of CuO nanorod arrays on copper and their application as a cathode in dye-sensitized solar cells," *Materials Chemistry and Physics*, vol. 93, pp. 35-40, 2005.

[12] X. Zhang, *et al.*, "High-power and high-energy-density flexible pseudocapacitor electrodes made from porous CuO nanobelts and single-walled carbon nanotubes," *ACS nano*, vol. 5, pp. 2013-2019, 2011.

[13] M. Muhibbullah and M. Ichimura, "Fabrication of copper oxide thin films by the drop chemical deposition technique," *Materials Research Bulletin*, vol. 47, pp. 1968-1972, 2012.

[14] L.-T. Lee, *et al.*, "Current mode atomic force microscopy (C-AFM) study for local electrical characterization of conjugated polymer blends," *Ambio*, vol. 41, pp. 135-137, 2012.

Shear Horizontal Surface Acoustic Wave COMSOL Modeling on Lithium Niobate Piezoelectric Substrate

S. T. Ten, U. Hashim, W.W. Liu, K.L.Foo, C.H. Voon,
H.Hisham and T. Nazwa
Institute of Nanoelectronic Engineering (INEE)
Universiti Malaysia Perlis (UniMAP)
01000 Kangar Perlis.

S. T. Ten, A. Sudin, M.S. Nur Humaira
Mechanization and Automation Centre,
MARDI, P.O.Box 12301 GPO, 50774 Kuala Lumpur.
stten@mardi.gov.my

Abstract— *Shear horizontal surface acoustic wave (SHSAW), one type of the Surface Acoustic Wave is most suitable for the liquid based application as SHSAW has the advantage of acoustic energy is not being radiated into liquid. Hence, SHSAW device has the potential to provide high-performance sensing platform in pathogen sensing research. Since 1960, there have been a lot of complicated theoretical models for the SAW devices development, such as from delta function model, equivalent circuit model, coupling-of-modes (COM) model, P-matrix model and Computer Simulation Technology Studio Suite (CST), finite element analysis (FEA) model. Unfortunately, these models have been developed more toward telecommunication application such as signal filters and resonators. However, SHSAW device development in this research is not meant for signal filter or resonators but used for surface sensing purpose through mass loading effects detection, therefore COMSOL Multiphysics was used to modeling the SHSAW for pathogen sensing development. Three resonant frequencies of 15 μm, 20 μm and 25 μm width electrodes were simulated in this paper as preliminary evaluation of the applicable of COMSOL Multiphysics in SHSAW research. The comparison has shown that the difference between theoretical calculation and simulation result is less than 10%.*

Keywords— *Acoustic Waves Based Sensors; Biosensors; E.coli; COMSOL; FEA model.*

I. INTRODUCTION

White and Voltmer [1] were the first to design and develop surface acoustic wave (SAW) based devices by adding the thin film interdigital transducer (IDT) on the surface of piezoelectric substrate. Since that, SAW based devices were used for telecommunication application and are key components in RF and IF stages of many electronic systems. SAW devices can generate and detect acoustic waves in the range of 100MHz to a few GHz using interdigital transducers (IDTs) on the surface of a piezoelectric crystal. In this process, electromechanical conversion is occurring, to generate mechanical wave propagation by transforming electrical input to mechanical wave and for detection, by transforming mechanical wave back to electrical signal. In this way, the acoustic energy is strongly confined at the surface of the device in the range of the acoustic wavelength, regardless of the thickness of the complete substrate [2]. From review paper by Ten [3], SAW sensors are very sensitive to changes in mass, density, viscosity on the guiding layer. The waves are potentially very sensitive towards any changes on the surface especially in the middle sensing area where is generally coated with a chemically and biologically selective layer for adsorption of analyte species. The binding binding of the targeted specimen in analyte with sensitive sensing area in SAW will cause as a mass loading, viscosity and conductivity changes. This research has been directed to detect patogen by using this concept and sensitivity of the device is the critical issue. Modelings is certainly a great help in speeding up the process of device parameters optimization. Many theoretical models have been developed to comprehend the underlying mechanism of the SAW propagations since SAW was discovered.

A number of analytical and numerical studies such as delta function model [4], equivalent network model [5], Green's function model [6] P-matrix model, coupling of mode method (COM) [7] and Computer Simulation Technology Studio Suite (CST) have been developed to analyze SAW based device. However, these models are more towards to SAW filter and resonator design as assumption and simplification is made for those models, thus not actually design for full-scale behavior prediction of SAW devices, unable to be used to simulate the mass loading effect as for SAW based sensor application. On the other hand, finite element analysis is undertaking full scale analysis of SAW devices [8-9]. For instant, Atashbar [10] used finite element simulation in hydrogen SAW gas sensor development. Thus, this paper presents the process of using the finite element analysis (FEA) modeling, COMSOL Multiphysics to obtain the resonant frequencies of three different IDT sizes on 64^0YX Lithium Niobate (LiNbO$_3$). Comparison was made to verify the applicable of COMSOL Multiphysics in modeling the SHSAW in the research.

II. THEORY

Surface acoustic waves can be generated on the free surface of an elastic solid (piezoelectric substrate). One the input side, the waves are generated from converting signal voltage variations into mechanical acoustic waves and the other side output is converting mechanical SAW vibrations back into output voltages. The propagation of the acoustic wave in a piezoelectric material is governed by the following coupled

---

Research is sponsored by Nano Fund from MOSTI headed by MARDI and INEE in UniMAP.

978-1-4799-5761-3/14 $31.00 © 2014 IEEE

electromechanical constitutive equations or known as piezoelectric constitutive relations [11] in stress-charge form:

$$[T] = [c][S] - [e^t][E] \tag{1}$$

$$[D] = [e][S] + [\varepsilon][E] \tag{2}$$

Where T = vector stress, c = vector elastic stiffness coefficient, S = vector strain, e = vector piezoelectric constant coefficients, E = vector electric field intensity, D = vector electrical displacement density, ε = vector permittivity.

The fundamental frequency of the SAW resonator mainly depends on the pitch of the IDTs, which is chosen to be equal to the SAW wavelength λ, and the sound speed of the piezoelectric layer v:

$$f_0 = v/\lambda \tag{3}$$

According to Zhang [12], when wave propagates in a piezoelectric material, material density, particle displacement, acceleration and electric potential are related to each other. COMSOL Multiphysics simulation has the ability in including all the consideration.

III. METHODOLOGY

A two port delay line SHSAW sensor (Fig. 1) was used in this study. Piezoelectric materials are anisotropic material, thus different orientation of a cut crystal or Euler angle will affect the parameters changing in piezoelectric constitutive relations and consequent different type of wave propagation produced. In order to obtain SHSAW, 64^0YX LiNbO$_3$ (with the Euler angle (180,26,180) is used as piezoelectric substrate. The material property of different cut crystal can be obtained by crystal rotation calculation matrices. The modified set of 64^0YX LiNbO$_3$ material properties including the c, e and ε matrices were obtained through the Crystal Rotation Calculator at zephrasoft.com (shown in Table 1). The thickness and width of substrate are 100µm and 122 µm are used for the three different electrode width simulations. The length of the device is depended on the number of pairs and width of the electrode and also the distance of the delay line. Therefore the length of the devices are different; for 15µm width electrode is 934 µm, for 20µm width electrode is 1244 and for 25µm width electrode is 1554 µm. The design parameters of the SHSAW device are electrode width, number of finger pairs, delay line and aperture (Fig.2 and Table 2). No damping effect was added in these studies.

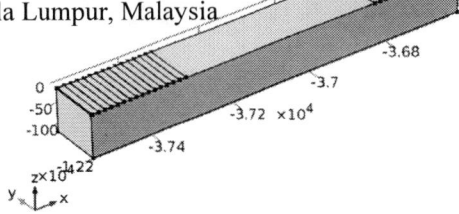

Fig.1. Schematic of a two-port delay line SAW sensor.

TABLE 1. DEFAULT COMSOL STRESS-CHARGE FORM 64^0YX LiNbO$_3$ MATERIAL PROPERTY

Description	Value
Elasticity matrix (Ordering: xx, yy, zz, yz, xz, xy), x10^{10} c [Pa]	{{20.30, 5.014, 7.786, -0.3127, 0, 0}, {5.014, 21.35, 7.964, -0.2921, 0, 0}, {7.786, 7.964, 22.52,-1.917, 0, 0}, {-0.3127, -0.2921, -1.917, 6.464, 0, 0}, {0, 0, 0, 0, 6.997, 1.145},{0, 0, 0, 0, 1.145, 6.503}}
Coupling matrix, e [Cm^{-2}]	{{0, 0, 0, 0, 2.230, -3.869}, {-2.335, -0.9857, 2.575, 3.123, 0, 0}, {-0.9162, -0.02306, 2.467, 1.041, 0, 0}}
Relative permittivity, ε	{{44.00, 0, 0}, {0, 41.12, 5.910}, {0, 5.910, 31.88}}

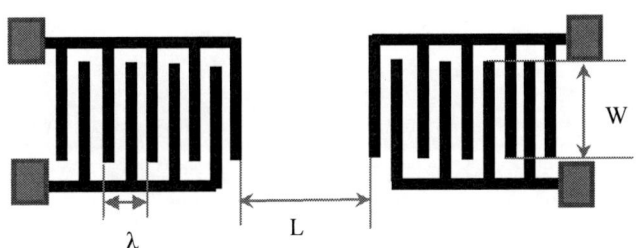

Fig.2. Design parameters of the SHSAW device

TABLE 2. IDT DESIGN PARAMETERS FOR 3 DIFFERENT ELECTRODE WIDTHS

Electrode width (µm)	Wavelength, λ (µm)	No. of finger pairs	Delay line, L (µm)	Aperture, W (µm)
15	60	4	480	120
20	80	4	640	120
25	100	4	800	120

IV. RESULTS AND DISCUSSION

COMSOL Multiphysics manages to show 2 and 3 dimension plot surface total displacement simulation (Fig. 3) and the signal output at vary frequency response (Fig. 4) of a two ports delay SHSAW device. The x,y and z are shown as the global axes and the indicator bar shows displacement in micrometer (Fig.3).

978-1-4799-5761-3/14 $31.00 © 2014 IEEE

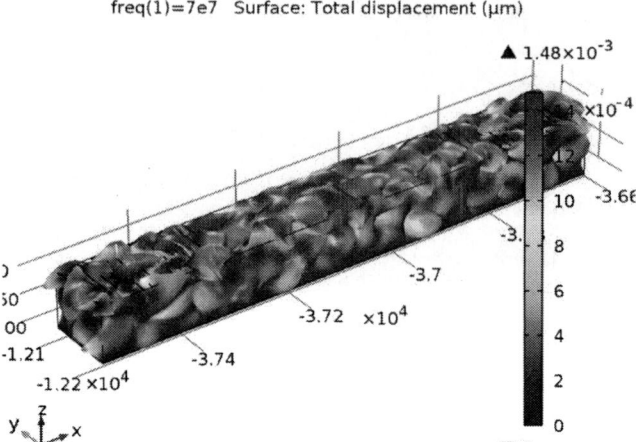

freq(1)=7e7 Surface: Total displacement (μm)

Fig.3. COMSOL simulation 3D plot surface to displacement of a two ports delay SAW sensor at vary frequency response.

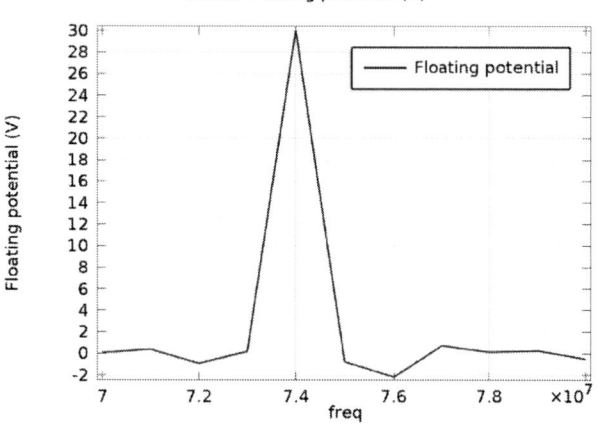

Global: Floating potential (V)

Fig.4. Frequency varying output signal obtained at receiver or output terminal.

According to Zhang[12], since the model SAW device has reciprocal and symmetric design for the two IDTs, i.e., S11=S22 (reflection coefficients) and S12=S21 (transmission coefficients), quantified the insertion loss (-20log|S12|), and from the electric potential values obtained from both transmitter and receiver IDTs, insertion loss can be obtained by the calculation:

$$\text{Insertion Loss (S21)} = 20 \times \log|V_{output} / V_{input}| \qquad (4)$$

V_{input} is the electric potential at the terminal input or input IDT sweeping through the setting frequency range and the V_{output} is the terminal output or receiver IDT where the voltage output is measured. The resonant frequencies of the different width size electrodes are indicated at the peak of the insertion loss varying frequency plotting (Fig.5)

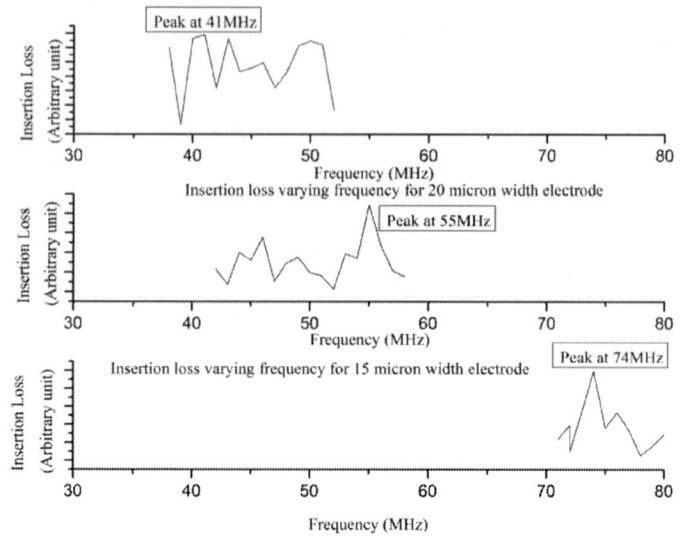

Fig.5. Frequency varying insertion loss signal of vary width size of electrodes.

In order to determine the applicable of COMSOL Multiphysics, comparison between calculated resonant frequency using (3) and simulated resonant frequency were made (Fig.6). The value of wave propagation velocity on the surface of 64^0YX LiNbO$_3$, v, is 4478m/s and the wavelength λ is four times the electrode width.

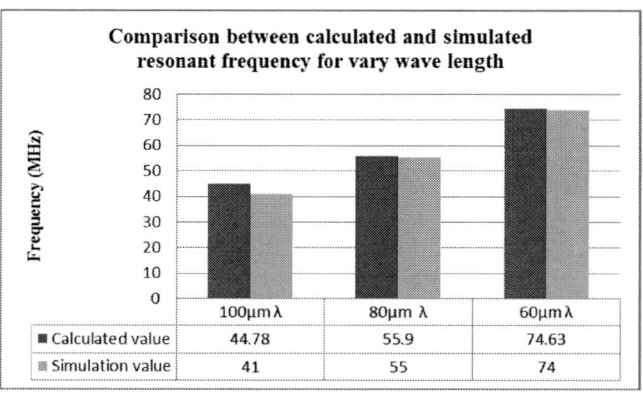

Comparison between calculated and simulated resonant frequency for vary wave length

	100μm λ	80μm λ	60μm λ
■ Calculated value	44.78	55.9	74.63
▨ Simulation value	41	55	74

Fig.6. Comparison between calculated and simulated resonant frequency for vary wavelength.

%, the largest difference was from the electrode width 25μm, and following by electrode width 20μm and the least difference is from 15μm one. The 25μm width electrode is contributed the largest mass effect as is compared to 20μm and 15μm one, moreover SHSAW device is very sensitive to mass loading effect, thus the result is showing the logical trend.

V. CONCLUSION

COMSOL Multiphysics can be used to simulate the SHSAW propagation on 64^0YX LiNbO$_3$ especially the resonant frequency from the results shown. The major constraint faced is large amount of computation physical memory is needed for high resolution data (simulate much smaller mesh element size) to achieve higher accuracy. For that, the future works are to further optimize the modeling by reduce the usage of computation physical memory and yet still can maintain the accuracy.

ACKNOWLEDGMENT

The authors are very grateful to all the supporting members in MARDI, Institute of Nano Electronic Engineering and School of Computer and Communication Engineering UniMAP.

REFERENCES

[1] R.M. White and F.W.Voltmer, "Direct piezoelectric coupling to surface eleastic waves", Appl. Phys. Lett., vol. 17, pp. 314-316, 1965.

[2] Grate, J.W. and G.C. Frye, Acoustic wave sensors. Sensors update, 2(1): p. 37-83, 1996.

[3] S.T.Ten, W.W.Liu, S. Ahmad, H.Uda, T.Nazwa, "Design and characteristic of CMOS fabricated acoustic waves based sensors for foodborne pathogen rapid detection," Proceedings in 2012 IEEE EMBS International Conference on Biomedical Engineering and Sciences ,P. 387-391, 2012.

[4] R. H. Tancrell, M. G. Holland, Acoustic surface wave filters, Proceedings of IEEE 59, 393-409, 1971.

[5] W. R. Smith, H. M. Gerals, J. H. Collins, T.M. Reeder, H. J. Shaw, Analysis of interdigitated surface wave transducers by use of equivalent circuit model, Transactions of Microwave Theory Tech. 17 856-864, 1969.

[6] D. Qiao, W. Liu, P. M. Smith, General Green's functions for SAW device analysis, IEEE transactions on Ultrasonics, Ferroelectrics and Frequency Control 46 (5), 1242-1253, 1999.

[7] E. Akcakaya, A new analysis of single phase unidirectional transducers, IEEE transactions on Ultrasonics, Ferroelectrics and Frequency Control 34, 45-52, 1987.

[8] G. Xu, Direct finite element analysis of frequency response of a Y-Z Lithium Niobate SAW filter, Smart Materials Structure 9, 973-980, 2000.

[9] G. Xu, Q. Jiang, A finite element analysis of second order effects on the frequency response of a SAW device, Journal of Intelligent Material Systems and Structures 12, 69-77, 2001.

[10] M. Z. Atashbar, B. J. Bazuin, M. Simpeh, S. Krishnamurthy, 3D FE simulation of hydrogen SAW gas sensor, Sensors and Actuators B 111-112, 213-218, 2005

[11] C.Campbell, Surface Acoustic Wave Devices for Mobile and Wireless Communications, Academic Press, 1998.

[12] Zhang, Guigen. Orientation of Piezoelectric Crystals and Acoustic Wave Propagation. Proc. Of the 2012 COMSOL Conference, Boston, 2012.

A Study on the Role of Solvent on Properties of Gd$_x$Zn$_{1-x}$O Films Synthesized by Sol-gel Method

C.A Norhidayah, S.A Kamaruddin, N.Nafarizal, M.Z
Sahdan
Microelectronic and Nanotechnology Research Center
Universiti Tun Hussein Onn Malaysia
86400 Batu Pahat, Johor
zainizno@gmail.com

A.R Nurulfadzilah, S.N.M Tawil
Department of Electrical and Electronic Engineering
Universiti Pertahanan Nasional Malaysia
57000 Kuala Lumpur, Malaysia

Abstract—The increasing demand for high-performance and low cost optoelectronic devices motivates many researchers to develop more efficient transparent conductive oxide (TCO) films. Among the popular TCOs, the past decade has seen the emergence of zinc oxide (ZnO) as one of the potential materials for the fabrication of transparent conductive electrodes. The aim of this work is to study the influence of different solvents on the properties of Gd$_x$Zn$_{1-x}$O (x $\leq$ 0.01) films synthesized by sol-gel spin coating technique. Consequently, three different solutions were prepared with different solvents [2-Methoxyethanol (2-ME), ethanol (EtOH) and isopropanol (IPA)]. The structural, surface roughness and optical properties were investigated using an X-ray diffractometer (XRD, PANanalytical xpert-pro), atomic force microscope (AFM, Park XE-100) and ultra violet-visible spectrophotometer (UV-Vis, Shimadzu UV 1800), respectively. As a result, all films were found to have polycrystalline with hexagonal wurtzite structure. In addition, AFM analysis revealed that the film synthesized using EtOH exhibiting the smallest surface roughness of about 4.12 nm compared with IPA and 2-ME of about 5.12 nm and 12.22 nm, respectively. Also, the optical transmittance spectra indicate that all films exhibit good transparency in the visible spectral range with an average transmission of approximately 97.5%, 97.2% and 96.1% for EtOH, IPA and 2-ME, respectively. The optical band gap energy (Eg) values was estimated to be around 3.26 ~ 3.30 eV using Tauc's model with the lowest value for IPA (3.26 eV) and highest for 2-ME (3.30 eV). In a nutshell, the solvents are playing a key role for controlling the growth and nucleation in the preparation of Gd$_x$Zn$_{1-x}$O solution and it strongly affects the properties of the film.

Keywords— Transparent Conductive Oxide (TCO); Zinc oxide (ZnO); Sol-gel; Spin Coating; Gadolinium(Gd)

I. INTRODUCTION

Transparent conducting oxide (TCO) films have been investigated extensively for optoelectronic devices, such as flat panel displays, solar cells and light-emitting diodes (LED) [1-3]. To develop more efficient TCO films, the films should have high transmittance in the visible region (400 to 800 nm), low resistivity, and high thermal or chemical stability [4, 5]. In general, indium tin oxide (ITO) has been widely used as TCO material because of its excellent electrical and optical properties. However, ITO has low stability, high toxicity and high cost fabrication, motivating effort to develop alternatives [4, 5].

In the past decade, zinc oxide (ZnO) has emerged as one of the promising candidates to replace ITO due to its low cost and excellent properties compared to ITO [4]. ZnO is a very interesting semiconducting material with a wide band gap of 3.37 eV and higher exciton binding energy (60 meV) at room temperature. It also has special properties such as long-term stability, relatively low material costs, simple processing due to its compatibility with wet chemical etching, biocompatibility, environmental friendliness, excellent radiation resistance and etc. In addition, ZnO is a unique material that exhibits excellent chemical, physical and optical properties that are useful for the advancement of technology in the field of optoelectronics, optics, spintronics, biomedical and solar cell applications [6, 7]. Accordingly, low resistivity and highly transparent of ZnO films are required for optoelectronic devices and other application. However, pure ZnO films have high resistivity due to its stoichiometric nature which could not produce large number of free charge carriers. In order to obtain suitable conductive ZnO films, it can be doped with a variety of elements. Doping ZnO with rare-earth or 4f transition elements is a popular technique for manipulating the optical properties of ZnO system. Typically, rare-earth materials are the noble candidates for optically active ions in semiconductor materials because they show a lot of absorption and fluorescence transitions in the almost visible region. These elements have additional significant features in contrast to other optically active ions including the wavelengths of their emission and absorption transitions are moderately unresponsive to host materials. Also, they have longer lifetimes of energy states, higher quantum efficiency and good color rendering index [8, 9], which lead to their excellent performance in optoelectronic devices.

Several deposition techniques such as chemical vapour deposition (CVD), radio-frequency sputtering (rf sputtering), spray pyrolysis and sol–gel method can be used to obtain doped and undoped ZnO films [10-12]. Among these techniques, the sol–gel method reflects a number of advantageous. It is a wet chemical process, which offers a cost effective and simple to coating. Moreover, it can perform well in atmospheric pressure without the need for expensive vacuum equipment and is capable to produce large area coating [13, 14]. However, in this technique there are some parameters that control the crystal growth [12]. One of these parameters is the type of solvent. The kinetics of nucleation and growth is expected to be strongly dependent on the properties of the solvent. In fact, during the growth of colloidal ZnO nanoparticles, the solvent is not only provide a medium for the reaction, but also act as a ligands to help and control the morphology and particle size of ZnO [15].

In this study, Gd had been chosen as dopants and is expected to improve the transmittance of ZnO films. Indeed, Gd has unique physical properties that are useful in improving the optical properties of the films. Gd-doped ZnO ($Gd_xZn_{1-x}O$) films were synthesized by a low cost sol–gel spin coating technique using different solvents. The effects of different solvents on the material properties were studied in details.

II. EXPERIMENTAL PROCEDURE

A. Preparation of Solution

$Gd_xZn_{1-x}O$ films were deposited on microscope glass substrates by sol-gel spin coating technique. As a starting material, zinc acetate dehydrate [Zn $(CH_3COO)_2 2HO$, Merck], gadolinium acetate dehydrate [$Gd(C_2H_3O_2)_3.xH_2O$ Sigma-Aldrich] and monoethanolamine [MEA, C_2H_7NO, Merck] were used as a precursor, dopant and stabilizer, respectively. In this work, three different solutions with different solvents namely 2-Methoxyethanol [2-ME, Sigma-Aldrich], ethanol [EtOH, J.T. Baker], and isopropanol [IPA, J.T Baker] were prepared. In a typical process, 0.4 M of zinc acetate dehydrate and 1 at. % of Gd was mixed in 30 ml of different solvent. The mixture solution was then vigorously stirred with a magnetic stirrer at 60 °C for 30 min. then after 30 min MEA was added drop by drop in this solution with constant stirring for 2h at 60 °C until it become clear and homogeneous. All solutions prepared were aged at room temperature for 24 h before deposition process.

B. Deposition of $Gd_xZn_{1-x}O$ films

Before deposition process, the glass substrates were cleaned in an ultrasonic bath for 10 min using acetone (CH_3COCH_3), ethanol, and deionized water, sequentially. Then, each of the solution was spin-coated using 3-step program (1000 r.p.m (5s), 3000 r.p.m (30s), 1000 r.p.m (5s)) by putting few drops on the well-cleaned glass substrates. Thereafter, the samples were preheated at 300 °C in hot plate for 5 min to evaporate the solvent. This process was repeated 10 times to obtain requires dense film of appropriate thickness. Finally, all samples were annealed in annealing furnace at 500°C for 1 hour to remove organic residue.

C. Characterization Technique

The crystallinity of the films were characterized using an X-ray diffractometer (XRD, PANanalytical xpert-pro) which was operated with CuKα (λ = 1.5406 A°) radiation. The surface topologies of the films were investigated using atomic force microscope (AFM, Park system, XE-100). Meanwhile, the thicknesses of the films were measured using a KL-Tenko surface profiler and the films were optically characterized using an ultra violet-visible spectrophotometer (UV-Vis, Shimadzu UV 1800).

III. RESULTS AND ANALYSIS

A. Structural Analysis

Fig. 1 (a-c) presents the XRD spectra of $Gd_xZn_{1-x}O$ films deposited on glass substrate using different solvent namely EtOH, IPA and 2-ME, respectively. These spectra indicate that the films have polycrystalline nature (JCPDS card file no. 01-079-2205) and a preferred orientation with the c-axis perpendicular to the substrates. The dominant diffraction peak for all films was (0 0 2). In addition to the (0 0 2) peak, other peaks such as (1 0 0), (1 0 1), (1 0 2) and (1 1 0) with comparatively lower intensities were also observed for all films. Also, no new peaks associated with the GdO or Gd clusters detected, indicating that the incorporation of gadolinium did not affect the wurtzite phase of ZnO. These observations indicate that the films do not have any phase segregation or secondary phase formation.

Based on Fig. 1(a-c), all films exhibit the highest intensity of XRD diffraction peak on (0 0 2) plane, therefore the calculation of the crystallite size was highlighted at (0 0 2) plane. The crystallite size was calculated using XRD data in accordance with Debye-Scherer's formula [12].

$$D = \frac{0.9\lambda}{\beta \cos\theta} \quad (1)$$

Where λ, θ and β is the X-ray wavelength (0.154 nm), Bragg's diffraction angle in degrees and full width at half maximum (FWHM) of the peak corresponding to the θ value in radians, respectively. The average crystallite sizes of the films are listed in Table 1. According to the results, the crystallite size of the $Gd_xZn_{1-x}O$ films ranges from 6.96 to 13.93 nm. The result shows that the films prepared using 2-ME produce the smallest crystallite size (6.96 nm). However, the films prepared using EtOH produce the largest crystallite size (13.93 nm). The largest crystallite size displayed by the film prepared using EtOH solvent can be attributed to higher viscosity of EtOH compared with other solvents. The results are in accordance with the AFM findings. Since the crystallite size is inversely proportional to the FWHM, we found that the crystallite size increased at as the FWHM decreased. The increase of the crystallite size has been indicated by the enhancement of the crystallinity and c-axis orientation of the films [13].

The dislocation density (δ), defined as the length of dislocation lines per unit volume of crystal structural. Its represents the amount of defects in the crystal which can be estimated from the following equation:

$$\delta = \frac{1}{D^2} \quad (2)$$

As observed from table 1, the film synthesized by EtOH has good crystallization because of their small dislocation density values which represent the amount of defects in the film. Besides, a larger dislocation density implies a higher hardness [16]. It is also known that above a certain crystallite size limit (~ 20 nm) the strength of materials increases with decreasing crystallite size.

B. Surface Topologies Analysis

Fig. 2(a-c) shows the three-dimensional (3-D) surface topology images of the samples. A scanning probe microscope was used to observe the surface topologies of the films in a scan size of 1 μm × 1 μm by non-contact mode. Based on Fig. 2(a) we can see that the surface is quite smooth with fine structure, and more uniform grains. However, for Fig. 2(b) and (c), they have non-uniform grains and relatively rough surface. From this, it was concluded that the films have a uniform grain distribution gives a smooth surface topology. Other than that, the wide variation in grain size and surface topology revealed the importance of solvent in the synthesis of this film. This may be due to different volatile solvent, which is more volatile solvent evaporates quickly during the spin coating process (called solvent quenching effect) leaves grains are randomly oriented on the surface. The root-mean-square average (RMS) roughness, as determined from the AFM measurements revealed that the lowest surface roughness (4.12 nm) has been observed for the most volatile solvent EtOH and the highest surface roughness (12.22 nm) for the less volatile 2-ME solvent.

C. Optical Analysis Analysis

Fig. 3 represents the optical transmittance spectra of $Gd_xZn_{1-x}O$ films deposited using different solvent recorded at room temperature in the wavelength region of 300 – 800 nm. As can be seen, the optical transmittance spectra indicate that all films exhibit good transparency in the visible spectral range with an average transmission of approximately 97.5%, 97.2% and 96.1% for EtOH, IPA and 2-ME, respectively. The highest transmittance obtained in EtOH may be due to the structural homogeneity and crystallinity as evidenced from Fig. 1. In addition, the steepness of the absorption edge is an identification of the good crystallinity of the films. By contrast, the EtOH-derived films exhibit sinusoidal behavior this indicates that the top film surface is smooth and uniform as evidenced from AFM image. However, a slight shift of transmission curves to lower wavelengths is observed for curves of 2-ME-derived films in comparison with EtOH and IPA films. This shift is ascribed to the increase in band gap energy.

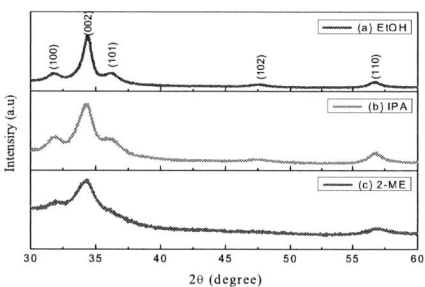

Fig. 1. XRD spectra of $Gd_xZn_{1-x}O$ films deposited using different solvents; (a) EtOH, (b) IPA and (c) 2-ME, respectively.

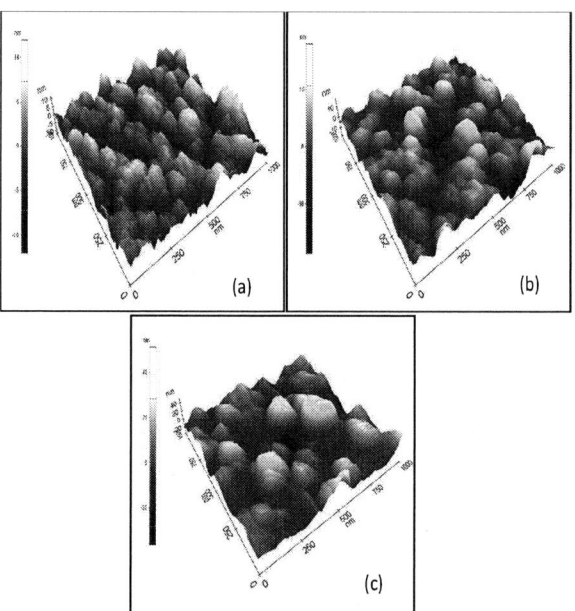

Fig. 2. AFM topologies image for $Gd_xZn_{1-x}O$ films deposited using different solvents; (a) EtOH, (b) IPA and (c) 2-ME, respectively.

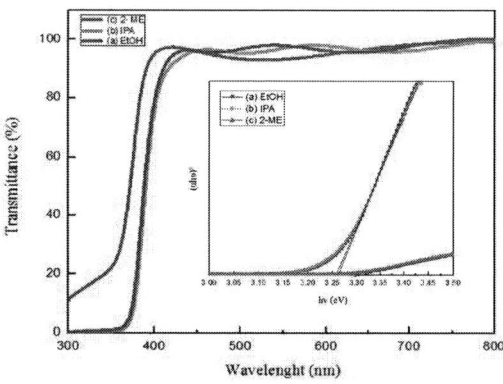

Fig.3. Transmittance spectra of $Gd_xZn_{1-x}O$ films deposited using different solvents; (a) EtOH, (b) IPA and (c) 2-ME, respectively. The insets show the plot of $(\alpha h v)^2$ vs. photon energy (hv) of $Gd_xZn_{1-x}O$ films.

TABLE I. EVALUATED STRUCTURAL PARAMETERS AND OPTICAL BAND GAP VALUE OF ALL FILMS.

Solvent	Plane	FHWM $(\beta)^o$	2θ	D (nm)	δ x10^{-3} (nm)$^{-2}$	Eg (eV)
EtOH	002	0.5904	34.390	13.93	5.153	2.27
IPA	002	0.8266	34.304	9.94	10.121	2.26
2-ME	002	1.1808	34.264	6.96	20.643	3.30

The absorption coefficient can be calculated based on the thickness of the films and the transmittance data using the Lambert's formula [11]:

$$\alpha = \frac{1}{t} \ln \frac{1}{T} \qquad (3)$$

Where t and T is the thickness of the films and transmittance data, respectively. The optical band gap (E_g) can be determined from the experimental spectra of the absorption coefficient (α) as a function of the photon energy ($h\nu$) using the equation given by caglar *et al.* [17]:

$$(\alpha h\nu)^2 = A(h\nu - Eg) \qquad (4)$$

Where h, ν and A is the Planck constant, frequency of the incident photon and energy-independent constant between 10^7 and 10^8 m^{-1}, respectively. Accordingly, the optical band-gap can be obtained by extrapolating the corresponding straight lines downwards to the photon energy axis in the Tauc's plot as shown in the inset of Fig. 3. As can be seen, the optical value for all films lies in the range of 3.26 to 3.30 eV with the lowest value for IPA (3.26 eV) and the highest value for 2-ME (3.30 eV). The variability in the band gap values in different solvents can be attributed to the structure and topologies of the film surface, which would change the inter-atomic bond of the films [12]. Thus, the use of different solvent affects the structure of the film network which causes the change in the optical properties of the films.

IV. CONCLUSION

In conclusion, Gd$_x$Zn$_{1-x}$O films have successfully deposited onto microscopes glass substrate by sol-gel spin coating technique. Consequently, we have studied the effect of three different solvents namely ethanol (EtOH), isopropanol (IPA) and 2-Methoxyethanol (2-ME), respectively on the properties of Gd$_x$Zn$_{1-x}$O (x≤0.01) films. The XRD result revealed that the formation of a hexagonal wurtzite structure of ZnO without a trace peaks for the cubic phase of GdO. The crystallite sizes of 6.96~13.93 nm were calculated from the XRD data. Meanwhile, the AFM studies show that the films using EtOH has a quite smooth with fine structure, and more uniform grains. However, for IPA and 2-ME, they have non-uniform grains and relatively rough surface. Also, the highest value of optical transparency was found for the sample using EtOH solvent and the lowest of optical transparency has been observed for the sample using 2-ME solvent. Based on Tauc's plot analysis, the optical band gap is found to change with the use of different solvents. The results show that the use of different solvent in the preparation of the solution affects the structure of Gd$_x$Zn$_{1-x}$O film which causes a change in the properties of the films.

ACKNOWLEDGMENT

A special thanks to Ministry of Higher Education Malaysia for the financial support through FGRS grant vote No. 0823 from Universiti Tun Hussein Onn Malaysia (UTHM) and FRGS/1/2013/SG06/UPNM/02/1 from Universiti Pertahanan Nasional Malaysia (UPNM)

REFERENCES

[1] Burrows, Paul E., Gordon L. Graff, Mark E. Gross, Peter M. Martin, Ming-Kun Shi, M. Hall, E. Mast, C. Bonham, W. Bennett, and Michael B. Sullivan, "Ultra barrier flexible substrates for flat panel displays," Displays 22, no. 2, 2001, pp 65-69.

[2] Koida, Takashi, Hitoshi Sai, Hajime Shibata, and Michio Kondo. "Trend of transparent conductive oxides for solar cells," In Active-Matrix Flatpanel Displays and Devices (AM-FPD), 2012 19th International Workshop on, IEEE, 2012, pp. 45-48.

[3] Puzzo, Daniel P., Michael G. Helander, Paul G. O'Brien, Zhibin Wang, Navid Soheilnia, Nazir Kherani, Zhenghong Lu, and Geoffrey A. Ozin. "Organic light-emitting diode microcavities from transparent conducting metal oxide photonic crystals," Nano letters 11, no. 4, 2011, pp. 1457-1462.

[4] Chen, Hanhong, Aurelien Du Pasquier, Gaurav Saraf, Jian Zhong, and Yicheng Lu. "Dye-sensitized solar cells using ZnO nanotips and Ga-doped ZnO films," Semiconductor Science and Technology 23, no. 4 2008, 045004.

[5] Santra, Sampurna, Pratik Ghosh, and Arindam Biswas. "Investigation of Low Loss Heavily Doped Plasmonic Semiconductors: Looking Beyond Lossy Conventional Noble Metals." International Journal of Emerging Technology and Advanced Engineering, Volume 4, Issue 4, 2014

[6] Sharma, Parmanand, Amita Gupta, Frank J. Owens, Akhisha Inoue, and K. Venkat Rao. "Room temperature spintronic material—Mn-doped ZnO revisited," Journal of magnetism and magnetic materials 282, 2004, pp. 115-121.

[7] Z. L. Wang, "Novel nanostructures of ZnO for nanoscale photonics, optoelectronics, piezoelectricity, and sensing." Applied Physics A 88, no. 1, 2007, pp. 7-15.

[8] Thirumalai, J., R. Chandramohan, R. Divakar, E. Mohandas, M. Sekar, and P. Parameswaran. "Eu3+ doped gadolinium oxysulfide (Gd2O2S) nanostructures—synthesis and optical and electronic properties." Nanotechnology 19, no. 39, 2008, pp. 395703.

[9] Krishnan, Rajagopalan, Jagannathan Thirumalai, Ithrees Basha Shameem Banu, and Anthuvan John Peter. "Rugby-ball-shaped Na0. 5La0. 5MoO4: Eu3+ 3D architectures: synthesis, characterization, and their luminescence behavior," Journal of Nanostructure in Chemistry 3, no. 1, 2013, pp. 1-5.

[10] Akgul, Guvenc, Funda Aksoy Akgul, Klaus Attenkofer, and Markus Winterer. "Structural properties of zinc oxide and titanium dioxide nanoparticles prepared by chemical vapor synthesis," Journal of Alloys and Compounds 554, 2013, pp.177-181.

[11] Jun, Min-Chul, and Jung-Hyuk Koh. "Effects of NIR annealing on the characteristics of al-doped ZnO thin films prepared by RF sputtering," Nanoscale research letters 7, no. 1, 2012, pp. 1-7.

[12] Foo, K. L., M. Kashif, U. Hashim, and Wei-Wen Liu. "Effect of different solvents on the structural and optical properties of zinc oxide thin films for optoelectronic applications," Ceramics International 40, no. 1, 2014, pp. 753-761.

[13] D. Bao, H. Gu, and A. Kuang, "Sol-gel-derived c-axis oriented ZnO thin films, Thin Solid Films", 312, 1998. no. 1-2, pp. 37–39,

[14] Z. Liu, Z. Jin, W. Li, and J. Qiu, "Preparation of ZnO porous thin films by sol-gel method using PEG template", Materials Letters, vol. 59, 2005, no. 28, pp. 3620–3625,

[15] Znaidi, Lamia. "Sol–gel-deposited ZnO thin films: a review." Materials Science and Engineering: B 174, no. 1 (2010): 18-30.

[16] John, Rita, and Rajaram Rajakumari. "Synthesis and Characterization of Rare Earth Ion Doped Nano ZnO." Nano-Micro Letters 4, no. 2 2012.

[17] Caglar, Mujdat, Saliha Ilican, and Yasemin Caglar. "Influence of dopant concentration on the optical properties of ZnO: In films by sol–gel method." Thin Solid Films 517, no. 17, 2009, pp. 5023-5028.

The Impact of Channel Doping in Junctionless Field Effect Transistor

B.S Lim, M.K. Md Arshad, Noraini Othman, M.F.M. Fathil, M.F. Fatin and U. Hashim

Institute of Nano Electronic Engineering (INEE),
Universiti Malaysia Perlis (UniMAP)
01000 Kangar, Perlis, Malaysia.
mohd.khairuddin@unimap.edu.my

Abstract—**In this paper, we present the simple approach in study the impact of channel doping on the operation of the junctionless transistor transistor in 25 nm gate lengths through 2D-TCAD Sentaurus simulation tools. We increase the channel doping up to the level of doping source and drain, thus creating the junctionless phenomena between source and drain. The transistor parameters such as threshold voltage, transconductance, subthreshold slope, drain-induced barrier lowering are extracted. The impacts of low and high drain voltages are also considered. The higher the doping concentration the larger drain current can be produced, however the drawback is larger subthreshold slope is also obtained due to wider channel preventing fully-depletion.**

Keywords— Junctionless transistor; Doping concentration level; Drain voltage

I. INTRODUCTION

In general, transistors are the fundamental parts of common electronic devices that exist in current electronic device while it contains semiconductor junctions. The most general type of junction is the p–n junction, which is making up by using the contact between p-type silicon that doped with impurities to generate an excess of holes while n-type silicon doped to create an excess of electrons. Since the invention of integrated circuit 60 years ago, the number of transistors on a single silicon microchip has been greatly increasing exponentially since 1970s, and it is increasing from a few hundred to reach several billion today. As a result, transistors are becoming smaller in size and it causes the increasingly difficult to produce a high quality junctions. Apart from that, it is very hard to change the level of doping concentration of a material over length that is shorter than 10 nm. Moving forward, continuous device miniaturization, such as decreasing the physical channel length and gate dielectric thickness, is becoming more difficult in semiconductor technology. Furthermore because of difficulties of doping profile to control short-channel effects of bulk silicon channel, novel non–planar structures are being look for [1].

As we thorough in the problem, nanometer regime (particular channel length about 10 nm), transistors becoming so tiny which will increasingly difficult to generate high-quality junctions especially for case very shallow junction.

Furthermore, if the junction create is too depth, it may causes the electric field between source and drain overlap or touch one to another, resulting tremendous current leakage. Current leak from junctions, upping power consumption and will cause overall inefficiency in the devices. However, creating a high quality of junctions is a major factor that will raise the costs in the chip manufacturing process. This can be shown as the gateways for current, they are the most important part in transistors, and manufacturing high quality junctions can be expensive and hardly to be controlled since it requires many procedures of lithography process.

Therefore, a new approach has been recommended recently [2] which it not need to form the junction and it may makes the process of manufacture become much simpler, even at small dimensions. This new kind of technique known as junctionless FET, which is made using uniformly and heavily doped silicon nanowires, and long channel devices can perform a better functionality compare to conventional transistors [3]. Figure 1 shows the longitudinal cross section of a junctionless FET.

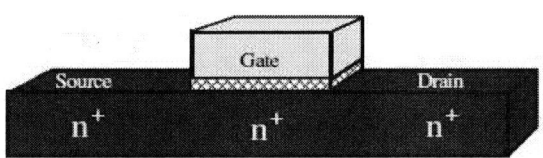

Figure 1 Longitudinal cross section of junctionless FET[3].

This type of transistor can be summarizing as no junctions and no doping concentration gradients, The doping concentration in the channel is same to that in the source and drain. Since the gradient of the doping concentration between source and channel or drain and channel become zero, so no diffusion will be carry on, which remove the high cost for ultrafast annealing techniques and allows the device to fabricate with shorter channels. Thus, many advantages can be obtain from this type of transistor such as near-ideal sub threshold slope, extremely low leakage currents, and less degradation of mobility with gate voltage and temperature than classical transistors[4].

In this study, we look into the impact of junction and junctionless transistor in 25 nm gate lengths. We increase the channel doping up to the level of doping of source and drain, thus creating the junctionless phenomena between source, channel and drain. Although we understand, the benefits of junctionless transistor requires small and narrow channel (i.e. 3D-Device simulation), but as a matter of our understanding purposes, this study is carried out by mean of 2D - TCAD Sentaurus device simulation.

II. SIMULATION PROCEDURE

Figure 2 illustrates the simulated device structure created by using structure editor in Sentaurus TCAD simulation. The simulated structure is obtained by a series command for the physical parameter which requires defining the material such as silicon thickness, buried oxide thickness, gate oxide, etc. In this simulation, the silicon, buried oxide and equivalent gate oxide thickness are 7 nm, 10 nm and 1.2 nm respectively. The drain voltage is varied into two different values i.e. 20 mV and 1 V. The gate length is fixed to 25 nm. The drain and source are doped with phosphorus with the doping concentration of $1x10^{14}cm^{-3}$ to $1x10^{18}cm^{-3}$ and bulk silicon substrate is fixed to $1x10^{14}cm^{-3}$. The doping in the channel is varied until it is identical to that in the source and drain, thus it form junctionless thus creating zero gradient of between source and channel or drain and no diffusion able to take place.

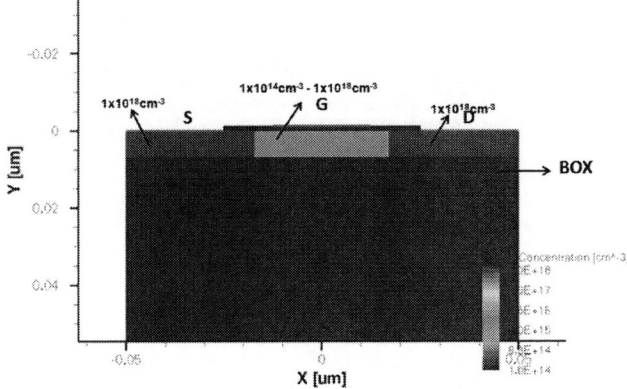

Figure 2 Device structure of the 2-D SOI junctionless

III. RESULTS AND DISCUSSION

Figure 3 illustrates the current-voltage characteristics (I_d-V_g) curve for 20 mV for several values of doping concentration i.e. $1x10^{14}$ cm^{-3} to $1x10^{18}$ cm^{-3}. One can see that, the current increases with the increase of channel doping. On the hand, the off state current (measured at $V_g = 0$ V) is also increased.

The result obtained is in correlation with general trend, i.e. the higher the doping concentration the larger drain current can be produced. Furthermore, the characteristics are

shifted towards smaller V_g values as doping concentration is increased.

Based on this result, the transistor parameters such as threshold voltage (V_{th}), transconductance (g_m), subthreshold slope (SS), and drain-induced barrier lowering (DIBL) are extracted which is presented next.

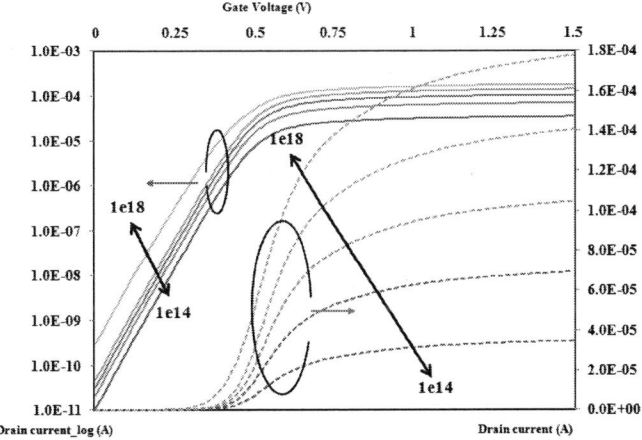

Figure 3 I_d-V_g characteristics

Figure 4 shows the transconductance (extracted from derivative of drain current, d I_d/dV$_g$) and Figure 5 shows the threshold voltage (extracted from derivative of transconductance, d g_m/dV$_g$) for several doping concentrations. While Table 1 shows the extracted parameters to estimate device performance of the devices, labeled with the relating doping concentrations. Threshold voltage is obtained with the maximum transconductance method at $V_d = 20$ mV. The threshold voltage becomes smaller as the doping concentration increased. On the other hand, the transconductance becomes slightly increase. This can be seen with I-V curves (Fig. 3) shifted on the left.

The subthreshold slope becomes larger with higher doping concentration. This may explain by the higher doping concentration, the more mobile carriers, and then stronger field needed for carrier depletion. Additionally, a larger depletion capacitance (or smaller depletion width) is the cause for a larger subthreshold slope [3]. Although, this is considered as drawback, but would like to admit that the junctionless FET may be a good alternative if compare to conventional MOSFET due to its subthreshold slope that obtained that is below 80mV/dec which is within the tolerance of short channel devices. [5]. Probably, the increase of SS with the increases of channel doping due to non-fully depletion in wider device i.e. this simulation is based on 2D-device simulation.

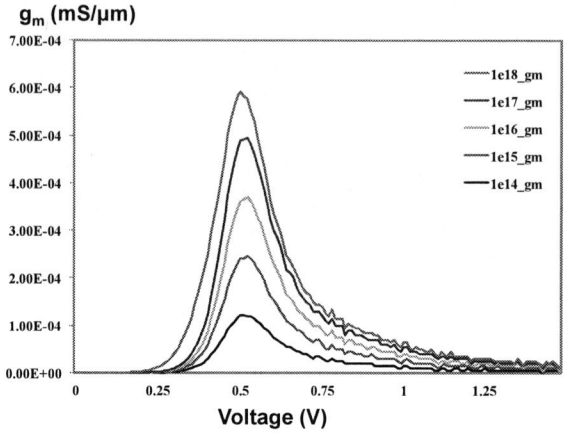

Figure 4 Transconductance

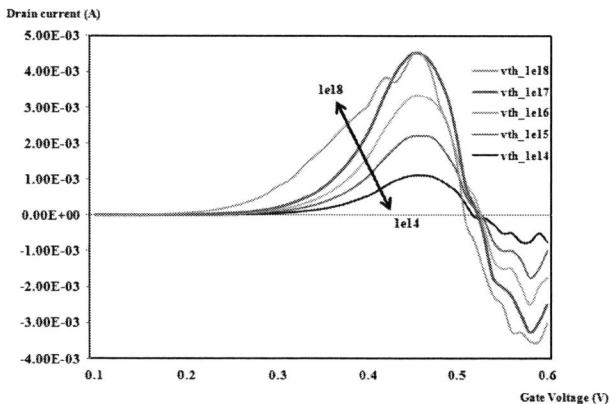

Figure 5 Threshold Voltage

Table 1 Extracted parameters providing an estimation of the performance of devices

Doping Concentration (at 20mV)	1e14	1e15	1e16	1e17	1e18
Max. g_m (mS/um)	0.122	0.125	0.125	0.130	0.135
Thereshold Voltage (V)	0.455	0.45	0.455	0.45	0.37
Subthreshold Slope (mV/decade)	50.7	55.5	58.3	65.6	79.8

IV. CONCLUSION

On the whole, the simulation of junctionless transistor has shown the impact of channel doping concentration on the operation of junctionless FET. The result shows that the trend of higher the doping concentration the larger drain current can be produced. Furthermore, the characteristics are shifted towards smaller V_g values as doping concentration is increased. On the other hand, the transconductance and threshold voltage becomes smaller as the doping concentration increased. In addition, the results proof that the junctionless transistor can be a good choice compare to conventional MOSFET from the shown characteristics. The subthreshold slopes produced are below 80mV/dec, able to shows the junctionless transistor is within with the tolerance of short channel devices. The 3D simulation is compulsory to investigate the junctionless transistor to achieve lower subthreshold slope

ACKNOWLEDGMENT

The authors are grateful to Microelectromechanical Systems (MEMS) lab for the equipment facilities and TCAD Sentaurus license during the simulation process. The author also would like to acknowledge all the team members in Institute of Nano Electronic Engineering (INEE) for their guidance and help.

REFERENCES

[1] S. R. Sahu, "Review of Junctionless transistor using CMOS technology and MOSFETs," *Int. J. Comput. Appl.*, vol. 1, no. 1, pp. 8–11, 2012.

[2] J. P. Colinge, C. Lee, and A. Afzalian, "Nanowire transistors without junctions," *Nat. Nanotechnol.*, vol. 5, pp. 225 – 229, 2010.

[3] C. Technology, "Performance estimation and Variability from Random Dopant Fluctuations in Multi-Gate Field Effect Transistors: a Simulation Study," 2010.

[4] T. Solankia and N. Parmar, "A Review paper : A Comprehensive study of Junctionless transistor," *Natl. Coference Recent Trends Eng. Technol.*, vol. 1, no. 5, pp. 1–5, 2011.

[5] C. Lee, A. Afzalian, R. Yan, N. Dehdashti, and J. Colinge, "Junctionless MuGFETs," *Appl. Phys. Lett.*, vol. 94, no. 5, pp. 3–4, 2009.

[6] M. Zabeli, N. Caka, M. Limani, and Q. Kabashi, "The impact of MOSFET's physical parameters on its threshold voltage," *WSEAS Conf. MINO*, pp. 54–58, 2007.

Experimental Analysis on SU8-Micromolding Structure of PDMS (Poly-Dimethylsiloxane) Based Microfluidic Channel

Marianah Masrie, Burhanuddin Yeop Majlis, *SMIEEE* and Jumril Yunas, *MIEEE*
Institute of Microengineering and Nanoelectronics (IMEN)
Universiti Kebangsaan Malaysia
Bangi Selangor, Malaysia
burhan@vlsi.eng.ukm.my

Abstract— This paper reports experimental study of a microfluidic channel fabrication for bioparticles detection system. The microfluidic channel is fabricated by standard MEMS soft photolithography process implementing negative photoresist SU-8 and poly-dimethylsiloxane (PDMS). Based on the characterization process in the fabrication, an optimum structure of PDMS microfluidic has been achieved. A proper UV exposure dosage can be identified through the observation of the SU-8 mold film thickness and sidewalls profile produced in the characterization process. From this, UV expose for 60 s with exposure energy at 156 mJ/cm^2 is considered as the optimal expose condition in this work. In addition, the difference between the PDMS microchannel pattern from the SU-8 mold are also observed. With these results, the produced structures can provide suitable channel formation for portable bioparticles detection purpose in lab on Chip applications.

Keywords—SU-8 mold, PDMS microfluidic, photolithography, Lab on a Chip

I. INTRODUCTION

There is a growing need for developing microfluidic devices in many area of biomedical engineering especially for bioparticle detection which should be able to fast detect, portable sensing by providing on-site laboratory analysis of biological and chemical samples[1, 2]. PDMS is one of the most widely used polymers in fabricating microfluidic devices [3]. PDMS based microfluidic structure offer biocompatibility, low cost and optically transparent for light wave between 235 nm to 1100 nm which is suitable for UV/VIS absorbance detection [4]. Typically, the PDMS microfluidic is fabricated using standard soft lithography through a replica molding process. This polymer can be replicated following various thickness, length and width of the mold. A deep microchannel patterns with various depth of 500, 250,75 and 15 μm for absorbance cell detection were · successfully fabricated [5]. For the deepest microchannel depth pattern from 500 μm, it was found that the vertical wall was out of shape due to limitation of UV light penetration while fabricating the mold. While Horiuchi [6] compared the PDMS transferred pattern from complicated SU-8 mold patterns by investigating the top and bottom width layer.

In addition, SU-8 is also widely used as the master mold high aspect ratio pattern of microfluidic structures. This negative photoresists polymer has unique property which is high viscosity that can produce vertical walls and specific film thickness with a proper spin coating in photolithography process ([7, 8]. In a previous works, characterization of SU-8 is one of the critical aspects that must be considered in order to achieve an optimum high aspect ratio. It was found that the process in photolithography process; pre-bake, exposure time, post bake time and developing time are important to produce high aspect ratio of SU-8 mold [9, 10]. Jin et. Al [11] identified a pre-bake time in photolithography process affecting the developed SU-8 mold. This problem can be solved by optimizing that process and therefore resulting desired aspect ratio. While increasing and decreasing of ultraviolet (UV) exposure time has been found to affect the SU-8 sidewall profiles [12]. Upon UV exposure at 365 nm, the cross linking of the SU-8 polymer is affected by acid concentration produced during photochemical reaction between SU-8 and UV light [13].

In this paper, we present the fabrication and characterization of the microfluidic channel for bioparticles detection by optimizing the PDMS micro molded structures. Standard photolithography process is used to pattern the SU-8 for the master mold. Therefore, the characterization is performed by implementing different UV exposure dosages during pattern transfer and the effect is studied for each of the samples. From this work, we are able to identify the optimum exposure energy in photolithography process for improving the profile quality of SU-8 mold.

978-1-4799-5761-3/14 $31.00 © 2014 IEEE

II. METHODOLOGY

The microfluidic master mold is based on SU-8 with film thickness of 80 μm and width of 100 μm, respectively. The microfluidic structure consist of 2 mm diameter inlet and outlet , 3 mm diameter detection chamber and the channel length is 15 mm. The molding structure is fabricated using standard photolithography process.

To create the microfluidic structure, A PDMS based polymers is poured on to fabricated SU-8 based master mold structure using soft lithography process, as shown in Fig.1.

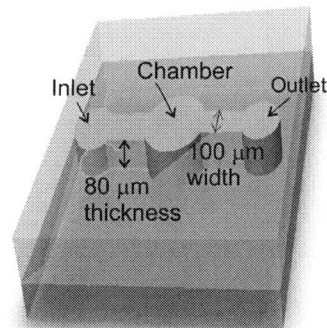

Fig. 1. PDMS based microfluidic structure

A. SU-8 master mold fabrication

The master mold with 80 μm film thickness was produced through standard photolithography process as shown in Table I. This film thickness can be achieved using a single coating process of SU-8 2075 photoresist from MicroChem [14]. In this work, a silicon wafer with size of 2.5x2.5 cm^2 was used as initial substrate sample of the mold master. The substrate was treated using standard cleaning procedure before proceed with the rest of the photolithography process.

To obtain SU-8 master mold structure with thickness at 80 μm, the deposited SU-8 on the silicon substrate was pre spun at 500 rpm for 10 s and ramped at 2500 rpm for another 20s. The mold was soft baked at 65°C for 5 minutes and continued bake at 95°C for another 10 minutes before exposing to UV light at 365 nm with 2.3mW/cm^2 light intensity (Karl Suss MJB 3) through a transparency mask. After exposure, it was baked again (65°C 2 min, 95°C, 8 min) and a visible mask pattern on the SU-8 film can be seen within 5 to 15 seconds during post exposure bake at 65°C. Then it was developed in SU-8 developer for 8 minutes. After the pattern appeared the mold was rinsed using isopropyl alcohol (IPA) and blow dried using nitrogen. As the final process, the mold was hard baked for 30 minutes at 150 °C.

B. Microchannel structure fabrication

The soft lithography process was initiated by mixing the PDMS elastomer base (Sylgard® 184 from Dow Corning) with curing agent, ratio of 10:1. The mixture was de-gased in a vacuum chamber until air bubble was completely removed. To obtain the PDMS pattern layer, the mixture was poured on the SU-8 master mold and cured in an oven for 1 hour at 60°C.

After peeling process, the microfluidic channel can be produced by sealing the pattern layer with flat surface of PDMS produced by the similar process. All process is illustrated in Table II.

TABLE I. PHOTOLITHOGRAPHY PROCESS FLOW AND SCHEMATIC PROCESS FOR 80μM FILM THICKNESS SU-8 MASTER MOLD

Step No.	Process Description	Schematic
1	Standard cleaning	Si wafer
2	Spin coat	SU-8 / Si wafer
3	Soft bake	
4	UV expose	UV 365 nm / SU-8 / Si wafer / Mask
5	Post exposure bake	
6	Develop, Rinse & Dry	
7	Hard bake (optional)	SU-8 master mold / Si wafer

TABLE II. SOFT LITHOGRAPHY PROCESS FLOW AND SCHEMATIC PROCESS FOR 80 μM DEPTH MICROFLUIDIC CHANNEL

Step No.	Process Description	Schematic
1	PDMS Preparation; Curing agent 10:1	
2	Mixture degas	
3	Mold on to master and cure	PDMS / Si wafer
4	Peel off	PDMS / Si wafer
5	Bond on to PDMS base layer	PDMS / PDMS

III. RESULTS AND DISCUSSION

The SU-8 molds with desired 80 μm thickness and 100 μm width, were successfully fabricated using standard photolithography process which was exposed with different exposure energy. The resist contains a photo-acid generator that absorbs photons from UV light. Then, strong acid is produced to cross-link the resist during post exposure bake [15]. The exposure energy, E (mJ/cm^2) can be obtained by multiplying the incident energy, I_o (mW/cm^2) with the exposure time, t (s) [16].

$$E = I_o t \qquad (1)$$

The intensity from 365 nm UV source is average at 2.6 mW/cm^2. This was measured using UVX digital ultraviolet intensity meter from Cole-Palmer. Based on different exposure time range in between 40 – 120 sec, the exposure energy, E obtained by equation (1) is presented in Table III.

TABLE III. EXPOSURE TIME WITH ASSOCIATE ENERGY

Exposure time t (sec)	Exposure energy P (mJ/cm^2)
40	104
50	130
60	156
80	208
120	312

Based on the characterization, only significant results were selected for comparison. Fig. 2 shows the cross sectional SEM images for three SU-8 mold samples and the equivalent PDMS microchannel structure. Exposure energy at 104 mJ/cm^2 is considered as low dosage as the whole structure of the SU-8 mold indicated in Fig.3(a) is deformed. This is possibly due to insufficient of acid formation in order to cross link the photoresist. As a result, the SU-8 can be easily removed by the developer especially at the bottom part during develop process. The PDMS microchannel structure also encounter problem such as deformation at the top part since it was demolded from the SU-8. The range of exposure dosage recommended by the Microchem is from 150 to 215 mJ/cm^2. As can be seen from Fig. 3, at 60 s UV exposure with exposure energy at 156 mJ/cm^2 is considered as the ideal expose in this work. The film thickness is measured at 79.539 μm which is very close to the desired thickness of 80 μm. The mold structure produced through this dosage was still has structural deformation due to the sidewall as depicted in Fig 3.(b). However, the quality of the mold was slightly improved as a near vertical wall structure was resulted. Obviously, when the exposure energy was increased to 312 mJ/cm^2, the structure at the top was expanded to 99.000 μm which is almost to the desired top width at 100 μm. Yet, there was a problem with the film thickness as it was decreased tremendously.

The analytical results (Fig.3) show that the top and bottom width (W_t and W_b) of SU-8 are important for the sidewall profile. The top width structures tend to expand as the exposure energy is increased. On the other hand, the bottom width dimensions are obviously deformed due to insufficient UV light penetration to the bottom part. Fig.3 also shows the effect of the exposure energy to SU-8 mold film thickness. As the energy increased, the film thickness is decreased moderately. At specific spin coat to produce 80 μm, (pre spin at 500 rpm for 10 s and spin at 2500 rpm for 20s) 156 mJ/cm^2 proportionate to 60 s UV exposure was preferred in this work.

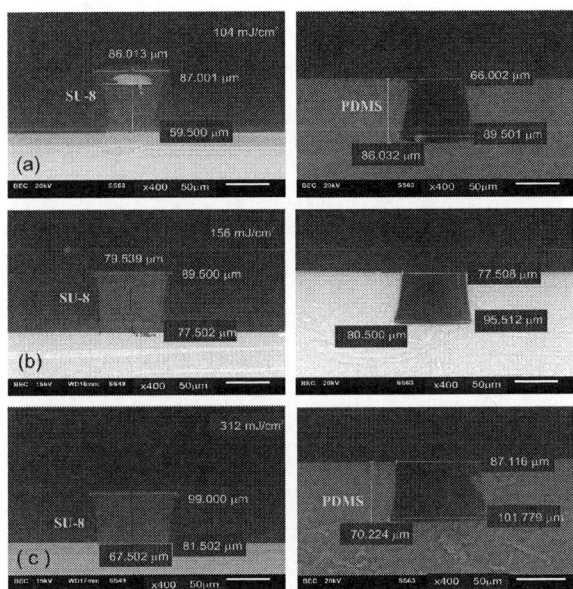

Fig.2. SEM images of SU-8 mold and the PDMS microchannel structure, (a) low expose, (b) proper expose and (c) over expose

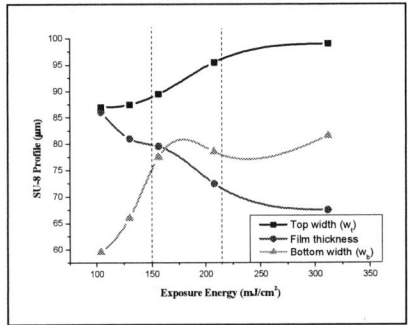

Fig.3. SU-8 structure profile at different exposure energy

The difference of SU-8 mold top and bottom width with the PDMS transferred pattern dimension were also investigated. Significant error is observed from the differences which are inconsistent error plotted in Fig.4. The average difference between SU-8 top width and the transferred pattern to PDMS is approximately to 4.5% with the highest and lowest error are at 6.8 % and 2.8%. While for the transferred pattern of bottom width to the PDMS to

PDMS is having large percent of error which is about 7.3% (Fig. 5). In addition, the differences between SU-8 mold thickness and transferred pattern to PDMS depth are plotted in Fig.6 having a percentage error about 1.5%. It is considered that these differences may be caused by inaccurate volume mixture of PDMS and curing agent and mechanical deformation during peel of process which can affect the stiffness of the cured PDMS.

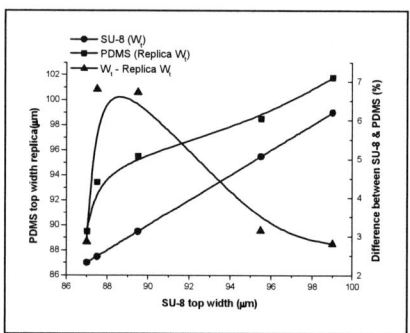

Fig. 4. SU-8 and PDMS transferred pattern top width

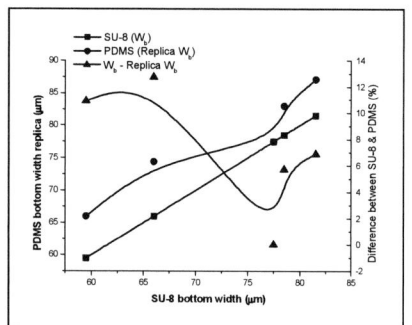

Fig. 5. SU-8 and PDMS transferred pattern bottom width

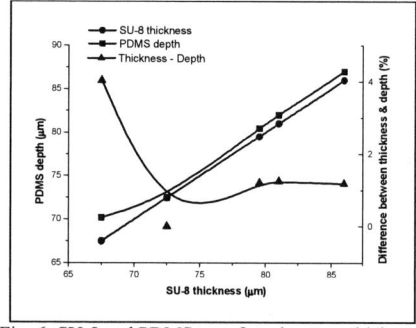

Fig. 6. SU-8 and PDMS transferred pattern thickness

IV. CONCLUSION

The SU-8 mold and the transferred pattern on PDMS layer were successfully fabricated and analyzed. The study was aimed to find optimum photolithography parameter for the fabrication of microfluidic channel for lab on chip application. From the characterization process, it was shown that variation of UV exposure energy could result significant profile for the SU-8 mold as low expose and over

expose can caused deformation to the mold. Finally, by optimizing the UV exposure parameters in photolithography process, the quality of SU-8 mold profile was improved where near vertical sidewalls of SU-8 mold was achieved. In addition, the specific UV dosage was identified to obtain 80 μm SU-8 film thickness.

Acknowledgment

This works is supported by Research Grant: UKM/NND/(1)/TD11-002 (Development of lab-on chip for peripheral blood stem cell isolation and rapid detection of tropical diseases from blood).

References

[1] D. A. Boehm, P. A. Gottlieb, and S. Z. Hua, "On-chip microfluidic biosensor for bacterial detection and identification," *Sensors and Actuators B: Chemical*, vol. 126, pp. 508-514, 2007.

[2] J. Johari, J. Yunas, and A. A. Hamzah, "Piezoelectric micropump with nanoliter per minute flow for drug delivery systems," *Sains Malaysiana*, vol. 40, pp. 275-281, 2011.

[3] A. R. Bahadorimehr, J. Yunas, and B. Y. Majlis, "Low cost fabrication of microfluidic microchannels for Lab-On-a-Chip applications," in *Electronic Devices, Systems and Applications (ICEDSA), 2010 Intl Conf on*, 2010, pp. 242-244.

[4] J. C. McDonald, D. C. Duffy, J. R. Anderson, D. T. Chiu, H. Wu, O. J. A. Schueller, *et al.*, "Fabrication of microfluidic systems in poly (dimethylsiloxane)," *Electrophoresis*, vol. 21, pp. 27-40, 2000.

[5] D. Onoshima, J. Wang, M. Aki, K. Arinaga, N. Kaji, M. Tokeshi, *et al.*, "A deep microfluidic absorbance detection cell replicated from a thickly stacked SU-8 dry film resist mold," *Analytical Methods*, vol. 4, pp. 4368-4372, 2012.

[6] T. Horiuchi and S. Shiratori, "Multi-Layer Thick Resist Process Using Liquid-Crystal-Display Projection Exposure and Vacuum Treatment Process," *Journal of Photopolymer Science and Technology*, vol. 25, pp. 455-460, 2012.

[7] J. Spratley, M. Ward, and P. Hall, "Bending characteristics of SU-8," *Micro & Nano Letters*, vol. 2, pp. 20-23, 2007.

[8] S. Natarajan, D. Chang-Yen, and B. Gale, "Large-area, high-aspect-ratio SU-8 molds for the fabrication of PDMS microfluidic devices," *Journal of Micromechanics and Microengineering*, vol. 18, p. 045021, 2008.

[9] J. Zhang, K. Tan, and H. Gong, "Characterization of the polymerization of SU-8 photoresist and its applications in micro-electro-mechanical systems (MEMS)," *Polymer testing*, vol. 20, pp. 693-701, 2001.

[10] J. Liu, B. Cai, J. Zhu, G. Ding, X. Zhao, C. Yang, *et al.*, "Process research of high aspect ratio microstructure using SU-8 resist," *Microsystem Technologies*, vol. 10, pp. 265-268, 2004/05/01 2004.

[11] P. Jin, K. C. Jiang, and N. Sun, "Microfabrication of ultra-thick SU-8 photoresist for microengines," in *Micromachining and Microfabrication*, 2003, pp. 105-110.

[12] M. B. Chan-Park, J. Zhang, Y. Yan, and C. Yue, "Fabrication of large SU-8 mold with high aspect ratio microchannels by UV exposure dose reduction," *Sensors and Actuators B: Chemical*, vol. 101, pp. 175-182, 2004.

[13] J. Melai, C. Salm, S. Smits, V. M. Blanco Carballo, J. Schmitz, and B. Hageluken, "Considerations on using SU-8 as a construction material for high aspect ratio structures," 2007.

[14] MicroChem. SU-8 2000 Permanent epoxy negative photoresist [Online]. Available: www.microchem.com

[15] R. Feng and R. J. Farris, "Influence of processing conditions on the thermal and mechanical properties of SU8 negative photoresist coatings," *Journal of Micromechanics and Microengineering*, vol. 13, p. 80, 2003.

[16] M. J. Madou, *Fundamentals of Microfabrication: The Science of Miniaturization*: Taylor & Francis, 2002.

Mechanical Characteristics of Porous Silicon Membrane for Filtration in Artificial Kidney

Norhafizah Burham, Azrul Azlan Hamzah, Burhanudin Yeop Majlis, *SMIEEE*

Institute of Microengineering and Nanoelectronics (IMEN)
Universiti Kebangsaan Malaysia
Bangi, Malaysia
fiezah@msn.com

Abstract—Silicon is a promising material due to it having reliable and desirable characteristics for making porous silicon membrane. Porous membrane is widely used in various applications especially in bioMEMS, Lab on Chip and MEMS. Normally, porous membrane functions as a part of filtration system that can be integrated with other systems to make a complete device. The porous silicon membrane is simulated using COMSOL 4.3a for mechanical verification. This work compares the simulation result of the silicon membrane design with theoretical calculation. This paper studies the effect of pressure across the silicon membrane based on the deflection and von Mises stress at the centre of silicon membrane. The maximum deflection and von Mises stress of different membrane thickness and pore shapes are compared against various levels of pressure applied on the silicon membrane surface. The 100 nm thin silicon membrane studied was found to be far superior to the 25 nm silicon thin membrane, being able to mechanically withstand the applied pressure up to 7.33 kPa (55 mmHg).

Keywords—Silicon Membrane; Pores; COMSOL ver. 4.3; Deflection; von Mises Stress

I. Introduction

Porous silicon membrane can be used as a filtration system that can be integrated with other components to make a complete device especially in bioMEMS application. However, this porous membrane has a limitation such as poor pore distribution, inadequate mechanical strength, low throughput, high cost and so on [1-4] during the fabrication and filtration process.

Porous systems are engineered to mimic the natural filtration system for use in implantable drug delivery system, bioartificial organ and other medical devices [5]. The desired membrane material used must possess biocompatibility and anti-biofouling behaviour to be implanted in biomedical devices. In addition, the material must have good physical and chemical stability to get better membrane performance with ordered pore structure. Compared to other materials, inorganic material has very good strength and chemical stability in contrast with organic and composite materials [6].

Silicon and other semiconductors are inorganic materials that have very good properties in terms of strength and chemical stability. That is the reason why porous silicon

became a multipurpose material which are greatly considered for many applications such photoelectronics [7], photonic crystal [8], optical filter [9], membrane and molecular sieve [3, 10].

The deflection and von Mises stress of porous silicon membrane is studied using Comsol 4.3(a) to verify the mechanical strength of the membrane by applying pressure on the membrane surface. Various parameters were investigated like how membrane thickness and pore shape influence membrane deflection and von Mises stress when pressure is applied. The simulation result will be used to predict the maximum deflection and membrane stress in fabricating the actual porous silicon membrane.

II. Material and Method

A. Modelling of porous silicon membrane

In order to design and optimize the deflection and stress on the silicon surface, Comsol Multiphysics software version 4.3(a) is used to design a single frame (15 μm x 15 μm) of silicon membrane with ordered pores as shown in Fig. 1. The uniform pores have been designed on the membrane surface to study the factors affecting the deflection of the membrane, especially the membrane thickness and pore shape. By applying an external pressure on membrane surface, the deflection and von Mises stress can be studied in detail.

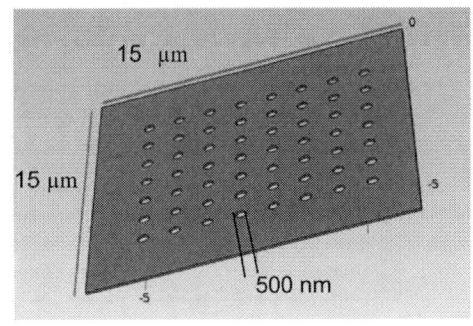

Fig. 1. Silicon membrane model

B. Simulation

In order to analyze the deflection performance, the theoretical result and simulation result are compared. A single frame of membrane without any pore was simulated in COMSOL ver. 4.3(a) and the deflection result compared with theoretical result. Thimoshenko equation [11] is used to define the deflection at the maximum point or centre of membrane.

$$w_{max} = \left(\frac{p_0 a^4}{\pi^4 D}\right)\left[\frac{1}{3+3\left(\frac{a}{b}\right)^4+2\left(\frac{a}{b}\right)^2}\right] \quad (1)$$

$$D = \frac{Et^3}{12(1-v^2)}$$

w_{max} = the deflection of the centre of the membrane

P_0 = pressure

a and b = width and length

t = membrane thickness

E = Young Modulus

v = Poisson ratio

Similarly the tensile stress σ_o in the middle of the membrane can be found using von Mises stress

$$\sigma_0 = \frac{P_0 a^2 b^4}{2t^2(a^4+b^4)} \quad (2)$$

Different membranes with thicknesses of 25 nm, 50 nm, 100 nm and 150 nm with pore size diameter of 500 nm were simulated. The pressure applied on the membrane surface area ranges from 15 mmHg up to 55 mmHg. The value of the membrane pressure is based on the kidney filtration membranes from 15 mmHg for the first membrane to 55 mmHg for the third membrane [12]. The pressure is converted to Pascal and ranges approximately from 1.99 kPa to 7.33 kPa.

The calculated and simulated values are plotted in Fig. 2. It could be seen that the calculated values closely approximate the simulated values for all thickness variations, even when the deflection values are beyond 0.03 μm.

Similarly, the stress is proportional to the square of the silicon membrane frame by equation 2. Fig. 3 compares the calculated and simulated values of stress for the silicon membrane. The calculated values closely approximate the simulated result for all thickness variations.

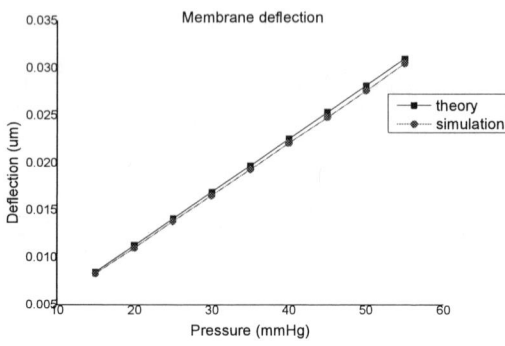

Fig. 2. Comparison between calculation and simulation membrane deflection

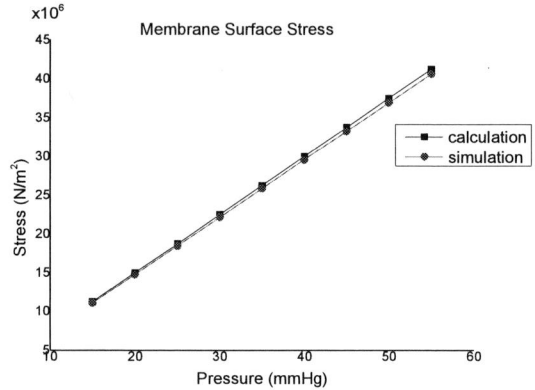

Fig. 3. Membrane von Mises stress comparison between calculation and simulation

III. Result and Discussion

A. Deflection of Different Membrane Thickness

The membrane thickness is an important parameter in studying the deflection and stress on the membrane surface. The deflection of the membrane is reduced and became approximately constant as the membrane thickness is increased. This indicates that thickness plays an important role in membrane deflection due to its capacity to withstand substantially higher pressure up to 55 mmHg with a maximum deflection of 1.6 μm.

The deflection of various membrane thicknesses when applied with pressure from 15 mmHg up to 55mmHg is plotted in Fig. 4. The porous membrane deflection is proportional to the pressure applied with various membrane thicknesses.

The characteristic lines get more linear by decreasing the membrane thickness. As expected, by increasing the pressure applied, the deflection increases linearly for thicker membrane. There is an opportunity to design a thin membrane, balancing the linear interrelationship between pressure increase and deflection desired to the specific application. However, in this artificial kidney application it is not suitable to have the thinner membrane. This is because the thin membrane can easily break and will damage the pore structure of the silicon membrane. So, the 100 nm membrane is more suitable in this application due to the consistency of the silicon membrane in withstanding higher pressure.

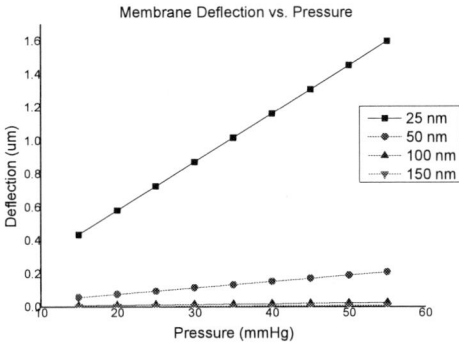

Fig. 4. Membrane deflection for various applied pressure at different values of membrane thickness

B. Stress with Different Membrane Thickness

The thin silicon membranes once again give linear characteristics for the von Mises stress as shown in Fig. 5. From the graph, the von Mises stress is proportional to the pressure applied. Again, the 25 nm membrane's stress increases rapidly with applied pressure and the 100 nm membrane's stress increases very slowly and is almost constant.

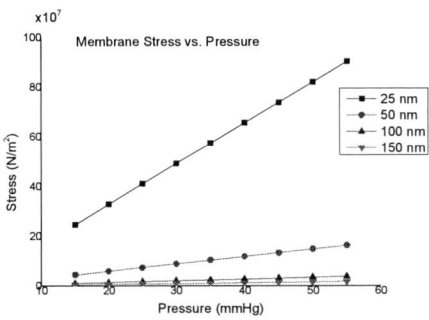

Fig. 5. von Mises stress for various applied pressure at different values of membrane thickness

C. Pore Shape

Various pore shapes were studied, namely circle pores, square pores and slit pores. The different pore shapes are studied to define which shape is more suitable for a filtration membrane. A kidney filtration system consisted of two types of pores which are the circle and slit pores. The first and second membrane used circle pores while the slit pores are used at the third membrane. The square pores shown in Fig. 6 are studied to make a comparison whether it can perform as good as the circle pores.

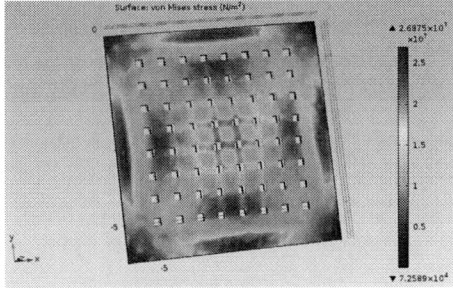

Fig. 6. Membrane stress for square pore shape

Fig. 7. Membrane stress for slit pore shape

The comparison between circle pores, square pores and slit pores deflection with applied pressure were analysed. The deflection result is shown in Fig 8. The circle pores has the lowest deflection compared to the slit and square pores. However the square pores deflection is not much different compared with the circle pores. By calculating the slope of the square and circle pores, it was found that they are approximately the same with a difference of only 0.0002.

The slit pore shape shows the highest linearity deflection when pressure is applied. By comparing the limitation of silicon membrane without pores, the suitable pressure applied to the membrane should be no more than 20 mmHg. Furthermore, by comparing the actual kidney filtration membrane, the slit a pore is used at the third membrane with 15 mmHg applied pressure.

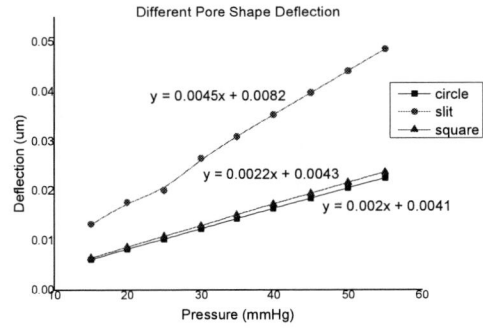

Fig. 8. Membrane deflection with various pore shapes

The von Mises stress on membrane surface of different pore shapes is shown in Fig. 9. Again, the slit pore has the highest surface stress. By comparing the maximum stress of silicon membrane without the pores in Fig. 2, the silicon membrane can withstand a surface stress up to 40 MPa. So, the maximum stress on the slit pores membrane must be 40 MPa. Fig. 9 shows the maximum pressure of the slit pore membrane is 20 mmHg which meet the requirement of no-pore silicon membrane. In addition, stress is related to the diameter of pore. It can be proved by calculating the stress of one circle pore and one slit pore using equation 2. By calculating this stress, it was found that the slit pores has the highest stress compared to the circle pores.

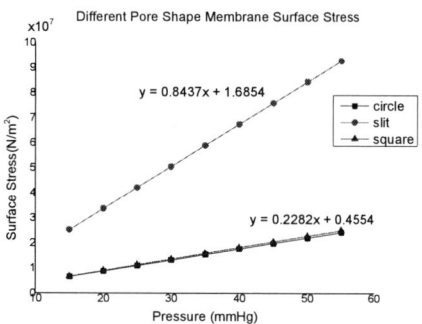

Fig. 9. Membrane stress with various pore shapes

The simulation result on the silicon membrane is used to estimate the deflection and stress of actual silicon membranes during fabrication process. This result is used to ensure the membrane can withstand various pressures. The fabrication of a membrane is shown in Fig. 10 after KOH etching process. The frame of the silicon membrane is 10.13 μm x 12.25 μm.

Fig. 10. One frame of silicon membrane

After the KOH etching process is completed, the pores will be produced on the membrane surface using electrochemical etch. Fig. 11 shows the silicon membrane with the pores.

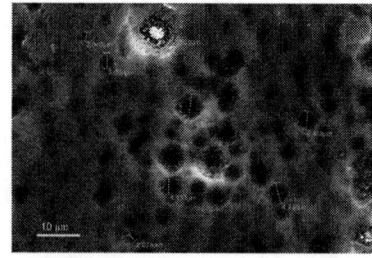

Fig. 11. Pores on silicon membrane surface

IV. CONCLUSION

As conclusion, the membrane deflection and stress depend on the thickness of the membrane. It can be shown by applying varying pressures on silicon membranes with different thicknesses. The thin silicon membrane has the highest linearity compared to the thick membrane. The 100 nm and 150 nm give a constant deflection when 15 mmHg to 55 mmHg of pressure is applied to the membrane. Again, for von Mises stress, it was found that the thin membrane has the highest linearity stress compared to the 100 nm thick membrane. As for different pore shapes, the slit pores give the highest deflection and von Mises stress compared to square pores and circle pores. By comparing different pore sizes, the pore shape limitation can be estimated when applying the pressure. Finally, this simulation result is used to estimate the actual deflection and von Mises stress of silicon membrane during fabrication in the future.

References

[1] S. Kuiper*, C.J.M.v.R., W. Nijdam, M.C. Elwenspoek: 'Development and applications of very high flux microfiltration membranes', Journal of Membrane Science, 1998, 150, pp. 8

[2] Hien D. Tong, H.V.J., Vishwas J. Gadgil, Cazimir G. Bostan, Erwin Berenschot, Cees J. M. van Rijn, and Miko Elwenspoek: 'Silicon Nitride Nanosieve Membrane', Journal of Nanoletters, 2004, 4, (2), pp. 5

[3] Christopher C. Striemer1, T.R.G., James L. McGrath2 & Philippe M. Fauchet1: 'Charge- and size-based separation of macromolecules using ultrathin silicon membranes', Nature PublishingGroup, 2007, 445, pp. 749-753

[4] Wessling, A.v.d.B.a.M.: 'Silicon for the perfect membrane', Nature PublishingGroup, 2007, 445, pp. 1

[5] Adiga, S.P., Jin, C., Curtiss, L.A., Monteiro-Riviere, N.A., and Narayan, R.J.: 'Nanoporous membranes for medical and biological applications', Wiley Interdisciplinary Reviews: Nanomedicine and Nanobiotechnology, 2009, 1, (5), pp. 568-581

[6] Thusu, T.: 'Inorganic Porous Membranes for Liquid Phase Separation', Separation and Purification Methods 2001, 30, (2)

[7] Lehmann V, G.U.: 'Porous Silicon Formation: A Quantum Wire Effect', Applied Physics Letter, 1991, 58, pp. 856-858

[8] Nicewarner-Pena SR, F.R., Reiss BD, He L, Pena DJ, Walton ID, Cromer R, Keating CD, Natan MJ: 'Submicrometer Metallic barcodes', Science, 2001, 294

[9] V.Lehmann, S.R.: 'MEMS techniques applied to the fabrication of anti-scatter grids for X-ray imaging', Sensors and Actuators A Physical, 2002, 95, pp. 202-207

[10] Hamzah, A.A., Abidin, H.E.Z., Majlis, B.Y., Nor, M.M., Ismardi, A., Sugandi, G., Tiong, T.Y., Dee, C.F., and Yunas, J.: 'Electrochemically deposited and etched membranes with precisely sized micropores for biological fluids microfiltration', Journal of Micromechanics Microengineering, 2013, 23, pp. 9

[11] Ugural, A.C.: 'Stresses in plates and shells', Library of Congress Cataloging, p.317, 1981

[12] Gerard J. Tortora, Bryan Derrickson, "Principles of Anatomy and Physiology", John Wiley & Sons Inc., 11 Edition, pp 1004-1006, 2006

Bonding Stress Analysis on Palladium Bond Pad Between Cu and Au FAB

Sauli. Z.*, Retnasamy, V.
School of Microelectronic Engineering, Universiti
Malaysia Perlis (UniMAP), Kampus Pauh Putra, Arau,
Perlis.
*Corresponding email: zaliman@unimap.edu.my

Mamat, H.
MIMOS Berhad, Technology Park Malaysia,
Kuala Lumpur.

Abstract–**This papers presents the simulation investigation of bond pad stress analysis during wire bonding contact stage. The contact of gold free air ball (FAB) and copper (FAB) on Palladium (Pd) bond pad were evaluated through simulation. The stress of wire FAB, Pd bond pad and copper lead frame during FAB contact were compared for both wire materials. Ansys 11 was used for the analysis. Outcome of the analysis exhibited that the contact force significantly influences the stress on the FAB, bond pad and lead frame. FAB wire material comparison showed that the stress of Cu wire FAB was larger compared to the Gold FAB.**

Keywords- **Copper FAB; Gold FAB; Pd bond pad; contact stage; Ansys**

I INTRODUCTION

Wire bonding is significant process where interconnections are provided to the integrated circuit during packaging assembly process. This process involves micronized wires being bonded from the lead frame and die. In general gold (Au) and copper (Cu) wires are utilized for the bonding process which is executed with the combination of ultrasonic energy, bond force and heat [1]. The operating life time of the wire bond packages are significantly dependent on the bonded wires as signals and electric current are passed through these wires. Therefore, sustaining the reliability of the wire bonds is extremely important. However, appraising the reliability of the wire bond is very tedious due to involvement of vast bonding parameters. Hence various researches utilize simulation as method to comprehend and distinguish the wire bonding process as it is cost and time effective approach when compared with experimental method. Mechanical response of wire bonding process on Cu/low-K structures was examined by Yeh,C.L et al. [2] through simulation. The impending deficiency on copper via and the bond pad for various elastic modulus of different material were investigated. The deformation of gold wire on aluminum bond pad bonding was investigated through simulation by Saiki, H et al. [3].The distributions of Al film after contact stage of gold wire were simulated in their study. The characteristics of copper wire bonding on aluminum bond pad were investigated by Hsu, H.C et al. [4] ,by the means of simulation. The incidences of aluminum bond pad compression during copper wire bonding

process were evaluated. The wire bond failure mechanism was evaluated by van Driel, W. D et al. [5], by using simulation. The relationship between wire and molding compound and the significance of interfacial delaminated were assessed. Stress on LED die during Cu wire bonding was scrutinized by Chen, Z et al. [6], through simulation. Their study was focused on evaluating the plastic deformation and thermal heating in the LED packages caused by the wire bonding process parameters. The significance of intermetallic compound growth for copper wire bonding process was investigated Yuan, C. et al. [7], through simulation. The effectiveness of the wire pull test on fine pitch pads was investigated by Sundararaman, V et al. [8], through simulation. The deformation of the wire bond shape under pull test condition was scrutinized. Consistencies of wire pull test ON Au wires bonded on the Cu/low-K wafer were scrutinized by Chang-Lin, Y et al. [9] through numerical simulation. The influences of material properties on the eminence of the wire bond were investigated. Hence, as demonstrated by various researches, simulation process is utilized in every part of wire bonding process to characterize, improvise and develop the wire bonding process. At this moment in time, the significant advancement in integrated circuit design requires commencement of advance bond pad material to improve the reliability of the device and to protect the structures below the bond pad layer. Palladium (Pd) is one of the proposed bond pad material [1]. Clauberg, H. et al [10], performed thorough assessment on NiPd and NiPdAu bond pads as an alternative to the existing Al bond pad. Anand, T. J. S. et al. [11] studied and examined the influence of high bonding temperatures and high temperature storage on IMC growth for copper wirebond on three types of bond pads ,ie. Al, AlSiCu and NiPdAu. Key outcome of their examination demonstrated that The NiPdAu bond pad exhibited gradual growth of the IMC compared to the other two bond pad materials.

Nevertheless, the adaption of the Pd bond pad to the wire bonding environment requires inceptive investigation to obtain proper bonding parameters. Thus, there is a need to attain baseline information regarding the bonding effect on Pd bond pad due to the minimal work done on stress analysis on Pd bond pad bonding to the best of the authors' knowledge.

Therefore, in this study, the contact of gold free air ball (FAB) and copper (FAB) on Pd bond pad were evaluated through simulation. Comparisons between gold (FAB) and copper (FAB) contact on the Pd bond pad were done. The stress of Au FAB, Cu FAB, Pd bond pad and lead frame during contact stage were assessed. The simulation was focused to inceptive contact stage between FAB and bond pad. Contact force ranging from 1N to 5N. Ansys 11 was used for the analysis.[12].

II METHODOLOGY

The general process step of wire bonding ball bond formation is explained in the following sequence. First a gold wire is threaded through the capillary and an Electronic Flame Off (EFO) spark is used to melt the wire which forms a gold ball at the end of the wire. Second, the wire is then pull back to position the ball against the base of the capillary. Third, the tool is lowered to the bond pad and the Au ball is pressed against it. Pressure and heat is applied to the bond pad and with applied ultrasonic energy, the ball bond is formed. Fourth, the ball is left welded to the surface and the tool is lifted up and forms a wire loop as it moved to the second bond position. Fifth, the bond pad is placed under the bonding tool and the tool is lowered to make a bond which is known as stitch bond. Finally the capillary is lifted up after forming the stitch bond and a wire clamp pulls and splits the wire free. Then tool is then raised up and the clamp feeds enough amount of wire through capillary to repeat the bonding process once again [1].

Thus in this work, the contact of gold free air ball (FAB) and copper (FAB) on Pd bond pad were evaluated through simulation.Prior to simulation, there are several assumption made to simplify the simulation process. There are stated as follow. The temperature of the FAB and bond pad was presumed to be the same. Next, bilinear rate dependent elastic plastic material was presumed for FAB and elastic plastic material property was presumed for bond pad and lead frame. Next, the pressure, heat and ultrasonic energy effect was neglected in this work due the requirement of advanced numerical techniques and simulation period. Hence, preface analysis of contact stage between the FAB and bond pad was only taken into consideration. Von Mises stress criterion was classified as the outcome of this work. A 3D model was developed using Ansys version 11[12] for the analysis. The 3D model constitutes of wire FAB, Pd bond pad and copper lead frame with the following design dimension. The diameter of wire was 25μm and the diameter of the FAB was 70 μm. The Pd bond and copper lead frame had a thickness of 1.5μm and 3μm respectively. The 3D FAB model is illustrated in Figure 1. The 3D model was designed and meshed with 199582 elements (Solid 187) with attained grid independence. The contact area between FAB, bondpad and lead frame was defined with Trage174 and Conta174 element. The 3D FAB Model is illustrated in Figure 1. Contact forces in the range of 1N to 5N were applied to the FAB in the negative Y direction. The contact of two FAB materials, copper and gold was compared and the results are attained in terms of von Mises stress. The lead frame bottom area together with sides of the bond pad was

constrained to oppose movement during FAB contact on bond pad. Material properties are displayed in Table 1.

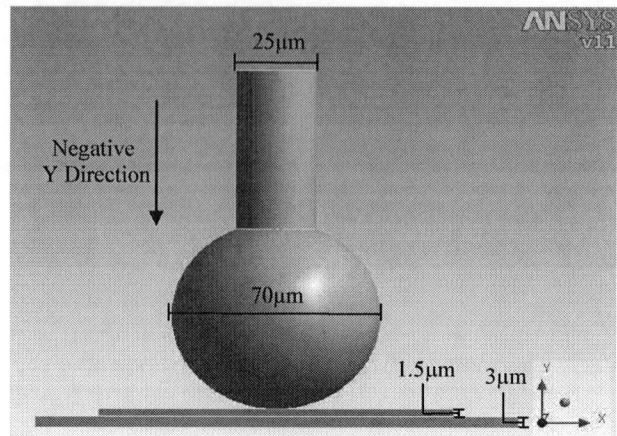

Fig. 1. 3D FAB Contact Model

TABLE1 MATERIAL PROPERTIES

Material	Density, ρ(g/cm3)	Young's modulus,E (Pa)	Poisson ratio, v
Gold	19.3	78 x 10⁹	0.44
Pd	12.023	121 x 10⁹	0.39
Copper	8.94	130 x 10⁹	0.34

III RESULTS AND DISCUSSION

Contact of gold free air ball (FAB) and copper (FAB) on Pd bond pad were evaluated through simulation.. Comparison between gold free air ball (FAB) and copper free air ball (FAB) contact on the Pd bond pad were done. The stress of Au FAB, Cu FAB, Pd bond pad and lead frame during contact stage were assessed. The simulation was focused to inceptive contact stage between FAB and bond pad for contact force varied from 1N to 5N. For all contact forces, the stress was evaluated. Results are discussed in terms of Von Mises stress and are summarized in Table 2. By referring the simulated result from Table 2, it was observed for both wire material, the FAB exhibited highest stress compared with Pd bond pad and lead frame. This was due to deformation of FAB upon contact on the bond pad. Thus, with the augmentation of contact force caused an increase in the deformation of FAB

TABLE 2: RESULTS

Force (N)	Ball Stress (MPa)		Bond Pad Stress(MPa)		Lead Frame Stress(MPa)	
	Au	Cu	Au	Cu	Au	Cu
1	11000	12700	4080	10300	3890	4210
2	14900	24900	5110	16600	5000	7390
3	17700	32300	8720	19500	6600	8910
4	18700	33000	12100	21500	8060	10300
5	20500	34500	13900	22400	9660	11700

The simulated results were compared with the work done by Yeh,C.L et al.(2005). In their work, the stress on Al bond during the contact stage of Au FAB was 558 GPa for the similar bonding conditions. In addition, the maximum stress attained by the copper via below the Al bond pad was 524 GPa. However, in this simulation work, the maximum stress of the Pd bond pad and Cu lead frame attained were 13900 MPa and 9660 MPa respectively for Au FAB at contact force of 5N. Thus, from this comparison, it can be seen that the Pd bond pad is much robust when compared to the Al bond pad. This observation is also supported by Clauberg, H. et al (2010). Results of their analysis suggested that the NiPdAu based bond pad has exceptional properties in terms of bond strength, bond parameter window, reliability and bond pad damage compared to the existing Al bond pad. In addition, throughout the contact FAB on the bondpad, stress was exerted on the wire FAB, bond pad and lead frame due to the contact force of the wire FAB as obtained by the simulated results. From the FAB material comparison, the stress exerted by Cu FAB was larger than that of the Au FAB. This is due to the higher stiffness of copper FAB compared to the gold FAB. (Harman,G. ,2010; Chen, Z.,2011). For this simulation work, the Young modulus of the copper FAB and gold FAB were 130 GPa and 78 GPa respectively. Thus it is suggested to control and utilize appropriate bonding process parameters for Cu wire bonding more attentively compared to the Au wire bonding. In addition suitable contact force are required to be utilized in order to avoid excessive stress distribution on the bond pad and lead frame during the bonding process as it may induce damage to the structures below the bond pad which will indirect reduce the reliability of the bonded device.

IV CONCLUSION

Contact of gold free air ball (FAB) and copper (FAB) on Pd bond pad was evaluated in this work through simulation. Comparison between gold (FAB) and copper (FAB) contact on the Pd bond pad were done. The stress of Au FAB, Cu FAB, Pd bond pad and lead frame during FAB contact were assessed. The simulation was focused to inceptive contact between FAB and bond pad with contact force ranging from 1N to 5N. Ansys version 11 was used for the analysis. The results obtained were based on Von Mises Stress. The simulated results indicated with augmentation of contact force, the stress on both wire FAB, Pd bond pad and copper lead frame amplified. As for the wire material comparison, copper FAB induces higher stress when compared with the gold FAB. This simulation work can be utilize a guideline for future work on Pd bond pad bonding where the effect of ultrasonic energy during wire bonding process and bonding parameter optimization on Pd bond pad can be studied.

ACKNOWLEDGEMENT

The authors acknowledge the support and thank the School of Microelectronic Engineering,Universiti Malaysia Perlis for the facilities. The authors appreciation are extended to the Ministry of Higher Education for the support given.

REFERENCES

[1] G. Harman, *Wire bonding in microelectronics*: McGraw-Hill, 2010.

[2] C.L. Yeh and Y.S. Lai, "Transient analysis of the contact stage of wirebonding on Cu/low-K wafers," *Microelectronics Reliability*, vol. 45, pp. 371-378, 2005.

[3] H. Saiki, Y. Marumo, H. Nishitake, T. Uemura, and T. Yotsumoto, "Deformation analysis of Au wire bonding," *Journal of Materials Processing Technology*, vol. 177, pp. 709-712, 2006.

[4] H.C. Hsu, W.Y. Chang, C.L. Yeh, and Y.S. Lai, "Characteristic of copper wire and transient analysis on wirebonding process," *Microelectronics Reliability*, vol. 51, pp. 179-186, 2011.

[5] W. D. van Driel, R. van Silfhout, and G. Q. Zhang, "On Wire Failures in Microelectronic Packages," *Device and Materials Reliability, IEEE Transactions on*, vol. 9, pp. 2-8, 2009.

[6] Z. Chen, Y. Liu, and S. Liu, "Modeling of copper wire bonding process on high power LEDs," *Microelectronics Reliability*, vol. 51, pp. 171-178, 2011.

[7] C. Yuan, E. Weltevreden, P. van dan Akker, R. Kregting, J. de Vreugd, and G. Q. Zhang, "FE modeling of Cu wire bond process and reliability," in *Thermal, Mechanical and Multi-Physics Simulation and Experiments in Microelectronics and Microsystems (EuroSimE), 2011 12th International Conference* ,pp. 1/5-5/5, 2011.

[8] V. Sundararaman, D. R. Edwards, W. E. Subido, and H. R. Test, "Wire pull on fine pitch pads: an obsolete test for first bond integrity," in *Electronic Components & Technology Conference, Proceedings. 50th*, pp. 416-420, 2000.

[9] C.L. Yeh, Y.S. Lai, and C.L. Kao, "Transient simulation of wire pull test on Cu/low-K wafers," *Advanced Packaging, IEEE Transactions on*, vol. 29, pp. 631-638, 2006.

[10] H. Clauberg, P. Backus, and B. Chylak, "Nickel–Pd bond pads for copper wire bonding," *Microelectronics Reliability*, vol. 51, pp. 75-80, 2011..

[11] T. Anand, C. K. Yau, Y. S. Leong, L. W. Keat, and H. M. Ting, "Microstructural and mechanical analysis of Cu and Au interconnect on various bond pads," *Current Applied Physics*, vol. 13, pp. 1674-1683, 2013.

[12] Sauli, Z., Retnasamy, V., Rahman, N. A., Norhaimi, W. M. W., Ramli, N., & Vairavan, R., "Shearing speed induced stress comparison on gold and copper ball interconnection", the Proceedings of International Conference on Computational Intelligence, Modelling and Simulation, 156-159, 2012.

PCB Depanelling Stress Distribution Simulation Analysis Using Designated Thru Holes

Retnasamy, V.*, Sauli. Z., Vairavan, R.
School of Microelectronic Engineering, Universiti
Malaysia Perlis (UniMAP), Kampus Pauh Putra, Arau,
Perlis.
*Corresponding email: vc.sundres@gmail.com

Mamat, H.
MIMOS Berhad, Technology Park Malaysia,
Kuala Lumpur.

Abstract– **This work demonstrates the evaluation of stress distribution of the printed circuit boards (PCB) during the depanelling process and technique to manage it were investigated. The stress distribution of the PCB were evaluated using 4 types of PCB geometry, one without hole, one with single front hole, one with single centric hole and one with three through holes. The holes were placed in various positions to scrutinize the stress distribution of the PCB. The PCB boards were displaced with heights in the range of 1cm till 5cm. Ansys ver 11 was utilized to perform the simulation. Key results showed that the hole structures assisted in managing the stress distribution during the arching process of the PCB subjected to its position on the PCB.**

Keywords- **Depanelling process ; PCB; single hole; centric hole; through hole ; Ansys**

I INTRODUCTION

The advancement of portable electronic devices such as handheld smartphones, laptops, USB flash drives, personal digital assistance and etc. are lashing towards compact size specifically on PCB assemblies is emerging in a rapid pace with appended utility and compact sized. Therefore, to gratify these necessities, the PCB is packed with complex circuitry and components. In general, the PCB provide power, heat dissipation pathways and support to surfaced mounted components [1].

There have been various studies conducted by researches on PCB with the aim of enhancing the performance of PCB. The approximation stresses on solder joints PCB during mechanical arching were examined by Wong et al. [2] to evaluate the ways of reducing the stress on the solder joints. In their study, the PCB were modeled as tri layer arrangement. The analysis of drop applause of multilayered PCB was done by Wang et al. [3] to evaluate performance of the PCB during drop impact. Comparison between three types of PCB and effects of material properties were done. Analysis on rigid flex printed circuit through taguchi method was done by Huang et al. [4]. The maximum strain plated through holes on the PCB were investigated through simulation. The ways to diminish vibration damage in PCB were investigated by Chomette et al. [5]. It was identified that the vibration damage relies on the profile of huge band excitation. The arching performance of woven fiber composite in a multi-layer PCB was studied by Li et al. [6] through simulation. It was indicated that the woven composite material properties obscure the arching stiffness analysis. Therefore, the consequence of woven fiber material properties during the arching process was investigated. The reaction of PCB during board level drop test was examined by Tong et al. [7]. The effects of acceleration, strains, and resistance on the distribution of solder joints on the PCB during drop test were investigated. The denouement of residual stress on the flexing capability of PCB was investigated by Jun et al. [8] through simulation. Three point arching test was utilized to evaluate the residual stress in the solder joints in various process conditions. Hence, it can be seen that efforts are being placed in characterizing the behavior of the PCB to enhance the reliability of electronic products.

During the manufacturing process, the assembly of the electronic devices is performed in hefty range. These electronic products are produced in batch form where the PCB are exposed to board level assembly. During this assembly process, the PCB are exposed to various assembly process condition such as mechanical arching during depaneling process, which is a process where the PCB are individualized. This depanelling process will induce stress distribution on the PCB and the components mounted on the PCB. This may induce damage on the solder joint of the component which will reduce the reliability of the electronic device. Therefore, method to manage the stress distribution on the PCB during the depanelling process has to be explored due to the minimal work done in this area to the best of the authors knowledge.

As a result, in this paper, the stress distribution of PCB during the depanelling process and technique to manage it were investigated. The stress distribution of the PCB were evaluated using 4 types of PCB geometry, one without hole, one with single front hole, one with single centric hole and one with three through holes. The holes were placed in various positions to scrutinize the stress distribution of the PCB. The stress distributions of all 4 PCB were compared. The analysis was done through simulation. The arching process was replicated by using displacement technique. The PCB were displaced with heights ranging from 1cm till 5cm. Ansys 11 was used for the simulation analysis.

978-1-4799-5761-3/14 $31.00 © 2014 IEEE

II METHODOLOGY

Ansys 11[9] was used for the analysis. 3 dimensional models were developed to resemble the PCB.In this work, 4 types of 3D model were developed for the analysis. The 3D model consisted of, one model without hole, one model with single front hole, one model with single centric hole and one model with 3 through holes. Figure.1 illustrates the 3D models which were utilized for the simulation.

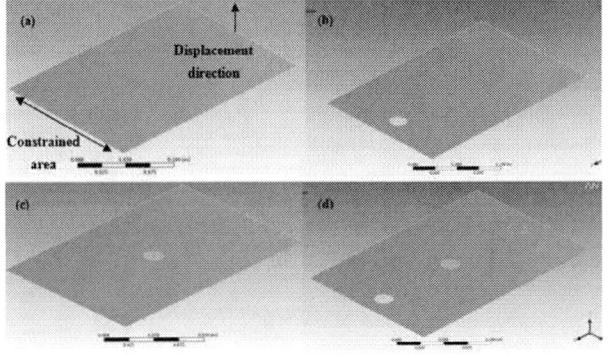

Fig. 1. The 3D model of PCB, (a) Without hole and (b) with single front hole (c)with single centric hole, (d) 3 through holes

All four models were developed using (Solid 186) elements. The 3D PCB models were meshed with 69867, 67520, 68872 and 69308 numbers of elements which were generated by the meshing option available in the simulation software for the PCB without hole, with 3 through holes, with single centric hole and with single hole respectively. The general dimension of all PCB were set to 175mm x 250mm with thickness of 1mm. The diameter of the holes used for the PCB models were 30 mm. For the single hole PCB model, the hole was place 15mm from the constrained area. As for single centric through hole PCB model, the hole was placed at the center. As for the through hole PCB model, the hole was placed at an equivalent distance from constrained area. As the designs of all four models are in three dimensions, The *XYZ*-coordinate system was set to the PCB 3D models. The displacement technique was used to replicate the arching motion which occurs during the depanelling process. The displacement loading was placed in the Y direction as shown in Figure 2. The displacement height was varied from 1cm to 5cm. Next constrained was applied to the PCB models as shown in Figure 2 to result in zero degree movement. Table 1 shows the material properties used.

TABLE1
MATERIAL PROPERTIES

Material	Density, ρ(kg/cm3)	Young's modulus,E (Pa)	Poisson ratio, v
FR4	1850	22×10^9	0.136

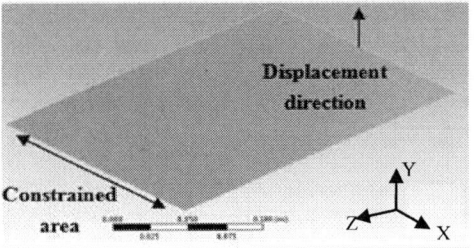

Fig. 2. Mechanical Arching Boundary conditions applied to the 3D model

III RESULTS AND DISCUSSION

The stress distribution of PCB during the arching process and technique to manage it were investigated. The stress distribution of the PCB were evaluated using 4 types of PCB geometry, one without hole, one with single front hole, one with single centric hole and one with three through holes. The holes were placed in various positions to scrutinize the stress distribution of the PCB.The stress distributions of all 4 PCB were compared.The simulated results are based on von Mises stress of stress distribution. The results of the simulation is summarized in Table 2 where PCB1 is PCB without hole, PCB 2 is PCB with single front hole, PCB 3 is PCB with single centric hole and PCB 4 is PCB with 3 through holes.

TABLE 2:SIMULATION RESULTS

Disp (cm)	PCB 1 (Pa)	PCB 2 (Pa)	PCB 3 (Pa)	PCB 4 (Pa)
1	534500	803420	529690	809220
2	1069000	1606800	1059400	1618400
3	1603500	2410300	1589100	2427600
4	2138000	3213700	2118800	3236900
5	2672500	4017100	2648500	4046100

In general, the simulated results exhibited that the stress distribution of the PCB are significantly influenced by the hole structure. From the simulated results, PCB with 3 through holes exhibited the highest stress distribution among the 4 types of PCB. This is followed by the PCB with single front hole which had the second highest stress distribution. As for the single centric hole PCB model, the stress distribution was the lowest among all 4 types of PCB models. Largest stress for all 4 PCB were obtained at the displacement height of 5cm In the course of the depaneling process, the mechanical arching causes a curvature bend on the PCB structure [10]. During the arching motion, stress occurs on the PCB structure. The stress is initiated on the obverse side of the PCB. This stress is then disseminated along the width and length of the PCB structure. The dissemination of stress will then induce damage to the surface mounted components. This can be visualized in

Figure 3(a). The red region near the constrained area of the PCB exhibits the maximum stress region. Hence when, the surface mounted components are placed in this region, the solder joint of the components will experience high stress and tension which may induce crack to the solder ball. Therefore, a hole structure was set close to the constrained area of the PCB. This was done to hub the stress on to the hole structure which function as a frail link as it could not ward off the conflicting stress. As a result, the stress distribution will be directed towards the hole structure and high stress region will build up around the boundary of the hole structure. Therefore, the red region along the width of the PCB is reduced and is focused to the periphery of the hole structure. This can be visualized in Figure 3(b). Hence, the stress on the solder joints can be reduced. However, by referring to Figure 3(c), it is observed that the centric hole structure assisted less in minimizing the high stress region, the red region near the constrained area of the PCB due to the position of the hole structure as it was further away from the constrained area. This explains why the single centric hole PCB modal exhibited the lowest stress distribution among the 4 models in this study. When the amount of hole structure is increase, the stress distribution is diverse and focused on the each hole structure as visualized in Figure.3(d). Among the three holes, the hole near the constrained area exhibited high stress region around it's perimeter when compared with the other two holes. Thus , this explains why the 3 through hole PCB model exhibited the highest stress distribution among the 4 types of PCB. Therefore, it can be seen that the hole structure has assisted in diverting the high stress region towards its periphery and hence the other part of the printed circuit

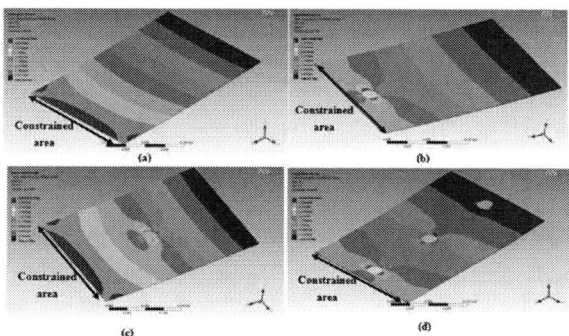

Fig. 3. Stress contour of the PCB plates. (a) without hole and (b) with single front hole (c) with single centric hole (d) with 3 through holes

board exhibits less stress. The positioning of the hole structure has also a noteworthy influence on the stress distribution as suggested by the simulated results. Hence, by positioning the hole structure at proper area of the PCB, the stress distribution can be manage which will reduce the damage on the solder joints of the surface mounted components.

IV CONCLUSION

In this paper, the stress distribution of PCB during the arching process and technique to manage it were investigated. The stress distribution of the PCB were evaluated using 4 types of PCB geometry, one without hole, one with single front hole, one with single centric hole and one with three through holes. The holes were placed in various positions to scrutinize the stress distribution of the PCB.The stress distributions of all 4 PCB were compared. The analysis was done through simulation. The arching process was replicated by using displacement technique. Key results of the analysis showed that the stress of the PCB augmented with increased displacement height. The 3 through hole PCB exhibited the highest stress among the 4 types of PCB geometry. It was also observed that the hole structure assisted in managing the stress distribution during the arching process of the PCB. The positioning of the hole structure also has significant effect in lowering the stress along the length and width of the PCB.

ACKNOWLEDGEMENT

The authors acknowledge the support and thank the School of Microelectronic Engineering,Universiti Malaysia Perlis for the facilities. The authors appreciation are extended to the Ministry of Higher Education for the support given.

REFERENCES

[1] K. R. Haberle and R. J. Graves, "Cycle time estimation for PCB assemblies," *Electronics Packaging Manufacturing, IEEE Transactions on,* vol. 24, pp. 188-194, 2001.

[2] E. Wong and C. Wong, "Approximate solutions for the stresses in the solder joints of a PCB subjected to mechanical arching," *International Journal of Mechanical Sciences,* vol. 51, pp. 152-158, 2009.

[3] Y. Wang, K. Low, H. Pang, K. H. Hoon, F. Che, and Y. Yong, "Modeling and simulation for a drop-impact analysis of multi-layered PCB," *Microelectronics Reliability,* vol. 46, pp. 558-573, 2006

[4] S. Huang, K. Yung, and B. Sun, "A finite element model and experimental analysis of PTH reliability in rigid-flex printed circuits using the Taguchi method," *International Journal of Fatigue,* vol. 40, pp. 84-96, 2012.

[5] B. Chomette, S. Chesne, D. Rémond, and L. Gaudiller, "Damage reduction of on-board structures using piezoelectric components and active modal control—Application to a PCB," *Mechanical Systems and Signal Processing,* vol. 24, pp. 352-364, 2010..

[6] L. Li, S. Kim, S. Song, T. Ku, W. Song, J. Kim, M. Chong, J. Park, and B. Kang, "Finite element modeling and simulation for arching analysis of multi-layer PCB using woven fiber composite," *Journal of Materials Processing Technology,* vol. 201, pp. 746-750, 2008.

[7] T. Y. Tee, J.e. Luan, E. Pek, C. T. Lim, and Z. Zhong, "Novel numerical and experimental analysis of dynamic distributions under board level drop test," in *Thermal and Mechanical Simulation and Experiments in Microelectronics and Microsystems, 2004. EuroSimE 2004. Proceedings of the 5th International Conference,* pp. 133-140, 2004.

[8] J. Wang, P. He, and F. Xiao, "The effect of residual stress on the flexing strength of PCB assembly," in *High Density Microsystem Design and Packaging and Component Failure Analysis, 2004. HDP'04. Proceeding of the Sixth IEEE CPMT Conference,* pp. 146-150, 2004.

[9] Z. Sauli, V. Retnasamy, K. Vengdasalam, M. H. Aziz, R. M. Hatta, and R. Vairavan, "Stress analysis on through holes in PCB," in *Computational Intelligence, Modelling and Simulation (CIMSiM), 2012 Fourth International Conference* ,pp. 144-147,2012.

[10] D. Lau, M. Tsang, S. R. Lee, J. Lo, L. Fu, J. Jin, and S. Liu, "Experimental testing and computational stress analysis of PCB for the

failure prediction of passive components under the depaneling load condition," in *Electronic Components and Technology Conference, 2005. Proceedings. 55th*, pp. 1783-1791, 2005

High Power LED Heat Dissipation Simulation Analysis via Heat Sink Fin Variation

Retnasamy, V.*, Sauli. Z., Vairavan, R., Taniselass, S.

School of Microelectronic Engineering, Universiti Malaysia Perlis (UniMAP), Kampus Pauh Putra, Arau, Perlis.*Corresponding email: vc.sundres@gmail.com

Mamat, H.

MIMOS Berhad, Technology Park Malaysia, Kuala Lumpur.

Abstract– In this work, significance of heat sink fins numbers on the heat dissipation of single chip high power LED package was addressed. The investigation was carried out through simulation by utilizing Ansys version 11. In this work, the heat sink fin numbers were increased from 4 fins to 6 fins, 8 fins, 10 fins and 12 fins respectively. The heat dissipation were evaluated in terms of operating junction temperature, thermal resistance and von Mises stress of the LED chip. Results of the analysis showed that with the increment of heat sink fins, the surface area of the heat sink also increases which in turn reduces the junction temperature of the GaN chip.

Keywords- heat sink; fin number; GaN chip ; high power LED; junction temperature;von Mises stress;Ansys

I INTRODUCTION

Revolution to the lighting industry, the high power light emitting diodes (LEDs) is the latest lighting technology utilized due to its vast advantage in terms of luminous efficiency, power saving and long operating life over the conventional lighting technology[1].Nevertheless, this revolution had created substantial amount of thermal management challenges as the heat created by the chip requires sufficient dissipation to limit the degradation in performance, reliability, efficiency and failure of the high power LED package [2]. Various works have been done to address the heat dissipation issue faced by the high power LEDs. Kim et al. [3] performed a thermal scrutiny on LED array with heat pipe cooling method and scrutinized the ascension of the junction temperature of LED array with varied boundary settings in terms of the effect of heat pipe, convection condition, and ambient temperature. Liu et al. [4] demonstrated a novel thermal management by fabricating silicon based thermoelectric cooling device integrated with the high power LED. Liu et al. [5] demonstrated improvement of the microject cooling system for thermal management application of high power LED arrays where the effectiveness of the cooling system with three various microject structures was investigated. Lu et al. [6] reported an improvement in thermal management of high brightness LED package by implementing a flat loop heat pipe cooling mechanism. Investigation was done to characterize the thermal efficiency of loop heat pipe in terms of start-up performance, temperature uniformity and thermal resistance

under various heat loads and incline angles. Deng et al. [7] proposed liquid metal as the coolant for high power light emitting diodes. In their study, the heat dissipation efficiency, thermal resistance under varied volume flow and instantaneous thermal shock on LED were scrutinized. Lin et al. [8] studied the heat transfer attributes of aluminum oscillating heat pipes. The heat pipes comprised of parallel and square based channels. Parametric analysis was done to obtain the preeminent design in terms of sizes, cross-section and number of turns.

One of the solutions for thermal challenge of high power LED is the utilization external cooling system such as heat sink which would enhance the overall heat dissipation of the package. However, the design of the heat sink such as fin numbers may play an influential part in lowering the operating junction temperature of the high power LED. Hence, this paper reports the investigation of the significance of heat sink design in terms of fins numbers on the heat dissipation of single chip high power LED package. The investigation was carried out through simulation by utilizing Ansys version 11. In this work, the heat sink fin numbers were increased from 4 fins to 6 fins, 8 fins, 10 fins and 12 fins respectively. The heat dissipation was evaluated in terms of operating junction temperature, thermal resistance and von Mises stress of the LED chip.

II METHODOLOGY

The significance of heat sink fins numbers on the heat dissipation of single chip high power LED package was investigated by using Ansys version 11[9] for the simulation. The analysis was done by designing five 3D single chip high power LED models. The basic packgage structure of the single chip high power LED consisted of GaN chip, sapphire substrate, Au-Sn die attach, copper rectangular heat slug, metal core printed circuit board, thermal interface material and aluminum heat sink. In this work, the heat sink fin numbers were increased from 4 fins to 6 fins, 8 fins, 10 fins and 12 fins respectively. The heat dissipation were evaluated in terms of operating junction temperature, thermal resistance and von Mises stress of the LED chip. The dimension of the 3D model is listed in Table 1 where l=length, w = width and h= height. The material properties of the 3D model is listed in Table 2 and the heat sink design variation in terms of fin numbers is listed

978-1-4799-5761-3/14 $31.00 © 2014 IEEE

in Table 3. Figure 1 illustrates the 3D LED package with varied heat sinks

TABLE 1:
3D MODEL DIMENSION

LED Structure	Dimension (mm)
GaN	l=1, w=1, h=0.25
Sapphire	l=1, w=1, h=0.25
Au–20Sn (Die Attach)	l=1, w=1, h=0.125
Copper (Heat slug)	l=5, w=5, h=1
MCPCB	l=8, w=6, h=0.25
TIM	l=8, w=6, h=0.125
Aluminum(Heat sink)	l=20, w=20, h=10.625

TABLE 2:
MATERIAL PROPERTIES

Material	Thermal conductivity k (W/m°C)
GaN	130
Sapphire	42
Au–20Sn (Die Attach)	57
Copper (Heat slug)	401
MCPCB	201
TIM	0.75
Aluminum(Heat sink)	237

TABLE 3:
HEAT SINK FIN NUMBER VARIATION

Number of fins	4	6	8	10	12
Fin width (mm)	2	2	2	1.8	1.5
Spacing (mm)	4	1.6	0.57	0.22	0.18

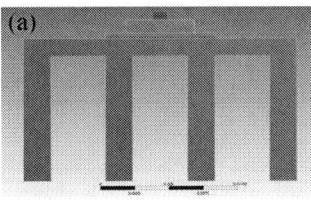

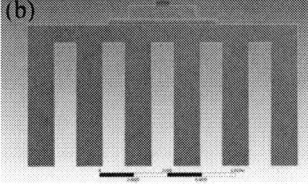

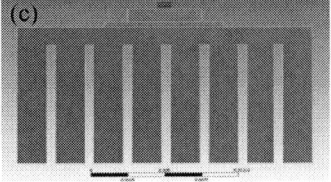

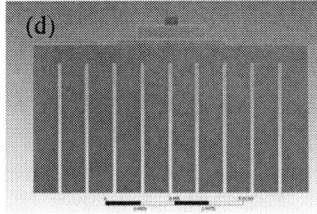

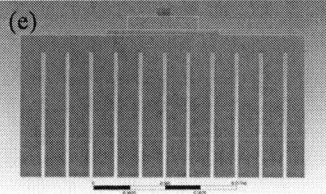

Fig. 1.The 3D model single chip high power LED package (a) 4 fins (b) 6 fins (c) 8 fins (d) 10 fins (e) 12 fins

The GaN chip served as the only plane of heat source where input power of 1 W was applied for the analysis. The simulation environment was set to natural convection condition of h= 5 W/m²C with ambient temperature of 25°C. The five 3D single chip high power LED models were meshed with 350k tetrahedral elements with achieved grid independence. The end time of the simulation was set at 10000s to achieve steady state output. The junction temperature and von Mises stress of the GaN chip with varied heat sinks were calculated by the Ansys software. In addition, Equation 1 was used to manually calculate the thermal resistance of the GaN chip with the simulated junction temperature. Equation 1 is defined as:

$$R_{JA} = \left(\frac{T_J - T_A}{P}\right) \tag{1}$$

Where R_{JA} is thermal resistance, T_j is the junction temperature, T_a is the ambient temperature, P is the input power [9]

III RESULTS AND DISCUSSION

This study presents the investigation on the significance of heat sink fins numbers on the heat dissipation of single chip high power LED package. The investigation was carried out through simulation by utilizing Ansys version 11. In this work, the heat sink fin numbers were increased from 4 fins to 6 fins, 8 fins, 10 fins and 12 fins respectively. The heat dissipation were evaluated in terms of operating junction temperature, thermal resistance and von Mises stress of the LED chip at input power of 1 W. Figure 2 shows the temperature gradient for the LED model with 4fin heat sink. From Figure 2, it can be observed that the junction temperature attained a steady state after 8000s with the junction temperature value of 108.71 °C. This approach was used to attained the junction temperature of the other LED design with varied heat sink fin number of 4 fins, 6 fins, 8 fins, 10 fins and 12 fins respectively and the end results

978-1-4799-5761-3/14 $31.00 © 2014 IEEE

after steady state are computed computed in Table 4 which compromises of junction temperature von Mises stress and thermal resistance of the of GaN chip .By referring to the results in Table 4, it observed that as the heat sink fin number increases from 4 fins to 12 fins with increment of 2 fins for each simulation run, the junction temperature of the GaN chip exhibited significant reduction. At input power of 1 W, the GaN chip recorded 108.71°C junction temperature with 4 fin heat sink and 72.32 °C junction temperature with 12 fin heat sink. Thus, the junction temperature recorded a temperature drop of 36.39 °C. This trend was also observed for the von Mises stress of the LED chip the GaN chip. For input power of 1 W, the GaN chip with 4 fin heat sink recorded 212.46 MPa von Mises stress and 115.65 MPa with 12 fin heat sink. After increasing the number of fins in the heat sink, the von Mises stress of the LED showed a drop of 96.81 MPa in stress. In addition, the thermal resistance of the GaN chip has also decreased. At input power of 1 W, the thermal resistance of GaN LED chip with 4 fin heat sink showed thermal resistance of 83.71 °C/W and 47.32 °C/W for the GaN LED chip with 12 fin heat sink . Thus, the thermal resistance showed total reduction of 36.39 °C.

The temperature contour of the LED model with 4 fin heat sink is illustrated in Figure 3. From the illustration, the red region of the contour signifies the max temperature of the LED package.

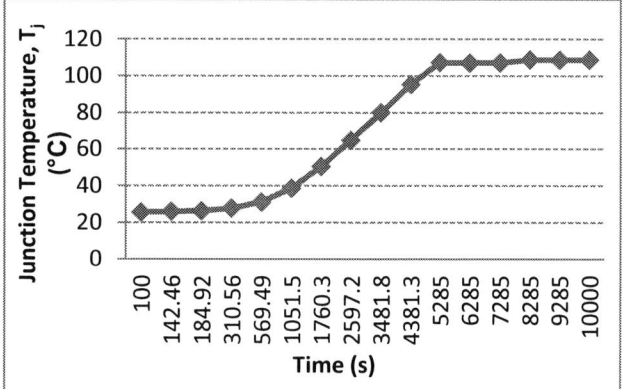

Fig. 2.Junction temperature gradient the LED model with 4fin heat sink

TABLE 4: SIMULATION RESULTS

Heat sink (Fins)	Junction Temperature (°C)	Von Mises Stress (MPa)	Thermal Resistance (°C/W)
4	108.71	212.46	83.71
6	93.23	173.76	68.23
8	83.45	148.48	58.45
10	76.88	128.91	51.88
12	72.32	115.65	47.32

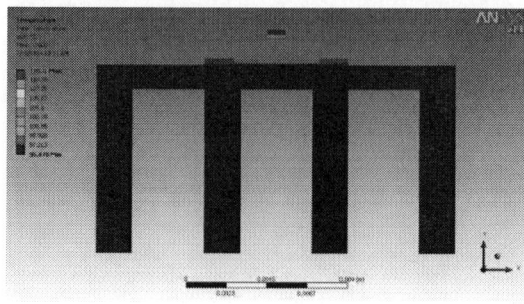

Fig.3.The temperature contour of the LED model with 4 Fin heat sink

Hence, by summing up the overall findings of this study, the surface area of the heat sink influences the heat dissipation of the LED. The overall reduction in terms of junction temperature, von Mises stress and thermal resistance is due to the increase of the overall surface area of the heat sink [10]. The surface area plays an important role in the heat dissiaption of the LED [1,2,5]. With the increment of heat sink fins, the surface area of the heat sink also increases. Thus heat generated by the LED chip has more surface area to be disseminated. The junction temperature of the GaN chip reduces with the increase surface area of heat sink, the von Mises stress and thermal resistance of the GaN has simultaneously decreased. Thus , the total surface area of the heat sink significantly influences the heat dissipation of the LED package. With the increase of the heat dissipation , the life time of the LED package can be extended [10]. Hence, this simulation work can be utilized as a guideline for fabricating the heat sink for actual application.

IV CONCLUSION

This paper reports the significance of heat sink fins numbers on the heat dissipation of single chip high power LED package. The investigation was carried out through simulation by utilizing Ansys version 11. In this work, the heat sink fin numbers were increased from 4 fins to 6 fins, 8 fins, 10 fins and 12 fins respectively. The heat dissipation were evaluated in terms of operating junction temperature, thermal resistance and von Mises stress of the LED chip. Results of the analysis showed that with the increment of heat sink fins, the surface area of the heat sink also increases which in turn reduces the junction temperature of the GaN chip.

ACKNOWLEDGEMENT

The authors acknowledge the support and thank the School of Microelectronic Engineering,Universiti Malaysia Perlis for the facilities. The authors appreciation are extended to the Ministry of Higher Education for the support given

REFERENCES

[1] A. Vipradas, A. Takawale, S. Tripathi, V. Swakul, A. Kaisare, and S. Tonapi, "A parametric study of a typical high power LED package to enhance overall thermal performance," in *Thermal and Thermomechanical Phenomena in Electronic Systems (ITherm), 2012 13th IEEE Intersociety Conference*, pp. 308-313, 2012.

[2] R. Vairavan, Z. Sauli, V. Retnasamy, N. Khalid, K. Anwar, and N. Abdullah, "Natural Heat Convection Analysis on Cylindrical Al Slug of LED," *Applied Mechanics and Materials,* vol. 487, pp. 536-539, 2014.

[3] L. Kim, J. H. Choi, S. H. Jang, and M. W. Shin, "Thermal analysis of LED array system with heat pipe," *Thermochimica Acta,* vol. 455, pp. 21-25, 2007.

[4] C. K. Liu, M.J. Dai, C.-. Yu, and S.L. Kuo, "High efficiency silicon-based high power LED package integrated with micro-thermoelectric device," in *Microsystems, Packaging, Assembly and Circuits Technology, 2007. IMPACT 2007. International*, pp. 29-33, 2007.

[5] S. Liu, J. Yang, Z. Gan, and X. Luo, "Structural optimization of a microjet based cooling system for high power LEDs," *International Journal of Thermal Sciences,* vol. 47, pp. 1086-1095, 2008.

[6] X.Y. Lu, T.C. Hua, M.J. Liu, and Y.X. Cheng, "Thermal analysis of loop heat pipe used for high-power LED," *Thermochimica Acta,* vol. 493, pp. 25-29, 2009..

[7] Y. Deng and J. Liu, "A liquid metal cooling system for the thermal management of high power LEDs," *International Communications in Heat and Mass Transfer,* vol. 37, pp. 788-791, 2010.

[8] Z. Lin, S. Wang, J. Huo, Y. Hu, J. Chen, W. Zhang, and E. Lee, "Heat transfer characteristics and LED heat sink application of aluminum plate oscillating heat pipes," *Applied Thermal Engineering,* vol. 31, pp. 2221-2229, 2011.

[9] R. Vairavan, V. Retnasamy, and Z. Sauli, "CuDia Slug Size Variation Analysis on Heat Dissipation of High Power LED," *Applied Mechanics and Materials,* vol. 487, pp. 33-36, 2014.

[10] L. Y. Bing, "On Thermal Structure Optimization of a Power LED Lighting," *Procedia Engineering,* vol. 29, pp. 2765-2769, 2012.

The Impact of Scaled Channel Length in Tunneling Field Effect Transistors (TFETs)

Nurul Huda Abdul Rahman[1], M. K. Md Arshad[2], Noraini Othman[1], M.F.M. Fathil[2], M.S. Nur Humaira[2] and U. Hashim[2]

[1]School of Microelectronic Engineering
[2]Institute of Nano Electronic Engineering (INEE),
Universiti Malaysia Perlis (UniMAP)
01000 Kangar, Perlis, Malaysia.
mohd.khairuddin@unimap.edu.my

Abstract— In this paper, we investigate the channel length (L_{CH}) of the silicon-on-insulator (SOI) n-type tunneling field effect transistor (NTFET) with the respect of device performance. 2D-device simulation was used for simulating the devices with 30 nm gate length of SOI NTFET with 10 nm thin buried oxide (t_{BOX}) and 7 nm thin silicon body (t_{si}).. The device performance, such as threshold voltage (V_{TH}), ON current (I_{ON}), OFF current (I_{OFF}), and subthreshold swing (SS) was extracted from the current-voltage characterizations of SOI NTFET. The longer the channel length, the lower the SS value and I_{ON}/I_{OFF} ratio obtained in these simulations. Unfortunately, the SS values obtained throughout these simulations were still higher than the typical SS value of TFET device which is supposed to be lower than 60 mV/decade. However, the SS values obtained was still lower compared to the SS value of the MOSFET. On the other hand, I_{ON}/I_{OFF} ratio still high which is better for switching operation of the devices.

Keywords— *Tunneling Field Effect Transistor (TFET), Metal Oxide Semiconductor Field Effect Transistor (MOSFET), and Silicon-On-Insulator (SOI)*

I. INTRODUCTION

MOSFET device is the most common transistor used in both digital and analog circuits. If the MOSFET device is scaled down to nanometer dimension size, there will produce a short channel effect of the device. Several problems associated with short channel effect such as reduced device performance, enhanced leakage current, punch through, and Drain Induced Barrier Lowering (DIBL). In addition, TFET has almost same fabrication process steps with standard CMOS processing where it does not require additional process steps or mask. Therefore, TFET becomes an alternative way instead of the MOSFET to overcome these problems because TFET does not affected by these problems.

TFET is suitable to replace the MOSFET because TFET can achieve higher I_{ON}/I_{OFF} ratio where it is better for low power device application. The researcher had concluded that the TFET can achieve higher I_{ON}/I_{OFF} ratio at lower operating voltages [1]–[4]. Guo et al. concluded that, "TFETs employ band-to-band tunneling (BTBT) as the carrier injection

mechanism, which can provide steeper than 60mV/decade of SS value at room temperature". There are several advantages TFET over MOSFET such as better immunity towards short channel effects, the SS value does not limited to 60mV/decade, enhanced the speed of the devices due to tunneling, higher ON/OFF current ratio to get faster switching operation, and suitable for low power applications due to lower leakage current [5]–[7]. In this work, we investigate the channel length (L_{CH}) of the silicon-on-insulator (SOI) n-type tunneling field effect transistor (NTFET) with the respect of device performance.

II. SIMULATED STRUCTURE

Fig. 2 a illustrates the cross section of a basic SOI NTFET device structure, simulated by using TCAD Sentarus® simulator tool software. The models that had been included in the simulation were Shockley-Read Hall recombination and hydrodynamic transport model. For mobility models we choose Enormal, PhuMob, and HighFieldSaturation models. In addition to that, the thin layer mobility had been included in these simulations since the silicon body thickness (t_{si}) is only 7 nm thick as shown in Zone 4 of Fig. 2 a.

The gate oxide thickness (t_{ox}) is 1.2 nm whereas the buried oxide thickness (t_{BOX}) is 10 nm as shown in Zone 1 and Zone 2 respectively. Hence, it shows that this SOI NTFET device is in nanometer dimension size. In addition, due to thin t_{BOX}, the back gate electrode can be used for dynamic V_{TH} control to optimize the tradeoff between high performance and low power or to compensate process-induced V_{TH} variations to improved parametric yield [8]. On the other hand, reverse back-gate biasing which is applies to TFET device can suppress short-channel effects [8]. This 2-D SOI NTFET devices have been simulated to analyze the device performance at two drain biases i.e. V_D = 20 mV and 1.0 V. The V_G were swept simultaneously from 0 to 3 V with step of 10 mV.

This NTFET device had an opposite doping type of the source (boron) and drain (phosphrous) where the drain had the same doping type with substrate. However, the drain had higher doping concentration compared to the substrate. The substrate and channel were lightly doped with n-type dopant while the source and the drain were heavily doped with p-type and n-type dopants respectively as shown in Fig. 2 a. The doping concentration in the source and the drain were 10^{20} cm^{-3} whereas the doping concentration in the substrate and the channel were 10^{16} cm^{-3}. Hence, the channel behavior was intrinsic (undoped). Channel length had been varied arbitrarily throughout this project to study the effect of channel length on device performance as will be explained next.

Fig. 2 b shows the total doping concentration of the different structures. The doping concentration measurements were taken across the a-b line as shown in Fig. 2 a. The channel length (L_{CH}) was various for each of the sample structure. There were structure A, structure B, and structure C with L_{CH} of 34 nm, 36 nm, and 38 nm respectively. The channel length started with 34 nm due to the gate length is 30 nm. Besides that, there must have at least 2 nm gap between the gate to the drain and the gate to the source. On the other hand, Fig. 2 b shows that the doping concentrations for each of the sample structure were constant in every region.

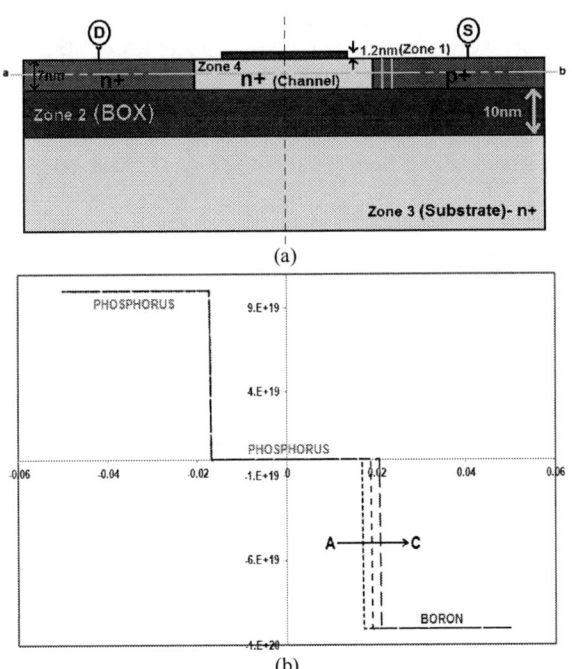

(a)

(b)

Fig. 2: (a) The simulated structure of SOI N-type TFET (NTFET) device structure with (b) the total doping concentration across the a-b line.

III. RESULTS AND DISCUSSION

A. Current-voltage (I_D-V_G) characteristics

Fig. 3 illustrates the current-voltage (I_D–V_G) characteristics of the different gate lengths in linear-linear (left) and log-linear (right) scale. Although the variation is quite small, but one sees that structure C produce highest I_D

while structure A produce lowest I_D.. The extraction value of the threshold voltage, drain induced barrier lowering (DIBL), on-to-off current ration (I_{ON}/I_{OFF}) and subthreshold slope (SS) will be discussed next.

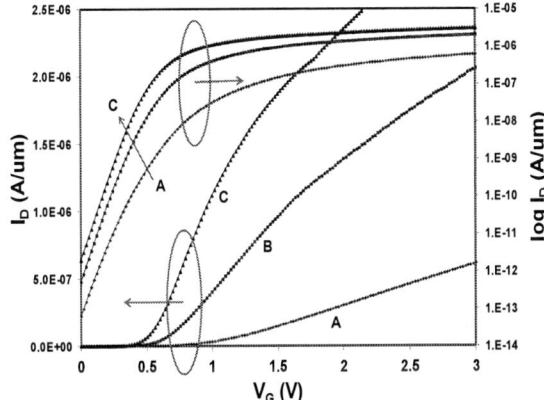

Fig. 3: The I_D-V_G characteristics of the SOI NTFET devices at $V_D = 20$mV and $V_G = 3$V.

B. Threshold voltage (V_{TH})

Fig. 4 shows the extracted results from I_D–V_G characteristics (Fig. 3). The V_{TH} value of this SOI N-TFET device is extracted by using constant current (CC) method. The CC method is determine by using the I_D–V_G characteristics shown in Figure 3 when the V_G corresponding to a given arbitrary constant I_D where the typically value for constant I_D is 10^{-7}A [9]. Structure A has the highest V_{TH} value compared to structure C which has the lowest V_{TH} value. On the other hand, the V_{TH} value of structure B is lies in between structure A and structure B. From Fig. 4, the V_{TH} value is decrease as the L_{CH} increase.

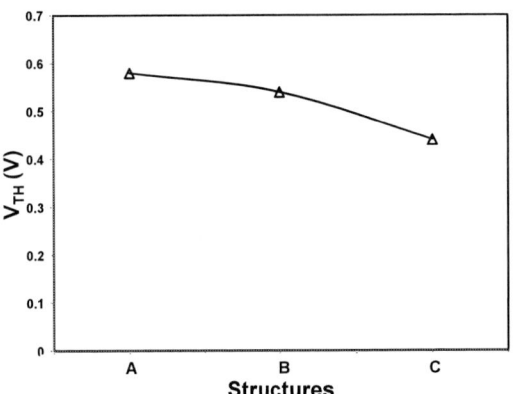

Fig. 4: The V_{TH} values for different structures of SOI NTFET devices.

C. I_{ON}/I_{OFF} ratio

The variation of I_{ON}/I_{OFF} ratio of the different structures can be seen in Fig. 5. Structure A has the highest I_{ON}/I_{OFF} ratio, while structure C has the lowest I_{ON}/I_{OFF} ratio, whereas structure B I_{ON}/I_{OFF} ratio lies in between. The I_{ON} values were extracted when V_G at 1.5V. On the other hand, the extracted values of I_{OFF} were extracted when V_G at 0V. The extracted value of I_{ON} and I_{OFF} slightly increases due to the availability of more charge carriers as the channel length increases. From the extracted results, the I_{OFF} values found throughout these simulations are still lower than the MOSFET which is less than 1 nA/μm. This is because the TFET operate in reverse-biased p-i-n diodes [10]. The smaller the value of I_{ON} and I_{OFF} generated by the devices, the better switching time of the devices. Hence, it will enhance the operating speed of the devices.

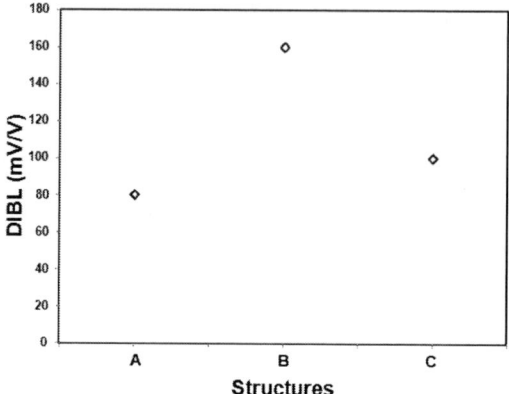

Fig. 6: The DIBL values for different structures of SOI NTFET devices at V_D = 20 mV and V_D = 1.0 V.

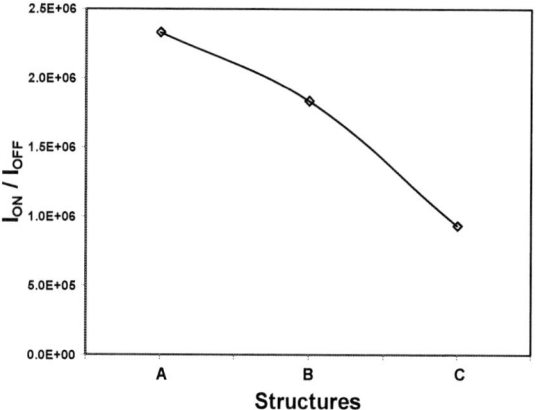

Fig. 5: The I_{ON}/I_{OFF} ratio versus the structures of SOI NTFET devices.

D. Drain Induced Barrier Lowering (DIBL)

Fig. 6 illustrates the DIBL values for different structures of SOI N-TFET devices. The DIBL values can be measured by determining the difference between the V_{TH} value of the substrate bias of V_D = 20 mV and substrate bias of V_D = 1.0 V of SOI N-TFET devices. Structure B has the highest DIBL value while structure A has the lowest DIBL value. Meanwhile, structure C lies between structure A and structure B. It can be seen that the increase in I_D associated with the changes of the DIBL. The different amount of DIBL between structure A and structure B is higher compared to the different amount of DIBL between structure B and structure C. Therefore, the quantity of the I_D increase between structure A and structure B is higher compared than the quantity of the I_D increase between structure B and structure C.

E. Subthreshold Slope (SS)

Fig. 7 illustrates the SS values of SOI NTFET devices at V_D = 20mV for different structures. Structure A has the highest SS value compared to structure C which has the lowest SS value. On the other hand, structure B lies in between structure A and structure C. The extracted SS values of the structures were in the range between 110-to-96 mV/decade, which is higher than 60 mV/decade for TFET device. We can find that, the desired SS values does not achieve throughout these simulations. Hence, the future works can be done to analyze this NTFET device to achieve less than 60 mV/decade. The dopants density and junction depth of the channel, source, and drain or t_{si} is very thin in this project, which could be the main parameter that affects the SS value of the NTFET device. This is because most of other research or work which had been introduced this TFET device designed very deep junction depth for their device [5], [10].

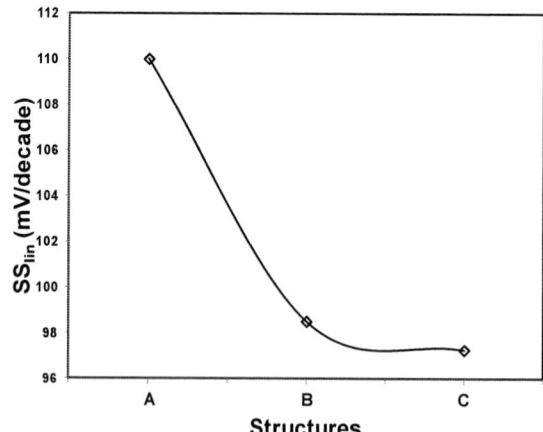

Fig. 7: The linear subthreshold slope (SS) for different structures of SOI NTFET devices.

Table 1 indicates that comparison between the fabricated and simulated devices of the SOI NTFET devices on device performances [10]. The fabricated of the SOI NTEFT device have 70 nm of L_G, 2 nm of t_{OX}, and 70 nm of t_{BOX} by Choi et al., whereas the simulated SOI NTFET device have 30 nm of L_G, 1.2 nm of t_{OX}, and 10 nm of t_{BOX} that has been used in this simulation. The fabricated device have better V_{TH} and SS values compared to the simulate device. However, the simulate device has better I_{ON} and I_{OFF} values. One can see that, the V_{TH} value, the SS value, I_{ON}/I_{OFF} ratio increases as the device size become smaller in term of L_G, t_{OX}, and t_{BOX}. The increases of the V_{TH} and the SS value is not favorable for low power device application, whereas the increases of the I_{ON}/I_{OFF} ratio are good for low power device application.

TABLE 1: THE DEVICE PERFORMANCES FOR FABRICATED AND SIMULATED DEVICES OF SOI NTFET DEVICES [10].

	V_{TH} (V)	I_{ON} (A)	I_{OFF} (A)	SS (mV/dec)
Fabricate Device	0.12	1.21μ	5.4 n	52.800
Simulate Device	0.52	1.87μ	2.0 p	97.238

IV. CONCLUSION

Overall, the I_{ON}/I_{OFF} ratio has been optimized by optimizing the L_{CH} of these NTFET devices. The I_{ON}/I_{OFF} ratio of these NTFET devices found throughout this project is high. This higher I_{ON}/I_{OFF} ratio is important to the semiconductor devices as it will affect the switching time for the device to operate. Besides that, this NTFET device is very suitable for lower power applications as the I_{OFF} obtained throughout this project was very low.

ACKNOWLEDGEMENT

The authors are grateful to the Institute of Nano Electronic Engineering (INEE) of University Malaysia Perlis (UniMAP) for the opportunity they give us to publish this paper. Special thanks to the Microelectronics School for providing licenses of TCAD Sentaurus which were used throughout this project. The author also would like to acknowledge all the team members in Institute of Nano Electronic Engineering (INEE) for their guidance and help.

REFERENCES

[1] R. Fritschi, K. Boucart, F. Casset, P. Ancey, and A. M. Ionescu, "Suspended-Gate Gate Drain Ground," pp. 8–10, 2005.

[2] G. Amaratunga and W. Reddick, "Silicon surface tunnel transistor," *Appl. Phys. Lett.*, vol. 67, no. 4, pp. 494–496, 1995.

[3] K. Gopalakrishnan, P. B. Griffin, and J. D. Plummer, "I-MOS: a novel semiconductor device with a subthreshold slope lower than kT/q," *Dig. Int. Electron Devices Meet.*, pp. 289–292, 2002.

[4] A. Guo, P. Matheu, and T. K. Liu, "SOI TFET Enhancement via Back Biasing," *IEEE Trans. Electron Devices*, vol. 58, no. 10, pp. 3283–3285, 2011.

[5] M. Aswathy, N. M. Biju, and R. Komaragiri, "Simulation Studies of a Tunnel Field Effect Transistor (TFET)," *Int. Conf. Adv. Comput. Commun.*, 2012.

[6] Woo Young Choi, "Comparative study of tunneling Fields effect transistors and Metal Oxide semiconductor Field Effect Transistors," *Jpn. J. Appl. Phys.*, vol. 49, no. 4, pp. 1–13, 2010.

[7] Christian Philipp Sandow, "Modelling, Fabrication and Characterization of silicon tunnel field effect transistor," RWTH Aachen, Germany.

[8] T.-J. K. LIU, "Through the Back Gate," *Advanced Substrate News*, 2008.[Online].Available: http://www.advancedsubstratenews.com/2008/05/through-the-back-gate/. [Accessed: 26-Dec-2013].

[9] F. J. Garc, M. Estrada, and Y. Yue, "A review of recent MOSFET threshold voltage extraction methods," *Microelectron. Reliab.*, vol. 42, pp. 583–596, 2002.

[10] W. Y. Choi, B. Park, J. D. Lee, and T. K. Liu, "Tunneling Field-Effect Transistors (TFETs) With Subthreshold Swing (SS) Less Than 60 mV/dec," *IEEE Electron Device Lett.*, vol. 28, no. 8, pp. 743–745, Aug. 2007.

Performance Evaluation of Dual-Channel Armchair Graphene Nanoribbon Field-Effect Transistor

Adila Syaidatul Azman, Zaharah Johari and Razali Ismail

Department of Electronics and Computer Engineering
Faculty of Electrical Engineering
Unversiti Teknologi Malaysia
81310 UTM Johor Bahru, Johor, Malaysia
dila.syaida@gmail.com

Abstract—**Graphene has become a potential successor to silicon in electronic devices. In this paper, the performance of dual-channel armchair graphene nanoribbon field-effect transistor (AGNR FET) is investigated. Both physical and electrical properties of dual-channel AGNR FET are simulated using Atomistic Tool Kit from Quantum Wise. Their band structures and transmission spectra are analyzed. Current-voltage characteristic is then extracted and the performance of single and dual-channel AGNR FETs is compared. From the simulation, it is found that dual-channel AGNR FET exhibits significant improvement in ON current over two fold. Results obtained will give insight in the implementation of dual-channel AGNR FET for performance enhancement in future electronic devices.**

Keywords—*armchair graphene nanoribbon, dual-channel, field-effect transistor.*

I. INTRODUCTION

Graphene has attracted numerous research attentions since it was discovered in 2004 [1]. Its unique structure makes it another successor to Silicon for future electronic devices. Graphene in its two-dimensional (2D) form do not have band gap that is essential in switching devices. As a solution, the 2D graphene is carved down into narrow width becoming a one-dimensional graphene nanoribbon (GNR). GNR is useful as switching devices as it has bandgap that allow devices to switch on and off. Therefore, GNR is of great interest to be used as the conducting channel in FET for channel length less than 10 nm to replace the conventional silicon material [2].

GNR may present in two types depending on its edges namely zigzag and armchair. Zigzag GNR (ZGNR) has intrinsic zero band gap due to its metallic behavior. However, it is necessary to induce bandgap in ZGNR by applying voltage to the electrodes and metal gate [3]. On the contrary, armchair GNR (AGNR) possesses semiconducting properties and its bandgap is tunable by its width [3]. Therefore, AGNR will be the focus of the study in this paper.

Optimized performance and ability to deliver sufficient drive current are desirable in electronic devices like transistor. Researchers have started to apply multiple conducting channels in FET. For example,a local back-gated aligned carbon nanotube (CNT) array transistor has been proven to exhibit current density of more than 40 μA/μm [4]. The practicality of implementing array channel in FET is also evident in array vertical nanowire transistor with gated surround structure where it exhibits improvement in current flowing [5]. The enhancement in array FET performance is promising and yet to be demonstrated using graphene. This is of great interest since graphene and CNT originated from the same source and the fabrication process of graphene is simpler compared to CNT. A number of researchers have demonstrated various ways of fabricating array graphene-based FET [6-8]. It is shown that the array graphene field-effect transistor (GFET) exhibits enhancement in on/off current ratio Obviously, the back-gated graphene FETs array based on the nano-wall graphene channel demonstrated typical field-effect behavior with linear and saturation regions [9] This observation is fascinating as saturation is a hot debate topic in graphene-based FET. This indicates that multiple conducting channels of graphene may play significant role in some applications.

Although the array-channel graphene-based device has been fabricated, additional works particularly at computational level need to be done. This is to clarify the relation between the numbers of conducting channel with the FET performance. This has not been reported in Ref. [6-8]. Therefore, in this paper the performance of dual-channel AGNR FET is investigated. Hence, performance comparisons using the single and dual-channel AGNR FETs are assessed. In the next section, details of the computational approach used in this study are described. Fig. 1 below shows the structure of single and dual-channel AGNR FETs.

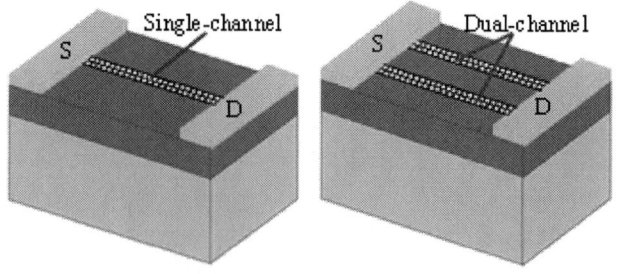

Fig. 1. Structures of single and dual-channel AGNR FET without the top-gate

978-1-4799-5761-3/14 $31.00 © 2014 IEEE

II. COMPUTATIONAL METHOD

The simulation is carried out using Atomistix Tool Kit (ATK) software from Quantum Wise version 13.8.1.

A. Model

Different configurations of AGNR are designed with different widths and lengths for closed system of single-channel and dual-channel AGNR FETs as shown in Fig. 2. The red lines indicate the source and drain electrodes and the blue lines indicate the metal regions which act as the gate of the AGNR FETs. The channel width and length of these configurations is extracted and tabulated in Table I. The AGNR sheet is terminated with hydrogen atoms to avoid end-effect by calculation. The optimized bond lengths of carbon-carbon and carbon-hydrogen are 1.42Å and 1.10Å, respectively.

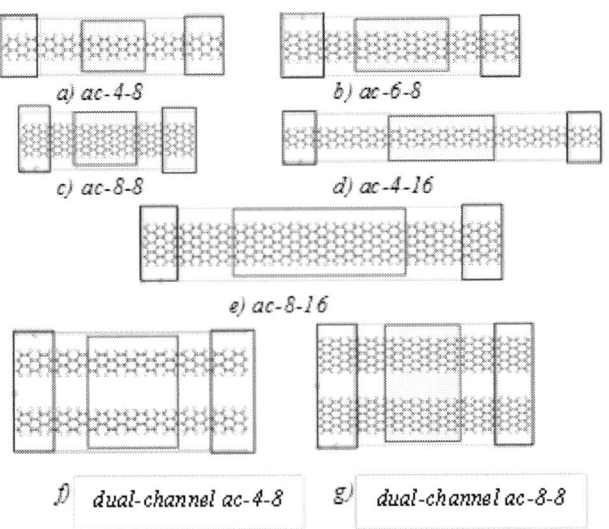

Fig. 2. Different configurations for closed systems of single and dual-channel AGNRFETs. The name is according to the dimension of the configuration, which is "ac-W-L", where ac stands for armchair, W is the width in terms of number of carbon atoms, and L is the length of the scattering region in terms of repetition of one pair of hydrogen atoms.

TABLE I: CHANNEL WIDTH AND LENGTH FOR EACH CONFIGURATION OF AGNR FET

Configuration	Z-Matrix	
	Channel Width (Å)	Channel Length (Å)
ac-4-8	5.59	32.68
ac-6-8	8.05	32.68
ac-8-8	10.51	32.68
ac-4-16	5.59	66.78
ac-8-16	10.51	66.78
Dual-channel ac-4-8	19.27	32.68
Dual-channel ac-8-8	29.12	32.68

B. Extended-Huckel and Non-Equilibrium Green's Function Methods

Self-consistent Extended-Huckel and Non-Equilibrium Green's Functional (NEGF) methods [10] are used in order to study the physical and electrical properties of AGNR FETs. In the first stage, the physical properties of AGNR such as band structures, transmission spectra and density of states are calculated using the Recursion calculator. In the second stage, similar methods are used to analyze the current versus voltage characteristics. To solve Poisson equation in the "electrode" calculations, Fast Fourier transform (FFT) and multigrid techniques are used [10].

III. RESULTS AND DISCUSSION

First, energy band gap was calculated for different configurations of AGNR. According to Hou *et. al.* [10] and Son *et al.* [11], the energy band gap decreases as width of the AGNR increases. From Fig. 3, it is clearly shown that the energy band gap of the AGNR decreases as the width increases. For narrowest AGNR, which is ac-4-8 with channel width = 5.59 Å, the energy band gap is approximately 2.25 eV. As we increase the channel width to 8.05 Å for ac-6-8 configuration, the energy band gap is reduced to 1.2 eV. Meanwhile, for ac-8-8 configuration, with channel width of 10.51 Å, the energy band gap is the smallest, which is 0.1 eV. In addition, for ac-4-16 configuration, the energy band gap is approximately 2.25 eV. On the other hand, the energy band gap for ac-8-16 configuration is approximately 0.08 eV. It is worth noting that the energy band gap is infinitesimal if the AGNR width is kept constant while varying the channel length.

The transmission spectra (TS) for single-channel and dual-channel AGNR FETs with ac-8-8 configuration are plotted in Fig. 4. It can be observed that the transmission coefficient is zero around the Fermi energy level for both single and dual-channel AGNR FETs. Additionally, the width of the zero transmission energy is approximately the same to the amount of band gap. This is in line with observation made by Hou *et al.* [10].

Comparing the TS for single and dual-channel AGNR FETs in Fig. 4, it is observed that beyond zero TS window the single and dual-channel AGNR FETs have transmission coefficient of 1 and 2 respectively. This indicates that there are only one π band appears at the top of the valence band and one π* band appears at the bottom of the conduction band in the single-channel AGNR FET [10]. Meanwhile, the transmission coefficient of 2 achieved in dual-channel AGNR FET indicates the present of two conducting channels. The increment by two fold in transmission coefficient as the number of conducting channel double may explain the enhancement in drain current in latter section.

To analyze the current-voltage characteristic, voltage is supplied to the electrodes. In this study, the voltage used at the drain (right electrode) is 2.0 V, while the source at the left electrode is grounded. The metal gate is biased with 1.0 V. Fig. 5(a) shows the difference in the drain current, I_D for single-channel AGNR FET configurations with two different values of channel length at $V_D = 1.98$ V, which are 32.68 Å and 66.78 Å. The I_D for channel length 32.68 Å is 0.1431 mA, while the I_D for channel length 66.78 Å is 0.1403 mA. This shows that the I_D changes are insignificant although the length of the AGNR is reduced to almost halves. On the contrary, considerable effect can be observed when the width of the AGNR is reduced to halve. This is evident in Fig. 5(b) where I_D is substantially decreased. This indicates that the width in AGNR plays an important role in FET performance.

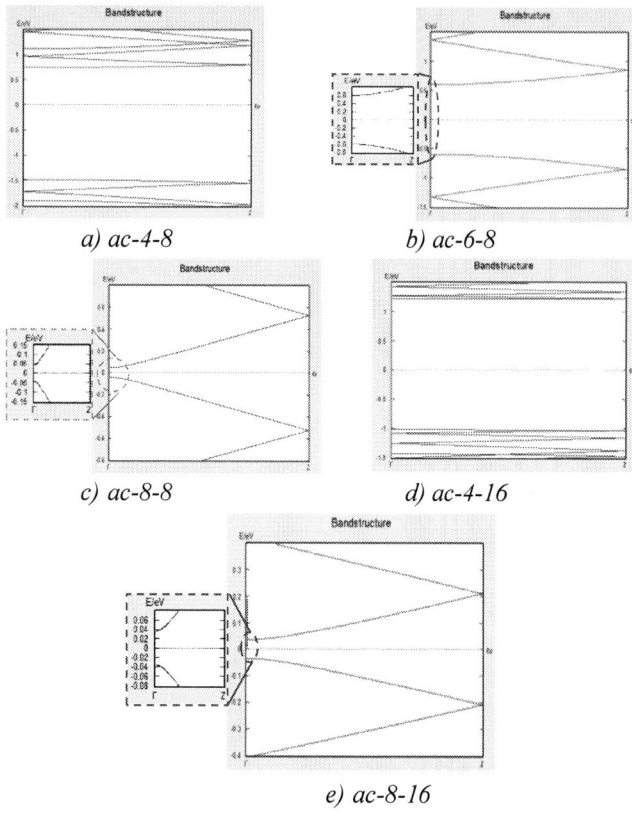

a) ac-4-8   b) ac-6-8

c) ac-8-8   d) ac-4-16

e) ac-8-16

Fig. 3. Band structures for different configurations of AGNR.

However, the I_D for the AGNR FET does not saturate. This could possibly due to the relatively long linear region of graphene, supported by the experiment that had been done by Liu *et al.* using graphene FETs array. They had proven that the I_D of graphene approximately saturates at $V_{DS} = 20$ V [9]. This is not attainable in this study since V_{DS} was swept up to only 2.0 V considering in actual case, graphene could not often withstand high voltage without breaking. Nevertheless, saturation region is desirable in devices like switches and amplifiers. Most importantly, saturated I_D is necessary for a fast switching transistor in high frequency electronic devices. The unsaturated behavior of I_D prohibits the device to switch on and off. Hence, further studies need to be done to investigate the optimized conditions for the AGNR FET to achieve saturation and to reduce the saturation voltage using more than one conducting channel.

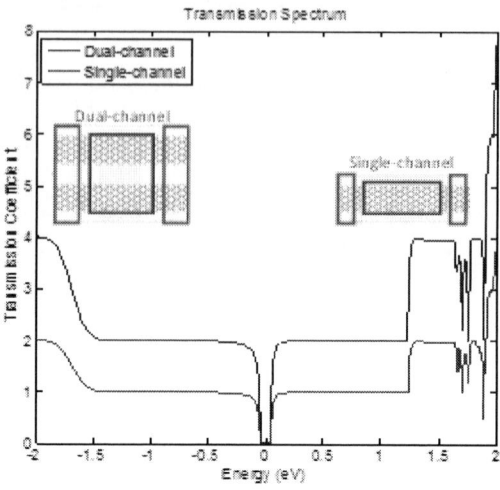

Fig. 4. Transmission spectra for single and dual-channel AGNR FETs for ac-8-8 configuration biased at 1.0V.

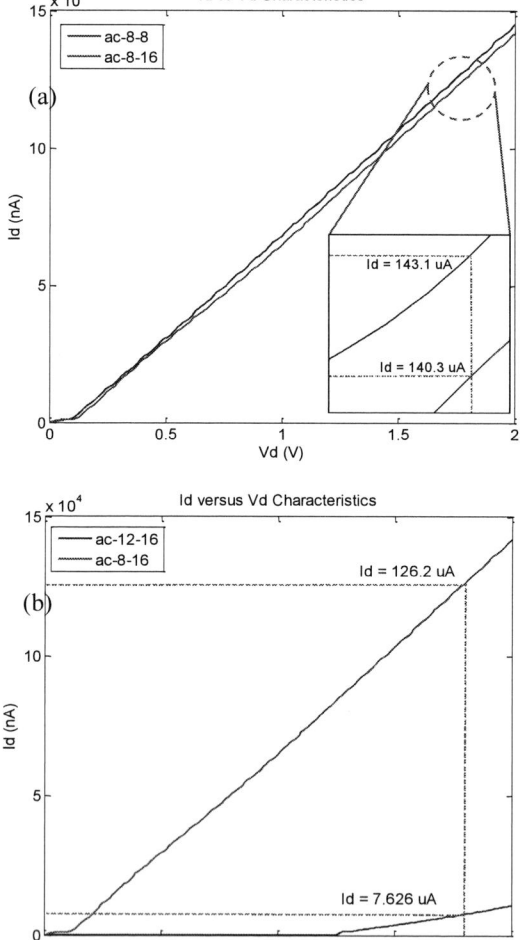

Fig. 5.   (a) I_D-V_D characteristic biased at 1.0V for single-channel AGNR FET W = 10.5083 Å for two different L. (b) I_D-V_D characteristic for single-channel AGNR FET configurations with L = 66.78 Å for two different W.

To evaluate the dual-channel AGNR FET performance, the current-voltage characteristics of single and dual-channel ac-8-8 configuration biased at 1.0V are compared in Fig. 6. From Fig. 6, it can be seen that the I_D for dual-channel AGNR FET is higher than the I_D of single-channel AGNR FET. The value of the drain current for single-channel AGNR FET is 0.1293 mA, while the I_D for dual AGNR FET is 0.2594 mA at V_D= 1.8 V which is approximately double the value of the I_D for single-channel AGNR FET. Since it has two conducting channels, the number of electrons flowing from the drain to the source is double the number of electrons flowing in the single-channel AGNR FET. As the current flowing is directly proportional to the number of electrons flowing from the drain to the source, therefore, it is obvious that the current flowing in dual-channel AGNR FET is double the current flowing in single-channel AGNR FET. To the best of the knowledge, there is no experimental data to compare with the simulation result present. Despite, the trend of the conductivity is similar to that achieved in Ref. [4] where the ON current linearly increase with number of CNT channels.

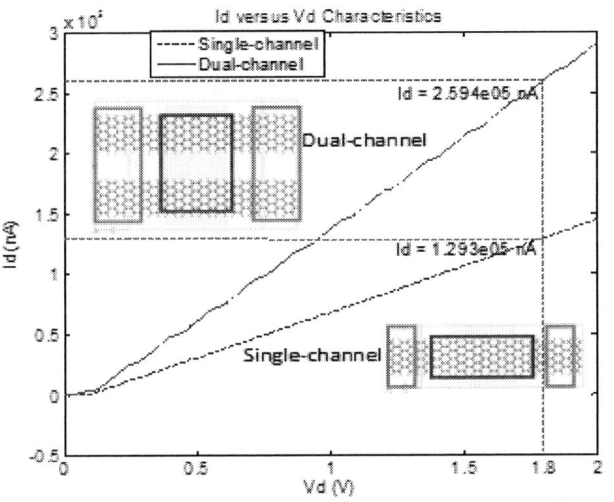

Fig. 6. Drain current versus drain voltage characteristics for single-channel and dual-channel armchair GNR FET ac-8-8 configurations.

IV. CONCLUSION

In this paper, the performance of dual-channel AGNR FET is studied using Extended-Huckel and NEGF methods. From the simulation, it is found that the dual-channel AGNR FET exhibits I_D improvement over two fold compared to single-channel AGNR FET. This shows feasibility of using multiple conducting channels in future electronic devices. The outcome of this study will stimulate experimental effort to confirm the finding.

ACKNOWLEDGMENT

Authors would like to acknowledge the financial support from Research University grant of the Ministry of Higher Education (MOHE), Malaysia under Project Q.J130000.2523.05H23. Also thanks to the Research Management Center (RMC) of Universiti Teknologi Malaysia

(UTM) for providing excellent research environment in which to complete this work.

REFERENCES

[1] K. S. Novoselov, A. K. Geim, S. Morozov, D. Jiang, Y. Zhang, S. Dubonos, *et al.*, "Electric field effect in atomically thin carbon films," *science,* vol. 306, pp. 666-669, 2004.

[2] R. Ismail, M. T. Ahmadi, and S. Anwar, *Advanced Nanoelectronics*: CRC Press, 2012.

[3] E. Kan, Z. Li, and J. Yang, "Graphene Nanoribbons: Geometric, Electronic, and Magnetic Properties," 2011.

[4] A. D. Franklin, A. Lin, H.-S. Wong, and Z. Chen, "Current scaling in aligned carbon nanotube array transistors with local bottom gating," *Electron Device Letters, IEEE,* vol. 31, pp. 644-646, 2010.

[5] A. Hellemans, "Ring around the nanowire [News]," *Spectrum, IEEE,* vol. 50, pp. 14-16, 2013.

[6] J. G. Son, M. Son, K. J. Moon, B. H. Lee, J. M. Myoung, M. S. Strano, *et al.*, "Sub-10 nm Graphene Nanoribbon Array Field-Effect Transistors Fabricated by Block Copolymer Lithography," *Advanced Materials,* vol. 25, pp. 4723-4728, 2013.

[7] G. Liu, Y. Wu, Y.-M. Lin, D. B. Farmer, J. A. Ott, J. Bruley, *et al.*, "Epitaxial graphene nanoribbon array fabrication using BCP-assisted nanolithography," *ACS nano,* vol. 6, pp. 6786-6792, 2012.

[8] X. Liang and S. Wi, "Transport Characteristics of Multichannel Transistors Made from Densely Aligned Sub-10 nm Half-Pitch Graphene Nanoribbons," *ACS nano,* vol. 6, pp. 9700-9710, 2012.

[9] H. Liu, B. Chen, X. Li, C. Lu, Y. Ding, and B. Lu, "Fabrication of patterned graphene FETs array," *Physics Procedia,* vol. 32, pp. 229-234, 2012.

[10] Z. Hou and M. Yee, "Electronic and transport properties of graphene nanoribbons," in *Nanotechnology, 2007. IEEE-NANO 2007. 7th IEEE Conference on,* 2007, pp. 554-557.

[11] Y.-W. Son, M. L. Cohen, and S. G. Louie, "Energy gaps in graphene nanoribbons," *Physical review letters,* vol. 97, p. 216803, 2006.

Novel Low-Voltage RF-MEMS Switch: Design and Simulation

Ma Li Ya, Norhayati Soin
Department of Electrical Engineering
Faculty of Engineering, University of Malaya
50603 Kuala Lumpur, Malaysia
maliya8445@gmail.com norhayatisoin@um.edu.my

Anis Nurashikin Nordin
Department of Electrical and Computer Engineering
International Islamic University Malaysia
53100 Kuala Lumpur, Malaysia
anisnn@iium.edu.my

Abstract—This paper presents a novel design of a low-voltage radio frequency (RF) micro-electromechanical system (MEMS) switch with its electro-mechanical and microwave characteristics' simulation, as well as its virtual fabrication process. The RF-MEMS switch employs a shunt capacitive structure. There are eight beams supporting a rectangular membrane to suspend over a coplanar waveguide (CPW) transmission line. The outer four crab-leg beams are used to bypass the signal line to the ground by the coupling capacitance and via when switch is actuated; and the inner four serpentine beams are used to supply the DC actuation voltage to the membrane. The developed RF-MEMS switch has a very low pull-in voltage of 3.53V; and the maximum von Mises stress under actuated condition is 13.1808MPa. The capacitance ratio of the switch is 218.5, with switch-on capacitance of 68fF. The switch's microwave characteristics are obtained by AWR Design Environment 10® simulation; its insertion loss and isolation is -1.7532dB and -18.394dB respectively at 20GHz; and at frequency of 60GHz, the insertion loss is -7.2499dB and isolation is -27.293dB. A simple low-cost three-mask process with photoresist (PR-S1800) as sacrifice layer to release the membrane is proposed in this design; and a virtual fabricated device is simulated using IntelliFab v8.7® software.

Keywords—*RF-MEMS; switch; low-voltage; electrostatic; virtual fabrication*

I. INTRODUCTION

Low-cost MEMS switches are prime candidates to replace the conventional GaAs FET and p-i-n diode switches in RF and microwave communications systems, mainly due to their low insertion loss, good isolation, linear characteristic and low power consumption [1]. Various RF-MEMS switches have been proposed and designed so far; normally they can be classified into two types, metal-contact switches and capacitive switches. Metal-contact switches, as their name implies, use metal-to-metal direct contact to achieve an ohmic contact between a switch membrane and its signal line [2], and work for DC or low-frequency circuit. Capacitive RF-MEMS switches generally are composed by a signal line, a thin dielectric layer on the signal line, and a suspended membrane. During actuation, a coupling capacitance exists between the signal line and the membrane and these structures are capable of switching up to 100GHz.

The most popular capacitive switches are usually constructed in shunt configuration and actuated by

electrostatic force. Compared with other actuation mechanisms, the biggest limitation of electrostatic capacitive switches is their high actuation voltage (or pull-in voltage). In [3], the fabricated CMOS RF-MEMS switch has an insertion loss less than 0.98dB and isolation of 12.4 to 17.9dB from 10 to 20 GHz, and capacitance ratio of 91; however, its pull-in voltage is around 68V. Jamie Yao et al. proposed a low-loss RF switch with insertion loss and isolation of 0.25dB and 35dB respectively at 35 GHz; but the actuation voltage is about 50V [4]. Various researchers are devoting their work to design RF-MEMS switches with low pull-in voltage, in order to make them compatible with most CMOS circuits. For example, there was an actuation voltage of 10V RF-MEMS switch proposed using robust single-crystal silicon [5]. In 2002, Devarajan et al. proposed three RF-MEMS switches with different hinge geometries, where the lowest pull-in voltage can be achieved is 9V [1]. In 2006, Dai et al. fabricated three RF-MEMS switches also with different beam designs by CMOS process and the smallest pull-in voltage is 7V [6]. Sergio has designed and fabricated a 9V low-voltage RF-MEMS switch by utilizing highly compliant serpentine spring folded suspensions together with large area capacitive actuators [7]. Based on this, it can be seen that in order to reduce the pull-in voltage of RF-MEMS switches, most researchers focus on the supporting beams' design; and lots of different shapes of beams and membranes have been developed, since the efficient spring constant of whole system can directly affect the pull-in voltage.

In this paper, a very low-voltage design and virtual fabrication of a RF-MEMS switch with a novel supporting beams' structure is presented. Section II details the design of the RF switch and the relationship of the switch's spring constant with pull-in voltage. A robust design with pull-in voltage of 3.53V is achieved in this work. In section III, the FEM (Finite Element Method) simulations based on the RF-MEMS switch's electro-mechanical properties and microwave characteristics have been conducted. Section IV illustrates the three mask fabrication process simulated using IntelliFab®.

II. RF-MEMS SWITCH DESIGN

The structure of the RF-MEMS switch and its dimension's parameters are illustrated in Fig. 1 and Table I. The switch mainly contains a coplanar waveguide (CPW) transmission line which consists of two ground lines (G) and a signal line

(S), a very thin silicon nitride (Si_3N_4) layer above the signal line, a big rectangular membrane suspended over the signal line, and eight folded beams to support the membrane. As shown in Fig. 1(c), the thickness of each layer and their gaps are set as: $t_{membrane}$=0.8μm; g=3μm; t_{Si3N4}=0.1μm; t_{CPW}=1μm; and g_{CPW}=4μm.

When a pull-in voltage is applied between the membrane and signal line, an electrostatic force is produced and pulls the membrane down to the signal line. Most of the RF signal is bypassed to the ground by an induced coupling capacitance between the membrane and signal line, which is called the switch-off state. On the contrary, without the applied DC voltage, the membrane is at up-state position; the RF signal can be propagated from an input port to an output port, which is called the switch-on state. This pull-in voltage can be calculated by (1) [8]; and the relationship between the pull-in voltage (Vp) and spring constant (k) can be plotted in Fig. 2.

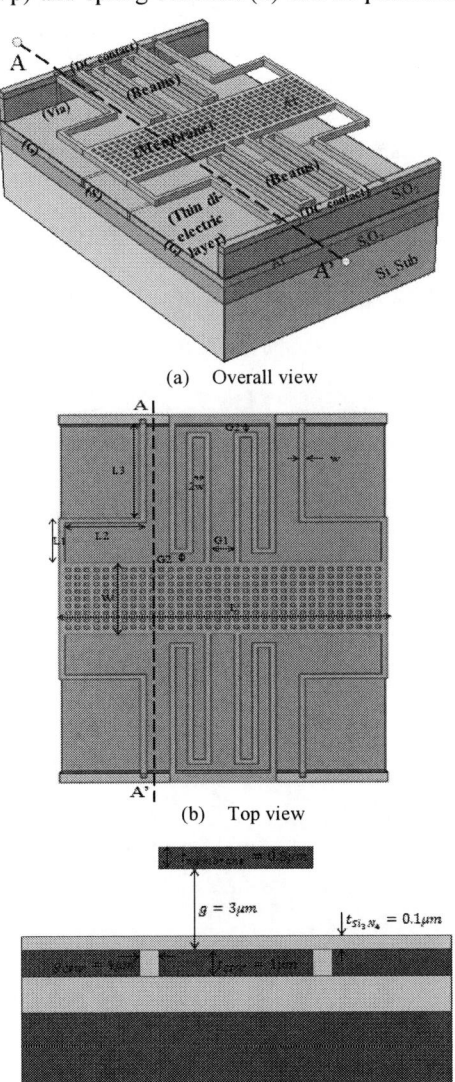

(a) Overall view

(b) Top view

(c) Cross-section view (A-A')

Fig. 1.   RF- MEMS capacitive switch design

TABLE I   Geometric Parameters of Membrane and Beams

Name	L	W	L1	L2	L3	G1	G2	w
Value (μm)	280	80	50	70	105	20	10	5

$$V_p = \sqrt{\frac{8kg_0^3}{27\varepsilon_0 A}} \qquad (1)$$

Where, k is the spring constant of the membrane; g_0 is the initial gap between the membrane and the signal line; ε_0 is the permittivity of air, 8.854×10-12F/m; and A is the area of the membrane, namely the product of the membrane's width and length, $W \times L$.

The highlight pull-in voltage value of 3.5V is the design objective of this work; and the corresponding value of spring constant should be around 0.3N/m. In order to design such low spring constant structure, the crab-leg beams and serpentine beams have been employed here, as shown in Fig. 1(a).

III.   FEM SIMULATION

The finite element method (FEM) simulation is utilized to estimate the mechanical and electric behaviors of the RF-MEMS switch using the software of IntelliSuite®, and the microwave characteristic by AWR Design Environment 10®. The model of the RF-MEMS switch is established in accordance with the dimensions in Fig.1 and Table I. The material of the switch is aluminum with a Young's modulus of 70GPa, Poisson's ratio of 0.33 and mass density of 2700kg/m³, which all follow the default setting in the materials' library of the software. In the boundary setting, the ends of eight folded beams and silicon substrate are fixed. A bias voltage is applied at the membrane and the DC potential of the signal line is set as 0V. Then a z-axis displacement of the membrane is produced. Fig. 3 illustrates the relationship between the applied bias voltage and the membrane's z-displacement of the switch. It can be seen that at voltage of 3.53V, the membrane almost touches on the lower Si_3N_4 layer and the z-direction displacement is around 2.9μm. The 3D z-direction displacement distribution and von Mises stress distribution of the membrane and beams, when the pull-in voltage of 3.53V applied, are displayed in Fig. 4. It clearly shows that the

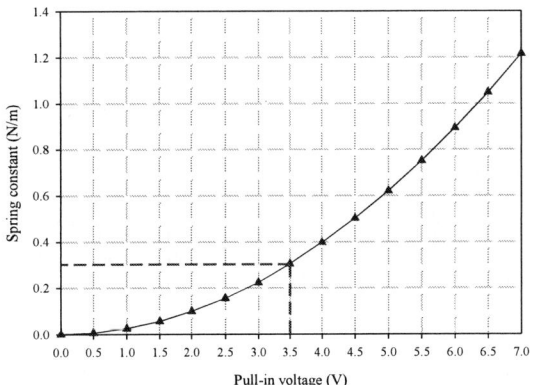

Fig. 2.   Relationship of the spring constant and pull-in voltage

IEEE-ICSE2014 Proc. 2014, Kuala Lumpur, Malaysia

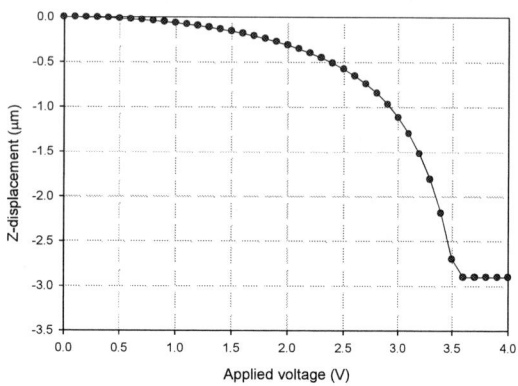

Fig. 3.   Relation of applied voltage and membrane displacement

membrane exhibits a uniform out-of-plane displacement of 2.9µm. The maximum von Mises stress of 13.2MPa is located at the end of the folded beams of the switch which is much lower than the yield strength of aluminum of 124MPa [8]. Therefore, the switch operates in the elastic range.

AWR Design Environment 10® has been used to simulate the microwave characteristics of the RF-MEMS switch. One

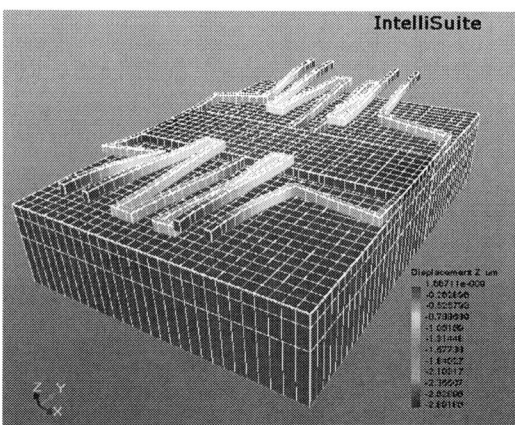

(a) Displacement distribution in z direction

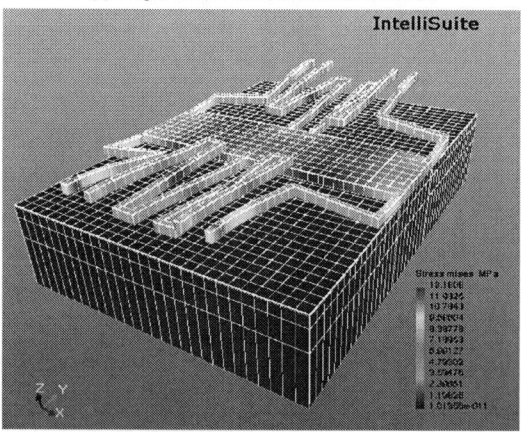

(b) von Mises stress distribution

Fig. 4.   Membrane displacement and stress distribution with the pull-in voltage applied.

of the modules in the software is Microwave office TM, which can be used to run 3D electromagnetic (EM) simulation based on finite element method to solve for port parameters and EM fields within arbitrary 3D structure.   Fig.5 (a) shows the simulated insertion loss and return loss of the RF-MEMS switch when it is not actuated and the membrane is at the up state; Fig.5 (b) displays the simulated return loss and isolation of the designed switch when it is actuated and the membrane is at down state. The simulated results expose that the RF-MEMS switch has an insertion loss of -1.7532dB and isolation of -18.394dB at 20GHz; and at the frequency of 60GHz, the insertion loss is -7.2499dB and isolation is -27.293dB. These two frequency ranges can be used to operate Military radar RF applications and wireless common link, respectively [9].

IV.   VIRTUAL FABRICATION PROCESS

IntelliFab, a process simulation module in IntelliSuite®, is specially designed for MEMS process modeling and virtual fabrication. With IntelliFab, a user can directly build a process traveller, simulate the process and visualize the fabricated 3D structure at the end of each step [10].

Fig. 6 lists the each step setting for the RF-MEMS switch's fabrication. There are totally three masks needed during the fabrication; one is for aluminum wet etch to obtain CPW transmission line; another one is for Si_3N_4 etch to form holes for via; and the third one is used to etch top aluminum layer to produce the membrane and beams. Fig. 7 presents the cross-section view of fabricated results by each step. In this fabrication process, using photoresist-S1800 as the sacrificial

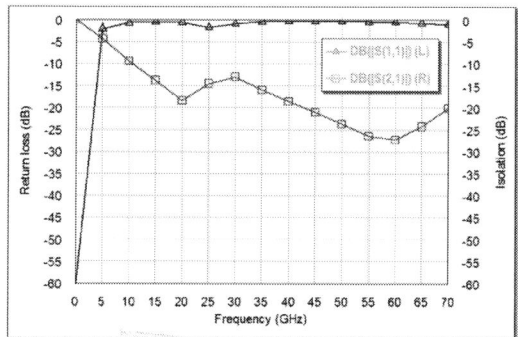

(a) Insertion loss and return loss at switch-on state

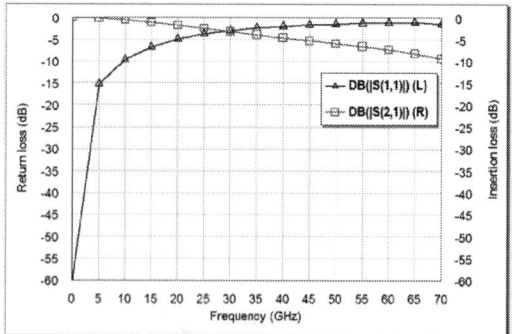

(b) Isolation and return loss at switch-off state

Fig. 5.   Microwave characteristics of the RF- MEMS switch

978-1-4799-5761-3/14 $31.00 © 2014 IEEE   144

#	*	☑ Type	Material	Process	Process ID	Process Option
1	☑	Definition	Si	Czochralski	100	
2	☑	Deposition	SiO2	PECVD	Ar	Conformal Deposition
3	☑	Deposition	Al	Sputter	Ar-Ambient	Conformal Deposition
4	☑	Deposition	PR-S1800	Spin	S1805	Conformal Deposition
5	☑	Exposure	UV	Contact	Suss	
6	☑	Etch	Al	RIE	Cl2_BCl3	Partial Etching
7	☑	Etch	PR-S1800	Wet	1112A	Partial Etching
8	☑	Deposition	Si3N4	PECVD	Ar	Planarization
9	☑	Deposition	PR-S1800	Spin	S1805	Conformal Deposition
10	☑	Exposure	UV	Contact	Suss	
11	☑	Etch	Si3N4	Wet	H3PO4	Partial Etching
12	☑	Deposition	Al	Sputter	Ar-Ambient	Planarization
13	☑	Deposition	PR-S1800	Spin	S1805	Conformal Deposition
14	☑	Exposure	UV	Contact	Suss	
15	☑	Etch	Al	RIE	Cl2_BCl3	Partial Etching
16	☑	Etch	PR-S1800	Wet	1112A	Partial Etching
17	☑	Etch	PR-S1800	Wet	1112A	Sacrifice

Fig. 6.   Virtual fabrication process steps

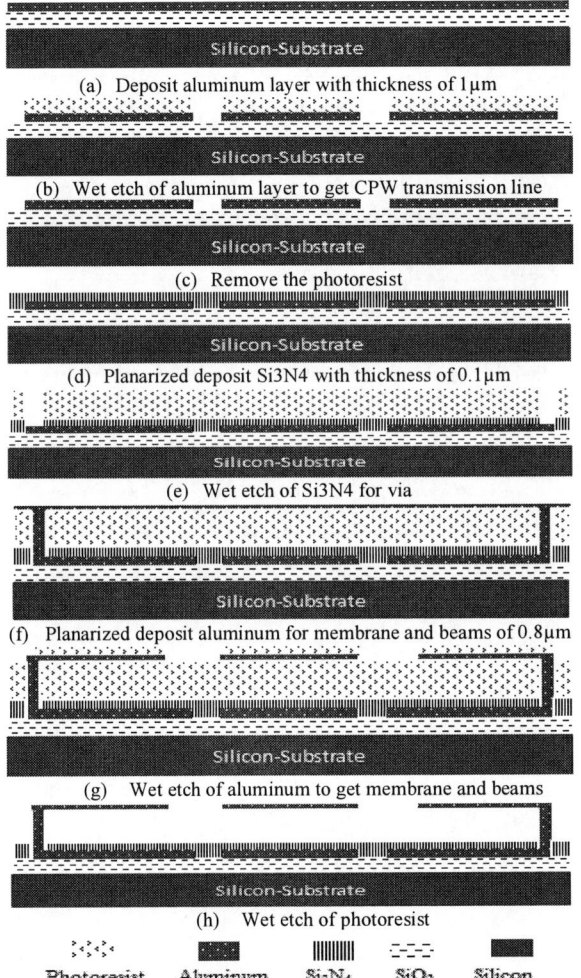

(a) Deposit aluminum layer with thickness of 1μm

(b) Wet etch of aluminum layer to get CPW transmission line

(c) Remove the photoresist

(d) Planarized deposit Si3N4 with thickness of 0.1μm

(e) Wet etch of Si3N4 for via

(f) Planarized deposit aluminum for membrane and beams of 0.8μm

(g) Wet etch of aluminum to get membrane and beams

(h) Wet etch of photoresist

Photoresist   Aluminum   Si₃N₄   SiO₂   Silicon

Fig. 7. Virtual fabrications' step for the RF-MEMS switch

layer to release the top membrane and beams is employed and the result shows that it not only simplifies the fabrication process, but it is also good to get a flat thin dielectric layer on the signal line, as shown in the final fabricated model of Fig. 8. Another important benefit to use the sacrificial layer to release the membrane and beams, is that there the designer does not need focus too much on the width of the beams and the sizes of the holes.

V.   CONCLUSION

A novel shunt capacitive RF-MEMS switch has been presented in this work. The electric and mechanical properties, as well as its virtual fabrication process have been illustrated also. The developed RF-MEMS switch is simulated by IntelliSuite® and shows a pull-in voltage of 3.53V and the maximum von Mises stress under actuated condition is 13.2MPa. The simulated insertion loss and isolation of the switch for frequency of 20GHz is -1.7532dB and -18.394dB respectively by AWR Design Environment®. At the frequency of 60GHz, the insertion loss is -7.2499dB and the isolation is -27.293dB. A three-mask process is employed in this switch's fabrication. By using photoresist-S1800 as the sacrifice layer to release the membrane and beams of the switch, a very thin and smooth dielectric layer on the signal line is obtained. Capacitance ratio of the design switch is around 218.5, with switch-on state capacitance of 68fF.

ACKNOWLEDGMENT

The authors would like to thank the financial support by the RACE fund (RACE 12-006-0006) and UM CR 004-2013 fund.

REFERENCES

[1] D. Balaraman, et al., "Low-cost low actuation voltage copper RF MEMS switches," in Microwave Symposium Digest, 2002 IEEE MTT-S International, 2002, pp. 1225-1228.

[2] J. H. Lau, Advanced MEMS packaging: McGraw-Hill, 2010.

[3] S. Fouladi and R. R. Mansour, "Capacitive RF MEMS switches fabricated in standard 0.35-μm CMOS technology," IEEE transactions on microwave theory and techniques, vol. 58, pp. 478-486, 2010.

[4] Z. J. Yao, et al., "Micromachined low-loss microwave switches," Microelectromechanical Systems, Journal of, vol. 8, pp. 129-134, 1999.

[5] J. M. Kim, et al., "Electrostatically driven low-voltage micromechanical RF switches using robust single-crystal silicon actuators," Journal of Micromechanics and Microengineering, vol. 20, p. 095007, 2010.

[6] C. L. Dai and J. H. Chen, "Low voltage actuated RF micromechanical switches fabricated using CMOS-MEMS technique," Microsystem technologies, vol. 12, pp. 1143-1151, 2006.

[7] S. P. Pacheco, et al., "Design of low actuation voltage RF MEMS switch," in Microwave Symposium Digest. 2000 IEEE MTT-S International, 2000, pp. 165-168.

[8] C. L. Dai and Y. L. Chen, "Modeling and manufacturing of micromechanical RF switch with inductors," Sensors, vol. 7, pp. 2660-2670, 2007.

[9] Z. Peng, "Dielectric charging of RF MEMS capacitive switch," Ph.D, Lehigh University, 2010.

[10] "IntelliFab Clean Room Tools - User Guide," ed: IntelliSense Software Corporation, 2012.

Three-Input Four-Output Voltage-Mode Multifunction Filter Using Two DDCCs

Montree Kumngern [1,3] and Usa Torteanchai [2,4]

[2] Department of Telecommunications Engineering, Faculty of Engineering,
King Mongkut's Institute of Technology Ladkrabang, Bangkok 10520, Thailand
[1] Civil Aviation Training Center, Bangkok 10900, Thailand
E-mail: [3] kkmontre@kmitl.ac.th, [4] usa@catc.or.th

Abstract—This paper presents a new voltage-mode universal filter with three-input and four-output using two differential difference current conveyors, two grounded capacitors and two resistors. The proposed filter is suitable for integrated circuit by using only grounded capacitor. Low-pass, band-pass, high-pass and band-stop voltage responses can be obtained simultaneously into one single topology and without component-matching condition requirements. The natural frequency and the quality factor can be controlled orthogonally by setting the passive components. PSPICE simulation results using 0.5 µm CMOS technology form MIETEC are included to confirm the theoretical analysis. The simulation results can be expressed that the proposed circuit agrees well with theory.

Keywords—universal filter, voltage-mode circuit, differential difference current conveyor

I. INTRODUCTION

Second-generation current conveyor (CCII) is important active building block for analog circuit design. CCII-based circuits provide the advantages such as wide bandwidth, high slew rate, low power consumption and simple circuitry [1]. Typically, conventional CCII is a single-ended device. Thus, the implementation of differential and floating input using conventional CCII is difficult. To solve this problem, a new differential difference current conveyor (DDCC) is proposed [2]. This device incorporates the advantages of the CCII and the differential difference amplifier such as high-input impedance and addition/subtraction voltage capability into a single device. In addition, if fully differential topology is needed, it can be served by fully differential current conveyor (FDCCII) [3]. Therefore, the applications and advantages in realizing various signal processing circuits, particularly from the frequency filters, using DDCC/FDCCII have been introduced in technical literature; see, for example [4]-[18]. Compared with DDCC, the structure of FDCCII is more complex, but FDCCII is suitable for fully differential filters. Thus, for the single-ended filter, DDCC is sufficient to support the realization. Many DDCC-based universal filters have been introduced [7]-[18]. These filters provide a good realization of low-pass (LP), high-pass (HP), band-pass (BP) and band-stop (BS) filtering functions. However, some of these structures do not exploit the full capability of the DDCC, where typically one of the two input terminals of the DDCC is not used.

In this paper, we attempt to realize a universal filter using full capability of DDCC, namely all terminals of DDCC are used. Thus, a new three-input four-output voltage-mode universal filter employing only two DDCCs, two capacitors, and two resistors is presented. The circuit can realize LP, BP, HP and BS transfer functions by suitably connecting the input and output terminals. The natural frequency and the quality factor can be controlled orthogonally by setting the value of passive components. The workability of the proposed circuit is confirmed by PSPICE simulators.

II. PROPOSED CIRCUIT

The circuit symbol of DDCC is Fig. 1. Its characteristic can be given by the following matrix equation:

$$\begin{pmatrix} V_x \\ I_{y1} \\ I_{y2} \\ I_{y3} \\ I_z \end{pmatrix} = \begin{pmatrix} 1 & -1 & 1 & 0 \\ 0 & 0 & 0 & 0 \\ 0 & 0 & 0 & 0 \\ 0 & 0 & 0 & 0 \\ 0 & 0 & 0 & 1 \end{pmatrix} \begin{pmatrix} V_{y1} \\ V_{y2} \\ V_{y3} \\ I_x \end{pmatrix}. \quad (1)$$

The CMOS implementation for DDCC is shown in Fig. 2 [2]. This schematic will be used for simulating in this paper.

The proposed multifunction biquadratic filter is shown in Fig. 3. The circuit is composed of two DDCCs, two grounded capacitors and two resistors. It should be noted that the use of grounded capacitor makes the proposed filter ideal for integrated circuit (IC) implementation [19], [20]. Using (1) and nodal analysis, the output signals V_{o1}, V_{o2}, V_{o3} and V_{o4} of Fig. 3 can be obtained as

$$V_{o1} = \left\{ \left(s^2 R_1 R_2 C_1 C_2 + s R_1 C_1 \right) V_{in1} + s R_2 C_2 V_{in2} \right. \\ \left. - s R_2 C_2 \right) V_{in3} \right\} / D(s) \quad (2)$$

$$V_{o2} = \left\{ s R_1 C_1 V_{in1} + \left(s R_2 C_2 + 1 \right) V_{in2} \right. \\ \left. - \left(s R_2 C_2 + 1 \right) V_{in3} \right\} / D(s) \quad (3)$$

$$V_{o3} = \left\{ \left(s R_1 C_1 + 1 \right) V_{in1} + V_{in2} - V_{in3} \right\} / D(s) \quad (4)$$

978-1-4799-5761-3/14 $31.00 © 2014 IEEE

IEEE-ICSE2014 Proc. 2014, Kuala Lumpur, Malaysia

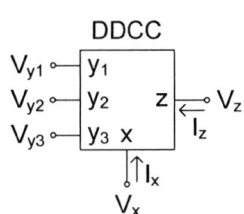

Figure 1. Circuit symbol of DDCC.

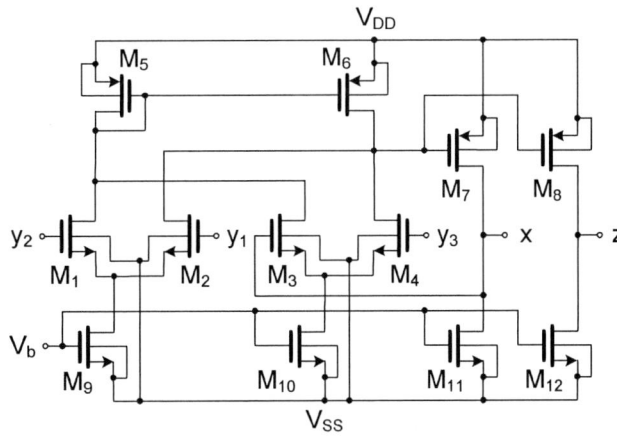

Figure 2. The CMOS implementation of DDCC.

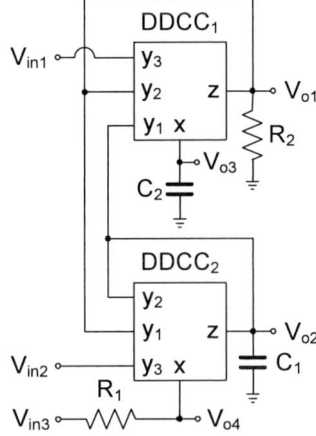

Figure 3. Proposed voltage-mode universal filter using DDCCs.

$$V_{o4} = \left\{ \left(s^2 R_1 R_2 C_1 C_2\right) V_{in1} + \left(s^2 R_1 R_2 C_1 C_2 + sR_1 C_1\right) V_{in2} + V_{in3} \right\} / D(s) \quad (5)$$

where $D(s) = s^2 R_1 R_2 C_1 C_2 + sR_1 C_1 + 1$. It is clearly seen from (2) to (5) that the filtering functions can be obtained by appropriately connecting the input and the output terminals. For example, HP, PF, BP and BS filters can be obtained, respectively, as

- HP: $V_{in1}=V_{in}$, $V_{in2}=V_{in3}=0$ and $V_{o4}=V_{out}$,
- LP: $V_{in2}=V_{in}$, $V_{in1}=V_{in3}=0$ and $V_{o3}=V_{out}$,
- BP: $V_{in1}=V_{in}$, $V_{in2}=V_{in3}=0$ and $V_{o2}=V_{out}$,
- BS: $V_{in1}=V_{in3}=V_{in}$, $V_{in2}=0$ and $V_{o4}=V_{out}$.

It should be noted that these filtering functions are obtained using only two DDCCs. This is the advantage of this active building block that provides the arithmetic operation capability of voltage signals and addition/subtraction of current signal. The natural frequency (ω_o) and the quality factor (Q) can be given by

$$\omega_o = \frac{1}{\sqrt{R_1 R_2 C_1 C_2}} \quad (6)$$

$$Q = \sqrt{\frac{R_1 C_1}{R_2 C_2}} \quad (7)$$

From (6) and (7), the parameter Q can be given by setting C_1/C_2 while parameter ω_o can be given by setting resistors R_1 and R_2.

III.   NON-IDEAL ANALYSIS

The ideal circuit performance so far has been based on the assumptions that the DDCC has no tracking errors and parasitic parameters. Thus, tracking errors and parasitic parameters of DDCC will be considered in this section. To consider the non-ideal effect of a DDCC, taking the non-idealities of the DDCCs into account, the relationship of the terminal voltages and currents can be rewritten as:

$$\begin{pmatrix} V_x \\ I_{y1} \\ I_{y2} \\ I_{y3} \\ I_z \end{pmatrix} = \begin{pmatrix} \alpha_{k1} & -\alpha_{k2} & \alpha_{k3} & 0 \\ 0 & 0 & 0 & 0 \\ 0 & 0 & 0 & 0 \\ 0 & 0 & 0 & 0 \\ 0 & 0 & 0 & \beta_k \end{pmatrix} \begin{pmatrix} V_{y1} \\ V_{y2} \\ V_{y3} \\ I_x \end{pmatrix} \quad (8)$$

where $\alpha_{k1}=1-\varepsilon_{k1v}$ and ε_{k1v} ($|\varepsilon_{k1v}|\ll1$) signifies the voltage tracking error from V_{y1} terminal to V_x terminal of the k-th DDCC, $\alpha_{k2}=1-\varepsilon_{k2v}$ and ε_{k2v} ($|\varepsilon_{k2v}|\ll1$) signifies the voltage tracking error from V_{y2} terminal to V_x terminal of the k-th DDCC, $\alpha_{k3}=1-\varepsilon_{k3v}$ and ε_{k3v} ($|\varepsilon_{k3v}|\ll1$) signifies the voltage tracking error from V_{y3} terminal to V_x terminal of the k-th DDCC and $\beta_k=1-\varepsilon_i$ and ε_i ($\varepsilon_i\ll1$) signifies the output current tracking error of the k-th DDCC. Re-analysis the proposed configuration of Fig. 3 with equation (8), the modified parameters ω_o and Q are obtained by

978-1-4799-5761-3/14 $31.00 © 2014 IEEE   147

$$\omega_o = \sqrt{\frac{\alpha_{12}\alpha_{21}\beta_1\beta_2}{R_1 R_2 C_1 C_2}} \qquad (9)$$

$$Q = \sqrt{\frac{\alpha_{12}\alpha_{21}\beta_2}{\beta_1} \frac{R_1 C_1}{R_2 C_2}} \qquad (10)$$

From (9) and (10), the tracking errors slightly change the parameters ω_o and Q. The active and passive sensitivities are tabulated in Table I.

TABLE I. SENSITIVITIES OF CIRCUIT COMPONENTS

X	$S_X^{\omega_{on1}}$	$S_X^{Q_{n1}}$
R_1	-0.5	0.5
R_2	-0.5	-0.5
C_1	-0.5	0.5
C_2	-0.5	-0.5
β_1	0.5	-0.5
β_2	0.5	0.5
α_{12}	0.5	0.5
α_{21}	0.5	0.5

IV. SIMULATION RESULTS

To prove the performance of the proposed filter, the filter in Fig. 3 was simulated using PSPICE simulators. The DDCC in Fig. 2 was implemented using 0.5 μm CMOS parameters from MIETEC. The transistor aspect ratios of DDCC were listed as Table II [21]. The supply voltages were ±2 V. The bias voltage V_B of DDCC was taken as -1.22 V.

TABLE II. TRANSISTORS ASPECT RATIO OF DDCC.

MOS transistor	W/L(μm/μm)
M_1–M_4	1.8/0.7
M_5–M_8	20/0.7
M_9–M_{10}	5.2/0.7
M_{11}–M_{12}	58/0.7

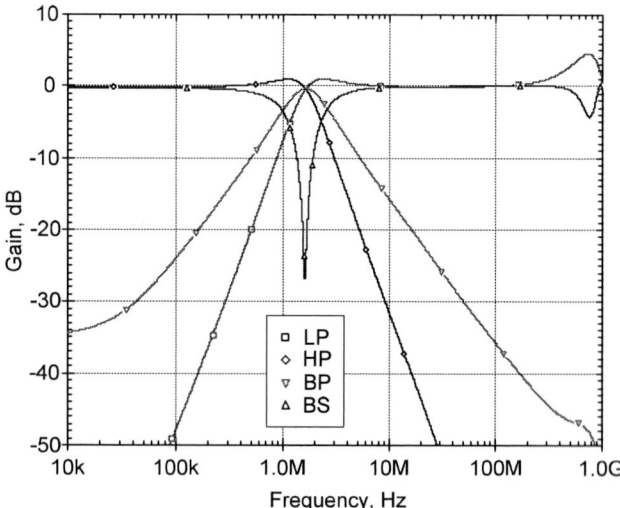

Figure 4. Simulated LP, HP, BP and BS filters.

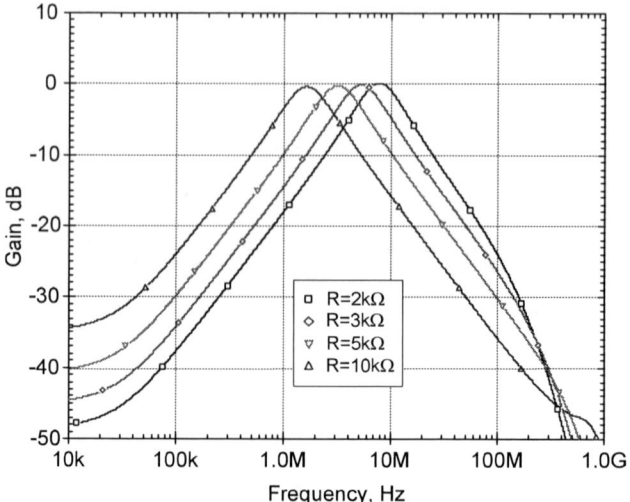

Figure 5. Simulated BP filter with different resistor.

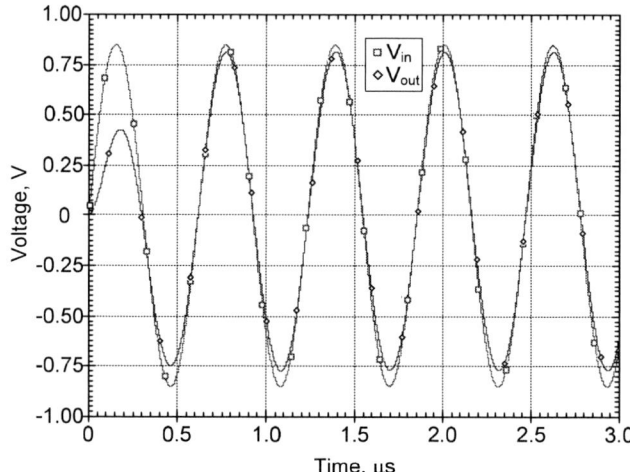

Figure 6. Transient response of the BP filter.

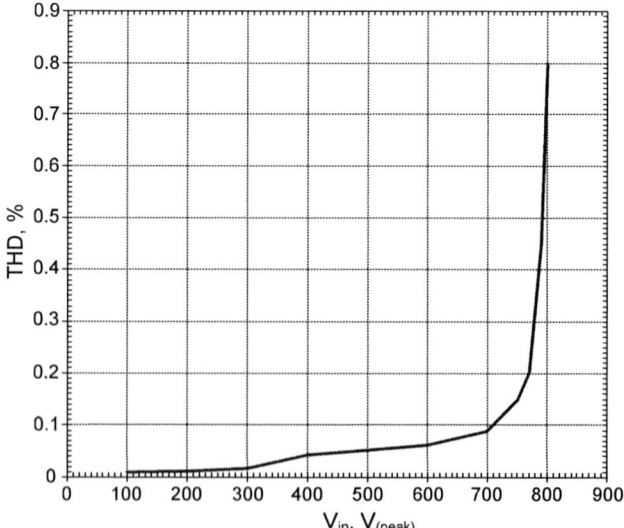

Figure 7. Simulated THD of the LP filter.

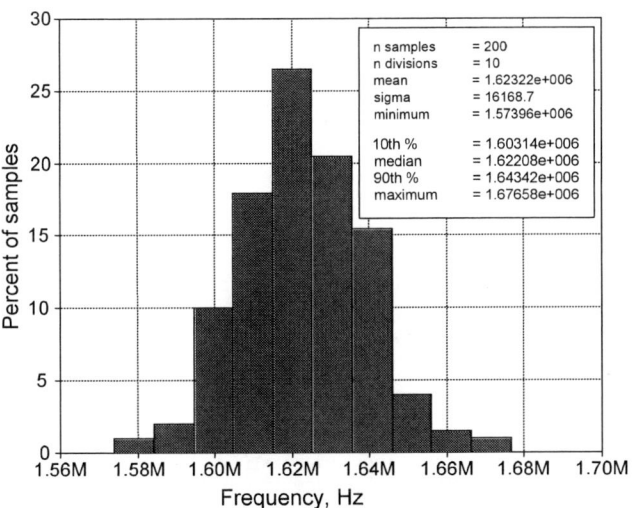

n samples	= 200
n divisions	= 10
mean	= 1.62322e+006
sigma	= 16168.7
minimum	= 1.57396e+006
10th %	= 1.60314e+006
median	= 1.62208e+006
90th %	= 1.64342e+006
maximum	= 1.67658e+006

Figure 8. Monte Carlo simulation for frequency response BP responses.

The proposed filter was designed using the capacitors $C_1 = C_2 = 10$ pF and the resistors $R_1 = R_2 = 10$ kΩ. This setting has been designed to obtain the LP, BP, HP and BS filters with $f_o = 1.59$ MHz and $Q = 1$. The simulated results for the LP, BS, HP and BP filter characteristics were shown in Fig. 4. The pole frequency of 1.62 MHz was obtained in this figure. Fig. 5 shows the simulated frequency responses of the BP filter when resistor R was varied (i.e., $R=R_1=R_2$). In order to test the dynamic range of the proposed filter, the input V_{in1} was applied by 850 mV (peak) of 1.62 MHz, the waveform of input and output were was in Fig. 6. It is evident from this figure that the amplitude of 850 mV peak can be applied without significant distortion. The Total Harmonic Distortion (THD) was 1 %. To show the linearity of the proposed filter, LP filter was tested by applying in-band frequency of 100 kHz, the THD with different amplitude was shown in Fig. 7. The proposed filter was also simulated by utilizing the Monte-Carlo statistical analysis. The Monte-Carlo analysis of the frequency characteristic with 1 % variations of the R_1, R_2, C_1 and C_2 was performed and the result of Monte Carlo analysis, using 200 runs, was shown in Fig. 8. The mean value was at 1.62 MHz with standard deviation σ = 16.16 kHz.

V. CONCLUSIONS

In this paper, a new voltage-mode multifunction filter using DDCCs is presented. The proposed structure employs two DDCCs, two grounded capacitors and two resistors. The use of grounded capacitor makes the proposed filter suitable for IC implementation. The proposed circuit provides LP, HP, BP and BS filters without component-matching conditions requirements. The proposed filter has been simulated by PSPICE simulators to confirm the workability and presented theory.

ACKNOWLEDGMENT

The authors would like to express sincere thank to the Telecommunications Research and Industrial Development Institute (TRIDI) of the office of National Telecommunications Commission of Thailand (NTC) for kindly supporting the equipments of the research laboratory.

REFERENCES

[1] C. Toumazou, F. J. Lidgey, D. G. Haigh, Analogue IC design: the current-mode approach, Peter Peregrinus, 1990.

[2] W. Chiu, S.–I. Liu, H.–W. Tsao, J.–J. Chen, "CMOS differential difference current conveyors and their applications," IEE Proceedings of Circuits, Devices and Systems, vol. 143, pp. 91-96, 1996.

[3] A. El-Adawy, A. M. Soliman, H. O. Elwan, "A novel fully differential current conveyor and applications for analog VLSI," IEEE Transactions on Circuits and Systems-II, vol. 47, pp. 306-313, 2000.

[4] C. M. Chang, B. M. Al-Hashimi, C. L. Wang, C. W. Hung, "Single fully differential current conveyor biquad filters," IEE Proceedings of Circuits, Devices and Systems, vol. 150, pp. 394-398, 2003.

[5] F. Gur, F. Anday, "Simulation of a novel current-mode universal filter using FDCCIIs," Analog Integrated Circuits and Signal Processing, vol. 60, pp. 231-236, 2009.

[6] H.-P. Chen, "Voltage-mode FDCCII-based universal filters," International Journal of Electronics and Communications, vol. 62, pp. 320-323, 2008.

[7] J.-W. Horng, W.-Y. Chiu, H.-Y. Wei, "Voltage-mode highpass, bandpass and lowpass filters using two DDCCs," International Journal of Electronics, vol. 99, pp. 461-464, 2004.

[8] H.-P. Chen, K.-H. Wu, "Voltage-mode DDCC-based multifunction filters," International Journal of Electronics, vol. 16, pp. 93-104, 2007.

[9] H.-P. Chen, "Universal voltage-mode filter using only plus-type DDCCs," Analog Integrated Circuits and Signal Processing, vol. 50, pp. 137-139, 2007.

[10] W.-Y. Chiu, J.-W. Horng, "High-input and low-output impedance voltage-mode universal biquadratic filtic using DDCCs," IEEE Transactions on Circuits and Systems-II, vol. 54, pp. 649-652, 2007.

[11] H.-P. Chen, S.-S. Shen, J.-P. Wang, "Electronically tunable versatile voltage-mode universal filter," International Journal of Electronics and Communications, vol. 62, pp. 316-319, 2008.

[12] H.-P. Chen, "Versatile universal voltage-mode filter employing DDCCs," International Journal of Electronics and Communications, vol. 63, pp. 78-82, 2009.

[13] H.-P. Chen, "Versatile multifunction universal voltage-mode biquadratic filter," International Journal of Electronics and Communications, vol. 64, pp. 983-987, 2010.

[14] J.-W. Horng, W.-Y. Chiu, "High input impedance DDCC-based voltage-mode universal biquadratic filter with three inputs and five output," Indian Journal of Engineering & Materials Sciences, vol. 18, pp. 183-190, 2011.

[15] J.-W. Horng, T.-Y. Chiu, Z.-Y. Jhao, "Tunable versatile high input impedance voltage-mode universal biquadratic filter based on DDCCs," Radioengineering, vol. 21, pp. 1260-1268, 2012.

[16] M. Kumngern, F. Khateb, K. Dejhan, P. Phasukkit, S. Tungjitkusolmun, "Voltage-mode multifunction biquadratic filters using new ultra-low-power differential difference current conveyors," Radioengineering, vol. 22, pp. 448-457, 2013.

[17] W.-Y. Chiu, J.-W. Horng, "High-input and low-output impedance voltage-mode universal biquadratic filter using DDCCs," IEEE Transactions on Circuits and Systems-II, vol. 54, pp. 649-652, 2007.

[18] J.-W. Horng, "High input impedance voltage-mode universal biquadratic filter with three inputs using DDCCs," Circuits System and Signal Processing, vol. 27, pp. 553-562, 2008.

[19] M. Bhusan, R. W. Newcomb, "Grounding of capacitors in integrated circuits," Electronics Letters, vol. 3, pp. 148-189, 1967.

[20] M. T. Abuelma'atti, "Grounded capacitor current-mode oscillator using single current follower," IEEE Transactions on Circuits and Systems-I, vol. 39, pp. 1018-1020, 1992.

[21] W. Tangsrirat, O. Channumsin, T. Pukkalanun, "Resistorless realization of electronically tunable voltage-mode SIFO-type universal filter," Microelectronics Journal, vol. 44, pp. 210-215, 2013.

Comparative study of subthreshold characteristics of different antimonide-based and nitride-based Dual Material Gate (DMG) HEMTs

Arman-Ur-Rashid*, Md. Aynal Hossain, Tanvir Rahman and Farseem M. Mohammedy
Department of Electrical and Electronic Engineering
Bangladesh University of Engineering and Technology (BUET)
Dhaka-1000, Bangladesh
*Email: arman0806161@yahoo.com

Abstract—As MOSFET technology reaching its limits, new devices like HEMTs (high electron mobility transistors) are gradually gaining more interests. HEMTs have high mobility due to reduce scattering for the spatially separated doped region. But as their dimensions reach the nano-scale region, different SCEs (short channel effects) and carrier transport inefficiency creep in. To solve these problems a new structure, DMG (dual material gate) HEMT has been proposed. In this paper we have shown comparison of subthreshold characteristics among different antimonide-based DMG HEMTs and between antimonide-based and nitride-based DMG HEMTs. We have found that as the band-offsets in the heterojunction become deeper, the effect of using DMG over SMG (single material gate) reduces in case of antimonide-based HEMTs. Our work also shows that DMG structure is more effective in reducing SCEs for antimonide-based HEMTs than nitride-based HEMTs, while nitride-based DMG HEMTs give more efficient carrier transport than its antimonide-based counterpart.

Keywords-Dual Material Gate (DMG),Single Material Gate (SMG), Short channel effect(SCEs), band-offset, antimonide-based HEMT

I. INTRODUCTION

Antimonide-based compound semiconductors are opening a new promising field for semiconductor industry as they can be used both as third generation infrared photo-detectors and in ultra-high speed, ultra-low power consuming integrated circuit fabrication [1]. On the other hand, nitride-based devices are well known for the optoelectronic devices in blue/green/UV spectral range as well as for high-voltage, high-power, and high-temperature electronics applications [2]. In the field device technology, HEMTs (high electron mobility transistors) are gaining vast interest due to their potential to replace MOSFETs. Inherently HEMTs are faster, low power consuming and easy to fabricate compare to MOSFETs [3]. So it is desirable to explore the field of antimonide-based and nitride-based HEMTs. Already it has been seen that antimonide-based HEMTs are easier to fabricate than nitride-based HEMTs [4]. But nitride-based HEMTs are proven for their high voltage, high power operation in microwave frequencies [5].

Scaling of device is a fundamental trend in semiconductor industry to make smaller, faster and less power consuming chips. HEMTs follow the same trend [6]. With the advancement of fabrication techniques and material science, even sub-micrometer range HEMTs with excellent speed performance are now reality [7]. But as gate length decreases below 300nm various short channel effects (SCEs) and poor carrier transport efficiency appear. So it becomes an interesting topic of research to solve these problems. Various structures like Dual Gate FET, Split-Gate (SG) FET etc. were proposed. But none can solve the short channel effects and improve carrier transport efficiency simultaneously. The ultimate solution was given by Long et al.[8] in his proposed Dual Material Gate (DMG) FET structure.

In DMG FET, two gates made of different materials of different workfunctions are joined laterally to create a single gate. The workfunction of one gate (M1) is chosen to be to be greater than the other (M2). Gate M1 works as control gate and gate M2 works as screening gate for drain voltage.It enhances the average electric field and inserts a step function in channel potential at the interface of two gates. The enhanced average electric field improves the carrier transport efficiency and the step function reduces the short channel effects. But as the scaling goes into deep sub-micron level, the need for an analytical model of subthreshold behavior of DMG HEMT emerges. S.P.Kumar et al. [9] first proposed an analytical model for the subthreshold behavior of nitride-based DMG FET. In our previous paper [10] the model of [9] was modified and an analytical model for antimonide-based HEMTs was developed.

In [10], the $Al_{0.7}Ga_{0.3}Sb/InAs$ combination was used for barrier and channel. Among the antimony-based material AlSb and GaSb are approximately lattice matched to InAs and these semiconductors are jointly known as $6.1A^o$ family. In this paper we also study the $Al_{0.7}Ga_{0.3}Sb/GaSb$ and $Al_{0.7}Ga_{0.3}Sb/In_{0.15}Ga_{0.85}Sb$ barrier/channel combinations and compare their results. We also derived the results of $Al_{0.2}Ga_{0.8}N/GaN$ DMG HEMT from the proposed model of [9] and compared those results with antimonide-based DMG HEMTs.

II. MODEL FORMULATION

Fig. 1 shows the cross-section of proposed DMG HEMT structure. Here, L1 and L2 are the lengths of gates with metal M1 and M2 respectively. L is the total length. The drain and source regions are doped uniformly with donor concentration $N_d^+ = 10^{25} m^{-3}$. The workfunctions of metal M1 is chosen $\varphi_1 = 5.3$ V, which is greater than the workfunctions of metal M2 $\varphi_2 = 4.1$ V. φ_1 is chosen greater than φ_2 to make the HEMT n-channel[11]. The gate voltage and drain voltage were chosen -0.2 V and 0.2 V respectively to make the barrier layers completely depleted and confined electrons in the heterointerface. The other parameters of model are listed in Table. I.

The details of the analytical model and simulation are given in [10] for antimonide-based DMG HEMTs and we followed the nitride-based DMG HEMT model from [9].

Table I. Lists of parameters

Parameter	Name of the parameter	Value
N_d	Doping in the barrier layer	10^{23} m^{-3}
L	Effective channel length	120nm
d_d	thickness of doped barrier layer	25nm
d_i	Thickness of spacer layer	5nm
ε_s	Dielectric constant of Al$_{0.7}$Ga$_{0.3}$Sb	13.1
T	Operating temperature	300K

In this paper we have chosen three different formations of antimonide-based DMG HEMTs to compare the effect of DMG structure within different antimonide-based HEMTs. The barrier layer is chosen to be Al$_{0.7}$Ga$_{0.3}$Sb which gives us the freedom of bandgap engineering with ternary AlGaSb (Al=1~0.7) where minimum mole fraction of Al required to make type I-straddling heterojunction is 0.7. Moreover it gives better hole blocking performance in valence band [12]. The first channel layer is InAs which will provide large conduction band offset, high peak electron velocity and high electronconcentration in InAs channel [4]. For our second HEMT's channel layer, we selected the Ga$_{0.85}$In$_{0.15}$Sb. It gives a shallower conduction band offset and its growth on GaSb substrate is well studied [13]. The third structure has GaSb as channel layer. It gives the shallowest band-offset in the

heterointerface among the chosen formations. In case of nitride-based DMG HEMT we followed the structure used in [9].Table II. gives the lists of various formations of DMG HEMTs used in our paper.

Table II. Formation of HEMTs

Barrier layer	Channel layer	Substrate
Al $_{0.7}$Ga$_{0.3}$Sb	InAs	GaSb
Al $_{0.7}$Ga$_{0.3}$Sb	Ga$_{0.85}$In $_{0.15}$Sb	GaSb
Al $_{0.7}$Ga$_{0.3}$Sb	GaSb	GaSb
Al $_{0.2}$Ga$_{0.8}$N	GaN	GaN

While forming HEMT structures, it is important to consider the strain in the heterojunctions [4]. Too much strain in forming the heterostructure degrades the mobility and thus hampers the main objective of using HEMT. Table III gives a percentage of strain for the various HEMTs in this paper.

Table III. Percentage strain in different heterojunctions

Heterostructure	Percentage of strain
InAs/ GaSb	0.6173
Ga$_{0.85}$In $_{0.15}$Sb/ GaSb	0.9426
Al $_{0.7}$Ga$_{0.3}$Sb/ GaSb	0.4544
Al $_{0.2}$Ga$_{0.8}$N /GaN	0.4829

III. RESULTS AND DISCUSSIONS

We have used ATLAS device simulator to simulate channel potential and electric field distribution along the channel to verify the model of [10]. Fig.2 shows electric field variation vs normalized channel length and the channel potential vs. normalized channel length for DMG Al$_{0.7}$Ga$_{0.3}$Sb/GaSb HEMT. Close match between proposed model in [10] and ATLAS's simulation is evident from the figure.

Fig.3 shows the channel potential variation vs normalized channel length for various combinations of barrier/channel of DMG and SMG HEMTs. It also highlights the difference of DMG HEMTs from SMG (single material gate) HEMTs. Clearly there is a step potential at the interface of two gates.

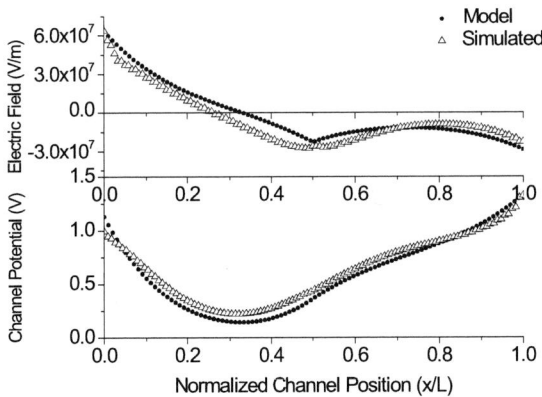

Figure.2: variation of channel potential and electric field vs the normalized channel length

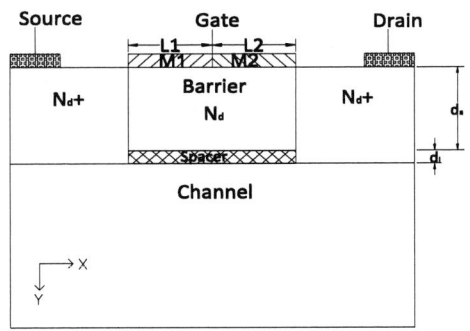

Figure.1: Cross sectional view of a DMG HEMT.

IEEE-ICSE2014 Proc. 2014, Kuala Lumpur, Malaysia

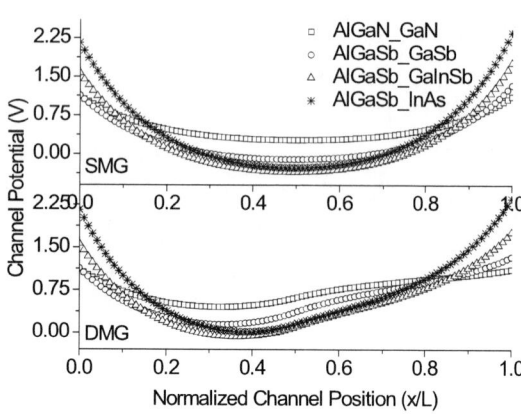

Figure. 3: variation of channel potential with channel length for different DMG and SMG HEMTs

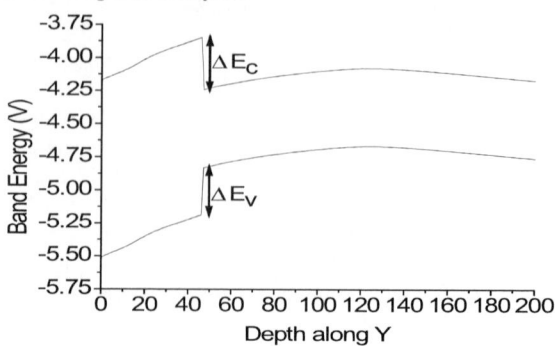

Figure.5: conduction band offsets and valenceband offsets at $Al_{0.7}Ga_{0.3}Sb$ /$Ga_{0.85}In_{0.15}Sb$ heteorjuntion.

Table IV. Magnitudes of potential-step for different HEMTSs at the interface of dual material gate

Material composition	Potential step at gates interface (V)
$Al_{0.2}Ga_{0.8}$ N /GaN	0.3331
$Al_{0.7}Ga_{0.3}Sb$ /InAs	0.4098
$Al_{0.7}Ga_{0.3}Sb$ /$Ga_{0.85}In_{0.15}Sb$	0.4472
$Al_{0.7}Ga_{0.3}Sb$ /GaSb	0.4529

Fig.4 highlights it. This step potential has a screening effect which reduces the drain induced barrier lowering (DIBL). As a result short channel effects are suppressed [9].

Now we listed the magnitudes step-function at the channel potential for different combinations of barrier/channel formation in Table IV. We can see from the result that all the antimonide-based DMG HEMTs shows greater step function at the heterointerface than nitride-based DMG HEMT. So if we use antimonide-based DMG HEMTs instead of nitride-based HEMTs short channel effects will be significantly lower.

But within the antimonide-based DMG HEMTs, the magnitude of step-function decrease as the band-offset in the heterointerface increases. The band-offset (highlighted in Fig.5) at the interface of barrier and channel is deepest for $Al_{0.7}Ga_{0.3}Sb$/InAs and shallowest for $Al_{0.7}Ga_{0.3}Sb$/GaSb. Table.V lists the conduction band and valence band offsets for different combination of barrier/channel layers. From these two tables we can say that effectiveness of DMG HEMTs over SMG HEMTs for suppressing short channel effects is more effective in formations where the band-offset is shallow.

Table V. Conduction and valence band offset at different heterojunction

Barrier layer	Channel layer	$\Delta Ec(eV)$	$\Delta Ev(eV)$
$Al_{0.7}Ga_{0.3}Sb$	InAs	1.2	-0.1
$Al_{0.7}Ga_{0.3}Sb$	$Ga_{0.85}In_{0.15}Sb$	0.3950	0.355
$Al_{0.7}Ga_{0.3}Sb$	GaSb	0.31	0.34
$Al_{0.2}Ga_{0.8}$ N	GaN	0.1960	0.1440

In Fig.6 we see the electric field variation vs the normalized channel length for both SMG and DMG HEMTs for different combinations of barrier/channel. Fig.7 highlights the difference between SMG and DMG structure. It is clearly visible from the Fig.7 that at the drain end the drain voltage has been screened by screening gate. Also DMG structure has an extra peak at the interface of two gates in its electric field profile. Both of these effects enhance the average electric field of the channel. Velocity of electrons is related with electric field distribution and so an enhanced average electric field leads to a better transport efficiency [9].

In Table VI. we listed the percentage change of electric field at the drain end. We can see that screening of gate voltage is

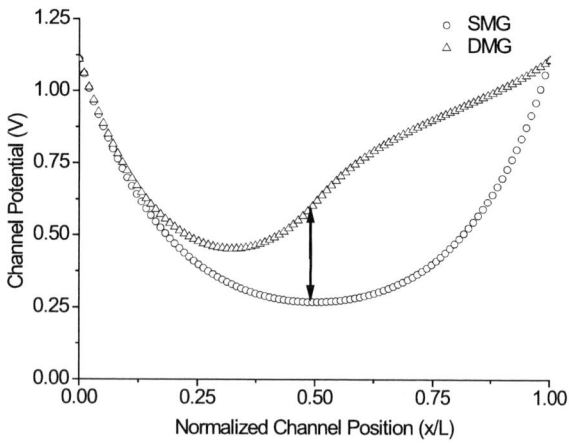

Figure4: potential step at the interface of two gates between SMG and DMG HEMTs. The figure has been producedfollowing [9] for $Al_{0.7}Ga_{0.8}$N/GaN HEMT.

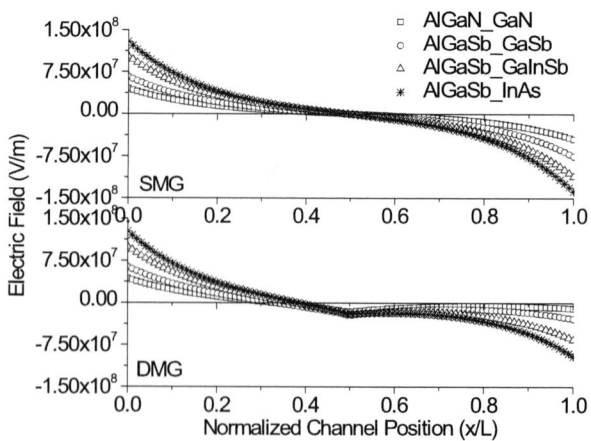

Figure 6: variation of electric field vs normalized channel length for various SMG and DMG HEMTs

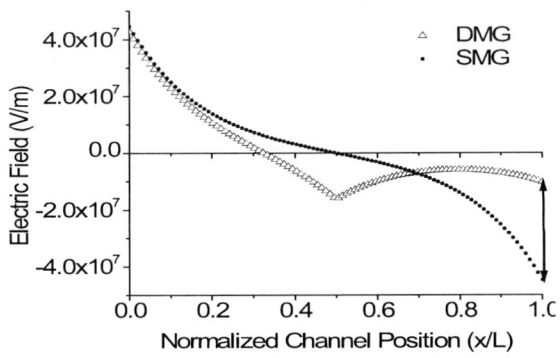

Figure 7: change in electric field at the drain end for DMG HEMTs. The figure has been produced following [9] for $Al_{0.2}Ga_{0.8}N/GaN$ HEMT

better in nitride-based DMG HEMTs than its antimonide-based counterparts. So nitride-based DMG HEMTs will show more efficient carrier transport.Amongantimonide-based HEMTs, as the band-offset increases, the improvement in the carrier transport efficiency decreases. So DMG structure is gives more efficient carrier transport in case of shallow band-offset formation.

Table VI. Percentage change of electric field at drain end

Hemt composition	Percentage change of electric field at drain end(%)
$Al_{0.2}Ga_{0.8}N$ /GaN	63.67
$Al_{0.7}Ga_{0.3}Sb$ /InAs	30.58
$Al_{0.7}Ga_{0.3}Sb$ /$Ga_{0.85}In_{0.15}Sb$	47.90
$Al_{0.7}Ga_{0.3}Sb$ /GaSb	61.96

IV. CONCLUSIONS

Both nitride-based and antimonide-based HEMTs are gaining vast interest for their unique abilities. So it is very important to search for better structures for these devices. DMG HEMT is a promising structure as it has the potential to solve the short channel effects and carrier transport efficiency related problems for deep sub-micron range of HEMTs. In our paper, we compared the effect of using DMG HEMTs between these two families. We found that, in case of suppressing SCEs DMG structure is more effective in antimonide-based DMG HEMTs than their nitride-based counterpart. But nitride-based DMG HEMTs give better carrier transport efficiency than antimonide-based DMG HEMTs. We also compared different combination barrier/channel for antimonide-based DMG HEMTs. We have seen that, DMG structure improves the performance better when the band-offset in the heterointerface of barrier/channel layer is shallow. As the depth of band-offset in barrier/channel heterointerface increases, the step function at the interface of two gates and screening of drain voltage decreases. So using DMG structure for antimonide-based HEMTs will be more meaningful in case of shallow band-offsets devices like $Al_{0.7}Ga_{0.3}Sb/GaSb$ DMG HEMT while performance will not improve much from normal SMG HEMTs if the band-offset is deep in the heterjunction like in $Al_{0.7}Ga_{0.3}Sb/InAs$ HEMT. So our work gives an assessment about effectiveness of Dual Material Gate (DMG) structure in different kind of HEMTs.

ACKNOWLEDGMENT

We would like to take this opportunity to thank Mr. Tonmoy Kumar Bhowmick for introducing us to this topic of research. We are also thankful for his various useful advices regarding our work. This work was supported under grant no. CASR#218, agenda97, held on 06/09/2009.

REFERENCES

[1] C. Liu, Y. Li and Y. Zeng, "Progress in Antimonide Based III-V Compound Semiconductors and Devices," *Engineering*, vol. 2 no. 8, pp. 617-624, 2010.

[2] I. P. Smorchkova, C. R. Elsass, J. P. Ibbetson, R. Vetury, B. Heying, P. Fini, E. Haus, S. P. DenBaars, J. S. Speck and U. K. Mishra, "Polarization-induced charge and electron mobility in AlGaN/GaN heterostructures grown by plasma-assisted molecular-beam epitaxy", *Journal of Applied Physics*, vol.86, pp. 4520-4526, 1999.

[3] Rashmi, Abhinav Kranti, S. Haldar, R.S. Gupta, "An accurate charge control model for spontaneous and piezoelectric polarization dependent two-dimensional electron gas sheet charge density of lattice-mismatched AlGaN/GaN HEMTs", *Solid-State Electronics*, Volume 46, Issue 5, Pages 621-630, May 2002.

[4] Brian R. Bennett, Richard Magno, J. Brad Boos, Walter Kruppa, Mario G. Ancona, "Antimonide-based compound semiconductors for electronic devices: A review", *Solid-State Electronics*, Volume 49, Issue 12, Pages 1875-1895,December 2005.

[5] O. Ambacher, J. Smart, J. R. Shealy, N. G. Weimann1, K. Chu, M. Murphy, W. J. Schaff, L. F. Eastman, R. Dimitrov, L. Wittmer, M. Stutzmann, W. Rieger and J. Hilsenbeck, "Two-dimensional electron gases induced by spontaneous and piezoelectric polarization charges in N- and Ga-face AlGaN/GaN heterostructures", *Journal of Applied Physics*, vol.85,pp. 3222-3233, 1999.

[6] Awano, Yuji; Kosugi, M.; Kosemura, Kinjiro; Mimura, T.; Abe, Masayuki, "Short-channel effects in subquarter-micrometer-gate HEMTs: simulation and experiment," *Electron Devices, IEEE Transactionson* , vol.36, no.10, pp.2260-2266, Oct 1989.

[7] Palacios, T.; Chakraborty, A.; Heikman, S.; Keller, S.; DenBaars, S.P.; Mishra, U.K., "AlGaN/GaN high electron mobility transistors with InGaN back-barriers," Electron Device Letters, IEEE , vol.27, no.1, pp.13-15, Jan. 2006.

[8] Wei Long; Ou, H.; Kuo, J.-M.; Chin, K.K., "Dual-material gate (DMG) field effect transistor," *Electron Devices, IEEE Transactions on* , vol.46, no.5, pp.865-870, May 1999.

[9] Sona P. Kumar, Anju Agrawal, Rishu Chaujar, Mridula Gupta, R.S. Gupta, " Analytical modeling and simulation of subthreshold behavior in nanoscale dual material gate AlGaN/GaN HEMT,"*Superlattices and Microstructures*, Volume 44, Issue 1, Pages 37-53,July 2008.

[10] Arman-Ur-Rashid, Md. Aynal Hossain, Tanvir Rahman, Farseem M. Mohammedy, "Analytical modeling and simulation of subthreshold behavior of dual material gate (DMG) $Al_{0.7}Ga_{0.3}Sb/InAs$ HEMT," accepted at the 3rd Int. Conf. on Informatics, Electronics & Vision (ICIEV), Dhaka, Bangladesh,2014.

[11] Chaudhry, A.; Kumar, M.J., "Controlling short-channel effects in deep-submicron SOI MOSFETs for improved reliability: a review," *Device and Materials Reliability, IEEE Transactions on* , vol.4, no.1, pp.99,109, March 2004.

[12] A.F.M Anwar, RT Webster, "Energy bandgap of $Al_xG_{1-x}As_{1-y}Sb_y$ and conduction band discontinuity of $Al_xGa_{1-x}As_{1-y}Sb_y$/InAs and $Al_xGa_{1-x}As_{1-y}Sb_y$/InGaAs heterostructures,"*Solid-State Electronics*, Volume 42, Issue 11, Pages 2101-2104, November 1998.

[13] Mohammedy, F.M.; Deen, M.J.; Thompson, D.A., "Extraction of Electron and Hole Ionization Coefficients From Metamorphically Grown InGaSb Diodes," *Electron Devices, IEEE Transactions on* , vol.56, no.3, pp.523,528, March 2009.

Characterization of Vertical Strained SiGe Impact Ionization MOSFET for Ultra-Sensitive Biosensor Application

Ismail Saad, Mohd. Zuhir H., Bun Seng C., Khairul A.M, Bablu Ghosh, N. Bolong

Nano Engineering & Material (NEMs) Research Group, Faculty of Engineering, Universiti Malaysia Sabah, 88999, Kota Kinabalu, Sabah, Malaysia.
ismail_s@ums.edu.my

Razali Ismail

Computational Nanoelectronics (CONE) Research Group, Faculty of Electrical Engineering, Universiti Teknologi Malaysia, 81310, Skudai, Johor, Malaysia.
zuhirhamzah@yahoo.com

Abstract—This paper venture into prospective ideas of finding viable solution of nanoelectronics device design by an assessment of incorporating vertical impact-ionization MOSFET (IMOS) with strained SiGe technology into a formation of an emerging device structure with elevated performance and reliable outcomes for future bio-based sensor application. Impact Ionization FET biosensors can be extremely promising for applications where ultra-high sensitivity and fast response is desirable. An ultra-low power with low Subthreshold Swing and high breakdown voltage are imperative for ultra-sensitive biosensor. Impact ionization MOSFET (IMOS) is expected to have a subthreshold swing (S) down to 20 mV/dec which is much lower compared to Conventional MOSFET (CMOS). This will eventually enhanced the switching behavior of the transistor and enhancing its electrical performance and response time particularly when scaled down into nanometre regime. However, vertical IMOS experience parasitic bipolar transistors (PBT) effect and low breakdown voltage. Parasitic Bipolar Transistor effect is a phenomenon where the MOSFET act as a minority carrier device like BJT instead of majority carrier device. This is not favorable for any power device or sensor. Dielectric Pocket (DP) is believed to be able to minimize the PBT effect while improving the performance of the device. Eventually, this device will prolong the increase density of transistor in a chip for future application of biosensor nanoelectronics.

Keywords—IMOS, Dielectric Pocket, VESIMOS, VESIMOS-DP, Parasitic Bipolar Effects, Biosensor

I. Introduction

Biosensors are central key component for modern society due to their wide applications in public healthcare, security, forensic industries, and environmental protection. Currently, enzyme-linked immunosorbent assay (ELISA) based on optical sensing technology is widely used as a medical diagnostic tool as well as a quality-control check in various industries. ELISA needs biomolecules labeling which requires bulky and expensive optical instrument to analyze the specimen. On the other hand, biosensors based on field-effect-transistors (FETs) [1–2] are highly attractive as they promise real-time label-free electrical detection, scalability,

inexpensive mass production, and on-chip integration of both sensor and measurement system.

However, the conventional FET (CFET) based biosensors exhibit some fundamental limitations on the sensitivity and response time [3–5]. Therefore, Impact ionization MOSFET (IMOS) is believed to be the potential candidates to solve some of the limitation leading to an ultra-sensitive and fast electrical biosensor [6]. The vertical IMOS device which is a planar-doped barrier MOSFET has been introduced and investigated [7]. The device does not suffer from either V_{TH} shifts or a change in subthreshold slopes with repeated measurements, mitigate hot electron damages [8–9] and capable functioning properly under high temperature [10]. Therefore, it was observed that the vertical concept of IMOS is better than the planar IMOS in terms of hot carrier effects and hence, device reliability. The vertical IMOS nevertheless suffer remarkable hysteresis and high supply voltage (V_{DS}) [11].

The concept of strained SiGe vertical IMOS was introduced as an attempt to bring down the supply voltages [12]. Both supply and threshold voltage reduce significantly when strained SiGe layer were integrated into the structure. Nevertheless, the device still experiencing low breakdown voltage causes by parasitic bipolar transistors (PBT) effect which affecting reliability of the device. An ultra-low power with low subthreshold swing and high breakdown voltage are imperative for ultra-sensitive and fast electrical biosensor. Dielectric Pocket (DP) is believed to be able to minimize the PBT effect while improving the performance of the device [13]. This paper will therefore investigate the characteristic of Vertical Strained-SiGe Impact Ionization MOSFET incorporating Dielectric Pocket (VESIMOS-DP) of finding viable solution for future biosensing applications.

II. Structure and Physical Models

The expansion of new approach towards high sensitive detection technique remains a major challenge in bio-based sensing field particularly on the electrical detection of biological species. It is broadly recognized that the sensor sensitivity will increase with the increasing of surface area per unit mass. Thus, high sensitivity at low biomolecule concentration is desirable. Response time is another vital parameter to determine the performance of biosensors. Response time is defined as the time required to obtain a desired sensitivity. Response time can be related to the subthreshold swing (S). Low S value implies low response time. In order to achieve lower S value, an application of strain or alternate materials/design is required for lowering the operating bias.

In general, electrical characteristics of a device are calculated by solving Poisson's equation and continuity equation numerically within the device defined meshes [15]. Poisson's equation relates variations in electrostatic potential to local charge densities. Continuity equations are then used to calculate the current densities of the electrons and holes. Continuity equations describe the electron and hole densities that evolve from transport processes and generation-recombination processes. Boltzmann transport framework is use to solve these two equations. The relationship between current density of holes and electrons and carrier concentration is exhibited during this self-consistent process.

Since VESIMOS-DP is an impact ionization based device, avalanche multiplication is the principal carrier injection mechanism for IMOS device operation. At sufficiently high electric field, the carrier is able to initiate the first impact ionization event to create an electron-hole pair. The resulting electron and hole pair created then begin a process of carrier multiplication known as avalanche breakdown. This helps to increase the current tremendously. When the electron travelled through a distance (dx), it will create an average of ($\alpha_n.dx$) electron-hole pairs in the process.

This increment in electron current density due to the electron and hole multiplication event [16] is given by:

$$\left.\frac{\delta J_n}{dx}\right|_n = \alpha_n J_n dx \qquad (1)$$

$$\left.\frac{\delta J_n}{dx}\right|_n = \alpha_n J_n dx \qquad (2)$$

Hence,

$$\frac{\delta J_p}{dx} = -\frac{\delta J_n}{dx} \qquad (3)$$

Impact ionization model of carrier generation mechanism has been employed in this study. II occurs in a sufficiently high electric field, under which the free carriers gain sufficient amount of energy to undergo collision with other free carriers and generate electron-hole pairs (EHP). When the generation rate of the free carriers is high enough, it would result in avalanche breakdown. The impact ionization model employed in this research study was Selberherr's [17] models, which is a local impact ionization model. The general II process can be described by the equation 6.

$$G = \alpha_n J_n + \alpha_p J_p \qquad (4)$$

Fig. 1 shows the detailed cross-sections schematic which simulated for the analysis of S value and suppressing the PBT effect of the VESIMOS transistors using Sentaurus package [14]. This structure comprises a source and a drain region with n+ doping, an intrinsic channel containing a highly doped δp+ layer (Boron = 4×10^{19}/cm^3) and two sided gates. The source/drain junction thickness was 200nm, the nitride spacer of 20nm and the polysilicon gate thickness of 100nm. The strain SiGe layer thickness is 20nm with Ge concentration of 30%. The thickness of DP layer is also 20nm.

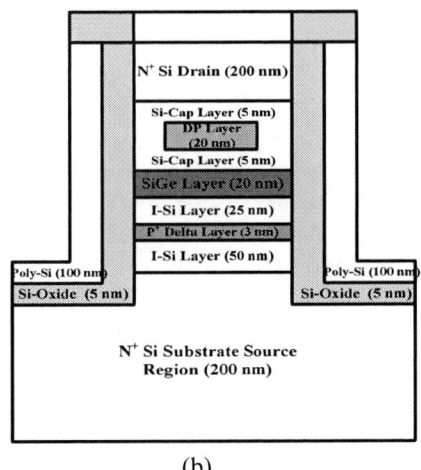

(a)   (b)

Fig. 1. Schematic cross-sections of vertical strained IMOS structure: (a) standard VESIMOS device, and (b) VESIMOS with DP device.

III. Profiles and Characterization

An ultra-low power with low S value and high breakdown voltage are imperative for ultra-sensitive and fast electrical biosensor. Fig. 2 shows the transfer characteristics ($I_{GS} - V_{GS}$) of VESIMOS-DP device with channel length L_g = 50nm and DP size = 60 nm and body doping $N_A = 4 \times 10^{19}$/cm^3.

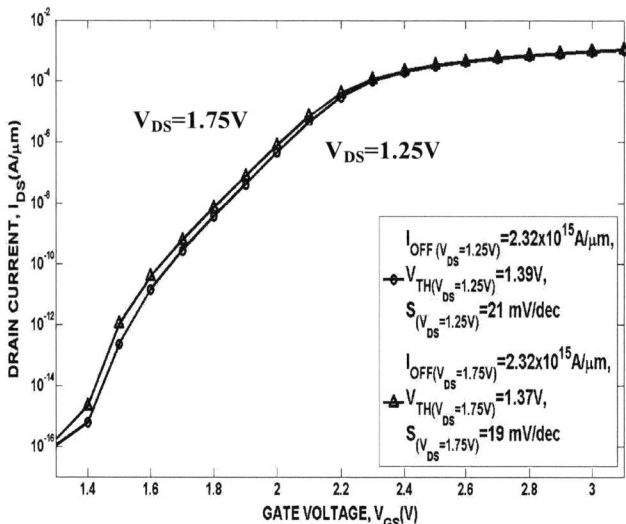

Fig. 2. Transfer Characteristics, I_{DS}-V_{GS} of VESIMOS – DP for $Si_{0.7}Ge_{0.3}$, S/D doping =2.0×10^{18}/cm^3, V_{DS}=1.75V

By using a linear extrapolation of transconductance g_m (V_{GS}) to zero, 1.39V threshold voltage (V_{TH}) is obtained at V_{DS} = 1.25V and reduced to 1.37V at V_{DS} = 1.75V. The moderately high V_{TH} is expected since the device has two potential barriers to get through which are DP layer and existing δp+ triangular potential barrier. Hence, require large amount of V_{GS} needed to collapse both barriers. The dependence of V_{TH} on V_{DS} that decreased from 1.39V at V_{DS} = 1.25V to 1.37V at V_{DS}=1.75V as the supply of drain voltage provides sufficient energy for the electrons to cross the δp+ potential barrier forming an ON current of the device. In the off-state operation mode, the transistors show a leakage current (I_{off}) which is independent of the V_{GS}, but increases with increasing of V_{DS}. A very low off-state leakage current I_{off} = 2.32 x 10^{-15} A/µm and good drive current I_{on} = 1.6×10^{-5} A/µm at V_{DS} = 1.25V was explicitly shown in fig. 2. It increases to I_{off} = 2.74 x 10^{-14} A/µm and I_{on} = 1.8×10^{-4} A/µm for V_{DS}=1.75V.

Subthreshold value is in direct proportion to the leakage currents. Thus, low S value indicates low leakage currents. A considerable low S value of VESIMOS-DP (S = 19 mV/dec) has justified incorporating DP layer at the drain end intrinsic region. This subthreshold voltage obtained, is much lower than the conventional MOSFET limit which is 60 mV/decade due to impact ionization mechanism of VESIMOS-DP device. It gives the VESIMOS-DP device fastest switching behavior and enhanced its electrical

performance and response time which is imperative for bio-based sensor characteristic. In addition, the output characteristic was also highlighted a very good drain current at different gate voltage with the increasing of drain voltage as shown in figure 3. It was happen due to the existence of strain SiGe at the channel region has enhanced the carrier transport in the VESIMOS-DP channel. Fig. 3 shows the superb performance of the output characteristics ($I_{DS} - V_{DS}$) for VESIMOS-DP device with L_g=50nm.

Fig. 3 show that the V_{DS} rise sharply then grew steadily before going into breakdown state. The device undergoes breakdown state above V_{DS} = 2.5V instead of entering bipolar mode for V_{GS} > 1.80V. In bipolar mode, the device acts as a Bipolar Junction Transistor (BJT) where n+ source of the device become an emitter, drain as a collector and δp+ layer as a base. The current flows between collector and emitter when base current which is created by holes generated during impact ionization process existed in δp+ layer region. This mechanism contributes to the lower subthreshold slope, but suffers from hysteresis [12] due to the PBT effect. However, with the presence of DP layer at the drain side region, the PBT effect has been suppressed for $V_{DS} \leq 2.5$V. For V_{DS} > 2.5V, the PBT effect has been minimized to a minimum level as the δp+ layer cannot contain with the surge of holes current due to higher II rate. Nevertheless, most of the device application operation voltage is below 2.5V which merit the incorporation of DP layer into the device.

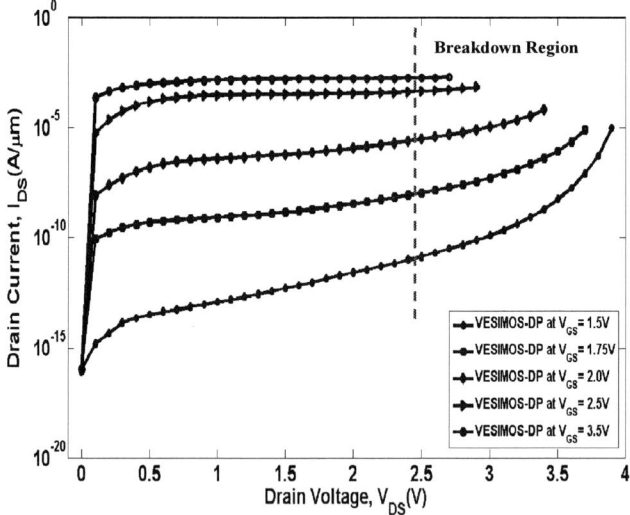

Fig. 3. Output characteristics of VESIMOS-DP device at different gate voltage.

Fig. 4 demonstrates the comparison of subthreshold characteristic for VESIMOS and VESIMOS-DP device taken at V_{DS}= 1.75V.

IEEE-ICSE2014 Proc. 2014, Kuala Lumpur, Malaysia

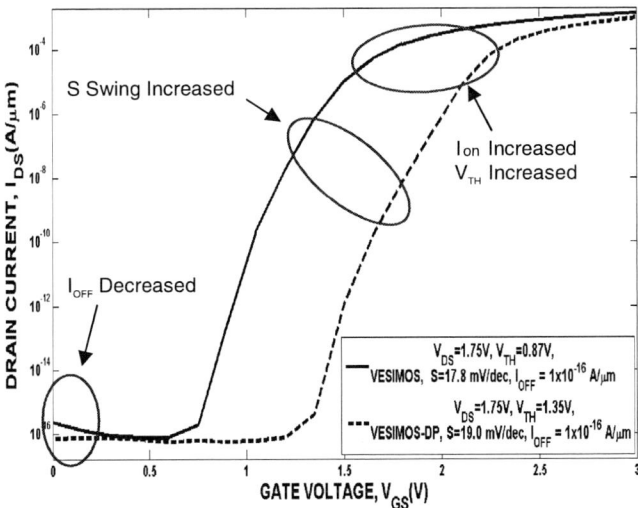

Fig. 4. Subthreshold characteristics of VESIMOS and VESIMOS-DP devices at V_{DS}=1.75V and $Si_{0.7}Ge_{0.3}$.

The off-state leakage current depicted to be lower by at least two decades in VESIMOS-DP device. A significant increase in S value and V_{TH} is visible in Fig. 5 for a VESIMOS-DP device when compared with VESIMOS. The subthreshold swing inclined from 17.8mV/dec to 19mV /dec and threshold voltage increase from 0.87V to 1.35V due to the presence of DP layer add another potential barrier for electron to get through from source to drain. Higher drive-on current is observed in VESIMOS-DP device which increased by a factor of 2 as compared to that only VESIMOS device. These results confirm that DP layer has significantly suppressed PBT effect by limiting the leakage current passing through drain region in bipolar mode.

IV. **Conclusion**

The investigation on the prospective ideas of finding viable solution of nanoelectronics device design by an assessment of incorporating vertical impact-ionization MOSFET (IMOS) with strained SiGe technology into a formation of an emerging device structure with elevated performance and reliable outcomes for future bio-based sensor application was successfully analyzed. Enhanced performance of vertical IMOS structure utilizing combined Strained SiGe + DP technology is studied based on process and device simulation tools. VESIMOS-DP shows excellent switching behavior with low S value of about 19mV/dec which is much lower compared to Conventional MOSFET (CMOS). This will eventually enhanced the switching behavior of the transistor and enhancing its electrical performance and response time. In addition, the vicinity of DP layer has been successfully suppressed the PBT effect by attaining high breakdown voltage. In many aspects, it is revealed that the incorporation of DP enhanced the electrical performance of VESIMOS for future development of biosensor nanoelectronic application.

Acknowledgment

The authors would like to acknowledge the financial support from FRGS fund (FRG253-TK-2-2013) of MOHE Malaysia. The author is thankful to the University Malaysia Sabah (UMS) for providing excellent research environment in which to complete this work.

References

[1] Yi Cui, Qingqiao Wei, *et al.*, "Nanowire Nanosensors for Highly Sensitive and Selective Detection of Biological and Chemical Species ", Science, vol. 293 (5533), 2001, pp. 1289-1292.

[2] M. Shinwaria, *et al.*, "Study of the electrolyte-insulator-semiconductor field-effect transistor (EISFET) with applications in biosensor design", Microelectronics Reliability, vol. 47 (12), 2007, pp. 2025–2057.

[3] Deblina Sarkar and Kaustav Banerjee, "Proposal for tunnel-field-effect-transistor as ultra-sensitive and label-free biosensors", Applied Physics Letter, vol. 100, 2012, pp. 143108.

[4] Deblina Sarkar and Kaustav Banerjee, "Fundamental limitations of conventional-FET biosensors: Quantummechanical-tunneling to the rescue", 70th Annual Device Research Conference, 2012, pp. 83 – 84.

[5] A. Moscatelli, P. Rodgers, *et al.*, "Biomolecular turn-ons Nature Nanotechnology", Appl.Phys. Letter, vol. 100 (7), 2012, pp. 275

[6] Gopalakrishnan K, Griffin PB, *et.al.*, "I-MOS: a novel semiconductor device with a subthreshold slope lower than kt/q", IEEE Electron Device Meeting, 2002, pp. 289–292.

[7] Abelein U, Born M, *et.al*, "A novel vertical impact ionization MOSFET (I-MOS) concept", 25th international conference on microelectronics (MIEL 2006), Pg. 127–129.

[8] Abelein U, Born M, *et.al*, "Improved reliability by reduction of hot-electron damage in the vertical impact-ionization MOSFET (I-MOS)", IEEE Electron Device Letter, vol. 28(1), 2007, pp. 65–67.

[9] Abelein U, Assmuth A, *et.al*, "Doping profile dependence of the vertical impact ionization MOSFET (IMOS) performance", Solid State Electron, vol. 51, 2007, pp. 1405–1411.

[10] Abelein U, Assmuth A, *et.al*, "Vertical 40 nm impact ionization MOSFET (I-MOS) for high temperature applications", 26th international conference on microelectronics (MIEL 2008), 2008, pp. 287–290.

[11] Kraus R and Jungemann C., "Investigation of the vertical IMOS-transistor by device simulation", International conference on ultimate integration of silicon, vol. 10, 2009, pp. 281–284.

[12] Dinh TV., *et.al*, "Investigation of the performance of strained-SiGe vertical IMOS-transistors", Solid State Device Research Conference, 2009, pp. 165–168.

[13] S.K. Jayanarayanan, *et.al*, "A Novel 50nm vertical MOSFET with a dielectric pocket", Solid-State Electronics, vol. 50, 2006, pp. 897-900.

[14] Sentaurus user Guide Device and Process Simulation Software, Sentaurus Inc, 2012.

[15] N.D. Jankovic and G.A. Armstrong,: Comparative analysis of the DC performance of DG MOSFETs on highly-doped and near-intrinsic silicon layers. Microelectronics Journal. vol. 35, 2004, pp. 647–653.

[16] McKay, "Avalanche Breakdown in Silicon", Phys Rev. 94, 1954, pp 877

[17] Selberherr S., "Analysis and Simulation of Semiconductor Devices", Springer-Verlag, Wien-New York, 1984.

Investigation on Iodine Flow Rate in MEH-PPV: I-MWCNT Nanocomposite Thin Film

M.S.P. Sarah, M. Mudasir, S.S. Shariffudin, H. Hashim, M.Rusop
Faculty of Electrical Engineering
Universiti Teknologi MARA
40450 Shah Alam, Malaysia
E-mail: puterisarah@salam.uitm.edu.my

Abstract—The effect of Iodine flow rate in nanocomposited MEH-PPV:I-MWCNT was investigated by means of electrical, optical and physical characterization. First, 120 mg of MWCNT was doped with 1g of Iodine using thermal chemical vapour deposition method (TCVD). The doping process was done for 1 hour with Iodine flow rate varied from 0.1, 0.3, 0.5, 0.7 and 0.9 *l*/min. On the other hand, 40 mg of MEH-PPV was stirred for 48 hours in tetrahydrofuran (THF). Next, the I-MWCNT was added to the solution to form nanocomposited MEH-PPV: I-MWCNT solution. The solution was then deposited on glass substrate using spin coating technique. The current-voltage (*I-V*) measurements were done in dark and under illumination. UV-Vis Spectrometer was used to measure the absorbance and transmittance. For physical properties, the characterizations were done using FESEM and Surface Profiler. From the *I-V* characteristic, thin film with Iodine flow rate 0.9 *l*/min gives the best result considering some response it gives towards light. Besides, the sample shows the highest photoconductivity with 2.82×10^{-3} S/cm. In optical properties, the thin film also gives value 0.89 unit of absorption spectra which is the highest value among other samples. The optimized flow rate will be used to fabricate an active layer of organic solar cell.

Keywords— MEH-PPV, I-MWCNT, nanocomposite, Iodine flow rate

I. Introduction

Conjugated polymer seldom used as an active layer in organic solar cells due to its low cost, simple processing and promising technology application [1]. The most common conjugated polymer used is Poly [2-methoxy-5-(2-ethyl-hexyloxy)-1, 4-phenylene vinylene] (MEH-PPV [2]. It behaves as a p-type semiconductor polymer; therefore it has low conductivity [3]. Despite that, it has good environmental stability, easy to control of its conductivity and can be manufacturer in large quantity with low cost [4].

However, MEH-PPV needs to be use with a conductive filler such as CNT due to its low absorption and conductivity [4]. With that, new form of nanocomposite was created to improve the physical structure and a novel behavior. Nanocomposites differ due to it has higher surface to voltage ratio of the reinforcing phase and also exceptionally high aspect ratio. CNT can be divided into two types which is single wall carbon nanotube (SWCNT) and multi wall carbon

nanotube (MWCNT) [5]. For this work, MWCNT has been used due to it has advantages compare to SWCNT in aspect of conductivity and absorbance. MWCNT are much cheaper compared to SWCNT. Therefore, they can be profitably used to make polymer or other composites enhancing their mechanical, thermal and electrical properties. Meanwhile, the reason MWCNT used is because of its low weight, high electrical conductivity, excellent strength and efficient thermal conductors [6].

Issues with MWCNT in MEH-PPV involved slight increment in conductivity and also dispersion properties issue [7]. Recent study [8] state that MWCNT has a saturated value that if too much MWCNT was added, it may wipe out. Iodine was chosen to be doped with MWCNT because recent study [9] shows that it can be effectively doped and exchange electron with Iodine. With the help of Iodine doping, the MWCNT can be well dispersed in organic solvent which usually a good solvent for most polymers even though doping Iodine and MWCNT was not clearly understood [10]. Besides, Iodine is one of the most useful atoms in the organic synthesis field. It has been used as the reactive agent of an aromatic π-electron system [11]. This is to give the corresponding cation radical. As a result, I-MWCNT can uniformly distribute in polymer matrices [12]. In this study, Iodine is the best candidate to alter and enhance the electrical and physical properties of MWCNT. Hence, the effect of different Iodine flow rate was examined.

II. METHODOLOGY

A. Cleaning Substrates

The process of substrates cleaning starts with immersion of the substrates in Acetone, Methanol and de-ionized (DI) water for 10 minutes each before it was dried with Nitrogen gas blower.

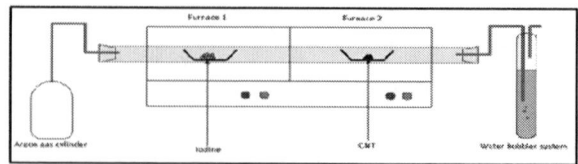

Fig. 1: Schematic diagram of thermal CVD system

B. Material Preparation

The MEH-PPV was first weighed before being mixed in tetrahydrofuran (THF). THF was used due to its porous surface and most suitable solvent for MEH-PPV [2] towards the application of organic solar cells. 40 mg of MEH-PPV were dissolved in 10 ml of THF and was stirred for 48 hours to guarantee MEH-PPV was well dissolves in solvent.

In this work MWCNT were manufactured by US Research Nanomaterial Inc. Before mixing the MWCNT with MEH-PPV, it was annealed 30 minutes at 450°C inside a chamber furnace to remove any unwanted particle and also to improve the physical structure of the MWCNT. The doping process was carried out using thermal chemical vapor deposition (CVD) as can be seen in Fig. 1. The Iodine and annealed MWCNT was weighed at 1g and 120 mg respectively. Then Iodine and MWCNT powder were placed onto an alumina boat before being inserted into the thermal CVD. Meanwhile, Argon gas was used as the carrier gas. For the doping process, Furnace 1 was filled with Iodine (in solid form) and heated at 100°C. On the other hand, the MWCNT powder was loaded into Furnace 2 at 800°C. Right after that, the quartz tube was purged with Argon gas for approximately 10 minutes to flush out the air inside the quartz tube and also to provide an inert environment during the doping process. During the process, different flow rate of iodine were varied which are 0.1 *l*/min, 0.3 *l*/min, 0.5 *l*/min, 0.7 *l*/min and 0.9 *l*/min. Once doping process finished, I-MWCNT was mixed with MEH-PPV solution and sonicated for 1 hour to make sure it was well dispersed.

Next step is the deposition of MEH-PPV:I-MWCNT solutions onto glass substrates by spin coating technique. The solutions were drop with 10 drops and spin at 2000 rpm. Then, thin film produced were dry using digital hot plate stirrer for 5 minutes at 50°C.

C. Characterization

There are three types of characterizations that took place; electrical, optical and physical properties. For electrical properties, samples are being analyzed using solar simulator to get the current-voltage (*I-V*) characteristics. Meanwhile, to measure optical properties, UV-Vis spectrometer JASCO/V-670 EX was used. From UV-Vis, transmittance and absorbance measurement can be obtained. Physical properties involved FESEM and thickness of the thin film.

III. RESULT AND DISCUSSION

A. Electrical Properties

The measured *I-V* characteristics of MEH-PPV: I-MWCNT nanocomposite thin film under illumination is shown in Fig. 2. *I-V* was measured from -0.5 V to 0.5 V. Low voltages was chosen due to avoiding samples' damage as MWCNT itself is a conductive filler and then doped with Iodine which make it more conductive [3]. The current changed linearly with the voltage which indicates ohmic behavior between the thin film and metal contact. At 0.5 V, for flow rate at 0.1 *l*/min, the current was so small

approximately 0 nA. The same readings occurred also for flow rate at 0.3 and 0.5 *l*/min. However, at 0.7 *l*/min, current starts to increase with nearly 20 nA and it increases further up at Iodine flow rate 0.9 *l*/min with current reading 40 nA. Increment in current might be due to combination of increased carrier donation by the MWCNT and increases in number of conductive path [4]. From the *I-V* measurement, resistivity, ρ and conductivity σ of the MEH-PPV: I-MWCNT nanocomposite thin film can be calculated using equation (1) and (2) respectively:

$$\rho = \left(\frac{V}{I} \right) \frac{wt}{l} \qquad (1)$$

$$\sigma = \frac{1}{\rho} \qquad (2)$$

where V is the supplied voltage, I is the measured current, w is the width, t is the thin film's thickness and l is the length between electrodes.

Fig. 3 shows the photoconductivity obtained from the *I-V* curve. By considering conductivity, thin film can be rate as it can actively transfer electron or behaves as resistance or conductance. Fig. 3 shows the photoconductivity of the nanocomposited MEH-PPV:I-MWCNT thin film. The photoconductivity was found to be increasing from 3.32×10^{-5} S/cm at 0.1 *l*/min up to 2.82×10^{-3} S/cm at 0.9 *l*/min of Iodine flow rate. It has been clearly observed that Iodine flow rate at 0.9 *l*/min has the highest conductivity. It was found, as Iodine flow rate increases, the MEH-PPV:I-MWCNT nanocomposites thin film are more conductive [5].

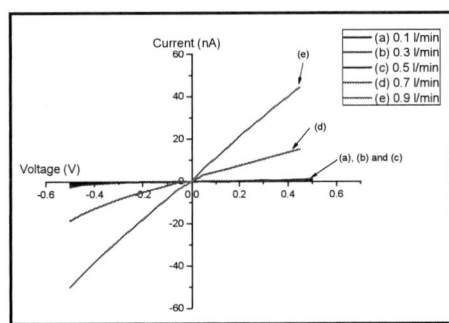

Fig. 2: *I-V* for MEH-PPV: I-MWCNT nanocomposite thin film under illumination

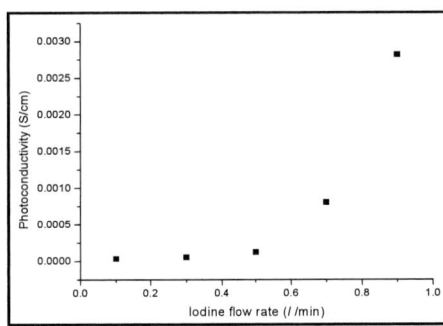

Fig. 3: Photoconductivity for MEH-PPV: I-MWCNT nanocomposite thin film

B. Optical Properties

The absorption spectra of MEH-PPV: I-MWCNT was presented in the range of wavelength from 350 nm to 800 nm. From Fig. 4, it can be observed that the absorbance value increases as Iodine flow rate being increased. The absorption spectra of 0.9 l/min show the highest absorption at average absorbance 0.89 at 550 nm. Also, the absorbance peaks tend to shift to longer wavelength. The red shift in the absorption peak may due to increase in the interfacial charge carrier concentration (charge density at the interface) which improves the extension of the band tails in the forbidden band gap of the organic compounds [6]. Therefore the optical absorption occurs at higher wavelength.

The absorption will influence the generation of electrons to excite to the correspond light at certain wavelength. With higher absorption, it will increases ability of electron to excite from highest occupied molecular orbital (HOMO) to lowest unoccupied molecular orbital LUMO. The absorption increases due to I-MWCNT absorb light in the infrared region (IR) of solar spectrum and semiconducting polymer absorb light in the ultra-violet (UV) and visible region [7]. Fig. 5 shows the transmittance spectra of nanocomposited MEH-PPV: I-MWCNT thin film for variation of Iodine flow rate from 0.1, 0.3, 0.5, 0.7 and 0.9 l/min in the range of wavelength from 350 nm to 800 nm. It demonstrates that transmittance of nanocomposited MEH-PPV: I-MWCNT thin film shows decreases in value with the increment in Iodine flow rate. As the Iodine flow rate increases, the transmittance decreased from 75, 25, 20, 13 and 8 T% respectively.

To estimate the value of optical band gap, tangent line should be drawn in the Tauc's plot touching most plots until intercepting the x-axis. The interception value is the optical band gap value and measured in eV. The result in Fig. 6 shows the optical band gap of 2.01, 2.12, 2.14, 2.14 and 1.60 eV for MEH-PPV: I-MWCNT with Iodine flow rate 0.1, 0.3, 0.5, 0.7 and 0.9 l/min respectively. From here, it is observed the optical band gap of the thin film can be tuned by varying the Iodine flow rate. Comparing the results with the optical band gap of pure MEH-PPV which is 2.20 eV [8], this work manages to reduce the optical band gap value in average of 0.20 eV. With the I-V characteristics being analysed in parallel with the optical band gap properties, the significant relationship between this parameter could be concluded; the smaller optical band gap ensure a higher conductivity of a thin film [9]. This will enhance the excitation of electrons from HOMO to LUMO once photon is absorbed [10].

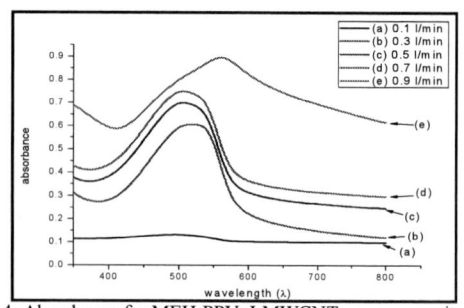

Fig. 4: Absorbance for MEH-PPV: I-MWCNT nanocomposite thin film

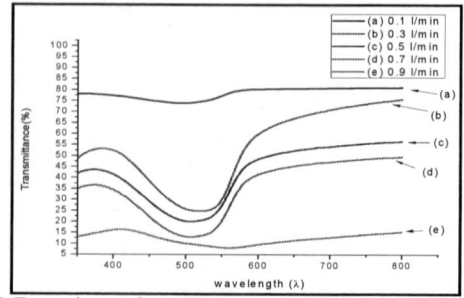

Fig. 5: Transmittance for MEH-PPV: I-MWCNT nanocomposite thin film

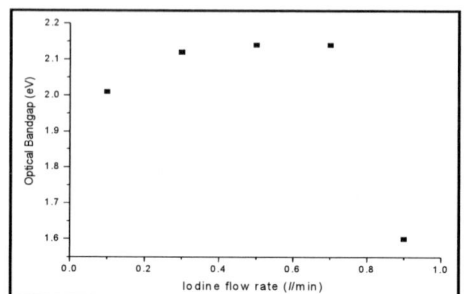

Fig. 6: Optical band gap for MEH-PPV: I-MWCNT nanocomposite thin film

C. Physical properties

Surface morphology was being observed by using field emission scanning microscopy (FESEM) with same magnification (5000x) at 5.0kV for all samples. From the figure, it can be assume that, higher flow rate will increase the growth rate and density of I-MWCNT [11]. Figure 7(a) shows the surface morphology of the lowest Iodine flow rate being applied which is 0.1 l/min. Part (a) of Fig. 7 depicts the FESEM images of only existence of nature polymer and small existence of I-MWCNT. Pores are also observed due to evaporation of solvent during film formation. Part (b) and (c) of Fig. 7 shows micrograph of MEH-PPV:I-MWCNT nanocomposite thin film appearing as non-smooth and agglomeration of I-MWCNT. Part (d) and (e) of Fig. 7 shows the morphology of MEH-PPV:I-MWCNT nanocomposite thin film at 0.7 and 0.9 l/min respectively. In the case of 0.9 l/min, I-MWCNT are clearly distinguishable in the image forming more than one layer compared to 0.7 l/min where only one layer image can be observed. The thin film's morphology changes with the different Iodine flow rate which could anyhow influence the charge carrier mobility and conductivity [12]. Anyhow, it can be said that increment of Iodine flow rate trigger the existence of I-MWCNT in the MEH-PPV matrix.

IV. CONCLUSION

The nanocomposited MEH-PPV: I-MWCNT thin film has been fabricated and studied. The effects of Iodine flow rate to the thin film were also examined. The characterization of electrical properties shows increment in current as the Iodine flow rate is increase. Thin film with highest Iodine flow rate, 0.9 l/min gives the best photoconductivity with 2.82×10^{-3} S/cm. In optical characterization, it can be said that as Iodine flow rate increases, absorbance also increases. Meanwhile, the optical band gap of thin film with 0.9 *l*/min shows the lowest value comparing with flow rate at 0.1, 0.3, 0.5 and 0.7 *l*/min.

Acknowledgment

This paper was made possible through the invaluable assistance and cooperation from various parties involved. The authors wish special appreciation to Faculty of Electrical Engineering. Special thanks also go to all members of NANO-ElecTronic Center (NET) and Nano-SciTech Center (NST), UiTM Shah Alam for their guidance, criticism and advices in completing this project.

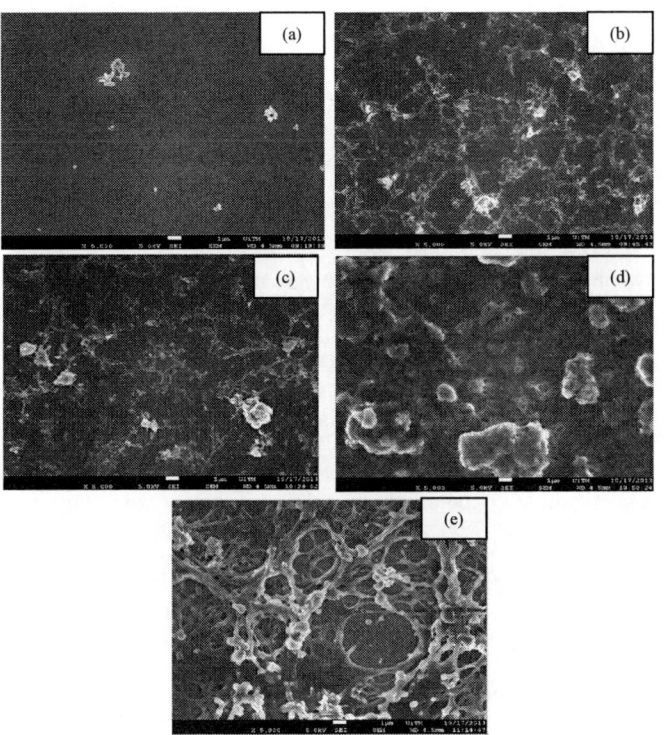

Fig. 7: FESEM images for MEH-PPV: I-MWCNT nanocomposite thin film at (a) 0.1 l/min (b) 0.3 l/min (c) 0.5 l/min (d) 0.7 l/min and (e) 0.9 l/min

References

[1] X. Yang, J. Loos, S. C. Veenstra, W. J. Verhees, M. M. Wienk, J. M. Kroon, *et al.*, "Nanoscale morphology of high-performance polymer solar cells," *Nano letters*, vol. 5, pp. 579-583, 2005.

[2] A. O. Sevim and S. Mutlu, "Post-fabrication electric field and thermal treatment of polymer light emitting diodes and their photovoltaic properties," *Organic Electronics*, vol. 10, pp. 18-26, 2009.

[3] A. Sertap Kavasoglu, F. Yakuphanoglu, N. Kavasoglu, O. Pakma, O. Birgi, and S. Oktik, "The analysis of the charge transport mechanism of n-Si/MEH-PPV device structure using forward bias I-V-T characteristics," *Journal of Alloys and Compounds*, vol. 492, pp. 421-426, 2009.

[4] M. Sarah, F. Zahid, and M. Rusop, "Investigation on IV for Different Heating Temperatures of Nanocomposited MEH-PPV: CNTs Organic Solar Cells," *International Journal of Photoenergy*, vol. 2012, 2012.

[5] Richard Elkins, Nathan Fierro, Erin Flanagan, Adam Haughton, Michael Kasser, Matthew Stair, *et al.*, "Utilizing Carbon Nanotubes to Improve Efficiency of Organic Solar Cells," 2006.

[6] A. M. K. Esawi and M. M. Farag, "Carbon nanotube reinforced composites: Potential and current challenges," *Materials & design*, vol. 28, pp. 2394-2401, 2007.

[7] L. Kumari, V. Prasad, and S. Subramanyam, "Effect of iodine incorporation on the electrical properties of amorphous conducting carbon films," *Carbon*, vol. 41, pp. 1841-1846, 2003.

[8] S. Xu, X. Zhu, H. Kotadia, H. Lu, S. H. Mannan, C. Bailey, *et al.*, "Remedies to control electromigration: Effects of CNT doped Sn-Ag-Cu interconnects," in *Electronic Components and Technology Conference (ECTC), 2012 IEEE 62nd*, 2012, pp. 1899-1904.

[9] W. Zhou, S. Xie, L. Sun, D. Tang, Y. Li, Z. Liu, *et al.*, "Raman scattering and thermogravimetric analysis of iodine-doped multiwall carbon nanotubes," *Applied Physics Letters*, vol. 80, p. 2553, 2002.

[10] T. Michel, L. Alvarez, J. L. Sauvajol, R. Almairac, R. Aznar, O. Mathon, *et al.*, "Structural selective charge transfer in iodine-doped carbon nanotubes," *Journal of Physics and Chemistry of Solids*, vol. 67, pp. 1190-1192, 2006/6// 2006.

[11] Y. Shibayama, H. Sato, T. Enoki, M. Endo, and N. Shindo, "Magnetic properties of activated carbon fibers and their iodine-doping effect," *Molecular Crystals and Liquid Crystals*, vol. 310, pp. 273-278, 1998.

[12] X.-M. Xie, Z.-L. Zhang, Y.-T. Liu, Q.-P. Feng, W. Zhao, and X.-Y. Ye, "Polymer assisted dispersion and alignment of carbon nanotubes," in *Nanotechnology, 2009. IEEE-NANO 2009. 9th IEEE Conference on*, 2009, pp. 123-125.

[13] N. Zayana and M. Rusop Mahmood, "Effect of Different Solvents on the Optical Properties and Surface Morphology of MEH-PPV Thin Films," *Advanced Materials Research*, vol. 667, pp. 402-406, 2013.

[14] L. Valentini, I. Armentano, P. Santilli, J. M. Kenny, L. Lozzi, and S. Santucci, "Electrical transport properties of conjugated polymer onto self-assembled aligned carbon nanotubes," *Diamond and Related Materials*, vol. 12, pp. 1524-1531, 2003.

[15] Y. Zhao, J. Wei, R. Vajtai, P. M. Ajayan, and E. V. Barrera, "Iodine doped carbon nanotube cables exceeding specific electrical conductivity of metals," *Scientific reports*, vol. 1, 2011.

[16] Y. A. Ismail, T. Soga, and T. Jimbo, "Features in optical absorption and photocurrent spectra of organic solar cells due to organic/organic interface," *Journal of Applied Physics*, vol. 109, p. 103109, 2011.

[17] E. Kymakis and G. Amaratunga, "Single-wall carbon nanotube/conjugated polymer photovoltaic devices," *Applied Physics Letters*, vol. 80, p. 112, 2002.

[18] N. Z. Yahya and M. Rusop, "Investigation on the optical and surface morphology of conjugated polymer MEH-PPV: ZnO nanocomposite thin films," *Journal of Nanomaterials*, vol. 2012, 2012.

[19] I. Khatri, T. Soga, T. Jimbo, S. Adhikari, H. R. Aryal, and M. Umeno, "Synthesis of single walled carbon nanotubes by ultrasonic spray pyrolysis method," *Diamond and Related Materials*, vol. 18, pp. 319-323, 2009/3// 2008.

[20] E. Bundgaard and F. C. Krebs, "Low band gap polymers for organic photovoltaics," *Solar Energy Materials and Solar Cells*, vol. 91, pp. 954-985, 2007.

[21] M. Kara, A. A. Teh, R. Ahmad, M. Rusop, and Z. Awang, "Gas flow rate and synthesis time dependence of MWCNT growth by chemical vapour deposition," in *RF and Microwave Conference (RFM), 2011 IEEE International*, 2011, pp. 359-362.

[22] Z. Zhuo, F. Zhang, J. Wang, J. Wang, X. Xu, Z. Xu, *et al.*, "Efficiency improvement of polymer solar cells by iodine doping," *Solid-State Electronics*, vol. 63, pp. 83-88.

Impact of Implantation Methods on Speed and Accuracy Trade-off in Calibrated TCAD Tool

Muhamad Amri Ismail

Technology Development Group
MIMOS Wafer Fab, MIMOS Berhad
Technology Park Malaysia, 57000 Kuala Lumpur, Malaysia
amris@mimos.my

Abstract— **Analytical-based and Monte Carlo-based are two methods available in TCAD for simulation of ion implantation step. This paper presents a selection of suitable implantation methods considering the speed and accuracy trade-off while fulfilling the calibrated TCAD requirements in MOSFET process and device simulations. Doping profiles from several device physicals such as channel, halo and source-drain structures are acquired to capture the impact of different implantation methods. The comparisons between measured and simulated doping profiles are presented to further investigate the trade-off as a function of energy levels and tilt angles. The best solution is proposed to obtain essentially calibrated TCAD simulation, without unnecessarily scarifying the simulation time.**

Keywords— *calibrated TCAD; analytical-based; Monte Carlo-based; MOSFET device; implantation methods*

I. INTRODUCTION

Technology Computer Aided Design (TCAD) offers various advantages with regards to continuous technology downscaling and exponentially growth of device densities. As an elegant tool to perform reliable process and device simulations, TCAD plays a very important role at every stage of technology developments. The availabilities of pre-silicon data such as process flows, device characteristics and SPICE model parameters are some of the benefits gained at the early stage of technology development while performing swift and numerous design of experiments (DOE) are the advanced capabilities gathered during the qualification stage[1]. Even after the qualification stage, TCAD is still relevant for technology improvements and statistical simulations such as design for manufacturing (DFM) and design for yield (DFY) analyses [2].

These advantages are certainly crucial in identifying related obstacles as well as reducing the technology development cost. Predictability is another advanced feature of TCAD where the simulation data are calibrated with the actual measurement data [3]. TCAD calibration could be divided into several levels namely 1-dimensional (1-D), 2-dimensional (2-D) and 3-dimensional (3-D) depends on the requirements and complexity of the process and device simulations. 1-D is the simplest calibration level where it deals with basic process parameters, 2-D is the intermediate level where it further

involves with final device characteristics and 3-D is the most complicated level which may requires an advanced processor to handle large experimental data [4]-[5].

Ion implantation is one of the standard process simulation embedded in any TCAD tool instead of diffusion and oxidation processes. In fact, the fabrication process is primarily starts with ion implantation to show the criticality of this step to the overall accuracy of TCAD calibration process. There are two types of implantation method usually furnished in commercial TCAD package known as analytical-based and Monte-Carlo-based [6]. These two methods have their own superiorities and drawbacks that further influence the time expenditure and precision of simulation data. In this paper, we propose an efficient way to select suitable implantation methods at different device structures without forfeiting the simulation cycle and accuracy especially for 1-D calibration. The proposed methodology focuses on implantation methods considering its holistic role to the whole fabrication process as it appears at the first and at various process steps. With respect to its usefulness in all calibration levels, 1-D calibration is selected rather than 2-D or 3-D where the concerns are mainly on the profiles of Secondary Ion Mass Spectrometry (SIMS) data.

II. SPEED AND ACCURACY TRADE-OFF

From ion implantation theories, several mechanisms such as ion channeling, implant doses, energies and tilt angles as well as accumulation of point defects are the most challenges part for accurate TCAD simulations. Channeling is generally related to the effect when the implant tilt angles are low while implant doses, energies and tilt angles are important for the compensation of variations in other process steps. Point defects are the phenomenon due to implant induced damage in the presence of high dose implant. A simple process simulation to consider for these implantation mechanisms in TCAD is through analytical-based method. The doping profile in analytical method is estimated using either a statistical Gaussian distribution function, Pearson function or advanced dual Pearson function [7]. For example in simple Gaussian distribution function, the doping profile, $N(x)$ is given by

$$N(x) = \frac{Q}{\sqrt{2\pi}\sigma_p} exp\left[-\frac{(x - R_p)^2}{2\sigma_p^2}\right]$$

(1)

where Q is the implant dose, R_p is the projected range and σ_p is the standard deviation of the projected range [8]. Projected range in (1) is referring to the characteristic of dopant distribution, where it is further represented by several implantation coefficients known as moments. Series of moments for particular implantation characteristics are tabulated in the form of lookup tables and that is the reason why simple analytical method is also known as table-based method.

In contrast, physical-based Monte Carlo uses more advanced approaches suitable for various dopants with consideration of every aspect in implantation issues from ion channeling to point defects analysis. Monte Carlo-based method uses classical binary collision theory to model the energy loss due to ion collision. In the binary collision theory, there are elastic and inelastic collision processes with regard to the interactions of the moving ion respectively with an atomic nucleus and the electrons. Then the energy loss, ΔE_n is modeled by

$$\frac{\Delta E_n}{E_0} = \frac{4M_1 M_2}{(M_1 + M_2)^2} cos^2(bl)$$

(2)

where E_0 is the kinetic energy before collision, M_1 and M_2 are the first and second mass particles respectively, b is the impact factor and I is the integral part [9]. In general, the energy loss model concerns on the separation between two mass particles and that is why physical-based Monte Carlo method is also known as particle-based. The integral part involves with scattering analysis where it is further based on number of particles used during Monte Carlo simulation.

Basically it depends to the user to select either to use simple analytical-based method or advanced Monte-Carlo based method since both are available in standard TCAD tool. In short, analytical-based method is simple in methodology by quick simulation on implantation step with certain accuracy but it is not detail out on the atomic level of the implantation mechanism. On the other hand, Monte Carlo-based method is more suitable for higher accuracy but it trade-off with simulation time since it needs to fit a huge number of iterations. For example, the default value for the number of particles in commercial TCAD simulator is 1000 to show that there are a lot of iterations involve in Monte Carlo-based method [9]. The advancement in computer hardware and processing has contributes in reducing the simulation time with Monte Carlo analysis but still suitable solution is required for efficient time management. For example, TCAD nowadays is applied to not only for a single device but also to the different device geometries for the same technology node with respect to device scalability. Therefore it is necessary to understand the criticality of Monte Carlo-based method at certain device

structures so that the simulation time is efficient for all device geometries.

III. EXPERIMENT

In order to investigate the impact of speed and accuracy trade-off in calibrated TCAD tool, a few wafers were fabricated for the collection of experimental data with respect to different implantation steps. The wafers were manufactured by MIMOS Wafer Fab using matured industry standard 0.35um CMOS process. Table 1 shows the summary of three wafers used in the analysis where each wafer contains different process flow and split conditions particularly for implantation steps of NMOS device. For wafer 1, after the oxidation step of initial well structure, P-well is formed using several Boron implantations, namely P-well implant, field implant and V_t-adjust implant. Wafer 2 is meant for NMOS halo structure where it contains the combination of Arsenic and Boron implants. Lightly doped drain (LDD) structure is created through Arsenic implant to reduce the peak electric field in channel region. Quad halo Boron implant with high tilt angle is required for better penetration of the dopants into the halo structure. Wafer 3 is specifically for N+ source-drain (S/D) structure where a high dose of Arsenic is implanted with some low tilt angles to build for low sheet and contact resistance values.

The doping profiles are then extracted from the fabricated wafers using SIMS instrument which later become the actual measured data. For the simulated data, Sentaurus Process (SProcess) from Synopsys TCAD tool is used to generate the related doping profiles [9]. With respect to the objective of this paper, both analytical-based and Monte Carlo-based are utilized during implant simulation steps of each split condition as per Table 1. The standard SProcess file needs to be associated with calibration file provided by Sentaurus Calibration Package for improvement in simulation accuracy, especially for Monte Carlo-based method [10]. The number of particles in Monte Carlo analysis used in this work is set at 6000 in order to ensure that the simulated result will be as smooth as possible with better fitting quality of the doping profiles.

TABLE I. PROCESS SPLITS

Wafer #	Device structures	Split conditions
1	P-well	Boron well implant Boron field implant Boron V_t-adjust implant
2	NMOS halo	Arsenic LDD implant Boron quad halo implant
3	N+ S/D	Arsenic high dose implant

IV. RESULT AND DISCUSSION

We used the result from TCAD simulation and compared with actual SIMS profiles for the analysis and proposal on suitable speed and accuracy trade-off. Since there are three wafers fabricated for the experimentation purpose, therefore three doping profiles have been analyzed accordingly. The comparison between measured and simulated data at P-well structure is depicted in Fig. 1. The result shows that Monte Carlo-based method is the best option for reproducing the actual measured profiles whereas analytical-based method fails to predict the profiles at deeper well structure. The same scenario goes to the doping profiles of NMOS halo structure where Monte Carlo-based displays more superiority compared to analytical-based method, as shown in Fig. 2. The common condition in both P-well and NMOS halo structures is the dopants are implanted at low energies. Another common condition that worth to be highlighted is the high tilt angles that used in both structures which required the precision of particle-based Monte Carlo for accurate simulation of implantation process.

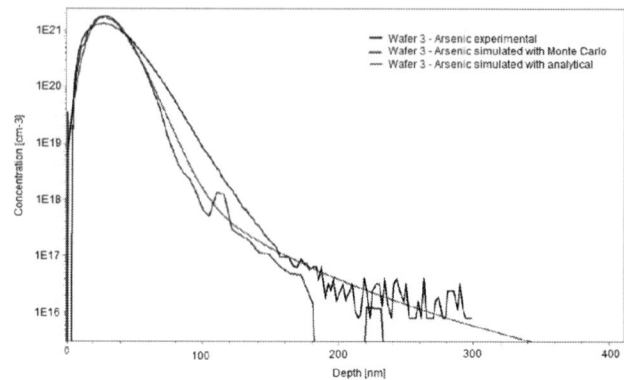

Fig. 3. Comparison of Arsenic SIMS profiles measured and simulated at N+ S/D structure

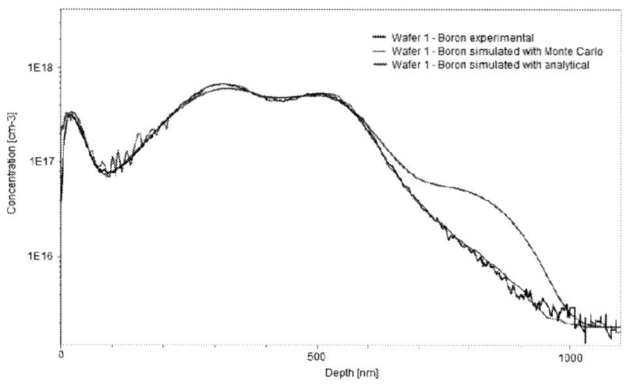

Fig. 1. Comparison of Boron SIMS profiles measured and simulated at P-well structure

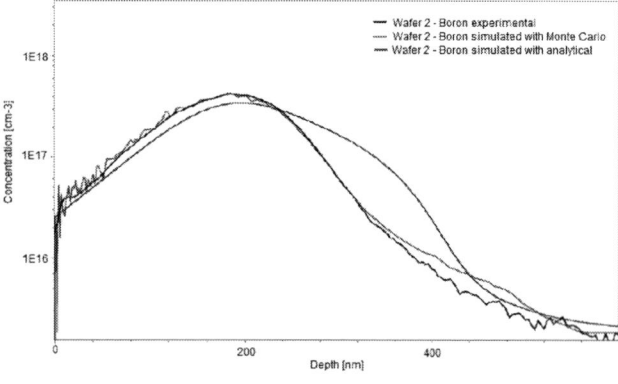

Fig. 2. Comparison of Boron SIMS profiles measured and simulated at NMOS halo structure

The doping profiles related to N+ S/D structure are depicted in Fig. 3. If previously analytical-based method suffers inaccuracy issues at P-well and NMOS halo structure, Fig. 3 shows that it could provide a qualitatively good agreement compared to actual SIMS profile. In other words, it is not necessarily required to proceed with slow Monte Carlo-based method in pursuing for calibrated TCAD process simulation when fast analytical-based method already met the requirement.

In this work, the difference between N+ S/D structure compared to P-well and NMOS halo structures are in term of energies and tilt angle values. The experimental result shows that the structure with low energies and high tilt angles will need to consider for the use of particle-based Monte Carlo method for TCAD calibration at high precision, as summarized in Table 2. From another standpoint, the structure with high energies and low tilt angles will certainly accurate even with simple analytical-based method as supported by experimental data without unnecessarily wasted the simulation time. These results are useful especially if the TCAD simulation needs to be applied to the several device geometries for the same technology node. By considering the impact of speed and accuracy trade-off, selection of suitable implantation methods during TCAD process simulation will guides for continuous computationally efficient works.

TABLE II.   SPEED AND ACCURACY TRADE-OFF

Implantation methods	Process conditions
Accurate Monte Carlo-based	Low energies High tilt angles
Simple analytical-based	High energies Low tilt angles

V.   CONCLUSION

This paper has presented an impact of different implantation methods available in TCAD process simulator with respect to speed and accuracy trade-off. The qualitatively good agreement between measured and simulated doping profiles proves that it is possible to achieve a well calibrated 1-D TCAD process simulation. This work shows that time-efficient analytical-based method could be used for simulation of implantation step with high energies and low tilt angles. Particularly for implantation step with low energies and high tilt angles, particle-based Monte Carlo method is the best solution to fulfill the accuracy requirements. Careful combination of analytical-based and Monte Carlo-based methods in the overall process flow is necessary to gain both benefits of computationally-efficient and highly calibrated TCAD tool.

REFERENCES

[1] R. W. Dutton, and A. J. Strojwas, "Perspective on technology and technology-driven CAD," IEEE Trans. on Computer-Aided Design of Integrated Circuits and Systems, vol. 19, no. 12, pp. 1544-1560, Dec. 2000.

[2] S. Tirumala, Y. Mahotin, X. Lin, V. Moroz, L. Smith, S. Krisnamurthy, L. Bomholt, and D. Pramanik, "Bringing manufacturing into design via process-dependent SPICE models," Proc. of IEEE International Symposium on Quality Electronic Designs, pp. 801-806, March 2006.

[3] C. K. Maiti, and T. K. Maiti, "Technology CAD of strained-engineered MOSFETs," in Strained-Engineered MOSFETs. CRC Press – Boca Raton, FL: 2013, pp. 206-208.

[4] S. K. Saha, "Introduction to technology computer aided design," in Technology Computer Aided Design: Simulation for VLSI MOSFET. CRC Press – Boca Raton, FL: 2013, pp. 1-44.

[5] A. N. Bhoj, R. V. Joshi, and N. K. Jha, "Efficient methodologies for 3-D TCAD modeling of emerging devices and circuits," IEEE Trans. on Computer-Aided Design of Integrated Circuits and Systems, vol. 32, no. 1, pp. 47-58, Jan. 2013.

[6] G. Hellings, and K. D. Meyer, "TCAD simulation and modeling of ion implants in Germanium," in High Mobility and Quantum Well Transistors. Springer - Netherlands: 2013, pp. 27-47.

[7] S. Li, and Y. Fu, "Advanced theory of TCAD process simulation," in 3D TCAD Simulation for Semiconductor Processes, Devices and Optoelectronics. Springer - New York: 2012, pp. 19-39.

[8] G. Hobler, and S. Selberherr, "Two-dimensional modeling of ion implantation induced point defects," IEEE Trans. on Computer-Aided Design of Integrated Circuits and Systems, vol.7, no. 2, pp. 174-180, Feb. 1988.

[9] Sentaurus Process User Guide, Synopsys Inc., Dec. 2010.

[10] Advanced Calibration User Guide, Synopsys Inc., Dec. 2010.

IEEE-ICSE2014 Proc. 2014, Kuala Lumpur, Malaysia

FDCCTA: The Active Building Block for Analog Signal Processing Applications

Montree Kumngern

Department of Telecommunications Engineering, Faculty of Engineering,
King Mongkut's Institute of Technology Ladkrabang, Bangkok 10520, Thailand
E-mail: kkmontre@kmitl.ac.th

Abstract—**This study presents a new active building block for analog signal processing applications, namely fully differential current conveyor transconductance amplifier. The proposed circuit is realized using differential difference current conveyors and transconductance amplifiers. Unlike, previous building blocks, both fully differential input and electronic tuning capability can be obtained into a single new device. The proposed building block is used to realize a universal biquadratic filter as an example application. PSPICE simulation results using 0.5 km CMOS technology from MIETEC are used to confirm the presented building block. The simulation results express that the proposed active building block can be used to realize analog signal processing circuits.**

Keywords–analog building block, fully differential current conveyor transconductance amplifier, differential difference current conveyor, transconductance amplifier

I. INTRODUCTION

In 1996, a new active device, differential difference current conveyor (DDCC) has been proposed [1]. This device has the advantages of both the second-generation current conveyor (CCII) and the differential difference amplifier (DDA) (such as high input impedance and arithmetic operation capability). The DDCC was developed next by including the transconductance amplifier (TA) at the output stage and called the differential difference current conveyor transconductance amplifier (DDCCTA) [2]. Therefore, this device has the DDCC as an input stage and has the TA as output stage that provided all the good properties of the DDCC and the function of electronic tuning of the TA.

Recently, a new active building block, the so-called fully differential second-generation current conveyor (FDCCII) has been proposed [3]. This device has the advantages of the fully DDA that use to improve the dynamic range in mixed-mode applications where fully differential signal processing is required and still has advantages of CCII. Several FDCCII-based circuits have been proposed; see, for example [4]-[8].

To increase the advantage of FDCCII, the transconductance amplifier is added into the output stage of FDCCII. Thus new version of FDCCII provides the electronic adjustable function at the output stage while still has good properties of FDCCII. This new device is called the fully differential current conveyor transconductance amplifier (FDCCTA). The use of FDCCTA-based circuits, passive resistors can be reduced which is ideal

for integrated circuit implementation. It can be expected that this active building block will be used to realize analog signal processing circuits in the future.

II. PROPOSED CIRCUIT

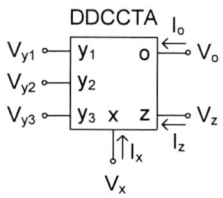

Figure 1.   Electrical symbol of DDCCTA [2].

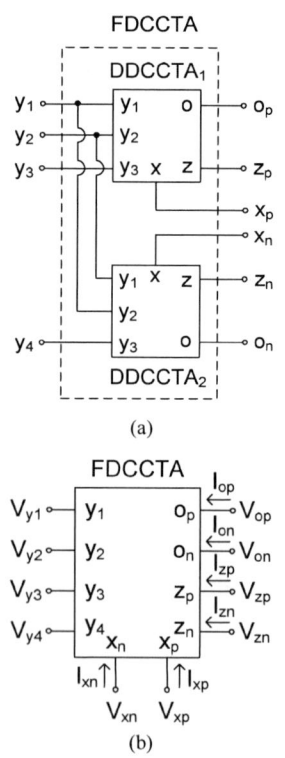

Figure 2.   Proposed FDCCTA: (a) realization of FDCCTA using two DDCCTAs, (b) electrical symbol of proposed FDCCTA.

978-1-4799-5761-3/14 $31.00 © 2014 IEEE   166

IEEE-ICSE2014 Proc. 2014, Kuala Lumpur, Malaysia

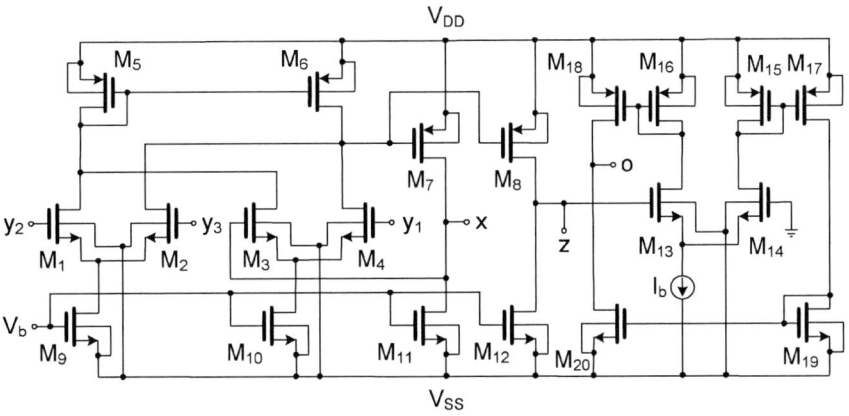

Figure 3. The CMOS implementation of DDCCTA.

Fig. 1 shows the electronic symbol of DDCCTA [2]. The port relations of an ideal DDCCTA can be described by

$$
\begin{pmatrix} I_{y1} \\ I_{y2} \\ I_{y3} \\ V_x \\ I_z \\ I_o \end{pmatrix} = \begin{pmatrix} 0 & 0 & 0 & 0 & 0 & 0 \\ 0 & 0 & 0 & 0 & 0 & 0 \\ 0 & 0 & 0 & 0 & 0 & 0 \\ 1 & -1 & 1 & 0 & 0 & 0 \\ 0 & 0 & 0 & 1 & 0 & 0 \\ 0 & 0 & 0 & 0 & g_m & 0 \end{pmatrix} \begin{pmatrix} V_{y1} \\ V_{y2} \\ V_{y3} \\ I_x \\ V_z \\ V_o \end{pmatrix}
\tag{1}
$$

This device is consisted of a DDCC and a transconductance amplifier (TA). The addition and subtraction circuit follows a differential difference input voltage $(V_{y1}-V_{y2}+V_{y3})$ to the output voltage at x-terminal V_x. If terminals x and z are connected by resistors, voltage V_x will transfer to the current at z-terminal and converts next to the voltage at z-terminal V_z. The voltage V_z will transfers to the current through the o-terminal I_o by the transconductance gain (g_m). From (1), $I_{y1} = I_{y2} = I_{y3} = 0$, it can express that high-input impedance is obtained. The CMOS implementation of DDCCTA is shown in Fig. 3. Transistors, M_1-M_{12}, are composed of a DDCC while transistors, M_{13}-M_{20}, are composed of a TA. The transconductance gain can be controlled by the bias current I_b.

The proposed FDCCTA is shown in Fig. 2. The circuit is consisted of two DDCCTAs. The port characteristics of an ideal FDCCTA can be given by

$$
\begin{pmatrix} V_{xp} \\ V_{xn} \\ I_{zp} \\ I_{zn} \\ I_{op} \\ I_{on} \end{pmatrix} = \begin{pmatrix} 0 & 0 & 0 & 0 & 1 & -1 & 1 & 0 \\ 0 & 0 & 0 & 0 & -1 & 1 & 0 & 1 \\ 1 & 0 & 0 & 0 & 0 & 0 & 0 & 0 \\ 0 & 1 & 0 & 0 & 0 & 0 & 0 & 0 \\ 0 & 0 & g_{mp} & 0 & 0 & 0 & 0 & 0 \\ 0 & 0 & 0 & g_{mn} & 0 & 0 & 0 & 0 \end{pmatrix} \begin{pmatrix} I_{xp} \\ I_{xn} \\ V_{zp} \\ V_{zn} \\ V_{y1} \\ V_{y2} \\ V_{y3} \\ V_{y4} \end{pmatrix}
\tag{2}
$$

The property of FDCCTA is similar to FDCCII, except the transconductance amplifiers as an output stage. There are two transconductance amplifiers at the output stage of FDCCTA. Two addition and subtraction circuits follow two differential difference input voltages $V_{y1}-V_{y2}+V_{y3}$ and $-V_{y1}+V_{y2}+V_{y4}$ to the output voltages at x_p-terminal and x_n-terminal V_{xp} and V_{xn}, respectively. If terminals x_p, x_n, z_p and z_n are connected by the impedances, voltages V_{xp} and V_{xn} will transfer to two currents I_{xp} and I_{xn}. These currents will convey to the currents at z_p- and z_n-terminals, and converted next to the voltages at z_p- and z_n-terminals V_{zp} and V_{zn}. The voltages V_{zp} and V_{zn} will transfers to the currents through the o_p-terminal (I_{op}) and the o_n-terminal (I_{on}) by the transconductance gains g_{mp} and g_{mn}, respectively. The transconductance gains g_{mp} and g_{mn} can be controlled independently by the bias currents. The CMOS implementation for completing of the proposed FDCCTA is shown in Fig. 4. From this figure, the bias currents I_{bp} and I_{bn} are used to control the transconductance gains g_{mp} and g_{mn}, respectively. Compared with FDCCII, FDCCTA provides the possibility of interactive electronic tuning of the parameters g_{mp} and g_{mn} that is valuable for analog circuit realizations.

III. FDCCTA-BASED UNIVERSAL FILTER

To confirm the operation of the proposed active building block, the proposed FDCCTA is used to realize a universal filter as shown in Fig. 5. The circuit employs only a FDCCTA and all-grounded passive components. It should be noted that the input voltages V_{in1} and V_{in2} of Fig. 5 is applied to the y_1 and y_4 terminals of the FDCCTA. Thus, the filter in Fig. 5 has the face of high-input impedance, which can be directly connected in cascade to realize high-order filters. It should be noted that the circuit uses grounded passive components which is ideal for integrated circuit implementation [9]. Routine analysis the filter in Fig. 5 yields the following voltage transfer functions:

$$
\frac{V_{o1}}{V_{in1}} = \frac{s^2 C_1 C_2 R_1}{s^2 C_1 C_2 R_1 + s C_2 + g_{mp}}
\tag{3}
$$

$$
\frac{V_{o2}}{V_{in1}} = \frac{s C_2}{s^2 C_1 C_2 R_1 + s C_2 + g_{mp}}
\tag{4}
$$

978-1-4799-5761-3/14 $31.00 © 2014 IEEE   167

IEEE-ICSE2014 Proc. 2014, Kuala Lumpur, Malaysia

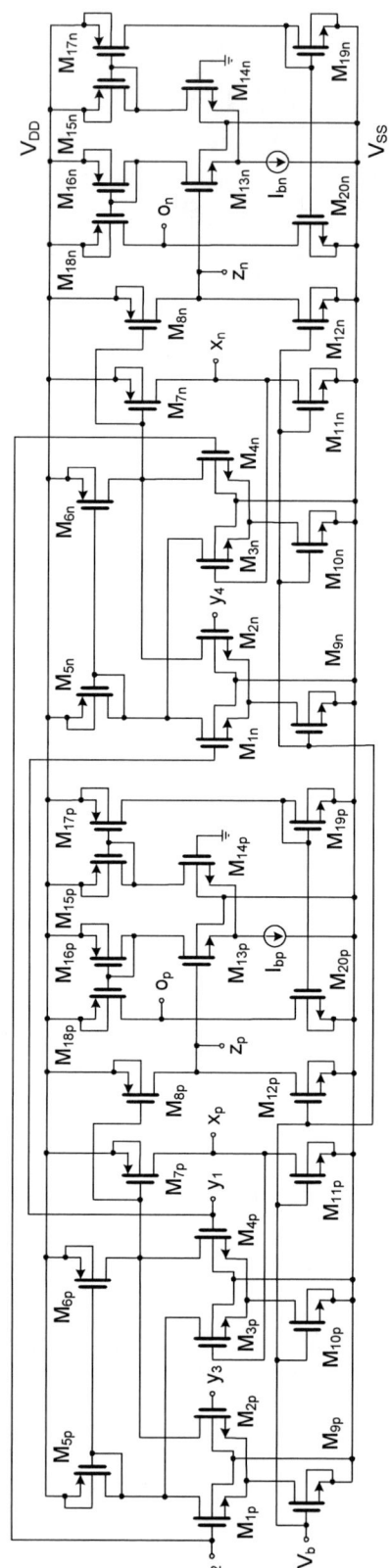

Figure 4.   The CMOS implementation of proposed FDCCTA.

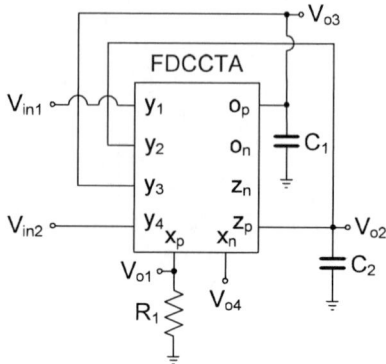

Figure 5.   Universal filter using FDCCTA.

$$\frac{V_{o3}}{V_{in1}} = \boxed{?}\,\frac{g_{mp}}{s^2 C_1 C_2 R_1 \boxed{?} sC_2 \boxed{?} g_{mp}} \tag{5}$$

$$\frac{V_{o4}}{V_{in1}} = \boxed{?}\,\frac{s^2 C_1 C_2 R_1 \boxed{?} g_{mp}}{s^2 C_1 C_2 R_1 \boxed{?} sC_2 \boxed{?} g_{mp}} \tag{6}$$

From (3)-(6), high-pass (HP), band-pass (BP), low-pass (LP) and band-stop (BS) filters can be respectively obtained. Moreover, if V_{in2} is connected to V_{o2}, all-pass (AP) filter can be obtained as

$$\frac{V_{o4}}{V_{in1}} = \frac{\boxed{?} s^2 C_1 C_2 R_1 \boxed{?} sC_2 \boxed{?} g_{mp}}{s^2 C_1 C_2 R_1 \boxed{?} sC_2 \boxed{?} g_{mp}} \tag{7}$$

Therefore, HP, BP, LP, BS and AP responses can be obtained by Fig. 5. In all cases, the natural frequency (ω_o) and the quality factor (Q) are calculated, respectively, as

$$\omega_o = \sqrt{\frac{g_{mp}}{C_1 C_2 R_1}} \qquad \boxed{?}\boxed{?}\boxed{?}$$

$$Q = \sqrt{\frac{C_1 R_1 g_{mp}}{C_2}} \qquad \boxed{?}\boxed{?}\boxed{?}$$

From (8)-(9), the parameter ω_o can be controlled by setting $g_{mp}=1/R_1$ and keeping C_1 and C_2 constant while the parameter Q can be given by C_1/C_2 and keeping g_{mp}/R_1 constant.

IV. SIMULATION RESULTS

TABLE I.   MOS TRANSISTOR ASPECT RATIOS OF FDCCTA IN FIG. 4.

MOS transistors	W/L(km/km)
M_{1p}–M_{4p}, M_{1n}–M_{4n}	1.8/0.7
M_{5p}–M_{8p}, M_{5n}–M_{8n}	20/0.7
M_{9p}–M_{10p}, M_{9n}–M_{10n}	5.2/0.7
M_{11p}–M_{12p}, M_{11n}–M_{12n}	58/0.7
M_{13p}–M_{20p}, M_{13n}–M_{20n}	4/0.7

978-1-4799-5761-3/14 $31.00 © 2014 IEEE   168

TABLE II. SIMULATED PERFORMANCES OF PROPOSED FDCCTA.

Parameters	Value
Technology	0.5 μm
Supply voltage	o2 V
Bias voltage (V_b)	-1.22 V
DC voltage range	o750 mV
DC current range	o145 μA
-3dB bandwidth voltage follower	≤ 1 GHz
-3dB bandwidth current follower	≤ 1 GHz
g_m ($I_{bp}=I_{bn}=1$ to 100μA)	23.86 μS to 0.18 mS
R_{yi}, C_{yi} (i=1,2,3,4)	>10 GΩ, 1.77 fF
R_{xp}, R_{xn}, L_{xp}, L_{xn}	52.8 Ω, 1.46 μH
R_{zp}, R_{zn}, L_{xp}, L_{xn}	529 kΩ, 17.4 fF
R_{op}, R_{on}, C_{op}, C_{on}	9.98 MΩ, 2.1 fF
Power consumption ($I_{bp}=I_{bn}=100$μA)	5 mW
Input noise	821 nV/√Hz

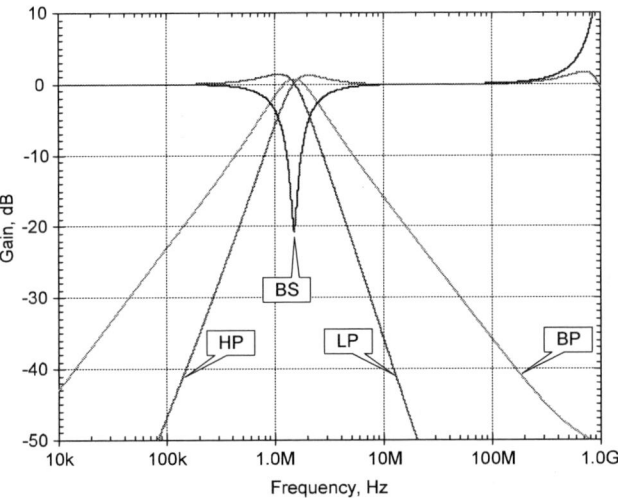

Figure 6. Simulated LP, BP, HP and BS of the proposed filter.

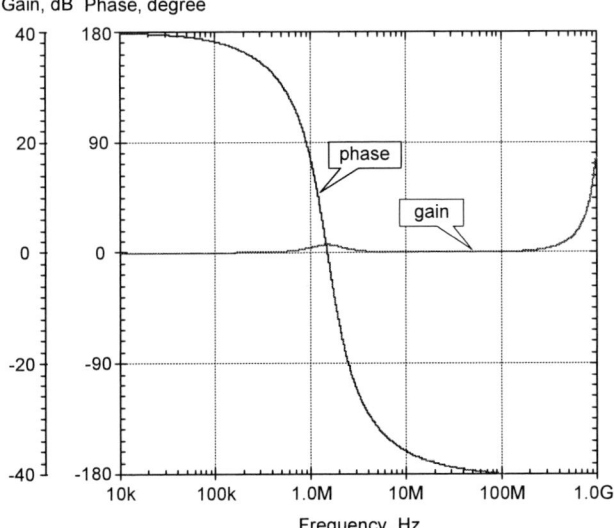

Figure 7. Simulated gain and phase responses of AP filter.

To verify the performance of the proposed circuit, Fig. 4 was simulated by PSPICE simulators using the 0.5μm from MIETEC. The transistor aspect ratios of FDCCTA were tabulated in Table I. The supply voltages were $V_{DD}=-V_{SS}= 2$ V. The bias voltage V_b was taken as -1.22 V. The simulated performances of the proposed FDCCTA are shown in Table II. To verify the workability of the proposed FDCCTA, the filter using the FDCCTA in Fig. 5 was simulated. As an example design, the capacitors $C_1 = C_2 = 10$ pF, the resistor $R_1 = 10$ kΩ and $I_{bp} = 16.5$ μA ($g_{mp} = 89.4$ μS) were given. Fig. 6 shows the simulated magnitude responses of the LP, HP, BP and BS filters. The magnitude and phase responses of AP filter were shown in Fig. 7. The frequency f_o of 1.49 MHz was expressed. Therefore, it can be confirmed using the filter in Fig. 5 that the proposed FDCCTA can be used to realize analog circuits.

V. CONCLUSION

In this paper, a new active building block, FDCCTA is proposed. This device has the properties of FDCCII and TA into a single active device. Therefore, the property of proposed FDCCTA is similar to FDCCII, except the output stage. The output stage of FDCCTA is the TA that can be provided electronically adjustable function which is valuable for analog and digital signal (mixed-signal) applications. The proposed FDCCTA is used to realize universal filter to verify the workability. The performance of the proposed FDCCTA and its application has been verified by PSPICE simulators.

REFERENCES

[1] W. Chiu, S.-I. Liu, H.-W. Tsao, J.-J. Chen, "CMOS differential difference current conveyors and their applications," IEE Proceeding–Circuits Devices and Systems, vol. 143, pp. 91-96, 1996.

[2] N. Pandey, S. K. Paul, "Differential difference current conveyor transconductance amplifier: a new analog building block for signal processing," Journal of Electrical, Computer and Engineering, 2011 (2011). Article ID 361384.

[3] A. A. El-Adwy, A. M. Soliman, H. O. Elwan, "A novel fully differential current conveyor and applications for analog VLSI," IEEE Transactions on Circuits Systems-II, vol.47, pp. 306-313, 2000.

[4] C.-M. Chang, B. M. Al-Hashimi, C. L. Wang, C. W. Hung, "Single fully differential current conveyor biquad filters," IEE Proceedings-Circuits, Devices and Systems, vol. 150, pp. 394-398, 2003.

[5] J.-W. Horng, C.-L. Hou, C.-M. Chang, H.-P. Chou, C.-T. Lin, Y.-H. Wen, "Quadrature oscillators with grounded capacitors and resistors using FDCCIIs," ETRI Journal, vol. 28, pp. 486-494, 2006.

[6] C.-M. Chang, A. M. Soliman, M. N. S. Swamy, "Analytical synthesis of low-sensitivity high-order voltage-mode DDCC and FDCCII-grounded R and C all-pass filter structures," IEEE Transactions on Circuits and Systems I: Regular Papers, vol. 54, pp. 1430-1443, 2007.

[7] H.-P. Chen, "Voltage-mode FDCCII-based universal filters," International Journal of Electronics and Communications, vol. 62, pp. 320-323, 2008.

[8] C.-N. Lee, C.-M. Chang, "Single FDCCII-based mixed-mode biquad filter with eight outputs," International Journal of Electronics and Communications, vol. 63, pp. 736-742, 2009.

[9] M. Bhusan, R. W. Newcomb, "Grounding of capacitors in integrated circuits," Electronics Letters, vol. 3, pp. 148-149, 1967.

Dependency of electrical characteristics on nano gap variation in pinch off lateral gate transistors

Farhad Larki, *MIEEE*, Arash Dehzangi*, *MIEEE*,
Sawal Hamid Md Ali, *MIEEE*, Azman Jalar, *MIEEE*,
Md. Shabiul Islam, *MIEEE* Burhanuddin Y. Majlis,
S*MIEEE*
Institute of Microengineering and Nanoelectronics
(IMEN), Universiti Kebangsaan Malaysia,
43600 Bangi, Selangor, Malaysia,
*Email: dehzangi@ukm.edu.my

Elias B Saion,
Department of Physics, University Putra Malaysia,
43400 Serdang, Selangor, Malaysia

Mohd Nizar Hamidon, *MIEEE*
Functional Devices Laboratory, Institute of Advanced
Technology, Universiti Putra Malaysia, 43300, Serdang,
Selangor, Malaysia

Sabar D. Hutagalung, *MIEEE*
Department of Physics, Faculty of Science, Jazan
University, Jazan,
Kingdom of Saudi Arabia

Abstract—the variation of electrical characteristics with nano size air gap variation between gates and channel of a pinch off lateral gate transistor were investigated using 3D Technology Computer Aided Design. It is found that smaller nanosize gaps which can be formed by approaching the lateral gates to the channel can improve the switching performance of the device significantly. Devices with different air gap demonstrate same *on* state current and maximum transconductance of 0.05 μS, however the on/off current ratio (I_{ON}/I_{OFF}) is varied by three orders of magnitude. The parameters such as electric field and band energy variation are investigated in order to explain the variation of electrical characteristics by air gap variation.

Keywords— Air Gap; Lateral gate (LG), TCAD Simulation, Junctionless Transistor (JLT)

I. INTRODUCTION

Transistors as one of the major elements of all modern digital logic devices need to be scaled down in order to improving the *on* state current, reducing the switching time and hence the logic delay. However, technical problem in fabrication process and new physical phenomena associated with scaling procedure of transistors introduces new challenges to the scaling process. Transistors with ultra-small size need ultra sharp doping concentration gradients in junctions which is really hard to be obtained in in new generation of transistors with small dimensions and ultra small channel length.

On the other hand, scaling of the gate oxide is hard to be attained due to the presence of gate leakage current, moreover scaling the gate length is hard to be demonstrated because of the short channel effect (SCE). The alternative techniques for fabrication and new device design have been introduced in the literature in order to overcome each of these obstacles. New class of transistors which known as Junctionless Transistors (JLTs) is introduced in order to overcome the problem of source/drain doping [1-2]. High-k gate dielectrics is used for

suppressing the direct tunneling current through gate oxides [3]. According to the new oxidation techniques the gate oxide thickness of the order of 1 nm can be achieved. An oxide thickness of 1.2 nm only a five atom thick oxide layer which means we are approaching a physical limit beyond which carrier tunneling current through the gate increases dramatically. Another issue associated with an excessively thin gate oxide is the loss of inversion charge, which leads to smaller gate capacitance and so smaller transconductance [4-5].

In order to obtain the optimal performance of transistors particularly from the circuit speed points of view still a lot of challenges are remained to be solved. Therefore, scaling is no longer the absolute solution for the circuit speed due to the interconnect delay and the difficulty of scaling transistors. In order to improve the circuit speed, reducing capacitance is more and more important [6].

Devices with nano air gap can be used as an option for reducing the capacitance between gate and channel due to reduce in parasitic fringe capacitance. The air gap structure also helps in reducing the sub threshold slope and provides significant power saving in both the standby state and during switching operation [7-8]. More application of nanosize air gap is reported in T-gate semiconducting carbon nanotube RF transistors [9], Tetracene single-crystal FETs [10], and organic FETs [11].

In this study, we report a comparison study of device called as Double Lateral Gate Junctionless Transistor (DGJLT), through 3D TCAD simulation. The device has double lateral gate and two symmetric nanosize air gaps which can modify the conduction properties of the channel. The transfer characteristics of device with two different gaps are calculated and the transconductance is extracted from its output. In order to deeply understand the effect of lateral gate and the impact of air gap variation different component of electric field along the carrier path, valence band edge energy, and recombination/generation rate of carriers are investigated.

II. METHODOLOGY

In this work, 3-D simulations of the DGJLTs were carried out by the Sentaurus 3-D device simulator [12]. Fig. 1 (a) and (b) present the isometric view and top view of the simulated DGJLT. The main parts of the device is labeled as LG for lateral gate with 200 nm width, air gap which has values of 50

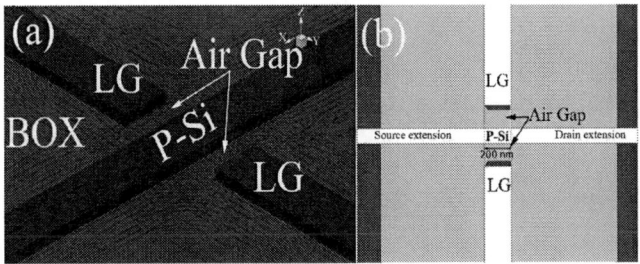

Fig. 1. The schematic of the simulated device (a) Isometric view and (b) Top view. The lateral gates, BOX, channel and air gap were labeled.

and 100 nm through the simulations, P-Si with 10^{15} (cm^{-3}) boron doped. All the active regions of the device have 100 nm thicknesses and all contacts work function is considered as 5.12 eV. The validation of the device with 100 nm gap with the experimental results already reported in [13-14].The simulations were carried out using hydrodynamic carrier transport model.

Since all the physical dimensions considered in this study are above 10 nm which consider as the critical border between quantum and classical regions [15-16], it can be assumed that the hydrodynamic model is the most accurate model for simulation of DGJLTs in the present work. Besides the fundamental equations, the simulator implemented the doping-dependent Masetti model for mobility and doping dependence Shockley–Read–Hall recombination-generation to account for leakage current.

III. RESULTS AND DISCUSSION

Transfer characteristics of devices with two different air gaps of 50 and 100 nm between lateral gates and channel are presented in Fig. 2 (a). It shows that the air gap is an important parameter in controlling the behavior of the device. As the gates get closer to the channel, carriers affect more significantly by the gates. This is in agreement with the model proposed [13, 17] in which explained that the pinch off effect of the device is mainly due to the electrostatic control of the carriers behavior by lateral gates effect. The *on* state current is same for two devices; however the *off* current is decrease by three orders of magnitude for device with smaller gap. Leakage current through gate and channel, which is more predominant in a solid insulator can be avoided by using nanosize air gaps [8].

The threshold voltage is approximately same for two devices, even though the device is biased with high gate voltage. In the

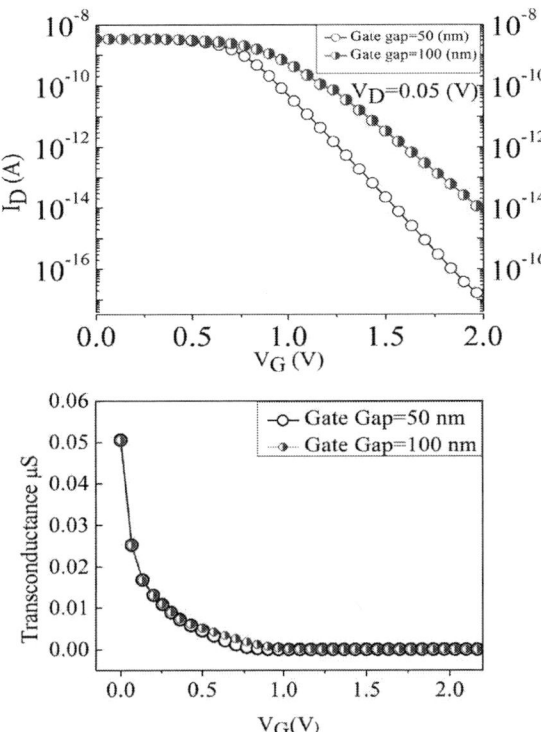

Fig. 2. Output characteristics comparison of two devices with 50 and 100 nm gate gap (a) Transfer characteristics and (b) Transconductance as a function of gate voltage.

devices with solid gate insulator such as a-Si TFTs at high gate voltage, the charge trapping in the insulator is the main cause of the threshold voltage shift [18]. It should be mention that the mechanism in low gate voltage is completely different in these devices. In low gate voltage the dominant mechanism is breaking bonds in the channel or the creation of new states with variation of the gate voltage [19]. In DGJLT, according to the air gap and just one interface with the BOX the amount of charge which can be trapped is negligible and the only source of threshold voltage shifts is breaking of the bonds. According to the physical dimensions of the channel and the crystallinity of the channel which kept untouched due to the fabrication of device with AFM nanolithography [17], the threshold voltage shift is approximately same for two devices. The transconductance ($g_m = \partial I_D / \partial V_G$) of the DGJLTs with two different gaps as a function of gate voltage is shown in Fig. 2(b). The overall trend is in accordance with pinch off devices which demonstrate the variation of device from *on* to *off* state. The maximum transconductance of 0.05 µS is calculated for both devices. Sharp decrease of transconductance presents the effect of positive gate voltage to deplete the channel. At low gate voltage the variation of the carriers and correspond current is not significant compare to the variation the gate voltage and consequently result in a decreasing value of the transconductance with increasing the gate voltage. However, with increasing the gate voltage above 0.5 V the carriers deplete more effectively and the rate of current variation is more comparing to the gate voltage variation. It should clearly

IEEE-ICSE2014 Proc. 2014, Kuala Lumpur, Malaysia

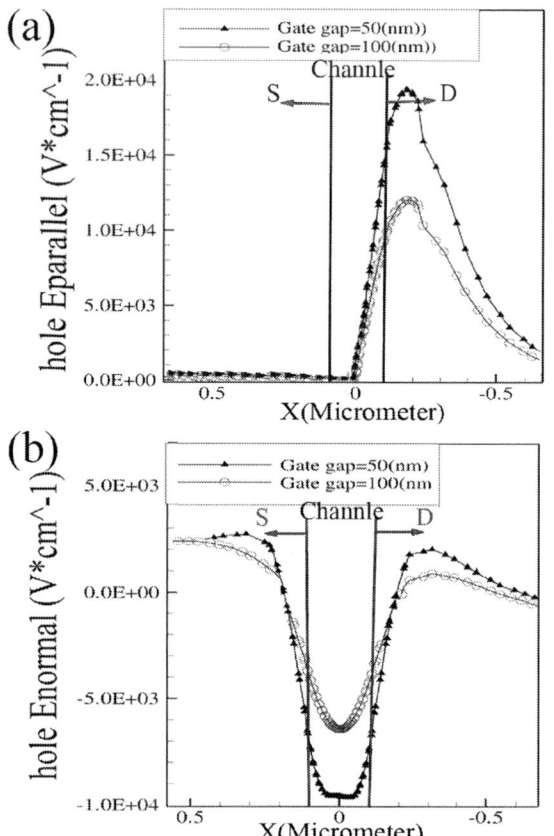

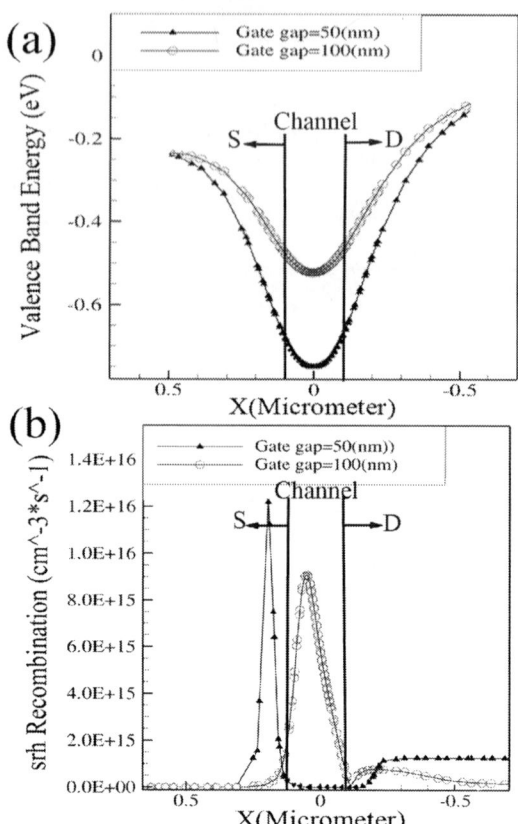

Fig. 3 Effect of gate gap variation of different components of the electric field, (a) Electric field exerted on holes in direction parallel to the current and (b) Electric field exerted on holes in direction perpendicular to the current. V_G=2.0 V and V_D=-1.0 V.

Fig. 4 The variation of Valence Band Edge Energy (a) and srh Recombination (b) of two devices with 50 and 100 nm gate gap along the channel

mention here that at high values of the gate voltage the carriers already depleted and the transconductance saturate. In order to demonstration more clear picture about the effect of air gap variation on device performance we now look at the variation of some parameters inside the device.

The electric field affected on the majority carriers in direction perpendicular and parallel to the channel is presented in Fig. 3 (a) and (b), respectively. Compared to the structure with 100 nm gap, normal and parallel electric field strength exhibited 18 and 25 % increase in the device with 50 nm gap. The appearance of the parallel electric field peak out of the gated area is a normal behavior of all junctionless transistors [20]. As a result of stronger electric field in the direction perpendicular to the current flow at smaller gate gap, it is expected that the arrangement of the carriers in the area under the gates can be more effectively modified in the device with smaller gap compare to the device with larger gap. According to the p-type channel, holes are the majority carriers and the lateral gates need to be biased with positive voltage in order to push the device to the *off* state. However, the electrons as the minority carriers attract to the gated area and as a consequence of the electron accumulation the area of higher potential energy forms in this area.

This area of higher potential energy works as a barrier which trap the holes which pass through the channel. The valence band edge energy variation along the channel axis for varying gate gap is shown in Fig. 4a. It clearly shows the formation of stronger potential well as the gate gap decreased from 100 nm to 50 nm. This potential well works as a trap for majority carriers (holes). As a consequence, the number of majority carriers decreases significantly. In this situation there is a possibility of inverting the channel particularly in the device with smaller gap since the number of minority carriers becomes much higher, however the area under the gates is not able to easily invert since the adjoining p type regions. Even if it were to invert, the reverse bias p-region contacting this inversion layer ($-V_D$), would limit the transport in this layer. These accumulated electrons in the channel recombine with the holes and increase the recombination rate of device. This is more significant in the device with 50 nm gap compared to the devices with 100 nm gate gap. The SRH recombination rate is presented in Fig. 4b confirm the later discussion about the carrier recombination. According to the previous discussions, it is expected that in the device with gate gap larger that a critical value the device simply begins to behave like a resistor and the gate completely loss control over the carriers and consequently very low switching speed can be expected. It is predicted that at very higher gate and drain voltages these devices also show an acceptable switching speed and leakage

978-1-4799-5761-3/14 $31.00 © 2014 IEEE

current value. However, for the next generation of transistors low power consumption is a crucial issue which needs to be considered.

IV. CONCLUSIONS

In this work, the impact of nanosize air gap variation on the output characteristics of DGJLT is investigated. The characteristics clearly show that the air gap is an important parameter which can effectively improve the performance of the device. The smaller gap results in lower *off* current, since the *on* current remains unaltered. The variation of the characteristics with air gap is also explained based on electric field component variation in the channel, valence band edge energy, and the rate of recombination of the carriers.

ACKNOWLEDGMENT

The authors gratefully acknowledge that this work was financially supported by Funding, Dana Pembangunan Penyelidikan, DPP-2013-061, Dana Impac Perdana, DIP-2012-16 , and AKU95 HiCoe.

.

REFERENCES

[1] Colinge, J.P., Lee, C.W., Afzalian, A., Akhavan, N.D., Yan, R., Ferain, I., Razavi, P., O'Neill, B., Blake, A., and White, M.: 'Nanowire transistors without junctions', Nat Nanotechnol, 2010, 5, (3), pp. 225-229

[2] Colinge, J., Kranti, A., Yan, R., Lee, C., Ferain, I., Yu, R., Dehdashti Akhavan, N., and Razavi, P.: 'Junctionless Nanowire Transistor (JNT): Properties and design guidelines', Solid-State Electron, 2011

[3] Hall, S., Buiu, O., Mitrovic, I.Z., Lu, Y., and Davey, W.M.: 'Review and perspective of high-k dielectrics on silicon', Journal of Telecommunications & Information Technology, 2007, 2007, (2)

[4] Sze, S.M.: 'Semiconductor devices: physics and technology' (John Wiley & Sons, 2008. 2008)

[5] Taur, Y.: 'CMOS design near the limit of scaling', IBM J. Res. Dev., 2002, 46, (2.3), pp. 213-222

[6] Park, J.M.: 'Improving CMOS Speed and Switching Power with Air-Gap Structures', 2011

[7] Sinha, N., Jones, T.S., Guo, Z., and Piazza, G.: 'Body-biased complementary logic implemented using AlN piezoelectric MEMS switches', J. Microelectromech. Syst., 2012, 21, (2), pp. 484-496

[8] Cha, S., Jang, J., Choi, Y., Amaratunga, G., Ho, G., Welland, M., Hasko, D., Kang, D.-J., and Kim, J.: 'High performance ZnO nanowire field effect transistor using self-aligned nanogap gate electrodes', Appl. Phys. Lett, 2006, 89, (26), pp. 263102

[9] Che, Y., Badmaev, A., Jooyaie, A., Wu, T., Zhang, J., Wang, C., Galatsis, K., Enaya, H.A., and Zhou, C.: 'Self-aligned T-gate high-purity semiconducting carbon nanotube RF transistors operated in quasi-ballistic transport and quantum capacitance regime', ACS nano, 2012, 6, (8), pp. 6936-6943

[10] Xia, Y., Kalihari, V., Frisbie, C.D., Oh, N.K., and Rogers, J.A.: 'Tetracene air-gap single-crystal field-effect transistors', Appl. Phys. Lett, 2007, 90, (16), pp. 162106

[11] Chen, Y., and Podzorov, V.: 'Bias Stress Effect in "Air-Gap" Organic Field-Effect Transistors', Adv. Mater, 2012, 24, (20), pp. 2679-2684

[12] Sentaurus, T.: 'User Guide,Version D-2010.03, March 2010– Synopsys, 2010', synopsys, 2010

[13] Larki, F., Dehzangi, A., Abedini, A., Abdullah, A.M., Saion, E., Hutagalung, S.D., Hamidon, M.N., and Hassan, J.: 'Pinch-off mechanism in double-lateral-gate junctionless transistors fabricated by scanning probe microscope based lithography', Beilstein J. Nanotechnol, 2012, 3, (1), pp. 817-823

[14] Larki, F., Dehzangi, A., Saion, E.B., Abedini, A., Hutagalung, S.D., Abdullah, A.M., and Hamidon, M.N.: 'Simulation of transport in laterally gated junctionless transistors fabricated by local anodization with an atomic force microscope', Phys. Status Solidi A-Appl. Mat, 2013, 210, (9), pp. 1914-1919

[15] Lee, C.W., Yun, S.R.N., Yu, C.G., Park, J.T., and Colinge, J.P.: 'Device design guidelines for nano-scale MuGFETs', Solid-State Electron, 2007, 51, (3), pp. 505-510

[16] Dehzangi, A., Larki, F., Hutagalung, S., Saion, E., Abdullah, A., Hamidon, M., Majlis, B., Kakooei, S., Navaseri, M., and Kharazmi, A.: 'Numerical investigation and comparison with experimental characterisation of side gate p-type junctionless silicon transistor in pinch-off state', Micro Nano Lett, 2012, 7, (9), pp. 981-985

[17] Dehzangi, A., Abdullah, A.M., Larki, F., Hutagalung, S.D., Saion, E.B., Hamidon, M.N., Hassan, J., and Gharayebi, Y.: 'Electrical property comparison and charge transmission in p-type double gate and single gate junctionless accumulation transistor fabricated by AFM nanolithography', Nanoscale Res. Lett, 2012, 7, (1), pp. 1-9

[18] Powell, M.: 'Charge trapping instabilities in amorphous silicon-silicon nitride thin-film transistors', Appl. Phys. Lett, 1983, 43, (6), pp. 597-599

[19] Deane, S., Wehrspohn, R., and Powell, M.: 'Unification of the time and temperature dependence of dangling-bond-defect creation and removal in amorphous-silicon thin-film transistors', Phys. Rev. B, 1998, 58, (19), pp. 12625

[20] Colinge, J.P., Ferain, I., Kranti, A., Lee, C.W., Akhavan, N.D., Razavi, P., Yan, R., and Yu, R.: 'Junctionless Nanowire Transistor: Complementary Metal-Oxide-Semiconductor Without Junctions', Sci. Adv. Mater, 2011, 3, (3), pp. 477-482

The Annealing Temperature Effect on the Structure and Electrical Properties of Titanium Dioxide (TiO$_2$) Film Deposited by Reactive RF Sputtering

S. Norhafiezah, RM Ayub, M. K. Md Arshad, A.H. Azman, M. F. Fatin, M.A. Farehanim and U. Hashim

Institute of Nano Electronic Engineering (INEE),
Universiti Malaysia Perlis (UniMAP)
01000 Kangar, Perlis, Malaysia.
mizfyza@gmail.com

Abstract—**Titanium dioxide (TiO$_2$) thin film is deposited using Reactive Radio Frequency (RF) sputtering on Si (100) wafer and annealed in N$_2$ for 2 hours at different temperatures i.e. 500 °C, 750 °C and 1100 °C. The TiO$_2$ peak is characterized using X-ray diffraction (XRD). At 500 °C and 750 °C, only anatase peak is observed with the grain size of 150.72 nm and 186.51 respectively. As the temperature increase to 1100 °C, both anatase and rutile structures start to grow but the grain size is reduced to 67.88 nm. The confirmation of grain and the surface roughness is determined by using atomic force microscopy (AFM). The grain sizes become larger from 66.58 nm to 86.01 nm as the temperature increase from 500 °C to 750 °C as well as the surface roughness (0.271 nm to 1.201 nm). However, at 1100 °C, grain size shows no significant different i.e. 84.41 nm (compared at 750 °C) and slightly higher surface roughness of 2.194 nm. Thus, the 1100 °C annealing temperature requires to attain rutile structure and the smaller particle size. The electrical properties of TiO$_2$ film annealed at 1100 °C shows small amount of current flow through the device thus will be suitable to be used in biosensor application.**

Keywords— Titanium dioxide; RF sputtering; Annealing temperature; Grain size; Electrical properties

I. INTRODUCTION

Titanium Dioxide can be presented in three basic of crystalline polymorphous phases, namely anatase and rutile (tetragonal structure) and brookite (orthorhombic structure). It has been reported that rutile structure is the most stable phase compared to anatase and brookite [1]. However, anatase structure will transform to rutile structure upon heating [2]. In general, the various deposition techniques will result in different TiO$_2$ structure, surface morphology and other properties of TiO$_2$ thin film. Therefore, the variation of temperature has become the most influential parameter in the formation of crystallites and also the grain sizes. It has been reported [3, 4] that the higher annealing temperature will produce larger grain size. It is also mentioned that for temperatures of more than 800 °C, additional rutile phases will be obtained [3]. It has been verified that by increasing the annealing temperature would improve the electrical sensitivity of TiO$_2$ thin film [5]. Hence, it is appropriate to use in

biosensor application that requires high sensitivity of detection [6].

Nowadays, titanium dioxide has emerged as a stringent semiconductor material for biosensor application, photocatalysis and cosmetic industry [7]. Since TiO$_2$ has some beneficial features such as non-toxic, less expensive, chemically stable and bio-compatible [8], TiO$_2$ has become one of the most favorable materials in bio-sensor application.

There are various deposition technique of the titania film such as sputtering, sol-gel, atomic layer deposition and spray pyrolysis [9, 10]. However, sputtering technique is the most widely used in wafer fabrication technology and more favorable due to its higher deposition rate and better uniformity [11].

In this paper, TiO$_2$ thin films have been deposited on Si (100) wafer using reactive RF sputtering method. The effect of annealing temperature to the TiO$_2$ particle size and the electrical characteristic was investigated with the aims to find the optimum annealing temperature in order to obtain a smaller grain size and higher electrical sensitivity for the biosensor application.

II. EXPERIMENTAL PROCEDURE

A. Deposition of TiO$_2$ film

The p-type silicon (100) wafer with 10 ohm/cm resistivity was first cleaned using Piranha, RCA-1 and RCA-2 solution to remove the residue, organic and metallic contamination, respectively. The wafer was immersed in diluted HF acid to remove the native oxide on the top of the wafer surface. The TiO$_2$ film was deposited by using Reactive Radio Frequency (RF) Sputtering technique (Penta Vacuum E303 Sputter CVD System) using a titanium target with 99.9% of purity and 3 inches of the diameter. Prior to fill up the chamber with the mixture of 50 sccm Argon (purity 99.99%) and 5 sccm Oxygen (99.99%), the machine chamber was evacuated to 5.5 × 10^{-5} torr. During the sputtering process, the substrate temperature, RF power and target-to-substrate distance were fixed to 50 °C, 100W and 14 cm, respectively. All the sample were annealed at 500 °C, 750 °C and 1100 °C for 2 hours using a furnace in N_2 atmosphere. The effect of annealing temperature variation with respect to the grain size and the electrical properties of TiO_2 film was investigated.

B. MOS Capacitor Fabrication Process

The aluminum (Al) metal was deposited on top and bottom of the wafer which act as an electrode and ohmic contact, respectively. The Metal-Oxide-Semiconductor (MOS) capacitor was fabricated using conventional lithography method including the pattern transfer and aluminum etching process. The cross section of the process flow was shown in Fig 1.

C. Characterization details

The crystal structure of TiO_2 film was determined using an X-Ray Diffractometer (Bruker D2 Phaser) with the 2θ scan range from $20°$ to $60°$. The crystallite size was calculated using Scherer equation:

$$D = 0.9 \lambda / \beta \cos \theta \qquad (1)$$

Where λ is the X-ray wavelength, β is full-width at half maximum (FWHM) and θ is the Bragg angle of the diffraction peak. The Atomic Force Microscopy SPA-400 was utilized to measure the surface roughness and the grain size of TiO_2 film with 2 μm scanning area and 2 Hz scanning speed. The electrical characterization was carried out using Keithley 6487 Pico-ammeter Source Measuring Unit (SMU) interfaced with Leios TMXpert software. The sweep voltage from 0V to 5V with steps of 0.5 V was applied to the SMUs to measure the current.

III.   RESULT AND DISCUSSION

A. Structural Study

Fig. 2 depicts the XRD spectra of TiO_2 film deposited on Si (100) wafer using reactive RF sputtering technique and annealed at 500 °C, 750 °C and 1100 °C for 2 hours. The diffraction peaks obtained were compared with the JCPDS data card 21-1272 for anatase [12] and 21-1276 for rutile [12] structure. It can clearly observed that for lower annealing temperature i.e. 500 °C and 750 °C only anatase structure appeared (004) in the crystal system. For 750 °C, we obtain very small peak corresponded to the titanium sub-oxide (Ti_2O_3) and titanium nitride (TiN). As the temperature increase to 1100 °C, both anatase (004 and 101) and rutile (110) structure were observed. However, due to the strong background of the peak, amorphous phases were obtained through-out the films. This study proved that the annealing temperature is strongly influential in the crystallographic direction of the titania films [13].

Table 1 shows the calculated grain size using (1). It can be seen that the particle size increase from 150.72 nm to 186.51 nm as the temperature increase from 500 °C to 750 °C. However, as the higher annealing temperature is applied (1100 °C), the grain size is reduced to 67.88 nm. This is thought to be due to the N_2 gas effect on the structure of TiO_2 film during the annealing process [14]. The nitrogen gas will restrict other

gases in the air from performing any additional bonding with titanium dioxide film, thus resulting denser structure.

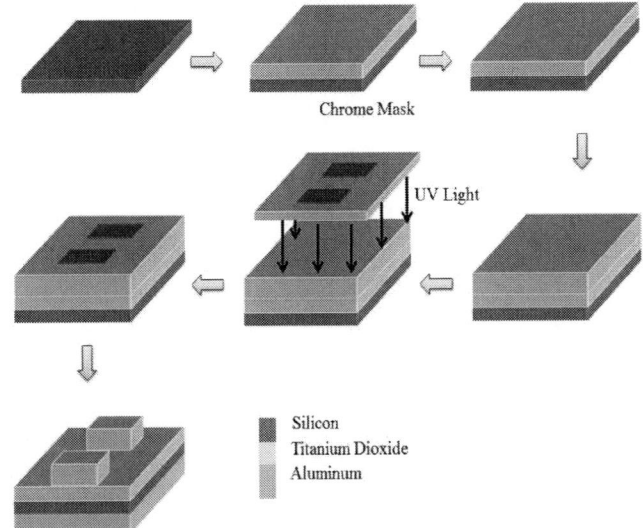

Fig. 1.   Process step of fabricating MOS capacitor

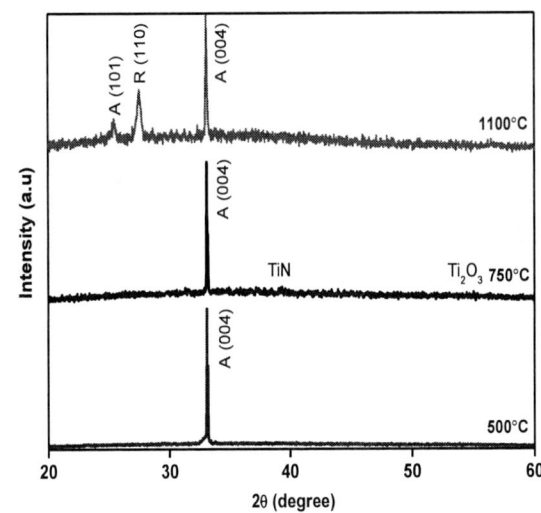

Fig. 2.   XRD pattern of TiO_2 film annealed at different temperature (A: anatase, R: rutile)

TABLE I.   ANNEALING TEMPERATURE, FWHM AND CALCULATED CRYSTALLITE SIZE

Annealing Temperature (°C)	FWHM (rad)	2θ (degree)	Grain Size (nm)
500	959.12×10^{-6}	33.01	150.72
750	743.14×10^{-6}	33.00	186.51
1100	2.13×10^{-3}	32.98	67.88

TABLE II. MEASURED VALUE OF ROUGHNESS AVERAGE, RMS AND GRAIN SIZE

Annealing Temperature (°C)	Roughness Average, Ra (nm)	Root Mean Square, RMS (nm)	Grain Size (nm)
500	0.271	0.344	66.58
750	1.201	1.506	86.01
1100	2.194	2.731	84.41

The AFM images of TiO_2 thin film annealed at 500 °C, 750 °C and 1100 °C for two hours were demonstrated in Fig. 3. Fig. 3 (a) shows the grains started to grow with a small particle and heterogeneous on the wafer surface. As the temperature increase up to 750 °C (Fig. 3 (b)), the structure was set to form an elongated granular shape of grains uniformly across the wafer. This result is in good agreement with [5]. However, as higher annealing temperature (1100 °C) applied, the structure become denser and more homogenous (Fig 3(c)). The grain size obtained because the annealing temperature that recondense the adatoms around the nucleation sites to form the grains [15] and the forming of crystalline structure as obtained in XRD results.

The formation of grains was due to the single crystal structure transformation to poly-crystalline during the annealing process. Low annealing temperature (500 °C) indicated 66.58 nm of nanoparticle size. As the temperature increase to 750 °C, the grain size was also increase to 86.01 nm. The AFM results have a good agreement with the calculated XRD result which is the particle size increased as the temperature increased from 500 °C to 750 °C. However, the higher temperature (1100 °C) indicates small grain size i.e. 84.41 nm compared to 750 °C. This shows that even at higher temperature, the rutile structure capable to yield TiO_2 film with a smaller grain size. The surface roughness, root-mean-square, and roughness thickness were increased respectively as the annealing temperature increased as shown in Table 2 which have good agreement with [16].

B. Electrical Characteristic

Fig. 4 depicts the current-voltage measurement of TiO_2 thin film deposited on silicon substrate which has been annealed at 500 °C, 750 °C and 1100 °C for 2 hours in N_2 gases. The current was decreased as the annealing temperature increased. The decrement of the current was due to the increment size and number of the particles growth on the silicon surface. This result has good agreement with [17]. Moreover, larger grain size will restrict the electron migration between the grain boundaries thus, only small amount of current can pass through the device. Therefore, the resistivity of TiO_2 nanoparticle was incresed considerably after the post-heating process. Larger particle size will improve the strcuture of the film become more denser and reduced the number of porosity [18].

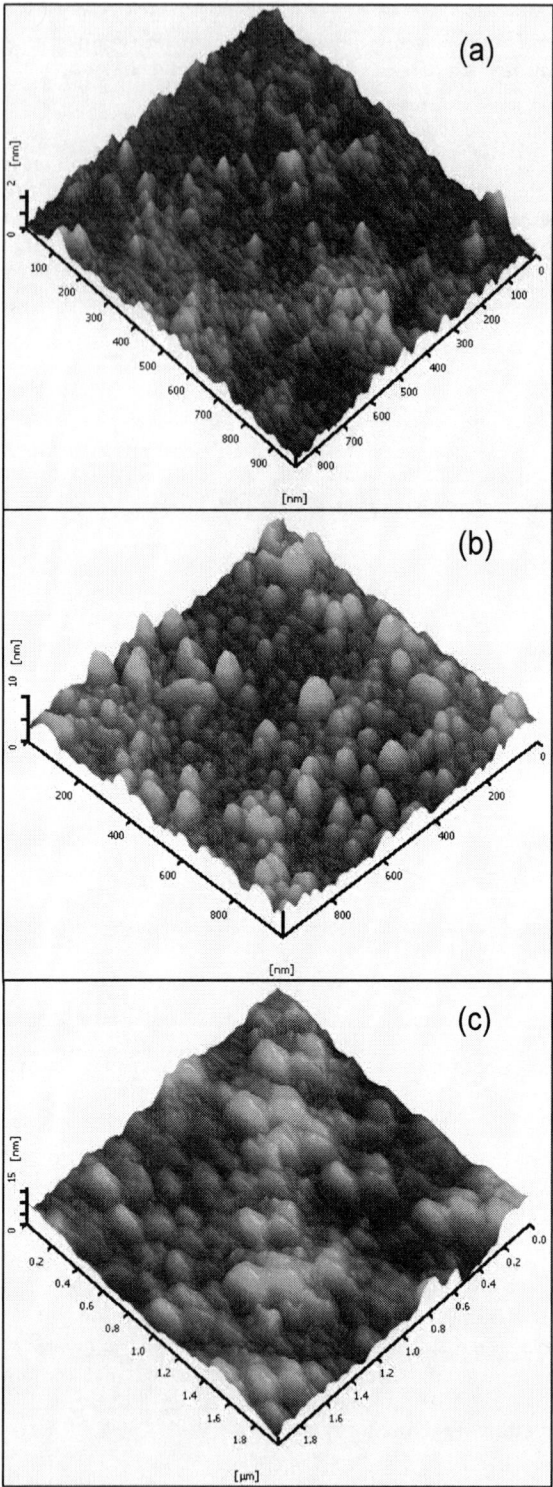

Fig. 3. 3D AFM images of TiO_2 film annealed at different temperature (a) 500 °C (b) 750 °C (c) 1100 °C

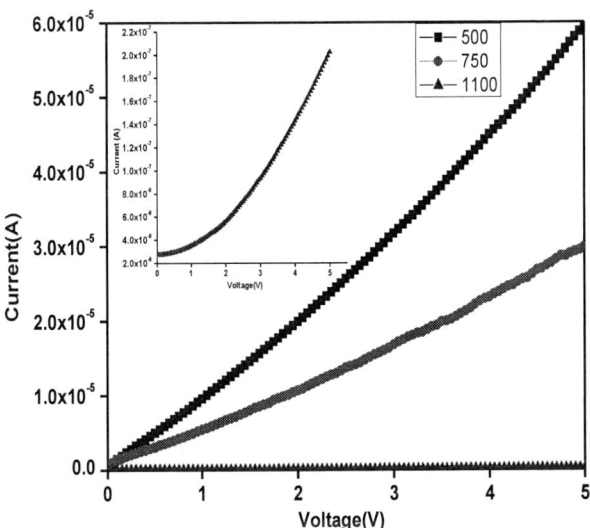

Fig. 4. I-V curve for TiO₂ annealed at 500 °C, 750 °C and 1100 °C. The inset graph is the close up version of film annealed at 1100 °C

IV. CONCLUSION

In this study, the transformations of amorphous to crystalline (anatase and rutile) phases were highly affected by the enhancement of annealing temperature from 500 °C to 1100 °C. Thus, the higher temperature will result in spherical shape of grain, but the rougher TiO₂ layer. The nanoparticle size was investigated using XRD and in line with AFM results. The high annealing temperature, i.e. 1100 °C will restrict the current flow throughout the TiO₂ surface due to the large size of nanoparticles and will result only small amount of current. Hence, in our case, the optimum annealing temperature was 1100 °C for the biosensor application due to its small particle size and high sensitivity current acquired during measurement.

ACKNOWLEDGMENT

The authors are grateful to the Department of Higher Education, Ministry of Higher Education, (KPT) for funding this research through the Fundamental Research Grant Scheme (FRGS) with the code number 9003-00360, entitled The Study of Electron Tunneling through Single/Multiple Layer Dielectric Thin Film. Special thanks to Advanced Materials Research Centre (AMREC) lab for the equipment facilities during the deposition process. The author also would like to acknowledge all the team members in Institute of Nano Electronic Engineering (INEE) for their guidance and help.

REFERENCES

[1] M. Ahmadi, "Study of Different Parameters in TiO2 Nanoparticles Formation," *J. Mater. Sci. Eng.*, vol. 5, pp. 87–93, 2011.

[2] L. Bedikyan, "Titanium Dioxide Thin Films: Preparation and Optical Properties," *J. Chem. Technol. Metall.*, vol. 6, no. 48, pp. 555–558, 2013.

[3] J. Ben Naceur, M. Gaidi, and F. Bousbih, "Annealing effects on microstructural and optical properties of Nanostructured-TiO 2 thin films prepared by sol–gel technique," *Curr. Appl. ...*, vol. 12, no. 2, pp. 422–428, Mar. 2012.

[4] M. M. Hasan, A. S. M. A. Haseeb, H. H. Masjuki, and R. Saidur, "Structural and Electrical Properties of TiO2 RF Sputtered Thin Films," *Mater. Sci. Eng.*, vol. 75, pp. 68–71, 2000.

[5] S. Nadzirah and U. Hashim, "Annealing effects on titanium dioxide films by Sol-Gel spin coating method," *RSM 2013 IEEE Reg. Symp. Micro Nanoelectron.*, pp. 159–162, Sep. 2013.

[6] A. Ansari, M. Alhoshan, M. Alsalhi, and A. Aldwayyan, "Nanostructured metal oxides based enzymatic electrochemical biosensors," no. February, 2010.

[7] A. Rasooly, "Biosensor technologies," *Editor. / Methods*, vol. 37, no. 1, pp. 1–3, Sep. 2005.

[8] C. Jianrong, M. Yuqing, H. Nongyue, and W. Xiaohua, "Nanotechnology and biosensors," vol. 22, pp. 505–518, 2004.

[9] M. Alam and D. Cameron, "Preparation and characterization of TiO2 thin films by sol-gel method," *J. sol-gel Sci. Technol.*, pp. 137–145, 2002.

[10] M. Yamagishi, S. Kuriki, P. K. Song, and Y. Shigesato, "Thin film TiO2 photocatalyst deposited by reactive magnetron sputtering," *Thin Solid Films*, vol. 442, no. 1–2, pp. 227–231, Oct. 2003.

[11] D. Kaczmarek, J. Domaradzki, D. Wojcieszak, and B. Gornicka, "XRD and AFM Studies of Nanocrystalline TiO2 Thin Films Prepared by Modified Magnetron Sputtering," *Int. Spring Semin. Electron. Technol.*, pp. 159–162, May 2008.

[12] L. Miao, P. Jin, K. Kaneko, a Terai, N. Nabatova-Gabain, and S. Tanemura, "Preparation and characterization of polycrystalline anatase and rutile TiO2 thin films by rf magnetron sputtering," *Appl. Surf. Sci.*, vol. 212–213, pp. 255–263, May 2003.

[13] S. Ben Amor, L. Guedri, G. Baud, M. Jacquet, and M. Ghedira, "Influence of the temperature on the properties of sputtered titanium oxide films," vol. 77, pp. 903–911, 2002.

[14] C. H. Heo, S.-B. Lee, and J.-H. Boo, "Deposition of TiO2 thin films using RF magnetron sputtering method and study of their surface characteristics," *Thin Solid Films*, vol. 475, no. 1–2, pp. 183–188, Mar. 2005.

[15] H. Xiao, *Introduction to Semiconductor Manufacturing Technology.* Prentice Hall, 2001, pp. 483–484.

[16] M. M. Hasan, a. S. M. a. Haseeb, R. Saidur, H. H. Masjuki, and M. Hamdi, "Influence of substrate and annealing temperatures on optical properties of RF-sputtered TiO2 thin films," *Opt. Mater. (Amst).*, vol. 32, no. 6, pp. 690–695, Apr. 2010.

[17] M. Ahmad and N. Rasheid, "Effect of Annealing Temperature on Titanium Dioxide Thin Films Prepared by Sol Gel Method," *Issues Phys.*, vol. 1017, pp. 109–113, 2008.

[18] D. Yoo, I. Kim, S. Kim, C. H. Hahn, C. Lee, and S. Cho, "Effects of annealing temperature and method on structural and optical properties of TiO2 films prepared by RF magnetron sputtering at room temperature," *Appl. Surf. Sci.*, vol. 253, no. 8, pp. 3888–3892, Feb. 2007.

Effect of Process Parameter Variability on the Threshold Voltage of Downscaled 22nm PMOS using Taguchi Method

Afifah Maheran A.H., Menon, P.S., S. Shaari,
Institute of Microengineering and Nanoelectronics (IMEN),
Universiti Kebangsaan Malaysia (UKM)
43600 Bangi, Selangor, Malaysia
susi@eng.ukm.my

I. Ahmad, Noor Faizah Z.A.
Centre for Micro and Nano Engineering(CeMNE), College
of Engineering, Universiti Tenaga Nasional (UNITEN)
43009 Kajang, Selangor, Malaysia

Kalaivani T.
Infrastructure University Kuala Lumpur, Unipark Suria,
Jalan Ikram-Uniten, 43000 Kajang, Selangor, Malaysia

P.R. Apte
Indian Institute of Technology at Bombay, Powai, Mumbai,
India

Abstract— This paper provides the enhancement of 22nm planar PMOS transistor technology through downscaling, design parameter simulation and optimization process. The scaled down device is optimized for its process parameter variability using Taguchi method. The aim is to find the best combination of fabrication parameters in order to achieve the target value of the threshold voltage (V_{th}). A combination of high permittivity material (high-k) and metal gate is utilized simultaneously in replacing the conventional SiO_2/Poly-Si technology. For this, Titanium dioxide (TiO_2) was used as the high-k material and tungsten silicide (WSi_x) was used as the metal gate. The simulation results show that the optimal threshold voltage (V_{th}) of -0.289 V ± 12.7% is achieved in accordance to the ITRS 2012 specifications. This provides a benchmark towards the fabrication of 22 nm planar PMOS in future work.

Keywords—22 nm PMOS, scaling down, high-k/metal gate, Taguchi Method

I. INTRODUCTION

Complementary Metal Oxide Semiconductor (CMOS) device are being scaled down extremely in each technology generation and it is now into nano-regime generation. This phenomenon happens due to market demands of low-cost-high-performance device which leads the technology to achieve a higher integration density and performance in a single chip. Thus, researchers and designers invented a new technology devices using high permittivity (high-k) materials as the gate dielectric material which is being combined with metal gates to cater the problems occurred in nano-scale regime. This invention only replaced the traditional silicon dioxide (SiO_2) and polycrystalline silicon (poly-Si) gate structure design respectively [1].

In our previous work, a planar PMOS device utilizing a 22 nm technology was successfully scaled down from 32 nm. It utilized the high-k/poly-Si gates where the process parameters variability are optimized to achieve the best threshold voltage [2]. In this research, the improvement in downscaling of the

device structure is taken into account in order to achieve a smaller dimensional device with a decent threshold voltage (V_{th}) by optimizing the process parameter variability using Taguchi method [3].

To date, the threshold voltage is being agreed world widely to be one of the most important parameter that affects the power consumption of a device. This parameter is acknowledged as an output that changes due to variability in semiconductor processes that largely impacts the device operation [4]. The process variation trends shows that the threshold voltage increases with decreasing technology node and eventually affects the CMOS electrical properties and performance [5]. The variability in fabrication processes of nano-scale devices must be taken into account during device design where otherwise the errors in the variability will contribute to the difficulty of getting a good V_{th} value.

Taguchi method is used to solve multiple process parameter optimization problems with minimum set of experiments, where it uses an orthogonal arrays to implement the variability process. This method is similar to fractional factorial designs but has additional Parameter Design and Tolerance Design [6]. With added noise factors called signal-to-noise ratio (SNR), process parameter variability using Taguchi method becomes more reliable. In this experiment, since the objective is to achieve a decent target value which is also called as nominal value, the SNR (Nominal-the-Best, NTB) analysis is utilized [7]. In addition, the analysis of variance (ANOVA) was executed in order to identify the most significant value of the process parameters. With the combination of the SNR and ANOVA analysis, the prediction of the best combination of the PMOS's process parameters can be achieved excellently.

As mentioned, in order to achieve the goal, four process parameters with two noise factors are chosen to implement the variablility process. By varying the process parameters with different levels constructed by referring to the L9 Orthogonal Array, the objective of this work in order to achieve the V_{th}

value as close possible to the ITRS 2012 projection for 22 nm planar PMOS device where the range is between -0.289 V ± 12.7% will succeed [8].

II. SCALING DOWN

The PMOS transistor can be scaled down using the downscale rule [9] of $\beta = \dfrac{1}{\alpha}$ where $\alpha = 32nm / 22nm$ since the downscaling is from 32 nm gate length to 22 nm. Hence $\beta = 0.6875$. This ratio value is used to downscale the device dimensions and the mesh settings.

III. MATERIALS AND METHODS

A. Fabrication using TCAD Simulation Tools

The 22nm planar PMOS transistor is fabricated virtually. The sample used in this experiment is a p-type silicon substrate with <100> orientation. N-wells are firstly created by developing a 200Å oxide screen on wafers followed by phosphorus doping with a dosage of 4.5×10^{11} atom/cm^2. It is then annealed. The next step is to produce a 130 Å thickness of Shallow Trench Isolator (STI). The function is to produce an isolation of neighbouring transistor. Then, the making of a trench structure took place by depositing a 1500Å nitride layer, followed by the grown of a sacrificial oxide layer (PSG) with temperature of 950°C. It is then etch to complete the trench structure. The gate oxide is grown and Boron Difluorite (BF$_2$) with dosage of 1.75×10^9 atom/cm^2 after the threshold-adjustment implantation is done.

Then the high-k material, TiO$_2$ (dielectric permittivity = 80) is deposited for a final thickness of 2 nm and this is followed by etching and adjusting the length to produce a 22 nm gate length. The deposition of WSi$_x$ metal is then took place with a thickness of 53 nm and etched to produce the metal gate structure [10]. Halo structure is implanted by phosphorus dose of 1.5×10^{10} atom/cm^2 and this is one of the process parameters that can be varied. Then spacers are formed followed by source-drain implantations, with boron implantation with dosage of 44.0×10^{13} atom/cm^2. The next process is the development of 0.015 µm Borophosphosilicate Glass (BPSG) layer followed by compensation implantation process with dose of 1.45×10^{11} atom/cm^2 using phosphorous dose. Lastly, aluminium layer is then deposited to form the metal contact for the source and drain.

The final procedure is the electrical characteristic measurement using ATLAS simulation module in order to optimize the best value of threshold voltage (V_{th}). Figure 1 shows the 22nm PMOS transistor with its net doping profile.

B. Taguchi Orthogonal Array Method

As mentioned in Section I, four process parameters and two noise factors were identified to form the L9 Taguchi orthogonal array and each value at different levels are tabled in Table I and Table II respectively.

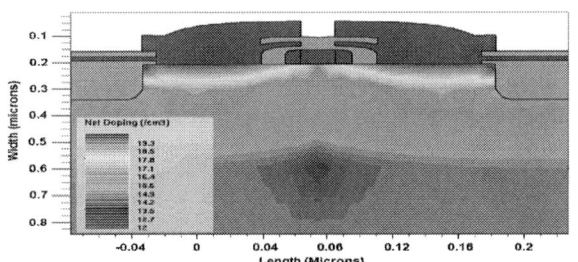

Fig. 1. The profile doping of the PMOS transistor

TABLE I. PROCESS PARAMETERS

Factor	Process Parameter	Unit	Level		
			1	2	3
A	Halo Implantation Dose (10^{10})	atom/cm^2	1.0 (A1)	1.5 (A2)	2.0 (A3)
B	Halo Tilting Angle	degree	35 (B1)	40 (B2)	45 (B3)
C	S/D Implantation Dose (10^{13})	atom/cm^2	44.0 (C1)	44.05 (C2)	44.10 (C3)
D	Compensation Implantation Dose (10^{11})	atom/cm^2	1.35 (C1)	1.4 (C2)	1.45 (C3)

TABLE II. NOISE FACTORS

Factor	Noise Factor	Unit	Level 1	Level 2
X	Sacrificial Oxide Layer Annealing Temperature	°C	905 (X1)	910 (X2)
Y	Gate Oxide Growth Annealing Temperature	°C	900 (Y1)	905 (Y2)

IV. RESULTS AND DISCUSSIONS

Implementing the L9 Orthogonal Array of Taguchi Method experiment [7], the results of the device's threshold voltage (V_{th}) are analysed. In order to get the best design of the target value (V_{th}) which is also known as the nominal value, there are important steps that must be considered. Firstly is to determine one process parameter that could be the adjustment factor so that the process parameters can be adjusted to be closer to the target value. Secondly, is to decide the best level for the other three process parameters. Finally, the selected levels from the analysis will be then simulated again in order to verify the predicted optimal design with respect to the noise factors.

A. Analysis for 22nm PMOS Device

The results of L9 array of 36 simulations for the V_{th} data is shown in Table III. Based on the results, the next step is to determine the significant control factors for the process parameters. A control factor is the factor that gives more efficiency to the device characteristics. In this case, the SNR analysis is used to identify the optimal process parameters in the experiment and the SNR for each level of the process parameter is computed. Regardless of the category of the performance characteristics, the SNR value increases or decreases proportionately to performance [11]. SNR of NTB quality characteristics is used to achieve the targeted V_{th} where the intended results needs to be exactly or closer to the target value. The SNR for NTB, η can be expressed as [7] $\eta = 10 \log_{10} \left[\dfrac{\mu^2}{\sigma^2} \right]$ where μ is the mean value and σ is the variance in the experiment. In order to choose the best level of the process parameters, one must choose the highest mean value of each process parameter. Table IV shows the SNR NTB mean values of the process parameters.

The highest mean value of each process parameter in designing the 22nm PMOS device is considered as the best level. From the table, it shows that the highest level was Level 3 with mean value of 84.59 dB for Halo implantation dose (Factor A). The second process parameter is Halo tilting Angle (Factor B) where level 2 has the highest value of 84.61 dB. While Factor C and Factor D which are the S/D Implantation dose and compensation implantation dose has Level 1 and Level 3 as the highest value of 84.53 dB and 85.11 dB respectively. At this point, the best level of the process parameters is A_3, B_2, C_1, and D_3.

B. Analysis of Variance (ANOVA)

The factor effect percentage on the SNR indicates the priority of a process parameter to reduce variation [12]. The result of ANOVA for the PMOS device is presented in Table V. From the table, the dominant factor and adjustment factor can be identified in order to find the best process parameter combination. The factor with highest percentage of contribution on SNR NTB will have a greater influence on the stability of V_{th} and this is also called the dominant factor. The factor with lowest percentage of SNR NTB and at the same time highest percentage on mean is considered as an adjustment factor.

The results of Table V apparently shows that Factor D (compensation implantation dose) has the most dominant impact on the resulting threshold voltage in a PMOS device and it can be identified as a dominant factor with percentage of 96.72%. The adjustment factor can be identified as the factor with the lowest percentage on the SNR NTB but the highest percentage on mean. And it is clearly shown that Factor A (Halo implantation dose) is the adjustment factor.

TABLE III. V_{TH} VALUES RESULT

Exp. No	V_{TH} (Volts)			
	X1Y1	X1Y2	X2Y1	X2Y2
1	- 0.28882	- 0.28885	- 0.28885	- 0.28887
2	- 0.29597	- 0.29599	- 0.29599	- 0.29601
3	- 0.30287	- 0.30289	- 0.30289	- 0.30291
4	- 0.31826	- 0.31828	- 0.31828	- 0.31830
5	- 0.29056	- 0.29058	- 0.29058	- 0.29061
6	- 0.29673	- 0.29675	- 0.29676	- 0.29677
7	- 0.31472	- 0.31474	- 0.31474	- 0.31476
8	- 0.31917	- 0.31919	- 0.31920	- 0.31922
9	- 0.28940	- 0.28942	- 0.28942	- 0.28945

TABLE IV.   S/N RATIO (NOMINAL-THE-BEST) VALUES

Factor	S/N Ratio (Mean)			Total mean S/N Ratio
	Level 1	Level 2	Level 3	
A	84.43	84.47	84.59	
B	84.56	84.61	84.31	84.49
C	84.53	84.46	84.49	
D	83.32	85.05	85.11	

TABLE V.   RESULT OF ANOVA

Factor	Factor Effect on S/N Ratio (Nominal-The-Best) (%)	Factor Effect on Mean (%)
A	0.64	16.55
B	2.50	13.97
C	0.14	0.34
D	96.72	69.14

C. Confirmation of Optimum Factor

Before finalizing the best combination of process parameters, the candidate of adjustment factor value must be establish first by adjusting the dosage value until V_{th} is close enough to the nominal value. Referring to the results in Section A and Section B, it can be concluded that the process parameters combination is $A_{(swept)}$, B_2, C_1, and D_3. Simulations with the value sweeps on Factor A from 1.0×10^{10} atom/cm^2 to 2.0×10^{10} atom/cm^2 are performed to determine the optimum parameter. Once the swept process is achieved to get the desired value range, the final combination of the process parameters can be determined. Taguchi method suggested the best combination of the process parameters for the device and it is shown in Table VI.

Based on Table VI, the final experiment is performed to verify the reliability of the Taguchi Method prediction. Then the last optimization step is to get the best value of the V_{th} by adding noise factors. Table VII show the final results of the experiment with added noise factor.

From the final results in Table VII, the mean is -0.30733 V and it is shown that the values are well within the targeted set by ITRS 2012 which is -0.289 V $\pm$ 12.7%. The results is in the range with reported by other researchers [10].

TABLE VI.   BEST COMBINATION OF THE PROCESS PARAMETERS

Factor	Process Parameter	Unit	Level	Best Value
A	Halo Implantation Dose	atom/cm^2	Sweep	1.0×10^{10}
B	Halo Tilting Angle	degree	2	40
C	S/D Implantation Dose	atom/cm^2	1	44.0×10^{13}
D	Compensation Implantation Dose	atom/cm^2	3	1.45×10^{11}

TABLE VII.   FINAL RESULTS OF V_{TH} WITH ADDED NOISE

$V_{TH}1$ (X1,Y1)	$V_{TH}2$ (X1,Y2)	$V_{TH}3$ (X2,Y1)	$V_{TH}4$ (X2,Y2)	V_{TH} (Mean)
-0.30731	-0.30732	-0.30733	-0.30735	-0.30733

V. CONCLUSIONS

As a conclusion, the optimum solution in achieving the best process parameter combinations in designing a 22nm PMOS device was positively predicted by Taguchi Method. Threshold voltage (V_{th}) is one of the main factors in determining the functionality of the nano-scaled device. In this research, the compensation implantation dose is identified as the dominant factor while Halo implantation dose is identified as the adjustment parameter for this device. Therefore it is proven that by using a combination of TiO_2/WSi_x as a high-k/metal gate planar structure, a fabrication's process parameter for 22nm PMOS transistor can be optimized in order to achieve the targeted result and the V_{th} value is well within the ITRS 2012 requirements of -0.289 V ± 12.7%. To the author's knowledge, this is the first time that such work has been reported in the literature.

ACKNOWLEDGMENT

The authors would like to thank Universiti Kebangsaan Malaysia (UKM) and the Malaysian Ministry of Education for sponsoring the publication of this work using grant no ERGS/1/2012/STG02/UKM/02/2. Universiti Teknikal Malaysia Melaka (UTeM) and Universiti Tenaga Nasional (UNITEN) are also acknowledged for moral and operational support throughout the project.

REFERENCES

[1] J. W. Sleight, I. Lauer, O. Dokumaci, D. M. Fried, D. Guo, B. Haran, S. Narasimha, C. Sheraw, D. Singh, M. Steigerwalt, X. Wang, P. Oldiges, D. Sadana, C. Y. Sung, W. Haensch, and M. Khare, "Challenges and Opportunities for High Performance 32 nm CMOS Technology," *IEEE IEDM*, pp. 1–4, 2006.

[2] A. H. Afifah Maheran, P. S. Menon, I. Ahmad, and Z. Yusoff, "Threshold Voltage Optimization in a 22nm High-k / Salicide PMOS Device," *IEEE Regional Symposium on Micro and Nano Electronics (RSM2013)*, pp. 134–137, 2013.

[3] H. A. Elgomati, B. Y. Majlis, A. M. A. Hamid, P. M. Susthitha, and I. Ahmad, "Modelling of Process Parameters for 32nm PMOS Transistor Using Taguchi Method," *Asia Modelling Symposium*, pp. 40–45, May 2012.

[4] M. H. Sulieman, "Threshold-voltage variations effects on the reliability of nano-scale CMOS logic gates," in *Nanotechnology, 2009. IEEE-NANO 2009. 9th IEEE Conference on*, 2009, pp. 744–747.

[5] K. Kuhn, C. Kenyon, A. Kornfeld, M. Liu, A. Maheshwari, W. Shih, S. Sivakumar, G. Taylor, P. VanDerVoorn, and K. Zawadzki, "Managing Process Variation in Intel's 45nm CMOS Technology," *Intel's Technology Journal*, vol. 12, no. 2, pp. 93–109, 2008.

[6] P. S. Menon, K. Kandiah, A. A. Ehsan, and S. Shaari. The development of a new responsivity prediction model for In(0.53)Ga(0.47)As interdigitated lateral PIN photodiode," *Journal of Optical Communications*, vol. 30, pp. 2-6, 2009.

[7] A. H. Afifah Maheran, P. S. Menon, I. Ahmad, S. Shaari, H. A. Elgomati, and F. Salehuddin, "Design and Optimization of 22 nm Gate Length High-k/Metal gate NMOS Transistor," *Journal of Physics: Conference Series*, vol. 431, pp. 1–9, Apr. 2013.

[8] "ITRS 2012. www.ITRS2012.net." .

[9] S.M. Sze, *Semiconductor Devices Physics and Technology*, 2nd ed. USA: John Wiley & Sons Inc., 2002, pp. 202–204.

[10] A. Khakifirooz and D. A. Antoniadis, "MOSFET Performance Scaling - Part II: Future Directions," *IEEE Transactions on Electron Devices*, vol. 55, no. 6, pp. 1401–1408, 2008.

[11] A. H. Afifah Maheran, P. S. Menon, I. Ahmad, and S. Shaari, "Effect of Halo structure variations on the threshold voltage of a 22nm gate length NMOS transistor," *Materials Science in Semiconductor Processing*, vol. 17, pp. 155–161, Jan. 2014.

[12] A.H. Afifah Maheran, P.S. Menon, I. Ahmad, S. Shaari., "Optimization of Process Parameters for Lower Leakage Current in 22nm n-type MOSFET Device using Taguchi Method," *Jurnal Teknologi*, vol. 68, no. 4, pp. 1–5, 2014.

An Efficient ROM Compression Technique for Linear- Interpolated Direct Digital Frequency Synthesizer

Qahtan Khalaf Omran[1], Mohammad Tariqul Islam[1], *Senior Member, IEEE*, [1]Department of Electrical, Electronic & System Engineering),Faculty of Engineering and Built Environment, University Kebangsaan Malaysia (UKM),43600 UKM, Bangi, Selangor, Malaysia.

Norbahiah Misran[1], *Member, IEEE* ,Mohammad Rashed Iqbal Faruque[2], *Member, IEEE*, [2]Space Science Center (ANGKASA), Research Center Building,Universiti Kebangsaan Malaysia (UKM), 43600 UKM, Bangi, Selangor, Malaysia.

Abstract— **A direct digital frequency synthesizer (DDFS) based on piecewise linear approximation method is presented in this paper. The proposed method allows sequential read access to memory cells per one clock cycle using time sharing. The output values will be momentarily stowed and read at a later time; thereby the slope is simply derived from these sinusoid points at successive phase angles. As a consequence, the DDFS only needs to store fewer coefficients and the hardware complexity is significantly shortened. The proposed DDFS has been analyzed using MATLAB Simulink and examined over entire Nyquist frequency band. The simulation results show a promising result of 84 dBc spurious free dynamic range (SFDR). The resultant low complexity architecture along with high spectral purity synthesized signal meets the specifications of recent portable battery-driven products.**

Keywords— DDS; Linear interpolation; DDFS

I. INTRODUCTION

Sinusoidal output direct digital frequency synthesis (DDFS) system is regarded as being the most suitable for portable low battery drain transceivers. It has better unique feature than its rival counterpart- indirect DPLL synthesizer, in that its output's frequency, phase and amplitude can be precisely and easily handled in digital form. Consequently, DDFS is capable of being incorporated with different digital modulations by using digital signal processing method.

The conventional DDFS structure given in Fig.1 incorporates three fundamental circuits: phase accumulator (PA), phase to sinusoid amplitude converter (PSAC), which is usually employs ROM as a sine mapper and linear digital to analog converter (DAC). FIW represents an input to the PA named the "frequency instruction word" ROM based DDFS create sine-wave output by indexing through sine lookup table (LUT); the LUT contains the sine amplitude values corresponding to each possible phase values. Periodically, the PA increments its register content by the input value (FIW) till it overflows and rolls up. Each overflow of PA coincides to exactly one period of synthesized sine wave and the frequency of generated signal is as follows:

$$f_{out} = \frac{FIW}{2^M} f_{clk} \qquad (1)$$

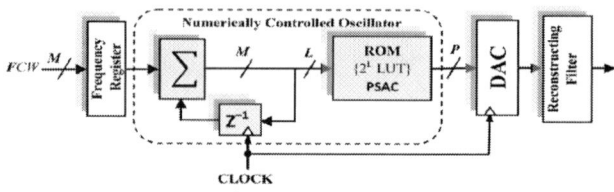

Fig. 1. DDFS basic structure

For experimental consideration, the DDFS could only generate frequencies up to 40% of the clock frequency; the constraint comes from the sampling theorem. It is obvious that whenever L increases, the size of ROM increases exponentially. The large ROM consumes high power and wide area which is degrading the performance of DDFS, taking into account that the upper bound of the SFDR based on phase truncation is given by

$$SFDR \approx 6.02L\text{-}3.92\text{dB} \qquad (2)$$

Which can be improved only with high phase resolution (i.e. large L) and that makes it difficult to obtain high spectral purity with small chip area and low power consumption. However, undesirable excessive power use of goods and services must be alleviated in portable communication equipment, and this makes it necessary to apply some compression techniques to scale down the ROM size while still retaining high spectral purity.

In pursuing such a destination, many attempts have been built and several techniques have been suggested. These methods include exploitation of trigonometric identities and approximation of the sine function. The alternative approaches completely discarded the need of a ROM by computing the samples of sine amplitude of the digital phase contents. The ROM elimination has been investigated in many researches, such as [1]. The ROM-less DDFS is based on a special phase conversion algorithm such as Taylors series evaluation [2], the second order parabolic approximation[3], ,the coordinate rotation digital computer (CORDIC) algorithm [4], the piecewise linear approximation [5] ,the Quasi-Linear Interpolation Method [6],the 8th —order even polynomial[7] and so on . However, it will be hard for any method that is based on high-order polynomials to meet the speed requirement of wireless communication applications. For instance, the hardware realization of Taylor's series' evaluation is complicated and the running speed is slow. The CORDIC method is complicated and has low efficiency due to their difficulties in computing the sine function [8]. This report investigates and offers an efficient DDFS architecture which is based on piecewise linear estimation. The proposed DDFS can simultaneously achieve an excellent spectral purity without additional circuitry.

II. PIECEWISE LINEAR APPROXIMATION BACKGROUND

Piecewise linear interpolation (LI) approximation is a well-known technique to generate a sinusoidal wave form with low hardware cost. In such a technique, the first sine quadrant is divided into $s = 2^N$ linear segments and each segment approximated by a linear polynomial of the form.

$$p_i(x) = C_i + M_i.(x - x_i) \ , x_i \le x \le x_{i+1} \quad 1 \le i \le S \quad (3)$$

Where M_i, C_i are the slope and initial amplitude coefficients of the ith linear segment respectively, and x represents the scaled phase ramp input in the interval [0, 1].

For a uniform segmentation, the segments' intervals are equal in length and the lower bound of each interval x_i is equal to $(i-1)/s$, hence the length of each segment's subinterval is 1/s [9-10]. Equation (3) can be written as follows

$$p_i(x) = C_i + M_i.(x - x_i), \quad \& \frac{i-1}{s} \le x \le \frac{i}{s} \quad (4)$$

Direct implementation of (4) involves two lookup ROMs The first lookup ROM utilizes for the storage of coefficients C_i and the second ROM for storage of coefficients M_i.

In this paper we add the coefficients M_i evaluation in the hardware, consequently the lookup ROM which is used for storing coefficients M_i is eliminated. To perform the derivation of the slope coefficients, in the following we introduce a new technique that allows accessing the memory twice at one clock cycle using *Time Sharing*. By this approach, we need not have a new complex coefficient calculations, the coefficients simply relate to a fewer conventional sine LUT points.

III. CIRCUIT REALIZATION USING TIME SHARING

In this section, for generating the sinusoidal function, we first exploit the quadrature wave symmetry and then employ the linear interpolation technique to estimate the first quarter of the sine wave. The approximation involves segmentation the phase interval by s subinterval, and each segment could be defined by two polynomial coefficients M_i and C_i. The coefficient M_i which represents the slope coefficient of its segment can be estimated from the sine function as follows.

$$M_i = \frac{1}{s}(sinx_{i+1} - sinx_i) \quad , \ 1 \le i \le s \quad (5)$$

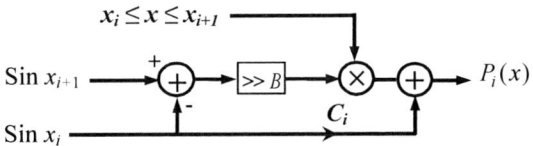

Fig.2 . Realization of the ith sine interpolation point, B the quantized segment length in bits.

Equation (5) can be easily realized by subtracting the Sin x_i at successive phase angles and then right shifted the result by B position using the binary (shift right) operation. By adding the initial coefficients C_i (which is equal to sin x_i points) to the resulted value, the realization of the whole $P_i(x)$ function is skilled.

Fig. 2 shows the $P_i(x)$ realization. It has been clearly seen that it requirement get two sequential sine points at the same time to behaviour the procedure of subtraction and extraction of the slope later. Nonetheless, the accessing of the memory is valid only once at a specific clock cycle. This study offers architecture of the pulse forming circuit which is doing the task of time sharing and the procedure itemized beneath to stimulate about this problem.

At each clock period, the content of address bus represents the instantaneous discrete phase sample. Since only a portion of sinusoid samples is reserved, the A MSB's bit of the L-2 phase accumulator output is used to select segment initial amplitudes Ci. The remaining B LSB's bits ($B = L$-2-A) which represent the (x_{i+1}- x_i) intervals are used to compute the interpolated sine samples.

- At the first time slot, Δt of clock period the value of phase register output advance by one before applied to the address bus ,consequently the output of the ROM is Sin xi+1 during this time slot.

- After Δt the address back propagate to pervious value, so the output of the ROM is sin x_i.

- The content of ROM at these successive time slots will be subtracted; hence, the gradient is simply derived and kept unchanged during the segment interval.

IV. MATLAB SIMULATION RESULTS

On the basis of the suggested PSAC architecture, we have designed the proposed DDFS with targeted SFDR of 84dBc. The proposed DDFS was analyzed using MATLAB Simulink with 15-bit phase accumulator and 32 segments. According to [10], architecture of first order approximation with 32 segments has worst case spur of -84 dBc. A MATLAB Simulink model of DDFS, where A is 5-bit length, B 8-bit length and P 13-bit amplitude resolution, yields to total memory size of ($2^5 \times 14 = 448$ bit) is shown in Figure 3. The compression ratio can be estimated in respect of uncompressing ROM excluding the sine quadrature summitry

978-1-4799-5761-3/14 $31.00 © 2014 IEEE

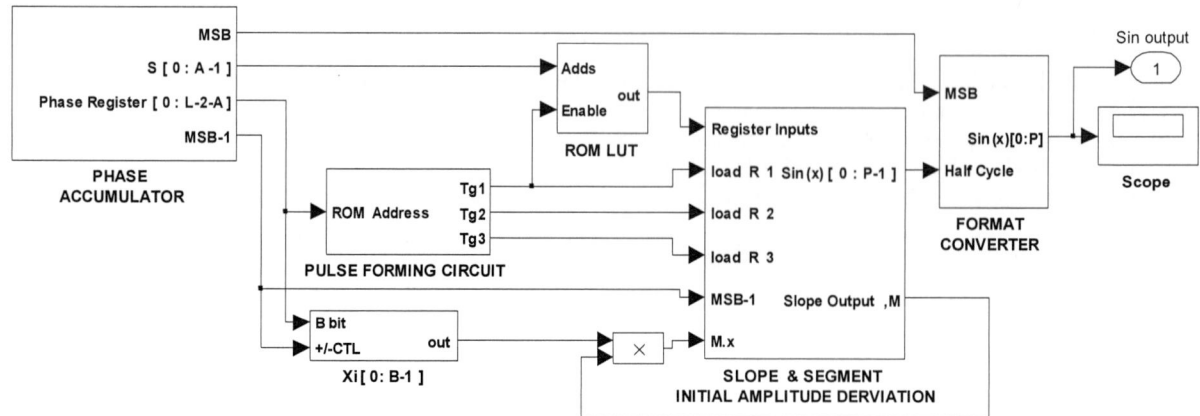

Fig. 3. MATLAB Simulink Model for proposed architecture

i.e. (The lookup ROM size is $2^{L-2} \times P$) and $P=L-1$, where L and P are the truncated phase resolution and the output amplitude resolution respectively. The resultant architecture showed a 256:1 ROM compression ratio. The ROM compression ratio is calculated as follows :$(2^{L-2} \times P) / 2^A \times P = (2^{13} \times 14) / (2^5 \times 14) = 256$ and if we include the amount of ROM compression due to the sine quadrature summitry, the compression ratio becomes: $(2^L \times P) / 2^A \times P = (2^{15} \times 14) / (2^5 \times 13) = 1102.77$. The spurious level for the DDFS are shown in Figures 4 and 5 for the output frequencies of 2%, and 38% of clock frequency, when the *FIW* is equal to 655, and 12452 respectively. The simulation results indicate 84 dBc of SFDR over the first Nyquist band; this is conforming to the theoretical SFDR upper bound estimated in [10] where:

$$SFDR = 24 \text{ dBc} + 20 \log S^2 \qquad (6)$$

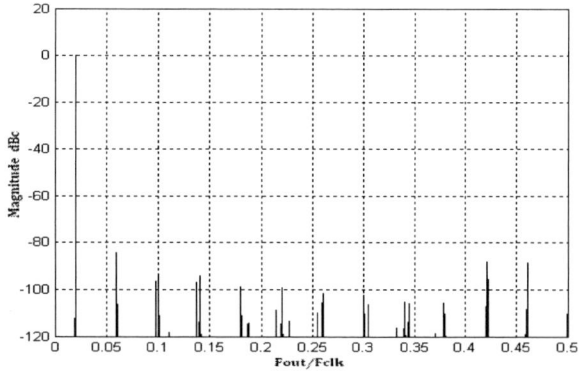

Fig. 4. DDFS output spectra at the first Nyquist band for $f_{out} = 0.02 \times f_{clk}$.

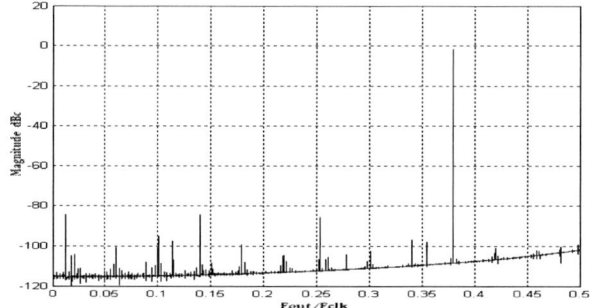

Fig. 5. DDFS output spectra at the first Nyquist band for $f_{out} = 0.38 \times f_{clk}$.

V. CONCLUSIONS

A single lookup ROM-based phase- to- sine mapper is introduced in this paper. The conventional linear interpolation based PSAC architecture has been developed. The proposed ROM elimination based on modified linear interpolation technique. Unlike many reported architectures, which used complex circuits to compute the sine samples, only 32 points from a standard sine LUT with less registers are sufficient. The complexity of DDFS architecture is significantly simplified by using an efficient phase to amplitude conversion architecture. It was shown that a ROM compression ratio of 1102.77:1 was achieved. The proposed DDFS has been investigated and tested over the entire Nyquist frequency range. The spurious free dynamic range of synthesized sinusoid achieved 84dBc which is adequate for more recent communications systems.

REFERENCES

[1] C. C. Wang, Y. L. Tseng, H. C. She, C. C. Li, and R. Hu, "A 13-Bit resolution ROM-less direct digital frequency synthesizer based on a trigonometric quadruple angle formula," IEEE Trans. Very Large Scale Integr. (VLSI) Syst., vol. 12, no. 9, pp. 895–900, Sep. 2004..

[2] Hai, U., Khan, M.N., Imran, M.S. &Rehan, M. (2005). Compressed ROM High Speed Direct Digital Frequency Synthesizer Architecture, *in*

Proc.of IEEE International Conference on Microelectronics (pp. 36–39).

[3] A. M. Sodagar and G. R. Lahiji, "A pipelined ROM-less architecture for sine-output direct digital frequency synthesizers using the second order parabolic approximation," IEEE Trans. Circuits Syst. II, Expr. Briefs, vol.48, no. 9, pp. 850–857, Sep. 2001.

[4] A. Madisetti, A. Y. Kwentus, and A. N. Willson, "A 100-MHz, 16-b, direct digital frequency synthesizer with a 100-dBc spurious-free dynamic range," IEEE J. Solid-State Circuits, vol. 34, no. 8, pp.1034–1043, Aug. 1999.

[5] Q. K. Omran, M. T. Islam, and N. Misran, "FPGA-Based implementation of a new phase-to-sine amplitude conversion architecture," Elektronika ir Elektrotechnika (Electronics and Electrical Engineering), vol. 19, no. 10, pp. 103–108, 2013.

[6] A. Ashrafi, R. Adhami, and A. Milenkovic, "A direct digital frequency synthesizer based on the Quasi-linear interpolation method," IEEE

Trans. Circuit Syst. I, Reg. Papers, vol. 57, no. 4, pp. 863–872, Apr.2010.

[7] Y. H. Chen and Y. A. Chau, "A direct digital frequency synthesizer based on a new form of polynomial approximations," IEEE Trans. Consum. Electron, vol. 56, no.2, pp. 436–440, May 2010.

[8] D. De Caro and A. G. M. Strollo, "High-performance direct digital frequency synthesizers using piecewise-polynomial approximation," IEEE Trans. Circuit Syst., vol. 52, pp. 324–336, Feb. 2005.

[9] D. De Caro , N. Petra, and A. G. M. Strollo , "Direct Digital Frequency Synthesizer Using No uniform Piecewise-Linear Approximation," IEEE Trans. Circuit Syst. I, Reg. Papers, vol. 58, pp. 2409– 2419, OCTOBER 2011.

[10] J. M. P. Langlois and D. Al-Khalili, "Novel approach to the design of direct digital frequency synthesizers based on linear interpolation, "IEEE Trans. Circuits Syst. II, Analog Digit. Signal Process. vol. 50, no. 9, pp. 567–578, Sep. 2003.

Modeling the Velocity Saturation Region of Graphene Nanoribbon Transistor

M. Hosseinghadiry, Razali Ismail*, F. Fotovvatikhah, M. Khaledian, M. Saeidmanesh

Faculty of Electrical Engineering, Universiti Teknologi Malaysia, 81310, UTM Johor Bahru, Johor, Malaysia.
(*E-mail: razali@fke.utm.my)

Abstract— A semi-analytical model for impact ionisation coefficient of graphene nanoribbon (GNR) is presented. The model is derived by calculating the probability of electrons reaching ionisation threshold energy E_t and the distance travelled by electron gaining E_t. In addition, ionisation threshold energy is semi-analytically modeled for GNR. We justify our assumptions using analytical modeling and comparison with simulation results. Gaussian simulator together with analytical modeling is used in order to calculate ionisation threshold energy and Kinetic Monte Carlo is employed to calculate ionisation coefficient and verify the analytical results. Finally, the ionization profile is presented using the proposed models and simulation is carried out. The results are compared with that of silicon.

Keywords— *Length of velocity saturation region, modeling, Graphene Nanoribbon Field Effect Transistor*

I. INTRODUCTION

Moore's law has predicted the trend of silicon technology over the last four decades [1]. Along the way, it has provided the trend of CMOS technology for today's global information society. It is thought that silicon will be the dominant technology for at least 10 more years [1]. However, there are some uncertainties regarding the behavior of the transistors under extreme low dimensions making silicon questionable for future circuit's technology [2]. Therefore, the end of silicon era has been predicted a number of times due to technical obstacles.

Short channel effect is one of the most concerning issues in the nanoscale devices. When the junction depth is comparable to the channel length, which is the case in nanoscale MOSFETs, barrier height decreases causing threshold voltage to reduce. In addition, when a high drain voltage is applied to a short channel device, the barrier is lowered even more, resulting in further decrease in the threshold voltage. This phenomenon is called drain-induced barrier lowering (DIBL).

Eventually, the device reaches the punch-through condition, when the gate is completely not able to control the current flow.

Graphene, which is one-atomic layer of graphite has been recently introduced as a material for the channel of field effect transistors (FET) in order to increase the speed and reduce the channel depth to avoid short channel effects. However, short channels results in long velocity saturation region which is a

negative factor in FET performance. Therefore the study of this factor in the graphene-based FETs is essential and of great importance. This factor can be determined using the model presented in the breakdown voltage of the single gate power Graphene Nanoribbon FET (GNRFET).

The length of the saturation region (L_d) can be obtained by solving the surface potential at $V(x=L_d)=V_S$, where V_S is the saturation voltage. Therefore, it is required to review the surface potential models first. Although there are many analytical and semi-analytical models for surface potential of silicon-based devices, there is a lack of research in modeling of this parameter for carbon-based transistors. Therefore, we briefly review the reported models for conventional silicon transistors.

Imam [3] presented a model for the threshold voltage of a typical double-gate MOSFET using a two dimensional surface potential model. Using proper boundary conditions, the potential along the x and y directions can be calculated. This model is considered as a 2D model.

The same approach as what Imam used was employed by Yang [4] to calculate breakdown voltage of double gate power MOSFET. After the surface potential model is obtained, the lateral electric field is modeled. Then using the ionization coefficient of silicon film and avalanche breakdown condition, the breakdown voltage is calculated and compared with experimental results.

A simple analytical expression of the 3D potential distribution along the channel of a lightly doped silicon trigate MOSFETs in weak inversion was derived in [5]. It was based on a perimeter-weighted approach of symmetric and asymmetric double-gate MOSFETs. The analytical solution was compared with the numerical solution of the 3D Poisson's using Flex PDE simulator. Finally using the model, sub-threshold slope and short channel effects are studied.

In conclusion, it can be said that surface potential models can be broadly classified into one-dimensional (1D), two-dimensional (2D), and three-dimensional models (3D). In this paper, 1D analytical models for surface potential, lateral electric field and length of saturation region are presented and the behavior of GNR transistors are studied in saturation region.

978-1-4799-5761-3/14 $31.00 © 2014 IEEE

II. SATURATION REGION MODELING OF GRAPHENE NANORIBBON TRANSISTORS

A. Selecting a Template (Heading 2)

Fig. 1 shows the cross section of the field effect transistor, which is used for modeling. Graphene nanoribbon with small bandgap is used in the channel and the effect of junctions is ignored. In this model, t_{OX} is the thickness of gate oxide, ε_{OX} is dielectric constant, t_G is the channel thickness, W and L are the channel length and width respectively.

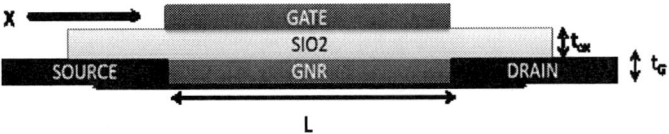

Fig. 1. Schematic cross section of a top-gated GNRFET [6]. Typical device parameters are; doping concentration N=5×10^{16} m^{-2}, t$_{OX}$=1 nm, t$_{GNR}$=0.39 nm, L = 20 nm, and w$_G$=5 nm [7].

Poisson equation is generally used to analytically study the surface potential [4, 8-10]. Applying Gauss law for the GNRFET presented in Fig. 1 and using the similar approach presented in [9,10], the 1-D Poison's equation for a GNR can be written as

$$\frac{d^2V(x)}{dx^2} - \frac{V(x)}{\lambda^2} = -\frac{qN}{\varepsilon_G} - \frac{V_G + V_{BI}}{\lambda^2} \quad (1)$$

where, $V(x)$ is the surface potential in the drain region (between saturation point and drain), $\varepsilon_G = 3 \times 10^{-11}$ F/m is GNR dielectric constant, $V_G = 0.1$ V is the gate voltage and q is the electronic charge [11]. The parameter $\lambda = (\varepsilon_G t_G t_{OX}/\varepsilon_{OX})^{-2}$ is the relevant length scale for potential variation [12]. V_{BI} is the built-in voltage which is written as [10].

$$V_{BI} = \frac{hv_F}{6qw_G} - V_T \ln\left(\frac{N}{n_i}\right) \quad (2)$$

Where, $v_F \sim 10^6$ m/s is the Fermi velocity, $V_T = K_BT/q$ is thermal voltage, n_i and N are the intrinsic and doping carrier concentration respectively.

$V(0) = V_S$, $V(\Delta L) = V_D$, $E(0) = E_S$, where V_S, V_D, ΔL, E_S are saturation voltage at onset of saturation region, drain voltage, length of saturation velocity region (LVSR) and saturation surface electric filed [9] respectively, are the required boundary conditions of equation (1). Assuming $A = -qN/\varepsilon_G - (V_G+V_{BI})/\lambda^2$ and solving equation (1) yields

$$V(x) = -\lambda^2 A + \left(\lambda^2 A + V_S\right)\cosh\left(x/\lambda\right) + \lambda E_S \sinh\left(x/\lambda\right) \quad (3)$$

Using $E(x) = -dV(x)/dx$, lateral electric field $E(x)$ could be written as

$$E(x) = \left(\lambda A + \frac{V_S}{\lambda}\right)\sinh\left(x/\lambda\right) + E_S \cosh\left(x/\lambda\right) \quad (4)$$

In order to calculate length of drain region, Eq. (3)can be numerically solved at $x = \Delta L$ giving

$$\Delta L \approx L - \frac{\lambda V_S \frac{1}{2}\exp\left(\Delta L/\lambda\right)}{V_D + \lambda^2 A - \frac{1}{2}\exp\left(\Delta L/\lambda\right)\left(\lambda^2 A + V_S\right)} \quad (5)$$

To calculate saturation voltage and electric field in the channel for simplicity, the electric field could be assumed linear [8, 9, 12]. Therefore, it can be seen that $d^2V(L_E)/dx^2 = -E_S/L_E$, where L_E is the effective channel length ($L_E = L - \Delta L$). Using Eq. (1), one can get

$$E_S = \frac{L_E}{\lambda^2 - L_E^2}\left(\frac{t_G t_{OX}qN}{\varepsilon_{OX}} + V_G + V_{BI}\right) \quad (6)$$

In addition, V_S can be computed by integrating over Eq.(6) considering $V(0)=0$ giving

$$V_S \approx \frac{t_G t_{OX}qN}{\varepsilon_{OX}} + V_G + V_{BI} \quad (7)$$

III. SIMULATION RESULTS AND DISCUSSION

Using the model, we examine the effect of different parameters on the length of saturation region. It is worth mentioning that the model was verified using MEDICI simulation tool applying silicon material into the model. At the moment there is no simulation tool with the ability to incorporate Graphene.

The effect of different drain voltages on surface electric field is studied and the results are shown in Fig. 2. Saturation voltage is considered to be 0.23 V based on the calculations. This figure shows the electric field profile follows an exponential form depending on the distance from the source.

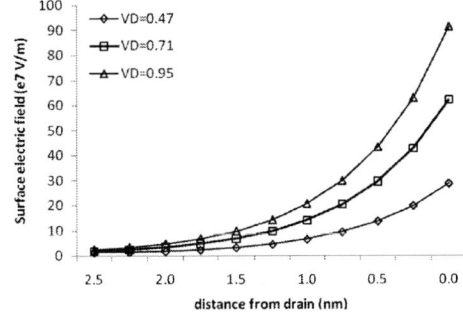

Fig. 2. Surface electric field distribution at different drain voltages at W$_G$=5nm (E$_G$=0.8 eV).

Fig. 3 shows the saturation electric field at different channel lengths and drain-source voltages. As can be seen, increasing V_D results in slight increase in E_S when t_{OX} is in the range of 1 ~ 5 nm. In addition, this figure shows that by increasing t_{OX} and L, saturation electric field increases and decreases respectively.

The length of drain region at different channel length and oxide thickness is shown in Fig. 4 and 5 respectively. In addition, the effect of drain voltage is shown as well. Higher drain voltage results in shorter effective channel length due to higher electric field and longer drain region. Thick oxide results in lower vertical electric field and longer drain region.

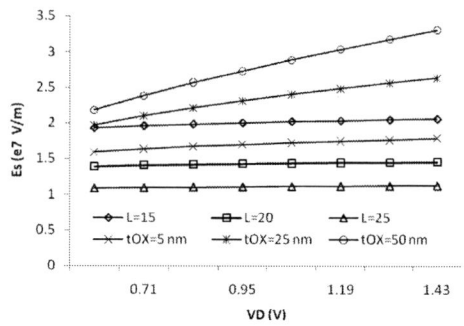

Fig 3. Saturation electric field at different drain voltages at W_G=5nm (E_G=0.8 eV).

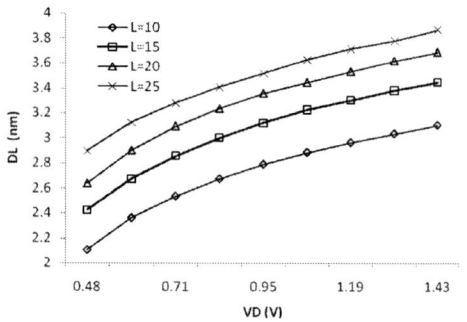

Fig.4. The length of velocity saturation region (LVSR) with different channel lengths and drain voltages at W_G=5nm (E_G=0.8 eV).

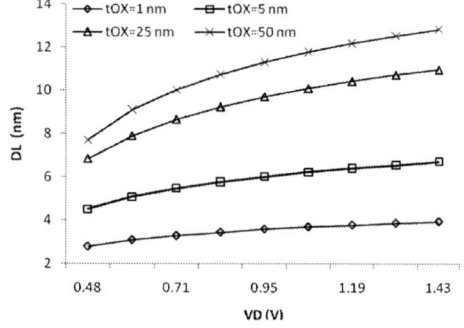

Fig. 5.The length of velocity saturation region (LVSR) vs. oxide thickness variations at different drain voltages at W_G=5nm (EG=0.8 eV).

IV. CONCLUSION

Surface potential, lateral electric field and the length of drain region of graphene based FET are studied analytically using a one-dimensional approach. It is seen that high electric field resulted at the drain region using short graphene-based channel which could be due to high ionization rate. The high electric field resulted in an extended drain region which makes the effective channel short. Using the model, the threshold and breakdown voltage could be further studied.

ACKNOWLEDGMENT

Authors would like to acknowledge the financial support from Research University grant of the Ministry of Higher Education (MOHE), Malaysia under Project Q.J130000.2523.05H23. Also thanks to the Research Management Center (RMC) of Universiti Teknologi Malaysia (UTM) for providing excellent research environment in which to complete this work.

REFERENCES

[1] M. C. Lemme, T. J. Echtermeyer, M. Baus, and H. Kurz, "A Graphene Field-Effect Device," *Electron Device Letters, IEEE,* vol. 28, pp. 282-284, 2007.

[2] M. H. Ghadiry, M. N. S., M. Rahmani, M. Ahmadi, and A. A. Manaf, "Modelling and simulation of saturation region in double gate Graphene nanoribbon transistors," *Semiconductors J.,* vol. 46, pp. 126-129, 2011.

[3] M. A. Imam, M. A. Osman, and A. A. Osman, "Threshold voltage model for deep-submicron fully depleted SOI MOSFETs with back gate substrate induced surface potential effects," *Microelectronics Reliability,* vol. 39, pp. 487-495, 1999.

[4] W. Yang, X. Cheng, Y. Yu, Z. Song, and D. Shen, "A novel analytical model for the breakdown voltage of thin-film SOI power MOSFETs," *Solid-State Electronics,* vol. 49, pp. 43-48, 2005.

[5] A. Tsormpatzoglou, C. A. Dimitriadis, R. Clerc, G. Pananakakis, and G. Ghibaudo, "Semianalytical Modeling of Short-Channel Effects in Lightly Doped Silicon Trigate MOSFETs," *Electron Devices, IEEE Transactions on,* vol. 55, pp. 2623-2631, 2008.

[6] I. Meric, M. Y. Han, A. F. Young, B. Ozilmaz, P. H. Kim, and K. L. Shepard, "Current saturation in zero-bandgap, topgated graphene field-effect transistors," *nature nanotechnology,* vol. 3 pp. 654-659, 2008.

[7] M.Cheli, P.Michetti, and G. Iannaccone, "Model and performance evaluation of field-effect transistors based on epitaxial graphene on SiC," *IEEE Trans. on electron devices,* vol. 57, 2010.

[8] D. Krizaj, G. Charitat, and S. Amon, "A new analytical model for determination of breakdown voltage of Resurf structures," *Solid-State Electronics,* vol. 39, pp. 1353-1358, 1996.

[9] M. El Banna and M. El Nokali, "A pseudo-two-dimensional analysis of short channel MOSFETs," *Solid-State Electronics,* vol. 31, pp. 269-274, 1988.

[10] Q. Zhang, T. Fang, H. Xing, A. Seabaugh, and D. Jena, "Graphene nanoribbon tunnel transistors," *IEEE Electron device letters,* vol. 29, pp. 1344-1346, 2008.

[11] T. Stauber, N. M. R. Peres, and F. Guinea, "Electronic transport in graphene: A semi-classical approach including midgap states," *Physical Review B,* vol. 76, p. 205423, 2007.

[12] H. Wong, "Drain breakdown in submicron MOSFETs: a review," *Microelectronics Reliability,* vol. 40, pp. 3-15, 2000.

Transistor Sizing Methodology for Low Noise Charge Sensitive Amplifier With Input Transistor Working in Moderate Inversion

N. Aimaier, R. M. Sidek, M. N. Hamidon, N. Sulaiman
Department of Electrical and Electronic Engineering
University Putra Malaysia
Seri Kembangan, Malaysia
mknurahmat@gmail.com, roslinams@upm.edu.my

Abstract— In this paper noise contribution of current source transistors and sizing methodology in charge sensitive amplifier for application in the front-end readout electronics is presented. In modern deep-submicron technologies, MOS transistor operating region tends to shift from strong inversion to moderate inversion, this makes traditional square-law MOS device modeling not applicable anymore. Thus a simplified EKV model, which is quite successful in all CMOS operating regions, has been adopted to develop a new analytical methodology to optimize geometry of current source transistors so that the noise contribution from these transistors is only a fraction of input transistor noise. A charge sensitive amplifier based on dual PMOS cascode structure is designed by adopting this current source transistor sizing methodology, and has been simulated using 130nm CMOS technology. The proposed methodology and noise contribution from current source transistors have been found in good agreement with simulation results using deep-submicron CMOS technology.

Keywords— *CSA; EKV model; noise optimization*

I. Introduction

Charge sensitive amplifier (CSA) is an important component in front-end readout (FER) electronic system (Fig. 1) for the application of nuclear physics and medical imaging. A CSA senses the electrical charge (Q_{in}) from the radiation detector (C_{det}) and convert the charge signal to voltage signal. The FER electronics should have high resolution in order to detect charge signals, thus the CSA should have minimum noise level. The input transistor in CSA is usually seen as the most dominant noise source in amplifier, and many works have been done to optimize input transistor width for minimum noise contribution [1], [2], [3]. In modern deep-submicron CMOS technologies, the noise from current source transistors cannot be neglected, and optimization methodology for current source transistors geometry was also described in [4]. However, the requirements for low power dissipation and properties of deep-submicron technologies make the input transistor works in moderate inversion [5], and the simple current source transistor noise optimization methodology is not valid anymore. This paper uses the simplified EKV model to develops an analytical methodology to find out an optimal

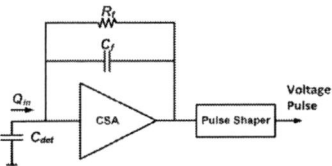

Fig. 1. Front-end readout electronic system

geometry for current source transistors, such that the noise contribution from these transistors results in only a fraction of input transistor noise with the input transistor working in moderate inversion. Based on this proposed methodology, a CSA with CR-RC pulse shaper are designed and simulated to verify the validity of the methodology.

The paper is organized as follows. In section II, the simplified EKV model and derivation of optimum input transistor width is described. The proposed current source transistor sizing methodology is presented in section III, followed by a CSA design and noise simulation results are presented in section IV, and section V provides conclusion.

II. Simplified EKV Model and Optimum Input Transistor Width

The design and optimization of input MOSFET is to achieve minimum noise in FER system. The input transistor noise components are channel thermal noise and flicker (1/f) noise. For a CSA followed by a fast shaper, the input transistor thermal and 1/f noise contribution to the FER system can be represented in terms of ENC (Equivalent Noise Charge) [6]:

$$ENC^2_{in(th)} = \frac{C_t^2}{q^2}\left(\frac{1}{T_P} a_w n\gamma \frac{4kT}{g_m} \right) \tag{1}$$

$$ENC^2_{in(f)} = \frac{C_t^2}{q^2}\left(\frac{K_f}{C_{ox}WL} \frac{(2\pi)^{\alpha_f}}{T_P^{1-\alpha_f}} Af(\alpha_f) \right) \tag{2}$$

where $C_t = C_{det} + C_f + C_g$, and C_{det} is the detector capacitance, C_f is the CSA feedback capacitance, C_g is the input transistor capacitance, q is unit electron charge, T_P is the shaper peaking time, a_w is the thermal excess noise factor, n is the sub-threshold slop coefficient, γ is dimensionless coefficient that ranges from 1/2 (weak inversion) to 2/3 (strong inversion), k is the Boltzmann constant, T is the absolute temperature, g_m is the transconductance of MOSFET transistor, K_f is the 1/f noise coefficient, C_{ox} is the gate oxide capacitance per unit area, W and L are channel width and length, respectively, α_f is the 1/f noise slope coefficient, Af (α_f) is the ENC coefficient that values depend on 1/f noise slope coefficient and shaper order.

In a low power and high speed FER system design, the contribution of 1/f noise is less than channel thermal noise in input transistor [5], and we made such assumption in the following analysis. Thus, the total noise from input transistor can be simplified as thermal noise. There is an optimum input transistor width that leads to minimum thermal noise with the input transistor working in strong inversion [5]:

$$\frac{\partial ENC^2_{in(th)}}{\partial W} = 0 \Rightarrow W_{SI} = \frac{C_{det} + C_f}{6C_{ov} + 2C_{ox}L} \quad (3)$$

For moderate inversion, the simple formula (3) is not accurate, since the equation $g_m = [2\mu C_{ox}(W/L)I_D]^{1/2}$, where I_D is the drain current, used to derive (3) is not valid in moderate inversion. Thus, we will introduce the simplified EKV model, which is quite successful in the moderate inversion. The EKV model uses a basic variable normalized forward current to indicate the inversion level of MOS transistor operating in saturation region, and equation for normalized forward current is given by [5]:

$$i_f = \frac{I_D}{2n\mu C_{ox}(W/L)V_T^2} \quad (4)$$

where V_T is the thermal voltage. It is commonly assumed that strong inversion takes place where $i_f > 10$ and weak inversion takes place in the region of $i_f < 0.1$[5]. In modern low power and deep-submicron design, the transistor usually works in moderate inversion duo to low drain current I_D and high gate oxide capacitance per unit area C_{ox}. The ratio of the transconductance to the drain current can be calculated as [5]:

$$g_m / I_D = f(i_f) / nV_T \quad (5)$$

where

$$f(i_f) = \frac{1}{\sqrt{i_f + 0.5\sqrt{i_f} + 1}} \quad (6)$$

There is a relationship exists between optimum input transistor width in the moderate inversion and strong inversion, and the relationship is given by [5]:

$$W_{opt} = W_{SI} \frac{1}{1 + A \times (W_{SI} / i_{noW})^m} \quad (7)$$

where

$$i_{noW} = W * i_f \quad (8)$$

A and m are constant and their values are $A = 0.25$, $m = 0.61$. The detailed derivation of optimum input transistor width in the moderate inversion can be found in the same paper.

III. The Proposed Current Source Transistor Sizing Methodology

The proposed methodology is to find an optimum current source transistor geometry that, the thermal and 1/f noise from current source transistor results in only a fraction of input transistor noise. The thermal and 1/f noise contributions from current source transistor calculated as [4]:

$$ENC^2_{cs(th)} = \frac{C_t^2}{q^2}\left(\frac{1}{T_P} a_w n\gamma \frac{4kT}{g_{m(in)}}\right)\left(\frac{g_{m(cs)}}{g_{m(in)}}\right) \quad (9)$$

$$ENC^2_{cs(f)} = \frac{C_t^2}{q^2}\left(\frac{K_{f(cs)}}{C_{ox}(WL)_{(cs)}}\frac{(2\pi)^{\alpha_{f(cs)}}}{T_P^{1-\alpha_{f(cs)}}} Af(\alpha_{f(cs)})\right)\left(\frac{g_{m(cs)}}{g_{m(in)}}\right)^2 \quad (10)$$

where subscript (in) and (cs) correspond to parameters of input and current source transistor, respectively. Since the total noise of input transistor is simplified as channel thermal noise in this paper, in order to minimize noise from current source transistor, the ratio of thermal and 1/f noise of current source transistor to the thermal noise of input transistor has to be as small as possible. Thus current source transistor thermal noise and 1/f noise scaling factor F_{th} and F_f are defined as below:

$$F_{th} = \frac{ENC^2_{cs(th)}}{ENC^2_{in(th)}} = \frac{g_{m(cs)}}{g_{m(in)}} = \frac{I_{D(cs)}\sqrt{i_{f(in)} + 0.5\sqrt{i_{f(in)}} + 1}}{I_{D(in)}\sqrt{i_{f(cs)} + 0.5\sqrt{i_{f(cs)}} + 1}} \quad (11)$$

$$F_f = \frac{ENC^2_{cs(f)}}{ENC^2_{in(th)}} = \frac{K_{f(cs)}(2\pi)^{\alpha_{f(cs)}} Af(\alpha_{f(cs)})T_p^{\alpha_{f(cs)}} g^2_{m(cs)}}{C_{ox}(WL)_{(cs)} a_{w(in)} n\gamma 4kT g_{m(in)}} \quad (12)$$

rearranging (11) and (12) by using (1), (5), (6), (9) and (10), an optimized current source transistor width and length relationship is derived as below:

$$(W/L)_{(cs)} = \frac{I_{D(cs)}}{2n\mu_{(cs)}C_{ox}V_T^2 i_{f(cs)}} \quad (13)$$

$$(WL)_{(cs)} = \frac{1}{F_f} \frac{K_{f(cs)}(2\pi)^{\alpha_{f(cs)}} Af(\alpha_{f(cs)}) T^{\alpha_{f(cs)}} I_{D(cs)}^2 f^2(i_{f(cs)})}{C_{ox} a_{w(in)} n^2 \gamma_{(in)} 4kT V_T I_{D(in)} f(i_{f(in)})} \quad (14)$$

IV. CSA Design Example and Noise Simulation

The proposed current source transistor sizing methodology has been adopted to a dual PMOS cascode amplifier (Fig. 2) [4]. The CSA has been designed using 130nm CMOS technology with 1.2V voltage supply. It is recommended that one has to avoid using minimum channel length in order to obtain the best noise performance [7], usually it is suggested to use 2~3 times of the minimum channel length, thus L_{min}=0.4μm is used as minimum channel length in the CSA design. For the low power consumption, a power constraint P_D=200μW is set for the CSA, which results in around 167uA drain current for input transistor. Table I and Table II list the MOSFET model parameters and the designed CSA specifications. Using (3) and (7), and with the parameters in Table I and Table II, an optimum input transistor geometry was calculated as $(W/L)_1$=74μm/0.4μm.

Based on (5) and (11), the thermal noise scaling factor F_{th} for M4 is difficult to achieve below 10% due to large amount of I_{D4}. However for M5, F_{th} can be easily remained below 10%, thus F_{th}=5% is setup for M5 in the following noise analysis. Interestingly, from hand calculation, 1/f noise scaling factor F_f for M5 is always remains below 5% as long as the

TABLE II. DESIGNED CSA SPECIFICATIONS

Parameters	Specification
Power consumption	200μW (V_{DD}=1.2V)
Input test charge (Q_{in})	10fC
Detector capacitance (C_{det})	1pF
Feedback capacitance of CSA (C_f)	100fF
Feedback resistance R_f	50MΩ
Open loop gain of the amplifier	65dB
Current source bias current for M4 (I_{D4})	155μA
Current source bias current for M5 (I_{D5})	12μA
Shaper peaking time	100ns

channel length is not lower than 0.4μm, this will be verified by further analysis from noise simulation. Next step is to calculate the geometries of current source transistors for different noise scaling factor. Based on (13) and (14), the calculated geometries of M4 and M5 are shown in Table III.

In order to verify this methodology, six amplifiers with current source transistors geometries corresponded to Table III are designed. A typical radiation detector readout channel with a CSA followed by a CR-RC pulse shaper was setup as shown in Fig. 3 [4]. To account for noise contribution is only from CSA, we made noise contributions from detector, CSA feedback resistor and shaper all equals to zero. The rms noise voltage has been measured at the shaper output, and the ENC can be calculated by following equation [4]:

$$ENC = \frac{V_{n(rms)}}{V_{s(out)}} \times \frac{Q_{in}}{q} \quad (15)$$

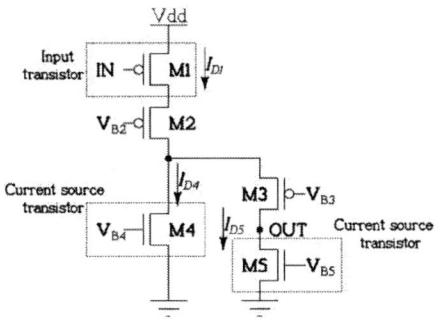

Fig. 2. The CSA based on the dual PMOS cascade structure

TABLE I. MOSFET PARAMETERS OF 130NM CMOS TECHNOLOGY USED FOR NOISE CALCULATIONS

Parameters	Specification
Transconductance K of PMOS (NMOS)	174 (405) μA/V²
Overlap capacitance C_{ov} of PMOS (NMOS)	0.432 (0.354) fF/μm
Gate oxide capacitance per unit area (C_{ox})	13.8 fF/μm²
1/f noise coefficient K_f of PMOS (NMOS)	2×10⁻²⁴ (1.5×10⁻²⁵)
Thermal excess noise a_w of PMOS (NMOS)	≈1 (1.2) [6]
ENC coefficient $Af(\alpha_f)$	0.6 [6]
1/f noise slope coefficient α_f of PMOS (NMOS)	1.2 (0.85) [7]

TABLE III. CALCULATED GEOMETRY OF CURRENT SOURCE TRANSISTORS FROM THEORETICAL NOSIE SCALING FACTOR

M5				M4					
F_{th}	W/L	W (μm)	L (μm)	F_{th}	F_f	W/L	W*L (μm²)	W (μm)	L (μm)
5%	1.96	1.2	0.6	15%	10%	1.33	1.08	1.2	0.9
5%	1.96	1.2	0.6	15%	20%	1.33	0.49	0.8	0.6
5%	1.96	1.2	0.6	15%	30%	1.33	0.34	0.7	0.5
5%	1.96	1.6	0.8	20%	10%	2.41	1.88	2.1	0.9
5%	1.96	1.6	0.8	20%	20%	2.41	0.91	1.5	0.6
5%	1.96	1.6	0.8	20%	30%	2.41	0.61	1.2	0.5

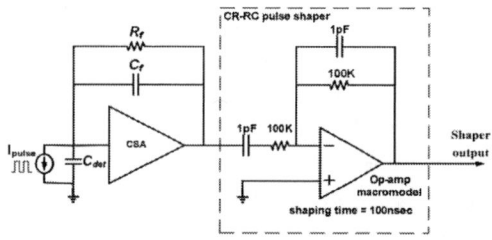

Fig. 3. The front-end readout channel noise simulation test bench [4]

where $V_{n(rms)}$ and $V_{s(out)}$ are rms noise voltage and signal amplitude at the shaper output, respectively, Q_{in} is the input charge from the detector. The calculated ENC of each transistor from noise simulation for designed six CSAs are shown in Fig. 4. In the figure, the left part is when F_{th} equals to 15% and the right part is when F_{th} equals to 20% for M4. The comparison of theoretical noise scaling and simulated results are summarized in Table IV.

It can be seen from the Table IV that, when F_{th}=15% for M4, simulated thermal noise scaling for current source transistor M5 is a little higher than theoretical scaling and simulated F_f for M5 is a little bit higher than theoretical 5% except when F_f=30% for M4. With the thermal noise scaling factor 20% for M4, the simulated F_f is higher than theoretical F_f for M4. These differences might be expected from not

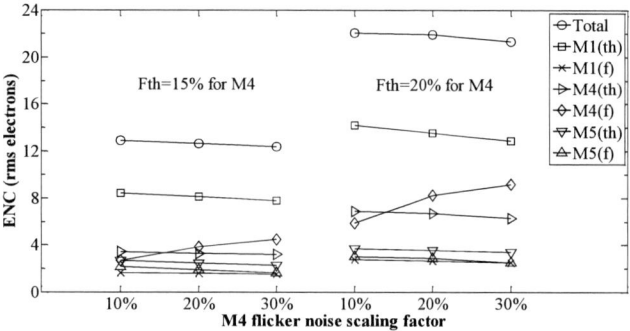

Fig. 4. Simulated ENC of M4 for flicker noise scaling factor F_f=10% to 30% and thermal noise scaling factor F_{th}=15% and F_{th}=20% with F_{th}=5% for M5

TABLE IV. COMPARISON OF THEORETICAL AND SIMULATED NOISE SCALING FACTOR

Theoretical scaling				Simulated noise scaling			
M5		M4		M5		M4	
F_{th}	F_f	F_{th}	F_f	F_{th}	F_f	F_{th}	F_f
5%	< 5%	15%	10%	9.9%	6.4%	16.5%	10.1%
5%	< 5%	15%	20%	9.0%	5.3%	16.4%	22.1%
5%	< 5%	15%	30%	8.5%	4.4%	16.9%	33.1%
5%	< 5%	20%	10%	6.7%	4.5%	23.5%	17.2%
5%	< 5%	20%	20%	6.8%	4.4%	24.6%	36.9%
5%	< 5%	20%	30%	6.9%	3.7%	24.0%	50.9%

considering the MOSFET second order effects, such as electron mobility μ and 1/f noise coefficient K_f, because EKV model assumes that electron mobility is constant in the inversion channel. However the effective mobility depends on gate bias voltage [8], and 1/f noise coefficient varies in different channel length and bias condition [7]. Despite some small differences, the simulated noise scaling factor in a good agreement with the theoretical noise scaling factor, and current source transistors M4 and M5 noise contribution is always a fraction of input transistor noise contribution.

v. Conclusion

The transistor sizing methodology for low noise CSA is developed with input transistor working in moderate inversion. As a complementation for current source transistor geometry optimization with input transistor working in strong inversion, this methodology also makes sure that current source transistors thermal noise and 1/f noise are always a fraction of input transistor noise. By adopting this methodology, a simplified EKV model was used, which is quite successful in MOSFET moderate inversion operating region, to extract the optimum current source transistors aspect ratio. Then six sets of CSAs were designed using 130nm CMOS technology to get noise contributions of each transistor from simulation. The simulation results show good agreement with current source transistor theoretical noise scaling factor. Thus, the proposed current source transistor sizing methodology can be used for designing low-noise CSA for the application of FER electronics.

References

[1] T. Noulis, S. Siskos, G. Sarrabayrouse, and L. Bary, "Advanced low-noise X-ray readout ASIC for radiation sensor interfaces," IEEE Trans. Circuits Syst. I: Regular Papers, vol. 55, pp. 1854-1862, February 2008.

[2] Y. Hu, G. Deptuch, R. Turchetta and C. Guo, "A low-noise, low-power CMOS SOI readout front-end for silicon detector leakage current compensation with capability," IEEE Trans. Circuits Syst. I: Fundamental Theory and Applications, vol. 48, August 2001.

[3] P. O'Connor and G. De Geronimo, "Prospects for charge sensitive amplifiers in scaled CMOS," Nuclear Instruments and Methods in Physics Research, vol. 480, pp. 713-725, March 2001.

[4] Ming-cheng Lin, and M. Syrzycki, "Current source transistor optimization methodology for noise optimized charge sensitive amplifier with fast shaper," 24th IEEE Canadian Conference on Electrical and Computer Engineering, CCECE 2011, Niagara Falls, ON, 2011

[5] P. Grybos, M. Idzik and P. Maj, "Noise optimization of charge amplifiers with MOS input transistors operating in moderate inversion region for short peaking times," IEEE Trans. Nuclear Science, vol. 54, pp. 555-560, June 2007

[6] G. De Geronimo, and P. O'Conner, "MOSFET optimization in deep submicron technology for charge amplifiers," IEEE Trans. Nuclear Science, vol. 52, pp. 3223-3232, December 2005.

[7] V. Re, M. Manghisoni, L. Ratti, V. Speziali, and G. Traversi, "Design criteria for low noise front-end electronics in the 0.13μm CMOS generation," Nuclear Instruments and Methods in Physics Research, vol. 568, pp. 343-349, July 2006.

[8] C. C. Enz, F. Krummenacher, and E. A. Vittoz, "An analytical MOS transistor model valid in all regions of operation and dedicated to low-voltage and low-current applications," Anlaog Integrated Circuits and Signal Processing, vol. 8, pp. 83-114, July 1995.

Fabrication of Interdigitated Microelectrodes for CuO Nanowires I-V Measurement

Tiong Teck Yaw[2], Chang Fu Dee[1], Azrul Azlan
Hamzah, Burhanuddin Yeop Majlis, Muhamad Mat
Salleh, Mohd Faizal
MEMs, Institute of Microengineeering and Nanoelectronics,
Universiti Kebangsaan Malaysia,
Bangi, Malaysia.
[1]cfdee@ukm.edu.my; [2]t_yawt05@yahoo.com

Saadah Abdul Rahman
Low Dimensional Materials Research Centre, Department
of Physics, University of Malaya,
Kuala Lumpur, Malaysia.

Abstract—**An interdigitated electrode consisted of 50 pairs of 5 µm wide microelectrodes, separated by 5 µm gap has been fabricated using single layer positive photoresist lift-off process. The effect of UV exposure time to electrode width and photoresist overcut angle were optimized for the fabrication of an array interdigitated electrodes. The fabricated interdigitated electrode has been used to measure the electrical properties of thick film and nanowires sample. The Cu_2O film and CuO nanowires have been contacted on the aluminium interdigitated electrode through the joule heating technique. Schottky behaviour based I-V characteristics of both Cu_2O film and CuO nanowires were observed at room temperature. The formation of schottky barrier on the sample has been discussed.**

Keywords—*Interdigitated microelectrodes; CuO; Cu_2O; Nanowires; Photolithography*

I. Introduction

Interdigitated electrodes (IDE) are widely used in chemical sensing [1], gas sensing [2] and biosensing [3] application. Stable and large area contact has makes it as a great contact electrode especially for nanomaterials. It has been used for dielectrophoresis align and contact on nanostructures for various sensors applications [4]–[8]. IDEs were widely fabricated through the used of photolithography method. For single layer lift-off method, negative photoresists with undercut side wall profile were usually selected for lift-off process in order for preparing IDEs. Cheaper positive resist with higher temperature resistance property also can be used for lift-off process. For positive resist, the best side wall profile that is suitable for lift-off without any surface treatment is 90° as this can prevent the formation of uneven electrode edge. In this experiment, we intended to perform the lift-off process using single layer positive resist with different overcut side wall profiles without additional surface modifications. The effects of UV exposure duration to the overcut angle of developed photoresist and the width of metal electrodes were studied. Suitable methods for the preparation of a clean, neat and clear electrode edge IDE using positive photoresist were detailed. The IDEs were used to measure I-V characteristic of Cu_2O thick film (TF) and CuO nanowires (NWs).

This work is supported by Fundamental Research Grant Scheme (FRGS/1/2013/SG06/UKM/02/1) and Funding for Higher Institutions' Centre Of Excellence (HICOE) AKU95

II. Methodology

Fig. 1 illustrated the processes involve in the fabrication of aluminium IDEs in this experiment. In this fabrication, n-type silicon wafer with orientation of 111 were used as the substrates. It has been clean with a standard wafer cleaning procedure using RCA 1 ($H_2O/NH_4OH/H_2O_2$ (5:1:1)) for organic clean, 15 minutes at 70°C followed by oxide strip process in H_2O:HF (50:1) for 15 seconds and RCA 2 ($H_2O/HCl/H_2O_2$ (6:1:1)) for ionic clean, 10 minutes at 70°C.

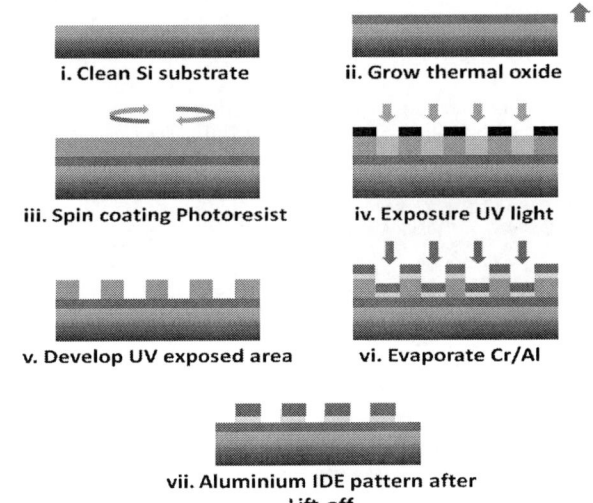

Fig.1 Process flow of the aluminium IDEs fabrication.

The thermal oxide (150 nm) was then grown on silicon substrates as an insulating layer under 1500 sccm oxygen flow at 1000°C for 3 hours. Subsequently, the substrates were spin coated with positive photoresist (Futurrex PR1-1000A) 1 µm thick at 3000 rpm for 40 seconds. The samples were pre-baked at 120°C for 120 seconds prior to the UV exposure. Next, the IDEs patterns were transfer from a chrome photomask to the samples with exposure to ultraviolet light (356 nm) under the intensity of 2.41 mW/cm². The exposure times were 10, 30, 50 and 70 seconds respectively. The UV exposed region for all samples were developed using RD 6 developer with duration fixed at 40 seconds. Cr/Al (10 nm/150 nm) contact electrodes were deposited on the samples using NanoFilm Sputtering System and Med 010 Balzers Union metal evaporator at 2.0×10^{-5} Torr respectively. Thin layer of chromium was deposited as an adhesion layer that enhanced the adhesion of aluminium electrodes on the substrates. Lift-off process has been done through two different ways: acetone spraying on the samples and combination of acetone spraying and sonication of samples in acetone. Those samples were observed by using Olympus STM 6 Microscope and JEOL JSM-6510LV Scanning Electron Microscope (SEM).

CuO NWs and Cu_2O TF were synthesized with thermal oxidation of the copper plate at 600°C, oxygen flow 1500 sccm, for 3 hours as reported [9]. During the process of CuO NWs growth with thermal oxidation, double layers comprise of Cu_2O TF and CuO NWs formed on the Cu substrate [10, 11]. The layer of CuO NWs and Cu_2O TF were taken off from the copper substrate and were placed on aluminium interdigitated electrode respectively for I-V characteristic measurement using Keithley 2400 power source. I-V measurements were done from -3 V to 3 V in dark at room temperature.

III. Results and Discussion

The side wall angles of the developed photoresist with different UV expose durations have been measured from the SEM images. The angles were measured by the yellow line in the resist side wall as shown in Fig 2 (a). Graph Fig. 2 (b) shows the effect of UV light exposure duration to photoresist side wall angle. Result shows that the photoresist side wall angle is approaching normal and getting smaller when it was exposed with higher intensity of UV light.

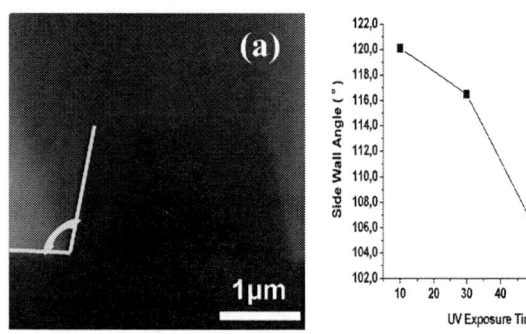

Fig. 2 (a) Overcut photoresist side wall angle. (b) Effect of UV exposure time to photoresist side wall angle graph.

The width of aluminium electrode was measured in SEM image as shown in Fig. 3 (a). Fig. 3 (b) shows the effect of UV light exposure duration to electrode's width. Samples which were exposed to higher dosage of UV light were found to have electrodes with wider width. In principle, photoresist exposed to higher UV dose will be developed faster. As we have fixed the photoresist development time for all samples at 40 seconds, the UV exposure time plays important role in deciding the electrode width and gap size from the standard width 5 µm in photomask. The gap became wider with longer UV exposure time and vice versa. As shown in graph Fig. 3 (b), the electrode's width is the same as in photomask

when it was exposed to UV light for 30 seconds. The UV light exposure dose for 30 seconds is 72.3 mJ/cm^2.

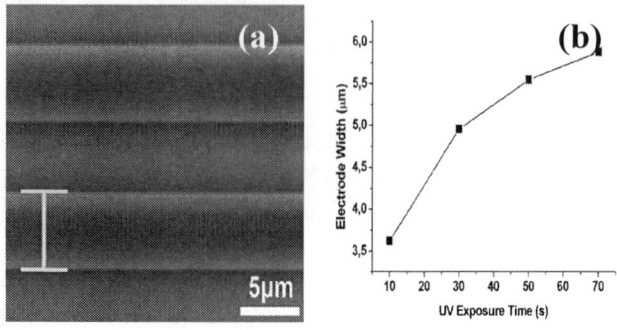

Fig. 3 (a) Electrode width measured after lift-off. (b) Effect of UV exposure time to electrode width.

Table I (a) shows the microscope images of patterned IDE photoresist after developed with RD 6 developer. All of the patterns were well developed and the gap width was found to be wider with higher exposure dose to the UV light. Table I (b) and (c) show the aluminium IDE patterns after lift-off by using acetone spray lift-off and acetone spray with sonication of sample in acetone lift-off method respectively. IDE patterns in Table I (b) show many regions with uncompleted photoresist lift-off especially for the patterns that were expose with 10 and 30 seconds of UV light. Yellow ellipse in the image in each sample indicates incomplete lift-off region of the IDE after acetone spray lift-off. As the samples were further lift-off by sonication in acetone, the lift-off process ended with clean, neat and clear electrode edge IDE as shown in Table I (c). These results show that the fabrication of IDE using positive resist with overcut profile and side wall angle higher than 90° can be done by using spray and sonication lift-off method. In this experiment, we found that when the sample is directly sonicated in acetone without prior acetone spraying, it suffers from aluminium debris contamination and it was relatively harder to remove the debris from the IDE chip.

SEM image of the fabricated array IDE chip was shown in Fig. 4 (a). A single IDE consisted of 50 pairs of 5 µm wide aluminium electrodes and separated by 5 µm gap was shown in Fig. 4 (b). The fabricated IDEs were clean from the aluminium debris and they have a sharp electrode edge after lift-off using acetone spraying followed by sonication in acetone.

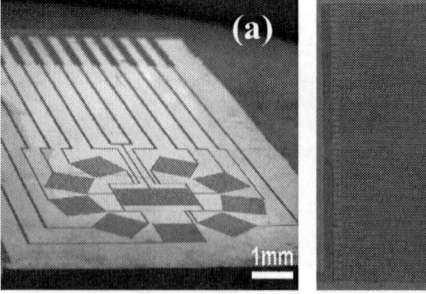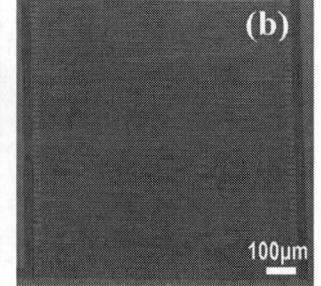

Fig. 4 (a) Array of IDE on chip. (b) Single IDE.

TABLE I. MICROSCOPE IMAGES OF IDE SAMPLES UNDER DIFFERENT UV EXPOSURE TIME AND AFTER TWO LIFT-OFF METHODS.

Expose Duration	(a) Photoresist IDE Pattern After Develop	(b) Al IDE Pattern After Spray Lift-off	(c) Al IDE Pattern After Spray + Ultrasonic Lift-off
10 s			
30 s			
50 s			
70 s			

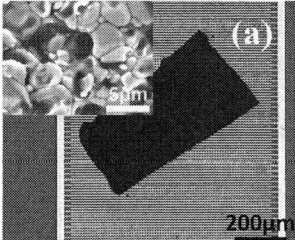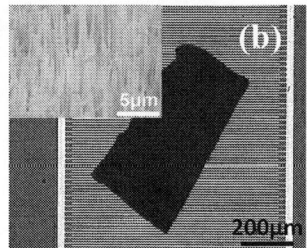

Fig. 5 (a) Cu_2O TF on IDE, inset: SEM image of the Cu_2O microstructure. (b) CuO NWs on IDE, inset: SEM image of CuO NWs.

A fabricated IDEs chip was used to measure I-V characteristic of the Cu_2O TF and CuO NWs by placing and contacting them on aluminium IDE separately as shown in Fig. 5 (a) and (b). The morphology of Cu_2O TF and CuO NWs were shown in the inset of Fig. 5 When the Cu_2O TF and CuO NWs were placed on IDE, they formed soft contacts on it. These soft contacts allow I-V characteristic to be measured on both sample, the results were labelled in red in Fig. 6 (a) and (b). The samples were measured again for second and third times. The second and third measurements shown in Fig. 6 (a) and (b) have better stability and lower resistance compared to the first measurement. It was found that after the first measurement the samples stick on the IDE well and cannot be removed easily from the IDE even with a pressurized air blow on them. This indicates a better quality of electrical contact on aluminium IDE have been formed after the first set of data has been collected. We believe, this phenomenon happened through the joule heating effect during the first measurement.

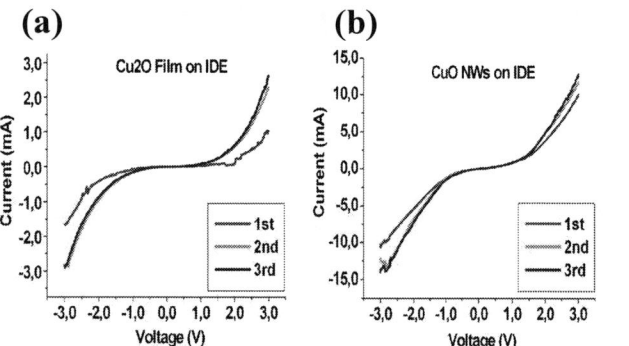

Fig. 6 I-V curve of (a) Cu_2O TF on IDE and (b) CuO NWs on IDE with three set of measurement.

Graphs in Fig. 6 indicate both Cu_2O and CuO sample exhibit schottky I-V characteristic when measured using aluminium IDE. The work function of Al, CuO and Cu_2O were 4.06-4.26 eV [12], 5.2-5.6 eV and 4.8-4.9 eV [13] respectively. Since Al has lower work function compared to Cu_2O TF and CuO NWs, theoretically it will form schottky contact with p-type Cu_2O TF and CuO NWs as displayed in Fig. 6 (a) and (b).

IV. Conclusion

In summary, a single layer Futurrex PR1-1000A photoresist has been optimized for patterning aluminium interdigitated electrode by lift-off technique. The width of the electrode can be decided with UV exposure duration. Positive photoresist with overcut side wall profile can be used for lift-off and very clean and neat patterns were retained after the process. The two steps lift-off process with acetone spraying followed by sonication in acetone has promise to a clean and standard IDE. Joule heating can be used to make better and more stable contact. The aluminium contacts for CuO NWs and Cu_2O TF show schottky behaviour.

Acknowledgment

Thanks to Ministry of Education Malaysia's (My Brain 15) schlolarship.

References

[1] A. E. Cohen and R. R. Kunz, "Large-area interdigitated array microelectrodes for electrochemical sensing," *Sensors Actuators B*, vol. 62, pp. 23–29, 2000.

[2] H. Wang, X. Wu, P. Dong, C. Wang, J. Wang, and Y. Liu, "Electrochemical Biosensor Based on Interdigitated Electrodes for Determination of Thyroid Stimulating Hormone," *Int. J. Electrochem. Sci.*, vol. 9, pp. 12–21, 2014.

[3] D. a. Brown, J.-H. Kim, H.-B. Lee, G. Fotouhi, K.-H. Lee, W. K. Liu, and J.-H. Chung, "Electric Field Guided Assembly of One-Dimensional Nanostructures for High Performance Sensors," *Sensors*, vol. 12, no. 5, pp. 5725–5751, Jan. 2012.

[4] J. Li, Y. Lu, Q. Ye, L. Delzeit, and M. Meyyappan, "A Gas Sensor Array Using Carbon Nanotubes and Microfabrication Technology," *Electrochem. Solid-State Lett.*, vol. 8, no. 11, p. H100, 2005.

[5] P. Hesketh and M. Gallivan, "The application of dielectrophoresis to nanowire sorting and assembly for sensors," in *Intelligent Control, 2005. Proceedings of the 2005 IEEE International Symposium on, Mediterrean Conference on Control and Automation*, 2005, no. 1, pp. 153 – 158.

[6] C. S. Lao, J. Liu, P. Gao, L. Zhang, D. Davidovic, R. Tummala, and Z. L. Wang, "ZnO nanobelt/nanowire schottky diodes formed by dielectrophoresis alignment across au electrodes," *Nano Lett.*, vol. 6, no. 2, pp. 263–266, 2006.

[7] G. W. Ho, "Gas Sensor with Nanostructured Oxide Semiconductor Materials," *Sci. Adv. Mater.*, vol. 3, no. 2, pp. 150–168, Apr. 2011.

[8] J. Suehiro, N. Nakagawa, S.-I. Hidaka, M. Ueda, K. Imasaka, M. Higashihata, T. Okada, and M. Hara, "Dielectrophoretic fabrication and characterization of a ZnO nanowire-based UV photosensor.," *Nanotechnology*, vol. 17, no. 10, pp. 2567–2573, May 2006.

[9] C.F. Dee, T.Y. Tiong, and M.M. Salleh, "A room temperature CuO nanowire sensor for organic volatile gases," in *proceeding of Recent Researches in Communications, Automation, Signal Processing, Nanotechnology, Astronomy and Nuclear Physics*, 2009, pp. 160–164.

[10] L. Yuan and G. Zhou, "Enhanced CuO Nanowire Formation by Thermal Oxidation of Roughened Copper," *J. Electrochem. Soc.*, vol. 159, no. 4, p. C205, 2012.

[11] G. Filipič and U. Cvelbar, "Copper oxide nanowires: a review of growth," *Nanotechnology*, vol. 23, no. 19, p. 194001, May 2012.

[12] S. Pookpanratana, R. France, R. Félix, R. Wilks, L. Weinhardt, T. Hofmann, L. T. Bismaths, S. Mulcahy, F. Kronast, T. D. Moustakas, M. Bär, and C. Heske, "Microstructure of vanadium-based contacts on n-type GaN," *J. Phys. D. Appl. Phys.*, vol. 45, no. 10, p. 105401, Mar. 2012.

[13] J.-D. Kwon, S.-H. Kwon, T.-H. Jung, K.-S. Nam, K.-B. Chung, D.-H. Kim, and J.-S. Park, "Controlled growth and properties of p-type cuprous oxide films by plasma-enhanced atomic layer deposition at low temperature," *Appl. Surf. Sci.*, vol. 285, pp. 373–379, Nov. 2013.

Spice Model Design for Carbon Nanotube Field Effect Transistor (CNTFET)

Soheli Farhana, AHM Zahirul Alam and Sheroz Khan
Department of Electrical and Computer Engineering, Faculty of Engineering
International Islamic University Malaysia, 53100 Kuala Lumpur, Malaysia

Abstract— In this paper, the design of SPICE circuit model has been discussed for Carbon Nanotube Field Effect Transistor (CNTFET) and predict device high frequency performance. An enhancement mode SPICE circuit model for nanotube transistor has been developed. A new CNTFET circuit can be developed by using this SPICE model. It can be also used to examine the performance benefits of the newly built transistor. The carbon nanotube field effect transistor SPICE model has been analyzed on the high frequency properties including the persuade of ballistic transport, kinetic inductance and quantum capacitance. The model enables device design and performance optimization for the future generation nano-electronics device modeling.

Keywords—carbon nanotube; transistor; quantum capacitance; spice model.

I. INTRODUCTION

Short channel considering dimension of device to avoid parasitic capacitance increase, economics in that devices and circuit can be processed at one time on a given wafer. A second benefit is the frequency capability of the active devices continues to increase, as intrinsic fT values increase with smaller dimensions while parasitic capacitances decrease. The channel carriers velocity is mostly responsible for the short-channel effect in the field effect transistor. The dimension scaling of the silicon made integrated circuit shows their physical limitation to be in nano scale [1]. Therefore CNTFET become popular in the next age integrity technology area due to its superior characteristics, such as semiconducting nanotube shows high carrier mobility [2], ballistic transport and nano dimensions. Single or multi layers of graphene forms a nanotube with the arrangement of honeycomb lattice that looks like a cylindrical shape. Carbon nanotube can be semiconducting or metallic depends on their roll direction which is known as chirality [3]. Low bias transport close to ballistic distances of a number of hundred nanometer can be found in nanotube where it shows very excellent performance in the nano-electronics field. Due to CNTFET's same I-V characteristics as MOSFET [4], therefore, scientists are encouraged to replace MOSFET with the CNTFET in the several applications.

CNTFETs are able to overcome most of the MOSFETs fundamental limitations as a promising new generation transistor. A ballistic or near-ballistic transport can be obtained with an intrinsic carbon nanotube (CNT) under low voltage bias to achieve the ultimate device performance [5]–

[6].

Intrinsic carrier concentration requires different sizes of CNT's electron affinities in order to the carrier continuity analysis. Electron affinities can be achieved from the CNT band-gap and graphite electron affinity. A technique to obtained the electron affinity was developed by subtracting the lowest band-gap of CNT's half value from the graphite electron affinity [7]. High electron mobility can be appeared in the carbon nanotubes. Small effective masses and low scattering rates cause the high mobility. The scattering rate increases due to electric fields increase. Few researchers are analyzed the low-field mobility [8]-[10] for the CNTFET model modeling purpose.

The remaining part of this article is arranged as: the CNTFET SPICE model is discussed in section II, that is used in this development. describes the I-V characteristics of CNTFET device is elaborated in Section III. Finally, Section IV is ended with the conclusion of this work.

II. SPICE MODEL FOR THE CARBON NANOTUBE TRANSISTOR

A. Sub-Bands Calculation, Δp

CNT sub-band's minima energy is elaborated as the following equations,

$$E(k_1, k_2) = V_{pp}\pi \sqrt{\frac{1 + 2\cos(2\pi k_1) + 2\cos(2\pi k_2)}{+ 2\cos(2\pi(k_1 - k_2))}} \quad (1)$$

where k_1 and k_2 depend on the n and m parameters,

$$k_1 = \frac{q}{N}\left(\frac{2n+m}{d_R}\right), k_2 = \frac{q}{N}\left(\frac{2m+n}{d_R}\right) \quad (2)$$

Now, $E(k_1, k_2)$ values are calculated from (1). Where m and n indicate the chiral vectors. The minima of the first, second, ..., p^{th} energy sub-bands are shown in Fig. 1. Considering a nanotube of diameter 1.95 nm for the analysis of the sub-band analysis of the single walled nanotube. From the analysis, the energies are obtained from the sub-bands are 0.4eV, 0.5eV, 0.7eV, 0.9eV, 1.1eV, 1.4eV and 1.5eV.

The authors would like to thank Ministry of Higher Education (MOHI) through FRGS research grant .

978-1-4799-5761-3/14 $31.00 © 2014 IEEE

IEEE-ICSE2014 Proc. 2014, Kuala Lumpur, Malaysia

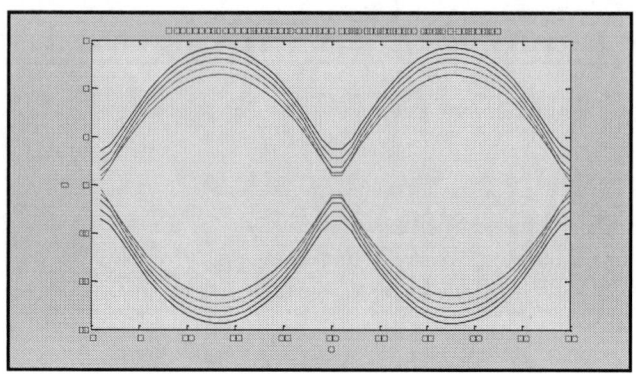

Fig. 1. Sub-band diagram for the Energy versus wave number

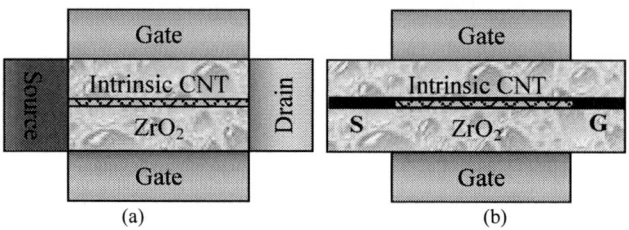

(a)   (b)

Fig. 2. Two kinds of CNTFETs: (a) Schottky barrier (SB) CNTFET (b) MOSFET-like CNTFET.

CNTFET consists with the different numbers of CNTs lies in its channel area. As shown in Fig. 2, there are basically two types of CNTFETs are available for research. They are MOSFET-like CNTFET and Schottky barrier CNTFET (SB-CNTFET). MOSFET-like CNTFET's channel is made of intrinsic semiconducting CNT. This transistor functionality depends on the theory of barrier-height modulation. Where, the channel is made of intrinsic semiconducting CNT in SB-CNTFET. The direct tunneling through the Schottky barrier (SB) at the source-channel junction is applied in the semiconducting nanotubes. The trans-conductance of this SB-CNTFET is controlled by the gate voltage due to the barrier-width is modulated by the application of gate voltage. A SPICE compatible model for CNTFET using SPICE circuit simulation is developed in this work where the parameters are verified by MATLAB simulation.

B. Model Illustrations

The flowchart to evaluate the drain current and the quantum capacitances is illustrated in Fig. 3. The aim of this SPICE model illustration is to analyze the drain current I_D with respect to the surface potential and quantum capacitance.

The key parameters used in this CNTFET model are the specific voltage ξ_i(S/D), the surface potential Ψ_S, the source (drain) Fermi level μS/D and the sub-band energy level Δp.

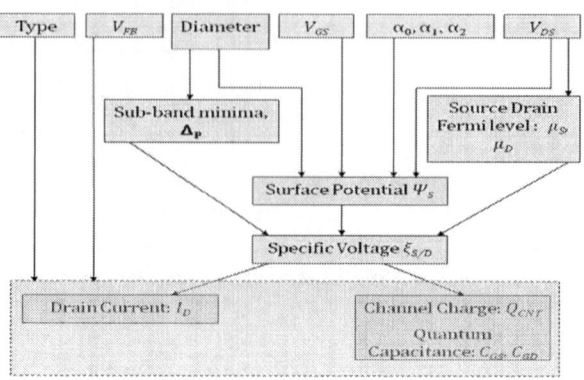

Fig. 3. CNTFET Parameters Illustration Through Flowchart.

Simple models can be developed for the MOSFET-like CNT transistor by considering that the gate-source voltage V_{GS} produces charges in the channel and also modulates the top of the source-drain energy band. When the source-drain barrier decreases, a ballistic current flows between the source and the drain. The voltage gate V_G lowers the channel potential with a certain amount Ψs and thus induces an accumulation of charges in the channel. This extra charge produces a voltage drop (V_G- Ψ_s) across the insulator as a result, which lowers the energy band by Ψ_s. therefore, by modifying V_G the band profile can be moved up or down. The total charge n_i in the CNT channel is the sum of the charges in the source and the drain, each of them being given by,

$$n_i = (1/2)\int_0^\infty D(E - E_c)f(E - E_c - \mu_i)dE \quad (3)$$

where i=S, D denotes the drain or source, respectively, D(E) is the DOS of the CNT, *f(E)* is the Farmi-Dirac distribution function, E_c is the bottom of the conduction band, and µi denote the energy levels of the source or drain.

The calculation of the DOS of CNT is represented and analyzed for simplified model [11], [15]-[22]. Here, a simplified model is used that does not account for the mixing of states. In this case the DOS near the Fermi level is, [12]

$$D_l(E) = D_0 E(E^2 - \Delta E_l^2)^{-1/2} \quad (4)$$

where $D_l(E) = 8/3\pi\gamma_0 a_{c-c}$ and ΔE_i is the minimum of the lth conduction subband (l= 1, 2, 3...), which can be written as,

$$\Delta E_l \cong (E_g/2)[6l - 3 - (-1)^l]/4 \quad (5)$$

For $\mu_S = 0$ and $\mu_D = -eV_{DS}$, denoting specific voltage,

$$\alpha_{i,l} = (\psi_S - \Delta_l - \mu_i)/k_B T \quad (6)$$

978-1-4799-5761-3/14 $31.00 © 2014 IEEE   198

We can determine the charge in the lth subband as,

$$n_{i,l} = \int_0^\infty (4k_B T / 3\pi\gamma_0 a_{C-C})\{1 + \exp[((z^2 + \Delta_l^2)^{1/2} \quad (7)$$
$$- \Delta_l)/k_B T - \alpha_i]\}^{-1/2} dz$$

The total charge on all subbands being,

$$n_{CNT} = \sum_l {}^{`}(n_{S,l} + n_{D,l}) \quad (8)$$

With $n_{i,l}$ given by (7). using a mean-value of the conduction band minimum for the first subband and dropping the contribution of other subbands, the above sum is finally reduced to two terms corresponding to the source and drain, respectively. The drain current I_D is given by,

$$I_D = (4ek_B T / h)$$
$$\times \sum_l \ln\{[1 + \exp(-\alpha_{S,i})] + \ln[1 + (-\alpha_{D,l})]\} \quad (9)$$

The basic characteristics of the transistor, $I_D(V_{GS}, V_{DS})$, is determined by repeating the above procedures for various (ψ_S, V_{DS}) points. The equivalent circuit of the CNTFET, depicted in Fig. 4, shows the CNTFET block together with the quantum capacitances and extrinsic resistances.

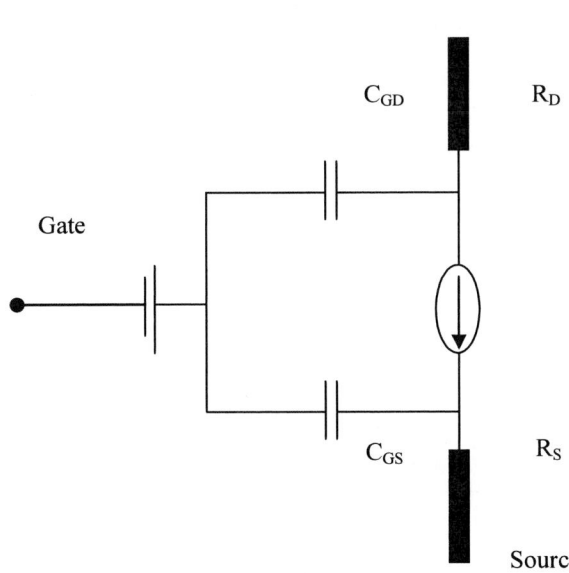

Fig. 4. The equivalent SPICE circuit of CNTFET.

III. RESULTS AND DISSCUSSION

A. CNTFET Characteristics

CNTFET's SPICE circuits model shown in Fig. 4 that has been simulated in PSPICE library. The whole calculation is then verified and simulated in MATLAB program. Fig. 5 shows the analysis of CNTFET barrier height when $\varphi_b = E_g/2$. The I-V characteristics of CNTFET is shown in Fig. 5. The results are obtained for 1.95 nm diameter CNT. In the analysis, the flat band voltage $V_{dd}/2$ is used where $V_d = V_{gs} = 0.4V$. From comparison table 1, we found a drive current $I_D = 720$ µA at 0.4 V and the ration of $I_{on}/I_{off} = 4\times10^3$. The maximum transconductance in the saturation region is gm = 1.8 mS is observed from the result..

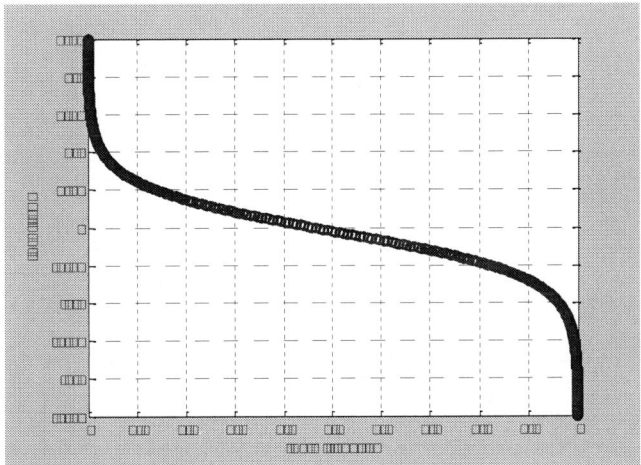

Fig. 5. CNTFET band diagram.

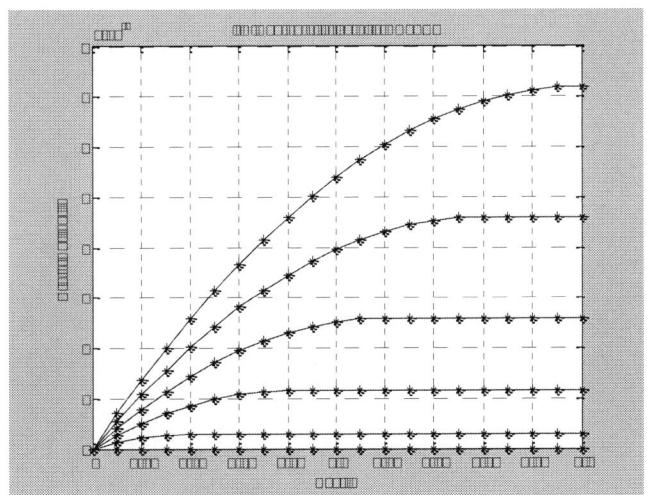

Fig. 6. I_D-V_{DS} characteristics of CNTFET with d = 1.95 nm.

TABLE I. COMPARISON BETWEEN CNTFET AND MOSFET

Parameter	CNTFET	MOSFET[14]
Channel length,L	14nm	32nm
Channel width,W	28nm	125nm
Channel area	$3.92*10^{-16}m^2$	$5.63*10^{-15}m^2$
Nanotube diameter	1.95nm	-
Chiral vector (n,m)	(25,0)	-
Transconductance,gm	1.8 mS	148µS
Gate to source capacitance	14aF	65aF
Gate to drain capacitance	14aF	37aF
Cutoff frequency	10THz	27.72 Ghz
Current density,$I_d$max	720µA	50µA
On-Off ratio	$4*10^3$	$9.54*10^5$
Subthreshold swing	73.5 mV/decade	113.67mV/decade
Drain-induced barrier lowering	54.73 mV/V	83.89mV/V

IV. CONCLUSION

A SPICE planar circuit model has been developed for CNTFET, which mainly focus on I-V characteristics of CNTFET. The developed model has been confirmed by the experimental data. A pair of analysis then analytically performed using different parameter of the CNTFET. The predisposition and possessions of these parameters on I-V characteristics have been calculated. Drain current-voltage curve from analytical equations also have been presented. Development of this model tools is essential for realizing the technology potential of the new devices.

REFERENCES

[1] International Technology Roadmap for Semiconductors (ITRS) reports, http://www.itrs.net/reports.html

[2] Hong Li, Chuan Xu, Navin Srivastava, and Kaustav Banerjee, "Carbon Nanomaterials for Next-Generation Interconnects and Passives: Physics, Status, and Prospects", IEEE Trans. Electron Devices, vol. 56, no. 9, Sep, 2009.

[3] R. Saito, G. Dresselhaus, and M. S. Dresselhaus, Physical Properties of Carbon Nanotubes. London: Imperial College Press, 1998.

[4] A. Javey, R. Tu, Farmer DB, Guo J, Gordon RG, Dai H: High performance n-type carbon nanotube field-effect transistors with chemically doped contacts. Nano Letters 2005, 5:345–348.

[5] A. Javey, J. Guo, D. B. Farmer, Q. Wang, D. Wang, R. G. Gordon, M. Lundstrom, and H. Dai, "Carbon nanotube field-effect transistors with integrated ohmic contacts and high-k gate dielectrics," Nano Lett., vol. 4, no. 3, pp. 447–450, 2004.

[6] D. Mann, A. Javey, J. Kong, Q. Wang, and H. Dai, "Ballistic transport in metallic nanotubes with reliable Pd ohmic contacts," Nano Lett., vol. 3,no. 11, pp. 1541–1544, 2003.

[7] A. Charlier, R. Setton, and M.-F. Charlier, "Energy component in a lattice of ions and dipoles: Application to the K(THF) C compounds," Phys. Rev. B, Condens. Matter, vol. 55, no. 23, pp. 15 537–43, 1997

[8] G. Pennington and N. Goldsman, "Semiclassical transport and phonon scattering on electrons in semiconducting carbon nanotubes," Phys. Rev. B, Condens. Matter, vol. 68, pp. 45 426-1–45 426-11, 2003.

[9] T. Durkop, S. A. Getty, E. Cobas, and M. S. Fuhrer, "Extraordinary mobility in semiconducting carbon nanotubes," Nano. Lett., vol. 4, pp. 35–9, 2004.

[10] T. Durkop, B. M. Kim, and M. S. Fuhrer, "Properties and applications of high-mobility semiconducting nanotubes," J. Phys. B, Condens. Matt., vol. 16, pp. R553–80, 2004.

[11] H. Hashempour, F. Lombardi, "Circuit-level modeling and detection of metallic carbon nanotube defects in carbon nanotube FETs". DATE07 2007.

[12] S. Lin, Y. Kim, F. Lombardi, Y. Lee, "A new SRAM cell design using CNTFETs", IEEE ISOCC 2008.

[13] S. Abdolahzadegan, P. Keshavarzian, K. Navi, "MVL current mode circuit design through carbon nanotube technology", European Journal of Scientific Research 2010, 152:163.

[14] A. Javey, R Tu, Farmer DB, Guo J, Gordon RG, Dai H: High performance n-type carbon nanotube field-effect transistors with chemically doped contacts. Nano Letters 2005, 5:345–348.

[15] S. Farhana, Z. Alam, S. Motakabber and S. Khan, "Development of a Spice Model for Carbon nanotube Field Effect Transistor (CNTFET)", 4th International Conference on Functional Materials & Devices 2013 (ICFMD - 2013), Penang, Malaysia, 8 - 11 April, 2013.

[16] S. Farhana, Z. Alam, S. Motakabber, and S. Khan. "Design and Development of a Simulator for Modelling Carbon Nanotube." In IOP Conference Series: Materials Science and Engineering, vol. 53, no. 1, p. 012049. IOP Publishing, 2013.

[17] S. Farhana, A. H. M. Alam, Sheroz Khan, and S. M. A. Motakabber. "Modeling of optimum chiral carbon nanotube using DFT", 13th IEEE Conference on Nanotechnology (IEEE-NANO), 2013 (2013): 853-857.

[18] S. Farhana, and A. H. M. Alam. "Environmental Friendly Device: Modeling of Carbon Nanotube with Optimum Chirality." Australian Journal of Basic & Applied Sciences 7, no. 8 (2013).

[19] S. Farhana, A. H. M. Zahirul Alam, Sheroz Khan, and S. M. A. Motakabber. "Modeling of small band-gap CNT for designing of faster switching CNTFET." In Business Engineering and Industrial Applications Colloquium (BEIAC), 2013 IEEE, pp. 589-592. IEEE, 2013.

[20] S. Farhana, A. H. M. Alam, and Sheroz Khan. "High frequency small signal modeling of CNTFET." IEEE Regional Symposium on Micro and Nanoelectronics (RSM), 2013 (2013): 364-367.

[21] S. Farhana, and AHM Zahirul Alam. "Modeling of a Carbon Nanotube Sensing Device." Middle-East Journal of Scientific Research 17, no. 10 (2013): 1475-1478.

[22] S. Farhana, A. H. M. Alam, Sheroz Khan, and S. M. A. Motakabber. "Investigation on carbon nanotube electronics structure." Pensee 75, no. 11 (2013): 123-129.

Study of Interdigitated Electrode Sensor for Lab-on-Chip Applications

Mohammadmahdi Vakilian, Burhanuddin Yeop Majlis
Institute of Microengineering and Nanoelectronics (IMEN)
Universiti Kebangsaan Malaysia (UKM)
43600 UKM Bangi, Selangor, Malaysia
Email: vakilian@eng.ukm.my

Abstract— In this paper, interdigitated electrodes (IDEs) are studied. IDEs are one of the most popular transducers, widely employed in various technological and analytical applications, especially in the field of biological and chemical sensors due to their low cost, ease of fabrication process and excellent sensitivity. The increasing demand for lab-on-chip (LoC) devices and for designing miniaturized sensors, it becomes essential to utilize of such transducers. In the current study, MEMS module of Comsol multiphysics software is used for the simulation of IDE sensor. The dependence of the IDE capacitance on the geometric parameters including IDE finger spacing, thickness, width, sensor spatial wavelength and sensitive layer thickness are discussed.

Keywords—interdigitated electrode; sensor; capacitance; lab-on-chip

I. Introduction

Interdigitated electrodes (IDEs) or interdigitated capacitors (IDCs) are the widely employed periodic electrodes configuration; since their introduction in the early 1970s, they have gained popularity and have been extensively used in various domains [1], including quality control of the curing processes, investigating the dielectric properties of the thin films [2], production of electronic parts, for instance, microwave filters [3], surface acoustic wave devices [4] and supercapacitor [5]. Apart from these, IDEs have also been broadly utilized in biological and chemical sensing, in which the changes in capacitance or impedance have been gauged based on the relationship between the analyte and a sensitive layer [6]. In the recent years, IDEs have been used in transducers for lab-on-chip (LoC) devices [7-9].

The miniaturization of electrodes and their low cost have made the planar interdigitated electrode arrays as popular framework for designing sensor device [10]. Uncomplicated and economical mass-fabrication procedure and the potential of using in a number of applications without substantial changes in the sensor design are the most significant advantages of these sensors [11]. Generally such sensors have been utilized for detecting capacitance, impedance, dielectric constant and mass conductivity in the biological medium [12]. Fundamentally, the structure includes two parallel coplanar electrodes, with periodical repetition of design (length, width,

electrode spacing). These sensors require the dimensioning to maximize the sensitivity of the biological measurement [13].

Traditionally, a capacitor comprises parallel plates, where the sensing and driving electrodes have been proximally located opposite to each other, with consistently distributed electric field. If the electrodes progressively open, then the electric field will not be constricted within a small area between the electrodes, however, it will be extended into a broader space, which generates a fringe field. However, if the electrodes open up to a coplanar plane, then the fringe field becomes prevalent between the electrodes, and this type of sensor is known as planar sensor. Fig. 1 illustrates the conversion of a parallel-plate capacitor to a planar sensor based on the fringe effect. The planar structure enables the examination of material under test (MUT) from only one side, which is especially beneficial when there is the restriction for accessing to both sides of the MUT [14, 15].

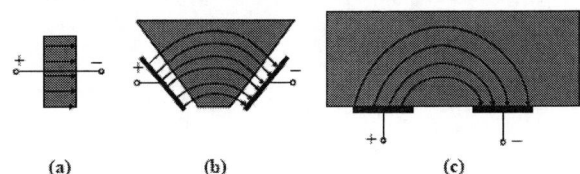

Fig. 1. Transition from a parallel-plate capacitor to a fringe field sensor. [14, 15] (a) a parallel-plate capacitor whose; (b) electrodes open up to provide; (c) one-sided access to the MUT

In this paper, an interdigitated electrode (IDE) sensor is simulated. The aim of this study is to evaluate the effects of the electrode geometry such as thickness, width, gap between electrodes, spatial wavelength and sensitive layer thickness on the overall capacitance.

II. Methodology

A layout and cross-section of an IDE with two comb electrodes have been illustrated in Fig. 2; where W is the width of the fingers and G is the width of the gaps between electrodes. Each comb electrode has been connected to a pre-set potential (+V/−V) and comprises a number of fingers with length L. The electrodes fabricated from gold have been deposited on a silicon substrate with relative permittivity of 11.9 then a sensitive layer made of SU-8 with relative permittivity of 3 has been deposited upon the electrodes. In this work, two materials with relative permittivities of 1 and 80 have been used as MUT. The reason behind this selection is to show the results with MUT relative permittivity (ε_2) less and greater than sensitive layer relative permittivity (ε_1). The spatial wavelength of electrodes (λ) has been defined [13, 16]:

$$\lambda = 2(W+G) \qquad (1)$$

The ratio of sensitive layer height (h) to electrodes spatial wavelength (λ) has been defined as layer adimensional thickness (r = h/λ). Another parameter is the metallization ratio (η) that has been defined [13, 16]:

$$\eta = \frac{W}{W+G} = \frac{2W}{\lambda} \qquad (2)$$

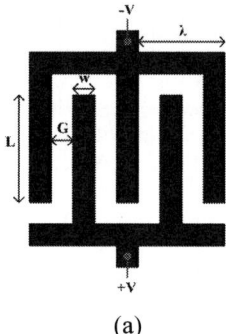

(a)

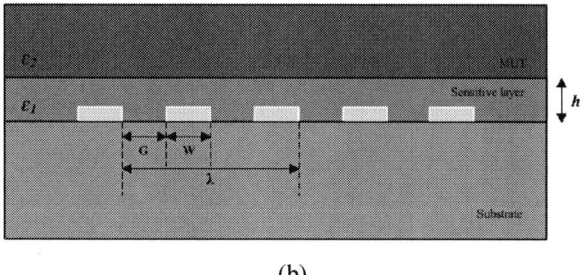

(b)

Fig. 2. (a) Layout of an IDE (b) cross section of an IDE sensor

In this research a 3D model of interdigitated electrode sensor was designed and simulated using electrostatics mode of MEMS module in Comsol multiphysics software. The effects of IDE geometry and sensitive layer thickness on the capacitance changes were investigated. In each finite element, for determining the capacitance following equations will be used [17]:

$$-\nabla \varepsilon_0 \varepsilon_r \nabla V = \rho \qquad (3)$$

$$\boldsymbol{D} = \varepsilon_0 \varepsilon_r \boldsymbol{E} \qquad (4)$$

where ρ is the charge density, E is the electric field and D is the electric displacement. The energy needed for charging a capacitor is equal to that of the electrostatic field, and is demonstrated by:

$$W_e = \int_\Omega (\boldsymbol{D}.\boldsymbol{E}) d\Omega \qquad (5)$$

The capacitance is calculated from the integral of the electric energy density by the following equation:

$$C = \frac{2}{V_p^2} \int_\Omega W_e \, d\Omega \qquad (6)$$

where V_p = 1 V is the value of the applied voltage in the port of the sensor.

III. RESULTS AND DISCUSSION

Fig. 3 shows the interdigitated electrode sensor simulated in Comsol multiphysics. In this section we have investigated the dependence of the capacitance of IDE on the geometric parameters including IDE finger spacing, thickness, width, sensor spatial wavelength and sensitive layer thickness. It is crucial to understand such features for evaluating and developing future IDE design.

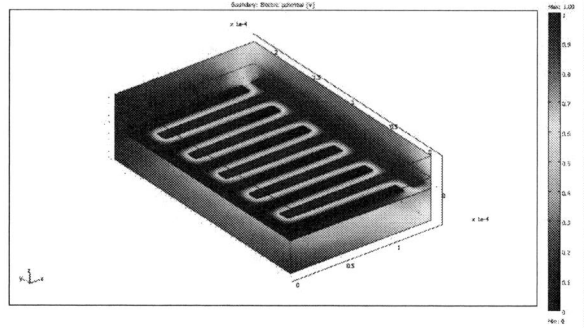

Fig. 3. Model of interdigitated electrode sensor

A. Dependence on IDE Finger Width

IDE electrodes with various widths have been computed and assessed. Fig. 4 illustrates the influence of various IDE finger widths on the total IDE capacitance. The experimental outcomes have exhibited that the overall IDE capacitance is strongly impacted by the width of IDE finger. The change of finger width from 10 μm to 25 μm has increased the overall capacitance by 50 %.

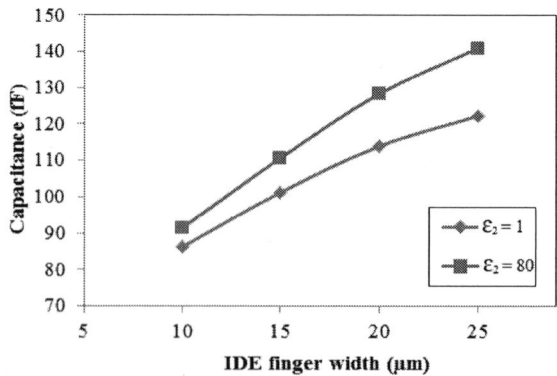

Fig. 4. Capacitance change over various IDE finger widths (L=100μm, G=10μm, T=500nm, ε_1=3, h=10μm)

B. Dependence on IDE Finger Thickness

For the purpose of investigating the importance of the finger thickness, it is essential to compute the capacitances of various thicknesses of IDE finger. As illustrated in Fig. 5, the effect of the thickness of finger on the IDE capacitance is not significant, and there will be minimal increase in the capacitance as a result of the increase in the fringing field capacitance.

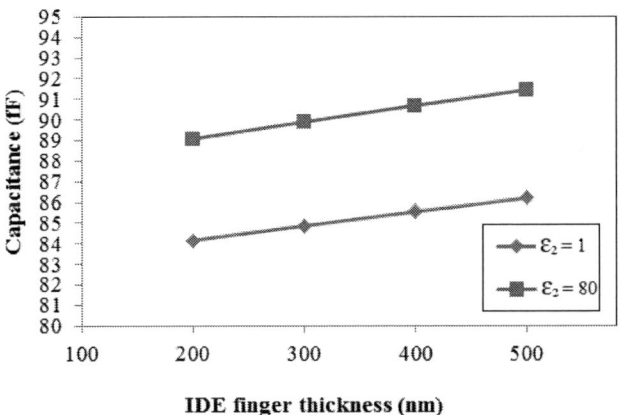

Fig. 5. Capacitance change over various IDE finger thickness (L=100μm, W=10μm, G=10μm, ε_1=3, h=10μm)

C. Dependence on IDE Finger Spacing

IDE electrodes with various spacing have been calculated and assessed. Fig. 6 illustrates the capacitance changes over various IDE finger spacing. The experimental results show a strong influence of IDE finger spacing on the overall capacitance.

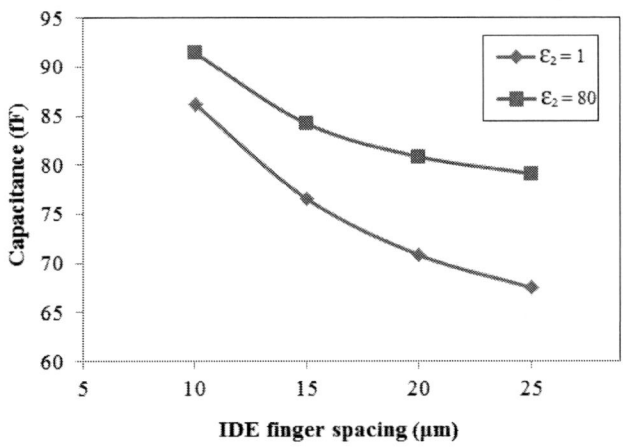

Fig.6. Capacitance change over various IDE finger spacing (L=100μm, W=10μm, T=500nm, ε_1=3, h=10μm)

D. Dependence on IDE Spatial Wavelength (λ) and Sensitive Layer Thickness (h)

The relevance among the overall IDE capacitance, thickness of sensitive layer (h) and spatial wavelength (λ) is a significant factor for designing the IDE sensor. As described in previous section r=h/λ. As depicted in Fig.7, the capacitance of IDE sensor has reached to an approximately fixed value for r ≥ 0.5. This indicates that the IDE sensor is insensitive to the distance from the IDE plane larger than half of the spatial wavelength (λ).

According to Fig.7, it is evident that, if the relative permittivity of the MUT is less than the relative permittivity of sensitive layer, the overall capacitance raises by increasing the value of r. On the other hand, the tendency for MUT relative permittivity larger than sensitive layer relative permittivity is the opposite; The reason behind this trend is that for MUT relative permittivity larger than the sensitive layer relative permittivity, the difference of capacitance component which is in series with the capacitance of MUT has a more significant role, while the internal capacitance overcomes the overall capacitance change in the other case.

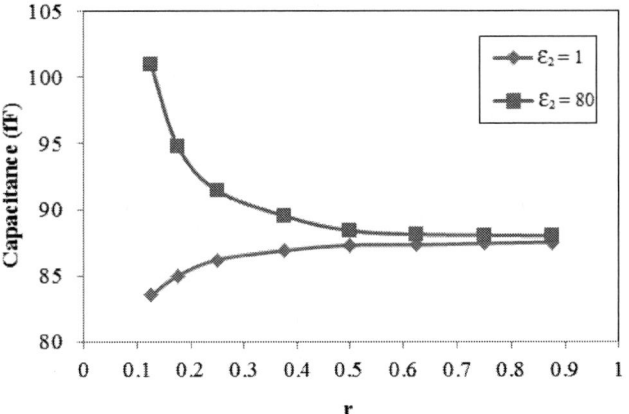

Fig.7. Capacitance as a function of the parameter r (L=100μm, W=10μm, G=10μm, T=500nm, ε_1=3)

IV. CONCLUSION

During this study, the focus has been on the interdigitated electrode (IDE) sensor. The structure of IDE definitely trumps over other structures of electrodes for designing the sensor device due to their low cost, ease of fabrication process and excellent sensitivity. An important characteristic of these sensors is the ability to utilize them over a broad range of applications without significant changes in the sensor design.

In this paper, MEMS module of Comsol multiphysics software was used for the design and simulation of IDE sensor. As it was shown electrode geometry is an important criteria for interdigitated sensor design. Simulation results showed that the effect of the IDE finger thickness on the total capacitance is not considerable. In addition, IDE finger width and spacing had strong effects on the overall capacitance. Also it could be seen that when the relative permittivity of the MUT was less than the relative permittivity of sensitive layer, the overall capacitance raised by increasing the value of r. The trend for MUT relative permittivity greater than sensitive layer relative permittivity was the opposite. The application of the results obtained from this research could be very useful in designing the future interdigitated sensors.

ACKNOWLEDGMENT

This work is supported by Ministry of Science, Technology and Innovation (MOSTI), Malaysia under the project NND/ND/(1)/TD11-002 (Development of Lab-on-Chip (LoC) for peripheral blood stem cell isolation and rapid detection of tropical diseases from blood).

REFERENCES

[1] R. Igreja and C. Dias, "Extension to the analytical model of the interdigital electrodes capacitance for a multi-layered structure," *Sensors and Actuators A: Physical,* vol. 172, pp. 392-399, 2011.

[2] Z. Chen, A. Sepúlveda, M. Ediger, and R. Richert, "Dielectric spectroscopy of thin films by dual-channel impedance measurements on differential interdigitated electrode arrays," *The European Physical Journal B,* vol. 85, pp. 1-5, 2012.

[3] B. H. Moeckly, L. S.-J. Peng, and G. M. Fischer, "Tunable HTS microwave filters using strontium titanate thin films," *Applied Superconductivity, IEEE Transactions on,* vol. 13, pp. 712-715, 2003.

[4] T.-D. Luong, V.-N. Phan, and N.-T. Nguyen, "High-throughput micromixers based on acoustic streaming induced by surface acoustic wave," *Microfluidics and nanofluidics,* vol. 10, pp. 619-625, 2011.

[5] H. E. Zainal Abidin, A. A. Hamzah, B. Y. Majlis, J. Yunas, N. Abdul Hamid, and U. Abidin, "Electrical characteristics of double stacked Ppy-PVA supercapacitor for powering biomedical MEMS devices," *Microelectronic Engineering,* vol. 111, pp. 374-378, 2013.

[6] R. Igreja and C. Dias, "Dielectric response of interdigital chemocapacitors: The role of the sensitive layer thickness," *Sensors and Actuators B: Chemical,* vol. 115, pp. 69-78, 2006.

[7] D. Dubuc, K. Grenier, M. Poupot, and J.-J. Fournie, "Broadband Microwave biosensing based on interdigitated capacitor for Lab-On-Chip applications," in *New Circuits and Systems Conference (NEWCAS), 2012 IEEE 10th International,* 2012, pp. 529-532.

[8] J. S. Shim, M. J. Rust, and C. H. Ahn, "A large area nano-gap interdigitated electrode array on a polymer substrate as a disposable nano-biosensor," *Journal of Micromechanics and Microengineering,* vol. 23, p. 035002, 2013.

[9] F. Tasdemir, S. Zihir, E. Ozeren, J. H. Niazi, A. Qureshi, S. S. Kallempudi, *et al.,* "A new lab-on-chip transmitter for the detection of proteins using RNA aptamers," in *Microwave Conference (EuMC), 2010 European,* 2010, pp. 489-492.

[10] S. C. Mukhopadhyay, "Sensing and instrumentation for a low cost intelligent sensing system," in *SICE-ICASE International Joint Conference,* 2006, pp. 1075-1080.

[11] M. Ibrahim, J. Claudel, D. Kourtiche, M. Nadi, F. Montaigne, and G. Lengaigne, "Optimization of planar interdigitated electrode array for bioimpedance spectroscopy restriction of the number of electrodes," in *Sensing Technology (ICST), 2011 Fifth International Conference on,* 2011, pp. 612-616.

[12] P. Geng, X. Zhang, W. Meng, Q. Wang, W. Zhang, L. Jin, *et al.,* "Self-assembled monolayers-based immunosensor for detection of Escherichia coli using electrochemical impedance spectroscopy," *Electrochimica Acta,* vol. 53, pp. 4663-4668, 2008.

[13] R. Igreja and C. Dias, "Analytical evaluation of the interdigital electrodes capacitance for a multi-layered structure," *Sensors and Actuators A: Physical,* vol. 112, pp. 291-301, 2004.

[14] X. Hu and W. Yang, "Planar capacitive sensors–designs and applications," *Sensor Review,* vol. 30, pp. 24-39, 2010.

[15] A. V. Mamishev, K. Sundara-Rajan, F. Yang, Y. Du, and M. Zahn, "Interdigital sensors and transducers," *Proceedings of the IEEE,* vol. 92, pp. 808-845, 2004.

[16] A. M. Sampson, E. T. K. Peterson, and I. Papautsky, "Interdigitated array microelectrode capacitive sensor for detection of paraffinophilic Mycobacteria," 2008.

[17] P. Oikonomou, G. Patsis, A. Botsialas, K. Manoli, D. Goustouridis, N. Pantazis, *et al.,* "Performance simulation, realization and evaluation of capacitive sensor arrays for the real time detection of volatile organic compounds," *Microelectronic Engineering,* vol. 88, pp. 2359-2363, 2011.

Simulation of Temperature Dependent Characterization of Charge Transport in Organic Transistor

Umar Faruk Shuib, Khairul Anuar Mohamad
Faculty of Engineering, Universiti Malaysia Sabah
Kota Kinabalu, Sabah, Malaysia
umarfarukshuib@gmail.com

Ismail Saad, Bablu Kumar Gosh, Tamer A. Tabet
Faculty of Engineering, Universiti Malaysia Sabah
Kota Kinabalu, Sabah, Malaysia

Abstract—In this paper, we present a simulation work in organic transistors using technology computer-aided design (TCAD) Sentaurus tool. A pentacene as an active semiconductor layer is used to simulate and design a top contact structure of p-channel organic transistor. The electrical characterization on these devices was investigated at temperature (T) ranging from 70 to 300 K to estimate the effects of Meyer-Neldel rule (MNR) for charge transport mobility (μ_{MN}). In the linear region of output characteristics for T = 70, 100, 140, 180, 220, 260 and 300 K, simulation showed maximum drain current (I_{ds}) for temperature of 70 K is 0.394 µA and for 300 K is 3.54 µA. The transfer characteristics exhibit threshold voltage values ranging from 18.3 to 24.6 V at 70 K to 300 K range, using IEEE1620 Standard. Then, the MNR model reveals low E_a has high μ_{MN} of 0.0093 cm^2 V^{-1} s^{-1} at T_{MN} = 208 K, while higher E_a values of 63.8 meV has lower μ_{MN} = 0.0034 cm^2 V^{-1} s^{-1}, which shows an inverse relationship between E_a and μ_{MN}.

Keywords—TCAD; organic transistor; charge transport mobility; linear mobility; Meyer-Neldel Rule model; temperature;

I. INTRODUCTION

Organic transistors have drawn much interest in organic material technology because of the capability for flat, flexible and inexpensive electronic devices as compared to inorganic semiconductors even despite the lower carrier mobility than conventional transistor as the construction can be done at room temperature, on inexpensive yet flexible plastic or other organic substrates [1]. Varied temperature research on organic transistor has been used to characterize charge transport. Multiple trapping and release (MTR) are the most common established model for charge transport in organic semiconductors because the exposed on thermally activated charge transport, which is activation energies. This method of the multiple trap and release model demanding on Arrhenius behavior for effective mobility and gaining advantage for model mobility temperature dependence in organic transistor. Moreover, an accurate extraction of transistor model parameters is benefits from modelling and circuit simulation and could be a time-saving task. Another issue to be taken into advantage is gaining high mobility device by variable temperature parameter. In this paper, the methods to extract the mobility were taken into on the linear region output characteristics of an organic transistor in the range from 70 K to 300 K.

This research was supported financially by the Ministry of Education Malaysia through Fundamental Research Grant Scheme (FRGS0306-TK-1/2012).

II. TCAD SIMULATION DESIGN

A schematic cross-section of the top contact organic transistors for the TCAD simulation using S-Device is shown in Fig. 1. The 200 nm thickness SiO$_2$ is used as an insulator and then 10 nm thickness for pentacene as an active layer. Gold is then as source and drain contacts with thickness 30 nm. The temperatures that are used for TCAD simulation are T = 70, 100, 140, 180, 220, 260 and 300 K. The device configuration summarized in Table I.

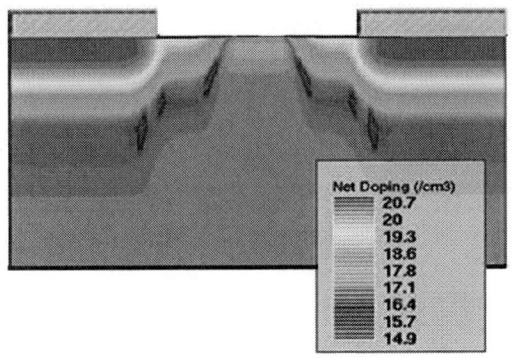

Fig. 1. Top contact organic transistor device structure.

TABLE I. DEVICE CONFIGURATION

Parameter	Value
Width	100 µm
Channel length	10 µm
Pentacene thickness	10 nm
Gold source and contact thickness	30 nm
n++ Si gate thickness	40 nm
SiO2 insulator thickness	200 nm

Poole-Frenkel like electric-field dependence, which is the inverse variation in activation energy against the square root of electric-field strength expressed in (1), has been employed for pentacene active channel [3].

$$\mu(E) = \mu_0 \exp\left[-\frac{\Delta}{kT} + \left(\frac{\beta}{kT} - \gamma\right)\sqrt{E}\right] \quad (1)$$

where $\mu(E)$ is the field dependent mobility, μ_0 is the zero field mobility, E is the electric field, Δ is the zero field activation energy, β is the electron Poole-Frenkel factor, γ is the fitting parameter, k is the Boltzman constant, and T is the temperature [3]. The setting parameters used for the TCAD simulation are summarized in Table II.

TABLE II.   PARAMETERS FOR TCAD SIMULATION

Parameter	Symbol	Value
Activation energy	Δ	0.1 eV
Poole-Frenkel factor	β	3.58×10^{-5} eV(cm/V)$^{1/2}$
Fitting parameter	γ	10^{-5} eV(cm/V)$^{1/2}$
Initial mobility	μ_0	0.03 cm^2/Vs

III. RESULT AND DISCUSSION

A. Linear Mobility Extraction Model

IEEE1620 Standard has suggested that drain current (I_{ds}) of an organic transistor would be calculated using classic theory of conventional non-organic transistors for linear region and saturation region as shown in (2) and (3), respectively [4]:

$$I_{ds} = \frac{WC_i\mu}{L}\left[\left(V_{gs} - V_{th}\right)V_{ds} - \frac{V_{ds}^2}{2}\right] \quad (2)$$

$$I_{ds} = \frac{WC_i\mu}{2L}\left(V_{gs} - V_{th}\right)^2 \quad (3)$$

where, V_{ds} is drain-source voltage, V_{gs} is gate-source voltage, W is the transistor channel width, L is the transistor channel length, μ is the field-effect mobility, V_{th} is the threshold voltage and C_i is the capacitance per unit area of the gate insulator. Equation (2) can be written as in (4) for $V_{ds} \ll (V_{gs} - V_{th})$,

$$I_{ds} = \frac{WC_i\mu}{L}\left(V_{gs} - V_{th}\right)V_{ds}, V_{gs} > V_{th} \quad (4)$$

as the transconductance is

$$g_m = \frac{\partial I_{ds}}{\partial V_{gs}} \quad (5)$$

Thus the relationship between (4) and (5), the mobility in the linear region can be described as

$$\mu_{linear} = \frac{Lg_m}{WC_iV_{ds}} \quad (6)$$

Fig. 2 shows the output characteristics of the device from TCAD simulation at 70 K and 300 K at gate voltages of -20 V, -40 V and -60 V on uniform mobility. It is observed that the behavior of the output characteristics is identical with the output characteristics of the conventional transistor, which both figure also shows linear and saturation region curves [3]. The maximum drain current, I_{ds} for temperature of 70 K is 0.394 µA as shown in Fig. 2(a) and the maximum I_{ds} for 300 K of temperature is 3.54 µA as shown in Fig. 2(b). It has been observed that I_{ds} is directly proportionate to temperature as the drain current will be increased as the temperature increase [5].

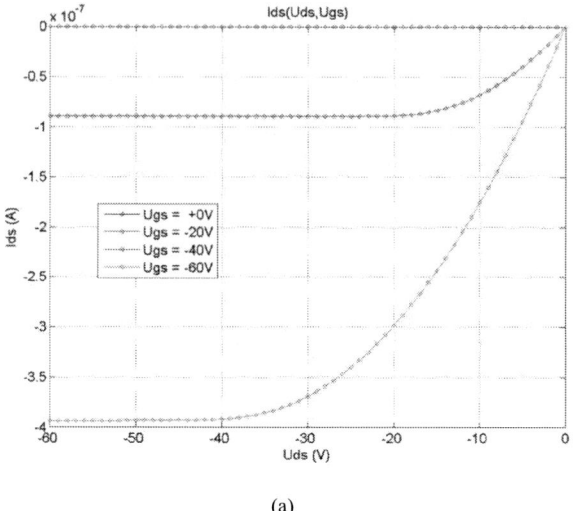

(a)

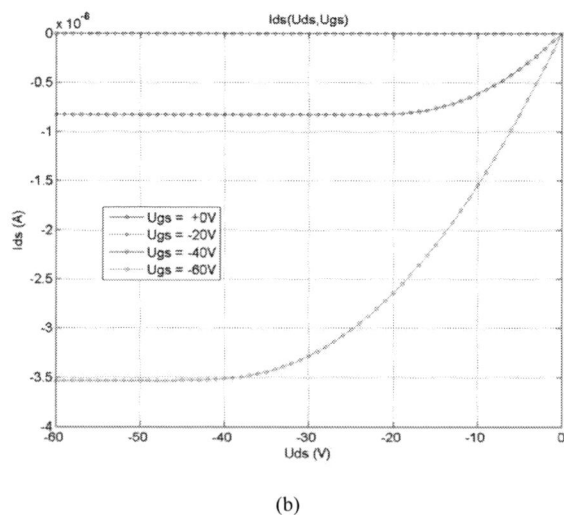

(b)

Fig. 2. The output characteristics of the OFET device shown at different gate voltages (a) at 70 K and (b) at 300 K.

In analytical model, threshold voltage is a fixed value which is calculated from physical specifications of the transistor. However, it has been shown that various parameters affect threshold voltage such as temperature and channel length, also different values for threshold voltage in linear and saturation regions of the transistor. In linear region, threshold voltage is extracted from I_{ds} vs $\sqrt{V_{gs}}$ curve [3]. In the simulation method, threshold voltage is not a given number, but it is a value which can be calculated from the results as it is done in the analytical model. In Fig. 3(a) is the transfer characteristics curve in the linear region for $T = 70, 100, 140, 180, 220, 260$ and 300 K shows threshold voltage value of 18.3 V to 24.6 V. Moreover, the values of threshold voltage for T at the saturation region are between 23.6 V and 38.1 V as obtained from Fig. 3(b).

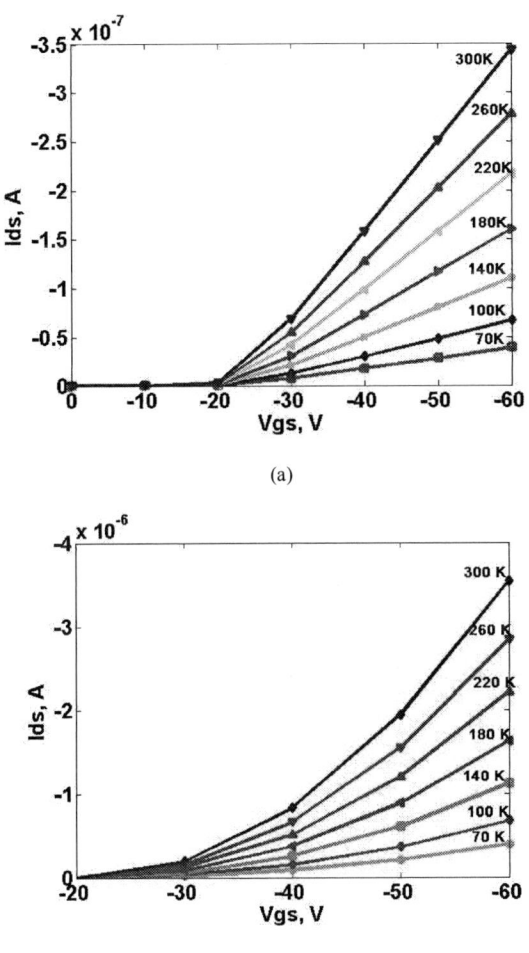

(a)

(b)

Fig. 3. The Transfer characteristics at different temperatures. (a) In the linear region for $V_{ds} = -2$ V and (b) in the saturation region for $V_{ds} = -60$ V.

B. Effective mobility towards Meyer-Neldel Rule

In order to study the behavior of thermally activated of organic transistors, the Meyer-Neldel Rule (MNR) model is widely used [6]. The model can describe the temperature dependence of the mobility that be described as

$$\mu_{linear} = \mu_0 \exp\left(\frac{-E_a}{kT}\right) \qquad (7)$$

where μ_0 is a prefactor that depends on E_a, as E_a is the variable activation energy, T is the absolute temperature, and k is Boltzmann constant. The relation between μ_0 and E_a is called the MNR as shown in (8)

$$\mu_0 = \mu_{MN} \exp\left(\frac{E_a}{kT_{MN}}\right) \qquad (8)$$

substituting (8) into (7) gives:

$$\mu_{linear} = \mu_{MN} \exp\left[-E_a\left(\frac{1}{kT} - \frac{1}{kT_{MN}}\right)\right] \qquad (9)$$

where μ_{MN} is the mobility which is independent to E_a when $T=T_{MN}$. A crossing point must be obtained (μ_{MN}, $1/T_{MN}$) if plotted μ versus $1/T$ for different V_{gs}.

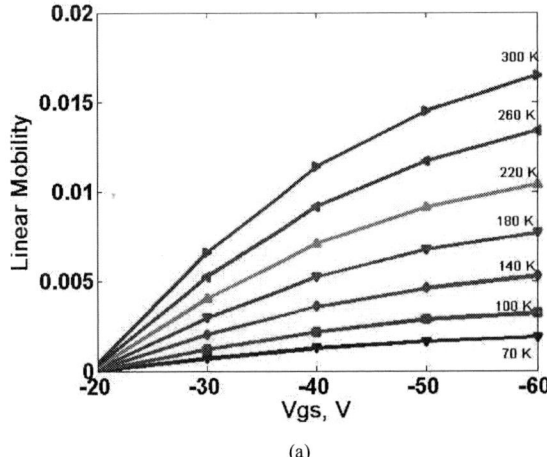

(a)

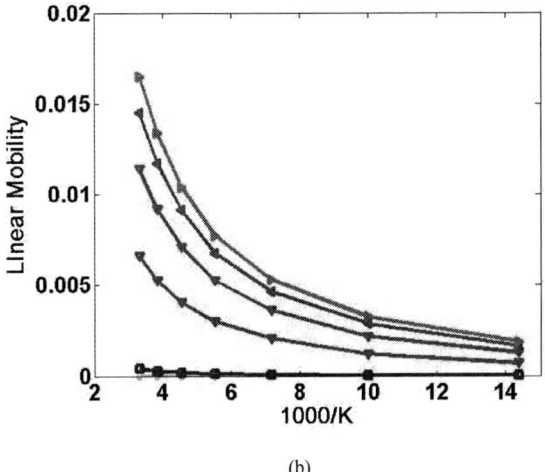

(b)

Fig. 4. (a) Mobility in the linear region versus V_{gs} at different temperatures from 70 K to 300 K. (b) Arrhenius plot of linear mobility at different gate voltages for every temperature.

In order to analyze data using the MNR describes in (9), Fig. 4(a) were converted into an Arrhenius plot as in Fig. 4(b) for different gate voltages between -10 V and -60 V. The activation energy can be calculated from the fit data point of linear mobility for each gate voltage from the slopes [6]. From the calculation, the value of $\mu_{MN} = 0.0093$ cm^2 V^{-1} s^{-1} and the $T_{MN} = 208$ K which corresponds to an $E_{MN} =1/ (k_B\ T_{MN}) = 55.8$ meV at V$_{gs}$ = -60 V. On the other hand, at V$_{gs}$ = -30 V, the $\mu_{MN} = 0.0034$ cm^2 V^{-1} s^{-1} and the $T_{MN} = 182$ K which corresponds to an $E_{MN} = 63.8$ meV, which can be explain that lower E$_a$ obtain lower mobility than higher E$_a$ value [2]. The Arrhenius plot of Fig. 4(b) can be fitted by two different lines for each V_{gs} depending on the temperature range of 70 K to 300 K with two E_a for each V_{gs} depending on the temperature range. The results are summarized in Fig. 5(a) shows that for range T from 140 − 70 K and 300 − 180 K has value of E$_a$ of 7.26 − 47.8 meV and 12.8–395 meV respectively. In context of variable temperature study, the E$_a$ value is directly proportionate with temperature as higher temperature obtains more trapping in the channel [2].

Finally the data of the fitting lines in Fig. 3(b) for certain temperature gives the prefactor μ_o as defined in (7) for each E_a. The data are summarized in Fig. 5(b). The minimum μ_o obtain is 2.47 x 10^{-4} cm^2 V^{-1} s^{-1} and the maximum μ_o is 3.56 x 10^{-3} cm^2 V^{-1} s^{-1}.

IV. CONCLUSION

Organic transistors were simulated using TCAD Sentaurus tool with Poole-Frenkel mobility model. The temperature dependence of mobility for different V_{gs} is to be found thermally activated with activation energies with the applied V_{gs} value by observations that can be explained as an inverse relationship between E$_a$ and μ_{MN}, which with higher E$_a$, gives lower μ_{MN} as stated in MNR model. Data collected assuming a trap energy of MNR model reveal low E$_a$ has high μ_{MN} of 0.0093 cm^2 V^{-1} s^{-1} at $T_{MN} = 208$ K, while higher E$_a$ values of 63.8 meV has lower $\mu_{MN} = 0.0034$ cm^2 V^{-1} s^{-1}. The free carrier mobility, $\mu_{0,}$ is found to be 2.47 x 10^{-4} cm^2 V^{-1} s^{-1} at minimum E$_a$ and 3.56 x 10^{-3} cm^2 V^{-1} s^{-1} at maximum E$_a$, reveals that lower temperature region has lower mobility than the higher temperature region. However, the effective mobility does not achieve the high mobility device benchmark as the maximum effective mobility is 0.016 cm^{2}V^{-1} s^{-1}.

ACKNOWLEDGMENT

This research was supported financially by the Ministry of Education Malaysia through Fundamental Research Grant Scheme (FRGS0306-TK-1/2012). We thank the reviewers and editor for their comments, which improved this manuscript.

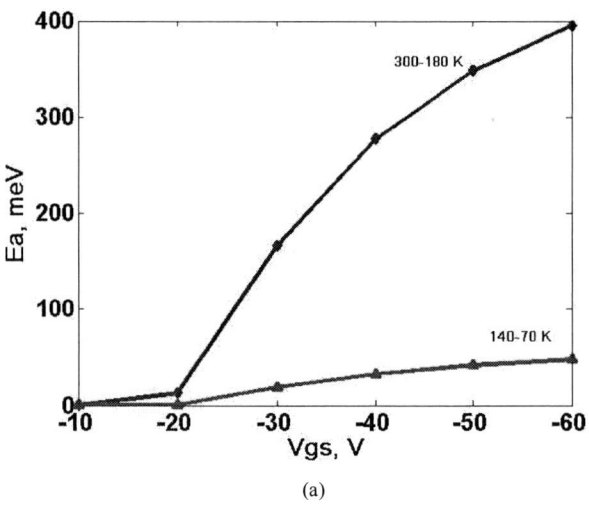

(a)

REFERENCES

[1] J. M. Shaw and P. F. Seidler, "Organic Electronics: Introduction," *IBM J. Res. Dev.*, vol. 45, no. 1, pp. 3-10, 2001.

[2] J. A. Letizia, J. Rivnay, A. Facchetti, M. A. Ratner, and T. J. Marks, "Variable temperature mobility analysis of n-channel, p-channel, and ambipolar organic field effect transistor" Adv. Funct. Mater. 2010, 20, 50-58.

[3] S. Locci, "Modeling of the physical and electrical characteristics of organic thin film transistors," Ph.D. in Electronic and Computer Engineering, Dept. of Electrical and Electronic Engineering, University of Cagliari. 2009. 12-16

[4] D. Gupta, M. Katiyar, and D. Gupta, "Mobility estimation incorporating the effects of contact resistance and gate voltage dependent mobility in top contact organic thin film transistors", Department of Materials & Metallurgical Engineering, Samtel Center for Display Technologies Indian Institute of Technology Kanpur.

[5] G. Horowitz, M.E. Hajlaoui and R. Hajlaoui, "Temperature and gate voltage dependence of hole mobility in polycrystalline oligothiophene thin film transistors" *Journal of Applied Physics*, v 87, pp. 4456-4460, 2000

[6] M. Ullah, T. B. Singh, G. J. Matt, C. Simbruner, G. Hernandz-Sosa, S. N. Sariciftci, and H. Sitter, " Temperature dependence of the charge transport in a C$_{60}$ based organic field effect transistor" Institute of Semiconductor and Solid State Physics, JKU Linz, Austria.

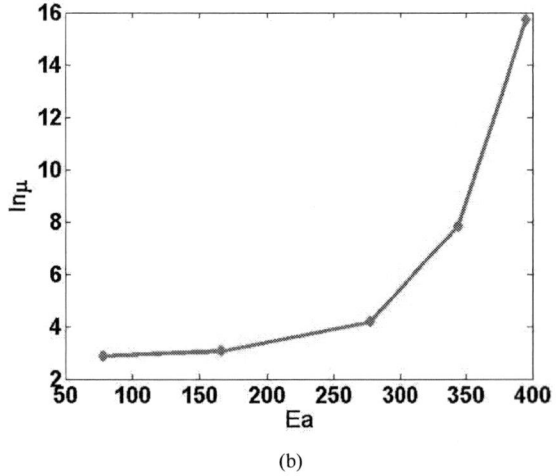

(b)

Fig. 5 (a) V_{gs} dependent E_a evaluated using the data from Fig. 3(a) in two different ranges of T. (b) Plot of μ_0 verses E_a using the data from Fig. 3(b).

Process-induced NBTI-imbalance of high-*k*/metal-gate deep-submicron CMOS

Y. Abdul Wahab, N. Soin, S. Shahabuddin
Department of Electrical Engineering
Faculty of Engineering, University of Malaya,
Kuala Lumpur, Malaysia.

H. Hussin
Center of Electronic Engineering Studies
Faculty of Electrical Engineering, Universiti
Teknologi MARA, Selangor, Malaysia.

Abstract—**Negative bias temperature instability (NBTI) is a critical reliability concern for deep-submicron high-k metal-gate p-MOSFETs. This paper reports the impact of aggressive junction-depth scaling with laser spike annealing (LSA) superactivation on NBTI-imbalance. The testbed device simulated in this work incorporates advanced process steps of shallow trench isolation (STI) with intrinsic stress, high-k metal gate, dipole charge, stress engineering with epi-SiC and epi-SiGe pockets and dual stress liner (DSL) on-state drain current improvements for gate-first CMOS processing. The results indicate that as the laser annealing decreases and dielectric thickness increases, the drain current would be further reduced. The increase of 100 ^{0}C in laser spike annealing temperature yields ~2 to 5 times leakage reduction. It is also investigated that thin HfO$_2$ layers with lower barrier energy leads to higher and faster occupation of the charged trap state during charging phase. Energy transition from E' center to neutral precursors state with lower LSA temperature leads to a lower occupation of the charged trap, thus responsible for faster neutralization. A correlation of the time dependency for the charged trap concentration, interfacial density and the drain current is observed with ~10% remarkable current degradation due to the trap accumulation. The measured ΔVth reduced for higher laser-induced high activation of the dopant thus is considered for the evaluation of the device performance.**

Keywords—*NBTI; LSA; superactivation; epi-SiGe;*

I. INTRODUCTION

The advanced of ultra-scaled MOSFETs in the deep-submicron regime with high-k dielectric and metal gate [1] demanded a continual target of design-in reliability to reduce the degrading effects by improving efficiency and accuracy [2] with aggressive junction-depth scaling for higher drive current with simultaneous stringent control of short–channel effects (SCE) [3]. Accurate prediction of lateral dopant distribution and activation has been a persistent concern for the diffusion less annealing, in addition to maximize device performance. Due to this integration of concerns, fast-ramp laser spike annealing (LSA) now receives considerable attention due to super activation of the dopants, low-resistance ultra-shallow S/D extension and silicide contact junctions; whereas the conventional spike-rapid thermal annealing (RTA) results in excessive dopant diffusion and limited electrical activation [4].

Advances in laser annealing process of Hf-based high-k metal-gate transistors with aggressive equivalent oxide thickness (EOT) scaling results significant reduction in gate

leakage current. This, however, comes with more than 30% increase in operating electric fields [5] and as a result, reliability issues such as hot carrier effect, oxide breakdown, and bias temperature instability (BTI) become critical concerns. Negative BTI (NBTI) is a PMOS-specific aging effect that increases the device's threshold voltage and reduces the carrier mobility over a period of time and under certain stress conditions. NBTI has recently drawn more attention due to the extensive role of ultrathin oxide devices.

In this work, we investigate the NBTI-induced performance on high-k/metal gate with the substantially improved short-channel performance with higher current capability of fast ramp laser spike annealing to achieve super activation. We also extensively analyze the impact of LSA variation of advanced thermal and diffusion process in the degradation process. Furthermore, we present a two-fold comprehensive understanding of NBTI degradation for effects of parametric variations of the barrier energies on the evolution of the trap states and the drain current change due to stress and the interfacial charge accumulation to evaluate the threshold voltage degradation.

II. METHODOLOGY OF GATE-FIRST ADVANCED-PROCESS

A. Device Description

The simulated samples used in this work are 6 Å EOT pMOS devices with shallow trench isolation (STI) and intrinsic stress, gate stack consisting of a thin SiO$_2$ interfacial oxide layer (IL), and with extensively varied geometry of HfO$_2$ high-k dielectric layer and AlN metal gate. A distinct process procedure of dipole charge at the high-k oxide interface to investigate the high-k degradation effect is also implemented. After the gate stack deposition, the LSA is performed under a static stress mode at elevated temperature to activate the S/D dopants and suppress transient-enhanced diffusion to describe the parameter degradation.

B. Laser Superactivation Numerical Method

The laser anneal model used can simulate the inhomogeneous thermal distribution [4,5] which declare in better accuracy for stress calculation as well as heat transfer delay in devices. The process temperature, heat boundary conditions during the ramp-up and ramp-down steps and thermal properties of the material are the significant parameters to solve the laser heat annealing equation. The proposed degradation analysis with the substantially improved short-

University Malaya High Impact Research Grant
(UM.C/HIR/MOHE/ENG/19) and (UMRG RP014B-13AET)

978-1-4799-5761-3/14 $31.00 © 2014 IEEE

channel performance with higher current capability of fast ramp laser spike annealing to achieve super activation incorporated with the defect maps generated in TCAD device simulation. The defects dispersion computed in time constants is primarily a property of the Si/SiO_2 interface. The modified scattering rate models emphasize the finite distribution of the activation energies for the Si/SiO_2, disregards the discrete Ea values. Therefore, single Ea values are substituted by a bond-dispersion energy distribution. Furthermore, the process flow incorporates with advanced stress engineering with epi-SiGe pockets; stress memorization, silicidation, and dual stress liner (DSL) to improve the on-state drain-current drive capability as evidenced by the experimental data.

III. RESULTS AND DISCUSSION

A. Input and Output Characteristics

Fig. 1 shows the variation of the drain current with gate to source voltage (Vgs) for different HfO_2 thicknesses of 32nm high-k PMOS stressed at the important formation of self-interstitial clusters in the temperature range from 600 ^{0}C to 900 ^{0}C during laser spike annealing. As the laser annealing decreases and dielectric thickness increases, the drain current would be further reduced leading to higher drain current degradation.

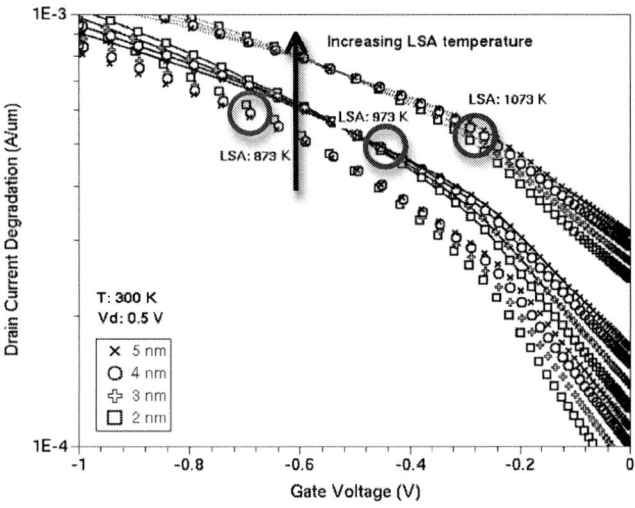

Fig. 1. Drain current degradation of different HfO_2 thickness range from 2nm to 5 nm of 32 nm high-k pMOSFET stressed at the important formation of self-interstitial clusters in the temperature range from 600 ^{0}C to 800 ^{0}C during laser spike annealing.

LSA devices show relatively higher on-state current and steeper sub threshold switching characteristics compared to thick dielectric layer devices. The high on-state current LSA-processed devices are attributed to low parasitic S/D resistance achieved by super activation of the dopants [2,3]. Fig. 2 shows the measured sub threshold characteristics for 22 nm and 32 nm gate length p-channel fabricated by the simulated process sequence at T_o= 600 ^{0}C and T_o + 200 ^{0}C laser temperatures with linearly scaled-y axis and on-current of low and high gate

voltages as presented in Fig. 2. It shows an improvement of approximately 35 % of the drain saturation current under full bias conditions of Vg 2 = 2*Vt at higher laser temperature. It is worth noting that this device performance significantly agreed with the dopant activation improvement revealed in later discussion of this work. Fig. 3 illustrates the Ioff–Ion of the effective leakage reduction due to evolution of the LSA performance ranging from T_o= 600 ^{0}C to T_o + 300 ^{0}C. The above analysis shows that the Ioff current is determined by the maximum allowed source/drain sub threshold leakage current. The maximum gate leakage current is related to the maximum source/drain leakage current at threshold and the increase of 100 ^{0}C in laser spike annealing temperature yields ~2 to 5 times leakage reduction as agreed by Hebb Jeff et al and Wang Yun et al. [6]-[8].

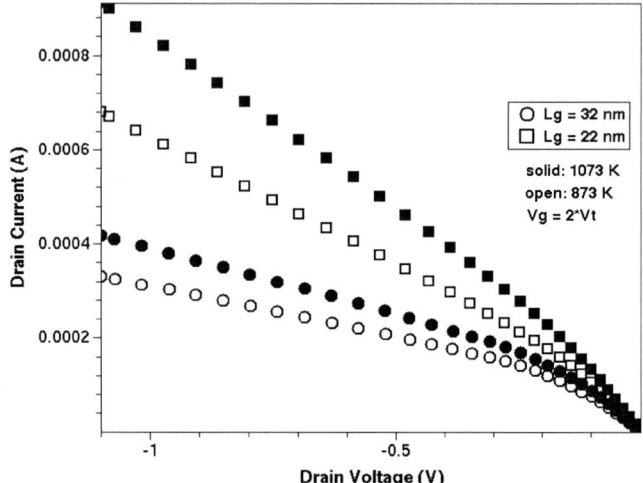

Fig. 2. Output characteristics of 22 nm and 32 nm physical gate length PMOS of linearly scaled-y axis: Laser anneal of low (open) and high (solid) under full bias conditions of Vg 2 = 2*Vt.

B. Trapping and Detrapping-NBTI

Fig. 4 presents the kinetics of the charged trap state at positively charged E' center (S2) at the lower barrier energy, Eb = 0.5 eV responsible for the transition from precursor state (S1) to (S2) state for all four different HfO_2 dielectric thicknesses. To further illustrate its significant finding, thin HfO_2 layers (2nm) with lower barrier energy leads to higher and faster occupation of the charged trap state during the stress phase, in agreement with Ribes et.al [9].

Therefore, NBTI demonstrated weak dependence on the geometry, based on oxide edge effects during discharging mechanism. This is due to missing lateral field, whereby a homogeneous damage of the active region can be seen [10]. Meanwhile a barrier energy of a transition from neutral E' center (S3) to state S1, Ea = 0.3 eV with lower LSA temperature leads to a lower occupation of the charged trap at stress phase and responsible for faster neutralization during relaxation, as shown in Fig. 5 (a) and (b).

IEEE-ICSE2014 Proc. 2014, Kuala Lumpur, Malaysia

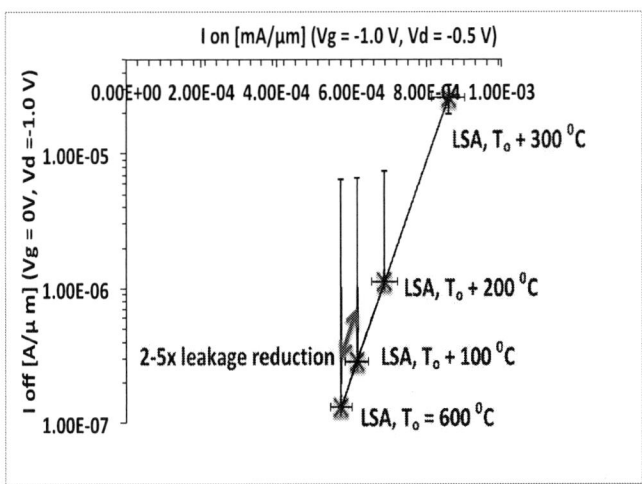

Fig. 3. Ioff-Ion evolutions for pMOSFETs with SiGe channel at 600 °C to 900 °C laser spike annealing.

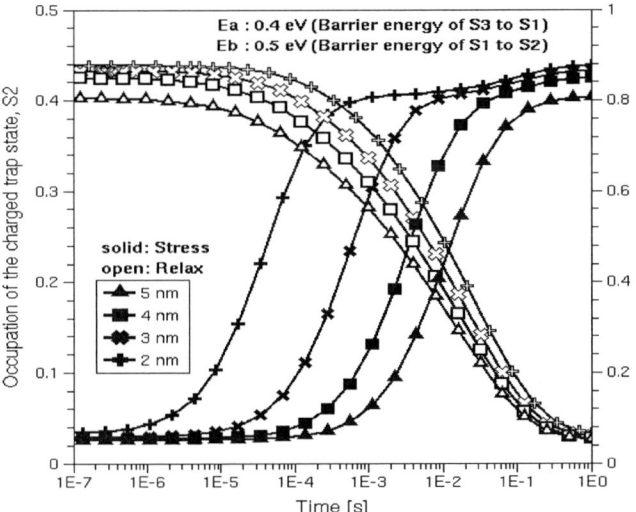

Fig. 4. Time dependency of the charged trap state S2 during the stress and relaxation stage for various thicknesses of HfO$_2$ dielectric layer.

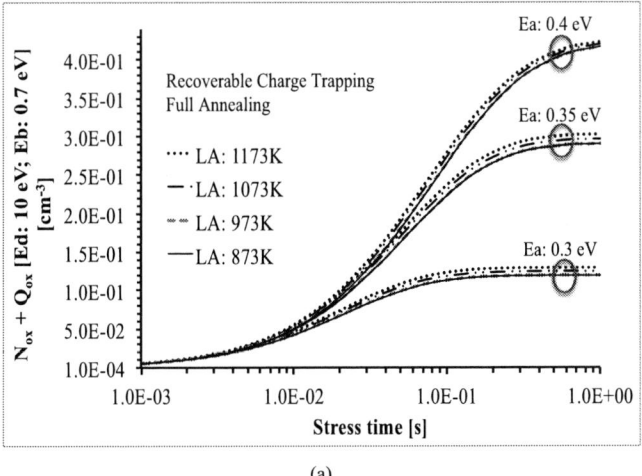

(a)

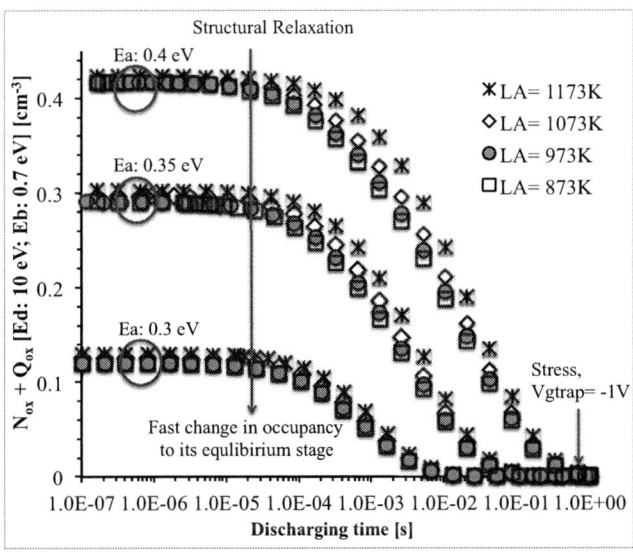

(b)

Fig. 5. Time dependency of the occupation of the charged trap of positively charged E' center for barrier energy of a transition from S3 to S1, Ea = 0.3, 0.35, and 0.4 eV and Eb =0.7 eV during (a) stress and (b) discharging phase

Fixed ΔVth activation energy Ea = 0.4 eV, a correlation of the time dependency for the charged trap concentration, interfacial density and the drain current is observed with ~10% current reduction due to the trap accumulation are shown in Fig. 6. With NBTI being a temperature-activated mechanism, the measured ΔVth reduced for higher LSA temperature as confirmed by Fig. 7.

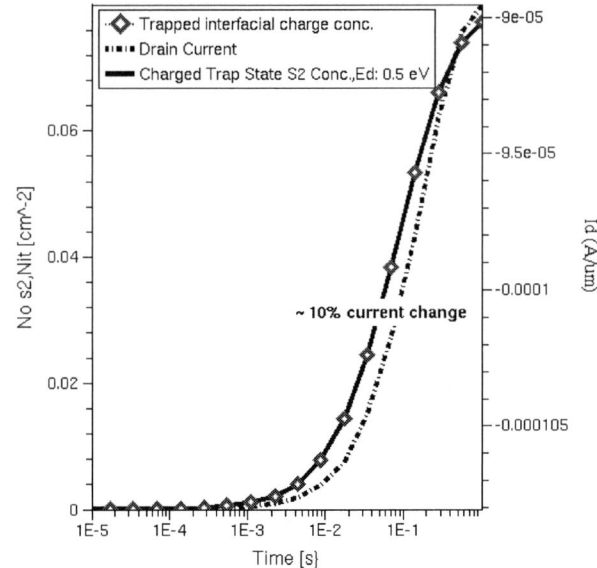

Fig. 6. Time dependency of the concentration of the charged trap at switching trap of positive E' state (S2), trapped interfacial charge concentration and the drain current at 873 K laser.

IEEE-ICSE2014 Proc. 2014, Kuala Lumpur, Malaysia

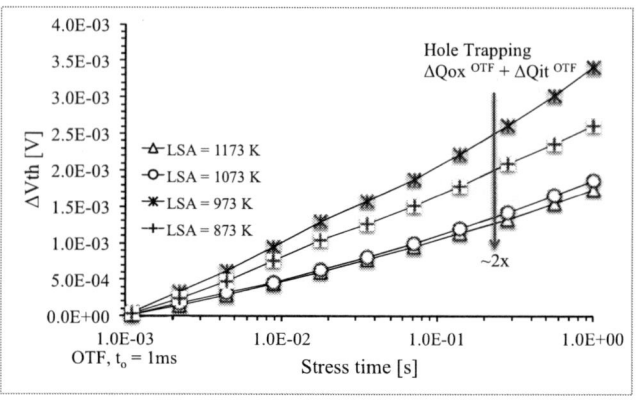

Fig. 7. Measured NBTI Δ Vth trends for different laser anneal temperatures at the super-threshold $|V_{Gstress}| = |V_{th0}|$ -0.5 V under RT.

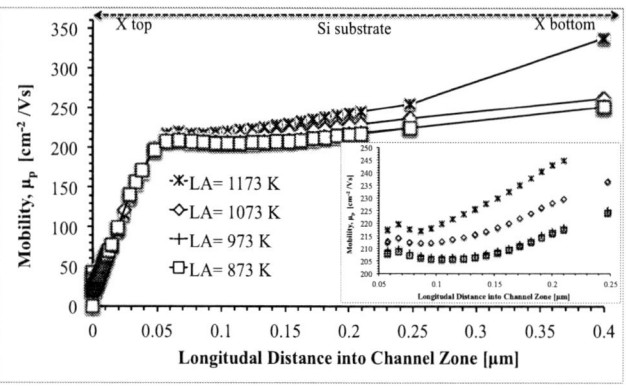

Fig. 8. Hole mobility as a function of the effective electric field for pMOSFETs with channel at 873 K to1173 K laser spike annealing.

This is due to high electrical field across the gate oxide with an elevated laser temperature leads to an electrochemical reaction in interface. The open dangling bonds act as interface states; trap the channel carriers for certain time duration. The trapped interface states shift the threshold voltage and reduce the channel hole mobility as in Fig. 8. Fig. 8 demonstrated a hole mobility as a function of the effective electric field for pMOSFETs with channel at 873 K to1173 K laser spike annealing. The peak hole mobility of the device with 900 ^{0}C (1173 K) lasers anneal is about 300 cm^2/(V·s), which is enhanced approximately 1.3 times higher compared to 800 ^{0}C (1073 K). During high laser anneal, the dangling bonds were passivized typically under hydrogen atmosphere and become an electrical inactive saturated bonds. The applied negative bias temperature stress tends to break these bonds within an electrochemical reaction [11], [12].

IV. CONCLUSION

We have reported that fast ramp laser spike annealing to achieve super activation with abrupt dopant profile enables the fabrication of high quality, improved short-channel performance with higher current capability. Excellent results are achieved for Ioff–Ion of the effective leakage reduction due to evolution of the laser performances. The LSA-induced NBTI-imbalance was highlighted, being particularly significant for drain current change in the threshold voltage degradation.

ACKNOWLEDGMENT

We would like to acknowledge the University Malaya High Impact Research Grant (UM.C/HIR/MOHE/ENG/19) and UMRG RP014B-13AET, which support this work.

REFERENCES

[1] Mistry K., et al. A 45 nm logic technology with high-k + metal gate transistors, strained silicon, 9 Cu interconnect layers, 193 nm dry patterning, and 100% Pb-free packaging. Int Electron Dev Meet Tech Dig, (2007) 247–50.

[2] ITRS 2008 (http://www.itrs.net).

[3] Hata, S. F., et al. "NBTI degradation effect on advanced-process 45nm high-k PMOSFETs with geometric and process variations." *Microelectronics Reliability* 50.9 (2010): 1283-1289.

[4] A. Pouydebasque., et.al "International Electron Devices Meeting 2005, IEDM Technical Digest, 2005, p. 679.

[5] Hard V, Dennis M, Parthasarathy C. NBTI degradation: from physical mechanisms to modeling. Microelectron Reliab 2005:1–23.

[6] T. Grasser, B. Kaczer, W. Goes, T. Aichinger, P. Hehenberger, M. Nelhiebel, "A Two-Stage   Model for Negative Bias Temperature Instability," *IEEE 47th Annual International Reliability   Physics Symposium,* pp. 33 - 44, April 2009.

[7] Hebb, J., Wang, Y., Chen, S., Shen, M., Zhou, S., Wang, X., & Owen, D. (2009, September). Expanded application space for laser spike annealing of CMOS devices. In *Advanced Thermal Processing of Semiconductors, 2009. RTP'09. 17th International Conference on* (pp. 1-7).

[8] Wang, Y., Chen, S., Shen, M., Wang, X., Zhou, S., Hebb, J., & Owen, D. (2008, May). Laser spike annealing for advanced CMOS devices. In *Junction Technology, 2008. IWJT'08. Extended Abstracts-2008 8th International workshop on* (pp. 126-130). IEEE.

[9] Ribes G, et al. Review on high-k dielectrics reliability issues. IEEE Trans Dev Mater Reliab2005;5(1): 5-19.

[10] Schlunder, C. (2009). Device reliability challenges for modern semiconductor circuit design–a review. *Advances in Radio Science*, 7(15), 201-211.

[11] Abdul Wahab, Y., A. F. Ahmad, H. Hussin, and N. Soin. "Reduction of annealed-induced wafer defects in dual-damascene copper interconnects." Microelectronics Reliability (2012).

[12] H Hussin, N Soin, MF Bukhori, YA Wahab, S Shahabuddin, New Simulation Method to Characterize the Recoverable Component of Dynamic Negative-Bias Temperature Instability in p-Channel Metal Oxide Semiconductor Field-Effect Transistors,Journal of Electronic Materials, 1-7, 2014.

978-1-4799-5761-3/14 $31.00 © 2014 IEEE

Synthesization of Nickel Nanoparticles Embedded in SU8 Polymer for Electromagnetic Actuator Membrane

Muzalifah Mohd Said, Jumril Yunas, Roer
Ekapawinto, Burhanuddin Yeop Majlis
Institute Microengineering and Nanoelectronics (IMEN)
Universiti Kebangsaan Malaysia
43600 UKM Bangi, Selangor, Malaysia
Email:jumrilyunas@ukm.my

Badariah Bais,
Faculty of Engineering and Built Environment
Universiti Kebangsaan Malaysia
43600 UKM Bangi, Selangor, Malaysia
Email: badariahbais@ukm.my

Abstract—A study of synthesization and magnetic properties of cured polymer composite on nickel (Ni) as an actuator membrane is prepared in this work. SU8 3050 polymer is used to mix with magnetic nanoparticles Ni to produce a hybrid material. The mixing process are done by using mechanical stirrer, ultrasonication and manually hand stirring. In this work an easy sol-gel method is selected to embed magnetic particles into the polymer matrix. Characterization of particle distribution is observed by metallurgical microscopy and scanning electron microscope (SEM). It is observed that the spreading of Ni particles inside the polymer chain are constant without clustering effects. Magnetization curves of each sample are traced at room temperature using vibration sample magnetometer (VSM). A good compromising magnetic properties with patternable high aspect ratio structures have been successfully fabricated using a standard MEMS process.

Keywords— actuator membrane, magnetic nanoparticles, hybrid material, magnetization curve.

I. INTRODUCTION

An actuator membrane is one of important parts in MEMS sensor and actuation system. Membrane for the actuating purpose is a membrane made from soft and elastic material which is able to have consistent bending and response to external stimulus such as heat, pressure, magnetic, electrostatic and etc. Research interest is to develop hybrid magnetic material that will be magnetically responsive, micro-machining ability and highly flexible for the electromagnetic actuator membrane. A reliable microstructure of polymer composites with magnetic properties is very desirable and gives promising potential.

There are several ways of doing material synthesization on magnetic nanoparticles such as sonochemical, sol-gel, sputtering, ball milling, mechanochemical, and polyol [1], [2]. Sol-gel method has been chosen as it proved by F.C Fonseca et.al that this method reduces the problem on material degradation because of oxidation of magnetic particles.

A study of synthesization process, magnetic properties of cured polymer nanocomposite and micromachining process of pattern transfer on polymer containing Ni particles are done in this work.

II. SYNTHESIZATION PROCESS

A. Mixing method

In this work, magnetic particles are embedded in polymer matrix using an effortless sol-gel method. The process of mixing the composite is prepared under UV-light protected condition. Ultrasonication, mechanical stirrer and manually hand stirring are the selected mixing techniques in order to attain homogeneous magnetic particles dispersion in SU8. The mixing time is fixed for 10 minutes each. The process of mixing are shown in Fig. 1.

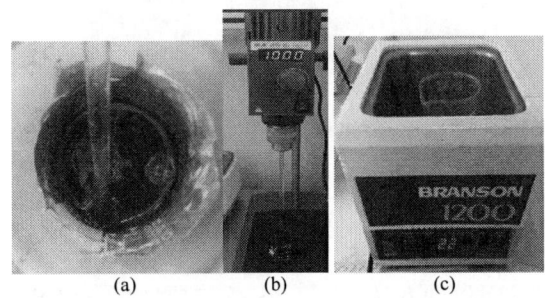

Fig. 1: Mixing process of composite (a) manually hand stirring, (b)mechanical stirrer machine, (c) ultrasonication.

B. Particles distribution

Characterization of particle distribution is observed by metallurgical microscopy and SEM. Fig. 2 shows the distribution of Ni particles in SU83050 for (a) 5% of Ni particles using manually hand stirring, (b) 4% using the mechanical stirrer and, (c) 3% using the ultrasonication stirring method .

The average size of Ni particles is 30nm. Black color indicates nickel particles. It is obviously observed that the spreading of Ni particles inside the polymer chain are constant, especially for mixing method of mechanical stirrer.

TABLE I.   PROCESS FORMULATION OF THE SAMPLES

Sample Number	SU8 Composite Formulation
1	1% Ni manually hand stirrer 3000rpm spin coat
2	2% Ni manually hand stirrer 3000rpm spin coat
3	4% Ni manually hand stirrer 3000rpm spin coat
4	5% Ni manually hand stirrer 3000rpm spin coat
5	2% Ni mechanical stirrer 1000rpm spin coat
6	2% Ni mechanical stirrer 2000rpm spin coat
7	2% Ni mechanical stirrer 3000rpm spin coat
8	4% Ni ultrasonic 3000rpm spin coat

The SEM image of the composite thickness after spin coating and curing process are shown in Fig. 3 below.

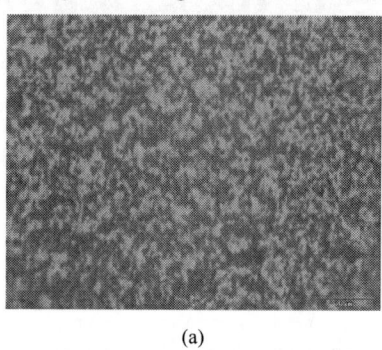

(a)

(b)

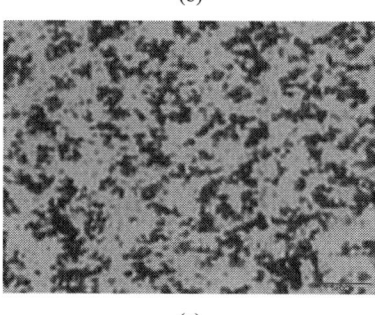

(c)

Fig. 2: Microscopic of Ni particles distribution in SU83050 using method :(a) 5% Ni manually hand stirring, (b) 4% Ni mechanical stirrer, and (c) ultrasonication .

C. Coating

After mixing process, the composite is spin coated on the silicon wafer with different   speeds for having various thickness. The spin coater speeds are a)500rpm for 10 seconds and 1000rpm for 20 seconds, b) 500rpm for 10 seconds and 2000rpm for 20 seconds, and c) 500rpm for 10 seconds and 3000rpm for 20 seconds.

Curing process of the composite is done as suggested by the polymer supplier. Study of magnetic properties of polymer nanocomposite are accomplished by using vibration sample magnetometer (VSM). Samples are mapped as in Table I.

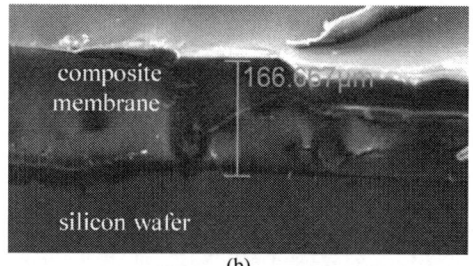

(b)

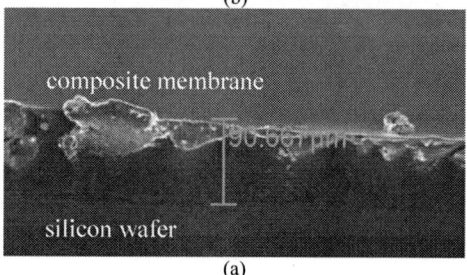

(a)

Fig. 3: SEM image on thickness of composite after spin coating and curing processes (a)1000rpm and (b)3000rpm.

The spin coater speed and residual composite thickness is as shown in Table II below.

TABLE II.   SPEED OF SPIN COATER MACHINE AND ITS CORRESPONDING MEMBRANE THICKNESS.

Spin Coater Speed (rpm)	Thickness of Membrane
1000	167μm
2000	140μm
3000	90μm

III.   MAGNETIC PROPERTIES OF POLYMER NANOCOMPOSITE

The magnetic hysteresis loop tells about the properties of magnet material such as retentivity/remanance, residual flux density, coercive force, permeability, and reluctance. The association between magnetization M, magnetic flux density B and permeability μ in magnetic material as stated by H.Sung et al [3] are;

$$B=\mu_0 M \tag{1}$$

978-1-4799-5761-3/14 $31.00 © 2014 IEEE

In B-H curve, by applying magnetic field H, B will revolve into (2).

$$B = \mu_0 (H+M) \qquad (2)$$

A. Strring Method

Magnetization curves of each sample are traced at room temperature using VSM . Fig. 4 to Fig. 7 show the hysteresis loops proved that the cured polymer composites do not block ferromagnetic performance of nanomagnetic particles immersed in it.

By referring to Fig. 4 shows magnetization curves of 2% Ni-SU8 nanocomposite is mixed using two different methods which are mechanical stirrer and manually hand stirring methods. Magnetization curve of the mechanical stirrer method gives a higher rententivity value which is 1.8408m emu which equal to 0.82T compare to hand stirring value which is 1.4947m emu and this is equal to 0.66T.

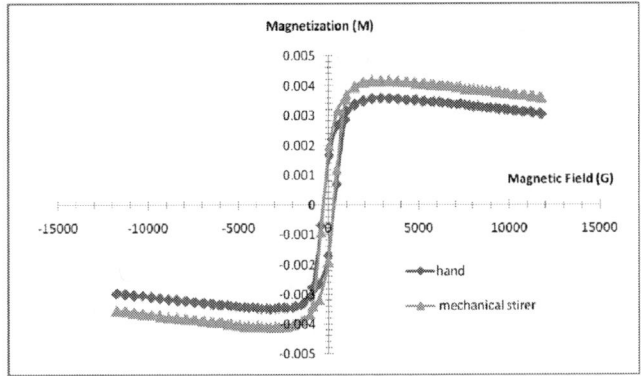

Fig. 4: Synthesization of composite using mechanical stirrer and manually hand stirring.

Graph in Fig. 5 shows hysteresis loops for 4 % Ni-SU83050 mixed using ultrasonication and manually hand stirring method. Magnetization curve of the ultrasonication mixing method gives extremely low rententivity values compare to the magnetization curve of manually hand stirring method.

Ultrasonication methods in this experiment are not directly reacting with the composite. Tab water is poured inside the ultrasonic machine and a beaker containing polymer nanocomposites is put in the water. The ultrasonic sound will vibrate the water and transmit very small vibrating wave to the composite inside the beaker. This is the reason why the ultrasonication method are not practically suitable for mixing method.

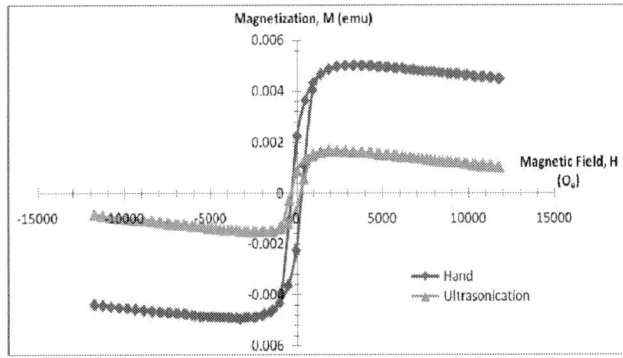

Fig. 5: Synthesization of composite using manually hand stirring and ultrasonication.

From above observation, manually hand mixing method gives a higher magnetization value compare to ultrasonication mixing method, and mechanical stirrer mixing method give the best curve result among others. This proved that using mechanical stirrer gives a higher rententivity value for the magnetic property of polymer nanocomposite. This is because of the evenly distributed Ni particles inside the SU8 matrix provide better magnetic orientation of the particles to react with each others to embrace better magnetization and reduce the experience magnetic hyperthermia cause of macrospin and coherent rotation which causes heat dissipation of magnetic nanoparticles [4][5].

B. Volume and Physical Size

Magnetization for sample 1, 2, 3 and 4 are shown in Fig. 6. The magnetization curves illustrate as a purpose of magnetic field for different percentages of Ni in SU83050. For 1% Ni shows remanance of 875.22µ emu which is equal to 0.4T. While for 5% Ni shows remanance of 2.7210m emu that is equal to 1.2T. Magnetization curves show small value as the composite hold less nickel. Coercivity value for all curves are about 300G as all composites contain the same size of magnetic material and conducted in the room temperature only. The coercivity is the value of force required to have zero magnetism in a material.

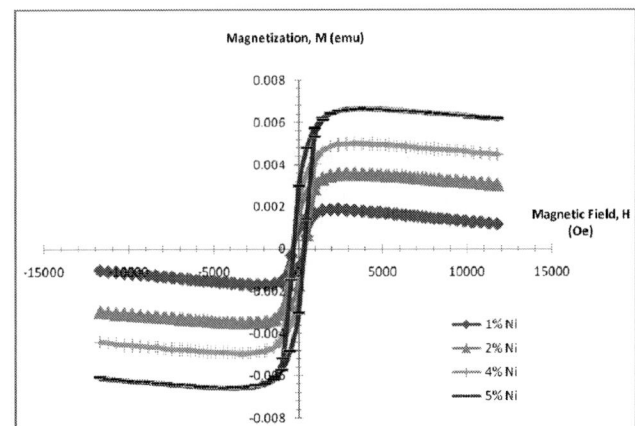

Fig. 6: The effects of nickel volume variation in SU83050.

For sample 5, 6 and 7, the composite is spin coated on the silicon wafer for different speed. Magnetization curves of all samples are shown in Fig. 7. The curves demonstrate magnetization variance for three different thickness of polymer nanocomposite membrane.

It is clearly noticed that thicker the composite membrane gives the higher magnetization curve value. For 167μm of membrane thickness shows rententivity of 2.2275m emu whereas 90μm membrane gives rententivity of 1.2152m emu. This is because, the thicker membrane composite will extend the volume of membrane, hence raise the magnetic particles restrain in the membrane. The more magnetic particles restrain, the magnetic strength of the structure will grow better. Therefore the magnetization for thicker membrane gives a higher magnetization rate.

Fig. 8: SEM image of fabricated nanocomposite cantilever on silicon wafer.

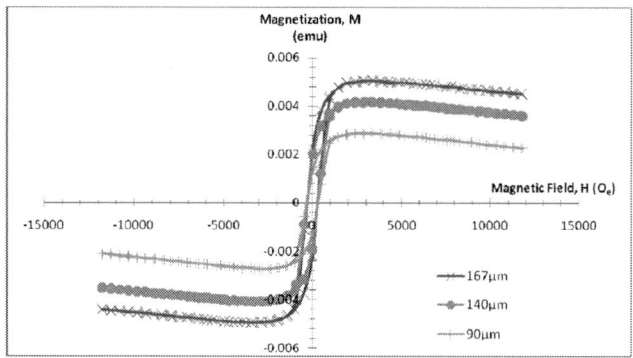

Fig. 7: Hysteresis loops of polymer composite for different thickness.

C. Pattern Transfer

Fig. 8 shows the SEM image of arrays of cantilever fabricated on the silicon wafer. Fabrication process of the polymer nanocomposite is done as recommended in the polymer data sheet provided by the supplier. This proved the potential of SU83050 polymer nanocomposite containing 2% Ni to be patterned and fabricated using standard MEMS process.

IV. CONCLUSION

In conclusion, after curing process of the hybrid material, the remanance value of Ni is around 1.2T. Changes in value depend on stirring process, thickness, and volume of Ni particles. The study has thriving of a pattern transfer experiment and measurement of hysteresis loops on membranes containing magnetic particles. The optimization of fabrication processes on a polymer nanocomposite need to be done in order to have good and precise MEMS structures. A successful of having patternable MEMS structures with consistent magnetic properties is very desirable for electromagnetic actuator especially to build a fluidic injector in Lab-On-Chip system.

ACKNOWLEDGMENT

This work is supported by the Institute of Microengineering and Nanoelectronic (IMEN), Universiti Kebangsaan Malaysia, and study funding by UTeM Malacca, Malaysia.

REFERENCES

[1] B. Cruz-franco, T. Gaudisson, S. Ammar, A. M. Bolarín-miró, F. S. De Jesús, F. Mazaleyrat, S. Nowak, G. Vázquez-victorio, R. Ortega-zempoalteca, R. Valenzuela, U. Autónoma, and M. De Reforma, "Magnetic Properties of Nanostructured Spinel Ferrites," *IEEE Trans. Magn.*, vol. 50, no. 4, pp. 2–7, 2014.

[2] F. C. Fonseca, G. F. Goya, and R. F. Jardim, "Superparamagnetism and magnetic properties of Ni nanoparticles embedded in SiO 2," *Phys. Rev. B*, vol. 66, pp. 1–5, 2002.

[3] H. W. F. Sung and C. Rudowicz, "A closer look at the hysteresis loop for ferromagnets - A survey of misconceptions and misinterpretations in textbooks," 1999.

[4] H. Mamiya and B. Jeyadevan, "Magnetic Hysteresis Loop in a Superparamagnetic State," *IEEE Trans. Magn.*, vol. 50, no. 1, pp. 1–4, Jan. 2014.

[5] A. M. Hyperthermia and R. Theories, "Hysteresis Loss-Induced Temperature in Ferromagnetic Nanoparticle," *IEEE Trans. Magn.*, vol. 50, no. 1, pp. 3–6, 2014.

Characterization of MEMS Structure on Silicon Wafer using KrF Excimer Laser Micromachining

M. Mazalan, S. Johari , B. P. Ng and Y. Wahab

Advanced Multi-disciplinary MEMS-based Integrated Electronics NCER Centre of Excellence (AMBIENCE)
School of Microelectronic Engineering, Universiti Malaysia Perlis (UniMAP)
Perlis, Malaysia
mazleemazalan@unimap.edu.my

Abstract— This paper presents preliminary parametric studies of KrF laser micromachining ablation effects on Silicon. Four parameters are studied, namely laser energy, pulse rate, number of laser pulses, and Rectangular Variable Aperture (RVA) in X and Y direction. At present, the study is focused on the production of microchannels using laser micromachine, in which its dimension is examined and measured. We found that the number of laser pulse is non-linearly proportional with the ablated channel width, with the etching rate of approximately 1 to 5 um for 50 laser pulses. This is similar with the measured depth of the microchannel. The changes in the measured channel width are most significant when the laser energy is increased. Some debris and recast can also be observed around the edge of the microchannel particularly during the variation of the laser pulse frequency. When varying the RVA, it is observed that the surfaces of the ablated microchannels are not smooth with a lot of debris accumulated at the channel edge and a few discolorations. Finally, a microcantilever structure is fabricated with the aim of demonstrating the capability of the laser micromachine.

Keywords— *Silicon (Si); Laser Micromachining*;

I. Introduction

The conventional method of realizing MEMS structures is using micro/nano fabrication technique which involves additive manufacturing process. In this process, successive layer of photoresist, oxide layer and metal layer, for example are deposited on a bare substrate. This is time consuming and costly as it involves several deposition and etching process as well as various chemicals. In addition, these methods are inherently restricted to two-dimensional (2D) fabrication and also to selective materials. Existing MEMS technology has been widely accomplished via microfabrication process that typically use lithography, with most MEMS devices applied in biomedical areas. For instance, Zee at al. [1] have developed a polymer-based gas sensors, which is used to help diagnose patients health since it is known that certain diseases cause the body to generate specific gases. The fabrication of the sensor arrays involved large volume of solvent which was required for deposition of the polymer. An additional removable mask was also needed to selectively deposit each polymer into a specific area. This technique also has difficulty forming sub-millimeter sensors due to poor adhesion to the

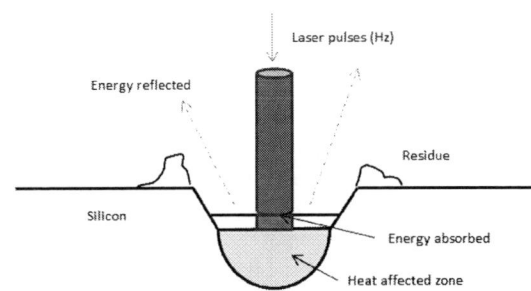

Fig 1. Laser-material interaction [3].

substrate when the mask was removed. In a different example, Wu et al [2] integrated four different processes namely DRIE etching, trench-filled molding, polysilicon layer deposition and bulk releasing in order to enhance the performance of several MEMS devices. While this work was capable of producing high performance devices, the need to combine four different processes is highly time and cost consuming.

To overcome these drawbacks, this research aims to characterize the use of laser micromachine as an alternative method for the development of MEMS devices towards biomedical application. The laser micromachining process involves laser-material interaction, where when laser radiation interacts with the targeted substrate/material (Fig 1), the substrate will absorb the laser energy. The absorbed laser energy is converted to lattice vibration energy (thermal) resulting in the temperature rise and heating of the targeted surface. This will eventually melt and vaporize the targeted material.

Laser micromachining processes propose distinctive capabilities in terms of materials flexibility, three dimensional devices processing, cost-efficient and less environmental impact than many existing technologies. By using a focused laser for direct patterns writing, the needs for photomasks can be avoided. The prototyping of new materials is also straightforward using laser etch process, compared to the effort and additional solvents required during deposition and etching process. In this paper, preliminary studies on the parameters of 248 nm KrF excimer laser micromachine ablation effects in producing microchannel on silicon wafer are presented. There are four parameters involve, namely laser energy, number of laser pulses, frequency and Rectangular Variable Aperture (RVA) in X and Y direction. Laser energy

978-1-4799-5761-3/14 $31.00 © 2014 IEEE

is directly related to the volume of the excimer laser gas and the   charging voltage, number of laser pulses (N) is the number of laser pulses exposed onto the materials surface leading to ablation effect, frequency is the number of laser pulses per second, and RVA is the size of the laser beam measured in x and y direction. The physical deformation and ablated wall side is observed and measured by using high power microscope and scanning electron microscope (SEM). The data collected is plotted to obtain the relationship between the laser parameters and the ablation effects. At present, the study is more focused on the dimension of the microchannel rather than the surface roughness. A demonstration of 3D-micromachining of a cantilever by using optimized parameters obtained in this characterization study is performed.

II. Results and Discussion

Initially, the effects of the number of laser pulses variation are observed by measuring the width and the depth of the microchannels using high power optical microscope. Five number of laser pulse of 50, 100, 150, 200 and 250 are considered. The relationships between the number of laser pulse and the width and depth of the microchannels are plotted in Fig. 2. For 50 laser pulses, the resulting width is measured to be 252.78 um. The width increases by 1 micron when the number of laser pulse is increased to 100. As the number of pulse is increased to 150, the measured width is 258.28 um, 5 micron exceeding the former width measurement. For 200 and 250 laser pulses, the measured width is 262.90 um and 263.34 um respectively. The number of laser pulse is non-linearly proportional with the ablated channel width, with the etching rate of approximately 1 to 5 um for 50 laser pulses. This is similar with the measured depth of the microchannel with the etch rate of 1 to 4 um. Fig 3 shows the top view of the microchannel with varying number of laser pulses. The edge of the microchannel is not really smooth. At present, the reason for this is still an ongoing investigation. The discoloration between the channel and the substrate is probably due to the heat diffusion into the surrounding materials.

We then investigated the effects of the laser energy variation on the width and depth of the microchannel. During this process, the number of laser pulses is set to 300. The minimum laser energy applied is 3.6 mJ resulting in the measured width of 234.30 um, 20 micron smaller than the width measured during variation of the laser pulse number. As the energy is doubled, the width increased by ~ 4 um. At the laser energy of 10.8 mJ, the measured width is 239.96 um. For the laser energy of 14.4 mJ and 18 mJ, the measured width is 245.30 um and 266.42 um in that order. The changes in the measured width are significant as the energy is increased, as plotted in the graph in Fig.4. This behaviour is expected, as higher laser powers accumulate more heat and the focused beam has energy exceeding the machining threshold as reported by [4]. In principle, all decomposition or material removal from a solid target is the consequence of an energy input into the target, resulting in overcoming the solid's binding energy or the machining threshold of the material. This can also be seen at the measured depth where the depth increases from 19.52 um to 48.3 um when the laser energy is

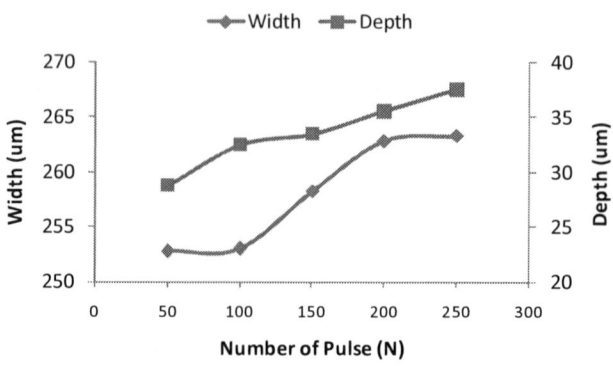

Fig 2. Measured width and depth as a function of the number of laser pulse.

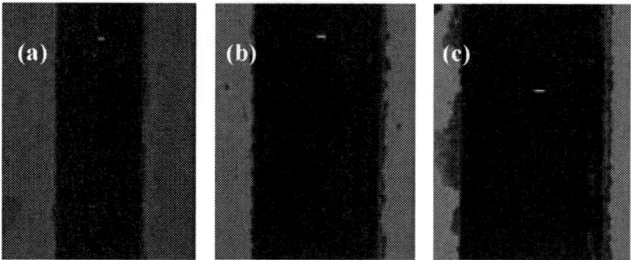

Fig 3. Optical micrographs of microchannel's top view for varying number of laser pulse of (a) 50, (b) 150 and (c) 250.

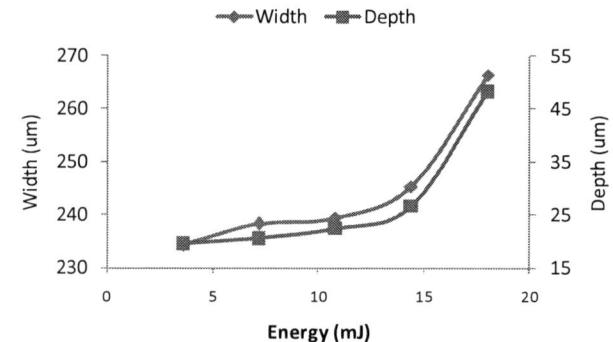

Fig 4. The dependance of the Si channel width and depth produced by varying laser energy.

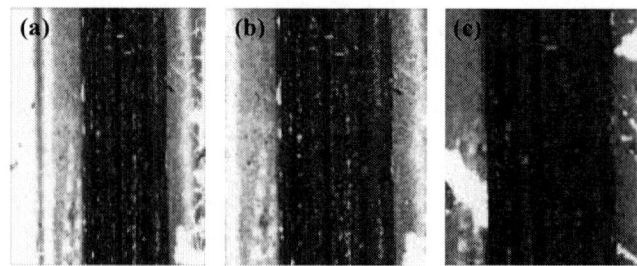

Fig 5. Optical images of top view of the microchannel captured at (a) 3.6 mJ, (b) 10.8 mJ and (c) 18 mJ.

increased from 3.6mJ to 18 mJ. The edge of the channels is also observed to be rough and jagged as observed in Fig 5. Fig 6 shows the dependence of the microchannel width and depth on the laser frequency (the number of laser pulses per second). At the frequency of 100 Hz, the width and the depth is measured to be 244.42 um and 43.15 um respectively. To our surprise, the measured width decreases by only 1 micron as the frequency is increased from 100 Hz to 500 Hz with the step of 100 Hz. The measured depth however fluctuates from 43.15 um to 36.3 um when the frequency is increased. Based on Fig 7, some debris and recast can be observed around the edge of the microchannel. Debris can be defined as materials produced during the vaporization of the targeted material which fall back onto the target surface and accumulated at the surrounding of the ablation area. According to [5], the debris can be reduced by cleaning the ablated area immediately after ablation or using gas jets. This however will increase the processing time of the laser micromachining. Another proposed method to eliminate debris is to deposit photoresist on the ablated area before machining. The resist has to be removed afterwards using photoresist developer.

The final parameter investigated in this study is the effect from the variation of the Rectangular Variable Aperture (RVA). RVA can also be referred as a simple beam shaping device, which contains horizontal and vertical blades that can be independently adjusted to form rectangular apertures in both x- and y- direction. Independent adjustments give the micromachine the advantageous ability to adjust the aperture center to the most intense portion of the beam. We varied the RVA in both x- and y- direction from 0.5 to 10 mm and measured the width and depth of the microchannel as plotted in Fig 8. The most significant changes observed are the width of the channel which increased from 37 um to 302.72 um when RVA-X is increased from 0.5 mm to 10 mm. The depth also doubled from 18 to 36 um for the minimum and maximum valued of RVA-X. For RVA-Y, the width and depth increases by 22% and 32% respectively as the RVA dimension is increased from 0.5mm to 10mm. The top view of the microchannel captured at different values of RVA-X and RVA-Y are depicted in Fig 9 and Fig 10. The surfaces of the ablated microchannels are not smooth with a lot of debris accumulated at the channel edge and a few discolorations.

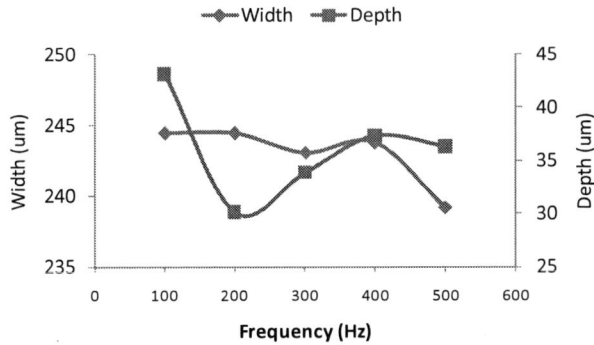

Fig 6. Cutting width and depth versus laser frequency.

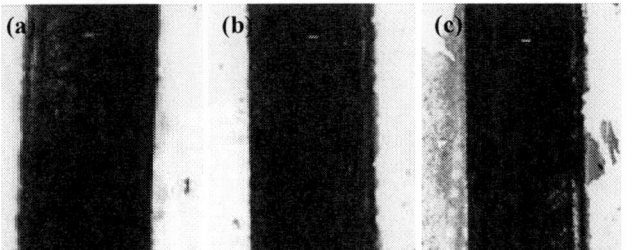

Fig 7. Optical images of microchannel produced by laser micromachined at laser frequency of (a) 100 Hz, (b) 300 Hz. and (c) 500 Hz.

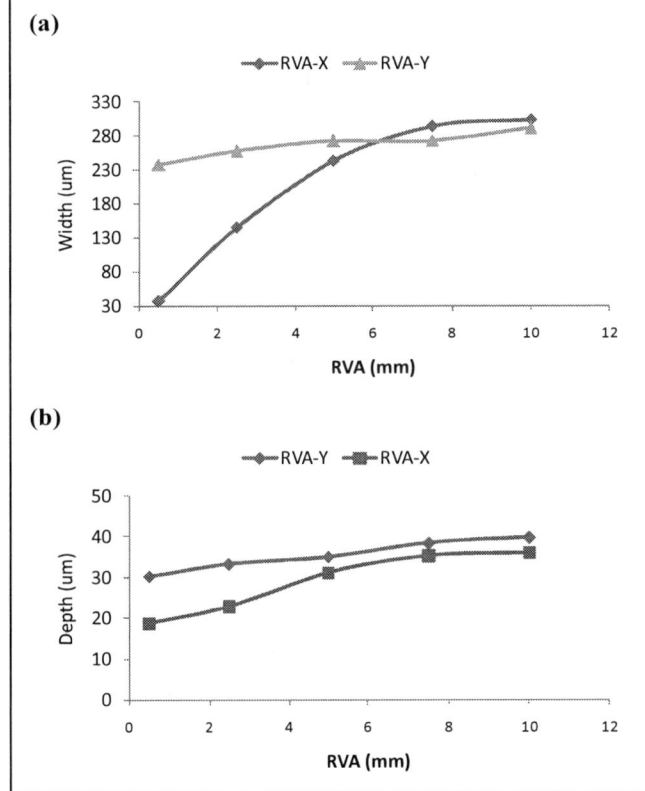

Fig 8. Results from varying RVA-X and RVA-Y: (a) The ablated channel width and (b) Measured channel depth.

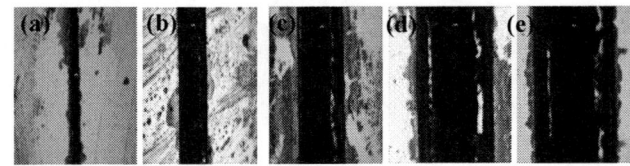

Fig 9. Optical images of microchannels produced by laser micromachined at RVA-X dimension of (a) 0.5 mm, (b) 2.5 mm, (c) 5 mm, (d) 7.5 mm and (e) 10 mm.

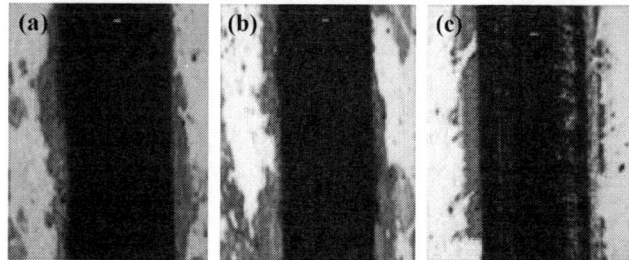

Fig. 10. Optical micrographs of microchannels produced by laser micromachined at RVA-X dimension of (a) 0.5 mm, (b) 5 mm and (c) 10 mm.

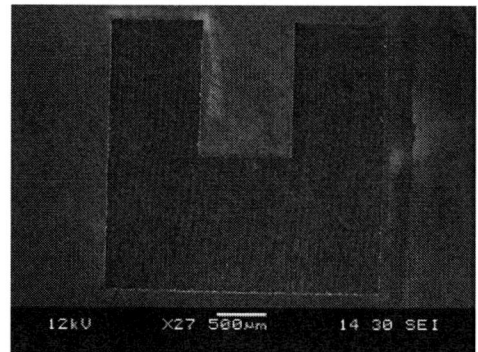

Fig. 11. SEM image of cantilever fabricated using laser micromachine.

III. Demonstration of Application

The characterization data collected and presented in the previous section have been used as a guideline to fabricate cantilever structures. A cantilever, which is a beam fixed firmly at only one end and left freely at other end is the most simplified MEMS based device. Microcantilevers have been typically applied for chemical, biological and physical sensing. The fabrication of this structure is presented in order to demonstrate the microfabrication capability of the laser micromachine. The cantilever structure was designed on a 9 mm^2 silicon substrate with the thickness of 115.60 um. The cantilever length and width was set to 1.5 mm and 1 mm respectively. During the fabrication, the laser energy was set to 18 mJ with 300 laser pulses. Laser frequency of 400Hz was also used with RVA-X and RVA-Y dimensions both set to 1 mm. The fabricated cantilever image was then observed and inspected using Scanning Electron Micrograph (SEM) as shown in Fig. 11.

IV. Conclusion

In this research, we have demonstrated the application of laser micromachine to fabricate silicon based cantilever. The fabrication parameters were obtained after investigations of four main laser micromachine parameters, which are laser energy, number of laser pulses, laser frequency and Rectangular Variable Aperture (RVA) in X and Y direction. The results indicate that the quality of the microchannels produced are dependent on all parameters. The most significant parameter is the RVA, where the larger the RVA, the more beam are focused on the targeted materials, resulting in more pronounced channel width. The laser frequency has the least effect on the microchannel width and depth produced. When varying each parameter, we observed some debris is deposited at the edge of the microchannel. We have found that one of the solutions to avoid any debris formation is to deposit photoresist on the ablated area before machining. The resist however need to be developed after the ablation process is complete. Future works include studying the surface roughness of the material after etching process. The ablation effects of laser micromachine on other materials such as glass, PDMS, PMMA and SU-8 photoresist will also be investigated.

Acknowledgment

The authors would like to thank the technicians at AMBIENCE laboratory and SoME cleanroom for their assistance. This work is supported by Research Grant UNIMAP FRGS 9003:00405.

References

[1] F. Zee and J.W. Judy, "MEMS chemical gas sensor using a polymer-based array", Proceedings of International Solid-State Sensors and Actuators Conference, Sendai, Japan, pp. 1169-1172.

[2] M. Wu and W. Fang, "Design and fabrication of MEMS devices using the integration of MUMPs, trench-refilled molding, DRIE and bulk silicon etching process", J. Micromech. Microeng, 15, pp. 1-8, 2004.

[3] A. F. M. Anuar, Y. Wahab, F. Hamzah, M. Najmi, S. Johari, M. Mazalan, N. I. M. Nor, and M. K. Md Arshad, " KrF Excimer Laser Micromachining of Silicon for Micro- Cantilever Applications", Proceedings of International Electronic Conference on Sensors and Applications, June 2014.

[4] Y. V. White, M. Parrish, X. Li, L. M. Davis and W. Hormeister, "Femtosecond micro- and nano-machining of materials for microfluidic applications", Proc. of SPIE Vol. 7039, 70390J, 2008.

J. Li and G. K. Ananthasuresh, "A quality study on the excimer laser micromachining of electro-thermal-compliant micro devices", J. Micromech. Microeng, 11, pp. 38-47, 2001.

A Study on the gas sensing effect on current-voltage characteristics of ZnO Nanostructures

W.H.Khoo and S.M.Sultan

Electronics and Computer Engineering Department
Faculty of Electrical Engineering, Universiti Teknologi Malaysia
81310 UTM Skudai
Johor Darul Takzim

—Current-voltage characteristics of ZnO nanostructures were studied under different gas ambient using nonequilibrium Green's function and density functional theory, DFT technique. It was found that I-V characteristics of ZnO nanostructures depend strongly on the type of gas molecules present. The sensitivity factor of more than 200% achieved with the presence of a single molecule of NO_2 gas. Meanwhile, there were no significant changes towards CO and NH_3 gas molecules. However, CO adsorption can significantly suppress the transmission spectrum of ZnO nanostructures. Under the same applied bias voltage, the current through ZnO nanostructures decreases with increasing CO concentrations.

—

I. Introduction

Nanomaterials such as quantum dots, nanoparticles, nanowires, nanotubes, nanogaps and nanoscale films [1-4], have received enormous attention due to their suitable properties for designing novel, nanoscale gas sensors. In addition, the high surface-to-volume ratios for nanomaterials allow a large amount of the atoms related to the gases to be located at or close to the surface, for detection. Among these nanomaterials, ZnO has attracted increasing attention due to its potential range of applications. Nanostructured ZnO possesses high surface area, good biocompatibility and chemical stability and is non-toxic. It also shows biomimetic and high electron communication features, making it suitable for potential applications in gas sensors [4]

For practical applications, chemical gas sensors with fast response, low detection limit, and large active region are highly desired. Recently, great advances have been made in ZnO nanostructures to detect the presence of hazardous and toxic gaseous molecules such as CO, NO_2 and NH_3[4]. The sensing mechanism is to detect the change in conductance of the semiconducting ZnO nanostructures induced by interactions between adsorbed molecules and the nanostructures. However, experimental data give inconsistent results and limited to a few gas molecules. In addition, there have been very few theoretical studies on the interaction of gas molecules with ZnO nanostructures such as nanowires. While most theoretical works on ZnO nanostructures focus on their binding energies [5], highest occupied molecular orbital-lowest unoccupied molecular orbital gaps [6], band structures and elasticity [7], there have been only a few reports on their electrical characteristics [8]. Thus this work attempts to study the effect of different gas molecules on the surface of the ZnO nanowires based on density functional theory (DFT) combined with nonequilibrium Green's function (NGF) technique.

II. Methodology

A ZnO nanostructure was modelled with the height of 0.56 Å and the length of 1.6 Å. The geometrical configuration for this study is composed of semi-infinite left and right electrodes separated by the scattering region (central region).Figure 1(a) shows the schematic diagram of ZnO nanowire before introducing the gas molecule. The Zn–O bond length is determined to be 1.8 Å which is in good agreement with other theoretical results [8]. In addition, an 1 : 1 Zn–O ratio were considered in constructing this structure. Figure 1(b) shows the adsorption of CO gas molecules on the surface of the ZnO nanowire along the scattering region. Other gas molecules were also tested which include NH_3 and NO_2.

The second part in this simulation study is to investigate the effect of the concentration density of gas molecules on the electrical characteristics of the ZnO nanostructure. In this part CO gas is selected. The adsorbed CO gas molecules on ZnO nanowire can be controlled by varying the number of CO molecules in the scattering region. The study starts with the lowest CO concentration, when only one CO molecule is included in the scattering region as shown in Fig.1 (b). The CO concentration is increased by adding CO molecules uniformly to the symmetric sites on the top and bottom of the NW as shown in Fig.1(c). Finally, the CO molecules increased to four symmetric sites as shown in Fig.1(d). Using these structures, the I-V characteristics have been studied using DFT and NEGF combined with the two-probe model (electrode-nanotube-electrode) which is implemented in ATOMISTIX TOOLKIT package.

978-1-4799-5761-3/14 $31.00 © 2014 IEEE

The I-V characteristics have been calculated with the Landauer formula [9],

$$I_D = \frac{2e^2}{h} \int T(E, V_D)[f_1(E) - f_r(E)]dE \quad (1)$$

where e is the electron charge, h is Planck's constant, and $f_1(E), f_r(E)$ represent the Fermi distribution functions at the left and right electrodes, respectively. $T(E, V_D)$ stands for the transmission coefficient as a function of the electron energy,E and bias voltage,V_D.

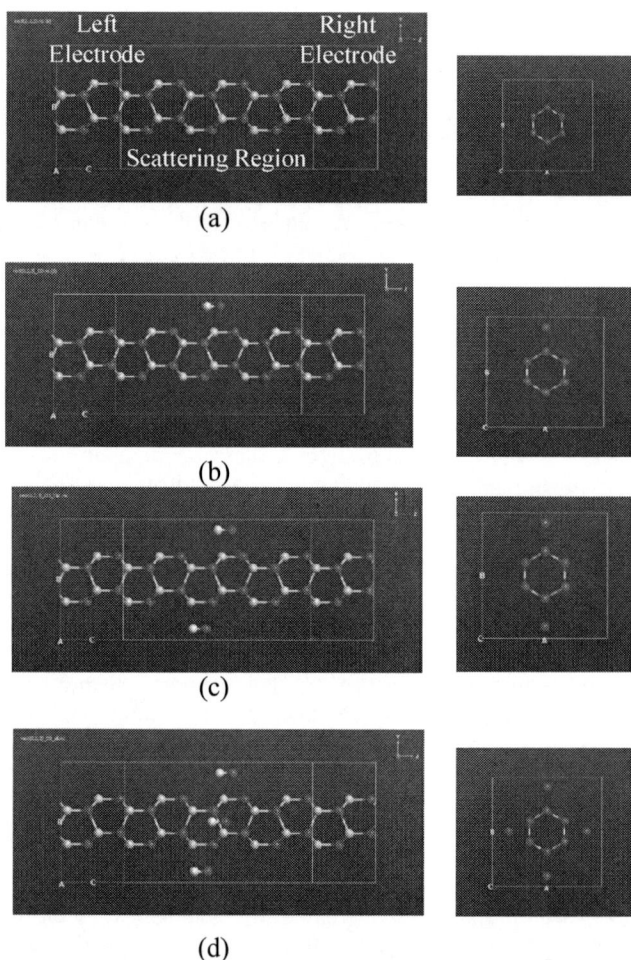

(a)

(b)

(c)

(d)

Fig.1. Schematic diagrams of ZnO nanowire sensors *(left)* side view *(right)* cross-sectional view of the nanowire structure. (a) Reference structure without any gas (b) One CO molecule (c) 2 CO molecules (d) 4 CO molecules adsorbed over the ZnO nanostructure.

III. Results and Discussion

Figure 2(a) shows the current voltage characteristics (I_D-V_D) of the ZnO nanostructures under different gas environment. From the plot, it is found that at a bias voltage of

2 V, the current level before gas exposure was 1 µA. Under NO_2 gas, current increases more than 3 times from 1 µA to 3.66 µA . This shows the ZnO has a good sensitivity towards NO_2 gases which in this case the sensitivity is more than 200% (S=$\Delta I/I_A$). However, there were no significant effect on a single molecule NH_3 and CO gases. This is an indication of less sensitivity of the ZnO nanostructures towards these gases.

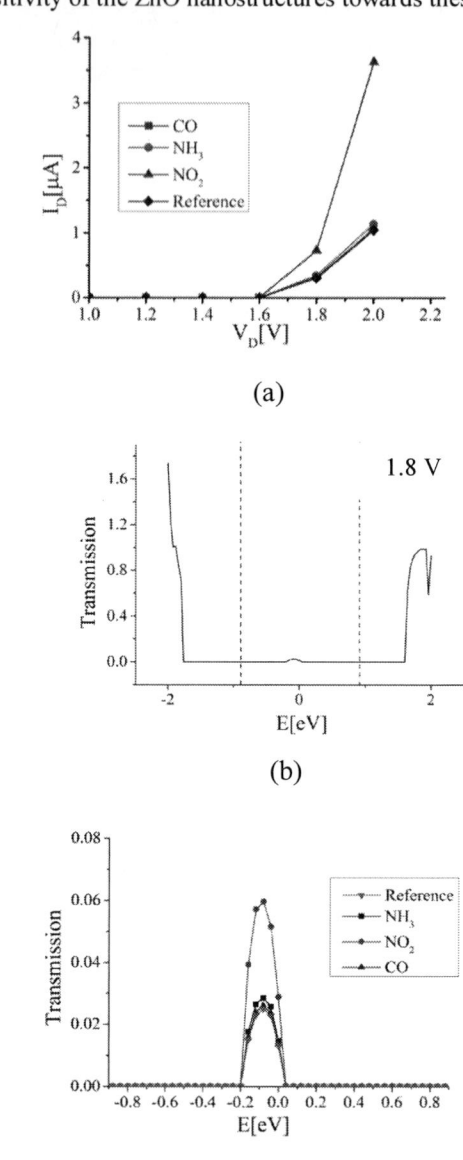

(a)

(b)

(c)

Fig.2. (a) Current-voltage characteristics of ZnO nanostructures under the influence of a single CO, NH_3, NO_2 molecules. The black plot is the reference device without any gas molecules. (b)Transmission spectrum of the reference device at a bias of 1.8 V. (c) Enlarged transmission spectra of ZnO nanostructure under different gas molecules.

To further study the results, transmission spectra have been calculated. Fig. 2 (b) shows the transmission spectrum of the reference device without the presence of any gas molecules at 1.8 V bias. The area between the two dotted lines represents the bias window, which is referred to the energy interval from the chemical potential of the left electrode to that of the right electrode. The transmission peaks refer to the conducting channels. The higher the transmission peaks in the bias window, the higher the current. Figure 2(c) shows the enlarged view of the transmission spectrum with the presence of different gas molecules. On comparing the transmission spectrum due to different gas molecules adsorption, it is found that under NO_2 gas molecule, the transmission peak enhanced at energy bias at -0.08 eV which lead to larger current values. However, there is no significant effect on the transmission peaks during the CO and NH_3 gases.

To further assess the effect of gas molecules concentration, CO gas molecule was considered at three different adsorption sites. Fig. 3(a) shows the electrical characteristics for different CO molecules concentration. It is observed that the current decreases with the increase of CO concentration density. Taking an example of bias voltage of 2 V, the current is about 1 µA with no CO adsorption, then decreases to 0.8 µA (2-CO molecule). The current decreases further to 0.6 µA at 4-CO gas molecule concentration.This result is associated to the scattering effects and the quantum interferences of localized impurity states. Similar trend can also be seen in Ref[10] on their SiCNT device.

Fig. 3(b) shows the transmission spectra of ZnO nanowire with different CO concentration at a bias voltage of 1.8 V. Reference refers to the pristime ZnO nanowires. When a single CO molecule is attached on ZnO nanostructure, there is no significant change on the transmission spectrum. However, when two CO molecules are attached, the transmission spectrum exhibits a small suppression at -0.08 eV. As the CO concentration increases, the transmission at this point is substantially suppressed.

With the increase of CO concentrations, the density of the impurity states increases significantly. Thus, the transmission decay can be due to scattering effects and the quantum interferences due to the proximity between the impurity states.

IV. Conclusions

In conclusion, using NEGF and DFT techniques from the ATOMISTIX Tool, a simple ZnO nanostructure model was designed. The electrical characteristics have been investigated for different gas molecules: CO, NH_3 and NO_2. It was found the ZnO nanostructure is sensitive by more than 200% towards NO_2 gas. Simulation with different concentration of CO gas molecules yield a decrease in current and suppression of transmission peak with increasing CO concentration. A maximum sensitivity factor of 66.7% is achieved.

(a)

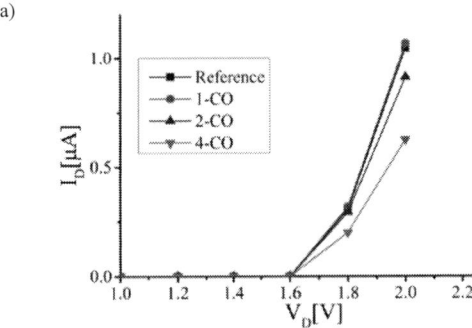

(b)

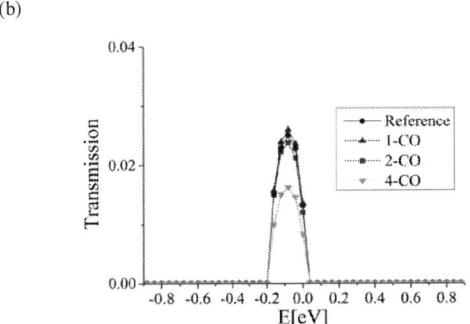

Fig.3. (a) Current-voltage characteristics of ZnO nanostructures with different CO molecules concentration. (b) Enlarged transmission spectra of ZnO nanostructure under different CO molecules concentration at bias of 1.8 V.

Acknowledgment

Authors appreciate Mr Wong King Kiat from Universiti Teknologi Malaysia for helping us on the technical aspects of the Quantum Wise software. Authors would like to acknowledge the financial support from Research University grant of Universiti Teknologi Malaysia under Projects Q.J130000.2723.00K83 and Ministry of Higher Education of Malaysia.

References

[1] X. Duan, Y. Huang, Y. Cui, J. Wang, and C. M. Lieber, "Indium phosphide nanowires as building blocks for nanoscale electronic and optoelectronic devices.," *Nature*, vol. 409, no. 6816, pp. 66-69, 2001.

[2] M. Curreli, R.Zhang , F.N.Ishikawa,, H.K.Chang, R.J.Cote, C.Zhou, M.E.Thompson, " Real-Time, Label-Free Detection of Biological Entities Using Nanowire-Based FETs," *IEEE Trans. on Electron Devices* ,vol. 7, no. 6, pp. 651-667,2008.

[3] B. Y. Xia and P. Yang, "Chemistry and Physics of Nanowires," *MRS Bulletin*, vol. 30, no. 5, pp. 351-352, 2003.

[4] M.M.Arafat, B.Dinan, S.A.Akbar and A.S.M.A Haseeb,"Gas Sensors Based on One Dimensional Nanostructured Metal-Oxides: A Review,"*Sensors*, vol.12,pp.7207-7258,2012.

[5] C.Li, W.Guo,Y.Kong, and H.Gao, "First-principles study on ZnO nanoclusters with hexagonal prism structures,"Appl. Phys. Lett. Vol. **90**,pp. 223102, 2007.

[6] B.Wen, R.Melnik, "Relative stability of nanosized wurtzite and graphitic ZnO from density functional theory," *Chem. Phys.Lett*, vol.466, pp.84,2008.

[7] Z.C.Tu and X.Hu, "Elasticity and piezoelectricity of zinc oxide crystals, single layers, and possible single-walled nanotubes," Phys. Rev. B , vol. 74,pp. 035434, 2006.

[8] Z.Yang, B.Wen, R.Melnik, S.Yao and T.Li, "Geometry dependent current-voltage characteristics of ZnO nanostructures:A combined nonequilibrium Green's function and density functional theory study,"*Appl.Phys.Letts,* vol.95, pp. 192101,2009.

[9] M. Büttiker, Y. Imey, R. Landauer, and S. Pinhas, "Generalized many-channel conductance formula with application to small rings,"*Phys.Rev.B.*,vol.31, pp.6207,1985.

[10] J.-M.Jia, S.-P.Ju, D.-N.Shi, and K.-f.Lin, "CO adsorption on a zigzag SiC nanotube: effects of concentration density and local torsion on transport,"*J.Nanopart Res*, vol.15,pp 1977,2013.

Memristive Behavior of HF-Etched Sputtered titania Thin Films

Z. Aznilinda

Faculty of Electrical Engineering
Universiti Teknologi MARA (UiTM)
81750 Masai, Malaysia
aznilinda@johor.uitm.edu.my

M. M. Ramly , NS Kamarozaman and S.H.Herman

NANO-Electronic Centre (NET)
Faculty of Electrical Engineering
Universiti Teknologi MARA (UiTM)
40450 Shah Alam, Malaysia

Abstract— **This paper demonstrates the fabrication method and reports the essential physical characterization of a memristive device with TiO_2 or titania as an active layer. The memristive device was fabricated on glass substrate. Titania thin films were grown in two layers by RF-magnetron sputtering technique onto the substrates. The first layer is a titania layer etched by 1% HF (Hydrofluoric acid) before the deposition of the second layer. The etching time was varied; for 5 seconds and 7 seconds. Current–voltage (I-V) curves of the samples were measured from the voltage loop ranging from 0V to -5V, -5V to 5V then back to 0V and also from -5V to 5V then back to -5V. It was proven that the HF-etch give an improvement in the memristive behavior when it is etched at 7 s.**

Keywords— *memristive behavior, memristor, TiO_2, titania, HF-etching.*

I. INTRODUCTION

Memristor is a two-terminal electronic device [1-2] that exhibit unique electrical characteristics that it retain a state of internal resistance based on the history of applied voltage and current. The resistance is depending on the magnitude and polarity of the voltage applied to it. When the voltage is turned off, the memristor remembers its most recent resistance until the next time its turned on [3]. Memristor was first theorized in 1971, known as the missing fourth fundamental passive circuit element predicted by Leon Chua[1]. Its properties could not be replicated by a combination of the other fundamental components.. It is an element that has the relation between charge and flux ($d\varphi$=Mdq) [1, 3] and which is evidently missing from the basic fundamental circuit element which contains the resistor, capacitor, and inductor. The name 'memristor' was connected to this device, because of the property of behaving like a non-linear resistor with memory.

The first memristor was experimentally demonstrated by Stanley Williams and team at Hewlett-Packard Laboratories (HP Lab) in 2008 [2] and was successfully found and structured in their nano-scale cross bar array that relates to Leon Chua's theory [3].

This work was partially supported by Ministry of Higher Education (MHOE) Malaysia under the Niche Research Grant Scheme (NRGS) (Project code: 600-RMI/NRGS 5/3 (7/2013)).

The active layer of memristor generally consists of one or two layers of thin films using numerous materials such as organic monolayer [4-5] amorphous silicon[6-7] ferroelectric material [8], and titanium dioxide (TiO_2) . TiO_2 or titania has the potential to produce good memristive behavior [2-3, 9-10] and widely used as the active layer in the memristive device. The memristive behavior in the device occurs due to the charge displacement in the active layer [1, 11-12] where the charge, originating from the oxygen vacancies in the active layer controls the conductivity and switching behavior of the device.

Memristive device has been successfully fabricated by few techniques such as sol gel [13] thermal oxidation [14], electron beam evaporation [9], and sputter deposition [11, 14-16]. Memristive device requires thickness of nano-scale in order to acquire its memristive behavior [1-2] that is below 10nm [2]. To control the thickness of the thin film to be at nano-scale, Radio frequency (RF) magnetron sputtering technique offers process parameters that can be controlled [17-18].

The main carrier in a memristive device is the oxygen vacancies also known as oxygen deficiency. Oxygen vacancies layer is required in the memristive device to act as the transporting element for the current to flow and creating a conduction layer. In order to create this layer, the titania layer is being treated to produce a deficient oxygen layer with oxygen vacancies known as TiO_{2-x}. The methods employed in this work in order to synthesize the deficient oxygen layer is the etching process.

The HF-etching process is used to structure the semiconductors with an extensive range of etching speeds and different anisotropic aspect. HF-etching process can also reduce electronic properties [19] and due to this we introduce etching process as one of the process in creating the oxygen vacancies. During the removal of the etched material, the atoms level from the material may also be removed. In this work, the surface of the first deposited titania layer was being etched in order to roughen the surface by 1% HF etching at 5 and 7 seconds. The difference in memristive behavior and its physical properties between both etching time are being discussed in this work.

II. METHODOLOGY

A. TiO₂ Thin Film Deposition by RF Magnetron Sputtering and Surface Treatment Etching Process

Silicon and glass substrates were cleaned by the standard cleaning method. Both substrates were coated with 60nm thick of platinum as its bottom electrode. A 99.99% TiO_2 target was used in an $Ar:O_2$ ambient of 50:1 at a working pressure of 5 x 10^{-3} Torr. The system was vacuumed to a pressure less than 5.0 x10^{-6} Torr in order to minimize the residual gas components before the deposition started. A pre-sputter cycle was run to remove contamination occurs on the target surface.

The first layer of TiO_2 thin film is grown by RF magnetron sputtering at sputtering power of 300 W for 15 minutes on silicon substrates with substrate temperature 250°C and biased at 20V. A surface treatment is then being done on the first layer before the deposition of the 2nd layer of TiO_2 which is also deposited using RF magnetron sputtering.

The first TiO_2 layer was exposed to surface treatment roughened by HF etching method for 5 and 7 seconds before the deposition of the 2nd layer TiO_2 thin films. It is etched using 1% HF solution. The second layer of TiO_2 is deposited for 5 minutes.

B. Characterization

The memristive behavior was measured using current-voltage (I-V) measurement system with two ways cycle by taking from the voltage loop ranging 0V to -5V, -5V to 5V then back to 0V and also from -5 to 5 V then back to -5V. The second cycle was measured to observe the effect on the sample when being applied by external bias for many times and at different cycle measurement.

The cross-section image, thicknesses and the composition of the thin film was measured using Field Emission Scanning Electron Microscope (FESEM, JEOL JSM 7600F) with EDS (Energy Dispersive X-ray Spectroscopy).

III. RESULT AND DISCUSSION

The memristive I-V behaviors are plotted in Figure 3. From the observation, sample with 7 seconds etching time shows better memristive behavior with a bow-tie alike curve that matches the electrical behavior reported for memristors [2-3]. This is due to more particles including oxygen ions are being removed at a longer etching time as discovered in TABLE II. The high oxygen vacancies results from the 7 seconds etching time provide more mobile ions for that conductivity to happen in the device.

It is also observed that the result with voltage bias from 0 V → -5 V → 5 V → 0 V, Figure 3(c) of 7 sec etching time, shows poorer memristive behavior with noise at the negative switching compared to the cycle of voltage bias from -5 V → 5 V → -5 V, Figure 3(d). This may possibly due to an electroforming path that have been created during the first biasing process so giving the next biasing cycle to exhibit a less noise hysteresis [20].

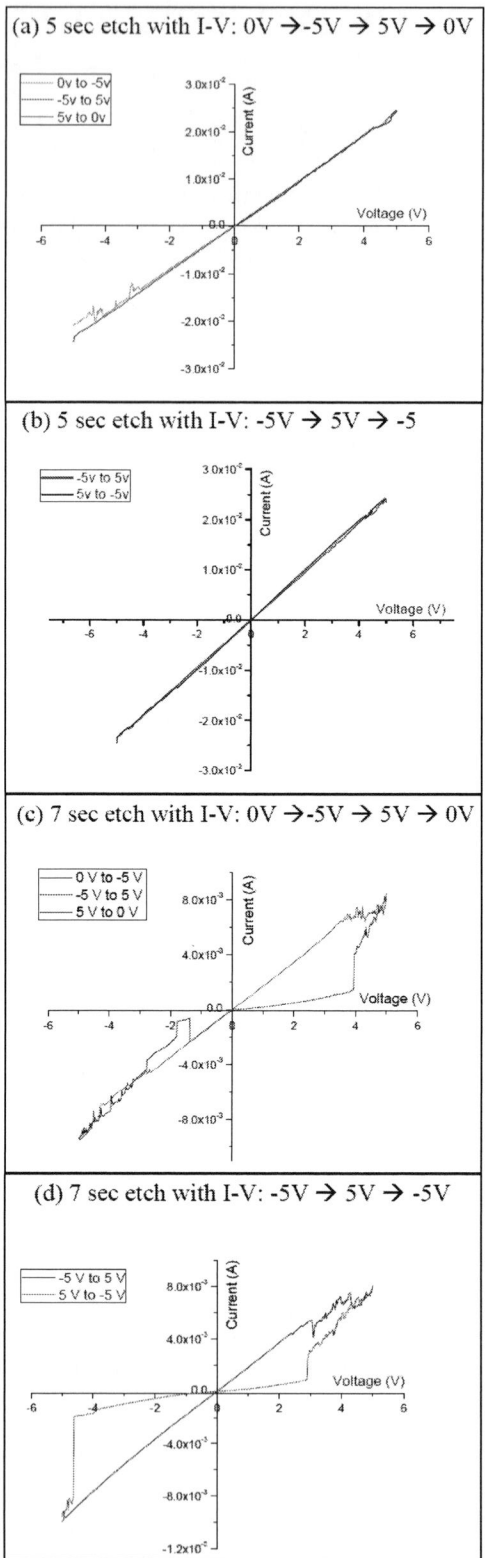

Fig. 3 Memristive switching curves experiential from the current-voltage (I-V) measurement system at different etching time.

However, comparing this sample, Figure 3(c) to other samples, it is observed that the switching cycles produced a similar result where the 0 V → -5 V started with HRS (High Resistance State) and then continue with LRS (Low Resistance State) when biased with -5 V to 5 V which shows a normal memristive switching behavior.

Fig. 4(a) and (b) shows the cross-section image and surface morphology image on the first layer of TiO_2 deposited on glass coated with platinum and etched for 7 seconds respectively. Columnar growth of TiO2 can be observed. The surface morphology shows bigger cluster size compare to the one being treated with plasma.

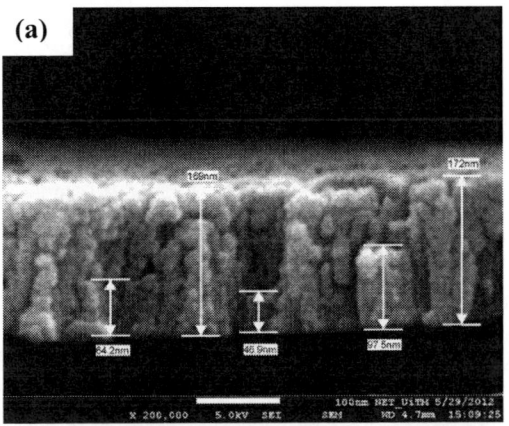

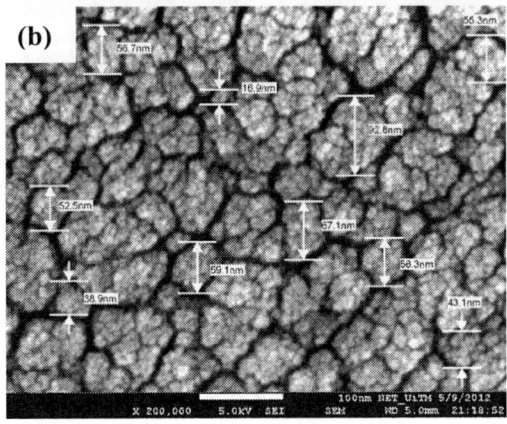

Fig.4 (a) Cross-section images of deposition of TiO_2 layers at 7 seconds etching time. (b) FE-SEM images of the particle structure on the TiO_2 thin film for 7 seconds etching time.

TABLE II.
THICKNESS AND OXYGEN CONCENTRATION OF TiO_2 THIN FILM ETCHED AT DIFFERENT ETHING TIME. METHOD.

Etching time (second)	Thickness layer 1 (nm)	Thickness Layer 2 (nm)	Oxygen concentration of etched titania layer (atomic weight %)
5	127	242	47.92
7	97.5	74.5	45.45

Table II shows the thickness of each layer by observing the FE-SEM cross-section image. It is observed that the thickness of the sputtered titania layer with shorter etching time is higher as compared to the longer etching time. The EDS data shows that the oxygen concentration in sample with 7 seconds etching period is lesser than 5 seconds etching period treatment with 45.45 wt% and 47.92 wt% respectively. During the etching process, the longer the sample exposed to the etching solution, the longer the titania surface react to the HF and dissolves in the HF solution. This has caused a bigger amount of etched atoms thus results in a thinner titania layer. During the etching process, it not only removes the Ti particles but also the O particles and result in oxygen vacancies and creating TiO_{2-x} layer. At the same time the active layer may become thinner and these both effects may improve the memristive behavior [21].

Figure 5 shows the surface topography images and the surface roughness of the titania thin film from the AFM system. The roughness of sample with 5 seconds etching time is higher compared to sample with 7 seconds etching time. Surface roughness are higher at 5 seconds etch may due to a too short etching time as longer etching time may help to smoothens a film surface [22-23]. Etching process with low solution concentration requires longer time to remove more particles from the sample [23]. Longer etching time may already meet the optimum or the adequate time for the etchant to remove a layer of the material and eventually gives a smooth surface. And longer HF-etching process may reduce the stress concentration on a surface thus giving a smoother surface [23].

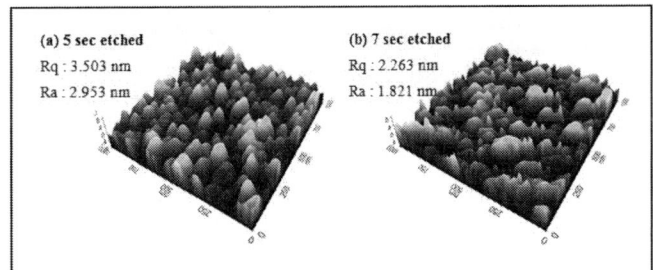

Fig.5 3--dimensional AFM images of sputtered titania layer etched at (a) 5 seconds (b) at 7 seconds

IV.   CONCLUSION

In this study, we have successfully fabricated a memristive device with two layers of TiO_2 by sputtering method and applying etching treatment process in order to create the active layer. by having first layer along with the removal of the oxygen. When employ this method onto a titania layer, it helps to remove the titania and also the oxygen atoms and provides oxygen vacancies and creating TiO_{2-x} layer. At the same time the active layer may become thinner and both effect may improve the memristive behavior. The method contributes to the creation of damage surface and difference in the memristive behavior. Having a higher damage surface contributes to better memristive behavior which may cause by a higher oxygen vacancies created at the interface.

ACKNOWLEDGMENT

The authors would like to thank the technichians and science officers at NET for their kind support in this research.

REFERENCES

[1] L. Chua, "Memristor-The missing circuit element," *Circuit Theory, IEEE Transactions on*, vol. 18, pp. 507-519, 1971.

[2] D. B. Strukov, *et al.*, "The missing memristor found," *Nature*, vol. 453, pp. 80-83, 2008.

[3] R. Williams, "How We Found The Missing Memristor," *Spectrum, IEEE*, vol. 45, pp. 28-35, 2008.

[4] J. J. Blackstock, *et al.*, "Internal Structure of a Molecular Junction Device: Chemical Reduction of PtO_2 by Ti Evaporation onto an Interceding Organic Monolayer," *The Journal of Physical Chemistry C*, vol. 111, pp. 16-20, 2007/01/01 2006.

[5] D. R. Stewart, *et al.*, "Molecule-Independent Electrical Switching in Pt/Organic Monolayer/Ti Devices," *Nano Letters*, vol. 4, pp. 133-136, 2004.

[6] D. Sacchetto, *et al.*, "Memristive devices fabricated with silicon nanowire schottky barrier transistors," in *Circuits and Systems (ISCAS), Proceedings of 2010 IEEE International Symposium on*, 2010, pp. 9-12.

[7] M. Haykel Ben Jamaa, *et al.*, "Fabrication of memristors with poly-crystalline silicon nanowires," in *Nanotechnology, 2009. IEEE-NANO 2009. 9th IEEE Conference on*, 2009, pp. 152-154.

[8] Y. Kaneko, *et al.*, "A novel ferroelectric memristor enabling NAND-type analog memory characteristics," in *Device Research Conference (DRC), 2010*, 2010, pp. 257-258.

[9] K. Miller, *et al.*, "Memristive Behavior in Thin Anodic Titania," *Electron Device Letters, IEEE*, vol. 31, pp. 737-739, 2010.

[10] J. J. Yang, *et al.*, "Memristive switching mechanism for metal//oxide//metal nanodevices," *Nat Nano*, vol. 3, pp. 429-433, 2008.

[11] T. Prodromakis, *et al.*, "Fabrication and electrical characteristics of memristors with TiO $2/TiO2+x$ active layers," 2010, pp. 1520-1522.

[12] T. Prodromakis, *et al.*, "Practical micro/nano fabrication implementations of memristive devices," in *Cellular Nanoscale Networks and Their Applications (CNNA), 2010 12th International Workshop on*, 2010, pp. 1-4.

[13] N. Gergel-Hackett, *et al.*, "Memristors With Flexible Electronic Applications," *Proceedings of the IEEE*, vol. PP, pp. 1-8, 2011.

[14] L. Ying-Tao, *et al.*, "A low-cost memristor based on titanium oxide," in *Solid-State and Integrated Circuit Technology (ICSICT), 2010 10th IEEE International Conference on*, 2010, pp. 1148-1150.

[15] H. Y. Jeong, *et al.*, "Interface-Engineered Amorphous TiO2-Based Resistive Memory Devices," *Advanced Functional Materials*, vol. 20, pp. 3912-3917, 2010.

[16] Z. Aznilinda, *et al.*, "Effect of Plasma Treatment on Memristive Behavior of Sputtered Titania," *Presented at the NANOSMAT conference in Sept2012 in Czech Republic*, 2012.

[17] P. B. Nair, *et al.*, "Effect of RF power and sputtering pressure on the structural and optical properties of TiO2 thin films prepared by RF magnetron sputtering," *Applied Surface Science*, vol. 257, pp. 10869-10875, 2011.

[18] P. Singh and D. Kaur, "Room temperature growth of nanocrystalline anatase TiO2 thin films by dc magnetron sputtering," *Physica B: Condensed Matter*, vol. 405, pp. 1258-1266, 2010.

[19] C. I. H. A. A. G. Baca, *Wet etching and photolithography of GaAs and related alloys, Dry etching of GaAs and related alloys* vol. 6. United Kingdom: EMIS processing, 2005.

[20] M. D. Pickett, *et al.*, "Switching dynamics in titanium dioxide memristive devices," *Journal of Applied Physics*, vol. 106, pp. 074508-074508-6, 2009.

[21] N. S. Kamarozaman, *et al.*, "Effect of TiO2 Seed Layer Thickness to the Growth of TiO2 Nanostructures by Immersion Method for Memristive Device Application," *Applied Mechanics and Materials*, vol. 393, pp. 63-67, 2013.

[22] C. Edstrom, "Wet Etching of Optical Thin Films," Chemistry, Jonkoping Institute of Technology, Jonkoping, 2010.

[23] S. Posritong, "Effect of Hydroflouric acid etching Followed by Unfilled Resin Application on The Biaxil Flexural Strength of A Glass-Based Ceramic," Master of Science in dentistry, Dentistry, Indiana University School of Dentistry, Indianapolis, 2012.

Investigation of the Inserted LT-AlGaN Interlayer in AlGaN/GaN/AlGaN DH-FET Strucutre on Si Substrates

Yu-Lin Hsiao, Chia-Ao Chang, and Edward Yi Chang

Department of Materials Science and Engineering
National Chiao-Tung University
1001 University Rd., Hsinchu, 30010 Taiwan
edc@mail.nctu.edu.tw

Abstract—A novel AlGaN/GaN/AlGaN double-heterostructure field effect transistor (DH-FET) structure with an inserted LT-AlGaN interlayer grown on 150 mm Si substrate has been studied. The DH-FET structure has been characterized by transmission electron microscopy (TEM), secondary ion mass spectrometry (SIMS) and X-ray diffraction (XRD). It is found that the inserted LT-AlGaN interlayer can further induce the compressive stress to compensate the tensile stress. Furthermore, the inserted LT-AlGaN interlayer acts as a dislocation filter to reduce threading dislocation propagation. These results indicate that the inserted LT-AlGaN interlayer plays an important role in the novel DH-FET structure.

Keywords—*AlGaN/GaN/AlGaN; double heterostructure FET (DH-FET); Interlayer*

I. INTRODUCTION

In the past decade, AlGaN/GaN single-heterostructure field effect transistors (SH-FETs) on Si substrates with high breakdown voltage and low specific on-resistance characteristics have been studied extensively. [1] These excellent performances make GaN devices attractive for high-power switching applications. Moreover, the fabrication cost of GaN grown on large diameter Si substrates is much cheaper than those of GaN-on-SiC and GaN-on-sapphire. Recently, 600 V GaN-on-Si commercial power devices are available and seem competitive to traditional Si-based power devices. Therefore, it is predictable that GaN-on-Si devices will have a great potential to improve power efficiency and become an emerging candidate for power electronic markets in the near future. [2]

The epitaxy growth of GaN-on-Si is a key issue for GaN power devices. Several methods as viewed from epitaxial structure to enhance the breakdown voltage have been proposed in the last few years. Increasing the epitaxial thickness to improve the crystalline quality of GaN growth can reduce the leakage current from the high dislocation density layer. In order to grow thick GaN layer, strain engineering is crucial to overcome the large lattice and thermal mismatches between GaN and Si. Thus, the insertion of the thin low temperature (LT) interlayer (AlN or AlGaN) into GaN layer induces compressive stress to compensate the large tensile stress during cooling down process. [3] A different approach to achieve higher breakdown voltage was based on AlGaN/GaN/AlGaN structure to build double heterostructure field effect transistors (DH-FETs), with improved electron confinement and higher breakdown field from the AlGaN back barrier. [4] However, studies on the combination of LT-AlGaN interlayer and DH-FET on large diameter Si substrates are still lacking.

In this paper, we propose a novel AlGaN/GaN/AlGaN double-heterostructure field effect transistor (DH-FET) structure with an inserted LT-AlGaN interlayer. Advantages of the LT-AlGaN interlayer and the DH-FET structure are presented.

II. EXPERIMENTS

Fig. 1 shows the schematic cross section of the proposed AlGaN/GaN/AlGaN DH-FET structure with the inserted LT-AlGaN interlayer. The proposed structure grown on 150 mm Si (111) substrate was carried out in a Thomas Swan metal organic chemical vapor deposition (MOCVD) system. Firstly, a transition layer (800 nm) was grown on the Si substrate, followed by a 1 μm AlGaN back barrier layer. Before the second AlGaN back barrier growth, a thin LT-AlGaN interlayer was inserted between two AlGaN back barrier layers. Then a 90 nm GaN layer was grown as a channel. Finally, a 30 nm AlGaN barrier layer was grown on the top.

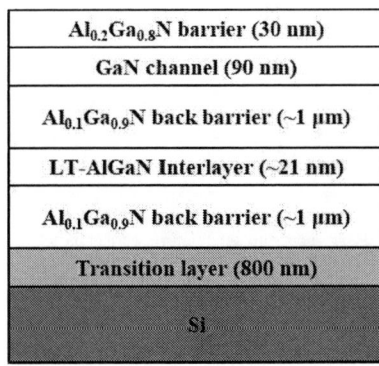

Fig. 1. Schematic cross section of the DH-FET structure with the inserted LT-AlGaN interlayer

A Laytec EpiCurve® TT system was used for in-situ measurement of the wafer temperature, reflectance and wafer curvature during epitaxial layer growth. Material analysis was characterized by transmission electron microscopy (TEM), secondary ion mass spectrometry (SIMS) and X-ray diffraction (XRD).

III. RESULTS AND DISCUSSION

Fig. 2 shows the transients of temperature and reflectance (upper figure) and the wafer curvature (lower figure) obtained with a Laytec EpiCurve® TT system during the growth of the DH-FET structure. The LT-AlGaN interlayer was introduced and the growth temperature is around 750 ℃. The in-situ wafer curvature measurement is essential for optimizing growth conditions and controlling the stress in order to obtain crack-free wafer. A negative curvature trace was observed after the LT-AlGaN interlayer was deposited. The compressive stress was induced to relax the tensile stress.

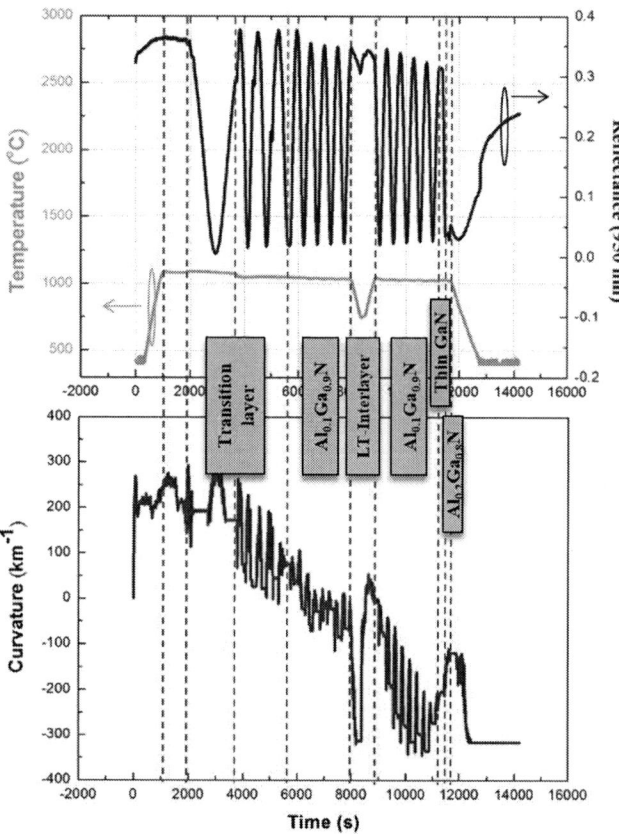

Fig. 2. In-situ measurements of temperature, reflectance and curvature during growth of DH-FET

In order to investigate the detail of each epi-layer, SIMS depth profile measurement was performed on the proposed DH-FET structure as shown in Fig. 3. From the depth profile, the increments of the Al- and Ga-related signals were observed near 1125 nm below the surface. It indicates that the

interlayer between two AlGaN back barrier layers is the LT-AlGaN interlayer. However, the Al signal is only for qualitative analysis not for quantitative analysis, it needs further characterization to investigate the Al composition in the LT-AlGaN interlayer.

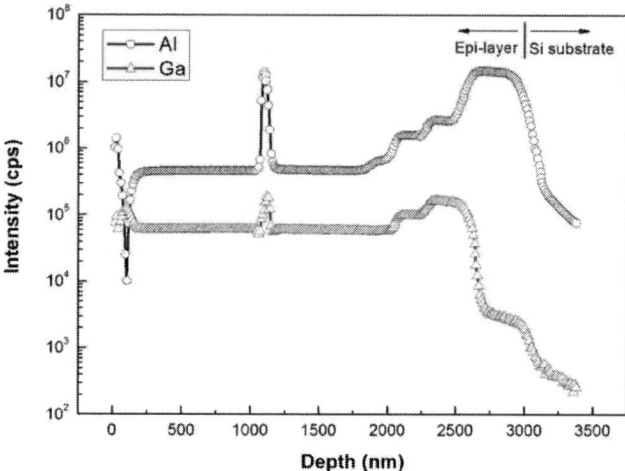

Fig. 3. SIMS depth profile of Al and GA for the DH-FET structure with the inserted LT-AlGaN interlayer

Fig. 4 shows XRD HRXRD ω-2θ scan for the DH-FET structure. The back barrier is $Al_{0.1}Ga_{0.9}N$ and the interlayer is $Al_{0.5}Ga_{0.5}N$. According to XRD rocking curves, the full-widths at half-maximum (FWHMs) of X-ray rocking curves for the $Al_{0.1}Ga_{0.9}N(002)$ ω-scan and $Al_{0.1}Ga_{0.9}N(102)$ ω-scan were 578.6 and 1103 arcsec, respectively.

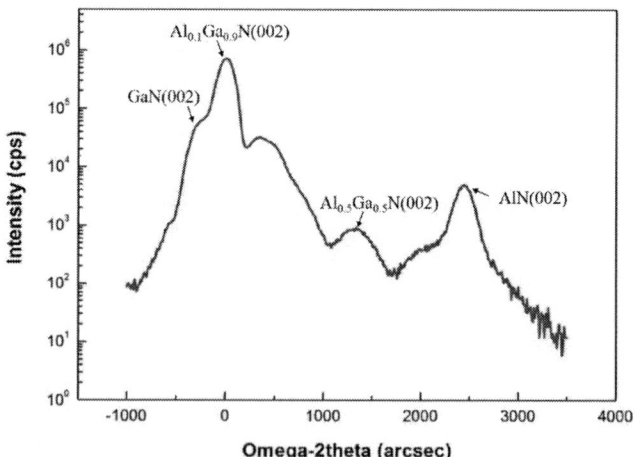

Fig. 4. XRD ω-2θ scan of DH-FET epitaxial structure. The inset shows the rocking curve of the $Al_{0.1}Ga_{0.9}N$ (002) peak. The Al composition of LT-AlGaN is 50%.

Fig. 5 shows the overview of the whole structure and the interfaces between LT-AlGaN interlayers by cross sectional TEM, respectively under bright field condition. Threading dislocation can be observed. Fig. 6 shows the interfaces between LT-AlGaN interlayer and AlGaN back barrier layers. The thickness of AlGaN interlayer is 21 nm.

Fig. 5. Cross section TEM image of the DH-FET structure

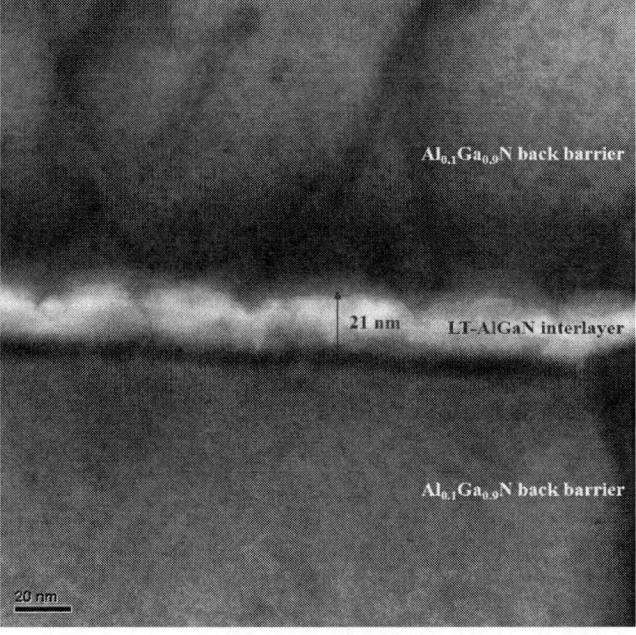

Fig. 6. Interfaces between LT-AlGaN interlayer and two AlGaN back barrier layers

IV. CONCLUSION

In this study, high quality AlGaN/GaN/AlGaN DH-FET structure with the inserted LT-AlGaN interlayer was successfully grown on 150 mm Si substrates. The DH-FET structure has been characterized by transmission electron microscopy (TEM), secondary ion mass spectrometry (SIMS) and X-ray diffraction (XRD). It is found that the LT-AlGaN

interlayer plays an important role in stress relaxation and dislocation reduction for the growth of the DH-FET structure.

ACKNOWLEDGMENT

This work was sponsored by the NCTU-UCB I-RiCE program, Ministry of Science and Technology, Taiwan, and TSMC, under Grant No. NSC 103-2911-I-009-302 and the Ministry of Economic Affairs, Taiwan, under Grant No. 102-EC-17-A-05-S1-154.

REFERENCES

[1] S. Selvaraj, A. Watanabe, A. Wakejima and T. Egawa, "1.4-kV breakdown voltage for AlGaN/GaN high-electron-mobility transistors on Silicon substrate," IEEE Electron Device Letter, vol. 33, no. 10, pp. 1375-1377, Oct. 2012.

[2] M. Su, C. Chen and S. Rajan, "Prospects for the Application of GaN Power Devices in Hybrid Electric Vehicle Drive System," Semiconductor Science Technolgy., vol. 28, no. 7, pp. 074012-9, 2013

[3] S. Fritze, P. Drechsel, P. Stauss, P. Rode, T. Markurt, T. Schulz, M. Albrecht, J. Bläsing, A. Dadgar, and A. Krost, "Role of low-temperature AlGaN interlayers in thick GaN on silicon by metalorganic vapor phase epitaxy," Journal of Applied Physics, vol. 111, pp. 124506-6, 2012.

[4] K. Cheng, H. Liang, M. Van Hove, K. Geens, B. De Jaeger, P. Srivastava, X. Kang, P. Favia, H. Bender, S. Decoutere, J. Dekoster, J. del Agua Borniquel, S. W. Jun, and H. Chung, "AlGaN/GaN/AlGaN double heterostructures grown on 200 mm Silicon (111) substrates with high electron mobility," Applied Physics Express, vol. 5, no. 1, pp. 011002-3, 2012.

Statistical Process Modelling For 32nm High-K/Metal Gate PMOS Device

Afifah Maheran A.H.[1], Noor Faizah Z. A.[2], P. S. Menon[1], I. Ahmad[2], P.R. Apte[3], Kalaivani T.[4], Salehuddin, F.[5]

[1]Institute of Microengineering and Nanoelectronics (IMEN), Universiti Kebangsaan Malaysia (UKM)
43600 Bangi, Selangor, Malaysia
[2]Centre for Micro and Nano Engineering (CeMNE), College of Engineering, Universiti Tenaga Nasional (UNITEN)
43009 Kajang, Selangor, Malaysia
[3]College of Engineering Pune, Shivaji Nagar, Pune, Maharashtra 411005, India
[4]Infrastructure University Kuala Lumpur, Unipark Suria, Jalan Ikram-Uniten, 43000 Kajang, Selangor Darul Ehsan, Malaysia
[5]Universiti Teknikal Malaysia Melaka, Hang Tuah Jaya, Melaka, Malaysia
susi@eng.ukm.my

Abstract— The evolution of MOSFET technology has been governed solely by device scaling, delivered an ever-increasing transistor density through Moore's Law. In this paper, the design, fabrication and characterization of 32nm $HfO_2/TiSi_2$ PMOS device is presented; replacing the conventional SiO_2 dielectric and Poly-Silicon. The fabrication and simulation of PMOS transistor is performed via Virtual Wafer Fabrication (VWF) Silvaco TCAD Tools namely ATHENA and ATLAS. Taguchi L9 Orthogonal method is then applied to this experiment for optimization of threshold voltage (V_{TH}) and leakage current (I_{OFF}). The simulation result shows that the optimal value of V_{TH} and I_{OFF} which are 0.1030075V and $3.4264075 \times 10^{-12}$A/um respectively are well within ITRS prediction.

Keywords—32nm PMOS, High-K dielectrics, Metal Gate Transistor, Silvaco, Taguchi Method.

I. INTRODUCTION

CMOS technology has been dominating the microelectronic industry for almost fifty decades when complex semiconductor and communication technologies starts being developed. Since then, the size of CMOS transistor has been shrinking continuously validating through Moore's Law where feature size of a transistor is scaled at a rate of approximately 0.7 times for every 18 months [1]-[3]. Design choices which takes into account transistor performance with technology optimization features along with ensuring robust product functionality has become the most highlighted aspect due to the high demand for smaller, faster, and cheaper technologies. However, the increasing of wafer fabrication process parameter variation has been perceived as one of the major roadblocks to further technology scaling, such as short-channel effects (SCE), drain induced barrier lowering (DIBL), and hot carrier effect which in turn, led to the innovation of high-K/metal gate devices [4].

High-K/metal gate technology is introduced to enhance the transistor performance, replacing SiO_2/Poly-Si and is being widely implemented nowadays. High-K dielectrics such as from Hf-based and Zr-based oxide have excellent electrical properties and thermal stability which will aid to achieve low leakage current while then being paired with metal gates to eliminate Poly-Si depletion. Previously, the self-aligned silicide (SALICIDE) technology has been used to reduce sheet resistance (R_s) of Poly-Si gates and now has been improved to metal silicide [5]. Metal such as tungsten and titanium is alloyed at the top layer of poly-Si to produce an ohmic contact between aluminium metal wire and poly-Si. This helps in producing better physical and electrical properties of a transistor [6].

Currently, the researchers face two conflicting design challenges. The first is to discover high performance design and implementation techniques that can meet the stringent speed constraints and secondly is to consider a low-power design approaches to prolong the operating time of devices. Since power and speed are usually trade-offs, an accurate threshold voltage and leakage current are crucial in order to produce a proper digital Integrated Circuit (IC). Hence, threshold voltage and leakage current are set as the focal point in this research where the control factors are chosen based on former researched parameters' that influenced the device performance the most [7]. Ion implantation such as Halo Implant and S/D Implant plays a vital role in solving scaling consequences as they are comparatively easy to control within a wafer. However, each factor will give different effects on the result as there is no linear relationship between each other [1], [4].

A 32nm $HfO_2/TiSi_2$ PMOS transistor is studied and presented in this paper through investigation and optimization of the variation effects of four process parameters on the transistor's performance. It is in line with the objective of this paper to find the best solution in meetings the ITRS prediction with threshold voltage of 0.103 V (+/- 12.7%) and leakage current below 150 nA/um [8]. Therefore, Taguchi Method is applied to reduce development time consumption besides ensuring the output is in the tolerable quality range. Taguchi method is used to solve multiple process parameter optimization problems with less number of experiments therefore being more robust [6]-[7]. This method is similar to fractional factorial designs but has additional Parameter Design and Tolerance Design [9]. The transistor modelling and analysis are simulated virtually via Silvaco ATHENA module and its electrical characteristics are validated through ATLAS module. The structure of the paper is organized as follows: Section II explores the experiment materials and methods, Section III discusses the performance analysis and the results; and Section IV concludes the paper.

II. EXPERIMENT DESCRIPTIONS

A. PMOS Virtual Fabrication Recipes

PMOS transistor is design virtually through ATHENA module. The fabrication processes is based on previous experiments [1, 4] except the introduction to technology of High-K/Metal Gate and the difference number of dopants. The summary of PMOS design is shown in Table I.

After metal 2 deposition, the PMOS transistor design is considered complete after the metallization and etching are performed for the electrode formation and the bonding pads are opened [1] and then will undergo the electrical characteristic simulation via ATLAS module to provide specific results.

Table 1: PMOS Fabrication Recipe

Process Step	NMOS Parameters
Silicon substrate	• <100> orientation
Retrograde well implantation	• 200Å oxide screen by 970°C, 20 min of dry oxygen • 4.5x10^{11} cm^{-3} Phosphorous • 30 min, 900 °C diffused in nitrogen • 36 min, dry oxygen

STI isolation	• 130Å stress buffer by 900°C, 25 min of dry oxygen
	• 1500Å Si$_3$N$_4$, applying LPCVD
	• 1.0 um photoresist deposition
	• 15 min annealing at,
	  » 900 °C (X1)
	  » 910 °C (X2)

Gate oxide	• diffused dry oxygen for 0.1 min, 815 °C

Vt adjust implant	• 1.8x10^{11} cm^{-3} Boron difluoride
	• 5 KeV implant energy, 7° tilt
	• 20 min annealing at,
	  » 800 °C (Y1)
	  » 795 °C (Y2)

High-K/Metal gate deposition	• 0.002 um HfO2 • 0.1 um TiSi2 • 17 min, 900 °C annealing
LDD implantation	• 5.23x10^{13} cm^{-3} Phosphor • 20° tilt
Sidewall spacer deposition	• 0.047 um Si$_3$N$_4$
S/D implantation	• 1.42x10^{14} cm^{-3} Boron • 10 KeV implant energy • 7° tilt
PMD deposition	• 0.3 um BPSG • 25 min, 850 °C annealing
Metal 1	• 0.04 um Aluminum
IMD deposition	• 0.05 um BPSG • 15 min, 950 °C annealing
Metal 2	• 0.12 um Aluminum

The completed doping profile of 32nm HfO$_2$/TiSi$_2$ PMOS transistor is shown is Fig. 1 where the figure indicates the tabulation of silicon, high-k/metal gate, silicon nitride and aluminium. Doping concentration is said to be one of the factors that will determine the electrical characterization of the transistor such that good doping concentration will ensure the transistor works effectively with perfect gate control and less leakage current [1].

B. Taguchi L9 Orthogonal Array Method

The control factors (CF) which are chosen based on previous research includes Halo Implantation, Halo Tilt Angle, S/D Implantation and Compensation Implantation. The noise factor (NF) parameter on the other hand includes Sacrificial Oxide Layer Temperature and V$_T$ Adjust Implant Temperature. However, in order to establish an impartial simulation circumstance, same control factors and noise factors are used to measure both V$_{TH}$ and I$_{OFF}$. It is simplify the results and strengthen the conclusions. The values for process parameters and noise factors at different levels are listed in Table II and Table III respectively. The L9 Orthogonal Array experimental layout can be seen at [6].

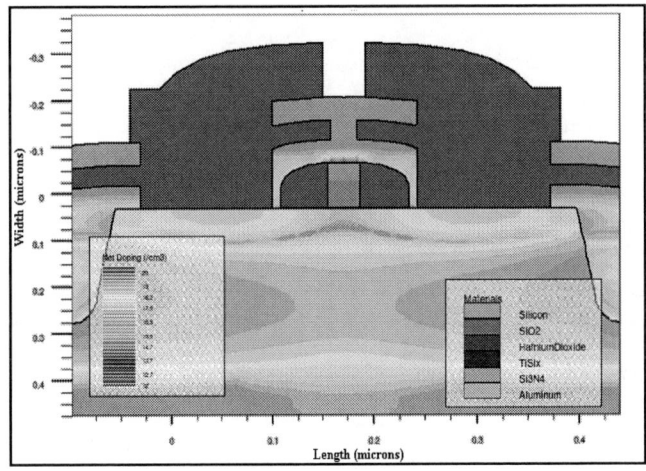

Fig. 1. The Doping Profile of 32nm PMOS Transistor

TABLE I.   CONTROL FACTORS AND THEIR LEVELS

Control Factor (CF)	Unit	L1	L2	L3
Halo Implant, (A)	atom/cm^3	5.230x10^{13}	5.234 x10^{13}	1.524 x10^{13}
Halo Tilt Angle, (B)	°	18	20	25
S/D Implant, (C)	atom/cm^3	1.40x10^{13}	1.41 x10^{13}	1.42 x10^{13}
Compensation Implant, (D)	atom/cm^3	0.98x10^{12}	1.0 x10^{12}	1.10 x10^{12}

TABLE II.   NOISE FACTORS AND THEIR LEVELS

Noise Factor (NF)	Unit	L1	L2
Sacrificial Oxide Growth Temperature (X)	°C	900	910
V$_T$ Adjust Implant Temperature (Y)	°C	800	795

III. PERFORMANCE ANALYSIS AND RESULTS

A. ATLAS Overlay Results

The results for electrical characteristic graphs of PMOS transistor which generated via ATLAS are as shown in Fig. 2 and Fig. 3 respectively where Fig. 2 shows the graph of Drain current (Id) vs. Drain voltage (Vd) at Vg=0.5 V, 1.1 V, 2.2 V and 3.3 V and Fig. 3 shows the graph of Drain current (Id) vs. Gate voltage (Vg) at Vd=0.1 V and 1.1 V.

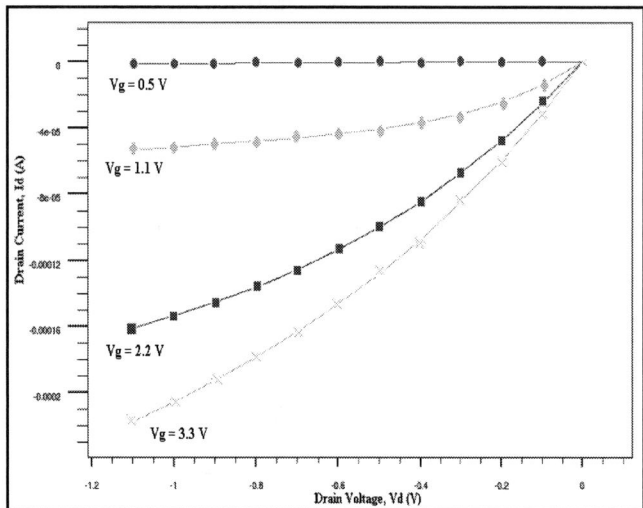

Fig. 2. Graph of Drain current (Id) versus Drain voltage (Vd)

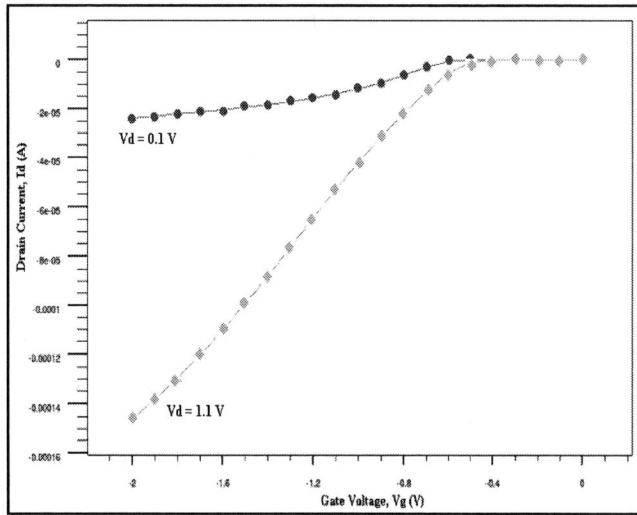

Fig. 3. Graph of Drain current (Id) versus Gate voltage (Vg)

B. V_{TH} and I_{OFF} Optimization

PMOS parameters are optimized after 36 simulations as per laid out in Table III. However, the details of the analysis can be comprehended at [1, 4, 5, 6]. This statistical approach is being practices to get the optimum V_{TH} as well as lowest possible I_{OFF} by first, examining all four control factors which will give an output nearest to the ITRS prediction. Then it is optimized using Taguchi analysis with different noise factor. In this research, threshold voltage of PMOS device belongs to the nominal-the-best quality characteristic while its leakage current belongs to the smaller-the-best quality characteristics. The experimental results of V_{TH} and I_{OFF} are as shown in Table IV and Table V respectively.

The next step is analysing the factor effect percentage on S/N Ratio as it indicates the priority of a process parameter to reduce variation. Larger S/N Ratio percentage give the better characteristic quality and greater influence on the device performance. In this experiment, the effect of each process parameter on the S/N Ratio at different levels can be separated out because the experimental design is orthogonal. At this stage, the priority of the process parameters with respect to the V_{TH} and I_{OFF} are scrutinized to determine the optimum combinations of the process parameters. The process is called analysis of variance (ANOVA). The results for S/N Ratio and ANOVA for V_{TH} and I_{OFF} are as shown in Table VI and Table VII respectively.

TABLE III. V_{TH} VALUES FOR PMOS DEVICE

Exp. No.	Threshold Voltage, V_{TH} (V)			
	X_1,Y_1	X_1,Y_2	X_2,Y_1	X_2,Y_2
1	0.0790799	0.079359	0.0787406	0.0790229
2	0.102321	0.102399	0.101938	0.102219
3	0.189397	0.189699	0.18902	0.18932
4	0.076547	0.0768266	0.0762139	0.0764938
5	0.0966123	0.0968917	0.0962289	0.0965094
6	0.199963	0.20027	0.199597	0.199898
7	0.0691116	0.069389	0.0687765	0.0690557
8	0.112411	0.112705	0.112029	0.112321
9	0.194086	0.194386	0.193714	0.19401

TABLE IV. I_{OFF} VALUES FOR PMOS DEVICE

Exp. No.	Leakage Current, I_{OFF} (pA/um)			
	X_1,Y_1	X_1,Y_2	X_2,Y_1	X_2,Y_2
1	73.5719	73.0653	74.2545	73.7427
2	40.1564	39.8847	40.5765	40.3017
3	4.93999	4.90833	4.98246	4.95049
4	78.3573	77.8171	79.0855	78.5397
5	47.4551	47.1333	47.9546	47.6291
6	3.82436	3.80009	3.8568	3.83231
7	98.8588	98.1756	99.7802	99.0899
8	29.7833	29.5833	30.0926	29.8903
9	4.45662	4.42821	4.49474	4.46604

TABLE V. S/N RESPONSE AND ANOVA FOR V_{TH}

CF	S/N Ratio (dB)			Factor Effect (%)	
	L1	L2	L3	Variance	Mean
A	53.54	34.72	52.62	30.81	1.85
B	49.45	52.39	39.03	13.48	94.81
C	35.17	53.57	52.12	28.64	0.5
D	52.59	35.47	52.81	27.07	2.85

TABLE VI.   S/N Response and ANOVA for I_off

CF	S/N Ratio (dB)			Factor Effect on Variance (%)
	L1	L2	L3	
A	212.23	212.30	212.53	0.015
B	201.62	208.29	227.16	98.635
C	213.83	212.34	210.89	1.219
D	212.04	212.11	212.91	0.131

Results show that Halo Implantation consumes highest influence on V_{TH} (30.81%). Thus, Halo Implantation is set to be the dominant factor. Adjustment factor is chosen based on lowest possible variance with highest mean impact. Hence, Halo Tilt Angle is set to be the adjustment factor. This parameter is then swept between 18° to 25° to get V_{TH} closer to the nominal value. For I_{OFF} analysis, Halo Tilt Angle contributes a major effect (98.64%) on variance and hence it is set as a dominant factor. The recommended process parameters for V_{TH} and I_{OFF} are A_1, $B_{2\ (swept)}$, C_2, D_3 and A_3, B_3, C_1, D_3 respectively. The best setting parameters which has been suggested by Taguchi analysis is as shown in Table VIII while Table IX shows the results of confirmation experiment at different noise level to verify the accuracy of Taguchi analysis prediction. The comparison between ITRS prediction and the simulation results of V_{TH} and I_{OFF} are as shown in Table X. From the table, it is clearly shown that PMOS design meets the ITRS standard with even lower targeted I_{OFF}.

TABLE VII.   Best Setting of Process Parameters

Control Factor	Units	Best Value	
		V_{TH}	I_{OFF}
Halo Implantation	atom/cm^3	5.23x10^{13}	5.24x10^{13}
Halo Tilt Angle	°	19.79	25
S/D Implantation	atom/cm^3	1.41x10^{13}	1.40x10^{13}
Compensation Implantation	atom/cm^3	1.1x10^{12}	1.1x10^{12}

TABLE VIII.   Results of Confirmation Experiment

Parameter	X_1,Y_1	X_1,Y_2	X_2,Y_1	X_2,Y_2	Mean
V_{TH} (V)	0.10306	0.103339	0.102679	0.102952	0.1030075
I_{OFF} (pA/um)	3.42287	3.40123	3.45169	3.42984	3.4264075

TABLE IX.   Simulation Results vs. ITRS Prediction

Parameter	ITRS Prediction	Simulation Results
V_{TH}	0.103 V	0.1030075 V
I_{OFF}	150 x 10^{-9} A/um	3.4264075x10^{-12} A/um

IV.   CONCLUSIONS

A comprehensive study on 32nm High-K/Metal Gate PMOS transistor was presented. The device simulations were implemented in Silvaco and optimized using Taguchi method. The optimum threshold voltage and low leakage current are reported to be well within ITRS prediction and thus proves that these experiments are not only possible but achievable taking into account various factors.

ACKNOWLEDGMENTS

The authors would like to express sincere gratitude to En. Muhaimin Aminuddin from ICMic Sdn. Bhd., Universiti Kebangsaan Malaysia (UKM) and Universiti Tenaga Nasional (UNITEN) for moral and operational support throughout the project. Special thanks to MMICare Association for supporting the conference's fee and expenditures via Gift Her With Life Fund (2014) Award.

REFERENCES

[1] F.Salehuddin, I.Ahamd, F.A.Hamid, A.Zaharim, "Influence of HALO and Source/Drain Implantation on Threshold Voltage in 45nm PMOS Device," *Australian Journal of Basic and Applied Sciences*, pp.55-61, 2011.

[2] Hazura, H. , Hanim, A.R. , Mardiana, B. , Menon, P.S, "An analysis of silicon waveguide phase modulation efficiency based on carrier depletion effect," *IEEE-ICSE Proc.*, pp.348-350, 2010.

[3] Kelin J. Kuhn, "CMOS Transistor Scaling Past 32nm and Implications on Variation", *Intel Corporation*, Portland Technology Development.

[4] H.A. Elgomathi, B.Y.Majlis, I.Ahmad, F. Salehuddin, A, Zaharim, F.A. Hamid, "Application of Taguchi Method in the Optimization of Process Variation for 32nm CMOS Technology," *Australian Journal of Basic and Applied Sciences*, pp.346-355, 2011.

[5] Thimas Skotnicki, "Materials and device structure for sub-32nm CMOS nodes", *Microelectronic Engineering*, pp.1845-1852, 2007.

[6] Afifah Maheran A.H., Menon P.S., I.Ahmad , H.A.Elgomati, B.Y. Majlis, F.Salehuddin, "Design and Optimization of 22nm NMOS Transistor," *Australian Journal of Basic and Applied Sciences*, pp.1-8, 2012

[7] A. H. Afifah Maheran, P. S. Menon, I. Ahmad, and S. Shaari, "Effect of Halo structure variations on the threshold voltage of a 22nm gate length NMOS transistor," *Materials Science in Semiconductor Processing*, vol. 17, pp. 155–161, Jan. 2014.

[8] "ITRS 2012. www.ITRS2012.net."

[9] P. S. Menon, K. Kandiah, A. A. Ehsan, and S. Shaari, "The development of a new responsivity prediction model for In(0.53)Ga(0.47)As interdigitated lateral PIN photodiode," *Journal of Optical Communications*, vol. 30, pp. 2-6, 2009.

Finite Element Modeling of Dielectrophoretic Microelectrodes Based on a Array and Ratchet Type

Muhamad Ramdzan Buyong, Norazreen Abd Aziz, Azrul Azlan Hamzah, M.F. Mohd Razip Wee,
Burhanuddin Yeop Majlis
Institute of Microengineering and Nanoelectronic (IMEN),
Universiti Kebangsaan Malaysia (UKM)
43600 Bangi, Selangor, Malaysia
Email: muhdramdzan@ukm.edu.my

Abstract—**This research describes an investigation of nonuniform electric field for dielectrophoretic forces (F_{DEP}) application in particles and cells manipulation. In an electro kinetics occurrence, a miniaturized array and ratchet type microelectrodes has been simulated. The study of optimal F_{DEP} behavior on the electric field distribution for both type microelectrodes was characterized and optimized by finite element method, (FEM). A set of array and ratchet type microelectrode are biased to generate asymmetric electric field distribution. Normalization of microelectrode simulation result shows that array and ratchet type produced a comparable electric field strength and direction. Deployment of additional dimension for array type electrode, three poles produced the highest of electric field strength of 7.513 e^7 V/m and displacement field direction of 2.758 e^{-3} C/m². Simulation results are used to design a higher sensitive and selective of a dielectrophoretic (DEP) microelectrode for selection, collection and processing of particle and cell using optimal F_{DEP} that determination advancement in the development of dielectrophoretic a lab-on-a-chip. Ultimately, the findings of this work is possible to contribute in medical sciences research for the enrichment of stem cell from bone narrow and peripheral blood form via integration DEP into a lab on a chip, (DLOC) concept application.**

Keywords— Microelectrode, dielectrophoretic, lab-on-a-chip,

I. INTRODUCTION

Nowadays, the characteristic of particles and cells in electric field particularly nonuniform fields is huge of interest to scientist of various clusters such as physics, chemistry, life science and engineering [1,2]. The capability to characterize and manipulate of particles and cells that suspended in a fluid by contactless technique has considerable prospective deployment in macro or nano biotechnological platforms for autonomous lab-on-chip devices. Moving forward of characterization for particle, which is lifeless objects not complicated as dealing with live cell. Living cell characterizations need to take further tread and stipulation. Therefore a F_{DEP} is the physical of an electro kinetics occurrence whereby neutral or charged particles and cells exposed to nonuniform electric field required further exploration for dielectrophoretic advances application.

The ultimate goal of this research is generating efficient nonuniform electric field in lab-on-a-chips. Consequently, the novelty and relevant of this study is enhanced nonuniform electric field in term strength and direction, afterward integrating DEP platform microelectrodes into microfluidics channel system. Thus, the entire objective is to configure integration of DEP into lab-on-a-chips.

The potential application in medical sciences of DLOC is deployment in the healthcare technology [3, 4] such as stem cell extraction devices. Previous work by R. Pethig, demonstrated capabilities the enrichment of haematopoietic stem cell from bone narrow and peripheral blood and adult stem cell from adipose tissues [4]. For that reasons, it required for precision technology in order to manipulate and characterize particles and cells that related to the strength and direction of F_{DEP} that produced by nonuniform electric field.

II. THEORY OF DIELECTROPHORETIC FORCES, F_{DEP}

Nonuniform electric field subject to time stationary (DC) or time varying (AC) electric field created a net force that engaged in the direction of locations with related to the polarization properties of the matter. Dielectrophoresis is the motion or particles and cells in a nonuniform electric field, was first observed by Pohl [5] and the latest by R. Pethig [3]. In an electric field *E,* a dielectric particles behaves as an effective dipole with dipole moment *p* proportional to the electric field, that is [7]

$$\mathbf{p} \alpha E \qquad (1)$$

The constant of proportional depends in general on the geometry of the dielectric particle. In the presence of an electric field gradient, the force on a dipole is given by [5]

$$\mathbf{F} = (\,\mathbf{p}.\nabla)E \qquad (2)$$

where P is the constant dipole moment vector, ∇ is del vector and *E* is the external electric field. Combining the two equations the (F_{DEP}) for homogenous particle suspended in electric field gradient is given by [6]

$$\mathbf{F}_{DEP} = 2\,\pi\,\varepsilon_{medium}\,R^3 CMF(\nabla E^2) \qquad (3)$$

where r is the radius of particle, ε_m is the permittivity of the suspending medium and ΔE is the gradient of the rms electric

978-1-4799-5761-3/14 $31.00 © 2014 IEEE

field. CMF is the real part of the Clausius-Mossotti factor, which is given by

$$CMF = \frac{(\varepsilon_{particle} - \varepsilon_{medium})}{(\varepsilon_{particle} + 2\varepsilon_{medium})} \qquad (4)$$

where $\varepsilon_{particle}$ and ε_{medium} are the complex permittivities of the particles or cells and medium.

The collective method to calculate F_{DEP} in term of strength force related to proportional response that depends to factor such as particles or cells that correlate to physical and biophysical properties respectively, medium permittivity and conductivity and sinusoidal input voltage and frequency signal applied [1]. In this work, according to Eq. (3) we know that sinusoidal input voltage of electric field is proportional to the strength of F_{DEP}. Thus, enhancement the strength of F_{DEP} can be further improved by increased sinusoidal input voltage. Alternative to that, the strength force of electric field of F_{DEP} related to microelectrode configuration geometry operating and effectiveness functioning approaches. Therefore, finite element modeling (FEM) was used towards to evaluate the theoretical of electric field force for DEP microelectrodes based on array and ratchet type. Consequently, the results can be scaled to the model by multiplying it with a correction factor without losing information about the particles or cells and medium properties.

III. MODELLING CONCEPTUAL DESIGN

Towards understanding the details of the electric field behavior, dielectrophoretic array microelectrodes are arranged in array of two columns and one row as shown in Fig 1.

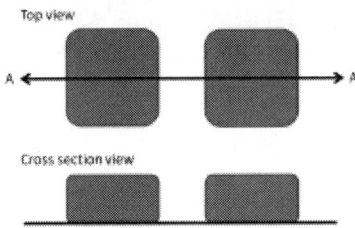

Fig. 1. Top and cross sectional view of DEP array for bottom microelectrode configuration.

The geometry and boundary condition of dielectrophoretic array type microelectrodes with particle used in the FEM model is illustrates in Fig 2. The sphere shape particle is modeled in two dimensional as a sphere with diameter of 8 µm and height from microelectrodes 2 µm. The array type of dielectrophoretic microelectrodes is modeled in geometry of a square or a regular quadrilateral. The length and width is 10 µm with a height of 2 µm. The distances of each electrode are variable spacing of 5 µm and 10 µm in arranged in an array. DEP array electrodes divided into two type configurations, which are three and two polarity microelectrodes. The three polarity microelectrodes consisted of P-P-N and P-N-N and for the two polarity microelectrode P-N. P is for symbolized for a positive node and N for a negative node polarity. The physical construction of the three polarity microelectrode is building for P-P-N, from left to right is a positive microelectrode polarity (highlight line) from the bottom and top microelectrode and a negative microelectrode polarity from bottom as shown in Fig. 2.

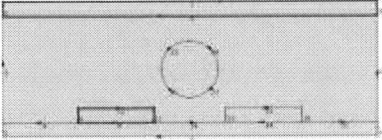

Fig. 2. DEP array, three poles microelectrode of P-P-N configuration.

Thus for P-N-N, from left to right is a positive microelectrode polarity (highlight line) from the bottom and a negative microelectrode polarity from the top and bottom as shown in Fig. 3.

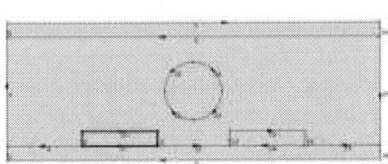

Fig. 3. DEP array, three poles microelectrode of P-N-N configuration.

The geometry and boundary condition of dielectrophoretic ratchet type microelectrodes is modeled in geometry of a triangle and a rectangle as shown in Fig 4. A triangle is an equilateral triangle with all side has the similar length of 10 µm with all angles measuring of 60˚. A rectangle length and width are 10 µm and 5 µm with a height of 2 µm. Similar to the ratchet type microelectrodes, the sphere shape particle is modeled into two dimensional as a sphere with diameter of 8 µm and height from microelectrodes 2 µm. The distances of triangle and a rectangle electrode with variable spacing of 5 µm and 10 µm in arranged in a sequence.

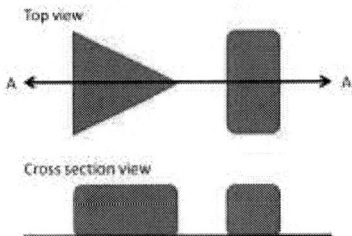

Fig. 4. Top and cross sectional view of DEP racthet for bottom microelectrode configuration.

The physical construction of dielectrophoretic ratchet type microelectrodes is building by P-N only. From left to right is a triangle microelectrodes connected to the positive node polarity (highlight line) with a rectangular connected to the negative node polarity. Both constructions are form on the bottom substrate only as shown in Fig 5.

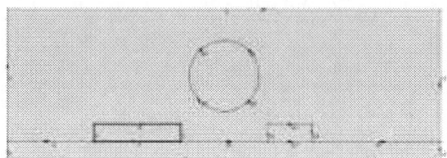

Fig. 5. DEP racthet, two poles microelectrode of P-N configuration.

IV. RESULTS AND DISCUSSIONS

Subsequent setting up the model as describe at previous section, the electric field is simulated approximately at the microelectrodes region in the AC/DC module. FEM results showed at Fig. 6 to 9, a closer observation on four parameters. Firstly is the surface for electric field V/m, second is the contour for displacement field C/m^2, third is the green cone for electric field direction and lastly is black arrow for electric displacement field direction in the solutions domains region the microelectrodes. Surface and contour proved that electric field strength and concentration is maximums close to the microelectrodes edge for all configurations. The green cone and black arrow indicate the direction of electric field gradient and electric displacement field, towards the microelectrodes from positive to negative source.

Investigation based on the discovery from reproduction model is the observation of the strength and direction of an electric field in the water as medium with suspending particle. Firstly we observed, the electric field lines constantly move towards from a positively charged microelectrode to a negatively charged microelectrode (green cone is the electric field direction and black arrow is the electric displacement field direction). Second, the electric field lines produced did not tangle with each other. Third, the electric field strength increased when the distance between positively and negatively charged microelectrode are reduced. Fourth, the electric field lines meet at the surface of a microelectrode, with the lines perpendicular to the surface of microelectrode. Finally, the electric field lines are biased when the electric field interacted with particles as FEM results showed at Fig. 6 to 9.

Further analysis based on numerical discovery from reproduction model as shown in Table 1 and plotted in Figure 10. It was found that, the electric and displacement field value are increased due to electrode spacing electrodes is decreased. In fact the electric and displacement field value for ratchet electrode type of P-N is slightly similar compared to array electrode type of P-N in 10 μm and 5 μm spacing. However the electric and displacement field value for array electrode type of P-P-N and P-N-N are higher compared to ratchet electrode P-N respectively in 10 μm and 5 μm spacing.

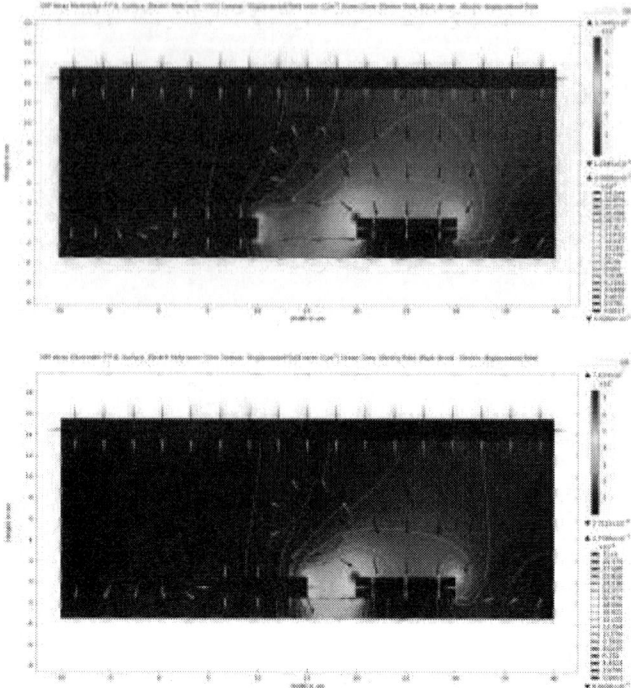

Fig. 6. DEP array, three poles microelectrode of P-P-N configuration with 10 μm (top) and 5μm (bottom) spacing.

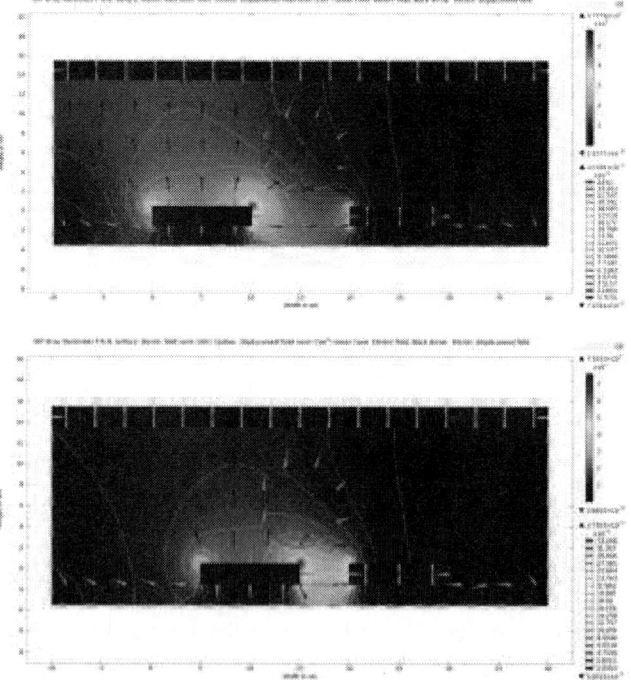

Fig. 7. DEP array, three poles microelectrode of P-N-N configuration with 10 μm (top) and 5μm (bottom) spacing.

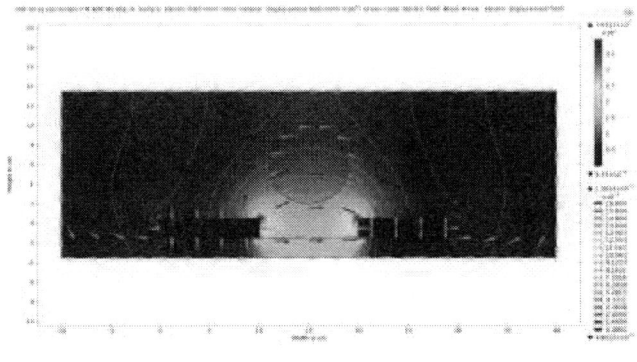

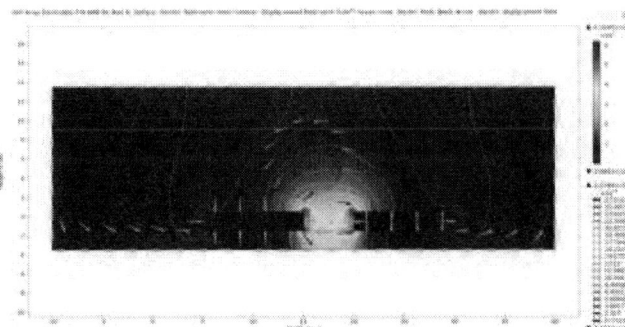

Fig. 8. DEP array, two poles microelectrode of P-N configuration with 10 μm (top) and 5μm (bottom) spacing.

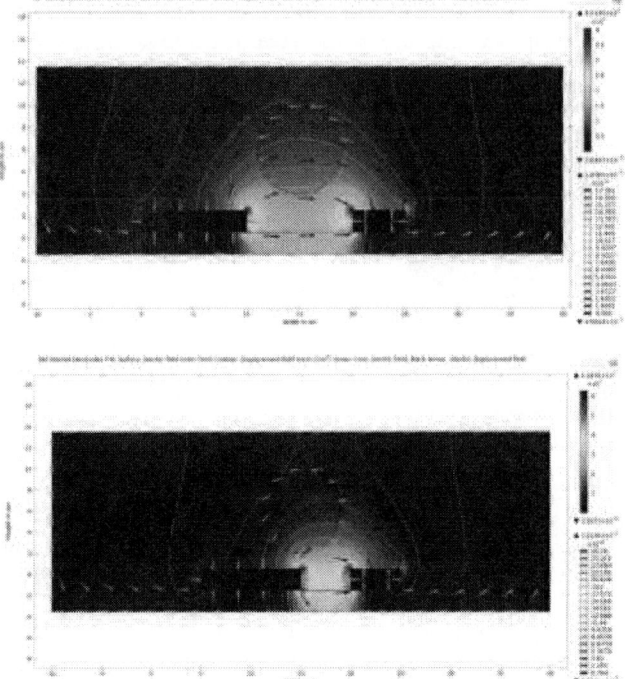

Fig. 9. DEP ratchet, two poles microelectrode of P-N configuration with 10 μm (top) and 5μm (bottom) spacing.

Table 1: Simulation result of electric and displacement field.

Electrode type	Array electrode type						Ratchet electrode	
Spacing μm	10			5			10	5
Arrangement number	P-P-N	P-N-N	P-N	P-P-N	P-N-N	P-N	P-N	P-N
Electric field, V/m E+7	5.7	5.8	4.0	7.4	7.5	6.2	4.1	6.3
Displacement field, C/m² E-3	2.1	2.0	1.4	2.8	2.6	2.3	1.4	2.2

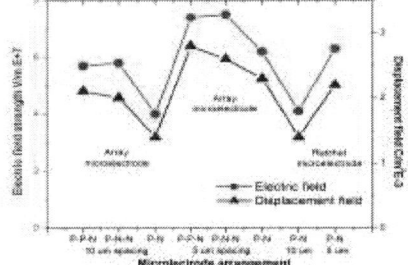

Fig. 10. Electric and displacement field of array and ratchet microelectrodes configuration.

V. CONCLUSION

We have characterized a finite element modeling for dielectrophoretic microelectrodes based on an array and ratchet type. Simulation result show that ratchet type give slight similar effective electric field distribution strengths compared to array type in two poles microelectrode arrangement. Deployment of additional poles for array type electrode, three poles gave highest of electric field and displacement field distribution. This finally allows the reliable and efficient of simultaneous analysis by integration of the three dimensional platforms into microfluidics system to produced DLOC devices for selection, collection or processing of particle and cell using optimal field of force.

REFERENCES

[1] T. Honegger, David Peyrade, Moving Pulsed Dielectrophoresis. Lab Chip 2013, 13, 1538 -1545.

[2] K. Khoshmanesh., S. Nahavandi., S. Baratchi., A. Mictchell., K. Kalantar-Zadeh, Dielectrophoretic Platforms for Bio-microfluidics Systems. Biosensors and Bioelectronics 26 (2011) 1800-1814.

[3] R. Pethig, Review Article Dielectrophoresis: Status of the Theory, Technology and Applications. Biomicrofludics 4, (2010) 022811-1 – 022811-35.

[4] R. Pethig., A. Menachery., S. Pells., P. De Sousa, Dielectrophoresis: A Review of Application for Stem Cell Research. Journal of Biomedic. and Biotech. (2010) 1-7.

[5] H. A. Pohl., 1978. Dielectrophoresis. Cambridge, UK: Cambridge University Press.

[6] K. F. Hoettges., Dielectrophoresis as a Cell Characterisation Tool. Microengineering in Biotech. Methods in Molecular Biology 583 (2010) 183-198.

[7] P. J. Burke., Nanodielectrophoresis: Electronic Nanotweezers. American Scientific Publisher (2003)

Dielectrophoretic Characterization of Array Type Microelectrodes

Muhamad Ramdzan Buyong, Norazreen Abd Aziz, Azrul Azlan Hamzah, Burhanuddin Yeop Majlis
Institute of Microengineering and Nanoelectronic (IMEN),
Universiti Kebangsaan Malaysia (UKM)
43600 Bangi, Selangor, Malaysia
Email: muhdramdzan@ukm.edu.my

Abstract—**This research describes an investigation of nonuniform electric field for dielectrophoretic forces, F_{DEP} application in particles manipulation. In an electrokinetics occurrence, a miniaturized array type of two poles microelectrodes has been simulated using engineered particle and tested using graphite metalloid particles. The particles can be attracted towards the regions of strong electric field depending upon the particles is more polarisable than the suspending medium. Dielectrophoresis offers the controllable, selective and accurate manipulation of target graphite metalloid particles. The surface area of graphite attracted to microelectrodes gradually increased starting from 4 seconds for 3412, 3845, and 3764 um², hence 4589, 4465 and 4739 um² at the 6 seconds mark and finally 5588, 5569 and 5644 um² after 8 seconds for three different test run respectively. Further study of optimal F_{DEP} behavior on the electric field distribution for three poles microelectrodes was characterized by finite element method, (FEM). The outcome, F_{DEP} response is further improved by additional poles microelectrodes from top side, instead of side by side in term the strength and direction of electric and displacement field. Ultimately, the findings of this work is possible to contribute in medical sciences research for the enrichment of stem cell from bone narrow and peripheral blood form via integration DEP into a lab on a chip, DLOC concept application.**

Keywords— Microelectrode, dielectrophoretic, lab-on-a-chip,

I. INTRODUCTION

Placing two planar electrodes and applying an alternating current voltage across them, an nonuniform electric field can be created. This field can be used to trap, dielectrophoretically particle. The ability to manipulate small objects is essential to experiments on the micron and nano scale level [1,2], especially in biology application. When a polarizable biological particle such as cell is exposed to an electric field, electrical charges are induced on the particle and medium interface. The induced charges make up dipoles aligned parallel or against to the applied field dependence frequency input applied that relative to biological particle and medium properties.

Dielectrophoretic, DEP the induced motion of polarizable biological particle in a nonuniform electric field, has been proven as a versatile mechanism to transport, accumulate, separate and characterize bio particles in microfluidic systems. The integration of DEP systems into the microfluidics enables the inexpensive, fast, highly sensitive, highly selective and label-free detection and analysis of target biological particle. Therefore the potential applications of this research in molecular research, forensic science, diseases research and medical discovery.

Based on the literature review, further potential application in medical sciences using DEP, where deployment is in the healthcare technology [3, 4] such as stem cell extraction devices. Previous work by R. Pethig, demonstrated capabilities the enrichment of haematopoietic stem cell from bone narrow and peripheral blood and adult stem cell from adipose tissues [4]. For that reasons, it required for precision technology in order to characterize and manipulate particles and cells that related to the strength and direction of F_{DEP} produced by nonuniform electric field.

The ultimate goal of this research is generating efficient nonuniform electric field in lab-on-a-chips. Consequently, the novelty and relevant of this study is enhanced nonuniform electric field in term of strength and direction using two poles microelectrodes in order to have capability attracted red particle and repelled green particle shown in Figure 1.

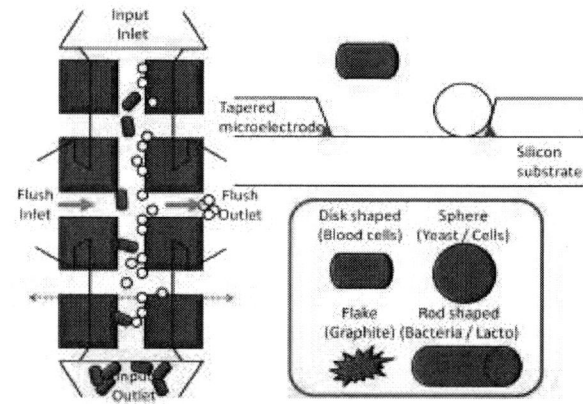

Fig. 1. Schematic of two poles microelectrode configuration in manipulation, attracted sphere particle towards microelectrodes and repelled oval particle from microelectrodes

II. THEORY OF DIELECTROPHORETIC FORCES, F_{DEP}

Nonuniform electric field subject to time stationary (DC) or time varying (AC) electric field created a net force that engaged in the direction of locations with related to the polarization properties of the matter. Dielectrophoresis is the

978-1-4799-5761-3/14 $31.00 © 2014 IEEE

motion or particles and cells in a nonuniform electric field, was first observed by Pohl [5] and the latest by R. Pethig [3]. In an electric field E, a dielectric particles behaves as an effective dipole with dipole moment p proportional to the electric field, that is [7]

$$p \propto E \qquad (1)$$

The constant of proportional depends in general on the geometry of the dielectric particle. In the presence of an electric field gradient, the force on a dipole is given by [5]

$$F = (p.\nabla)E \qquad (2)$$

where P is the constant dipole moment vector, ∇ is del vector and E is the external electric field. Combining the two equations the (F_{DEP}) for homogenous particle suspended in electric field gradient is given by [6]

$$F_{DEP} = 2\pi\varepsilon_{medium}R^3 CMF(\nabla E^2) \qquad (3)$$

where r is the radius of particle, ε_m is the permittivity of the suspending medium and ΔE is the gradient of the rms electric field. CMF is the real part of the Clausius-Mossotti factor, which is given by

$$CMF = \frac{(\varepsilon_{particle} - \varepsilon_{medium})}{(\varepsilon_{particle} + 2\varepsilon_{medium})} \qquad (4)$$

where $\varepsilon_{particle}$ and ε_{medium} are the complex permittivities of the particles or cells and medium.

The collective method to calculate F_{DEP} in term of strength force related to proportional response that depends to factor such as particles or cells that correlate to physical and biophysical properties respectively, medium permittivity and conductivity. sinusoidal input voltage and frequency signal applied [1] shown in Fig. 2

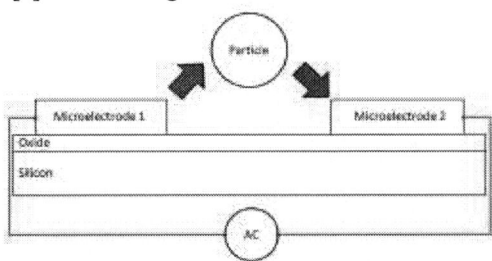

Fig. 2. Schematic of two poles microelectrode array type of P-N configuration.

In this work, according to Eq. (3) we know that sinusoidal input voltage of electric field is proportional to the strength of F_{DEP}. Thus, enhancement the strength of F_{DEP} can be further improved by increased sinusoidal input voltage. Alternative to that, the strength force of electric field of F_{DEP} related to

microelectrode configuration geometry operating and effectiveness functioning approaches. Therefore, finite element modeling (FEM) was used towards to evaluate the theoretical of electric field and displacement field for DEP microelectrodes shown in Fig. 3.

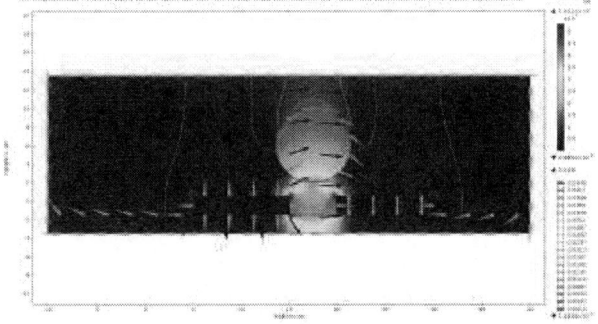

Fig. 3. COMSOL FEM of two poles microelectrode array type of P-N configuration.

III. EXPERIMENTAL PROCEDURE

Investigation via experimental work for two poles microelectrode array type of P-N, (P for positive polarity and N for negative polarity) configurations was developed using graphite powder size of < 20 µm with mass of 0.05 gram blended with deionised water with quantity of 100 ul. Utilizing precision syringe to produce a microliter droplet of 50 µl ± 5%, the mixture of graphite metalloid particle and deionised water is dispensed to the center of the four quadrant microelectrode shown in Figure 4.

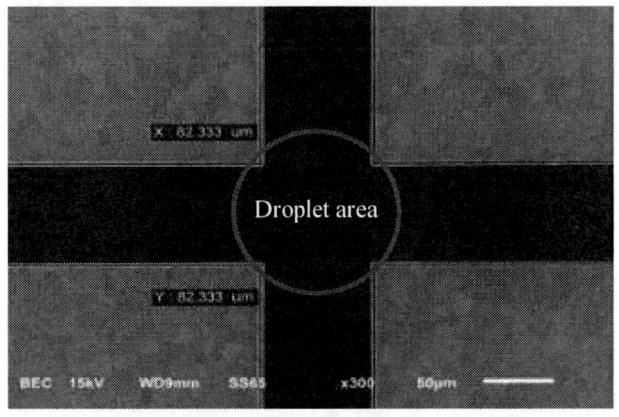

Fig. 4. SEM image of the two poles microelectrode with spacing of 80 µm.

The experimental work for characterization of DEP microelectrode is use micromanipulator bench station with standard prober system. The quadrant microelectrode is separated into four zones, the top left and right corner of the microelectrode is connected to the positive polarity of source node while the bottom left and right corner of the microelectrode is connected to the ground node. Three test run using similar input for 10 Vp-p voltage at 1.5 MHz frequency was applied to the four quadrant microelectrode for a period of 8 seconds per test run. The signal input is directly connected to the prober and the graphite metalloid particle movements is visualized using the microscope as shown in Fig. 5.

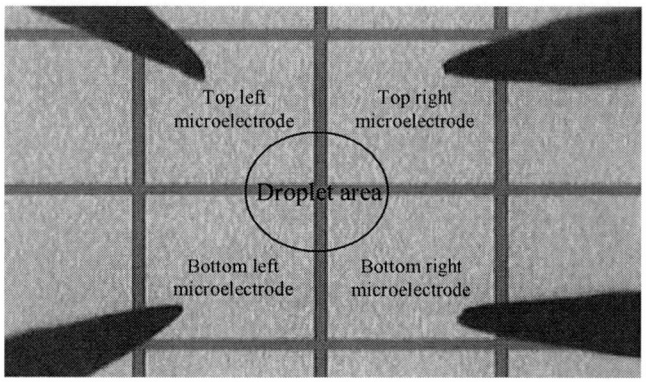

Fig. 5. Probing of four quadrant dielectrophoretic microelectrode.

IV. RESULTS AND DISCUSSIONS

Further experimental work set-up includes sinusoidal signal generator, micromanipulator bench system and microscope with colour video camera to quantify DEP trapping response. Then images were compiled and subsequently analyzed to measure graphite surface area of graphite trapped at microelectrode within specified areas showed in Fig. 6 to 8. From the experimental point of view, graphite intact and trapped at the edge of microelectrode indicated as positive dielectrophoresis. On the other hand, still several graphite un-trap bypasses through two poles microelectrodes.

The surface area of graphite attracted to microelectrodes gradually increased starting from 4 seconds for 3412, 3845, and 3764 um2, hence 4589, 4465 and 4739 um2 at the 6 seconds mark and finally 5588, 5569 and 5644 um2 after 8 seconds for three different test run respectively and plotted in Figure 9.

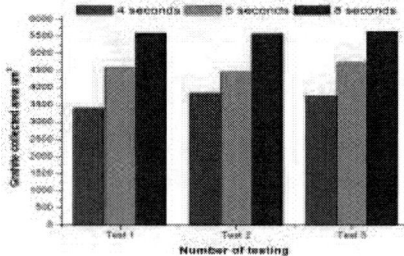

Fig. 9. Comparison of graphite collected area for 3 different number of testing.

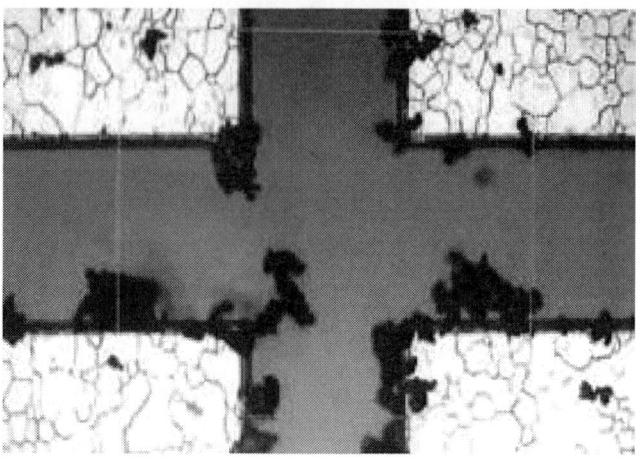

Fig. 6. Surface area of graphite attracted to microelectrodes edges at 4 seconds mark.

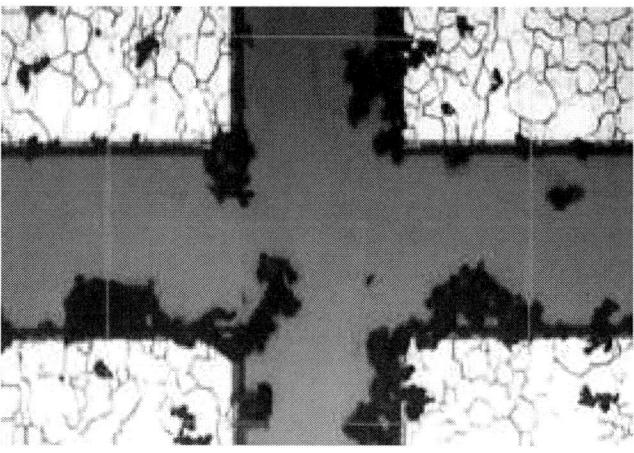

Fig. 7. Surface area of graphite attracted to microelectrodes edges at 6 seconds mark.

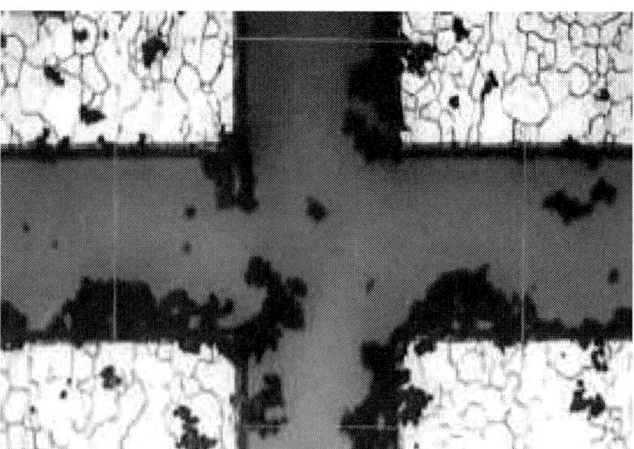

Fig. 8. Surface area of graphite attracted to microelectrodes edges at 8 seconds mark.

In order to achieve highly sensitive and selective characterization of targeted particle, three dimensional DEP platforms is required, shown in Figure 10. Additional microelectrodes from top side increase the strength of the nonuniform electric field relative to increasing F_{DEP}. Based on the simulation result we can enhance the strength and direction of nonuniform electric field values when we increased the number of poles of the microelectrodes from two to three-poles. Consequently, in order to achieve a highly efficient characterization of targeted graphite metalloid particles, the three poles microelectrodes DEP platform is proposed. The simulation result demonstrated the significant advantages of having additional microelectrodes at the top side. In fact, analysis from the simulation proved that additional microelectrodes on the top side increase the nonuniform electric field from top side in order increases the F_{DEP}. According to simulations result, the average ratio of electric and displacement field on two and three poles microelectrode approximately is 1:1.35. In this sense three poles microelectrode F_{DEP} was 1.35 stronger than two poles microelectrode F_{DEP} that using same input voltage and frequency value. Positively, the additional microelectrode from the top side angle is the catalyst to increase the strength and direction of nonuniform electric field.

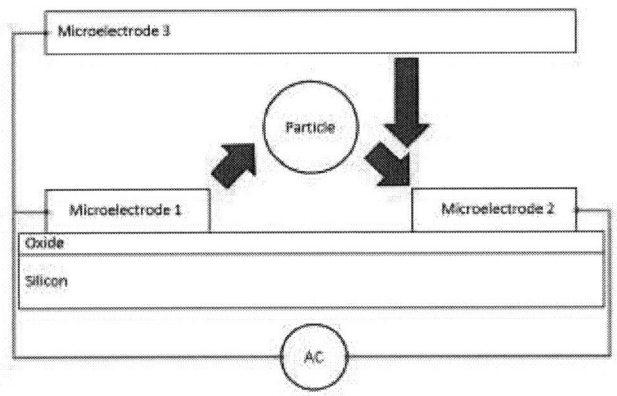

Fig. 10. Schematic of three poles microelectrode array type of P-P-N configuration.

Collectively, the novelty and relevance of this work is to increased number of poles of electric field microelectrode, thus electric and displacement field value is increased. The simulation result demonstrated the significant advantages of having additional microelectrodes at the top side. In fact, analysis from the simulation proved that additional microelectrodes on the top side increase the nonuniform electric field from top side in order increases the F_{DEP} as shown in Figure 11. This enable to achieve highly selective and sensitive characterization of targeted particles and cells.

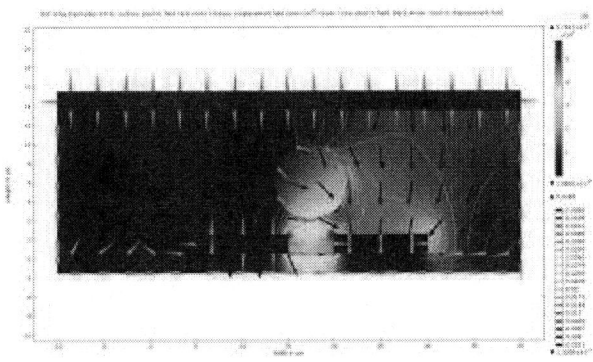

Fig. 11. COMSOL FEM of three poles microelectrode array type of P-P-N configuration.

V. CONCLUSION

We have characterized via experimental work and a finite element modeling for two poles dielectrophoretic microelectrodes. Both result of two poles dielectrophoretic microelectrodes be able to further improved the performance to have higher effective electric field distribution strengths. Deployment of additional poles microelectrode, three poles dielectrophoretic microelectrodes gave highest of electric field and displacement field distribution via simulation Collectively, the novelty and relevancy of this work is to do characterization of two and three poles DEP microelectrode platforms. Finally allows the reliable and efficient of simultaneous analysis by integration of the three dimensional platforms into microfluidics system for selection, collection or processing of particle and cell using optimal field of force.

REFERENCES

[1] T. Honegger, David Peyrade, Moving Pulsed Dielectrophoresis. Lab Chip 2013, 13, 1538 -1545.

[2] K. Khoshmanesh., S. Nahavandi., S. Baratchi., A. Mictchell., K. Kalantar-Zadeh, Dielectrophoretic Platforms for Bio-microfluidics Systems. Biosensors and Bioelectronics 26 (2011) 1800-1814.

[3] R. Pethig, Review Article Dielectrophoresis: Status of the Theory, Technology and Applications. Biomicrofludics 4, (2010) 022811-1 – 022811-35.

[4] R. Pethig., A. Menachery., S. Pells., P. De Sousa, Dielectrophoresis: A Review of Application for Stem Cell Research. Journal of Biomedic. and Biotech. (2010) 1-7.

[5] H. A. Pohl., 1978. Dielectrophoresis. Cambridge, UK: Cambridge University Press.

[6] K. F. Hoettges., Dielectrophoresis as a Cell Characterisation Tool. Microengineering in Biotech. Methods in Molecular Biology 583 (2010) 183-198.

[7] P. J. Burke., Nanodielectrophoresis: Electronic Nanotweezers. American Scientific Publisher (2003)

Structural damage of Si-implanted in the In$_{0.53}$Ga$_{0.47}$As thin film

Muhammad Zulkhairi Roslan, , Dilla D. Berhanuddin, Mohd Ambri Mohamed, M.F. Mohd Razip Wee, Farhad Larki, Burhanuddin Yeop Majlis

Institute of Microengineering and Nanoelectronics (IMEN)
Universiti Kebangsaan Malaysia (UKM),
(National University of Malaysia)
43600 UKM, Bangi, Selangor, D.E.,
Malaysia

mf.mohdrazipwee@gmail.com

Abstract—**Damage profiling of implanted ions in semiconductor's layer is crucial in order to accurately estimate the ion distribution and concentration in the target substrates. It also gives the predicted number of vacancies and interstitials after the collision events. This is particularly important prior to the ion implantation so as to reduce the defect formation and damage to the target's lattice which subsequently degrade the performance of the device. In this paper, we studied the optimized energy and range of ions implanted silicon in In$_{0.53}$Ga$_{0.47}$As film by utilizing the Stopping Range of Ions in Matter (SRIM) simulation. The effects of implantation energy in different thickness are also discussed based on creation of phonons, vacancies and ionization.**

Keywords—Silicon implantation; implantation energy; ;depth profiles;SRIM

I. Introduction

Ion implantation is the critical innovation that catalyzed the integrated circuit industry in the 1970's until its exponential growth in the following decades. With the ion implantation technique, a specific amounts of n-type and p-type dopant can be implanted with an outstanding control and repeatability thus enable to create the source and drain region in the CMOS device However, several parameters in the ion implantation process such as ion energy and dose, dose rate, implantation temperature, time and annealing method need to be handled precisely as they contribute to the changes in lattice structure of targeted substrates, thus affecting their electrical properties. The excessive value of these parameters can lead to a critical lattice defects (Stacking faults, micro-twins and impurities) that will damage the crystalline structure and subsequently the performance of the device. This problematic has been partially resolved by introducing an annealing step at high temperature usually higher than 700° C in order to obtain a homogeneous dopant and impurity redistribution.

Since the matured silicon technology for CMOS is approaching its theoretical limits InGaAs has become a very promising candidate for the highly scaled device due to its high electron mobility, high injection velocity and unique band alignment[[1],[2, 3].Furthermore, the technology of compound semiconductor like InGaAs are approaching to be compatible with conventional CMOS process. However, there are a few challenges need to be overcome before the realization of an efficient and commercially produced InGaAs device. Low leakage and thermally stable gate dielectric with low interface state density, as well as defect free junctions with low external or access resistance are some major problems that need to be resolved before it can be integrated in the commercialization stage[4].

The dopants for InGaAs consists of carbon, Beryllium and iron [5] but Silicon is a common material use as dopants for InGaAs substrates during the ion implantation process. It acts amphoterically in InGaAs and forming Si$^+_{Ga}$ or Si$^-_{As}$ defects [6]. Thus, the net carrier concentration is determined by the fraction of Si atoms occupying gallium sites compared with those occupying arsenic sites or contributing to the acceptor-like defect complexes.

In this work, a numerical simulation of silicon implantation in the In$_{0.53}$Ga$_{0.47}$As as a target material has been investigated by the Stopping Range of Ions in Matter (SRIM).Different parameters such as energy of implantation and channel thickness have been varied in order to fully investigate and obtain the optimized penetration depth and minimized the lattice defects caused by the Si ions. During implantation, the main stopping mechanisms such as the phonons and ionization for different thickness at optimized energy are investigated.

II. Simulated structures and models

Using the TRIM software, users define the ion and the target. For the ion, Si$^+$ is selected with the angle is been set at 0°, perpendicular to the target layer. Meanwhile, the target data is constructed by the three layered structure; oxide/channel/buffer layer. For the oxide layer, Al$_2$O$_3$ is chose to be paired with the In$_{0.53}$Ga$_{0.47}$As's channel and buffer layer as it offers a high band gap ($\approx$9 eV), a high breakdown electric field (5-30 MV/cm), and a satisfactory result in term of

978-1-4799-5761-3/14 $31.00 © 2014 IEEE

equivalent oxide thickness (EOT) with high thermal stability (up to at least 1000°C)[2][7]. However, extra precaution needs to be taken when defining a compound target layer, such as $In_{0.53}Ga_{0.47}As$. The atomic stoichiometry needs to be correctly defined; as here we define 0.53 for Indium, 0.47 for Gallium and 1.00 for Arsenic atoms. Then, the density of this $In_{0.53}Ga_{0.47}As$'s channel and buffer layer is defined at $5.50 g/cm^3$. For the first set of simulation, a 2/10/50 nm of oxide/channel/buffer layer is established. In order to investigate the maximum penetration of Si+, different energies have been used. The same method is applied to the second and the third set of simulation, with only the channel thickness is modified to 20nm for the second set and 30nm for the third. Total number of ions for this simulation is fixed at 10000 which are sufficient for the simulation to be considerably accurate.

In our simulation, two types of the calculations have been employed; quick calculation of damages and the full calculation of damages. The different between these two is the way it calculates the damage. The full calculation will follow every re-coil until its energy drops below the lowest displacement energy of any target atom. Hence all collisional damage to the target is analyzed. [8] So, the quick calculation is used to find the optimize energy for the chosen thickness of channel layer. Once found, the full calculation is employed to fully investigate the implantation process.

III. Results and discussions

During the first step of the simulation, structures with 2 nm Al_2O_3 as the oxide layer and $In_{0.53}Ga_{0.47}As$ with different thickness of 10, 20, and 30 nm were implanted by Si ions with various energies of 5 keV to 30 keV. The optimized Si energies for implantation of the $In_{0.53}Ga_{0.47}As$ layer with 10, 20 and 30nm were 7, 14, and 24 keV, respectively. It was observed that Si with higher energies than the above mentioned threshold, traverse the layer and the peak of the ions concentration shift toward the buffer layer. Fig.1 (a-c) presents the distribution of the Si in different depth of the target, for constant oxide thickness and various $In_{0.53}Ga_{0.47}As$ layer thicknesses in optimized incident energy. Since the Ordinate unit is in (Atoms/cm3)/ (Atoms/cm2), when it is multiplied by an implantation dose (ions/cm2), the dopants concentration (atoms/cm3) vs. depth can be obtained. A 50 nm buffer layer analogous to the channel at the bottom was added in order to catch most of the ions which penetrate to the first layer. Practically, this buffer layer can be used from the same or different materials as the channel with different doping concentration. For our case, we used the same material. The nature of damage induced by Si+ ions implanted into the structure of $In_{0.53}Ga_{0.47}As$ film is another parameter of interest in this study. At room temperature the target damage which is mainly appear by self-annealing, disappears because the lattice atoms have adequate energy to allow simple target damage to re grow back into its original crystalline form. In our simulation due to the limitation of SRIM software, the damage which is calculated is that which would happen for an

implantation at $0°$ K. Practically, the implantation is realized with nitrogen cooling at the target area to minimize in situ annealing [7]. The basic damage types is completely analogue to the room temperature, however the quantity of final damage will be varied.

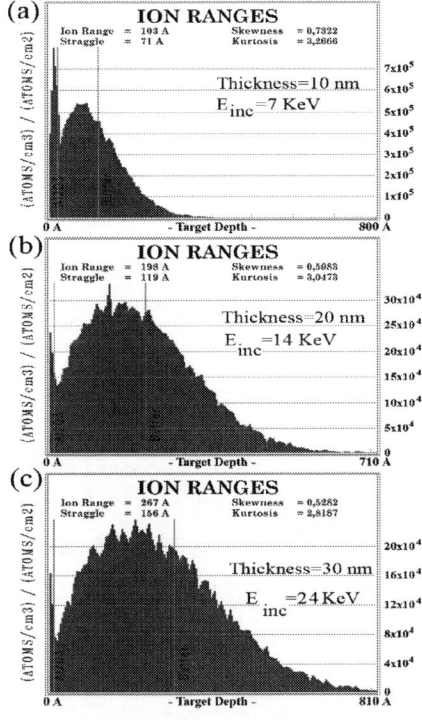

Fig. 1 The distribution of the Si in different depth of the target for constant oxide thickness and various $In_{0.53}Ga_{0.47}As$ thicknesses in optimized incident energy.

Based from ion ranges in Fig 2, the lateral distribution was extracted to get some idea of the straggle. The values of projected and radial straggle need also to be considered if the implantation is used for the FET application. . Short channel effect is one of the problem that could happen if the lateral straggle was not optimized that could affect the device's performance [9].

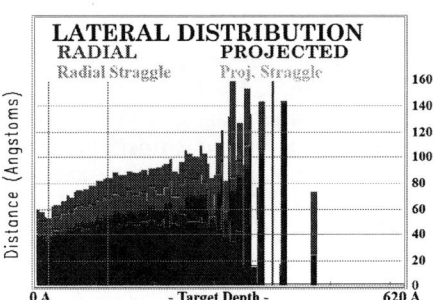

Fig. 2 The lateral, projected and radial distribution of the Si in different depth of the target for constant oxide thickness and 10nm $In_{0.53}Ga_{0.47}As$ thicknesses in optimized incident energy.

Fig. 3 illustrates the vacancy distribution after implantation, for $In_{0.53}Ga_{0.47}As$ damaged by Si ions. Vacancies are an empty lattice site which normally created with the displacement of the atoms during the collision with the energetic incident ion.

it, and each has different displacement energy, then the displacement energy will change as each atom of the cascade hits different target atoms. Table. 1 summarizes the important parameters related to the damage created in the crystal

Table. I

Variation of calculated parameters related to the energy loss as a function of incident energy in different $In_{0.53}Ga_{0.47}As$ thickness.

| | Vacancies /Ion | % Energy loss | | | | | |
| | | Ionization | | Vacancies | | Phonons | |
		Ions	Recoils	Ions	Recoils	Ions	Recoils
10 nm, 7keV	265	20.22	12.63	0.54	4.74	2.27	59.59
20 nm, 15keV	265	20.32	12.57	0.54	4.7	2.27	59.55
30 nm, 25keV	437	23.69	12.13	0.41	4.67	1.67	57.43

structure during the implantation process. It shows that by increasing the incident dose in thicker layers the number of vacancies created per ion increases. The energy loss information in Table 1 allows dividing the incident ion energy into various types, including Ionization, vacancies and phonons. When the target is implant by the incident ions the electrons of the target absorb some energy from the moving ions and also from the recoil atoms. This energy will be released in the form of heat if the target is a metal, or as phonons if the target is an insulator. This energy loss to the target electrons is known as Ionization. Fig.4 (a-c) shows ionization from the incident ions and also from recoiling target atoms. The ionization as a result of ions increases as the thickness increase, however ionization due to the recoils show insignificance variation with thickness

Fig. 3 The damage (vacancy) profile as a function of target depth in (a) 10 nm, (b) 20 nm, and (c) 30 nm of Si-implanted $In_{0.53}Ga_{0.47}As$ layer.

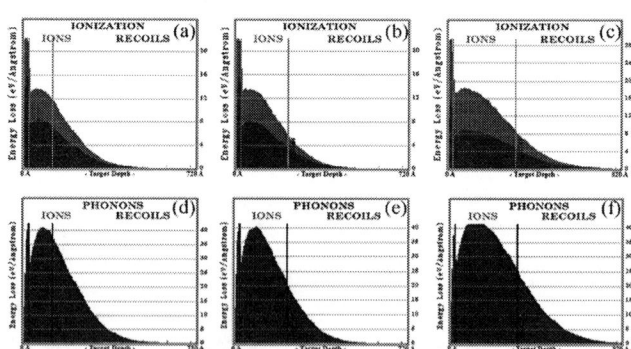

Fig. 4 Ionization (a-c) and phonons (d-f) variation in different depth for 10, 20, and 30 nm $In_{0.53}Ga_{0.47}As$.

The knocked out atom from the lattice of the target will be displace from its original position. The minimum energy required to knock a target atom far enough away from its lattice site so that it will not immediately return is called as displacement energy (E_{dis}). Since the atomic number of the incident ion (Si) and target atom is different, in order for a vacancy to be created the value of E_i (incident ion energy) and E_t (the target atoms energy in the lattice) must be greater than displacement energy ($E_i > E_{dis}$ and $E_t > E_{dis}$). There are more possibility with respect to the incident and knocked out atom energy. If $E_i < E_{dis}$ and $E_t > E_{dis}$ then the incident atom becomes a stopped interstitial atom in the target crystal structure. Finally, if $E_i < E_{dis}$ and $E_t < E_{dis}$, then the incident atom will be an interstitial and E_i+E_t is released as phonons. It should be mentioned that since target is a compound semiconductor ($In_{0.53}Ga_{0.47}As$) which had different elements in

The phonons in different depth of the $In_{0.53}Ga_{0.47}As$ due to the ions and recoils are presented in Fig. 4 (e-f). It is clear that majority of the energy from recoils are loss to become phonons compared to ions. The presence of phonons indicates that a lot of energy is losing into heat. In the structure with 10 nm $In_{0.53}Ga_{0.47}As$ thickness the ions generate phonons with 3.17% of their incident energy of 7 keV, and the recoiling atoms contribute an additional 63.51% = 66.68% (total). Total respectively. Multiplying these values by the total ion energy, we can obtain that 4.4 keV from 7 keV, 9.4 keV from 15 keV, and 15.4 keV from 26 keV of phonons add to the target structure.Fig 5 (a-c) present the vacancy created in different layers and due to the different elements in the structure. The maximum vacancies are created in O sites of the oxide layer and in the As sites of target. The In and Ga sites vacancy is approximately half of that in As.

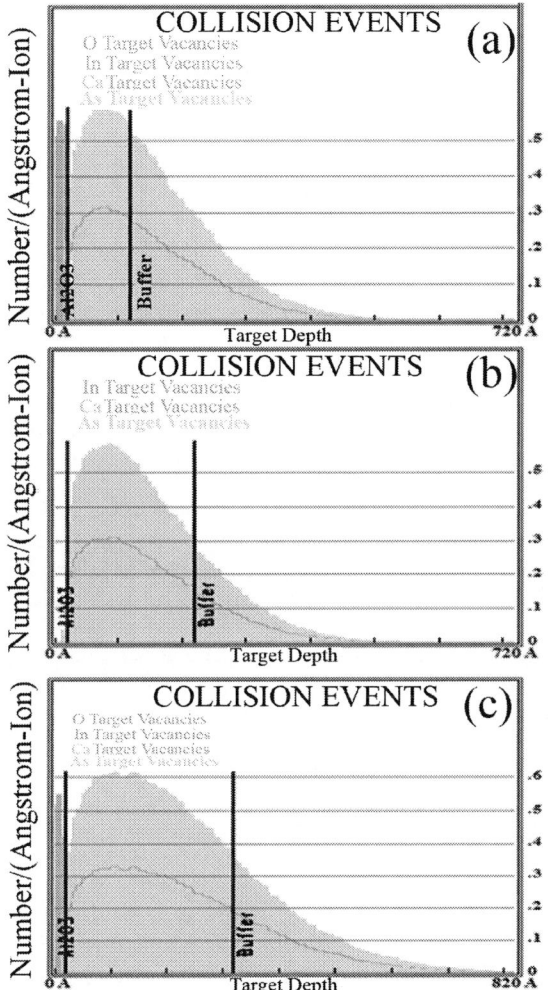

Fig. 5 Vacancies created in different layers and due to the different elements in the structure in different depth for (a)10, (b)20, and(c)30 nm $In_{0.53}Ga_{0.47}As$.

energy corresponds to 20 and 30 nm is 62.71 and 59.31%,

IV. Conclusion

In this work, the optimized energy and range of Si+ implanted in $In_{0.53}Ga_{0.47}As$ was studied. We found that for 10nm, 20nm and 30nm of thickness, the optimized energy is 7keV, 15keV and 24keV, respectively. It is also found that the most important sources of energy loss come from phonons and ionization process. These two processes contribute almost 97% of energy loss compared to the vacancies process. The optimized parameters obtained in this work can be used as the reference for implantation of thin film $In_{0.53}Ga_{0.47}As$ with ultra-thin layer Al_2O_3 for FET device application

Acknowledgment

The authors acknowledge support for this work provided by Universiti Kebangsaan Malaysia (grant project HICOE-AKU-95). Muhammad Zulkhairi Roslan thanks Prof Abdelkrim Khelif and Franck Chollet, from FEMTO-ST for their great support and assistance.

References

[1] [1] M. Egard, L. Ohlsson, M. Ärlelid, K. M. Persson, B. M. Borg, F. Lenrick, et al., "High-frequency performance of self-aligned gate-last surface channel In 0.53Ga0.47As MOSFET," IEEE Electron Device Letters, vol. 33, pp. 369-371, // 2012.

[2] [2] Y. Xuan, Y. Q. Wu, and P. D. Ye, "High-performance inversion-type enhancement-mode InGaAs MOSFET with maximum drain current exceeding 1 A/mm," IEEE Electron Device Letters, vol. 29, pp. 294-296, 2008.

[3] [3] V. Djara, K. Cherkaoui, M. Schmidt, S. Monaghan, É. O'Connor, I. M. Povey, et al., "Impact of forming gas annealing on the performance of surface-channel In 0.53Ga 0.47As MOSFETs with an ALD Al2O3 gate dielectric," IEEE Transactions on Electron Devices, vol. 59, pp. 1084-1090, 2012.

[4] [4] P. C. McIntyre, Y. Oshima, E. Kim, and K. C. Saraswat, "Interface studies of ALD-grown metal oxide insulators on Ge and III-V semiconductors (Invited Paper)," Microelectronic Engineering, vol. 86, pp. 1536-1539, 2009.

[5] [5] S. Pearton, "Ion implantation for isolation of III-V semiconductors," Materials Science Reports, vol. 4, pp. 313-363, 1990.

[6] [6] J. Wagner, H. Seelewind, and W. Jantz, Dopant incorporation in Si-implanted and thermally annealed GaAs.

[7] [7] E. J. Kim, E. Chagarov, J. Cagnon, Y. Yuan, A. C. Kummel, P. M. Asbeck, et al., "Atomically abrupt and unpinned Al2O3/In0.53Ga0.47As interfaces: Experiment and simulation," Journal of Applied Physics, vol. 106, 2009.

[8] [8] J. F. Ziegler, J. P. Biersack, and M. D. Ziegler, SRIM, the stopping and range of ions in matter:2008.

[9] [9] M. F. M. R. Wee, A. Dehzangi, S. Bollaert, N. Wichmann, and B. Y. Majlis, "Gate Length Variation Effect on Performance of Gate-First Self-Aligned In0. 53Ga0. 47As MOSFET," PloS one, vol. 8, p. e82731, 2013..

Corner compensation mask design on (MEMS) accelerometer structure

Norliana Yusof, Abdullah C.W. Noorakma
Faculty of Design Arts and Engineering Technology
Universiti Sultan Zainal Abidin,
21300 Kuala Terengganu, Terengganu, Malaysia
norliana@unisza.edu.my cwnoorakma@unisza.edu.my

NorhayatiSoin
Department of Electrical Engineering
Faculty of Engineering, University of Malaya
50603 Kuala Lumpur, Malaysia
norhayatisoin@um.edu.my

Abstract—**This paper presents the analysis effect of etching temperature and KOH concentration on convex corner undercutting of (MEMS) accelerometer structure. The Intellisuite CAD simulation software was used for the simulation analysis. From the analysis it was found that the optimum etching condition for this convex corner was at 25 wt% KOH concentration and 63 °C etching temperature. Different types of compensation mask corners were designed which are corner, square and triangle in order to study the undercutting phenomena. In this case, the square corner compensation mask was chosen as it shown the most suitable compensation mask for this design of accelerometer. The etching simulation was continued with square corner compensation mask etched in the optimized temperature and KOH concentration and it indicated that the square corner compensation mask is the most suitable mask to solve the convex corner undercutting for this accelerometer structure.**

Keywords—Convex Corner Undercutting; Accelerometer; Intellisuite CAD simulation; Compensation mask; Etching

I. INTRODUCTION

There are two approaches of etching techniques for bulk micromachining which are wet and dry [1]. Among both techniques, anisotropic wet etching was the popular method to fabricate MEMS accelerometer in recent technologies. Anisotropic wet etching has received much attention due it greater performance compared to dry etching. Such dry etching technique produces vertical materials removals that conform to the mask design which may include the curved features and it depends on the crystallographic alignment between the mask and the accelerometer structure. On the other hand, the wet etching was easy to use and low cost, but also can provide smooth surface [2, 3].

In wet bulk micromachining it usually carried out using anisotropic etchants like KOH (Potassium Hydroxide), TMAH (Tetra Methyl Ammonium Hydroxide) and EDP (Ethylene Diamine Pyrocatecol). Among these etchants, KOH solution has the advantages of simplicity, easy to handle, low cost and homogeneous etching rate of the (1 0 0) crystal plane[4].

KOH shows strong anisotropy, and shows large values of etch rate ratios among orientations of about (100). It means that high controllability can be expected in etched profiles, while suppressing mask undercut. KOH solutions are less toxic than other etchants, hence are easy to process[5]. These are the main reasons why KOH is widely used for fabricating silicon microstructures in the industry. Nevertheless, the critical problem normally occurred during the etching process of MEMS rectangular/square corners was the deformation of the edges due to undercutting. Corner erosion leads to deformed rectangular structures, which will subsequently influence the device performance [6].This effect consequently caused a low performance of MEMS sensors especially in the fabrication of acceleration sensors where total symmetry and perfect 90° convex corners on the proof mass are mandatory for good device prediction and specification.

From previous studies, the most compensation structures were carried out on (100) silicon substrate and various convex corner compensation structures have been develop to prevent the undercut of wet etching. However, these structures will occupy some in plane spaces especially for the case of deep silicon etching. In order to prevent the undercut issue, the compensation mask's patterns such as triangle, square and bar have been developed to protect the desired right angled corner. As a result they have become the three classical compensation methods that are still used nowadays. Nevertheless, each individual compensation structure is independent and common principle of constructing a compensation pattern is not clearly put forward [7].

Other than corner compensating by mask, the undercutting issue could also be improved by controlling etching condition such as temperature and KOH concentration. According to previous research by N. Soin, the convex corner undercutting phenomena is significantly reduced at low etching temperature and high KOH concentration respectively [8].

In this study, we demonstrate the effect of temperature and KOH concentration to the convex corner during etching and proposed MEMS accelerometer structure with different shaped appearances of corner compensation masks.

II. METHODOLOGY

A. Masks Design

The developed design approach is modified based on the research work done by R.Mukhiya [9]. The researchers has chosen <100> square, <100> thin bar and <100> wide bar structure as their corner compensation structure. However, in this simulation, we have modified the corner compensation structure as shown in Fig. 2 respectively. Intellisuite CAD simulation software has been used for the simulation analysis in this study. The dimensions of masks design used in the process of realization of accelerometer structure are shown in Fig. 1. The structure is etched from both top and bottom directions of silicon wafer. Among various corner compensation structures, for investigation we have chosen bar, square and triangle structures. Square compensation structure is the easiest to design, analyze and most space efficient [5]. The corner compensation structures with design parameters are shown in Fig. 2.

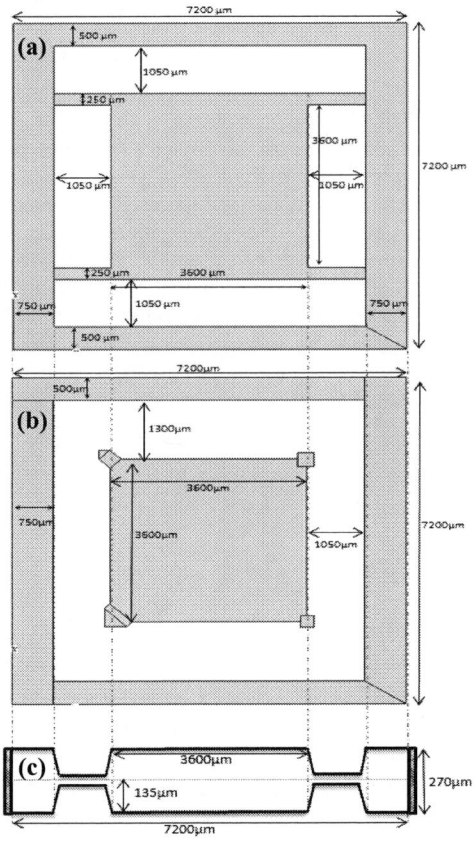

Fig. 1. (a) Top mask (b) Bottom mask (c) cross section view of complete accelerometer structure

B. Wet Etching Simulation Process

The wet etching simulation was carried out using KOH etchant of <100> silicon wafer. During the simulation, the final etching time was 4.5 hours in order to achieve nearly perfect 90° of convex corners. Etching rate is very temperature dependent. In this simulation, the range of the KOH concentration and the etching temperature was chosen to be from 10 wt% to 45 wt% and 30°C to 85 °C respectively.

III. SIMULATION RESULTS AND ANALYSIS

A. Determination of optimum KOH temperature on undercutting convex corner

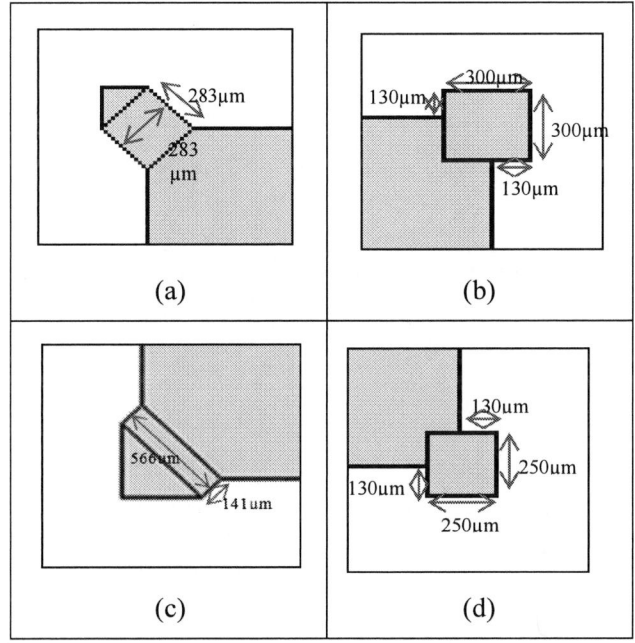

Fig. 2. Corner compensation structures (a) corner (b) square 1 (c) triangle (d) square 2.

The variation of etching temperature and etching rates as a function of etching depth is shown in Table I. As can be seen in Fig. 3 the etching temperature increased with the increased of etching rate. However, both trends were only seen in the crystal plane (100) and (110) meanwhile there is no significant increment in etching depth for crystal plane in (111). At the beginning of the temperature until 50 °C the etching depth was slightly increased. The etching depth was noticeable increased sharply at 50 °C onwards. The major increment was seen at 80 °C.

The etching simulation performed in this simulation has shown that most efficient etching temperature might occur from 62 °C to 64 °C. In order to study the optimum etching temperature, four type of corner compensation mask was designed and tested in the etching parameter study as shown in Fig. 5. It can be seen that at temperature of 62 °C, the convex corner started to etch nearly complete. As the temperature reached at 63 °C, the convex corner was shown completely etched.

The temperature was continually increased and it has been seen that the convex corner started to undercut and deformed. Hence from structure analysis it can be concluded that the optimum temperature was at 63 °C.

TABLE. I VARIATION OF ETCHING TEMPERATURE AS FUNCTION OF ETCHING DEPTH

Etching Temp.(°C)	Etching Rate (µm/hour)			Etching depth		
	{100}	{110}	{111}	{100}	{110}	{111}
30	3.33	5	0.04	14.985	22.5	0.18
35	5.12	7.705	0.07	23.04	34.673	0.315
40	6.91	10.41	0.1	31.095	46.845	0.45
45	10.29	15.57	0.16	46.305	70.065	0.72
50	13.67	20.73	0.22	61.515	93.285	0.99
55	19.825	30.165	0.355	89.2125	135.765	1.598
60	25.98	39.6	0.49	116.91	198.2	2.205
65	36.765	56.23	0.765	165.443	253.035	3.443
70	47.55	72.86	1.04	213.975	327.87	4.68
75	65.83	101.175	1.575	292.235	455.288	7.088
80	84.11	129.49	2.11	378.495	582.705	9.495
85	114.14	176.22	3.11	513.63	792.99	13.995

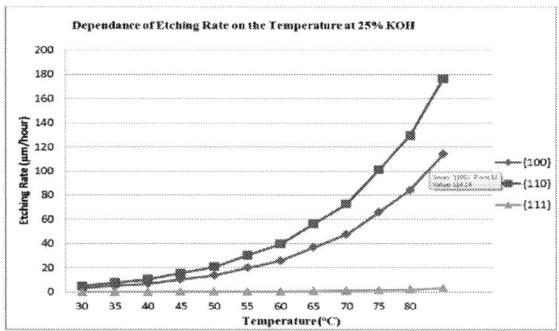

Fig. 3. Dependence of etching rate with temperature at 25 wt% KOH concentration

B. Determination of optimum KOH concentration on undercutting convex corner

The variation of etching concentration as function of etching rate is shown in Table II. The temperature has been set at 63 °C as the previous simulation shown 63 °C is the optimum temperature. The dependence of etching rate on the KOH concentration is presented in Fig. 5.

The analysis indicates that the most efficient KOH concentration might occur at KOH concentration from 20 to 30 wt%. Therefore in order to confirm that, the analysis was extended with concentration from 20 to 30 wt% to etch four types of corner compensation masks. The description of convex corner undercutting dependence on KOH concentration was illustrated in Fig. 6. Four different masks corner have been etched at three different KOH concentrations for about 4.5 hours.

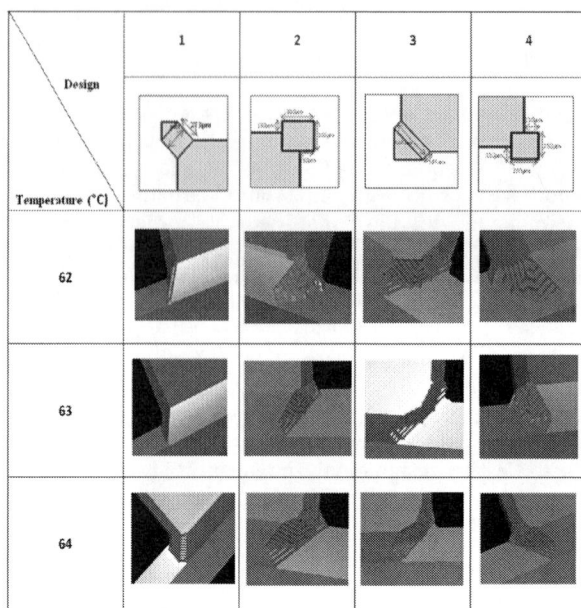

Fig. 4 Etched morphology of convex structure at 4.5 hours etching time at different etching temperature.

As the concentration reached at 20 wt%, the convex corner was nearly complete been etched. Finally at concentration 30 wt%, it can been seen that the structure deformed due to undercutting issue. Thus, it can be conclude that the most optimum concentration was at 25 wt%. The previous study performed by N. Soin also shown that the most ideal KOH concentration was around 30 wt% and slightly greater than our result [8].

TABLE II. VARIATION OF ETCHING CONCENTRATION AS FUNCTION OF ETCHING DEPTH

KOH Conc. (%wt)	Etching Rate (µm/hour)			Etching depth		
	{100}	{110}	{111}	{100}	{110}	{111}
10	31.58	48.248	0.639	142.11	217.125	2.88
15	33.25	50.801	0.671	149.625	228.6	3.015
20	33.39	51.016	0.678	150.255	229.59	3.06
25	32.451	49.578	0.655	146.025	223.11	2.97
30	30.566	46.696	0.616	137.565	210.15	2.79
35	27.93	42.767	0.564	125.685	192.465	2.52
40	24.908	38.056	0.506	112.095	171.27	2.295
45	21.458	32.784	0.438	96.57	147.51	1.98

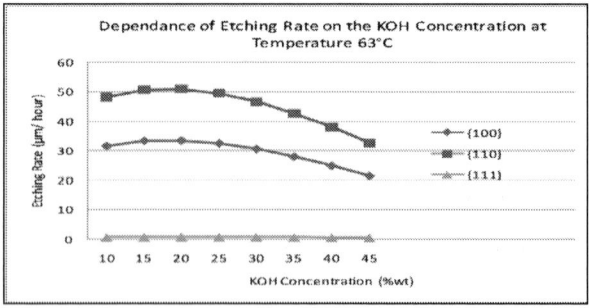

Fig. 5. Dependence of etching rate on KOH concentration at temperature 63 °C.

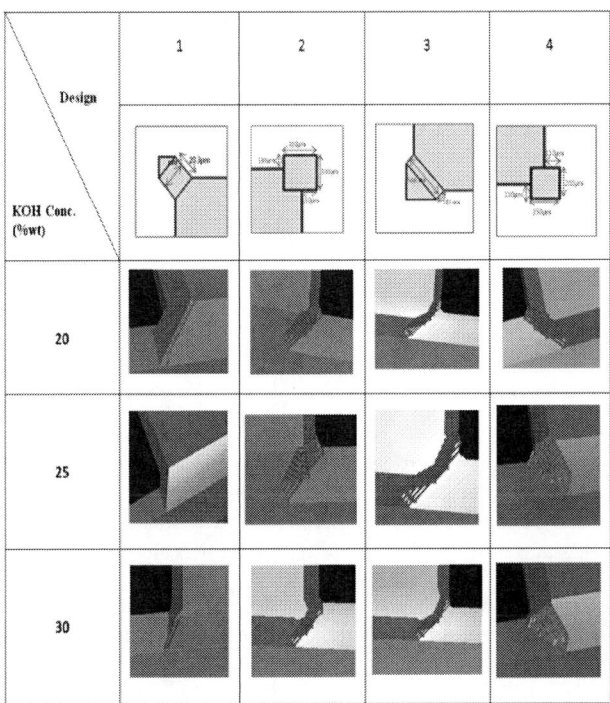

Fig. 6. Etched morphology of convex corner structure at 4.5 hours etching time at difference KOH concentration.

C. Determination of the best shape of compensation mask

Early data was interpreted in this study and it was identified that the suitable etching temperature was at 63 °C and 25 wt% KOH concentration. Based on four types of convex corner compensation mask, it was found that the best design was the square corner. Therefore, the simulation was extended to etch the accelerometer structure with the square corner compensation mask. The result was displayed in Fig. 7. The compensation mask had covered the accelerometer structure successfully at etching temperature 63 °C in 25 wt% KOH concentration.

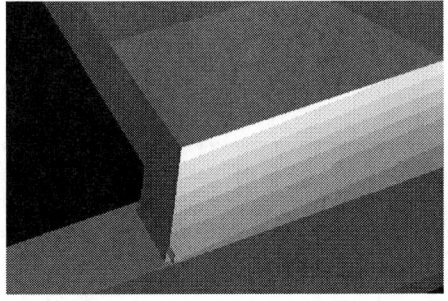

Fig. 7. Etched accelerometer structure with square corner compensation mask

IV. CONCLUSION

Etching simulations of convex corner of accelerometer structure by KOH anisotropic etching have been performed. It was observed that the etching parameter such as temperature and concentration of KOH solution have influenced the corner undercutting phenomena. The simulation result was identified that the optimum etching temperature was at 63 °C and the KOH concentration at 25 wt%. Based on the optimum etching parameter applied to different type of compensation mask, it was shown that the suitable corner compensation mask was square corner pattern.

ACKNOWLEDGEMENT

The authors would like to thank the financial support by the RAGS fund (RAGS/2013/UNISZA/SG02/1).

REFERENCES

[1] J. Judy, "Microelectromechanical systems (MEMS): fabrication, design and applications," Smart Mater. Struct., vol. 10, no. 6, pp. 1115, 2001.

[2] M.A. Hines , "In search of perfection: understanding the highly defect selective chemistry of anisotropic etching," Annu. Rev. Phys. Chem. Vol. 54, pp. 29 – 56, 2003.

[3] G.S. Higashi , Y.J. Chabal , G.W. Trucks , K. Raghavachari , Ideal hydrogen termination of the Si (111) surface , Appl. Phys. Lett. Vol. 56, 656 – 5818, 1990.

[4] J. Han, S. Lu, Q. Li, X. Li, and J. Wang, "Anisotropic wet etching silicon tips of small opening angle in KOH solution with the additions of I2/KI," Sensors Actuators A Phys., vol. 152, no. 1, pp. 75–79, May 2009.

[5] J.-B. Waldner, Nanocomputers and Swarm Intelligence. John Wiley & Sons, 2013.

[6] X.-P. Wu and W. H. Ko, "Compensating corner undercutting in anisotropic etching of (100) silicon," Sensors and Actuators, vol. 18, no. 2, pp. 207–215, Jun. 1989.

[7] Fan W, Zhang D.A simple approach to convex corner compensation in anisotropic KOH etching on (100) silicon wafer. J. Micromech Microeng, 2006, 16:1951.

[8] N. Soin, B.Y. Majlis, "Development of perfect silicon corrugated diaphragm using anisotropic etching," Microelectronic Eng., vol.83, pp. 1438-1441, 2006.

[9] R. Mukhiya, a. Bagolini, T. K. Bhattacharyya, L.Lorenzelli, and M. Zen, "Experimental study and analysis of corner compensation structures for CMOS compatible bulk micromachining using 25wt% TMAH," Microelectronics J., vol. 42, no. 1, pp. 127–134, Jan. 2011.

Electronic state transition in cooperatively interacting point-defects in semiconductor crystals

Mohd Ambri Mohamed and B. Y. Majlis
Institute of Microengineering and Nanoelectronics, The
National University of Malaysia, 43600 UKM Bangi,
Selangor, Malaysia.
ambri@ukm.edu.my

M. H. Ani,
Department of Manufacturing and Materials,
Kulliyyah of Engineering, International Islamic University,
Kuala Lumpur, Malaysia
mhanafi@iium.edu.my

Abstract— Electron state transition of deep level point defects in a semiconductor crystal was studied. Low-temperature grown GaAs produced excess antisite As (As_{Ga}) which produces localized spin when doped with Be. A nearly abrupt decrease of 1.7% of the resistance is detected at a temperature around 4 K which is consistent with abrupt decrease of magnetization. These observations are explained as a result of cooperative transition of electron states of As_{Ga} defects. First-principal calculations of the electron state of an As_{Ga} atom with a shallow acceptor Be show that at the transition an As_{Ga}^+ ion is displaced to the interstitial site and becomes a neutral atom and finally results in formation of a hole producing enhancement in conductivity.

Keywords— LT-GaAs; antisite As; Molecular Beam Epitaxy

I. INTRODUCTION

In the past three decades, innovations in materials growth technologies for semiconductor crystals have been the key element for the development of new electronic and optical devices. The properties of a semiconductor crystal can be tailored based on the location of the atomic species in the crystal lattice. For example, by utilizing a molecular beam epitaxy (MBE) growth technique, the energy levels of semiconductors and wave functions of electrons in semiconductors can be engineered [1,2].

The formation of point defects in a semiconductor crystal demonstrate unusual electronic properties as represented by EL2 defects in semi-insulating GaAs and DX centers in III-V compound semiconductors [3]. Antisite As (As_{Ga}) in semi-insulating GaAs bulk crystal known as EL2 defects are among the most studied defects experimentally and theoretically. These defects result in formation of two different states known as normal state and metastable state. At normal state, antisite As formed at substitutional site. However, antisite As is displaced to interstitial site when illuminated with light at low temperatures, hence results in electronically inactive state. This phenomenon occurs even at relatively low defect concentration of 10^{16} / cm^3.

In the event of high concentration of deep level point defects, there is a possibility that these defects interact cooperatively on each other through lattice strains. These defects display the electronic configuration different from the sp^3 bond scheme by large lattice distortions, hence could

results in changes of magnetic, electron transport and optical properties. To investigate this possibility, we grew GaAs at low temperature to create high concentration of native point defects and at the same time doped Be to create localized spin (As_{Ga}^+) in the semiconductor crystals.

From samples with high concentration of As_{Ga}^+ ions, we found nearly abrupt changes of the magnetization at low temperatures around 4 K [4]. These changes suggest a possibility that a transition of electronic states of As_{Ga} defects similar to that of EL2 defects cooperatively occurs in Be-doped LT-GaAs layers. Based on the temperature-dependence of the electrical resistance, we observed a nearly abrupt decrease of the electrical resistance at a temperature corresponding to that of the change in the magnetization [4]. With first-principle calculations of the electron states, these observations are explained as a result of cooperative displacements of As_{Ga}^+ ions to interstitial sites which lead to the formation of holes in the valence band. The significance of this study is that they have shown that a sudden change in the electron transport property of a semiconductor crystal is a result of a cooperative transition of deep-level point defects.

II. EXPERIMENTAL

It is difficult to measure the electron transport property with a lift-off sample because of the requirement of making contacts on both sides of a lift-off sample [4]. The lift-off sample is too fragile to handle with thickness of merely few microns. The resistance cannot be measured in the lateral direction of a LT-GaAs layer because of extremely high resistivity of the sample at low temperature [5]. Unlike the sample used in magnetization measurement, the sample needs to stay intact with the substrate for the investigation of the electron transport property. We reproduce a sample used in magnetization measurement [4] with a different structure, which is a bi-layer structure consisting of a Be-doped LT-GaAs layer as illustrated in Fig. 1. By using this structure, we successfully measured the temperature-dependence of the resistance of a LT-GaAs even at low temperature. For this sample, a Be-doped high-temperature (HT)-GaAs layer was first grown at 520°C to a thickness of 3 μm with a Be concentration of 1.5×10^{19} cm^{-3} on a 150 nm-thick-non-doped GaAs buffer layer. The resistivity of this HT-GaAs layer was 0.006 Ωcm and remained nearly constant at low temperatures. The sample with the HT-GaAs layer was removed from the MBE system and part of its surface was covered with a Ta sheet to allow the partial growth of LT layer. The sample was returned to the main chamber and a buffer Be-doped GaAs layer was grown at 580°C after the desorption of the oxide layer at 600°C. The substrate temperature was subsequently reduced to the growth temperature of Be-doped LT-GaAs layer. The LT layer was grown to a thickness of 1.5 μm on the uncovered surface of the HT-GaAs layer.

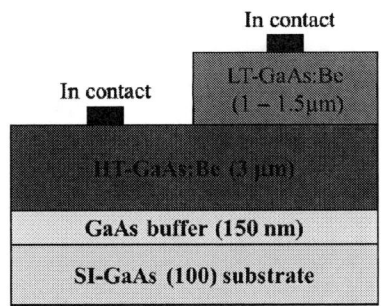

Fig. 1. Schematic structure of sample used in electrical transport measurement

A rectangular 3×8 mm² sample, half of which had a Be-doped LT-GaAs layer and the other half having a Be-doped HT-GaAs layer, was cut and two Indium contacts were made with a separation of 4 mm. Indium was used as the electrodes in this study because a good adhesion to the GaAs layers that can be sustained for repeated measurements over a wide temperature range. Electrodes were formed by annealing at 200°C in a purged nitrogen atmosphere for 7 min; an annealing at temperatures higher than 300°C results in the redistribution of As_{Ga} atoms in the LT-GaAs layer [6]. Two Indium contacts were used for both current and voltage measurements. Hence, the contacts resistance was included in the measured resistance. By preparing a van der Pauw sample of the Be doped HT-GaAs layer, we confirmed that the contact resistance became negligible in comparison to the resistance of the LT-GaAs in the measurement temperature range with the annealing condition described above. Physical Property Measurement System, from Quantum Design was used for electron transport measurements.

III. RESULTS AND DISCUSSIONS

There are a few earlier studies on the electron transport properties of LT-GaAs layers at low temperatures [7-8], but no measurement of the electrical transport properties of a LT-GaAs layer below 80 K has been carried out due to the thermally assisted hopping conduction which leads to an extremely high resistance at low temperatures. In our earlier measurements of a single LT-GaAs layer on a SI-GaAs substrate with electrode spacing of 4 mm in the lateral direction, the resistance increased rapidly with lowering the temperature and exceeds 7 MΩ at 120 K as shown in Fig. 2.

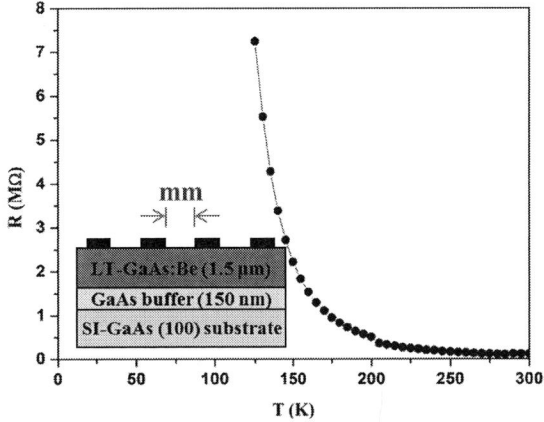

Fig. 2. Temperature-dependence of electrical resistance measured in a sample with lateral direction.

By using the bi-layer structure as shown in Fig. 1, we successfully measured the temperature-dependence of the resistance of a LT-GaAs layer to as low as 2 K, as shown in Fig. 3. Measurements were carried out on cooling the samples in and out of magnetic field. Temperature dependence of the resistance from 300 to 2 K is shown in the inset. The resistance at room temperature is attributed to that of the HT-GaAs layer because the resistance of Be-doped LT-GaAs layer is negligible at room temperature. In the high temperature regime between 300 and 100 K, the resistance exponentially increases with the lowering of the temperatures as expected from hopping conduction. The apparent decrease in the activation energy for hopping conduction at temperatures below 100 K is attributed to their variable-range hopping [9] or parallel conduction through oval defects [10]. A pronounced decrease of the resistance was observed around 4K. There is a remarkable difference between the results in the absence and presence of the magnetic field as seen in Fig. 3. The origin of this difference is not known at the present stage.

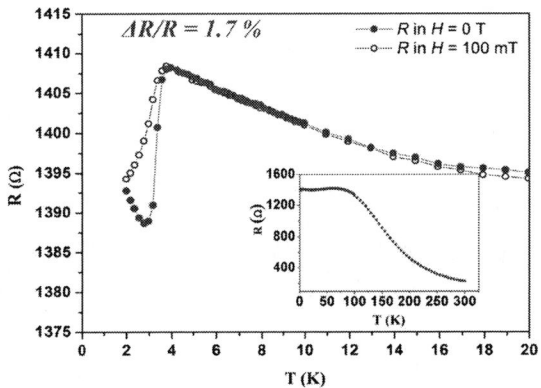

Fig. 3. Temperature-dependence of electrical resistance in a partially grown Be-doped LT-GaAs

The nearly abrupt change of the resistance in Fig. 3 occurs at a temperature close to that at which the change of the magnetization was observed in the previous study [4], implying that both changes result from the same origin. Metal In undergoes a superconductivity transition at 3.4 K in the absence of magnetic field. The observed decrease of the resistance cannot be attributed to it because the resistance starts to decrease at 3.8 K and also occurs even in a magnetic field of 100 mT, which is greater than the critical magnetic field [11]. The abrupt change in the magnetization of a Be

temperatures corresponding to those of the change of the magnetization has been observed. These results are explained as a result of cooperative displacements of $AsGa+$ ions to interstitial sites which lead to the formation of holes in the valence band. The occurrence of an abrupt change in the electron transport property which originates from a cooperative transition of deep-level point defects may have its future device applications.

ACKNOWLEDGMENT

This work is supported in part by Ministry of Education Malaysia (HICOE-AKU-95) and MEXT, Japan. MAM also thanks P. T Lam and N. Otsuka of JAIST for their support and assistance in the DFT calculations.

REFERENCES

[1] L. Esaki, IEEE Joural of Quantum Electronics, QE-22,1611, 1986.

[2] H. Sasaki, Proc. Int. Symp. Foundations of Quantum Mechanics, Tokyo, 94, 1983.

[3] Imperfections in III/IV Materials, Semiconductors and Semimetals, Vol. 38, eds. E. R. Weber, (Academic Press, New York, 1993)

[4] Mohd Ambri Mohamed, Pham Tien Lam, K. W. Bae, and N. Otsuka, "Cooperative transition of Electronic States of Antisite As Defects in Be-Doped Low-Temperature-Grown GaAs layers," J. Appl. Phys., 110, 123716 , 2011.

[5] F. Shimogishi, K. Mukai, S. Fukushima, and N. Otsuka, "Hopping conduction in GaAs layers grown by molecular-beam epitaxy at low temperatures," Phys. Rev. B, 65, 165311, 2002.

[6] M. R. Melloch, J. M. Woodall, E. S. Harmon, N. Otsuka, F. H. Pollak, D.D. Nolte, R. M. Feenstra, and M. A. Lutz, "Low-Temperature Grown III-V Materials," Annu. Rev. Mater. Sci., vol. 25, pp. 547-600, 1995.

[7] K. W. Bae, Mohd Ambri Mohamed, D. W. Jung, and N. Otsuka, "Direct Exchange Interaction of Localized Spins Associated With Unpaired sp Electrons in Be-Doped Low-Temperature-Grown GaAs layers," J. Appl. Phys., 109, 073918, 2011.

[8] S. Fukushima, T. Obata, and N. Otsuka, "Electrical properties of nearly stoichiometric GaAs grown by molecular beam epitaxy at low temperature," J. Appl. Phys., 89, 380, 2001.

[9] N. F. MOTT and W. D. TWOSE, Adv. Phys. 10, 107, 1961.

[10] S. Matteson and H. D. Shih, Appl. Phys. Lett. 48, 47, 1986

[11] C. Kittel, "Introduction to Solid State Physics, " 8th ed. (Wiley, New York, 2005), p. 261.

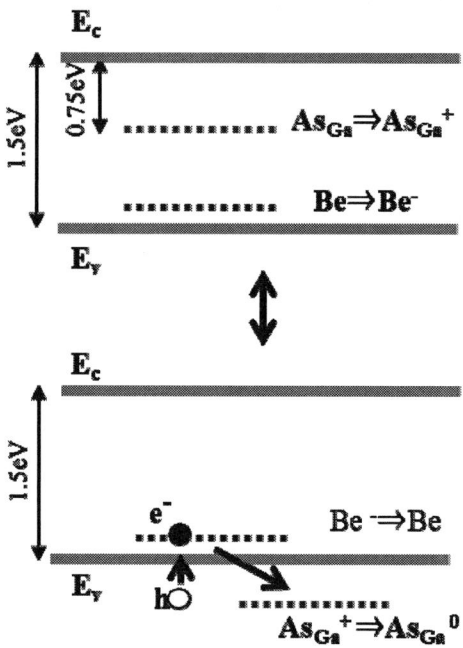

Fig. 4. Energy band diagram of LT-GaAs in the presence of Be shallow acceptor in normal (top) and displaced state (bottom)

doped LT-GaAs layer at a low temperature resulted from the cooperative transition of $AsGa$ defects which was similar to the transition of EL2 defects in GaAs. With the observation of the change in the resistance in the present study we can now substantiate the above hypothesis with the aid of band structure as shown in Fig. 4. These results indicate that the transition from a normal to the displaced state results in the anihilation of a localized spin and generation of a hole in the valence band. The latter change promotes in a transition from hopping conduction via $AsGa$ defects to the conduction by holes in the valence band. Experimental results shown in Fig. 3 are, therefore, can be explained by the transition of $AsGa$ defects from the antisite to the interstitial site of the crystal lattice.

IV. CONCLUSION

In conclusion, we have studied electron transport properties of Be-doped LT-GaAs layers at cryogenic temperatures. A nearly abrupt decrease of 1.7% of the electrical resistance at

Effects of Material and Membrane Structure on Maximum Temperature of Microheater for Gas Sensor Applications

Mimiwaty Mohd Noor, Gandi Sugandi, Mohd Faizal Aziz and Burhanuddin Yeop Majlis, *SMIEEE*
Institute of Microengineering and Nanoelectronics (IMEN)
Universiti Kebangsaan Malaysia, 43600 Bangi
Selangor, Malaysia
Email: burhan@eng.ukm.my

Abstract— The material selection for membrane is important in designing a microheater. A membrane is used as an insulator layer to prevent heat dissipation from the microheater to the substrate. At the same time, the thermal characteristic of the microheater is influenced by the insulator layer. A study on the effects of material and membrane structure on the maximum temperature of the microheater for gas sensor applications has been carried out using Heat Transfer Module of COMSOL 4.2. Three different membrane materials namely silicon nitride (Si_3N_4), silicon dioxide (SiO_2) and polyimide and two types of membrane structures namely full-membrane and bridge-membrane have been chosen for the study. Their effects on the microheater temperature are presented. The resistive meander type of microheater is used in this study. The heater material is platinum. The thickness and the area of the heater are 2 µm and 600 µm x 680 µm respectively. The thickness of each membrane is 5 µm. The area of the full-membrane and the bridge-membrane are 2500 µm x 2500 µm and 850 µm x 850 µm respectively.

Keywords— *Gas sensor; microheater; membrane; polyimide*

I. INTRODUCTION

The microheater is an important part of the gas sensor as it functions to provide a specific temperature that is required by the sensor. The operating temperature for a gas sensor is dependent upon the type of the target gases and the sensing material used. The operating temperature for a commercial Figaro Taguchi semiconductor gas sensor is 400 °C with the heater power consumption of 300-830 mW [1].

The gas sensor consists of a microheater, a sensing electrode and sensing material. At the specific temperature required by the sensor, the resistance of the sensing material changes in the presence of the target gas. The change of resistance of the sensing material is then detected by the sensing electrode and subsequently property changes in the electrode give output to the sensor of which target gases are present.

As the sensitivity of the gas sensor is influenced by the temperature, the microheater becomes one of the important parts of the gas sensor. The microheater is required to be able to work at a low power consumption, have a fast thermal response time, low thermal mass and able to provide uniform temperature to the active area of the sensor. Thermal characteristics of the microheater can be optimized by

controlling the heat loss and the microheater configuration as well as the selection of membrane material and structure.

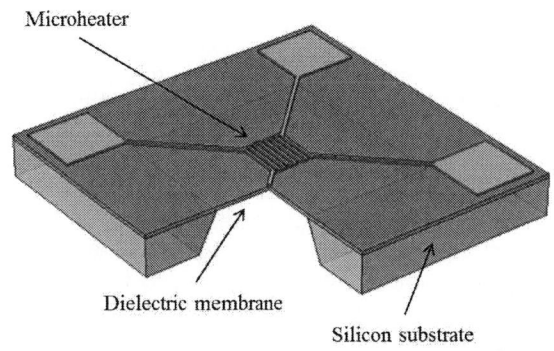

Fig.1. Typical structure of a microheater.

Figure 1 shows the typical structure of a microheater. Platinum is a common material used for microheaters because of its stable temperature coefficient of resistance and accuracy. Various heater geometries have been reported in the literature [2]. Generally the heater is designed in the shape of a meander or spiral. As for the membrane, materials with low thermal conductivity such as silicon nitride and silicon dioxide have been generally used as an insulator layer to prevent the heat dissipation from the microheater to the substrate. Both materials are applicable for high temperature of microheater up to 700 °C [3]. The use of polyimide as a membrane has also been reported recently and this material is applicable for the microheater temperature as high as 400 °C [4, 5]. The silicon underneath the membrane is generally removed allowing it to be suspended and thermally isolated from the substrate. In the fabrication process, this can be done by using the KOH wet etching process from the back side of the wafer [6].

In this work, we study the effect of material and membrane structure on the maximum temperature of the microheater by using Heat Transfer module of COMSOL 4.2. Platinum microheater with the meander shape has been chosen in this study. Three different membrane materials namely silicon nitride, silicon dioxide and polyimide and two types of membrane structure namely full-membrane and bridge-membrane have been chosen in order to perform the study. Their effect on the microheater temperature is discussed.

II. DESIGN AND SIMULATION

A. Design

The meander microheater geometry with the thickness of 2 µm has been used in this study. The material of the microheater is platinum with the area of 600 µm x 680 µm. The width as well as the gap of the microheater is 40 µm each. Dimensions of the microheater are shown in Figure 2.

Two types of membrane structure have been designed in this study namely the full-membrane structure of which the membrane is connected to the substrate along its entire perimeter and the bridge-membrane structure which the active membrane is supported by four beams as shown in Fig. 3a and 3b respectively. The thickness of the membrane is fixed at 5 µm. The area of the full-membrane and bridge-membrane are 2.5 mm x 2.5 mm and 850 µm x 850 µm respectively. The microheater is surrounded by an air block of 5000 µm x 5000 µm x 2000 µm.

B. Simulation

Three different types of membrane material have been chosen in this study. The properties of the three membranes are shown in Table I. The microheater has been simulated at same input power that is 250 mW. Only conduction and convection loss have been considered in this study.

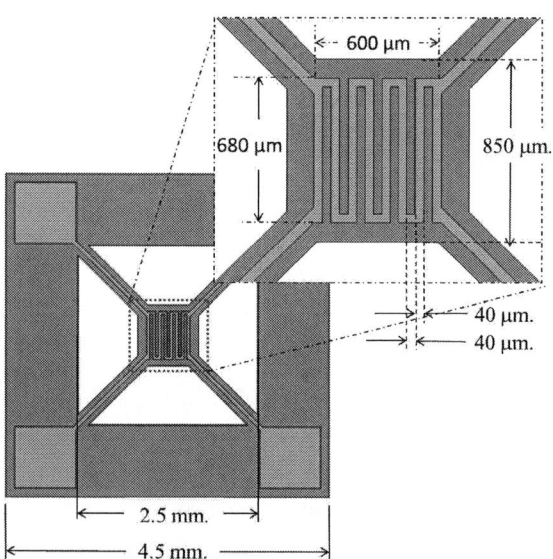

Fig.2. Dimension of the microheater.

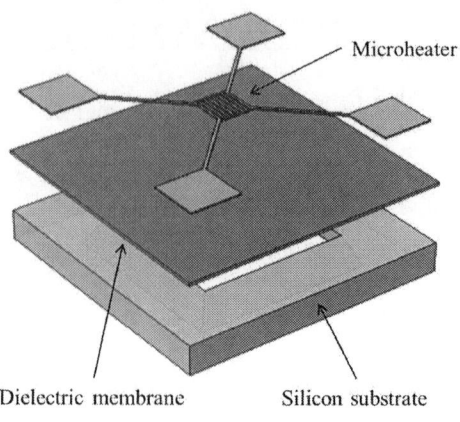

(a)

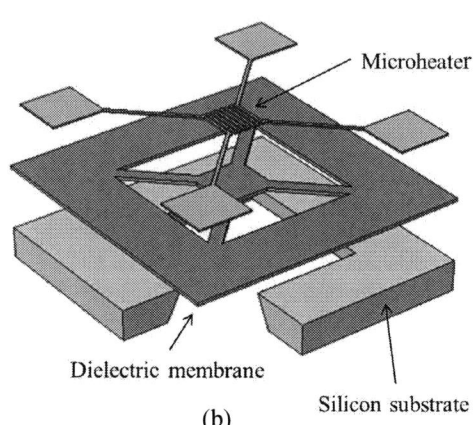

(b)

Fig.3. Schematic diagram of microheater layer (a) full-membrane structure (b) bridge-membrane structure.

TABLE I   MATERIAL PROPERTIES OF THE MEMBRANE USED FOR THE SIMULATIONS

Material	Si3N4	SiO2	Polyimide
Heat capacity at constant pressure (J/(kg*K))	700	730	1100
Thermal conductivity (W/m*K)	20	1.4	0.15
Density (kg/m^3)	3100	2200	1610
Relative permittivity	9.7	4.2	3.3
Young's modulus (pa)	250×10^9	70×10^9	2.7×10^9
Poisson's ratio	0.23	0.17	0.33

III. RESULTS AND DISCUSSION

The simulation results of the microheater on different types of membrane material and structure are summarized in Table II. At the same input power of 250 mW, polyimide shows the highest temperature of the microheater on both types of full-membrane and bridge-membrane structure with the temperature of 233.17 °C and 234.11 °C respectively. This is followed by the microheater on the silicon dioxide membrane that shows the temperature of 185.53 °C and 209.67 °C on the full-membrane and bridge-membrane structure respectively. The lowest temperature is shown by the microheater on the silicon nitride membrane which is the temperature are 63.94 °C and 146.96 °C on the full-membrane and bridge-membrane structure respectively. Polyimide with the lowest thermal conductivity has been able to minimize the conductive heat transfer from the heater to the membrane and subsequently reducing the sensor thermal loss compared to the other two materials. For this reason, the highest temperature of the microheater can be seen on the polyimide membrane, followed by silicon dioxide and silicon nitride membrane respectively.

Figure 4 shows the simulation results of the temperature distribution of the microheater on the polyimide membrane for full and bridge types of membrane structure.

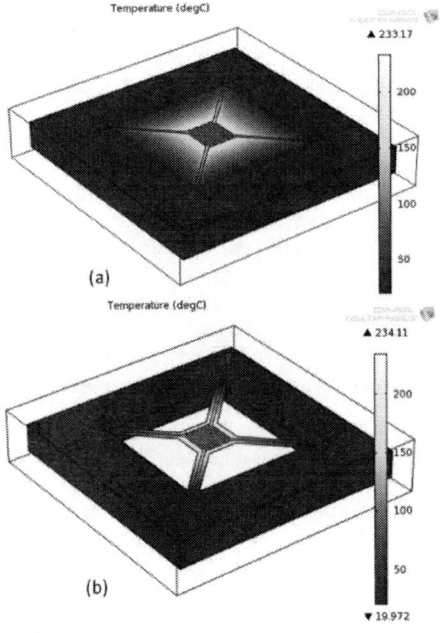

Fig.4.   Temperature distribution of the microheater on the polyimide membrane (a) full-membrane structure (b) bridge-membrane structure.

The microheaters on the bridge-membrane structure show a higher temperature than the microheaters on the full-membrane structure. Temperature of the microheater on silicon nitride, silicon dioxide and polyimide membranes has increased about 56.5 %, 11.5 % and 0.4 % respectively on the bridge membrane structure. Silicon nitride has the highest value of thermal conductivity thus contributes to the highest thermal loss over the whole area of the membrane. Reducing the silicon nitride membrane area has significantly increased the microheater temperature. Silicon dioxide has a lower thermal conductivity as compared to silicon nitride. Reducing the membrane area resulted in a lesser effect on the increment of

microheater temperature compared to the increment of microheater temperature on the silicon nitride membrane. Polyimide has the smallest value of thermal conductivity. Hence, reducing the membrane area does not give a significant effect on the increment of the microheater temperature. Bridge-membrane structure provides a smaller membrane area compared to the full-membrane structure. Reducing the membrane area has resulted in the heat dissipation from the heater to the membrane to reduce. The structure has significantly improved the temperature of microheater on the high thermal conductivity silicon nitride membrane.

TABLE II   TEMPERATURE OF THE MICRO HAEATER ON DIFFERENT TYPES OF MEMBRANE MATERIAL AND STRUCTURE AT THE POWER INPUT OF 250 MW.

Material	Temperature (°C)		Temperature increment of the microheater on the bridge-membrane (%)
	Full-membrane	*Bridge-membrane*	
Si3N4	63.94	146.96	56.5
SiO2	185.53	209.67	11.5
Polyimide	233.17	234.11	0.4

Figure 5 shows the heat distribution profile of the microheaters at the input power of 250 mW. As shown in Fig. 5a and 5b, the micro heater on silicon nitride membrane has the greatest area of temperature flat zone distributed over the membrane as compared to the other two membranes due to its high thermal conductivity characteristic. Although the silicon nitride membrane has improved the temperature distribution of the microheater, a higher power input is required for the microheater to reach the same temperature that has been reached by the microheater on silicon dioxide and polyimide membrane at the input power of 250 mW.

Figure 6 shows the thermal response time of the microheaters. For the full-membrane structure as shown in Fig. 6a, temperature of the micro heater on high thermal conductivity silicon nitride membrane is the lowest resulting in the fastest response time that is about 0.1 second for the micro heater to reach the maximum temperature of 63.94 °C. However, the temperature is not significant for metal-oxide gas sensor application at low power consumption. Microheater on the silicon dioxide and polyimide took about 0.35 seconds and 0.45 seconds to reach the maximum temperature of 185.53 °C and 233.17 °C respectively. For the bridge-membrane structure as shown in Fig. 6b, both microheaters on the silicon nitride and silicon dioxide membrane took about 0.25 seconds and 0.30 seconds to reach the maximum temperature of 146.9 °C and 209.67 °C respectively while it was documented at 0.30 seconds for the microheater on polyimide membrane to reach the maximum temperature of 234.11 °C. The temperature of all microheaters on bridge-membrane structure is higher than microheaters on full-membrane structure yet this structure has resulted in a faster response time for the microheaters to reach the maximum temperature except for the microheater on the silicon nitride membrane. Microheater on the silicon nitride membrane experienced 56 % temperature increment on the bridge-membrane structure resulted in longer response time for

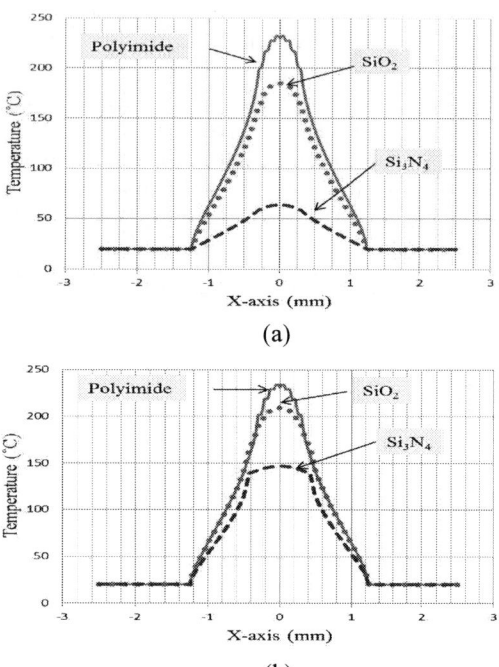

(a)

(b)

Fig. 5. Heat distribution profile of the microheaters at the power input of 250 mW (a) full-membrane structure (b)bridge-membrane structure.

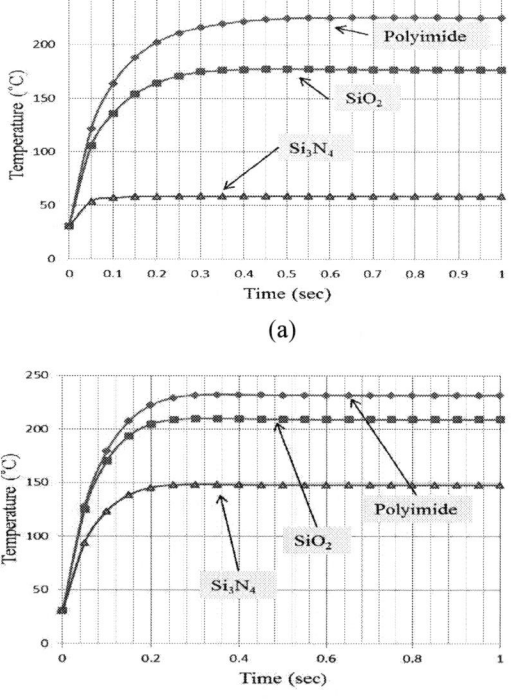

(a)

(b)

Fig.6. Comparison of thermal response time of the microheaters (a) full-membrane structure (b)bridge-membrane structure.

the microheater to reach the maximum temperature as compared to the microheaters on silicon dioxide and polyimide membrane. The bridge-membrane structure reduced the heat loss of the microheater to the membrane due to its small area of membrane. It also resulted in a fast response time of the micro heater to reach the maximum temperature.

IV. DISCUSSION

A study on the effect of material and membrane structure on maximum temperature of microheater for gas sensor applications has been presented. Three different materials namely silicon nitride, silicon dioxide and polyimide and two types of membrane structure namely full-membrane and bridge-membrane were chosen to perform the study. Polyimide has the lowest thermal conductivity minimizing the heat dissipation from the heater to the membrane resulting in the microheater to reach the highest temperature. Although the polyimide membrane provided the least in thermal loss, its application is limited at temperatures up to 400 °C only as compared to silicon oxide and silicon nitride that are applicable in the higher range of temperature (400 °C - 700 °C). Silicon nitride distributed the most uniform temperature of microheater over the membrane. However, a higher power input is required for the microheater to reach the same temperature that has been achieved by the microheater on silicon dioxide and polyimide membrane at the input power of 250 mW. The bridge-membrane structure that has a smaller area compared to the full-membrane structure reduced the microheater heat loss to the membrane and subsequently improved the temperature performance and response time of the micro heater for all three membrane materials. The structure can provide significant results for use in metal-oxide gas sensor devices with a relatively low power consumption.

ACKNOWLEDGMENT

This work is supported by the Research University Grant DIP-2012-16.

REFERENCES

[1] Figaro Taguchi Gas Sensor, TGS816 Data-sheet Gas Sensor Products, http://www.figaro.co.jp/en/product/docs/tgs816_product_infomation_rev 01.pdf [September 17, 2013]

[2] O. Sidek et al "Effect of Heater Geometry on the High Temperature Distribution on a MEMS Micro-hotplate," 3rd Asia Symposium on Quality Electronic Design, pp.100-104, 2011.

[3] Jin-Ho Yoon and Jung-Sik kim, "Study on the MEMS-type gas sensor for detecting a nitrogen oxide gas," Solid-State Ionics, vol.192, pp. 668-671, 2011.

[4] D. Briand, S. Colin, J. Courbat, S. RAible, J. Kappler and N. F.de Rooij,"Integration of MOX gas sensors on polyimide hotplates", Sensors and Actuators B, 130, pp. 430-435, 2008.

[5] Mimiwaty M. Noor, Gandi Sugandi and Burhanuddin Yeop Majlis, "Micro-heater Filament of Polyimide Membrane for Gas Sensor Applications," Proceeding of IEEE International Conference on Semiconductor & Electronics (ICSE2012), pp. 360-362, 2012.

[6] Mimiwaty M. Noor, Badariah Bais and Burhanuddin Yeop Majlis. "The effects of temperature and KOH concentration on silicon etching rate and membrane surface roughness," Proceeding of IEEE International Conference on Semiconductor & Electronics (ICSE2002), pp. 524-528, 2002.

FEM analysis of wavelength effects in piezoelectric substrate

Norazreen Abd Aziz, Badariah Bais, Muhamad
Ramdzan Buyong and Burhanuddin Yeop Majlis
Institute of Microengineering and Nanoelectronics (IMEN),
Universiti Kebangsaan Malaysia, Bangi, Malaysia
burhan@ukm.my

Anis Nurashikin Nordin
ECE Department, Kulliyyah of Engineering, International
Islamic University Malaysia (IIUM), Gombak, Malaysia
anis.nordin@gmail.com

Abstract— In this paper, we discussed simulation of several annular surface acoustic wave (A-SAW) devices using various wavelengths to identify its effects on the focusing properties and to analyze the propagation of Rayleigh waves in piezoelectric substrate. By choosing Y-cut Z propagating Lithium Niobate as the substrate and aluminum electrodes as the IDT, we modeled the A-SAW devices using Comsol Multiphysics. We used 8 pairs of annular electrodes with thickness of 1 μm with three different design's wavelength of 100 μm, 150 μm and 200 μm, respectively. To minimize the computational time in determining the optimum frequency i.e. resonant frequency of the device, only one pair of electrode for each design is simulated under eigenfrequency analysis in 2D piezoelectric (pzd) module. To understand the Rayleigh waves behavior, simulation of the whole device structure was done under frequency domain analysis in 2D-axisymmetric piezoelectric module. From the simulation results, it can be observed that SAW displacement profiles, electric potential field and operating frequency are significantly influenced by the wavelength. The formation of focused acoustic waves at the center of A-SAW device suits them in biosensing and microfluidic actuation applications that require detection or manipulation of localized variations.

Keywords—surface acoustic wave, Rayleigh wave, Lithium Niobate, Comsol Mutiphysics.

I. INTRODUCTION

Surface acoustic waves (SAWs) are sound waves propagating along the surface of semi-infinite solids. Interdigital electrodes (IDT) are comb-shaped structure fabricated over the piezoelectric substrate. When an AC signals applied to the input of the electrodes, the resulting electric fields between adjacent fingers produce mechanical deformation in the piezoelectric crystal. At a resonant frequency, conversion of electrical energy to mechanical energy from each electrode add constructively which produces dynamic strain and initiate elastic waves to travel along the surface.

A typical SAW device action can be achieved in two ways i.e. one port SAW having several IDTs finger with reflectors and two ports SAW which uses two groups of IDTs, one act as the transmitter and another as the receiver. Fig. 1 shows the schematic layout of one and two port SAW devices. The reflectors in one port device act to contain the outward propagating waves within the device. On the other hand, the input energy from the transmitter group in two port SAW device will propagate across the delay line between the two transducers, and detected at the receiver electrodes. SAW devices are widely used in mobile phone technology [1], wireless communication [2], telecommunication [3] and recently, an extensive work focus on biosensor [4], microfluidic [5], microtransport [6] and actuation [7] applications.

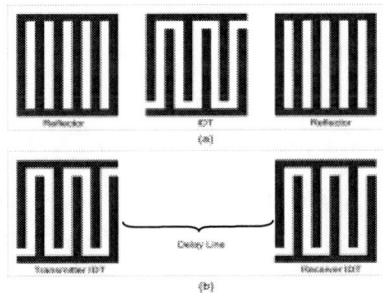

Fig. 1. (a) Schematic layout of one port conventional SAW device (b) Schematic layout of two port conventional SAW device

In this work, we developed a focusing SAW device adopting annular interdigital electrodes to generate high intensity acoustic fields. From literature, focusing SAW device provides acoustic waves emission with better intensity, compression ratio width and confined spot compared to conventional electrodes [8]. These phenomena suits for acoustic streaming, actuating, transporting or dispensing small volume of liquid. Realizing these potential applications, comprehensive understanding of device's performance is needed. This paper discuss on the effects of device wavelengths obtained by varying the electrodes spacing of the focusing SAW device in lithium niobate medium. Lithium niobate is preferred as a medium for waves propagation due to its high piezoelectric coupling properties. Evaluation of the acoustic wave generated including operating frequency of device, acoustic displacement field and its focusing properties is also discussed. The analysis is supported by finite element simulations model of piezoelectric module in Comsol Multiphysics.

978-1-4799-5761-3/14 $31.00 © 2014 IEEE

II. Theory and Device Structure

The displacement of acoustic waves on a surface of piezoelectric substrate i.e. Rayleigh elastic wave has components in the surface-normal and surface-parallel directions with respect to the direction of the wave propagation. The particle in the upper surface takes elliptical path having both the components. Fig. 2 shows the coordinates system of Rayleigh waves propagation.

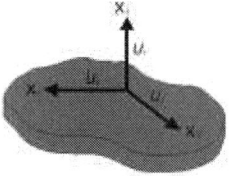

Fig. 2. Coordinates system of Rayleigh waves propagation in piezoelectric substrate

$$u_1 = Z_1 e^{(-b_1 x_3)} + Z_2 e^{(-b_2 x_3)} e^{jk(x_1 - ct)}$$

$$u_2 = (-b_1 jk) Z_1 e^{(-b_1 x_3)} + (jk/b_2) Z_2 e^{(-b_2 x_3)} e^{jk(x_1 - ct)}$$

$$u_3 = 0$$

$$b_1 = k(1 - c^{2/v_l^2})^{1/2}$$

$$b_2 = k(1 - c^2/v_t^2)^{1/2}$$

where Z_1 and Z_2 are amplitude meanwhile v_l and v_t are longitudinal and transverse wave velocities, respectively, k is the wave number and c is the velocity of SAW. The propagation of SAW depends on the substrate material properties, the crystal cut and the structure of electrodes. Meanwhile, the wavelength of A-SAW device, λ is determined by the electrodes' finger spacing, p and they are related by $\lambda = 4p$. The associated frequency, f of the waves propagating with velocity, v is given by $f = v / \lambda$. Since the geometry of electrodes is in the shape of comb-circular, the emission of acoustic waves follows the shape of the transducer and confined only in the region of the innermost electrodes [9]. Fig. 3 shows the top layout of the annular electrodes structures that is used in this study. By solving the above equations, FEM can generate the displacement profiles and voltages at each node of the devices.

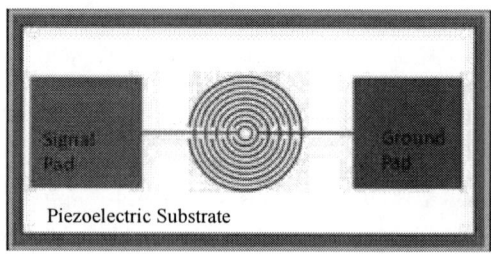

Fig. 3. Top schematic layout of annular electrodes on piezoelectric substrate

III. Finite Element Modeling

Applying alternating voltage to the annular IDT patterned on piezoelectric substrate generates surface acoustic waves. The IDT geometry dictates the SAW wavelength along with its resonant frequency. Therefore, the spacing between electrodes is modified to simulate the A-SAW devices operating in different frequency ranges and generating focused acoustic waves with varying wavelengths. In order to evaluate the effects of wavelength on the Rayleigh wave propagation characteristics and its focusing properties of A-SAW, we simulated 8 finger pairs A-SAW devices with varying finger spacing of 25 μm, 37.5 μm and 50 μm. 2D eigenfrequency analysis was run in early stage of the study to identify the operating frequency and the potential distribution underneath the metal electrodes. Neglecting the edge effect and to simplify the design, we reduced the model geometry to one pair of annular IDT in 2D Piezoelectric Device (pzd) Comsol simulation as shown in Fig. 4(a). The elastic, permittivity, stress constants of the Y cut Z propagating lithium niobate substrate are taken from [10]. Aluminum is chosen as the electrode with values of Poisson ratio, density, Young's modulus taken from the material library in Comsol. Results from the eigenfrequency analysis are used as a guideline for frequency range setting in 2D axisymmetric frequency domain analysis. Since the IDTs are periodic in nature and alternately in positive and negative potentials arrangement, it was drawn in frequency domain under 2D axisymmetric Piezoelectric (pzd) module also. The 2D axisymmetric layout as shown in Fig. 4 (b) is sufficient to model the A-SAW device as a whole since it will be later extruded in 3D for the harmonic analysis.

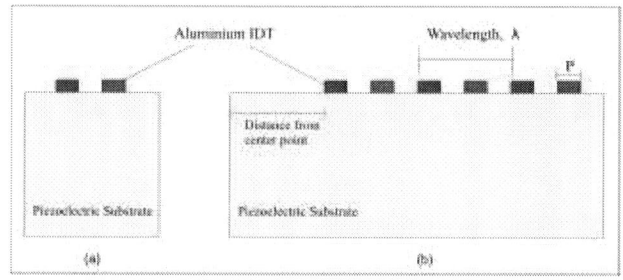

Fig. 4. (a) 2D layout for eigenfrequency analysis (b) 2D axisymmetric layout for harmonic analysis

IV. Results And Discussions

We used an eigenfrequency analysis to identify the resonant frequency and electrical potential distribution of the A-SAW. Fig. 5 shows the displacement profiles and potential distribution of one pair aluminum electrodes. From Fig. 5 (a), (c) and (e), device with wavelengths of 100 μm, 150 μm and 200 μm produced maximum displacement of 3.4 nm, 3.6 nm and 3.5 nm, respectively. Conversion of mechanical deformation to electrical formation produces an electric field in the piezoelectric substrate. Formation of electric field is shown in Fig. 5(b), (d) and (f). All three devices produced almost the same amount of electric potential and displacement regardless of their wavelengths. However, the penetration depth of acoustic waves in piezoelectric is greater as we increase the gap of the metal electrodes. Table 1 shows the comparison of theoretical and simulation data extracted from eigenfrequency analysis. The variation of phase velocity and resonant frequency simulated in this work is less than 10% from the theoretical value.

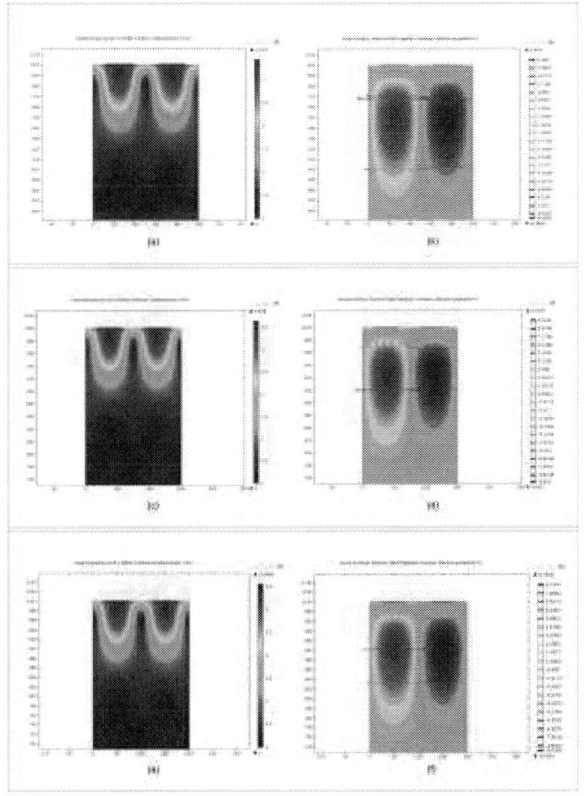

Fig. 5. (a) Deformed shape plot of the resonance mode of λ = 100 μm (b) Electric potential distribution at resonance mode of λ = 100μm (c) Deformed shape plot of the resonance mode of λ = 150μm (d) Electric potential distribution at resonanace mode of λ = 150μm (e) Deformed shape plot of the resonance mode of λ = 200μm (f) Electric potential distribution at resonance mode of λ = 200μm

TABLE I.   THEORY AND DATA SIMULATION COMPARISONS

Lambda (μm)	Resonant Frequency (MHz)		Phase Velocity		Variation (%)
	Theory	FEM	Theory	FEM	
100	34.88	37.9	3488	3790	8.6
150	23.25	25.3	3488	3795	8.8
200	17.44	18.9	3488	3780	8.3

We further analyzed the SAW emission's characteristic of the whole device under frequency domain analysis. From the susceptance versus frequency graph as

shown in Fig. 6, it is clear to see the resonant frequency for three devices simulated in this work agrees well with the resonant frequency obtained in the eigenfrequency analysis. Smaller periodic gap produces higher displacement and higher susceptance value.

Observation on the displacement profiles at the center of the A-SAW device over certain range of frequencies is also done in frequency domain study as shown in Fig. 7. From Fig. 7, device with wavelength of 100 μm shows maximum displacements amplitude of 24 nm at a frequency of 37.58 MHz. Meanwhile, devices with 150 μm and 200 μm show maximum displacements amplitude of 15 nm with the operating frequency of 18.48 MHz and 24.82 MHz, respectively. It was also found that the resonant frequency obtained from the harmonic analysis is closer to the theoretical value compared to the one obtained in eigenfrequency analysis. This indicates that emulating the whole device will produce more accurate results. Therefore, smaller wavelength produces higher operating frequency and displacement amplitude which can be achieved by designing device with smaller electrode spacing.

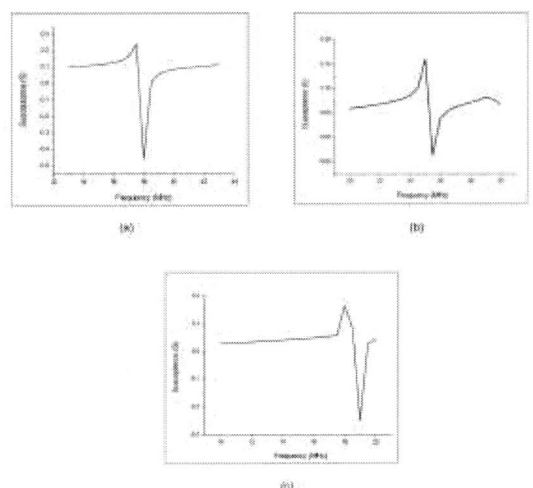

Fig. 6. Susceptance versus frequency for A-SAW device with wavelength of (a) 100 μm (b) 150 μm (c) 200 μm

Since the emission of the acoustic waves is confined at the center of electrodes, radius of the innermost electrodes for all devices is set to be 50 μm. Fig. 8 shows the displacement profiles of the three devices. From Fig. 8, the graphs show propagation of mechanical wave underneath metal electrodes finally converges at the center of the three devices. It can also be observed that smaller wavelength with steeper slope generates more intense focused wave. Extrusion from 2D z-displacement profiles along the horizontal axis of the device proves that annular periodic electrodes produces focused wave at the center of A-SAW device. Fig. 9 shows the 3D displacement contour of A-SAW device with wavelength of 100 μm.

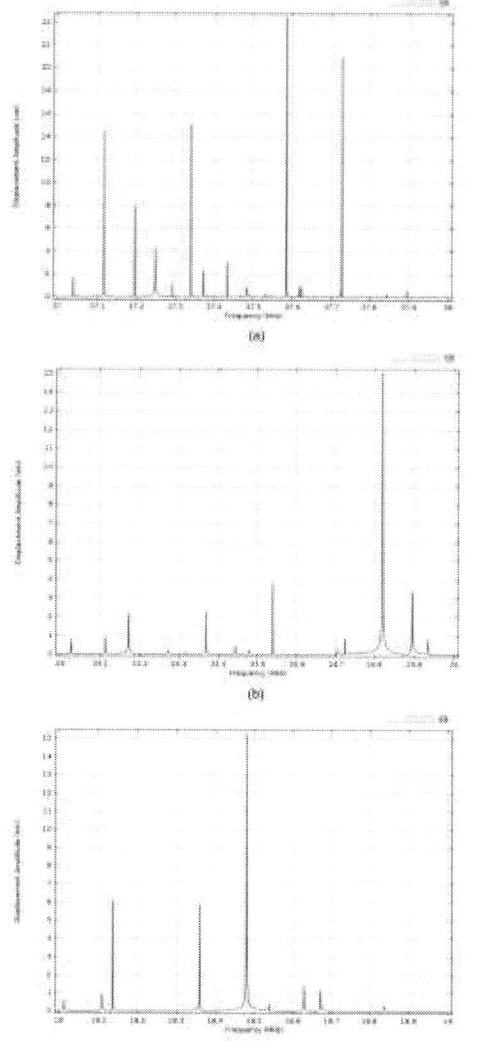

Fig. 7. Harmonic response of A-SAW device with wavelength of (a) 100 μm (b) 150 μm (c) 200 μm

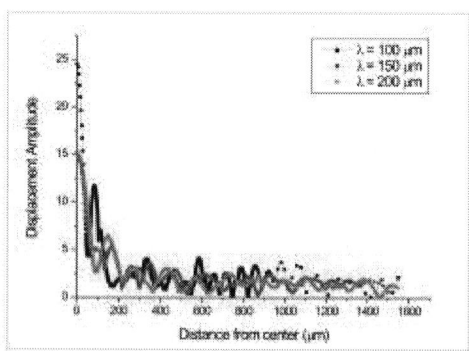

Fig. 8. Displacement profiles of A-SAW devices

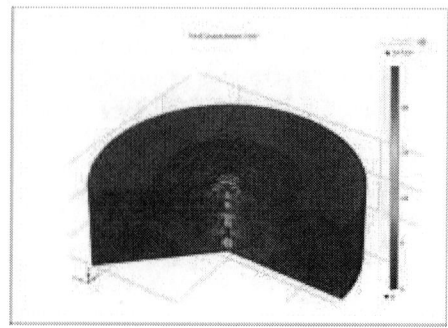

Fig. 9. Displacement profiles of A-SAW device with wavelength of 100 μm

V. CONCLUSION

A-SAW devices with 25 μm, 37.5 μm and 50 μm finger spacing are simulated using piezoelectric module of Comsol Multiphysics. Our simulation results indicate that device with smaller electrode's gap i.e. 25 μm produce higher operating frequency, greater amplitude displacement and better focusing ability compared to the other devices. These findings lay the groundwork for further investigation on the effect of other geometrical parameters such thickness of electrodes, number of finger pair and radius of innermost electrodes on SAW displacement profiles, electric potential field and operating frequency.

REFERENCES

[1] M. Kadota, "6E-5 High Performance and Miniature Surface Acoustic Wave Devices with Excellent Temperature Stability Using High Density Metal Electrodes," *2007 IEEE Ultrason. Symp. Proc.*, no. 1, pp. 496–506, Oct. 2007.

[2] A. C. Tikka, S. F. Al-Sarawi, and D. Abbott, "Finite element modelling of SAW correlator," vol. 6799, pp. 679915–679915–9, Dec. 2007.

[3] S. Ballandras, a. Reinhardt, V. Laude, a. Soufyane, S. Camou, W. Daniau, T. Pastureaud, W. Steichen, R. Lardat, M. Solal, and P. Ventura, "Simulations of surface acoustic wave devices built on stratified media using a mixed finite element/boundary integral formulation," *J. Appl. Phys.*, vol. 96, no. 12, p. 7731, 2004.

[4] Q. Cai, R. Wang, L. Wu, L. Nie, and S. Yao, "Surface Acoustic Wave Enzyme Sensor Applied to the Kinetic Assay of Acid P hosphatase," vol. 120, pp. 2833–2836.

[5] Y. Bourquin, J. Reboud, R. Wilson, and J. M. Cooper, "Tuneable surface acoustic waves for fluid and particle manipulations on disposable chips.," *Lab Chip*, vol. 10, no. 15, pp. 1898–901, Aug. 2010.

[6] L. Y. Yeo and J. R. Friend, "Ultrafast microfluidics using surface acoustic waves," *Biomicrofluidics*, vol. 3, no. 1, p. 12002, 2009.

[7] U. Demirci and G. Montesano, "Single cell epitaxy by acoustic picolitre droplets.," *Lab Chip*, vol. 7, no. 9, pp. 1139–1145, Sep. 2007.

[8] S. K. Sankaranarayanan and V. R. Bhethanabotla, "Numerical analysis of wave generation and propagation in a focused surface acoustic wave device for potential microfluidics applications," *IEEE Trans. Ultrason. Ferroelectr. Freq. Control*, vol. 56, no. 3, pp. 631–643, 2009.

[9] V. Laude, D. Gérard, N. Khelfaoui, C. F. Jerez-Hanckes, S. Benchabane, and A. Khelif, "Subwavelength focusing of surface acoustic waves generated by an annular interdigital transducer," *Appl. Phys. Lett.*, vol. 92, no. 9, p. 094104, 2008.

[10] Y. Rao and G. Zhang, "3D Modeling of a Surface-Acoustic-Wave Based Sensor," 2007.

Performance Analysis of Zinc Oxide Piezoelectric MEMS Energy Harvester

Umi Milhana Jamain, Nur Hidayah Ibrahim, Rosminazuin Ab Rahim

Kulliyyah of Engineering
International Islamic University Malaysia
Kuala Lumpur, Malaysia
umimilhanajamain@gmail.com

Abstract— **This paper presents the design and analysis of MEMS piezoelectric energy harvester. Zinc oxide (ZnO) MEMS piezoelectric energy harvester has been utilized as piezoelectrically active cantilever for mechanical to electrical transduction. A COMSOL Multiphysics model was used which provide accurate information on the frequency, stress and voltage output of a ZnO piezoelectric energy harvester. Few design parameters have been studied which are rectangular cantilever, triangular cantilever, rectangular cantilever with proof mass and using different types of piezoelectric materials. The effects of varying geometrical dimensions of the device were also investigated. From simulation results, it was found out that ZnO piezoelectric energy harvester with the length of 150 µm, width 50 µm and thickness of 4 µm generates 9.9184 V electric potential under the resonance frequency of 0.71 MHz and 1 µN/m² mechanical force applied.**

Keywords— *MEMS; piezoelectric energy harvester; ZnO; COMSOL Multi Physics*

I. INTRODUCTION

Energy harvesting is the process by which energy is derived from external sources (e.g. solar power, thermal energy, wind energy, salinity gradients, and kinetic energy), captured, and stored for small, wireless autonomous devices, like those used in wearable electronics and wireless sensor networks. Energy harvesters provide a small amount of power for low-energy electronics. Energy from mechanical vibrations can be harvested by the use of piezoelectric, electromagnetic, and electrostatic [1].

When pressure (stress) is applied to a material, it creates a strain or deformation in the material. The capability of the piezoelectric thin film in generating an electrical output in response to mechanical energy or vibration has given a significant impact in our daily lives. Piezoelectric thin film has been widely used in various MEMS applications such as surface acoustic wave (SAW) resonators, pressure sensors, biomedical and energy harvesting [2-4]. In energy harvesting application, a piezoelectric energy micro-generator typically harvests mechanical energy or vibrations and converts it to electrical energy through piezoelectric effect. Different piezoelectric materials can affect the performance of the energy harvester due to different piezoelectric constants. Examples of piezoelectric materials include lead zirconate titanate (PZT), zinc oxide (ZnO) and aluminum nitrate (AlN).

This work is supported by Ministry of Education, Malaysia under the project RAGS 12-008-0008.

Compare to AlN, PZT material has higher piezoelectric coefficient and high dielectric constant [5].

Many researches on MEMS device utilizing ZnO and AlN as the piezoelectric transducers have been reported [5-7]. In this paper, an investigation on the utilization of both ZnO and AlN piezoelectric materials as the transducer for energy harvester will be carried out.

II. THEORETICAL

In determining the resonance frequency of the cantilever beam during bending. Stoney's equation which relates cantilever end deflection δ to applied force σ is referred:

$$\delta = \frac{3\sigma(1-v)}{E\left(\frac{L}{t}\right)^2} \qquad (1)$$

where v is Poisson's ratio, E is Young's Modulus, L is the cantilever's length and t is the cantilever's thickness. The cantilever's spring constant, k can be expressed as

$$k = \frac{3EI}{L^3} \qquad (2)$$

where I is moment of inertia of the cantilever and the equation of moment of inertia is as given as

$$I = \frac{bh^{\wedge}3}{12} \qquad (3)$$

where b is width of the cantilever and h is the thickness. In order to identify the resonant frequency, the following equation is referred

$$f(res) = \frac{0.32\sqrt{k}}{\sqrt{m}} \qquad (4)$$

where mass, $m = \rho.h.L.w$. In this case, ρ is the resistivity, h is the cantilever's height, L is the cantilever's length and w is the cantilever's width. It shows that the resonance frequency is a function of spring constant and cantilever's mass.

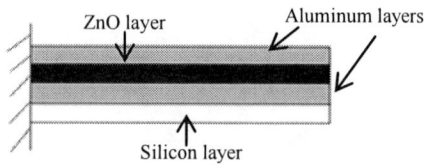

Fig. 1. Cantilever structure which comprises of ZnO layer sandwiched between Al layers with bottom Si supporting layer

III. SIMULATION WORKS

Simulation works of piezoelectric energy harvester comprises of two main analyses which are: stationary analysis and eigenfrequency analysis.

A. Cantilever Structure

As shown in Fig. 1, the piezoelectric energy harvester is made of ZnO cantilever sandwiched between Al layers which serves as the top and bottom electrodes, respectively. The bottom layer is made of Silicon (Si) substrate as the supporting layer. The structure's dimension is 150 μm length, 50 μm width with 1 μm thickness for each layer.

B. Materials Properties

The properties of Aluminum, ZnO and Silicon used in the ZnO piezoelectric energy harvester are shown in Table 1.

C. Simulation Settings

The boundary conditions are set by fixing one end of the cantilever while the other end is set free. A force of 1×10^{-6} N/m^2 is applied on the upper face of the cantilever. In meshing, the boundaries in the geometry are discretized into triangular or quadrilateral boundary elements. The geometry edges are discretized into edge elements. Tetrahedral meshed has been used in this design. The element size is set as normal and the complete mesh consists of 28620 elements.

D. COMSOL Analysis

Few analyses have been carried out in order to identify the effects of the contributing parameters to the performance of the piezoelectric energy harvester. Stationary analysis solves for stationary displacements or a steady-state condition. All loads and constraints are constant. While eigenfrequency

TABLE 1.   MATERIAL PROPERTIES

	Materials		
Material Properties	***ZnO***	***Al***	***Silicon***
Young Modulus, E (Pa)	6.83×10^9	70×10^9	170×10^9
Density, ρ (kg/m^3)	5680	2700	2329
Poisson's Ratio	0.2	0.33	0.28

analysis is used in order to find the first 6 modes of frequency of the cantilever beam and their corresponding shape of deformation. The first mode is considered as the excitation frequency which acts as resonance frequency and therefore gives maximum electrical potential.

IV. SIMULATION RESULTS

Simulation of the piezoelectric energy harvester can be divided into few stages: stationary analysis, eigenfrequency analysis and the utilization of both analyses in determining the resonance frequency and resultant output voltage in response to different contributing factors.

A. Stationary Analysis

When force is applied from the downward direction (-z) of the cantilever, the cantilever vibrates resulting in the deflection of the ZnO cantilever. The stationary analysis shows that the beam is deflected downward, in (−z) direction, which is in the same direction of the applied external load. As illustrated in Fig. 2, the maximum displacement is observed at the beam's free end at a value of $5.85 \times 10^{-9} \mu m$. It shows that the displacement is maximum at the free end and zero at fixed end. The deflected ZnO cantilever generates an electric potential. As shown in Fig. 3, the electric potential generated is high at the fixed end of the cantilever with maximum value of 9.91V.

Fig. 2. Simulation capture of cantilever's displacement at fixed applied force of 1×10^{-6} N/m^2 with maximum value of 5.85×10^{-9} μm was observed at the fixed end.

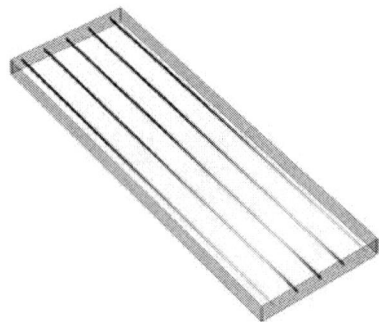

Fig. 3. Simulation capture of generated electric potential at fixed applied force of 1×10^{-6} N/m^2 with maximum value of 9.91 V was observed at the fixed end.

TABLE 2. FREQUENCY MODE AND SHAPE OF DEFORMATION OF CANTILEVER

Shape of deformation	Frequency (MHz)	Mode
	0.17	1
	1.02	2
	1.08	3
	1.92	4
	3.11	5
	3.29	6

B. Eigenfrequency Analysis

The eigenfrequency analysis solves for the natural frequencies of a structure and its corresponding mode shapes. The values of the modes of frequency and their corresponding deformation shapes are illustrated in the Table 2.

The first mode frequency is chosen as the resonance frequency of the cantilever since its bending behaviour is suitable for the energy harvester application. Fig. 4 shows the graph plot of the frequency versus total displacement of the cantilever. Maximum displacement was observed at resonance frequency of 0.171 MHz.

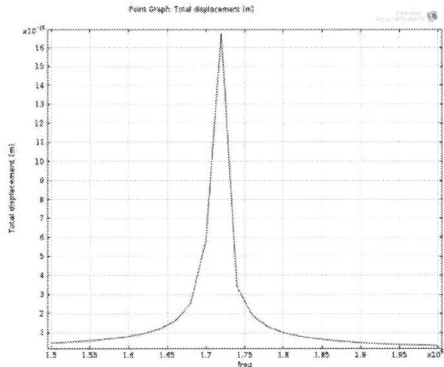

Fig. 4. Graph of total displacement versus frequency for the simulated cantilever.

TABLE 3. COMPARISON BETWEEN ANALYTICAL AND SIMULATION RESONANCE FREQUENCY

Model	Frequency (MHz)	
	Analytical	Simulation
Rectangular	0.177	0.171

Table 3 shows the comparison of analytical resonance frequency obtained from (1) and the simulated resonance frequency from COMSOL. The simulated value is more or less equal to the analytical value which indicates the validity of the simulation works.

C. Effects of Different Piezoelectric Materials

An investigation on different piezoelectric materials has been carried out in order to understand the effect of piezoelectric materials to the energy harvester's performance. In this analysis, different piezoelectric materials which are ZnO, PZT and AlN have been used for the same rectangular-shaped cantilever model. A force of 1×10^{-6} N/m^2 has been applied to all three models. The models are analyzed using stationary and eigenfrequency analyses to observe the resultant displacement, electric potential and resonance frequency.

Table 4 shows the cantilever's performance for all the three models. Using ZnO piezoelectric material, both resonance frequency and generated electric potential are at maximum with values of 0.17 MHz and 9.91 V, respectively compare to the other materials. It shows that ZnO is the best candidate in generating maximum output power for the piezoelectric energy harvester application.

TABLE 4. THE CANTILEVER PERFORMANCE FOR DIFFERENT PIEZOELECTRIC MATERIALS

Piezoelectric Materials	Displacement (µm)	Resonance frequency (MHz)	Electric potential (V)
ZnO	5.85×10^{-9}	0.17	9.91
PZT	1.08×10^{-10}	0.15	9.01
AlN	8.66×10^{-11}	0.20	9.62

D. Effects of Different Geometrical Dimensions

For this analysis, the variation of length was applied to the cantilever. In this investigation, the ZnO piezoelectric material is used with an applied force of 1×10^{-6} N/m^2 is maintained for all analyses.

Fig. 5 shows the results of the displacement for variation of length of the cantilever. As the length increases, the displacement also increases. The length varies from 100 µm to 150 µm. However, for cantilever's width variation, it was observed that the bigger the width, the smaller the displacement of the cantilever as shown in Fig. 6. Therefore, cantilever's dimension with proper length to width ratio should be considered in order to get maximum performance of the device.

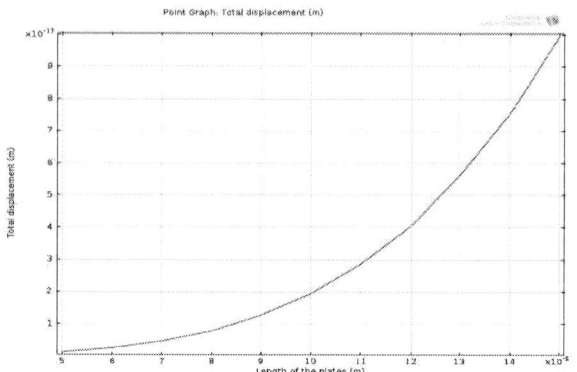

Fig. 5. Graph of the variation of length of the cantilever versus total displacement under stationary analysis.

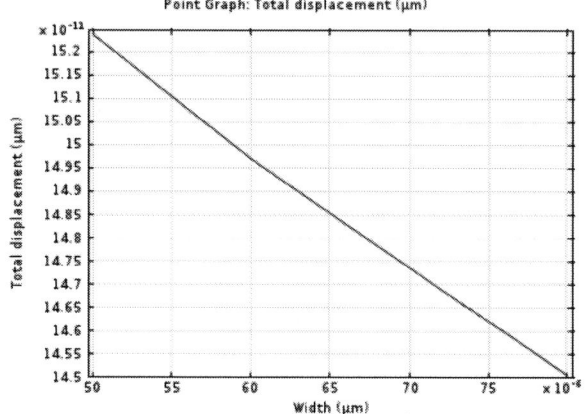

Fig. 6. Graph of the variation of width of the cantilever versus total displacement under stationary analysis.

V. Conclusions

The stationary and eigenfrequency analyses were used in analyzing the performance of piezoelectric energy harvester. The consistency in the value of the applied load, geometry, and boundary conditions allowed us to compare the obtained results from the stationary analysis for the three different types of piezoelectric material models.

Additionally, the comparison reveals that the piezoelectric material ZnO has the highest direct piezoelectricity relationship, in which, when the load of $1x10^{-6}$ N/m^2 was applied on the beam's free end, the model that consists of ZnO material gave the highest electric potential of 9.91 V. In fact, this is due to the direct relationship in the piezoelectric materials between the strain and electrical potential, in which, if one has increased, the other will increase as well. The eigenfrequency analysis has helped us in determining the resonance frequency of the cantilever. As the length of the cantilever in increase, the displacement of the cantilever also increases which contributes to high electric potential.

References

[1] A.Aini , A.N Nordin, H.Salleh ,"A comparative study on MEMS piezoelectric microgenerators," Microsyst Technol, 2010.

[2] D. L. DeVoe and A. P. Pisano A P 2001 "Surface micromachined piezoelectric accelerometers (PiXLs)", J. Microelectromech. Syst 10, pp. 180-6, 2001.

[3] R. Ruby, "Review and comparison of bulk acoustic wave FBAR, SMR technology", Proc. IEEE Ultrasonics Symp., vol 7, pp. 1029-40, 2007.

[4] P. Muralt P, "Piezoelectric micromachined ultrasonic transducers based on PZT thin films", IEEE Trans. Ultrason. Ferroelectr. Freq. Control 52, 2276-88, 2006.

[5] A. Kuoni, R. Holzherr, M. Boillat, N. F. Rooij, "Polyimide membrane with ZnO piezoelectric thin film pressure transducers as a differential pressure liquid flow sensor", J. Micromech. Microeng. 12, 2003.

[6] P Ivaldi, J Abergel, M H Matheny, L G Villanueva, R B Karabalin, M L Roukes, P Andreucci, S Hentz and E Defaÿ, "50 nm thick AlN film-based piezoelectric cantilevers for gravimetric detection", J. Micromech. Microeng. 21, 2011.

[7] B. Kumar, S. W. Kim, "Energy harvesting based on semiconducting piezoelectric ZnO nanostructures", Nano Energy 1(3), pp. 342-355, 2012

978-1-4799-5761-3/14 $31.00 © 2014 IEEE

Channel length effect on the saturation current and the threshold voltages of CNTFET

Abu Hanifah Muhamad Ali and M. H. Ani,
Department of Manufacturing and Materials,
Kulliyyah of Engineering, International Islamic University,
Kuala Lumpur, Malaysia
mhanafi@iium.edu.my

Mohd Ambri Mohamed
Institute of Microengineering and Nanoelectronics, The
National University of Malaysia, 43600 UKM Bangi,
Selangor, Malaysia.
ambri@ukm.edu.my

Abstract— Carbon nanotubes (CNTs) are such promising material in future microelectronic devices due to their great property in conductivity, mechanical strength and light weight. Field effect transistor (FET) has already come to its most maximum efficiency because of their reduced size leads to decrease in their capability in conducting electric. It is interested to embed ballistic electron transfer capability of CNTs on FET. In this study, direct growth method of CNTs was employed to attach it on FET electrodes with various terminal gaps. The results show that CNTFET has successfully fabricated, with averaged saturation currents always lowest at the channel size of 15μm. While their highest measured threshold voltage value is 4.291 V at 15 μm gap. This phenomenon is attributed to the change of CNTs' chirality, which apparently changes the metallic type of CNTs to the semiconducting CNTs.

Keywords—CNTFET; Direct Growth Method; Transconductance; Chirality

I. INTRODUCTION

In 1998, Tans et. al. has successfully demonstrated the functionality of CNTFET at room temperature [1]. This is an important step in microelectronic engineering. However, mass-production of CNTFET is still far reached due to technical challenges to deposit and correctly align the CNT onto FET devices. Random networks of SWCNTs offer an alternative to parallel alignment. Several methods for forming random networks for CNTFETs, such as direct growth [2], solution dropping [3], and various transfer printing techniques [4–6] were reported and some high-quality electrical properties of the resulting FETs were also demonstrated.

Direct growth of CNTFETs using CVD has attracted much attention because it has many advantages compared to other methods, i.e arc discharge and laser ablation [7]. The CVD has several important advantages such as relatively uncomplicated and easily available raw material. The technique can also be used to produce vertically aligned nanotubes, fibers, monoliths, foams and powders.

In this study, CNTFETs were fabricated using direct growth CVD method. The performance of the CNTFETs was evaluated using standard semiconductor analyzer.

II. EXPERIMENTAL

A. FET fabrication

Semiconductor grade Si/SiO_2 substrate was immersed in de-ionized water with ultrasonic agitation for 10 minutes. The sample then was cleaned with ethanol and subsequently etched by plasma process. In order to increase adhesiveness between resist and substrate, it was baked on 110° C hotplate for 5 minutes and spin coated using hexamethyldisilan (HMDS). Then, it was baked again to enhance the reaction between HMDS and –OH group from SiO_2 surface.

Lift of layer (LOL-2000) was applied on the substrate before the substrate was spin coated with photoresist, which was baked at 110° C for 3 minutes. The resist was exposed to UV light, which then positive photo resist was used to print the UV light through the photomask. Exposed photoresist was cleaned with solvent, rinse with purified water, dried on hotplate and undergoes plasma etched to remove unwanted residues. After that, the electrode made up of Mo and Co, was deposited onto substrate using sputtering method. The substrate was submerged into remover (1165) for 12 hours, and rinsed with ethanol.

The thickness of SiO_2 layer was 400 nm, while the channel width was 250 μm. The channel length of the FET was varied from 10, 15 and 20 μm. The capacitance value of the SiO_2 was defined as $1 \times 10^{-3} F/cm^3$.

B. CNT deposition

In order to grow the CNT directly, the samples was placed in CVD reactor which using ethanol as carbon source. 50 ml of ethanol was allocated in carbon feedstock. The reactor was heated to 900° C and supplied with $Ar-H_2$ gas with internal pressure of 400 Pa. After the condition stabilized, the flow of $Ar-H_2$ gas was then terminated and ethanol vapor was supplied into the chamber immediately. This is the point where deposition of CNT has begun, and the process was continued for 30 minutes. After deposition process, the sample was characterized using FE-SEM and semiconductor analyzer.

III. RESULTS AND DISCUSSION

FE-SEM image in Figure 1 shows drain and source terminals are connected by the CNTs network. Growth of CNTs was successfully done by direct growth method with

978-1-4799-5761-3/14 $31.00 © 2014 IEEE

Cobalt as the catalyst. The right side is the source terminal while the left one is the drain terminal. The nucleation of CNTs has begun in both electrodes due to the existence of the catalyst. The catalyst is sputtered into the electrodes during the fabrication processes. The device would be functioning if the threshold voltages are managed to be overcome. The threshold referred to the amount of the voltage need to be supply to the gate terminals so that the enhancement layer could be produce between source and drain terminals.

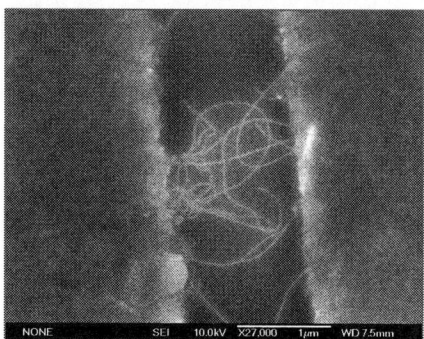

Fig. 1. The CNTs bridging the drain and source terminals.

Through the semiconductor analyzer, the output and transfer curves were successfully determined. Above all the information obtained from the output curves, it was understood that the channel length significantly affect the saturation current as shown in Figure 2. The saturation current of the 10 μm channel length was always higher compared to the 15 μm regardless of their gate voltages (V_{gs}). However, the saturation current increase when the gate voltages increased from the 15 μm to 20 μm.

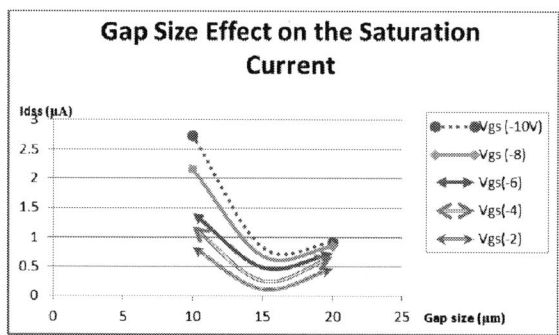

Fig. 2. The channel length effect to the saturation current.

The changed of the trend at 15 μm not only occurred at the output curve but the transfer curves as well. From transfer curves the most significance measurement that being affected by the channel length is threshold voltages as shown in Table 1. The threshold voltages were determined from the transfer curves (I-Vgs) generated from the semiconductor analyzer, where the threshold points were determined from where

current starts to switch from off to on state. The threshold voltages increased from 1.58 V to the 4.30 V. However, threshold reduced to 2.76 V when the channel length is being increased to 20 μm. Negative sign was ignored, considering the device is p-type CNTFET.

TABLE I.   THE THRESHOLD VOLTAGES CORRESPONDING TO THE CHANNEL LENGTH OF THE CNTFETs

Channel length (μm)	10	15	20
Threshold voltage (V)	-1.58	-4.30	-2.76

The positive sign and the negative sign of the measurements is not the main concern since it not really explained the channel length effect. The main issues that deal with the channel length effect are the energy barrier and the resistance to produce the conductive layer. The higher the saturation current basically explained the lower the energy barrier of the devices to allow the electron to pass through it. In this case 15 μm shown the highest energy barrier compared to 10 μm and 20 μm channel length. The increment of the energy barrier basically due to the increment of the band gaps values. The band gaps values must be referred to the band gaps of the CNTs since it was the material for the conductive channel in the devices. Because of that, the changes of the band gaps values of CNTs structure most likely due to the changes of structure in CNTs and the structure of the CNTs is always related to their chirality. It could be said that the changes of chirality occurred between 15 μm and 20 μm. The change of the chirality between the regions is supported by the threshold voltages since their changes of trends occurred between 15 μm and 20 μm as well. The threshold voltage is amount of the voltages needs in order to produce conductive layer. Normally the metallic type materials including the metallic type CNTs has lower band gap and lower threshold voltages. However, their conductivity was easily affected when their travel distance increased. The travel distance (channel length) effect is more susceptible to CNTs because it has very small diameter. The effect was reduced when the channel length exceeds 15 μm. It most probably due to changes from metallic type to the semiconductor of CNTs as the channel path increased. The change to semiconducting CNTs as the channel length increase is attributed to the diminishing of catalyst effects. The ethanols which act as a carbon source deposited more pure carbon without inherit any metallic properties from cobalt since it is growing further away from the electrode. The semiconducting CNTs are more consistent in conductivity and less affected by the increasing in channel length. On top of that, the thermal effect also affects the performances of the devices. The thermal source come from the voltages supply increased the internal resistance in metallic CNTs compared to semiconducting CNTs. Thus the longer the metallic CNTs will cause greater resistance. However, when the chirality of the CNTs changes due to the reduction of the catalyst effect and their own tendency to be semiconducting CNTs, it directly changes the CNTFET devices performances. In the near future, an in-situ study will be performed during direct growth of CNTs to clarify the mechanism of the changes of CNTs chirality during the growth of CNTs.

IV. CONCLUSION

Single Wall Carbon Nanotubes has been successfully grown onto the FET devices. The devices were differentiated by varying their channel lengths. The saturated currents have almost quadratic trend, lowest at 15 μm. Threshold current was determined to be dependent to channel length respectively. The change of CNTs chirality was critical in between length of 15 to 20 μm, where CNTs change from metallic to semiconducting properties.

ACKNOWLEDGMENT

This research was supported by the Institute of Microengineering and Nanoelectronics, The National University of Malaysia and International Islamic University Malaysia. It was sponsored by Ministry of Education, Malaysia through grant RAGS13-002-0065

REFERENCES

[1] S.J. Tans, A.R.M. Verschueren, C. Dekker, Nature 393 (1998) 49.

[2] E.S. Snow, J.P. Novak, P.M. Campbell, D. Park, Appl. Phys. Lett. 82 (2003) 2145.

[3] M. Shiraishi, T. Takenobu, T. Iwai, Y. Iwasa, H. Kataura, M. Ata, Chem. Phys. Lett. 394 (2004) 110.

[4] M.A. Meitl, Y. Zhou, A. Gaur, S. Jeon, M.L. Usrey, M.S. Strano, J.A. Roger, Nano Lett. 4 (2004) 1643.

[5] S.-H. Hur, D.-Y. Khang, C. Kocabas, J.A. Rogers, Appl. Phys. Lett. 85 (2004) 5730.

[6] S.H. Hur, O.O. Park, J.A. Rogers, Appl. Phys. Lett. 86 (2005) 243502.

[7] N.M. Mubaraka, E.C. Abdullah, N.S. Jayakumar, J.N. Sahu, Journal of Industrial and Engineering Chemistry 20 (2014) 1186–1197.

Polydimethlsiloxane (PDMS) Microchannel with Trapping Chamber for BioMEMS Applications

Ummikalsom Abidin, Burhanuddin Yeop Majlis, *SMIEEE* and Jumril Yunas, *MIEEE*
Institute of Microengineering and Nanoelectronics (IMEN)
Universiti Kebangsaan Malaysia (UKM)
43600 UKM Bangi, Selangor, Malaysia
Email: burhan@ukm.edu.my

Abstract—**This study presents fabrication and simulation of a microchannel for BioMEMS applications. The basic construction of this microfluidics channel consists of an inlet and an outlet, a microchannel for transporting continuous fluid flow and a trapping chamber as mean of trapping and separating the intended biological cells. The microchannel is constructed using polydimethlsiloxane (PDMS) using replica molding technique from SU-8 mold. The flow characteristics and the pressure drop experienced by microchannel have been modeled and simulated using Finite Element Analysis (FEA). The simulation results revealed a linear relationship of velocity magnitude and pressure drop with volumetric flow rate in the range of 0.5 to 1000 μL/min. Furthermore, the velocity streamlines indicated a laminar fluid characteristic is maintained in the microchannel flow at maximum volumetric flow rate of 1000 μL/min. Pressure drop is a vital parameter in a microchannel design due to the bonding limit between the PDMS microchannel and its substrate. In this work, a maximum pressure drop of 14.8 kPa is developed at maximum volumetric flow rate of 1000 μL/min. The pressure drop is in a safe limit for PDMS microchannel bonded with PDMS layer substrate operation.**

Keywords—microchannel; BioMEMS; microfluidics; polydimethylsiloxane (PDMS); Finite Element Analysis (FEA)

I. INTRODUCTION

Bio-Micro Electro Mechanical Systems (BioMEMS), Micro Total Analysis System (μTAS) and Lab-on-chip (LoC) are the related field of research utilizing fluid transport phenomena in microscopic scale volume for chemical and biological analysis on a single chip. Microscopic channel or microchannel is defined as very small channel of dimension in the range of micron to millimeter. The microchannel is able to handle fluid flow of picoliter, nanoliter and microliter in volume [1]. The study of fluid in this volume classification is known as microfluidics. Currently, microchannels are widely used to transport biological particles ranging from DNA, proteins, and cells in BioMEMS applications.

Microchannel has been fabricated from different type of materials ranging from silicon, glass, metals and polymers. Various fabrication processes i.e. micromolding, surface and bulk micromachining have been utilized in the microchannel production. Among all the materials and fabrication techniques, polydimethylsiloxane (PDMS) polymer and micromolding or replica molding technique have been widely

chosen for microchannel fabrication [2]. PDMS is a biocompatible material which is nontoxic to cells. PDMS is permeable to gases but impermeable to water [3]. Gas permeability within the PDMS will maintain living cells viability in the microchannel for biological studies. PDMS has optical transparency down to 230 nm wavelength, hence make any optical detection methods possible. The primary advantage of using PDMS as microchannel material is due to its ease of fabrication using replica molding technique. The replica molding technique enables simple, fast and economical way of fabricating microchannel where disposable issue is critical in handling biological samples. PDMS is also known to have a good reversible bonding with many surfaces i.e. silicon, glass and PDMS layer. Reversible bonding of PDMS can be realized using technique such as plasma cleaner and corona discharge to activate the PDMS surface.

This work presents microfabrication of PDMS microchannel using replica molding technique with UV-cured SU-8 permanent negative photoresist as a mold. The PDMS consisted of an inlet and outlet with a trapping chamber. Trapping chamber is used for extensively bioparticles separator and microincubator. A chamber is A FEA analysis revealed the velocity magnitude and field inside the microchannel at a wide range of volumetric flow rate. Furthermore, pressure drop values across the microchannel are also computed in determining the safe operation limit of the selected volumetric flow rate.

II. THEORETICAL

Fluid flow is governed by well known Navier-Stokes equation which is

$$\rho\left(\frac{\partial V}{\partial t} + V \cdot \nabla V\right) = \rho g - \nabla p + \mu \nabla^2 V \tag{1}$$

where, ρ is the fluid density, μ is the fluid dynamic viscosity, t is time, g is the gravitational acceleration and V is the fluid velocity vector. The Navier-Stokes Equation is further simplified for flow in *x*-direction with assumptions of steady, incompressible and Newtonian fluid flow with no other forces involved to be

$$\frac{1}{\rho}\frac{dp}{dx} = \frac{\mu}{\rho}\frac{d^2 v_x}{dz^2} \tag{2}$$

In a fully developed cylindrical or rectangular microchannel flow, the flow profile is described as parabolic laminar. Hagen-Poiseuille flow is the pressure driven flow in microchannels which shows pressure drop and volumetric flow rate correlation. For rectangular microchannel, with height of, h and width of w, equation (2) can be further simplified to give

$$\Delta p = \frac{12 \mu L Q}{w h^3} / \left[1 - 0.63 \tanh\left(\frac{h}{w}\right) \right] \qquad (3)$$

where, Δp is the pressure drop between the two ends of the channel, L is the total length of channel, and Q is the volumetric flow rate across the channel [4]. In this work, the velocity magnitude and pressure drop data are obtained from the simulation using FEA.

III. MICROCHANNEL FABRICATION

Micro molding or replica molding technique using cured SU-8 photoresist mold is used to fabricate PDMS microchannel of various geometries and designs. In this work SU-8 3050 photoresist mold is fabricated on silicon substrate by photolithography technique.

A. SU-8 Photoresist Mold

In this work SU-8 3050 permanent epoxy negative photoresist from Microchem Company is used. The SU-8 3050 has kinematic viscosity and density of 12,000 mm²/s and 1153 kg/m³ respectively [5]. SU-8 3000 series is intended for thick, thermally and chemically stable structure in micromachining applications. Prior to lithography process, a silicon substrate is first cleaned with standard silicon cleaning process. Silicon substrate dehydration process at 200 degree Celsius for 10 minutes is performed on an aluminum topped hotplate to ensure SU-8 processing reliability. The selection of spin speed for the desired SU-8 mold thickness is available from the manufacturer's datasheet. In this work the recommended spin speed for 67 μm thickness SU-8 mold is 2000 revolution per minute (rpm) for 30 second (MicroChem Spin Speed vs. Thickness for SU-8 3000 resists at 23 °C Japan and Asia). The details process parameter of SU-8 mold microfabrication is as shown in Table 1.

TABLE I.   PARAMETERS FOR SU-8 MOLD FABRICATION

Parameters	Value	Unit
Thickness	67	μm
Spin Speed	2000	rpm
Soft-bake Time at 95°C	24.8	mins
UV-Exposure Time	96	s
Post-Bake Time at 65°C at 95°C	1 4	mins
Development Time	5	mins
Hard Bake Time at 150°C	30	mins

B. Polydimethlsiloxane (PDMS) Microchannel

PDMS is a biocompatible polymer materials for BioMEMS applications. In this work PDMS replica molding technique from cured SU-8 mold has been performed to fabricate microchannel with trapping chamber. The Sylgard 184 Silicone Kit (Dow Corning, USA) contained PDMS oligomer and crosslinking pre-polymer agent. The PDMS oligomer and crosslinking agent is mixed in a ratio of 10:1 in weight ratio. The mixture is then thoroughly mix with disposable plastic pipette tip for two minutes until mixture is milky due to air bubbles. To remove the air bubbles from the mixture a degassing process is performed. The mixture is degassed under vacuum condition in preen plasma machine for 30 minutes or until free bubbles mixture is obtained. After the degassing process, the PDMS mixture is ready to be poured on the SU-8 mold for replica molding. The uncured PDMS which has been poured onto the SU-8 mold is then baked in oven at 65 degree Celsius for 1 hour for curing or crosslinking the polymer. Peeling process is performed on the way to detach the PDMS microchannel pattern layer from the SU-8 mold. To complete the device, the PDMS pattern layer is bonded to the PDMS tubing layer treated with corona discharge. The results of the microfabrication are discussed in results and discussion section.

IV. MODELING AND SIMULATION

In this work the microchannel is defined with one inlet and outlet. A chamber of square-shaped of 10.0 mm x 10.5 mm is designed as a main domain for trapping bioparticles flowing in the continuos microchannel flow. The model geometry used in this FEA is as shown in Fig. 1. Water at 25 degree Celsius with density, ρ of 997.13 kg/m³ and dynamic viscosity, μ of 8.91 x 10⁻³ Pa.s is used as the medium in this preliminary study. A three dimensional (3D) analysis of x-, y- and z-component of geometry and physics are involved. The physics used to solve this problem is single phase laminar fluid flow and the stationary study is selected. The fluid flow is considered Newtonian, incompressible and steady in the analysis. The flowing flow in the microchannel is drives by the pressure drop prescribed between the inlet and the outlet.

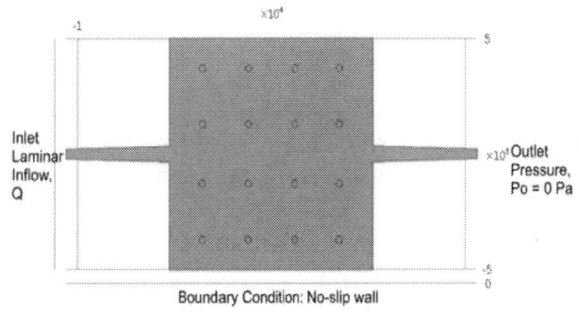

Fig. 1. FEA model geometry with boundary conditions used

The mesh element used in this study is physics controlled of unstructured triangular mesh. In this finite element analysis, normal meshing consisted 578,468 elements and 866,557 degree of freedoms are solved. The computational solution time is 5 minutes which consumed about 3.5 Gigabytes computer memory. The FEA computed the fluid's velocity components $V = (u, v, w)$ in the x, y and z directions and its pressure p in the region defined by the geometry and boundary conditions. To validate the simulation results, a comparison between the Hagen-Poiseuille theoretical equation and simulation result is performed as given in Table 2.

TABLE II.   VALIDATION OF THEORETICAL AND SIMULATION RESULTS OF A STRAIGHT MICROCHANNEL OF 21,750 µM X 462 µM X 67 µM

	Theoretical	Simulation	Error
Velocity magnitude	891.0 µm/s	923.8 µm/s	4.8 %
Pressure drop	34.5 mPa	35.6 mPa	3.4 %

V.   RESULTS AND DISCUSSION

The SU-8 mold and the PDMS microchannel are successfully fabricated by replica molding technique explained in the microfabrication section. The data of water velocity magnitude, velocity field and the pressure drop across the microchannel have been obtained from the FEA simulation. The discussion of the results is in the following sub-section.

A.   Microfabrication Results

The SU-8 mold fabricated onto the silicon wafer has a well define pattern and adhered strongly onto the silicon substrate as shown in Fig. 2. A SU-8 material is an epoxy based photoresist of two steps cross-linking process. Upon UV exposure, a strong acid forms and then become a catalyst for thermally driven epoxy cross linking during the post exposure baking phase. Fig. 2 (b) and (c) show the Scanning Electron Microscopy (SEM) results of the cured SU-8 mold fabricated in this work. The microchannel mold height of 73 µm is obtained from the SEM result. From the manufacturer datasheet, at a spin speed of 2000 rpm, a 67 µm structure height should be achieved. The reason of this discrepancy is due to the SU-8 solvent evaporation as a result aging and frequent opening and closing the SU-8 resist bottle in the laboratory. The solvent evaporation increased the SU-8 resist viscosity hence produced a thicker structure. Non- vertical sidewalls were also observed for this SU-8 structure. From the technical datasheet, the manufacturer's recommended on using long pass filter during UV exposure in order to obtained vertical sidewalls structure. The filter is use to eliminate UV exposure below 350 nm. However no filter is used in this work during UV exposure.

The PDMS microchannel replicates the SU-8 mold pattern by replica molding technique. The cured PDMS layer is clear and optically transparent. To complete the microchannel as BioMEMS device, a cured PDMS layer with inlet and outlet tubing is bonded to the pattern layer. The bonding is performed by treated the PDMS surface with corona discharge. The complete PDMS microchannel device which is ready for testing is as shown in Fig. 3.

Fig. 2.   (a) SU-8 mold on a silicon substrate (b) SEM of the SU-8 mold thickness (c) SEM of the SU-8 mold pillar diameter

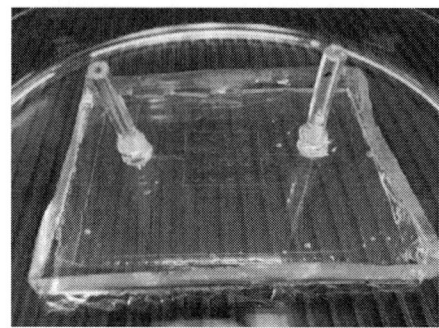

Fig. 3.   PDMS microchannel bonded with PDMS tubing (Tygon®) layer

B.   Simulation Results

In this study, simulation of water flows in a microchannel with trapping chamber is discussed. The graph of maximum velocity magnitude and pressure drop at 0.5, 1.0, 5.0, 10, 25, 50, 100, 500 and 1000 µL/min is shown in Fig. 4.

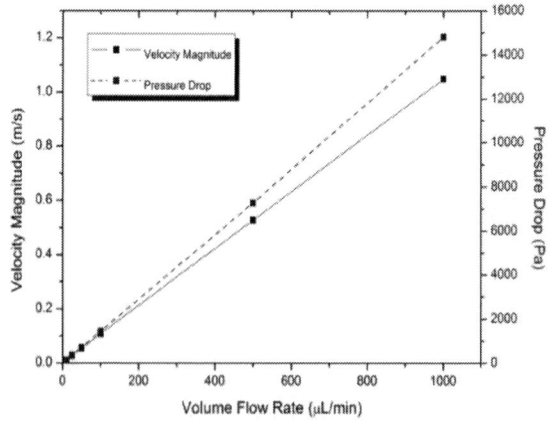

Fig. 4.   Graph of maximum velocity magnitude and pressure drop at 0.5, 1.0, 5.0, 10, 25, 50, 100, 500 and 1000 µL/min

The velocity magnitude varies linearly with the inlet laminar volume flow rate. The minimum and maximum velocity magnitude of 5.392×10^{-4} m/s and 1.049 m/s is obtained at flow rate of 0.5 and 1000 µL/min respectively. The velocity magnitude at volumetric flow rate of 1000 µL/min is shown in Fig. 5. The simulation result clearly shows the velocity magnitude of water flow in the microchannel. The maximum water velocity is observed at the inlet and outlet area. The reason of this high velocity magnitude is due to the design of the smallest cross-sectional area at the inlet and outlet. This results agree with the mass continuity equation of constant density fluid which is defined as $Q = V_{ave}/A$ where, Q is the constant volumetric flow rate, V is average velocity and A is the cross-sectional area. A significant decreased of water velocity magnitude is observed at the largest cross-sectional area of the microchannel which is the trapping chamber. The zero value of velocity magnitude at the microchannel walls confirmed the wall no-slip condition. Fig. 6 shows the streamlines of velocity field at maximum volumetric flow rate of 1000 µL/min. The streamlines indicated the flow is in laminar regime where the Reynolds number is less than 2300 in a continuous liquid flow.

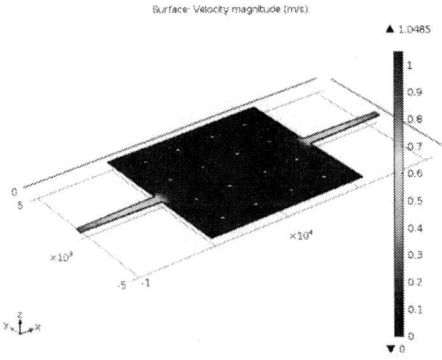

Fig. 5. Velocity magnitude of the microchannel at volumetric flow rate of 1000 µL/min

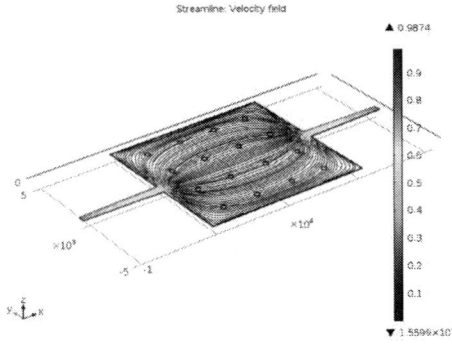

Fig. 6. Streamlines velocity profile of the microchannel at volumetric flow rate of 1000 µL/min

C. Pressure Drop

The graph of pressure drop across the microchannel at different volume flow rate is shown in Fig. 7. The minimum and maximum pressure drop of 7.17 Pa and 14.81 kPa have been observed at volume flow rate of 0.5 and 1000 µL/min respectively. The pressure drop value across the microchannel conveys important information on the safe limit of the PDMS microchannel bonding. PDMS to PDMS surface treated with corona discharge has bonding strength of 60 kPa in liquid flow [6]. As in this work, the bonding strength of PDMS-PDMS surface treated with corona discharge is able to withstand maximum flow rate of 1000 µL/min.

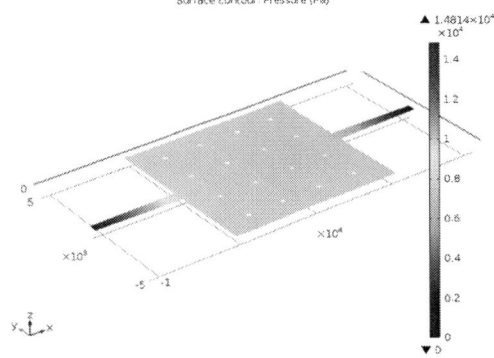

Fig. 7. Pressure drop across the microchannel at volumetric flow rate of 1000 µL/min

VI. CONCLUSION

Fabrication of PDMS microchannel by replica molding technique using SU-8 is successfully demonstrated. FEA results shown the velocity magnitude, velocity profile and pressure drop across the microchannel at different volume flow rate. The fluid flow parameters data are important to know in designing functional and reliable microchannel for the intended BioMEMS applications.

ACKNOWLEDGEMENT

We would like to thank Ministry of Higher Education of Malaysia and Universiti Kebangsaan Malaysia for the research grant under the project NND/ND/(1)/TD11-002

REFERENCES

[1] L. Y. Yeo, H.-C. Chang, P. P. Y. Chan, and J. R. Friend, "Microfluidic devices for bioapplications.," *Small*, vol. 7, no. 1, pp. 12–48, Jan. 2011.

[2] G. M. Whitesides, "The origins and the future of microfluidics.," *Nature*, vol. 442, no. 7101, pp. 368–73, Jul. 2006. I.S. Jacobs and C.P. Bean, "Fine particles, thin films and exchange anisotropy," in Magnetism, vol. III, G.T. Rado and H. Suhl, Eds. New York: Academic, 1963, pp. 271-350.

[3] S. K. Sia and G. M. Whitesides, "Microfluidic devices fabricated in poly(dimethylsiloxane) for biological studies.," *Electrophoresis*, vol. 24, no. 21, pp. 3563–76, Nov. 2003.

[4] I. Papautsky and A. B. Frazier, "A Review of Laminar Single-Phase Flow in Microchannels," pp. 1–9, 2001.

[5] Microchem Technical Datasheet, "SU-8 2000 Permanent Epoxy Negative Photoresist Processing Guidelines for SU8-2025, SU8-2035, SU8-2050, and SU8-2075," www.microchem.com.

[6] K.-S. Koh, J. Chin, J. Chia, and C.-L. Chiang, "Quantitative Studies on PDMS-PDMS Interface Bonding with Piranha Solution and its Swelling Effect," *Micromachines*, vol. 3, no. 4, pp. 427–441, May 2011.

Influence of Gadolinium Doping on The Crystalline Structure and Optical Properties of Zinc Oxide Thin Films

N. Sarip, F. Mahmud, M.Z. Sahdan

Microelectronics and Nanotechnology-Shamsuddin
Research Centre (MiNT-SRC)
Universiti Tun Hussein Onn Malaysia
86400 Parit Raja, Batu Pahat, Johor, Malaysia
zainizno@gmail.com

S.N.M. Tawil

Department of Electrical and Electronic Engineering
Universiti Pertahanan Nasional Malaysia
Kem Sungai Besi, 57000 Kuala Lumpur, Malaysia

Abstract—**This report represents an experimental investigation on the structural and optical modification of ZnO thin films towards the Gd dopant content, whereby the desired films were deposited on glass substrates by chemical solution method. The prepared samples were characterized by X-ray diffraction (XRD), field-emission scanning electron microscopy (FE-SEM), and surface profiler. The optical properties were studied by UV-Visible spectroscopy technique. The influence of Gd dopant on structural and optical properties of the prepared thin films were investigated and discussed based on the structure modification and band gap of undoped and Gd-doped ZnO thin films.**

Keywords☐ chemical solution deposition; doping; gadolinium; zinc oxide

I. Introduction

Zinc oxide (ZnO) is well known in the scientific community to possess a lot of advantages in its nature, making it as one of the most prominent material for a variety of electronic and opto-electronic applications [1]. However on the contrary, doped ZnO can hold some very different properties compared to that of pure ZnO. Group III elements, i.e. Ga, Al, and In can be used as *n*-type dopant in ZnO, have been generally recognized to influence the optical and also the electrical properties of the materials [2]. Other elements of rare earth group, i.e. Ce, Er, Eu, La, Tb, Tm, Yb, and Dy as *p*-type dopant to ZnO, have also attracted significant attention among the community due to their unique optical properties which give to intense emission peaks in the visible and near infrared range [3]. Hence, the electrical conductivity, type of conduction and band gap range, including the magnetic characteristics of the nanomaterial can be manipulated through doping. Therefore, doping effect can enhance the present properties of ZnO and provides space for new applications possible.

Presently, ZnO-based opto-electronic devices have been reported to appear as doping of ZnO began to become accessible. The facile addition of metal precursor to the solution may not be as straightforward as it seems since the metal atom might not fuse in the ZnO structure, producing a secondary phase denoted to the existence of foreign element [4]. Fabrication of stable and device-quality *p*-type ZnO, however, has not been realize regardless of a large number of publications stating successful demonstration of *p*-type [5, and reference therein]. Numerous reports on impurities doped with ZnO have been driven to minimize the resistivity and to boost the opacity of the films [6, and reference therein]. Gadolinium (Gd) metal of the lanthanum group element was selected as dopant in the work reported here based on the the the Gd_2O_3 physical properties in optical application in amendment to both the electrical and optical characteristics of the ZnO thin films.

This work is expected to provide the science and research information necessary towards the realization of the new enhanced functionality devices through the integration of photonic, electronic and magnetic properties in a single device that will be beneficial for spin light-emitting diodes (LEDs), novel microprocesses and sensitive biological and chemical sensors. Our motivation on ZnO thin film is, therefore, to examine the structural and optical properties obtained through a feasible, toxic free, chemical solution deposition method by manipulating the Gd doping content. Comparative studies concerning the characterization of pure and Gd-doped ZnO thin films were also review in this paper.

II. Experimental Setup

A. Materials

Zinc acetate dehydrate, Monoethanolamine were purchased from Sigma-Aldrich. Gadolinium nitrate hexahydrate was procured from Acros Organics. Other chemicals were employed directly. Deionized water collected from the Merck Millipore RiOs 3 system was used throughout all experimentations. Gd–doped ZnO thin films were synthesized as stated in former reports [7,8].

B. Synthesis of pure and Gd-doped ZnO thin films

Gd–doped ZnO thin films were synthesized with addition of different stoichiometric percentile of gadolinium nitrate with respects to the principle solution. Undoped ZnO thin film was deposited for comparison with the Gd–doped thin films, in effort to examine the doping effect at various content values in relative to structural, optical and electrical properties.

C. Apparatus

The shapes and micrographical structures of the ZnO nanorods thin films were studied using a field-emission scanning electron microscope (FE-SEM, JEOL JSM-7600F) along with the X-ray diffractometer (XRD, Bruker D8 Advance). The obtained images were used to calculate the average crystallite size of the ZnO thin films grown on the substrate. UV-Visible spectra were recorded on a Shimadzu UV 1800 spectrophotometer whereas the thickness of the samples was measured with the utilization of the Alpha Step-IQ Surface Profiler (KLA Tencor).

III. Results and Discussion

A. Structural characterization

The diffraction spectra of the deposited undoped and Gd-doped ZnO thin films was shown in Fig. 1. The (1 0 0), (0 0 2), (1 0 1), (1 0 2), (1 1 0) peaks were observed in entirety. The results were in consent to the JCPDS indexed of ZnO hexagonal wurtzite structure (JCPDS No. 36–1451, a = 3.24982 A, c = 5.20661 A) with surprisingly no indication peaks of Gd_2O_3 phases could be observed in the XRD pattern. The dominant (0 0 2) reflex for entire films in the XRD spectra suggesting that the crystalline particle grew perpendicularly to the substrate. The small peak shifts corresponding to the sample of Gd–doped films as compared with undoped ZnO film confirms that gadolinium atoms doping deformed the original crystalline arrangement that arises from the lattice strain [9]. The undoped ZnO film obtained the highest level of c-orientation. Additionally, the average particles diameter decreased as the increase in x value (Table 1). The average crystallite size was estimated from Debye–Scherer relation [10]:

$$D = 0.9\lambda/\beta cos\theta \qquad (1)$$

where D is the average diameter of the crystallite, λ is the X-ray wavelength (CuKα, 0.154 nm), θ and β are the Bragg diffraction angle and the peak width in radians at half-maxima, correspondingly.

The texture coefficient (TC) corresponds to the individual plane texture, whose variation to standard indicates the favourable plane growth [11]. According to Illican et al., the value of $TC_{(002)}$ corresponds to a value of one implies to randomly oriented crystallite in the film, while higher values indicate the abundance of grains oriented in (0 0 2) direction. The variation of the highest $TC_{(002)}$ was evident for 2.0 at. % Gd content.

Fig. 3a–d show the FESEM micrographs of undoped and Gd–doped ZnO nanorods thin films with x values of 1.0, 2.0 and 4.0 at. %. Literally, well defined hexagonal facet nanorods, typically having diameter 27.6 nm was observed for

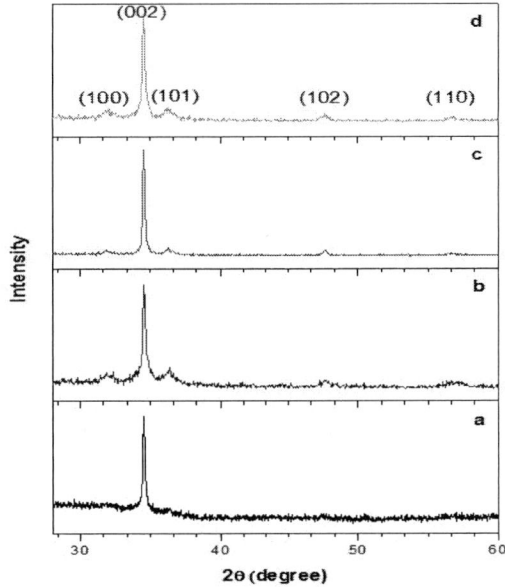

Fig. 1 XRD patterns of Gd–doped ZnO thin films at Gd content of x = 0 at. % (a), x = 1.0 at. % (b), x = 2.0 at. % (c) and x = 4.0 at. % (d), respectively with their orientation marked in parenthesis.

TABLE I.   COMPARISON BETWEEN GD-DOPING ZNO THIN FILMS WITH DIFFERENT X VALUES.

Gd content, x (at. %)	Size[a], D (nm)	Ave. diameter[b], d (nm)	Band gap energy, E_g[c] (eV)	Texture coefficient[d], $TC_{(002)}$ (%)
0.0	37.84	27.6	3.20	42.68
1.0	35.20	23.4	3.21	81.97
2.0	22.31	23.1	3.23	92.89
3.0	36.30	27.5	3.25	86.17

[a] Measured by XRD.
[b] Average nanorods diameter determined by FESEM.
[c] Measured from the $(\alpha h\nu)^2$ versus photon energy, $(h\nu)$ graph.
[d]Texture coefficient (TC) calculated from XRD spectrum data. The preferred orientation represents the texture of (0 0 2) plane was determined using the Harris method of $TC_{(002)} = (I_{(002)}/I_{0(002)})/\Sigma_i (I_{(002)}/I_{0(002)})$ x 100%, where $I_{(002)}$ is the measured relative intensity of (0 0 2) plane and $I_{0(002)}$ is the standard intensity of (0 0 2) plane taken from [11].

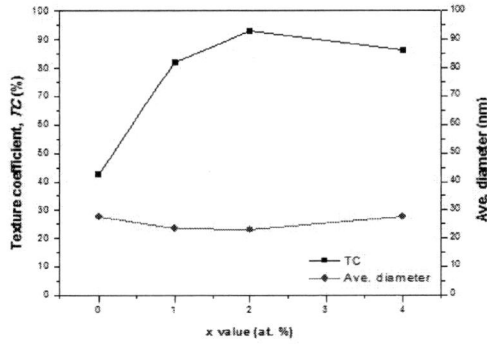

Fig. 2 Variation of texture coefficient and particle diameter with Gd content.

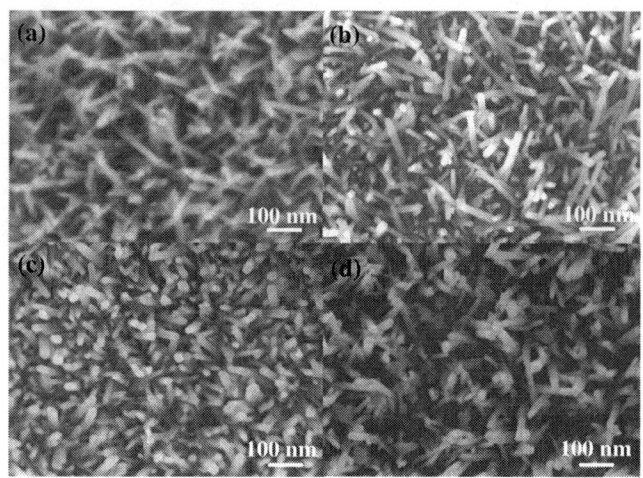

Fig. 3 FE–SEM images of $Zn_{1-x}Gd_xO$ thin films at (a) x = 0, (b) x = 1.0 at. %, (c) x = 2.0 at. % and (d) x = 4.0 at. % attained at 100 000 X magnification.

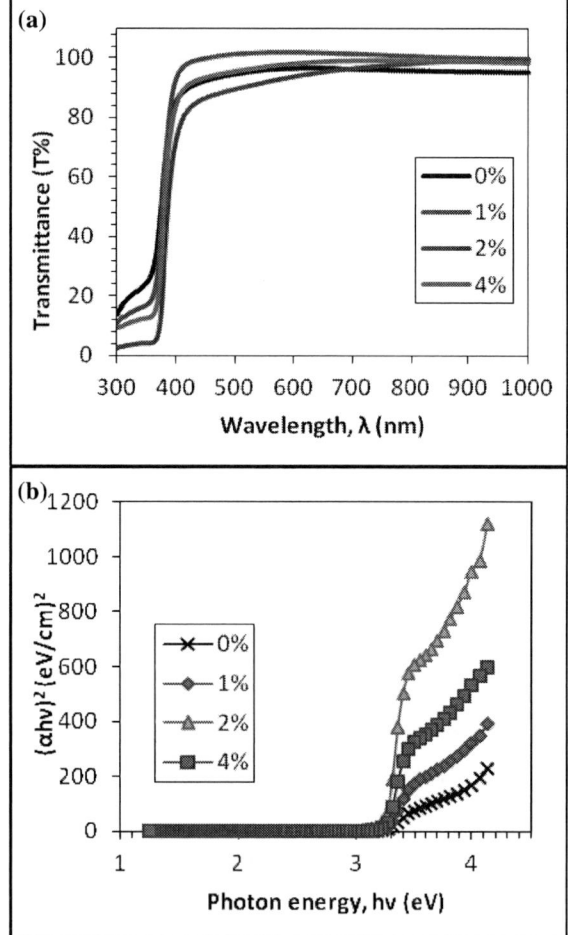

Fig. 4 (a) Variation of transmittance (%T) with wavelength; (b) plot of $(\alpha hv)^2$ as a function of photon energy of undoped and Gd-doped ZnO thin films.

undoped ZnO (Fig. 3a). Such nanorods were arranged in similar array, covering entire substrate surface resulting in the appearance of single formation as depicted by the XRD spectra. Addition of Gd (1 at. %) causes crystal to orient arbitrarily denoted to the appearance of significant (1 0 0), (1 0 1), (1 0 2) and (1 1 0) peaks reading on the XRD spectra over all doped ZnO samples. With increased Gd incorporation, diameter of the nanorods shown a decreasing trend with the exception of 4.0 at. % as plotted in Fig. 2. In relative, Fig. 3b–d present evidence of spatially packed nanorods distribution across the films. This was attributed to the increase orientation of Gd–doped ZnO nanorods by increasing Gd content as supported by the XRD spectra analysis

B. Optical and electrical characterizations

Optical properties of Gd-doped ZnO with x values varied from 0 at. % to 4.0 at. % were further explored by measuring their UV–Vis transmittance spectra (Fig. 4a). All films showed high transmittance in the visible range of correspondingly above 82% regardless of doping concentration. The absorption spectra of the Gd-doped ZnO with different x values were analyzed to discover the significance of Gd doping content to the band gap enlargement. As ZnO is direct band gap semiconductor, its band gap originated from the equation [12]:

$$\alpha hv = A. \ (hv - E_g)^{1/2} \qquad (2)$$

where α, E_g and A are absorption coefficient, band gap energy and constant, respectively. The optical band gap (E_g) of ZnO thin films can be determined from a linear line extrapolation in a plot of $(\alpha hv)^2$ versus hv (Fig. 4b) in accord to the above given formula. Band gap values of 3.20 eV, 3.21 eV, 3.23 eV and 3.25 eV were recorded correspondingly for 0 at. %, 1.0 at. %, 2.0 at. % and 4.0 at. % Gd contents.

IV. Conclusion

Both undoped and Gd-doped ZnO films were prepared on glass substrates by chemical solution deposition method. The XRD patterns of ZnO thin films showed polycrystalline wurtzite with a preferential (0 0 2) orientation with a slight peak deviation in the Gd-doped ZnO films from the undoped ZnO film. The grown samples were highly transparent in the visible region and have a broad absorption edge in the near-ultraviolet region with 2.0 at. % Gd surpassed the rest of the films. Increased doping content results in increased trend of grain size. The absorption edge analysis showed that the optical band gap energy, E_g, for the ZnO thin film was 3.20 eV and Gd-doped ZnO thin films were 3.21 eV, 3.23 eV and 3.25 eV for the 1.0, 2.0 and 4.0 at. % of Gd content, respectively. These results suggest that 2.0 at. % of Gd-doped ZnO thin film prepared demonstrates the high-quality c-axis orientation with 92% of TC and highest optical absorbance measurement near UV range.

Acknowledgment

This work are financially supported by The Ministry of Higher Education Malaysia through the Exploratory Research Grant Scheme (E001) and Fundamental Research Grant Scheme (FRGS/1/2013/SG06/UPNM/02/1).

References

[1] Wang Z.L, "Ten years' venturing in ZnO nanostructures: from discovery to scientific understanding and to technology application," Chinese Sci Bull, vol. 54, pp. 4021-4034, 2009.

[2] Z. Lin, S. Guangjie, S. Shitao, Q. Xiujuan, and H. Sihuizhi, "Development on transparent conductive ZnO thin films doped with various impurity elements," Rare Metals, vol. 30, pp. 175-182, 2011.

[3] R. Zamiri, A.F. Lemos, A. Reblo, H.A. Ahangar, and J.M.F. Ferreira, "Effects of rare-earth (Er, La and Yb) doping on morphology and structure properties of ZnO nanostructures prepared by wet chemical method." Ceramic International, vol. 40, pp. 532-529, 2014.

[4] A. Chandra Bose, and N. Rajeswari Yogamalar, "Synthesis, dopant study and device fabrication of Zinc Oxide nanostructures: mini review," Prog. in Nanotechnology and Nanomaterials, vol. 2, pp. 1-20, 2013.

[5] U. Ozgur, D. Hofstetter, and H. Morkoc, "ZnO devices and applications: a review of current status and future prospects," Proceeding of the IEEE, vol. 98, pp. 1255-1268, 2010.

[6] S. Kumar, P. Kaur, C.L. Chen, R. Thangavel, C.L. Dong, Y.K. Ho, J.F. Lee, T.S. Chan, T.K. Chen, B.H. Mok, S.M. Rao, and M.K. Lou, "Structural, optical and magnetic characterization of Ru doped ZnO nanorods," J. of Alloys and Compounds, vol.588, pp. 705-709, 2014.

[7] N. Sarip, C.A. Norhidayah, S.A. Kamaruddin, F. Mahmud, S.N.M. Tawil, M.Z. Sahdan, "Synthesis and Characterization of Zinc Oxide Nanostructures by Different Sonication Period," Advanced Materials Research, vol. 925, pp. 110-114, 2014.

[8] S.A. Kamaruddin, M.Z. Sahdan, K.Y. Chan, M. Rusop,and H. Saim, "Zinc oxide microrods prepared by sol-gel immerse technique," Microelectronics International vol. 27, pp. 166-169, 2010.

[9] B.D. Cullity, and S.R. Stock. Elements of X-ray diffraction, 3rd. ed., Upper Saddle River, NJ : Prentice Hall, 2001.

[10] H. Ghayour, H.R. Rezaie, Sh. Mirdamadi, and A.A. Nourbakhash, "The effect of seed layer thickness on alignment and morphology of ZnO nanorods," Vacuum, vol. 86, pp. 101-105, 2011.

[11] S. Illican, M. Caglar, and Y. Caglar, "Determination of the thickness and optical constants of transparent indium-doped ZnO thin films by the envelope method," Material Science-Poland, vol. 25, pp. 709-718, 2007.

[12] Y. Liu, K. Ai, Q. Yuan, and L. Lu, "Fluorescence-enhanced gadolinium-doped zinc oxide quantum dots for magnetic resonance and fluorescence imaging" Biomaterials, vol. 32, pp. 1185-1192, 2011.

Structural and Electronic Properties of Li doped Decahedral Ag$_{12}$Li$_1$ Bimetallic Nanoclusters

Shaikat Debnath
Department of Electrical Engineering
University of Malaya
Kuala Lumpur, Malaysia
shaikat.debnath@gmail.com

Suhana Mohd Said
Department of Electrical Engineering
University of Malaya
Kuala Lumpur, Malaysia
smsaid@um.edu.my

Abstract— This paper is focused on the quantum modeling of the electronic and structural properties of nanoclusters. Two doping positions for Lithium (Li) metal doped in a 13-atom Decahedron Silver (Ag), the centrally-doped (c-doped) and surface doped (s-doped) in Ag$_{12}$Li$_1$ bimetallic nanoclusters were modeled. Decahedral Ag$_{13}$ and centrally/core doped(c-doped) and surface/shell doped (s-doped) Ag$_{12}$Li$_1$ bimetallic clusters have been investigated in the framework of density functional theory (DFT) calculations. Both of the c-doped and s-doped clusters exhibited higher Binding Energy (Eb) and Ionization Potential (IP) than the decahedral Ag$_{13}$. The s-doped Ag$_{12}$Li$_1$ was found with higher affinity for electrons compared to Ag$_{13}$ and c-doped Ag12Li$_1$. The geometric and energetic considerations here indicate that photoelectron spectroscopy might be useful to characterize these species experimentally. Potential applications include nano-devices and catalysis.

Keywords— DFT calculation, Bimetallic cluster, Silver-Lithium nanoalloy

I. INTRODUCTION

Bimetallic nanoclusters, which are the aggregation of two types of metal atoms of nanoparticle size, constitute very attractive properties and a subject of technological interest because of its distinct physical and chemical properties. [1] They exhibit particular and distinctive structural, [2] electronic, [3] optical [4] and magnetic [5] properties, which depend not only on their size but also on their chemical composition and atomic arrangement. [6] Such special features of bimetallic nanoclusters, along with the facility to tune their physical, electronic and chemical properties in molecular level by bimetallization [5] has led to a technological interest in catalysis [6, 7] and also in the application and development of new nano-devices for electronics. [8] [9]

This work is concerned with one atom Li doped in center (c-doped) and surface (s-doped) of 13-atom Ag$_{12}$Li$_1$

bimetallic clusters. Silver nanoparticles have unique optical, electrical and thermal properties that have made them an outstanding plasmonic metal finding application in various fields from photovoltaic to biological and chemical sensors. [10] On the contrary, Lithium (Li) being the elemental free-electron Group IA metal, has been identified as excellent potential for plasmonic nanometals. It has been reported [11, 12] that Lithium has good absorption efficiency (Q$_{abs}$) with very low inter-band transition losses at optical frequencies. In fact, these transition losses are comparable to that of silver along with its property of exhibiting the strongest free-electron-like-behavior, which result in very prominent Surface Plasmon Resonance (SPR) in visible-UV range. [13] Thus, elemental free electron metal like Lithium's addition into the Ag nanocluster might introduce some significant property-changes in the Silver nanocluster which will be an interesting study to explore. But in our current work, optical properties of the Ag$_{12}$Li$_1$ bimetallic cluster will not be discussed; rather it will be strictly limited only to the structural and electronic properties of the Ag$_{12}$Li$_1$ structures.

On a separate note, the 13-atoms cluster is the smallest core-shell structure (about 0.6 nm of diameter) of Silver, with either icosahedral or, decahedral symmetry, which possesses a specific symmetry where a central atom is surrounded by 12-atoms closed shell. [14] In geometry, a dodecahedron is any polyhedron with twelve flat faces, but usually a regular dodecahedron is meant: a Platonic solid. It is composed of 12 regular pentagonal faces, with three meeting at each vertex, and is represented by the Schläfli symbol {5,3}. It has 20 vertices, 30 edges and 160 diagonals. So, the core-shell 13-atom clusters can be considered as small enough to be treated with first-principles (DFT) methods, which can be considered as the quantum models for large nanoparticles for which such calculations cannot be done.

On a separate note, it is worth mentioning that the electronic properties and optical response of silver cluster lies between that of alkali metals and transition metals. So, it is quite challenging to clarify the properties of silver nanocluster due to its close lying *d electrons*, which quench

978-1-4799-5761-3/14 $31.00 © 2014 IEEE

the oscillator strengths by screening the *s*-electrons and get partially involved in the excitations. On the contrary, the absorption spectra of alkali metal clusters are explained as collective oscillation of *s valence electrons* and the shift of absorption energies of very small particles, compared to the larger ones explained by a spill out (extension of the electronic wave functions out of the "classical volume" of the cluster) of these *s-electrons*. [15]

On the other hand, density functional theory (DFT) calculations have been found satisfactory in explaining the structural and electronic properties of different bimetallic nanoclusters. [14, 16, 17] Till date, reports have been found on photoionization spectroscopy of metal dimers of Ag-Li [18] but no report has been found on any DFT based or, any other theoretical study on the structural and electronic properties of Ag-Li alloys in nano-scale.

In this paper, a theoretical DFT calculation has been presented to achieving the structural and electronic properties of one atom Li doped in center (c-doped) and surface (s-doped) of 13-atom Decahedral (Dh) $Ag_{12}Li_1$ bimetallic clusters, where the theoretically achieved values of Binding energy and HOMO-LUMO gap, Ionization potential and Electron affinity have been used to explain the electronic and chemical features of the Li-doped bimetallic clusters.

II. COMPUTATIONAL METHOD

In this study, we present the density functional spin-polarized calculations using DMol3 of Accelrys Inc. code. [19, 20] It was used to investigate the geometrical and energetic stabilities and optical properties of Ag_{13} with Decahedral (Dh) symmetry and c-doped and s-doped $Ag_{12}Li_1$ bimetallic structures, based on previous papers. [16, 17] All geometrical structures of the neutral clusters were optimized by Generalized Gradient Approximation (GGA) and Perdew-Burke-Ernzerhof (PBE) [21] exchange-correlation functional in DFT level. To deal with the electrons of both the decahedral Ag_{13} and the bimetallic structures, the DFT Semi-core Pseudo-potentials was used. The orbital cut off range and Fermi smearing were selected as 5.0 A^0 and 0.001 Ha respectively. The self-consistent-field (SCF) procedures were performed with the aim of obtaining well converged geometrical and electronic structures with a convergence criterion of 10^{-6} a.u. The energy, maximum force and maximum displacement convergence were set as 10^{-6} Ha, 0.002 Ha/A^0 and 0.005 A^0 respectively.

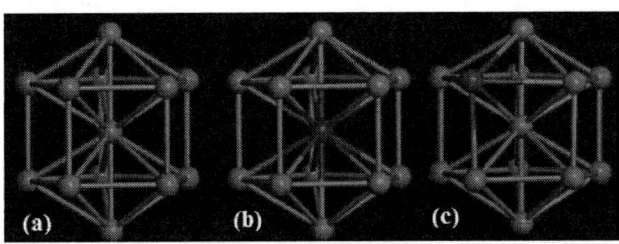

Figure1. (a) Decahedral Ag_{13}, (b) c-doped $Ag_{12}Li_1$, (c) s-doped $Ag_{12}Li_1$ [Li atoms are in magenta color]

Table 1: The geometrical symmetry, spin, binding energy and HOMO-LUMO gaps for Ag_{13}, c-doped and s-doped $Ag_{12}Li_1$ clusters.

Cluster	Symmetry	Spin	E_b (eV)	IP (eV)	EA (eV)	HOMO LUMO Gap (eV)
Ag_{13}	D_h	1	19.19	5.61	2.08	0.22
$Ag_{12}Li_1$ (c-doped)	D_{5h}	1	19.79	5.71	2.20	0.20
$Ag_{12}Li_1$ (s-doped)	C_S	1	21.06	6.97	1.05	0.20

III. STRUCTURAL AND ELECTRONIC PROPERTIES

Figure 1 shows the structural and electronic motifs of decahedral Ag_{13} with D_{5h} symmetry, c-doped $Ag_{12}Li_1$ with D_{5h} symmetry and s-doped $Ag_{12}Li_1$ with C_S symmetry. Nonlocal functional Generalized Gradient Approximation (GGA) with PBE exchange function has been used for the geometrical optimization of the above mentioned clusters. Total energy and binding energy of all the clusters were checked to determine the ground state of all the clusters with different spin multiplicity from singlet to octet. It has been found that spin-doublet provides the most stable structure for the 13-atom decahedral Ag_{13} structure alike the previous reports. [16, 17]

From the Figure-1, it is evident that doping of Lithium has provided some significant change in the cluster structure as it can be seen that, decahedral Ag_{13} had D_{5h} symmetry, whereas as a result of doping lithium into the surface has changed that D_{5h} symmetry to C_S symmetry, whereas the c-doping didn't bring any changes in the symmetry.

For the current study of electronic properties of the clusters, Table-1 is put above with Binding energy (E_b) and HOMO-LUMO gaps of the Ag_{13}, c-doped and s-doped $Ag_{12}Li_1$ clusters.

Binding energy is one of the very crucial elements to study the stability of the clusters. Here we have defined Binding energy in the following way:

$$E_b = -E_{cluster} + nE_{Ag} + mE_X$$

Here, n=12, m=1 for $Ag_{12}Li_1$, n=13, m=0 for Ag_{13}

From the data of Table-1, it is evident that both c-doped and s-doped $Ag_{12}Li_1$ clusters have higher binding energy than the decahedral Ag_{13} which means, both lithium doped structures are more stable than the decahedral Ag_{13}. The large difference in the electronegativity of Ag and Li might have created some ionic bonding in the bimetallic cluster, which may be the reason for the higher binding energy, which eventually provides the higher stability. Other thing is, as the binding energy is higher for the s-doped $Ag_{12}Li_1$ than the c-doped $Ag_{12}Li_1$, it can be said that Li atom tends to occupy the surface position in the 13-atom $Ag_{12}Li_1$ bimetallic cluster rather than the core-position.

Apart from binding energy, the relative strain-index is another way to predict about the stability of the clusters. This index number is a value between 0 and 1, which tends towards 1.0 in the limit of extremely large clusters and it is zero if there is no binding at all. These values indicate the presence of strain in the clusters and give an indication of the relative stability of the cluster. Relative strain-index is defined as following way:

Relative Strain-index = Binding Energy/ Sum of cohesive energies of the cluster atoms

The cohesive energy of Ag and Lithium are 2.95 eV and 1.63 eV respectively [22]. Using the binding energy values from Table-1 and the above mentioned values of cohesive energies, the relative strain-index of Ag_{13}, c-doped $Ag_{12}Li_1$ and s-doped $Ag_{12}Li_1$ clusters are found 0.50, 0.53 and 0.57 respectively. So, it can be said that, relative strain index-wise, s-doped $Ag_{12}Li_1$ is the most stable structure among the three clusters. Thus, it is apparent that the relative strain-index data have followed and verified the trend of binding energy data of the clusters.

The energy gap between HOMO and LUMO is considered as a useful quantity in case of determining the electronic stability of the clusters. It is reported that magic clusters with high symmetry and closer electronic shell usually have large HOMO-LUMO gaps. [23] From Table-1, it is seen that both of the doped bimetallic clusters have very similar HOMO-LUMO gaps as decahedral Ag_{13}, as Ag_{13} has its HOMO-LUMO gap of 0.22 eV, whereas both the c-doped and s-doped $Ag_{12}Li_1$ structures are found with slightly lower HOMO-LUMO gaps of 0.20 eV. On the other hand, it is question-worthy that how efficient the HOMO-LUMO gaps can be regarding the stability prediction of metal clusters. It is because of the fact that metals clusters are natually good electrical conductors with very low gap in between their HOMO and LUMO orbitals.

In order to study the chemical reactivity of the clusters, Ionization Potential (IP) and Electron Affinity (EA) are needed to be considered with importance. Here, IP is defined as the energy needed to for removal of an electron from the cluster, whereas EA is defined as the energy needed to attach an electron into the cluster. Here in our calculations, we have used the following ways to calculate IP and EA:

$$IP = -E_{cluster} + E_{cluster}^{+}$$

$$EA = -E_{cluster} + E_{cluster}^{-}$$

From the Table-1, we can see that both the doped cluster have higher IP than Ag_{13} cluster. Here, it needs to be mentioned that the experimental values of ionization potential (IP) and electron affinity (EA) of Lithium in bulk

structures are 5.39 [24] [25] and 0.62 eV [26] respectively. The higher electro-positivity of the doped atom might be the reason for higher IP of the bimetallic clusters compared with the decahedral Ag_{13}. On the other hand, s-doped $Ag_{12}Li_1$ is found with lower EA than Ag_{13}. This is due to the fact that a high electropositive atom in the surface of the bimetallic cluster will have more affinity for electrons, thus lowering the EA of the s-doped $Ag_{12}Li_1$ bimetallic cluster. Thus, it can be concluded that when Lithium is doped into the surface of the Ag_{13} cluster, the resultant s-doped $Ag_{12}Li_1$ bimetallic cluster shows more affinity for electrons and it needs more energy to extract electron from the s-doped bimetallic cluster compared to c-doped $Ag_{12}Li_1$ and decahedral Ag_{13}.

Overall, it can be concluded that, the c-doped and s-doped $Ag_{12}Li_1$ clusters exhibit better electronic and chemical stability than decahedral Ag_{13}. So, for the applications of opto-electronics where stability of the nano-material is crucial, instead of Ag_{13}, lithium doped $Ag_{12}Li_1$ cluster can provide some attractive features as the alternative material.

IV. CONCLUSION

The structural and electronic properties of decahedral Ag_{13} and c-doped and s-doped $Ag_{12}Li_1$ bimetallic nanoclusters have been investigated using DFT. It is found that the Li atom prefers to tend to the surface position in the bimetallic cluster rather than the core-position. Both the c-doped and s-doped $Ag_{12}Li_1$ clusters exhibited higher binding energy, i.e. higher electronic and chemical stability compared to monometallic decahedral Ag_{13} cluster. It is also found that, s-doped $Ag_{12}Li_1$ has the highest affinity for the electrons compared to the other two structures. Thus, it can be concluded that the c-doped and s-doped $Ag_{12}Li_1$ bimetallic clusters have attractive structural and electronic properties, which are important for potential applications in the field of nano-devices and catalysis.

ACKNOWLEDGEMENT

The authors wish to acknowledge the University of Malaya–Ministry of Higher Education Grant UM.C/625/1/HIR/MOHE/ENG/29, University of Malaya Research Grant (UMRG) RP014C-13AET and Malaysian Ministry of Science and Technology's Science Fund (06/01/03/SF0831) for the financial support

REFERENCE

[1] E. E. Zhurkin, T. Van Hoof, and M. Hou, "Nanoscale alloys and core-shell materials: Model predictions of the nanostructure and mechanical properties," *Physical Review B*, vol. 75, p. 224102, 06/04/ 2007.

[2] J. Chen, D. Wang, J. Xi, L. Au, A. Siekkinen, A. Warsen, *et al.*, "Immuno Gold Nanocages with Tailored Optical Properties for Targeted Photothermal

Destruction of Cancer Cells," *Nano Letters,* vol. 7, pp. 1318-1322, 2007/05/01 2007.

[3] P. A. Derosa, J. M. Seminario, and P. B. Balbuena, "Properties of Small Bimetallic Ni−Cu Clusters," *The Journal of Physical Chemistry A,* vol. 105, pp. 7917-7925, 2001/08/01 2001.

[4] A. M. Schwartzberg and J. Z. Zhang, "Novel Optical Properties and Emerging Applications of Metal Nanostructures†," *The Journal of Physical Chemistry C,* vol. 112, pp. 10323-10337, 2008/07/01 2008.

[5] E. Janssens, S. Neukermans, H. M. T. Nguyen, M. T. Nguyen, and P. Lievens, "Quenching of the Magnetic Moment of a Transition Metal Dopant in Silver Clusters," *Physical Review Letters,* vol. 94, p. 113401, 03/21/ 2005.

[6] K. Kaizuka, H. Miyamura, and S. Kobayashi, "Remarkable Effect of Bimetallic Nanocluster Catalysts for Aerobic Oxidation of Alcohols: Combining Metals Changes the Activities and the Reaction Pathways to Aldehydes/Carboxylic Acids or Esters," *Journal of the American Chemical Society,* vol. 132, pp. 15096-15098, 2010/11/03 2010.

[7] N. C. Barnard, S. G. R. Brown, F. Devred, J. W. Bakker, B. E. Nieuwenhuys, and N. J. Adkins, "A quantitative investigation of the structure of Raney-Ni catalyst material using both computer simulation and experimental measurements," *Journal of Catalysis,* vol. 281, pp. 300-308, 7/25/ 2011.

[8] R. Ferrando, J. Jellinek, and R. L. Johnston, "Nanoalloys: From Theory to Applications of Alloy Clusters and Nanoparticles," *Chemical Reviews,* vol. 108, pp. 845-910, 2008/03/01 2008.

[9] S. Duan and R. Wang, "Bimetallic nanostructures with magnetic and noble metals and their physicochemical applications," *Progress in Natural Science: Materials International,* vol. 23, pp. 113-126, 4// 2013.

[10] A. K. Yu, A. A. Kudrinskiy, A. Y. Olenin, and G. V. Lisichkin, "Synthesis and properties of silver nanoparticles: advances and prospects," *Russian Chemical Reviews,* vol. 77, p. 233, 2008.

[11] M. G. Blaber, M. D. Arnold, and M. J. Ford, "Search for the Ideal Plasmonic Nanoshell: The Effects of Surface Scattering and Alternatives to Gold and Silver," *The Journal of Physical Chemistry C,* vol. 113, pp. 3041-3045, 2009/02/26 2009.

[12] M. G. Blaber, M. D. Arnold, and M. J. Ford, "A review of the optical properties of alloys and intermetallics for plasmonics," *Journal of Physics: Condensed Matter,* vol. 22, p. 143201, 2010.

[13] P. R. West, S. Ishii, G. V. Naik, N. K. Emani, V. M. Shalaev, and A. Boltasseva, "Searching for better plasmonic materials," *Laser & Photonics Reviews,* vol. 4, pp. 795-808, 2010.

[14] M. Harb, F. Rabilloud, and D. Simon, "Structural, electronic, magnetic and optical properties of icosahedral silver-nickel nanoclusters," *Physical Chemistry Chemical Physics,* vol. 12, pp. 4246-4254, 2010.

[15] M. Harb, F. Rabilloud, D. Simon, A. Rydlo, S. Lecoultre, F. Conus, *et al.*, "Optical absorption of small silver clusters: Agn, (n=4−22)," *The Journal of Chemical Physics,* vol. 129, pp. -, 2008.

[16] W. Ma and F. Chen, "Optical and electronic properties of Cu doped Ag clusters," *Journal of Alloys and Compounds,* vol. 541, pp. 79-83, 11/15/ 2012.

[17] Y. Rao, Y. Lei, X. Cui, Z. Liu, and F. Chen, "Optical and magnetic properties of Cu-doped 13-atom Ag nanoclusters," *Journal of Alloys and Compounds,* vol. 565, pp. 50-55, 7/15/ 2013.

[18] J. S. Pilgrim and M. A. Duncan, "Photoionization spectroscopy of AgLi," *Chemical Physics Letters,* vol. 232, pp. 335-340, 1/20/ 1995.

[19] B. Delley, "An all-electron numerical method for solving the local density functional for polyatomic molecules," *The Journal of Chemical Physics,* vol. 92, pp. 508-517, 1990.

[20] B. Delley, "From molecules to solids with the DMol3 approach," *The Journal of Chemical Physics,* vol. 113, pp. 7756-7764, 2000.

[21] J. P. Perdew, K. Burke, and M. Ernzerhof, "Generalized Gradient Approximation Made Simple," *Physical Review Letters,* vol. 77, pp. 3865-3868, 10/28/ 1996.

[22] C. Kittel, "Introduction to Solid State Physics," 2005.

[23] H. Yildirim, A. Kara, and T. S. Rahman, "Tailoring Electronic Structure Through Alloying: The AgnCu34−n (n = 0–34) Nanoparticle Family," *The Journal of Physical Chemistry C,* vol. 116, pp. 281-291, 2012/01/12 2011.

[24] J. E. Sansonetti and W. C. Martin, "Handbook of basic atomic spectroscopic data," *Journal of Physical and Chemical Reference Data,* vol. 34, pp. 1559-2259, 2005.

[25] D. R. Lide, *CRC Handbook of Chemistry and Physics, 88th Edition*: Taylor & Francis, 2007.

[26] T. Andersen, H. Haugen, and H. Hotop, "Binding energies in atomic negative ions: III," *Journal of Physical and Chemical Reference Data,* vol. 28, pp. 1511-1533, 1999.

Dispersion Characteristics of Twisted Clad Chiral Nihility Fibers

N. Iqbal, M.A. Baqir and P.K. Choudhury[*], *Senior Member, IEEE*

Institute of Microengineering and Nanoelectronics
Universiti Kebangsaan Malaysia
43600 UKM Bangi, Selangor, Malaysia
[*]E-mail: pankaj@ukm.my

Abstract—**Electromagnetic behavior of twisted clad optical fiber structure is investigated under the situation of core section being composed of chiral nihility metamaterial. Twists in the fiber clad are introduced in the form of sheath helical conductor loadings at the core-clad interface. The dispersion relation for the fiber structure is deduced and analyzed (considering low-order sustained hybrid modes) in respect of the effect on the dispersion characteristics due to the alterations in helix pitch angle. It have been found that the fiber structure generally shows decrease in normalized propagation constant with the increase normalized frequency parameter. Also, the dispersion behavior is greatly affected due to the variations in the angle of twists in the helix pitch.**

Keywords—*Twisted clad fibers; complex mediums.*

I. INTRODUCTION

Metamaterials can be designed to achieve prescribed electromagnetic properties [1]. These have been proved to be of great potentials in emerging technologies, e.g. sub-wavelength focusing, cloaking, broadband antennas, oscillators etc. [2,3]. Within the context, chiral metamaterials constitute the subclass in which the electromagnetic behavior can be tailored by varying the chirality. Reference [2] states the phenomenon of negative refraction/reflection, as achievable by suitably adjusting the chirality. These also exhibit circular dichroism – the feature that is used in many applications. For example, circular dichroism is used in sugar industries to measure the concentration of sugar solutions. Uniaxial anisotropic chiral metamaterial guides bounded with DB mediums [4], and those with twisted clad DB mediums have also been reported in the literature [5].

In the context of chirality, Lakhtakia reported the concept of nihility medium in which the permittivity ε and the permeability μ of a medium simultaneously become zero [6], and the wave propagation remains forbidden. Later Tretyakov *et al.* introduced the concept of nihility in chiral metamaterials [7], and defined its constitutive parameters, such as $\varepsilon \to 0$, $\mu \to 0$ and $\kappa \neq 0$. The phenomenon of negative refraction/reflection can also be achieved by these mediums [7]. One of the interesting characteristics of these has been – when a chiral nihility medium is backed with conductor, power through these mediums becomes zero [8].

In the present communication, we put efforts to study the dispersion behavior of chiral nihility core twisted clad optical fiber corresponding to a few low-order propagation modes. In the fiber structure, windings of conducting sheath helix are introduced at the core-clad interface, which serves the purpose to bring in twists in the fiber structure. We aim to observe the effects on the dispersion features due to alterations in helix pitch angle. It has been found that the helical turns impose great impact on the dispersion behavior, as observed in other communication [9,10]. The results reflect that the normalized propagation constant generally exhibits a decreasing tendency with the increase in normalized frequency parameter. Apart from this, some other features in respect of mode degeneracy are also observed. To the best of our knowledge, it would be the first investigation of helical clad optical fiber with the core as chiral nihility medium.

II. ANALYTICAL TREATMENT

Figure 1 illustrates the schematic of optical fiber having the core/clad parametric values as $0 < \rho \leq a$ and $\rho \geq a$, respectively. We essentially use cylindrical polar coordinates (ρ, ϕ, z) for the analysis. For the sake of simplicity, the clad section is assumed to be infinitely extended, and the time t- and the axis z-harmonic waves propagate along the z-direction of the guide. Further, the core section of fiber is assumed to be a chiral nihility medium with the chirality parameter as κ, and the clad has the refractive index as n_{cl} with the assumption that $\kappa > n_{cl}$. Moreover, a conducting sheath helical wrap is introduced at the core-clad interface with ψ as the characteristic pitch angle of helix.

We now consider the propagation of time harmonic electromagnetic waves through core/clad section of fiber. The constitutive relations in an isotropic chiral medium can be written as [11]

$$\mathbf{D} = \varepsilon \mathbf{E} - j\kappa \sqrt{\varepsilon_0 \mu_0} \mathbf{H} \tag{1a}$$

$$\mathbf{B} = \mu \mathbf{H} + j\kappa \sqrt{\varepsilon_0 \mu_0} \mathbf{E} \tag{1b}$$

By using chiral nihility conditions, the constitutive relations become

$$\mathbf{D} = -j\kappa \sqrt{\varepsilon_0 \mu_0} \mathbf{H} \tag{2a}$$

$$\mathbf{B} = j\kappa \sqrt{\varepsilon_0 \mu_0} \mathbf{E} \tag{2b}$$

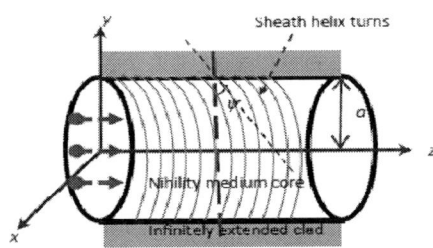

Fig. 1. Twisted clad optical fiber.

In above equations, ε_0 and μ_0 are, respectively, the permittivity and the permeability of free-space, and ε and μ are those corresponding to medium. Also, κ is the chirality parameter. Now, in chiral/chiral nihility metamaterials, two modes – the left-circularly polarized (LCP_) and the right-circularly polarized (RCP_+) – propagate, and the associated wave numbers are

$$k_\pm = \pm \kappa k_0 \tag{3}$$

Throughout discussion, the subscript plus (+) corresponds to the right-circularly polarized wave and the subscript minus (−) to the left-circularly polarized one. The wave impedance of chiral nihility medium can be written as

$$\eta_1 = \lim_{\substack{\varepsilon \to 0 \\ \mu \to 0}} \sqrt{\frac{\mu}{\varepsilon}} \tag{4}$$

The electric/magnetic fields through chiral nihility metamaterial [7] can be written as

$$\mathbf{E} = \mathbf{E}_+ + \mathbf{E}_- \tag{5a}$$
$$\mathbf{H} = \frac{j}{\eta_1}(\mathbf{E}_+ - \mathbf{E}_-) \tag{5b}$$

In chiral nihility metamaterial, the longitudinal components of electric/magnetic fields can be written as

$$E_{+z1} = AJ_m(k_{r+}\rho)e^{jm\varphi} \tag{6a}$$
$$E_{-z1} = BJ_m(k_{r-}\rho)e^{jm\varphi} \tag{6b}$$

with

$$k_{r\pm} = k_0\sqrt{\kappa^2 - (\beta/k_0)^2} = k_r \tag{7}$$

k_0 and k_r respectively, being the free-space and the chiral nihility medium wave numbers, β is the propagation constant in medium, and m is the azimuthal mode index that can take discrete values.

The total longitudinal electromagnetic field is the sum of the right- and the left-circularly polarized waves, which provides

$$E_z = (A+B)J_m(k_r\rho)e^{jm\varphi} \tag{8a}$$
$$H_z = \frac{j}{\eta_1}(A-B)J_m(k_r\rho)e^{jm\varphi} \tag{8b}$$

Now, the transverse field components through the chiral nihility core can be achieved by incorporating the axial field components in Maxwell's equations, as the procedure adopted in [12]. The finally obtained transverse field components are

$$E_{\rho 1} = (A+B)\left[\frac{jm\kappa k_0}{k_r^2\rho}J_m(k_r\rho) - \frac{j\beta}{k_r}J'_m(k_r\rho)\right]e^{jm\varphi} \tag{9a}$$
$$H_{\rho 1} = \frac{j}{\eta_1}(A-B)\left[\frac{jm\kappa k_0}{k_r^2\rho}J_m(k_r\rho) - \frac{j\beta}{k_r}J'_m(k_r\rho)\right]e^{jm\varphi} \tag{9b}$$

$$E_{\varphi 1} = (A+B)\left[\frac{m\beta}{k_r^2\rho}J_m(k_r\rho) - \frac{\kappa k_0}{k_r}J'_m(k_r\rho)\right]e^{jm\varphi} \tag{9c}$$
$$H_{\rho 1} = \frac{j}{\eta_1}(A-B)\left[\frac{m\beta}{k_r^2\rho}J_m(k_r\rho) - \frac{\kappa k_0}{k_r}J'_m(k_r\rho)\right]e^{jm\varphi} \tag{9d}$$

In similar fashion, as adopted for core region, the axial components of electric/magnetic fields through the clad region can be written as

$$E_{z2} = (C+D)K_m(\tau_r\rho)e^{jm\varphi} \tag{10a}$$
$$H_{z2} = \frac{j}{\eta_2}(C-D)K_m(\tau_r\rho)e^{jm\varphi} \tag{10b}$$

and the transverse field components in fiber clad will have the forms as

$$E_{\rho 2} = -\frac{j}{\tau_r^2}\Big[\beta\tau_r(C+D)K_m(\tau_r\rho) + \frac{j\omega\mu m}{\rho}\beta(C-D)K'_m(\tau_r\rho)\Big]e^{jm\varphi} \tag{11a}$$
$$H_{\rho 2} = -\frac{j}{\eta_2\tau_r^2}\Big[\frac{j\omega\varepsilon m}{\rho}\beta(C+D)K_m(\tau_r\rho) - j\beta\tau(C-D)K'_m(\tau_r\rho)\Big]e^{jm\varphi} \tag{11b}$$
$$E_{\varphi 2} = \frac{j}{\tau_r^2}\Big[\frac{j\beta m}{\rho}(C+D)K_m(\tau_r\rho) - j\omega\mu\tau(C-D)K'_m(\tau_r\rho)\Big]e^{jm\varphi} \tag{11c}$$
$$H_{\varphi 2} = -\frac{j}{\eta_2\tau_r^2}\Big[\omega\varepsilon\tau(C-D)K_m(\tau_r\rho) - \frac{j\beta m}{\rho}(C+D)K'_m(\tau_r\rho)\Big]e^{jm\varphi} \tag{11d}$$

In eqs. (8), (9), (10) and (11), $J_m(\cdot)$ and $K_m(\cdot)$ are, respectively, Bessel functions of the first and the second kinds, and the prime represents their derivatives with respect to the argument. Also, the parameter τ_r is defined as

$$\tau_r = \sqrt{\beta^2 - n_{cl}^2} \tag{12}$$

Now, upon implementing the boundary conditions at the core-clad interface, the dispersion relation for fiber structure can be deduced. In the case of fibers loaded with conducting sheath helix structure, the boundary conditions require that the tangential components of electric field become zero in the direction of conductivity of helix, and also, the tangential components of electric and magnetic fields remain continuous at core-clad interface [9]. Implementing these boundary conditions, through the use of the set of eqs. (8)–(11), the dispersion relation for the fiber structure under consideration may be deduced, after some rigorous mathematical steps, as follows:

$$\begin{aligned}&\{a^2k_0^2\kappa^2k_r^4\alpha^2\sin(2\psi) + 4ak_r^2\kappa k_0\,\alpha\delta J_m(ak_r) + \\ &4\Delta J_m^2(ak_r)\} \times \{2m\beta\eta_2 k_r\gamma\cos^2\psi K_m(a\tau_r) + \\ &a\tau_r^2k_r^2\eta_2\tau_r\gamma^2\cos^2\psi + 4j\tau_r\sin\psi K_m(a\tau_r) \times \\ &(m\beta\cos\psi - ja\eta_2\tau_r^2\sin\psi)\} = 0\end{aligned} \tag{13}$$

where

$$\alpha = J_m(ak_r) - J_{m+1}(ak_r) \tag{14}$$
$$\gamma = K_{m-1}(ak_r) + K_{m+1}(ak_r) \tag{15}$$
$$\delta = -m\beta\sin(2\psi) + ak_r^2\cos(2\psi) \tag{16}$$
$$\Delta = m^2\beta^2\sin(2\psi) - 2am\beta k_r^2\cos(2\psi) - a^2k_r^4\sin(2\psi) \tag{17}$$

In these equations, $\eta_2\ (= \sqrt{\mu/\varepsilon})$ is the impedance of clad section.

III. RESULTS AND DISCUSSION

We now investigate the dispersion behavior of fiber structure under consideration for which the clad refractive index is taken to be $n_{cl} = 1.45$, and the operating wavelength as $\lambda_p = 1.55$ μm. Considering variations in core chirality and pitch angle, the values of normalized propagation constant β/k_0 are plotted against the normalized frequency parameter $V = ak_0\sqrt{\kappa^2 - n_{cl}^2}$; these plots are shown in figs. 2 and 3 corresponding to the helix pitch angles as $0°$ and $45°$, respectively. Two different values of chirality parameter κ, viz. 1.55 and 1.75, are taken into account. The values of propagation constant β corresponding to each mode are basically obtained from the dispersion relation eq. (13).

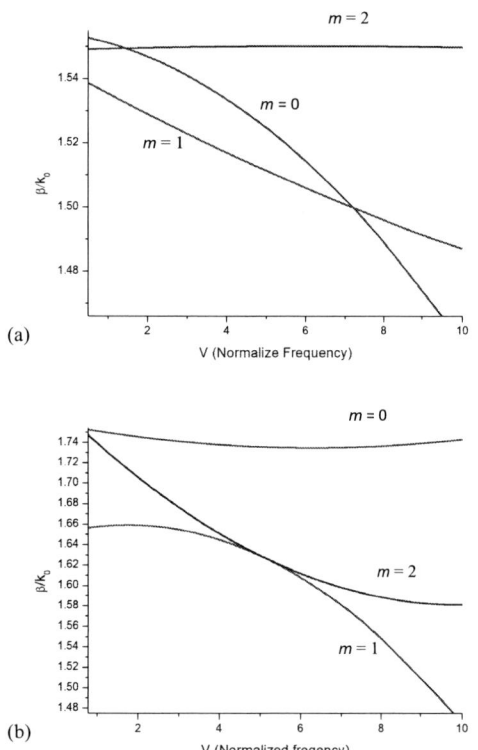

(a)

(b)

Fig. 2. Plots of β/k_0 vs. V corresponding to $\psi = 0°$ and with $\kappa = 1.55$ (a) and $\kappa = 1.75$ (b).

Figures 2a and 2b, respectively, correspond to the situations when the chirality parameter is assumed to be 1.55 and 1.75, and the helical windings are oriented perpendicular to the direction of wave propagation, i.e. $\psi = 0°$. The dispersion plots are made for meridianal ($m = 0$), and two low-order skew ($m = 1$ and $m = 2$) modes. Comparing these two figures, we observe that the variation in chirality parameter brings in drastic changes in the dispersion behavior. When the chirality is assumed as $\kappa = 1.55$, the effective refractive index corresponding to the $m = 2$ mode remains almost constant upon varying the normalized frequency parameter V, as shown in fig. 2a. The increase of core chirality to $\kappa = 1.75$ alters the dispersion behavior

corresponding to this mode (i.e. $m = 2$), and we observe from fig. 2b that the normalized propagation constant decreases with increasing V. We further notice from fig. 2b that, corresponding to $m = 0$, the normalized propagation constant remains almost the same with increasing normalized frequency – the feature which is not seen in fig. 2a. Otherwise, for the other two modes, the β/k_0 parameter generally exhibits decrease with increasing V, as observed in figs. 2a and 2b.

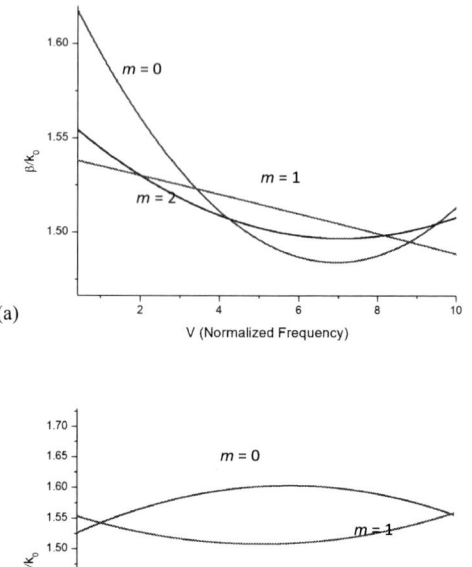

(a)

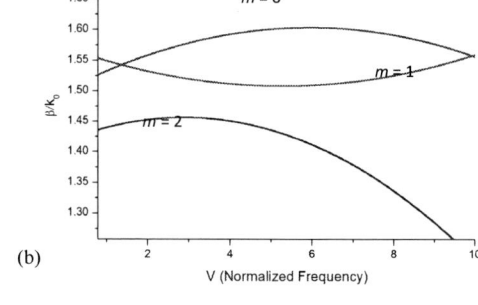

(b)

Fig. 3. Plots of β/k_0 vs. V corresponding to $\psi = 45°$ and with $\kappa = 1.55$ (a) and $\kappa = 1.75$ (b).

A comparison of figs. 3a and 3b too shows variations in the dispersion behavior of fiber with changes in chirality parameter, when the helix pitch angle assumes 45°. However, the effect due to alterations in pitch may be clearly viewed upon comparing figs. 2a and 3a, and also, figs. 2b and 3b. We observe that an increase in helix pitch lifts up the constant nature of normalized propagation constant with increasing V corresponding to any of the modes – the feature noticed in figs. 2. In figs. 3 too we observe that the normalized propagation constant generally decreases with the increase in V. Furthermore, we notice that the higher-order modes behave more differently than the low-order ones upon alterations in helix pitch, which evidently indicates the impact of introducing conducting sheath helix geometry in such chiral nihility fibers. Moreover, certain modes exhibit conditions close to degeneracy, and this feature remains useful for optical applications.

IV. CONCLUSION

From the foregoing discussion, it can be concluded that the dispersion characteristics of twisted clad optical fibers with chiral nihility core exhibit interesting features, which may essentially be altered upon either changing the helix pitch angle or the chirality parameter. It remains interesting owing to the observed decrease in normalized propagation constant corresponding to certain modes with increasing normalized frequency parameter – which is generally not found in the case of dielectric fibers. Furthermore, certain situations correspond to nearly constant normalized propagation constant with changing normalized frequency, and in some cases, conditions close to degeneracy are also noticed. All these features account for useful optical applications of such complex structured optical fiber made by amalgamating chiral nihility and chiral mediums wherein a conducting sheath helix is introduced at the core-clad interface.

ACKNOWLEDGEMENTS

The authors are thankful to the Ministry of Higher Education (Malaysia) for providing financial support to the work. Also, they sincerely acknowledge the constant encouragement and help received from Prof. B.Y. Majlis, the director of Institute of Microeng. and Nanoelectronics (Universiti Kebangsaan Malaysia).

REFERENCES

[1] V.G. Veselago, "The electrodynamics of substances with simultaneously negative value of ε and μ," Sov. Phys. Usp., vol. 10, pp. 509–14, 1968.

[2] J.B. Pendry, "A chiral route to negative refraction," Science, vol. 306, pp. 1353–1355, 2004.

[3] D. Rainwater, A. Kerkhoff, K. Melin, J.C. Soric, G. Moreno, and A. Alù, "Experimental verification of three-dimensional plasmonic cloaking in free space," New J. Phys., vol. 14, pp. 013054.1–013054.13, 2012.

[4] M.A. Baqir and P.K. Choudhury, "Propagation through uniaxial anisotropic chiral waveguide under DB-boundary conditions," J. Electromagn. Waves and Appl., vol. 27, pp.783–793, 2013.

[5] M.A. Baqir and P.K. Choudhury, "Flux density through guide with microstructured twisted clad DB medium," J. Nanomaterials, vol. 2014, Article ID: 629651, 2014.

[6] A. Lakhtakia, "An electromagnetic trinity from 'negative permittivity' and 'negative permeability'," Int. J. Infrared and Millimeter Waves, vol. 23, pp. 813–818, 2002.

[7] S. Tretyakov, I. Nefedov, A. Sihvola, S. Maslovski, and C. Simovski, "Waves and energy in chiral nihility," J. Electromagn. Waves and Appl., vol. 17, pp. 695–706, 2003.

[8] Q. Cheng, T.J. Cui, and C. Zhang, "Waves in planar waveguide containing chiral nihility metamaterial," Opt. Commun., vol. 276, pp. 317–321, 2007.

[9] M. Ghasemi and P.K. Choudhury, "Propagation through complex structured liquid crystal optical fibers," J. Nanophotonics, vol. 8, pp. 083997–083997, 2014.

[10] P.K. Choudhury, "Liquid crystal optical fibers for sensing applications," Proc. SPIE, vol. 8818, pp. 88180E-1–88180E-10, 2013.

[11] S. Tretyakov, A. Sihvola, and L. Jylha, "Backward-wave regime and negative refraction in chiral composites," Phot. and Nanostructures, vol. 3, pp. 107–115, 2005.

[12] J. Dong, "Exotic characteristics of power in the chiral nihility fiber," Prog. In Electromagn. Res., vol. 99, pp. 163–178, 2009.

IEEE-ICSE2014 Proc. 2014, Kuala Lumpur, Malaysia

A 48GHz-78GHz MMIC Sub-Harmonic Pumped Image Rejection Mixer

Shengzhou Zhang

Institute of VLSI
Zhejiang University
Key Lab. of RF Circuits and Systems, MOE.
Hangzhou Dianzi University
Hangzhou, China
szzhang@zju.edu.cn

Lingling Sun[*], Jincai Wen[*], Jun Liu

Key Lab. of RF Circuits and Systems, MOE.
Hangzhou Dianzi University
Hangzhou, China
sunll@hdu.edu.cn, jcwen@hdu.edu.cn

Abstract—The paper presents the design of a 48-78GHz sub-harmonic pumped image rejection mixer (SHIRM) based on a 3μm GaAs technology. The SHIRM contains two identical 2nd APDP-Based SHP mixers, a modified quadrature-phase RF Lange coupler and an in-phase LO Wilkinson power divider. The quasi-lumped topology is utilized with the compact size of 1.7×1.6 mm². The conversion gain of -14±1dB and the minimum image rejection radio of 20dB over the frequency range 48-78GHz covering the whole V-band are simulated, with fixed IF of 1GHz and the LO input power of 12dBm. The 3dB bandwidth is superior to 30GHz. The LO-to-RF and 2LO-to-RF isolations are superior to 25dB and 70dB, respectively. The 2nd harmonic component of LO signal is suppressed by the symmetrical structure.

Keywords—*image rejection mixer; sub-harmonic pumped mixer; APDP (anti-parallel diode pair); monolithic microwave integrated circuit (MMIC); mHEMT; quasi-lumped topology*

I. Introduction

In the 60GHz band wireless communication, V-band has attracted tremendous interests, because these frequency spectrums are hardly used and the relatively numerous free license can be obtained [1-2]. Image rejection mixers (IRMs) play an important role in V-band wireless communication systems. The phase-cancellation technique is utilized to distinguish the desired RF input signals and the undesired image signals resulting from the down-converted products. Therefore, the frequency spacing between the desired RF inputs and image signals could be ignored. It could save the preselection cost of extra mixers and filters for down-conversion [3]. In addition, due to the no dc power and half of frequency requirement of the normal local oscillator (LO), the passive sub-harmonic pumped (SHP) mixers are more effective than the fundamental mixers for mm-wave/THz transceivers. They gain great research efforts for high frequency applications [4]. Up to now, the published V-band IRMs are mainly focused on MMIC research and practical applications not to expand bandwidth [5-7], not to mention the whole V-band (50-75GHz).

In this work, a 48GHz-78GHz subharmonic pumped image rejection mixer (SHIRM) has been designed and simulated. To achieve a high broadband performance, the RF divider employs a modified quadrature-phase Lange coupler. The LO splitter utilizes an in-phase Wilkinson power divider and the sub-harmonic pumped (SHP) mixers uses the APDP configuration and the quasi-lumped topology. The simulation results exhibit the maximum conversion gain of 13dB and the minimum image rejection radio of 20dB at the RF bandwidth of 48-78GH.The designed MMIC mixer chip size is as compact as 1.7×1.6 mm².

II. Circuit Design

A. Configuration of the SHPIRM

The configuration of the present broadband image-rejection mixer is shown in Fig.1. It is achieved including two 2nd APDP-Based SHP mixers, a LO in-phase Wilkinson power divider and a RF quadrature-phase Lange coupler. The 2nd APDP-Based SHP mixers employ the λ/4@ LO open stub and the λ/2@ RF short stub with the quasi-lumped topology to enhance the port isolations and reduce the chip size [8]. Since IF frequency is far away from LO and RF frequencies, the IF quadruture combiner will occupy large areas. Therefore, with the off-chip lumped elements quadrature power splitter (LQPS), the IF quadruture combiner can be extracted from the chip [9].

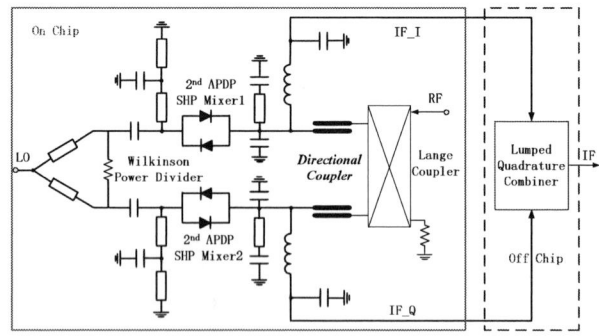

Fig. 1. Schematic of the broadband SHPIRM

The image rejection performance is realized through the symmetrical quadratic structure with two identical SHP mixers on the path. For the down-conversion, the quadrature-phase RF signals and the in-phase LO signals are pumped to the both 2nd APDP-Based SHP mixers, respectively. The major components

This work is supported by the Major State Basic Research Development Program of China (973 Program) under Grant 2010CB327403 and the National Natural Science Foundation of China (61331006, 60906015)

978-1-4799-5761-3/14 $31.00 © 2014 IEEE

of the down-converted signals are $f_{RF}\pm2f_{LO}$. The in-phase LSB signal ($f_{RF}-2f_{LO}$) is the desired signal and extracted out at the IF port. The out-phase USB signal ($f_{RF}+2f_{LO}$) would be filtered and the image signal ($2f_{LO}-f_{RF}$) can be canceled at the IF port via a lumped quadrature hybrid combining the two IF I, Q signals.

TABLE I.   SIGNALS PHASE OF THE SHPIRM

	Down Conversion						
	RF	2LO	USB	LSB	IF-Comb	USB	LSB
SHP Mixer1	$\pi/2$	0	$\pi/2$	$-\pi/2$	$\pi/2$	π	0
SHP Mixer2	0	0	0	0	0	0	0

B. Wilkinson Power Divider

The Wilkinson power divider (WPD) could accurately offer two in-phase and equiamplitude signals with a low insertion loss. Compared to the lumped/semi-lumped elements structure, the distribute WPD occupies a larger chip area, but owns better reliability, input/output reflection coefficients (S11, S22) and isolation ratio (S32) performance [10]. Moreover, multi-stage WPD performs a considerable higher bandwidth than one stage structure, but the insertion loss will increase [11]. This work chooses a single stage distribute structure WPD with micro-strip line. Fig.2 shows the simulation results. With the LO inputs frequency of 25-35GHz, the insertion loss (S21, S31) is better than -3.25dB and the amplitude imbalance almost equal to zero. The isolation (S32) and the return loss of three ports are superior to -17dB and -18dB, respectively.

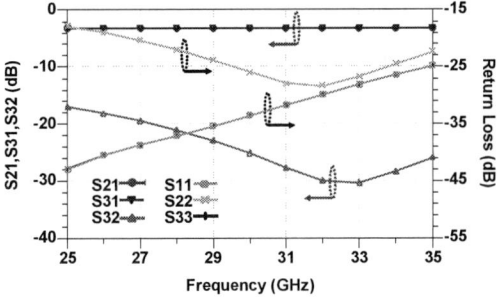

Fig. 2.   S-parameter of the single stage distribute WPD

C. Lange Coupler

For the broadband SHPIRM, the two 2nd APDP-Based SHP mixers are identical and the WPD provides two broadband well-balanced LO signals. Hence, the image rejection ratio (IRR) is mainly determined by the imbalances of the RF divider. The Lange coupler and the branch line coupler are both commonly used as quadrature-phase couplers. The branch line coupler has narrow band performance, especially when reduced in size [8]. Due to its compact structure, the Lange coupler is utilized in this design and shown in Fig.3.

The structure of the lateral-coupled structure Lange coupler is indeed inhomogeneous and asymmetric. It will prohibit the occurrence of odd- and even-modes and cause the amplitude and phase imbalances, especially at high frequencies. How to compensate the imbalances of odd- and even-mode phase velocities is the main consideration in the design. The key dimensions (L, W, and S) can be calculated based on the computed values of the odd- and even-mode impedances and effective dielectric constants using the assumption of coupled micro-strip lines [12,13].

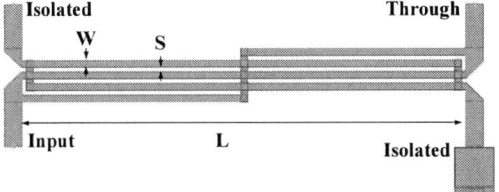

Fig. 3.   The layout of the lateral-coupled Lange coupler

The 3μm GaAs technology provides one 2.5μm thick metal layer and one 1.5μm thin layer. The thin metal layer which is the first metal shows better precision than the thick metal layer. Especially at high frequencies, the skin effect is more obvious. Thus, this design adopts the thin metal layer as the signal line. Moreover, to avoid the parasitic parameters of the via whole, the terminate resistance and expand the bandwidth, which greatly influence the input/output reflection coefficients and the output phase imbalance [10], a MIM capacitance is connected in series with the terminate port of the Lange coupler.

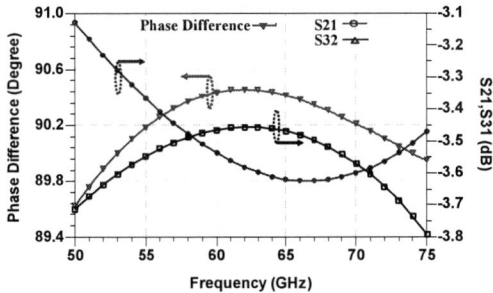

Fig. 4.   Phase imbalance and Amplitude of the Lange Coupler

After the optimization executed by Agilent Momentum simulation, the key dimensions L, W and S are 500μm, 8μm and 5μm, respectively. Eventually, as shown in Fig. 4, for RF inputs frequency over the whole V-band (50-75GHz), the amplitude and phase imbalance are less than 0.5dB and 0.6°. The insertion loss (S21, S31) is better than -3.8dB.

III.   Simulation Results of the Broadband SHPIRM

All the simulations are based on the commercial software Agilent ADS and full wave electromagnetic (EM) simulator. The Wilkinson power divider and the Lange coupler were designed. Furthermore, the performance of SHPIRM was optimized using HP/EEsof harmonic balance analysis. The layout of SHPIRM is showed in Fig. 5, with chip size of 1.7×1.6 mm^2.

For the LSB down-conversion mode, the influence of LO power to the SHPIRM performances are investigated. The Fig.6 and Fig.7 show the dependence of conversion gain and

IRR versus RF frequency with different LO input power, where fixed IF of 1GHz. The simulation results show that the conversion gain is sensitivity to the LO power. On the contrary, the IRR is not sensitivity in case of different LO power from 9dBm to 13dBm. When the LO power is above 12dBm, the conversion gain nearly saturates. The best conversion gain and IRR for the whole V-band occur at the LO power of 12dBm. Under the condition of the LO power fixed at 12dBm, the conversion gain is -14±1dB and the image rejection ratio is better than 20dB at the frequency range of 48-78GHz covering the whole V-band with fixed IF of 1GHz. The 3dB bandwidth is superior to 30GHz.

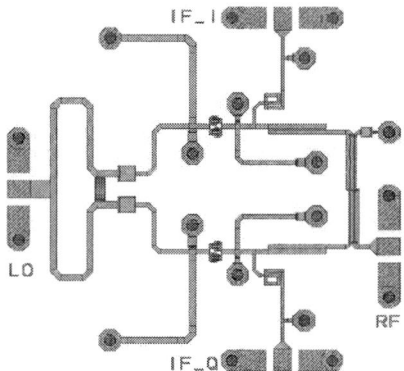

Fig. 5.   The layout of the SHPIRM

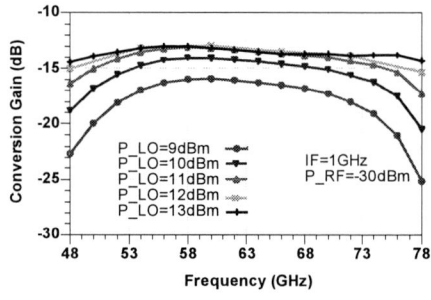

Fig. 6.   Conversion gain vs. RF frequency with different LO power

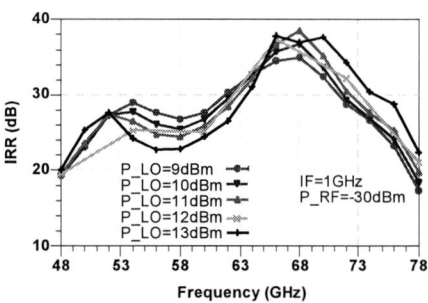

Fig. 7.   IRR vs. RF frequency with different LO power

Fig.8 and Fig.9 show the simulation results of the dependence of the conversion gain and IRR versus RF frequency with different IF frequencies, where fixed the LO power of 12dBm. When the IF frequency increases, the imbalance of the Lange coupler deteriorates, resulting in the

IRR decrease and the conversion gain also decline. The minimum conversion gain and IRR are 16.5dB and 20dB, regarding the IF frequency varying from DC to 2GHz and RF bandwidth of 48GHz to 78GHz.

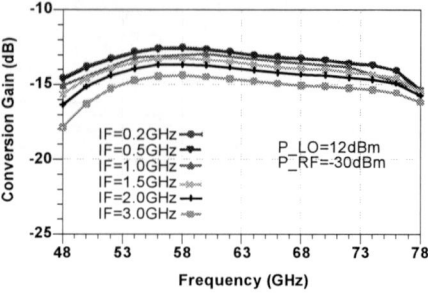

Fig. 8.   Conversion Gain and vs. RF frequency with different IF freq.

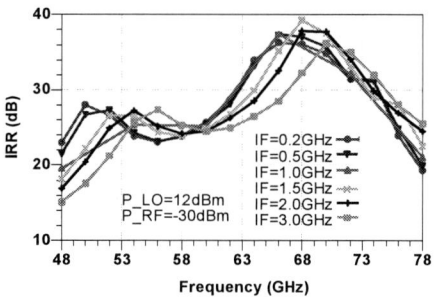

Fig. 9.   IRR and vs. RF frequency with different IF freq.

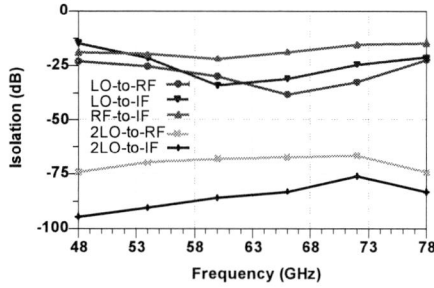

Fig. 10. Port isolations vs. RF frequency

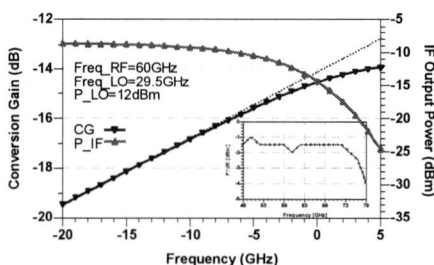

Fig. 11.   IF output power vs. RF input power

Meanwhile, the simulation results of port isolations versus RF frequency are given in Fig.10. The LO-to-RF isolation versus RF frequency for the whole V-band is better than 25dB. The SHPIRM has a completely symmetrical structure, suppressing the second harmonic component of LO signal and enhancing the 2LO to RF and 2LO to IF isolations. Compared to the single end SHP mixer, all isolations of SHPIRM are all improved. With RF fixed at 60GHz and LO fixed at 29.5GHz, the input/output characteristic is shown Fig. 11. The P1dB (1dB compression) can be obtained for RF input power of -2dBm. For the whole V-band the P1dB varies between -1 and -2.5dBm. CG is the conversion gain.

TABLE II. COMPARISON OF THE REPORTED IRMS

Ref	Diode/HEMT mixer	CG*(dB)	IRR*(dB)	RF Freq. (GHz)	3dB bandwidth (GHz)	LO Power (dBm)	On/Off Chip IF	Year
[12]	Diode	-14 ~ -10	>20	24-27.5	2.5	10	OFF	2004
[13]	Resistive	-16 ~ -13	>19	57-66	9	8	OFF	2005
[14]	Resistive	-20 ~ -10	4-30	53-70	12	NA	ON	2007
[15]	Resistive	-29.5 ~ -21	15-34	30-40	8	12	OFF	2007
[16]	Diode	-14 ~ -10	20-29	29-36	5	11	OFF	2009
[17]	Resistive	-21 ~ -9	25	70-100	15	4	OFF	2009
[18]	Diode	-14 ~ -12	24-37	71-86	15	14	OFF	2012
This Work	Diode	-15 ~ -13	20-38	48-78	30	12	OFF	

(CG*: Conversion Gain, IRR*: Image rejection ratio)

Table II summarizes the previously reported SHPIRM fabricated on InP, GaAs process. Compared to the reported mixers, the present mixer has the 3dB bandwidth of 30GHz. In the utilized 3μm GaAs process, the diode is the conventional interdigital capacitor configuration instead of mushroom gate. Thus, the conversion gain is not significant.

IV. Conclusion

A 48GHz-78GHz MMIC APDP-Based sub-harmonic pumped image rejection mixer has been designed and simulated in this paper. The SHPIRM consists of the LO in-phase Wilkinson power divider, a RF quadrature-phase Lange coupler, and two 2nd APDP-Based SHP mixers. Duo to its large size, the IF quadrature combiner is realized with off-chip lumped elements. The designed SHPIRM exhibits the maximum conversion gain is -13dB and the minimum image rejection ratio (IRR) is better than 20dB over the frequency range 48-78GHz covering the whole V-band (50-75GHz), with fixed IF of 1GHz. The present SHPIRM with a completely symmetrical structure suppresses the harmonic component of LO signal. This present SHPIRM is suitable for the prevalent V-band receiver and transmitter module applications.

References

[1] S. E. Gunnarsson et al., "60 GHz Single-Chip Front-End MMICs and Systems for Multi-Gb/s Wireless Communication," IEEE Journal of Solid-State Circuits, vol. 42, no. 5, pp. 1143–1157, May 2007.

[2] K. Hettak et al., "Highly Integrated 60 GHz SSB MMIC Mixer With No DC Power Consumption Based on Subharmonic LO and CPW Circuits in GaAs pHEMT Technology," IEEE 7th European Microwave Integrated Circuits Conference (EuMIC), pp. 536–539, October 2012.

[3] Cochrane, J. B. and F. A. Marki. 'ThinFilm Mixers Team Up to Block Out Image Noise," Microwaves, vol. 16, no. 3, March 1977.

[4] M. V. Schneider and W. W. Snell, Jr.,"Harmonically pumped stripline down-converter,"IEEE Trans. Microw. Theory Tech., vol. MTT-23, no.3, pp. 271–275, Mar. 1975

[5] C. Zelley, et al., "A 60 GHz Integrated Sub-harmonic receiver MMIC," in IEEE GaAs IC Symp. Tech. Dig., pp. 175-178, November 2000.

[6] O. Vaudescal,et al. "A highly integrated MMIC chipset for 60 GHz broadband wireless applications,"presented at the IEEE MTT-S Int. Microwave Symp., Seattle, WA, vol.3, pp.1729 –1732, June 2002.

[7] S. Kishimoto, K. Maruhashi, M. Ito, Y. Hamada, K. Ohata, "60-GHz-band intentional LO-leakage APDP mixer for SSB self-heterodyne transmitter module", Microwave Symp, vol.1, pp.183-186, June 2004.

[8] H. Okazaki, and Y. Yamaguchi, "Wide-Band SSB Sub-harmonically Pumped Mixer MMIC," IEEE Transactions on Microwave Theory and Techniques, vol. 45, no. 12, pp. 2375-2379, August 1997.

[9] Sten E. Gunnarsson, Dan Kuylenstierna, and Herbert Zirath, "A 60 GHz MMIC pHEMT Image Reject Mixer with Integrated Ultra Wideband IF Hybrid and 30 dB of Image Rejection Ratio," Asia Pacific Microwave Conference (APMC), pp.4-7, December 2005.

[10] Y. Xu, Y. Guo, R. Xu, B. Yan, Y. Wu, "An High Performance Ka-band MMIC Fourth Harmonic Image Rejection Diode Mixer," Journal of Electromagnetic Waves and Applications, vol. 22, no. 17-18, pp. 2355-2364, April 2008.

[11] Andreas Wentzel, Viswanathan Subramanian, et al., "Novel Broadband Wilkinson Power Combiner," Proceedings of the 36th European Microwave Conference, pp. 212-215, September 2006.

[12] Moon-Que Lee, Seong-Mo Moon, et al., "Sub Harmonically Pumped Image Rejection Mixer for Ka-band Applications", in Proc. Gallium Arsenide Appl. Symp. (GAAS), pp. 151-154, October 2004.

[13] M. Varonen, et al., "Resistive HEMT Mixers for 60-GHz Broad-Band Telecommunication," IEEE Trans. Microw. Theory Tech., vol. 53, no. 4, pp. 1322-1329, April 2005.

[14] S. E. Gunnarsson, D. Kuylenstierna, and H. Zirath, "Analysis and Design of Millimeter-Wave FET-Based Image Reject Mixers," Microwave Theory and Techniques, IEEE Transactions on, vol.55, no.10, pp.2065-2074, October 2007.

[15] Yang, T., Z. Yang, Y. You, and C. Li, "Small size Ka-band monolithic fourth harmonic image rejection mixer," Microw. Opt. Technol. Lett., vol. 49, no. 3, pp.505-507, March 2007.

[16] Pramod K. Singh, Sarbani Basu, and Yeong-Her Wang, "Integrated Compact Broad Ka-band Sub-harmonic Single Sideband Up-converter MMIC," Progress in Electromagnetic Research, pp.305-307, May 2009.

[17] M Gavell, M Ferndahl, SE Gunnarsson, M Abbasi, H Zirath, "An Image Reject Mixer for High-Speed E-band (71-76, 81-86 GHz) Wireless Communication", IEEE Compound Semiconductor Integrated Circuit Symposium (CISC), pp.1-4, October 2009.

[18] Andy Dearn, Liam Devlin, James Nelson, "A Sub-harmonic E-band IRM/SSB Realized on a Low Cost PHEMT Process", Proceedings of the 7th European Microwave Integrated Circuits Conference (EuMIC), pp.282-284, October 2012.

Gold Nanoplates as Sensing Material for Plasmonic Sensor of Formic Acid

Marlia binti Morsin

Faculty of Electrical and Electronic Engineering,
Universiti Tun Hussien Onn Malaysia,
86400, Parit Raja, Batu Pahat, Johor, Malaysia
marlia@uthm.edu.my

Muhamad Mat Salleh and Akrajas Ali Umar

Institute of Microengineering and Nanoelectronics (IMEN),
Universiti Kebangsaan Malaysia,
43600 UKM Bangi, Selangor, Malaysia
mms@ukm.edu.my, akrajas@ukm.edu.my

Abstract— The gold (Au) nanoplates were grown using seed mediated growth method on the substrate surface. This process consists of two-steps process; seeding and growth of the nanocrystals. The cation polymer namely poly – l – lysine (PLL) was deposited on the substrate to obtain high yield nanoplates before immersed into seed solution. The morphology image of Au nanoplates sample by FESEM shows the yield can be estimated up to 63 % over all area of substrate surface. The optical absorption has two peaks that are associated with transverse (t-SPR) and longitudinal LSPR (l-SPR) respectively. The gold nanoplates sample is tested into formic acid and it was found that their intensities and peak position were changed with the change of surrounding medium. Consequently, using an optical sensor system, variation concentration of formic acid solutions from 1 % to 50 % can be detected. The higher concentration of formic acids will damage the Au nanoplates samples.

Keywords— *Gold nanoplates; Localized surface plasmon resonance (LSPR) ; Optical Sensor; Formic Acid*

I. INTRODUCTION

The Localized Surface Plasmon Resonance (LSPR) of gold nanostructures gains great attention among researchers particularly in biomedical sensing [1] referring to its sensitive characteristics to the dielectric surrounding medium. The LSPR properties are very dependent on the shape and size of gold nanostructures such as sphericals, rods, cubes and plates. Normally, gold nanospherical were synthesized and used in sensing applications [2-5]. The sensing parameter is based on the change in LSPR peak.

The typical technique for synthesis of gold nanostructures is wet chemical method like seed mediated growth method [6]. This technique has two process; i.e. nucleation and followed by growth of the metal nanoparticles. This method offers simple preparation and low cost compared to top down approach such as lithography and electron beam. Moreover, we can produce large amount of nanocrystals with variety of size and shapes. These advantages are very beneficial in LSPR or well known as plasmonic sensor.

A colourless chemicals namely formic acid (CH_2O_2) has been used as a preservative and antibacterial agent in livestock feed [7]. In the poultry industry, it is sometimes added to feed

to kill E. coli bacteria. It is also functioned as a coagulant in rubber manufacturing process and in various cleaning product. It has been found that ingesting highly concentrated amounts of formic acid can result in kidney and liver damage. Unfortunately, this chemical is also used in food preparation process as a preservative to help protect against food deterioration caused by micro-organisms [8].

This paper reports the development of high yield of gold (Au) nanoplates to detect the presence of formic acids in plasmonic sensors. This chemical is a permitted food additive but very harmful in large amounts

II. METHODOLOGY

Gold (Au) nanoplates were grown on quartz substrate using seed mediated growth method [9] with modification. The chemicals used for the synthesis were hydrogen tetrachloroaurate ($HAuCl_4.3H_2O$), trisodium citrate ($C_6H_5Na_3O_7$), sodium tetraborohydride ($NaBH_4$), poly-l-lysine (PLL), ascorbic acid (AA), cethyltrimethy ammonium boromide (CTAB) and poly (vinyl pyrrolidone) (PVP). All these chemicals were purchased from Sigma-Aldrich except trisodium citrate, sodium tetraborohydride and ascorbic acid which were obtained from Wako Pure Chemical Ltd. The solutions of these chemicals were prepared using deionized (DI) water obtained from pure lab UHQ ELGA with resistivity around 18.2 MΩcm.

This process began by immersing the quartz substrate into 5 % poly-l-lysine (PLL) for 30 minutes to impose a positive charge on the substrate surface. Then, two types of solutions were prepared namely seed solution and growth solution. The seed solution was prepared by mixing 0.5 ml of 0.01 M HAuCl4 with 2.0 ml of 0.01 M trisodium citrate and 18 ml DI water. After that, 0.5 ml of 0.1 M cold aqua NaBH4 was added into the solution. The substrate was immersed into seed solution for 1 hour at room temperature. The steps were repeated twice. After that, the substrate was annealed at 150 °C for 1 hour using vacuum oven to strengthen the gold nanoseeds on the surface. After this treatment, the substrate was immersed for 5 hours into growth solution. The growth solution was prepared by mixing 0.5 ml of 0.01 M $HAuCl_4$, 10 ml of 1 mM PVP, 8 ml of 0.1M CTAB and 2 ml DI water with addition of

978-1-4799-5761-3/14 $31.00 © 2014 IEEE

0.1 ml of 0.1 M ascorbic acid. Finally, the substrate was annealed at 100 ºC for 1 hour to remove the surfactant. The change of the substrate color after each process is shown in Fig. 1.

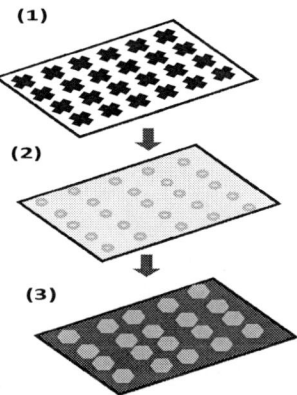

Fig. 1. The change of substrate color after (1) PLL treatment (2) Seeding Process and (3) Growth Process of Au nanoplates

The prepared sample of Au nanoplates was used as sensing material in the optical plasmonic sensor to detect formic acid. This system was setup with a sensor chamber, a light source, and a spectrometer. The duplex fiber optical probe system is used as light beam route. One of the fiber arms was used as light beam transmitter toward Au nanoplates sample and the other fiber arm was used to transmit the reflected light to the spectrometer. The signal was analyze using OOIBase32 software as spectrum analyzer tool. The optical responses of the Au nanoplates sample were recorded as the absorption spectra of light at first when the sample immersed in deionized (DI) water and then immersed in formic acid solution as an analyte. The analyte chemical was purchased from HmBG Chemicals.

III. RESULTS AND DISCUSSION

A. Structural Characterizations

The gold (Au) nanoplates have been successfully grown on the quartz substrate and the final substrate color was violet [10] indicates the generation of Au nanoplates on the surface as shown in Fig. 2.

The formation of Au nanoplates on the quartz substrate was confirmed by X-ray diffraction (XRD). The XRD result is shown in Fig. 3 and has been compared to the JCPDS-004-0784 file for gold. The spectrum shows two sharp peaks with higher peak can be indexed as the (111) - crystallographic planes of face - centered cubic (fcc) gold nanocrystals at 38.175 °. This plane was oriented in parallel to the surface of the substrate, same as Au nanoplates formation. Another peak observed at 44.3 ° with fcc gold crystals planes (200). The crystalline size of this sample is 264.8 Å with Full Width at Half Maximum (FWHM) is 0.307.

Fig. 2. The formation of the Au nanoplates on the substrate surface.

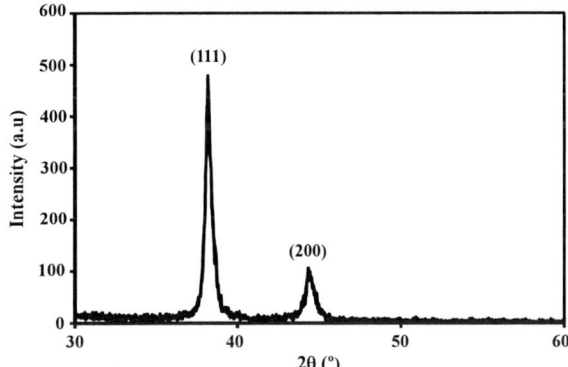

Fig. 3. The XRD of the Au nanoplates

Then, the Au nanoplates sample was analyzed using FESEM Zeiss Supra 55VP. Fig. 4 shows the morphology image of Au nanoplates sample by FESEM. From the image, the presence of two groups of Au nanoplates sizes can be observed. First, the Au nanoplates with edge length more than 150 nm and the other are the Au nanoplates with smaller edge length (less than 50 nm). Over all area, the yield of big nanoplates with PLL treatment can be estimated to be 63 % on the surface area. The plate shapes are flat rod, truncated hexagonal, asymmetric hexagonal, symmetric hexagonal and triangular. Besides, spherical Au nanoparticles and irregular Au shapes were also observed.

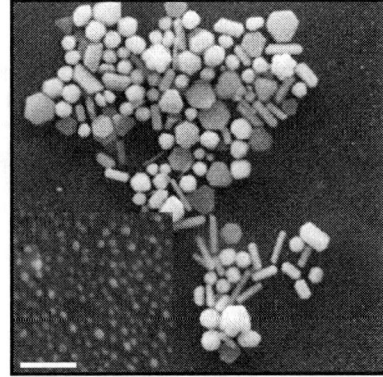

Fig. 4. : FESEM image of gold nanoplates grown on the substrate using seed mediated growth method. Scale bar is 1 μm.

B. Plasmonic Response Study

The plasmonic property of the Au nanoplates samples was studied from the optical absorption spectra in three different media; air, deionized (DI) water and formic acid (FA) that were measured using our plasmonic sensor system setup. The sample of Au nanoplates was immersed into the water and the spectrum was recorded. The results are shown in Fig. 5.

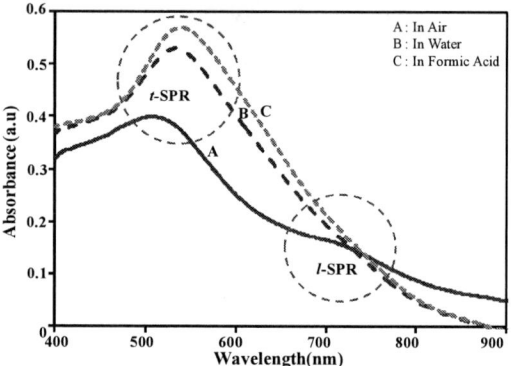

Fig. 5. The spectra of gold nanoplates in 3 different medium (A) air (B) deionized (DI) water and (C) 1% formic acid.

The absorption spectrum of the sample in air has two absorption bands where peaks are located at 530 nm and 710 nm. These peaks are associated with transverse and longitudinal LSPR respectively. The longitudinal LSPR peak disappeared for the sample in the water and formic acid medium. The presence of water and formic acid molecules in the surrounding medium will influence the resonance of the Au nanoplates samples, thus attenuate the resonance in horizontal direction which is parallel with the surface.

Also, it was found that the LSPR peak positions of the spectrum were red shifted and the intensity was changed with the change of surrounding medium. The changes of the spectra can be related with the change of the refractive index (RI) of the medium [11]. In this case, the analyte medium (FA) is higher than RI of air and water (RI_{FA} = 1.37, RI_{water} = 1.33, RI_{air} = 1).Hence, the difference between peak position and intensity of the LSPR spectra gold nanoplates sample in formic acid solution and water is used as the sensing sensitivity parameters.

Next, the sensitivity of Au nanoplates towards various concentrations of formic acid solutions was tested and recorded. The formic acid solutions were diluted with DI water for seven different concentrations; 1 %, 3 %, 5 %, 7 %, 10 %, 30 % and 50 %. The results for various concentration of formic acid solution are shown in Fig. 6.

As well, it appears that the sensing responses of t-SPR are linear with increasing of the formic acid concentrations. It was observed for larger concentration of formic acid (50 %), the peak position of the spectrum was blue shifted even though the intensity was still increase. Furthermore, the concentration of formic acid higher than 50 % will damage the Au nanoplates samples by looking the violet color fades on the sample surface.

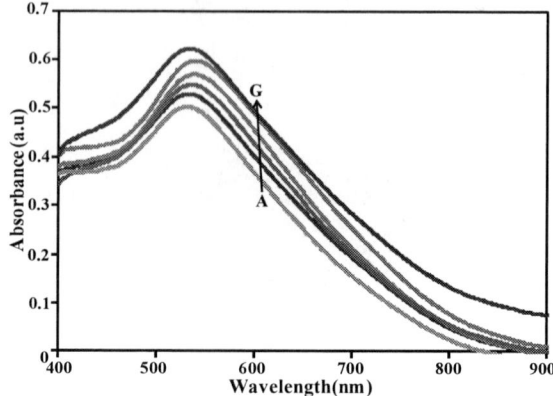

Fig. 6. The plasmonic responses of Au nanoplates with variations of formic acid concentrations; (A) 1 % (B) 3 % (C) 5 % (D) 7 % (E) 10 %, (F) 30 % and (G) 50 %.

IV. CONCLUSION

The gold (Au) nanoplates were successfully grown on the substrate surface by seed mediated growth method. The optical absorption spectra of Au nanoplates samples in air, deionized water and formic acid demonstrated the LSPR property is very sensitive with the change of surrounding medium. Consequently, using an optical sensor system, various concentrations of formic acid solutions as low as 1 % can be detected.

ACKNOWLEDGMENT

The first author would like to acknowledge Universiti Tun Hussien Onn for SLAB - MOHE fellowship.

This work was supported by Ministry of Education Malaysia under ERGS/1/2012/STG02/UKM/01/1 grant.

Also thanks for Universiti Kebangsaan Malaysia for all the facilities and equipment provided for this project, i.e.; FESEM, XRD and UV-Vis

REFERENCES

[1] K.M. Mayer and J.H. Hafner, "Localized Surface Plasmon Resonance Sensors", *Chemical Reviews* vol. 111 (6), pp. 3828-3857, 2011

[2] M.H. Tu, T. Sun and K.T.V. Grattan, "Optimization of the Gold – Nanoparticle - based Optical Fibre Surface Plasmon Resonance (SPR) – based Sensors", *Sensors and Actuators B: Chemical* vol. 164 (1), pp. 43 – 53, 2012

[3] Marlia Morsin, M.M. Salleh, A.A Umar, " Detection of Boric Acid using Localized Surface Plasmon Resonanace Sensor of Gold Nanoparticles, *The 14th International Meeting on Chemical Sensor*, pp .1418 - 1412, 2012

[4] S. Nengsih, A.A. Umar, M.M. Salleh and M. Yahaya, " Detection of Formaldehyde in Water: A Shape-Effect on the Plasmonic Sensing Properties of the Gold Nanoparticles", *Sensors* vol. 12(8), pp. 10309-10325, 2012

[5] Marlia Morsin, M.M. Salleh, A.A Umar,Burhanuddin Yeop Majlis, " High sensitivity localized surface plasmon resonance sensor of gold nanoparticles: Surface density effect for detection of boric acid "*Semiconductor Electronics (ICSE), 2012 10th IEEE International Conference on*, pp .352 - 356, 2012

[6] A.A Umar, M. Oyama, "A cast seed-mediated growth method for preparing gold nanoparticle-attached indium tin oxide surfaces", *Applied Surface Science* vol. 253,pp. 2019-2022, 2006

[7] T. J. Humphrey and D. G. Lanning, "The vertical transmission of salmonellas and formic acid treatmentof chicken feed; A possible strategy for control", *Epidem. Inf.* Vol. 100, pp. 43-49, 1988

[8] Bill Ruzicka, "Formic Acid Use Handbook And Manual Of Treatments", *MiteGone Enterprises Int.*,pp 1 – 83, 2011

[9] A.A Umar, M. Oyama, M.M. Salleh and B.Y. Majlis, "Formation of Gold Nanoplates on Indium Tin Oxide Surface: Two-Dimensional Crystal Growth from Gold Nanoseed Particles in the Presence of Poly(vinylpyrrolidone)", *Crystal Growth & Design* Vol. 9 (6), 2835-2840, 2009

[10] A. Miranda, E. Malheiro, E. Skiba, P. Quaresma, P. A. Carvalho, P. Eaton, B. de Castro, J. A. Shelnuttcd and E. Pereira, "One-pot synthesis of triangular gold nanoplates allowing broad and fine tuning of edge length", *Nanoscale*, Vol. 2, 2209–2216 , 2010

[11] J.N. Anker, W.P. Hall O,. Lyandres, N. C. Shah, J. Zhao and R.P.V. Duyne, "Biosensing with Plasmonic Nanosensors", *Nature Materials* Vol.7,442-453,2008

IEEE-ICSE2014 Proc. 2014, Kuala Lumpur, Malaysia

Design of 40GHz Multilayer End Coupled Band Pass Filter using LTCC Technology

Zulkifli Ambak[1], Azmi Ibrahim[1],Hizamel Mohd Hizan[1],Ahmad Ismat Abdul Rahim [1],Mohd Zulfadli Mohamed Yusoff [1]

Communication Technology
TM Research and Development Sdn Bhd
Lingkaran Teknokrat Timur
63000,Cyberjaya,Selangor
zulkifliambak@tmrnd.com.my

Razali Ngah [2] and Young Chul Lee [3]

Wireless Communcation Center [2]
Faculty of Electrical Engineering
Universiti Teknologi Malaysia
81310,Skudai,Johor

Division of Marine Electronics and Communication Engineering [3]
Mokpo national Maritime University
(MMU),Mokpo,Jeonnam, South of Korea.

Abstract— **Multilayer end coupled band-pass filters (BPFs) using Low Temperature co-fired Ceramic (LTCC) technology have been presented as a parts of 40 GHz Remote Antenna Unit Downlink for Radio over fiber (RoF) system applications. In order to control the precise distance between two adjacent resonators, the resonators was vertically located on different layers. The BPFs were designed and fabricated by using Ferro A6S tape. Its dielectric constant and loss tangent are 5.8 and 0.002, respectively, at 40GHz. The overall size of the fabricated filter was 10.5 x 2.5 x 0.77 mm3. The measured insertion and return loss were 3.17dB and 18.53dB at 42GHz, respectively.**

Keywords—BPF; LTCC; RAU; RoF

I.

LTCC as a ceramic multilayer technology has a great potential to satisfy the demand for higher data rate at millimetre-wave (mm-wave) applications. LTCC has an advantages to miniaturize the mm-wave circuitry especially passive module like filter and antenna using low loss material systems,for example Ferro A6S material which has a dielectric constant of 5.9 and loss tangents of 0.002 [1].In order to realize the remote antenna unit for mm-wave RoF, the system requires low loss device, compact and high advanced packaging via LTCC multilayer technology.

Figure 1 show the mm-wave RoF system concept while Figure 2 show a block diagram of the 40GHz RAU downlink for RoF system. Generally, the mm-wave RoF application is of growing interest in the application of high date rate wireless communication [2]-[4]. Wireless networks based on RoF technologies is a promising cost-effective solution to meet the ever increasing user bandwidth and wireless demands especially to connect the unconnected area due to

geographical, building block, hacking issue, buried obstacle, etc.

This paper focuses on design of 40GHz band pass filter as a key component for development of the 40GHz RAU downlink for mm-wave RoF system.

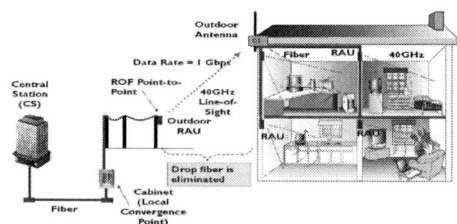

Fig.1. mm-wave RoF system concepts

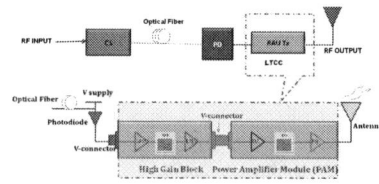

Fig.2. Block Diagram of the 40GHz RAU downlink for RoF system

There have been several reports on mm-wave filters. For example, 40GHz BPF with laminated RT/Duroid RT5880 was presented in [5] while using LTCC in [6]-[8]. The hairpin filter in [5] employs 127μm Rogers RT/Duroid 5880 with dimension of size 9.095mm x 2.810mm including GSG pad. The filter in [6] designed by Xu Ziqiang et al, employs LTCC Ferro A6M substrate with size of 11.7mm x 6.6mm. The filter performance in [6] shows 1.6dB measured insertion loss larger than simulated result and center frequency shifted by 0.6GHz. In this paper, multilayer end coupled BPF using LTCC

978-1-4799-5761-3/14 $31.00 © 2014 IEEE   294

technology was designed and fabricated. The simulation results by CST and then compared with the measurement results of the fabricated filter. The design for this mm-wave BPF was compact in size compared with filter in [5] and [6]. The overall size of this fabricated filter was 10.5 mm x 2.5 mm x 0.77mm.

II.

Some of the indispensable and most important components to realize the RAU downlink are mm-wave BPF. In order to meet the specification for RAU downlink for mm-wave RoF applications, keeping BPF structures to minimum size and weight is also very important. The band pass filter is a key component of the RAU downlink for RoF system to keep out of the sensitive signal.

A. Design Specification

Table I shows the design specification for multilayer end coupled BPF using LTCC technology to meet the requirement of the Broadband Fixed Wireless Access in the 40GHz Band.

TABLE I.   DESIGN SPECIFICATION

No	LTCC MULTILAYER END COUPLED BPF		
	Items	Targeted	Unit
1	Frequency range	40.5-43.5	GHz
2	Center Frequency	42	GHz
3	Pass band width	3	GHz
4	Insertion Loss (S21)	< 3	dB
5	Return Loss (S11)	>10	dB
6	Impedance	50	Ω

B. Design and Simulation

The multilayer end coupled band pass filter was designed and fabricated by using in-house TMRND's LTCC process. This LTCC process was discussed in detail in section III. In this work, the multilayer end coupled BPF was designed using 8 layers LTCC Ferro A6S tapes systems with relative dielectric constant and a loss tangent of 5.8 and 0.002 respectively. The thickness of single layers was 96μm. It consist of the two layers are required for end coupled resonator BPF as shown in Figure 4 which divided into 4 segments include input port, coupling with spacing gap and resonators. Additional 6 layers are used as GND planes. The total dimension of proposed band pass filter was 10.5mm x 2.5mm x 0.77 mm.

Figure 3 and 4 show the design layout at top view and cross section for the end coupled BPF respectively. There are 3 samples were investigated and analyzed in order to check the filter performances in term of insertion loss and size.The performance of the 3 LTCC BPF samples with Ground Signal Ground (GSG) ports were investigated. All the finding results will be analyzed.

Figure 5 shows the simulation result for the end coupled BPF. The center frequency was 42GHz with insertion loss, S21 less than 3dB in the pass band of 40.5 to 43.5, and return loss more than 10dB. Given that f_2 was the upper cut-off frequency while f_1 was the lower cut off frequency and f_c was the center frequency. Therefore, the Fractional Bandwidth (FBW) was calculated using (1)

$$FBW = \frac{f_2 - f_1}{f_c} \dots\dots\dots\dots\dots\dots\dots\dots\dots(1)$$

From the calculation, the filter has FBW of 7.14%, hence this filter can be categories as narrowband filter because of FBW less than 20%.

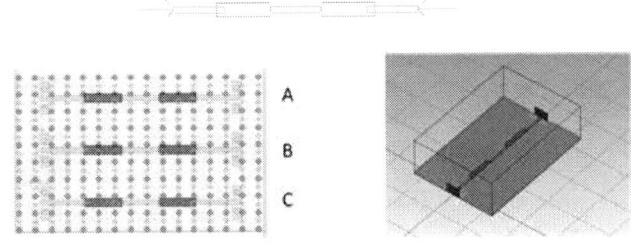

Figure 3. Top view of the 40GHz LTCC End coupled band pass filter.

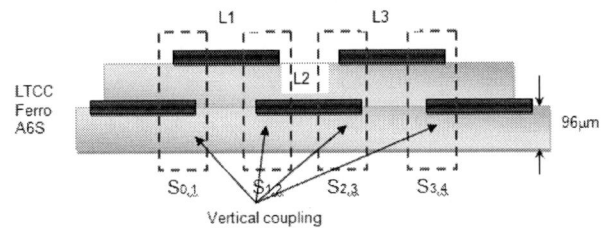

Figure 4. Cross section view of LTCC BPF consisting of end coupled fabricated on LTCC.

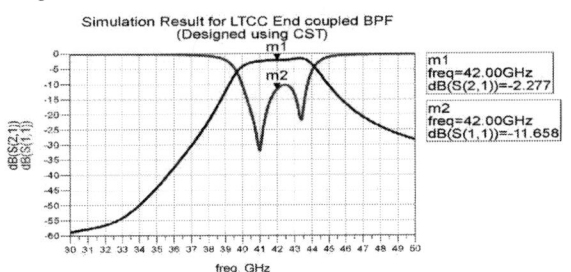

Figure 5. Simulation results of the LTCC End coupled BP

III

Figure 6 shows the capabilities and flow diagram of in-house TMRND LTCC Fabrication process as reported in [9].The most important LTCC process was screen printing and firing.The shrinkage factor need to be consider during design stage. In this works, Ferro A6S was selected as substrate. The shrinkage for Ferro A6S was about 15% in the x-and y directions and about 24% in the z-direction. All the LTCC process, need to checked carefully to avoid misallignment and error.

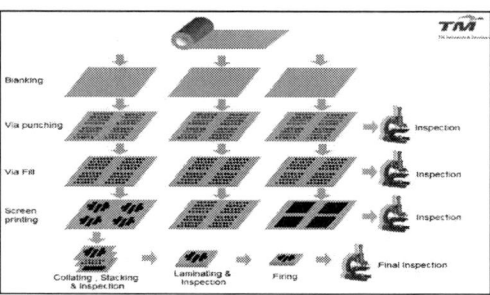

Figure 6. TMRND's in-house LTCC Process Flow

Table II shows the material parameter and characteristic for Ferro A6S.The measured dielectric constant, εr was 5.8 at 40GHz using T-resonator method.

TABLE II.   FERRO A6S MATERIAL PARAMETER AND CHARACTERISTIC

No	TMRND's IN-HOUSE LTCC CAPABILITIES		
	Parameters	Ferro A6S	Unit
1	Dielectric constant, εr	5.8 ± 0.2	
2	Dissipation factor, tan δ	0.002	
3	Substrate height,H	96	μm
4	Conductor Thickness	10	μm
5	Shrinkage XY	15-16	%
6	Shrinkage Z	20-25	Ω

IV

The multilayer end coupled band pass filter was designed and fabricated on the substrate of Ferro A6S green sheet by using linear arrays of silver via-hole having the diameter of 200μm and via- to-via pitch of 600μm. Fig. 7 shows fabricated multilayer end coupled BPF with 3 samples

mark as sample A,B and C. The detail physical dimensions of the LTCC BPF are listed in Table III.The dimensions for fabricated samples was slightly different compared with designed. In order to control the precise distance between two adjacent resonators, they are vertically located on different layers. For example, sample C shows the overlap S0,1 at 200μm vertically located on different layers. All the designs complied with in-house TMRND's LTCC design guidelines with minimum spacing at 100μm between lines.

TABLE III.   DESIGNED AND FABRICATED THE MULTILAYER END COUPLED BPF DIMENSION

Item	DESIGN AND FABRICATED LTCC BPF DIMENSION				
	Designed	Fabricated (in μm)			Remarks
	(in μm)	A	B	C	
S 0,1	210	160	200	200	Overlap
S 1,2	30	30	25	30	Spacing
S 2,3	30	30	25	30	Spacing
S 3,4	210	160	200	200	Overlap
L1	1410	1446	1418	1486	
L2	1343	1333	1359	1370	
L3	1410	1446	1418	1486	

The measurement was done by using R&S ZVA50 network analyzer and Cascade Microtech 450μm probe tips. Figure 8, 9 and 10 shows the measurement results compared with simulation results. It can be seen from the measured result that the insertion loss, S21 was slightly increased while return loss, S11 was better than 10dB compared with simulated result. The sample C was selected because of lowest insertion loss, S21 at 3.17dB and highest return loss at 18.53dB compared with sample A and B. However, the measured insertion loss at center frequency was 0.89dB larger than simulated results, and the different might due to shrinkages of the conductor layer which resulting in increase of spacing and gap of each filter section. Additionally, the increased insertion loss might be caused roughness of the fired circuits. Furthermore, the measured result has good agreement compared with simulated result. The simulated and measured responses of the design were summarized in Table IV.

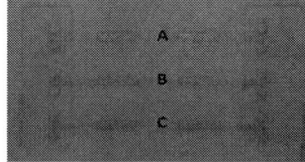

Fig. 7. Photograph of the fabricated LTCC end coupled BPF with conductor Gold

TABLE IV.   SIMULATED AND MEASURED RESULTS FOR THE MULTILAYER END COUPLED BPF

Parameter	SIMULATED AND MEASURED RESULTS					Unit
	Spec	Simulated	Measured			
			A	B	C	
Center Frequency	42	42	42	42	42	GHz
Pass band width	3	4.5	3.5	4.3	3.5	
Insertion Loss (S21)	< 3	2.28	5.95	3.93	3.17	dB
Return Loss (S11)	> 10	11.66	20.15	16.03	18.53	dB

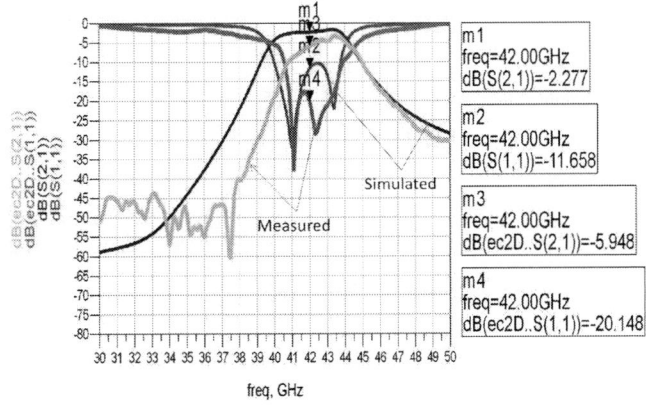

Fig. 8. Measured and simulated results for LTCC BPF (Sample A)

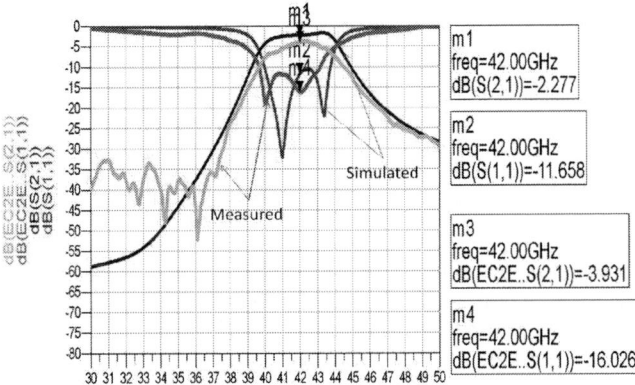

Fig. 9. Measured and simulated results for LTCC BPF (Sample B)

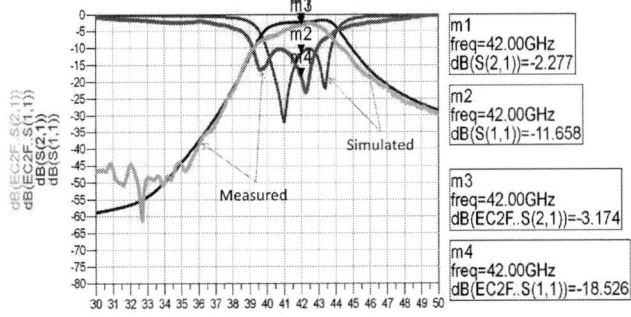

Fig.10. Measured and simulated results for LTCC BPF (Sample C)

V.

The multilayer end coupled band pass filter using LTCC technology with 8 layers LTCC Ferro A6S has been designed an fabricated. The measured insertion loss of fabricated end coupled band pass filter for sample C was better compared to sample A and B. Its measured results was 3.17dB and return loss was 18.53dB at center frequency of 42GHz which is in good agreement with simulated results.As summarized, the variations between the measured and designed dimensions were defined and measured relatively to the guide wavelength at the design center frequency. Therefore, need to optimize the design in the next stage for further improvement. The overall size of the fabricated multilayer end coupled band pass filter is 10.5 mm x 2.5 mm x 0.77 mm.

REFERENCES

[1] http://www.ferro.com
[2] Syamsuri Yaakob, Muhammad Zamzuri Abdul Kadir, Sevia Mahdaliza Idrus, Norhakimah Md Samsuri, Romli Mohamad and Nazif Emran Farid, "On the carrier generation and HD signal transmission using the millimeter-wave radio over fiber system", Optik Elsevier, Volume 124, Issue 23, December 2013, Pages 6172–6177.
[3]Luca pergola,"LTCC based RF Frontends for WLAN Applications and Radio over Fiber Systems," P.H.D thesis ,ETH Zurich,2007
[4] H. B. Kim, A. Wolisz, and A. A. Description, "A Radio over Fiber based Wireless Access Network Architecture for Rural Areas." Proceeding 14th IST Mobile and Wireless Communication Summit,Dresden, Germany,June 2005.

[5] Azah Syafiah Mohd Marzuki, Surati Selamat, Amran Naemat,Khaidir Khalil, Azlinda Tee Md Azlan Tee and Young Chul Lee, "40GHz planar filter on Duroid substrate", RFM , Penang, Dec 2013.
[6] Xu Ziqiang,Shi Yu,Zeng Zhiyi,Liao Jiaxuan and Li Tian, "34GHz Bandpass filter for Low Temperature Co-fired Ceramic System in Package Application", Chinese Journal of Mechanical Enginnering, Vol XX,No X, 2011.
 [7] Min-Soo Kang ,Bong-Su Kim, andMyung-Sun Song, "End Coupled Stripline BPF Using LTCC in Millimeter-Wave," 24th International Conference on Microwave and Millimeter Wave Technology Proceedings,2004.
[8]Y.H Cho,Y.C Lee,J.W Lee,M.s.song and C.s Park, "A Fully Embedded LTCC Stripline Parallel Coupled BPF for 40GHz BMWS Application",IEEE,2004.
[9] Z.Ambak, M.R.Saad ; R.Alias, A.Ibrahim,S.M Shapee,Zulfadli M. Yusof, M.R.Yahya, "Design of Interdigital Band Pass Filter for WLAN applications using LTCC technology",IEEE(APACE),2010

Effect of Different Tunnel diodes on The Efficiency of Multi-junction III-V Solar cells

Hung-Wei Yu, Hong-Quan Nguyen, Chen-Chen Chung, Ching-Hsiang Hsu, Chih-Jen Hsiao and Edward Yi Chang

Department of Materials Science and Engineering, National Chiao Tung University,
1001 University Road, Hsinchu 300, Taiwan

Abstract—InGaP/GaAs dual-junction solar cells with different tunnel diodes (TDs) grown on misoriented GaAs substrates are investigated. It is found that the solar cells with P^{++}-AlGaAs/N^{++}-GaAs TDs grown on 10° off GaAs substrates show a higher external quantum efficiency (EQE) but also generate a higher peak current density (J_{peak}) than the solar cells with P^{++}-GaAs/N^{++}-InGaP TDs grown on 10°off GaAs substrates. Furthermore, smooth surface (rms roughness: 1.54 Å) and sharp interface for the GaAs/$Al_{0.3}Ga_{0.7}As$ TDs were obtained when the (100) tilted 10° off toward [111] GaAs substrate was used. The conversion efficiency of InGaP/GaAs dual-junction solar cell with N^{++}–GaAs/P^{++}–AlGaAs TD grown on the (100) tilted 10° off toward (111) GaAs substrate is close to 20%.

Keywords—*N^{++}-GaAs/P^{++}-AlGaAs tunnel diodes, misoriented GaAs substrates, InGaP/GaAs dual junction solar cells*

I. INTRODUCTION

TUNNEL diode (TD), which must be both highly transparent and heavily doped, acts as a series connection between tandem solar cells. It is a very important component for III-V multijunction solar cells, as III-V solar cell devices have to be operated at higher concentration ratios. The current density of a tunnel diode is composed of three components as shown in equation (1):

$$J= J_{tunnel} +J_{excess} + J_{thermal} \tag{1}$$

These components are J_{tunnel}, the band-to-band tunneling current density, J_{excess}, the excess current density, and $J_{thermal}$, the minority-carrier diffusion current density or thermal current density. If, at higher concentrations, the peak current density (J_{peak}) in the TD is lower than the short-circuit current density (J_{sc}), the overall performance of the III-V tandem solar cell will be dramatically affected. Currently, for III–V multijunction solar cell, wide band gap materials with a low electrical resistivity, such as AlGaAs and InGaP, are adopted as TD materials because of their lower optical absorption. In fact, the increase in the band gap of TD materials made from III–V semiconductors results in a decrease in tunneling current density. This implies that the reduction in impurities in TDs, such as oxygen atoms that act as nonradiative traps, is very important when higher band gap materials are used.

The AlGaAs epitaxial layer is a very important material for many high-speed electronic and optoelectronic devices [1]-[2], because the difference in the lattice parameters between GaAs and AlGaAs is very small, which prevents the generation of undesirable interface states. However, the AlGaAs epitaxial layer as a TD material is known to have a serious oxygen incorporation problem as compared with InGaP epitaxial layer for III–V multijunction solar cell application. It has been reported that the reduction of these impurities from an AlGaAs epi-layer can be achieved using shorter growth interruption [3], liquid metal bubblers [4], and (311) oriented GaAs substrates [5]-[6]. On the contrary, InGaP TDs show a lower interface recombination rate than AlGaAs TDs when electrons move away from the TD interface.

In a previous work, we proved that the 10° off misorientation GaAs substrates not only reduce the oxygen-impurity content in the N^{++}-GaAs/P^{++}-AlGaAs TD, but also produce high-quality TD with a smooth surface and a sharp interface [7]. Although the 10° off misoriented GaAs substrate can efficiently reduce nonradiative traps in the N^{++}-GaAs/P^{++}-AlGaAs TD, a dual-junction solar cell grown on 10° off misorientation GaAs substrates using a N^{++}-GaAs/P^{++}-AlGaAs TD has not yet been reported.

II. EXPERIMENT

We report the results of comparison between InGaP/GaAs dual-junction solar cells with a N^{++}-InGaP/P^{++}-GaAs TD (sample A) and those with a N^{++}-GaAs/P^{++}-AlGaAs TD (sample B), both grown on 10° off misorientation GaAs substrates using a metal organic chemical vapor deposition (MOCVD: EMCORE D180) system. Trimethylgallium (TMG), trimethylaluminum (TMAl), and trimethylindium (TMIn) were used as group III sources, whereas arsine (AsH$_3$) and phosphine (PH$_3$) were used as group V sources. The precursors for p-type and n-type dopants were carbon-tetrabromide (CBr$_4$) and dimethyl-telluride (DMTe), respectively. For GaAs/AlGaAs TDs, higher doping levels can be accomplished in AlGaAs epi-layers with carbon doping than in GaAs. The GaAs epitaxial layer used for the n-type side in the TDs was found to have lower optical absorption because of the generation of the Burstein-Moss shift [8]. The band gap of n-type GaAs in the TDs slightly increases with the increase in doping concentration, resulting in more light passing through the TD to the GaAs cell. For an InGaP/GaAs TD, carbon-doped InGaP was not used in this study because carbon is very difficult to embed in InP-based materials by MOCVD. All films in this study were grown at a low pressure of 40 torr and a hydrogen flow rate of 28000 sccm.

978-1-4799-5761-3/14 $31.00 © 2014 IEEE

III. RESULTS AND DISCUSSION

Figure 1 (a) and 1 (b) show the layer structures of the InGaP/GaAs dual-junction solar cells with different TD materials. The framed GaAs bottom cell consists of a 0.1 μm InGaP back surface field (BSF) layer, a 3 μm GaAs base layer, a 0.1 μm GaAs emitter layer, and a 0.05 μm InGaP window layer. The InGaP top cell consists of a 0.03 μm AlInP BSF layer, a 0.5 μm InGaP base layer, a 0.05 μm InGaP emitter layer, and a 0.03 μm AlInP window layer. A multilayer metal consists of Ni (6 nm)/Ge (50 nm)/Au (100 nm)/Ni (45 nm)/Au (250 nm) was used as the front contact, and a Ti (50 nm)/Pt (60 nm)/Au (250 nm) layer was used as the back contact in this study. The antireflection coating is a single layer of Si_3N_4 with a thickness of 75 nm.

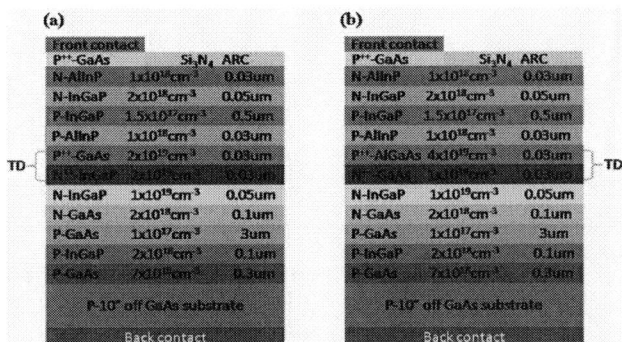

Fig. 1 Structures of InGaP/GaAs dual-junction solar cells with different tunnel diode materials: (a) the cell design with P++-GaAs/N++-InGaP tunnel diode (TD), (b) the cell design with P++-AlGaAs/N++-GaAs TD.

Figure 2 shows the current-voltage (I-V) characteristics of the InGaP/GaAs dual-junction solar cells with different TD materials. Table I shows the performance comparison of samples A and B measured at one sun, including open-circuit voltage (V_{oc}), short-circuit current density (J_{sc}), Fill Factor (FF), and conversion efficiency (Eff.). From the experimental results, J_{sc} of sample A is smaller than those of samples B. J_{sc} generated by a III-V multijunction solar cell is dependent on the incident light and can be simulated using the following equation :

$$J_{sc} = \int_0^\infty \eta_c(E)\big(1 - R(E)\big)a(E)b_s(E)dE \qquad (2)$$

where $\eta_c(E)$ is the probability of electrons collected under illumination; $R(E)$ is the probability of photon reflection; $a(E)$ is the probability of absorption of a photon of energy E; and $b_s(E)$ is the incident spectral photon flux density. Equation (2) shows that a low quantum efficiency (QE) may lead to a reduction in J_{sc} for a solar cell device operated under illumination.

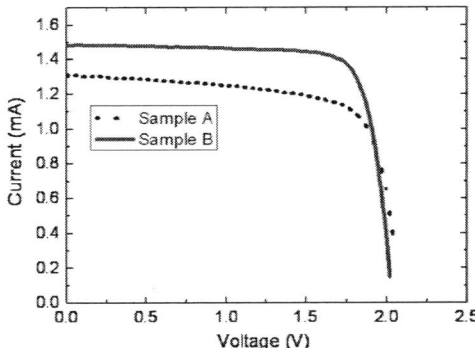

Fig. 2 I-V characteristics of InGaP/GaAs dual-junction solar cells measured at one sun. Sample A: the cell design with a P++-GaAs/N++-InGaP TD grown on 10° off misorientation GaAs substrates; sample B: the cell design with a P++-AlGaAs/N++-GaAs TD grown on 10° off misorientation GaAs substrates

Table I. I-V characteristics of the InGaP/GaAs dual-junction solar cells with different tunnel diodes grown on misoriented GaAs substrates.

Tunnel junction	V_{oc} (V)	I_{sc} (mA/cm^2)	FF (%)	Eff. (%)
Sample A	2.07	10.5	72.0%	15.78
Sample B	2.03	11.9	78.9%	19.24

The characteristics of the dual-junction solar cells with different TD materials, such as P++-GaAs/N++-InGaP and P++-AlGaAs/N++-GaAs, grown on 10° off GaAs substrates are investigated on the basis of the spectral responses of the EQE of the dual-junction solar cells, as shown in Fig. 3. The non-square shape of the spectral responses of the InGaP top cell is observed owing to higher reflection losses (10-50%) in the range from 300 to 500 nm. From the experimental results, it is found that the total EQE of the dual-junction solar cells with the P++-AlGaAs/N++-GaAs TD (sample B) is higher than that with the P++-GaAs/N++-InGaP TD (sample A) at the same misorientation angle (10° off). The maximum EQEs of sample B are 82% for the InGaP top cell and 85% for the GaAs bottom cell. An increase in the EQE of the GaAs bottom cell of sample B is confirmed, and this is due to the smaller absorption losses in the P++-AlGaAs/N++-GaAs TD than in sample A.

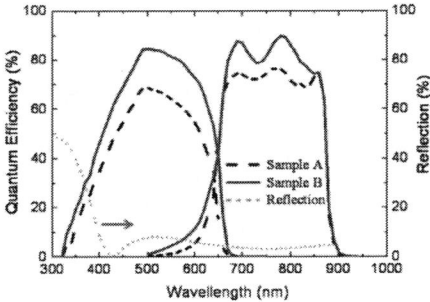

Fig. 3 EQE of InGaP/GaAs dual-junction solar cells. Sample A: the cell design with a P++-GaAs/N++-InGaP TD grown on 10° off misorientation GaAs substrates; sample B: the cell design with a P++-

AlGaAs/N^{++}-GaAs TD grown on 10° off misorientation GaAs substrates

In order to determine the reason for the inferior epitaxial quality of sample A, photoluminescence (PL) using a 532 nm laser source was employed to determine the band gap and epitaxial quality. It is found that the band gap of the InGaP cell of sample A is 1.92 eV, which is larger than those of samples B (1.82 eV), as shown in Fig. 4. It is suggested that a transition of an ordering/disordering InGaP epitaxial layer may occur in sample A when the higher substrate misorientation angle (10° off) was adopted [9]-[10]. The poor epitaxial quality of sample A is attributed to crystal defects associated with disordered structures in the matrix, which further decreases QE and J_{sc} of III-V solar cells devices. Experiments also confirmed that InGaP/GaAs dual-junction solar cells with a P^{++}-AlGaAs/N^{++}-GaAs TD grown on 10° off GaAs substrate (sample B) do not generate a disordering InGaP epitaxial layer during material growth, and thus results in superior I-V characteristics. Besides, the slope of sample A is lower than those of samples B, as shown in Fig. 2, suggesting that the series resistance (Rs) of sample A is higher than those of the others. Higher series resistance also leads to lower FF and J_{sc} of III-V solar cells devices.

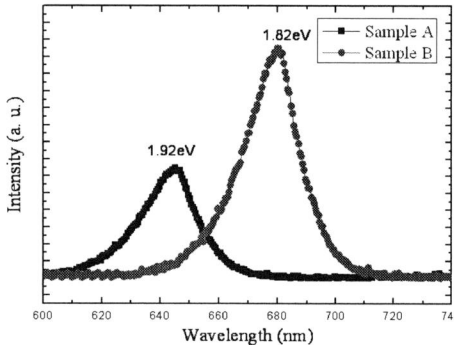

Fig. 4 PL of InGaP/GaAs dual-junction solar cells. Sample A: the cell design with a P^{++}-GaAs/N^{++}-InGaP TD grown on 10° off misorientation GaAs substrates; sample B: the cell design with a P^{++}-AlGaAs/N^{++}-GaAs TD grown on 10° off misorientation GaAs substrates.

IV. CONCLUSION

In conclusion, we have demonstrated that, compared with the cell design with a P^{++}-GaAs/N^{++}-InGaP TD (sample A), InGaP/GaAs dual junction solar cells with a P^{++}-AlGaAs/N^{++}-GaAs TD grown on 10° off GaAs substrates (sample B) exhibit superior photovoltaic conversion efficiency when operated at one sun. The cell design with a P^{++}-GaAs/N^{++}-InGaP TD grown on the 10° off GaAs substrate generates a disordering InGaP epitaxial layer during material growth, as shown in Fig. 4. This implies that the higher substrate

misorientation angle, such as 10° off, may result in a disordering InGaP epitaxial layer. However, the disordering InGaP structure did not appear in sample B. It is believed that the cell design with a P^{++}-AlGaAs/N^{++}-GaAs TD grown on the 10° off GaAs substrate can suppress the generation of the disordering InGaP structure during material growth. Besides, higher Rs for sample A also lead to the degradation of FF and J_{sc} of III-V solar cells, as shown in Fig. 2. These results suggest that InGaP/GaAs dual-junction solar cells with a P^{++}-AlGaAs/N^{++}-GaAs TD grown on 10° off misorientation GaAs substrates not only produce a high EQE but also generate a higher peak current density (J_{peak}) than that with a P^{++}-GaAs/N^{++}-InGaP TD grown on the 10° off GaAs substrate.

ACKNOWLEDGMENT

This work was supported by the NCTU-UCB I-RiCE program, National Science Council, Taiwan, under Grant No. NSC. 103-2911-I-009-302.

REFERENCES

[1] L. Ouattara, A. Mikkelsen, N. Skold, J. Eriksson, T. Knaapen, E. CÄavar, W. Seifert, L. Samuelson, and E. Lundgren, "GaAs/AlGaAs nanowire heterostructures studied by scanning tunneling microscopy," Nano Lett., Vol. 7, no. 9, pp. 2859-2864, June 2007.

[2] K. Takahashi, S. Yamada, Y. Minagawa, and T. Unno, "Characteristics of Al$_{0.36}$Ga$_{0.64}$As/GaAs tandem solar cell with pp-n-n structural AlGaAs solar cells," Sol. Energy Mate. & Sol. Cells, Vol. 66, no. 1-4, pp. 517-524, February 2001.

[3] Y. Kadoya, H. Noge, T. Someya, and H. Sakaki, "Effect of oxygen incorporation at AlGaAs/GaAs interfaces on the electrical properties of two-dimensional electron gas," Appl. Phys. Lett., Vol. 70, no. 5, pp. 595-597, November 1996.

[4] J. R. shealey and J. M. Woodall, "A new technique for gettering oxygen and moisture from gases used in semiconductor processing," Appl. Phys. Lett., Vol. 41, no. 1, pp. 88-89, April 1982.

[5] K. Tamamura, J. Ogawa, K. Akimoto, Y. Mori, and C. Kojima, "Carbon incorporation in metalorganic chemical vapor deposition (Al,Ga)As films grown on (100), (311)A, and (311)B oriented GaAs substrates," Appl. Phys. Lett., Vol. 50, no. 17, pp. 1149-1151, February 1987.

[6] T. E. Lamas, A. A. Ouivy, C. S. Sergio, G. M. Gusev, and J. C. Portal, "High mobility of a three-dimensional hole gas in parabolic quantum wells grown on GaAs(311)A substrates," J. Appl. Phys., Vol. 97, no. 7, pp.076107-1-076107-3, March 2005.

[7] H. W. Yu, E. Y. Chang, H. Q. Nguyen, J. T. Chang, C. C. Chung, C. I. Kuo, Y. Y. Wong, and W. C. Wang, "Effect of substrate misorientation on the material properties of GaAs/Al$_{0.3}$Ga$_{0.7}$As tunnel diodes," Appl. Phys. Lett., Vol. 97, no. 23, pp. 231903-1-231903-3, December 2010.

[8] B. E. Sernelius, "Band gap shifts in heavily doped n-type GaAs," Phys. Rev. B, Vol. 33, no. 12, pp. 8582-8586, December 1986.

[9] H. Murata, I. H. Ho, T. C. Hsu, and G. B. Stringfellow, "Surface photoabsorption studies of the chemical structure of GaInP grown by organometallic vapor phase epitaxy," Appl. Phys. Lett., Vol. 67, no. 25, pp. 3747-3749, October 1995.

[10] L. C. Su, I. H. Ho, and G. B. Stringfellow, "Kinetically controlled order/disorder structure in GaInP," Appl. Phys. Lett., Vol. 65, no. 6, pp. 749-751, May 1994.

Design and optimization of a Mach-Zehnder Interferometer (MZI) for optical modulators

Hanim Abdul Razak, Hazura Haroon
Faculty of Electronics and Computer Engineering (FKEKK)
Universiti Teknikal Malaysia Melaka (UTeM)
Melaka, Malaysia
hanim@utem.edu.my

P. Susthitha Menon, Sahbudin Shaari, Norhana Arsad*
Institute of Microengineering and Nanoelectronics (IMEN)
*Faculty of Engineering and Built Environment (FKAB)
Universiti Kebangsaan Malaysia (UKM)
Bangi, Malaysia

Abstract— This paper presents an investigation of the most influential parameters in designing a Mach Zehnder Interferometer (MZI) on Silicon-On-Insulator (SOI) using Taguchi method. The symmetric Y-junction is employed in designing the MZI. Meanwhile, parameters affecting the design of a MZI such as the area of the waveguide, offset in X direction (lx), offset in Z direction (lz) and the arm length were considered and varied for optimization purposes. Design and optimization of passive MZI structure is crucial prior to implementation of active devices.

Keywords— *MZI; Opti BPM; Taguchi method; SOI*

I. INTRODUCTION

Integrated optical devices on silicon have been commonly used in communications, interconnects, chemical and biochemical applications. The possibility of manufacturing the integrated optical circuit by utilizing the standard silicon fabrication process and the prospect of integrating optical and electrical functions to form a smart system are the main reason behind this phenomenon. Besides, these devices feature miniaturization, mechanical stability and freedom from electromagnetic interference (EMI) [1][1, 2]. In particular, Silicon-On-Insulator is becoming a favourite material for development of optical devices as the structure provides high refractive index contrast between silicon and the buried oxide layer (Δn~3.0)[2], [3].

The Mach Zehnder interferometer (MZI) is the most popular interference structure due to its simplicity both in design and fabrication, with the existence of the reference arm for compensation in common-mode effect [4]. Various applications of waveguide devices based on MZI configurations such as switches/modulators [5], [6], multi/demultiplexers [7], [8], splitters [9], [10], and sensors [11], [12], are possible and have been widely investigated. Therefore, the design and optimization of passive MZI structure is crucial prior to implementation of the abovementioned devices.

Configurations of integrated MZI usually comprises of a two back-to-back 3-dB couplers connected by a pair of widely separated waveguides. The symmetric Y-junction, directional coupler or multi-mode interferometer (MMI) can be utilized for this purpose. Design of a passive MZI based on tapered

MMI has been experimentally executed by Thompson et al with the extinction ratio (ER) of 32dB [9]. In this paper, MZI with symmetric Y-junctions on SOI is investigated and optimized.

Four parameters affecting the design of the MZI are considered and the most influential parameter will be determined using Taguchi method. Taguchi method offers a systematic application for designing and analyzing the product quality at the design stage. The optimized product utilizing Taguchi method is robust to the variation of experimental conditions and other noise factors. In addition, the designers will be able to determine which factors are affecting the product design the most and vice versa. Meanwhile, the signal-to-noise ratio (SNR) analysis allows the designer to find out the optimal parametric combinations of the product[13], [14]. This paper presents an investigation of the most influential parameter in designing an MZI based on Y-junctions by evaluating the extinction ratio of the devices. The design utilizes the finite difference beam propagating method (BPM) software.

II. THEORY AND MODELING

The modeling of the MZI was performed using a Finite Difference BPM which solves Maxwell's equations by using finite differences in place of partial derivatives. BPM is a step-by-step method of simulating the passage of light through any waveguiding medium. The design and simulation of the device were done in three dimension (3D) isotropic module of OptiBPM 9.0. The 3D dimensions are X-direction (vertical) for transverse direction, Y-direction for depth and Z-direction (horizontal) for propagation. The power in the output waveguide of the MZI for each experiment for a range of wavelength 1540-1610 nm, where TE polarized light was collected for determination of the extinction ratio for performance evaluation. The power in the output waveguide in the device was defined by producing power overlap of the current BPM field with the fundamental mode of the studied waveguide in the simulator.

Parameters affecting the design of a MZI such as the area of the waveguide, offset in the X direction (lx), offset in the Z direction (lz) and the arm length were considered and varied

Universiti Teknikal Malaysia Melaka (UTeM)

978-1-4799-5761-3/14 $31.00 © 2014 IEEE

for optimization purposes. The values of the parameters were varied according to Table 1. Meanwhile, Fig. 1 and Fig. 2 depict the schematic diagram of the MZI and the rib waveguide structure. The experiments were simulated utilizing the L9 orthogonal array in Taguchi method which requires three values of control factors with significant effect to the power output of the device. Therefore, the three level values in Table 1 were selected according to the L9 orthogonal array. Values of these levels were selected based on previous works. Taguchi method eliminates the necessity to run a large number of experiments by introducing a specially designed set array of experiments [13], [14].

Extinction ratio (ER) is a crucial parameter in analyzing the performance of an optical device. It is the ratio of two power levels and it is given by:

$$ER(dB) = 10 \log \frac{P_1}{P_2} \qquad (1)$$

where P1 is the input power and P2 is the output power.

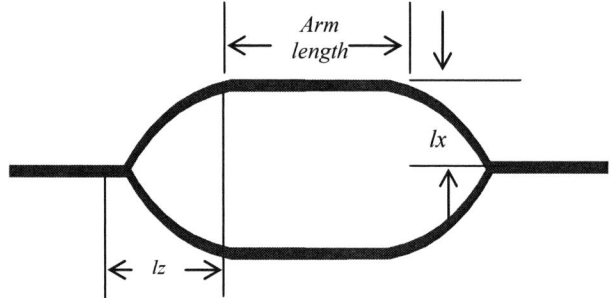

Fig. 1. Schematic diagram of MZI (top view).

TABLE 1 CONTROL FACTORS AND THEIR LEVELS

Symbol	Control factors	Level 1	Level 2	Level 3
A	Offset in X direction, lx (µm)	40	45	50
B	Offset in Z direction, lz (µm)	3000	5000	10000
C	Rib area (µm²)	2.0	2.5	3.3
D	Arm length (µm)	1000	2000	3000

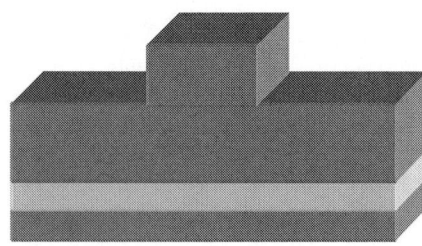

Fig. 2. Structure of rib waveguide (cross section of MZI).

III. RESULTS

The output light field propagation of the device is depicted in Figure 3. The light field distribution is at its best if there is no reflection occurring at the boundary and that the boundary condition is homogeneous [19]. Figure 4 shows the optical light propagation simulated through the device. A well confined light through the device ensures a high output power at the output arms.

Table 2 shows the combinations of experiment for different levels of control factors and the calculated extinction ratios. In order to determine the extinction ratio, the power output waveguide for each experiment were varied for a range of wavelength (1540-1610nm). The determined ER for each experiment were analyzed and processed using Taguchi method in order to predict the optimal design and to identify the most influential parameter in designing the device. In this paper, SNR of larger-the-best quality characteristics, η is performed. A larger SNR value represents better performance characteristics. SNR for 'larger the better' is given by:

$$\eta = -10 Log_{10} \left[\frac{1}{n} \Sigma \left(\frac{1}{Y_1^2} + \frac{1}{Y_2^2} + \ldots + \frac{1}{Y_n^2} \right) \right] \qquad (2)$$

where n is the number of tests in each experiment and Yi represents the experimental value of the output power [14].

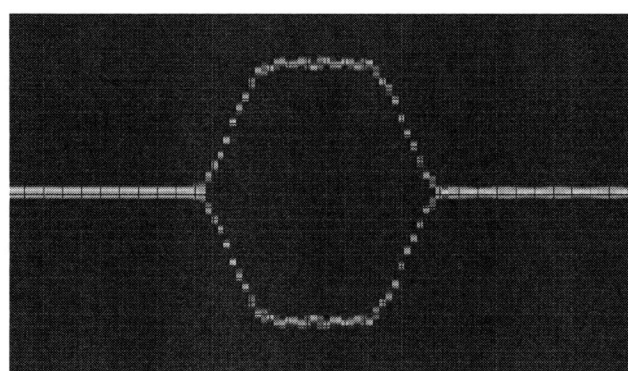

Fig. 3 Simulated optical light propagation through the MZI

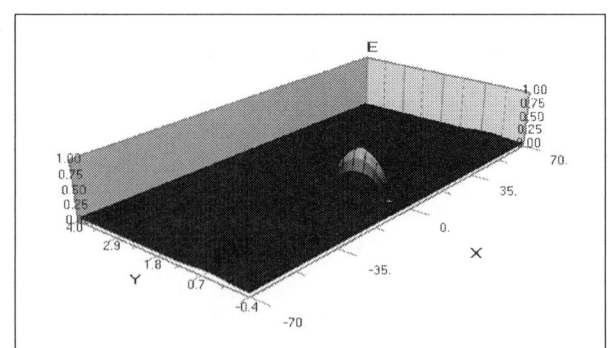

Fig. 4 Simulated output light field distribution at the end of propagation

The calculated SNR values are displayed in Table 2. Meanwhile, Fig. 5 shows the standardized effects of the ER of the device. The graph compares the relative magnitude and the statistical significance of all the main factors. It can be deduced that control factor C (rib area) has the most influence on the ER of the MZI by 49% followed by control factor D (arm length) by 20% then control factor B (lz) with 19% whereas the least contributing factor in the performance of the device is control factor A (lx) with 12%. In this analysis, larger rib area improves the ER of the device due to the fact that a larger rib area can propagate a higher intensity of light. Nevertheless, appropriate arm length must be selected in designing the MZI. Control factor B, lz determines the splitting angle between the MZI arms, therefore, smaller value of lz results in smaller splitting angle and thus reduces loss. Smaller value of lx are also desirable as shorter lx reduces the length of S-bend and reduces loss at the S-bend area.

In order to predict the optimum design for the device, Fig. 6 can be analyzed by selecting the highest level for each control factor. In this case, the optimum design is A3B1C3D3. The predicted optimum design was simulated and the SNR for the optimized design was obtained. To validate the conclusion drawn during the analysis, it is best to run a confirmation experiment. The SNR value obtained for the confirmation run is 13.1. Therefore, upon optimization, the SNR improves by 17.69%, in comparison to the highest SNR value prior to optimization (experiment 7) from Table 2.

TABLE 2 EXPERIMENTAL LAYOUT FOR L9 ORTHOGONAL ARRAY WITH DETERMINED EXTINCTION RATIOS AND SNRS

Experiment	Levels of control factors				Extinction	SNR (larger the better)
	A	B	C	D	Ratio (dB)	
1	1	1	1	1	1.53	3.69
2	1	2	2	2	1.16	1.29
3	1	3	3	3	2.76	8.82
4	2	3	3	3	2.50	7.96
5	2	2	3	1	2.42	7.68
6	2	3	1	2	0.15	-16.48
7	3	1	3	2	3.60	11.13
8	3	2	1	3	1.20	1.58
9	3	3	2	1	1.87	5.44

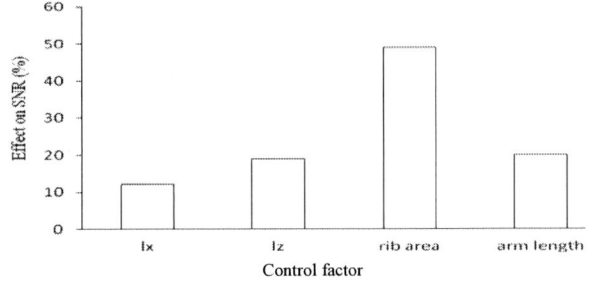

Fig. 5 Percentage of control factors effect on the device performance

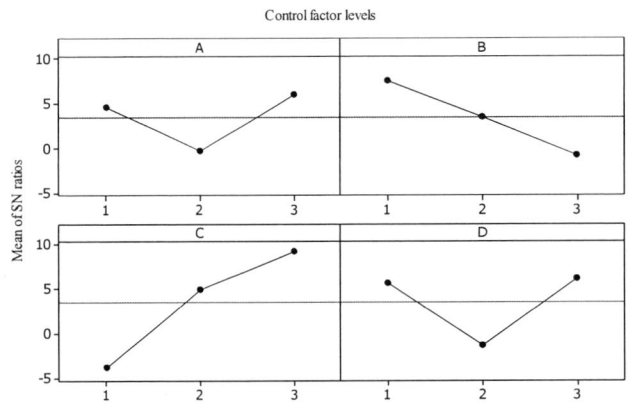

Fig. 6 Factor effect plot

IV. CONCLUSION

We have demonstrated and analyzed the effect of MZI design parameters on the ER of the device. Substrate area of the waveguide, offset in X direction (lx), offset in Z direction (lz) and the arm length were varied and the ER was extracted and compared. Using Taguchi method, the rib area was found to be the most influential design parameter. The finding of this work can be laid as a guidance in designing other active devices based on MZI.

ACKNOWLEDGMENT

The authors would like to thank Dr. Fauziyah Salehuddin and Prof. P. R. Apte for inputs on Taguchi's method. Universiti Kebangsaan Malaysia is acknowledged for the initial start-up of this research using grants GUP-2012-012 and Industri-2011-015. Universiti Teknikal Malaysia Melaka is acknowledged for funding the publication of this paper under grant PJP/2013/FKEKK(38B)/S01249.

REFERENCES

[1] B. Jalali and S. Fathpour, "Silicon Photonics," *J. Light. Technol.*, vol. 24, no. 12, pp. 4600–4615, Dec. 2006.

[2] A. Liu, L. Liao, D. Rubin, J. Basak, Y. Chetrit, H. Nguyen, D. W. Kim, A. Barkai, R. Jones, N. Elek, R. Cohen, N. Izhaky, and M. Paniccia, "Silicon photonic integration for high-speed applications," 2008, vol. 6898, p. 68980D–68980D–10.

[3] R. Soref, "The Achievements and Challenges of Silicon Photonics," *Adv. Opt. Technol.*, vol. 2008, pp. 1–7, 2008.

[4] R. W. Chuang, M.-T. Hsu, Y.-C. Chang, Y.-J. Lee, and S.-H. Chou, "Integrated multimode interference coupler-based Mach–Zehnder interferometric modulator fabricated on a silicon-on-insulator substrate," *IET Optoelectron.*, vol. 6, no. 3, p. 147, 2012.

[5] H. Hazura, A.R . Hanim, B. Mardiana, and P. S. Menon, "An analysis of silicon waveguide phase modulation efficiency based on

carrier depletion effect," *2010 IEEE Int. Conf. Semicond. Electron.*, pp. 348–350, Jun. 2010.

[6] A. Hanim, B. Mardiana, H. Hazura, and S. Saari, "On the modulation phase efficiency of a silicon p-i-n diode optical modulator," *Int. Conf. Photonics 2010*, pp. 1–3, Jul. 2010.

[7] M. Heid, S. Spalter, G. Mohs, A. Farber, and H. Melchior, "160 Gbit/s demultiplexing based on a monothically integrated Mach Zehnder Interferometer," in *IEEE 27th European Conference on Optical Communication*, 2001, pp. 82–83.

[8] J. . Viens, C. L. Callender, J. P. Noad, and L. Eldada, "Compact Wide-Band Polymer Wavelength-Division Multiplexers," *IEEE Photonics Technol. Lett.*, vol. 12, no. 8, pp. 1010–1012, 2000.

[9] D. J. Thomson, Y. Hu, G. T. Reed, and J. Fedeli, "Low Loss MMI Couplers for High Performance MZI Modulators," vol. 22, no. 20, pp. 1485–1487, 2010.

[10] S. K. Raghuwanshi, V. Kumar, D. Chack, and S. Kumar, "Propagation Study of Y-Junction Optical Splitter Using BPM,"

2012 Int. Conf. Commun. Syst. Netw. Technol., pp. 625–629, May 2012.

[11] D. a May-Arrioja, P. LiKamWa, J. J. Sánchez-Mondragón, R. J. Selvas-Aguilar, and I. Torres-Gomez, "A reconfigurable multimode interference splitter for sensing applications," *Meas. Sci. Technol.*, vol. 18, no. 10, pp. 3241–3246, Oct. 2007.

[12] F. Prieto, B. Sepulveda, A. Calle, and A. Liobera, "An integrated optical interferometric nanodevice based on silicon technology for biosensor applications," *Nanotechnology*, vol. 14, pp. 907–912, 2003.

[13] M. S.Phadke, *Quality engineering using robust design.* Prentice Hall, 1989.

[14] F. Salehuddin, I. Ahmad, F. A. Hamid, and A. Zaharim, "Application of Taguchi Method in Optimization of Gate Oxide and Silicide Thickness for 45nm NMOS Device," *Int. J. Eng. Technol. IJET-IJENS*, vol. 9, no. 10, 2009.

An Ultra-low Power and Area Efficient 10 bit Digital to Analog Converter Architecture

Iffa Binti Sharuddin and L. Lee
Faculty of Engineering
Multimedia University, Cyberjaya Campus
Selangor, Malaysia
iffa1305@gmail.com, linilee@mmu.edu.my

Abstract—**An ultra-low power and area efficient successive approximation register (SAR) analog-to-digital converter (ADC) is presented. To achieve ultra-low power performance, a digital-to-analog converter (DAC) architecture is proposed that combined a 4-bit thermometer coded and a 6-bit C-2C array to form a 10-bit DAC. Thereby, power consumption and area of the design are drastically reduced by virtue of lower switching activity and smaller size capacitor array. Add on to that, the architecture also has better linearity. The proposed 10-bit DAC is designed and simulated in a 0.18 μm CMOS process. Simulation results show that it only consumed 1.74 nW at 1.5 V power supply.**

Keywords—analog-to-digital converter; successive approximation

I. Introduction

Analog-to-Digital Converter (ADC) is the key to mixed signal System-on-Chips (SOCs) in that it provides the interface between the analog world and digital processing. There are overwhelming type of ADCs exist in the market today. The most popular ADC architectures available today include the successive approximations register (SAR) ADC, flash ADC, pipelined ADC and sigma-delta ($\Sigma\Delta$) ADC. SAR ADC is very suitable for large-scale wireless sensor networks since it has advantages in term of speed, resolution and power consumption. These characteristics are also highly sought after in the physiological applications. SAR ADC has a numbers of advantages compared to other Analog-to-Digital Converter (ADC) design. SAR ADC consists of one comparator in the whole system and it can achieve low power consumption in the range of nano-Watt [1]. It has advantages of zero-cycle latency, low latency time, high accuracy, low power and easy to use.

This paper proposes an ultra-low power digital-to-analog converter (DAC) that is designed in 0.18 μm CMOS technology. The proposed 10-bit DAC consists of a 4-bit thermometer coded DAC and a 6-bit C-2C array, combined by using segmentation. The size of the C-2C DAC increases linearly with the resolution, thus occupying less area than a binary-weighted DAC. Moreover, the capacitor sizes are fixed and small, so it has the same capacitive load, thus easing the switch design as well as the layout, since the entire topology is composed of replicas of unit C and 2C capacitor block. There

is no need for charging or discharging large capacitors, making the C-2C DAC faster than a binary-weighted capacitor. On the other hand, thermometer coded DAC has good linearity and less glitches [2]. To improve the linearity of the DAC, thermometer coded DAC is used in the MSB part. Combining these two designs, an ultra-low power 10-bit DAC has been designed.

II. Design Methodology

The proposed SAR ADC based on the C-2C array is used instead of the binary weighted capacitor (BWC) [3]. Fig. 1 shows the proposed DAC circuit. The drawbacks of the binary-weighted capacitor array are large area and high power consumption [4]. The total capacitor value increases exponentially with the number of bits and the minimum value of unit-capacitor is restricted by kT/C noise. Add on to that, the settling time of DAC increases in the binary weighted capacitor. To solve this problem, C-2C array DAC has been used whereby the values of the capacitors are drastically reduced. This type of DAC can achieve higher speed and less power. In this context, the power consumption increases linearly while in BWC, the power rises exponentially. Unit of C-2C can generate accurate reference capacitance but with the cost of large output impedance and error equivalent to input capacitance of comparator being introduced [4]. The objective of this proposed DAC design is to have 10 bit resolution achieving ultra-low power consumption at nano-Watts range.

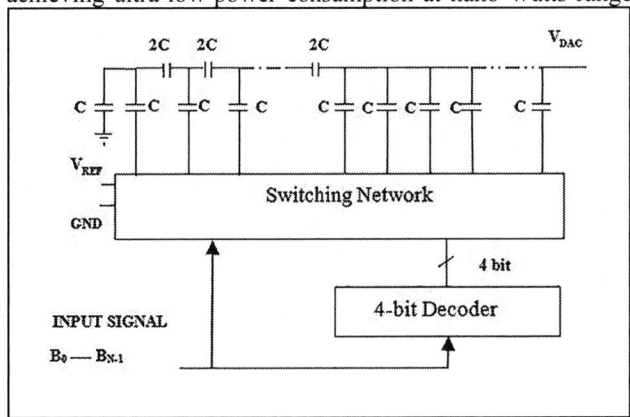

Fig. 1. Combination of C-2C array and thermometer coded DAC

However, the linearity of the C-2C architecture is affected by parasitic capacitances on the interconnecting nodes of the C-2C DAC. This cause the resolution of the architecture is limited to 4 or 6 bits only [5]. Also, monotonicity cannot be guaranteed for these high resolution C-2C DAC. The solution is to design the first MSB bits in thermometer coded architecture and the lower bits in the C-2C array architecture. So, the first 6 bit LSB is implemented by using C-2C and the last 4-bit MSB is realized by using thermometer coded DAC. The thermometer coded DAC is well known for its monotonic transfer, low glitch noise and good linearity.

III. Circuit Description

A. Design of Thermometer Coded DAC

Fig. 2 shows the example of 3-bit thermometer coded [2]. As shown in the figure, all unit capacitor is being controlled by the signal from the binary-to-thermometer decoder. Each switch circuit that is connected to each unit capacitor will response to the signal from the decoder and produce the desired output. When the digital input increases by 1LSB, one capacitor is charged. The analog output increases as the digital input increases. With this way, monotonicity is guaranteed. Thermometer codes have more bits than their counterparts especially when a bit is large, however they have advantages of changing one bit at a time to avoid glitches [6]. These are the advantages of thermometer-coded (or unary-weighted capacitor (UWC)) based architecture.

All Boolean algebraic equation of the decoder can be constructed by using DeMorgan's Law. The decoder would be designed and implemented by using minimum number of logic gates based on the law. The combination of logic gates of NAND, NOR, and NOT are used to implement the decoder. There are 15 outputs with the first combination 0000 produces no output [7]. The total logic gates used in this decoder is reduced to 39 gates with the use of DeMorgan's Law compared to normal Boolean algebraic equation of the decoder.

Total capacitor of thermometer coded DAC can be calculated by using (1). For a 4-bit design, thermometer coded DAC has a total of 15 unit capacitors.

$$\text{Total capacitance} = C_{unit}(2^N-1) \qquad (1)$$

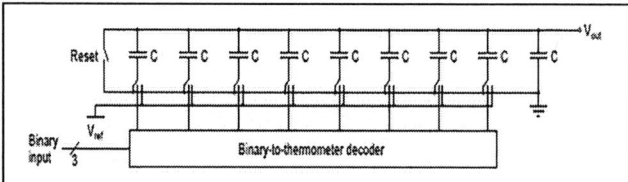

Fig. 2. Example of 3-bit thermometer coded DAC [2]

B. Design of C-2C Array

C-2C array architecture is preferred compared to binary weighted capacitor as it has more relax criteria in capacitor matching since it is composed of unit capacitances, C and 2C

only. C-2C array provides higher speed and consumes lesser power than binary weighted capacitor. Fig. 3 shows the C-2C array circuit designed. The drawback of this design is that it has high output impedance and this will introduce error equivalent to input capacitance of comparator [4]. This drawback can be mitigated by using other architecture in the MSB part, for example thermometer coded DAC. The total capacitors used for 6-bit C-2C array DAC design is equaled to 17 unit capacitors based on (2) below.

$$\text{Total capacitance} = C_{unit}(3N-1) \qquad (2)$$

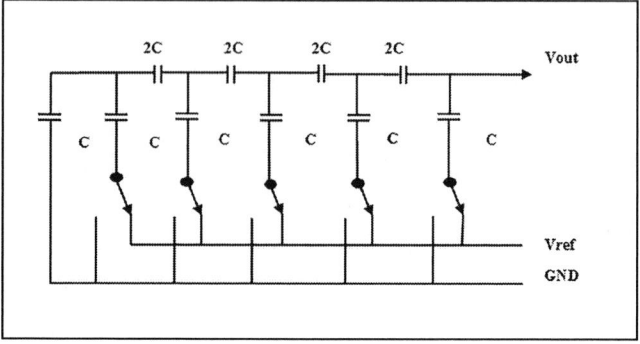

Fig. 3. 6-bit C-2C array DAC circuit design

C. Design of Switch

The combination of complementary NMOS and PMOS switches were chosen for Vref switching. CMOS transmission gate (TG) has the lowest on-resistance and can pass full range of signal. NMOS switch is used for ground. The dummy switch technique is added to the switch design to reduce charge injection and clock feed through effect. As shown in Fig. 4, minimum sized PMOS and NMOS devices have been employed for all switches design of the DAC.

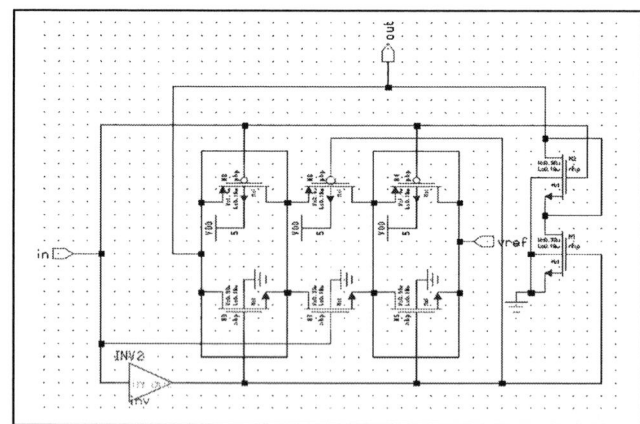

Fig. 4. Dummy switch for one bit DAC

IV. Proposed Design

For 10-bit DAC, we combined both architectures; thermometer coded and C-2C array by using segmentation. In order to have monotonic transfer function, thermometer coded DAC is used for MSB part while C-2C array is used for LSB part. To have balance capacitor on both MSB and LSB part, we choose MSB part to be equaled to 6 bit while the LSB part is set equaled to 4 bit. Therefore, 15 similar capacitors are selected with each size of C_{unit} and it is split with 17 capacitor sizes of C and 2C. Both architectures were combined by using attenuation capacitor, C_A technique. To have balance weight capacitor on both sides, the value for C_A is set equaled to 2C. So, the total of the capacitor in this proposed architecture is equaled to $34C_{unit}$. Verification through simulation was performed and desired output is obtained as shown in Section V.

V. Performance Result

Input stimulus in the form of binary coded 10-bit data is being fed to the converter. The total of 1024 combination of digital signal is converted into the correspondent "staircase" voltage levels. The minimum size of unit capacitor, C_{unit} equivalent to 1 fF is used. Hence, the total capacitance of the proposed architecture is equaled to 34 fF. Based on simulation, the power consumption of the DAC is equaled 1.74 nW only at the voltage supply of 1.5V.

A. Test bench for 10 bit DAC

Fig. 5 is the schematic circuit for the proposed 10-bit DAC design. The power supply (VDD) of the DAC is equaled to 1.5V. The voltage reference (Vref) is equaled to VDD. The period of the clock is equaled to 125 ns which is approximately equivalent to 8 MHz

B. DAC performance

Fig. 6 shows individual steps of DAC outputs along with the input digital signal b1 − b3. The calculated value of ideal V_{LSB} is equaled to 1.464 mV by using (3). The measured step value of individual bit is equal to 1.47 mV as calculated using (4). The value of the measured individual DAC step is almost equal to ideal V_{LSB} value and this shows that monotonicity of the 10-bit DAC is guaranteed.

$$V_{LSB}=V_{REF}/2^N \qquad (3)$$

$$\text{Step Value } [n] = V[n] − [n−1] \qquad (4)$$

Fig. 7 shows the DAC cycling through all of its steps. Good linearity of the DAC output voltage level was observed as in Fig. 7.

C. Switch Performance

Fig. 8 shows the comparison of glitch at mid code transition with and without the use of dummy switch in the DAC design. The dot line represents the switch without the dummy switch and the bold line as with dummy switch. As shown in the Fig. 8, the amplitude of the glitch is greatly reduced when the dummy switch in Fig. 4 is employed for each bit of DAC. However, the glitch is the main limitation for this work as the work to improve the glitches is still in progress.

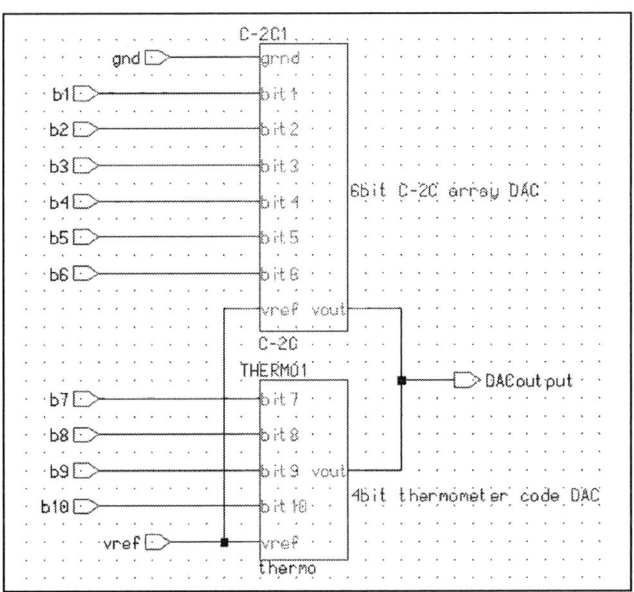

Fig. 5. Schematic circuit for proposed 10-bit DAC design

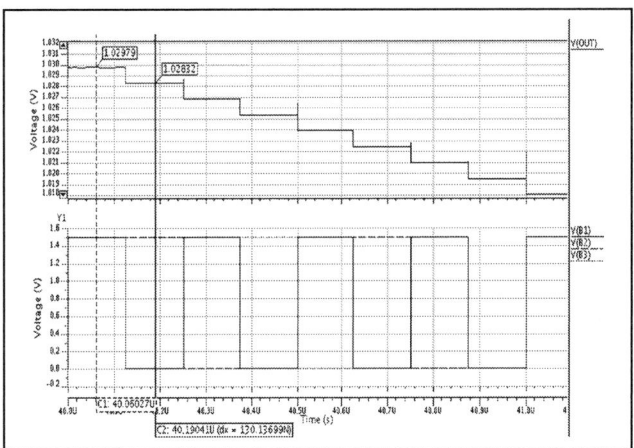

Fig. 6. Individual DAC step response

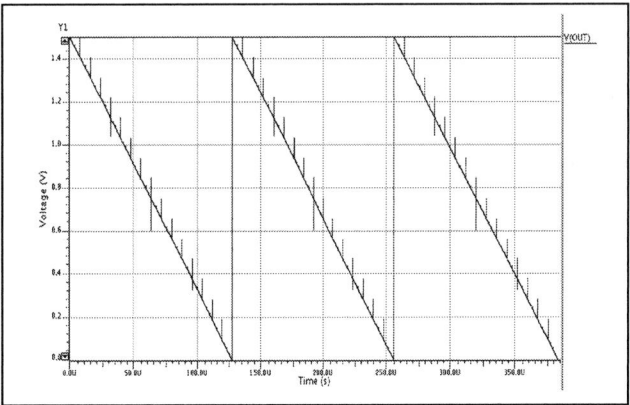

Fig. 7. DAC ramp response

TABLE II.	COMPARISON OF AREA CONSUMPTION
DAC Architecture	Area Consumption
Charge Redistribution [8]	$2^N A_{unit}$
Split Capacitor [9]	$2^N A_{unit}$
Segmentation (BWC and UWC) [3]	$2^{NLSB} A_{unit} + (2^{NMSB} - 1) A_{unit}$
Capacitor Based Hybrid (C-2C and BWC) [11]	$(3N_{LSB} + 1) A_{unit} + 2^{NMSB} A_{unit}$
This work	$(3N_{LSB} + 1) A_{unit} + (2^{NMSB} - 1) A_{unit}$

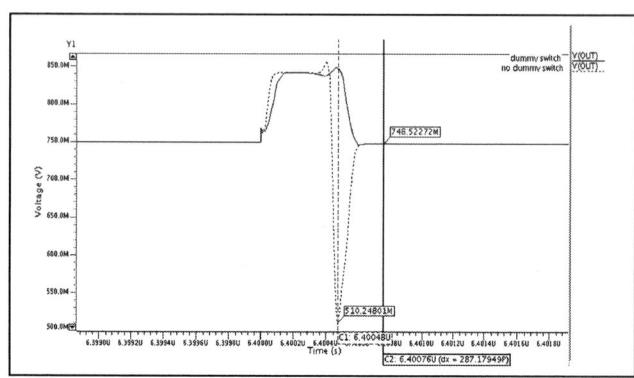

Fig. 8. Glitch at mid transition

VI. Comparison with Other DACs

In this section the present work is compared with the other published DAC designs. All compared designs with relevant parameters are presented in Table I and Table II.

In Table I, the conventional charge redistribution technique for the DAC part is presented [8]. The major limitation of this architecture is that the total array rises exponentially with the resolution. Splitting method in [9] had been proposed to overcome this problem. Although the split method has ultra-low power consumption, they are very sensitive to parasitic capacitance [10]. Other than that, segmentation technique which combined binary weighted capacitor with unary weighted capacitor has been used [3]. A hybrid DAC that combined two different architectures such as C-2C ladder and binary weighted capacitors has been proposed in [11]. This hybrid was designed to have advantages of each DAC. As for this proposed work, the combined architecture of C-2C and UWC has shown significant reduction in terms of power with only 1.74 nW consumption at 1.5V power supply. All the DAC compared in the Table I is based on pre layout simulation except for DAC presented in [8]. From Table II, this work has the lowest total capacitor among other DAC architectures. This architecture has advantages in terms of area consumption and low power dissipation.

TABLE I.   COMPARISON OF DACS PERFORMANCE

Ref.	DAC Architecture	Tech [μm]	Supply Voltage [V]	N-bit	Fs (kS/s)	Pd [nW]
[8]	Charge Redistribution	0.13	1.0	10	1	33
[9]	Split Capacitor	0.065	0.7	10	1	0.75
[3]	Segmentation (BWC and UWC)	0.13	1.0	10	100	80
[11]	Capacitor-based Hybrid (C-2C and BWC)	0.18	1.0	8	10	60
This Work	Sub-DAC (C-2C and UWC)	0.18	1.5	10	8	1.74

VII. Conclusion

In this paper a 10-bit SAR ADC has been designed by using the segmentation technique of combining a 4-bit thermometer coded DAC with a 6-bit C-2C array DAC. Utilizing the benefits of the architecture; good linearity and area efficient, the design has been simulated in 0.18μm CMOS technology. The total power consumption reported is equaled to 1.74 nWatts. This ultra-low power consumption DAC is highly desirable especially in physiological applications.

References

[1] Pouya Kamalinejad, Shahriar Mirabbasi and Victor C.M. Leung, "An Ultra-Low-Power SAR ADC with an Area-Efficient DAC Architecture," *IEEE Int. Symp. Circuits and Systems*, May 2011, pp. 13-16.

[2] Boschker, Frank B, " Design of a 12bit 500Ms/s standalone charge redistribution Digital-to-Analog Converter", M.S. thesis, University of Twent, 2008.

[3] R. Lotfi, R. Majidi, M. Maymandi-nejad, and W.A. Serdijn, "An ultra-low-power 10-Bit 100-kS/s successive-approximation analog-to-digital converter," *IEEE Int. Symp. Circuits and Systems*, pp.1117-1120, 24-27 May 2009.

[4] Hoonki Kim, Young-Jae Min, Yonghwan Kim and Soowon Kim, "A low power onsumption 10-bit rail-to-rail SAR ADC using a C-2C capacitor array," *IEEE Int. Conf. Electron Devices and Solid-State Circuits*, pp.1-4, 8-10 Dec. 2008.

[5] H. Balasubramaniam, W. Galjan, W.H. Krautschneider and H. Neubauer, "12-bit hybrid C2C DAC based SAR ADC with floating voltage shield," *3rd Int. Conf. Signals, Circuits and Systems*, pp.1-5, 6-8 Nov. 2009.

[6] Yang Lin, "Design of a 8-bit CMOS Unit-Element Current-Steering Digital-to-Analog Converter", M.S. thesis, University of Maine, 2008.

[7] J. Huynh, B. Ngo, M. Pham and L. He, "Design of a 10 bit TSMC 0.25 μm CMOS digital to analog converter," Sixth Int. Symp. Quality of Electric Design, pp.187-192, 21-23 March 2005.

[8] Dai Zhang, A. Bhide and A. Alvandpour, "A 53-nW 9.1-ENOB 1-kS/s SAR ADC in 0.13- μm CMOS for Medical Implant Devices," *IEEE Journal of Solid-State Circuits*, vol.47, no.7, pp.1585-1593, July 2012.

[9] Dai Zhang and A. Alvandpour, "A 3-nW 9.1-ENOB SAR ADC at 0.7 V and 1 kS/s," *Proc. the ESSCIRC*, pp.369-372, 17-21 Sept. 2012.

[10] A. Rodriguez-Perez, M. Delgado-Restituto and F. Medeiro, "A low-voltage low-power successive approximation reconfigurable ADC based on SC techniques," *Ph.D. Research in Microelectronics and Electronics Conf., PRIME 2009*, pp.8-11, 12-17 July 2009.

[11] Sang-Hyun Cho, Chang-Kyo Lee and Jong-In Song, "Design of a 1-Volt and μ-power SARADC for Sensor Network Application,". *IEEE Int. Symp. Circuits and Systems*, pp.3852-3855, 27-30 May 2007.

Implementation of Low Power Compressed ROM for Direct Digital Frequency Synthesizer

Salah Hasan Alkurwy[1], Sawal Hamid Md Ali[2]

[1,2] Department of Electrical, Electronic & System Engineering, Faculty of Engineering & Built Environment University Kebangsaan Malaysia (UKM), 43600 UKM, Bangi, Selangor, Malaysia. E-mail: mr.salah65@yahoo.com E-mail: sawal@eng.ukm.my

Md. Shabiul Islam[3]

[3]Insitute of Microengineering and Nanoelectronics (IMEN), University Kebangsaan Malaysia (UKM), 43600 UKM, Bangi, Selangor, Malaysia. E-mail: shabiul@ukm.edu.my

Abstract—**A Low Power Compressed ROM Look-up table has been presented in this paper to achieve low power consumption of the direct digital frequency synthesizer as well small core size. The quarter wave symmetry technique is used to store only one quarter of the sine wave. The suggested 12-bit compressed ROM designed consists of three 4-bit sub-ROMs based on an angular decomposition technique and trigonometric identity. Exploiting the advantages of sine-cosine symmetrical attributes together with XOR logic gates, one sub-ROM block can be removed from the design. These techniques, compressed the ROM into 368 bits. The ROM compressed ratio is 534.2:1, with only two adders, two multipliers, and XOR-gates with high frequency resolution of 0.029 Hz.**

Keywords; direct digital frequency synthesizer (DDFS); ROM Look-up table (ROM LUT), phase accumulator (PA).

I. Introduction

DDFS or direct digital synthesis is a system that generates time-varying digital waveform and later converts it to an analog signal. The first DDFS designed by Tierney in 1971[1], consists of a phase accumulator(PA) to generate 0π to 2π digital values, phase-to-sinusoid amplitude converter or ROM LUT to store the addressing data for phase mapping, Digital-to-analog converter (DAC) to convert the sine amplitude into analog sinusoidal signal, and LPF to remove the unwanted harmonic signals.

The PA is a counter that linearly increases (sawtooth) with each timing clock based on the N-bit input, to the extent of $2^N - 1$, and resets back to zero. The PA increments the LUT address that corresponds to the increment of the angle. The frequency control word (FCW) controls the incremental step in the PA and, therefore, controls the address reading of LUT. The output frequency of the DDFS f_{out} can be defined as follows:

$$f_{out} = (M \times f_{clk})/2^N \qquad (1)$$

Where M is the frequency tuning word, f_{clk} is the clock frequency, and N is the N-bit PA. The frequency resolution of the DDFS is determined by the clock frequency f_{clk} and the N-bit PA, as shown in (2):

$$f_{Res} = f_{clk} / 2^N \qquad (2)$$

The phase-to-amplitude converter, also called ROM, is a sine LUT to convert the phase output of the PA to the corresponding amplitude of the phase in the LUT. The ROM LUT contains a memory address of one cycle (0 to 2π) of the waveform to be generated. The size of the ROM is 2^N words, and the waveform purity depends on the N-bit. However, the high accuracy requires a large ROM size.

A simple technique to compress the ROM is to use the quarter symmetry of the sine wave to store only a quarter (0: $\pi/2$) of the sine waveform and two MSBs of the phase output values to reconstruct the full sine wave. Multiple methods and techniques were used to compress the quarter ROM.

The angular decomposition technique was suggested in the first DDFS system designed by Tierney [1], wherein complex multiplications were conducted on the ROM data because of the exponential sum of values. The complex multiplications used in this approach made it undesirable for an application. Two segmented angles, coarse (A) and fine (B), based on trigonometric identities and with simple multiplication were used by Hutchinson [2], for quarter ROM design. The method used by Sunderland in [3], wherein the quarter ROM LUT was partitioned into three sub-ROMs (A, B, and C) based on the trigonometric relation. The Sunderland approach was improved in [4], based on the sine-phase difference algorithm.

The polynomial approximation method was used in the DDFS design to optimize the SFDR. In [5], De Caro proposed a dual-slope technique to optimize the piecewise linear approximation, for phase to sine mapping. This technique improved spectral purity and ROM size. The author suggested another DDFS architecture in [6] based on a piecewise linear approximation technique with non-uniform segment length to the input of three groups of multiplexers, to maximize the SFDR and reduce the size of the ROM coefficients.

This paper presents a low power compressed ROM design for 32-bit DDFS system design, based angular decomposition technique and trigonometric identity.

II. Compressed ROM Design

Quarter-wave symmetry technique is must use in ROM look-up table design to save 75% of the ROM size. This technique is implemented by storing only one quarter (0: $\pi/2$) of the sine waveform and two most significant bits (MSB) from the phase accumulator are used to reconstruct the full sine wave. The angular decomposition technique used to approximate the quarter phase angle, and thus reduce the ROM size.

The basic concept of the angular decomposition technique; is to partition the quarter ROM into three sub-ROMs (A, B, and C), such that A < ($\pi/2$), B < ($\pi/2$) * (1/2A), and C < ($\pi/2$) * (1/2A +B).

According to the trigonometric relation, the sine wave function given in:

$$Sin\ (A + B + C) = Sin\ (A + B)Cos\ C +$$
$$+Cos\ A\ Cos\ B\ Sin\ C - Sin\ A\ SinB\ Sin\ C \qquad (1)$$

Where

$$Sin\ (A + B) = Sin\ A\ Cos\ B + Cos\ A\ Sin\ B \qquad (2)$$

The equation for cos B calculation is given by

$$Cos\ B = Cos((\pi/2 \times [0 : 2^B - 1)]/2^B) \times (1/2^B)) \quad (3)$$

Using three equal 4-bit sub-ROMs in the proposed design make the cos B values ($cos\ B = cos\ (0.046019) = 0.999999677 \cong 1$).

Therefore, equation (2) can be reduced as given by (4)

$$Sin\ (A + B) = Sin\ A + Cos\ A\ Sin\ B \qquad (4)$$

With the relative values of A, B, and C, equation 1 may be re-written as

$$Sin\ (A + B + C) = Sin\ A + Cos\ A\ Sin\ B + Cos\ A Sin\ C \quad (5)$$

As shown in equation 5, the compressed ROM look-up table, involves four sub-ROMs, for Sin A, Cos A, Sin B and Sin C. The ROM size of the conventional ROM compressed to 544 bit. Two multiplier adders and two adders need to achieve the required ROM design.

The block circuit diagram of the compressed ROM LUT design based angular decomposition technique (equation 5), is shown in Fig. 1.

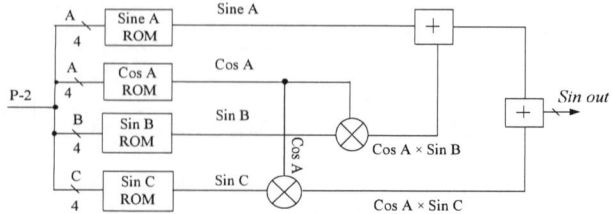

Fig. 1. The architecture of conventional compressed ROM look-up table design based angular decomposition technique

The calculated results of (sin-$cos\ A$) shows that they are inversely symmetry. Based on this, $cos\ (A)$ can be obtained by complementing the sin (A) values and this is achieved by connecting the $sin\ (A)$ output and the high V_{CC} to the XOR logic gate inputs. In this way, only one addressing sub-ROM is needed for $sin\ A$ and $cos\ A$ values, and the new ROM circuit compressed to 368 bit with Two multiplier adders and two adders .

The architecture of new compressed ROM LUT design after hardware reduction is shown in Fig. 2.

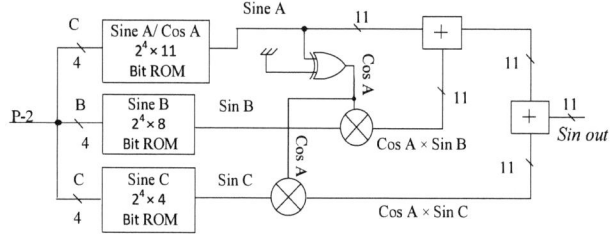

Fig. 2. The architecture of new compressed ROM

III. DDFS Architecture

The phase accumulator designed with 32-bit, in the presented DDFS system, to achieve high frequency resolution, whereas truncates to reasonable 14-bit phase out, due the ROM size reduction. The 32-bit PA together with new compressed ROM LUT circuit, combined with digital-to-analog block to achieve the desired sine wave output. The proposed DDFS, which consists of PA. Compressed ROM LUT and DAC using the wave symmetry technique is shown in Fig. 3.

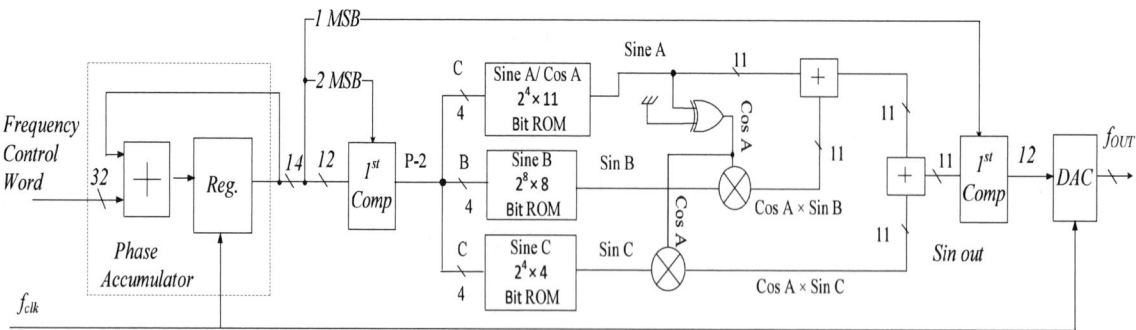

Fig. 3   The direct digital frequency synthesizer (DDFS) architecture

IV. Result and Discussion

The proposed DDFS design was coded in Verilog HDL, successfully simulated in ALTERA Quartus II software, and implemented with a Cyclone III FPGA kit board, as shown in Fig. 4

The simulation result shows that the 07FFFFFF hexadecimal format of the frequency word input FCW produce the PA (sawtooth) phase output. The Q2 half cycle (o –π) waveform shows at the compressed ROM output and the full sine wave (o –2π) is shown as a sine wave output of the proposed DDFS system.

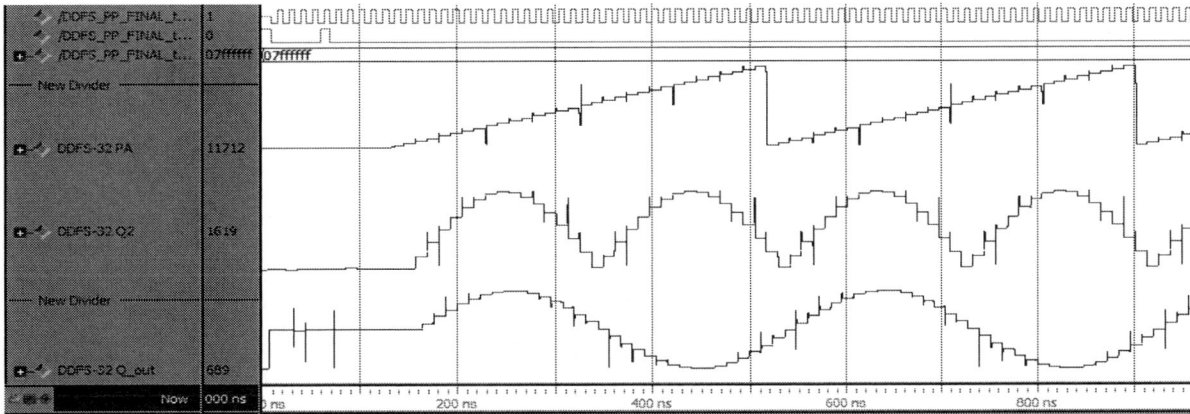

Fig. 4.   The gate level simulation of the direct digital frequency synthesizer (DDFS) system design

The 32-bit DDFS has been implemented on the Cyclone III FPGA kit board. The truncated 14 bits required 196608 ($2^{14} \times 12$) to draw the quarter sine wave. The applied of the quarter-wave symmetry and an angular decomposition technique based on trigonometric identity, reduce the ROM size to only 368 bits, with 534.2:1 ratio.

The measured DDFS output waveforms and spectra for different clock frequencies illustrates the purity of the sine wave output as shown in Fig. 5, for (FCW = 2^{28}-1 = 0FFFFFFF) in hexadecimal format, and F_{clk} = 125 MHz (Cyclone III kit board clock frequency). The output frequency that was calculated is F_{out} = ($FCW / 2^{N}$) × F_{clk} = (2^{28}-1 / 2^{32}) × 125*10^6 = 7.812499 MHz. The measured output frequency from oscilloscope is 7.813 MHz and this is closely matched with the calculation result.

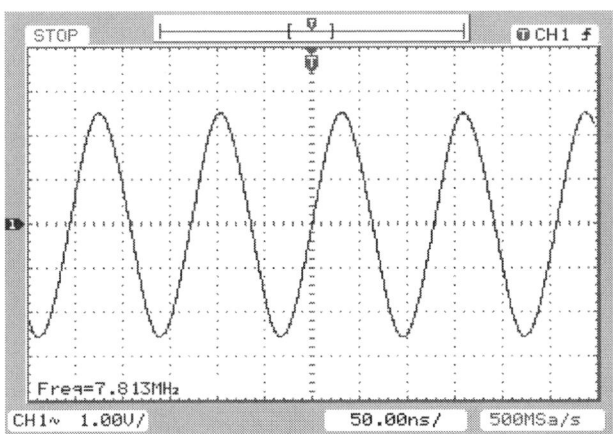

Fig. 5   sine wave output signal of the DDFS

Table 1 shows the comparison of the ROM size used in the proposed DDFS system design with some other parameters. The table shows that the proposed design has the smallest ROM size.

TABLE I. COMPARISON BETWEEN ROM SIZES WITH PREVIOUS DDFS WORKS

Ref.	PA (bit)	Truncated phase (bit)	Amplitude phase (bit)	ROM size (bit)	Truncation Ratio
[3]	20	14	12	3328	59.1 : 1
[4]	31	15	14	3072	149.3 : 1
[7]	28	14	12	832	236 : 1
[8]	24	14	12	480	409.6 : 1
[9]	32	14	12	2176	90.35 : 1
[10]	15	15	15	1216	404 : 1
[11]	32	16	14	1664	551.3 : 1
[12]	24	14	12	672	292.5 : 1
This work	32	14	12	368	534.2 : 1

V. Conclusion

A 32-bit DDFS system design presents in this paper. The quarter wave symmetry and angular decomposition technique based on trigonometric identity were used to reduce to compress the ROM. Exploiting the advantages of sine-cosine symmetrical attributes together with XOR logic gates, one sub-ROM block can be removed from the proposed ROM design, and the ROM compressed to only 368 with a 534.2:1 ratio. The proposed system was successfully simulated in ALTERA Quartus II software. The PA and compressed ROM look-up table have been implemented on the Cyclone III FPGA kit board and connect to the DAC circuit. The DDFS system performance has been measured and verified using oscilloscope.

References

[1] J. Tierney, C. M. Rader, and B. Gold, "A digital frequency synthesizer," *Audio and Electroacoustics, IEEE Transactions on,* vol. 19, pp. 48-57, 1971.

[2] H. B. H, Jr. (1975) Contemporary frequency synthesis techniques. *IEEE Press.* pp. 25–45.

[3] D. A. Sunderland, R. A. Strauch, S. S. Wharfield, H. T. Peterson, and C. R. Cole, "CMOS/SOS frequency synthesizer LSI circuit for spread spectrum communications," *Solid-State Circuits, IEEE Journal of,* vol. 19, pp. 497-506, 1984.

[4] H. T. Nicholas, H. Samueli, and B. Kim, "The optimization of direct digital frequency synthesizer performance in the presence of finite word length effects," in *Frequency Control Symposium, 1988., Proceedings of the 42nd Annual,* 1988, pp. 357-363.

[5] D. De Caro and A. G. M. Strollo, "High-performance direct digital frequency synthesizers in 0.25 μm CMOS using dual-slope approximation," *Solid-State Circuits, IEEE Journal of,* vol. 40, pp. 2220-2227, 2005.

[6] D. De Caro, N. Petra, and A. G. M. Strollo, "Direct Digital Frequency Synthesizer Using Nonuniform Piecewise-Linear Approximation," *Circuits and Systems I: Regular Papers, IEEE Transactions on,* vol. 58, pp. 2409-2419, 2011.

[7] F. Curticapean and J. Niittylahti, "A hardware efficient direct digital frequency synthesizer," in *Electronics, Circuits and Systems, 2001. ICECS 2001. The 8th IEEE International Conference on,* 2001, pp. 51-54 vol.1.

[8] D. De Caro and A. G. M. Strollo, "High-performance direct digital frequency synthesizers using piecewise-polynomial approximation," *Circuits and Systems I: Regular Papers, IEEE Transactions on,* vol. 52, pp. 324-337, 2005.

[9] Y. Byung-Do, C. Jang-Hong, H. Seon-Ho, K. Lee-Sup, and Y. Hyun-Kyu, "An 800-MHz low-power direct digital frequency synthesizer with an on-chip D/a converter," *Solid-State Circuits, IEEE Journal of,* vol. 39, pp. 761-774, 2004.

[10] L. S. J. Chimakurthy, M. Ghosh, F. F. Dai, and R. C. Jaeger, "A novel DDS using nonlinear ROM addressing with improved compression ratio and quantization noise," *Ultrasonics, Ferroelectrics and Frequency Control, IEEE Transactions on,* vol. 53, pp. 274-283, 2006.

[11] F. Babak and P. Keshavarzi, "A Novel DDFS Based on Trigonometric Approximation with a Scaling Block," in *Information Technology: New Generations, 2009. ITNG '09. Sixth International Conference on,* 2009, pp. 102-106.

[12] D. De Caro, N. Petra, and A. G. M. Strollo, "Reducing Lookup-Table Size in Direct Digital Frequency Synthesizers Using Optimized Multipartite Table Method," *Circuits and Systems I: Regular Papers, IEEE Transactions on,* vol. 55, pp. 2116-2127, 2008.

Adhesion Enhancement for Electroless Plating on Mold Compound for EMI Shielding with Industrial Test Compliance

Tai, Min Fee[1,2]*, Lee, Swee Kah[3], Goh, Soon Lock[3], Kok, Swee Leong[1]**, Kenichiroh.Mukai[4], Tafadwa Magaya[4]

[1]Faculty of Electronic and Computer Engineering, Universiti Teknikal Malaysia Melaka, 76100 Malaysia
[2]Atotech (Malaysia), Tingkat 4B-2, Wisma Fiamma 20, Jalan 7A/62A, Bandar Manjalara, 52200 Kuala Lumpur
[3]Infineon Technology, Free Trade Zone, Batu Berendam Melaka, Malaysia
[4]Atotech USA Inc., 369 Inverness Pkwy, #350, Englewood CO 80112
*min-fee.tai@atotech.com, **sweeleong@utem.edu.my

Abstract – In the manufacturing process, metal capping for EMI shielding is done during the integrated circuit (IC) assembly process, which hinders the attempt of reducing the size of electronic device and also incurs higher cost of assembly. Therefore direct deposition of metal on IC mold compound is desirable. Conventional metal plating techniques, however fail tape test. This paper studies the condition of plating metal directly onto surface of mold compound with the enhancement of novel non-etching adhesion promoter CovaBond MR[TM]. By plating direct onto mold compound, the shielding capping task can be done in array form before the die saw process in IC manufacturing (before IC assembly), which reduce the thickness and dimension of chip and improves design flexibility of circuit board as well as reduce the manufacturing cost. The industrial test results in this paper have proven the performance of the enhanced metal plating technique.

Key words: EMI shielding, electroless NiP, mold compound, reliability, CovaBond MR[TM]

I. INTRODUCTION

Conventionally Electromagnetic Interference (EMI) shielding is accomplished by metal can on chip as shown in Fig. 1(a). Fig. 1(b) shows direct conformal metal plating on mold compound, which is able to create a thin and smaller dimension EMI shielding solution [1]. This provides higher flexibility on board design by reducing the need of grouping on certain area, improve chip density and reduce the total thickness of the board. In order to increase the performance of EMI shielding, a double shielding can be established as shown in Fig. 1(c), by combining plating method and metal can, which further enhance the EMI shielding capability.

Conformal shielding on mold compound has been done in market by laser roughening follow by silver paste, or using sputtering technology. The cost of conformal shielding could reduce much lower if it is done by electroless plating. However, electroless plating method is not desirable due to difficult to plate on non-conductive surface, and its poor adhesion performance.

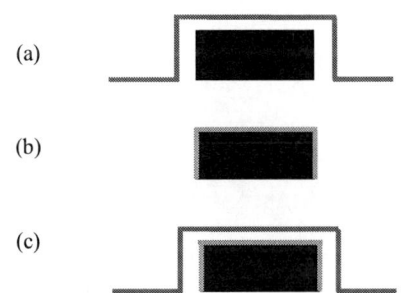

Fig. 1: Metal capping techniques: (a) Conventional metal can on device, (b) Conformal plating on mold compound helps to reduce space and height, (c) Conformal plating on mold compound combines with metal can to enhance EMI shielding effect.

II. ADHESION ENHANCEMENT AND PLATING TECHNIQUES

The key challenge to plate metal directly on mold compound is to provide adequate adhesion between metal and dielectric. Although technique of plating on plastic are quite common in fabrication process, but the same technique cannot be direct apply into mold compound. One of the reasons is the fact that mold compound is not purely organic resin instead it has high percentage of silicon oxide (normally 60-90%), which known as glass filler. Another reason is aging effect; whereby in electronic industry, aging simulation during qualification work is very critical. Device need to undergo extreme change of temperature and humidity. Normally delamination issue tends to arise thus fail in industrial standard test requirement.

Surface roughness is important to ensure the integrity of the bond between the mold and EMI shielding material. Ideal etching method should offer a better micro-roughness with high relative surface area increase (RSAI) after surface treatment. High quality micro-roughening can provide high contact area for metal plating, and reduce chemical attack between plating layer in subsequent process in manufacturing process.

978-1-4799-5761-3/14 $31.00 © 2014 IEEE

One of the surface roughing techniques is using permanganate etch which can produce acceptable roughness on polymer material economically. Fig. 2(a) shows surface of mold compound where resin has been etched off by permanganate and glass filler appear on surface. However, due to resin contain in mold compound is low, permanganate etch alone is not sufficient to achieve the required roughness.

Another method we tried by performing etching on glass filler by using fluoride base chemical. The purpose of fluoride base chemical is able to attack and dissolve the glass filler, hence create roughness on the glass filler layer as shown in Fig. 2(b). In this paper, a novel non-etching adhesion promoter using CovaBond MR™ from Atotech [3] is evaluated to compare to the existing techniques. CovaBond MR™ is a unique wet chemical process that enables deposition of metals (e.g Cu and Ni) with good adhesion onto difficult to plate substrates such as molding resin. It employs a combination of mechanical anchoring and chemical bonding, thus enabling direct metallization with very good adhesion.

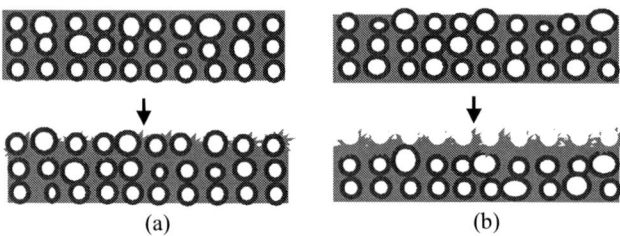

(a)   (b)

Fig. 2: Surface roughing process by, (a) permanganate etches and (b) fluoride etch.

Plating on mold compound starts with cleaning and create roughness on surface, follow by activation to enable surface to be conductive and works as catalyze to electroless nickel-phosphorous (NiP) plating etch as shown in Fig. 3.

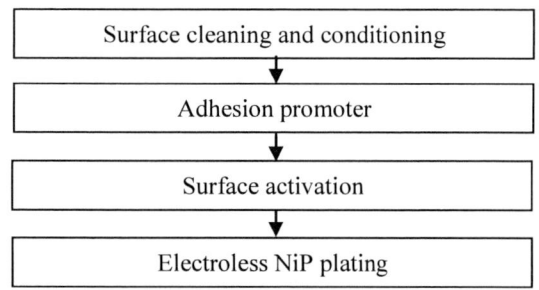

Fig. 3: Process Flow for Nickel-Phosphorous Plating on Mold Compound

III. ADHESION MECHANISM

Due to of the high glass filler content and the nature of the molding resin itself, the surface of the mold compound is normally quite rough, as shown in Figure 4.

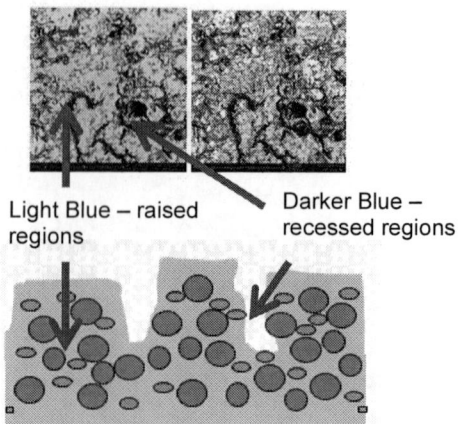

Fig 4: Structure Of mold compound

The adhesion promotion process starts with spraying the substrate with CovaBond MR™ which excellently wets & fills the surface topography, including recessed areas. Later the solvent evaporates; condensation leads to a thin CovaBond MR™ film. Solvent penetrates into substrate, inducing the resin surface to swell, thus increasing the available area for permanganate etching. CovaBond MR™ film imparts added selectivity for the permanganate attack on the resin, thus increasing the available surface area for adhesion. Particularly the recessed regions are subject to softening and swelling, while the raised regions are protected by the CovaBond MR™ film serving as a temporary and local etch resist.

Process followed by permanganate etching as illustrated in Figure 5(a) and 5(b). CovaBond MR™ film completed etched from surface within ≤ 2min. Dense matrix on top of raised regions retard the etching process and induce a planarization effect. Permanganate preferentially etches in susceptible CovaBond MR™ soaked recessed regions causing rougher trenches. Higher RSAI and retention of distinct recessed area/higher level area structure leads to increased adhesion. Within 5 to 15 min of permanganate etching, glass fillers get partially exposed, providing additional anchor points for electroless copper adhesion.

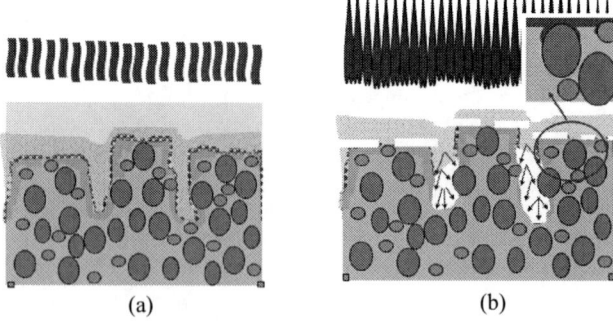

(a)   (b)

Fig 5: Covabond MR™, (a) coating and (b) oxidizing process.

IV. EXPERIMENTAL RESULTS

Fig. 6(a) shows the surface of a mold compound of a SEM photo under a magnification of 1000 times, before it is going through the surface treatment. It can be seen the surface is not severely rugged compared to Fig. 6(b) and Fig. 6(c), which are being treated with permanganate etch and fluoride etch respectively. CovaBond MR[TM] combines mechanical anchoring with chemical adhesion leading to better adhesion strength for similar surface roughness of permanganate etches as shown in Fig. 7.

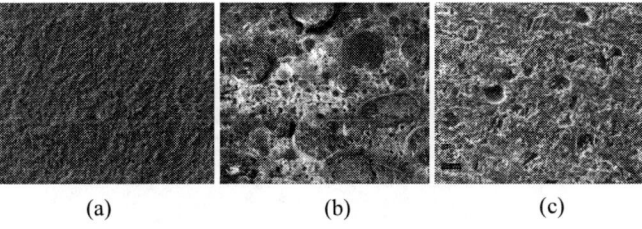

| (a) | (b) | (c) |

Fig. 6: Surface of mold compound at a magnification of 1 k scanning electron microscope, (a) before surface treatment, (b) after permanganate etches on resin, (c) after fluoride etches on glass filler.

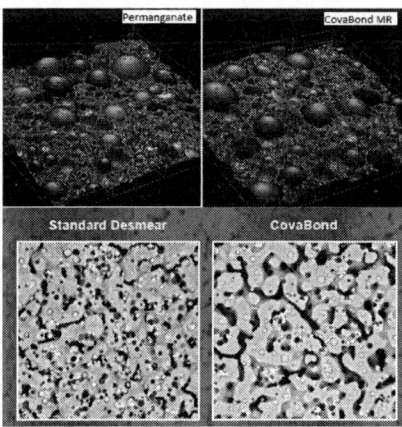

Fig. 7: CovaBond MR[TM] combines mechanical anchoring with chemical adhesion leading to better adhesion strength for similar surface roughness of permanganate etches

In order to understand the adhesion capability, samples are treated with different adhesion promoter, and follow by autocatalytic nickel-phosphorous (NiP) plating for 5 um thickness. Tape test based on industrial standard was carried out to test on samples and the results are shown in Table 1 with the quantified measurement of peel strength.

The tensile strength tester as discussed in was being used for the measurement. As NiP is fragile, copper was plated on top of mold compound. The results are shown in Table 2.

The experiment test summarize that CovaBond MR[TM] gives significant adhesion performance compare to others adhesion promoter. We target peel strength of above 4N/cm for industrial buy off. This reading is benchmark to peel strength requirement in the IC substrate manufacturing.

Table 1: Result of different adhesion promoter versus peel strength

No.	Test condition	Peel strength test [6]
1.	Without treatment of adhesion promoter	No test result available as sample has already blistered
2.	Permanganate etch	< 1.5 N/cm
3.	Fluoride etch	< 2.0 N/cm
4	CovaBond MR[TM]	4-5 N/cm

Table 2: Results of different adhesion promoter versus tape test.

Test condition	Before tape test	After tape test
Without treatment of adhesion promoter	Blister found immediate after NiP plating. Tape test fail.	
Permanganate etch	No blister after NiP plating; but the entire NiP peels off after tape test.	
Fluoride etch	  No blister after NiP plating, tape test condition improved; only minor peel off.	
Covabond MR[TM]	No blister after NiP plating. Tape test pass.	

V. INDUSTRIAL RELIABILITY TESTS

The purpose of semiconductor device reliability testing is primarily to ensure that shipped devices, after assembly and adjustment by the customer, exhibit the desired lifetime, functionality and performance in the hand of end users. In order to comply to the electronic manufacturing requirement, some standard reliability tests have been performed. In this section, all of the aging tests will be conducted on samples produced only with CovaBond MR ^{TM}.

Tests were conducted on chip Mound on Board (MOB) as the experimental reference. Table 3 shows the results of preconditioning samples with standard Moisture Sensitivity Test Level 1 (MSL1), then followed by Pressure Cooker Test at 96 hours (MSL1 + PCT96), Pressure Cooker Test 192 hours (MSL1 + PCT192), Thermal Cycling of 500x (MSL1+ TC500), 1000x (MSL1 + TC1000) and 2000x (MSL1 + TC2000).

Table 3: Results of electrical test after varies type of aging conditions.

No.	Test condition	Electrical Resistance ,ohm			
		Average	Min	Max	Standard deviation
1	MOB (reference)	0.952	0.7	1.2	0.146
2	MSL1:Precondition	0.965	0.7	1.2	0.129
3	MSL1+ PCT96	0.978	0.8	1.2	0.112
4	MSL1 + PCT192	1.072	0.9	1.2	0.091
5	MSL1 + TC500	0.953	0.8	1.2	0.091
6	MSL1 + TC1000	0.919	0.7	1.2	0.141
7	MSL1 + TC2000	0.722	0.4	1.2	0.159

Electrical test was carried out to verify the consistency of device performance to investigate the fatigue of samples. From the electrical tests, the experimental results show that there is no significant change of electrical resistance, which indicates that no separation and minor crack occur. However, for the test on TC2000, the sample resulted in a noticeable drop of resistance, which shows that heat treatment of the NiP affecting the resistivity. It has been reported that, at a temperature as low as 150°C, resistivity decreases commonly due to the fact that it releases physically adsorbed hydrogen, which therefore increases the conductivity [4].

After the aging and electrical tests, tape test were repeated, as presented in Table 4. There is no peel off of nickel flake after all aging test. It shows that the adhesion between nickel-phosphorous and mold compound is remained strong.

On cross section investigation as shown in Fig. 8, there is no separation found between NiP and mold compound, which confirmed that Covabond MR ^{TM} adhesion is able to withstand with the entire reliability test.

Table 4: Tape test after electrical test and after varies type of aging conditions.

Test condition	Tape test result
MSL1+ PCT96	Pass
MSL1 + PCT192	Pass
MSL1 + TC500	Pass
MSL1 + TC1000	Pass
MSL1 + TC2000	Pass

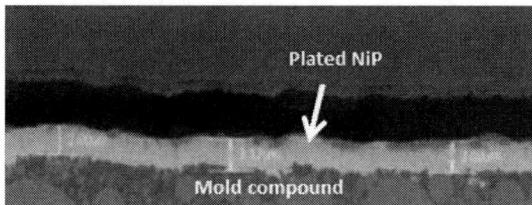

Fig. 8. Cross section view between mold compound and plated metal after TC 1000, no delamination noticed.

VI. CONCLUSION

From the experiment results, they show that CovaBond MR ^{TM} enhances adhesion of electroless nickel to epoxy mold, and the process meets the high level criteria of reliability test on electronic manufacturing requirement. This conformal plating on mold compound provides alternative solution for IC packaging industry especially useful for EMI shielding capability requirement of an IC.

Acknowledgement

The authors would like to acknowledge the support of this work by Advanced Sensors and Embedded Systems (ASECs) Research Group, Universiti Teknikal Malaysia Melaka (UTeM) and also the industrial partners, Atotech Inc. and Infineon Technologies (M) Sdn. Bhd. in sharing the information, facilities and materials.

References

[1] K.H.Liao, A.C.H.Chan, C.H.Shen, I.C.Lin and H.W. Huang. Novel EMI shielding methodology on highly integration SiP module. 2nd IEEE CPMT Symposium, Japan, 2012.

[2] W.J.Zhang, Q.Lu, T. Hang, M.Li and D.L.Mao. Electroless plating copper cones on lead frame to improve the adhesion with epoxy molding compound. 13th Int. Conf. on Electronic Packaging Technology and High Density Packaging (ICEPT-HDP), 2012.

[3] CovaBond MRTM- Enabling Improve Adhesion of Electroless Copper to Smooth Substrate, Molecular Interface Technology (MIT), Atotech Berlin, 2012.

[4] Ron Parkinson, Properties and applications of electroless nickel, Nickel Development Institute, Jun 2001.

IEEE-ICSE2014 Proc. 2014, Kuala Lumpur, Malaysia

A Low Power 0.18μm CMOS Technology Integrating Dual-Slope Analog-to Digital Converter

Ili Shairah Abdul Halim, Nor Syazwana Mohd Yusof, Siti Lailatul Mohd Hassan
Faculty of Electrical Engineering
Universiti Teknologi MARA
Shah Alam, 40450, Selangor, Malaysia
e-mail: shairah@salam.uitm.edu.my, nsyazwana@yahoo.com

Abstract— **In this paper, a 4-bit integrating dual slope analog-to digital converter (DS-ADC) is designed which consumes low power and simplicity but slow conversion time. The design utilizing a Silvaco Electronic Design Automation (SEDA) tools with an advanced 0.18μm CMOS Technology using 1.8V power supply. This integrating dual slope ADC contains five main components, which are switching, integrator, comparator, control logic and counter at which the integrator is realized with a two-stage operational amplifier (op-amp) that provides sufficient gain, ICMR and low power dissipation. Simulation confirms that the proposed DS-ADC architecture shows a power efficiency of 2.4739mW with 1.06μs conversion time.**

Keywords- Dual Slope ADC, Two stage op-amp, Integrator, Comparator, Control Logic

I. INTRODUCTION

A dual slope integrating ADC (DS-ADC) provides a high accuracy, high resolution, inexpensive, good noise rejection, and ideal for digitizing low bandwidth signal. They are commonly found in slow-speed, low sensitive to passive device variation applications such in digital multimeter, panel meter and even medical instrumentations. DS-ADC integrates an unknown input voltage (V_{in}) for a fixed amount of time (T_{int}), then "de-integrates" (T_{deint}) using a known reference voltage (V_{ref}) for a variable amount of time. Operational amplifier is the key analog component of DS-ADC and the most widely used circuit approach for the implementation of MOS op-amp is the two stage configuration which provides good common mode range, output swing, voltage gain and CMRR in a simple circuit [5-6]. There are many types of analog-to-digital converters which can be classified according to the concept on which they were designed. For example, there are Charge-Coupled ADC, Digital-Ramp ADC, Successive Approximation ADC, Voltage-to-Frequency ADC, Integrating ADC, Delta-Sigma ADC, Flash ADC and some of these converters require the use of digital-to-analog converters (DACs), analog comparators and some logic modules. Integrating ADCs are commonly designed either in single or dual slope by which the single slope approaches is less accurate and make measurement repeatability quite difficult to attain [1-2]. The accuracy of a single slope ADC depends on the resistance and capacitor used. Thus to eliminate the dependency of resistance and capacitor, there is a need of a Dual slope ADC and this ADC requires no DAC as illustrated in Fig. 1 [1-3]. Fig. 1 clarified on the behavior of the DS-ADC. The first integration is of fixed length, dictated by the counter, in which the sample-and-held signal is integrated, resulting in the first slope.

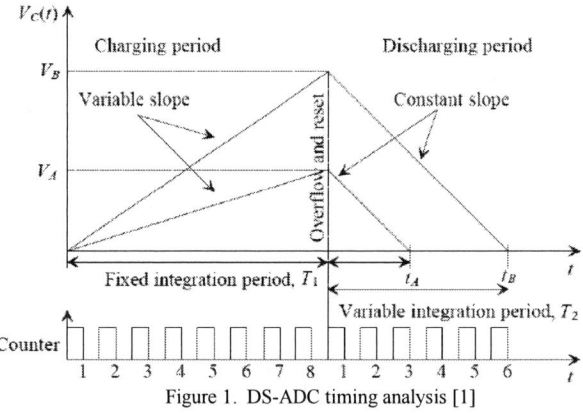

Figure 1. DS-ADC timing analysis [1]

After the counter overflows and is reset, the reference voltage is connected to the input of the integrator. Since V_{in} was negative and the reference voltage is positive, the inverting integrator output will begin discharging back down to zero at a constant slope. A counter again measures the amount of time for the integrator to discharge, thus generating the digital output. Notice that the first slope varies according to the value of the input signal, while the second slope, dependent only on V_{ref}, is constant. Similarly, the time required to generate the first slope is constant, since it is limited by the size of the counter. However, the discharging period is variable and results in the digital representation of the input voltage [1]. Primary emphasis is also placed on CMOS amplifiers because of their more widespread use and that they are the key elements of most analog subsystems, particularly in converters which then making the performance of many systems is strongly influenced by op-amp performance [6-7]. The proposed design will lead to a better performance of integrating ADC in terms of its power, error rejection and circuit simplicity in comparison to other ADC architectures and it is specially design in an advanced 0.18μm CMOS technology and a two stage op-amp as a key for increasing system performances. The design is wholly employed by Silvaco EDA Tools beginning from schematic, testbenches, simulation and verification.

II. METHODOLOGY

A. Block Diagram

Fig. 2 illustrates the each module that integrates the DS-ADC as a whole complete system.

978-1-4799-5761-3/14 $31.00 © 2014 IEEE   317

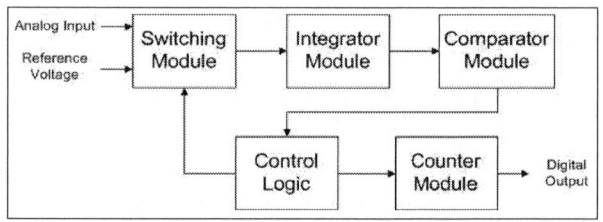

Figure 2. DS-ADC Block Diagram

B. DS-ADC Design

i. Complete System

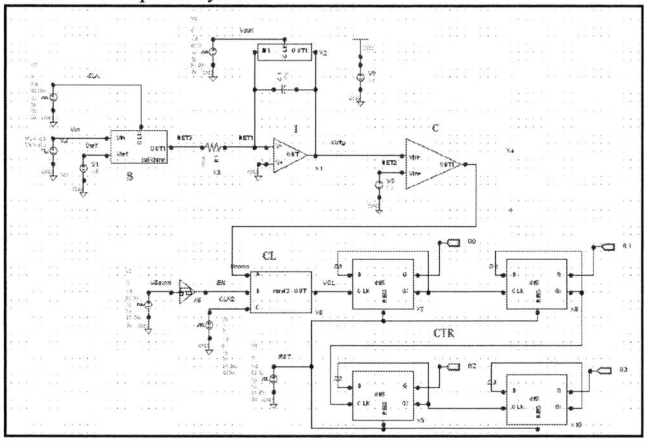

Figure 3. DS-ADC complete system

As shown in above Fig. 3, the whole complete system of DS-ADC is built by combining the modules of switching-S, integrator-I, comparator-C, control logic-CL and counter – CTR. The system is powered by 0.9V analog signal and 1.8V of reference voltage and V_{DD} at 25ºC temperature.

III. RESULTS AND DISCUSSIONS

A. Two Stage Op-amp

Table I pointed out the specification results according to the circuit in Fig. 4.

TABLE I. TWO-STAGE OP-AMP DESIGN SPECIFICATION

Parameter	Specification
Input Offset Voltage	5.4mV
Gain	45.43dB
CMRR	20dB
ICMR	0.6 – 1.72V
Slew Rate	0.09V/µs
Power Dissipation	58µW

B. Integrator

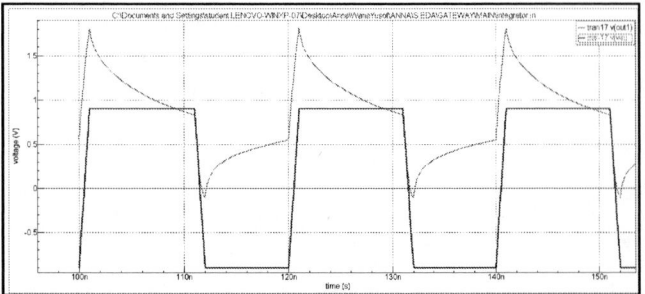

Figure 4. Integrator simulation

Fig. 4 is the transient simulation of an integrator. The negative feedback of the op-amp ensures that the inverting input will be held at 0 volts due to the virtual ground concept. If the input voltage is exactly 0 volt, there will be no current through the resistor, therefore no charging of the capacitor, and therefore the output voltage will not change.

TABLE II. VARIOUS INTEGRATOR VALUES

Resistor, R1 (Ω)	Capacitor, C1 (pF)	V_{charge} (V)	$V_{discharge}$ (V)
100	0.1	1.8	-1.8
100	1	2.0	-2.0
1k	10	1.5	-1.5
10k	10	1.2	-1.2

Table II clarified the different values of RC that make up the integrator and the corresponding output of charging and discharging values. It cannot be guaranteed what voltage will be at the output with respect to ground in this condition, but it can be said that the output voltage will be constant. However, if a constant, positive voltage (0.9V) is applied to the input, the op-amp output will fall (discharging) at a linear rate, in an attempt to produce the changing voltage across the capacitor necessary to maintain the current established by the voltage difference across the resistor. Conversely, a constant, negative voltage (-0.9V) at the input results in a linear, rising voltage (charging) at the output. The output voltage rate-of-change will be proportional to the value of the input voltage.

C. Comparator

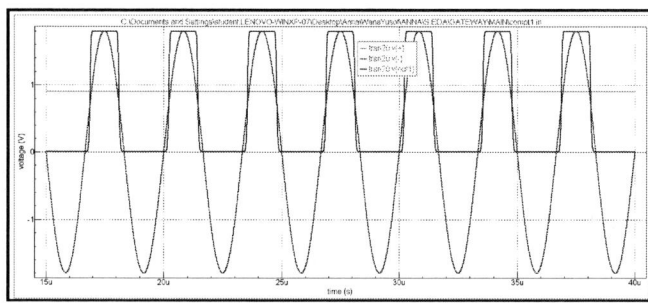

Figure 5. Comparator simulation

As mentioned earlier, a comparator basically compares between the two voltages that are applied on the amplifier terminals. The comparator circuit is simulated with 0.9V as a reference voltage (V+) and a sinusoidal 1.8V into the inverting input (V-) that produced a binary output of '1' as long as the inverting input potential is higher than that of reference voltage as described in Fig. 5.

D. Switching Module

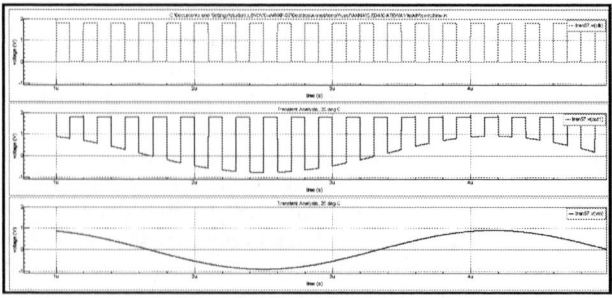

Figure 6. Switching simulations

When the input voltage, V3 as in Fig. 6, is the higher supply rail, the output, V_{out}, goes to V2 (the supply voltage). The NMOS device at the bottom turns on and the PMOS device on top shuts off. When the input goes to low, the output goes to V1 (reference voltage) turning on the PMOS and turning off the NMOS. Table III summarized the operation of the switching circuit.

TABLE III. INVERTER SWITCH

Clock, CLK	Output Voltage, V_{out}
1	V_{in}
0	V_{ref}

E. Control Logic

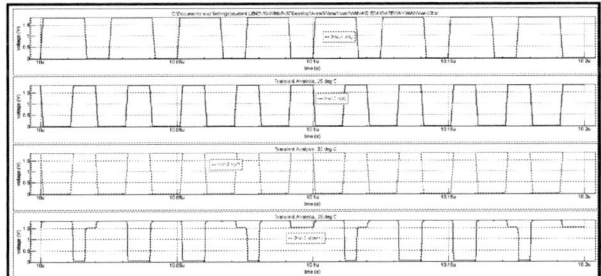

Figure 7. Control logic simulation

Since the control logic module is designed on a simple 3-input NAND gate, the output produced is easily achieved. A high logic of '1' would always true if either input of the gate is '0' whereas logic '0' at the output when all 3-inputs are at high logic as proved in Fig. 7.

F. Counter

Configurations of a single bit counter with 'reset' element functioning as resetting or clearing the output, means a binary '0' will be presented at the Q output. The proposed DS-ADC is a 4-bit system that is realized through a ripple counter stated that only the first flip-flop is clocked by an external clock and all subsequent flip-flops are clocked by the output of the preceding flip-flop as in both Fig. of 8 (a) and (b).

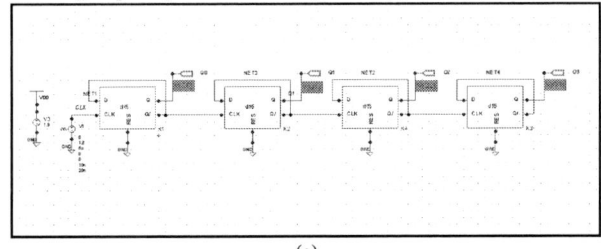

(a)

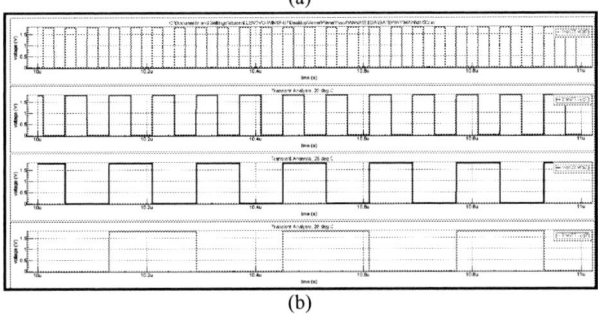

(b)

Figure 8. Counter simulation (a) 4-bit ripple counter (b) 4-bit ripple counter simulation

G. Complete System

The dual-slope operation differs from the single-slope in that V_{intg} is now compared to ground and two voltages; V_{ref} and V_{in} are both integrated. Initially a negative input is connected to the integrator, ramping V_{intg} until the counter overflows.

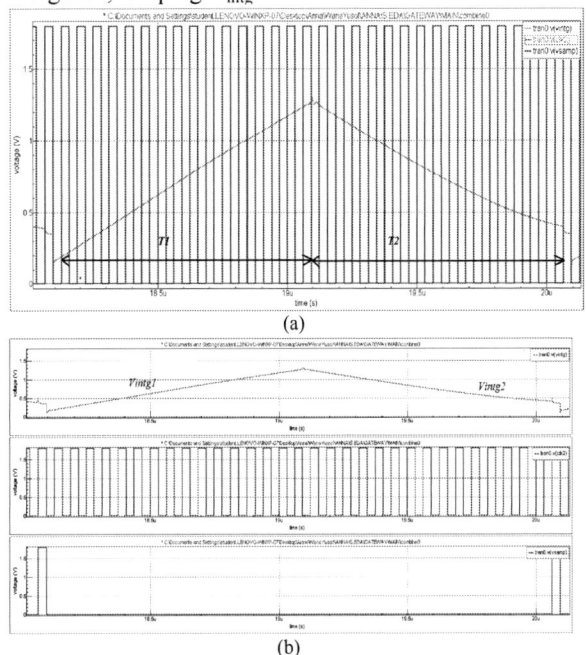

(a)

(b)

Figure 9. Integration slope of DS-ADC

Since a negative value is being integrated on the inverting input, the integrator output will always positive and greater than zero, so the counter will continue until it overflows, which happens at 2^N clock cycles (= T1). (3) gives a value for V_{intg}. Looking at the slope ($V_{intg1}/T1$), the dependence on V_{in} is clear and since V_{in} is a variable, so is the slope.

$$V_{intg1} = \frac{-\int_0^{T1} \frac{Vin}{R} dt \cdot T1}{C} = \frac{Vin.T1}{RC} \qquad (3)$$

After overflow switches, V_{ref} is integrated; at the same time, the control logic triggers the reset to clear the counter. The counter now begins checking how many clock cycles are needed to ramp V_{intg} down to zero. (4) gives the value of V_{intg} for this time period, T2. Note, (4) is negative since V_{ref} is on the inverting input; this means that the integrator is now ramping down.

$$V_{intg2} = \frac{1}{C} \int_0^{T2} \frac{Vref}{R} dt = \frac{-Vref.T2}{RC} \qquad (4)$$

Again, looking at the slope ($V_{intg2}/T2$), it is clear that it will be constant since V_{ref} and RC do not change. The value of V_{intg} after T2 will be the sum of V_{intg1} and V_{intg2} and as the illustration shows in Fig. 9 (a) and (b), it should be zero.

$$V_{intg1} + V_{intg2} = 0;$$
$$0 = \frac{Vin.T1}{RC} - \frac{Vref.T2}{RC};$$
$$\frac{Vin.T1}{RC} = \frac{Vref.T2}{RC} \rightarrow \frac{Vin}{Vref} = \frac{T2}{T1} \qquad (5)$$

T2/T1 gives the digital output of the input voltage. (5) is revealing in that there is no dependence on RC. Essentially, errors in the slope of V_{intg} (determined by RC) will cancel out since the same integrator is used to calculate T1 and T2. Also, since the same clock is used to measure these times, errors in the clock such as jitter should also cancel [8]. All these calculation and result are summarized in below Table IV, referring to the complete circuit of Fig. 3.

TABLE IV.   INTEGRATION ANALYSIS

Parameter	Value
V_{in} (V2)	-0.9V
V_{ref} (V1)	1.8V
Clock cycles	$2^N = 16$
R1	100kΩ
C1	10pF
V_{intg1}	1.3V
V_{intg2}	-1.3V
T1	305.6 μs
T2	286.5μs

TABLE V.   DS-ADC CORRESPONDING OUTPUT

Decimal	Bit	Ideal	Actual
1	0001	0.1125	0.1401
2	0010	0.2250	0.2835
3	0011	0.3375	0.3932
4	0100	0.4500	0.4963
5	0101	0.5625	0.6101
6	0110	0.6750	0.6762
7	0111	0.7875	0.8430
8	1000	0.9000	0.9903
9	1001	1.0125	1.0630
10	1010	1.1250	1.1736
11	1011	1.2375	1.2750
12	1100	1.3500	1.3723
13	1101	1.4625	1.4609
14	1110	1.5750	1.5430
15	1111	1.6875	1.6771

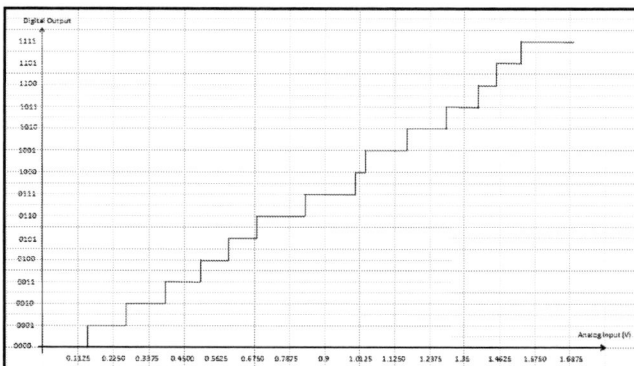

Figure 10.   DS-ADC transfer curve

As tabulated in Table V, the least significant bit (LSB) refers to the rightmost bit in the digital input word. The LSB defines the smallest possible change in the analog output voltage. The LSB will always be denoted as D_0. One LSB can be defined as:

$$1LSB = V_{ref}/2^N \qquad (6)$$

The value of 1 LSB for this ADC can be calculated using (6) and is the ideal step width (1/16). Therefore, with V_{ref} = 1.8V, 1 LSB = 0.1125V. Based on Fig.10, the digital output 4-bit DS-

ADC is plotted versus the analog input, V_{in}. Since the input signal is a continuous signal and the output is discrete, the transfer curve of the ADC resembles that of a staircase. Another fact to observe is that the *2N* quantization levels correspond to the digital output codes 0 to 15. Thus, the maximum output of the ADC will be 1111 (2^N - 1) [1]. The overall characteristics of conversion time, power and others of the DS-ADC design are tabulated in TABLE VI.

TABLE VI.   DS-ADC CHARACTERISTICS

Integrating DS-ADC	
Technology	0.18μm
Temperature	25°C
Analog Input, V_{in}	0.9V,10kHz frequency
Reference Voltage, V_{ref}	1.8V
Conversion Time	1.06μs
Power Consumption	2.47392mW

IV.   CONCLUSION

A Dual Slope ADC is one valuable ADC in signal processing for its high noise immunity, low cost and low architecture but slow in speed. In this paper, a 4-bit DS-ADC has been designed and simulated in 0.18μm CMOS technology by EDA Tools. The design does not utilize any DC external biasing circuit and only need to apply V_{dd} because it consumes an independent current and voltages sources. The total power consumption of this design is only 2.474mW with 1.06μs conversion time. It is suggested as future work to enhance the DS-ADC transfer curve to be as accurate as the ideal ADC, perhaps by optimize the two stage op-amp and an extra sample-hold circuit.

REFERENCES

[1] R. Jacob baker, Harry W. Li & David E. Boyce, "CMOS Circuit Design, Layout, and Simulation", Department of Electrical Engineering Microelectronics Research Center, The University of Idaho, 1998.

[2] Hasan Krad, "A Dual-Slope Integration Based Analog-To-Digital Convertor", Department of Computer Science And Engineering, College of Engineering, Qatar University, P.O. Box 2713, Doha, Qatar, American J. of Engineering And Applied Sciences 2 (4): 743-749, 2009, Issn 1941-7020, © 2009 Science Publications.

[3] Nidhi Agrawal, "Design of 10-Bit Dual Slope Adc for Isfet Based Ph Meter", Department of Electronics & Communication Engineering, Thapar University, Patiala, India, July 2010.

[4] Marina Fekry M. Bestauros, "Design and Implementation of A Dual Slope Adc for PTAT Temperature Sensors Used In Hdrc® Vision Applications", Institute for Microelectronic, University Of Stuttgart, 2010.

[5] Ecen4827/5827 Lecture notes, "Two Stage Op-Amp: Dc Solution (Part 1)", Department of Electrical and Computer Engineering, University of Colorado.

[6] Paul R. Gray and Robert G. Meyer, "MOS Op-Amp Design- A Tutorial Overview", IEEE Journal of Solid-State Circuits, Vol.SC-17, NO.6: 969-982, December 1982.

[7] Measurement Computing, "Analog to Digital Conversion", Norton, MA 02766, info@mccdaq.com.

[8] Antonio Ablea, "Advanced Analog IC Design", Boise State University, May 19, 2008.

COMPARATIVE STUDY ON 8T SRAM WITH DIFFERENT TYPE OF SENSE AMPLIFIER

Siti Lailatul Mohd Hassan, Idalailah Dayah, Ili Shairah Abdul Halim
Faculty of Electrical Engineering
UniversitiTeknologi Mara
40450 Shah Alam
Selangor Darul Ehsan

ABSTRACT- This paper presents the comparative study on 8T SRAM with different type of sense amplifier. These sense amplifiers are voltage-mode sense amplifier (VMSA) and current-mode sense amplifier (CMSA). The first objective of this research is to design the 8T SRAM and sense amplifier and the next objective is to identify which design has better performance in term of power, speed, stability and area. Sense amplifiers are one of the most critical circuits in the periphery of CMOS memories that plays an important role to reduce the overall sensing delay and voltage. Earlier voltage mode sense amplifiers sense the voltage difference at bit and bit lines bar but as the memory size increase the bit line and date line capacitances increases. The tools used for simulation is SILVACO EDA Gateway and SILVACO EDA Expert for layout using 0.18um technology. The results show that the CMSA has higher speed with lower delay, and low power dissipation than VMSA. But, VMSA has smaller cell area and higher noise margin than CMSA. CMSA is more suitable for high speed performance and low power circuitry and VMSA is best suited for stability and smaller design.

Index Terms – Voltage mode sense amplifier, Current mode sense amplifier, 8T SRAM

I. INTRODUCTION

Technology and supply voltage scaling continues to improve the logic circuit delay with each technology generation. However, the speed of the overall circuit is increasingly limited by the signal delay over long interconnects and heavily loaded bit-lines due to increased capacitance and resistance[1].

Static Random Access Memory (SRAM) is a type of semiconductor volatile memory (RAM) which keeps its data until the power is turns OFF. SRAM will store the binary logic bits '1' or '0' [2]. It consists of an array of memory cells along with the row and column circuitry. SRAM has design to fill two needs that are to provide direct interface with CPU at speeds not achievable by DRAMs and to replace DRAMs in systems that require very low power consumption. The stability and area of SRAM need to be concern in designing SRAM cell. SRAM cell must be able to write and read data and keep it as long as the power is applied. The main challenge in designing SRAM cell is to ensure that the circuitry holding the state is weak enough to be overpowered during a write, and still strong enough to be not disturbed during read operation.

SRAM represents a large portion of the chip, and it is expected to increase in the future in both portable devices and high-performance processors. To achieve longer battery life and higher reliability for portable application, low-power SRAM array is a necessity [3].

Sense amplifiers are used to translate small differential voltage to a full logic signal that can be further used by a digital logic. The need for increased memory capacity, higher speed, and lower power consumption has defined a new operating environment for future sense amplifiers. Sense amplifiers are mainly used to read the contents of SRAM and DRAM cells [4] Sense amplifier is one of the important peripheral circuits in memory as it strongly influences the memory access times [3].

This sense amplifier can be operated in voltage, current and charge mode but we operate them in current-mode because they present a low impedance to the inputs and respond to the differential current rather than to the voltage between the inputs, this can reduce interconnect delay in long wires there by providing speed improvement. The current mode sense amplifier reduces the bit line swing during read operation as compared to voltage mode sensing technique. It proves that current sensing technique would be faster than voltage mode due to the low impedance termination of the current mode. It shows that current sensing is relatively insensitive to the bit line capacitance [5]. This gives the motivation to use current mode sensing in the bit lines in SRAM.

In this paper, we explore the design of the 8T SRAM connected with different type of sense amplifier which is voltage mode sense amplifier and current mode sense amplifier by using 0.18um technology process and we compare the various designs based on delay, power dissipation and cell area. The two sense amplifier topologies compared are current mode sense amplifier (CMSA) and voltage mode sense amplifier (VMSA).

II. CIRCUIT DESCRIPTION

A. 8T SRAM

The 8T SRAM circuit consists of two cross coupled inverters which is M1, M2, M3 and M4. There are two extra NMOS transistors connected to the cross-coupled inverters where each of them pulls down the path of the cross-coupled inverters so that it can achieve the leakage power reduction. In addition, both drain inverters is connected to the NMOS transistors so that it connected to the write-word line (WWL) at their respective gate terminals and the drain is connected to the BitLine (BL) and BitLine_Bar (/BL). The write-word line is used to select the cell while the bit lines are used to perform write and read operations on the cell [8]. Both BitLine and BitLine_Bar are used to improve the speed of write and read operations.

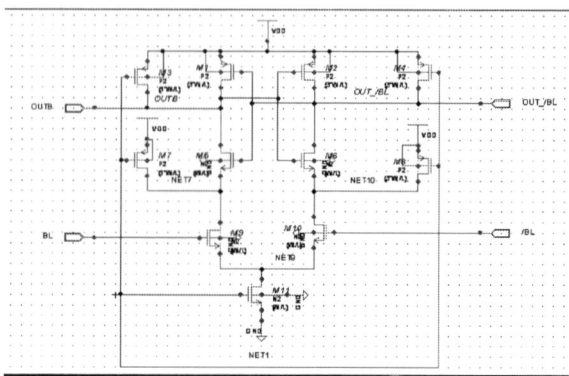

Fig 2: CMSA schematic circuit with 8T SRAM

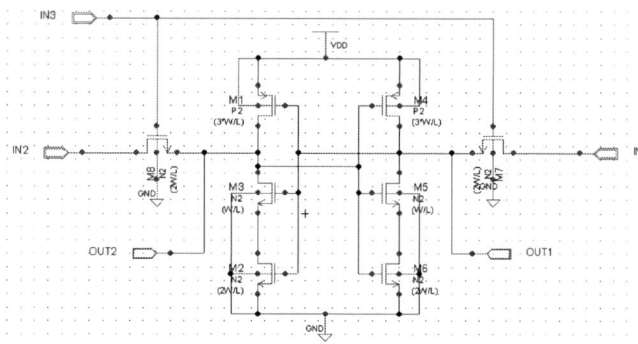

Fig.1: Schematic 8T SRAM

i. Write and Read operation

To get the write and read operation, the input BL, BL_Bar and WWL plays an important role. Writing a value into the 8T SRAM cell is done by forcing the BL and BL_Bar high while keep the other low. To get the write operation high '1', input WWL and BL must be high '1' and the BL_Bar is low '0'. Whereas, to get the output Write low '0', BL is set to low '0' and the BL_Bar is set to high '1'. Thus the output go low '0'. Before reading, both bitlines are pre-charged high and SRAM cell is selected [10]. For read operation, the sense amplifier will take charge as sense amplifier are used to sense which line is being pulled down to perform the read operation of the stored data.

B. Current Mode Sense Amplifier

Figure 2 shows the CMSA schematic circuit. The M9 and M10 are connected to BL and BL_Bar which drives the transistors gate. Transistors M3, M4, M7 and M8 are the pre-charged transistors where the M1, M5 and M2 and M6 are the inverter pair cross coupled that resolves the BL differential voltage. Traditionally, this topology has been used because the memory bitlines are driving high impedance (gate) and full discharge of array bitline due to timing mismatch is not a concern [7]. The output of OutB and Out_/Bl is connected into inverters that convert the voltage differential at their inputs on BL and BL_Bar to a full swing at the output.

This sense amplifier design is based on the current differential that produced by M9 and M10 in the two pull down branches of this sense amplifier. During the read operation, when the BL or BL_Bar is low depending on the data stored in the cell. Gate M11 is connected to input pin Sense Enable (SE), and when it's set to low, it will automatically turned ON and all the precharged transistors simultaneously turned OFF. Since the gate voltage of M9 and M10 differs by the generated bitline differential, their channel currents are unequal. Thus the current that flow in M9 and M10 is unequal and it makes the voltage in either OutB or Out_/BL fall faster other than the other node. Therefore, this voltage is resolved by the cross-coupled inverters.

C. Voltage Mode Sense Amplifier

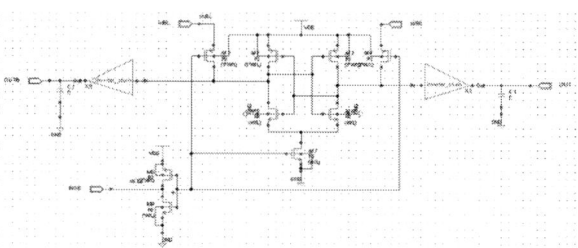

Fig 3: VMSA schematic circuit with 8T SRAM

Figure 3 above shows the schematic of voltage mode sense amplifier. There are also two cross-coupled inverters in this design M1, M5, and M2, M6 that resolve the differential voltage on the bitlines to a full swing at the output OUTB and OUT. The obvious advantage of this topology over the CMSA is the lower number of transistors needed which means faster access and smaller footprint [7].

The operation of this circuit is directly based on the voltage differential developed on its internal nodes by the input bitlines inBL and in/BL. The output of the circuit is connected with inverter chain so that it gives the good output swing waveform. When the wordline in 8T SRAM is ON, M7 will turn OFF and M3 and M4 turn ON. The gate terminal M7 is connected to the input of sense enable (SE) so that the cross-couples inverter amplifies the differential voltage to its output swing.

III. DESIGN METHODOLOGY

A. Schematic and Layout

The simulation of this design is by using Gateway SILVACO EDA Tool using the process technology of 0.18um. The first step in designing the circuit is to design the 8T SRAM circuit. To decide the best width and length for the design, we must run the simulation and check the voltage transfer curve (VTC) either it is symmetrical or not. The sizing calculations are based on the ratio of inverter which is 3:1. Based on VTC waveform, Vm point is symmetry at 2.4471V and is approximately to 2.5V, therefore the sizing of W/L in the design is acceptable. The sizing of the PMOS and NMOS is shown in Table1 below. The sizing of PMOS is 3 times larger than NMOS. Next step is to design the sense amplifier for current mode and voltage mode. The sizing of these sense amplifiers follows 8T design sizing. All schematic design must comply the DRC for simulation purpose.

The software used to design the layout is Expert SILVACO EDA Tool. The process technology used is 0.18um which is from SILTERRA. After the process of layout design, the nest step is LVS checking using Guardian LVS to ensure schematic and layout are both compatible with each other.

TABLE I: TRANSISTORS SIZING

TRANSISTOR	SIZING W/L (um)
PMOS	3*W/L = 60/1
NMOS	W/L = 20/1

IV. SIMULATION RESULT AND DISCUSSIONS

A. Speed of operation

CMSA is well known for the high-speed circuit for SRAM. Based on Table 2, we can see the different value of delay between CMSA and VMSA when V_{DD} is varied and the capacitance load is set to 5pF. When V_{DD}= 2V, the delay of CMSA is 44.89% smaller than the VMSA where VMSA is 12.44nS and the CMSA is 6.86nS. As the value of V_{DD} increases, the delay decreases. However VMSA still has larger delay compared to the CMSA. Thus, CMSA has the better speed compared to VMSA.

TABLE II: DELAY BY USING CL= 5pF

	VMSA	CMSA	%
V_{DD} (V)	Delay (nS)	Delay (nS)	
2	12.44	6.86	44.89
2.5	5.39	4.91	8.91
3	3.94	3.71	5.84
3.3	3.49	3.32	4.87
4	3.43	2.75	19.83
4.5	3.02	2.50	17.22
5	2.75	2.31	16.00

B. Current and Power Dissipation

The current dissipation (I_D) output value is shows in Table 3, the V_{DD} is varied from 2V to 5V and the load capacitance is set to 5pF. Current dissipation is measured at the output of the sense amplifier. Results show that CMSA has very small amount of current dissipation. For V_{DD}=2V, I_D for CMSA is 0.97387uA and VMSA is 5.77075uA which is 83.12% larger than CMSA and the trend continues up to V_{DD}=5V. CMSA has the best current dissipation where its current dissipation compared to VMSA.

TABLE IIII: CURRENT DISSIPATION BY USING CL= 5pF

	VMSA	CMSA	%
V_{DD} (V)	I_D (uA)	I_D (uA)	
2	5.77	0.97	83.12
2.5	4.05	0.27	93.29
3	18.66	2.89	84.52
3.3	62.48	4.04	93.54
4	75.86	3.42	95.49
4.5	117.69	7.33	93.77
5	115.38	6.32	94.52

Table 4 shows the power dissipation for both sense amplifiers. CMSA has the less power dissipation compared to VMSA as CMSA also has low current dissipation.

TABLE IV: POWER DISSIPATION BY USING CL= 5pF

	VMSA	CMSA	%
V_{DD} (V)	Power Dissipation (uW)	Power Dissipation (uW)	
2	11.54	1.95	83.10
2.5	10.13	0.68	93.29
3	55.99	8.67	84.52
3.3	206.18	13.32	93.54
4	303.42	13.68	95.49
4.5	529.62	32.97	93.77
5	576.88	31.63	94.52

C. Noise Margin

Table 5 shows the noise margin of both CMSA and VMSA sense amplifiers. It is good to have higher noise margin as noise margin is the amount of noise that a circuit can withstand. Therefore, from the results, VMSA has better noise margin compare to CMSA.

TABLE V: Noise Margin for VMSA and CMSA

VDD (V)		VMSA	CMSA	%
4.5	NML	0.1087	0.0997	8.28
	NMH	2.1979	2.2024	-0.21
5	NML	0.2107	0.1108	47.41
	NMH	2.4975	2.2961	8.06

D. Area

The transistor area for the VMSA design (39.015x 89.17um) is 30.98% less compared to the CMSA design (67.6 x 114.645 um). Since the VMSA is known of its simplicity of circuit, it has the small area of layout compared to CMSA.

V. CONCLUSION

This paper compares two sense amplifier using 8T SRAM. From the results obtain, CMSA is found to have less delay, thus higher speed. It also has small current and power dissipation compared to VMSA. In term of noise margin, VMSA has better noise margin than CMSA. VMSA also has smaller cell area than CMSA. It can be concluded that the CMSA is best suited for applications where low-voltage, low-power and high-speed and VMSA is best suited for smaller design area and stability.

ACKNOWLEDGEMENT

The author would like to acknowledge with gratitude, UiTM Research Management Institute (RMI) and Faculty of Electrical Engineering, UiTM for supporting this work under Excellent Fund (Research Intensive Faculty) code (82/2012).

REFERENCES

[1] N. Dist, "Analysis of New Current Mode Sense Amplifier," pp. 1–6.

[2] J. Zhu, N. Bai, and J. Wu, "A Review of Sense Amplifiers for Static Random Access Memory," vol. 30, no. 1, 2013.

[3] A. V Gayatri, "Efficient Current Mode Sense Amplifier for Low Power SRAM," vol. 1, no. 2, pp. 147–153, 2011.

[4] N. Verma, S. Member, and A. P. Chandrakasan, "A 256 kb 65 nm 8T Subthreshold SRAM Employing Sense-Amplifier Redundancy," vol. 43, no. 1, pp. 141–149, 2008.

[5] J. Wang and H. Lee, "New Current-Mode Sense Amplifier for Low-Voltage Low-Power SRAM Design," pp. 1–5.

[6] J. Singh and B. Raj, "SRAM Cells for Embedded Systems," 2014.

[7] B. Mohammad, P. Dadabhoy, K. Lin, and P. Bassett, "Comparative study of current mode and voltage mode sense amplifier used for 28nm SRAM," *2012 24th Int. Conf. Microelectron.*, no. Icm, pp. 1–6, Dec. 2012.

[8] P. Y. Chiang, "Sense amplifier power and delay characterization for operation under low-Vdd and low-voltage clock swing," *2009 IEEE Int. Symp. Circuits Syst.*, pp. 181–184, May 2009.

[9] A.-T. Do, Z.-H. Kong, K.-S. Yeo, and J. Y. S. Low, "Design and Sensitivity Analysis of a New Current-Mode Sense Amplifier for Low-Power SRAM," *Ieee Trans. Very Large Scale Integr. Vlsi Syst.*, vol. 19, no. 2, pp. 196–204, 2011.

[10] S. K. Jain and P. Agarwal, "A low leakage and SNM free SRAM cell design in deep sub micron CMOS technology," *19th Int. Conf. VLSI Des. held jointly with 5th Int. Conf. Embed. Syst. Des.*, p. 4 pp., 2006.

[11] G. V. Kristovski and Y. L. Pogrebnoy, "New sense amplifier for small-swing CMOS logic circuits," *IEEE Trans. Circuits Syst. II Analog Digit. Signal Process.*, vol. 47, no. 6, pp. 573–576, Jun. 2000.

[13] M. Sinha, S. Hsu, a Alvandpour, W. Burleson, R. Krishnamurthy, and S. Borkar, "High-performance and low-voltage sense-amplifier techniques for sub-90nm SRAM," *IEEE Int. Syst. SOC Conf. 2003 Proc.*, pp. 113–116, 2003.

Discrete Event Simulation Modeling for Semiconductor Fabrication Operation

MA Chik[1],SR AB Rahim[2] AZ MD Rejab[3] K Ibrahim[4]
Fab Operations Department
Silterra Malaysia Sdn Bhd
Kulim, Kedah, Malaysia
mohd_azizi@silterra.com[1]

U. Hashim
Institute of Nanoelectronic Engineering
Universiti Malaysia Perlis
Kangar, Perlis, Malaysia
uda@unimap.edu

Abstract – This research is to investigate simulation modeling method for semiconductor fabrication factories (FAB) that known for complex manufacturing operation from various the literatures. This paper covers literatures from various publications since past 10 years. The significant simulation model used in common and semiconductor fabrication will be selected for evaluation in semiconductor fabrication manufacturing operation scenario with high mixed. Depends to the products mixed configuration, cycle time to complete semiconductor fabrication will take 60 to 90 days. The longer period needed due to wafer process required 300 to 1000 steps, process re-entrance to similar equipment more than 30% of the total steps further increase the complexity with many configuration due to setup changes. The simulation also needs to configure for process queue time, process dedication, several of difference for equipment configuration, capability and capacity. Primary simulation techniques reviewed in this analysis are discrete event, petri-net, gaming, virtual, intelligent, monte carlo and hybrid to understand individual strength and common usage in the market. This research summarized the highest usage, most uses and compatibility in similar operation for semiconductor fabrication. In summary, the research concluded DES is the most suitable technique for simulating FAB operation because of its nature of queuing and leaving concept that fits and resulted to 95% accuracy for WIP forecasting.

Keywords— Semiconductor Fabrication (Wafer Fab), Discrete Event Simulation (DES)

I. INTRODUCTION

Competitiveness of wafer technology makes the manufacturing execution became more complex and competitive. This is because the increasing number in processing steps in newer product, various cycles times requirements at respective processing step, fast turn around demand by customer, and cost competitiveness ([1] [2] [13]. Process to fabricate a chip on a wafer required 300 to 1000 steps process re-entrance to similar equipment more than 30% of the total steps, cycle time can be up to 90 days depends to complexity and priority given for a product with minimum of 60 equipment. Notice the ratio between number of steps and equipment is almost 5 times, which is due to the wafer fabrication process is not like a single line assembly process in normal factory instead of re-entrance

back to the same process [2] [3]. Fig. 1, describe the re-entrance process that generally apply for semiconductor fabrication. Normally the product mixes for wafer fabrication is based on Aluminum and Copper layers.

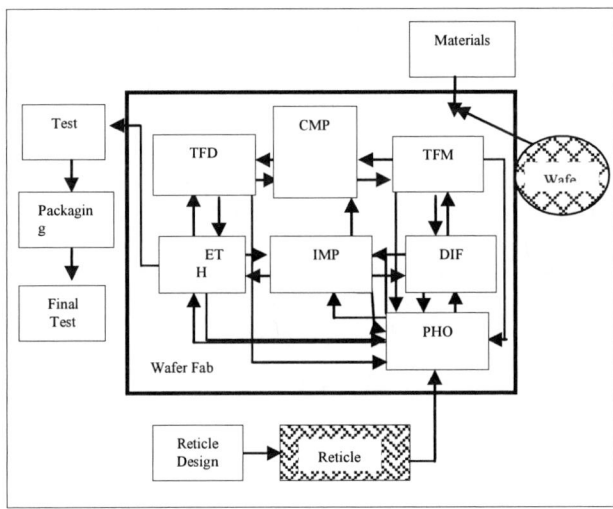

Fig. 1. Example of re-entrance process in semiconductor Fabrication

The fabrication of product used list of gasses like oxidation, nitride ammonia, resists, and several of chemicals for unnecessary layers removal to develop layers for the transistors and insulators and also metal interconnect. In the overall process, ion implantation, diffusion process except alloy and some of the process that prior contact masking in photolithography process, which step that is at about 55% of the total process which toward front of the overall wafer process for generic CMOS devices. Others process like passivation, CMP is been done after contact masking and toward to the end process. In the front processes, a product wafer will only visited to the respective tool in those area only which typical consists of 55% of the overall wafer fabrication process. However when the product enter the back end process, the challenge to dispatch lot are greater. This is due tool in Photolithography and Etch where the processes are made in the tool that shared front end and also

978-1-4799-5761-3/14 $31.00 © 2014 IEEE

at the backend, where more lot need to compete to use the tool for processing. For wafer dispatching, consideration of the process method for the respective tool which include single wafer processing, batch, cascading queue time, and same group of incoming thickness are built into the model. The overall process for manufacturing process required simulation model to verify impact of changes due to its complexity. The development of simulation will not be an easy task. This paper will analyze and conclude the right approach for simulation model for semiconductor fabrication.

II. LITERATURE REVIEW

Many simulation modeling approaches have been developed over the years by researchers from different fields. These approaches differ in terms of model detail intensity, representation, and underlying assumptions. These approaches can be loosely classified into eight types: discrete event simulation (DES), hybrid simulation (HS), Monte Carlo simulation (MCS), intelligent simulation (IS), Petri net (PN), simulation gaming (SG), virtual simulation (VS), and distributed simulation (DS). The review conducted based on published manuscripts on simulation in ten years indicates that most of the simulation manuscripts were published with DES types [4] [6] [9]. Fig. 2 shows the summary of the simulation model based on the summary of the literature. Further analysis conducted in this paper on recent publication revised the data from Jahagirian further. A quick review from eight type simulation practices shows that DES and hybrid shows significant usages compared others for practically used. Data for DS hardly available in this research and it may require more effort to review the usage. In this research, additional three simulations that scored next highest after Hybrid included for further verification in semiconductor fabrication potential used.

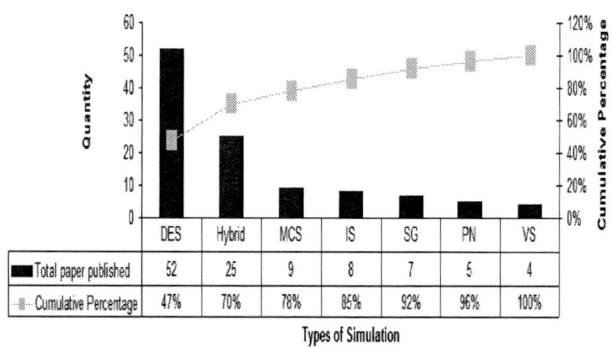

Fig. 2. Summary of Simulation Types found in literature [4] [5]

DES is probably the most commonly utilized simulation method in operational research. DES models a system as a sequence of discrete events; the entities (e.g., products) are believed to shift to different states as time passes. Entities join the system and "visit" particular states (not necessarily only one state) before leaving the system. DES systems refer to links of queues and servers in the system. DES models are dynamic, that is, the movement of time plays a

decisive role [4] [5] [6]. Many researchers employ DES to solve complex re-entrant manufacturing and operation problems include used of DES to simulate manufacturing scheduling in a coffee factory [7] [8] [9]. A large number of end products (more than 300 mixes), sporadic demand, and limited shelf life make manual production scheduling impractical. An integrated system that incorporated DES into the scheduling process was developed to address the problem. The researchers found that the simulation model helped evaluate/modify the schedule until a satisfactory schedule was determined. The model captured the overall effect of the schedule and helped improve the throughput of the system that perform an iterative process with an enterprise resource planning (ERP) system and a DES program to simulate scheduling systems [9] [10]. The researchers configured the arrival time of the product at a new process, which was later incorporated into the ERP system and regarded as the fixed time of scheduling.

HS involves the combination of various simulation techniques to solve a problem. HS models are commonly utilized in simulations to solve various manufacturing problems [4] found that 10 % of simulations are considered hybrid. The best known example of such an approach is the combination of DES and system dynamic (SD). Research on this particular combination has focused on the concept of "enterprise modeling and simulation" where the impact of production decisions is evaluated with DES models and is investigated based on enterprise-level performance measures.

MCS is one of the earliest simulation techniques developed; however, it has played a trivial role in manufacturing and business domains. Its usage is mainly limited to "static" problems or numerical problems with a stochastic nature such as those in property valuation and risk management [5] [6] [11]. B. Scibilia (2011) adopted MCS to perform optimization of the CMP process in fab. Y.H. Kwak and L. Ingall (2007) presented a review of the applications of MCS and their relevance to risk management and analysis in project management. The researchers reported that MCS is applied in many sectors such as semiconductor, geophysical, marine, and aerospace[11] [12]. Results from literature review shows used of DES most in the semiconductor fabrication.

III. METHODOLOGY

This paper will provide an assessments or evaluation of existing simulation techniques in semiconductor fabrication with final objective to forecast how many WIP (wafer) will completed all the process in next 30 days [12] [13]. During this process configuration of below steps needed.

1. Configure starts plan, lots, process flow that contains steps and recipes at all equipment.

2. Integration with existing database or flat files for WIP, lots status, current cycle time, remaining cycle time till completion, hold release until, active until, equipment down until, probability of potential success, reworks, qualification prior executing production lot.

3. To input process dedication for equipment, and equipment model need like single processing, batching, minimum and maximum batching, batch interval and wait no longer that x period.

4. Simulate 90 days period with variable changes in between (during running).

5. Provide results for product need to wait for process and during process for each step for with 95% accuracy.

In this analysis, the simulation model needs to perform logical operation similar as shown in Fig. 3. The simulation requirement as describe in Fig. 4.

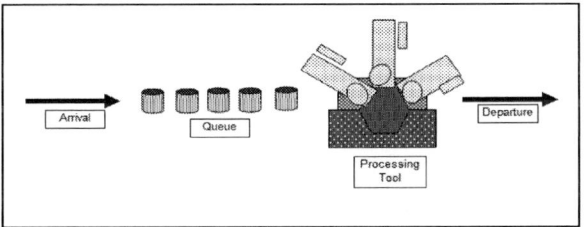

Fig. 3. Example of manufacturing operation in semiconductor fabrication

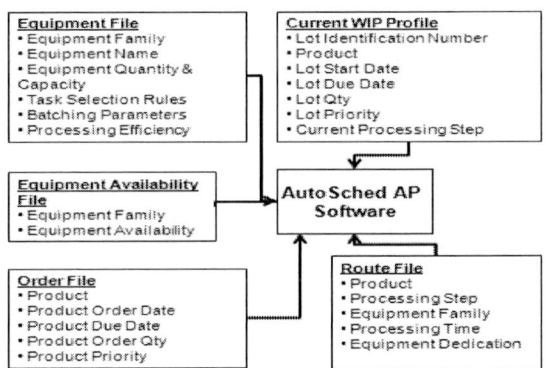

Fig. 4. Example requirement input to do simulation for semiconductor fabrication [12].

IV. DATA ANALYSIS

The process to develop simulation model utilized the simulation techniques learn from literature is performed. Lesson learned during the process is plotted in the table 1. Table 1, summarized that DES, is the most suitable simulation techniques for semiconductor fabrication case study.

TABLE I SUMMARIZED THE EVALUATION IN THE BULLET NUMBERING 1 TO 5.

No	High=5, Medium=3, Low=1			Types of Simulation Models						
	Category			DES	Hybrid	MCS	IS	SG	PN	VS
1	Ability to configure starts plan, lots, process flow that contains steps and recipes for each equipment			5	5	3	3	1	3	1
2	Ease of integration with existing database or flat files for WIP, lots status, current cycle time, remaining cycle time till completion, hold release until, active until, equipment down until, probability of potential success, reworks, qualification prior executing production lot.			5	5	3	3	1	1	1
3	Ability to consider process dedication for equipment, and equipment model need like single processing, batching, minimum and maximum batching, batch interval and wait no longer that x period.			5	3	5	3	5	5	1
4	Ability to simulate 90 days period with variable changes in between (during running).			5	3	3	3	3	1	3
5	Ability to process large data and resulted for detail data analysis			5	3	5	5	5	3	5
	Total			25	19	19	17	15	13	11

To understand which simulation model can be easily used in the real application, a simple number has been allocated into the category to summarize the process difficulty. The categories are High, Medium and Low. High means less effort to be used with the category specified and less time spending based on actual works been done. This review is being done among the researchers and engineers that involved in this project.

Five points will be given for High. Medium means moderate of time needed to spend to use this approach include time for modification will take no more than two days. Low means, not favorable due to its difficulty of using it. A same group of engineers evaluate the simulation type and summarized with the following result. In this analysis, DES score highest which is 25, versus others that score 19 to 11.

Further analysis to validate accuracy is performed using pair t-test. The accuracy equation in this analysis was based on

$$1 - (100\% \times |\,(Forecast-Actual)/Actual)|) \qquad (1)$$

Table 3 shows the accuracy measurement for WIP arrival forecast based on 95% accuracy.

TABLE III, ACCURACY SUMMARY BASED ON PAIR T-TEST VALUE

Simulation Model	P value for Pair T-Test for 95% Forecast Accuracy		
	95% CI	98% CI	99% CI
Discrete event simulation (DES)	0.052	0.052	0.052
Petri-net (PN)	0.047	0.047	0.047
Simulation gaming (SG)	0.042	0.042	0.042
Virtual simulation (VS)	0.049		
Intelligent simulation (IS)	0.039	Significant below 0.03	
Monte carlo simulation (MCS)	0.024		
Hybrid simulation (Hybrid)	0.031		

For t-test approach in this analysis, the research is attempt to get P value more than 0.05, which means that there is no significant difference between actual versus forecast. Accuracy in DES is further validated with absolute percentage error (APE) [14] [15]. Table 4 shows example of 5 runs of simulation forecast against the actual where WIP is based on difference month. Accuracy of the run is 96.7%.

TABLE IV DES ACCURACY VALIDATION WITH APE

No of Simulation Forecast	Forecast	Forecast Y...	Actual	APE	APEn...
A	200	Y_A	214	6.54%	$=\|(Y_A-ActualA)/(ActualA)\|$
B	231	Y_B	222	4.05%	$=\|(Y_B-ActualB)/(ActualB)\|$
C	224	Y_C	215	4.19%	$=\|(Y_C-ActualC)/(ActualC)\|$
D	290	Y_D	287	1.05%	$=\|(Y_D-ActualD)/(ActualD)\|$
E	289	Y_E	291	0.69%	$=\|(Y_E-ActualE)/(ActualE)\|$
n		Y_n	Actualn		$=\|(Y_n-Actualn)/(Actualn)\|$
MAPE	3.30%				$(\frac{1}{n}\sum_{I-1}^{n}\|(Y_n - Actual_l)/Actual_l\|)*100$
MedAPE	4.05%				$\sum_{n-1}^{n}\frac{1}{n}\sum_{I-1}^{n}\|(Y_n - Actual_l)/Actual_l\|$
SKEW	0.16				$\frac{n}{(n-1)(n-2)}\sum_{I-1}^{n}\|(Y_e - Actual_l)/Actual_l\|)$
Accuracy	96.70%				$1-(\frac{1}{n}\sum_{I-1}^{n}\|(Y_n - Actual_l)/Actual_l\|)*100$

V. CONCLUSION

In this paper, an analysis to validate which simulation techniques is most suitable for semiconductor fabrication manufacturing or operation was performed. The analysis is based requirement to develop simulation model from real semiconductor fabrication. A summary of survey table shared and accuracy performance is done with pair t-test value. The results are summarized in the survey table 2, shows that DES approach is the easiest method to perform simulation model for semiconductor fabrication. Most is due the existing logical approach that designed in the technique. Table 3, summarize that DES model has p value higher compared to other simulation techniques. Table 4 shows that the DES accuracy shows in the in forecast accuracy of 96.7% using absolute percentage error (APE). Its concluded that DES technique achieved higher accuracy for forecasting purposes. The company has decided to use DES technique in the simulation model. It will also be useful for manufacturing students to learn various of the simulation techniques and apply some of them in the classroom to help them shorten the learning curved when the work in the industry later. A new knowledge gains from simulation in gamming that potentially be used for smart sampling techniques for industry problem in future work.

MENTZER, ACKNOWLEDGMENT

The Authors would like to thank Silterra Management, Manufacturing Systems Department that helps to provide data and support for the process that results in understanding of which method is practical to be used in the industry. The authors also would like to thank Collaborative Research in Engineering, Science & Technology Center (CREST) for providing computer servers, where simulation programmed and analysis are being done.

REFERENCES

[1] MA Chik, CY Ve , P Balakrishna, U Hashim, I Ahmad and B Mohamad, "A study for Optimum Productivity in 0.16μm Products in Wafer Fabrication Facility," IEEE International Conference on Semiconductor Electronics (ICSE), 2010, pp 377-380.

[2] MA Chik, K Ibrahim, MH Saidin, FM Yusof, G devandran and U Hashim. Development of Capacity Indices for Semiconductor Fabrication", 10th IEEE International Conference on Semiconductor Electronics 2010,pp 649-653.

[3] MA Chik, YE Teck,and LK Wee, A study for optimum productivity yield in 0.16μm mixed of wafer fabrication facility," IEEE International Conference on Semiconductor Electronics (ICSE), 2010, pp 377-380.

[4] M Jahangirian, T Eldabi, A. Naseer, L Stergiouslas and & T Young, ," Simulation in manufacturing and business: A review", European Journal of Operational Research, 20(9),2009, pp 1-13.

[5] W. Lenders and C. Baier, "Genetic algorithms for the variable ordering problem of binary decision diagrams", 5th International Advances in Grid and Pervasive Computing Conference Proceedings, Heidelberg, 2005, pp 1-20.

[6] L. Shuo, J. Zhao, L. Yin and X. Wang, "An algorithm of task scheduling in survivability", 3rd International Artificial Intelligence and Computational Intelligence Conference Proceedings, Heidelberg, 2011, pp. 85-92.

[7] O. Engin, G. Ceran and M. K. Yilmaz, "An efficient genetic algorithm for hybrid flow shop scheduling with multiprocessor task problems", Journal of Applied Soft Computing, 11(3), 2011,pp 3056-3065.

[8] T. Heb, J. Seifert and P. Schegner, "Comparison of Static and Dynamic Simulation for Combined Heat and Power Micro Units", 17th PSCC Proceedings, Stockholm, 2011, pp 200-207.

[9] M. Jahangirian, T. Eldabi, A. Naseer, L. K. Stergiouslas and T. Young, "Simulation in manufacturing and business: A review", European Journal of Operational Research, 20(9),2009, pp 1-13.

[10] W. Chan, Y. J. Son and C. Macal, "Agent-based simulation tutorial - simulation of emergent behavior and differences between agent-based simulation and discrete- event simulation", Winter Simulation Conference Proceedings, 2010, pp 135-150.

[11] N. Ward "Predictability as a key Compnent of Productivity", Applied Materials Automation Software Symposium (Singapore), Keynote Speaker, 2012.

[12] SR AB. Ibrahim, I Ahmad, MA Chik, AZ Md. Rejab and U. Hashim, U. "Development of Manufacturing Simulation Model for Semiconductor Fabrication". World Academy of Science, Engineering and Technology (WASET), 2013.

[13] K Ibrahim, MA Chik and U Hashim. "Managing Demand Variabiliy to Achieve Optimum Cost and revenue in Wafer foundry". 11th Asia Pacific Industrial Engineering & Management Systems Conference, 2010.

[14] RJ Hyndman and AB Koehler, "Another look at measures of forecast accuracy ".International Journal of Forecasting 22, 2006, pp 679-688.

[15] JT Mentzer, and K Kahn, Forecasting technique familiarity, satisfaction, usage and application, Journal of Forecasting, 1995, 14, pp 465-476.

Horrendous Capacity Cost of Semiconductor Wafer Manufacturing

K Ibrahim[1] , MA Chik[2]
Operations Department
Silterra Malaysia Sdn Bhd
Kulim, Kedah, Malaysia.
Kader_ibrahim@silterra.com[1], Mohd_azizi@silterra.com[2]

U. Hashim
Director
Institute of Nanoelectronic Engineering
Universiti Malaysia Perlis, Kangar, Perlis, Malaysia.
uda@unimap.edu.

Abstract— **Semiconductor wafer manufacturing, being one of the most advanced and complex process, commands high level of utilization of the available tools to ensure maximum productivity. It is also very important to keep the operations as lean as possible to ensure cost effectiveness. In this research we will show that the cost additional capacity is outrages. This is the main reason more and more companies are opting out of fab owners club. Others are scaling down and going fables and fab-light. Capacity utilization and capacity maximization is the key to a successful fab. Fabs continuously look for ways to increase the capacity by improving productivity. Beyond certain productivity level, fabs must spend on purchasing tools. Semiconductor tools are expensive and in many cases there will be a need to spend in the support infrastructure. The escalating cost really brings out the creativity and innovation among the fab engineers. This paper discusses what actions are taken to address or mitigate this issue. The research is based on some available data from SilTerra Malaysia Sdn Bhd wafer fab in Kulim.**

Keywords— semiconductor, manufacturing, capacity, cost.

I. INTRODUCTION

Capacity of a semiconductor wafer manufacturing facility is monitored closely by everyone in the fab to ensure maximum and optimum utilization. This is to ensure the fabs can recover the huge investment in capital expenditure. The capacity is increasing at an unbelievable rate and will be reaching 278 million 200 mm wafers by 2020 [1]. The cost of adding capacity has increased exponentially. A typical 300mm fab with a run rate of 30,000 wafers per month will cost US$7 billion compared to about US$100 million in 1980 [2]. Figure 1 shows how the cost has increased over the years, GlobalFoundries spent more than US$10 billion on the latest 300mm wafer fab in New York [3]. The cost of wafer fabrication plant is increasing at a rate that is unimaginable. This is due to the fact that wafer fabrication is a capital intensive game. Only those with a lot of cash and reserves can play this game.

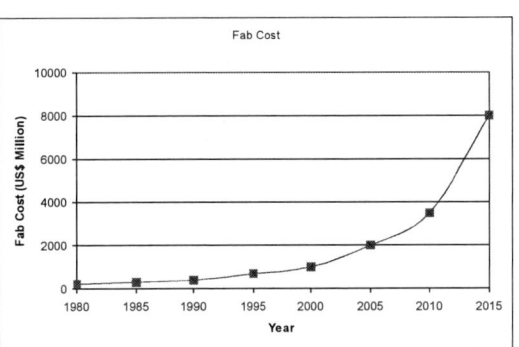

Figure 1: Fab cost from 1980 to 2015 [2].

As such more and more traditional fab owners are opting to exit the fab owners club to become fables or fab-light. The fab owner's club member is reducing drastically even though the number of fabs in the world may be increasing. The introduction of the 450mm wafer fab will make this game even more interesting and eventually only a few players will be left to play the game due to the huge investment needed. It takes 10 years for the new wafer size to reach 35% worldwide silicon surface area consumption [4] [9]. Based on the 200mm and 300mm wafer evolution, 450mm wafer evolution will start and by 2025 represent 35% of world silicon that is being consumed. This again will drive the wafer fabrication capital spending through the roof. Already today a 300mm wafer fab will cost about US$10 billion. A 450mm wafer fab will easily be US$15 billion based on the previous wafer size change transformation. This will definitely reduce the number of players in this field. Only those with a lot of cash reserve will continue to invest in this business and reap the benefits. The entry barrier to this business becomes extremely high that one day, the very few players will monopolize the business and start taking advantage of this situation.

II. CAPACITY COST

The capacity cost that we will discuss here is the fixed cost incurred due to investment in building, infrastructure, machines and tools [6]. The cost of building a wafer has been steadily increasing and there is no sign it will stop. Fabs that want to increase capacity have a few options. They can increase the number of tools in the current clean-room utilizing the readily available utilities and infrastructure. This is possible if there is

978-1-4799-5761-3/14 $31.00 © 2014 IEEE

available space in the clean-room. Fabs can also explore adding an annex to the existing building and create and extension. This section is usually referred to by a suffix. If the main fab is called fab 4, this section will be called fab 4A. This alternative is explored when there is still available support infrastructure from facilities. In this option, there is also licensing deals that can be of advantage to the fabs. Many of the required software are based on fab license or site license. Some fabs also go out and buy other fabs to add capacity. This is an easy option to acquire an entity that is already operational. The advantage here is time and cost. The other more costly alternative is building a brand new fab. Even in this option, many fabs resort to installing used equipments where possible to reduce investment cost.

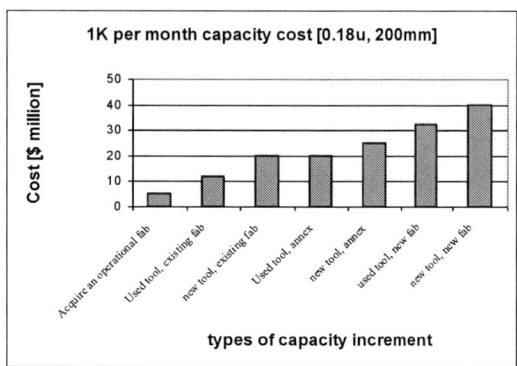

Figure 2: Fab capacity increment cost for 200mm wafer fab.

Figure 2 shows the cost involved in increasing the capacity employing various methods. Although acquiring another fab seems like a logical solution, it usually is the most difficult to execute due to many hurdles. This method needs huge upfront investment to acquire the fab and there is no way of increasing the capacity slowly at a pace of 1K/month increments. With the latest advancement in technology, there is almost no new 200mm wafer fab being built. Potential investors are buying used fabs and converting them to suit their needs. Those in the very high technology race are building 300mm fabs.

Fabs are facing challenges on how to increase the capacity at the lowest possible cost with the lowest risk [9]. There are situation where tools are acquired cheap in the used market and fabs spent a lot of money in refurbishing, rebuilding and in many cases end up with other licensing issues. In going back to the OEM (original equipment manufacturer), fabs end up forking out a lot of money to get the license legalized. One of the reason is because OEM themselves are also suppliers of used tools. When a fab bought their tool through other means, this denies them the business opportunity. Nowadays the costs of adding tools are increasing because fabs are going back to the OEM to reduce risk and uncertainty.

Fabs that are chasing the technology are also looking at many aspects before deciding which way to go. Moving to the next wafer size will give immediate 30% savings in the cost per silicon area [5]. The wafer cost will continue to escalate and eventually fabs need continuous strategy in cost reduction. Newer fabs are fully automated and this reduces the direct labour drastically. More than 50% of head count can be reduced and this will significantly reduce cost and drastically improves productivity. Cost structures in US, Europe and Japan shows that more than 30% of wafer manufacturing cost are attributed to salary and benefits [7]. Fabs started moving to Asia to reduce this impact, but now the salary cost in Asia is also catching up fast. Moving to the next wafer size and full automation is now a necessity and inevitable.

Wafer or silicon manufacturing cost can be reduced by moving to the next wafer size, but this comes with a huge price. 50% increase in capital expenditure is expected. Even within same wafer size, fabs are experiencing increase in cost over the years due to more and more automation, finer geometry requirements and more sophistication in the tools.

III. ALTERNATIVE CAPACITY

Over the years wafer fabs have become creative in adding capacity. It started with sharing capacity between fabs in the same campus. In Singapore, fab A diverts some of the WIP in certain bottleneck process to be processed in fab B that has similar capability and unutilized capacity. This is managed through automation where the number of wafers that will travel inter-fab will be decided every morning. Overall this fab could better utilize the resources available across fabs. In Taiwan similar situation happens and on top of that WIP is also transported to fabs outside the campus through proper logistic planning. At a given time of the day, a bunch of lots at process that needs help are transported to different fabs that may be outside the campus. There are situation where lots are shipped to a location that is far during crisis. This happens when critical equipment is down for a prolonged period and the fab has another wafer manufacturing location overseas or even very far away. In order to get some of the critical lots to customers, fabs are left with no choice but to resort this method.

In certain fabs, bottleneck processes are outsourced to other fabs where such process exist and are available. In this case, it is a business transaction between 2 entities and for each process step a fee is charged. This is usually a short term measure until the fabs increase their bottleneck process capacity.

There are also fabs that do outsourcing where certain orders are sent to other fabs with pre determined business arrangement. In some fabs this activity is called as order desk where orders are distributed to various fabs around the world that have had pre qualification done. In this arrangement wafers may be manufactured in any fab within the company or even in a fab outside the company. This is transparent to the customer, where customer still receives the ordered good as per the agreement. Again this is a short term measure where fabs buy time before additional capacity is available in house. Customers also do not prefer their wafers to be fabricated elsewhere where they have less visibility and control [8] [9]. Customers will also demand reduction in ASP (average selling price) [9].

Since adding capacity in wafer fabrication is a big decision and it is not easy to get shareholder approval, funding and financing, there is always delay in decision making. This is true for smaller companies where cash reserves are limited. The cost of building a clean-room is also escalating over time due to various factors [9]. Figure 3 shows the escalating cost of building a clean-room for 300mm fabs from various fabs across the world. The obvious reasons are the technology nodes advancement that needs a stricter clean-room environment.

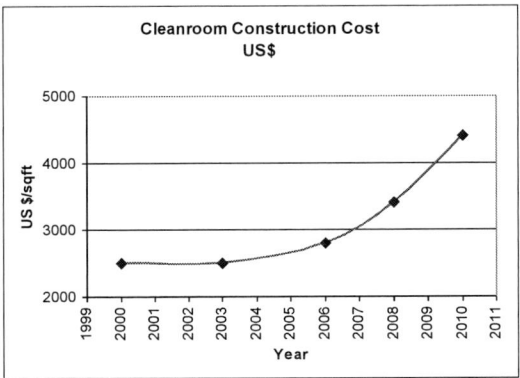

Figure 3: Fab clean-room construction cost.

IV. DISCUSSION

Wafer fabs are facing trying times where adding capacity means survival. Due to business expansion, every fab will have to add capacity to ensure they are in a position to support the customers who are also expanding in their business. Once fabs cannot provide the capacity needed, then customers will start looking elsewhere. Loosing customers is inevitable. Getting customers and keeping them is an extremely difficult task in wafer fabrication. On the other side fabs are struggling to add capacity to satisfy the customer's capacity needs. This is prevalent mainly for the smaller foundries with revenues below US$500 million per year. These smaller foundries are also struggling to reduce cost. The bigger fabs has the economies of scale advantage and they are always ahead. The ridiculous cost of increasing capacity is another factor that is stopping these fabs from expanding. Many fabs are for sale or looking for partners for consolidation due to this pressure. The solution to the escalating cost of adding capacity is to be creative and bold to explore new business models. Some of the ideas are like collaboration among smaller fabs to utilize idling capacity, outsource bottleneck processes to other fabs that have excess capacity and getting customers to consign the tools. There are many other ideas that can be brought to the table for discussion. This can only happen if all related parties have an open mind to explore various models.

V. CONCLUSION

Wafer foundries needs to find its own niche to survive or face extinction. With fierce competition among the foundries, foundries are looking for ways to fulfill customer demand without spending huge capital expenditures. Each fab must evaluate their strength and weakness before deciding which path to select. The intense competition and product introduction in the consumer market is always forcing fabs to consolidate, sell or adopt a new product strategy. It is inevitable that there will be fewer foundries in the future as more mergers, acquisition and partnership take place.

ACKNOWLEDGMENT

All the data discussed and presented were obtained from database at SilTerra Malaysia Sdn Bhd. We are deeply indebted to SilTerra in allowing us access to many of the proprietary information. I would like to also thank Prof Uda Hashim and Mohd Azizi Chik for their discussion and encouragement in completing this paper.

REFERENCES

[1] anysilicon, "Foundry Ranking by Capacity 2013-2014", http://anysilicon.com/foundry-ranking-capacity-2013-2014, accessed on 22nd May 2014.

[2] S.W. Jones, "Integrated Circuit Economics", IC Knowledge LLC, 2010.

[3] D. Reisinger, "GlobalFoundries' NY chip plant to get up to $10B boost" 3rd Jan 2014, CNET .

[4] IC Knowledge LLC, "White paper- Forecasting the 450mm Ramp Up", 2014.

[5] N. Mokhoff, "Semi industry fab costs limit industry growth", EE Times 3rd Oct 2012.

[6] Y-C Chou, C-T Cheng and Yi-Yu Liang, "Analysis Of Capacity Strategies in The Uncertain Environment of Semiconductor Manufacturing", Proceedings of the Fifth Asia Pacific Industrial Engineering and Management Systems Conference 2004.

[7] K. Ibrahim, Silterra internal proprietary report, unpublished, 2014

[8] MA Chik, K Ibrahim, MH Saidin, FM Yusof,G devandran, and U Hashim. Development of Capacity Indices for Semiconductor Fabrication." 10th IEEE Internation Conference on Semiconductor Electronics. 2010. pp 649-653.

[9] K Ibrahim, MA Chik and U Hashim. Managing Demand Variabiliy to Achieve Optimum Cost and revenue in Wafer Foundry. 11th Asia Pacific Industrial Engineering & Management Systems Conference 2010.

Low power and low voltage SRAM design for LDPC codes hardware applications

Rosalind Deena Kumari Selvam, C. Senthilpari, Lee Lini
Faculty of Computing & Informatics,
Faculty of Engineering
Multimedia University
Cyberjaya, Malaysia
rosalind@mmu.edu.my

*Abstract:*The Low Voltage Low Power (LVLP) 8T, 11T, 13T and ZA SRAM cell is designed using the dynamic logic SRAM cell. The SRAM cells are implemented using pass transistor logic technique, which is mainly focused on read and write operation. The circuits are designed by using DSCH2 circuit editor and their layouts are generated by MICROWIND3 layout editor. The Layout Versus Simulation (LVS) design has been verified using BSIM 4 with 65nm technology and with a corresponding voltage of 0.7V respectively. The simulated SRAM layouts are verified and analyzed. The SRAM 8T gives power dissipation of 0.145 microwatts, propagation delay of 37.2 pico seconds, area of 14 x 8 micrometers and a throughput of 4.037 nano seconds.
Keywords— power dissipation, delay, throughput, SRAM cell

I. INTRODUCTION

Lower power operation has become of crucial importance in VLSI Design. One of the ways of obtaining power reduction is by lowering the power supply and has been seen to be good and effective. One of the essentials of IC design techniques is to lower the power of memory circuits with a minimum tradeoff on its performance. The VLSI industry is constantly striving towards achieving high density, high speed and low power devices in the CMOS technology. As the size of the transistor is being reduced to about 70% of its earlier version using new technology, the density of these devices on chip has been increased and also reduction in delay time has been obtained to satisfy the demand of high performance.

Different memory circuits like the SRAM, covers a considerable area in the design of digital ICs. Arun Ramnath Ramani and Ken Choi have [1] shown that it is possible to push the design of low power SRAMs into the sub threshold region and then compared it with various parameters like speed, power consumption and average power delay product. Yashwant Singh and D. Bhoolchandani have [2] focused on design ofSRAM cell with dynamic Vt and dynamic standby voltage to mitigate the leakage power dissipation. Simulation results show significant reduction in power dissipation in standby mode of SRAM cell. An8T-CDC column-decoupled SRAM was designed using a half-select free design by Rajiv V. Joshi et al [3] which enabled enhancement in voltage scaling capabilities, and there was a 30%–40% power reduction in comparison to standard 6T techniques. A 10T SRAM cell circuit was designed by Takahiko et al [4]considering the static noise margin(SNM) .This circuit was

ratio-less and prevented the loss of information. Kumkum Verma et al [5] proposed a design of 6T SRAM cell at 180 nm using two types of architecture namely bank partitioning and matrix array architecture which focused on optimizing power and delay. They found in order to lower the power in SRAMs it was necessary to reduce the capacity bit and word lines. By this the power dissipation reduces by 78% and speed by 23% at the cost of 6.93% more transistor count.[6].A study of SRAM cell based on tunneling field effect transistors(TFET) was made by Xuebei Yang and KarthikMohanram. They proposed a new 6T TFET SRAM circuit utilizing ground lowering read assist (RA). This proved to have better performance and reliability. It also occupied about 10 to 15% less area and consuming at least 4 orders of magnitude of lower static power making it a good choice for low power high density SRAM applications.

Our proposed dynamic design is a tradeoff between existing challenges highlighted by the authors and gives a better performance in terms of power dissipation, propagationdelay and speed.In the present paper, a modified SRAM cell is proposed to implement LDPC decoders, which is used in DSP algorithms and cryptography. The proposed SRAM cell is compared with other different type of SRAM cell designs. These SRAM cells have been designed using CAD tools such as DSCH2 for logic design and Microwind 3 for the layout design and timing simulation. The SRAM cell is simulated and the results are compared with other published results in terms of power dissipation, propagation delay, and total chip area. The results obtained from simulation shows that the proposed circuit performs better in terms of propagation delay and throughput.

II. Design Method

The design of the circuit used follows the dynamic CMOS logic, which is also known as Pre-charge-Evaluate logic. This design allows reducing the number of transistors used to implement any logic function substantially. In this method the output node capacitance is first pre-charged and then the output voltage is evaluated based on the applied inputs.Both these operations are driven by a clock signal which drives one NMOS and one PMOS transistor in each dynamic stage.

978-1-4799-5761-3/14 $31.00 © 2014 IEEE

When the clock signal is low, the PMOS precharge transistor starts to conduct while the complementary NMOS transistor is off. The output capacitor is charged up through the conducting PMOS transistor to a logic high level of $V_{out} = V_{DD}$. The input voltages at this stage have no affect on the output as one of the transistors is turned OFF. When the clock signal becomes high, the precharge transistor turns off and the other transistor which is the complementary one is turned on. The output node voltage now changes eitherhigh or low depending on the input voltage levels. If the input signals create a path to conduct between the output node and the ground then the output capacitance will start to discharge to 0.

This paper is mainly focused on the design of SRAM cell using dynamic logic. A fundamental NMOS dynamic logic circuit comprises of an NMOS pass transistor which drives the gate of another NMOS transistor.A periodic clock signal drives the pass transistor and it helps to charge up or charge down a capacitance depending on the input signal. This leads to the concept of logic 0 when the clock is low (0) and the capacitance is in the charge down mode or logic 1 when the clock input is high (1) and the capacitor is in the charge up mode. The output of the inverter will have a value logic 1 or logic 0 depending on the voltage at the capacitance.

Logic '1' transfer: If the voltage at the node is assumed to be 0 initially, anda logic 1 level is applied to the input terminal. The clock signal at the gate will now increase from 0 to V_{DD} at $t = 0$. Now the pass transistor will start to conduct as soon as the clock becomes active. The pass transistor will work in the saturation state and charges up the capacitor. The pass transistor will turn off when the voltage at the node equals the maximum voltage and the gate source voltage will be equal to threshold voltage.

Logic '0' transfer: If it is assumed that the voltage at the node is 1 initially and logic 0 is applied at the input terminal. The clock signal at the gate of the pass transistor changes from 0 to V_{DD}. The pass transistor starts to conduct as soon as the clock signal becomes active and the current flows in the opposing direction of as during charge up of the capacitor. The pass transistor operating in the linear region discharges the capacitor and the source voltage now becomes 0.

Our design uses a 65nm technology size for 8T, 11T, 13T and ZA circuits using the logic 0 and logic 1 transition for inputs of WL= 1 and the BL input as '0' and '1'. The circuits below show the basic setup for the simulation. According to dynamic logic technique and pass transistor logic, the 8T, 11T, 13T and ZA circuits are shown in Fig. 2(a), Fig. 2(b), Fig. 2(c) and Fig. 2(d) respectively.

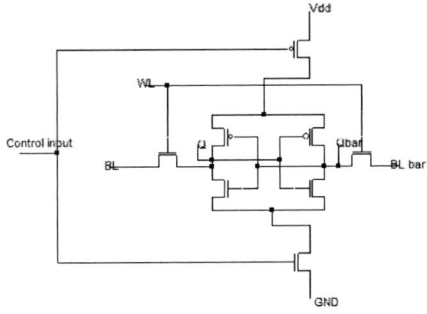

Fig. 2(a): 8T SRAM

The circuits shown above in Fig.2 (a), 2(b), 2(c) and 2 (d) have been designed using the dynamic logic. The operation of the circuit is divided by the control input into two distinct phases that are the pre-charge and evaluate intervals.

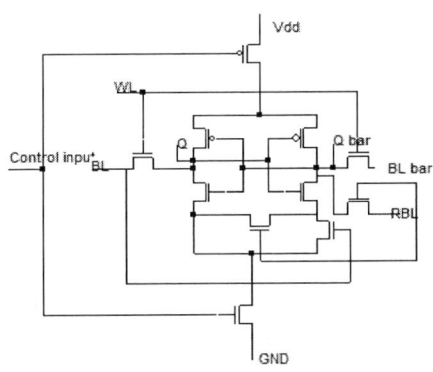

Fig. 2 (b): 11T SRAM

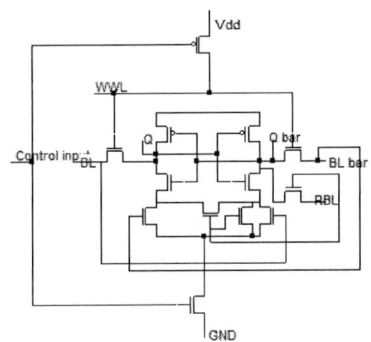

Fig. 2(c) 13T SRAM

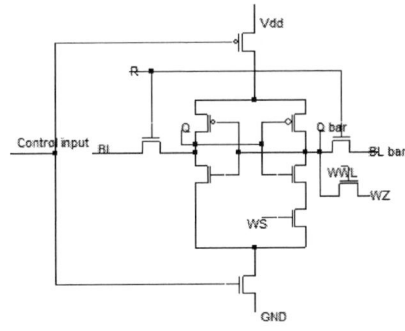

Fig. 2 (d) ZA SRAM

Timing Diagrams:

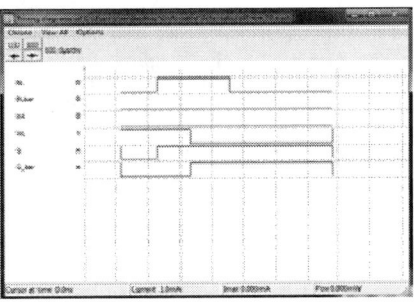

Fig. 3(a) 8T

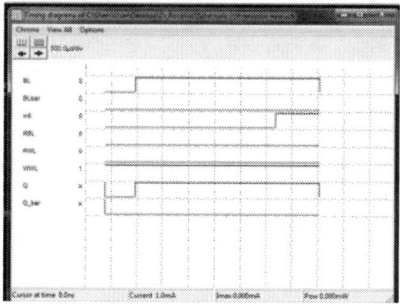

Fig. 3(b) Icular stack.

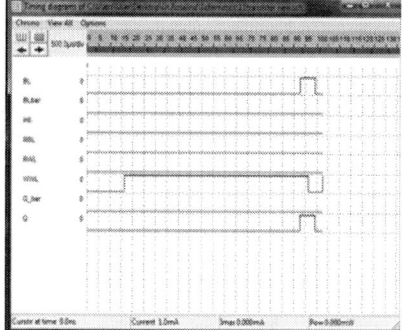

Fig. 3(c)13T

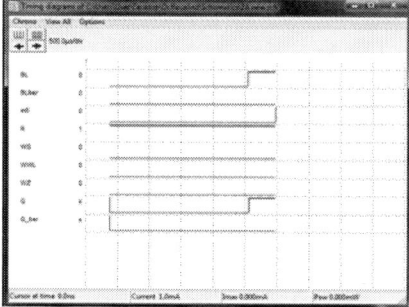

Fig. 3(d) ZA

A condition of 0 at the control input defines the pre-charge where the PMOS is conducting while the NMOS is cutoff as is seen in the timing diagram in Fig. 3(a), 3(b), 3(c) and 3(d) respectively. The switching of the control input between 0 and 1 and the effect on the circuits of 8T, 11T, 13T and ZA is seen in the timing diagrams of Fig.3 (a), 3(b), 3(c) and 3(d).In the part III of the paper the results obtained using the above techniques achieve low power dissipation, less delay and better speed

III. RESULTS AND DISCUSSION

A VLSI circuit can be determined as a complex maze of paths that influence on the whole circuit function. When the designis not appropriately simulated it will ruin the whole circuit operation. In order to ensure the rightness of the circuit there are a few ways to minimize these effects. One of the methods to correct the fault is by putting the analysis data on the certain probe points and follows the data flow through the circuit. The output points later can be observed to determine whether the circuit has handled the data in an appropriate manner. The observation of these data as they flow through the computer representation of a circuit is called the simulation and the selection of input data is known as a test-vector generation.Our SRAM cell is designed in context with

dynamic logic; the results are evaluated in terms of logic 0 and logic 1 for pre evaluation charges. The outputs and leakage are evaluated according to pre evaluation charge logic.

Table 1 shows the Logic 0 simulation result of dynamic 8T SRAM cell, which was measured with WL =1 and BL=0 for the SRAM cell. The function of the SRAM cell is designed by NMOS tree. When the input is 1 the current that passes through the drain is low, so the output drops the voltage of V_{DD}, in line with the CMOS 65nm technology.According to NFET standards, the output voltage and the dissipation is less compared with other existing results. The parameters components resulting from the layout are analyzed using BSIM4 analyzer. The output parameters values of delay and falltime are made perfect due to dynamic logic. The dynamic logic cells are acting as a push-pull device, which maintains the flow of charges in a regular manner. So, our dynamic logic based 8T SRAM cells give lower delay than other logic. The area A can be calculated from SRAM 8T whole cell which includes input pads and output pads. Any memory logic will need to place a bit of data at the correct location of the memory cell. The proper selection of the cell location will be done if the logic is given sufficient energy. This can be calculated from ½ CV^2. The throughput of the memory logic always depends on the number of stages of stack. Our 8T SRAM cell has proper load capacitance which consumes sufficient energy to place the logic in the part

Table I: Logic 0 simulation result of dynamic SRAM cell

Type	V_O (V)	I_D (mA)	P_D(W)	Propagation Delay	PDP	A=WxH	Through put
8T	0.7	0.06	0.145 x10⁻⁶	37.16 x10⁻¹²	5.388 x10⁻¹⁸	14x8μm	4.372 ns
11T	0.69	0.06	0.101 x10⁻⁶	8.70 x10⁻¹¹	8.787x10⁻¹⁸	17x8μm	4.087ns
13T	0.67	0.25	1 x10⁻³	8.41 x10⁻¹⁰	8.41 x10⁻¹³	20x8μm	6.841ns
ZA	0.67	0.03	6 x10⁻⁹	8.50 x10⁻¹⁰	5.1 x10⁻¹⁸	16x8μm	8.85ns
8T	0.7	0.06	0.145 x10⁻⁶	3.72 x10⁻¹¹	5.388 x10⁻¹⁸	14x8μm	4.037ns
11T	0.68	0.06	0.101 x10⁻⁶	7.00 x10⁻¹¹	7.07 x10⁻¹⁸	17x8μm	4.07ns
13T	0.69	0.25	1 x10⁻³	9.25 x10⁻¹⁰	9.25 x10⁻¹³	20x8μm	6.925ns
ZA	0.66	0.03	606 x10⁻⁹	8.50 x10⁻¹⁰	5.151 x10⁻¹⁶	16x8μm	8.85ns

Table 2 shows the Logic 1simulation result of dynamic SRAM cell, which was measured with WL =1 and BL=0 and WL= 1 and BL=1for the SRAM cell. The NMOSFET tree determines the function of the SRAM cell here. When the input is 1 the output node voltage remains at the logic 1 high level ordrops to a logic low level depending on the input voltage levels.The output voltage drops to its correct logic level after a certain time delay.The voltage falls to a value almost where the output voltage equals Vdd. The parameters of components resulting from the layout are analyzed using BSIM4 analyzer. The output parameter values of delay and fall time are made almost to the best value because of dynamic logic.

Table II: Logic 1 simulation result of dynamic SRAM cell

Type	V_O (V)	I_D (mA)	P_D(W)	Propagation Delay	PDP	A=WxH	Through put
8T	0.56	0.056	7.817 x10⁻⁶	3.72 x10⁻¹¹	2.907x10⁻¹⁶	14x8μm	4.0372ns
11T	0.66	0.056	2.341 x10⁻⁶	3.60 x10⁻¹¹	8.427 x10⁻¹⁷	17x8μm	4.036ns
13T	0.68	0.249	7 x10⁻⁶	9.09 x10⁻¹⁰	6.363 x10⁻¹⁵	20x8μm	6.909ns
ZA	0.67	0.249	906 x10⁻⁹	9.00 x10⁻¹⁰	8.154 x10⁻¹⁵	16x8μm	8.9ns
8T	0.56	0.056	7.817 x10⁻⁶	3.72 x10⁻¹¹	2.907 x10⁻¹⁶	14x8μm	4.0372ns
11T	0.54	0.056	9.061 x10⁻⁶	1.04 x10⁻¹⁰	9.423 x10⁻¹⁶	17x8μm	4.104ns
13T	0.68	0.249	7 x10⁻⁶	9.42 x10⁻¹⁰	6.594 x10⁻¹⁵	20x8μm	6.942ns
ZA	0.67	0.249	1.306 x10⁻⁶	9.84 x10⁻¹⁰	1.285 x10⁻¹⁵	16x8μm	8.984ns

978-1-4799-5761-3/14 $31.00 © 2014 IEEE

Table III: Comparison of power dissipation, delay and PDP

Parameter	Proposed circuit	Ref [1]	Ref [2]
Power Dissipation	0.145×10^{-6}	-	0.0687×10^{-6}
Delay	37.2×10^{-12}	1018.3×10^{-12}	-
PDP	2.907×10^{-16}	1.347×10^{-18}	-

From the 8T SRAM cell simulation, the results obtained and its comparison with other existing circuits is shown in Table 3.From the table it can be noted that our SRAM 8T cell compared with Arun et al. Ref [1] and Yashwant Singh et al Ref [2] SRAM circuits. Our proposed dynamic SRAM circuit has less critical path, due to lower switching threshold.So our circuit has given 96.34% improvement in terms of delay. But our circuit consumes more power when compared with reference [1] and reference [2] due to logic transition in dynamic cell.It is seen that for logic 0there is a small change in the power dissipated than during the logic 1 where some significant difference can be observed. The simulation of voltage versus time using Micro wind 2 shows the performance of the circuit for the 8T,11T, 13T and ZA circuits for the conditions of logic 1 control input and inputs write logic WL=1 and bit logic BL=0. From the simulation diagram the output power obtained for an input voltage of 0.7V can be observed. The propagation delay in terms of rise time t_r and fall time $t_f$is noted. It can be seen that the rise time and fall time are small values hence making the propagation delay to be a minimum. The output voltage is observed to be almost reaching V_{DD}. This proves that our proposed circuit is better designed in terms of performance than other existing circuits. Fig 4 (a) (b) (c) (d) Voltage Vs time for 8T, 11T, 13T and ZA circuits respectively.

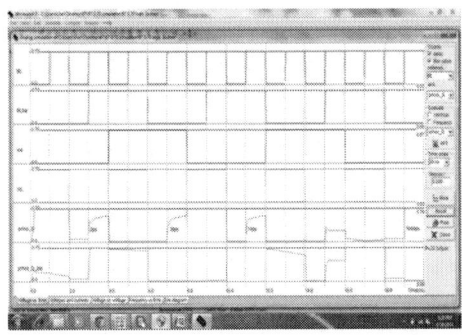

Fig 4(a) 8T

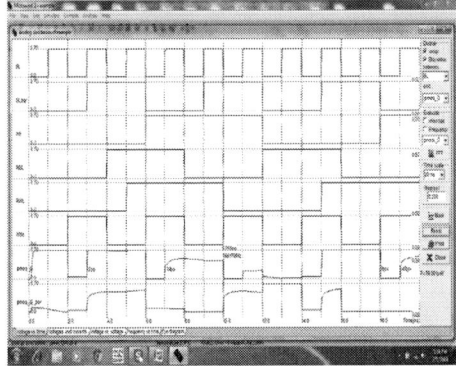

Fig 4(b) 11T

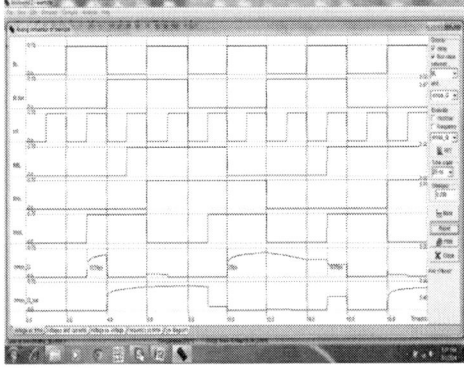

Fig 4 (c) 13T

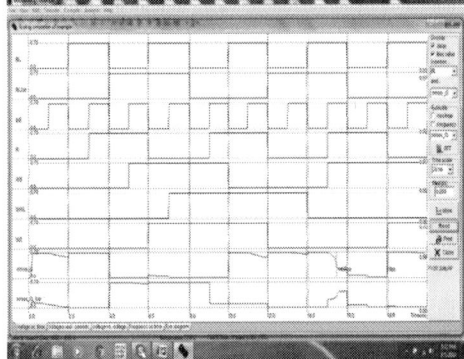

Fig 4 (d) ZA

CONCLUSION

This design has been implemented with the low voltage low power application where it tries to reduce the power consumption on the SRAM cell circuit. The data or output from this simulation does not give the much expected result. This is due to the many problems faced when dealing with the simulation process.This shows that during the designing process, we need to follow all aspects stated in the manual as a guide line in order to design a better circuit. According to dynamic Design concept, this SRAM cell achieved less consumption of power, lower delay, high speed and high throughput than existing SRAM cell.

REFERENCES

[1] Arun Ramnath Ramani and Ken Choi , 'A Novel 9T SRAM Design in Sub-Threshold Region' International IEEE conference on Electro/Information Technology (EIT),2011

[2] Yashwant Singh and D. Boolchandani, 'SRAM Design for Nanoscale Technology with Dynamic V_{th} and Dynamic Standby Voltage for Leakage Reduction'International IEEE conference on Signal Processing and Communication, 2013.

[3] Rajiv V. Joshi,, Rouwaida Kanj, and Vinod Ramadurai, 'A Novel Column-Decoupled 8T Cell for Low-Power Differential and Domino-Based SRAM Design', IEEETransactions on Very large scale integration (vlsi) systems, vol. 19, no. 5, may 2011.

[4] Takahiko Saito, Hitoshi Okamura, Hiromasa Yamamoto and Kazuyuki Nakamura, 'A Ratio-less 10-Transistor Cell and Static Column Retention Loop Structure for Fully Digital SRAM Design, IEEE Memory Workshop (IMW) 2012,

[5] Kumkum Verma Sanjay Kumar Jaiswa lDheeraj Jain Vijendra Maurya, 'Design & Analysis of 1-Kb 6T SRAM Using Different Architecture'2012 Fourth International Conference on Computational Intelligence and Communication Networks.

[6] Xuebei Yang and Kartik Mohanram, 'Robust 6T Si tunneling transistor SRAM design 'Design', Automation and Test in Europe IEEE Conference 2011

Gold-Free Cu-Metallized III-V Solar Cell

Ching-Hsiang Hsu[a], Hsun-Jui Chang[a], Hung-Wei Yu[b], Hong-Quan Nguyen[b], Jer-Shen Ma[a] and Edward Yi Chang[a,b,,c], IEEE Fellow

[a]Institute of Photonic System and Institute of Lighting and Energy Photonics
[b]Department of Materials Science and Engineering and
[c]Department of Electronic Engineering
[a,b,c]National Chiao Tung University
[a,b,c]No.1001 University Road, Hsinchu 30010 Taiwan

Abstract —Au-free, fully Cu-metallized InGaP/InGaAs/Ge triple-junction solar cells using Pd/Ge/Cu as front contact and Pt/Ti/Pt/Cu/Cr as back contact were fabricated and the results are reported for the first time. From the specific contact resistance measurement, these Cu-metallized ohmic contacts have low contact resistance in the order of 10^{-6} Ω-cm^2. AES and TEM results clearly show the formation mechanisms of the Cu-metallization ohmic structures, for Pd/Ge/Cu contact, it was due to the formation of Ge diffusion into the GaAs layer, and for the Pt/Ti/Pt/Cu/Cr contact, it was due to high work function of Pt layer, these copper metallized ohmic contacts were quite stable even after 310 $^{\circ}$C annealing. The I-V curves of the Cu-metallized InGaP/InGaAs/Ge triple-junction solar cells showed similar electrical characteristics to the solar cells with Au-metallized triple junction solar cell. Overall, the Pd/Ge/Cu and Pt/Ti/Pt/Cu ohmic contacts have been successfully applied to the InGaP/InGaAs/Ge triple-junction solar cells and demonstrated excellent performance.

Keywords—Copper-metallization; III-V solar cell; low-cost

I. INTRODUCTION

Solar cell is one of the fastest growing industries in the renewable energy field, since the solar energy could be converted into electricity directly by photovoltaic technology without the environmental pollution and the solar power will not be exhausted in the near future. [1] Among the different types of solar cells, the InGaP/InGaAs/Ge triple-junction solar cell which has the highest conversion efficiency because the wide-bandgap of InGaP/InGaAs/Ge triple-junction solar cells could absorb about 97% of the solar spectrum. In addition, the III-V solar cells are suitable for thin film form, resistant to radiation damage, and can be used with light concentrator. However, the In-GaP/InGaAs/Ge triple-junction solar cells are currently used in high performance applications due to the high manufacture cost. In order to improve the III-V solar cell efficiency to increase the market share, several technologies have been developed for the InGaP/InGaAs/Ge triple-junction solar cells, which include the enhancement of the conversion efficiency by the improvement of the anti-reflection layer (sub-wavelength structure), [2] new epitaxy structure design (lattice match; current match), [3] and operating with concentrator technology. On the other hand, due to the high cost of Ge substrate, Ge/Si substrate is proposed to replace the conventional p-type germanium (p-Ge) substrate as substrates. [4] In this paper, Cu is used to replace the conventional gold-based ohmic contact in the III-V solar cell for cost reduction.

The focus of this paper is to provide a new alloyed system on Pd/Ge/Cu and Pt/Ti/Pt/Cu for n-type and p-type ohmic contacts were used respectively to fabricate the Cu-metalized InGaP/InGaAs/Ge triple-junction solar cells. The Cu-based alloy has been used in metallization for the silicon based very-large scale integration because of its lower electrical resistivity, higher electro migration resistance, and lower cost. [5] However, there are only a few reports on the copper metallization of GaAs devices like metal semiconductor field-effect transistors, high electron mobility transistors have been reported but not for III-V solar cell since Cu diffuses very fast into the operated areas and is an deep acceptor for III-V based material. [6-7] So an effective barrier between Cu and the active region of the device is needed if Cu is to used be for the metallization of III-V based devices.

In the III-V solar cell structure, Pd/Ge/Cu and Pt/Ti/Pt/Cu metals are used for the n-type and p-type ohmic contacts, respectively. In n-type ohmic contact system, Germanium is used as the diffusion barrier that formed Cu_3Ge with Copper to prevent Cu diffuse into the GaAs substrate. In the other hand, the Platinum is used as the diffusion barrier for p-type ohmic contact system because it has a high melting point. [8] Here, we are reporting for the first time about fabrication and electrical performance of the Au-free, fully Cu-metallized InGaP/InGaAs/Ge multi-junction solar cells.

II. EXPERIMENTAL PROCEDURE

The InGaP/InGaAs/Ge multi-junction solar cells that layer structure consists of (from bottom to top) a p-type Ge substrate (200 um, ~10^{18} cm^{-3}), an InGaAs junction, an In-GaP junction, and an n-type GaAs capping layer (500 nm solar cell devices were fabricated using a standard lithography process. The wafer was cleaned with ACE and IPA solutions, and followed, the Pd/Ge/Cu and Pt/Ti/Pt/Cu metals were sequentially deposited using an e-gun evaporator that used for the front side and backside electrodes contacts, respectively, then to form alloy after annealing. Silicon nitride (75 nm) of AR coating was grown by PECVD. The structure of the InGaP/InGaAs/Ge multi-junction solar cells in this study is shown in Figure 1. And the conversion efficiency of the solar cell devices was measured using a solar simulation (AM1.5, 25 °C).

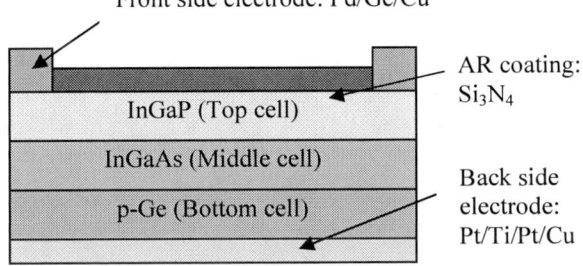

Figure 1 Cross section of Au-free fully Cu metallized In-GaP/InGaAs/Ge multi-junction solar cell.

III. RESULT AND DISCUSSION

A. Cu-Metallization Anaysis

The specific contact resistance of the n-type GaAs and p-type Ge ohmic contact extracted from the TLM as a function of annealing temperature 220 °C ~ 310 °C. The lowest specific contact resistance of the alloyed Pt/Ti/Pt/Cu is 6.9 x 10^{-6} Ω-cm^2 at 250 °C, and Pd/Ge/Cu is 4.4 x 10^{-6} Ω-cm^2 at 310 °C. The result also indicated the resistance of conventional Au/Ge/Ni/Au is 1.4 x 10^{-6} Ω-cm^2, and Ti/Pt/Au is 4.2 x 10^{-6} Ω-cm^2. The TEM image and the EDX profiles of the Pd/Ge/Cu ohmic metal structure after 250 °C annealing were shown in Figure 2 From Figure 2(a) and 2(b), it could be observed the Cu$_3$Ge compound started to form with vertical grain boundary. Figure 2(c) indicated that there was no Cu atom diffused into the GaAs substrate near the Pd/GaAs interface after 250 °C annealing. Figure 3 is the Auger electron spectroscopy (AES) depth profiles of the Ti/Pt/Cu ohmic structure after annealing, as estimation for the figure, the Pt can effectively prevent Cu diffusion into the substrate when annealing temperature up to 350°C.

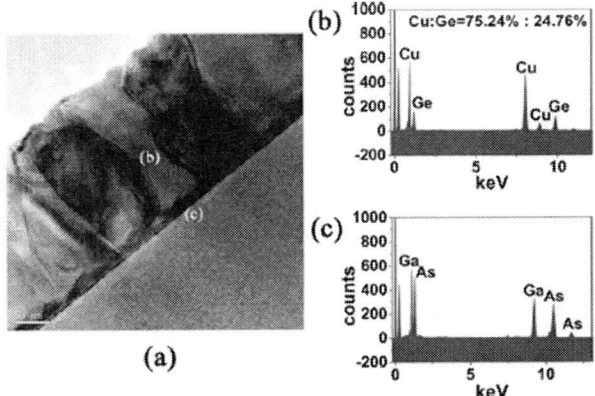

Figure 2 (a)The TEM image of the Pd /Ge /Cu Ohmic contact. (b) The EDX profiles of the Cu$_3$Ge compound grain. (c)The EDX profiles of the GaAs substrate near the Pd/GaAs interface after annealing at 250 °C.

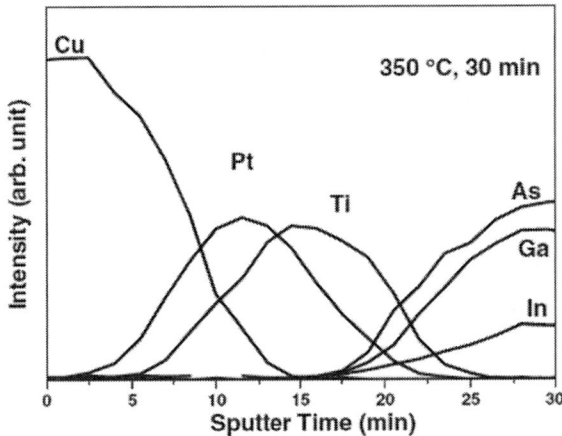

Figure 3 AES depth profiles of the InGaAs/Ti/Pt/Cu sample after 350 °C for 30 min annealing.

B. Efficiency of Cu-metallized Solar Cell

Figure 4 is the I-V curves of the novel Cu-metallized solar cell compare with the conventional Au-metallized solar cell. The fill factor (FF) of Cu-metallized solar cell and Au-metallized solar cell is 79.88%, 84.64%, respectively, and conversion efficiency is 23.78%, 24.06%, respectively. The FF value of copper contact system decreasing was due to R_{SH} value raising by instable dicing process, which not caused by Pd/Ge/Cu ohmic structure (Figure 4). When he yield rate of dicing could be improved in the future, the FF value and efficiency of Cu-metallized solar cell will be increased.

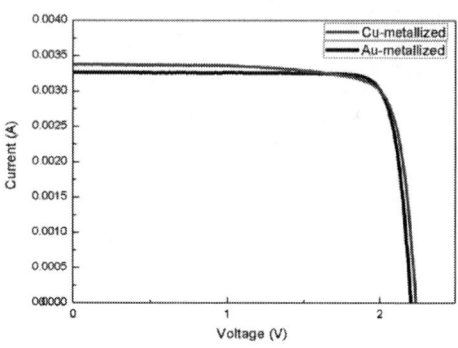

Figure 4 Conversion efficiency of the solar cell with Cu and Au metallization.

IV. CONCLUSIONS

An Au-free, fully Cu-metallized solar cell using Pd/Ge/Cu as front contact and Pt/Ti/Pt/Cu/Cr as back contact were fabricated and the results are reported for the first time. From the specific contact resistance measurement, these Cu-metallized ohmic contacts have low contact resistance in the order of 10^{-6} Ω-cm^2. TEM and AES results clearly show the formation mechanisms of the Cu-metallization ohmic structures, for Pd/Ge/Cu contact, it was due to the formation of Ge diffusion into the GaAs layer, and for the Pt/Ti/Pt/Cu/Cr contact, it was due to high work function of Pt layer, these copper metallized ohmic contacts were quite stable even after 310 $^{\circ}$C annealing. The I-V curves of the Cu-metallized InGaP/InGaAs/Ge triple-junction solar cells showed similar electrical characteristics to the solar cells with Au-metallized triple junction solar cell. Overall, the Pd/Ge/Cu and Pt/Ti/Pt/Cu ohmic contacts have been successfully applied to the InGaP/InGaAs/Ge triple-junction solar cells and demonstrated excellent performance of stability.

ACKNOWLEDGMENT

This work was sponsored by the NCTU-UCB I-RiCE program, Ministry of Science and Technology, Taiwan, and TSMC, under Grant No. NSC 103-2911-I-009-302.

REFERENCES

[1] K. Nishiokaa, T. Takamotob, et al., "Evaluation of InGaP/InGaAs/Ge triple-junction solar cells and optimization of solar cell's structure focusing on series resistance for high-efficiency concentrator photovoltaic systems," *Sol. Energ. Mat. Sol. C.,* vol. 90, no. 9, pp. 1308-1321, May 2006.

[2] J. W. Leema, J. S. Yu, et al., "Efficiency improvement of III–V GaAs solar cells using biomimetic TiO2 subwavelength structures with wide-angle and broadband antireflection properties," *Sol. Energ. Mat. Sol. C.,* vol. 127, pp. 43-49, Aug. 2014.

[3] D. C. Lawa, R. R. King, *et al.,* "Future technology pathways of terrestrial III–V multijunction solar cells for concentrator photovoltaic systems," *Sol. Energ. Mat. Sol. C.,* vol. 94, no. 8, pp. 1314-1318, Aug. 2010.

[4] H. W. Yu, E. Y. Chang, *et al.,* "Effect of graded-temperature arsenic prelayer on quality of GaAs on Ge/Si substrates by metalorganic vapor phase epitaxy," *Appl. Phys. Lett,* vol. 99, 171908, Oct. 2011.

[5] A. G. Baca, F. Ren, *et al.,* "A surey of ohmic contacts to III-V compound semiconductors," *Thin solid films,* vol. 308, pp. 599-606, 1997.

[6] E. D. Marshall1, B. Zhang, *et al.,* "Nonalloyed ohmic contacts to n- GaAs by solid- phase epitaxy of Ge," *J. Appl. Phys.,* vol. 62, no. 942 Mar. 1987.

[7] K. A. Darling, R. K. Guduru, *et al.,* "Thermal stability, mechanical and electrical properties," *Intermetallics* vol.16, pp. 378-383, Feb 2008.

[8] W. Macherzyński, M. Wośko, *et al.,* "Fabrication of ohmic contact based on platinum to p-type compositionally graded AlGaAs layers," 2nd National Conference on Nanotechnology, Journal of Physics: Conference Series 146, 012034, 2009.

On-Wafer Scattering Parameter Characterization of Differential Four-Port Networks LNA using Two-Port Vector Network Analyzer

Maizan Muhamad[1,2], Norhayati Soin[2], Harikrishnan Ramiah[2], Norlaili Mohd Noh[3], Chong Wee Keat[2]

[1]maizan@salam.uitm.edu.my

[1]Faculty of Electrical Engineering, University Teknologi MARA, Shah Alam, Malaysia.
[2]Faculty of Engineering, University Of Malaya, Kuala Lumpur, Malaysia.
[3]School of Electrical and Electronic Engineering, University Sains Malaysia.

Abstract—**This paper presents a technique that enables very accurate measurement for S-parameter of differential low noise amplifier by means of a standard two-port vector network analyzer (VNA). This technique involves by terminating two ports at one time while another two ports are measured. Accurate characterization of a two port device requires a four-port vector network analyzer, which might be not easily available. Thus, it is a common practice to terminate the two of the four ports to be used which the conventional/standard two port VNA. Even though the above approach is applicable but the reliability and conformity of the test method is still limited and uncertain. For verification, the measurement using four-port VNA have been conducted to test the devices S-parameters are accurately similar with the two port network. The fabricated on-wafer differential LNA structure was tested and measured with normal two-port VNA and also four-port VNA. By using this technique, there is no need to purchase a four-port VNA. By using this technique, any multi-port circuit network can be measured. The LNA has been implemented in RF 0.13um CMOS process. The differential LNA shows the measured performance in term of gain is equal to 17.4 dB. This give the percentage difference of 0.63 compared with measured using four-port VNA. The circuit consume only 9 mW power while dissipating 7.59mA from a 1.8 V supply. Generally, the measured results of the on-wafer fabricated differential LNA show good agreement for both set up.**

Keywords—differential; S Parameter; Two-port; LNA; WLAN

I. Introduction

The rapid development of wireless communication industry has driven RF integrated circuit based CMOS technology an attractive package for radio transceiver front-end circuits for implementation of various wireless communication systems. LNA is the first block of receiver system [1]. Basically, a differential LNA is usually better than single-ended LNA. It's prominent properties such as rejection to parasitic couplings, immunity to common-mode noise and also the ability to increase dynamic range and many more makes it preferred over single ended [2].

Even though the differential structure must be conserved inside the chip, sometimes there are event where the input signal is single-ended such as radio-frequency image

filters and intermediate frequency filters in a RF receiver. Moreover, there are many other circuits that call for differential signals in order to perform their purpose. In these conditions, there is a need for a stage that be able to transform single-ended signal to differential signals [2]–[5].

An important issue that should be taken into consideration is the linearity of the low noise amplifier. Linearity is important to get less distortion output. Nevertheless, there is always an exchange between gain, linearity and etc. A differential LNA is better to be implemented in a mixed-mode design where a single-ended LNA is prone to be affected by substrate noise and other interferers on chip of the receiving block [6].

This paper presents the conversion of a normal two-port s-parameters and four-port s-parameters measurement data. On-wafer fabricated differential LNA structure is tested with two-port and four-port VNA respectively. The relationship data is presented as well. Even though the standard conversion method could possibly be used a normal two-port VNA, a mix-mode measurement system is also carried out in order to test the accuracy of the set up [7].

II. Methodology

A. Differential LNA Circuit Design

For differential LNA, the circuit adopt cascode inductively degenerated topology which is the most commonly used topology. For the sake of simplicity, the biasing circuit of differential LNA is not shown in Figure below.

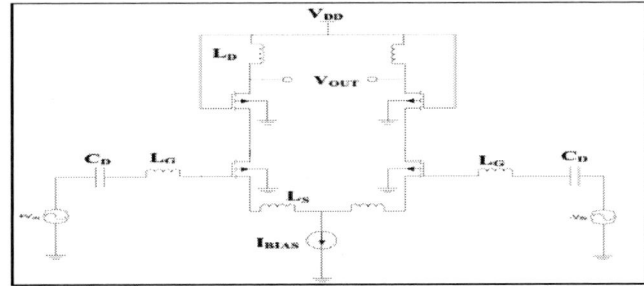

Fig. 1. Inductively Degenerated Differential LNA

B. S-Parameter of Differential LNA

The definitions of s-parameters have been expressed in terms of the input and output power waves of any electrical network [8]. Differential LNA circuit can be characterized based on its response to differential and common mode signal. This include whichever mode of transformation[9]. S-parameters is the most appropriate technique to characterize the nature of RF signal [8]. Fig. 2 shows an LNA device under test (DUT) which has been constructed as a basic four-port network and has been established in [10] . Equation (1) is the standard S-parameter illustration of this circuit network.

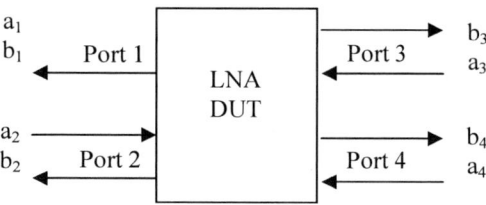

Fig. 2. Structure of a four-port network

$$\begin{bmatrix} b_1 \\ b_2 \\ b_3 \\ b_4 \end{bmatrix} = \begin{bmatrix} S_{11} & S_{12} & S_{13} & S_{14} \\ S_{21} & S_{22} & S_{23} & S_{24} \\ S_{31} & S_{32} & S_{33} & S_{34} \\ S_{41} & S_{42} & S_{43} & S_{44} \end{bmatrix} \begin{bmatrix} a_1 \\ a_2 \\ a_3 \\ a_4 \end{bmatrix} \qquad (1)$$

The relationship between two port network and four port network has been analyzed in [11] .The four by four differential S-matrix is briefly summarized as stated in equation (2) and has been given in [8]. This is done by manipulating the definitions of differential voltages and currents into equation of power waves. Which is S_{DD11} refers to the differential-mode return loss at differential port 1, S_{DD12} the differential insertion loss from differential port 2 to differential port 1, S_{DD21} the differential insertion loss from port 1 to port 2 and S_{DD22} the differential return loss at port 2. Another 12 S-parameters are described as DC referring to the differential to common-mode conversion, CD referring to the common-mode to differential-mode conversion and CC referring to the pure common-mode parameters [12].

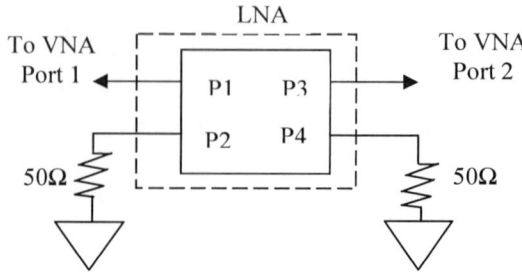

Fig. 3. Differential LNA terminated with single-ended 50Ω at each port.

Fig. 3 shows inexpensive technique of characterizing differential LNA networks with two-port VNA by utilizing a method of measuring the s-parameters of two ports at one time while the other two ports are terminated with 50 Ω.

C. Measurement and analysis

The differential LNA circuit is design for 0.13µm RF CMOS technology and simulated using Cadence Virtuoso. It has been fabricated using process design kit provided by Silterra (M) Sdn Bhd. Figure 4 shows the microphotograph of differential LNA. The on-wafer test and measurement is conducted on a cascade's probe station. Impedance standard substrate (ISS) is used for Wafer calibration. Details design and performance of the differential LNA can be obtained here [13].

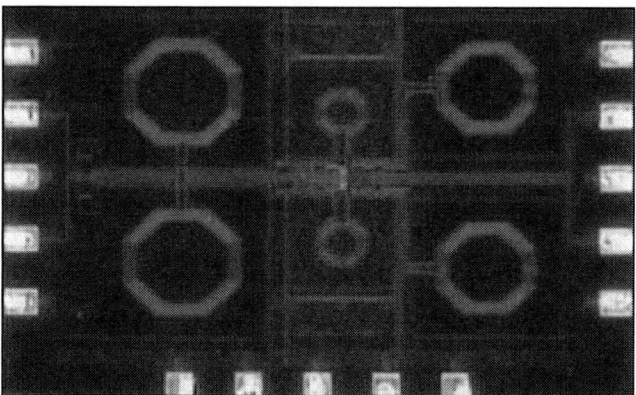

Fig. 4. Microphotograph of a differential LNA

$$\begin{bmatrix} s_{DD11} & S_{DD12} & S_{DC11} & S_{DC12} \\ S_{DD21} & S_{DD22} & S_{DC21} & S_{DC22} \\ S_{CD11} & S_{CD12} & S_{CC11} & S_{CC12} \\ S_{CD21} & S_{CD22} & S_{CC21} & S_{CC22} \end{bmatrix} = \frac{1}{2} \begin{bmatrix} (S_{11}-S_{12}-S_{21}+S_{22}) & (S_{13}-S_{14}-S_{23}+S_{24}) & (S_{11}+S_{12}-S_{21}-S_{22}) & (S_{13}+S_{14}-S_{23}-S_{24}) \\ (S_{31}-S_{32}-S_{41}+S_{42}) & (S_{33}-S_{34}-S_{43}+S_{44}) & (S_{33}+S_{32}-S_{41}-S_{42}) & (S_{33}+S_{34}-S_{43}-S_{44}) \\ (S_{11}-S_{12}+S_{21}-S_{22}) & (S_{13}-S_{14}+S_{23}-S_{24}) & (S_{11}+S_{12}+S_{21}+S_{22}) & (S_{13}+S_{14}+S_{23}+S_{24}) \\ (S_{31}-S_{32}+S_{41}-S_{42}) & (S_{33}-S_{34}+S_{43}-S_{44}) & (S_{31}+S_{32}+S_{41}+S_{42}) & (S_{33}+S_{34}+S_{43}+S_{44}) \end{bmatrix} \qquad (2)$$

By using standard four-port VNA as in Fig. 2 and two-port VNA as in Fig. 3, measurement of the s-parameter is carried out. Both measurement were analyzed and compared. Four-port s-parameters data can be taken directly from the measurement.

However by using two-port VNA to measurement differential LNA, six different measurement set ups are essential in order to calculate the relation between mixed-mode. In each set up, two ports of the LNA DUT are connected to any port of VNA, while the other remaining two ports are terminated to 50Ω loads. Fig. 3 shows one of the six set up that need to be done. For example, it shows s-parameters measurement of P1 (port 1) and P3 (port 3). With this set up, the S-parameters of P1P2, P1P3, P1P4, P3P2 and P3P4 can be measured respectively. This measurement will take 24 s-parameters data in which the s-parameters data of S_{11}, S_{22}, S_{33} and S_{44} are taken and measured three times.

TABLE I.   TWO-PORT SET UP FOR S-PARAMETERS MEASUREMENT

Port	Port 1	Port 2	Port 3	Port 4
(P1P3)	Input signal	Terminated with 50Ω	Input signal	Terminated with 50Ω
(P2P3)	Terminated with 50Ω	Input signal	Input signal	Terminated with 50Ω
(P2P4)	Terminated with 50Ω	Input signal	Terminated with 50Ω	Input signal
(P1P4)	Input signal	Terminated with 50Ω	Terminated with 50Ω	Input signal
(P3P4)	Terminated with 50Ω	Terminated with 50Ω	Input signal	Input signal
(P1P2)	Input signal	Input signal	Terminated with 50Ω	Terminated with 50Ω

Table I show the set up for two-port S-parameters differential LNA measurement. There are six different measurement that been carried out for this on wafer differential LNA.

III. Results and Analysis

Fig. 6 shows the comparison of S-parameters of the designed differential LNA that measured using four-port and two-port VNA.

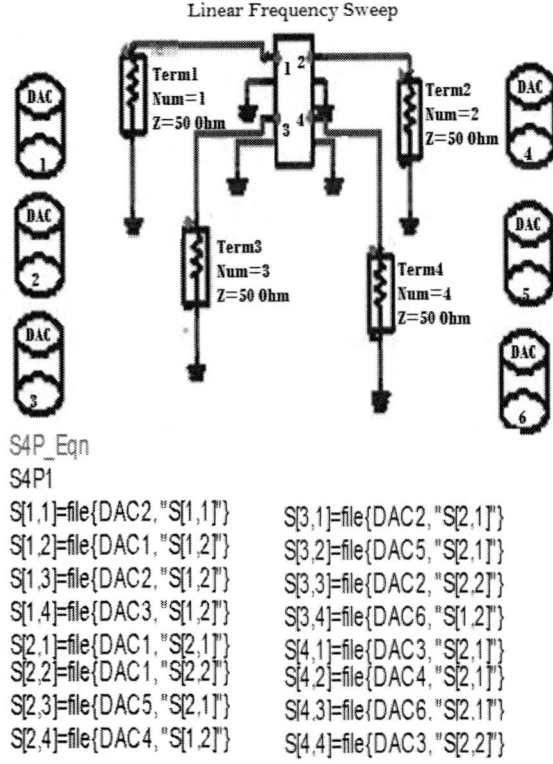

Fig.5. Measured using two-port data transform into mixed-mode using ADS

The measured four ports S-parameters data is transformed into differential mixed-mode base on Equation (2) by using Agilent's ADS software. It is more useful and convenient to avoid mistakes during data analysis and processing.

Fig. 5 illustrates equation based linear component and data file component that been used to process and select 16 s-parameters from six different set up measurement data files. The ADS data display window will showed the mixed-mode S-parameters when applying transformation formula in the data file component, DAC.

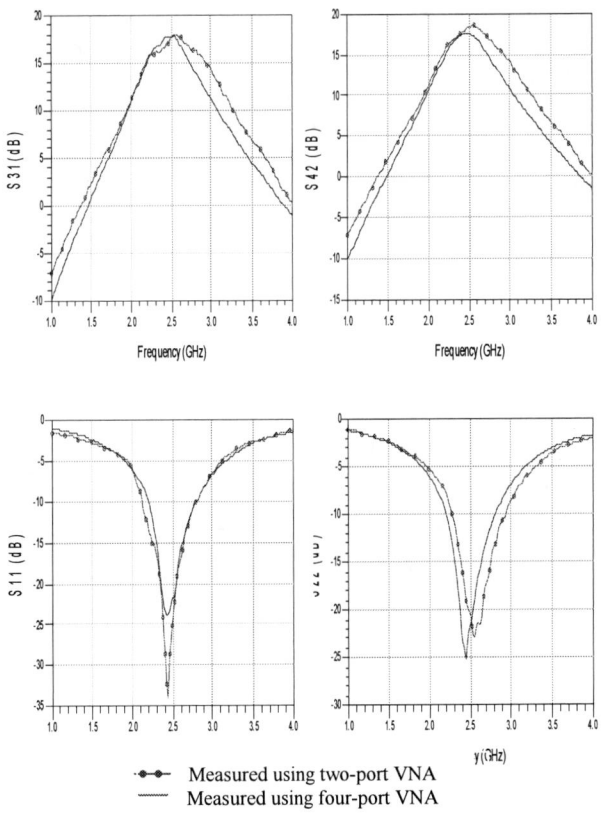

Fig.6. Comparison of the measured S Parameters for differential LNA

978-1-4799-5761-3/14 $31.00 © 2014 IEEE   341

Base from the curve, it shows that the LNA obtained good S parameters performance and mostly the peak performance occurs at the center operating frequency which is at 2.4GHz. The gain of the differential LNA which are S_{31} and S_{41} manage to achieve 17.28dB and 17.29dB respectively. S_{11} and S_{22} is the reverse isolation of the differential LNA. As depict in Fig. 6, both measurement results show good agreement for each S-parameters. Base from the graph, the reverse isolation parameter S_{11} taken using two-port is better compared to the other one. S_{11} reading at center frequency of 2.4GHz is -27.5dB.

This shows that the input impedance matching for this differential LNA is good. There is slightly frequency shift at S_{22} parameters which cause the both reading not properly overlapped. This might be due to the GSGSG probe not properly touch the wafer during measurement and some parasitic that present during the measurement. However this does not affect the overall reading and performance of the LNA since the requirement specify by the WLAN is still fulfill.

TABLE II.   COMPARISON OF S-PARAMETERS PERFORMANCE

S-Parameters (dB)	VNA Set Up		% Differences
	Two-port	Four-port	
S_{31}	17.12	17.28	0.93
S_{42}	17.40	17.29	0.63
S_{11}	-27.50	-24.50	11.5
S_{22}	-21.00	-22.00	4.65

Table II shows comparison of measured S-parameters performance of differential LNA taken at center frequency of 2.4GHz. This differential LNA follow requirement specify for WLAN / IEEE 802.11b/g standard.

IV. Conclusion

A technique of on wafer S-parameters characterization of differential LNA using Two-Port VNA is presented. This technique achieves to give measurement data as precise as four-port VNA equipment. Comparison for both measurement data is presented and show small percentage difference between the two. This difference might be due to probe placement position and planarity variation. However this small difference still can be justified and show good agreement between the two equipments set up.

A new and simple approach for data transforming from four-port data to mixed-mode S-parameters data is also introduced. This was done by using equation-based linear component (S4_eqn) and data file component (DAC) from Agilent's ADS tool. This method makes it easier to process several data more convenient and accurate.

Acknowledgment

The authors wish to thank MOHE for the FRGS grant 600-RMI/ST/FRGS 5/3/Fst (122/2010) and University of Malaya for PPP fund. The author is grateful for UiTM for the scholarship and RMI UiTM for the management of the FRGS fund for this project.

References

[1] T. H. Lee, "The Design of CMOS Radio-Frequency Integrated Circuits," *Cambridge Univ. Press*, 2004.

[2] M. Robens, R. Wunderlich, and S. Heinen, "Differential Noise Figure De-Embedding: A Comparison of Available Approaches," *IEEE Trans. Microw. Theory Tech.*, vol. 59, no. 5, pp. 1397–1407, May 2011.

[3] K. Vaz and M. Caggiano, "Measurement technique for the extraction of differential S-parameters from single-ended S-parameters," *27th Int. Spring Semin. Electron. Technol. Meet. Challenges Electron. Technol. Progress, 2004.*, vol. 2, no. I, pp. 313–317.

[4] C. Yin-Cheng, L. Shuw-Guann, L. Hsien-Yuan, C. Hwann-Kaeo, and J. Ying-Zong, "On-wafer differential noise figure and large signal measurements of low-noise amplifier," in *Microwave Conference, 2009. EuMC 2009. European*, 2009, pp. 699–702.

[5] Y. Chang, S. Lin, H. Liao, H. Chiou, and Y. Juang, "On-Wafer Differential Noise Figure and Large Signal Measurements of Low-Noise Amplifier," no. October, pp. 699–702, 2009.

[6] W. Fan, A. C. W. Lu, L. L. Wai, and B. K. Lok, "Mixed-Mode S-Parameter Characterisation of Differential Struc- tures," no. 1, pp. 66–70.

[7] W. Fan, A. Lu, L. L. Wai, and B. K. Lok, "Mixed-mode S-parameter characterization of differential structures," in *Electronics Packaging Technology, 2003 5th Conference (EPTC 2003)*, 2003, pp. 533–537.

[8] K. Vaz and M. Caggiano, "Scattering parameter characterization of differential four-port networks using a two-port vector network analyzer," *Proc. Electron. Components Technol. 2005. ECTC '05.*, vol. 2, pp. 1846–1853, 2005.

[9] M. Caggiano, "Error reducing techniques for the scattering parameter characterization of differential networks using a two-port network analyzer," *28th Int. Spring Semin. Electron. Technol. Meet. Challenges Electron. Technol. Progress, 2005.*, pp. 342–347, 2005.

[10] D. E. Bockelman and W. R. Eisenstadt, "Combined differential and common-mode scattering parameters: theory and simulation," *IEEE Trans. Microw. Theory Tech.*, vol. 43, pp. 1530–1539, 1995.

[11] S. Ling, W. Zhigong, and G. Jianjun, "A method for on-wafer S-parameter measurement of a differential amplifier by using two-port network analyzer," in *Asia-Pacific Microwave Conference Proceedings, APMC*, 2005, vol. 5.

[12] H. Ka Mun, K. Vaz, and M. Caggiano, "Scattering parameter characterization of differential four-port networks using a two-port vector network analyzer," in *Electronic Components and Technology Conference, 2005. Proceedings. 55th*, 2005, pp. 1846–1853 Vol. 2.

[13] M. Muhamad, N. Soin, H. Ramiah, N. M. Noh, and W. K. Chong, "Design of CMOS Differential LNA at 2 . 4GHz," in *IEEE International Conference of Electron Devices and Solid-State Circuits (EDSSC)*, 2013, pp. 1–2.

Strain Rate Effect on Micromechanical Properties of SnAgCu Solder Wire

I. Abdullah[1], R. Ismail[1] and A. Jalar[1], *Member, IEEE*
[1]Institute of Micro Engineering and Nanoelectronics (IMEN)
Universiti Kebangsaan Malaysia,
43600 Bangi, Selangor, Malaysia.
Email: eizhan@eng.ukm.my

Abstract— Dislocation behavior was occurs when an eutectic solder alloy of SnAgCu experiencing different strain at room temperature that require the further analysis in order to relate the physical and microstructure changes towards the mechanical performance of lead free solder. In this study, nanoindentation technique was applied to determine the hardness and modulus on six variant of strain (0.00015 mms^{-1}, 0.0015 mms^{-1}, 0.015 mms^{-1}, 0.15 mms^{-1}, 1.5 mms^{-1} and 15 mms^{-1}) after tensile test. The *P-h* curves and the micromechanical parameter namely hardness and residual modulus through nanoindentation test were conducted. The analysis were obtained strain rate sensitivity (*m*) and stress exponent (*n*) from dwell time in order to determine the mechanism of grains. The *P-h* curve result showed the pop-in event at the ranges of 100 nm to 300 nm. The micromechanical properties were show the increment of values at high strain rates. The dominated discontinuity local will occurrence the pop-in event and will activating dislocation distribution.

Keywords— *SnAgCu, nanoindentation, dislocation, P-h curve, pop-in event, strain rate.*

1. INTRODUCTION

Eutectic solder alloys are widely used in the microelectronics industry [1,2]. Deformation of material is divide by two namely twinning and dislocation. Dislocation of grain boundary was occurrence by some load on impact such strain and compression at constant temperature. However, the dislocation distribution of grain boundary would be the issue lack the hardness parameter [3]. In the correlation of mechanical properties and microstructure, the hardness which is defined as the indentation load divided by the contact area of the indentation made, they can be a sensitive enough parameter to represent the hardening potential of a grain boundary, however, it need a further research. In many report where an increase in the measured hardness was observed, the grain boundary hardening was attribute to segregation of impurity atoms [4]. For example, doped zone-refined metal such as Sn, Ag and Pb, tin segregation in alpha Ag-Sn alloy and Cu-Sn. Wang and Ngan (2004) were solve this issue with proper heat treatment procedure were followed to achieve enough segregation of impurities at grain boundary [5].

From modelling or experimental works, nanoindentation technique is frequently used to measure the properties material for solder alloys and other material at small volumes. In fact,

Gomez and Basaran (2006) reported the nanoindentation technique has gained popularity as an experimental tool to determine the material properties for specimens available in small volumes. On the result of nanoindentation on rate independent materials have shown a strong dependence of hardness on penetration depth by loading rates [6].

Most of the studies on the lead free deformation behaviour are the combination of bulk material testing with strain, strain rate. Therefore the analysis of micromechanical properties effect of solely the changes strain rates towards the deformation behaviour of lead free solder is crucially needed. To clarify the microstructure behaviour, nanoindentation test was carried in order to determine the micromechanical properties of Sn-3.0Ag-0.5Cu related to grain boundary with strain rate effect through tensile test. The deformation mechanisms was obtained measuring the stress rate sensitivity and stress exponent from the dwell time analysis results obtained.

2. SAMPEL PREPARATION AND METHODS

The eutectic of Sn-3.0Ag-0.5Cu (SAC305) solder wire was provided by local industry, Red Ring Solder (M) Sdn Bhd was selected in order to determine the micro-mechanical properties. These specimen were used for the present work. The selected specimen was cut into 10 cm of length and the acrylic tapes were attached on the top and the bottom of SAC solder wire before being used in tensile machine as shown in Fig 1.

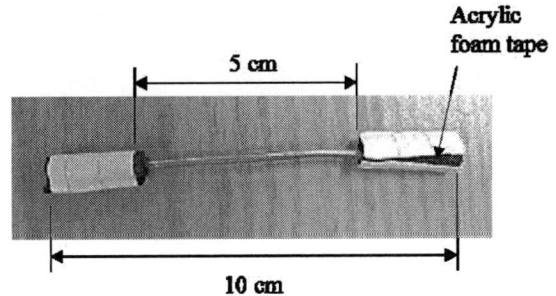

Fig. 1 SAC305 solder wire specimen for tensile test

The tensile test were performed on these specimen using INSTRON™ Microtester machine at room temperature with six variant of strain rates with the value of 0.00015 mms⁻¹, 0.0015 mms⁻¹, 0.015 mms⁻¹, 0.15 mms⁻¹, 1.5 mms⁻¹ and 15 mms⁻¹. These solder wire with length of 5mm was cut far end from the rupture area in order to prepared these samples for nanoindentation test. The nanoindentation samples were prepared by resin mounting the chosen 5mm length of SAC305 solder wire. Later, the wet grinding was performed on the sample using 600, 800 and 1200 grit of abrasive papers followed by polishing with 3 μm, 1 μm and 0.25 μm of diamond suspension on silk cloths. In order to determine the micro-mechanical parameters, the indentation were carried out by a Nanotest™ (Micro Materials) machine with a Berkovich tip (three-sided pyramidal) of ~50 nm indented area. The indentation were performed using loading-holding-unloading cycle's mode. A constant rate of 0.1 mN/s was applied to cross section of SAC305 surface until the maximum load (10 mN) was reached, then held to dwelling in 185 seconds at maximum load until unloading. Four location on cross section of SAC305 surface were to get the average values. All of indentation data were analysed based on the Oliver and Pharr theories [7].

3. RESULTS AND DISCUSSION

The indentation on SAC305 solder wire samples were carried out after tensile strain rate performed. Fig. 2 depicts the stress-strain curves on six variant of strain rates. From the indentation, indentation load (P) and penetration depth (h) are acquired from indentation penetration into cross section of SAC305 solder wire surface. The variations of P-h curve as-received and six variant of strain rates on SAC305 solder wire as depicted in Fig. 3. The loading of indentation was consist elastic and plastic deformation. Initial loading during the indentation has resulted in elastic deformation. When the depth of penetration increases, the plastic deformation of the material had started occur. The initial plastic deformation was triggered by initial changes at elastic region of P-h curve and also stress-strain curves (in Fig. 2). In addition, the shape of P-h curve is be related to mechanical properties of this material.

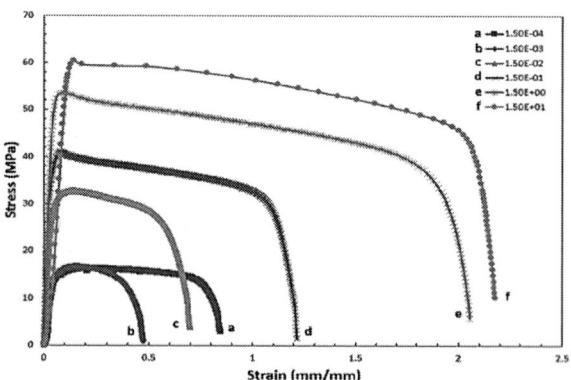

Fig. 2   Stress-strain curves of SAC305 solder wire from strain rate tensile test at room temperature (298.15 K).

According to Fig. 3, the loading section on P-h curve of indentation for different strain rates have glitches and ladder shape compared to as-received SAC305 solder wire. This phenomenon is called a pop-in or displacement burst [4]. The pop-in was occurrence at initial elastic region that the range from 100 nm to 300 nm. This occurrence of local discontinuity was happen during loading on strain rates. The pop-in phenomenon also related to local discontinuity during indentation and microstructure change effect from strain rate [8,9]. The initial pop-in may reveal the shift the perfect elastic to plastic deformation and would be initiate dislocation among of grain boundaries of this materials [10]. The pop-in less occur during low strain rates otherwise it often happens when the strain rate was increased.

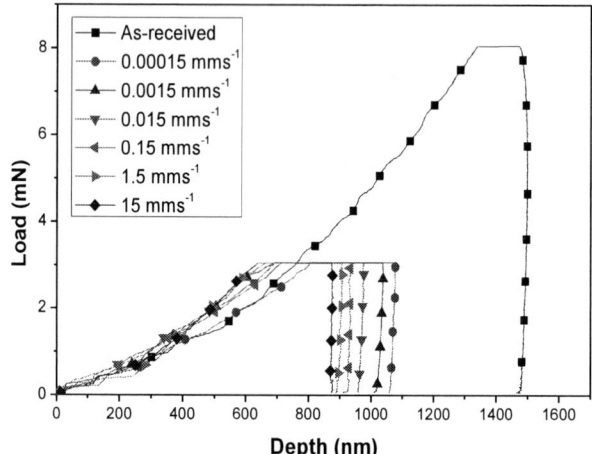

Fig. 3   The variations of P-h curves as-received and six variant of strain rates on SAC305 solder wire

Strain rate also affected to micromechanics properties through indentation. Fig 4(a) and (b) depict the strain rates effect to hardness and reduced modulus (elastic). According to Fig. 4, it was showed that the low strain rate (0.00015 mms⁻¹) has acquired small values (0.131 GPa and 42.933 GPa) for both of hardness and reduced modulus compared to high strain rate. As viewed, the both of values of as-received SAC305 solder wire have small value rather than six variant of strain rate due to the grain boundary have not deformed while unloading. These three of low strain have acquired the closely values of each other compared to the high strain rates. The significant of improvement in the hardness of strain rate 0.00015 mms⁻¹ to 15 mms⁻¹ is 41.2 percent. This is due to strain rate has changed the size and shape of the grains on the surface of this solder wire material [11]. In this alloy material, the hardness is strongly depends on the presence of defect in the vicinity of indentation and the specimen resistance to plastic deformation. In parallel to hardness changes, the values of reduced modulus are closely related to the elastic and plastic work that has changed the microstructure properties in terms of the grains changes through the dislocations formation [12].

To determine the mechanism of deformation of indentation, the determination of the stress exponent, n was conducted on indentation in the eutectic of SnAgCu using a constant load. Goodall and Clyne (2006) has stated that nanoindentation deformation analysis using constant load takes the maximum load and the maximum load applied during nanoindentation to compute the value of n [13]. Mahmudi et al. 2001 was describe the deformation mechanism according to n value which acquired from researchers as shown in Table 1[14]. m used in the determination of strain hardening occurs at the eutectic of SnAgCu which m is high value indicates a high strain hardening and less plastic deformation occurrence [15,16]. Table 2 shows the strain rate sensitivity, m and stress exponent, n in SAC305 solder wire for different strain rates at room temperature. The value of n for strain rate 0.00015 mms^{-1} is reduced from 7.5 to 4.9 for strain rate 0.0015 mms^{-1}. Then the value of n was increased from 5.6 to 6.8 at strain rates of 0.015 mms^{-1} and 0.15 mms^{-1}. But the value of n has dropped back on strain rate range of 1.5 mms^{-1} to 15 mms^{-1} is from 5.6 to 3.9.

Table 1 Description of deformation mechanism according to stress exponent, n

Stress exponent (n)	Deformation mechanism
1	diffusion
2	Grain boundary sliding
4 until 6	Dislocation in climb
Over than 8	Dislocation movement / Dislocation in climb

Table 2 n and m value acquired from the indentation at the cross section of SAC305 solder alloy wire after tensile test at room temperature (25 ° C)

Strain Rates (mms^{-1})	Stress Exponent (n)	Strain rate sensivity (m)
0.00015	7.5	0.13
0.0015	4.9	0.21
0.015	5.6	0.18
0.15	6.8	0.15
1.5	5.6	0.18
15	3.9	0.26

According to Table 2, it was showed that the strain rate effect on the n value which is determine the behaviour of the SAC305 solder alloy wire microstructure after tensile conducted. The value of stress exponent, n were show in the range of 0.00015 mms^{-1} to 15 mms^{-1} is between 3 to 8 (according to Table 1). According to the n value, there are a number of activities in microstructure activities which involves the grains where the occurrence of initial movement of dislocation at low strain rates and the dislocation would climb on a nearby the grains. This caused is a pop-in event in the SAC305 microstructure. Therefore, this phenomenon related between microstructural changes and indirectly alter the properties of hardness and modulus during strains performed.

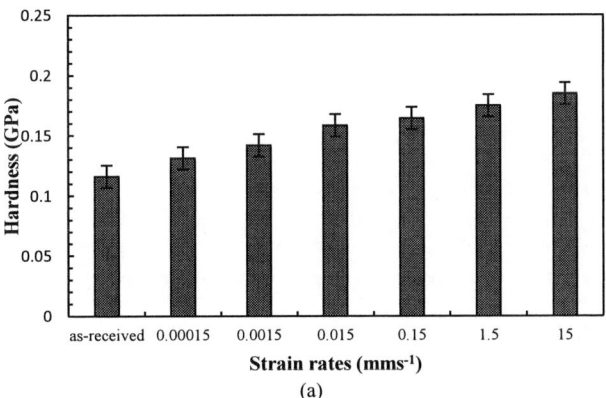

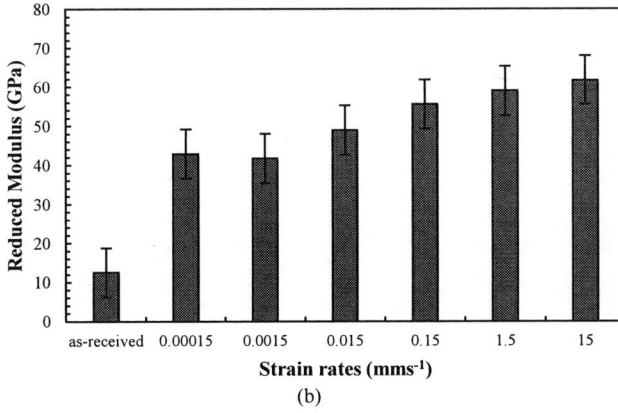

Fig. 4   Strain rates effect to hardness (a) and reduced modulus (b) of SAC305 solder wire

4. CONCLUSION

The indentation on SAC305 solder wire with six variant strain rate were performed. The P-h curve were show the initial local discontinuity was occur at elastic region and the pop-in event may revealed during indentation penetrate at the range of 100 nm to 300 nm. Therefore, the stress exponent played important role to clarify to the microstructure behaviour would started initial the dislocation movement which causes the occurrence of pop-in phenomenon.

ACKNOWLEDGEMENTS

This work has been sponsored by National University of Malaysia and under research university grant (UKM-OUP-NBT-29-143/2011, OUP-2012-120 and DIP-2012-14). The author would like to thank Ministry of Higher Education for a scholarship.

REFERENCES

[1]   J.H.L. Pang and B.S. Xiong, Lead-Free Solder Alloy, 28, 830–840, 2005
[2]   X. Deng, N. Chawla and M. Koopman, "Deformation behaviour of (Cu, Ag)–Sn intermetallic by nanoindentation", Acta Materialia 52, pp. 4291 – 4303, 2004.

[3]   F. Sanzoz and K.D. Stevenson, "Relationship between hardness and dislocation processes in a nanocrystalline metal at the atomic scale", Physical Review B 83, pp. 224101-1 – 224101-8, 2011.

[4]   T. Watanabe, S. Kitamura and S. Karashima, "Grain-boundary hardening and segregation in alpha-iron-tin alloy", Acta Metall. 28, 455, 1980.

[5]   M.G. Wang and A.H.W. Ngan, "Indentation strain burst phenomenon induced by grain boundaries in niobium", Journal of Materials Research vol. 19(8), pp. 2478 – 2486, 2004.

[6]   J. Gomez and C. Basaran, "Nanoindentation of Pb/Sn solder alloys; experimental and finite element simulation results", *International Journal of Solids and Structures* 43, pp. 1505–1527, 2006.

[7]   W.C. Oliver and G.M. Pharr, "An improvement technique for determining hardness and elastic modulus using load and displacement sensing indentation experiments", J. Mater Res 7, pp. 1564 -1583, 1992.

[8]   W.Y.W. Yusof, R. Ismail, A. Jalar, N.K. Othman and I.A. Rahman, "Microstructural evolution and mechanical properties of gamma-irradiated Au ball bonds", Material Characterization 01, pp. 1-22, 2014

[9]   C.A. Schuh, "Nanoindentation studies of material", Materials Today vol. 9(5), pp. 32-40, 2006.

[10]   S.R. Jian, C.J. Chan and T.C. Lin, "Berkovich nanoindentation on AlN thin film", Nanoscale Research Letter 5, pp. 935 -940, 2010.

[11]   D.Q. Yu, J. Zhao and L. Wang, "Improvement on the microstructure stability, mechanical and wetting properties of Sn–Ag–Cu lead-free solder with the addition of rare earth elements", Journal of Alloys and Compounds vol. 376 (1-2), pp. 170 – 175, 2004.

[12]   M.A. Meyers, A. Mishra and D.J. Benson, "Mechanical properties of nanocrystalline materials", Progress in Materials Science 51, pp. 427 – 556, 2006.

[13]   R. Goodall and T.W. Clyne, "A critical appraisal of The axtraction of creep parameters from nanoindentation data obtained at room temperature", Acta Materialia vol. 54(20), pp. 5489–5499, 2006.

[14]   R. Mahmudi, R. Rouminna and B. Raeisinia, "Investigation of stress exponent in the power-law creep of Pb-Sb alloys", Materials Science and Engineering: A 382(1-2): 15-22, 2004.

[15]   B.M. Kim, C.J. Lee and J. M. Lee, "Estimations of work hardening exponents of engineering metals using residual indentation profiles of nanoindentation", Journal of Mechanical Science and Technology 24, pp. 73 – 76, 2010.

[16]   M.N. Zulkifli, A. Jalar, S. Abdullah, N.K. Othman and M.A.A. Hamid, "Nanoindentation creep analysis of gold ball bond", IEEE 10th ICSE Proceeding, pp. 755-759, 2013.

Epitaxial Lift-Off of Large-Area GaAs Multi-Junction Solar Cells For High Efficiency Clean and Portable Energy Power Generation

N. Pan
MicroLink Devices, Inc.
6457 W. Howard St., Niles, IL 60714
United States
+1 847-588-3001, npan@mldevices.com

Abstract—Epitaxial lift-off (ELO) is a processing technique that enables thin epitaxial layers grown on GaAs substrate to be "peeled off" from the host substrate. The ELO process offers several key advantages for both performance enhancement and cost reduction of III-V electronic and optoelectronic devices. The epitaxial films can be transferred to new support substrates that are thin, flexible, lightweight, and with higher thermal conductivity than the original growth substrate. The GaAs substrate can be reused multiple times without any performance degradation. We have developed an ELO process capable of lifting off large areas of semiconductor material from substrates up to 6 inches in diameter without any degradation of material quality or performance characteristics. Solar cells with a 33% efficiency under 1-sun AM 1.5 and 31% efficiency at 1-sun under AMO were demonstrated using ELO. Production of this technology for fabricating thin, flexible large-area multi-junction solar cell arrays with very high efficiency has been established. Light weight portable solar sheets for remote charging have been produced. Other applications include electric-powered, unmanned aerial vehicles (UAVs), space satellites, and terrestrial solar concentrator systems.

Keywords—*Solar Cells, GaAs, High efficiency, Solar Panels, Lightweight*

I. INTRODUCTION

ELO has been previously developed for small area devices [1-2] for improvements in the device performance. This technology remained in the laboratory for many years after the first demonstration. MicroLink is the first company to commercialize this technology for large area inverted metamorphic (IMM) triple junction solar cells. The ELO technology provides unique device properties (flexibility, lightweight, high efficiency, low cost) which have not been previously realized [3]. The potential of ELO for other high power devices is very significant. Fig. 1 shows the flexibility of a 4-inch GaAs solar cell that was developed using ELO technology.

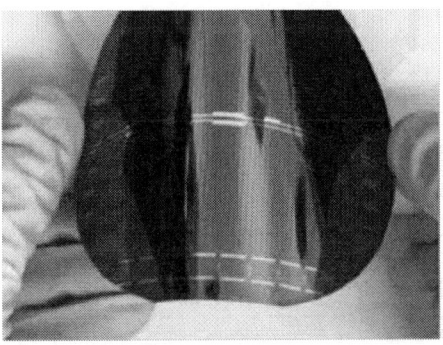

Fig. 1. 4-inch GaAs ELO foil attached to a thin and flexible metal backing is shown. The wafer contains two large-area (20-cm2) solar cells.

II. EXPERIMENTAL

All epitaxial structures were grown by metalorganic chemical vapor deposition (MOCVD) at 100 mbar using arsine (AsH3), phosphine (PH3), trimethylindium (TMI), trimethylgallium (TMG) as precursors and using a V/III ratio >50. Inverted metamorphic multijunction (IMM) InGaP/GaAs/InGaAs structures were grown on GaAs substrates. Figure 2 and 3 show schematics that outline the basic ELO process. The first layer deposited on the substrate is a thin high composition AlGaAs release layer. The solar cell epitaxial layers are then deposited, followed by application of a thick (1-2 mil) flexible alloy metal carrier layer. The wafer is then immersed in a concentrated HF-acid chemistry, which selectively dissolves the release layer (the etch selectivity relative to the GaAs epitaxial structure is greater than 1E5). The solar cell epitaxial layers are thereby completely separated from the GaAs substrate. The ELO process requires approximately 12 hours to complete, but is amenable to batch processing, enabling scaling up of the process to lift off hundreds of substrates within a 24-hour period. The released ELO foils are next mounted to a temporary, rigid carrier in order to carry out solar cell device processing (evaporation and lift off Ti-Au grid metallization, wet etch isolation, evaporation of TiO2-Al2O3 antireflection coating, device singulation). The completed solar cells are then removed from the temporary carrier (Fig. 2).

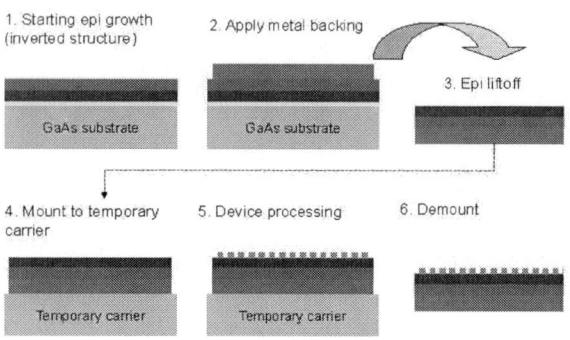

Fig. 2. Schematic of solar cell fabrication process using epitaxial lift-off structures is shown.

III. RESULTS

The ELO material that has been lifted off is of high quality and comparable to that of epitaxial material on the original substrate. The cross-section transmission electron micrograph showed the interface of the ELO layer and the metal handle substrate. No delamination, cracking, threading dislocations, or voids were observed in the TEM study. We have successfully fabricated large-area triple-junction IMM devices on ELO material. Fig. 3 shows I-V characteristics measured at NASA Glenn under AM0 spectrum for small (1-cm2) and large (20-cm2) devices. Similar performance is observed for small- and large-area devices, with a typical efficiency of 29%. The ability to fabricate large-area devices with high efficiency requires material that is highly uniform and free of cracks or defects. We have fabricated flexible solar sheets consisting of arrays of interconnected large-area ELO solar cells, as shown in Figure 4. The cells are interconnected using welded Ag-based foil ribbons and laminated between transparent sheets of polymer-based materials. The composite structure showed a high degree of flexibility and could be conformally attached to a curved structure, such as the wing of a solar-powered plane. Fig. 5 shows a typical 12 W GaAs solar sheet product that is used for remote charging. The unique feature of this product is the lightweight and compact size. This product has the highest power density that is ever produced in any solar cell technology. Larger output power panels are possible as the number of cells is increased. MicroLink is currently producing 20 W, 40 W, and 100 W panels.

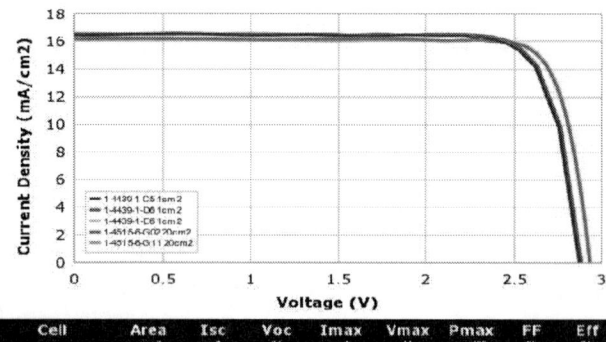

Cell	Area cm²	Isc mA	Voc V	Imax mA	Vmax V	Pmax mW	FF %	Eff %
1-4515-6-G02	20.1	322.88	2.93	311.63	2.57	800.69	84.6	29.1
1-4515-6-G11	20.1	322.09	2.93	311.45	2.59	805.88	85.3	29.3
1-4439-1-C5	1.0	16.56	2.87	15.93	2.46	39.23	82.5	28.7
1-4439-1-D6	1.0	16.68	2.88	16.14	2.47	39.93	83.0	29.2
1-4439-1-E6	1.0	16.70	2.88	16.06	2.47	39.69	82.5	29.0

Fig. 3. Large area I-V characteristics of 20 cm2 large area solar cells are shown.

Fig. 4. Flexible solar sheet consisting of an array of 30 interconnected, large-area GaAs ELO solar cells is shown.

Fig 5. A 12 W solar sheet product made from ELO GaAs solar cells is shown.

After the ELO process has been performed, the original substrate can be restored to epi-ready quality using a short repolishing step. The ability to reuse the substrate enabled significant cost reduction, particularly when the value of the substrate represents a significant fraction of the value of the final devices. Fig. 6 shows a histogram of triple-junction IMM cell efficiencies (1-cm2 active area) measured from a collection of 25 substrates that have been processed through four epitaxial growths and three repolishing cycles. The first epitaxial growth on the original GaAs substrate is labeled "prime". The average cell efficiency on the three repolished device populations is comparable to that of cells grown on the original, prime GaAs wafers, demonstrating high quality material. The number of possible reuses is limited primarily by the amount of material removed during repolishing. During this campaign the amount of polishing each cycle was approximately 40 μm, however we have shown that as little of 10 μm of polishing can be utilized. It is possible to achieve more than 10 reuses.

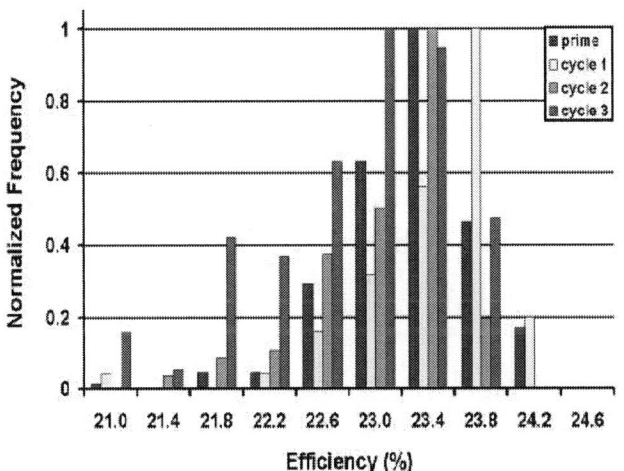

Fig. 6. Histogram of solar cell efficiencies (uncoated) showing device performance through three successive substrate reuse cycles is shown.

IV. CONCLUSION

MicroLink Devices has developed a wafer-scale epitaxial lift-off process technology for structures grown on GaAs substrates. The process is scalable for high production throughput as well as large-diameter (6-inch) substrates. The lifted-off semiconductor material is of high quality and has low defect density, enabling the fabrication of large-area (20-cm2)

multi-junction solar cell devices. We have also successfully integrated large-area ELO cells into lightweight and flexible solar arrays. Finally, we have demonstrated the ability to carry out multiple reuses of the original substrates. Epitaxial lift-off is an enabling process technology for cost reduction and improved device performance not only of multi-junction solar cells, but also potentially for a wide range of III-V electronic and optoelectronic device applications such as lasers, LEDs, photodetectors, and HBTs.

ACKNOWLEDGMENT

The authors gratefully acknowledge support from the AFRL Space Vehicles Directorate, the Department of Energy through the Solar America Initiative (SAI) Incubator program, and DARPA. The authors also acknowledge the excellent support provided by the NREL Measurements and Characterization group and from NASA Glenn for solar cell efficiency measurements.

REFERENCES

[1] M. Konagai, M. Sugimoto, and K. Takahashi, "High efficiency GaAs thin film solar cells by peeled film technology," J. Cryst. Growth 45, p. 277, 1978.

[2] E. Yablonovitch, D. M. Hwang, T. J Gmitter, L.T. Florez, J. P. Harbison, "Van der Waals bonding of GaAs epitaxial liftoff films onto arbitrary substrates," Appl. Phys. Lett. 56, p. 2419, 1990.

[3] R. Tatavarti, G. Hillier, A. Dzankovic, F. Tuminello, G. Martin, R. Navaratnarajah, D. P. Vu, and. N. Pan, "Lightweight, low cost GaAs solar cells on 4-inch epitaxial liftoff (ELO) wafers," in Proc. of 33rd IEEE PVSC conference, San Diego, 2008.

Memristor Applied In Delay Locked Loop For High Lock Speed And Wide Frequency Range

Siti Musliha Ajmal Binti Mokhtar
Faculty of Electrical Engineering
Universiti Teknologi MARA (UiTM)
Selangor, Malaysia

Wan Fazlida Hanim Abdullah
Faculty of Electrical Engineering
Universiti Teknologi Mara (UiTM)
Selangor, Malaysia

Abstract— Locking speed and operating frequency range of DLL is limited to the delays of its VCDL. Since memristor resistance changes according to external bias, memristor can be used as programmable resistor. In this paper, memristor is applied to VCDL as variable resistor where the memristor resistance regulates total delay of the VCDL. VCDL consists of current starved inverter as the delay unit. A voltage to current converter (VCC) is used to convert control voltage from capacitor, Vc to control current, Ic. The control current, Ic with memristor resistance generates new voltage, Vr that regulates the delay. Compared to conventional DLL, the proposed DLL design offers one more control parameter which is Vr. By applying memristor to the DLL, the control voltage Vc can be further manipulated. Simulation results show that proposed DLL with memristor has higher locking speed and can lock higher input frequency compared to conventional DLL without memristor. Therefore, higher locking speed and wide operating frequency range are both achievable by proposed DLL with memristor.

Keywords— *memristor; delay locked loop; memristor resistance; memristor model; programmable resistor*

I. INTRODUCTION

Phase locked loop (PLL) and delay-locked loop (DLL) are most commonly used for the clock synchronization in [1]–[3] and clock generation in SoC processor [4]. In [5], DLL is more preferable for clock synchronization and clock generation if there is no frequency multiplication. This is because DLL is using the voltage-controlled delay line (VCDL) that does not accumulate noise thus has better jitter performance than PLL [6], [7]. DLLs are generally more stable and easier to design with smaller area and faster locking time than the PLL. A DLL is a system that control its output signal's phase to always match with the input signal. A conventional analog DLL consists of phase detector, charge pump, low pass filter and voltage controlled delay line, (VCDL). The block diagram is shown in Fig 1. The DLL adjusts the delay of the delay line until the inputs to the phase detector are in phase. The drawback of conventional DLL is that it has finite delay range that limits operating frequency range. As reported in [8], the highest operating frequency of a DLL is limited by the minimal delay of a single delay unit while the lowest operating frequency is restricted by the length of delay line. Thus, designers always try to improve DLL performance through the VCDL. A good DLL will have wide frequency range, small area consumption and good jitter performance.

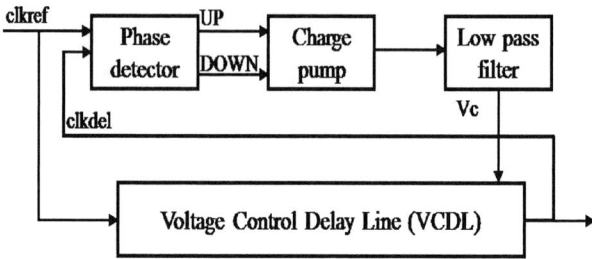

Fig. 1 Conventional analog DLL blocks

The theory of memristor is originally reported by Leon Chua in 1971 [9-10]. However, memristor remains as theory only until the first physical memristor model is realized by Stan William and his group from HP lab [11-12]. Among promising memristor application is implementation in CMOS integrated circuit as variable resistor. According to Leon Chua in [10], memristor is a resistor with varying resistance, where the resistance changes according to the total amount of charge that flow through over its entire history. Since memristor resistance changes according to supply voltage, memristor can be used as programmable resistor. Several programmable analog circuits demonstrating memristor based programming of threshold, gain, and frequency are proposed in work [13]. Another example of memristor-based programmable resistance is reported in work [14]. In this work, memristor is applied to VCDL as variable resistor where the memristor resistance regulates total delay of the VCDL. A voltage to current converter (VCC) is used to convert control voltage from capacitor, Vc to control current, Ic. The control current, Ic with memristor resistance generates new voltage, Vr that regulates the delay. Therefore, the delays are adjustable due to memristor variable resistance.

This paper is divided into 5 sessions including the introduction and conclusion. Session 2 explains on how memristor regulate delays of the voltage control delay line from one constant bias value which is Vc. The idea to control memristor resistance is using pulse signal that increases/decreases the resistance according to signal polarity and pulse width. Session 3 describes on the proposed DLL applied with memristor. A comparison between conventional and proposed DLL performance is presented. Session 2 and 3 also include design simulation results and discussions.

978-1-4799-5761-3/14 $31.00 © 2014 IEEE

II. MEMRISTOR APPLIED TO VCDL

Proposed memristor applied to VCDL that consists of current starved inverter as delay unit is shown in Fig 2. The gate voltage of M3 and M4 control the amount of current flow to middle inverter that has parasitic capacitor, C_{LOAD}. If the gate voltage is big, than the current flows is also large and otherwise. Thus the delay is determined by size of C_{LOAD} and amount of (dis)charging current, I_{DS} as shown in equation (1).

$$t_{delay} = C_{LOAD} \frac{V_{TN}}{I_{DS}} \qquad (1)$$

As shown in above equation, delay is regulated by controlling I_{DS} since C_{LOAD} is constant value when transistor in saturation region. Here, a memristor is applied to control gate voltage of M1 and M2 thus controls the current flows to M3 and M4 as shown in Fig 2. Assuming circuit is in saturation region with no channel length modulation, the I_{DS} is shown below.

$$I_{DS} = \frac{1}{2} \mu_n C_{OX} \frac{W}{L} \left(V_{GS} - V_{TH} \right)^2 \qquad (2)$$

$$V_{GS} = I R_{mem} \qquad (3)$$

Here, I is external current source to present constant current value. In next session, this external current source is replaced with voltage to current converter to convert current from Vc. Equation (2) and (3) is inserted in equation (1) to roughly estimate the delay.

$$t_{delay} = C_{LOAD} \frac{V_{TN}}{\frac{1}{2} \mu_n C_{OX} \frac{W}{L} \left(I R_{mem} - V_{TH} \right)^2} \qquad (4)$$

A. Pulse Signal Supply to Control Memristor Resistance

Pulse signal is used in order to control the memristor resistance. Positive or negative pulse signal that applied to memristor will increase or decrease the resistance gradually, until completely saturated at either R_{OFF} or R_{ON}. According to [14], memristor resistance change grade depends on the number of pulse, pulse amplitude, and pulse frequency. High number of pulse, high pulse amplitude value and low pulse frequency result in higher resistance.

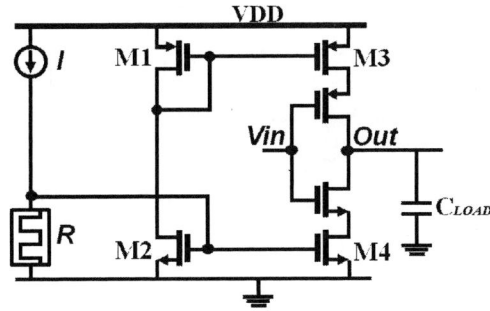

Fig. 2. Memristor applied with current starved inverter as delay unit for VCDL

Since memristor is not yet available in market due to the cost and technical difficulties in fabrication nanoscale device, few papers have proposed SPICE macro model for simulating memristor. In this work, to demonstrate memristor SPICE model behaviour, a SPICE model of memristor is adopted from [11]. The parameters are based on the actual fabricated memristor in our lab which are R_{ON}=100Ω, R_{OFF} =16kΩ, R_{INIT}=100, D=10N, uv=10F and p=10. Fig 3 and Fig 4 show simulation results for memristor resistance changes due to voltage supply polarity and number of pulse. In Fig 3, when negative voltage is applied to memristor, resistance increases until saturated at R_{OFF}. In contrast, when positive voltage is applied to memristor, resistance decreases until saturated at R_{ON}. In Fig 4, 1kHz pulse input voltage is applied to memristor. For small step resistance change, the pulse amplitude must be smaller which is achieved by high frequency pulse signal. Thus, high frequency of pulse signal input voltage is applied to memristor to control the resistance. The resistance changes 10 Ω/1ms pulse width.

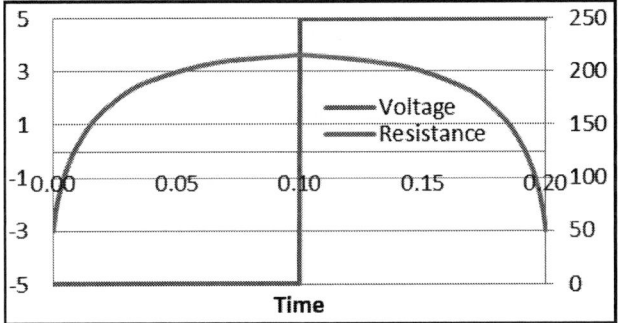

Fig 3. SPICE simulation results of memristor resistance and voltage versus time. R_{ON}=50Ω, R_{OFF} =220Ω, D=19.8nm, p=11.

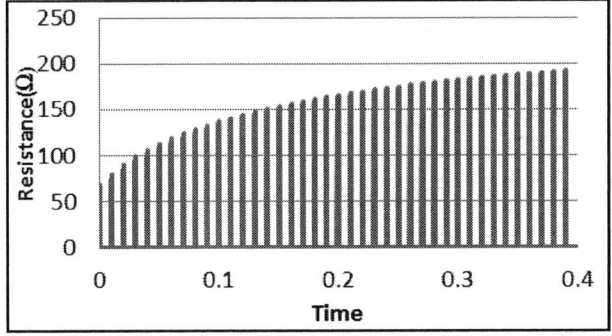

Fig 4. SPICE simulation results of memristor resistance versus time for pulse input voltage with 5V amplitude. Pulse frequency is 1kHz.

B. Simulation Result For Proposed Delay Element with Memristor

Fig 5 shows maximum and minimum delays for delay unit. However, current memristor model cannot work with circuit yet, thus memristor is replaced with resistor. Resistor value is based on memristor resistance value. According to equation (3), minimum delay is obtained when resistance highest while maximum delay is obtained when resistance lowest.

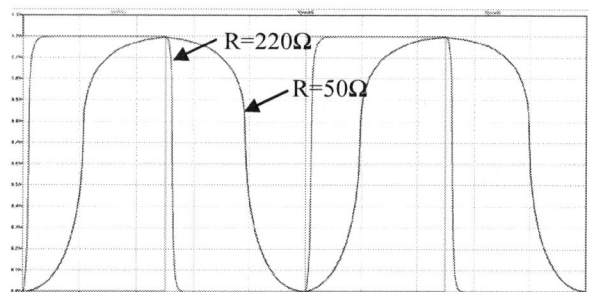

Fig 5. SPICE simulation results for maximum and minimum delays for delay element.

III. DLL WITH MEMRISTOR

Block diagrams for proposed DLL is shown in Fig 6. Notice that one block of voltage to current converter (VCC) is added up after low pass filter. Here, control voltage, Vc accumulated by low pass filter (capacitor) is converted by VCC to control current, Ic. However, the control current, Ic is usually in uA. Since resistance of fabricated memristor is even below than 1kΩ, Ic is multiplied to 100 times via current mirror in order to generate large enough Vr to turn on the transistor. This Vr will regulate the delay of delay line. The delay calculation is same with equation (3), however, in proposed DLL current source I is replaced by Ic. Schematic for VCC is shown in Fig 7. Delay line of VCDL consists of 8 delay units and 2 buffers to shape up the delayed output.

From equation (3) we can determine the delay of one delay unit. If the delay line is using N numbers of delay unit, thus total delays from delay line will be $N \times t_{delay}$.

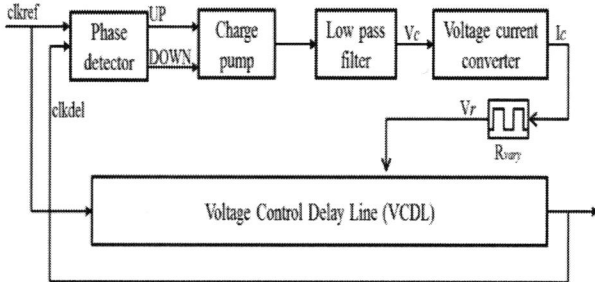

Fig 6. Proposed DLL with memristor

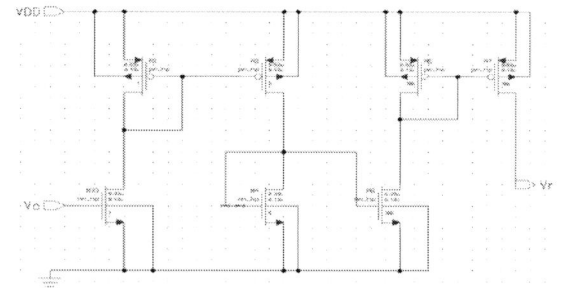

Fig 7. Schematic of voltage current converter

A. Simulation Result For Proposed DLL with Memristor

The circuit design and simulation for proposed DLL is using Mentor graphics in 0.13 process technology. Input frequency is 1GHz and simulation result is shown in Fig 9. When clkref leads clkdel, UP signal is generated thus current is flowing to capacitor thus increase Vc. Otherwise, when clkref lags clkdel, DOWN signal is generated thus current is flowing out from capacitor thus Vc is discharged to ground. The basic block diagram of charge pump is shown in Fig 8 and operation summarization is shown in table 1.

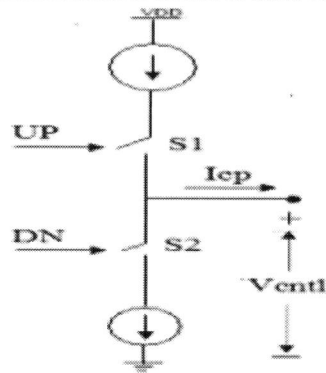

Fig 8. Basic block diagram of charge pump.

Table 1. Summary of PFD and charge pump operation.

Operations	PFD output	switch	Current	Vcntl
Clkref leads clkdel	UP signal high DOWN signal low	S1 on S2 off	Flows into capacitor	Increase
Clkdel leads clkref	UP signal low DOWN signal high	S1 off S2 on	Flows out of capacitor	Decrease
Clkdel equal clkref	UP signal low DOWN signal low	S1 off S2 of	No current flowing	Constant

As explained in session 2, current memristor model cannot work together with circuit yet, thus memristor is replaced with resistor. Resistor value is varied based on fabricated memristor resistance value which is from 50Ω (R_{ON}) to 220Ω (R_{OFF}) by 50Ω increment. As we can see in Fig 9, when resistance changes, value of Vr changes and this results in different rising time for the output clock. As resistance increases, Vr increases thus rising time is smaller and otherwise.

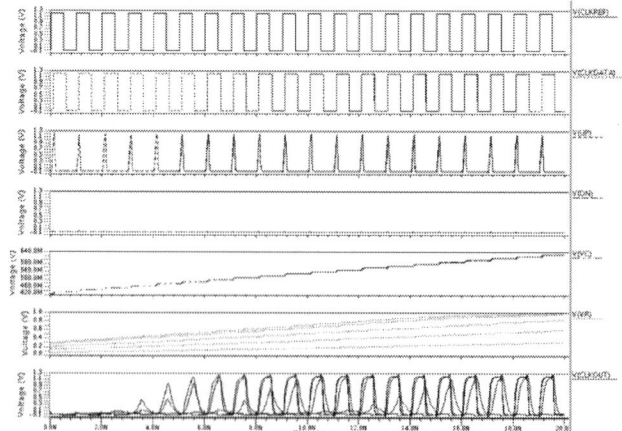

Fig 9. Simulation result for clkref leads clkdel for input frequency 1GHz and resistance from 50Ω to 220Ω

978-1-4799-5761-3/14 $31.00 © 2014 IEEE

Comparison between proposed DLL with memristor and conventional DLL without memristor is shown in Fig 10 and Fig 11. In Fig 10, input signal frequency is 1GHz with memristor resistance applied is 220Ω. Since Vr is larger than Vc, the locking time for proposed DLL is higher than conventional DLL. In Fig 11, the proposed DLL managed to locked clkout with clkref when the input signal frequency is 160MHz with memristor resistance is 70Ω. However, conventional DLL cannot lock the signal. This is because Vr is much lower than Vc, due to low resistance. By applying memristor to the DLL, the control voltage Vc can be further manipulated. Therefore, higher locking speed and wide operating frequency range are both achievable.

Table 2. Summary of conventional and proposed DLL comparison.

	DLL without memristor	DLL with memristor
Locking time (ns)	8	6
Frequency range	500 MHz~1 GHz	160 MHz ~ 2 GHz

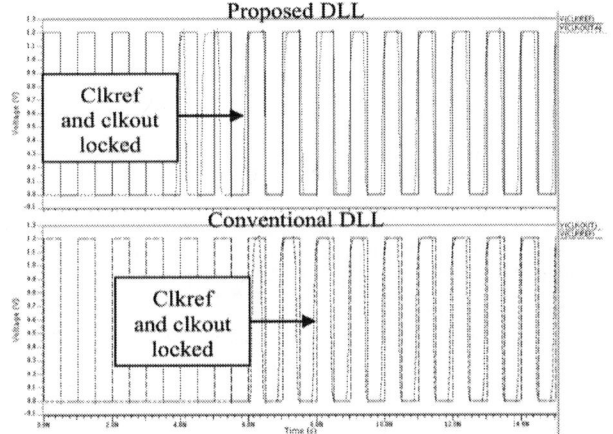

Fig 10. Comparison between proposed DLL with memristor and conventional DLL without memristor for input frequency 1GHz and resistance at 220 Ω

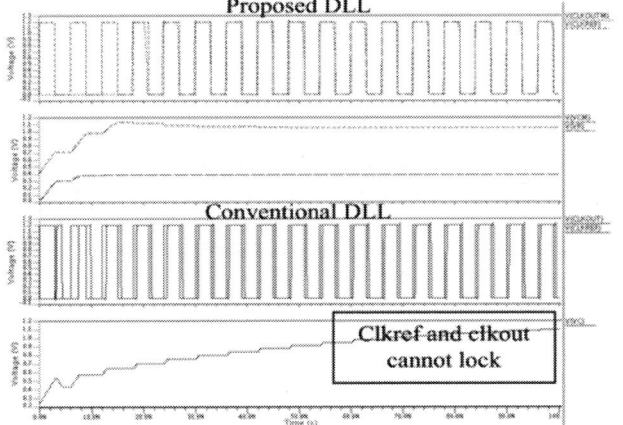

Fig 11. Comparison between proposed DLL with memristor and conventional DLL without memristor for input frequency 160MHz and resistance at 70 Ω

IV. CONCLUSION

In this work, memristor is applied as variable resistor with current starved inverter to regulate delays of VCDL. The memristor resistance is controlled by polarity of pulse signal supply. A voltage to current converter (VCC) is used to convert control voltage, Vc to control current, Ic. The control current, Ic with memristor resistance generates new voltage, Vr that regulate the delay of delay line. Therefore, the delays are adjustable compared to conventional DLL without memristor design thus proposed DLL design with memristor offers wider operating frequency range.

ACKNOWLEDGMENT

The authors would like to thank the Ministry of Science, Technology and Innovation (MOSTI) and the Research Management Institute of Universiti Teknologi MARA (UiTM) for providing the financial support under the Science Fund grant (100-RMI/SF 16/6/2 (31/2012).

REFERENCES

[1] R. B. Watson, Jr. and R. B. Iknaian, "Clock buffer chip with multiple target automatic skewcompensation," IEEE J. Solid-State Circuits, vol. 30, pp. 1267–1276, Nov. 1995.

[2] C. H. Kim et al., "A 64-Mbit, 640-Mbyte/s bidirectional data strobed, double-data-rate SDRAM with a 40-mW DLL for a 256-Mbyte memory system," IEEE J. Solid-State Circuits, vol. 33, pp. 1703–1710, Nov. 1998.

[3] Y. Moon, J. Choi, K. Lee, D. K. Jeong, and M. K. Kim, "An all-analog multiphase delay-locked loop using a replica delay line for wide-range operation and low-jitter performance," IEEE J. Solid-State Circuits, vol. 35, pp. 377–384, Mar. 2000.

[4] C. Kim, I.-C. Hwang, and S.-M. Kang, "A low-power small-area 7.28-ps-jitter 1-GHz DLL-based clock generator," IEEE J. Solid-State Circuits, vol. 37, pp. 1414–1420, Dec. 2002.

[5] K. H. Chen, and Y. L. Lo, "A fast-lock DLL with power-on reset circuit," International Symposium on Circuits and Systems, ISCAS,,04, pp. 23–26, May. 2004.

[6] H. H. Chang, J. W. Lin and S. I. Liu, "A fast locking and low jitter delay-locked loop using DHDL," IEEE J. Solid-State Circuits, vol. 38, no. 2, pp. 343–346, Feb. 2003.

[7] L. K. Soh, M. S. Sulaiman, and Z. Yusoff, "Fast-lock dual charge pump analog DLL using improved phase frequency detector," in Proc. International Symposium on VLSI Design, Automation and Test (VLSI-DAT), pp. 1–5, Jun. 2007.

[8] H. Chang, "A Wide-Range and Fast-Locking All-Digital Cycle-Controlled Delay-Locked Loop," IEEE Journal Of Solid-State Circuits, Vol. 40, No. 3, March 2005

[9] L.O. Chua, "Memristor-the missing circuit element," IEEE Trans. on Circuit Theory, vol. 18, pp. 507-519, 1971.

[10] S. M. K. Chua, L.O., "Memristive devices and systems," Proceedings of the IEEE, vol. 64, no. 2, pp. 209 – 223, Feb. 1976.

[11] D.B. Strukov, G.S. Snider, D.R. Stewart and R.S. Williams, "The missing memristor found," Nature (London), vol. 453, pp. 80-83, 2008.

[12] R. S. William, "How we found the missing memristor," IEEE Spectrum, p. 1-11,2008

[13] Y.V. Pershin, M.D. Ventra, "Practical Approach To Programmable Analog Circuit With Memristors," IEEE Transactions on Circuits and Systems, vol. 57, Issue 8, pp. 1857-1864, 2010

[14] S. Shin, K. Kim and S. Kang, "Memristor-based Fine Resolution Programmable Resistance and Its Applications", Int. Conference on Communications, Circuits and Systems, pp. 948–951,2009

Photoluminescence from Localized States in GaAsBi Epilayers

A. R. Mohmad, B. Y. Majlis

Institute of Microengineering and Nanoelectronics,
Universiti Kebangsaan Malaysia
43600 Bangi, Malaysia
armohmad@ukm.edu.my

F. Bastiman, C. J. Hunter, R. D. Richards, J. S. Ng,
J. P. R. David

Department of Electronic and Electrical Engineering
University of Sheffield
S1 3JD Sheffield, United Kingdom

Abstract— The structural and optical properties of $GaAs_{1-x}Bi_x$ samples with x = 0.048 and 0.06 were investigated by high resolution X-ray diffraction (HR-XRD) and photoluminescence (PL). The HR-XRD results show that both samples have a smooth and coherent $GaAs_{1-x}Bi_x$/GaAs interface. The PL peak energy shows an S-shape behavior with temperature (10 K to room temperature) which is an evidence of carrier localizations. An Arrhenius analysis reveals that the thermal quenching activation energy is 30 and 19 meV for the $GaAs_{0.952}Bi_{0.048}$ and $GaAs_{0.94}Bi_{0.06}$ sample, respectively. It was found that the activation energy is not necessarily originated from alloy fluctuations but may also due to CuPt ordering-induced band gap fluctuations. The power dependent PL indicates that the localized energy states are continuously present and located up to 40 - 47 meV from the valence band.

Keywords—photoluminescence; localization effects; activation energy; GaAsBi alloy;

I. INTRODUCTION

Bismuth (Bi) containing semiconductors, particularly $GaAs_{1-x}Bi_x$ is a promising material system for various applications including lasers [1], photo-detectors [2], solar cells [3] and spintronic devices [4]. This is mainly due to the large reduction of band gap with the incorporation of a relatively small amount of Bi and the large spin-orbit splitting. It was reported that the incorporation of Bi into GaAs reduces the band gap by reducing the energy of the conduction band minimum as well as increasing the energy of the valence band maximum [5,6]. The former reduces linearly by 23 - 28 meV/%Bi as predicted by the virtual crystal approximation while the latter increases by 64 meV/%Bi but becomes non-linear for x > 0.06 [5,6]. The increase of the valence band minimum is the result of anti-crossing interaction between the Bi level and the valence band edge of GaAs [7].

Despite the potential of $GaAs_{1-x}Bi_x$ alloy, the growth of this material system is not straightforward. The growth temperature of $GaAs_{1-x}Bi_x$ has to be lowered to $\leq$ 400 °C in order to reduce the surface segregation of Bi and hence incorporate significant amount of Bi into GaAs [8]. The growth temperature is significantly lower compared to the optimum growth temperature of GaAs which is ~ 590 °C. Besides, low temperature photoluminescence (PL) emission from $GaAs_{1-x}Bi_x$ samples was observed to be affected by carrier localizations [9-11]. A Monte Carlo analysis carried out by Imhof *et al* shows

that localization effects induced by alloy composition fluctuations and Bi clusters has a characteristic energy scale of 45 and 11 meV, respectively [11]. Besides, for $GaAs_{1-x}Bi_x$ sample with x = 0.026, localized energy states located up to ~ 90 meV above the valence band was observed [10]. In this report, power and temperature dependent PL will be used to investigate carrier localization effects in $GaAs_{1-x}Bi_x$ samples. This work uses two $GaAs_{1-x}Bi_x$ samples with x = 0.048 and 0.06.

II. EXPERIMENTAL DETAILS

The $GaAs_{1-x}Bi_x$ samples reported in this study were grown on semi-insulating GaAs (100) substrate in an Omicron molecular beam epitaxy – scanning tunneling microscopy (MBE - STM) system. After the substrate was degassed and the oxide layer was removed, a 80 nm of GaAs buffer was grown at 590 °C at a rate of 0.16 µm/h. Then, the growth temperature was reduced to 400 °C and a 25 nm of $GaAs_{1-x}Bi_x$ epilayer was grown, followed by a 50 nm of GaAs cap. During the growth of the $GaAs_{1-x}Bi_x$ epilayer, the Bi cell temperature was varied between 520 and 530 °C while other growth parameters were fixed.

The structural and optical properties of the as-grown samples were characterized by high resolution X-ray diffraction (HR-XRD) and PL. The X-ray source in the HR-XRD system was generated by accelerating the electrons with 50 keV of potential difference before hitting a copper target. The (004) θ-2θ scans were carried out and the measured spectra were fitted using RADS Mercury software. For PL measurements, a green 532 nm diode pumped solid state laser was used as the excitation source. A monochromator was used to disperse the PL in wavelengths and then detected by a liquid nitrogen cooled germanium detector. The spectral response of our PL setup is shown in [12]. The sample was mounted inside a closed-cycle helium cryostat system which consists of a gas compressor, a temperature controller, a vacuum pump and the cryostat itself. The compressor provides a high pressure helium gas whose expansion creates a refrigeration effect down to 10 K. The temperature inside the cryostat is controlled by the temperature controller as well as a heater and a temperature sensor inside the cryostat.

III. RESULTS AND DISCUSSIONS

A. High Resolution X-Ray Diffractions Results

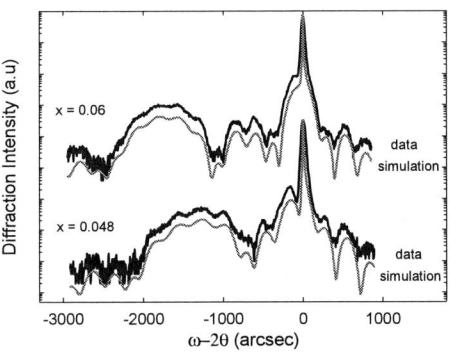

Fig. 1. HR-XRD spectra (black) and simulations (grey) for GaAsBi samples with $x = 0.06$ and 0.048.

Fig. 1 shows the HR-XRD spectra of two GaAs$_{1-x}$Bi$_x$ samples grown with different Bi cell temperature. For each sample, two main peaks can be observed; a sharp and high intensity peak located at 0 arcsec and a broad and low intensity peak located at negative peak splitting. The former originates from the GaAs layer (and the substrate) while the later originates from the GaAs$_{1-x}$Bi$_x$ epilayer. A broad and weak peak was observed because the GaAs$_{1-x}$Bi$_x$ layer is significantly thinner compared to the substrate. The simulation results nicely fit with the measured spectra and yielded $x = 0.048$ and 0.06 for sample grown with Bi cell temperature of 520 and 530 °C, respectively. Besides, the presence of well-defined fringes in the measured spectra indicates that both samples have a smooth and coherent interface.

B. Photoluminescence Results

Fig. 2 show the temperature dependent PL of the GaAs$_{1-x}$Bi$_x$ samples with excitation power, $P_{ex} = 50$ mW. A relatively low excitation was used to avoid the saturation of the localized states while ensuring measurable PL from both samples at room temperature (RT). The results show that, at RT, the PL peak energy of sample GaAs$_{0.952}$Bi$_{0.048}$ and GaAs$_{0.94}$Bi$_{0.06}$ is 1.11 and 1.04 eV, respectively. A band gap reduction of 70 meV was obtained by increasing the Bi concentration from 0.048 to 0.06. The band gap of GaAs$_{1-x}$Bi$_x$ reported in this work is consistent with the theoretical values predicted by tight binding analysis in [6]. For both samples, the PL peak energy shows an S-shape dependence with temperature which is a well-known artifact of carrier localization effects. The localization may be attributed to Bi composition fluctuations and Bi clusterings [11]. Similar observations also have been reported by other groups [10,11].

The S-shape behavior consists of three processes, (i) PL peak energy redshift between 10-80 K, (ii) PL peak energy blueshift between 80-140 K and (iii) another PL peak energy redshift for temperatures from 140 K to RT. Process (I) occurred due to the movement of excitons to other localized states with lower energy. As the temperature increases from 80 to ~ 140 K,

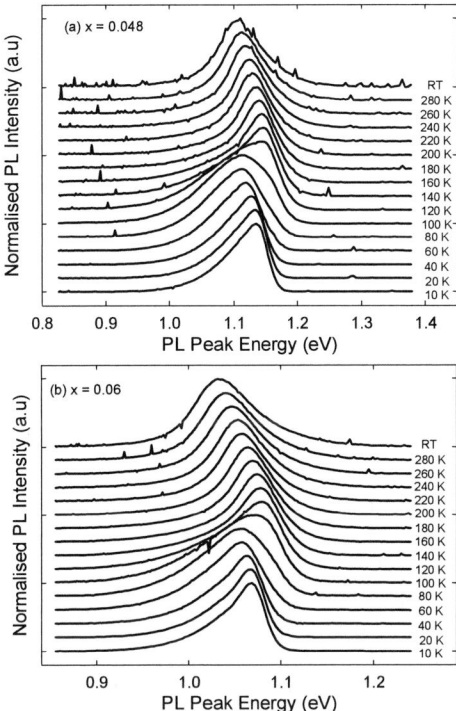

Fig. 2. Temperature dependent PL of GaAsBi samples with (a) $x = 0.048$ and (b) $x = 0.06$. The PL spectra were vertically shifted for clarity.

excitons started to possess sufficient energy to hop out of the localized states and become delocalized. For temperatures > 140 K, majority of the carriers are delocalized and the PL peak energy show a second redshift due to the dependence of the band gap with temperature. These processes are illustrated in Fig. 3. One of the methods to estimate the magnitude of the localization potential is by calculating the thermal energy required for the carriers to hop out of the localized states [13]. Since the carriers in both samples become delocalized at ~ 140 K, the thermal energy and the localization potential is approximately $k_B \cdot 140$ K = 12 meV.

Next, the integrated PL (IPL) intensity of the GaAs$_{1-x}$Bi$_x$ samples was plotted versus the inverse of temperature, known as an Arrhenius plot as shown in Fig. 4 (a). At low temperatures (10 – 40 K), the IPL intensity is almost unchanged but decreases exponentially at high temperatures. At 10 K, the IPL intensity of sample GaAs$_{0.952}$Bi$_{0.048}$ is ~ 3 times higher compared to the GaAs$_{0.94}$Bi$_{0.06}$ sample. However, when the temperature is increased from 10 K to RT, the IPL intensity of the GaAs$_{0.952}$Bi$_{0.048}$ sample reduces by almost three orders of magnitude while for sample GaAs$_{0.94}$Bi$_{0.06}$, the IPL reduces by only two orders of magnitude. The thermal quenching of the IPL intensity can be described by,

$$ I(T) = I_o/[1 + A \cdot exp\ (-E_a/k_BT)] , \qquad (1) $$

IEEE-ICSE2014 Proc. 2014, Kuala Lumpur, Malaysia

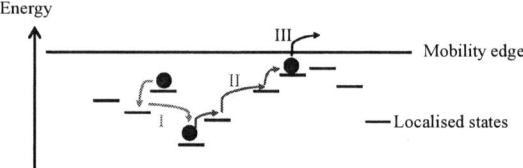

Fig. 3. Illustration of the hopping of carriers between nearby localized states due to the increase of temperatures.

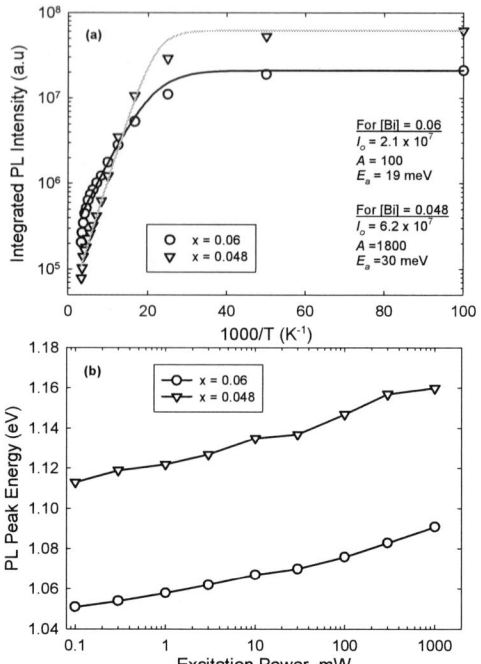

Fig. 4. (a) Arrhenius plot of the GaAsBi samples. The solid lines refer to the fittings using (1), (b) Variation of PL peak energy with excitation power

where $I(T)$ is the IPL intensity at temperature T, I_o is the IPL intensity at 10 K, A is the fitting parameter and E_a is the activation energy. The best fit using (1) yielded $A = 1800$, $E_a = 30$ meV for the GaAs$_{0.952}$Bi$_{0.048}$ sample and $A = 100$, $E_a = 19$ meV for the GaAs$_{0.94}$Bi$_{0.06}$ sample. However, the IPL intensity of the GaAs$_{0.94}$Bi$_{0.06}$ sample deviates from the fitting at high temperatures. This is probably due to the loss of carriers to non-radiative recombination centers which become increasingly dominant at high temperatures.

The 19 meV of activation energy for the GaAs$_{0.94}$Bi$_{0.06}$ sample is consistent with the localization potential estimated from Fig. 2 which is 12 meV. However, this is not the case for the GaAs$_{0.952}$Bi$_{0.048}$ sample in which the value of E_a is much larger (30 meV) compared to the 12 meV of localization potential. The result suggests that for this material system, the thermal quenching activation energy is not necessarily originated from alloy composition fluctuations. Recently, Norman *et al* have carried out a Transmission Electron Microscopy (TEM) study and observed the presence of CuPt$_B$-(111) ordering in GaAs$_{1-x}$Bi$_x$ (for x up to 0.1) which was attributed to the (2 × 1) surface reconstruction during GaAs$_{1-x}$Bi$_x$ growth [14]. It is possible that the E_a for the GaAs$_{0.952}$Bi$_{0.048}$ sample could represent a barrier between the ordered regions (which have a smaller band gap) and the disordered regions (which have a larger band gap) which contains the non-radiative centers. A similar observation was reported in InGaP material system in which the E_a increases with the degree of CuPt ordering [13].

Power dependent PL measurements were also carried out at 10 K as shown in Fig. 4 (b). As P_{ex} is increased from 0.1 to 1000 mW, the PL peak energy blue-shifts by 47 and 40 meV for the GaAs$_{0.952}$Bi$_{0.048}$ and GaAs$_{0.94}$Bi$_{0.06}$ samples, respectively. This observation verifies the presence of localization effects in the samples and consistent with the S-shape behavior shown in Fig. 2. The PL peak energy blue-shifts with P_{ex} due to the saturation of the localized states at high excitations. The amounts of blue-shift for both samples (47 and 40 meV) are larger compared to the 12 meV of localization potential estimated from Fig. 2 and the E_a deduced from Fig. 4 (a). This indicates the presence of deep localized states in the samples and that the states are continuously present and located up to 40 - 47 meV from the valence band.

IV. CONCLUSION

Temperature and power dependent PL measurements were carried out for GaAs$_{1-x}$Bi$_x$ samples with $x = 0.048$ and 0.06. The PL emissions from both samples show the presence of localization effects. A stronger localization was observed for the GaAs$_{0.952}$Bi$_{0.048}$ sample compared to the GaAs$_{0.94}$Bi$_{0.06}$ based on the larger thermal quenching activation energy and the larger amounts of PL peak energy blueshift with excitation. It was found that the localized energy states are continuously present up to 40 - 47 meV above the valence band minimum.

ACKNOWLEDGMENT

This work was supported by the Research University Grant ICONIC-2013-007 (Ministry of Education Malaysia), UK TSB, the Royal Society (J. S. Ng) and doctoral fellowship from the University of Sheffield-EPSRC (F. Bastiman).

REFERENCES

[1] P. Ludewig, N. Knaub, N. Hossain, S. Reinhard, L. Nattermann, I.P. Marko, S.R. Jin, K. Hild, S. Chatterjee, W. Stolz, S.J. Sweeney, and K. Volz, "Electrical injection Ga(AsBi)/(AlGa)As single quantum well laser", Appl. Phys. Lett., vol. 102, pp. 242115, 2013.

[2] C.J. Hunter, F. Bastiman, A.R. Mohmad, R. Richards, J.S. Ng, S.J. Sweeney, and J.P.R. David, "Absorption Characteristics of GaAsBi\GaAs Diodes in the Near-Infrared", IEEE Photon. Tech. Lett., vol. 24, pp. 2191, 2012.

[3] R.N. Kini, L. Bhusal, A.J. Ptak, R. France, and A. Mascarenhas, "Electron Hall mobility in GaAsBi", J. Appl. Phys., vol. 106, pp. 043705, 2009.

978-1-4799-5761-3/14 $31.00 © 2014 IEEE   356

[4] B. Fluegel, S. Francoeur, A. Mascarenhas, S. Tixier, E.C. Young, and T. Tiedje, "Giant spin-orbit bowing in GaAs$_{1-x}$Bi$_x$", Phys. Rev. Lett., vol. 97, pp. 067205, 2006.

[5] A.R. Mohmad, F. Bastiman, C.J. Hunter, J.S. Ng, S.J. Sweeney, J.P.R. David, and B.Y. Majlis, "Localization effects and band gap of GaAsBi alloys", Phys. Status Solidi B, in press.

[6] M. Usman, C.A. Broderick, A. Lindsay, and E.P. O'Reilly, "Tight-binding analysis of the electronic structure of dilute bismide alloys of GaP and GaAs", Phys. Rev. B, vol. 84, pp. 245202, 2011.

[7] K. Alberi, O.D. Dubon, W. Walukiewicz, K.M. Yu, K. Bertulis, and A. Krotkus, "Valence band anticrossing in GaBi$_x$As$_{1-x}$", Appl. Phys. Lett., vol. 91, pp. 051909, 2007.

[8] F. Bastiman, A.R. Mohmad, J.S. Ng, J.P.R. David, and S.J. Sweeney, "Non-stoichiometric GaAsBi/GaAs (100) molecular beam epitaxy growth", J. Cryst. Growth, vol. 338, pp. 57, 2012.

[9] A.R. Mohmad, F. Bastiman, J.S. Ng, S.J. Sweeney, and J.P.R. David, "Photoluminescence investigation of high quality GaAsBi on GaAs", Appl. Phys. Lett., vol. 98, pp. 122107, 2011.

[10] M. Yoshimoto, M. Itoh, Y. Tominaga, and K. Oe, "Quantitative estimation of density of Bi-induced localized states in GaAsBi grown by molecular beam epitaxy", J. Cryst. Growth, vol. 378, pp. 73, 2013.

[11] S. Imhof, A. Thranhardt, A. Chernikov, M. Koch, N.S. Koster, K. Kolata, S. Chatterjee, S.W. Koch, X. Lu, S.R. Johnson, D.A. Beaton, T. Tiedje, and O. Rubel, "Clustering effects in Ga(AsBi)", Appl. Phys. Lett., vol. 96, pp. 131115, 2010.

[12] A.R. Mohmad, F. Bastiman, J.S. Ng, S.J. Sweeney, J.P.R. David, "Room temperature photoluminescence intensity enhancement in GaAsBi alloys", Phys. Status Solidi C, vol 9, pp. 259.

[13] K.L. Teo, J.S. Colton, P.Y. Yu, E.R. Weber, M.F. Li, W. Liu, K. Uchida, H. Tokunaga, N. Akutsu, and K. Matsumoto, "An analysis of temperature dependent photoluminescence line shapes in InGaN", Appl. Phys. Lett., vol. 73, pp. 1697, 1998.

[14] A.G. Norman, R. France, and A.J. Ptak, "Atomic ordering and phase separation in MBE GaAsBi", J. Vac. Sci. and Technol. B, vol. 29, pp. 03C121, 2011.

Growth Parameters Optimization of GaN High Electron Mobility Transistor Structure on Silicon Carbide Substrate

Y. Y. Wong*, S. C. Huang, W. C. Huang, F. Lumbantoruan, Y. S. Chiu, H. C. Wang, H. W. Yu, E. Y. Chang, *IEEE Fellow*

Compound Semiconductor Device Laboratory
National Chiao Tung University
1001 Ta Hsueh Rd., Hsinchu, 30010 Taiwan
*Email: yuenyee98@nctu.edu.tw

Abstract - High electron mobility transistors heterostructures of AlGaN/GaN were grown by metalorganic chemical vapor deposition system on silicon carbide substrate. The growth parameters such as AlN buffer thickness, AlN spacer growth time and Al content in AlGaN barrier layer were optimized. Moreover, the effects of chamber pressure and V/III ratio at the initial growth state of GaN on film crystal quality were also investigated. The optimized AlGaN/GaN heterostructure has AlN buffer thickness of 120 nm, AlN spacer growth time of 10 s and Al content of 28% in the barrier layer. Furthermore, as a result of using higher chamber pressure and lower V/III, both the GaN crystal quality and electron mobility in the AlGaN/GaN were also significantly improved. After the growth parameter optimization, the GaN (002) and (102) planes exhibited X-ray rocking curve widths of 209 arcsec and 273 arcsec, respectively. Besides, the AlGaN/GaN structure also has an electron mobility of 1832 cm^2/V-s and a sheet electron density of 1.08×10^{13} cm-2, which yielding a sheet resistance of 316 Ω/sq.

Keywords- AlGaN/GaN, High electron mobility transistors, SiC

I. INTRODUCTION

AlGaN/GaN high electron mobility transistor (HEMT) has received huge attention for future high frequency and high power applications. Due to the GaN material strengths such as large energy bandgap (3.42eV), high sheet carrier density and high electron mobility, the GaN HEMT devices have clear advantages compared to the conventional Si- and GaAs-based technologies as summarized in Fig. 1. However, due to the lack of large-size commercial-grade native substrate, GaN materials are usually grown on a foreign substrate include Si, sapphire and silicon carbide (SiC). Among these substrate, SiC is the best choice in term of device performance and reliability. GaN with best crystal quality can be achieved on the SiC substrate as a result of relatively small lattice constant mismatch and thermal expansion coefficient mismatch (Table 1). Moreover, the high thermal conductivity of SiC substrate is also necessary for better heat dissipation during high power and high frequency operations of GaN devices. High performances of GaN devices are frequently demonstrated on SiC substrate [1-3].

Despite the advantages of SiC as compared to other substrates, good quality GaN material cannot be achieved with addressing several growth issues. First of all, due to poor surface wetting, good quality GaN film is difficult to grow directly on SiC [4]. Three dimensional islands will form at the initial growth stage and jeopardizing the GaN film quality. In order to overcome this issue, AlN is used as a buffer layer [5]. The AlN is a good wetting agent on SiC and form good coverage on it. Furthermore, the AlN has an intermediate lattice constant between that of GaN and SiC. Therefore, it can effectively reduce the lattice strain of GaN. Another critical issue of GaN grown on SiC is the crack of GaN film as a result of thermal expansion coefficient mismatch (Table I). The AlN buffer is also useful to mitigate this problem.

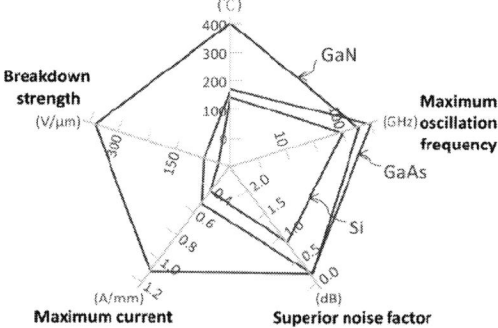

Fig. 1. GaN device strength as compared to that of Si and GaAs. (source: GaN Systems)

TABLE I. COMPARISON OF SUBSTRATES MATERIAL PROPERTIES FOR GAN GROWTH

Substrate	Si	Sapphire	SiC
Thermal conductivity (W/cm/K)	1.3	0.3	4.9
[a]Lattice constant mismatch (%)	+16	-16	-3.5
[a]Thermal expansion coefficient mismatch (%)	-54	+39	-12

[a](+): larger than that of GaN; (-) smaller than that of GaN

In this study, the growth parameters of AlGaN/GaN HEMT structure on SiC substrate would be optimized. In particular, the effects of AlN buffer thickness and AlN spacer thickness as well as the Al content in AlGaN barrier layer on the AlGaN/GaN structure would be discussed. Besides, the GaN crystal quality was also further improved by changing the initial growth condition of the GaN layer include chamber pressure and V/III ratio.

II. EXPERIMENTAL

AlGaN/GaN HEMT structures were grown by low-pressure metal-organic chemical vapor deposition (MOCVD, Thomas Swan) system on 2-inch semi-insulating (SI) SiC substrate. The main goal was to achieve high quality HEMT structure by optimizing the growth parameters. Figure 2 illustrated the schematic diagram of the AlGaN/GaN HEMT structure grown on SiC in this study. The GaN layer used in this study has a fixed thickness of 2.2 μm. In-situ reflectance (EpiTT, LayTec) was employed to monitor the growth behavior. After the growth, high resolution X-ray diffraction (HRXRD, Bede D1) was used to check the crystal quality of the growth films. Besides, the Al content and the AlGaN barrier thickness were also determined by HRXRD, using Ω-2Θ and X-ray reflectivity (XRR) scan modes, respectively. On the other hand, the thickness and surface morphology of AlN buffer layers were determined using scanning electron microscopy (SEM, Hitachi S-4700) and atomic force microscope (AFM, Digital Instrument, D3100), respectively. After the material characterization, Hall effect measurements were performed using the van der Pauw configuration to evaluate the basic electric properties of the AlGaN/GaN HEMT structure.

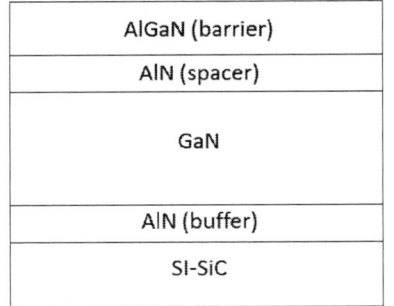

Figure 2. Schematic diagram of AlGaN/GaN HEMT structure on SiC substrate.

III. RESULTS AND DISCUSSION

A. The effect of AlN buffer thickness

GaN layers with increasing AlN thickness from 90 nm to 350 nm were prepared. The crystal quality of GaN film was determined by performing x-ray rocking curve (XRC) scans on both the GaN (002) and (102) planes. Figure 3 (a) shows the XRC widths of GaN films grown using different AlN buffer thicknesses. It can be observed that both GaN (002) and (102) plane widths decreased by increasing the buffer thickness from

90 to 120 nm. However, the XRC widths became broadened again when the AlN thickness larger than 120 nm. At the same time, cracks were also started to form on GaN films once the AlN thickness larger than 120 nm (Figure 3 (b). Another set of samples was also prepared by growing only AlN buffer layer grown on the SiC substrate. AFM images revealed that AlN buffer layers had a similar surface morphology despite having different thicknesses. Figure 3 (c) reveals the typical surface morphology of the bare AlN buffer with a root-mean-square (rms) roughness of approximately 0.4 nm. The results suggested the crystal quality of GaN film was improved by using thicker AlN buffer layer. However, once the buffer thickness greater than a critical thickness, the degradation of GaN film might be caused by the formation of cracks. These cracks could have initiated at the AlN due to the lattice constant mismatch between AlN and SiC substrate.

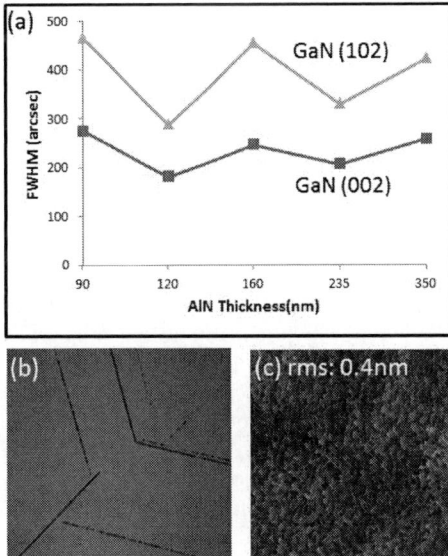

Figure 3. (a) FWHM of GaN (002) and (102) planes rocking curve widths as a function of AlN buffer thickness. (b) Optical microscope image on GaN film with AlN buffer layer thicker than 120 nm. (c) AFM of typical bare AlN buffer surface

B. The effects of AlN spacer layer

For AlGaN/GaN HEMT, the device's electrical performances are heavily depended on the electron properties confined within the 2-dimensional electron gas (2DEG) channel [6]. Therefore, AlGaN/GaN interface with good sharpness and abruptness is required to improve the electron mobility in the 2DEG channel. In the current study, a thin AlN spacer layer was inserted between the AlGaN and GaN layers in order to enhance the heterostructure interface. The thin spacer thickness was represented by deposition time (0 ~ 15 s). Hall measurements were performed on AlGaN/GaN HEMT prepared with different AlN spacer thicknesses and the electron mobilities of these samples are plotted in Figure 4. The electron mobility increased with the increased of AlN spacer growth time for 0 to 10 s which implied that the AlGaN/GaN interface was improved with the AlN spacer. However, for sample with AlN spacer growth at 15 s, the electron mobility started to degrade. A similar trend was also observed elsewhere [7, 8] .The AlN has a lattice constant smaller than either the GaN or the AlGaN. It is therefore, tensile strain would be induced onto the thin AlN spacer layer. Misfit dislocations (MDs) or micro-cracks would be generated in the spacer layer if the AlN thickness was in excess of a critical value. We believed that the mobility degradation was attributed to the above reasons.

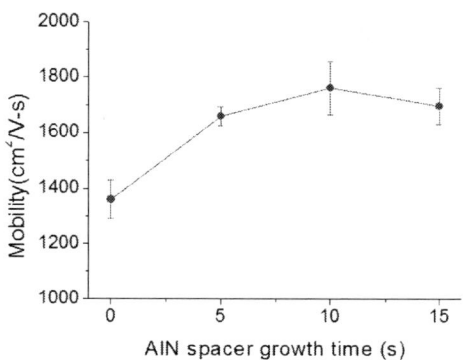

Figure 4. Hall mobility of AlGaN/GaN structure as a function of AlN spacer layer growth times.

C. The effect of Al content in AlGaN barrier layer

The electrons in the 2DEG channel of an AlGaN/GaN structure are induced by polarization field and surface states [9, 10]. Moreover, the strength of the polarization field increased with the increase of Al content and/or the barrier layer thickness. In the current work, the electron density in the 2DEG channel was altered by changing the Al content in the AlGaN barrier layer. The Al content was increased from 24 to 29% while the barrier thickness was kept at 23 nm for all samples. As shown in Table 2, the electron density in the 2DEG channel for sample A to D increased monotonically with the Al content. In contrast, the electron mobility increased from 1668 to 1875 cm^2/V-s for Al content increased from 24 to 25 %. After that, the electron mobility decreased to 1762 and 1677 cm^2/V-s for Al content equaled to 28 and 29%, respectively. As a result, the sheet resistance was decreased gradually. The same phenomenon was also observed in Ref. [11]. Sample C, with 28% of Al in the AlGaN, exhibited an electron concentration of 1×10^{13} cm^{-2} and electron mobility of 1762 cm^2/V-s, yielding a relatively low sheet resistance of 352 Ω/sq will be used for the subsequent experiment.

TABLE II. . HALL MEASUREMENT RESULTS OF ALGAN/GAN HEMT STRUCTURES WITH DIFFERENT AL CONTENT IN ALGAN BARRIER.

Sample	Al content in AlGaN (%)	R_s (Ω/sq)	Mobility (cm^2/V-s)	N_s (10^{13} cm^{-2})
A	24	522	1668	0.74
B	25	374	1875	0.91
C	28	352	1762	1.01
D	29	333	1677	1.12

D. The effect of GaN initial growth pressure and V/III ratio.

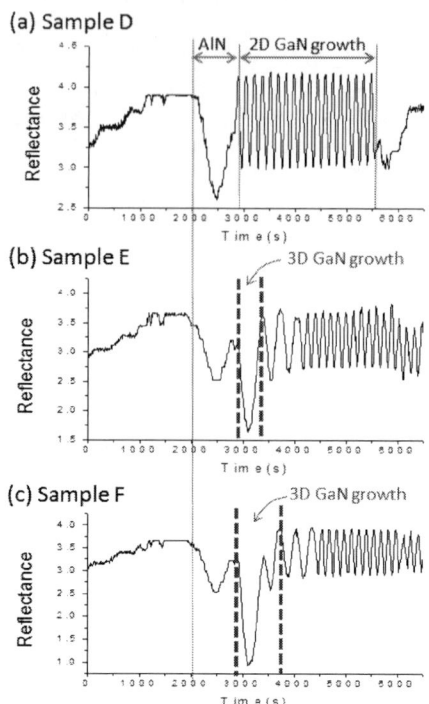

Fig. 5. Reflectance curves for sample (a) D, (b) E and (c) F. The dash lines indicate the coalescence time of GaN layer.

Our previous study reveals that the threading dislocation (TD) especially the edge-type TD will form scattering centers and reduce the electron mobility in the 2DEG channel [12]. Therefore, the crystal quality of GaN needs to be improved in order to further improve the electron mobility in the AlGaN/GaN HEMT structure. To achieve this, we adjusted the initial growth pressure and V/III ratio of the GaN layer. This is because higher chamber pressure [13] and lower V/III ratio [14] can result in longer coalescence time at the initial growth state of the GaN layer which is useful in edge TD annihilation. Figure 5 shows the reflectance curves of samples prepared with different chamber pressures and V/III ratios in the early growth state of GaN films. The growth pressure and V/III ratio of GaN film for sample D were 200 torr and 1227 respectively. 2-dimensional growth mode commenced immediately on the AlN buffer layer (Figure 5 (a)). For sample E, an additional layer of GaN grown with higher chamber pressure of 500 torr was inserted. Besides, the V/III ratio of this layer was 1761. As a result of higher growth pressure, the initial growth rate of GaN was significantly reduced and a 3-dimensional growth mode was clearly observed as shown in Figure 5 (b). The dash line indicates the 3D-to-2D growth transition which is also frequently referred as the coalescence time. In order to further increase the coalescence time, the V/III ratio of the initial GaN layer was reduced to 1408 for sample F (but maintain the chamber pressure at 500 torr). As can be seen in Figure 5 (c), the coalescence time was double (~1000 s) as compared to that of sample E. XRC and Hall measurement results of sample D, E and F are shown in Figure 6. As a result of delaying the coalescence time, the crystal quality of the GaN film was improved as suggested by the XRC results (Figure 6 (a)). The XRC widths of both GaN (002) and (102) planes were reduced with the coalescence time. Moreover, the electron mobility of the AlGaN/GaN structure was also improved from 1762 to about 1832 cm^2/V-s (Figure 6 (b)). Besides, the sheet resistance was reduced to about 316 Ω/sq as the sheet electron density was maintained at approximately 1×10^{13} cm^{-2} for these samples.

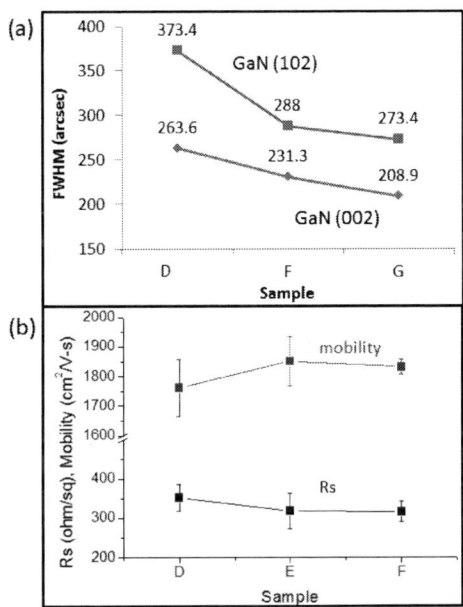

Fig. 6 (a) XRC widths of GaN (002) and (102) planes and (b) sheet resistance (Rs) and electron mobility of sample D, E and F

IV. CONCLUSIONS

AlGaN/GaN heterostructures for HEMT applications have been grown on the SiC substrate. The effects of AlN buffer thickness, AlN spacer growth time and Al content in AlGaN barrier layers were investigated. Besides, the influences of chamber pressure and V/III ratio during the initial growth state on the GaN crystal quality were also investigated. In this study, the optimized AlN buffer layer thickness was 120 nm. A growth time of 10-s was also most beneficial for the AlGaN/GaN interface. For AlGaN barrier layer, the 28% of Al content was used to achieve relatively higher electron mobility and higher sheet carrier density. Finally, an initial GaN layer with higher chamber pressure and lower V/III ratio was also successfully used to further improve the GaN crystal quality and the electron mobility in the AlGaN/GaN heterostructure.

ACKNOWLEDGMENT

This work was supported by the Taiwanese Ministry of Science and Technology under Research Grants Nos. 101-2221-E-009-173-MY2 and 102-2221-E-009-095- MY2

REFERENCES

[1] Y. F. Wu, A. Saxler, M. Moore, R. P. Smith, S. Sheppard, P. M. Chavarkar, *et al.*, "30-W/mm GaN HEMTs by field plate optimization," *IEEE Electron Device Letters*, vol. 25, pp. 117-119, 2004.

[2] C. Yu-Sheng, C. Jui-Chien, L. Tai-Ming, C. Yu-Ting, L. Chung-Yu, C. Chia-Ta, *et al.*, "RF characteristics of AlGaN/GaN HEMTs under different temperatures," in *Semiconductor Electronics (ICSE), 2012 10th IEEE International Conference on*, Kuala Lumpur, 2012, pp. 411-413.

[3] J. W. Chung, W. E. Hoke, E. M. Chumbes, and T. Palacios, "AlGaN/GaN HEMT With 300-GHz fmax," *Electron Device Lett., IEEE*, vol. 31, pp. 195-197, 2010.

[4] H. Lahrèche, M. Leroux, M. Laügt, M. Vaille, B. Beaumont, and P. Gibart, "Buffer free direct growth of GaN on 6H-SiC by metalorganic vapor phase epitaxy," *J. Appl. Phys.*, vol. 87, pp. 577-583, 2000.

[5] L. Liu and J. H. Edgar, "Substrates for gallium nitride epitaxy," *Materials Science and Engineering: R: Reports*, vol. 37, pp. 61-127, 2002.

[6] R. Tulek, A. Ilgaz, S. Gokden, A. Teke, M. K. Ozturk, M. Kasap, *et al.*, "Comparison of the transport properties of high quality AlGaN/AlN/GaN and AlInN/AlN/GaN two-dimensional electron gas heterostructures," *J. Appl. Phys.*, vol. 105, p. 013707, 2009.

[7] S. Niraj Man, L. Yiming, and C. Edward Yi, "Simulation study on electrical characteristic of AlGaN/GaN high electron mobility transistors with AlN spacer layer," *Jpn. J. Appl. Phys.*, vol. 53, p. 04EF08, 2014.

[8] A. Teke, S. Gökden, R. Túlek, J. H. Leach, Q. Fan, J. Xie, *et al.*, "The effect of AlN interlayer thicknesses on scattering processes in lattice-matched AlInN/GaN two-dimensional electron gas heterostructures," *New J. Phys.*, vol. 11, p. 063031, 2009.

[9] J. P. Ibbetson, P. T. Fini, K. D. Ness, S. P. DenBaars, J. S. Speck, and U. K. Mishra, "Polarization effects, surface states, and the source of electrons in AlGaN/GaN heterostructure field effect transistors," *Appl. Phys. Lett.*, vol. 77, pp. 250-252, 2000.

[10] M. Miyoshi, Y. Kuraoka, M. Tanaka, and T. Egawa, "Metalorganic Chemical Vapor Deposition and Material Characterization of Lattice-Matched InAlN/GaN Two-Dimensional Electron Gas Heterostructures," *Appl. Phys. Express*, vol. 1, p. 081102, 2008.

[11] H. Tang and J. B. Webb, "Molecular Beam Epitaxy for III-N Materials," in *III-Nitride Semiconductor Materials*, Z. C. Feng, Ed., ed: Imperial College Press, 2006, pp. 117-160.

[12] Y.-Y. Wong, E. Y. Chang, T.-H. Yang, J.-R. Chang, J.-T. Ku, M. K. Hudait, *et al.*, "The Roles of Threading Dislocations on Electrical Properties of AlGaN/GaN Heterostructure Grown by MBE," *J. Electrochem. Soc.*, vol. 157, pp. H746-H749, 2010.

[13] P. Caban, W. Strupinski, J. Szmidt, M. Wojcik, J. Gaca, O. Kelekci, *et al.*, "Effect of growth pressure on coalescence thickness and crystal quality of GaN deposited on 4H–SiC," *J. Crystal Growth*, vol. 315, pp. 168-173, 1/15/ 2011.

[14] S. Figge, T. Böttcher, S. Einfeldt, and D. Hommel, "In situ and ex situ evaluation of the film coalescence for GaN growth on GaN nucleation layers," *J. Crystal Growth*, vol. 221, pp. 262-266, 12// 2000.

Gap in pagination due to unavailable paper.

Pages 362-365

Characterisation of Nickel Germanide formed on Amorphous and Crystalline Germanium

Fahid Algahtani, Elena Pirogova
Electrical and Computer Engineering ,
RMIT University,
Melbourne, VIC, Australia
s3276519@student.rmit.edu.au

Anthony Holland
Centre of Technology,
RMIT University,
Ho Chi Minh City, Vietnam
anthony.holland@rmit.edu.au

Mark Blackford
Institute of Material Engineering ,
Australia Nuclear Science and Technology Organisation ,
Sydney, NSW, Australia

Jeffrey C McCallum, Brett C Johnson[4]
School of Physics,
University of Melbourne,
Melbourne, VIC, Australia

Abstract—**Nickel germanide formed on amorphous and crystalline germanium is investigated for material and electrical properties. The crystal quality of films formed is poorer for the germanides formed on amorphous germanium with a slight increase in sheet resistance. The grains of NiGe formed on amorphous germanium show a growth that is hexagonal like, extending into the substrate further than germanides grains formed on crystalline germanium. The NiGe formed on crystalline germanium has a much more uniform thickness and uniform grain size and shape. Hollow cone illumination shows that some recrystallisation of the amorphised region of germanium does occur.**

Keywords—Nickel germanide and crystalline germanide

I. INTRODUCTION

The use germanium to make semiconductors devices has been of significant interest recently. The high electron and hole mobilities of germanium gives it an advantage over silicon [1]. These desirable characteristics of germanium have led to it being pursued in CMOS technology [1]. This has been allowed to happened through the introduction of high permittivity (high-κ) materials instead of silicon dioxide, that are applied as gate dielectric materials in standard processing of CMOS and hence alternative semiconductor materials are an option as these new high-κ are deposited and not formed from the semiconductor like silicon dioxide on silicon.. In considering new materials to use for semiconductor devices, engineers are thus opting for germanium due to its attractiveness of high mobilities and compatibility with CMOS technology [2,3]. There are still challenges to be overcome before the full potential of germanium MOS devices can be realised. One of the challenges is the getting reliable ohmic contacts with shallow formation of junctions. Germanides (e.g. NiGe) are considered suitable for germanium contact technology. Notably the use of pre-amorphisation implant can be one strategy that can lead to achieving shallow doping profiles that have high

electrical activation of n-type dopants. Amorphisation is achieved by using Ge implantation, prior to dopant incorporation [4]. Shallow junction are more difficult to form on germanium crystalline substrates compared to silicon and hence the pre-amorphisation step is used before dopant implantation [4]. Germanides e.g. NiGe form good ohmic contacts to crystalline Ge and here we investigate the formation of NiGe on amorphised Ge.

The amorphous layer is normally crystallised at relatively low temperatures ~400°C and happens via solid-phase epitaxial crystallization that results to achieving significant improvement in the dopant activation and formation of germanide on amorphous germanium (a-Ge). Experiments have found that the formation of NiGe (on a-Ge) lowers the re-crystallisation temperature of amorphous germanium [5].

In this work, characterization of NiGe to amorphous and crystalline Ge is presented and results that demonstrate the sensitivity of the polycrystalline structure of the formed germanide to the crystal quality of the germanium used in the germanides formation are also presented and discussed.

II. FORMATION OF NICKEL GERMANIDE ON AMORPHOUS GERMANIUM

This study investigated Nickel germanide formation on amorphous germanium using three successive Ge^+ doses of $1x10^{15}cm^2$ at 550keV, 1MeV and 2MeV into n-type crystalline Ge, which had a resistivity $0.5\Omega cm$,. Samples were analysed using transmission electron microscopy (TEM) and by sheet resistance measurements. TEM samples were prepared by mechanically grinding slices cut from the substrates and then ion milling using 4kV Ar ions to get electron transmission. Sheet resistance measurements were performed using a standard four-point probe setup. a-Ge layers were formed on several 1cm x 1cm samples and the amorphous layer is clearly identified as indicated by cross-section TEM (XTEM), approximately 1.3μm thick as seen in Fig. 1(a). a-Ge and c-Ge samples were coated with 50nm of Ni using e-beam evaporation and NiGe was formed using various heat treatments in vacuum. NiGe is suitable for forming ohmic contacts to Ge because of its low resistivity that is desirable; in addition it forms at relatively low temperature. Heat treatments were undertaken in vacuum and also in a small glass annealing furnace with a steady flow of nitrogen gas. After heat treatment samples were prepared for TEM analysis. Fig.1(b) is a XTEM image which clearly shows the abrupt junction between c-Ge and polycrystalline NiGe. In Fig 2 (a) the NiGe grains can be clearly seen and the pointed bottoms with hexagonal shape is apparent. Fig. 2(b) is a XTEM image of the same sample area at higher magnification.

III. RESULTS AND DISCUSSION

NiGe formed at 400C for 30 minutes in a vacuum oven, from 50nm of Ni on a- and c-Ge was used for TEM analysis.

Depths of grains were found to narrow near boundaries of grain for the NiGe on a-Ge. The depth variation was from 60 to 160nm consideration of many grains. Fig. 3 is a high magnification XTEM image of these grains showing that the sidewalls of the grains are relatively smooth, but the bottom of the grains has a varied roughness of grains. Some grains show a sharp pointed structure, hexagonal like protruding into the substrate. Sharper grain boundaries between NiGe and the substrate and between grains were observed for NiGe on c-Ge. This is shown in Fig. 4.

Fig. 5(a) and (b) are XTEM images formed using hollow cone illumination and more clearly show that individual grain of nickel germanides formed. Fig. 5(a) also shows that there is improved crystallinity of the "amorphous" Ge nearer the bottom of the NiGe layer and this is not apparent in the standard XTEM images of Fig. 2(a) and (b). Fig. 5(b) shows that hollow cone illumination clearly identifies the geometry of individual grains of NiGe.

Low energy electrons for EDS in a SEM produced stoichometries of almost 50% Ni and 50% Ge which conforms to the EDS data of the separate grains in the TEM sample found in NiGe on a- and c-Ge. The resistance of sheet values of the NiGe films was obtained through a combination of subsequent measurements; the final results showed that NiGe on a-Ge has on average of 10 to 20% higher value but with considerable variation. The resistivity of NiGe on c-Ge is about 18μΩcm. The c-Ge material was found to be easily probed electrically but good ohmic contacts using NiGe had to

be made on the recrystallised (polycrystalline) Ge layer; this made it possible to take the measurements on it.

Overall results show that NiGe on crystalline and pre-amorphised germanium forms relatively easily but that further investigation is required to limit the protrusion of NiGe grains formed on a-Ge, into the Ge substrate. Despite the dramatic difference in structure of the NiGe grains on a-Ge and c-Ge there is relatively little difference in sheet resistance.

IV. CONCLUSION

Cross-sectional Transmission Electron Microscopy clearly shows a distinction between crystalline and amorphous germanium regions in image mode and also shows the distinct boundary between crystalline germanium and polycrystalline nickel germanide. When nickel germanide is formed on amorphous germanium the polycrystalline germanide grains are rougher at the bottom (compared with grains formed on crystalline germanium) and the boundary between geramanide and germanium is not distinct. Hollow cone illumination is used to demonstrate that the grain dimensions are much more varied for germanide formed on amorphous germanium. Despite the variation in grain sizes and in the variation in grain depth, the germanide layers have resistivities (determined from sheet resistances values and average layer thickness) that vary by only 10%.

ACKNOWLEDGMENT

The authors take this opportunity to sincerely acknowledge the Australian Institute of Nuclear Science and Engineering (AINSE) for supporting this project.

REFERENCES

[1] C. Claeys and E. Simoen. *Germanium-based Technologies from Materials to Devices,* Elsevier: New Yory, 2007.

[2] Y. Bin et al., "Low-Contact-Resistivity Nickel Germanide Contacts on n+Ge with Phosphorus/Antimony Co-Doping and Schottky Barrier Height Lowering", *Silicon-Germanium Technology and Device Meeting (ISTDM)*, 2012, pp. 1-2.

[3]. R. R. Leiten et al., "Ohmic contact formation on n-type Ge", *Appl. Phys. Letters*, 92 2008, 022106.

[4]. Y. L. Chao et al., "Dopant activation in bulk germanium and germanium on insulator", *MRS Proc. 829*, 2005, B9.18.1.

[5] J. H. Park et al., "Metal-induced dopant (boron and phosphorus) activation process in amorphous germanium for monolithic three-dimensional integration", *J. Appl. Phys.* 106, 2009, 074510.

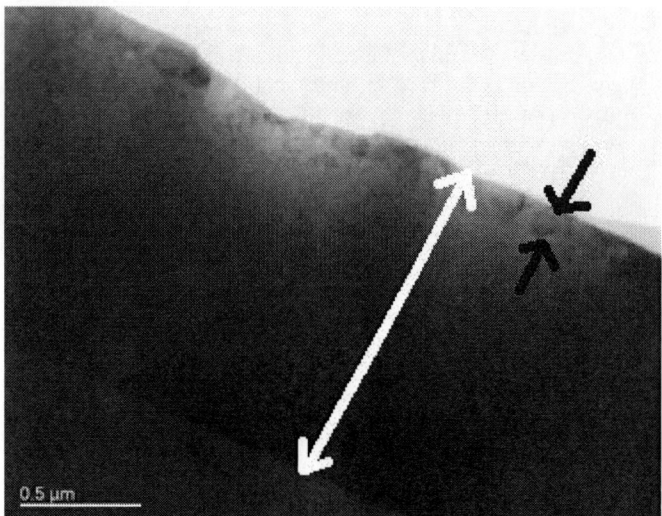

Figure 1 (a) XTEM image showing the amorphous Ge layer (white arrows), approximately 1.3μm resulting from ion implantation of Ge with three successive doses of Ge ions of $1x10^{15}cm^2$ at 550keV, 1MeV and 2MeV. Also shown is the region where NiGe was formed at the surface.

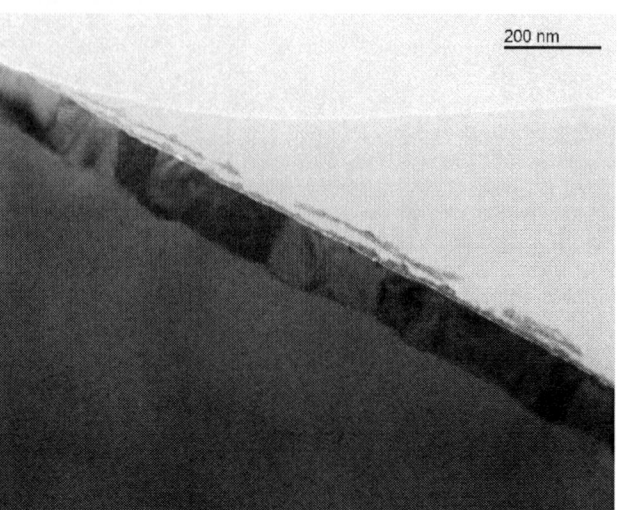

Figure 1 (b) XTEM image showing NiGe grains formed on crystalline Ge. (The residue above the polycrystalline NiGe layer is glue residue from TEM sample preparation).

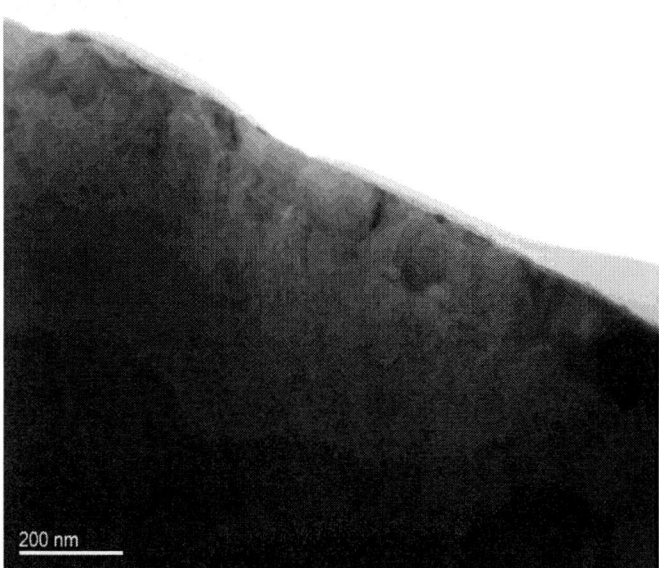

Figure 2 (a) XTEM image showing NiGe grains. The bottom of the NiGe layer is rough

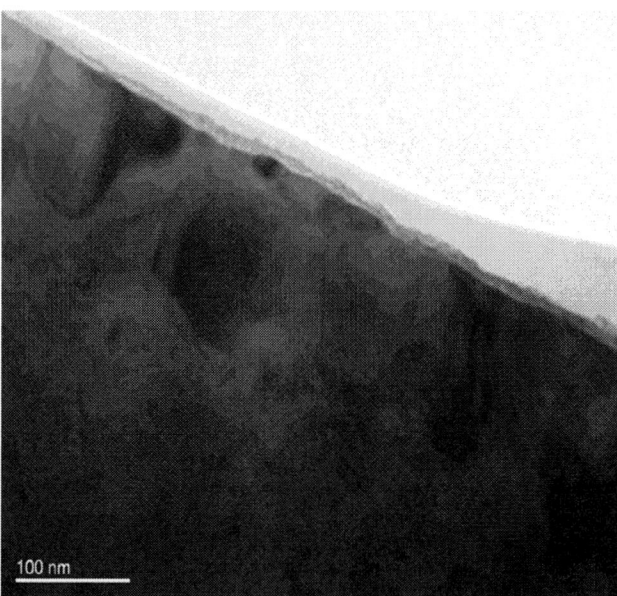

Figure 2 (b) XTEM image in Figure 2(a) but at higher magnification. The variety of geometry of grains is apparent

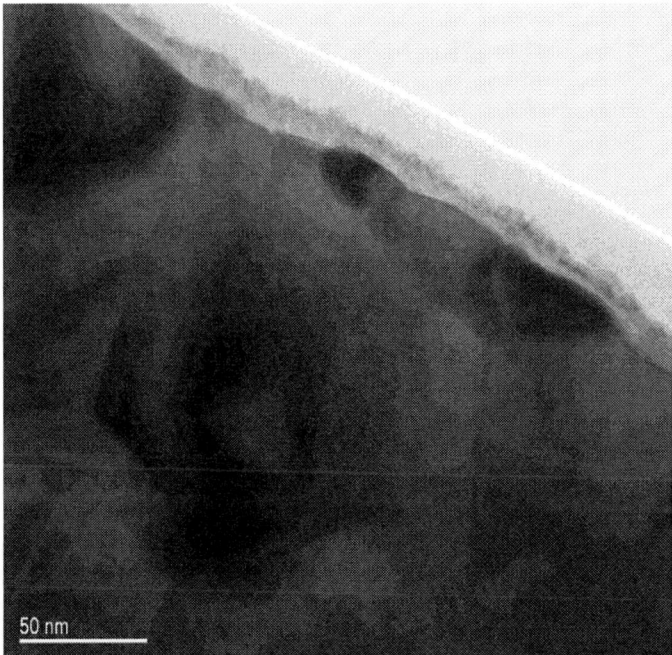

Figure 3. High magnification XTEM image of NiGe grains. The smooth sidewalls and rough bottom of the grains is apparent.

Figure 4. High magnification XTEM image of NiGe grains formed on crystalline Ge (with no amorphisation steps).

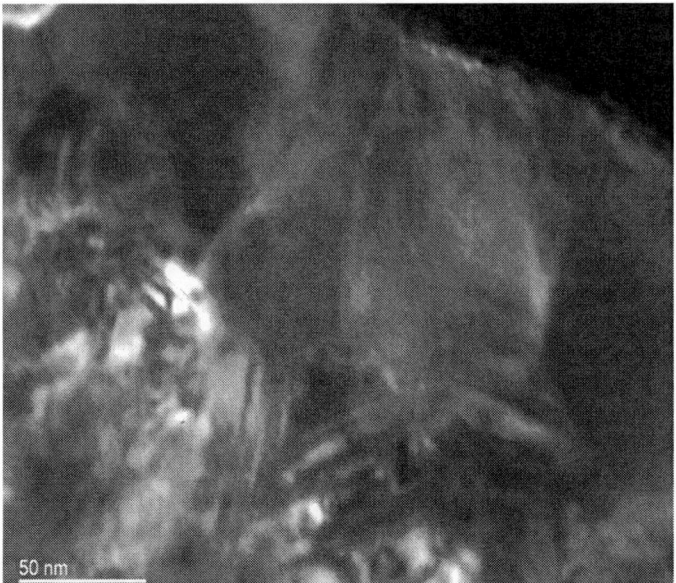

Figure 5 (a) XTEM image obtained using hollow cone illumination (HCI). There is evidence of recrystallization in the implanted a-Ge region below the NiGe grains.

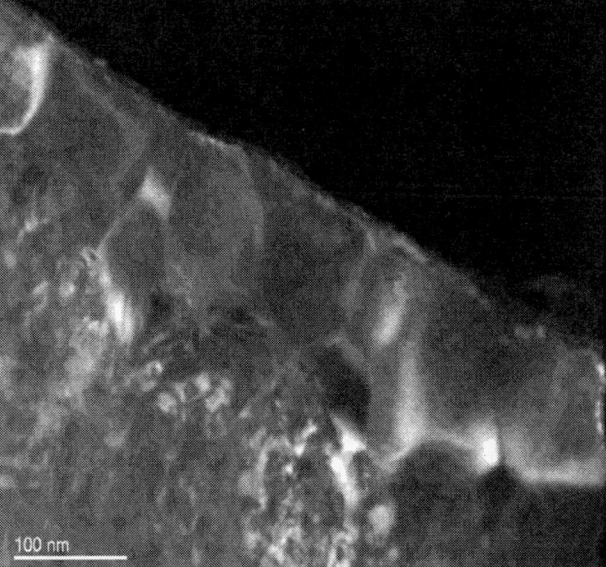

Figure 5(b) XTEM image more clearly showing the geometry of the NiGe grains formed on a-Ge.

A Field-Effect Device Based on an Exfoliated Thin Film of Few-Layer Graphene

S. R. Kasjoo, M. M. Ramli, and M. R. Zakaria
School of Microelectronic Engineering,
Universiti Malaysia Perlis (UniMAP)
02600 Arau, Perlis, Malaysia
shahrirrizal@unimap.edu.my

M. K. Md Arshad, R. Mat Ayub, R. A. Rahim,
and U. Hashim
Institute of Nano Engineering
Universiti Malaysia Perlis (UniMAP)
01000 Kangar, Perlis, Malaysia

Abstract—In this work, a back-gated field-effect device based on an exfoliated thin film of few-layer graphene (FLG) has been fabricated and some of its properties were characterized. The estimated hole mobility of the FLG film, extracted from the device transconductance, was approximately 843 $cm^2V^{-1}s^{-1}$ which was lower than the typical reported values. The reasons for the lower mobility were briefly discussed in terms of charged impurity density, and contact resistance between FLG film and metal. The use of mechanical exfoliation method in producing thin films of FLG, which is cheap, fast and simple, can also be exploited in the development of other graphene-based devices.

Keywords—charged impurity density, contact resistance; field-effect device; graphene; mobility; transconductance

I. INTRODUCTION

Graphene is the name given to a single layer carbon atoms densely packed into a honeycomb structure. Interest in graphene has been increased rapidly since the demonstration of the electric field effect in graphene films by K.S Novoselov et al in 2004 [1]. In the report, a very simple way to isolate graphene by peeling it off from graphite with Scotch tape, also known as the mechanical exfoliation method, has been introduced. The exfoliated graphene layers were then transferred onto a silicon dioxide (SiO_2) layer grown on a silicon substrate, and they can be identified easily under optical microscope by investigating the colour differences. Almost at the same period of time, C. Berger et al have reported an alternative way of producing large graphene flakes by thermal decomposition on the surface of silicon carbide, SiC [2]. Since then, many unique properties of graphene were discovered [3 – 5], and many techniques of producing this material have been developed [6 – 7], and thus graphene becomes one of the most valuable materials on earth nowadays.

In this paper, a back-gated field-effect device has been fabricated using an exfoliated thin film of few-layer graphene (FLG) on Si/SiO$_2$ substrate. The device was then characterized in order to study its properties. More details on the fabrication process and characterization of this device are presented in the next few sections. Suggestions to improve the performance of the field-effect devices are also discussed.

II. DEVICE FABRICATION

In the fabrication process, a 90-nm grown SiO_2 on an n-type silicon substrate with resistivity of 6.0 – 9.0 ohm-cm has been utilized. At this oxide thickness, mono-layer graphene and FLG films have the most contrast image under optical microscope [4]. The n-type silicon substrate served as a back-gate for the field-effect device.

As mentioned earlier, the FLG film used in this work was produced by implementing mechanical exfoliation method. A sticky (Scotch) tape was used to exfoliate FLG from graphite. The tape with flakes of FLG (after exfoliation) was then placed onto the Si/SiO$_2$ substrate and it was pressed gently so that the flakes were able to make a strong bonding with the SiO$_2$ layer. Using a fine optical microscope, the FLG flakes with single and multiple graphene layers can be easily identified. Different number of graphene layers will reflect different corresponded colors under microscope, hence based on these colors, thickness of FLG films can be estimated as shown in Fig. 1(a).

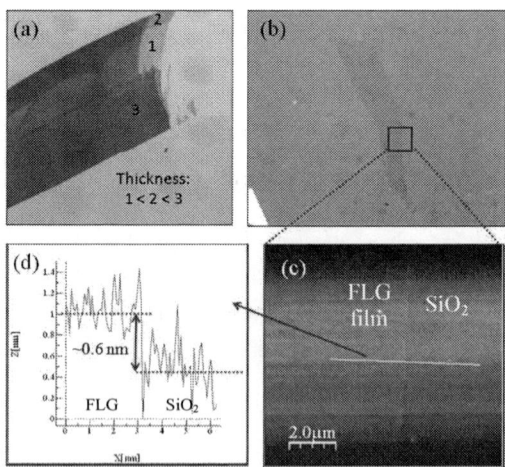

Fig. 1. (a) Optical image of the FLG films on top of a Si/SiO$_2$ substrate. Different thicknesses of the films yielded different contrasts and colors. (b) Optical image of a bi-layer graphene on a Si/SiO$_2$ substrate. (c) AFM image of the edge of the FLG film [shown in (b)]. (d) Corresponded profile [obtained from (c)] showing that the FLG layer has a thickness of 0.6 nm.

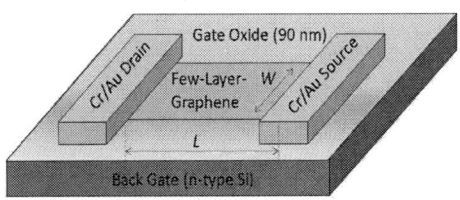

Fig. 2. Illustration of the fabricated back-gated field-effect device by utilizing an exfoliated thin film of FLG as the device channel which has length and width of L and W, respectively. Cr/Au was used to form ohmic contacts on the FLG film. The thickness of the gate oxide was 90 nm, and n-type Si served as the back-gate of the device.

Fig. 1(b) shows the optical image of the FLG film that has been utilized in the fabrication of the back-gated field-effect device. Based on the atomic-force microscope (AFM) image and the corresponded cross-section profile, as shown in Figs. 1(c) and 1(d), respectively, the estimated thickness of the FLG film was approximately 0.6 nm. The surface roughness of the Si/SiO$_2$ substrate was ~0.4 nm. By assuming a mono-layer graphene has a thickness of ~0.335 nm [8], the FLG film used in this work was believed to be a bi-layer graphene.

The formation of ohmic contacts (source and drain terminals) on the graphene was conducted using a standard metal lift-off process. The metals used were the thermal evaporation of a ~5 nm of Cr, followed by 100 nm of Au and annealing process. Chromium was chosen because it was found to interact more strongly with a clean mono-layer graphene [9]. This would help to reduce the contact resistance between Cr/Au and FLG film. A metal contact was then formed to the substrate as the back-gate contact.

Fig. 2 shows the illustration of the fabricated back-gated field-effect device in which the FLG film was the device channel with approximately rectangular in shape (for simplification). The length, L, and width, W, of this channel were ~13 m and ~10 m, respectively.

III. RESULTS AND DISCUSSION

Fig. 3(a) shows the I_D-V_{DS} characteristics of the fabricated device (I_D is the drain current and V_{DS} is the drain-source voltage), as a function of the applied back-gated bias, V_{BG}. As can be seen, the I_D-V_{DS} curves are linear, indicating the formation of ohmic contacts between Cr/Au and FLG film.

A standard I_D-V_{BG} measurement was also performed with V_{DS} fixed at 0.01 V. The result obtained, as shown in Fig. 3(b), indicates that the dominating conduction is only due to holes and not due to electrons. However, if V_{BG} is swept beyond 30 V, the presence of electron carriers might be observed (i.e., ambipolar behavior).

From the I_D-V_{BG} curve, the hole mobility, μ_h, can be estimated using the following equation [6],

$$\mu_h = \frac{g_m L}{C_{ox} W V_{DS}}, \tag{1}$$

where g_m is the transconductance defined as $g_m = \Delta I_D / \Delta V_{BG}$ and C_{ox} is the gate oxide capacitance per unit area. The value of g_m

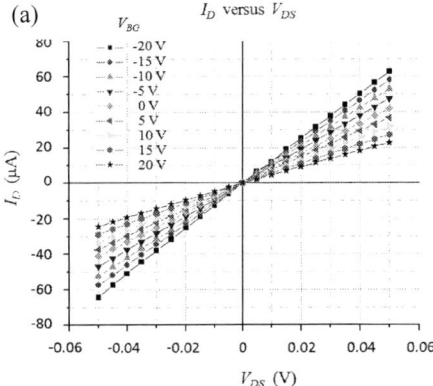

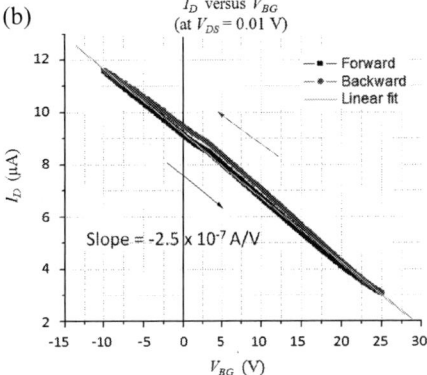

Fig. 3. (a) I_D as a function of V_{DS} with the variation of V_{BG} from -20 V to 20 V. (b) I_D as a function of V_{BG} measured at $V_{DS} = 0.01$ V. A small hysteresis was observed and the magnitude value of the slope obtained (i.e., transconductance, g_m) was -2.5 $\times$ 10^{-7} A/V.

can be obtained from the slope of the I_D-V_{BG} curve which has a magnitude of 2.5 $\times$ 10^{-7} A/V. The value of C_{ox} can be calculated using the following equation, $C_{ox} = (\varepsilon_{ox}\varepsilon_0)/t_{ox} = 3.75 \times 10^{-8}$ Fcm^{-2}, where ε_{ox} is the permittivity of the gate oxide ($\varepsilon_{ox} = 3.9$ for a typical SiO$_2$), ε_0 is the permittivity of free space (8.85 $\times$ 10^{-12} F/m), and t_{ox} is the thickness of the gate oxide (90 nm).

From (1), the estimated μ_h was ~843 cm^{2}V^{-1}s^{-1}. This value is very much lower than the typical reported values whereby $\mu_h > 2,000$ cm^{2}V^{-1}s^{-1} [1, 3 – 6]. In field-effect devices, the mobility of an FLG is very sensitive to the charged impurity density at the interfaces between gate oxide and substrate, and between gate oxide and FLG film [4]. A small hysteresis observed in Fig. 3(b) indicates the presence of significant charged impurities in the gate oxide. Moreover, a possible contamination of the FLG film due to improper handling or preparation of materials during the experiment (especially in the formation of ohmic contacts) can increase the contact resistance between metal and FLG, which might limit the performance of the device [6, 10]. This has to be considered in the estimation of μ_h. Therefore, we believe that the degraded value of μ_h in this work is strongly due to the combination of these two main factors.

There are several ways to improve the mobility and performance of the FLG-based field-effect devices. One of them is by replacing SiO_2 with single-crystal epitaxial $Pb(Zr_{0.2}Ti_{0.8})O_3$ (PZT) [11] or Boron-Nitride [12] films as the gate oxide. The reduction of the charged impurity density in the gate oxide is another key to achieve high mobility [4]. This can be done by a proper handling and preparation of materials during the experiments, and also by optimizing the oxidation process in order to produce gate oxide with minimum defects and impurities.

IV. CONCLUSION

In summary, we have demonstrated a working back-gated field-effect device based on a bi-layer graphene which can be used in many applications. For example, due to the high mobility of graphene, the device is capable to operate at terahertz (THz) frequencies which can be implemented in the THz imaging applications [13]. We have also shown that the mobility of the FLG film, extracted from the device transconductance, can be deteriorated not only due to the charged impurities in the gate oxide of the device but also because of the contamination in the FLG film. In addition, the utilization of mechanical exfoliation method in producing FLG thin films, which is cheap, fast and simple, can also be used in the development of other possible graphene-based devices such as self-switching diodes [14, 15], ballistic rectifiers [16] and geometric diodes [17]. This shows the impressive potential of graphene for future electronic devices.

ACKNOWLEDGMENT

This work was partially supported by the Ministry of Education (Malaysia). The authors would like to thank to all staffs in the Microelectronics and Nanostructures Group from the School of Electrical and Electronic Engineering, University of Manchester, United Kingdom for their technical help.

REFERENCES

[1] K. S. Novoselov, A. K. Geim, S. V. Morozov et al., "Electric field effect in atomically thin carbon films," Science, vol. 306, pp. 666 – 669, October 2004.

[2] C. Berger, Z. Song, T. Li et al., "Ultrathin epitaxial graphite: 2D electron gas properties and route toward graphene-based nanoelectronics," J. Phys. Chem. B, vol. 108, pp. 19912 – 19916, December 2004.

[3] M. C. Lemme, T. J. Echtermeyer, M. Baus, and H. Kurz, " A graphene field-effect device," IEEE Electron Device Lett., vol. 28, pp. 282 – 284, April 2007.

[4] K. Nagashio, T. Nishimura, K. Kita, and A. Toriumi, "Mobility variations in mono- and multi-layer graphene films," Appl. Phys. Express, vol. 2, pp. 025003-1 – 025003-3, January 2009.

[5] F. Schwierz, "Graphene transistors," Nature Nanotechnology, vol. 5, pp. 487 – 496, July 2010.

[6] X. Liang, Z. Fu, and S. Y. Chou, "Graphene transistors fabricated via transfer-printing in device active-areas on large wafer," Nano Lett., vol. 7, pp. 3840 – 3844, November 2007.

[7] K. S. Subrahmanyam, L. S. Panchakarla, A. Govindaraj, and C. N. R. Rao, "Simple method of preparing graphene flakes by an arc-discharge method," J. Phys. Chem. C, vol. 113, pp. 4257 – 4259, February 2009.

[8] T. L. Burnett, R. Yakimova, and O. Kazakova, "Identification of epitaxial graphene domains and adsorbed species in ambient conditions using quantified topography measurements," J. Appl. Phys., vol. 112, pp. 054308-1 – 054308-7, September 2012.

[9] R. Zan, U. Bangert, Q. Ramasse, and K. S. Novoselov, "Metal-graphene interaction studied via atomic resolution scanning transmission electron microscopy," Nano Lett., vol. 11, pp. 1087 – 1092, January 2011.

[10] A. Venugopal, L. Colombo, and E. M. Vogel, "Contact resistance in few and multilayer graphene devices," Appl. Phys. Lett., vol. 96, pp. 013512-1 – 013512-3, January 2010.

[11] X. Hong, A. Posadas, K. Zou, C. H. Ahn, and J. Zhu, "High-mobility few-layer graphene field effect transistors fabricated on epitaxial ferroelectric gate oxides," Phys. Rev. Lett., vol. 102, pp. 136808-1 – 136808-4, April 2009.

[12] I. Meric, C. R. Dean, N. Petrone, et al., "Graphene field-effect transistors based on boron-nitride dielectrics," Proc. of the IEEE, vol. 101, pp. 1609 – 1619, July 2013.

[13] L. Vicarelli, M. S. Vitiello, D. Coquillat, et al., "Graphene field-effect transistors as room-temperature terahertz detectors," Nature Materials, vol. 11, pp. 865 – 871, October 2012.

[14] F. Al-Dirini, F. M. Hossain, A. Nirmalathas, and E. Skafidas, "All-graphene planar self-switching MISFEDs, metal-insulator-semiconductor field-effect diodes," Scientific Reports, vol.4, pp. 3983-1 – 3983-8, February 2014.

[15] S. R. Kasjoo, and A. M. Song, "Terahertz detection using nanorectifiers," IEEE Electron Device Lett., vol. 34, pp. 1554 – 1556, December 2013.

[16] A. M. Song, A. Lorke, A. Kriele, J. P. Kotthaus, W. Wegscheider, and M. Bichler, "Nonlinear electron transport in an asymmetric microjunction: A ballistic rectifier," Phys. Rev. Lett., vol. 80, pp. 3831 – 3834, April 1998.

[17] Z. Zhu, S. Joshi, S. Grover, and G. Moddel, "Graphene geometric diodes for terahertz rectennas," J. Phys. D: Appl. Phys., vol. 46, pp. 185101-1 – 185101-6, April 2013.

Effects of Annealing Temperature on Dielectric Property of Nano-Films Lead Titanate Prepared on ITO Glass by Spin Coating Method

*Z. Nurbaya[1], I.H.H. Affendi[1], N.A. Azhar[1], M.H Wahid[2], and M. Rusop[1]
[1]NANO-ElecTronic Centre (NET), Faculty of Electrical Engineering,
[2]Faculty of Applied Sciences, Universiti Teknologi Malaysia (UiTM), 40450 Shah Alam, Selangor, Malaysia

*Z. Nurbaya
[3]Razak School of Engineering and Advanced Technology, Universiti Teknologi Malaysia (UTM), 54100 Kuala Lumpur, Malaysia
*email: nurbayazainal@gmail.com

Abstract—**Thin films capacitor is one of the most potential devices for storage application. This study managed to perform high performance thin films fabrication through low cost spin coating method and prepared at low annealing temperatures. On top of that, the role of substrate is also being discussed in order to improve the dielectric and ferroelectric property of the nano-thickness thin films. As in our study, ITO coated glass substrate is very suitable compared to ordinary soda lime glass substrate. It was found that the optimal dielectric permittivity, ε is about 150 for the thin films prepared by 500°C annealing temperature measured at 1 kHz frequency. A 199-nm thin film was obtained by two cycles of coating/heating process. Field emission scanning electron microscopy (FESEM) and atomic force microscopy (AFM) showed the surface of these films to be high densification with low surface roughness (ra=0.318 nm). It is believed that high electrical properties are attributed to the microstructural property by which homogenous interfacial layer between films and electrode base.**

Keywords—*lead titanate; dielectric property; annealing temperature; thin films processing;*

I. INTRODUCTION

Lead based ceramic with perovskite structure, i.e. lead titanate ($PbTiO_3$) offers wide range of nanoelectronic device capabilities and significantly flexible in material modification. Such modification includes A and B-site of perovskite ABO_3 structure improves electrical properties which in relation with the microstructural growth [1]–[5]. This is typically, the origin of lead titanate itself however less investigation made compared to the outstanding lead zirconium titanate (PZT) that offers more determination on multiferroic property. Therefore, we convinced that the improvement of $PbTiO_3$ must be acknowledged first by looking at certain factors like thin films processing method, starting materials, extra lead effect, and might also temperature depencies. Fabrication of the films can be made by using metal organic chemical vapor deposition (MOCVD), sputtering deposition, and also chemical solution deposition (CSD) [6]–[8]. Preparation of the thin film ferroelectric ceramic by MOCVD and sputter method is facing difficulties in stoichiometry control because of multi composite oxide. On the other hand, CSD offers great flexibility and much more concern on thin films homogeneity that can be easily obtained at low cost films production, such spin coating, hydrothermally coating, and dip coating [9]–[11].

In this study, we prepare the thin films by spin coating method owing to its credibility in covering large and complex area. Besides that, this method is a promising method to obtain nanoscaled thickness films with high densification films persistent for nanodielectric preparation based capacitor device applications. Moreover, the thin films preparations must be crucially monitored when applying heat during annealing treatment to avoid films crack and lead discrepancy as well. Thus, low synthesis temperatures are much more considerable to improve the electrical properties of $PbTiO_3$ thin films. In this study, we investigate the interesting electrical characteristic of $PbTiO_3$ thin films at influenced of low synthesis temperature of 400 to 600°C applied for annealing treatment.

II. METHODOLOGY

The $PbTiO_3$ precursor solutions were made through simple polymerization process at open atmosphere environment. The starting materials involved alkoxide material, Titanium tetraisopropoxide, (TTiP, 97%) and lead acetate trihydrate (PbAc, 99.9%) with the solvent, 2-methoxyethanol (2-MeOH, Sigma Aldrich, 99.9%). A clean beaker was prepared for mixturing PbAc powders in 100 ml solvent at 0.4molL^{-1} concentration. The Pb precursor was stirred at 100°C for about 1 hour and the heat was reduced to 60°C during the addition of TTiP liquids. The solution was continuously stirred for 2 hours. The process at this stage was handled in very quick movement so that the alkoxide solution will not vaporized quickly with humidity at room temperature. Then, the prepared solutions were added with 10wt% of PbAc powder which avoids lead compensation during annealing treatment. Notably, the solution did not include any stabilizer like either diethanolamine (DEA) or glacial acetic acid (GAA) that is purposely used for crack free solutions. After all the mixturing process which took about 4 hours, it was then completed by ageing process 24 hours at room temperature at slow rate stirring speed.

PbTiO$_3$ films processing using ITO coated glass as substrates are much favorable owing to its homogenous layer between films and electrode, thus reducing the issues of interfacial layer. The cleaned 4.0 cm^2 areas of each ITO substrates were prepared in environment atmosphere. The spinning program was set for 3000 rpm, 20 s and pre-spin at 2000 rpm, 10 s. Hot furnace was pre heated about 200°C, 10 minutes for drying process. The coating and drying thin films process were repeatedly done for five times thus to ensure uniform films as well as high dense films. Annealing treatment is compulsory for ceramic films processing due to its high melting point. Due to the low melting point of glass substrate, the films were annealed under 600°C in range of 400 to 600°C, for 30 minutes. The prepared films were cooled rapidly at room temperature to stop the heating treatment and yet promotes high crystalline of the films.

The prepared spin coated PbTiO$_3$/ITO glass thin films were characterized for electrical properties. So, the films were coated with Al metal as top electrode by using thermal evaporation method at 30 nm thickness under 0.09cm^2 area of steel mask. Electrical measurements were conducted as for investigation on the dielectric and I-V characteristic of the PbTiO$_3$ thin films at different annealing treatments. There should have impedance spectroscopy (Agilent 4294A) and I-V 2-point probe measurement (BUKOH KEIKI, CEP 2000).

III. RESULTS AND DISCUSSIONS

A. Dielectric Property

The plot of capacitance versus frequency (C-f) in range of 40 Hz to 110 kHz was measured at different thin films annealing temperature as shown in Fig. 1 (a) while the loss tangent versus frequency plot as shown in Fig. 1(b). The result shows the trend of dielectric performance with slightly difference to each other. It seems that the films prepared above and below than 500°C temperature indicates low capacitance values measured at 1 kHz frequency. They are 5.89, 5.99, 6.23, 6.19, and 6.01 nF for the entire films respectively. Thus, it shows that the optimal parameter is the films prepared at 500 °C and successfully obtained slightly lower tangent loss, 1.4% than others.

Dielectric constant values of each films can be obtained according to (1), however this study focused the real part of relative permittivity without concerning the factor of interfacial layer issues due to fact that the layer just at sub-10 nm thickness.

$$k = \frac{Cd}{\varepsilon_o A} \qquad (1)$$

Where, C is capacitance measurement of the films, d is the films thickness, ε_o is air permittivity, and the A is the area of metal contact which is 0.09cm^2. The constant values of dielectric material, in this study, entire PbTiO$_3$ films were generally obtained the constant in range of 900 Hz to 9 kHz upon dropping drastically above than 10 kHz frequency. This suggests that the films performs space charge polarization which is normally happen accumulation of charges at the interfacial layer between the films and metal contact [12].

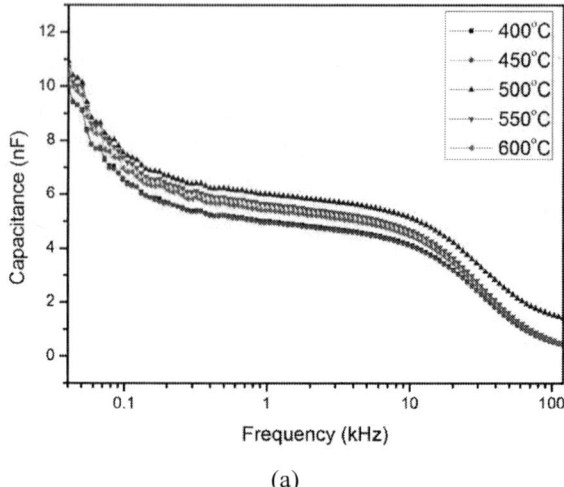

(a)

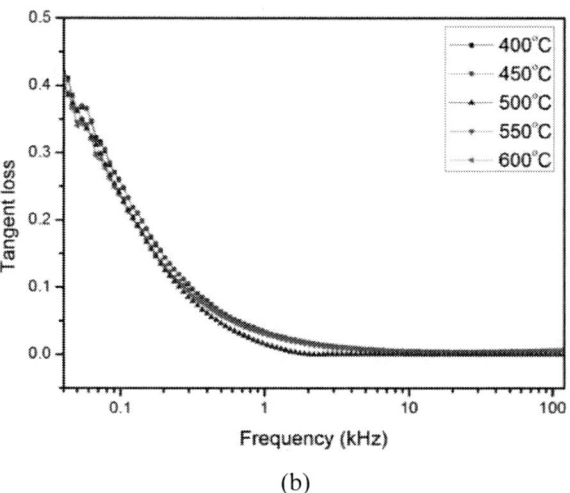

(b)

Fig. 1, Plot of lead titanate thin films dielectric property varied at different annealing temperatures, (a) C-f plot and (b) *tangent loss-f* plot

dielectric material, in this study, entire PbTiO$_3$ films were generally obtained the constant in range of 900 Hz to 9 kHz upon dropping drastically above than 10 kHz frequency. This suggests that the films performs space charge polarization which is normally happen accumulation of charges at the interfacial layer between the films and metal contact [12].

The tangent loss that represents the dissipation factor study of the films precisely for AC signals implementation. Interpretation of the tangent loss results shown in the Fig. 1(b) can be best described as in (2) and (3).

$$\varepsilon_r = \varepsilon_r' - j\varepsilon_r'' \qquad (2)$$

$$\tan(\delta) = \frac{\varepsilon''}{\varepsilon'} \qquad (3)$$

Where ε_r is dielectric constant, ε_r' is relative permittivity (i.e. determination of polarization), ε_r'' is energy loss permitted by dielectric medium, and tan (δ) is the level of how lossy the material is for ac signal. Dielectric constant value, k for the entire films are 103, 103, 150, 137, and 136 respectively for annealing temperature in this study. The dissipation factor by mean is the tangent loss acquired from the result in Fig. 1(b) of the entire films was 4.5, 4.5, 2.4, 4.4, and 4.3%. Thus, the films prepared by lowest annealing temperature gained such a highest percentage of dissipation factor compared to the films prepared above than 500˚C annealing temperature with just 2.4%. This must be taken into consideration as this factor may leads to the leakage current initiation. In addition, the reason behind high dissipation factor is due to the weakening of bonding energy which may have been consisted to the lowest annealing temperature applied during crystallization process [13].

B. I-V Characteristics

Current-voltage measurement of PbTiO$_3$ thin films in this

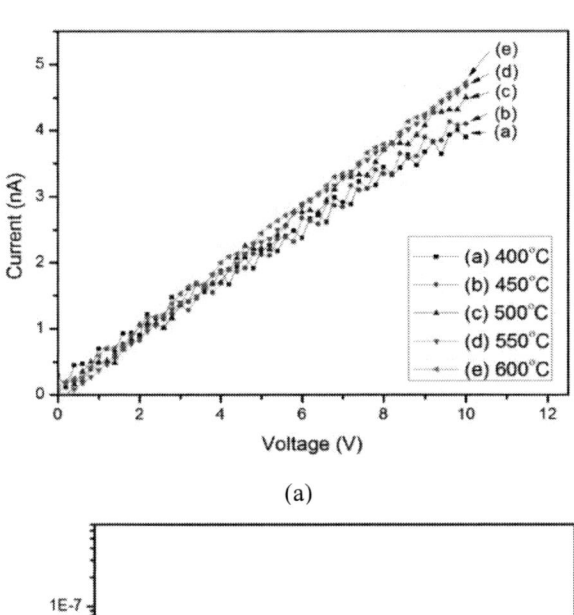

(a)

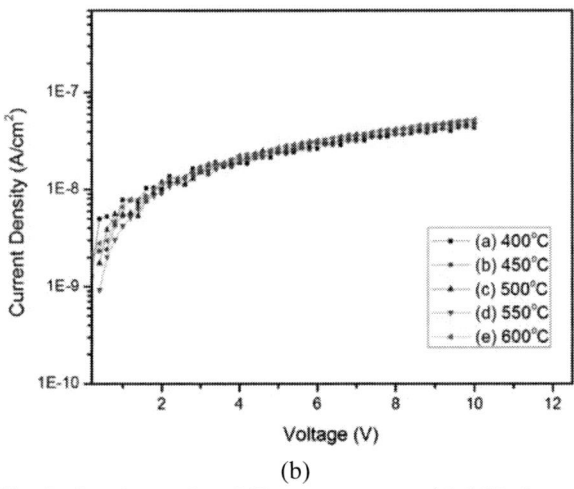

(b)

Fig. 2, 2-point probes I-V measurement, (a) I-V plot and (b) J-V plot.

study is to investigate leakage current density plus the resistivity of the films according to the results shown in Fig. 2(a), while Fig. 2(b).

The result satisfies that the entire films with 164, 166, 199, 194, 200 nm thicknesses in respective way of annealing temperatures shows ohmic behavior which is truly denotes the behavior of insulator material. Literally, the films prepared by 400, 450, and 500˚C were having noises distraction and needed smoothing manipulation to decrease the noise factor. In fact, higher annealing temperatures were measured to have such smooth results. It is may be attributed to the factor of the films uniformity which has been one of the consideration to promotes linear current-voltage relationship.

Regardless of that characteristic, the trend of I-V plot resulted in relation to the increase of annealing temperature. The currents are in range 10^{-9} A, which specifically about $3.89\text{x}10^{-9}$, $4.10\text{x}10^{-9}$, $4.50\text{x}10^{-9}$, $4.67\text{x}10^{-9}$, and $4.72\text{x}10^{-9}$ A for films prepared at 400, 450, 500, 550, and 600˚C respectively measured at 10 V. This finding is consistent to the previous reports that focused on the I-V characteristic of polycrystalline films of which were prepared by high annealing temperature is actually having high leakage current behavior rather than amorphous films [14]. Nevertheless, the plot of current density by means of the current division according to the metal contact area, 0.09cm^2 as shown in Fig. 2(b) were constantly unites result of entire films that approaching to one direction that begin at 2 V.

The issue of leakage current in single layer films has been improved by adding another layer to make multilayer configurations based films device. Jia *et al* had studied the homogenous multilayer configuration developed from different structural property including amorphous, polycrystalline, and microcrystalline which significantly improved the leakage current behavior for capacitor applications [15]. Others also reported different metal contact to be used as either top or bottom electrode [16], [17].

IV. CONCLUSIONS

In short, low synthesis temperature offers significant improvement of electrical properties. The preparation of PbTiO$_3$ thin films were succeeded even at such less than 200nm thickness making the thin films more potentials than conventional ceramic dielectric for high performance capacitor. Therefore, the films prepared by 500˚C annealing temperature could be the optimal annealing temperature for the single layer PbTiO$_3$ thin films.

The films thickness about 199 nm achieved highest dielectric constant, k=150 and performed adequate low dissipation factor about 2.4%. Furthermore, the I-V characteristic shows in a good relation with the processing temperature under range of ~10^{-9} A that is certainly denotes for high performance thin films capacitor applications.

ACKNOWLEDGMENT

The author would like to thank Research Management Institute (RMI), UiTM through the project 600-RMI/DANA 5/3/PSI (325/2013) and Ministry of Higher Education (MOHE) for financial support.

REFERENCES

[1] E. Dogheche, B. Jaber, and D. Re, "Optical waveguiding properties of … Pb, LaTiO$_3$/Al$_2$O$_3$ planar structures," J. Electroceramics, vol. 2:2, pp. 105–111, 1998.

[2] F.Z. Fadil, T. Lamcharfi, F. Abdi, and M. Aillerie, "Synthesis and characterization of magnesium doped lead titanate," Cryst. Res. Technol., vol. 46, pp. 368–372, Apr. 2011.

[3] H.S. Hwang, Y. Park, and W.S. Choi, "Electrical properties of Pb(Zr,Ti)O$_3$ films prepared on ITO glass," Microelectron. Eng., vol. 85, pp. 2456–2458, Dec. 2008.

[4] D.S.L. Pontes, E.R. Leite, F.M. Pontes, E. Longo, and J.A. Varela, "Microstructural, dielectric and ferroelectric properties of calcium modified lead titanate thin films derived by chemical processes," J. Eur. Ceram. Soc., vol. 21, pp. 1107–1114, 2001.

[5] Y.J. Yu, F.P. Wang, and L.C. Zhao, "Effects of rare earth Eu doping on ferroelectric properties of," Microelectron. Eng., vol. 66, pp. 726–732, 2003.

[6] S. Kim, M. Jun, and S. Hwang, "Preparation of undoped lead titanate ceramics via sol – gel processing," J. Am. Ceram. Soc., vol. 96, pp. 289–296, 1999.

[7] A. Pignolet, P.E. Schmid, L.Wang, and F. Levy, "Structure and electrical properties of sputtered lead titanate thin films," J. Phys. D Appl. Phys, vol. 619, pp. 58–61, 1991.

[8] C.H. Wang and D.J. Choi, "Effect of the Pb/Ti source ratio on the crystallization of PbTiO$_3$ thin films grown by metalorganic chemical vapor deposition at low temperature of 400˚C," J. Am. Ceram. Soc., vol. 13, pp. 207–213, 2001.

[9] T. Morita and Y. Cho, "Hydrothermally deposited PbTiO$_3$ epitaxial thin film," J. Korean Phys. Soc., vol. 46, pp. 10–14, 2005.

[10] R.A. Bakar and M. Rusop, "Electrical properties of sol-gel derived lead titanate thin films by dip coating technique," 2010 Int. Conf. Electron. Devices, Syst. Appl., pp. 415–417, Apr. 2010.

[11] D.G. Wang, C.Z. Chen, J. Ma, and T.H. Liu, "Lead-based titanate ferroelectric thin films fabricated by a sol-gel technique," Appl. Surf. Sci., vol. 255, pp. 1637–1645, 2008.

[12] M. Osada and T. Sasaki, "A- and B-Site modified perovskite nanosheets and their integrations into high- k dielectric thin films," Appl. Ceram. Technol., vol. 9, pp. 29–36, 2012.

[13] X.T. Li, W.L. Huo, W.J. Weng, G.R. Han, and P.Y. Du, "Effect of Mg doping on dielectric properties," J. Electroceramics, vol. 21, pp. 128–131, 2008.

[14] F. Chiu, J. Wang, J.Y. Lee, and S.C. Wu, "Leakage currents in amorphous Ta$_2$O$_5$ thin films," J. Appl. Phys., vol. 81, pp. 6911–6915, 1997.

[15] Q.X. Jia, L.H. Changb, and W.A. Andersonb, "Low leakage current thin film capacitors construction using a multilayer," Thin Solid Films, vol. 259, pp. 264–269, 1995.

[16] T. Ohno, T. Matsuda, T. Nukina, N. Sakamoto, N. Wakiya, S. Tokuda, and H. Suzuki, "Effect of the electrode structure on the electrical properties of alkoxide derived ferroelectric thin film," Mater. Lett., vol. 64, pp. 1742–1744, Aug. 2010.

[17] G. Han, J. Ryu, W.H. Yoon, J.J. Choi, B.D. Hahn, J.W. Kim, D.S. Park, and S. Priya, "Experimental investigation on the effect of top electrode diameter in PZT thick films," Mater. Lett., vol. 65, pp. 2193–2196, Jul. 2011.

[18] J. Moon, J.A. Kerchner, J. Lebleu, A.A. Morrone, and J.H. Adair, "Oriented lead titanate film growth at lower temperatures by the sol-gel method on particle-seeded substrates," J. Am. Ceram. Soc., vol. 80, pp. 2613–2623, 1997.

Surface Functionalization of Multiwalled Carbon Nanotube for Biosensor Device Application

M.F. Fatin, A. Rahim Ruslinda, M. K. Md Arshad, U. Hashim, S. Norhafizah, M. A. Farehanim
Institute of Nano Electronic Engineering
Universiti Malaysia Perlis
Kangar, Malaysia
nabilah.faudzi@gmail.com

Abstract— **CNT has been widely known for its excellent electrical and thermal conductivities. However, these desirable properties are limited for application due to the inertness of CNT. This work emphasize on a simple and practical method of MWCNT functionalization by sonication in mild acid condition. A mixture of nitric acid and sulfuric acid in ratio of 1:3 were demonstrated as an oxidizing agent to introduce carboxyl functional group on MWCNTs surface. The chemical bonding and surface morphology were examined using FTIR and SEM respectively and indicates a successful functionalization of carboxyl functional group. This work will be further developed for application of functionalized MWCNT for biosensor device.**

Keywords— *surface functionalization, mild acid, carboxyl group*

I. INTRODUCTION

Carbon nanotube (CNT) is an inert compound that has very powerful properties. Since the discovery of carbon nanotube in 1991 by a Japanese researcher, Sumio Iijima, it draws a great interest from researchers all around the globe [1]. There are a few types of carbon nanotubes like multiwalled carbon nanotube (MWCNT) which has first discovered, single walled carbon nanotube (SWCNT) which has been discovered two years later in 1993[2]. CNT structure varies with its synthesis method. SWCNT can be either metallic or semiconductor depends on its diameter and chirality while MWCNT exist as metallic compound [3]. Different angle of graphene folding contribute to different structure like armchair and zigzag. CNT has a tensile strength about 100 times stronger compared to steel and excellent thermal and electrical conductivities [4]. With the length range from hundreds nanometer to several micrometer and diameter several nanometer depends on its type, it produces a very large surface area. The long hollow tube structure makes the entire surface accessible for reaction. Due to its desirable properties, several methods have been developed to overcome the inertness by surface functionalization through chemical and electrical treatment of carbon nanotube.

Non-covalent functionalization usually will be in the form of absorption or adsorption. It is not preferable in some application like sensor because non-covalent bond is not strong enough to bind two compounds. The bond can break in case there are any compounds present that have higher affinity to the functional group. A device will not be reusable because non-covalent bond also tend to break after several washings. However, this type of functionalization will preserve the electrical and chemical properties of CNT because there are no changes on sp^2 orbitals of carbon. Non-covalent functionalization includes Van der Waals force, π-π stacking and electrostatic interaction.

Covalent modification on the surface is involving formation of strong bond. This type of functionalization is preferable to produce a reusable biosensor. Chemical oxidation treatment usually applied along with sonication or thermal treatment to create defects on the CNT surface and introduce the oxygenated species such as carboxyl functional group [5]. Covalent bond will change the state of carbon from sp^2 to sp^3 orbital thus changing its properties. In this paper, the effect of acid oxidation towards MWCNT surface was investigated. A device has been fabricated for the biosensor application of functionalized MWCNT.

II. EXPERIMENTAL

A. Chemicals and Instrument

MWCNT was purchased from Fibermax Composites (Greece) with diameter of 10 – 40 nm and length in the range of 1 – 25 μm. Nitric acid and Sulfuric acid were both from Merck Chemicals. The chemicals used in this experiment were purchased from Sigma-Aldrich. The water used is deionized water. Fourier Transform Infrared Spectroscopy (FTIR) was used to determine the chemical bond present in samples and Field Emission Scanning Electron Microscope (FESEM) was used to investigate the surface morphology.

B. Surface Functionalization

10 mg of raw MWCNT was weighed and diluted in 100 ml of 6M HNO_3 : H_2SO_4 mixture (1 : 3). The mixture was sonicated for 4 hours at 40°C. The ultrasonically agitated solution was filtered and the residue was rinsed for several times until the pH of the filtrate become neutral. The residue obtained was dried in vacuum oven at 80°C overnight. The raw and functionalized MWCNT were characterized by using FTIR and FESEM to investigate the changes in surface and bindings before and after chemical treatment.

978-1-4799-5761-3/14 $31.00 © 2014 IEEE

C. Surface Characterization Procedure

FTIR was used to detect the bindings that present in the MWCNT structure from the vibration of atom during exposure to IR radiation. Dry functionalized MWCNT was mixed with Potassium bromide (KBr) to form KBr pellet and exposed to IR source. The analysis was done in controlled humidity condition to avoid interference from water molecule in air. Raw and functionalized MWCNT was dispersed in isopropanol and dropped on silicon wafer for FESEM analysis. Samples were ultrasonically dispersed in isopropanol to a concentration of 1 μg/ml, dropped on silicon wafer and let dry overnight.

D. Device Fabrication and Characterization

The 1 cm × 1 cm SGFET was fabricated by using oxidation of silicon wafer to form 4000Å of Silicon dioxide (SiO2) layer. Source and drain electrode was deposited on SiO_2 surface by chemical vapor deposition of Nickel and Gold (Ni/Au) layer. A 300 μm channel was formed on the middle by photolithography and wet etching process. The functionalized MWCNT was dropped on the channel to increase the efficiency of the device as shown in Fig. 1. This device was electrically characterized to measure current-voltage (I-V) potential to be used as solution gate field effect transistor (SGFET) for biosensor application.

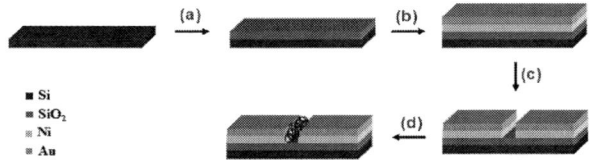

Fig. 1: Device fabrication process (a) Oxidation, (b) Ni/Au deposition, (c) Ni/Au channel formation, (d) Functionalized MWCNT deposition

III. RESULTS AND DISCUSSION

Based on previous study, acid oxidation treatment will result in deformities on MWCNT surface and introduction of oxygenated species at the defect sites [6,7]. Oxygenated species that may produce are carboxyl, hydroxyl, lactones and phenols [8]. However, carboxyl group is the most abundant functional group produced after acid oxidation.

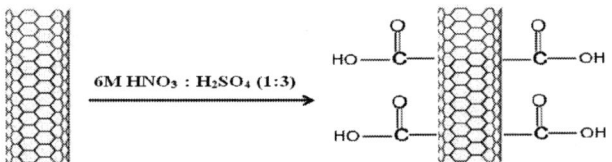

Fig. 2: MWCNT-COOH functionalization by acid oxidation

Fig. 3 shows the FTIR spectra of raw and functionalized MWCNT by using KBr pellet method in the region of 4000-400cm^{-1}. Usually, carboxyl functionalized CNT will shows a few important peaks which are peaks at 3500-3200cm^{-1} (O-H stretch), 1760-1690cm^{-1} (C=O stretch) and 1320-1000cm^{-1} (C-O stretch). CNT usually will produce peak at 1680-1640cm^{-1} which indicates -C=C- of alkenes and 1500-1400cm^{-1} which

indicates C-C stretch (in ring). [9,10,11]. O-H stretching present in the raw MWCNT spectra could be due to partial oxidation on MWCNT surface during purification by the manufacturer [12]. The functionalized MWCNT shows present of C=O stretching vibration at 1714.5cm^{-1} which arose from carboxyl functional group.

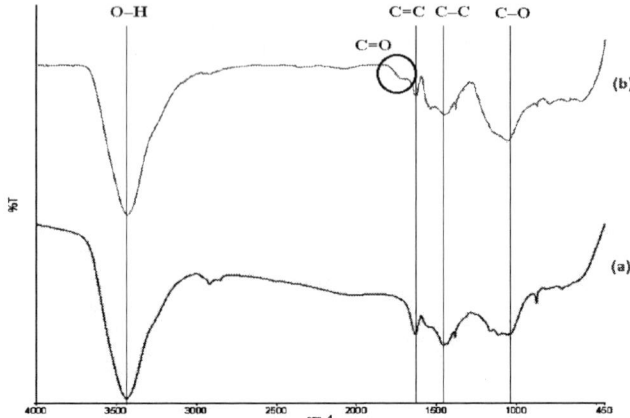

Fig. 3: FTIR spectra of raw MWCNT (a) and functionalized MWCNT (b). After functionalization, there is a peak (in circle) representing C=O stretch of carboxyl group.

Fig. 4 shows the FESEM image of MWCNT. Functionalized MWCNT shows the same worm-like bundling structure which resembles the raw MWCNT. This indicates that after chemical treatment with acid, the structure of long, hollow tube is still preserved. Usage of low concentration of acid and low temperature reduces the defect sites formed on MWCNT [13]. The raw MWCNT shows presence of bundling due to Van der Waals interactions between them while after functionalization, groovy CNT wall can be observed due to presence of carboxyl group at the defect sites.

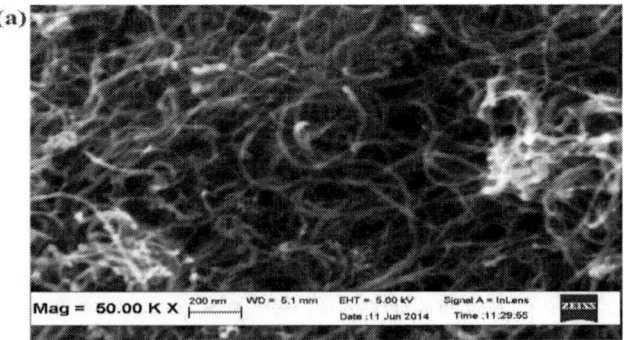

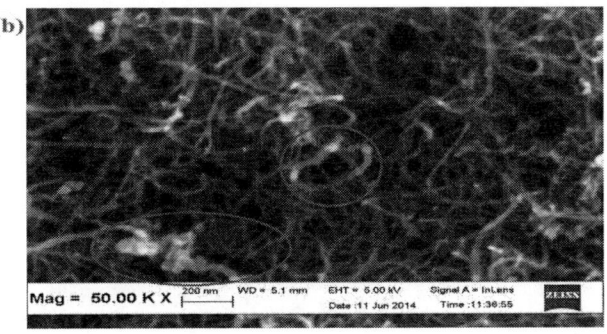

Fig. 4: FESEM image of (a) raw MWCNT and (b) HNO₃:H₂SO₄ treated MWCNT for 4 hours

The functionalized MWCNT has been dropped on fabricated device as shown in Fig. 5(a) in order to observe the behaviour against deionized water (DIW) and Phosphate buffer solution (PBS). Fig. 5(b) shows the current-voltage (I-V) characteristics of fabricated device. It was notified that the fabricated sensor showed lower resistance in PBS as compare to DIW. This may be attributed to more conductive nature of PBS due to the presence of more ions as compared to deionized water. From the I-V curve we can conclude that result after functionalization, MWCNT remains as conductive and can be used as an amplifier in biological compound detection.

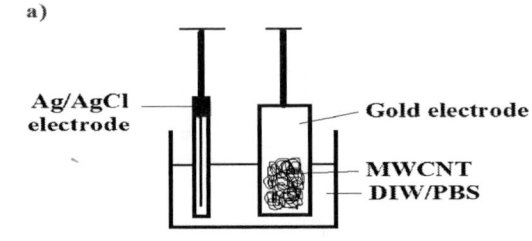

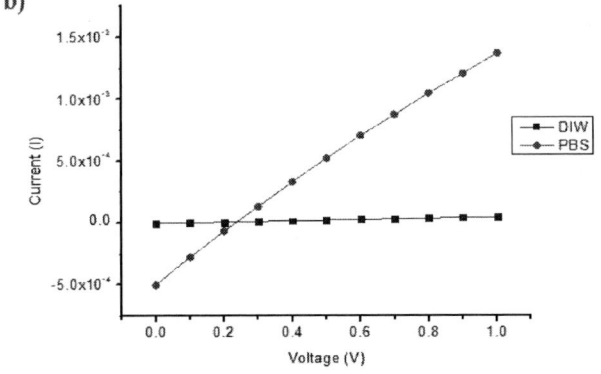

Fig. 5: (a) The electrical measurement setup (b) Electrical characterization of functionalized CNT in DIW and PBS

IV. Conclusion

Surface functionalization of MWCNT using mild acid oxidation through sonication method has demonstrated a successful introduction of carboxyl group on MWCNT surface. Though, functionalization method needs to be improvised to promote high density of functional group on MWCNT surface. The functionalized MWCNT will be further used on fabricated FET device for biosensor application to enhance the detection signal in biological reaction.

Acknowledgment

This work was funded by Fundamental Research Grant Scheme (9003-00375) and supported by Institute of Nano Electronic Engineering (INEE), Universiti Malaysia Perlis.

References

[1] S. Iijima, "Helical microtubules of graphitic carbon," Nature, 354, 56-58, 1991.

[2] S. Iijima, T. Ichihashi, "Single-shell carbon nanotubes of 1-nm diameter," Nature, 363, 603-605, 1993.

[3] W. Yang, K. R. Ratinac, S. P. Ringer, P. Thordarson, J. J. Gooding, F. Braet, "Carbon nanomaterials in biosensors: Should you use nanotubes or graphene?," Biosensors and Carbon Nanomaterials, 49, 2114-2138, 2010.

[4] K. A. Wepasnick, B. A. Smith, J. L. Bitter, D. H. Fairbrother, "Chemical and structural characterization of carbon nanotube surfaces" Anal Bioanal Chem, 396, 1003–1014, 2010.

[5] V. Datsyuk, M. Kalyva, K. Papagelis, J. Parthenios, D. Tasis, A. Siokou, I. Kallitsis, C. Galiotis, "Chemical oxidation of multiwalled carbon nanotubes," Carbon, 46, 833-840, 2008.

[6] X. Jiang, J. Gu, X. Bai, L. Lin, Y. Zhang, "The influence of acid treatment on multi-walled carbon nanotubes," Pigment & Resin Technology, 38, 165-173, 2009.

[7] F. Aviles, J. V. Cauich-Rodri´guez, L. Moo-Tah, A. May-Pat, R. Vargas-Coronado, "Evaluation of mild acid oxidation treatments for MWCNT functionalization," Carbon, 47, 2970-2975, 2009.

[8] V. Likodimos, T. A. Steriotis, S. K. Papageorgiou, G. E. Romanos, R. R. N. Marques, R. P. Rocha, J. L. Faria, M. F. R. Pereira, J. L. Figueiredo, A. M. T. Silva, P. Falaras, "Controlled surface functionalization of multiwall carbon nanotubes by HNO₃ hydrothermal oxidation," Carbon, 69, 311-326, 2014.

[9] D. A. Skoog, F. J. Holler, S. R. Crouch, Principles of Instrumental Analysis, Thomson Brooks/Cole, 6th Edition, 2007.

[10] H. Kitamura, M. Sekido, H. Takeuchi, M. Ohno, "The method for surface functionalization of single-walled carbon nanotubes with fuming nitric acid," Carbon, 49, 3851-3856, 2011.

[11] S. Pilehvar, J. A. Rather, F. Dardenne, J. Robbens, R. Blust, K. D. Wael, "Carbon nanotubes based electrochemical aptasensing platform for the detection of hydroxylated polychlorinated biphenyl in human blood serum," Biosensors and Bioelectronics, 54, 78–84, 2013.

[12] D. S. Ahmed, A. J. Haider, M. R. Mohammad, "Comparison of functionalization of multiwalled carbon nanotubes treated by oil olive and nitric acid and their characterization," Energy Procedia, 36, 1111-1118, 2013

[13] N. A. Buang, F. Fadila , Z. A. Majid, S. Shahir, "Characteristic of mild acid functionalized multiwalled carbon nanotubes towards high dispersion with low structural defects," Digest Journal of Nanomaterials and Biostructures, 7, 33-39, 2012.

Fabrication of SAW Device By Using Zno Thin Film as a Sensing Area

S.L.Lai, M.R. Zakaria
School of Microelectronic Engineering
Universiti Malaysia Perlis (UniMAP),
Arau, Perlis, Malaysia
leng_lai90@gmail.com, rosydizakaria@yahoo.com

U. Hashim, K.L Foo, R. Haarindra Prasad
Institute of Nano Electronic Engineering
Universiti Malaysia Perlis (UniMAP),
Kangar, Perlis Malaysia
uda@unimap.edu.my

Abstract— The limitation of human five main senses to detect the Nano-scale organism has introduced the idea for researcher in development several kinds of devices or sensors. The Surface Acoustic Wave (SAW) devices which wisely use in telecommunication field at the early time had been integrated into biosensor which able to detect the microorganism in aqueous state. Typically, a SAW sensor will have three elements which are transducers, sensing area and electronic signal processor. Zinc Oxide (ZnO) thin film use as piezoelectric substrate while metallic material as Inter Digital Transduces (IDTs). The new device structure which replaces the sensing area by using ZnO metal oxide layer is conducted in this study. Hence, the purpose is to investigate the behavior of various sizes of ZnO metal oxide layer as sensing area for the SAW device. Conventional lithography and Therefore, ZnO thin film layer by Sol-gel method is applicable use as SAW device. ZnO Sol-gel coating method were used to fabricate the SAW device. The SAW device with ZnO metal oxide layer shows the attractive result. The electrical characteristic and frequency response were used to assess the behaviour of the SAW device. X-ray Diffraction (XRD) and Atomic Force Microscope (AFM) was used to evaluate the ZnO structures.

Keywords— *ZnO Sol-gel ZnO thin film , SAW Biosensor device*

I. INTRODUCTION

Since the end of the 18th century, ZnO material has become attracted within the research community [1]. The large band gap of 3.30 eV allows ZnO material endorses some specific behavior in term of their physical and chemical properties[3]. The outstanding properties of the ZnO lead to the development for several applications such as piezoelectric devices, SAW devices, chemical sensor as well as biosensor [2,3]. ZnO used as sensing material on the sensor which can achieve excellent sensitivity, even though there is a small sample for detection [3,4]. SAW-base biosensors have higher mass sensitivities and allow a higher frequency range from several 100MHz to a few GHz [5]. SAW device is sensitive to small change of mass loading. The change of mass loading effect the surface conductivity of SAW device thus it modified the waveform and result as output. A typical SAW sensor consists of a pair of IDTs and a sensing area between the IDTs. An IDT use to generate surface acoustic wave as the input another IDT use to transduce the mechanical wave into electric signals. The wave travels along the sensing area which initially coated with some material for selective detection. Any reaction takes place at the sensing area will result the modification of the traveling wave due to the change of propagation velocity[6].

There are several types of method to deposit ZnO thin films on the SAW device as piezoelectric substrate likes sputtering, Chemical Vapor Deposition (CVD) and Metal Organic Chemical Vapor Deposition (MOCVD), laser ablation, Molecular Beam Epitaxy (MBE), Sol-gel and Filtered Cathodic Vacuum Arc (FCVA) [7]. Recently, deposition by ZnO powder has been studied through ultrasonic spray pyrolysis [8], microwave-assisted synthesis[9] and sol-gel direct precipitation [10]. Among the deposition methods, sol-gel method is commonly used in the study purpose due to its relative low cost as compared to other methods [4].

The aim of this research is to fabricate a ZnO based SAW sensor with ZnO thin film as sensing area. The surface morphology of ZnO film was detected by using AFM while XRD was used to determine the crystalline structure orientation of ZnO thin film.

II. EXPERIMENTAL PROCEDURE

A. Mask Preparation

For mask design, AutoCAD software was used to design the mask for the conventional fabrication process. Two masks design is needed in order to fabrication difference material layer on the substrate. The first mask is for IDT mainly to deposit the aluminum IDT electrodes and the second mask is for sensing area development which uses to allocate the ZnO thin film on the sensor. The IDT parameters are shown in Table 1. The mask design is illustrated in Fig. 1.

TABLE I.   MASK DESIGN PARAMETERS

Criteria	Parameters
Finger width (um)	300
Finger length (um)	4500
Number of IDT finger	10
Aperture (um)	4200
Size of bond pad (um)	2000 x 2000
Size of sensing area (um)	4800 x 3000

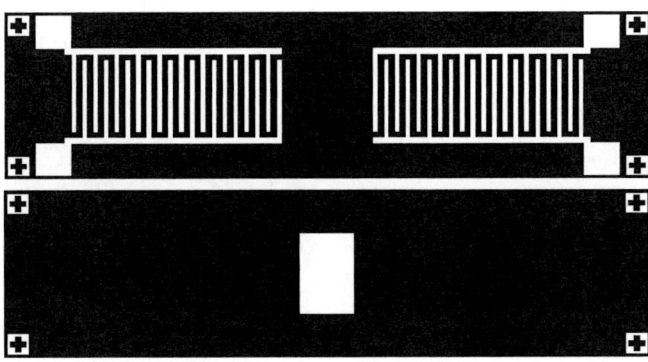

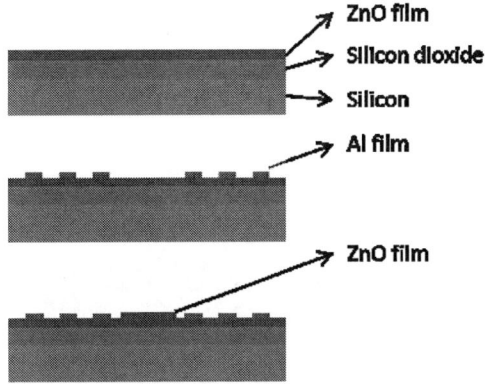

Fig. 1. Mask design for SAW sensor fabrication process (a) first mask for the IDT, (b) second mask for the ZnO thin film

Fig. 2. Process steps to fabricate SAW device

B Substrate Preparation

The fabrication process begins with samples dicing. The wafer is cut into desire samples size by using diamond scriber. Cleaning process based on the standard cleaning procedure was set up in order to remove the contamination on the silicon surface. The silicon dioxide was growth in the wet oxidation furnace for 2 hours with the thickness around 4000Å. After that, the sample was coated with a few layers of ZnO Sol-gel to form a piezoelectric substrate. The sample is then annealed for 2 hours at 500°C. The ZnO Sol-gel was prepared by solute the Zinc Acetate powder in the Ethanol solvent and stir with 1000 rpm at 60°C for about 20 minutes. After that, add the monoethanolamine (MEA) slowly until the solution becomes clear. Lastly the solution is left at room temperature for ageing more than 24 hours.

C Formation Of IDT

For this stage, etching process by lift-off technique is invoked due to the poor selectivity of the removing agent between ZnO film and Al films. Pattern of first mask was transferred on the sample which initially coated with a layer photoresist with 1700 rpm and bake for 90 seconds at 90°C. The photoresist used as the hard mask to protect the ZnO film. The opening regions of the photoresist mask enable the Al (about 500 Å) to deposit directly on the ZnO film through thermal evaporator. Afterwards, acetone used as removing agent to lift-off the Al film, which deposited on the photoresist surface.

D Formation Of Zno Thin Film On Sensing Area

The sample was coated with a layer of photoresist with spin coater at 1700 rpm and bake for 90 seconds at 90 °C. The design of second mask was transferred on the photoresist layer. The opening region at the center of the sample is the location where to coat the ZnO thin film. Besides that, another purpose of the photoresist is to protect the entire surface of the sample. Repeat the coating process by using Sol-gel method for few layers. Lastly, the mask layer is stripped by using acetone and leaving the ZnO film on the sensing area of the device. The general process flow of fabrication is demonstrated in Fig. 2.

III. RESULTS AND DISCUSSION

The results are obtained from the measurement of SAW device in term of optical, physical and electrical result will be explained in details in this part.

The XRD is one of important characterization process to confirm the measured material is existed on the sample. The crystal quality and orientation of the prepared ZnO thin film was characterized by using XRD. The result for XRD as shown in Fig. 3. All the diffraction peaks of the XRD result were competing with the data base of the EVA program base on hexagonal lattice of ZnO, [ZnO: PDF 00-036-1451]. In XRD result, only high crystalline structure shows sharp and narrow diffraction peak [11]. From the figure, the highest diffraction peak is at 34.5°, which mean the ZnO have preferential crystal growth on (002) plane [12]. This crystal orientation show that the ZnO thin film growth with a hexagonal wurtzite structure. Besides that, the refraction peaks of (100), (101), (102) and (110) do exist and match from the database when the 2Theta vary from 30° to 60°.

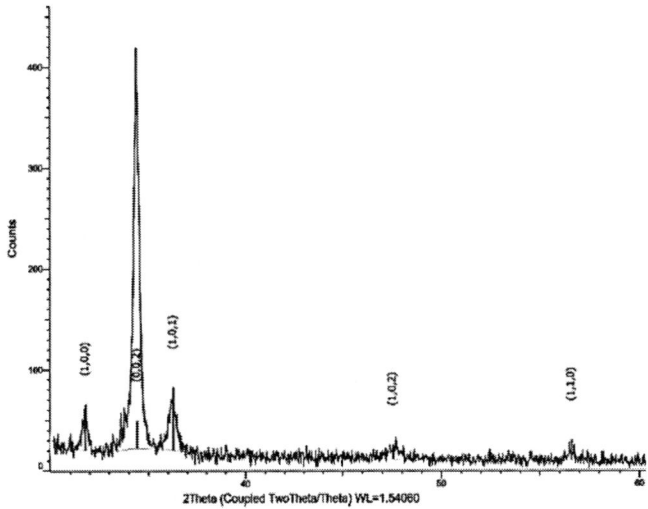

Fig. 3. XRD pattern for ZnO thin film

The morphology surface of ZnO thin film prepared by using sol-gel method before anneals and after was observed through AFM. Fig. 4(a) illustrates the 3-D view of ZnO thin film before annealing, while, Fig. 4(b) shows the 3-D view of ZnO thin film, which has undergoes annealing process at 500ºC for two hours.

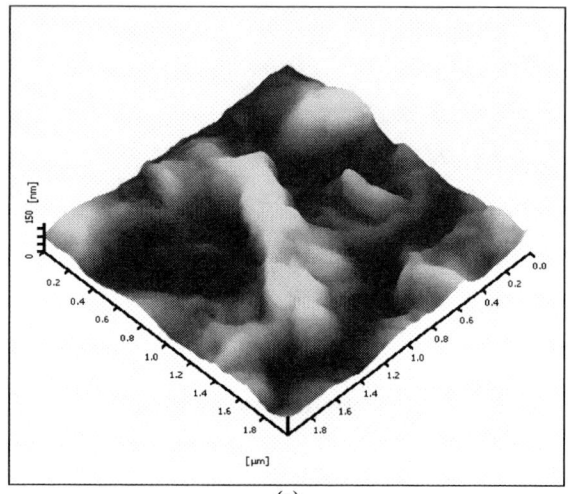

(a)

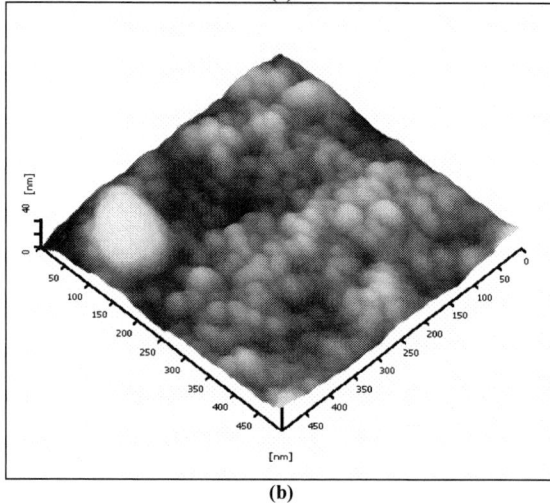

(b)

Fig. 4. AFM image of ZnO thin film (a) before annealing process (b) after annealing process

After the annealing process, the ZnO thin film produced smaller grain size and uniformly distributed. From the result obtain the value for the root mean square and average roughness has reduced about 50% after annealing process. The root mean square of the ZnO thin film is directly proportional to the average surface roughness; therefore the grain size of the ZnO thin film after annealing is becoming much smaller. The characteristic of the ZnO thin film before and after the annealing process has been shown in Table 2.

TABLE II. SURFACE CHARACTERISTIC OF ZNO THIN FILM UNDER AFM

Characteristic	Before annealing process (nm)	After annealing process (nm)
Root mean square (rms)	9.69	4.55
Peak to valley height(P-V)	50.98	42.91
Average roughness (Ra)	7.8	3.35

As shown in Fig. 5, I-V characteristic analysis is conducted by using Semiconductor Parameter Analyzer (SPA) in order to determine the resistance of the IDT device. The resistance value of this typical IDT design is 108.45kΩ. Whereas, examination of C-V curve for the IDT was conducted and the result is shown in Fig. 6. The capacitance value obtained is an approximately 424 PF. The central frequency for this typical SAW device can be calculated by using the $v = f\lambda$ [13], where v is the acoustic velocity of piezoelectric which is 3954 m/s, f is the central frequency and λ is the wavelength. The value λ of can be calculated from the IDT design parameters in which the $\lambda = 4d$, d is the finger width of the IDT. Therefore the λ is1.2 mm and the central frequency is 3.295 MHz.

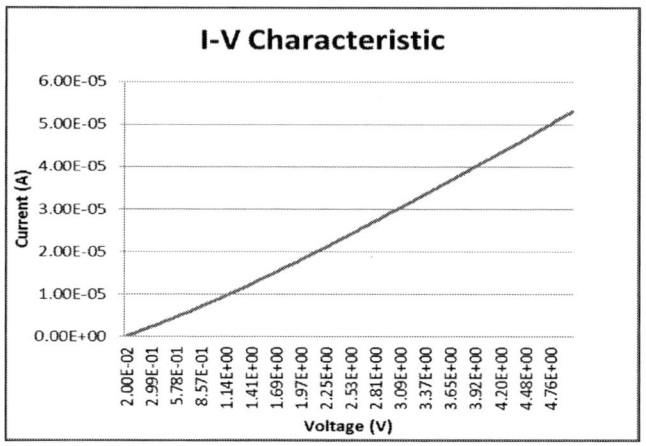

Fig. 5. I-V curve

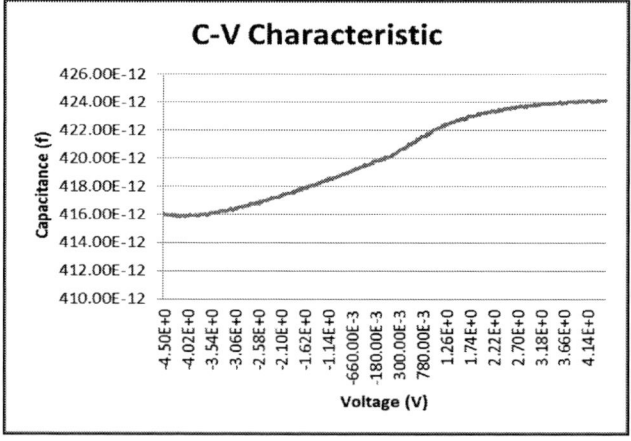

Fig. 6. C-V curve

IV. CONCLUSION

ZnO thin film coat on the SiO_2 surface by using Sol-gel method has been validated by using XRD and give the highest peak on crystalline structure (002). The grain size of ZnO thin film has become smaller after annealing process. The IV and CV characteristic display the electrical behavior of the IDT. The central frequency of this typical fabricated SAW device is 3.295MHz.

978-1-4799-5761-3/14 $31.00 © 2014 IEEE

ACKNOWLEDGMENT

The authors wish to express gratitude to Universiti Malaysia Perlis (UniMAP) for giving chance to finish the research in Cleanromm of School of Micro-E, and Institute of Nano Electronic Engineering (INEE).Thus, appreciation also goes to all lectures and technicians at both place.

REFERENCES

[1] A. A. M. Ralib, A. N. Nordin, and U. Hashim, "Finite element modeling of SAW resonator in CMOS technology for single and double interdigitated electrode (IDT) structure," *RSM 2013 IEEE Reg. Symp. Micro Nanoelectron.*, pp. 1–4, Sep. 2013.

[2] M. J. Choi, K. E. McBean, P. H. R. Ng, A. M. McDonagh, P. J. Maynard, C. Lennard, and C. Roux, "An evaluation of nanostructured zinc oxide as a fluorescent powder for fingerprint detection," *J. Mater. Sci.*, vol. 43, no. 2, pp. 732–737, Oct. 2007.

[3] R. H. Prasad, U. Hashim, K. L. Foo, and M. Shafiq, "Study the Optical Characteristic of ZnO Nanostructure Through Annealing at Various Time Period," vol. 832, pp. 68–72, 2014.

[4] a. Y. Oral, "Evaluation of Microstructures of ZnO Nanocrystals Fabricated by Different Methods," *Acta Phys. Pol. A*, vol. 123, no. 2, pp. 169–170, Feb. 2013.

[5] K. Länge, B. E. Rapp, and M. Rapp, "Surface acoustic wave biosensors: a review.," *Anal. Bioanal. Chem.*, vol. 391, no. 5, pp. 1509–19, Jul. 2008.

[6] T. J. Giffney, Y. H. Ng, and K. C. Aw, "A Surface Acoustic Wave Ethanol Sensor with Zinc Oxide Nanorods," vol. 2012, pp. 1–5, 2012.

[7] Y. Q. Fu, J. K. Luo, X. Y. Du, a. J. Flewitt, Y. Li, G. H. Markx, a. J. Walton, and W. I. Milne, "Recent developments on ZnO films for acoustic wave based bio-sensing and microfluidic applications: a review," *Sensors Actuators B Chem.*, vol. 143, no. 2, pp. 606–619, Jan. 2010.

[8] P. Singh, A. Kumar, and D. K. Ã, "Growth and characterization of ZnO nanocrystalline thin films and nanopowder via low-cost ultrasonic spray pyrolysis," vol. 306, pp. 303–310, 2007.

[9] Y. Liu, Y. Hu, M. Zhou, H. Qian, and X. Hu, "Applied Catalysis B : Environmental Microwave-assisted non-aqueous route to deposit well-dispersed ZnO nanocrystals on reduced graphene oxide sheets with improved photoactivity for the decolorization of dyes under visible light," *"Applied Catal. B, Environ.*, vol. 125, pp. 425–431, 2012.

[10] C. Chen, P. Liu, and C. Lu, "Synthesis and characterization of nano-sized ZnO powders by direct precipitation method," vol. 144, no. 3, pp. 509–513, 2008.

[11] K. L. Foo, M. Kashif, U. Hashim, and M. E. Ali, "Fabrication and Characterization of ZnO Thin Films by Sol-Gel Spin Coating Method for the Determination of Phosphate Buffer Saline Concentration," pp. 1–5, 2013.

[12] K. L. Foo, U. Hashim, H. Prasad, and M. Kashif, "Fabrication and characterization of IDE ZnO thin films using sol-gel method for PBS solution measurement," *2012 10th IEEE Int. Conf. Semicond. Electron.*, pp. 736–739, Sep. 2012.

[13] M. R. Zakaria, U. Hashim, R. M. Ayub, and T. Adam, "Design and Fabrication of IDT SAW by Using Conventional Lithography Technique Institute of Nano Electronic Engineering (INEE)," vol. 18, no. 9, pp. 1281–1285, 2013.

Fabrication of a CMOS-Compatible Surface Acoustic Wave Device for Application in Pathogen Sensing

S. T. Ten[1,2], U. Hashim[1], F.Malek[3], W.W. Liu[1],
K.L.Foo[1], C.H. Voon[1], F. H. Wee[3], Y.S. Lee[3], and
H.Hisham[1]
[1]Institute of Nanoelectronic Engineering (INEE)
[3]School of Computer and Communication Engineering,
Universiti Malaysia Perlis (UniMAP)
01000 Kangar Perlis.

S. T. Ten[2], A. Sudin[2], M.S. Nur Humaira[2],
[2]Mechanization and Automation Centre,
MARDI, P.O.Box 12301 GPO, 50774 Kuala Lumpur.
stten@mardi.gov.my

Abstract— **Surface acoustic waves (SAW) devices have been initially developed and used for the high-volume low-cost TV component. Due to the ultra-sensitivity to the surface perturbation, SAW based devices have been modified to be sensors. Initially, SAW based sensors were developed for gas detection and recently they have been moving towards biological detection. Shear horizontal surface acoustic wave (SHSAW), one of the SAW based type is most suitable for the application in liquid based condition. Ultra sensitive pathogen sensors are needed to improve the food security and quality of life as well. Hence, the ultra-high sensitive biosensor has been designed in this research towards the low concentration pathogen detection. One of the main SHSAW components is the interdigital transducer (IDT). Currently, there are variety of techniques to fabricate the accurate size electrodes, Electron beam lithography and X-ray lithography. However, these methods are very costly. Therefore, this paper is presenting the more economical fabrication process which is using complementary metal-oxide-semiconductor method to fabricate inter-digital transducers on 64^0 YX LiNbO$_3$ piezoelectric substrate. Comparisons were made for the theoretical calculation and fabricated measurement resonant frequency of the 3μm, 8 μm, 12 μm, 25 μm and 20μm width IDTs and the differences obtained are less than 5%.**

Keywords—Acoustic Waves Based Sensor; Biosensors; E.coli; Resonant Frequency; Wave Length; IDT.

I. INTRODUCTION

The first acoustic based sensor was made from modified quartz crystal for chemical sensing by adding a sorptive film on the crystal. In 1950s, it was further analyzed and improved [1-2]. In the late 1970s, Wohltjen and Dessy had advanced acoustic based sensor to the surface acoustic wave delay line design, which the acoustic waves are generated and detected with comb like structure electrodes, called inter-digital transducer (IDT) [3]. On the other side, the first biosensor had been introduced by Professor Leland C Clark Jnr. in year

1956, after that the number of research on this field had been increased from years to years [4]. According to the modern dentition, "a biosensor is an analytical device comprising a biological or biologically derived sensing element either integrated within or intimately associated with a physicochemical transducer" [5]. Transduction elements can be optical, electrochemical, calorimetric, magnetic, or mass sensitive. CMOS fabricated acoustic waves based sensors typically surface acoustic waves (SAW) sensors are able to process the outputs which can produce the information about the ultra-sensitive mass loading, properties of the bound materials and the changes of viscoelastic effects. According to Ivnitski[6], recent advances in bio sensing technologies are using electrochemical, piezoelectric, optical, acoustic and thermal biosensor. SAW based sensor is one of the piezoelectric type sensors. In this paper, Shear horizontal surface acoustic wave (SHSAW) device, one of the SAW based type was fabricated by using the complementary metal-oxide-semiconductor (CMOS) method. SHSAW, one of the SAW based types is most suitable for the liquid based application as it has the advantage of acoustic energy is not being radiated into liquid. The piezoelectric substrate 64^0 YX LiNbO$_3$ was used in this study as it is has the fast propagation wave velocity.

II. MATERIAL AND METHODS

A. Theory

From review paper by Ten [7], SAW sensors are very sensitive to changes in mass, density, viscosity on the guiding layer and these are the advantages as biosensors. The acoustic energy is strongly confined at the surface of the device in the range of the acoustic wavelength, regardless of the thickness of the complete substrate [8]. For this reason, the waves are potentially very sensitive towards any changes on the surface, such as a mass loading, viscosity and conductivity changes. With this principle, any changes on the surface such as mass loading caused by the binding of the targeted specimen in analyte with sensitive sensing area in SAW will be detected.

Therefore, if ssDNA pathogen is immobilized on the sensitive layer of SAW sensor and whenever even very low

---

Research is sponsored by Nano Fund from MOSTI headed by MARDI and INEE in UniMAP.

density targeted ssDNA pathogen, in the analyte will be hybridized on the sensitive surface of the SAW sensor, then the mass loading effect can be discriminated by using both the phase changes or synchronous frequency shifting in the surface acoustic wave inside the guiding layer between input and output IDTs [9]. This is the future work of this research as firstly, the IDTs have to be successfully fabricated and then only the mass loading effect experiments can be continued. Thus the accuracy of IDT fabrication is very critical task. The important parameters in determining the synchronous frequency of the device is the pitch of fingers of the IDTs. For simplicity, the most common IDT design which utilizes 1:1 interdigitation and equal spacing between all fingers on both sides was used. The pitch of the fingers, p, is the spacing between two fingers on the same side of the electrode. The consecutive fingers (alternating sides of the IDT) are always at equal but opposite voltage assuming a sinusoidal (AC) signal, consecutive fingers mark the location of maximal strain alternating between tension and compression. As such, the wavelength of the wave which is transduced by the piezoelectric substrate will be equal to pitch, p. The following relationship that describes the synchronous frequency, fo of the device:

$$fo = vp / p \qquad (1)$$

where vp is the propagation velocity of a wave on the substrate. It is important to note that vp is a material property, and that, as a result, the synchronous frequency is both determined by material selection and a design parameter. At the lower end of the frequency sensor of a SAW sensor, the necessary pitch of the IDT would be quite large. On the other hand, at the higher end of the frequency range, the necessary pitch of IDT would be quite small, limited by the minimum feature size resolvable by current photolithography techniques. Therefore, if the typical frequency ranges of SAW device are 10MHz to 3GHz, then the corresponding to pitches of approximately 1μm to 300μm. The IDT patents were designed by using AutoCAD software. The entire mask was printed onto a chrome glass surface. Currently, there are variety of techniques to fabricate the sub-micron electrodes is quite challenging using optical lithography, for instance Ultra-Violet (UV) and deep UV lithography, but by useful techniques such as off axis illumination, SAW resonators have been fabricated at 2.45 GHz [10] and 3.15 GHz [11]. Electron beam (e-beam) lithography has been used to fabricate SAW resonators at 10 GHz with electrode widths below 100 nm [12]. However, fabrication using e-beam lithography is an extremely slow serial writing process with low throughput which is suitable only for prototyping and not mass production. Well defined structures with excellent sidewall quality and sub-micron features, around 400 nm are currently possible using X-ray lithography (XRL) [13-14]. However, if the sensitivity is enough for the range of synchronous frequency under 1GHz than the more economical fabrication method is needed to consider such as CMOS method.

B. Fabrication Process

In the IDT deposition fabrication process flow (Fig.1), the piezoelectric substrate material, 64^0YX LiNbO3 is used. First process is the Physical Vapor Deposition (PVD), thermal evaporator process; deposit the metal aluminum onto the surface of the single polished LiNbO3 substrate, for the thickness about 300nm by using Auto 306 Vacuum Coater, PVD machine. The process is then followed by coating positive photoresist (PR1-2000A) by using spin coating machine (Model WS-650MZ-23NPP). The third process is pattern transfer by exposing UV light (MIDAS, MDA-400M) under the designed chrome mask (Fig.2).

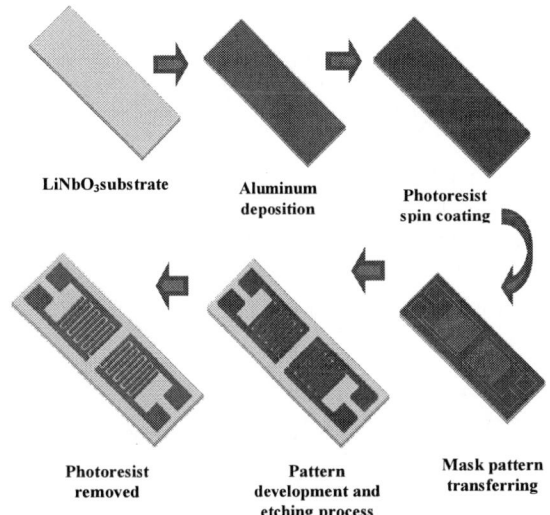

Fig.1. IDT deposition fabrication process flow

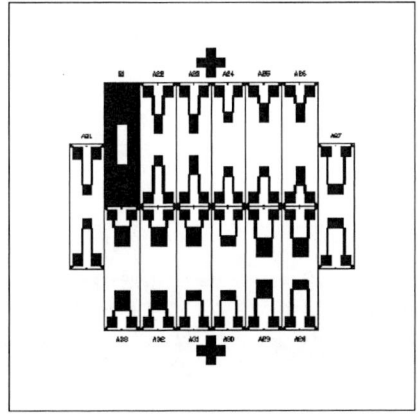

Fig.2. IDT pattern design on chrome glass surface

The forth process is the development process, removing the photoresist by Resist Developer RD6 to expose the part to be etched way. Consequently, the aluminum deposited area which is not cover by PR1-2000A will be etched away by Aluminum Etch 80-15-23-2 solution. The final process is washing away the remaining PR1-2000A by dripping Dimethyl Ketone onto the etched substrate in the spinning machine.

III. RESULTS AND DISCUSSION

The pattern transferring process using $LiNbO_3$ as substrate (Fig.3) testing has been completed. The optimum processes and parameters setting results were found as shown in Table 1:

TABLE 1. OPTIMUM PROCESSES AND PARAMETERS

Process Type	Optimum Parameters
i) Photoresist (PR1-2000a) coating.	
a) Spinning step 1	Ramp up 1000 rpm with acceleration 500 for 10 seconds
b) Spinning step 2	Spinning speed 4500 rpm with acceleration 12000 for 30 seconds.
c) Spinning step 3	Ramp down 0 rpm with acceleration 1200 for 5 seconds.
ii. Soft bake	105^0C for 60 seconds
iii. UV expose	6 to 8 seconds (for width IDT less than 10µm).
	10 seconds (for width IDT more than 10 µm)
iv. Bake	105^0C for 60 seconds
v. Development	Dip in RD 6 for 21 seconds. After that, rinse it with deionized water (DI) quickly and blow it gently to dry (do not spin to dry as the IDT photoresist pattern will be damaged). Repeat the process if the UV exposed photoresist area is not full removed.
vi. Etching	Etch the unwanted aluminium layer with Aluminium Etch, the duration is depended on the metal deposited thickness during the PVD process. It can be done by observing the colour changes.
vii. Stripping photoresist	Using acetone with spinning speed of 4000 rpm.

Fig. 3. IDTs on $LiNbO_3$ substrate.

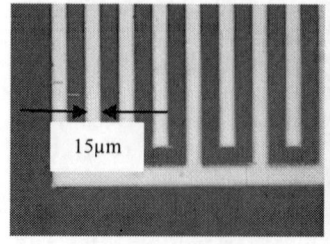 

Fig. 4. Appropriate parameters setting conditions

Fig. 5. Under etch condition

Fig. 6. Over etch condition

Fig. 7. Effect of contamination on the chrome mask surface

During the fabrication process, the tedious requirement for the above parameters settings will produce good IDTs deposition on $LiNbO_3$ substrate (Fig. 3). Optimum process and appropriate parameters setting will able to produce sharp and accurate size IDTs (Fig.4). On the other hand, not proper photoresist coating process and too short dipping process in RD 6 will cause the under etch result (Fig.5), however, over time dipping in RD6 or Aluminum Etch will cause the effect of over etched condition (Fig. 6). Thus, the IDT surface changing condition during the RD6 and Aluminum Etch dipping process must be observed cautiously. The defect of unwanted pattern transfer (Fig.7) is caused by the chrome mask surface condition, thus before the photolithography process, it is advisable to check the cleanness of chrome mask surface under high power microscope and make sure it is free of any contamination. Besides that, it is advisable to check the aluminum deposition condition before photolithography process as well, as it will affect the quality of fabricated IDTs as well. Five different IDT pitch sizes were fabricated; 12 µm (3 µm width IDT finger and 3 µm spacing between IDT fingers), 32 µm (8 µm width IDT finger and 8 µm spacing between IDT fingers), 48 µm (12 µm width IDT finger and 12 µm spacing between IDT fingers), 60 µm (15 µm width IDT finger and 15 µm spacing between IDT fingers) and 60 µm (20 µm width IDT finger and 20 µm spacing between IDT fingers). The synchronous frequencies of fabricated IDTs (Fig. 8) were measured by Agilent E8362B, network analyzer. As for the theoretical synchronous frequency calculation of the five different IDT pitch sizes were calculated based on (1), the value of vp for 64^0YX LiNbO3 is 4478m/s. Both calculated and measured synchronous frequencies of the five different IDT pitch sizes are compared in Table 2. The results clearly show that the difference of the calculated and measured synchronous frequencies is less than 4.5%.

TABLE 2. COMPARISON OF IDT CALCULATION AND FABRICATION

IDT Pitch (μm)	Calculation (Mhz)	Fabrication (Mhz)	% Error
12	373.167	384.750	3.10
32	139.938	144.900	3.55
48	93.291	97.525	4.53
60	74.633	72.100	3.39
80	55.975	58.3875	4.31

IV. CONCLUSION

In this paper, we present the method for the more economical method of aluminum IDT fabrication on 64^0YX LiNbO$_3$ substrate including the optimum processes and parameters setting. Besides that precaution steps are mentioned to produce good IDT deposition. From the result obtained, it has a very minor difference from the calculation one, less than 4.5%. Thus, aluminum IDT fabrication on 64^0YX LiNbO$_3$ can be done in more economical method, CMOS method if tolerance of 5% is allowed.

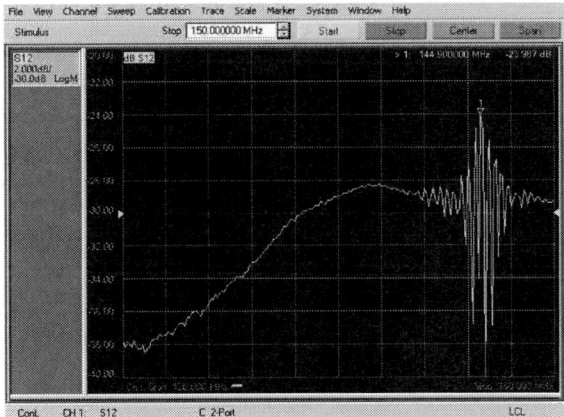

Fig.8. Insertion loss spectra of 32μm pitch on 640YX LiNbO$_3$

ACKNOWLEDGMENT

The authors are very grateful to all the supporting members in MARDI, Institute of Nano Electronic Engineering and School of Computer and Communication Engineering UniMAP.

REFERENCES

[1] Sauerbrey, *G. Z. Phys.*, 155, 206 (1959).

[2] King, Jr., W. H. *Anal. Chem.*, 36, 1735 (1964).

[3] Wohltjen, H. and Dessy, R. Anal. Chem., 51, 1458 (1979)

[4] Ngeontae, W., W. Janrungroatsakul, et al. (2009), "Novel potentiometric approach in glucose biosensor using silver nanoparticles as redox marker," Sensors and Actuators B: Chemical 137(1): p. 320-326, 2009.

[5] Turner, A.P.F., Biosensors--sense and sensitivity. Science, 290(5495): p. 1315-1317, 2000.

[6] Ivnitski, D., et al. , Biosensors for detection of pathogenic bacteria. Biosensors and Bioelectronics, 14(7): p. 599-624, 1999.

[7] S.T.Ten, W.W.Liu, S. Ahmad, H.Uda, T.Nazwa, "Design and characteristic of CMOS fabricated acoustic waves based sensors for foodborne pathogen rapid detection," Proceedings in 2012 IEEE EMBS International Conference on Biomedical Engineering and Sciences ,P. 387-391, 2012.

[8] Grate, J.W. and G.C. Frye, Acoustic wave sensors. Sensors update, 2(1): p. 37-83, 1996.

[9] Perpeet, M., et al., SAW Sensor System for Marker-Free Molecular Interaction Analysis. Analytical letters, 39(8): p. 1747-1757, 2006.

[10] U. Knauer, J. Machui, and Clemens C. W. Ruppel, Design, Fabrication, and Application of GHz SAW Devices, in IEEE MTT-S International Microwave Symposium Digest, pp. 1821–1824, 2001.

[11] A. Springer, F. Hollerweger, R.Weigel, S. Berek, R. Thomas,W. Ruile, Clemens C. W. Ruppel, and M. Guglielmi, Design and Performance of a SAW Ladder-Type Filter at 3.15 GHz Using SAW Mass-Production Technology, IEEE Transactions on Microwave Theory and Techniques, vol. 47, no. 12, pp. 2312–2316, 1999.

[12] H. Odagawa and K. Yamanouchi, 10 GHz Range Extremely Low-Loss Ladder Type Surface Acoustic Wave Filter, in Proceedings of the IEEE Ultrasonics Symposium, pp. 103–106, 1998.

[13] S. Achenbach, T. Mappes, R. Fettig, J. Kando, and J. Mohr, Process Conditions for the Fabrication of Sub-Wavelength Scale Structures by X-ray Lithography in PMMA Films, in Proceedings of SPIE- The International Society for Optical Engineering, no. v 5450, pp. 86–94, 2004.

[14] N. Yoshioka, A. Sakai, H. Morimoto, k. Hosono, Y. Watakabe, and S. Wadaka, Fabrication of Surface Acoustic Wave Devices by Using X-ray Lithography, Journal of Vacuum Science and Technology B. Microelectronics Processing and Phenomena, vol. 7, pp. 1688–1691, Nov-Dec 1989.

Preparation Of Zinc Oxide Piezoelectric Substrate for SAW Biosensor Device

[1]M. R. Zakaria, [1]F. Hamzah, [1]M. F. Omar, [2]U. Hashim, [2]R. Mat Ayub, [2]M. A. Farehanim, 2A Wesam Al-Mufti

[1]School of Microelectronic Engineering
[2]Institute of Nano Electronic Engineering
Universiti Malaysia Perlis (UniMAP), Arau, Perlis, Malaysia
rosydizakaria@yahoo.com, pamek.may13@gmail.com, uda@unimap.edu.my, Ramzan@unimap.edu.my

Abstract— To develop the SAW device, piezoelectric substrate is one of important factor that improve performance device. The material used such as ITO (Indium Thin Oxide) or LiTaO3 (Lithium Tantalate) extremely affect the cost as well as the preparation methods. To overcome this problem, the material known as ZnO (Zinc Oxide) has been chosen as the thin film layer of piezoelectric in developing the biosensor. These ZnO thin film acts as piezoelectric layer that made contact with the sensing element before the signal convert from a biological signal to an electrical signal by the transducer. Sol-gel deposition technique was used and zinc acetate as precursor compound and 2-Methoxyethanol as solvent to form the sol-gel. This sol-gel then be spin coat on the SiO_2/Si substrate which will later to be anneal with oxygen to make the thin film more dense, thinner and has smoother surface quality and higher resistivity. The surface morphology, electrical and optical characterization of the thin films will be studied using Scanning Electron Microscopy (SEM) and Atomic Force Microscopy (AFM), Four Point Probe and UV-Visible Spectropohotometer

Keywords—ZnO piezoelectric, Zno Sol-Gel, SAW Biosensor.

I. INTRODUCTION

In Biosensor technologies, substrate is one of the important factors that affecting to performance of sensing mechanism [1,2]. In development of Surface Acoustic Wave, there are three important parts in the SAW Biosensor device which is, substrate piezoelectric, sensing area and InterDigital Transducers (IDT). SAW device performance was influenced significantly by the piezoelectric layer [3]. The properties depend on the crystal-cut angles and the direction of SAW propagation. Properties that usually had impacts on the SAW device performance is the SAW velocity, electromechanical coupling factor, temperature, stability and propagation loss. A high SAW velocity will allows fabrication of high frequency SAW device nonetheless reduction of SAW velocity will produce smaller devices [3,4].

The most common use piezoelectric layer for acoustic wave sensor is quartz due to its low temperature coefficient comparable to the other piezoelectric material. There are several crystal orientation cut for quartz to minimize temperature coefficient such as ST cut for SAW device and AT and BT cuts for resonator. These cuts are widely available since it has been characterized in the past years and widely obtainable. Lithium niobate ($LiNbO_3$) on the other hand has a large electromechanical coupling coefficient (k^2) that allow two port acoustic wave device to use fewer finger pairs than quartz. Nevertheless $LiNbO_3$ is sensitive to temperature changes due to large temperature

coefficient and need a stable temperature environment to operate. Other material like silicon exhibits both piezo- and non-piezoelectric behavior while GaAs has very small k^2 [5].

Electromechanical coupling factor k^2 is a measure of interaction between electrical and mechanical fields in piezoelectric material or device. Materials with high electromechanical coupling will allow for smaller device size or reduce insertion loss [6]. Substrate or thin film for SAW devices with small temperature coefficient delay and high electromechanical coupling factor is preferred due to the temperature stability at the resonance frequency is a direct function temperature coefficient and acoustic wave velocity of SAW device. Wave scattering at crystalline defects and surface abnormalities results in acoustic wave transmission losses that significantly affect the insertion loss of the SAW device. Thus materials with small propagation loss are desirable because propagation loss is directly reflected in the increased insertion loss of devices. Roughly speaking, materials with higher SAW velocities usually possess propagation loss smaller than that of materials having smaller SAW velocities [7]. Cost constrained and integration issues in electronics integration for bulk material that were commonly used for piezoelectric such as quartz, lithium tantalate (LiTaO3), lithium niobate (LiNbO3) and sapphire (Al2O3) is the main limitation in the development of SAW device [8].

Zinc oxide (ZnO) is an inorganic compound that has a wide bandgap (3.3 eV), large excitation binding (60 meV), large electromechanical coupling factor and low dielectric constant. It has a unique characteristic that exhibit properties of semiconductors, piezoelectric and pyroelectric materials [7,9]. The thin film of ZnO is able to function in a very wide area of application such as a n-type semiconductor (II–VI), photoconductor, piezoelectric, birefrigence, acousto-optical, transparency in the infrared region and opto-electric [9]. This allows ZnO to be used in a device such as filter, resonator, solar cell, light emitter and sensor. So far ZnO thin film is widely used for mass consumer in a form of SAW filter for communications. Since ZnO is a nontoxic material and bio-safe, it can be safely used without toxic effect in biomedical application to immobilize and modify biomolecule [8,10,11]

Piezoelectric material can be deposited or grow in thin film form such as ZnO and AlN were used as overlays on silicon to fabricate acoustic wave device. The thin film that was grown or deposited must crystal oriented in a single and piezoelectrically active crystal orientation. This allows for

978-1-4799-5761-3/14 $31.00 © 2014 IEEE

fabrication of acoustic wave device that able to operate at much higher frequency 100Mhz – 1Ghz [10].

There are several deposition methods possible for ZnO thin films such as sputtering, pyrolysis, thermal-plasma, sol-gel and CVD. Each property of the thin film will be depended with the deposition technique. In this project ZnO thin film formation by sol-gel method will be used [12]. Sol-gel is a technique to form inorganic solid (metal or semi-metal oxide) through polymerization of precursor compound (metal salts, metal alkoxides, metal acetates metal acetylacetonates) solution through at a much lower temperature compare to high temperature chemistry. Sol-gel can be classified in aqueous and non-aqueous sol-gel process where either water or organic solvent is used. It consist of two main process step which is hydrolysis of the molecular precursor to form sol-soluble hydroxylated oligomers (dispersed tiny particle in liquid) and condensation in order to form a hydrate oxide hydrogel.

Although sol-gel technique enable creation of high quality transparent conductive oxide and low in cost, however the process require significantly longer time due to repetative coating and drying to achieve desired thickness. This hinder the usage in of sol-gel in industry [11,13].

II. METHODOLOGY

The fabrication process begins from the Silicon wafer (100) was cleaned by using a standard wet cleaning process. In order to improve resistivity of the substrate, the silicon oxide (SiO_2) was deposited with a thickness 3000 μm. The ZnO solution then was prepared by using sol-gel technique. In preparing the sol gel solution, several important parameters such as the ratio, chemical solvent, stabilizer and the main precursor need to be identified first. These parameters are very important in order to prepare fine and excellent Sol Gel. Therefore, some calculation will be carried out to determine the ratio of the chemical involved. Firstly, the quantity of the solvent and the main precursor which is Zinc Acetate Dehydrate (ZAD) are calculated using the equation (1) below:

$$Mass = MCV \qquad (1)$$

Where, M represents the molar mass of the Zinc Acetate Dehydrate while C is the concentration used. V is the solvent volume for 2-Methoxynethanol. These three parameters will be used to determine the mass of the Zinc Acetate Dehydrate used for the preparation of the sol gel solution. The molar mass of Zinc Acetate Dehydrate usually fix. However, the concentration and solvent volume are flexible depending on the desired total amount of the sol gel that needs to be prepared. Note that the ratio of the morality of ZAD to the Stabilizer is 1:1. The molarity of ZAD is 0.2 Mol/L, so the morality for the stabilizer is also 0.2 Mol/L. Equation 1 will be used to calculate the mass of the stabilizer used. The molar mass of the MEA and Acetic Acid are fixed while the concentration and the solvent volume are varied depending on the desired total amount of the sol gel that needs to be prepared. Both of this calculation will be carried out before the next step of the sol gel preparation.

Next, the mass of the ZAD powder will be solute in the solvent by putting the solution on the hot plate with the temperature 60 ^{0}C and the spin speed around 1000 rpm. These temperature and spin speeds are added to make sure that the ZAD powder is perfectly dissolved in the solvent. In the meantime, the stabilizer will be added drop by drop in order to obtain the fully transparence solution.

The final steps in preparing the sol gel solution are the transparence solution obtained will be kept at room temperature for 24 hours for aging purposes before it can be used for the deposition of the ZnO thin films. The final step is the sample will be annealed using annealing furnace with temperature 300^{0}C and 500^{0}C for 2 hours. The steps will be repeated using different samples to vary the number of layers of ZnO thin film for characterizing purpose. Fig. 1 shows the steps for deposit zinc oxide onto the silicon substrate.

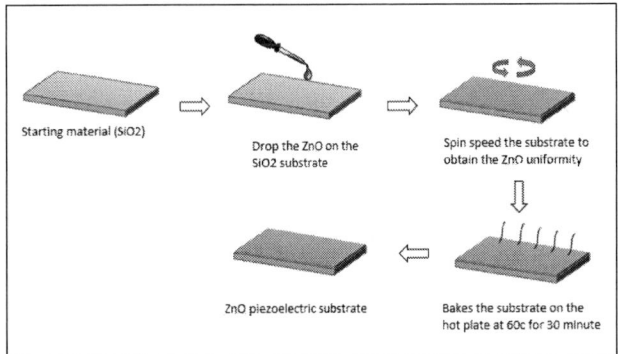

Fig. 1. Zinc Oxide deposition process steps

Metallization process then continued to form IDT on the Piezoelectric substrate. The process detail has been discussed in previous paper [14].

III. RESULT AND DISCUSSION

The result obtained from several characterization is very important in order to identify the parameter involved to improve the reliability and sensitivity of the biosensor. The methods are Physical, Electrical and Optical Characterizations.

A. . Physical Characterization

In physical characterization, several experiments have been conducted involving the usage of Atomic Force Microscopy (AFM) and Scanning Electron Microscopy (SEM). Result representing a 2-Methoxyethanol as the solvent shows in Fig. 2(a) and (b) for 3 layers and 5 layers after undergo to the 500 ^{0}C temperature. The scanning scales for the entire sample were taken from 0.5μm × 0.5μm to 1μm × 1μm and the images obtained were composed to 512 × 617 pixel dimensions. From the observation, the 2-Methoxyethanol, image shows the homogenous surface structure with the thickness 40 nm for 3-layers and 60nm for 5-layers thin films.

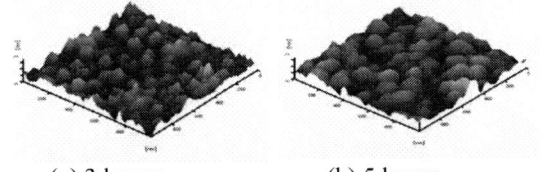

(a) 3 layers   (b) 5 layers

Fig. 2. The 3-D image of 2-Methoxyethanol annealed at 500^{0}C for (a) 3 layers, (b) 5 Layers

The grain sizes and root mean square (rms) were observed using the AFM surface analysis as shown in the Table I. The RMS value shows approximately ~4nm and the value also increase with the increasing of the annealing temperature. [15]. From the result, the average roughness (R_a) indicate that, the surface of 2-Methoxyethanol surface has a highest peak at 8.4 nm. It's shows that the solvent strongly influence the surface roughness of the thin films.

TABLE I.   PARAMETER OF ZnO THIN FILMS ANNEALED WITH DIFFERENT ANNEALING TEMPERATURE

Solvent	Annealing Temperature (^{0}C)	Mean Diameter of Grain Size (nm)	Root Mean Square (RMS) (nm)	Average Roughness, R_a (nm)
2-Methoxyethanol	300	5.07	1.701	1.314

In this study, SEM is used for the purpose of identifying the surface characteristic of the ZnO thin films. Fig. 3 shows the SEM images of the samples that undergo different layer coating. All the results taken were measured by the average applied voltage 15kV and the beam diameter around 1µm while the magnification range applied ranging from 10000X to 50000X. However, the selected result was applied with the 10000X in order to observe the difference surface morphology of the thin films. Image (a) and (b) from Fig. 3 representing the thin films deposited by using 2-Methoxyethanol as a solvent. In both cases, the annealing temperature is important in order to produce a homogenous surface structure. In addition, the particular structure becomes larger and denser as the annealing temperature increased. These results also strengthen on the effect of the annealing temperature to the morphological structure of the ZnO thin films that state the similar achievement [16]. In detail, the increasing in annealing temperature resulting in the higher of the packing density of the ZnO thin films which may be due to the disappearing gap between particles [17].

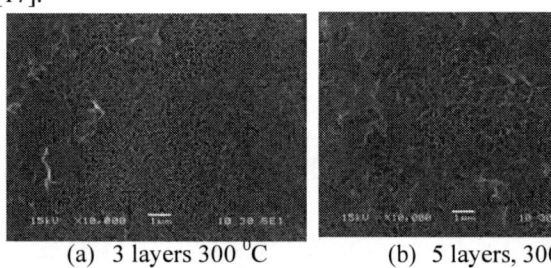

(a) 3 layers 300 ^{0}C   (b) 5 layers, 300^{0}C

Fig. 3.   SEM images of ZnO thin films after annealed 300^{0}C (a) ZnO film with 3 Layers, (b) ZnO film with 5 Layer

B. Optical Characterization

UV-VIS Spectrophotometer is equipment for measuring the optical transmittance or absorbance of a sample with respect to the wavelength of the electromagnetic radiation. Fig. 4 presents the ZnO thin film transmittance spectra heat for 2-Methoxyethanol. The range of the wavelength is taken from 300 nm to 800 nm. From the graph, the potential cut-off wavelength takes place from 370 nm – 385 nm. Both graphs exhibited sharp absorption edge at the cut-off wavelength range. These phenomena happen because at the cut-off wavelength, the ability of the electron at the valence band to move from the valence band to the conduction band increased resulting in an extreme drop off wavelength. The result obtained also proved that at the range of UV region (200nm – 400nm), the transmittance wavelength will decrease dramatically resulting in sharp absorption edge [18]. As can be seen from both graphs, ZnO thin film with 2-Methoxyethanol with 5 layers generated great optical transmittance which is 45.2% with thin film annealed at 300^{0}C. However, a ZnO thin film with 2-Methoxyethanol with 3 layes as a solvent shows the lowest optical transmittance at the annealing temperature 500^{0}C which is 3.7%. This is due to the fact that the thin film unable to absorb great optical transmittance because of high surface roughness. This result strengthens the theoretical explanation about the surface roughness discussed from the previous section and the effect of the surface roughness with the ability to absorb great optical transmittance [19].

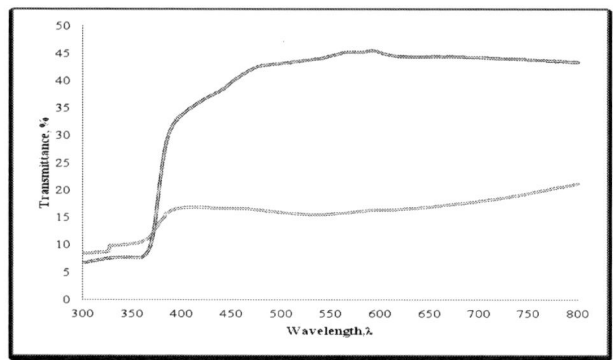

Fig. 4.   Zno thin films transmittance spectra annealed at 300^{0}C

C. Electrical Characterization

Electrical characterization focuses on the resistivity and (Current vs Voltage) I-V measurements of the ZnO thin films and SAW device. Four point probe equipment is used to measure the resistivity because of its capability to directly measure the sheet resistance of any semiconductor material. From Fig.5 observed that the values of sheet resistivity for total layer cases were dramatically increased when 2-Methoxyethanol annealed at 300^{0}C. From the result, the lowest resistivity value with 3.433MΩ/cm, 92.706M Ω/cm and 150.095M when the layer was increased from 1 to 5 layers, respectively. From previous studies, the lowest value of resistivity is due to the fact that the career concentration increases thus decrease the resistivity [19]. The career concentration is the ability of the electron moves from the conduction band to the valence band. In terms of thickness, carrier concentration increase due to the disruption of oxygen in the grain area, thus traps the carrier's ability to move to the valence band resulting in the increasing of the resistivity. From the graph, the ZnO thin films with 5 layers annealed at 300 ^{0}C produce greater resistivity is about 150.095 MΩ/cm.

TABLE II.   AVERAGE SHEET RESISTIVITY WITH DIFFERENT ANNEALING TEMPERATURE

Annealing Temperature	2-Methoxyethanol (Ω/cm)		
	1 layer	3 layers	5 layers
300^{0}C	3.422M	92.706M	150.095M

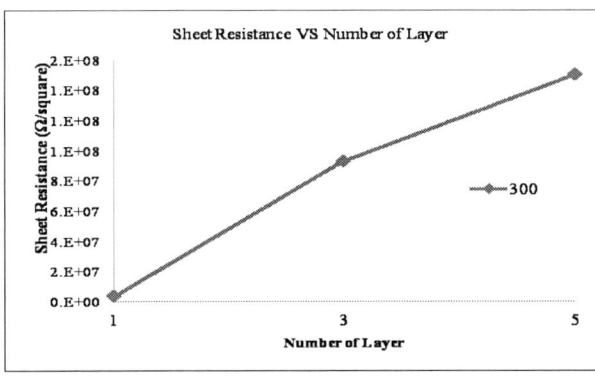

Fig. 5.   2-Methoxyethanol ZnO thin films at 300 C for various layers

From the Fig. 6 can be seen that there is a slight difference between the result of I-V curve for the IDT that fabricated on a piezoelectric substrate (ZnO) and without piezoelectric substrate (ZnO). The result shows that the current of the IDT that fabricate on piezoelectric substrate is much lower compared to the IDT without piezoelectric. This shows that by using piezoelectric substrate, it will affect the device where the sensitivity of the device will be increased because of the resistance is getting low. Besides that, piezoelectric substrate wills also improving device stability where SAW that propagates on piezoelectric substrate is more stabilize and it has low insertion loss. This shows that there are many advantages by using piezoelectric substrate in biosensor fabrication.

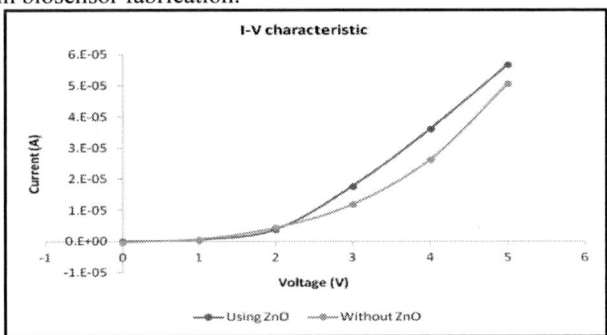

Fig. 6.   The I-V curve of IDT design with and without ZnO

CONCLUSION

From the research, the ZnO sol gel solution was prepared by using 2-Methoxyethanol as a solvent. The prepared sol gel was successfully coated on the silicon substrate to produce as a piezoelectric substrate. On the physical surface after undergo to AFM and SEM found that, the sample with many ZnO layers indicates that the provide higher surface roughness as compare to 1 layer 2-Methoxyethanol. The grain size of the thin film at a number of 5 layers was increased. Same phenomena were identified on the surface structure where the surface becomes denser due to the fact that disappearing gap between particles resulting in the increase in size of the particle as the post annealing temperature increase. From the IV result also shows that the result with ZnO Piezoelectric has better compared to the substrate without ZnO. From this result, we confidently the piezoelectric substrate is one of the important factors that affecting to SAW device performance.

ACKNOWLEDGMENT

The authors wish to thank University Malaysia Perlis (UniMAP), for giving the opportunities to do this research in the Micro & Nano Fabrication Cleanroom.

REFERENCES

[1]   K. Länge, B. E. Rapp, and M. Rapp, "Surface acoustic wave biosensors: a review.," Anal. Bioanal. Chem., vol. 391, no. 5, pp. 1509–19, Jul. 2008.

[2]   O. Tigli and M. E. Zaghloul, "Surface acoustic wave (SAW) biosensors," 2010 53rd IEEE Int. Midwest Symp. Circuits Syst., pp. 77–80, Aug. 2010.

[3]   D. A. Powell, K. Kalantar-zadeh, W. Wlodarski, and S. J. Ippolito, "Layered Surface Acoustic Wave Chemical and Bio-Sensors," vol. X, no. Section 2, 2006.

[4]   Y. Hur, et al., "Development of an SH-SAW sensor for the detection of DNA hybridization", Sensors and Actuators A: Physical, 2005. 120(2): p. 462-467.

[5]   M.-I. Rocha-Gaso, C. March-Iborra, A. Montoya-Baides, and A. Arnau-Vives, "Surface generated acoustic wave biosensors for the detection of pathogens: a review.," Sensors (Basel)., vol. 9, no. 7, pp. 5740–69, Jan. 2009.

[6]   K. Uchino, "Introduction to Piezoelectric Actuators and Transducers Kenji Uchino , International Center for Actuators and Transducers , Penn State University," no. 5, pp. 0–40.

[7]   Y. Q. Fu, J. K. Luo, X. Y. Du, a. J. Flewitt, Y. Li, G. H. Markx, a. J. Walton, and W. I. Milne, "Recent developments on ZnO films for acoustic wave based bio-sensing and microfluidic applications: a review," Sensors Actuators B Chem., vol. 143, no. 2, pp. 606–619, Jan. 2010.

[8]   M. P. Santos and E. J. P. Santos, "Pre-heating temperature dependence of the c-axis orientation of ZnO thin films," Thin Solid Films, vol. 516, no. 18, pp. 6210–6214, Jul. 2008.

[9]   Ballantine, D. S. Jr. White, R. M. Martin, S. J. Ricco, A. J. Frye, G. C. Zellars, E. T. Wohltjen, H., Acoustic Wave Sensors - Theory, Design, and Physico-Chemical Applications, Elsevier, 1997.

[10] R. N. Gayen, K. Sarkar, S. Hussain, R. Bhar, and A. K. Pal, "ZnO films prepared by modified sol-gel technique," vol. 49, no. July, pp. 470–477, 2011.

[11] Chien-Yie Tsay, K.-S. F., Yu-Wu Wang, Chi-Jung Chang, and C.-K. L. Yung-Kuan Tseng (2010). "Transparent semiconductor zinc oxide thin films deposited on glass substrates by sol–gel process." Ceramics International: 1791-1795.

[12] S. Shariffudin, m. S., and s. H. Herman "effect of film thickness on structural, electrical, and optical properties of sol-gel deposited layer-by-layer zno nanoparticles." transactions on electrical and electronic materials 13(2): 102 - 105. (2012).

[13] M.R. Zakaria, U. Hashim, M. H. I. Mohd Amin, "Design and fabrication of IDT saw by using conventional lithography technique," 2013, Middle-East Journal of Scientific Research 18 (9): 1281-1285,

[14] M.R. Khanlary, S. I. (2011). "Structural and optical properties of ZnO thin films prepared by sol–gel method." Micro & Nano Letters 6(9): 767-769.

[15] Urai Seetawan, S. J., Tosawat Seetawan, Chanipat Euvananont, Chabaipon Junin, and P. C. Chanchana Thanachayanont, Vittaya Amornkitbamrung (2011). "Effect of annealing temperature on the crystallography, particle

[16] Hongxia Li, J. W., Hong Liu, Huaijin Zhang, Xia Li (2004). "Zinc oxide films prepared by sol–gel method." Journal of Crystal Growth.

[17] Chien-Yie Tsay, K.-S. F., Yu-Wu Wang, Chi-Jung Chang, and C.-K. L. Yung-Kuan Tseng (2010). "Transparent semiconductor zinc oxide thin films deposited on glass substrates by sol–gel process." Ceramics International: 1791-1795.

[18] Shakti, P. S. G. a. N. "Structural and Optical Properties of Sol-gel Prepared ZnO Thin Film." Applied Physics Research 2: 19 - 28. (2010).

[19] M. F. Malek, N. Z., M. Z. Sahdan, M.H. Mamat, Z. Khusaimi and M. Rusop (2010). "Electrical Properties of ZnO Thin Films Prepared by Sol-gel Technique." 384 - 387.

Controlling growth rate of ultra-thin Silicon Dioxide layer by incorporating Nitrogen gas during dry thermal oxidation

A.H. Azman, RM Ayub, M. K. Md Arshad, S. Norhafiezah, M.F.M. Fathil, M.Z. Kamarudin, M. Nurfaiz, U. Hashim.

Institute of Nano Electronic Engineering,
Universiti Malaysia Perlis (UniMAP)
01000 Kangar, Perlis, Malaysia.
azmanhassan88@gmail.com

Abstract—**The continuing trend toward miniaturization of silicon devices is enforcing development of ultra-thin dielectrics. While the thermally grown SiO_2 has been used as a gate dielectric ever since the decade of silicon device began, it appears that the electrical and physical properties of pure SiO_2 are not good enough to provide acceptable for ultra-thin gate dielectric film. There are many available methods to control the ultra-thin film; In this paper we show a simple but promising method that incorporated nitrogen as a second gas in the dry oxidation process, on which the growth rate can be controlled. This method produce surface protective layers against impurity penetration, good interfacial characteristics and strengthens the oxide structure, which directly related to improvement the gate dielectric quality.**

Keywords— thermal dry oxidation, SiO_2, gate oxide, nitrogen gas, oxide growth rate, ultra-thin film.

I. INTRODUCTION

It has been a decade, the silicon dioxide (SiO_2) as the gate dielectric have been extensive study, research and development for the MOS semiconductor industry for enhancing performance at lower cost. Thinner and thinner oxide layer requires for sub-nano MOS device since the oxide plays an active role and its electrical quality should be preserved [1]. One major trend in the semiconductor industry is to shrink the device dimensions and scale down the gate dielectric [2]. The gate dielectric or the SiO_2 film must be scaled down to sub-nano range to get the result for fabricating more device per wafer (i.e., increase the device density) and thus reducing the cost per chip [3]. Nevertheless, the main reason of scaling the gate dielectric is to enhance the drive current and reduce the short channel effect due to scaling the gate length. However, to obtain the ultra-thin SiO_2 film becomes more challenges and the requirement of low leakage current is more stringent than ever [4]. The growth rate of this SiO_2 film is a major problem for growing a sub-nano gate dielectric, where in the initial stage of dry oxidation, there is a non-linear oxide growth and it can't be controlled [5].

Thermal Oxide or SiO_2 is one of the "building block" films used in making both simple and complex semiconductor device. It can be an excellent dielectric (insulating) thin film. It is also found on the device as a "gate oxide," this is a very thin thermal oxide located over the gate or active region of the individual transistors. Silicon dioxide is a grown oxide not CVD (chemical vapor deposition) and as such it has a higher integrity than most CVD oxide films and so far has demonstrated higher uniformities, less defects and higher dielectric strength than deposited Oxide thin films. For reference, the target index of refraction number for thermal oxide is 1.462. Thermal oxide is normally grown in a diffusion furnace (either vertical or horizontal) or it can also be grown in a Rapid Thermal Processor. RTP systems are normally used for thin dry Thermal Oxides or Implant annealing on devices where tight thermal budgets are an issue. Thermal oxide is grown at high temperatures from 800°C - 1200°C using either a "Wet" or "Dry" growth method . Dry thermal oxide is much the same ways as the wet thermal oxide process except instead of pyrogenic steam as the oxidation gas, only pure Oxygen is used. A dry oxidation produces a more uniform and denser thermal oxide with even higher dielectric strength than wet oxide. The major different in the growth of wet and dry oxide is the growth rate, dry oxide grows much slower than wet oxide. For this reason dry oxides normally used for thin gate and capacitor oxides where high uniformity and high dielectric strength are needed. Nitrogen incorporation into the dry oxidation is one of promising methods and many more techniques have been proposed. This method can produce surface protective layers against impurity penetration, producing good interfacial characteristics strengthen the oxide structure and thereby improve the gate dielectric quality, it also acts as an oxidant diffusion barrier that reduce the oxidation rate significantly. The advantage of this process is that the process is essentially hydrogen free [6]. It appears that nitrogen neutralizes growth sites at the oxide-silicon interface, which significantly slows down the oxidation process when N_2 is used as an oxidizing ambient [7].

This paper, we investigate and control the growth rate of dry oxidation incorporated with nitrogen gas for ultra-thin film of silicon dioxide (SiO_2) on top of a silicon wafer. We look into the dry oxidation process for a better quality gate dielectric and at slower growth rate since the normal growth rates are not good enough to produce a good quality dielectric layer in the case of ultra-thin gate dielectric. Basically the growth rate for is measured using as in Equation 1:

$$Growth\ rate = \frac{average\ thickness\ (\text{Å})}{process\ time\ (sec)} \qquad (1)$$

In our case, we consider the average thickness (**Å**) is the average oxide thickness at least at 9 points measured on the wafer surface while the process time for growth of SiO_2 is at 300 sec.

II. EXPERIMENTAL DETAIL

A. Starting material

4-inch silicon wafer with p-type (100) with resistivity between 1-20 Ohm-cm range is used in this experiment. The wafer is then cleaned by standard cleaning 1 (RCA1), standard cleaning 2 (RCA2) and Buffer-Oxide Etch (BOE) to remove the organic residues, metal ions and the native oxide on the silicon wafer [8], [9].

B. Equipment

The PID-controlled, three-zone horizontal quartz tube furnace (MDL 906 MODULAB) was designed to accommodate 4-inch wafers in quartz wafer boats. The furnace has a maximum operating temperature of 1200°C with power consumption approximately 8 kW. It has an associated gas supply system with two flow meters i.e. one for high-purity oxygen and another one for high-purity nitrogen. The system have been plumbed so that the furnace can be purged with nitrogen while the unit is heating to operating temperature, and then can flow either dry or wet oxygen under controlled conditions. For the film thickness measurement, we use the Filmetrics F20-UV. The thickness and refractive index can be measured in less than a second.

C. Process

For this experiment, the dry oxidation process is carried out to grow silicon dioxide (SiO_2) at temperature of 900°C for about 300 second. The nitrogen gas is set at three flow rates i.e. 1 standard liter-per-minute (slm), 2 standard liter-per-minute (slm) and 5 standard liter-per-minute (slm) while the oxygen is set at 10 standard liter-per-minute (slm). The oxide thickness that was measured using Filmetric F20-UV dielectric analyzer is shown in Table 1. The measurement was done at nine separate locations randomly on the wafer as shown in Fig. 1.

TABLE 1. Process parameters to control the SiO_2 growth rate.

Furnace temperature (°C)	Oxygen gas rate (slm)	Nitrogen gas rate (slm)	Time (sec)
900	10	1	300
900	10	2	300
900	10	5	300

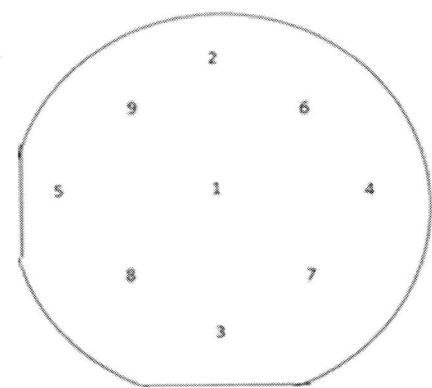

Fig. 1. *Thickness point on the silicon wafer surface.*

III. RESULT

A. The thickness of dry oxidation process incorporating with N_2 gas

In Fig. 2 shows the SiO_2 thickness for three differences of N_2 gas flow rates. One can see from the graph:

- Firstly, the thickness is significantly reduced with the increases of incorporating nitrogen gas in furnace. At 1, 2 and 5 slm nitrogen the average measured thicknesses are 224.122, 187.522 and 133.967 Å respectively. With the incorporation of N_2, the growth rate is slower compared with pure O_2. This N_2 is used as an oxidizing ambient, its acts as an oxidant diffusion barrier that reduce the oxidation rate significantly and that nitrogen neutralizes growth sites at the oxide-silicon interface, which significantly slows down the oxidation process. When using only pure oxygen, the oxygen directly interact with silicon to form the SiO_2 layer with no other gases that can disturb the oxygen reaction with the silicon layer.

- Secondly, non-uniformity is 6.85 %, 14.43 % and 10.42 % for 1, 2 and 5 slm respectively. The non-uniformity for the three parameters are not follow the trend that more nitrogen gases were added the more uniformity can be reach. In this result, the 1 slm nitrogen produces the best uniformity between the two other experiments. The other result should be uniform respectively. The worse uniformity is shown at 2 slm parameter where only 85.57% uniformity were establish. This maybe because of the defect or a native oxide that have encounter the silicon wafer. This native oxide grows on the surface of silicon wafers during wafer transport from the wafer-cleaning equipment to the thermal oxide growth equipment. Silicon wafers must be isolated from the ambient air during wafer transport [10]. This will give arise the non-uniformity because at certain point on the wafer is already have this native oxide. When the new oxide layer are grown, it will grow on top of the native oxide and will became thicker layer. The precaution ways to control this constrains are have to do in a clean room facility that have a very good ventilation and a proper ways to fabricate as in the industry. This non-uniformity is extracted based on Equation 2:

$$Non-Uniformity\ (\%) = \left(\frac{(Max\ Value - Min\ Value)(Å)}{(Average\ (Å)\ X\ 2)}\right) X100 \quad (2)$$

In our case, we consider the max value (Å) is the maximum oxide thickness value at any 9 points measured on the wafer surface, min value (Å) is the minimum oxide thickness value at any 9 points measured on the wafer surface. While average (Å) is the average oxide thickness for the experiment at least at 9 points measured on the wafer surface.

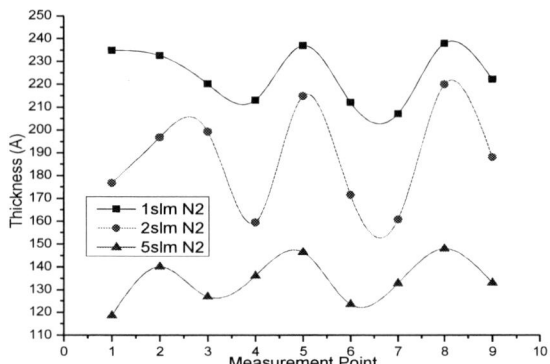

Fig. 2. Thickness result of three different dry oxidation processes incorporating of nitrogen gas.

B. The growth rate of dry oxidation incorporation N_2 gas.

In Fig. 3 show the growth rate of three different dry oxidation process that incorporation with nitrogen gas.

Firstly in Fig 3, it shown that the growth rate is significantly reduced with the increases of incorporating nitrogen gas in furnace. At 1, 2 and 5 slm nitrogen the growth rate measured are 0.74707, 0.62507 and 0.446 Å/sec respectively. The growth rate were calculated based on Equation 1. It is showed that all the three sample have a significant different result. This experiment shown that when more nitrogen gases flow during dry oxidation process there are more surface protective layer created that will make the growth rate became slower because of the oxidant barrier that reduce the oxidation rate significantly and that nitrogen neutralizes growth sites at the oxide-silicon interface, which significantly slows down the oxidation process when N_2 gas is used as an oxidizing ambient. When more nitrogen gas were added with the oxygen, the reaction that give to oxygen is the nitrogen will disturb the oxygen interaction with the silicon to form the SiO_2 layer. That is the cause why the growth rate is slower. When too much nitrogen is added the nitrogen will form one layer of nitride between the SiO_2 and silicon layer. This is

called the nitridation layer. This nitridation layer are not been recounted in this paper.

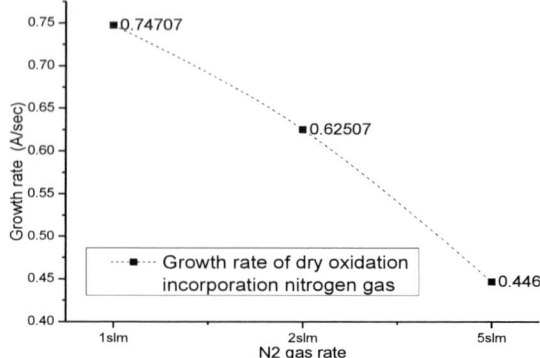

Fig. 3. Growth rate of three different dry oxidation process incorporation of nitrogen gas.

IV. CONCLUSION

In this continuing trend toward the miniaturization of silicon device is enforcing the development of ultra-thin gate dielectric. The thermally grown oxide has been used as the gate dielectric layer ever since the era of silicon device began. In this new era the ultra-thin gate dielectric became the most important gate dielectric layer to minimize the structure, thus to get a good quality of this oxide layer. This good quality gate dielectric also has to take into account to get a good device. This paper has shown one way to control the growth rate to produce an ultra-thin oxide layer by using a normal thermal dry oxidation process. The growth rate of this dry oxidation can be controlled to slow down the growth rate of thermal oxidation with the oxidation process that incorporated with the nitrogen gas. The incorporation of N_2 gas, the growth rate is slower compared with pure oxygen. This is because, only pure oxygen are used, the oxygen will directly interact with silicon wafer to form the SiO_2 layer with no other gases that can disturb the oxygen gas reaction with the silicon wafer. This is why the growth rate are faster than the incorporated with nitrogen gases. The higher the nitrogen gas rate, the slower the growth rate. With this method, the oxidation growth rate can be controlled to get a desired thickness of the oxidation for ultra-thin dielectric layer in the future electronic device.

ACKNOWLEDGMENT

The author would like to thank the Department of Higher Education, Ministry of Higher Education, (KPT) for funding this research through the Fundamental Research Grant Scheme (FRGS) with the code number 9003-00360, titled The Study of Electron Tunneling through Single / Multiple Layer Dielectric Thin Film.

REFERENCES

[1] L. Fonseca, F. Campabadal, B. Garrido, and J. Samitier, "A reliability comparison of RTO and furnace thin SiO 2 layers : effect of the oxidation temperature," vol. 40, pp. 61–75, 1998.

[2] H. Tseng and D. Ph, *The Progress and Challenges of Applying High-k / Metal-Gated Devices to Advanced CMOS Technologies*, no. January. 2010.

[3] A. J. Bauer and E. P. Burte, "4 nm GATE DIELECTRICS PREPARED BY RTP LOW PRESSURE OXIDATION IN O 2 AND N 2 O ATMOSPHERE," vol. 38, no. 2, pp. 213–216, 1998.

[4] R. M. Ayub, S. Norhafizah, A. H. Azman, U. Hashim, and T. Adam, "O RIGINAL A RTICLES Correlation between Oxide Breakdown Field and Effective Concentration in Ultra-Thin Tunnel Oxide of FG Flash Memory," vol. 9, no. 6, pp. 3451–3455, 2013.

[5] Ch. Hollauer, "The Deal-Grove Model," *Modeling of Thermal Oxidation and Stress Effects*. [Online]. Available: http://www.iue.tuwien.ac.at/phd/hollauer/node16.html.

[6] J. Ahn, W. Ting, D. Kwong, and S. Member, "Furnace Nitridation of Thermal SiO , in Pure N , O Ambient for ULSI MOS Applications," vol. 13, no. 2, pp. 12–14, 1992.

[7] S. Dimitrijev, H. B. Harrison, D. Sweatman, and A. D. Formulation, "bxtension of the Deal-Grove Oxidation Model to Include the Effects of Nitrogen," vol. 43, no. 2, 1996.

[8] M. Bachman, "RCA-1 Silicon Wafer Cleaning," 1999.

[9] M. Bachman, "RCA-2 Silicon Wafer Cleaning," 2002.

[10] K. Saga, H. Kuniyasu, and T. Hattori, "Influence of Ambient Oxygen and Moisture on the Growth of Native Oxides on Silicon Surfaces," pp. 1–3.

Fabrication and Characterization of Undoped Polysilicon Nanowire for pH sensor

C.C. Yee, M. K. Md Arshad*, M. Nuzaihan M. N, M.F.M. Fathil and U. Hashim

Institute of Nano Electronic Engineering (INEE),
Universiti Malaysia Perlis (UniMAP),
01000 Kangar, Perlis, Malaysia.
*Corresponding Email: mohd.khairuddin@unimap.edu.my

Abstract— **Polysilicon has great benefit in application of pH sensor due to the unique properties and easiness to use top-down approach. In this paper, we present fabrication and characterization of undoped polysilicon nanowire (NW) for pH sensor application. The fabrication processes steps involve were photolithography, etching, deposition and oxidation. 3-aminopropyltriethoxysilane or APTES were used to enhance the sensitivity of polysilicon layer as well as able to provide surface modification by undergoing protonation and deprotonation process. Surface analysis using SEM were used for surface morphology analysis. Different types of pH solution provide different resistivity and conductivity towards polysilicon surface. In addition, voltage, current, conductance against pH level are characterized and compared. Alkaline solution has the higher current as compared to acidic. This was due to the polysilicon layer contains more holes which are easily being attracted by – SiO⁻ to the surface and hence, forming a strong channel from source to drain. Results obtain reveal a linearity of pH measurement with a corresponding sensitivity of 4.65 nS/pH.**

Keywords— *polysilicon; nanowire; RIE; LPCVD; pH buffer.*

I. INTRODUCTION

Electronic devices are in the top hierarchy in the nanotechnology trend. It is very important as it has the potential to change every part of the human life. To achieve the nano scale, which ranges from 1 nm to 100 nm, advance machines and technologies are needed such as Reactive Ion Etching (RIE) and Low Pressure Chemical Vapour Deposition (LPCVD). By achieving reduction in size of a device, several significant factors will be reduced as well, such as size, weight and power consumption. Nano structures are able to work in handheld devices where sensing applications include medical, biochemical and food inspection. Nano structures are suitable as chemical and biological sensing element due to theirs high surface-to-volume ratios and quantum confinement effects [1].

Silicon was chosen as a material in semiconductor because of its high resistance to sustain high temperatures. Besides that, silicon has a tight bond (4 valence bond) which allows atoms to attach to one another easily. In the process of doping, silicon is easily doped with either n-type (phosphorus) or p-type (boron) elements. The polysilicon which are also know as polycrystalline silicon is formed from silica stone which has a purity of 99.99 %. Polysilicon bond strength is strong even when it is in the nanoscale. Apart from having superior electrical characteristics and excellent device stability [2], the use of polysilicon is promising for charged (bio)chemical species detection [3].

There are two approaches that can be used to fabricate nanowire which are the top-down approach and the bottom-up approach. The top-down approach was used in this fabrication due to it being a more conventional and advance method to produce a nano structure and to improve nanowire uniformity and sensitivity [4]. The top-down method starts with a large structure followed by patterning and etching the smaller features, hence producing a nanostructure [5]. Top-down approach usually uses RIE etching, which are costly to avoid any unsuspected etching profile [3], [5], [6]. However, in bottom-up approach, atoms assemble one at a time in predetermined order or metal catalytic growth [3]. It also provides atoms or molecules in the structure to be held covalently which are strong and reliable. The polysilicon nanowire in this project was fabricated using the top-down approach.

pH measurement devices are widely used in teaching laboratories as well as in industries. pH is a very important parameter that needs to be measured and controlled. Most liquids need to be tested to determine their nature which is, either acid or alkaline. The acidity or alkalinity of an aqueous solution is determined by the relative number of hydrogen ions (H^+) or hydroxyl ions (OH^-) present in them [7]. pH represent the hydrogen ion (in moles) per liter per liquid.

The operation principle of a nanowire sensor is that there are charges below the surface of the polysilicon layer which are causes by chemical reaction with the solution [8]. Hence, this causes a channel to be formed between the source and drain. 3-aminopropyltriethoxysilane(APTES) is able to modify the surface of the polysilicon or silicon oxide to provide a surface that can undergo protonation and deprotonation [9]. APTES contain $C_9H_{23}NO_3Si$ which are able to form covalent bonds with the polysilicon surface. In this paper, we demonstrate an undoped polysilicon wire used to detect the performance of the device in changes of pH sampling.

II. EXPERIMENTAL PROCEDURE

A. Device Fabrications

A bare p-type wafer of 4 inches was prepared and cleaned, followed by oxidation process. Silicon oxide was grown on top of the wafer with thickness about 178.98 nm. It was "estimated that for every 1 μm of SiO_2 grown, about 0.46 μm of silicon is consumed" [10]. After the oxidation process, polysilicon was deposited on top of the silicon oxide layer by using Low Pressure Chemical Vapor Deposition (LPCVD). The thickness of polysilicon layer was about 79.55 nm at 400 °C with 80 sccm of silane gases supplied. Complete deposition of polysilicon layer is as shown in Figure 1a. The complete 2 layered of wafer was cut into 12 dies using a diamond saw. Fabrication methods for all dies were the same. Next, the dies were taken to photolithography process. About 1 ml of photoresist was dripped on each die. Static coating was used so that the photoresist was thick enough to allow development process to be carried out on the wire. The chrome mask pattern that was designed was transferred to the surface of the wafer by UV light. Exposure time was 10s using the Mask Aligner. Then the die was dipped into the developer to remove the photoresist. After 3 s of developing, the die was taken to hard bake for 90 s to harden the photoresist. Developing process must be controlled due to the very small size of the wire. This factor makes the wire more prone to be over-developed over the area which will bring about resist profile problems [6]. The develop-bake process was repeated until unwanted photoresist was removed and the wire was clearly seen under the High Power Microscopy (HPM). The end of developing process is as shown in Figure 1b. Reactive Ion Etching (RIE) was the next step to remove parts of the polysilicon layer that were not blocked by the photoresist. The RIE recipes used was 50 sccm of SF 6, 30 sccm of O_2, 1.0 Pa of pressure, 250 V bias and etch for 10 s. The device after RIE and removal of photoresist is as shown in Figure 1c. After that, each die underwent metallization. Aluminum (Al) was deposited by using the thermal evaporator machine. After metallization, all dies were taken to photolithography process again for transferring pad image which was designed in different parts of the first chrome mask. About 1 ml of photoresist was dripped on the die followed by expose to UV light for 10 s. After transferring the image, each die was dipped into the developer solution for about 90 s to remove the softened photoresist. After that, each die was dipped into Al etchant to remove excessive Al that was not cover by photoresist. Then, it was cleaned with acetone followed by Deionised water (DI water). This step was to remove the photoresist residue on top of the Al pad. The complete device is as shown in Figure 1d. Finally, the device was taken for characterization by dripping solutions of different pH which were 2, 4, 7, 10 and 12 by using the Microprobe system with picoampere.

B. Sample Preparations

The device was cleaned with acetone and DI water before characterization process starts. 3-aminopropyltriethoxysilane (APTES) solution with 2 % of ethanol was dripped onto the wire surface for 3 hr so that there are surface modifications on polysilicon. Then, the device was rinsed several times in ethanol before baked at 100 °C for 10 min. The various pH buffer solutions were tested with litmus paper before characterization started to confirm the solution pH value as in Figure 2a. The litmus paper was dipped into the solution as shown in Figure 2b. After that, the colour change in the litmus paper was compared with the reference colour on the box as in Figure 2c. The device was probed on the probing station with the source and the drain connected as in Figure 2d. First characterization was done without any solution (i.e. air). Results obtained on air will be used to compare with other pH solutions. Every drop of solution was fixed to about 0.50 ml by using a single-channel micropipette on to the wires as shown in Figure 3. Next, DI water was dripped to test, followed by pH7, pH 10, pH 12, pH 4, and pH 2. In between before changing the testing solution, acetone and DI water was recommended to be used to clean the device surface due to the acetone properties that are able to clean the device surface.

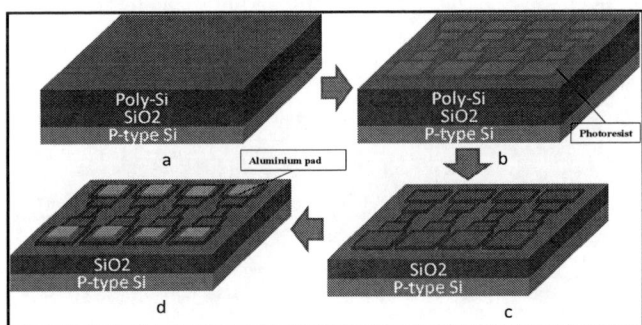

Figure 1: Summary of the process flow for a fabrication of polysilicon nanowire. a) After deposition of polysilicon and SiO_2. b) After developing of photoresist. c) After RIE and removal of photoresist with acetone. d) After etching of Al and formation of pad.

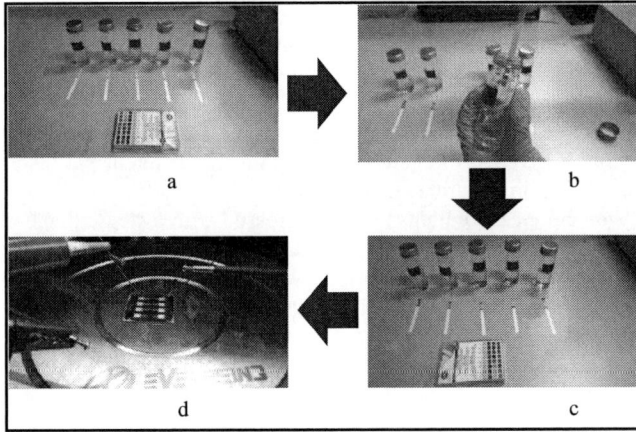

Figure 2: Process flow for preparation of pH buffer solution and characterization. a) Litmus paper was prepared. b) Litmus paper was immersed into each solution. c) Litmus paper changes color. d) Probe was connected to the source and drain of the device.

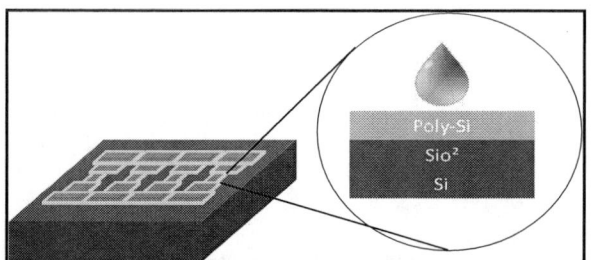

Figure 3: Schematic diagram for dripping of pH buffer solution on top of the modified polysilicon surface

III. RESULTS AND DISCUSSION

A. Surface Analysis

The normal developed resist was taken under High Power Microscope (HPM) was as shown in Figure 4a. Besides normal developed, there were several resist problem occur if development process not properly controlled [6]. Visual inspection was taken once the device was completely fabricated. Surface topology image was taken using Scanning Electron Microscopy (SEM) as shown in Figure 4b. Further reduction of the width can be achieved by using trimming process through deposition of silicon nitride as masking layer [5].

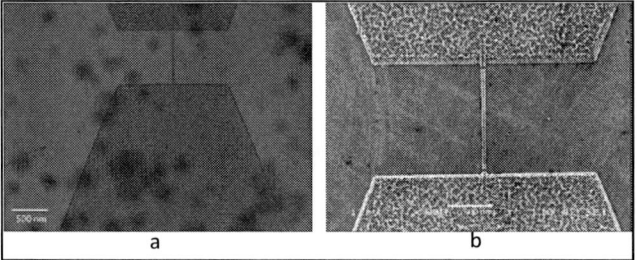

Figure 4: Images captured by using a) HPM and b) SEM.

B. Electrical Characterization

Once the device was completely fabricated, testing and measuring process starts. Before APTES, the device does not react to any pH buffer solutions and the results obtained were all the same. After APTES, the device was tested with air, pH2, pH4, pH7, pH10 and pH 12. The results were used to plot a Current-Voltage (I-V) graph for the different pH tested as shown in Figure 5. Drain voltage, V_{ds} varied from 0 V to 1 V as the measurements taken progressed with a step voltage of 0.05 V. Further increase in voltage exceeding 1V will causes the device to enter the breakdown region. From the results obtained, it can be seen that every pH and air overlaps each other at 0 V to 0.3 V. This indicated that no reaction occurred on the electrons and holes due to the small voltage supplied. pH buffer solution starts to deviate from each other above 0.3 V. It increased with the current, I_{ds}. pH 12 has the steepest graph compared to the others. It has a current of 481 nA at 1 V while pH 2 has a current of 397 nA at 1 V. The trends are in agreement with the previous reported findings [1]. This shows that there is a mixed surface functionality of the modified polysilicon nanowire between acidic and alkaline. APTES that are covalently linking on the polysilicon surface results a deprotonation and protonation of the $-NH_3$ and $-SiOH$ groups [9]. For alkaline solution, a high concentration of OH^- will caused the $-SiOH$ groups to deprotonate to $-SiO^-$ [9][1]. However for acidic solution, a high concentration of H^+ will caused the $-NH_2$ to protonate to $-NH_3^+$ [9][1]. These mechanisms are clearly shown in Figure 6. As $-SiO^-$ formed on the surface of undoped polysilicon, large amounts of holes are attracted to it. Highest current was obtained for pH 12 shown in the graph in Figure 5 proved that the amounts of holes are more than the amounts of electron in the polysilicon structure. At 1 V, pH 2 and pH 4 overlaps each other. This is because the amount of electrons attracted by $-NH_3^+$ groups is about the same. Air has the highest resistance because there were no chemical reaction that took place on the surface of the polysilicon and thus, no formation of the channel. Current obtained for air is relatively small which is only about 50.1 nA at 1 V.

Conductance of each pH buffer solution was measured and a graph was plotted, as shown in Figure 7. The conductance increases with the pH value. The conductance starts off with at 234.2 nS for pH 2 and followed by a slight increase for pH 4 at 235.9 nS. As explained before, the amount of electrons attracted is about the same for pH 2 and pH 4, hence forming an approximately same channel. However, there is a sharp increase in conductance after pH 4. The gradient of the slope is about 4.65 nS/pH. It can be increased with more effort in controlling polysilicon surface with reduction of wire's surface-to-volume ratio. Further improvement on sensitivity can be achieved with smaller nanowire dimension. As previously mentioned, the deposition of a layer of silicon nitride (Si_3N_4) as a masking layer is expected to improve the wire's trimming process [5].

The Conductance-Voltage (C-V) relationship was plotted as shown in Figure 8. It was clearly shown that as voltage, V_{ds} increases, the conductance also increases. Different voltage levels will cause an increase in conductance value due to the formation of strong channel from the source to drain. As for pH 12, more holes are attracted to the surface by $-SiO^-$ which causes a stronger hole channel to be formed. On the other hand, weaker electron channel are formed for pH 2. All buffer solution showed an increase in conductance except for air. As the voltage increase from 0.25V to 1V, the conductance of air is decreases due to the increase of the resistance. Without any solution that causes the holes and electron to take effect, the current that flows from the source to drain is less. Neither holes nor electron arranged themselves to make a channel for the current to pass through. The pH 7 buffer solution is in neutral state which means the amount of holes and electron attracted to the surface are approximately the same. Thus, the curve of pH 7 in Figure 8 is in between pH 10 and pH 4.

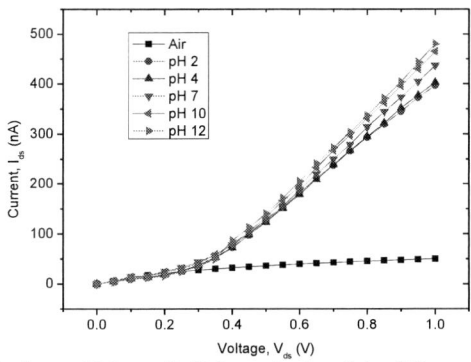

Figure 5: Current-Voltage (I_{ds}-V_{ds}) graph plotted in different pH buffer solution including air of a p-type wafer.

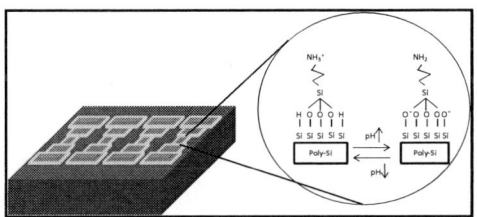

Figure 6: Schematic diagram of APTES-modified surface mechanisms after dipped with acidic or alkaline. Chemically boding with pH solution can easily be formed due to the superior effect of APTES.

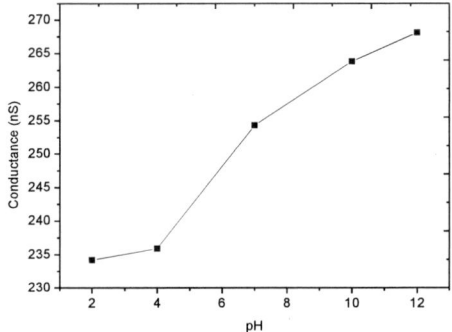

Figure 7: Graph plotted with conductance against pH 2, 4, 7, 10 and 12.

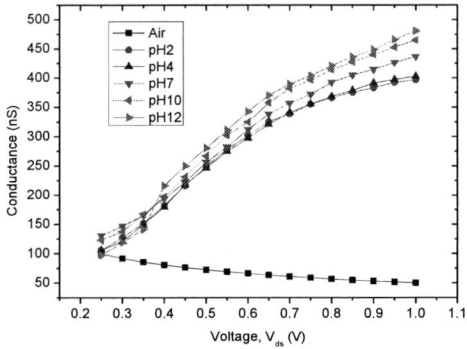

Figure 8: Effect of different pH buffer solution on undoped polysilicon NW. Conductance-Voltage graph are plotted to show difference in conductance against pH with increasing voltage.

IV. CONCLUSION

We have successfully demonstrated the fabrication and characterization of polysilicon nanowire as a pH sensor. Simple techniques such as conventional lithography method, and wet etching were applied. Current, voltage and conductance were measured when the wire were exposed to different pH buffer solution. Solution with pH 12 has the higher current as compared to pH 2 for a p-type wafer due to the large amount of holes that were attracted by $-$ SiO to the surface of an undoped polysilicon. In this work, the sensitivity of the device is 4.65nS/pH. However, the sensitivity can be improved by using APTES, thus making it suitable for bio-sensor applications. Further improvement on sensitivity might be achieved with depositing Si_3N_4 as masking layer.

ACKNOWLEDGMENT

We are grateful for the discussions that we have with our collaborators at the Institute of Nano Electronic Engineering (INEE) at University Malaysia Perlis. This work was supported by INEE at UniMAP, through the Nano Technology project.

REFERENCES

[1] S. Fatimah, A. Rahman, N. A. Yusof, U. Hashim, and M. N. Nor, "Design and Fabrication of Silicon Nanowire based Sensor," *Int. J. Electrochem. Sci.*, vol. 8, pp. 10946–10960, 2013.

[2] Y.-L. W. Chun-Yu Wu, Po-Yen Hsu, Chao-Lung Wang, Ta-Chuan Liao, Huang-Chung Cheng, "Polysilicon Nanowires Biosensors for pH Measurement and DNA Detection Utillizing High-k Dielectric Sensing Membrane," *Int. J. Electrochem. Soc.*, vol. 1, p. 1, 2012.

[3] F. Demami, R. Rogel, A. Salaün, and L. Pichon, "Electrical properties of polysilicon nanowires for devices applications," *Phys. Status Solidi*, vol. 3, pp. 1–4, 2011.

[4] M.-C. Chen, H.-Y. Chen, C.-Y. Lin, C.-H. Chien, T.-F. Hsieh, J.-T. Horng, J.-T. Qiu, C.-C. Huang, C.-H. Ho, and F.-L. Yang, "A CMOS-compatible poly-Si nanowire device with hybrid sensor/memory characteristics for System-on-Chip applications.," *Sensors (Basel).*, vol. 12, no. 4, pp. 3952–63, Jan. 2012.

[5] U. Hashim, S. A. B. Ariffin, and T. Adam, "Fabrication of Polysilicon Nanowires using Trimming Technique," *J. Appl. Sci. Res.*, vol. 8, no. 4, pp. 2175–2186, 2012.

[6] M. N. M. Nuzaihan, U. Hashim, T. Nazwa, and a. R. Ruslinda, "Fabrication of poly-silicon microwire using conventional photolithography technique: Positive resist mask vs aluminium hard mask," *RSM 2013 IEEE Reg. Symp. Micro Nanoelectron.*, pp. 211–214, Sep. 2013.

[7] A. D. Sheet, "The Theory of pH Measurement," pp. 1–8, 2010.

[8] W.-K. Ho, Y.-Y. Ho, Z.-R. Lin, C.-C. Hsu, and C.-L. Dai, "The pH sensor with the poly-Silicon nanowire," *2013 Conf. Lasers Electro-Optics Pacific Rim*, pp. 1–2, Jun. 2013.

[9] Y. Cui, Q. Wei, H. Park, and C. M. Lieber, "Nanowire nanosensors for highly sensitive and selective detection of biological and chemical species.," *Am. Assoc. Adv. Sci.*, vol. 293, no. 5533, pp. 1289–92, Aug. 2001.

[10] N. Taib, U. Hashim, A. Saifullah, T. S. Dhahi, and J. K. Setar, "Polysilicon Nanogap capacitive biosensors for the pH detection," *2011 IEEE Reg. Symp. Micro Nano Electron.*, vol. 1, pp. 250–252, 2011.

Annealing Temperature Effect on Nanostructured TiO$_2$ Films

[1]I.H.H. Affendi, [1]M.S.P. Sarah, [1,2]M. Rusop
[1]NANO-ElecTronic Centre (NET),
[2]NANO-SciTech Centre (NST), Institute of Science,
Universiti Teknologi MARA,
Shah Alam, Selangor, Malaysia
E-mail: irma.affendi@gmail.com, @ rusop@salam.uitm.edu.my

Abstract— The synthesization of TiO$_2$ sol-gel, by using titanium dioxide nano powder as precursor. In the production of nanostructured TiO$_2$, the annealing temperature for the film is differed to clarify the use of different temperature in annealing the film, the best temperature to get a good mobility in the current flow can be found. There are 7 samples with different drying temperature and different annealing temperature. The one with the highest IV characterization will be fabricated as thin film in the organic solar cell as the metal oxide film. As further discovered that the as-deposited without annealing sample have a quite thick film that it could not be characterized by Atomic Force Microscopy (AFM). The highest point of the current at 10V in the IV graph is 500°C annealing temperature of 6.06E-9 A which then makes it the highest in conductivity at 3.37E-6 Sm^{-1}. The current-voltage (I-V) measurement is used to study the electrical resistivity behaviour, hence the conductivity of the film to suit organic solar cell application.

Keywords—TiO$_2$ film; annealing temperature; drying temperature; sol-gel; IV characteristics;

I. INTRODUCTION

The titanium dioxide or titania (TiO$_2$) commercial exploitation started in the 20th century and it is rapidly increasing after the discovery of the sulphate processing in 1916 and then came the chloride process in 1958 [1].

Nowadays TiO$_2$ is still widely used as pigment, especially in paintings for white paints, anti reflection coatings for windows, anti-microbial coatings in hospitals, writings on foods e.g. the M&M chocolate, cosmetics and even now they use this TiO$_2$ sol-gel in organic solar cell as the n-type material. As in organic solar cell, TiO$_2$ is the inorganic material that works to produce an electron - hole pair when in contact with sun rays. Actually the uses of solar radiation (sun light) for energy is an ancient concept, but because of the recent concerns about the environmental impact and limitation of fossil fuels have made solar energy an increasing popularity.

There are many ways in depositing TiO$_2$ thin film such as thermal oxidation [2], chemical vapor deposition (CVD) [3], plasma oxidation [4], pulse laser deposition [5] and sol-gel process [6]. Sol gel spin coating method was chosen because it is simple and cheap since it processed from solutions and the process was easy to conduct. Since the spin coating technique has a simple operation, cheap because it uses solution and it is easy to run the process. Still, there are a few disadvantages of spin coating which might be quite a problem, especially on wastage for a big manufacturing company since for a typical spin coating processes utilize only 2-5% of the material dispensed onto the substrate [7], while the remaining 95% - 98% is flung off into the coating bowl and disposed. Not only the prices of the raw material increased substantially, but disposal costs are increasing as well [8].

In this investigation, the solution was based on previous research of the 0.01 molarity and based on the 3 deposition time thickness. The solution of TiO$_2$ is made from TiO$_2$ nano powder that acts as a precursor. Then the solution was deposited on the 2.5cm x 2.5cm microscope glass as the substrate by dropping the solution to the substrate [9]. In this experiment, the sol-gel technique is different from since in this process the TiO$_2$ solution was dropped on the substrate and let dried, whether on heat or on room temperature and then annealed to decrease the material wastage.

Then the development of the thin film characteristics was measured for the thickness by Surface Profiler (SP), Elemental-Dispersive X-ray (EDX) to measure the elemental composition and the electrical properties were measured by using current-voltage (IV) measurement and have been recorded as table and figures.

II. METHODOLOGY

A. Materials

The TiO$_2$ solution was prepared using all chemicals in the form as purchased and packaged by Sigma Aldrich, without further alterations or purification. Absolute ethanol (100%)

acts as the solvent, TiO_2 powder (99.5%) act as a precursor, Polyethylene Glycol (PEG) act as a stabilizer and Triton-X-100 as a surfactant and it may also act as a wetting agent.

B. Preparation of glass substrate

The microscope glass slide was used as the substrate in this experiment. The glass substrate was cut in 2.5cm x 2.5cm squares. Then, it is treated with acetone, methanol and distilled water in an ultrasonic bath for 10 minutes each and the sequence of acetone, methanol and distilled water was repeated 2 times. Lastly, the substrate was blow with Argon gas for drying purpose and to clean the substrate from impurities.

C. Synthesis of pure TiO_2 solution by sol-gel method.

The preparation of TiO_2 film was done using a sol-gel technique, TiO_2 nano powder was weighted as in the molarity calculation for 0.01M. Absolute ethanol was poured into a small 25ml beaker with enclosure and magnetic stirrer to be mixed with the weighted TiO_2 nano powder. 30minutes mixing amount of time to make sure that TiO_2 nano powder and absolute ethanol was mixed well. Then Triton X-100 was added and stirred for 2 hours. After the PEG was added and sonicated for another 30 minutes. The aging time was recorded at 24 hours time.

D. TiO_2 Films

The TiO2 film was spin coated onto the glass substrate that was cut 2.5cm x 2.5cm squares by dropping method. The dropping was done in approximately 5 seconds. The TiO_2 film was done in 3 layers due to previous experiment on layer deposition. Partly every layer will be heated at 80°C for 10 minutes. The purpose of heating the TiO_2 film is to vaporize the solvent or to volatile the material and to remove the contaminations. The process was repeated by the number of layers of the TiO_2 film. As prepared, part of the sample was annealed at 400°C, 425°C, 450°C, 475°C, 500°C in 1hour to restructure the lattice through atomic deformation and another part is as deposited and not annealed. The temperature 400 is chosen due to the fact that the films annealed below 400 are amorphous [10].

E. Characterization

The elemental composition of the film was measured by EDX to see if there are any impurities in the TiO_2 film formation. Then the IV measurement was used to see the resistivity and the conductivity of the thin film and surface profiler (SP) was used to measure thin film thickness.

III. RESULT AND DISCUSSION

TABLE I. THICKNESS BY SP AND IV CHARACTERIZATION AT 10V

Annealing Temperature (°C)	Thickness of TiO_2 film (μm)	IV measurement at 10V (A)
Dry 24°C 0°C Anneal	31.927	9.76E-10
Dry 80°C 0°C Anneal	25.276	5.73E-09
Dry 80°C 400°C Anneal	14.202	5.55E-09
Dry 80°C 425°C Anneal	12.468	5.79E-09
Dry 80°C 450°C Anneal	10.935	5.83E-09
Dry 80°C 475°C Anneal	8.795	5.81E-09
Dry 80°C 500°C Anneal	6.297	6.06E-09

In Table 1, the thickness was measured by SP and has been done several times after film depositions' done to get the average thickness of the film and this goes to all film characterized. Since the film is not uniformly built, the thickness is not uniform in every part of the film because of the sample's square shape and the depositing method by spin coating. Then for the current measurement, it is measured by the two point probe IV measurer. IV was measured several times to get the best average result for each film, this happens because of the non-uniformity of the film.

In terms of thickness in this decade, it is likely to have a thin film which 500°C would be favored by having 6.297μm thickness. Then it is the roughness that caught the attention when 500°C has the smoothest surface out of others at 3.57nm. Next row in the table is the IV characterization at 10V, as seen the IV characterization is increasing with the annealing temperature and this goes to the conductivity and the resistivity that will be clarified next.

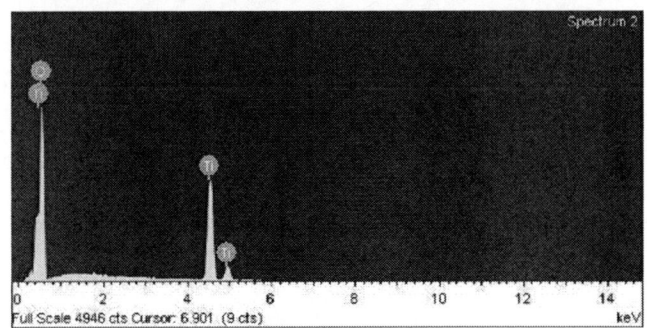

Fig. 1. EDX image of the TiO_2 film.

This Fig. 1 above shows the EDX image of the film. In the EDX image, the composition of the film can be seen and personalize that the film contains only Ti and O. In EDX analysis the existing properties of micro contaminant could vary from electron beam energy during EDX analysis [10]. This is due to the extremely short wavelength of the electron beam with relative strong energy. While 1 and 30 keV acceleration voltages are applied, the wavelength of the electron beam is 0.3876 and 0.0698 Å, respectively (11). So, the characterization has to be done by many instruments in order to get an accurate answer for the chemical part of the review. As for this investigation the focus is to get the best IV characterization.

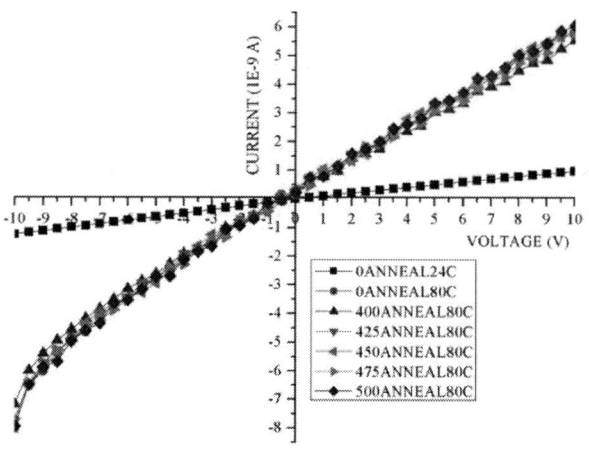

Fig. 2.   IV of dried film without annealing in room temperature and in 80°C, and dried in 80°C then anneal in 400°C, 425°C, 450°C, 475°C and 500°C.

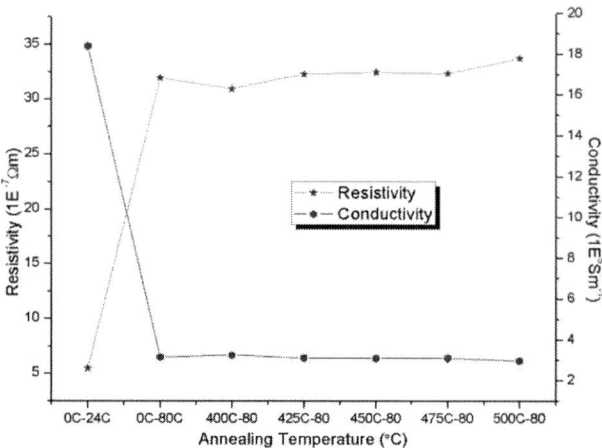

Fig. 3.   Graph of the conductivity and resistivity of nanostructured TiO_2 based on the IV measurement done for all sample.

Fig.2 above is the current-voltage (IV) characteristics of TiO_2 thin film with different annealing temperature. The IV-measurement was done in a 24°C temperature room, using the two-point probe method [12]. Metal contact was deposited by sputter coater to make the electrode in order to measure the IV since gold was used as the metal contact is because of gold is the standard metal contact used as the finishing electrode onto samples. This is due to it forms a good ohmic contact with materials [13]. The metal contact that was deposited is a 15nm thin gold film with a thickness that was deposited in 1 minute. IV measurement starts at 10V to -10V. Before the IV measurement could and would become 10V to -10V, we have tried whether the film is tough enough to stand the 20V difference and in the same time having a linear graph. It starts with 0V to 3V, then upgraded until the graph is not linear and

the film starts failing. Planted on the graph is the tough film and the highest IV measurement are at 500°C and the lowest is the as deposited film that really isolated from the other.

Based on the IV, resistivity and conductivity were calculated and was recorded in Fig. 3 graph which have a quite high resistivity and for conductivity, the graph is inverse with the resistivity. The one that have high resistivity and conductivity is again the 500°C. Since it has the thinnest film from others, this as a result of high annealing temperature means lots more of the solvents and impurities dispersed leaving a purely TiO2 film on glass substrate.

IV.   CONCLUSION

Sol-gel dropping-before-spinning technique (spin-coating) was chosen as the deposition process because of its simple and cheaper method in order to produce TiO_2 nanostructured film. The thickness of the films produced was decreasing by the increasing of annealing temperature and this was supported by the SP reading of each sample. The film was then measured by the two point probe for IV characterization and indeed the thickness supported by IV characteristics when compared with other films, the best film is the one annealed at 500°C with thickness of 6.297μm and IV at 10V is the highest of 6.06E-09A.

ACKNOWLEDGMENT

This work was supported by the Research Management Institute (RMI), NANO-Electronic Center (NET) and NANO-SciTech Centre of Universiti Teknologi MARA (UiTM).

REFERENCES

[1]   Reck, E.; Richards, M. Titanium dioxide — Manufacture, environment and life cycle analysis: The tioxide experience. Surf. Coat. Int., Part B: Coat. Trans., **1977**, 80, 568-572.

[2]   I. Saeki, Okushi, H. Konno, R. Furuichi, J. Electrochem. Soc. 143 (1996) 2226.

[3]   S.A. Campbell, H.S. Kim, D.C. Gilmer, B. He, T. Ma, W.L. Gladfelter IBM J. Res. Dev. 43 (1999) 383.

[4]   J. Clerk Maxwell, A Treatise on Electricity and Magnetism, 3rd Ed., vol. 2. Oxford: Clarendon, 1892, pp. 68-73.

[5]   Tinoco J C, Estrada M and Romero G 2003 Microelectron. Reliab. 43 895.

[6]   Z. Wang, U. Helmersson, P.O. Kall, Thin Solid Films 405 (2002) 50.

[7]   Haaland, P; McKibben, J.; and Paradi, M. "Fundamental Constraints on Thin Film Coatings for Flat-Panel Display Manufacturing," SID, p. 79.

[8]   Snodgrass, T. Newquist, C. "Extrusion Coating of Polymers for Next t-Generation Large-Area FPD Manufacturing", DISPLAY MANUFACTURING TECHNOLOGY CONFERENCE; 1; 40-44, 1994

[9]   Snodgrass, T. and Newquist, C. "Extrusion Coating of Polymers for Next Generation Large Area FPD Manufacturing," SID, p. 40.

[10] Chien-Yi Huang, "Chemical characterization of failures and process materials for microelectronics assembly.", Microelectronics International, Volume 26, Number 3, pg 41–48, 2009

[11] Lee, R.E. (1993), Scanning Electron Microscopy and X-ray Microanalysis, PTR Prentice-Hall, Englewood Cliffs, NJ.

[12] S..K..M Maarof, S. Abdullah, M. Rusop, "Synthesizationof Nanostructured Titanium Dioxide at Low Molarity of Solgel Process", MOIME 2013, Material Science and Engineering 46, 2013

[13] A.Y. Orala, et al., The preparation of copper (II) oxide thin films and the study of their microstructures and optical properties, Materials Chemistry and Physics, 83 (2004) 5.

Fabrication and Characterization of Polysilicon for DNA Detection

Y. M. Ang, *M. K. Md. Arshad, K. L. Foo, M. Nuzaihan Md. N., A.H. Azman U. Hashim
Institute of Nano Electronic Engineering (INEE),
Universiti Malaysia Perlis (UniMAP)
01000 Kangar, Perlis, Malaysia.
mohd.khairuddin@unimap.edu.my

Abstract — **We present the fabrication and electrical characterization of polysilicon and their properties with application in biomolecule sensors for DNA detection. Conventional photolithography technique was used to fabricate the DNA detection structure for two different wafer substrate i.e. N- and P-type. The fabrication processes involve of deposition, etching and oxidation to achieve the final structure. Surface modification, immobilization and hybridization were executed prior to electrical characterization by using cyclic voltammetry. It was observed that the modified surface with APTES achieved the highest current for both p- and n-type wafer with changes from 0.52 µA to 3.32 µA and from 0.57 µA to 2.52 µA respectively. Moreover, redox current of hybridization is observed approximately 22 % and 10 % larger than immobilized electrode for p- and n-type wafer.**

Keywords—Polysilicon thin film; DNA detection; Surface modification; immobilization; hybridization; Cyclic Voltammetry

I. INTRODUCTION

The completion of human genome study has encouraged the development of sensors for DNA diagnostics and forensic medicine. For example, with the biological properties, detection of cancer can be identified by the genetic changes during tumorigenesist. Moreover, infectious disease which is the first cause of death in worldwide are detectable by genetic. The changing of gene in DNA can be sensed. With the discovery of DNA, the cases of illness, disability and death are significantly reduces [1].

DNA (deoxyribonucleic acid) is an essential biological material and it is the genetic material of world living cells as well as viruses. In structure of DNA, there are four choices of nucleotides: Adenine, Guanine, Thymine and Cytosine. Adenine (A) and Guanine (G) are the purine bases whereas Thymine (T) and Cytosine (C) are pyrimidine bases. The bases DNA is from two polynucleotide strands with an alpha double helix. The genetic code is based on one of the strands. DNA is a genetic that enable to self-replicating or it can make copy of itself and because of the replication, the DNA allows the genetic information to pass devotedly to next generation.

During sensing of DNA, DNA hybridization plays an important role in identifies the cause of disease and detects the present of bacteria or virus. DNA detection is good environment monitoring and food safety [2]. The conventional sensing methods are poor because the time taken of processing is long. Thus, develop a sensitive, selective and simple to use DNA sensors are requesting to overcome the conventional sensor [3].

A biochemical receptor is needed to be attached to a sensor to translate the result into a measurable signal [4]. However, there are some challenges while the main challenge is the sensitivity and the specificity of the transducer and the receptor which mostly related to immobilization of the specific sensing element. Transducer in biosensor is used to detect DNA hybridization as well as piezoelectric, optical and electrochemical. Well, there is a lot of techniques has been developed for identification of DNA. PCR (Polymerase Chain Reaction) is one of the techniques which make large quantities of copies of a gene. Other examples are fluorescence based optical DNA detection, nanoparticle-based electrical DNA detection, redox-active reporter molecules, enzymes as electrical transducers, electrochemical methods, Surface Plasmon Resonance (SPR), surface stress measurement and Quartz Crystal Microbalance (QCM) [5],[6],[7].

In this work, polysilicon (Poly-Si) is used as the detection material to sense pork DNA. This is because polysilicon allows for highly sensitive and simple detection method. Polysilicon also has been study by many researchers in biosensor field as detect virus, glucose and others [8],[9].

II. EXPERIMENTAL PROCEDURE

A. Fabrication Preparation

In this experiment, the N-type (resistivity 0-100 ohm.cm, thickness 500+/-25 µm) and P-type (resistivity 1-20 ohm.cm, thickness 475-575 µm) wafer with diameter of 100 mm were used as a substrate.

Firstly, the wafers were cleaned using Buffered Oxide Etch (BOE). Next, a thick layer of oxide was grown in furnace using wet oxidation technique for 60 minutes. The average of 189.5 nm and 198.7 nm thickness of oxide layer was obtained for N-type and P-type wafers respectively using filmetric thin film analyzer. After that, polysilicon was deposited on the oxide layer using Low Pressure Physical Vapour Deposition

(LPCVD) technique at 650°C with 80 sccm silane gases supplied. A thickness of average 170 nm was obtained for both type of wafer by using filmetric thin film analyzer. Following next, gold was deposited on polysilicon which acts as a contact. Titanium was used as promoter to improve adhesion of gold on polysilicon. Length of 0.2 cm of titanium and length of 0.5 cm of gold were used in the process. Physical Vapour Deposition (PVD) technique was applied to deposit gold at 1050°C.

Then, pattern transfer process started with coated positive photoresist and soft baked for 90 seconds at 90°C to improve the adhesion of the photoresist. Next, expose the wafers under mask aligner machine for 10 seconds. 60 seconds of developing time was needed to form the desired structure. After that, hard bake was executed for 60 seconds with hot plate at 100°C to strengthen the resist. The fabrication process was followed by etching technique. Gold thin film was etched by aqua regia for 100 seconds with the ratio 3:1 mixture of hydrochloric acid HCl and nitric acid HNO_3. The etching process went further with removed titanium by mixture of ammonia (NH_4OH): hydrogen peroxide (H_2O_2): deionized water (DIW) with ratio 1:2:5. The time taken was 60 seconds. Strip photoresist with acetone was done after etching. The sample preparation process was ended with cutting the wafer into 1.5 cm x 0.5 cm sample size. Fig. 1 details out the fabrication process steps as explained above.

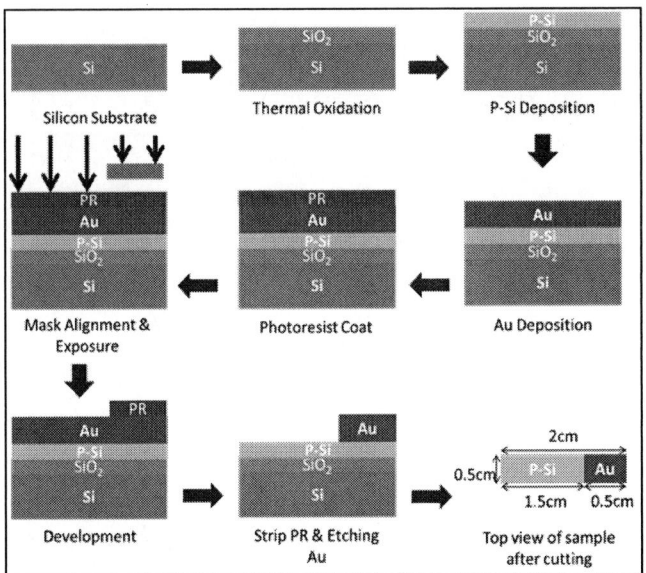

Fig. 1. Fabrication Process Steps

B. Characterization Preparation

Characterization processes for this experiment included surface modification, immobilization and hybridization. Fig. 2 shows the characterization process of the sample. Prior to DNA detection, two types of surface modification such as 3-Aminopropyltriethoxysilane (APTES) and gold nanoparticles (GNPs) was coated on the surface of polysilicon. The chemical reaction among polysilicon, APTES and GNPs were illustrated in Fig. 3.Surface modification began with coated

the polysilicon surface with 10µl of APTES which prepared by mixing of APTES: Ethanol: DIW with ratio 2:93:5 and incubated at room temperature for two hours. After that, the samples were washed by ethanol for 3 times and gently dry with blower to form APTES/Poly-Signet, APTES/Poly-Si electrodes were modified with GNPs to aid for immbolization of thiol-modified ssDNA probes onto each and also enhance the sensitivity of the DNA detection. The electrodes were immersed into GNPs solution for 30 seconds and dry on a hot plate at 150°C for 20 minutes. The step was repeated for 3 times to improve the reaction between APTES and GNPs.

After that, immobilization occurred with 2.5 µl of 10 µmol/L of probe DNA in phosphate-buffered saline (PBS) (pH 6.8, 50 mmol/L NaCl) was dropped on the modified electrodes surface to form gold-thiolate bond. 2.5 µl of ruthenium complex was added right after the probe DNA. The electrodes were then incubated at room temperature for 5 hours. Then, they were rinsed with PBS solution and DIW. Next, 2.5 µl of target DNA (10 µmol/L) and 2.5µl of ruthenium complex were dropped on the probe-treated surface to form double strained DNA (dsDNA) with known as hybridization. The electrodes with target DNA were incubating at room temperature for 4 hours and rinsed with PBS solution as well as DIW.

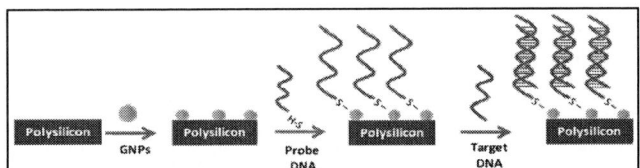

Fig. 2. Polysilicon thin films surface modification for DNA immobilization and hybridization process

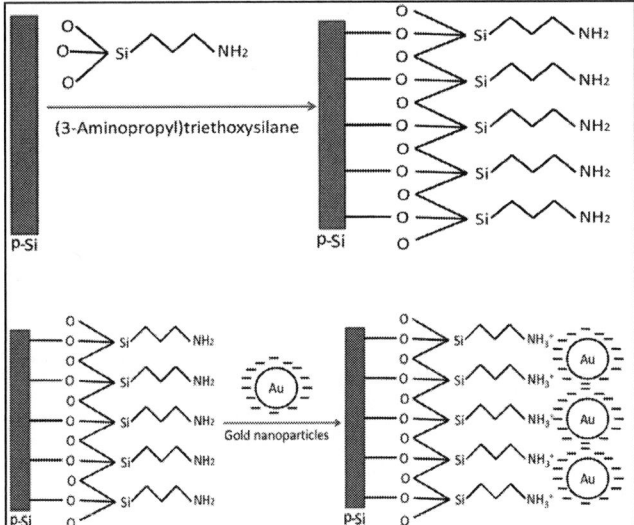

Fig. 3. Surface modification of Poly-Si with APTES and GNPs

C. Electrical Characterization Preparation

For electrical characterization, we only consider Cyclic Voltammetry (C-V) as a method to determine the DNA immobilization and hybridization detection due to easy access to the equipment. In this measurement, APTES/GNPs-modified polysilicon surface was acted as working electrode, platinum rod as counter electrode and Ag/AgCl as reference electrode. To prepare the electrolyte, 15 µl of 25 µmol/L of ruthenium complex was added into 60 ml of PBS buffer solution (pH 6.8) [10]. N-type and P-type modified wafer were used in C-V test. The measurement was conducted through potential range from 0.0 V to 2.0 V at 50 mV/s scan rate.

III. RESULT AND DISCUSSION

In this research, cyclic voltammetry was used to obtain information about redox potential and electrochemical reaction rates of analyte solution. The voltage was measured between working electrode and reference electrode, while current is measured between counter electrode and working electrode. The current will increase if the voltage is increased toward the electrochemical reduction potential of the analyte. However, the current decreases only if the increasing voltage past the reduction potential [11].

Fig. 4 shows the cyclic voltammograms of DNA detection using GNPs/APTES/Poly-Si electrode for P-type wafer. In this graph, initialize current obtained at 0 V was 0.57 µA. The bare electrode (without any surface modification) yielded least of redox reaction (i.e. 1.12 µA) while modified surface yielded the most redox reaction (i.e. 2.52 µA) when measured at 2 V. There was enhancement on the redox current after hybridization occurred as compared to that obtained from immobilization. We believe this might be due to huge amount of ruthenium interaction by hybridization DNA. For target DNA detection, the oxidation peak current for GNPs/APTES/Poly-Si electrode was increased from 1.22 µA (immobilization) to 1.49 µA (hybridization). The results demonstrated that hybridization had occurred with redox current ~22 % larger in comparison to the immobilized electrodes. The redox current is extracted based on formula given in Eq. 1.

$$\text{Redox Current, \%} = \frac{\text{Hybridization} - \text{Immobilization}}{\text{Immobilization}} \times 100\% \quad (1)$$

For N-type wafer, the trend of cyclic voltammograms was shown in Fig. 5. The initialized current measured at 0 V was 0.52 µA. The electrode without any surface modification yielded least of redox reaction as well with reading 1.67 µA at 2 V. The modified surface provided the most of redox reaction with reading 3.32 µA at 2 V. The oxidation peak current for GNPs/APTES/Poly-Si electrode went up from 2.02 µA (immobilization) to 2.22 µA (hybridization). This increment is caused by interaction among large amount of ruthenium produced by hybridization DNA. Based on Eq. 1, the redox current of hybridization was ~10 % greater than immobilized electrodes.

Comparing both type of wafer in term of cyclic voltammograms, they were in same trend which modified surface had the most redox current and bare electrode had the least of redox current. However, N-type wafer has the higher redox current in all the measurements as illustrated in Table 1. It markedly showed that N-type wafer has better in term of conductivity. Well, increment of redox current in hybridization electrodes indicated the ruthenium complex accumulated more in dsDNA with respect to ssDNA [12]. Our study showed that the accumulation of ruthenium complex was more on the electrode with dsDNA than the electrode with ssDNA. The reason might be due to the strong association of ruthenium complex (with positively charged) with the phosphate backbone (with negatively charged) and double helix of dsDNA. Resultant the accumulation of ruthenium complex was higher in the electrode with dsDNA as compared to the electrode with ssDNA, which was due to the present of unpaired electron [13].

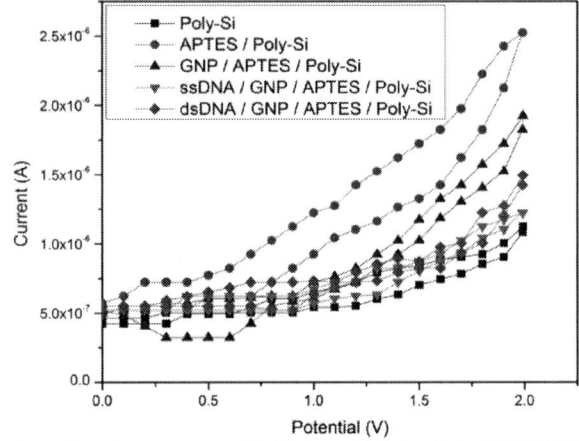

Fig. 4. Cyclic voltammograms of DNA detection using GNPs/APTES/Poly-Si electrode for P-type wafer.

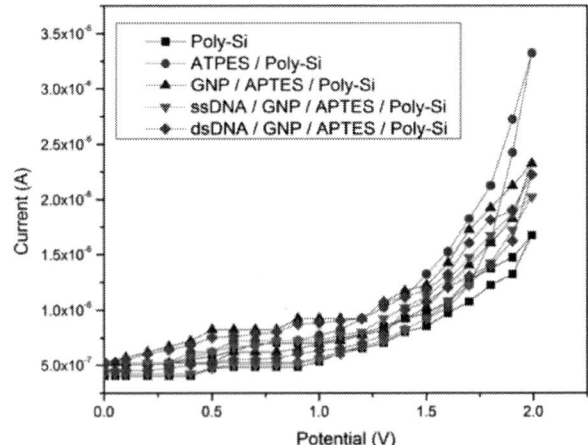

Fig. 5. Cyclic voltammograms of DNA detection using GNPs/APTES/Poly-Si electrode for N-type wafer.

TABLE I. REDOX CURRENT OF P-TYPE AND N-TYPE WITH VARIABLES SURFACE AT 2V

Surface \ Wafer	P-type (µA)	N-type (µA)
Poly-Si	1.12	1.67
APTES/Poly-Si	2.52	3.32
GNPs/APTES/Poly-Si	1.92	2.32
ssDNA/GNPs/APTES/Poly-Si	1.22	2.02
dsDNA/GNPs/APTES/Poly-Si	1.49	2.22

IV. CONCLUSION

In this study, we presented the fabrication and characterization of polysilicon for DNA detection for P- and N-type of wafers. It was found that DNA immobilization and hybridization on modified polysilicon surface for both type of wafer were successfully electrochemically investigated using cyclic voltammetry, suggesting its applications in low-cost bio diagnostics, forensic testing, food analysis and environmental monitoring. We demonstrated the thiol-modified ssDNA probes were successfully immobilized on the GNP/APTES-modified polysilicon surface for both type of wafer. DNA hybridization was detected voltammetrically with the present of ruthenium complex and measured sensitivity was up to ~22 % (P-type) and ~10 % (N-type) larger than immobilization. Both graphs showed the bare electrode gave least of redox reaction while modified surface (APTES/Poly-Si) surface gave the most redox reaction.

ACKNOWLEDGEMENT

The authors are grateful to the School of Microelectronic UniMAP and Institute of Nano Electronic Engineering UniMAP (INEE) for providing the opportunities and equipment facilities for the research. Appreciation goes to all the members in laboratory for their guidance and help. Without their guidance and ever encouraging advice, this research work would have not been possible.

REFERENCES

[1] M. E. Ali, S. Mustafa, U. Hashim, Y. B. Che Man, and K. L. Foo, "Nanobioprobe for the Determination of Pork Adulteration in Burger Formulations," *J. Nanomater.*, vol. 2012, pp. 1–7, 2012.

[2] M. E. Ali, U. Hashim, S. Mustafa, Y. B. Che Man, T. S. Dhahi, M. Kashif, M. K. Uddin, and S. B. Abd Hamid, "Analysis of pork adulteration in commercial meatballs targeting porcine-specific mitochondrial cytochrome b gene by TaqMan probe real-time polymerase chain reaction.," *Meat Sci.*, vol. 91, no. 4, pp. 454–9, Aug. 2012.

[3] U. Rant, K. Arinaga, S. Scherer, E. Pringsheim, S. Fujita, N. Yokoyama, M. Tornow, and G. Abstreiter, "Switchable DNA interfaces for the highly sensitive detection of label-free DNA targets.," *Proc. Natl. Acad. Sci. U. S. A.*, vol. 104, no. 44, pp. 17364–9, Oct. 2007.

[4] E. Palecek, M. Fojta, M. Tomschik, and J. Wang, "Electrochemical biosensors for DNA hybridization and DNA damage.," *Biosens. Bioelectron.*, vol. 13, no. 6, pp. 621–8, Sep. 1998.

[5] M. E. Ali, M. Kashif, K. Uddin, U. Hashim, S. Mustafa, and Y. Bin Che Man, "Species Authentication Methods in Foods and Feeds: the Present, Past, and Future of Halal Forensics," *Food Anal. Methods*, vol. 5, no. 5, pp. 935–955, Jan. 2012.

[6] M. E. Ali, U. Hashim, S. Mustafa, Y. B. C. Man, M. H. M. Yusop, M. F. Bari, K. N. Islam, and M. F. Hasan, "Nanoparticle sensor for label free detection of swine DNA in mixed biological samples.," *Nanotechnology*, vol. 22, no. 19, p. 195503, May 2011.

[7] K. L. Foo, "Fabrication and Characterization of ZnO Nanostructures for DNA Detection," Universiti Malaysia Perlis, 2013.

[8] A. Rahman and S. Fatimah, "Design and Fabrication of Silicon Nanowire based Sensor.," *Int. J. Electrochem. Sci.*, vol. 8, pp. 10946–10960, 2013.

[9] P. Yen, C. Huang, Y. Huang, H. Hsueh, C. Lin, C. Ho, F. Yang, H. Tsai, H. Liao, Y. Juang, C. Wang, S. Lu, and C. Lin, "The Polysilicon Nanowire based Biomolecular Sensor," pp. 2–4.

[10] K. L. Foo, U. Hashim, H. Prasad, and M. Kashif, "Fabrication and characterization of IDE ZnO thin films using sol-gel method for PBS solution measurement," *2012 10th IEEE Int. Conf. Semicond. Electron.*, pp. 736–739, Sep. 2012.

[11] D. Grieshaber, R. MacKenzie, J. Vörös, and E. Reimhult, "Electrochemical biosensors-Sensor principles and architectures," *Sensors*, pp. 1400–1458, 2008.

[12] V. Reddy, T. S. Ramulu, B. Sinha, J. Lim, R. Hoque, and J. Lee, "Electrochemical Detection of Single Nucleotide Polymorphism in Short DNA Sequences Related To Cattle Fatty Acid Binding Protein 4 Gene," vol. 7, pp. 11058–11067, 2012.

[13] Kelley, S. O., Jackson, N. M., Hill, M. G., & Barton, and J.K., *Long-Range Electron Transfer through DNA Films*, Angewandte. 1999, pp. 941–945.

Atomic Force Microscope base nanolithography for reproducible micro and nanofabrication

Arash Dehzangi[*], *MIEEE*, Farhad Larki, *MIEEE*,
Burhanuddin Y. Majlis, S*MIEEE,* Zainab Kazemi,
MohammadMahdi Ariannejad*,*
Institute of Microengineering and Nanoelectronics
(IMEN), Universiti Kebangsaan Malaysia,
43600 Bangi, Selangor, Malaysia,
[*]Email: dehzangi@ukm.edu.my

A Makarimi Abdullah,
Malaysian Institute of Marine Engineering Technology,
University of Kuala Lumpur, Malaysia,

Mahmood Goodarz Nasery, Manizheh Navasery,
Elias B Saion, Halimah Mohamed. K, Nasrin Khalilzadeh
Department of Physics, University Putra Malaysia,
43400 Serdang,
Selangor, Malaysia,

Sabar D. Hutagalung, *MIEEE,*
Faculty of Science, Department of Physics,
Jazan University, Jazan, Kingdom of Saudi Arabia,

Abstract— Atomic force microscopy nanolithography (AFM) is a strong fabrication method for micro and nano structure due to its high spatial resolution and positioning abilities. Mixing AFM nanolithography with advantage of silicon-on-insulator (SOI) technology provides the opportunity to achieve more reliable Si nanostructures. In this letter, we try to investigate the reproducibility of AFM base nanolithography for fabrication of the micro/nano structures. In this matter local anodic oxidation (LAO) procedure applied to pattern a silicon nanostructure on p-type (10^{15} cm^{-3}) SOI using AFM base nanolithography. Then chemical etching is applied, as potassium hydroxide (saturated with isopropyl alcohol) and hydrofluoric etching for removing of Si and oxide layer, respectively. All parameters contributed in fabrication process were optimized and the final results revealed a good potential for using AFM base nanolithography in order to get a reproducible method of fabrication.

Keywords— Local anodic oxidation, Silicon-on-insulator, Nanolithography; Atomic force microscope

I. INTRODUCTION AND REVIEW OF LITERATURE

The major techniques for fabrication of semiconductor devices are based on different lithographical procedures by means of top/down approach. In recent years, these conventional techniques are exceedingly developed. Nevertheless, some serious issues limit the applicability of these conventional techniques, which can be listed as costly operation, poor accessibility, multiple-step procedures and low accessibility for micro/nanoscale fabrication. In this matter uunconventional methods appear to have a good potential to make more applicable and low-cost techniques for fabrication of micro and nanostructures. Lately, some approaches such as soft lithography [1], nano-imprint lithography (NIL) [2], and atomic-force-microscope (AFM) base nanolithography [3, 4] appeared as extra options for nanofabrication. AFM base nanolithography techniques can be a proper technique for fabricating of nano and microstructurs [5, 6].

According to different AFM nanolithography application Xie.et al [7] classified these techniques into two main categories regarding to the principal of operational, such as force assisted and biased assisted AFM nanolithography. In Table I these two main groups are illustrated according to their operational basis, interactions of tip and sample in each technique, different pattern formation modes and a few examples of each group.

For force-assisted AFM base nanolithography, as a mechanical tip–surface interaction, a massive force is applied on AFM tip in order to perform patterning. In biased-assisted AFM nanolithography, in order to create a confined electric field, an electric bias is applied on AFM probe. In this case, the gap regularly has few nanometers, therefore a slight bias of some volts can be able to create a huge electric field (10^8-10^{10} V/m). In biased-assisted AFM base nanolithography for silicon, tip induced local anodic oxidation (LAO) in the air is considered as a crucial process. For the first time, LAO was reported in 1994 [8], performed on silicon-on-insulator (SOI) substrate, and it showed a reliable application in comparison with other scanning tunneling microscope method [9].

TABLE I. Comparison of two main category for AFM base nanolithography techniques

	Force-assisted AFM nanolithography	Bias-assisted AFM nanolithography
Operational principal	Huge force is applied on tip	Bias is applied on tip
Tip- surface interaction	Largely mechanical	Tip is like a nanoscale electrode to provide physical effect
Pattern formation modes	Patterns are shaped by probe using scratching, or pulling/pushing the surface (as well as atoms or molecules)	For high electric field, electrochemical or even explosive gas discharge procedures can be started to ease formation of patterns
Typical examples	Dip-pen nanolithography (DPN)	AFM nano-oxidation with electrostatic attraction

Supported by UKM (DIP-2012-16) and USM (304/PBAHAN/6039035), Malaysia

Rrecently, several works were reported in order to enhance the AFM-LAO nanolithography method [10, 11]. During LAO, an electric field is applied on AFM tip under ambient humidity in order to oxidize the specific surface of the substrate.

The presence of water meniscus in the gap between is necessary to facilitate the oxidation process on silicon substrate. For repetitive performance of AFM nanolithography technique as reproducible method for fabrication of micro/nanoscale devices a comprehensive understanding and control of the oxidation mechanism is required. In fact, a thorough understanding of the LAO procedure is complicated, since the motion of AFM cantilever or explicit spatial details between the tip and surface can be hard to observe [12]. In this matter, a qualitative mechanism was proposed [13] in three steps: (1) depassivation of the surface using high electric field, (2) oxyanion formation assisted by the presence of water meniscus and (3) oxyanions diffusion. In Table II, some basic models proposed for AFM-LAO are summarized.

Amongst different methods for etching process wet etching technique is a regular method to realize the silicon structure in AFM base nanolithography [14]. Solution of potassium hydroxide (KOH), which is saturated by isopropyl alcohol (IPA), is a renowned etchant for micro/nanolithography, where micro or nanostructure is defined by accurate crystalline plain regarding to anisotropic etching. It is important to note that the controllability on the cross section of silicon structure can be a good advantage of AFM base nanolithography. However, the reproducibility of such nanolithography is an important issue for further studies. In previous works [15-17], we investigated impact of experimental parameters variation on fabrication of micro and nanostructure on p-type (100) SOI using AFM base nanolithography and KOH+IPA wet etching.

In present work we performed optimized fabrication result in order to study the reproducibility of AFM base nanolithography. We try to suggest possible potential of AFM base nanolithography to generate multiple silicon micro and nanostructures. This may be used as a guide for upcoming experimental works about AFM base nanolithography, including to the previous ones to complete previous work.

TABLE II. Various local oxidation models by AFM

Oxidation model	Description
Cebrera-Mott model	Thickness of the oxide is controlled by diffusion restricted electric field [18]
Power- law model	AFM-LAO oxidation is merely realized for voltage exceeding a threshold voltage, and oxidation kinetics adjusts with power law [19]
Log kinetic model	Height of oxide is related to log(1/t) and linear performance amid 1/h vs log v [20]
Space charge model	- Different space charge dependence in process of oxidation (as a function of substrate doping). - Space charge is related to fast drop of initial growth rate - Direct and indirect pathway reaction (Alberty-Miller scheme) [21]

II. FABRICATION PROCESS

The schematic view for the fabrication process is shown in *Fig. 1* LAO process were performed by AFM machine (SPI3800N/ 4000, Chiba, Japan) in contact mode. It was carried out on B-doped (10^{15} cm^{-3}) SOI wafer p-type (100), prepared by using Unibond™ (UnibondInternational Ltd. UK), by a top silicon layer of 100 nm and with 145 nm buried oxide (BOX) thickness (the resistivity of 13.5–22.5 Ω cm) (Soitech).

All the parameters for fabrication process were optimized according to our previous experiments. It is important to mention that a right choice for preparation of a sample must be given to the sample prior to start the LAO process, particularly for SOI with a thin layer of silicon on the top. For sample preparation, all SOI samples were imposed by ultrasonic energy (5 min) in an ultrasonic bath. Then a modified and optimized RCA method was performed on samples to reach the best possible output. For LAO procedure, Cr/Pt coated AFM tips (Budgetsensors Cont-G) were used where the force constant and resonance frequency were preserved as 0.2 N m^{-1} and 13 kHz respectively.

Conductive Cr /Pt coated AFM tip showed acceptable oxide layer quality, for the purpose of fabrication. Room humidity (RH) was manageable from 60%-80% with the accuracy of 1%, where the optimized condition for the oxide thickness and quality was found to be in the range between 65-68% of the RH% at room temperature. Higher RH% beyond 70% leaded to the shadow effect [15] when the rate of the produced oxide thickness was unchanged.

After the oxidation part by AFM-LAO two distinct stages of wet chemical etching were conducted in order to achieve micro/nanostructures. As the etchant, KOH solution (System, with KOH pellets of 56.11 g/mol) was saturated by IPA (Merck, molar weight of 60.10 g/mol,).

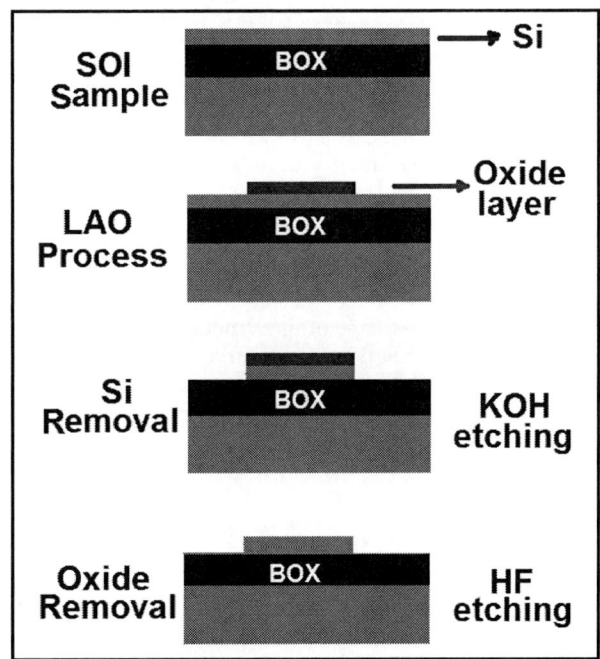

Fig. 1. Schematic of the fabrication procedure on SOI sample

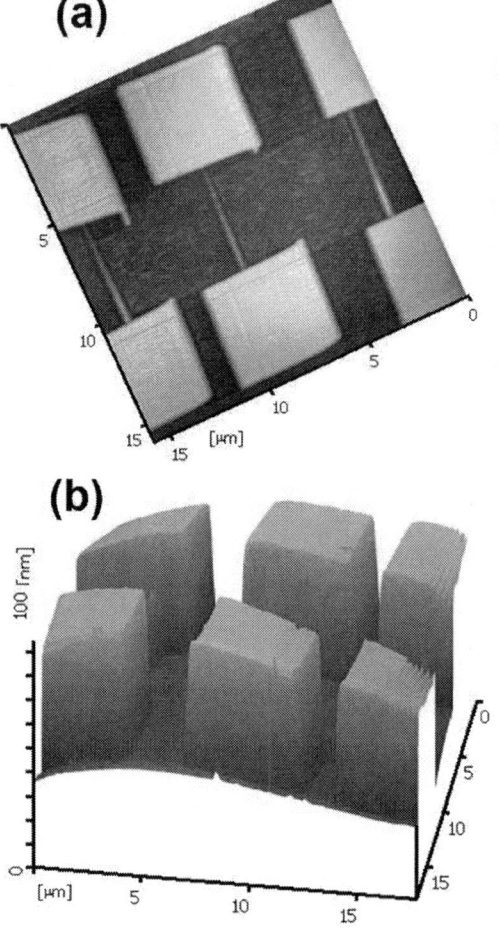

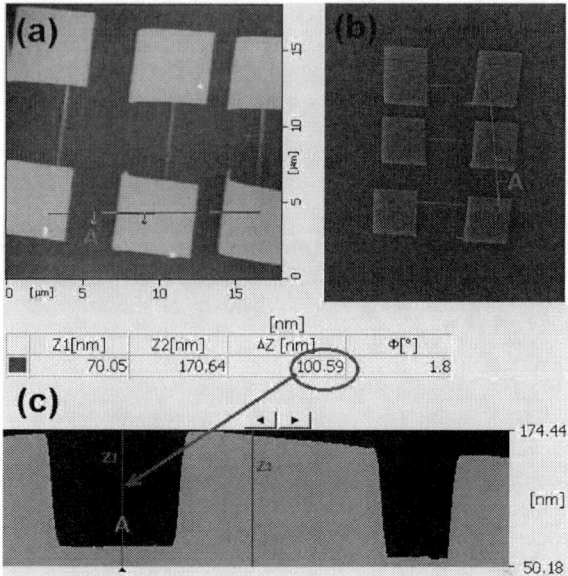

Fig. 3. (a) AFM image, (b) SEM image of microstructures, (c) AFM profile image for structures posterior of wet etching

Fig. 2. (a) AFM image the oxide pattern on SOI, (b) AFM image for structures posterior of wet etching

The BOX sheet acted as an etch-stop layer during the etching progression. The IPA mixed with KOH etchant can lead to the surface enhancement after etching progression [14]. All KOH+IPA wet etching procedures were performed in a sealed glass beaker, where the temperature was kept constant.

The second step of etching method and final stage step in fabrication process is oxide removal. The etchant is hydrofluoric acid (HF, JT-Baker) in aqueous solution. Through HF etching procedure, the oxide mask along with native oxide were eliminated. The shape of the structures fabricated in this work was simple square pads to check the reliability of the fabrication method by AFM base nanolithography.

III. RESULTS AND DISCUSSIONS

Fig. 2a shows AFM image for several oxide patterns on SOI surface after optimized LAO process. At room temperature using the best optimization, the constant voltage of 8V on AFM tip with the writing speed of 0.6μm/s was applied to run the oxidation process on the surface of the SOI sample. The silicon oxide shaped by LAO acts as etching mask on top of SOI substrate, where unoxidized silicon layer is considered to be removed by etching process (*Fig. 1*).

In all LAO process, optimized parameters were used to study the possibility of fabrication of multi structures. The oxide layers patterned on SOI wafer had good shape with the random thickness of 3-4 nm which can be adequate oxide layer range to perform as the mask on silicon surface. As it can be seen the same quality of the oxide patterns were extracted by using the optimized parameters. This can be a sign of LAO potential as a suggestion for reproducible method in micro and nano fabrication.

The optimized parameters for KOH+IPA etching were the solution of 30%wt. KOH with 10%vol. IPA at 63°C, immersed for 22 seconds and stirred at the rate of 600rpm. During etching, stirring of the solution can guarantee the uniformity for the final results of rtching.

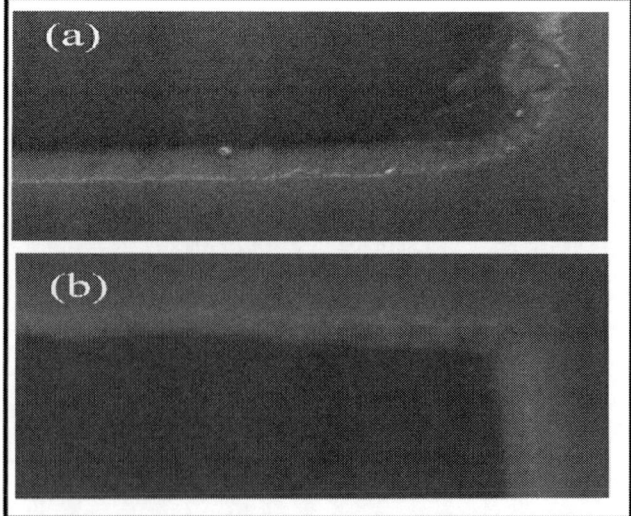

Fig. 4. (a) (b) SEM image of microstructures after etching process

For removing the oxide layer, the HF solution by the H_2O/HF ratio of 100:1 for 16 s to 18 s was applied to the structures. In *Fig. 2b*, the AFM image of the silicon structures after the optimized KOH etching process are shown. It can be seen that all structures has acceptable uniform shape. *Fig. 3a,b* shows AFM and SEM image for the silicon structure after the etching process.

As it is demonstrated in *Fig. 3c*, AFM profile image of the same structures reveals the same depth for all the structures, which is equal with the thickness of the highest silicon layer of the SOI (100 nm). This means that all undesired silicon around the oxide masks was successfully removed. *Fig. 4a,b* demonstrates SEM images for two different samples after the optimized etching by a KOH+IPA solution.

As a conclusion, the AFM base nanolithography was examined to get reproducible method to fabricate micro/nano multi-structures on SOI substrate by using the optimized parameters during the fabrication procedure.

It is worth noting that during the fabrication process, some minor parameters are needed to be considered and controlled to get proper result, such as deformation or geometry of AFM tip, order of oxidation, hydrophobicity, doping concentration and even the dielectric constant of silicon as the substrate material.

REFERENCES

[1] D. Qin, Y. Xia, and G. M. Whitesides, "Soft lithography for micro- and nanoscale patterning," *Nature protocols*, vol. 5, pp. 491-502, 2010.

[2] L. J. Guo, "Nanoimprint lithography: methods and material requirements," *Advanced Materials*, vol. 19, pp. 495-513, 2007.

[3] Y. Zhang, S. Santhanam, J. Liu, and G. Fedder, "Active CMOS-MEMS AFM-like conductive probes for field-emission assisted nano-scale fabrication," 2010, pp. 336-339.

[4] A. Malshe, K. Rajurkar, K. Virwani, C. Taylor, D. Bourell, G. Levy, *et al.*, "Tip-based nanomanufacturing by electrical, chemical, mechanical and thermal processes," *CIRP Annals-Manufacturing Technology*, vol. 59, pp. 628-651, 2010.

[5] M. A. Cook, C. K. Chan, P. Jorgensen, T. Ketela, D. So, M. Tyers, *et al.*, "Systematic validation and atomic force microscopy of non-covalent short oligonucleotide barcode microarrays," *PloS one*, vol. 3, p. e1546, 2008.

[6] A. Dehzangi, F. Larki, B. Y. Majlis, M. N. Hamidon, P. S. Menon, A. Jalar, *et al.*, "Numerical investigation of channel width variation in junctionless transistors performance," in *Micro and Nanoelectronics (RSM), 2013 IEEE Regional Symposium on*, 2013, pp. 101-104.

[7] C. Ju and P. J. Hesketh, "High index plane selectivity of silicon anisotropic etching in aqueous potassium hydroxide and cesium hydroxide," *Thin solid films*, vol. 215, pp. 58-64, 1992.

[8] E. Snow and P. Campbell, "Fabrication of Si nanostructures with an atomic force microscope," *Applied Physics Letters*, vol. 64, pp. 1932-1934, 1994.

[9] Y. K. Choi, D. Ha, T. J. King, and J. Bokor, "Investigation of gate-induced drain leakage (GIDL) current in thin body devices: single-gate ultra-thin body, symmetrical double-gate, and asymmetrical double-gate MOSFETs," *Japanese journal of applied physics* vol. 42, pp. 2073-2076, 2003.

[10] G. Pennelli, "Top down fabrication of long silicon nanowire devices by means of lateral oxidation," *Microelectronic engineering*, vol. 86, pp. 2139-2143, 2009.

[11] J. Martinez, R. V. Martínez, and R. Garcia, "Silicon nanowire transistors with a channel width of 4 nm fabricated by atomic force microscope nanolithography," *Nano Letters*, vol. 8, pp. 3636-3639, 2008.

[12] H. Seidel, L. Csepregi, A. Heuberger, and H. Baumgärtel, "Anisotropic etching of crystalline silicon in alkaline solutions I. Orientation dependence and behavior of passivation layers," *Journal of the Electrochemical Society*, vol. 137, pp. 3612-3626, 1990.

[13] B. Brennan, M. Milojevic, C. L. Hinkle, F. S. Aguirre-Tostado, G. Hughes, and R. M. Wallace, "Optimisation of the ammonium sulphide (NH4)2S passivation process on In0.53Ga0.47As," *Applied surface science*, vol. 257, pp. 4082-4090, 2011.

[14] A. Dehzangi, F. Larki, B. Y. Majlis, M. G. Naseri, M. Navasery, A. M. Abdullah, *et al.*, "Impact of KOH Etching on Nanostructure Fabricated by Local Anodic Oxidation Method," *Int. J. Electrochem. Sci*, vol. 8, pp. 8084-8096, 2013.

[15] A. Dehzangi, F. Larki, S. D. Hutagalung, M. Goodarz Naseri, B. Y. Majlis, M. Navasery, *et al.*, "Impact of Parameter Variation in Fabrication of Nanostructure by Atomic Force Microscopy Nanolithography," *PloS one*, vol. 8, p. e65409, 2013.

[16] A. Dehzangi, F. Larki, J. Hassan, S. D. Hutagalung, E. B. Saion, M. N. Hamidon, *et al.*, "Fabrication of p-Type Double Gate and Single Gate Junctionless Silicon Nanowire Transistor by Atomic Force Microscopy Nanolithography," *Nano Hybrids*, vol. 3, pp. 93-113, 2013.

[17] A. Dehzangi, A. M. Abdullah, F. Larki, S. D. Hutagalung, E. B. Saion, M. M. N. Hamidon, *et al.*, "Electrical property comparison and charge transmission in p-type double gate and single gate junctionless accumulation transistor fabricated by AFM nanolithography," *Nanoscale research letters*, vol. 7, p. 381, 2012.

[18] A. J. G. De Abajo, A. G. Pattantyus-Abraham, N. Zabala, A. Rivacoba, M. O. Wolf, and P. M. Echenique, "Cherenkov effect as a probe of photonic nanostructures," *Physical Review Letters*, vol. 91, pp. 143902/1-143902/4, 2003.

[19] F. J. García De Abajo, A. Rivacoba, N. Zabala, and N. Yamamoto, "Boundary effects in Cherenkov radiation," *Physical Review B - Condensed Matter and Materials Physics*, vol. 69, pp. 155420-1-155420-12, 2004.

[20] D. Stievenard, P. Fontaine, and E. Dubois, "Nanooxidation using a scanning probe microscope: An analytical model based on field induced oxidation," *Applied Physics Letters*, vol. 70, p. 3272, 1997.

[21] J. Dagata, F. Perez-Murano, C. Martin, H. Kuramochi, and H. Yokoyama, "Current, charge, and capacitance during scanning probe oxidation of silicon. I. Maximum charge density and lateral diffusion," *Journal of applied physics*, vol. 96, p. 2386, 2004.

Effects of Ethanol in Oxalic Acid on the Synthesis of Porous Anodic Alumina

N.H.M. Saleh[a], B.Y.Lim

School of Materials Engineering,
Universiti Malaysia Perlis (UniMAP),
Taman Muhibbah, 02600 Jejawi,
Perlis, Malaysia
[a]nadiahusnasaleh@yahoo.com,

C.H.Voon[b], S.T.Ten, M.N.Derman, K.L. Foo, M. K. Md Arshad, U.Hashim

Institute of Nanoelectric Engineering,
Universiti Malaysia Perlis (UniMAP),
Jalan Kangar-Alor Setar, Seriab 01000, Kangar,
Perlis, Malaysia
[b]chvoon@unimap.edu.my

Abstract— **Porous anodic alumina is a self-organizing porous material suitable as a template for obtaining nanostructured semiconductor materials. However, low temperature is generally used for the synthesis of porous anodic alumina. In this study, porous anodic alumina films were synthesized by a simple one-step anodizing technique at constant potential 40V using different volume percentage of ethanol in 0.5M oxalic acid at the temperatures of 25 °C. The current versus time transient was recorded by using Keithley sourcemeter. The morphology of the samples was viewed by a scanning electron microscopy. The current versus time transient decreased with the volume percent of ethanol, indicating reduction of growth rate of porous anodic alumina. Porous anodic alumina formed in oxalic acid without ethanol exhibit a complicated structure with irregular pore size and pore shape. Increasing volume percent of ethanol in the oxalic acid improved the pore size and shape. This is probably due to the cooling effect of the ethanol and prolonged time for pore organization. Typical morphology of porous anodic alumina can be formed by anodizing at 40 V at room temperature of 25 °C in 0.5 M oxalic acid with the addition of minimum of 30 volume percent of ethanol.**

Keywords—anodizing; porous anodic alumina; nanoporous; ethanol; oxalic acid

I. INTRODUCTION

Porous anodic alumina (PAA) is a self-assembly porous material that is formed by the anodizing of aluminum and its alloys in acidic electrolytes. It is widely used as template for the synthesis of semiconductor nanomaterials. The applications of these semiconductor nanomaterials in electronic industries are progressively widened [1-3]. Therefore, a great variation of nanodot arrays, nanoporous materials, nanowires, and nanotubes were synthesized [3-5]. For the past few years, efforts have been made by the scientific society for the development of method for the large scale synthesis of semiconductor nanomaterials with simple and low cost processing steps. Numerous research methods have been suggested for the synthesis of semiconductor nanomaterials. In general, synthesis method using porous anodic alumina as template is the most widely used method for the synthesis of semiconductor nanomaterials [6].

Well-ordered porous anodic alumina is one of the most widely used templates for the synthesis of semiconductor

nanomaterials. However, well-ordered porous anodic alumina formed only in narrow processing window. This is due to the regularity of porous anodic alumina is strongly dependent on the processing parameters. Most of the reported studies were focused on highly pure aluminum. Only few studies on the aluminum alloy were reported [7,8]. The cost of synthesis of porous anodic alumina can be reduced if aluminum alloys were used as alternatives for the synthesis of porous anodic alumina.

For oxalic acid, well-ordered porous anodic alumina with hexagonal pores structure can be formed at 40V. It is well known that a great amount of Joule heats are generated at the pores bottom during the anodizing process, which inhibits the formation of well-ordered porous anodic alumina [6]. In order to facilitate the dissipation of the heat that developed at the pores bottom of the sample, ethanol may be added as coolant to the electrolyte. Ethanol is added for its lower evaporation temperature of 78.4 °C, comparing to water to decrease the temperature at the oxide/metal interface.

The aim of this paper is to study the effect of addition of ethanol into oxalic acid on the synthesis of porous anodic alumina. A comparative study on the microstructure of porous anodic alumina from AA6061 that were anodized in oxalic acid with different volume percentage of ethanol was therefore reported in this paper. The current versus time transient were also studied in terms of its current behavior of each sample.

II. METHODS AND MATERIALS

A. Samples preparation

The AA 6061 samples were cut into 10mm x 50mm size and the surfaces were smoothened using SiC grinding paper up to 1500 grit to remove scratches formed on the surface during sectioning. The samples were then polished on polishing cloth using 0.1 μm diamond paste to achieve mirror-like surface. Samples were degreased by rinsing by using ethanol and followed by distilled water.

B. Anodizing

One step anodizing was employed in this study, in which the details were described elsewhere [10]. For the anodizing, the cathode was Pb plate. Cathode and anode were separated by 40mm. For this study, all samples were anodized under the same combination of processing parameters. Anodizing was performed at fixed voltage of 40V at 25 °C for 60 minutes. The concentration of oxalic acid was 0.5M with different mixture of volume percentage of ethanol (10%, 20%, 30%, and 40%). Each sample was sectioned into 10mm x 10mm in area of exposure to the electrolyte using polytetrafluoroethylene (PTFE) tape. The samples were rinsed thoroughly and dried in air. The anodized samples was viewed under scanning electron microscope (SEM).

III. RESULTS AND DISCUSSION

A. Current versus time transient

Fig.1 shows the current versus time transient for porous anodic alumina anodized in oxalic acid with increasing volume percent (vol %) of ethanol. Generally, the current versus time transients are typical for anodizing of aluminum and its alloy. Once the power supply was turned on, there is a current increase followed by a sudden decrease in a current to a minimum value. The current then rises gradually to a steady-state value and maintained until the power supply was turned off. This is due to the fact that at the initial stage, the current increase rapidly because of the low electrical resistance to the clean surface. When it reaches the maximum current, the non-conductive barrier layer started to grow and thickened which causes the current to decrease to a minimum value. However, the thickness of the barrier layer at the oxide/metal interface is not regular and pores were found on the alumina surface. Electric field concentrated at these pores where the alumina film is thinner [9]. Field-assisted oxide dissolution mechanism takes place at these pores and the current increases gradually [9]. The steady-state current is attained when the rate of oxide dissolution at the pores bottom equals the rate of oxide formation at the metal/oxide interface, indicating the end of pore rearranging process [10]. As the current versus time transient were typical for all anodizing process in oxalic acid, with and without ethanol, the addition of ethanol into the oxalic acid did not change the pore initiation and steady state pore growth.

Although the addition of ethanol did not change the shape of the current versus time transient, it can be observed from Fig. 1 that the steady state current decreased with the vol % of ethanol. It may be probably due to the weak chemical dissolution that was induced when ethanol was added into the oxalic acid that inhibit the movement of ions (O^{2-} / OH^- / Al^{3+}) [11]. This reduction of migration of ions further led to the lowering of the steady state current.

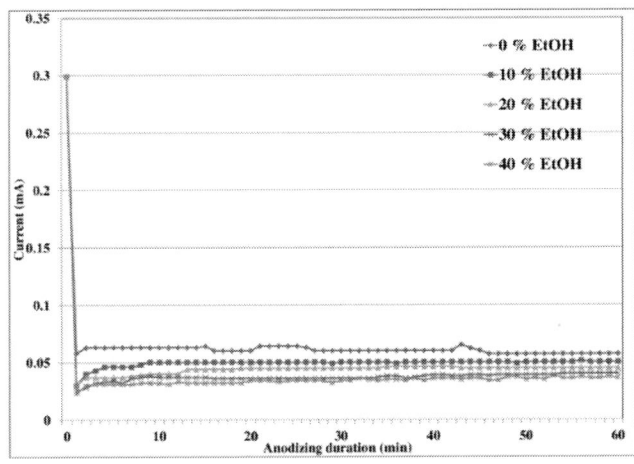

Fig. 1: Current versus time transient

B. Morphology of the porous anodic alumina

Fig.2 shows SEM top-views of porous structures for AA6061 plate anodized in 0.5M oxalic acid with increasing volume percentage (vol %). It can be observed that addition of ethanol into the oxalic acid affect the morphologies of the porous anodic alumina. As can be seen from Fig. 2(a), anodizing of AA 6061 plate at 25 °C in oxalic acid without addition of ethanol resulted in the emergence of porous anodic alumina without porous structure. The surface of the anodic alumina was rough. It is believed that this is due to the rough initial surface of AA 6061 plate. Addition of 10 vol % of ethanol in to the oxalic acid resulted in a flat surface, as can be observed in Fig. 2(b), although no porous structure was formed.

However, as in Fig. 2(c), addition of 20 vol % ethanol into the oxalic acid resulted in the formation of typical porous structures on the surface of AA6061 alloy. The surface was flat and the pores were randomly distributed over the surface. Some pores are merged together, as in the red circles in the Fig. 2 (c). As the vol % of ethanol was increased to 30 vol % as in Fig. 2(d), the image still showed the typical porous structure of alumina but with slightly different from Fig. 2(c). No pores merge together as observed when 20 vol% of ethanol was added into the oxalic acid. Besides, the pore distance and the pore size are uniform. Meanwhile, in Fig. 2(e), in which 40 vol % of ethanol was added, similar morphology was observed as on the surface of porous anodic alumina formed when 30 vol % of ethanol was added. The relatively rough surface of porous anodic alumina formed in 0.5 M oxalic acid with the addition of 40 vol % of ethanol is believed to be originated from the rough initial AA6061 surface.

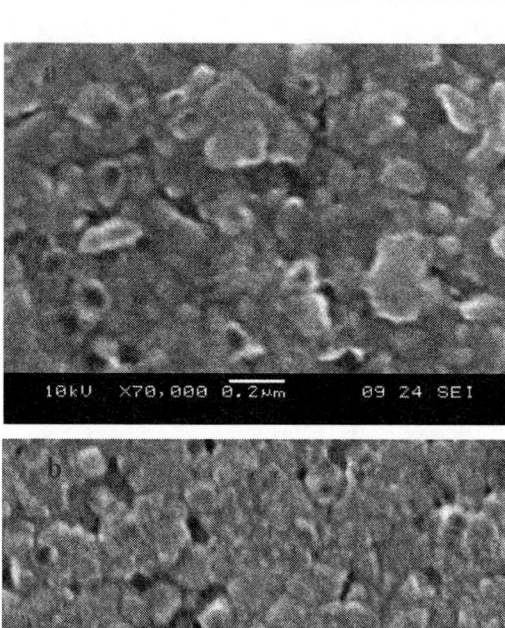

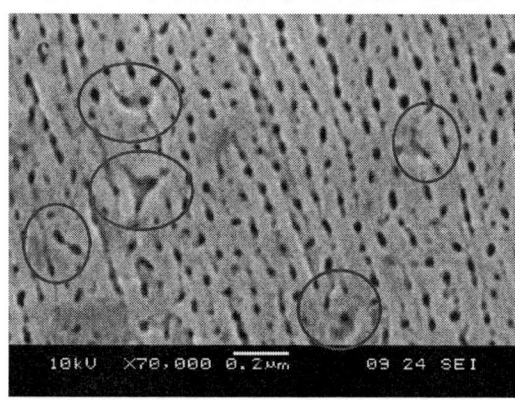

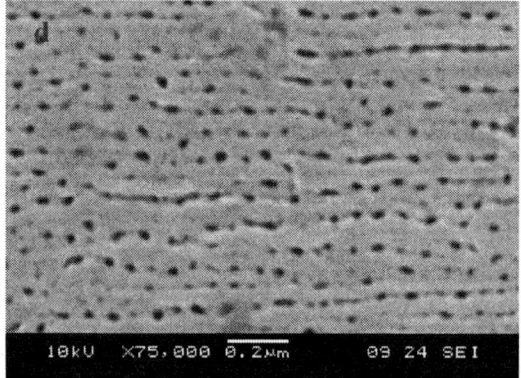

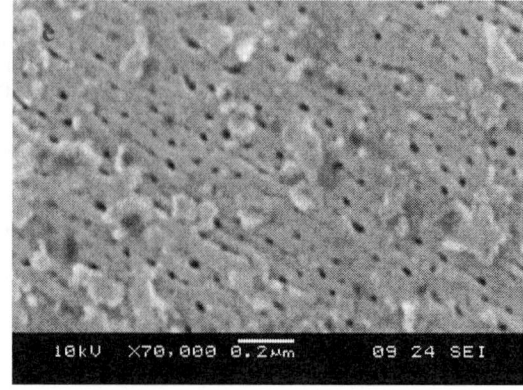

Fig. 2: Top-view SEM images of porous anodic alumina anodized in 0.5M oxalic acid with (a) 0% EtOH (b) 10% EtOH (c) 20% EtOH (d) 30% EtOH and (e) 40% EtOH

From observation, the addition of ethanol up to 30 vol % into 0.5 M oxalic acid affects significantly the pore growth. It can be observed from the above SEM images as the volume percentage of ethanol in the oxalic acid increases to 30 vol %; the porous anodic alumina became more evident. Typical porous anodic alumina morphology was obtained when a minimum of 30 vol % of ethanol was added into 0.5 M oxalic acid. Further increase of ethanol to 40 vol % did not further improve the morphology of the porous anodic alumina. Improvement of morphology of porous anodic alumina when up to 30 vol % of ethanol was added into the oxalic acid is probably due to the cooling effect of the low freezing point of ethanol (-114°C) [12]. Previous research has shown that local heating occurred at the pore bottoms during the anodizing process and causes the temperature of the pore bottom about 21°C higher the temperature of electrolyte. This high temperature at the pore bottoms would lead to the high growth rate of the porous anodic alumina with high Pilling–Bedworth ratio. This may subsequently cause the formation of anodic alumina structure without pores. The addition of ethanol probably led to the heat dissipation at the pore bottoms and enabled the formation of well-ordered porous anodic alumina. Ethanol was also used by Zaraska et. al [6] in their study of high field anodizing. From their finding, during anodizing, the cooling effect of ethanol was obvious in removal the heat dissipated at the pores bottom through the ethanol vaporization [6].

IV. CONCLUSIONS

In conclusion, the effect of addition of ethanol into oxalic acid on the synthesis of porous anodic alumina was successfully conducted. It is found that current decreased with the vol % of ethanol in 0.5 M oxalic acid from 0 vol % to 40 vol %. Anodic alumina without porous structure was formed in 0.5 M oxalic acid without ethanol and with the addition of 10 vol % of ethanol while typical porous structure of anodic alumina was formed in 0.5 M oxalic acid with 20, 30 and 40 vol % of ethanol. Porous anodic alumina can be formed with the addition of minimum of 30 vol % of ethanol into 0.5 M oxalic acid at room temperature of 25 °C and anodizing voltage of 40 V. With increasing of ethanol content from 20 to

40 vol %, the uniformity of pores arrangement and pore shape increased.

REFERENCES

[1] F.Y.Li, L.Zhang, R.M.Metzger, "On the growth of highly ordered pores in anodized aluminum oxide," Chem Mater, vol 10, pp. 2470–2480, 1998.

[2] J.M. Montero-Moreno, M. Sarret, C. Müller, "Influence of the aluminum surface on the final results of a two-step anodizing," Surf Coat Tech,vol 201, pp. 6352-6357, 2007

[3] X.Wang and G. R. Han, "Fabrication and characterization of anodic aluminum oxide template," Microelectron Eng, vol 66, pp. 166-170, 2003

[4] T. Maiyalagan, B. Viswanathan and U. V. Varadaraju, "Fabrication and characterization of uniform TiO_2 nanotube arrays by sol–gel template method," Bull Mater Sci, Vol. 29, pp. 705–708, 2006.

[5] J. Sarkar, G. Khan, and A. Basumallick, "Nanowires: properties, applications and synthesis via porous anodic aluminium oxide template," Bull Mater Sci, vol 30, pp. 271-290, 2007

[6] L. Zaraska, G.D. Sulka, and M Jaskuła, "The effect of n-alcohols on porous anodic alumina formed by self-organized two-step anodizing of aluminum in phosphoric acid," Surf Coat Tech,vol 204, pp. 1729-1737, 2010

[7] C.H.Voon, M.N.Derman, U.Hashim, "Effect of manganese content on the fabrication of porous anodic alumina," J Nano Met, vol 2012, Article ID 752926, pp. 1-9, 2012

[8] C.H.Voon, M.N.Derman, U.Hashim, K.R.Ahmad, K.L.Foo, "Effect of temperature of oxalic acid on the fabrication of porous anodic alumina from Al-Mn alloys," J Nano Met, vol 2013, Article ID 167047, pp.1-8, 2013

[9] C.H.Voon., M.N.Derman. , U.Hashim , K.R. Ahmad , and L.N.Ho ," A simple one-step anodising method for the synthesis of ordered porous anodic alumina, " J Exe Nanosci, vol 9, pp. 106–112, 2012

[10] M. A. Biason Gomes, S. Onofre, S. Juanto, L. O. de S. Bulhões, "Anodization of niobium in sulphuric acid media," J Appl Electrochem, vol 21, pp. 1023–1026. 1991

[11] X.Y. Han, W.Z. Shen, "Improved two-step anodization technique for ordered porous anodic aluminum membranes," J Electroanal Chem, vol 655, pp. 56-64, 2011

[12] Y. Li, M. Zheng, L. Ma and W. Shen, "Fabrication of highly ordered nanoporous alumina films by stable high-field anodization," Nanotechnology, vol 17, pp. 5101-5105, 2006.

Performance of Inverted Organic Solar Cell using Different Metal Electrodes

M. S. Alias, S.A. Kamaruddin, N. Nafarizal, M.Z. Sahdan

Microelectronic and Nanotechnology Research Center
Universiti Tun Hussein Onn Malaysia, 86400 Batu Pahat
Johor, Malaysia
zainizan@uthm.edu.my

Abstract—Organic solar cell is one of the fastest improving solar cells nowadays with improved power conversion efficiency approaching 10%. Here, we explore the performance of bulk heterojunction solar cell based on poly(3-hexyl thiophene) [P3HT] and [6,6]-phenyl-C61-butyric acid methyl ester [PCBM] by introducing ZnO nanoparticles buffer layer and appropriately tuning its energy level alignment by using different metal electrodes. The devices performance using two different high work function metal electrodes namely gold and platinum was investigated. The open circuit voltage (V_{oc}) was obviously changed using different metal electrodes. The device with Platinum electrode shows higher V_{oc} (0.2535 V) than the device with gold electrode by a factor of ~2. However, the efficiency was slightly lower than the gold device.

Keywords—*organic solar cell; P3HT:PCBM; ZnO thickness; open circuit voltage.*

I. INTRODUCTION

The finite supply of fossil fuels underscores the necessity of searching for renewable energy choices since it is computed to be diminished in the 35 years projection. Harvesting energy directly from sunlight using photovoltaic cell is a promising way to address the growing global energy needs. In recent years, organic solar cells (OSC) has drawn a lot of attention as means to be the next-generation solar cells due to their rapid increased in power conversion efficiency (PCE). This Organic solar cell is made from organic layer sandwiched between two different metal electrodes. With power conversion efficiency approaching 10 %, it is been projected to replace the traditional silicon solar cell in a few years more.

Nonetheless it still suffers from degradation, poor environmental stability, short lifetime and low power conversion efficiency mainly due to degradation at the active layer/electrode interface [1]. In order to reduce this degradation, the so called inverted organic solar cell has been introduced where the polarity of the device is reversed [2]. This inverted device gives new device architecture by using metal oxide as an electron extraction layer between electrode and the active layer and thus preventing the active layer from the unwanted oxidation and provides superior ambient stability [3]. In such structure, the metal electrodes can be less air sensitive, high work function metal such as silver, gold and platinum.

However, due to high exciton binding energy and their subsequent charge transport, OSC devices are highly sensitive to the morphology and thickness of the electron extraction layer (buffer layer). Although there are many studies elucidates the importance of thickness and morphology of the organic layer, there are limited studies on the morphology and thickness of buffer layer [4]. The processing of this buffer layer can result in different morphologies and therefore it is vital to study the role of morphology and thickness of this layer in order to maximize the device performance. Despite, the device performance also influenced by the open circuit voltage value (V_{oc}) in which limited primarily by the subtraction LUMO of electron acceptor and HOMO of electron donor [5]. Therefore in this work, the morphology and thickness of the buffer layer by incorporating zinc oxide (ZnO) nanoparticles is studied. Furthermore, the effect of different metal electrodes on the performance of inverted OSC device was further elucidated.

II. EXPERIMENTAL DETAILS

ITO substrate was cleaned in an ultrasonic bath using acetone for 3 minutes to remove any residual organic compound on the substrate. Then ITO was further cleaned with de-ionized (DI) water. After that, the substrate was dried using nitrogen gas purge. The ZnO layer was processed by sol-gel method using 0.4 M zinc acetate dehydrate as precursor and mixed with monoethanolamine [H_2N-CH_2CH_2OH] as stabilizer. The ZnO sol-gel solution was stirred overnight at 60°C for 24 hours. The solution was deposited onto ITO substrate using a spin coating technique and preheated at 250°C for 3 minutes before cooled down on alumina plate. The difference on ZnO buffer layer thicknesses were simply controlled using layer of deposition which varied from 53 nm to 642 nm. The organic layer P3HT and PCBM with 1:1 weight ratio were blend in 1ml of 1-2diclorobenzene solvent and stirred for 24 hours at room temperature. Then the blend was spin coated onto the ZnO/ITO substrate and preheated at 60°C for 5 minutes which producing thickness approximately 150 nm. The inverted organic solar cell device structure was completed by depositing gold electrode using DC sputter coater. In order to study the effect of different metal electrode, a device with platinum electrode was fabricated after the optimization of ZnO buffer layer thickness.

All the completed devices were annealed at 150°C for 5 minutes.

III. RESULTS AND DISCUSSION

Fig. 1 shows the FESEM images of nanoparticles ZnO buffer layer on ITO substrate with different thicknesses ranging from 53 nm to 642 nm (typical spin coating thin film thickness). The thicknesses were chosen because it can be varied easily by controlling the deposition layer. From Fig. 1(a), we can see the film particles with thickness of 53 nm are quite uneven and almost unnoticed even when we magnified the image up to 300, 000 times. At 131 nm, the film is rather porous and uniformed. As we further increase the thicknesses, the films nanoparticles are relatively well ordered and uniformed. The individual diameter of the ZnO nanoparticles observed are around 20-50 nm. This shows that we can obtained a uniform nanoparticles of ZnO buffer layer even with lesser layer of deposition. Figure 2 shows J-V curve of the device measured under AM 1.5 simulated solar radiations. The device performances are summarized in Table 1 with varied ZnO thicknesses ranging from 53 nm to 642 nm. From the table, we can see the device short circuit current density (J_{sc}) value enlarge in parallel with the increasing of ZnO thickness in the range of 53 nm to 131 nm but decrease again in the range of 224 nm to 642 nm. The J_{sc} value decrease is due to the burden of electron transport with thicker buffer layer. On the other hand, the value of open circuit voltage (V_{oc}) increases as we further increase the thickness of the ZnO buffer layer. The increasing of V_{oc} is because of the efficient electrons extraction and holes blocking capability of a thicker buffer layer.

Thicker ZnO buffer layer shows higher V_{oc} value. The device with 642 nm buffer layer has the highest V_{oc} value of 0.3939 V where it is slightly higher than the other devices by factor of ~3. The efficiency recorded was poor compared to the device with 131 nm buffer layer. The best performance recorded was the device with 131 nm achieving 0.0432 % with J_{sc} and V_{oc} value of 1.59871 mA/cm^2 and 0.1193 V respectively. Besides, there are no significant changes in the devices Fill Factor (FF). However, the efficiency is still low and only exhibits 0.0432%. One known possibilities to improve the V_{oc} of the device is by appropriately tuning the energy level alignment at the electrode/organic interfaces [6]. Nonetheless, the value of V_{oc} is limited primarily by the subtraction LUMO of electron acceptor and HOMO of electron donor [5]. Simultaneously, it is also influenced by the contacts between active layer and electrode [5].

TABLE I.   DEVICE PERFORMANCE WITH DIFFERENT THICKNESS

ZnO thickness (nm)	V_{oc} (V)	J_{sc} (mA/cm^2)	FF (%)	PCE (%)
53	0.1106	0.28711	19.72	0.0067
131	0.1193	1.59871	20.85	0.0432
224	0.1767	0.75469	24.55	0.0356
642	0.3939	0.28923	19.08	0.0236

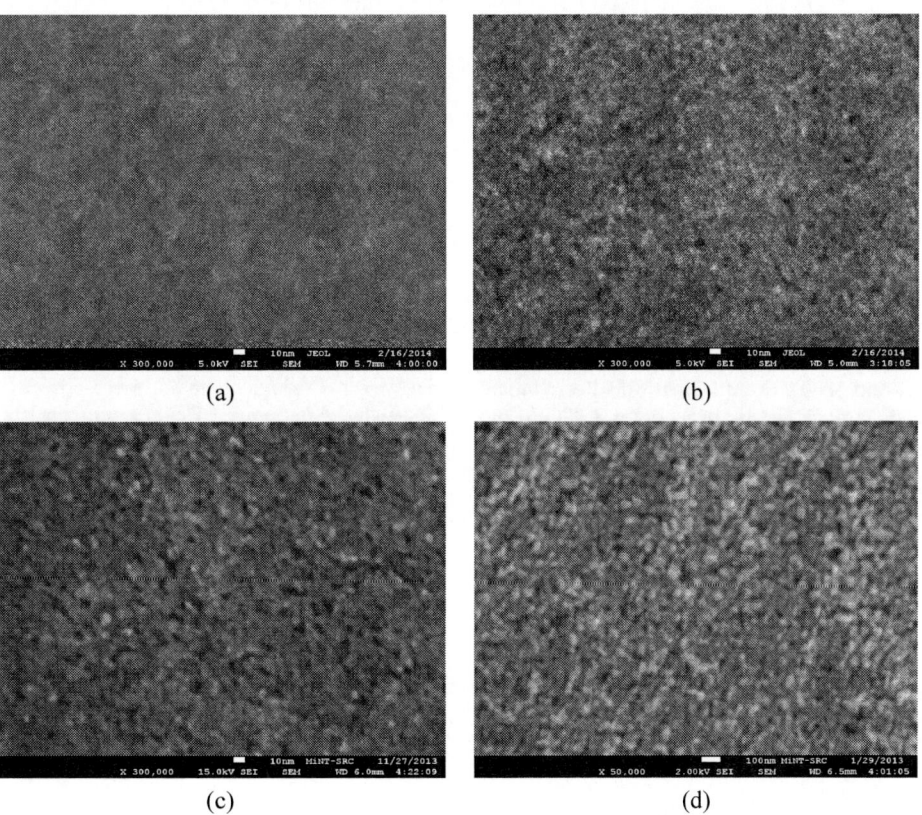

Fig. 1. FESEM images of ZnO layer with different thickness: (a) 53 nm, (b) 131 nm, (c) 224 nm and (d) 642 nm.

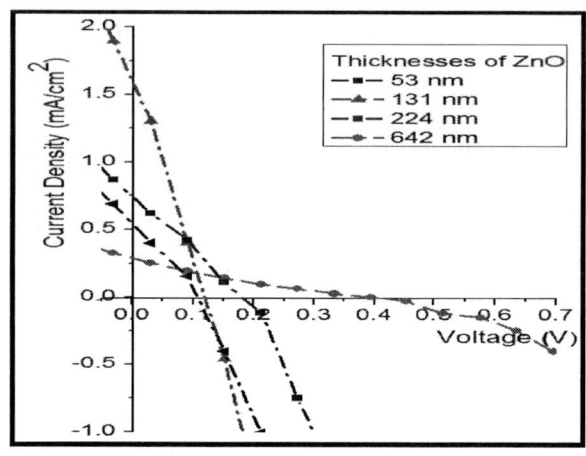

Fig. 2. *J-V* characteristic of the inverted organic solar cell with Au metal electrode.

TABLE II.   DEVICE PERFORMANCE WITH DIFFERENT ELECTRODE

Metal electrode	V_{oc} (V)	J_{sc} (mA/cm²)	FF (%)	PCE (%)
Au	0.1193	1.59871	20.85	0.0432
Pt	0.2535	0.4323	22.64	0.0276

Well said, a higher work function metal electrode and a buffer layer between active layer and metal electrode could improve V_{oc} of the device. With 131 nm ZnO thickness, a high work function metal electrode, Pt were used which giving the device configuration of ITO/ZnO/P3HT:PCBM:Pt. Table 2 shows the performance of the device with Pt and Au electrode. The device with Pt electrode shows a higher V_{oc} compared to Au electrode however the J_{sc} value shows otherwise. The J_{sc} value decreased from 1.59871 mA/cm² to 0.4323 mA/cm². This explains the reason of the low power conversion efficiency of the device with Pt electrode. The device with Pt electrode only exhibits 0.0276 %, a decreased value by a factor of ~2 from the device with Au. The low J_{sc} value is due to poor absorption of the device with Pt electrode.

Fig. 3 shows the energy level alignment of the inverted organic solar cell device. The energy-level alignment refers to the offset between the HOMO or LUMO levels and an electrode Fermi level. These offsets are labeled as the difference between donor HOMO and electrode Fermi level (ΔHOMO) and the difference between acceptor LUMO and electrode's Fermi level (ΔLUMO). In order to have a zero voltage drop at the interface (maximum V_{oc}), both HOMO and LUMO level should be in perfect alignment with the anode and cathode Fermi level. At the anode interface, ΔHOMO should be zero and at the cathode interface, ΔLUMO should also be zero. In other words, the electrode Fermi level should be close to donor HOMO. In both Au and Pt case, a minimum ΔHOMO and ΔLUMO is achieved.

From Figure 3, the value of ΔHOMO offset for Pt device is approximately 0.1 eV in which slightly lower than Au device (0.3 eV). This explains why the Pt electrode device V_{oc} value

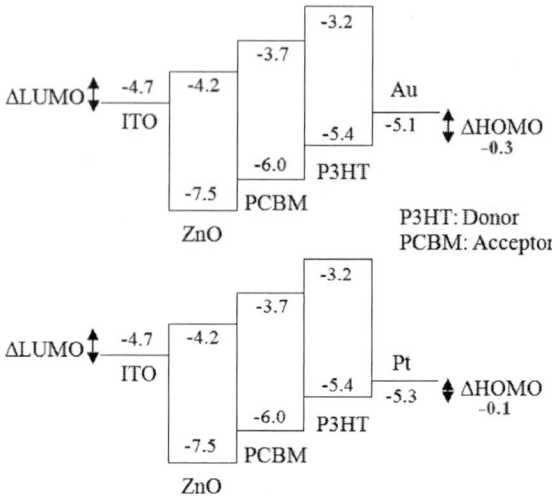

Fig. 3. Energy level diagram of the inverted organic solar cell device.

boosts by a factor of ~2. The Pt electrode Fermi level is closer to donor HOMO than Au electrode. The V_{oc} enhanced from 0.1193 V to 0.2535 V. This proves with appropriate tuning the energy level could optimize V_{oc} value of the device. However, the device with Au electrode gives better power conversion efficiency than Pt electrode although its V_{oc} value is slightly lower. In all cases, Au electrode is recommended as metal electrode rather than Pt electrode which minimize the absorption of the organic layer. Besides, the used metal oxides anode buffer layer can also boost the device V_{oc}. Metal oxides have the advantage over metal electrodes because they can achieve much higher work function than metals such as TiO_2 and Cu_2O. Therefore, with appropriate energy level alignment, we can achieve the ideal V_{oc} value for organic solar cell and consequently improve the power conversion efficiency for daily applications.

IV.   CONCLUSION

In summary, the different thicknesses of ZnO buffer layer affect the performance of the organic solar cell devices. The optimum ZnO thickness lies in between 53 nm to 131 nm. A thicker buffer layer device gives higher V_{oc} however it suffers from low power conversion efficiency. With Pt electrode, a boost V_{oc} value by a factor of ~2 was successfully tuned. However, Pt device shows lower power conversion efficiency than the Au device. Results suggest, higher work function metal electrode could reduce the absorption of the device. As conclusion, the optimization of ZnO buffer layer thickness leads to optimum power conversion efficiency of 0.0432 % with Au metal electrode.

ACKNOWLEDGMENT

The authors would like to thank Ministry of Education Malaysia for the financial support through FRGS grant Vote No. 1059. We also acknowledged Universiti Tun Hussein Onn Malaysia (UTHM) for the technical support.

REFERENCES

[1] M. Jørgensen, *et al.*, "Stability/degradation of polymer solar cells," *Solar Energy Materials and Solar Cells,* vol. 92, pp. 686-714, 2008.

[2] M. Glatthaar, *et al.*, "Organic solar cells using inverted layer sequence," *Thin Solid Films,* vol. 491, pp. 298-300, 2005.

[3] F. C. Krebs, *et al.*, "A roll-to-roll process to flexible polymer solar cells: model studies, manufacture and operational stability studies," *Journal of Materials Chemistry,* vol. 19, pp. 5442-5451, 2009.

[4] S. K. Gupta, *et al.*, "Understanding the role of thickness and morphology of the constituent layers on the performance of inverted organic solar cells," *Solar Energy Materials and Solar Cells,* vol. 116, pp. 135-143, 2013.

[5] W. Cai, *et al.*, "Polymer solar cells: Recent development and possible routes for improvement in the performance," *Solar Energy Materials and Solar Cells,* vol. 94, pp. 114-127, 2010.

[6] E. Bundgaard and F. C. Krebs, "Low band gap polymers for organic photovoltaics," *Solar Energy Materials and Solar Cells,* vol. 91, pp. 954-985, 2007.

Synthesis of Zinc Oxide Thin Film by Anodizing

C.H.Voon[1,a], N.Tukemon @ Tukiman[2,b], B.Y.Lim[2], U.Hashim[1], S.T.Ten[1], M.N.Derman[1], K.L. Foo[1], M. K. Md Arshad[1]

[1]Institute of Nanoelectric Engineering,
Universiti Malaysia Perlis (UniMAP),
Jalan Kangar-Alor Setar, Seriab 01000, Kangar,
Perlis, Malaysia
[2]School Of Materials Engineering,
Universiti Malaysia Perlis (UniMAP),
Taman Muhibbah, 02600 Jejawi,
Perlis, Malaysia
[a]chvoon@unimap.edu.my, [b]norazwani6446@gmail.com

Abstract— In this work, 99% purity zinc plates were anodized at potentiostatic conditions in distilled water at 25°C. The anodizing voltages were varied from 10 V to 30 V. The morphologies of the ZnO thin film were investigated by using scanning electron microscope (SEM) and the compositions were confirmed through the characterization using X-ray diffraction (XRD). The current versus time transients during the anodizing were recorded. SEM images show that thread-like nanostructures were present on the surface of Zn plates anodized at 25 V and 30 V. XRD patterns revealed that ZnO were present on all Zn plates anodized at anodizing voltages ranged from 10 V to 30 V. Current recorded during the anodizing process increased with the anodizing voltages due to the formation of thread-like nanostructures with lower electrical resistance.

Keywords— anodizing; zinc oxide; thin film; anodizing voltages; nanostructures

I. INTRODUCTION

Zinc Oxide (ZnO) is a well-known semiconductor that has wide direct band gap (3.37eV) at room temperature and large free excitation binding energy (60meV). This material is also well known for its high thermal stability, highly crystalline hexagonal structure, and high mechanical strength. These properties make ZnO a suitable electronic material in various fields as bio-molecule sensor, ultraviolet (UV) detector, light-emitting diode, chemical and gas sensor, solar cell, and optoelectronic.

ZnO have divergent structures, which the appearance are much abundant than any known nanomaterials including carbon nanotubes [1]. ZnO nanostructures can easily be synthesized in various forms or shapes, including nanorods, nanotubes, nanowires, nanobelts, nanoneedles etc. under specific growth condition [2]. For example, it was reported that ZnO nanowires were fabricated on Si (100) substrates by oxidation of metallic Zn powder at 600°C. The synthesized of materials have been found to have rod like, belt like, wire like and needle like morphologies [3]. A second-order self-assembly were applied to the fabrication of ZnO nanorods array [4]. The nucleation site were defined by patterning Au nanodots catalyst with self-organized array of anodized aluminium oxide nanopores. The nanorods were vertically aligned with respect to the hexagonal facets of gallium nitride (GaN) and have constant diameter of 60 nm and 400 nm average length.

ZnO thin films have been fabricated by cetyltrimethylmmonium bromide (CTAB)-assisted by hydrothermal method [5]. By using CTAB as surfactant, the morphology of ZnO can be changed in hydrothermal method. However, the products formed in aqueous solution are commonly lack in term of size and shape control [6]. The synthesizing of ZnO thin films using electrodeposition technique was also reported [7]. But, this method has relatively restrained number of elements that can be deposited from the aqueous electrolytes and requires the substrate to be conductive [8].

Anodizing is an electrolytic conversion process to form an oxide layer on certain metal to increase the thickness and regulate property of the natural oxide layer on the surface of metal parts [9-11]. Among the various type of methods, anodizing technique accommodate an easy route of synthesizing of thin films because of its simplicity, affordable cost setup from an economical view [14]. It was observed that the ZnO thin films can be produced from anodizing method. For example, Yamaguchi et al. observed that photocatalytic of ZnO were fabricated by anodizing zinc plate using acetaldehyde solution as the electrolyte [15]. Besides, Yam et al. studied the preparation of ZnO thin films by using acidified ethanolic anodizing [14]. While anodizing provides a low cost fabrication route of ZnO nanostructure thin film, this method require the use of chemicals that may lead to environmental issues. Therefore, a low cost alternative is required for the fabrication of ZnO thin film with simplified route and with minimum chemical use.

In this study, ZnO thin film was formed by anodizing Zn plate in distilled water. This technique provides several advantages in terms of reliability, repeatability, low temperature, and minimal chemical use. The aim of this current study is to investigate the feasibility of synthesis of ZnO thin film by anodizing of Zn plate in distilled water. Since anodizing process is an electrolytic process, the reactions are strongly dependant on the anodizing voltage. In this study, the anodizing of Zn plate was performed in distilled water of at room temperature with anodizing voltage ranged from 10 to 30 V. To the best of our knowledge, no published literature has been reported on the study of synthesis of ZnO thin film by anodizing using distilled water as electrolyte.

II. METHOD AND MATERIAL

A. Samples preparation

Zn plates (99% purity, thickness 1mm) with dimensions of 3 cm x 1.5 cm were cut and used in the anodizing process. The zinc plate were mechanically grinded by using silicon carbide paper and polished by using 6 μm and 1 μm alumina paste to mirror finishes. The Zn plate were then ultrasonically cleaned with ethanol and rinsed with distilled water.

B. Anodizing and characterization

For the anodizing process, zinc plate was used as anode while Pb plate with the same dimensions as cathode. Keithley SourceMeter was used as power supply. The anodizing was performed in distilled water of 25°C. 500 ml of fresh distilled water was used for the anodizing of each Zn plate. The distance between Pb plate and Zn plate was fixed to 40 mm. The schematic diagram of experiment set up is shown in Figure 1. The voltage applied was different ranging from 10 V to 30 V. The anodizing process was conducted under potentionstatic conditions. The duration of the anodizing was fixed for 120 hours. As anodized Zn plates were rinsed with distilled water and dried in air. The current during anodizing was recorded by using Keithley SourceMeter. Morphologies of the anodized Zn plates were examined by using scanning electron microscope (SEM). The composition of the anodized Zn plates was determined by using XRD.

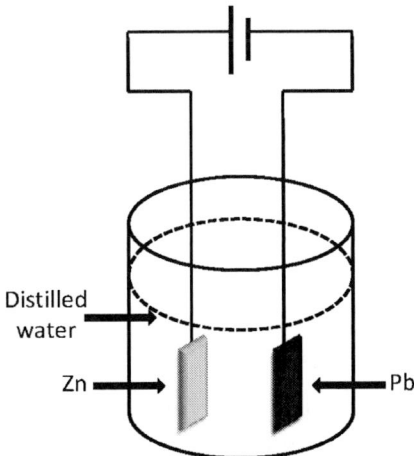

Fig. 1. Schematic diagram of experiment set up

III. RESULTS AND DISCUSSION

A. Morphology and composition of zinc oxide thin film

Fig. 2. shows the SEM images of Zn plates anodized in distilled water of 25 °C at voltages ranged from 10 V to 30 V. The obtained SEM images showed that the morphologies of the as anodized Zn plates were significantly influenced by the anodizing voltages. It can be seen that for Zn plates anodized at voltage 10, 15 and 25 V, no nanostructure was formed. The surfaces were relatively flat and with parallel scratches which were believed to be originated from the mechanical grinding and polishing process. However, thread-like nanostructures were observed on the surface of Zn plate anodized at 25 V. These nanostructures were randomly distributed over the surface of the anodized Zn plate. These nanostructures did not cover the entire surface such that the flat surface at the bottom of the nanostructures can be seen. Anodizing of Zn plate in 30 V resulted in entire surface covered by these thread-like nanostructures. These nanostructures were thick and randomly distributed over the entire surface.

The results were in good consistence with the observation reported by Amitha et. al. in which the Zn plates were anodized in deionized water [16]. They reported that morphologies of the ZnO thin film were significantly dependant on the anodizing voltage. However, no nanostructure was reported in their findings. Instead, wall-like structure was formed at 1 V, which changed gradually to well-defined porous structure as the voltage increased to 9 V [16].

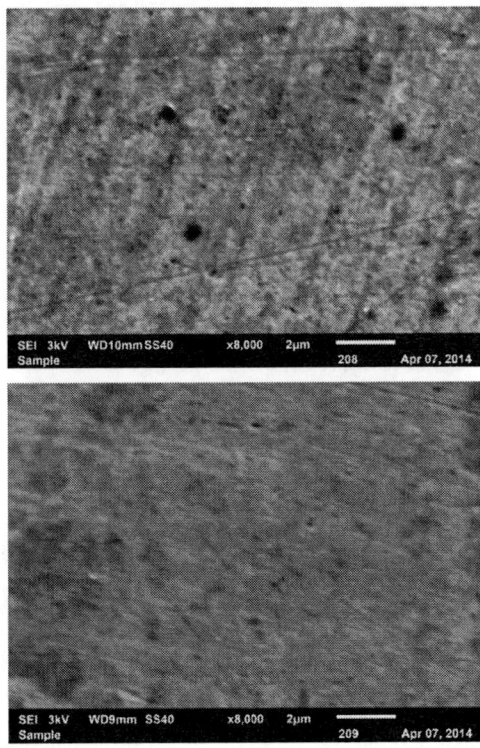

Fig. 2. SEM images of Zn plates anodized at voltages of (a) 10 V and (b) 15 V

IEEE-ICSE2014 Proc. 2014, Kuala Lumpur, Malaysia

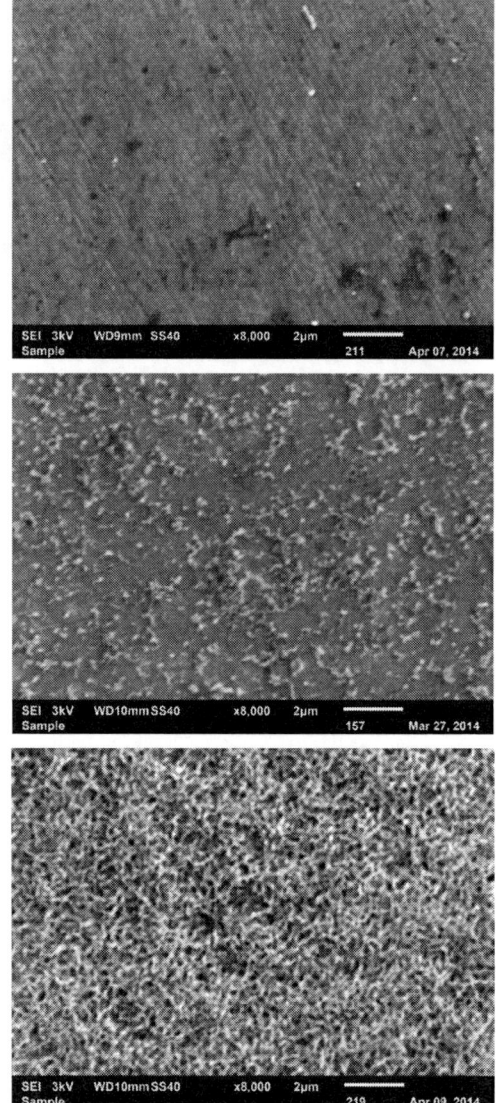

Fig. 2. SEM images of Zn plates anodized at voltages of (c) 20 V, (d) 25 V and (e) 30 V (cont)

To identify the composition of the thread-like nanostructure, the anodized Zn plates were characterized by using XRD. The XRD patterns of Zn plates anodized at voltages ranged from 10 V to 30 V are showed in Fig. 3. It can be observed that peaks attributed to Zn and ZnO were present in these XRD patterns. The presence of Zn peaks were due to the Zn plates which were used as substrates in this study. These XRD patterns also show that ZnO present in all Zn plates anodized at anodizing voltages ranged from 10 V to 30 V. Although the morphologies of the anodized Zn plates were different, as shown in Fig. 2., ZnO thin film were formed in all anodized Zn plates.

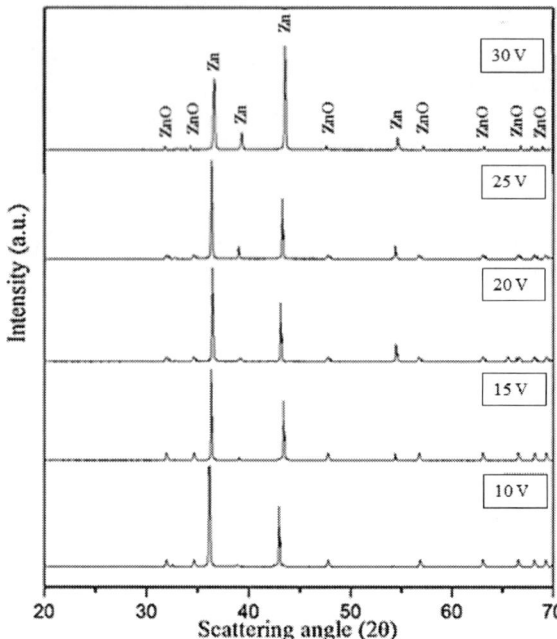

Fig. 3. XRD pattern of Zn plates anodized at voltages ranged from 10 V to 30 V

B. Current versus time Transient

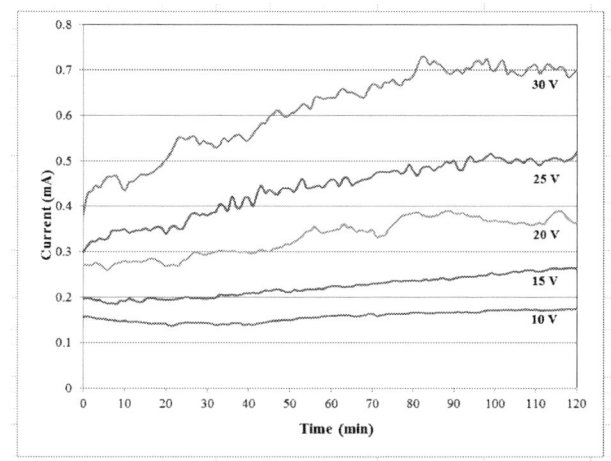

Fig. 4. Current versus time transient for Zn plates anodized at 10 V, 15 V, 20 V, 25 V and 30 V

To further understand the effect of the anodizing voltage on the formation of ZnO on the surfaces of Zn plates, the current versus time transients were studied. Fig. 4. shows the variation of current versus time for Zn plates anodized in distilled water of 2 °C with increasing anodizing voltage from 10 V to 30 V. It can be observed that the shape of current versus time transients for different voltages were similar. The values of currents were minimum at the commencement of the anodizing and increased with anodizing duration. However, the current increased with the increasing anodizing voltage. It is believed that this is related to the formation of ZnO thin

978-1-4799-5761-3/14 $31.00 © 2014 IEEE   422

film with different morphologies. It can be seen from Fig. 4. that the currents were relatively low for the formation of ZnO thin film without the thread-like nanostructures at 10 V and 15 V while currents were higher comparatively for the formation of ZnO thin film with thread-like nanostructures at 25 V and 30 V. It is believed that the thread-like nanostructure have lower electrical resistance which resulted in higher currents during the anodizing process at 25 V and 30 V.

IV. CONCLUSIONS

ZnO thin films were successfully synthesized by anodizing of Zn plates in distilled water at 25 °C at anodizing voltages ranged from 10 V to 30V. The morphologies of the ZnO thin films were significantly dependant on the anodizing voltages. ZnO thin films without thread-like nanostructures were formed at 10 V and 15 V while ZnO thin films with thread-like nanostructures were formed at 25 V and 30 V. XRD patterns show that ZnO were formed on all Zn plates anodized at anodizing voltages ranged from 10 V to 30V. The current versus time transient increased with the increasing anodizing voltage.

REFERENCES

[1] J. Wang and L. Gao, "Hydrothermal synthesis and photoluminescence properties of ZnO nanowires," Solid State Commun. China, vol 132, pp.269-271, October 2004.

[2] K. S. Babu, R. A. Reddy, R. A. Sujatha and K. V. Reddy, "Synthesis and optical characterization of porous ZnO," J. Adv. Ceramics. India, vol 2, pp. 260-265, May 2013.

[3] A. Sekar, S. H. Kim, A. Umar and Y. B. Hanh, "Catalyst-free synthesis of ZnO nanowires on Si by oxidation of Zn powders," J. Cryst. Growth. Korea, vol 280, pp. 509-515, April 2005.

[4] H. Chik, J. Liang, S. G. Cloutier, N. Kouklin and J. M. Xu, "Periodic array of uniform ZnO nanorods by second-order self-assembly," Appl. Phys. Lett. Rhode Island, vol 84, pp. 3376-3378, April 2004.

[5] R. S. Yadav, A. C. Pandey and S.S. Sanjey, "ZnO porous structures synthesized by CTAB-assisted hydrothermal process," Struct. Chem. India, vol 18, pp. 1001-1004, October 2007.

[6] D. Ramimoghadam, Mohd. Zubir Bin Hussein and Yun Hin Taufiq-Yap, "The effect of sodium dodecyl sulfate (SDS) and cetyltrimethylammonium bromide (CTAB) in the properties of ZnO synthesized by hydrothermal method," Int. J. Mol. Sci. India, vol 13, pp. 13275-13293, October 2007.

[7] L. Junwei, L. Zhifeng, E. Lei and Z. Zhichen, "Effects of potential and temperature on the electrodeposited porous zinc oxide films," J. Wuhan. Univ. Technol. China, vol 26, pp. 47-51, February 2011.

[8] A. J. Yin, J. Li, W. Jian, A. J. Bennet and J. M. Xu, "Fabrication of highly ordered metallic nanowire arrays by electrodeposition, "Appl. Phys. Lett. Rhode Island, vol 79, pp. 1039-1041, August 2001.

[9] C.H.Voon, M.N.Derman, U.Hashim, "Effect of manganese content on the fabrication of porous anodic alumina," J Nano Met, vol 2012, Article ID 752926, pp. 1-9, 2012.

[10] C.H.Voon, M.N.Derman, U.Hashim, K.R.Ahmad, K.L.Foo, "Effect of temperature of oxalic acid on the fabrication of porous anodic alumina from Al-Mn alloys," vol 2013, Article ID 167047, pp.1-8, 2013

[11] C.H.Voon, M.N.Derman, U.Hashim, K.R. Ahmad, and L.N.Ho, "A simple one-step anodising method for the synthesis of ordered porous anodic alumina," J Exe Nanosci, pp. 106–112, 2012

[12] L. S. Chuah, Z. Hassan and S. K. Mohd Bakhori, "Nanoporous ZnO prepared by electrochemical anodization deposition," Third International Conference on Smart Materials and Nanotechnology in Engineering, Malaysia, vol 8409, pp. 84092E-1 – 84092E-6, 2011.

[13] Y. Yamaguchi, M. Yamazaki, S. Yoshihara and T. Shirakashi, "Photocatalytic ZnO films prepared by anodizing," J. Electroanal. Chem. Japan, vol 442, p. 1-3, April 1998.

[14] S. W. Ng, F. K. Yam, L. L. Low, K. P. Beh, M. F. Muatapha, E. N. Sota, S. S. Tneh and Z. Hassan, "Self-assembled ZnO nanostripes prepared by acidified ethanolic anodization," Optoelectron. Adv. Mat. Malaysia, vol 5, pp. 89-91, January 2011.

[15] A. Ravanbakshah, F. Rashni, M. H. Sohi and R. K. Nekoui, "Synthesis of nanostructured zinc oxide thin films by anodic oxidation," Adv. Mater. Res. Iran, vol 829, pp. 347-351, November 2013.

[16] S. Amitha and N. K. Karuna, "Synthesis of zinc oxide porous structures by anodization with water as electrolyte," Appl. Phys. A. India, vol 129, pp. 151-157, June 2012.

Properties of Boron Doped Amorphous Carbon Films by Carbon Palm Oil for Carbon Based Solar Cell Applications

A. Ishak [1,1,*]

[1]NANO - ElecTronic Centre (NET), Faculty of Electrical Engineering, Universiti Teknologi MARA (UiTM), Shah Alam, Selangor, Malaysia
[2]Faculty of Electrical Engineering,
UiTM Sarawak, Kampus Kota Samarahan, Jalan Meranek Sarawak
Email: [*]ishak@sarawak.uitm.edu.my

M. F. Malek[1], and M. Rusop[3,4,**]

[3]NANO - ElecTronic Centre (NET), Faculty of Electrical Engineering, Universiti Teknologi MARA (UiTM), Shah Alam, Selangor, Malaysia
[4]NANO- SciTech Centre (NST), Institute of science, Universiti Teknologi MARA (UiTM), Shah Alam, Selangor, Malaysia
Email: [**]rusop@salam.uitm.edu.my

Abstract—The boron doped amorphous carbon were prepared by in-situ mixing of hydrocarbon palm oil and boron dopant and carrier gas, argon in the chamber by using bias assisted pyrolysis-CVD. The effect of substrate bias on the thickness, electrical and electronic properties of a-C:B film were investigated. The fabricated solar cell with the configuration of Au/p-C:B/n-Si/Au achieved conversion efficiency (η) of 0.453% at applied bias voltage of -20 V. This result showed the successful interstitial doping of boron in the amorphous carbon films deposited by this method and palm oil precursor as confirmed by the fill factor, open circuit voltage, and current density results.

Keywords—amorphous carbon; palm-oil; negative bias; boron; pyrolysis- CVD; solar cell

I. INTRODUCTION

The deposition of carbon films from the gas phase leads to materials of widely different structure and composition. The films can be crystalline, contain domains of any of these phases in more or less ordered arrangements, or even incorporate a considerable amount of hydrogen or other heteroatoms. Within the carbon films, amorphous carbon attained high interest due to its electronic and optical properties and chemical inertness [1-3].

In the solar cell applications, amorphous carbon was used as a semiconductor material thin film. The carbon precursors can be synthesized to form varieties of allotrope carbon such as amorphous carbon, carbon nanotube, fullerene, etc. using different method and precursors [4-6]. The conduction type of carbon films can be controlled by depositing dopants [7-9]. Boron is widely used as a p-type impurity in Si semiconductor and carbon films. Boron-doped films are employed in fabricating heterojunction solar cells using different precursors and techniques, such as chemical vapor deposition (CVD), ion implantation, and pulsed laser deposition. Different precursors and techniques result in different conversion efficiencies [10-12]. In this paper, the properties of boron doped amorphous carbon films at different negative bias are reported.

II. EXPERIMENTAL

Boron doped amorphous carbon (a-C:B) were deposited by using bias-assisted pyrolysis-CVD. Substrates (glass and n-silicon) were cleaned with acetone, methanol and rinse with deionizer water (DI) water for 15 min, respectively. The excess oxide layers of n-type silicon substrates were continued by the etching process with diluted hydrofluoric acid (10%) solution for about 3 min before rinsing in DI water. Substrates were blown with nitrogen gas. The cleaned of glass and silicon substrate was attached inside the chamber.

A liquid of palm oil precursor was heated in the bottle outside the chamber at around 150°C by using hot platter (Stuart CB162). The deposition conditions were set as follow: deposition temperature, 350°C, the vapor flow rate, 114 mL/min, argon flow rate, 200 mL/min and amount of boron, 1.5g, and negative bias varied from 0 and -20 V, duration time 1h deposition.

In the measurement of solar cell device, both sided of silicon (buttom and top) is deposited with approximately 60nm and 12nm of gold, respectively. Solar simulator (Bukuh Keiki EP200), surface profiler (Veeco Dektak 150), JASCO UV-VIS/NIR Spectrophotometer (V-670 EX) and field emission scanning electron microscopic (FESEM, ZEISS Supra 40VP) to characterize the electronic, thickness of film, optical, and surface morphology, respectively.

III. RESULT AND DISCUSSION

Fig. 1 shows the thickness of a-C:B film at 0 V and -20 V. The ion energy (E) is proportional to substrate bias voltage (V_b) [8]:

978-1-4799-5761-3/14 $31.00 © 2014 IEEE

$$E=(kV_b)/P^{0.5} \qquad (1)$$

where P is discharge pressure and k is a constant. According to formula (1), the energy increases with the increase of substrate bias from 0 to -20 V. It observe that the a-C:B film thickness increase from 150 nm to 220 nm. The change of thickness in the a-C:B films deposited with different substrate bias can be explained by the sub-plantation model [15, 16]. If the energy of an impinging species (E_{im}) lowers than penetration threshold (Ep), it cannot penetrate the surface, so it just stick to surface, and remain in their lowest energy state [15], at this case the thickness is low. When E_{im} is higher than E_p, the species have a probability to penetrate the surface, and enter a subsurface interstitial site leading to an increase of the thickness.

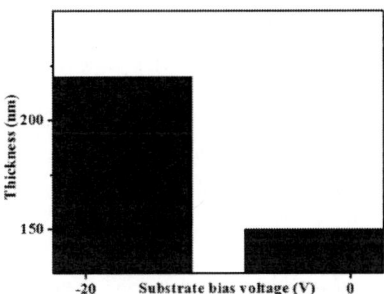

Fig. 1. The thickness of a-C:B films at 0 V and -20

AFM micrographs of the a-C:B films are given in Fig. 2. It can be seen that the average particle size on the film surface becomes smaller with increasing substrate bias from 0 to -20 V. At the bias voltage 0 V, the a-C:B film shows high peaks grain particles on the a-C:B film surface. When the bias voltage increased to -20 V the grain peaks particles of a-C:B film became low and dense, and the surface roughness decreased significantly from 7.89 nm to 1.05 nm.

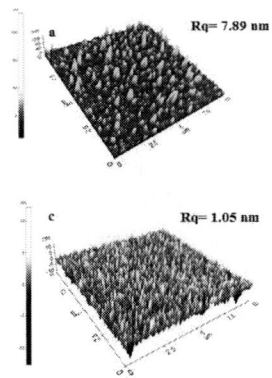

Fig. 2. AFM images of a-C:B film at (a) 0 V and (b) -20 V

For 0 V, the depositing particles undergo several collisions before they can reach the growing interface. Thus, the particles arrive following trajectories with different angles with respect to the growing surface. According to sub-plantation model, the growth of the film is a surface growth processes when E_{im} is lower than E_p. Surface diffusion on the surface causes the growth of the large clusters of sp^2 component leading to substantial roughening [21, 22]. When E_{im} is higher than E_p (for -20 V), the growth of the film is internal growth process. The surface morphology remains relatively unaffected by the penetration of the species, which leads to the smoother surface [22, 23].

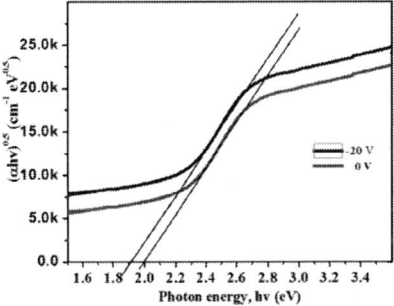

Fig. 3. Optical band gaps of boron doped amorphous carbon at 0V and -20V

The E_g of the thin films is obtained from the extrapolation of the linear part of the curve at the α=0, using the Tauc relation. The estimated E_g of a-C:B films at 0 V and -20 V are approximately 2.0 and 1.8 eV, respectively as shown in Fig. 3. The optical band gap decreases as the applied of negative bias voltage which might be sucssfullly intersial boron doping [10-12]. This value of the E_g was a good agreement why the surface roughness at -20 V has finer and denser than 0 V.

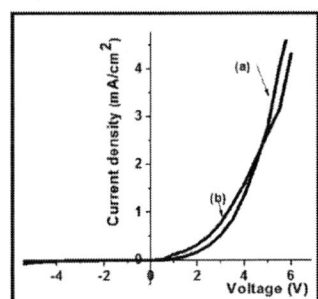

Fig. 4. The current-voltage characteristics of Au/a-C:B/n-Si/Au at (a) 0 V and (b) -20 V under dark measurement

The current density–voltage (J–V) characteristics of Au/a-C:B/n-Si/Au solar cells in dark environment are shown in Fig. 4. The Au/a-C:B/n-Si/Au solar cells display rectifying characteristic, indicating the formation of layers between the a-C:B film and silicon. The a-C:B layer acted as a p-type semiconductor with respect to n-silicon substrate, thus forming the rectifying characteristics.

TABLE I.

THE ELECTRONIC PROPERTIES OF Au/a-C:B/n-Si/Au SOLAR CELL FABRICATED AT 0 V AND -20 V

Negative bias (V)	Open-circuit voltage (V_{OC})	Current density (mA/cm²)	Fill factor (FF)	Conversion efficiency (%)

0	0.164	1.774	0.251	0.0381
-20	0.345	5.661	0.253	0.453

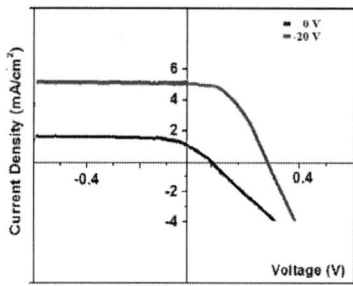

Fig. 5.The current-voltage characteristics of Au/a-C:B/n-Si/Au heterojunction solar cell device under illumination

The electronic properties of Au/a-C:B/n-Si/Au solar cell, including its open Voc, Jsc, FF, and efficiency are presented in Table. 1.The low Voc and Jsc values are attributed to the low built-in voltage, which are caused by the high amount of defect in the a-C:B films. The electronic properties of Au/a-C:B/n-Si/Au solar cells remarkably improved by the increase in the bias of the substrate. The improvement of conversion efficiency can be attributed to the successful boron incorporation, which increases the number of excess carriers and therefore increases the current density of the a-C:B film. The reduction of defects increases the built-in voltage, thereby prolonging the lifetime of excess carriers and providing wide diffusion. The energy-conversion efficiency achieved in this study (0.453% at -20 V) is higher than previously reported [24,25].

CONCLUSION

The change in the thickness, surface morphology, optical and electronic properties of the a-C:B films synthesized palm oil precursor were presented. The results show that with substrate bias of -20 V, the thickness increase and surface roughness becomes lower while the film become less transparency. These results strongly suggest that increasing the substrate bias produced more energetic ions, and these ions led to surface modifications of a-C:B films during the film deposition. The I–V measurement using the solar simulator revealed the formation of p-n abrupt junction of Au/a-C:B/n-Si/Au device. The conversion efficiency achieved to 0.453% at -20 V.

ACKNOWLEDGMENT

The authors thank to Ministry of Higher Education (MOHE) Malaysia for the scholarship, and Research Management Institute (RMI) Universiti Teknologi MARA (UiTM) for the facilities and the financial support.

REFERENCES

[1] Atthapon Srifa, Kajornsak Faungnawakij, Vorranutch Itthibenchapong, Nawin Viriya-empikul, Tawatchai Charinpanitkul, Suttichai Assabumrungrat, "Production of bio-hydrogenated diesel by catalytic hydrotreating of palm oil over $NoS_2/\gamma-Al_2O_3$ catalyst," Bioresource Technology, vol. 158, pp. 81-90, 2014.

[2] Jalal Rauhi, M. Alimanesh, S. Mahmud, C.H Raymond, M. Rusop, " A novel method for synthesis of well-aligned hexagonal ZnO nanostructures in field emission applications," Materials letters, vol. 125, pp. 147-150, 2014.

[3] Nugroho Dewayanto, Ruzinah Isha, Mohd Ridzuan Nordin, "Use of palm oil decanter cake as a new substrate for the production of bio-oil by vacuum pyrolysisA green precursor for carbon nanotube syhthesis," Energy Conversion and Management, vol. 86, pp. 226-232, 2014.

[4] A. Filomena Pinto, F.T. Varela, M. Gonçalves, R. Neto André, P. Costa, B. Mendes, "Production of bio-hydrocarbons by hydrotreating of pomace oil," Fuel, vol.116 pp. 84-93, 2014.

[5] Jalal Rauhi, M. Alimanesh, R. Dalvand, C.H Raymond, M. Rusop, "Optical properties of well-alligned ZnO nanostructure arrays synthesized by electric field-assisted chemical bath deposition," Ceramic international, vol.40, pp.11193-11198, 2014.

[6] A. Mallikarjuna Reddy, A. Sivasangkar Reddy, P. Sreedhara Reddy, "Effect of substrate bias voltage on the physical properties of dc reactive magnetron sputtered NiO thin films," J. Material Chemistry and Physics, vol. 125, pp. 434-439, 2011.

[7] Hongwei Zhang, Zhanjun Zhang, Jingjian Li, Shengmin Cai, "Effect of direct current bias voltages on supported bilayer lipid membranes on a glassy carbon electrode," Electrocheministry Communications, vol. 9, pp. 605-609, 2007.

[8] A. Bubenzer, B. Dischler, G. Brandt, P. Koids, "rf-plasma deposited amorphous hydrogenated hard carbon films: preparation, properties, and applications, Journal of Applied Physics, vol. 54, pp. 4590, 1983.

[9] M. Rusop, S. Abdullah, J. Podder, T. Soga, and T. Jimbo, "Effect of Gas Pressure on the Boron-Doped Hydrogenated Amorphous Carbon Thin Films Grown by Radio Frequency Plasma-Enhanced Chemical Vapor Deposition," Surface Review and Letters, vol. 13, pp. 7-12, 2006.

[10] J. Podder, M. Rusop, T. Soga, and T. Jimbo, "Boron doped amorphous carbon thin films grown by rf PECVD under different partial pressure," Diamond and related materials, vol. 14, pp. 1799-1804, 2005.

[11] M. Rusop, T. Soga, and T. Jimbo, "Properties of an nC: P/p-Si carbon-based photovoltaic cell grown by radio frequency plasma-enhanced chemical vapor deposition at room temperature," Solar energy materials and solar cells, vol. 90, pp. 291-300, 2006.

[12] M. Rusop, T. Soga, and T. Jimbo, "Properties of an nC: P/p-Si carbon-based photovoltaic cell grown by radio frequency plasma-enhanced chemical vapor deposition at room temperature," Solar energy materials and solar cells, vol. 90, pp. 291-300, 2006.

[13] S. M. Mominuzzaman, K. M. Krishna, T. Soga, T. Jimbo, and M. Umeno, "Raman spectra of ion beam sputtered amorphous carbon thin films deposited from camphoric carbon," Carbon, vol. 38, pp. 127-131, 2000.

[14] X. Tian, M. Rusop, Y. Hayashi, T. Soga, T. Jimbo, and M. Umeno, "A photovoltaic cell from p-type boron-doped amorphous carbon film," Solar energy materials and solar cells, vol. 77, pp. 105-112, 2003.

[15] J. Lascovich, R. Giorgi, and S. Scaglione, "Evaluation of the sp2/sp ratio in amorphous carbon structure by XPS and XAES," Applied surface science, vol. 47, pp. 17-21, 1991.

[16] N. Dwivedi, S. Kumar, J. D. Carey, R. Tripathi, H. K. Malik, and M. Dalai, "Influence of silver incorporation on the structural and electrical properties of diamond-like carbon thin films," ACS applied materials & interfaces, vol. 5, pp. 2725-2732, 2013.

[17] H. Dai, Y. Zhang, Z. Chen, and F. Zhai, "Investigating the structural and physical properties of hydrogenated amorphous carbon films fabricated by middle frequency pulsed unbalanced magnetron sputtering," Physica B: Condensed Matter, vol. 438, pp. 34-40, 2014.

[18] N. Dwivedi, S. Kumar, J. Carey, and H. K. Malik, "Photoconductivity and characterization of nitrogen incorporated hydrogenated amorphous carbon thin films," Journal of Applied Physics, vol. 112, p. 113706, 2012.

[19] J. G. Buijnsters and L. Vázquez, "Influence of external bias on the surface morphology of aC: H films grown by electron cyclotron resonance chemical vapor deposition," Surface and Coatings Technology, vol. 201, pp. 8950-8954, 2007.

[20] T. S. Cale and V. Mahadev, "Feature scale transport and reaction during low-pressure deposition processes," Thin films, vol. 22, pp. 175-276, 1996.

[21] P. Kelly and R. Arnell, "Magnetron sputtering: a review of recent developments and applications," Vacuum, vol. 56, pp. 159-172, 2000.

[22] X. Peng, Z. Barber, and T. Clyne, "Surface roughness of diamond-like carbon films prepared using various techniques," Surface and Coatings Technology, vol. 138, pp. 23-32, 2001.

[23] Y. Lifshitz, G. Lempert, E. Grossman, I. Avigal, C. Uzan-Saguy, R. Kalish, et al., "Growth mechanisms of DLC films from C+ ions: experimental studies," Diamond and related materials, vol. 4, pp. 318-323, 1995.

[24] Y. Hayashi, S. Ishikawa, T. Soga, M. Umeno, and T. Jimbo, "Photovoltaic characteristics of boron-doped hydrogenated amorphous carbon on n-Si substrate prepared by r.f. plasma-enhanced CVD using trimethlboron," Diamond and Related Materials, vol. 12, pp. 687-690, 2003.

[25] D. Kamaruzaman, N. Ahmad, I. Annuar, and M. Rusop, "Semiconducting Properties of Nanostructured Amorphous Carbon Thin Films Incorporated with Iodine by Thermal Chemical Vapor Deposition," Japanese Journal of Applied Physics, vol. 52, p. 11NL02, 2013.

Optical performance of MEH-PPV/ZnO Nanocomposite at Different Weight Percent for OLED Applications

N. E. A. Azhar[1], S. S. Shariffudin[1], Z. Nurbaya[1], I. H. H. Affendi[1], A. K. Shafura[1]

[1]NANO-Electronic Centre (NET), Faculty of Electrical Engineering, Universiti Teknologi MARA, 40450 Shah Alam, Selangor, Malaysia
Email: najwaezira@yahoo.com

M.Rusop[1, 2]

[2]NANO-SciTech Centre (NST), Institute of Science, Universiti Teknologi MARA, 40450 Shah Alam, Selangor, Malaysia

Abstract—**Nanocomposite based on zinc oxide (ZnO) nanostructures and poly [2-methoxy-5(2' –ethylhexyloxy)-phenylene vinylene) (MEH-PPV) of various weight percent have been obtained using sol-gel method. The substrates were deposit at 0.1 wt% to 0.4 wt% of ZnO with pure MEH-PPV to investigate the concentration effect of MEH-PPV/ZnO nanocomposite. The structural properties were characterized using FESEM and AFM to obtain the morphology of nanocomposite. From the AFM, it was found that the roughness is more uniform. The optical properties were obtained using ultraviolet-visible spectrometer (UV-Vis). It was found that the transmittance band increased with decreased of weight percent of ZnO nanostructures. For photoluminescence (PL) spectra shows that 0.4 wt% of ZnO at visible emission is due to emission characterisitic of PPV backbone which is arise from the relaxtion of excited π-electron to the ground state. This study will provide better performance and suitable for optoelectronic device especially OLEDs application.**

Keywords—Nanocomposite MEH-PPV/ZnO; weight percent; sol-gel method; structural and optical properties

I. INTRODUCTION

Nowadays, the developments of optoelectronic are widely used due to environment-friendly and good performance device. In 1987, Tang and Vanslyke look forward on structure organic light-emitting diodes (OLEDs) thin film with high brightness driving voltage and high electroluminescence (EL) showed good potential as solid state lighting source. There are several advantages using organic material in organic devices such as simple processing, large area, low cost and environment-friendly that attract for several applications [1].

An inorganic-organic nanocomposite have been introduced for optoelectronic device especially in organic light emitting diodes (OLEDs) because of its efficient to produce stable and high performance device [2]. During operation, degradations can occur in different parts of OLED such as chemical and photo-oxidation of the active layer, thermal instabilities, and diffusion of metal from electrodes [1]. In order to improve the stability of the organic device, an inorganic nanostructures which is oxide semiconductors are required to form composite.

In theoretically, by optimizing parameter such as concentration is necessary to control the functionalities of the composite. The preparation of nanocomposites is important to obtain high quality of material.

In this study, we investigated the nanocomposite that made by adding the different amount of nanotetrapods ZnO in MEH-PPV. Hence, the effects of MEH-PPV/ZnO nanocomposite on structural and optical properties were prepared using sol-gel method.

II. EXPERIMENTAL DETAILS

A. Preparation of Substrates

The Indium Tin Oxide (ITO) coated glass substrates were cut with size 2cm x 2 cm. The substrates were cleaned with acetone, methanol and DI water in the ultrasonic cleaner stage for 10 minutes for each. Then, the substrates were dried using a blower to remove the dust onto the substrates.

B. Preparation of ZnO Nanotetrapods

ZnO nanotetrapods were prepared and grown using thermal-CVD. 1g of zinc powder (99.9% purity; Sigma-Aldrich) was placed in an alumina boat at the centre of the tube of furnace 1 and furnace 2. The zinc powder was spread in the boat at furnace 1 as indicate in Fig. 1. Both ends of the quartz tube were encapsulated with silicon stopper. Argon gas was fed into the tube for 15 minutes to eliminate all other gases. The temperature of furnace 1 was set at 750°C while furnace 2 was set to 500°C. When the temperature reached the target temperature, oxygen gas was supplied into the tube with flow rate of 5 sccm. The growth time was 30 minutes under a constant flow of 100 sccm of argon gas with pressure of 1bars. Lastly, the furnace cooled down to room temperature [3].

C. Preparation of MEH-PPV/ZnO Nanocomposite

Poly[2-methoxy-5(2'–ethylhexyloxy)-phenylene vinylene) (MEH-PPV) were dissolved in toluene with 5mg/ml. The MEH-PPV solutions were stirred for 48 hours at room temperature [4]. After 48 hours, the ZnO powders were added into MEH-PPV solution and sonicated for 30 minutes. The nanocomposites were stirred for 1 hour before spin coat method.

The MEH-PPV/ZnO nanocomposites were formed by spin-coating method. The nanocomposites were prepared for pure MEH-PPV and 0.1 wt% to 0.4 wt%.The nanocomposites were spun at 2000 rpm with 60 seconds and dried onto hot plate at 50°C for 5 minutes.

D. Characterization Technique

The MEH-PPV/ZnO nanocomposite has been measured by thickness using surface profiler (VEECO DEETAK 150). The synthesization of ZnO has been measured by using field emission scanning electron microscope (JEOL JSM-J600F) and atomic force microscopy (AFM), JASCO UV-Vis-NIR spectrophotometer (V670 EX) for surface morphology and roughness respectively. For the optical properties, photoluminescence (PL) and transmittance spectra for MEH-PPV/ZnO nanocomposite were characterized using FlouroMax3 Horiba Jobin Vyon and UV-Vis spectrometer (JASCO UV-Vis) respectively.

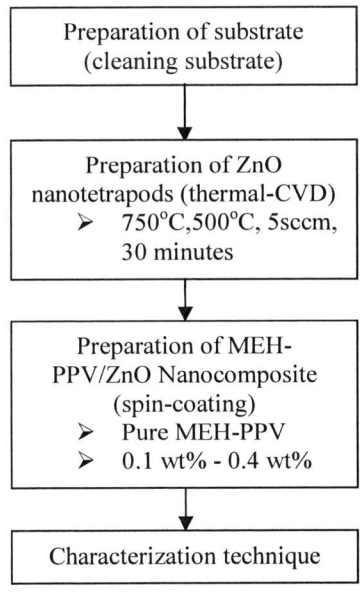

Fig. 1. Flow Chart of MEH-PPV/ ZnO Nanocomposite Process

III. RESULT AND DISCUSSION

A. Structural Properties

The thickness of MEH-PPV/ZnO nanocomposite was measured at different weight percent of ZnO using surface profiler. The value thickness of nanocomposite is shown in Table I.

The minimum thickness of MEH-PPV/ZnO nanocomposite is 44.7 nm for pure MEH-PPV while the highest thickness of films is 172.1 nm for 0.4 wt%. When the weight percent increase the thickness of MEH-PPV/ZnO nanocomposite increased. From the Table I, it can be concluded that the diameter size also increased with the weight percent. It can be proven by morphologies and AFM images shown in Fig. 2 and Fig. 3 respectively.

TABLE I.   THICKNESS OF MEH-PPV/ZNO NANOCOMPOSITE

MEH-PPV/ZnO Nanocomposite at different weight percent (wt%)	Film Thickness(nm)
Pure MEH-PPV	44.7
0.1	67.2
0.2	90.4
0.3	104.2
0.4	172.1

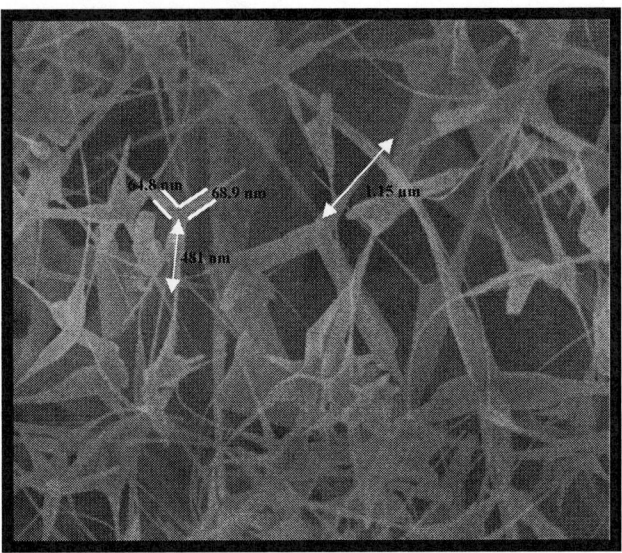

Fig. 2. FE-SEM images of ZnO nanotetrapods deposited by thermal-CVD

Surface morphology of nanotetrapods ZnO by thermal-CVD was obtained using field emission scanning electron microscopy (FE-SEM). Figure 2 shows the synthesis of nanotetrapods ZnO was deposited at 750°C, 5 sccm with 30 minutes deposition. The FESEM images representative the surface morphologies of the nanotetrapods at higher magnification which is 30k magnification with 100 nm. Diameters of the nanoneedles were in the range 64 nm to 89 nm. While the lengths of the nanoneedles 481 nm to 1.15 μm. Based on FESEM image can be seen that the nanoneedles were interconnected to each other [3]. When the weight percent increase, it can be seen that the diameter of nanotetrapods become more uniform hexagonal. When zinc reacted with oxygen, it can attribute to the thermodynamic equilibrium and kinetically controlled progress. Hence, the aggregation of the sample and the polymer clustering are visible in the range of nanometer for MEH-PPV only [5, 6].

The atomic force microscope (AFM) images in Fig. 3 with scan size 10 μm x 10 μm indicate that the nanocomposite were prepared at (a) pure MEH-PPV and (b) MEH-PPV/ZnO nanocomposite. From Fig. 3 shows the surface roughness of pure MEH-PPV and MEH-PPV/ZnO nanocomposite were ra=4.07 nm and ra=201.00 nm respectively. The surface roughness of MEH-PPV/ZnO nanocomposite was higher than pure MEH-PPV. It can be seen that the surface of pure MEH-PPV is more uniform compared to nanocomposite [6]. The surface roughness of the films might be affected the structures of ZnO. This is good agreement with the thickness of MEH-PPV/ZnO nanocomposite.

B. Optical Properties

The optical transmittances of nanocomposites were measured using UV-Vis-NIR spectrometer shown in Fig. 4. From Fig. 4 shows the transmittance spectra deposited at different weight percent concentration in the wavelength range between 300 to 600 nm.

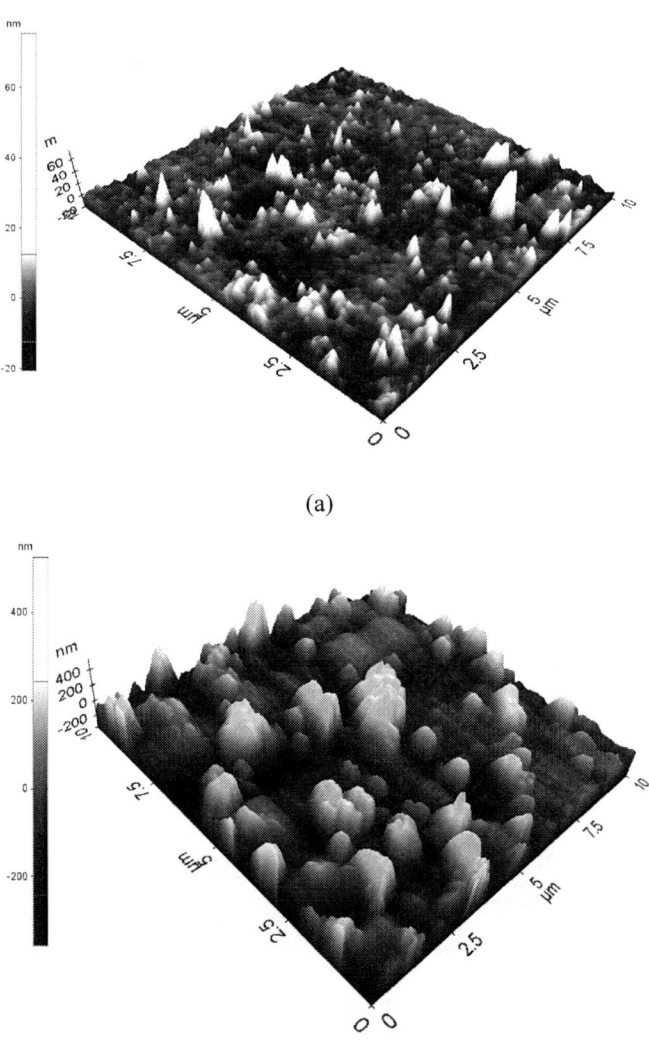

(a)

(b)

Fig. 3. AFM images of MEH-PPV/ZnO nanocomposite deposited by sol-gel spin coating method

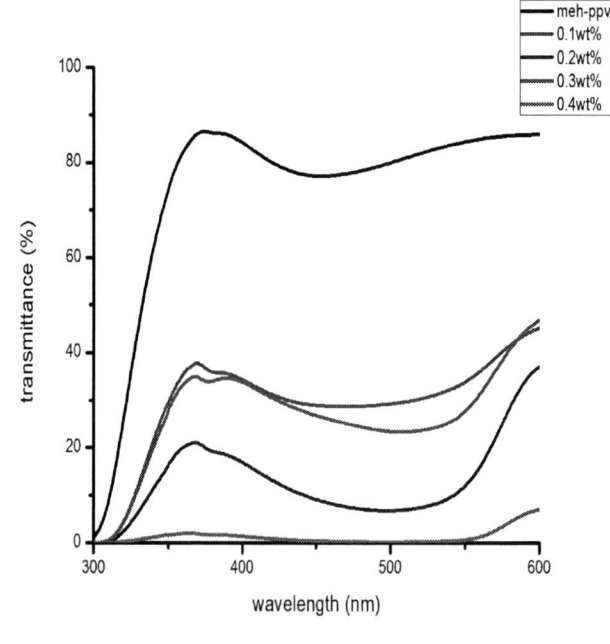

Fig. 4. Optical transmittance spectra of pure MEH-PPV and MEH-PPV/ZnO nanocomposite

For pure MEH-PPV sample was found to be transparent which is >80% in the visible-NIR range. The highest transmittance was obtained for sample deposited at pure MEH-PPV in 350 to 400 nm region while the lowest transmittance is 0.4 wt%. The reduction of transmittance properties is due to the increases of nanotetrapods ZnO concentration in the composite [7]. In contrast, the sample deposited at 0.4 wt% exhibit high absorption properties due to the low transmission. This is agreeing with thickness, AFM images and surface morphology above.

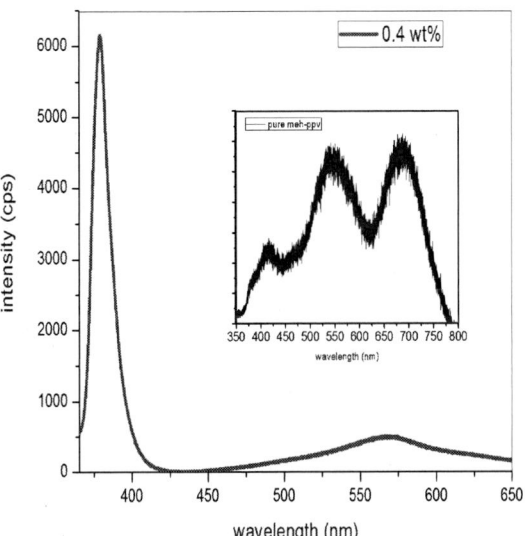

Fig. 5. Photoluminescence of MEH-PPV/ZnO nanocomposite deposited at pure MEH-PPV and nanocomposite 0.4 wt% of ZnO

The photoluminescence (PL) of MEH-PPV/ZnO nanocomposite were characterized at room temperature in the wavelength range of 325 nm to 800 nm. The PL spectra for MEH-PPV/ZnO were evaporated at pure MEH-PPV and nanocomposite 0.4 wt% of ZnO shown in Fig. 5. The MEH-PPV/ZnO nanocomposite were characterized under exciton of He-Cd laser (325 nm). The PL spectra consist of narrow peak in a near UV region (380 nm) and a broad peak of 571 nm in the visible region (yellow-orange band) for sample 0.4 wt% while the visible region for pure MEH-PPV at 687 nm (red emission). The peak at 571 nm is due to emission characterisitc of PPV backbone which is arise from the relaxtion of excited π-electron to the ground state [8]. While the peak at 687 nm is due to the interchain state [8, 9]. The higest emission was found at 0.4 wt%. The highest emission is due to high-energy transfer from particle to the polymer.

IV. CONCLUSION

In conclusion, this paper is successful characterized at different weight percent concentration using sol-gel spin coating method. The FESEM image shows that the homogeneity of nanotetrapods are good condition, however the aggregation of nanocomposite based on AFM images show the independent pattern when the weight percent concentration was varied. The UV-Vis transmittance spectrum for pure MEH-PPV is high. The PL spectra show that the emission peak slightly yellow-orange shift as ZnO concentration increased.

ACKNOWLEDGMENT

This work was supported by Research Management Institute (RMI) through the project 600-RMI/DANA 5/3/CIFI (139/2013) for financial, NANO-Electronic Center (NET) and NANO-SciTech Centre of Universiti Teknologi MARA, (UiTM) Malaysia.

REFERENCES

[1] T.-P. Nguyen, "Polymer-based nanocomposites for organic optoelectronic devices. A review," *Surf. Coatings Technol.*, vol. 206, no. 4, pp. 742–752, Nov. 2011.

[2] I. Musa, F. Massuyeau, E. Faulques, and T.-P. Nguyen, "Investigations of optical properties of MEH-PPV/ZnO nanocomposites by photoluminescence spectroscopy," *Synth. Met.*, vol. 162, no. 19–20, pp. 1756–1761, Nov. 2012.

[3] S. S. Shariffudin, S. H. Herman, and M. Rusop, "Self-Catalyzed Thermal Chemical Vapor Deposited ZnO Nanotetrapods," *Adv. Mater. Res.*, vol. 832, pp. 670–674, Nov. 2013.

[4] S. S. Shariffudin, M. H. Mamat, M. Rusop, U. T. Mara, and S. Alam, "Electrical & Optical Properties of Nanocomposite MEH-PPV / ZnO Thin Film," pp. 380–383, 2010.

[5] B. J. Schwartz, "Conjugated polymers as molecular materials: how chain conformation and film morphology influence energy transfer and interchain interactions.," *Annu. Rev. Phys. Chem.*, vol. 54, no. 3, pp. 141–72, Jan. 2003.

[6] N. Juhari, W. H. a. Majid, and a. I. Zainol, "The SEM & AFM Images of MEH-PPV Films below CLA Region," *Procedia Eng.*, vol. 53, pp. 354–361, Jan. 2013.

[7] S. Paper, M. H. Ibrahim, and A. S. Hamzahc, "The Optical Properties of Nanocomposite MEH-PPV:ZnO Thin Films," pp. 3–4, 2010.

[8] N. Z. Yahya and M. Rusop, "Investigation on the Optical and Surface Morphology of Conjugated Polymer MEH-PPV:ZnO Nanocomposite Thin Films," *J. Nanomater.*, vol. 2012, pp. 1–4, 2012.

[9] R. Traiphol, N. Charoenthai, T. Srikhirin, T. Kerdcharoen, T. Osotchan, and T. Maturos, "Chain organization and photophysics of conjugated polymer in poor solvents: Aggregates, agglomerates and collapsed coils," *Polymer (Guildf).*, vol. 48, no. 3, pp. 813–826, Jan. 2007.

Glass Etching for Cost-effective Microchannels Fabrication

N.M. Salih, N. Nafarizal, C.F. Soon, M.Z. Sahdan
Microelectronics and Nanotechnology Centre,
University Tun Hussein Onn Malaysia, 86400 Parit Raja,
Batu Pahat, Johor, Malaysia
zainizno@gmail.com

A. Tijjani, U. Hashim
Insitute Nano Electronic Engineering (INEE),
Universiti Malaysia Perlis (UNIMAP), 01000 Kangar
Perlis, Malaysia

Abstract— The available fabrication of microfluidic in industry involves with complex and expensive procedures. It is crucial to find another alternative which can reduce the cost production, simplify the process, and promote to a rapid production cycle. In this paper, a cost-effective approach of glass etching procedure for microchannels fabrication is described. The procedure involves with masking preparation using positive photoresist PR1-4000A, photolithography technique, and effective wet chemical etching process. From the experimental, good photoresist mask thickness and adherence was achieved. Then, acceptable etching rate, photoresist resistant, and surface roughness were obtained with suitable hydrofluoric acid (HF), hydrochloric acid (HCl), and de-ionized (DI) water composition. The presented simple fabrication process is suitable for fast prototyping and manufacturing disposable microfluidic devices. With the optimized process, a glass-based microchannels with etching depth up to 70µm was produced.

Keywords□ photolithography, wet etching, microchannels, microfluidic

I. INTRODUCTION

In micro-electromechanical systems, glass and silicon micro-machining is commonly used for microfluidic devices fabrication due to its fidelity in achieving small feature resolution, surface stability and solvent compatibility. Glass is a material of choice in microfluidics due to its beneficial optical properties, minimum chemical reaction, and the strength of anodic bonding which allows an excellent resistance to high pressures [1]. Due to the low cost and simplicity, commercially available microscopic slides have been used in this paper.

Various masking techniques have been implemented to make microchannels using glass substrate. Different mask layers for chemically glass etching have been reported using different materials such as Cr, Cr/Au, polysilicon to deposit a layer as a mask on glass in order to make an open region for wet chemical etchant by different deposition methods [2]. Approaching a low cost fabrication, we are using photolithographic technique to deposit the photoresist mask layer and HF-based wet etching method in this paper. HF-based methods usually result in a rough surface, but the

combination which consisting of hydrofluoric acid (HF), hydrochloric acid (HCl), and de-ionized (DI) water is expected to provides a smoother etched surface [3].

II. METHODOLOGY

The experiment involves two important procedures; Photolithography process and wet chemical etching as shown in Fig. 1.

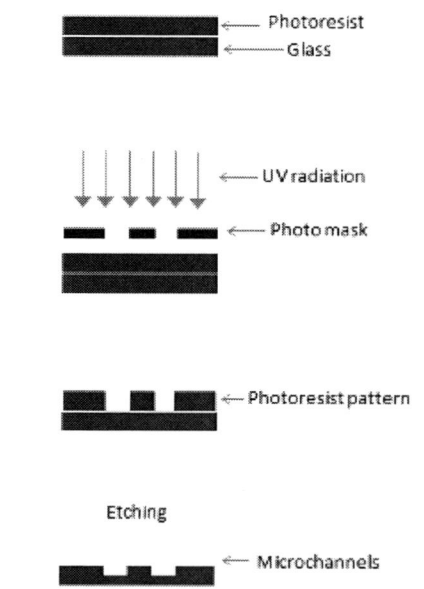

Fig. 1. Photolithography and wet etching process flow diagram

A. Photolithography

The glass slides were cleaned using ethanol and DI water. The cleaned glass substrates were then spin-coated with positive photoresist PR-4000A and exposed to UV radiation using a mask aligner machine together with photomask transparency. Then, the development of the photoresist was accomplished by immersing the exposed substrate into the RD6 developer and post baked on hot plate.

B. Wet Etching

In this study, the etching was performed using liquid chemicals. Etching solution contains a mixture with suitable concentration of DI water, HF, and HCl. The substrate with developed photoresist was immersed into the etching solution for an appropriate amount of time. After etching, the photoresist was removed from the glass surface using acetone, rinsed with DI water, and dried using nitrogen gas. The etching of the glass substrate was analyzed for different time and chemical etchant composition/ratio (HF: HCl: DI). .

III. RESULT AND DISCUSSION

Photoresist masking and etchant solution are the most important aspects in glass etching for microchannels fabrication. From the experiment, a well-developed photoresist masking with microchannels pattern as shown in Fig. 2 was produced. Proper photoresist coating, baking, UV-exposure, and development steps are important in preparing photoresist masking for chemical etching.

In photoresist coating, the spin speed plays a major role in achieving the desirable thickness and uniformity. From previous study, it was reported that a slow coating will produce a thick photoresist layer but, resulting in poor adhesion with the glass substrate [4]. Then, faster spin speed on the other hand will produce very thin photoresist layer and a low resistance to erosion [4]. In this work, an appropriate photoresist coating was achieved using coating speed combination of 300 rpm for 5 seconds, followed by 500 rpm for 20 seconds. The thickness of the photoresist was measured at 10-12μm which is believed able to withstand the attacks of the etchant solution up to 2 hours [5].

Besides photoresist thickness, good adhesion of photoresist on the glass substrate also crucial for well-defined photolithographic masking shape and etch structure. Fig. 3 shows the microscopic image of the microchannel's inlet with 3mm diameter. From the microscopic image in Fig. 3, it shows the masking pattern was developed with no major deformity on the channel's wall. It is believed that proper pre and post-baking promotes better adhesion of the photoresist on the glass surface [5]. Based on the experimental, after photoresist spin coating on glass, the substrate was pre-baked at 90 °C for 10 min resulting moisture-free surface. Then, immediately after exposure the resist was heated at 100°C for 3 min for adhesion promotion and post-baked at 130°C for 90 min in order to harden the photoresist. With a good photoresist thickness, adhesion, and hardness, it will prevent the photoresist from peeling off during etching.

In this study, different mixtures of HF, HCl, and DI water were studied in order to achieve acceptable etch rate, better photoresist resistant, and smooth microchannels surface. Based on the previous reports, the photoresist mask can resist only for a short time, which is less than 5 min, in highly concentrated HF solutions [3]. Due to this issue, DI water was added to dilute the etchant solution and improve the photoresist resistant up to more than 2 hours. Furthermore, surface roughness and channel geometry after etching will affect the performance of microfluidic devices [6]. It was reported that HF-based etchants usually result in a rough surface [3]. Subsequently, HCl was added into the etchant solution. By adding HCl, the insoluble products were dissolved during etching process and a smoother etched surface was achieved. From the analysis, it was observed that the etched glass substrate in Fig. 5 shows smoother surface compare to etched glass substrate in Fig. 4.

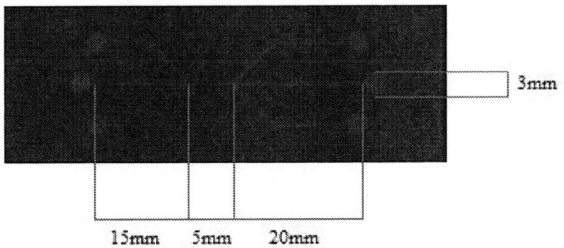

Fig. 2. Photoresist microchannels pattern on glass substrate

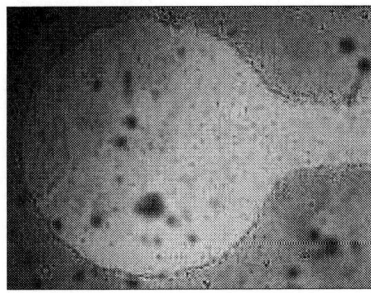

Fig. 3. Microscopic image of developed photoresist pattern (high power microscope, 10x magnification)

Fig. 4. Surface profiler image of etched glass substrate with diluted HF etchant (HF: DI, 1:4)

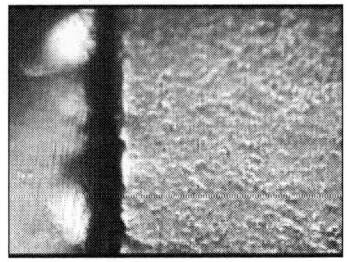

Fig. 5. Surface profiler image of etched glass substrate with etchant composition of HF: HCl: DI (1:2:2)

Fig. 4 shows the surface profiler image of etched glass substrate using diluted HF etchant (HF: DI, 1:4) while Fig. 5 shows the surface profiler image of etched glass substrate using combination of HF: HCl: DI (1:2:2). From the result, we had indicates that the combination of HF with HCl produces better surface roughness. After that, different etchant composition (HF: HCl: DI water) was investigated in this study (as shown in TABLE I). From the table, it was observed that higher concentration of HF in the etchant will produce lower etching rate but, higher resistance time of the photoresist mask. The etchant composition of 1:2:4 (HF: HCl: DI water) was selected due to acceptable etching depth, smooth etched surface, and high photoresist resistance time. From the experiment, the glass substrate was successfully etched with desired microchannels pattern as shown in Fig. 6. The etched glass substrate was produced with same dimension as the photoresist masking in Fig. 2.

The microscopic image in Fig. 7 shows that the etched pattern on the glass substrate has no major deformity which can lead to leakage and prevent fluid flow. However, further improvement can be done for better geometry shape and etching uniformity. Further analysis, the depth of the etched glass substrate was measured using a surface profilometer (Alpha-step IQ, KLA Tencor). The measurement was done on three different locations and the average reading was calculated. The etching of the glass substrate was analyzed for different immersing time in order to determine the average etching rate (as shown in Fig. 8). From the graph in Fig. 8, the average etching rate was calculated at 0.65µm/minute. From this work, the glass substrate was successfully etched up to 70µm in 2 hours.

TABLE I.   STUDY OF DIFFERENT ETCHANT COMPOSITION

Etchant ratio	Etching depth in 1hour (µm)	Surface roughness	Photoresist surviving time (min)
1: 2: 2	50.20	rough	60
1: 2: 3	47.58	rough	90
1: 2: 4	38.50	smooth	150
1: 2: 5	25.31	smooth	230

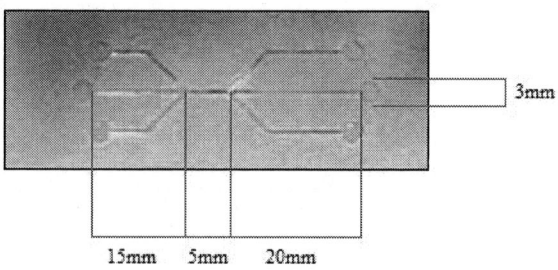

Fig. 6.   Microchannel etched pattern on glass substrate

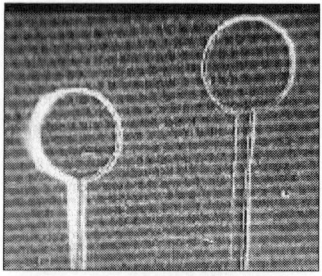

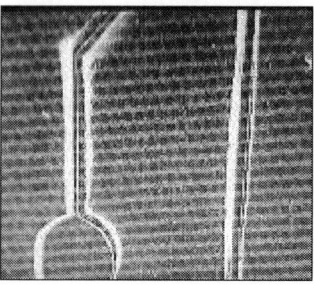

Fig. 7.   Microscopic image of etched pattern on glass substrate (low power microscope, 5x magnification)

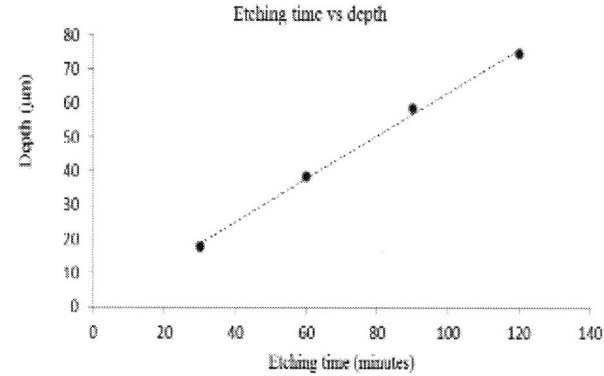

Fig. 8.   Etching time versus depth analysis graph

The etching rate obtained in this study is suitable for the microchannels fabrication. However, the study by C. Iliescu et al. shows that deeper etching of glass substrate was achieved using ultrasonic bath for the wet chemical etching process [7]. However, etching in ultrasonic bath is expected to decrease the photoresist resistant time as the adherence of the photoresist on the glass substrate will be affected [7].

IV.   CONCLUSION

This study had demonstrated a cost-effective microchannels fabrication using fast and less expensive glass etching approach. The experiment was performed using microscope slide glass, positive photoresist, photolithography technique, and wet etching processes. This work had indicates a proper photoresist mask preparation with acceptable thickness of 10-12µm and good adherence on glass substrate. Then, the wet etching process was investigated for better performance using suitable HF: HCl: DI water composition/ratio (1:2:4). From the result obtained, an optimum etching rate of 0.65µm/minute had produced a proper etched microchannels with etching depth up to 70µm with smoother surface and no major deformity on the geometry structure. It is expected that further improvement could be done in order to achieve better glass etching performance for effective microchannels fabrication.

ACKNOWLEDGMENT

The authors would like to thank Ministry of Higher Education Malaysia for the financial support through MTUN-COE grant Vote No. C020. We also acknowledged Universiti Tun Hussein Onn Malaysia (UTHM) and University Malaysia Perlis (UNIMAP) for technical support.

REFERENCES

[1] M. Stjernstrom and J. Roeraade, "Method for fabrication of microfluidic systems in glass", J. Micromech. Microeng. 8, pp. 33-38, 1998.

[2] F. E. H. Tay, C. Iliescu, J. Jing, J. Miao, "Defect–free wet etching through pyrex glass using Cr/Au mask", Microsyst Technol, vol. 12, pp. 935-939, 2006.

[3] G. A. C. M. Spierings, "Wet chemical etching of silicate glasses in hydrofluoric acid based solutions", J. Mater. Sci., vol. 28, pp. 6261-6273, 1993.

[4] G. Wang, Q. Li, Y. Dong., J. Chen, Fj. Gao, and S. Zhang, "Microfluidic Chips Etching Process on Soda-lime Glass", Proceedings of 2012 International Conference on Mechanical Engineering and Material Science (MEMS 2012), pp. 471-474, 2012.

[5] A. Bahadorimehr and B. Y. Majlis, "Fabrication of Glass-based Microfluidic Devices with Photoresist as Mask", Electronics and Electrical Engineering, Kaunas: Technologija, No. 10 (116), pp. 45–48, 2011.

[6] C. Kleinstreuer and J. Koo, "Computational analysis of wall roughness effects for liquid flow in micro-conduits", J. Fluids Eng., vol. 126, no. 1, pp. 1-9, 2004.

[7] C. Iliescu, J. Jing, F.E.H. Tay, J. Miao, and T. Sun, "Characterization of masking layers for deep etching of glass in an improved HF/HCl solution", Surface and Coatings Technology, vol. 198, pp. 314-318, 2005

Fabrication of SONOS Flash Memory Device By Using Engineered Tunnel Barrier Technique

M. R. Zakaria, Shahrir R. Kasjoo, A. F. Mahyidin
School of Microelectronic Engineering
Universiti Malaysia Perlis (UniMAP),
Arau, Perlis, Malaysia
rosydizakaria@yahoo.com, shahrirrizal@unimap.edu.my

A Wesam Al-Mufti, R. Mat Ayub, U. Hashim
Institute of Nano Electronic Engineering
Universiti Malaysia Perlis (UniMAP),
Kangar, Perlis Malaysia
Ramzan@unimap.edu.my, uda@unimap.edu.my

Abstract— **Flash memory is a device that used as a tool to store data electrically without external power supply. The charge-trap such as SONOS structure is the most widely used in flash memory technology fabrication due to the advantages of this device in term of scaling and performance characteristic. Conventional Flash memory with thickness 5nm single oxide shows good performance, but suffer leakage current and data retention. To overcome this problem, a SONOS flash memory was fabricated by using techniques that known as Engineered Tunnel Barrier to replace the conventional single oxide used in conventional flash memory. In this project, the total equivalent thickness oxide for all experiments is set at the 8nm to compare the performances. Thus, it will result in a faster write and erase speed. The analysis results will determine the most preferred structure that improved the programming characteristic.**

Keywords—Engineered Tunnel Barrier, SONOS, Flash Memory

I. INTRODUCTION

Flash memory is a one of device famously used as a storage medium. The ability of Flash memory to keep data even power is disconnected, high data retention and also in term of its scaling is the most reason for its popularity [1,2]. In fact, the Flash memory is able to challenge with hard disk as conventional memory storage and will replace it in the future. Flash memory is widely used in memory cards for digital system and may replace the traditional hard disk for data storage in computer in the future because of its small size and cheapness. Moreover, it also has smaller in size and faster time than a traditional hard drive memory [3,4]. Flash memory is a nonvolatile memory family memory and in a solid state, which means that are no moving parts. It's reliability and long-term of persistent storage device depending on its material and device structure. Flash memory structures can divided into two main classes, charge-trapping (CT) and floating gate (FG). The structure of a both flash memory cell is similar like a basic MOS transistor as shown in Fig. 1 [4,5].

Scalability is the most limitation problem in flash memories. Beside that, the non-scalability of the tunnel barrier and electron leak are parts of its disadvantages[6,7]. Therefore, the charge trap flash memory has been developed as an alternative to scaling or thinning floating gate method. These memory need some improvement in term of its structure because the conventional Flash memory generally has limitation in their programming performance. Tunnel oxide below than 5 nm will improve the device performance, but it will cause serious current leakage problems and can reduce the device data retention.

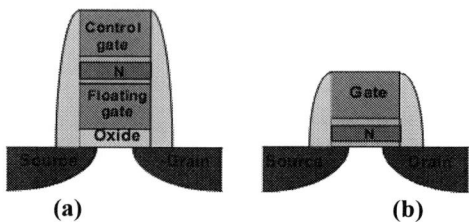

Fig. 1. Conventional Flash memory device
(a) Floating Gate Flash Memory (b) Charge Trap Flash Memory

There are two types of Charge Trap Flash, the SONOS and the Nano-crystal Dots (NC). The SONOS Flash memory device is a promising alternative to floating gate device for minimizing the scaling thining. The SONOS architecture consists of stacks (Si/ SiO_2/ Si_3N_4/ SiO_2/Si). The charges are stored and trapped inside the Si_3N_4 nitride layer since this material is discrete and can prevent the leakage current [8,9,10]. The SONOS device offers improved performance with a small cell size, single-level with low voltage, fast write/erase, improved memory retention, increased endurance, and radiation hardness. Fig. 2 shows the structure of techniques based on charge-trap concept compares the standard structure MOS been introduced to overcome the floating gate problem. This concept, called as an Engineered Tunnel Barrier-SONOS (ETB-SONOS). Recently low barrier, the good tunnel dielectrics like nitride as the high- κ materials have been demonstrated to be memory devices deposition materials as shown in Fig. 2. There are many advantages of using high- κ dielectric as additional dielectric such as low barrier results, larger tunnelling current and improves programming speed and voltage [11]. High- κ constant reduces the charging energy because the deep trap energy level can be obtained with the high- κ material [12,13]. The method that uses to prevent the leakage data is by covered the charge trap layer with IPD. The charge that store in the nitride able to store for years without needing any external supply. The overall retention time of data storage and programming performance are determined by the quality and the thickness of IPD. After the program voltage is removed, the data still remained in the charge trap layer after program bias is supplied at the device. When the barrier is made relatively thick, the longer time and higher voltage are needed to erase or write the charge trap data but it can increase the charge retention times. When the barrier is thin, the charge leakage at the charge trap will reduce the retention time, but will reduce the program and erasing process time [14].

978-1-4799-5761-3/14 $31.00 © 2014 IEEE

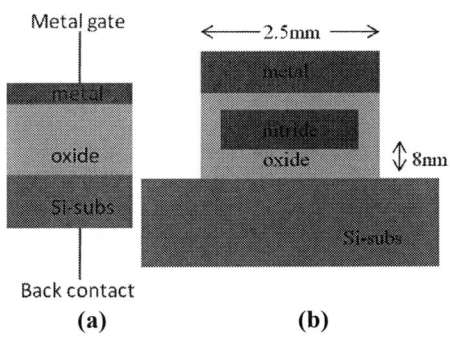

(a)   **(b)**

Fig. 2. Capacitor architecture. a) MOS Capacitor.(b) Conventional Flash Capacitor

This research is focused on finding the best thickness barrier for SONOS Flash Memory device performance in term of its programming tunneling and programming voltage. To achieve this goal, the extension study and experiments based of Charge-Trap Flash memory is designed and fabricated using SONOS charge trap concept to optimize the programming characteristic of the device. In this paper, the investigation was focused on the popular scaling effects using a thin tunnel oxide and engineering tunnel barrier. While Fowler-Nodheim Techniques tunneling mechanism through fabrication modeling was used to study the effect on device performance.

II. Modeling of the Flash Memory.

Fig. 3 illustrates the construction of the structure of conventional Flash SONOS. The basic structure of SONOS is same as a standard MOSFET device. In the first half of the research, the conventional SONOS flash with single tunnel oxide was fabricated to optimize the single tunnel oxide and the effect to device performance was studied. After that, the combination stack SiO_2 with high-κ dielectric such as Silicon Nitride was fabricated to study the effect of engineered tunnel barrier compare single tunnel oxide. We fabricate the SONOS with the asymmetric (3-Layer) and symmetric layer (2-Layer) with same EOT 8nm. The cross-section of SONOS flash memory for single oxide and ETB-SONOS as shows in Fig. 3, respectively.

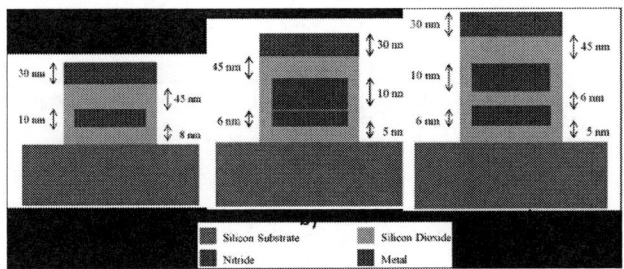

Fig. 3. SONOS Flash Memory. (a) Conventional SONOS. (b) 2-Layer ETB-SONOS (c) 3-Layer ETB- SONOS

III. Result And Discussion

Electrical characterizations and testing were carried out using a Dielectric Analyzer with mini probe station to produce a capacitive voltage (CV) and current voltage (IV) graph. Dielectric analyzer having the option to measure the electronic properties like conductivity, capacitance, inductance and impedance of device by varying the frequency, time and AC/DC voltage or current. The most important property of the Flash capacitor is that its capacitance changes with an applied DC voltage. As a result, the modes of operation of the Flash capacitor change as a function of the applied voltage. Fig. 4 shown a 10 kHz high frequency CV curve of conventional Flash capacitor. As a DC sweep voltage is applied to the gate, it causes the device to pass through accumulation, depletion, and inversion regions. From the result, oxide capacitance for Flash memory for conventional tunnel oxide is 17.96 nF. This C_{ox} is dependent on the thickness and area.

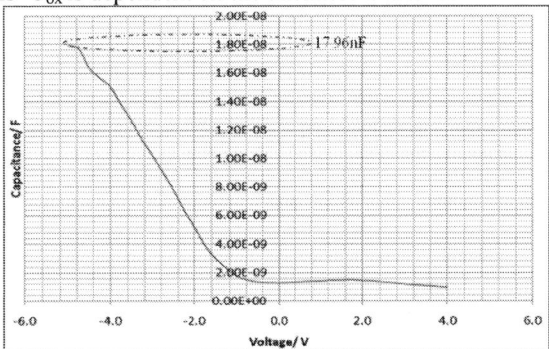

Fig. 4. Conventional Flash CV in 6.25μm² gate area.

Oxide thickness (t_{ox}) measurement was calculated from standard formula oxide capacitance (Cox) discuss in [15] and gained from dielectric analyzer measurement by referring gate area (A) in 6.25μm² and capacitance value from CV Fig. 4.

TABLE I.   Calculation Result of SiO_2 Thickness In Conventional Flash by CV Graph

Expected Parameter	Gate Area (A), μm²	Capacitance of Oxide (Cox), (nF)	Calculated of Oxide Thickness (t_{ox}), nm
Conventional Flash Device 8nm Fabrication	6.25	17.96	11.98

There are two tunnel barrier multi layer techniques that have been fabricated, asymmetric shows in Fig. 3(b) and symmetric in Fig 3(c). The effects of tunnel oxide thicknesses in these techniques also give direct impact on the device performance. The characterizations of these Flash capacitor involve capacitance voltage (CV) measurement at room temperature condition using dielectric parameter has compared with conventional SONOS in Fig. 5. The characterizations are done by manually set the electrical parameter of CV characteristic and connect the mini probe between metal and back side metal to the dielectric analyzer. The result for ETB Flash capacitor is shown in Figure 5(a) for 2-layers and Fig. 5(b) for 3-layers ETB Flash respectively. The figure shows the relationship between capacitance and control gate voltage (V_{cg}) in inversion and depletion region for high frequency capacitance. The voltage threshold (V_{th}) voltage is the minimum voltage that requires to make the device ON. The observation from CV graph, Vth also decreases due to the change of the charge in the depletion region. The slope of 8 nm conventional tunnel oxide was shown that the value of V_{th} at low applied voltage condition. An addition, the slope of 8 nm conventional tunnel oxide was achieved lowest capacitance oxide is around 17.96 nF compared with 5 nm and 3 nm of ETB tunnel oxide that give 37.98 nF and 62.13 nF. The analysis shows that C_{ox} will increase when the tunnel barrier scaling down from 8 nm to 3 nm of oxide thickness. The effect of the oxide thickness is clearly reflected in the Cox characteristic

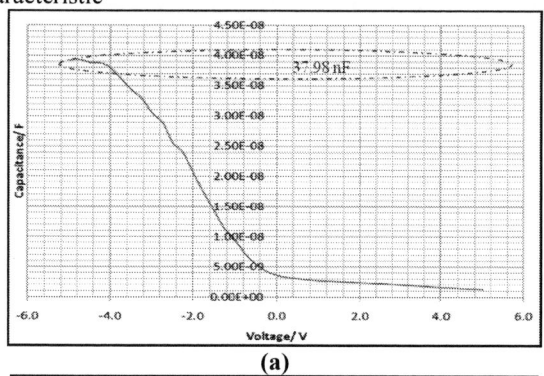

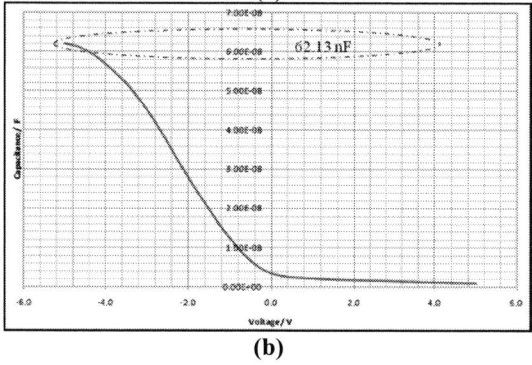

Fig. 5. ETB Flash Capacitor. a) Asymmetric Flash Capacitor, b) Symmetric Flash Capacitor

The oxide thickness calculations based on CV graph measurements can be very accurate by extracted from the strong inversion region, where the capacitance measured reflects the oxide capacitance. Table II shows the calculated result of the SiO_2 thickness using an equation of standard capacitance. From graph and calculation result, the SiO_2 deposition recipe achieved for first oxide layer of tunnel barrier in Flash conventional and ETB tunnel barrier by dry thermal oxidation at 850°C around 1 to 5 minutes growth time. The calculation of oxide thickness in symmetric and asymmetric stack structure is shown in Table II.

TABLE II.   CALCULATION RESULT OF SiO_2 THICKNESS IN ETB FLASH BY CV GRAPH

Expected Parameter	Gate Area (A), μm^2	Capacitance of Oxide (Cox), nF
Asymmetric Flash Device 8nm Fabrication	6.25	37.98
Symmetric Flash Device 8nm Fabrication	6.25	62.13

A. Effect of Single Tunnel SONOS to Programming Tunneling

The I-V curve is a simpler way to understand and determine the basic performance parameters of flash memory cell device and also can specify the behavior of tunnelling current in a semiconductor electrical circuit. Fowler-Nordheim tunnelling mechanism is mainly used to allow write and erase process on the flash memory device by giving the electric field on control gate and substrate. To show the behavior of FN tunnelling current and voltage of the conventional flash cell, the graph of the IV is plotted in Fig. 6. The behavior of Fowler-Nordheim tunneling is dependent on the voltage across the barrier and it will increase exponentially with the applied voltage [16]. From the figure for 8nm of single tunnel barrier thickness it produces 3.6 mA/cm^2 of tunnelling current when it sets the voltage to 7V.

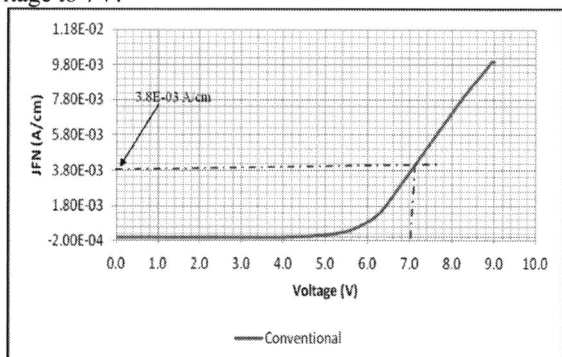

Fig. 6. Conventional Flash IV Characteristic in 8nm of EOT

B. Effect of Engineered Tunnel Barrier (ETB) on Programming Tunneling

The design with an asymmetrical (2-Layers) and symmetrical (3-Layers) tunnel barrier thicknesses of 8nm EOT have been processed for comparison with conventional tunnel oxide thickness. Fig. 7 shows the asymmetric and the symmetric tunnel barrier with similar EOT will affect the gate current with a variation of V_{CG}. From Figure, the slope approximately same in the depletion region. The tunneling voltage for asymmetrical and symmetrical was set at 6.2 V and 3.5 V respectively. As observed, the FN tunnelling current is producing at 3.6E-03 A/cm^2. The requirement of control gate voltage to produce the same tunnelling current for the device shows that the device performance is highly improved when changing the asymmetric (2-Layers) tunnel barrier to symmetric (3-Layers) tunnel barrier. This is because of the effect from using high-κ and low-κ combination in tunnel barrier such as effective mass and barrier height in energy band profiling.

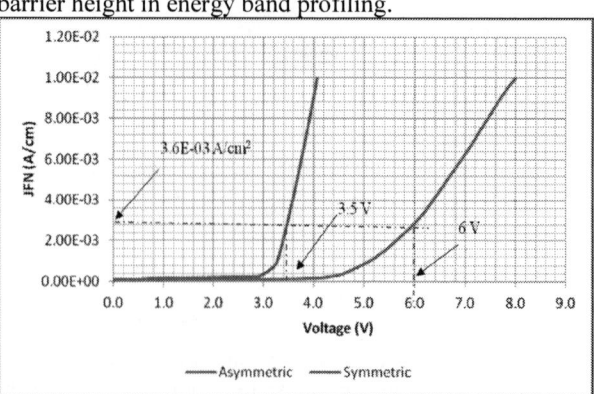

Fig. 7. Current-Voltage (IV) for conventional tunnel barrier and ETB Flash in 8nm of EOT

C. Programming Voltage Characteristic

For analysis the programming voltage, conventional tunnel oxide graph and ETB graph are plot in the same Fig. 8. The log-Current-Voltage (IV) with 8nm of EOT was plotting with green, red and blue for conventional, 2-Layered and 3-Layer, respectively. When the value of tunnelling current was set to 2.0E-03 A/cm^2, the programming voltage show decreases compared ETB and also conventional single oxide. The voltage for conventional is 6.6 V, while for 2-layer and 3-Layer are 5.6 V and 6.6 V. it shows that, the ETB SONOS is better than conventional SONOS in term of programming voltage. The effect of combination high-κ and oxide will cause the tunneling current is lower and enhanced the flow of an electrons through the barrier will be much easier. Addition the tunneling current through asymmetrical and symmetrical tunnel barrier was larger than that through the conventional tunnel oxide. In other words, the smaller programming voltage is achieved.

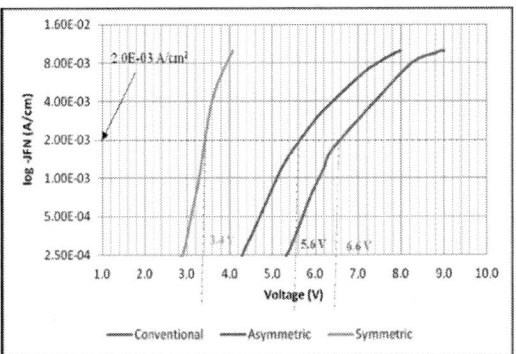

Fig. 8. Current-Voltage (IV) for conventional tunnel barrier and ETB Flash in 8nm of EOT

The results for programming tunneling and programming voltage as shows in Table III. When the tunneling current was set to the 2.0 E-3 A/cm^2, the programming voltage is decreased compared single tunnel oxide and engineered tunnel barrier. This phenomenon refers to conventional charge trap Flash memory that is related to thickness and dielectric parameter of tunnel oxide. Furthermore, these analyses also shown that the tunneling current of ETB is higher than conventional tunnel oxide while the programming voltage was decreased in achieved same tunneling current.

CONCLUSION

The Flash memory device by using technique charge Trap SONOS has successfully fabricated and it's result performance shows as predictable as expected. Several highlighted issues have been investigated in details and some of the problems has solved by using the proposed method. The device performance is highly improved when changing single tunnel oxide to ETB structure. This is because of the effects from using high-κ and oxide combination in tunnel barrier such as improve the dielectric parameter, effective mass and barrier height in energy band diagram. From the comparison of tunnel barrier, the effect on thickness of tunnel barrier shows the main factor that affected to the tunneling current and programming voltage. ETB technique was found to be higher than conventional tunnel oxide in tunneling current while the programming voltage was decreased ~3V. As a conclusion, the advantages of using high- κ with ETB is an alternative to improve device performance.

ACKNOWLEDGMENT

The authors wish to thank University Malaysia Perlis (UniMAP), for giving the opportunities to do this research in the Micro & Nano Fabrication Cleanroom in School of Microelectronic Engineering and Institute of Nanoelectronic Engineering (INEE).

REFERENCES

[1] Masuoko, F., Momodomi, M. (1987). New Ultra High Density EPROM and Flash EEPROM with NAND structure Cell. Electron Device Meeting, 1987 International, 25-26.

[2] Paul Hasler, Bradley A. Minch2, and Chris Diorio3, "floating-gate devices: they are not just for digital memories anymore" IEEEG. Groeseneken, H.E. Maes, J. Van Houdt, and J. S. Witters, "Basic of Nonvolatile Semiconductor Memory Devices" pp 10.

[3] D. Khang, S.M. Sze, "A floating gate and its application to memory devices", Bell Syst. Tech. J., Vol. 46 (1967)

[4] Paolo Pavan, Luca Larcher, Andrea Marmiroli, "Floating Gate Devices: Operation and Compact Modeling" pp 7-10,

[5] Wiliam D. Brown, Joe E. Brewer, "Nonvolatile Semiconductor Memory Technology" IEEE Press, university of Arkansas, pp 1-10. (1997)

[6] Casperson, J. D., Bell, L.D., Atwater, H.A. . (2002). Materials issues for layered tunnel barrier stucture. J. Appl. Phys., 92(1), 261-267.

[7] Julie D. casperson, L.Douglas Bell, Harry A.Atwater, "Material issues for layered tunnel barrier structures" Journal of Applied physics, Vol 92, num 1 (1 july 2002).

[8] Liu, Y. (2007). Advanced Tunnel Dielectrics for Nonvolatile Memory Technology. Yale University.

[9] Govoreanu, B., Blomme, P., Rosmeulen, M., Van Houdt, J., & De Meyer, K. (2003). VARIOT: a novel multilayer tunnel barrier concept for low-voltage nonvolatile memory devices. Electron Device Letters, IEEE, 24(2), 99-101.

[10] Govoreanu, B., Blomme, P., Rosmeulen, M., Van Houdt, J., & De Meyer, K. (2003). VARIOT: a novel multilayer tunnel barrier concept for low-voltage nonvolatile memory devices. Electron Device Letters, IEEE, 24(2)

[11] Prince, B. (2005, 10-10 Nov. 2005). Trends in scaled and nanotechnology memories. Paper presented at the Non-Volatile Memory Technology Symposium, 2005.

[12] Wiliam D. Brown, Joe E. Brewer, "Nonvolatile Semiconductor Memory Technology" IEEE Press, university of Arkansas, pp 10. (1997)

[13] Gehring, A., & Selberherr, S. Modeling of tunneling current and gate dielectric reliability for nonvolatile memory devices. Device and Materials Reliability, IEEE Transactions on, 4(3), 306-319. (2004).

[14] Gehring, A., & Selberherr, S. (2004). Modeling of tunneling current and gate dielectric reliability for nonvolatile memory devices. Device and Materials Reliability, IEEE Transactions on,

[15] MR Zakaria, U Hashim, RM Ayub, Z. Zailan A simulation study of the effect engineered tunnel barrier to the floating gate flash memory devices - Semiconductor Electronics (ICSE), 2012 10th IEEE 2012

[16] G. Eason, B. Noble, and I. N. Sneddon, "On certain integrals of Lipschitz-Hankel type involving products of Bessel functions," Phil. Trans. Roy. Soc. London, vol. A247, pp. 529–551, April 1955.

Integrated Constant Voltage Constant Current Readout Interfacing Circuit for EGFET Electrochemical Sensing

H. Guliga
Faculty of Electrical Engineering
Universiti Teknologi MARA
40450, Shah Alam Selangor
hasliadiguliga@yahoo.com

W. F. H. Abdullah
Faculty of Electrical Engineering
Universiti Teknologi MARA
40450, Shah Alam Selangor
wanfaz@salam.uitm.edu.my

S. H. Herman
Faculty of Electrical Engineering
Universiti Teknologi MARA
40450, Shah Alam Selangor
Hana1617@salam.uitm.edu.my

Abstract— Integrated constant voltage constant current (CVCC) readout interfacing for EGFET electrochemical sensing is presented. Development of integrated CVCC readout circuit is lead to single chip smart sensor system with the low cost production. CVCC technique is commonly used in investigation of ionic activities solution. The design architecture of CVCC has been targeted to Silterra 0.13um technology with the size of area include the IO pads design is 873μm x 811μm. the functionality of sub modules in this design included the three stage operational amplifier, current source and voltage reference circuit have been proved via the simulation results. Furthermore, the minimum input voltage from the ion sensing membrane for this readout circuit is approximately 0.73V.

Keywords — CVCC Readout circuit;   EGFET; three stage operational amplifier

I.   INTRODUCTION

Advancement in microsystem has led to development and research activities in sensor application and consumer electronic. Electrochemical sensor based on MOSFET technology was demonstrated by Piet Bergveld [1]. Ion sensitive field effect transistor (ISFET) that employs the basic structure of MOSFET was introduced to replace the ion sensitive electrode (ISE). This is because, ISFET has fast response time, mass manufacturing, micro size, robustness and potential in on chip integration[2-3]. Polysilicon that form the gate of MOSFET is removed and replaced with the structure of ion sensitive membrane material such as tin oxide(SnO_2)[4], silicon nitrate (Si_3N_4)[5], Titanium dioxide (TiO_2)[6] and aluminum oxide(Al_2O_3)[7]. This ISFET will immerged into the electrolyte solutions. The principle of ISFET based ion ionic exchange between ion sensitive membrane and electrolyte solution will produce the potential different that depends on the ion concentration. The variation of potential different will lead to the modulation of threshold voltage which affected to the drain source current.

Unfortunately, since ISFET is immerged into electrolyte solution, it will produce poor thermal sensitivity. Besides that, it is also sensitive to the light effect. In 1982's, the studies of the field effect electrochemical sensor were promote new level. Consequently, 1983's, J. Spiegel was introduce the extended gate field effect transistor (EGFET) to replace the traditional ISFET[8].The EGFET are introduced more advantages compare to traditional ISFET such as low cost, simple structure, easy to package, long term stability, insensitive to light and temperature and also disposable gate[9-11]. Also, the EGFET electrochemical sensor also promises the possibility lab on chip integration due to same silicon platform.

The Potential lab on chip integration of both electrochemical sensor and semiconductor technology has led to the miniaturization and mass-production. Readout interfacing circuit (ROIC) is plays important role in order to monitor the variation of ionic activities in electrolyte solution accordingly to real time. It is reported that from the previous literature review, many researchers work in areas of ISFET/EGFET ROIC design. They investigate and develop the different architectures of ROIC. Constant voltage constant current (CVCC) readout technique is recently used for ISFET due to accuracy and precision measurement[12-13]. The ability of CVCC readout interfacing circuit is to reduce the effect of temperature and drifting issue. Therefore, it will more suitable and useful for precision agriculture application. Currently, most of ROIC based on the CVCC technique was developed in discrete level since most researchers are focus on the characterization of ion sensing behavior. The interfacing between discrete devices will introduce more noise due to the impedance matching, increase in cost and area of design. Furthermore, another area of concern is power consumption. Available of the ROIC for ISFET/EGFET electrochemical sensor use dual supply. Advantage using dual supply is that, it can cover the output between positive and negative site. However, the power consumption will increase when the number of power supply increased.

978-1-4799-5761-3/14 $31.00 © 2014 IEEE   440

In this work, integrated CVCC readout interfacing circuit was introduced on this paper. Integration between sub-modules of CVCC technique on same silicon platform has been proved to reduce the cost and area of the design. Furthermore, single supply instead dual supply has been used in design to reduce the power consumption. The interfacing between the sensor and integrated readout circuit will introduce the smart sensor and system and create new potential market that benefit to the investment in world's technology

This paper divided into four sections. In section 2 briefly describes the EGFET sensor development. The proposed integrated readout interfacing circuit that implementation of the CVCC technique was described in section 3. The simulation result of integrated CVCC readout interfacing circuit included sub modules presented in section 4. Finally, in section 5, conclusion has been summarized of design.

II. EGFET DEVICE AS ELECTROCHEMICAL SENSING

A cross-section of EGFET structure is shown in fig. 1. The EGFET structure shows the ion sensitive membrane is separated from the commercial MOSFET by using the metal wire. Only the ion sensitive membranes will be immersed into electrolyte solution. The electrical potential exist between the ion sensitive membrane and electrolyte due to ion concentration of PH solution will create gate potential of the MOSFET. Based in the theory of MOSFET[14], the channel resistance value between the drain and source of MOSFET is based on different value of gate potential. Thus, the different level of PH solution has a different ion concentration that can be represented by the channel resistance between source and drain of EGFET.

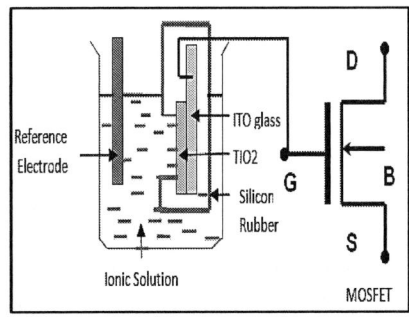

Fig. 1. basic cross-section of Extended Gate Field Effect Transistor (EGFET)

III. ELECTROCHEMICAL SENSOR READOUT INTERFACING CIRCUIT

In order to obtain measurement signal from ion sensitive membrane, EGFET will be interfaced with analog circuit. In this paper, constant voltage constant current (CVCC) readout interfacing circuit was employed for measuring the electrical response from the sensing membrane as shown in fig. 2. The electrochemical potential is monitored based on the output of readout interfacing circuit. The op-amp, A2 is form as follower amplifier to make sure that the output voltage is equal to the source potential of EGFET. The concept of CVCC is to ensure that the EGFET is active region by applied the constant bias voltage and constant current to the EGFET under isothermal point. The isothermal point of this sensing FET is I_{DS}, and V_{DS} are set at 100μA and 0.5V respectively. The current source is used to ensures that bias resistor and op-amp, A1 provide a constant voltage across between the drain and source of EGFET. Furthermore, the current sink ensures a constant current drawn between the drain-source. The electrochemical potential of electrolyte due to the ionic activity is monitored by source voltage of EGFET which is directly to the output of readout interfacing circuit.

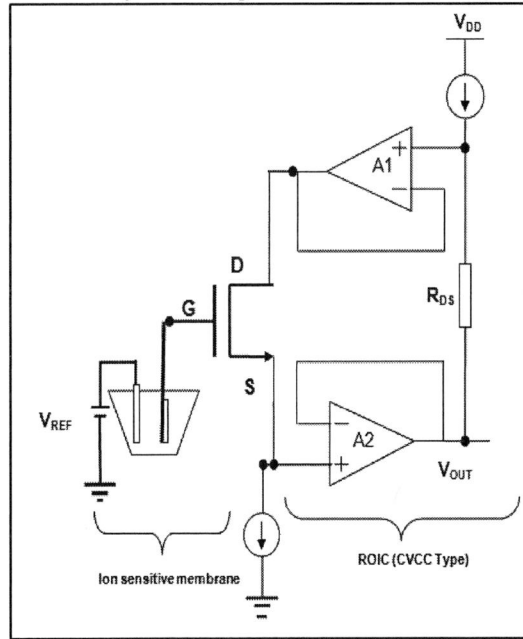

Fig. 2. CVCC Readout Interfacing Circuit Architecture for Electrochemical Sensing

In this work, the integrated CVCC readout interfacing circuit is developed by using with 0.13um technology silterra file. The integrated readout circuit consist of three sub modules; (i) op amps, (ii) current source/sink and (iii) biasing circuit. The three stage operational amplifier is implemented for this design due to the high common mode rejection ratio(CMRR), and high open loop gain with lowering power supply voltage[14] Due to operating drain-source current is small, high-swing current source and sink (figure 3) is implemented for this design rather than using single transistor mirror because it can reach maximum signal swing or very low minimum voltage requirement to reach the stable output current[14-15]. The current is controlled by external voltage (V_{BIAS}) or provided by the internal biasing circuit.

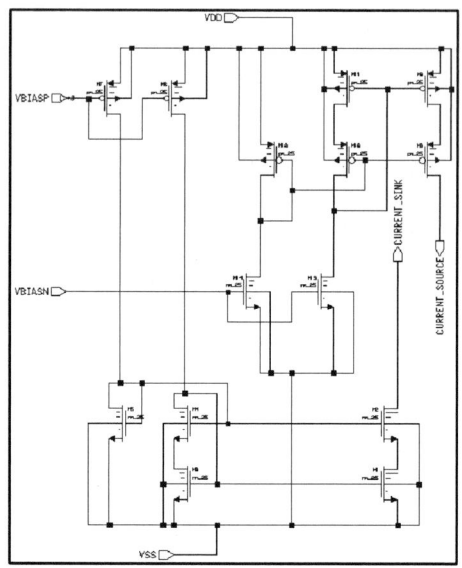

Fig. 3. High swing constant current source and sink architecture

IV. RESULT AND DISCUSSION

In this work, the integrated CVCC readout interfacing circuit is simulated using silterra 0.13um process technology. Table 1 shows comparison characteristic between schematic and layout simulation of three stage op-amp. There are different values between schematic and layout design due to existing of parasitic in layout design. Overall, the three stage op-amp of this project achieved all the specification requirements for both schematic and layout results simulation.

TABLE 1. COMPARISON CHARACTERISTIC THREE STAGE OP AMP SCHEMATIC AND LAYOUT

Specification	Schematic	Layout
VDD(V)	2.5	2.5
Gain (dB)	135	133
GBW(MHz)	15	15
Settling time(V/us)	9.925	19.04
PM Cc = 2pF	55.35^O	58.76^O
SR(V/µS)	Pos. : 0.1812 Neg. : -0.0394	Pos. : 0.157V Neg. : -0.0176
ICMR(V)	1.148 ~ 2.457	1.148 ~ 2.457
CMRR(dB)	125	-
Output Swing(mV)	0.280 ~ 2.433	0.38 ~ 2.432
Offset (mV)	6.82	5.69
Power dissipation (µW)	679.19 ~ 695.7	679.19 ~ 695.7

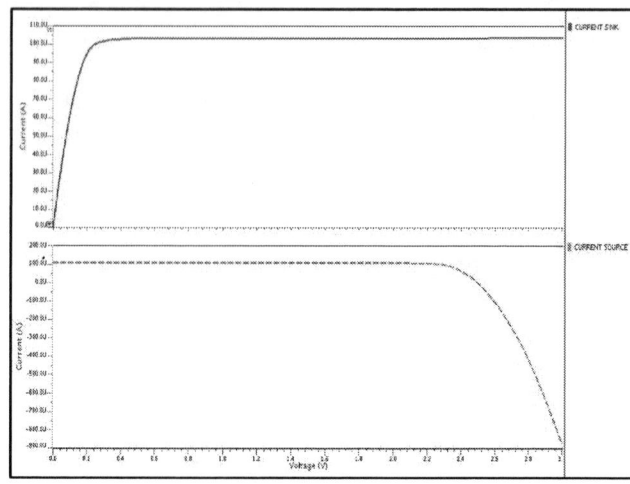

Fig. 4. Simulation of high swing cascode current source and sink result

Fig.4. shows the simulation of high swing cascode current source and sink. Red and blue lines represents as current sink and source respectively. It shows that the circuit in fig. 3 produces 100µA constant output current of sources and sinks. Moreover, the wide range operating of this circuit is approximately 0.2V~2.35V.

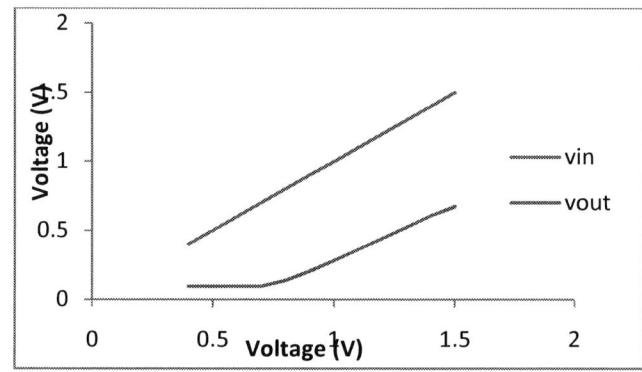

Fig. 5. Output simulation of integrated CVCC readout interfacing circuit

Fig. 5 shows the probe measurement on simulation between input versus output of single stage integrated CVCC readout circuit. The blue line is represents as input which is connected to the sensing membrane (voltage source). Also the red line is output of integrated CVCC readout circuit. Based on the previous explanation, the different ion concentration in electrolyte solution will produce a different electrical potential. Therefore in simulation process, this electrical potential will be replaced with DC voltage. The DC voltage will be swept at particular range is represent as varies of PH values. In this case, the DC voltage will be swept from 0.4V to 1.5V. It shows that, the minimum voltage input of the sensing FET for this design is approximately 0.73V

IEEE-ICSE2014 Proc. 2014, Kuala Lumpur, Malaysia

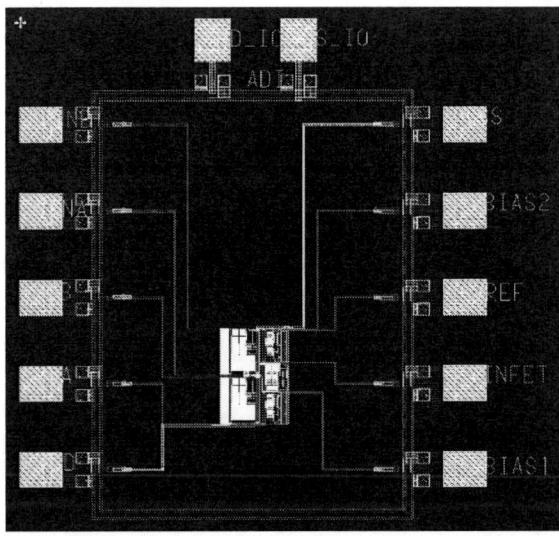

Fig. 6. layout design of integrated CVCC readout interfacing circuit include IO Pads design

Fig 6 shows the layout design of single stage integrated CVCC(constant voltage constant current) readout circuit include the I/O pads design. The size core of layout design without the I/O pads is approximately 118μm x 172μm but size layout of design will increase by including the I/O pads design approximately 873μm x 811μm

V. CONCLUSION

The integrated constant voltage constant current (CVCC) readout interfacing circuit for EGFET electrochemical sensing has been developed and presented. The design architecture has been designed by using Silterra 0.13um technology as shown in fig. 6. The architecture of this design consists of three sub modules included three stage op-amp, biasing circuit, and high swing current source/sink. The results based on simulation shows the constant current at 100um within 0.2V~2.35V ranges are successfully developed. Also, the result shows that requirement minimum voltage input from the ion sensing membrane for this readout circuit is approximately 0.73V.

ACKNOWLEDGEMENT

This work is partially supported by the Ministry of Higher Education Malaysia under the Fundamental Research Grant Scheme (Project Code:600-RMI/FRGS 5/3 (21/2013). The Author is a student registered with Universiti Teknologi MARA.

REFERENCES

[1] P. Bergveld, "Development of an ion-sensitive solid-state device for neurophysiological measurements." *Biomedical engineering, IEEE Transactions on* 1, pp. 70-71, 1970

[2] T. Nobpakoon, "A New Method for Current Differential ISFET/REFET Readout Circuit," *Int. J. Inf. Electron. Eng.*, vol. 3, no. 2, pp. 141–143, 2013.

[3] J. Chiang, J. Chou, and Y. Chen, "Study of the pH-ISFET and EnFET for Biosensor Applications," *Journal of Medical and Biological Engineering.* vol. 21, no. 3, pp. 135–146, 2001.

[4] L. Chi, J. Chou, W. Chung, T. Sun, and S. Hsiung, "Study on extended gate field effect transistor with tin oxide sensing membrane" , *Materials Chemistry and Physics* , vol. 63, pp. 19–23, 2000

[5] M. N. Niu., X. F. Ding,., & , Q. Y. Tong, "Effect of two types of surface sites on the characteristics of Si_3N_4-gate pH-ISFETs". *Sensors and Actuators B: Chemical*, vol. *37*, pp.13-17, 1996

[6] I. E. Fets, J. Chou, and C. Chen, "Fabrication and Application of Ruthenium-Doped Titanium Dioxide Films as Electrode Material for ion sensitive extended gate FETs," *Sensors Journal, IEEE* , vol. 9, no. 3, pp. 277–284, 2009.

[7] J.C. Chou, & C.Y. Weng. "Sensitivity and hysteresis effect in Al_2O_3 gate pH-ISFET," *Materials chemistry and physics*, vol. *71*, pp. 120-124, 2001

[8] J. Van der Spiegel, I. Lauks, P. Chan, D. Babic, "The extended gate chemical sensitive field effect transistor as multi-species microprobe", *Sensors and Actuators*, Vol. 4, PP. 291-298, 1983

[9] J. Chen, J. Chou, T. Sun, and S. Hsiung, "Portable urea biosensor based on the extended-gate field effect transistor," *Sensors and Actuators B: Chemical,* vol. 91, pp. 180–186, 2003.

[10] L. Chi, J. Chou, W. Chung, T. Sun, and S. Hsiung, "Study on extended gate field effect transistor with tin oxide sensing membrane" , *Materials Chemistry and Physics* , vol. 63, pp. 19–23, 2000

[11] M. A. Rosdan, S. H. Herman, W. F. H. Abdullah, N. S. Kamarozaman, and M. I. Syono, "Sputtered Titanium Dioxide Thin Film for Extended-Gate FET Sensor Application", *Micro and Nanoelectronics (RSM), 2013 IEEE Regional Symposium on.* pp. 219-222,

[12] M. Gotoh , S. Oda, I. Shimizu, A. Seki, E. Tamiya, & I. Karube (1989). "Construction of amorphous silicon ISFET," *Sensors and Actuators*, vol. *16* pp. 55-65, 1989

[13] W. F. H. Abdullah, M. Othman, & M. A. M. Ali, "Chemical field-effect transistor with constant-voltage constant-current drain-source readout circuit," In *Research and Development (SCOReD), 2009 IEEE Student Conference on* pp. 219-221. IEEE

[14] R. J. Baker. CMOS: circuit design, layout, and simulation. Wiley-IEEE Press. Vol. 18

Effect of Indium Concentration on Optical and Electrical Properties of In Doped ZnO Thin Films for Gas Sensing Application

M.Hannas[1,2,3,4], Azrif Manut[1,2], S.H.Herman[1,2], M.Rusop[1,2]

[1]NANO-Electronic Centre (NET)
[2]Faculty of Electrical Engineering, Universiti Teknologi MARA (UiTM) Shah Alam Selangor, MALAYSIA
[3]Institute Micro and Nanoelectronic (IMEN), National University of Malaysia (UKM) Bangi Selangor, Malaysia
[4]e-mail: hannas@ukm.my

Abstract—The deposition of Indium doped ZnO thin films using sol gel spin coating technique will be discussed in this paper. The concentrations of indium doping of the thin films were varied from 1.0 - 5.0 at. % indium. The effect of various doping concentrations 1.0 - 5.0 at. % indium on electrical, optical properties and gas sensing application were studied. The electrical properties were analyzed using I-V measurement (CEP2400). The optical transmittance and optical band gap were characterized by ultraviolet visible (UVVIS) spectrophotometer. Gold (Au) was used as a metal contact using electron beam thermal evaporator (ULVAC). The optical result show that the indium doped ZnO thin films revealed higher than 90% of transmittance value. The highest electrical resistivity was found $1.95E+04$ Ωcm at. 2% indium concentration. The optimum result of Indium doped ZnO thin films will be used for gas sensing application and gas sensitivity of the thin films will be extensively discussed in this paper.

Keywords: Indium doped ZnO thin film; Doping Concentration; Gas sensor; Sensitivity.

I. INTRODUCTION

Semiconducting material such as Ga_2O_3 [1], SnO_2 [2], Co_3O_4, etc., were extensively discussed for gas sensing detection. However, the temperature of gas sensing is quite high (~200–400 °C). Zinc oxide has widely used in semiconducting materials. Nowadays, there are many applications using zinc oxide based materials such as piezoelectric, electronic, electrical and optical properties. For example light emitting diodes (LED's), UV sensor, transparent electrodes, laser system, gas sensor,etc. ZnO has many good features such as wide range of band gap energy (3.37eV), good electron mobility (~60 meV) and very stable in n-type semiconductor [3]. There are many types of ZnO growth such as nanorods, nanobelts, nanowires, nanoflakes, nanorings and nanoflowers [4, 5]. Moreover, ZnO have many potential benefits such as low cost, non-toxic, excellent stability and very wide application in micro and nanoelectronics devices.

The deposited nanostructured ZnO thin films can be improved to have high conductivity, high electron mobility, high transparency and high porosity by doping process such as Sn, Al or In for electrical and optical performance [6].

There are many techniques to prepare nanostructured indium doped ZnO thin films such as RF magnetron sputtering system [7], reactive evaporation, chemical vapor deposition system

(CVD) [8], spray pyrolysis system, pulsed laser deposition system (PLD) [9], molecular beam deposition system (MBE) [10], and sol-gel spin coating process [11]. Sol gel spin coating techniques is one of the simplest techniques compared to others deposition equipment. There are many benefits and advantages of using sol gel spin coating techniques such as low operating costs, high porous thin films, capability of high volume deposition operation and easy to combine with other chemicals materials for doping process [12].

Recently, many researcher focused on the characterization of indium doped ZnO for UV sensor and solar cells applications. Moreover, Satis et al. reported on H_2S gas sensor based on indium-doped ZnO thin films. The thin films were deposited by spray pyrolysis technique with different indium concentration. The 1, 3 and 5 at. % Indium doped ZnO thin films have been reported, 3 at. % of indium doped ZnO thin films were obtained high response to H_2S gas [13]. In order to study high sensitivity of the indium doped ZnO thin film based gas sensing application, we characterized the indium doped thin film with electrical and optical properties using sol gel spin coating method. In this paper, we focused on the sensitivity of indium ZnO thin film using different doping concentration (1 to 5 at%) with different N_2 concentration (20 to 220 ml/min) using the optimum thin films among the doping concentration.

The characterization of the thin films based on electrical and optical properties were analyzed using spin coating method. The optimum result among the thin films will be selected for gas sensing application. Moreover, in this study inZnO thin films based gas sensors applications (different concentration oxygen and nitrogen gas) are very limited reports in literature.

Compare to others deposition method, sol-gel spin coating deposition is extensively used and very easy technique to deposit high uniformity of ZnO thin films. The effect of indium concentration on optical and electrical properties of indium doped ZnO thin films based gas sensing were investigated using simple gas chamber.

II. EXPERIMENTAL

The preparation of the InZnO thin films using sol gel spin coating techniques will be explained in details. Zinc acetate dehydrate [$Zn(CH_3COO)2\_2H_2O$] and indium (III) chloride ($InCl_3$) were used as a starting material. 2-Methox-yethanol was used as a solvent and MEA (monoethanolamine) [$NH_2C_2H_4OH$] as the stabilizer. Zinc acetate (0.4M) molar ratio to MEA was fixed at 1.0 %. In addition, the variation (at. 1,2,3,4 and 5%) was used for molar ratio of zinc acetate dehydrate to indium chloride. The solutions were mixed and stirred at 80 °C for 1 hour. Then, the solutions were aged 24 hour at room temperature. The precursor solution was deposited on glass substrates by spin coating system with 3000 rpm spin speed for 60 second. After the spin coating, the glass substrate of the thin film was dried at 150 °C using mini furnace. The spin coating process was prepared for 10 times to obtain the thickness in range 100-250 nm. Lastly, the samples were annealed at 500 °C to obtain good crystalline and high porosity of the indium doped ZnO thin films. The characterization of optical transmittances of indium ZnO thin films were analyzed by UV VIS spectrometer with range from 200 to 800 nm (wavelength). The electrical properties were measured by using I-V measurement KEITHLEY source meter (K2400). Gold (Au) target 99.99% of purity was used as a metal contact using electron beam thermal evaporator (ULVAC) with 60nm of the thickness. The sensitivity of the indium doped ZnO thin films were tested using oxygen and nitrogen gases with different concentration from 20-220 mL/min in gas chamber system at room temperature. The sensitivity of the thin films were analyzed using $S = (Ra - Rg)/Ra$ which is Ra is for resistance in the air and Rg is the sensor resistance by the N_2 gas.

III. RESULTS AND DISCUSSION

A. Transmittance and Optical Band Gap

Fig.1 shows the optical transmittance spectra of the nanostructured InZnO thin films in the wavelength of 200-800 nm with different doping concentration. The transmittance of the thin films show more than 90%.. The optical transmittance obviously difference in the near ultraviolet (200–380nm) range, where the transmittance increases with the increasing doping concentration due to the starting of the fundamental absorption in the UV range.

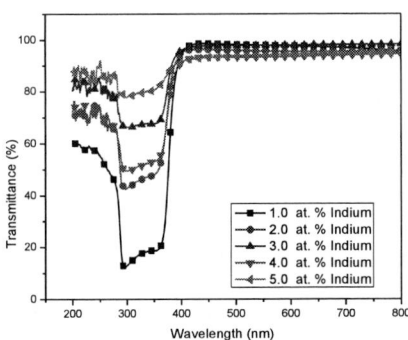

Fig. 1 The optical transmittance spectra of indium doped ZnO thin films at various doping concentration.

Fig.2 shows plot of $(\alpha h v)^2$ vs photon energy hv. The calculated values of optical band gap decreases from 3.26 to 3.10 eV with increasing the doping concentration. In the literature, effect of indium doping will increased the optical band gap of the thin films. This phenomenon showed that the

In^{3+} doping convinced the Burstein–Moss effect and band gap widening in nanostructured ZnO.

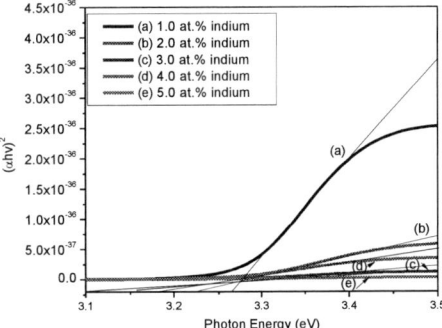

Fig.2 shows plot of $(\alpha h v)2$ vs photon energy hv.

B. Electrical Properties of the Indium Doped ZnO Thin Film.

Fig. 3 shows the current-voltage curve of indium doped ZnO thin films. The current-voltage characteristics have been investigated using Keithley 2400 at room temperatures. The voltage was applied from -10 to 10V. All of the indium doped ZnO thin film exhibits good Ohmic performance using 60nm thickness of Au metal contact.

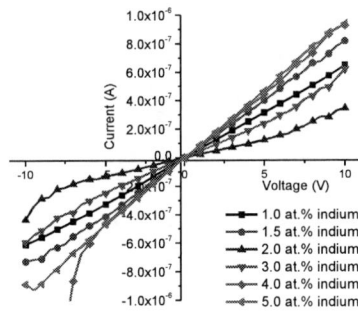

Fig.3 Current-Voltage curve of indium doped ZnO thin films with different doping concentration at 500 °C annealing temperature.

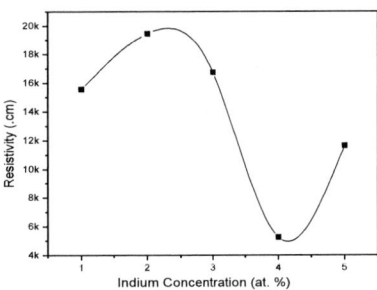

Fig.4 Resistivity of indium doped ZnO thin films with different doping concentration.

Fig. 4 shows the electrical resistivity of indium doped ZnO thin film decreases when increasing the doping concentration of the thin film. The decreasing value of electrical resistivity due to the loss of absorbed oxygen increases the donor concentration of the thin film surface and cause increases the thin film conductance.

The highest electrical resistivity was found 1.95E+04 Ωcm at. 2% indium concentration. The low resistivity was obtained 5.26E+03 Ωcm at. 4% indium concentration.

C. Gas Response measurement and Sensor Sensitivity

The IZO thin films deposited by sol-gel spin coating were analyzes with the different doping concentration for finding their optimum gas response. The gas response characterization was measured by linking through the high quality steel electrodes in 8cm x 8cm x 8cm of box chamber. Keithley K2400 source meter was used to supply DC power and current source. The gas response of indium doped ZnO thin films were tested at different test gases and different gas concentration. The sensitivity of gas sensor (S) was calculated as the resistance of the thin film in the ambient air (R_a) and R_g is the tested gas resistance (N_2 gas and O_2 gas). The gas sensor sensitivity, S, defined as follows:

$$S = \frac{Ra - Rg}{Ra} \qquad (3)$$

The sensitivity of indium doped ZnO thin film to N_2 at 220ml/min concentration of the 1,2,3,4 and 5at.% indium doped ZnO based sensor have been studied. Fig. 5 shows the variation of resistance measured in air (R_a) for IZO thin films deposited at different doping concentration at 220ml/min of N_2 gas. The enhanced sensing response will be found if either the value of Ra is lower or Rg is higher (Eq.(3)). The study on the selectivity of indium doped ZnO thin film based nitrogen gas sensor was carried out by monitoring the change in electrical resistance to different indium doping concentration including 1-5 at.% at 220 ml/min (N_2 gas). Fig. 5 shows that 5 at.% of indium doped ZnO thin film exhibit much higher increase in the sensor resistance after exposed to N_2 gas. Fig. 6 shows the histogram of gas response of different indium doping concentration thin film at a fixed concentration 220 ml.min (N_2 gas). For indium doping at 1,2,3,4 and 5 at.%, the sensitivities of the sensor were 9.8, 10, 5.4, 6.1 and 12.5 respectively. The higher response of the thin film suggested to be due to the sensitivity of the metal oxide semiconductor of the sensor determined by the interaction between the sensor surface and target gas [14]. The indium doped ZnO morphology provided a high surface area to interact N_2 molecules resulting in increased response.

Fig. 7 shows the response indium doped thin film as a function of N_2 concentration at 5 at. %. The response increased from 0.82% to 12.5%, as the N_2 concentration increased from 20 to 220 ml/min. The curve is almost linearly increasing in nature. The surface area, gas absorption and desorption kinetics play the main role with the penetrating constant originally increasing with increasing N_2 concentration until it reaches nearly saturation at very high N_2 concentration [15].

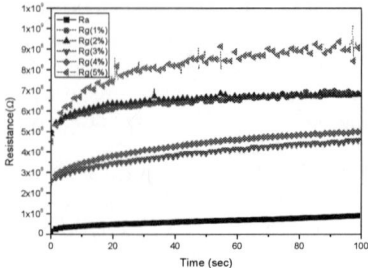

Fig.5 Variation of sensor resistance in presence of 220 ml/min N_2 gas with time for the indium doped ZnO thin films deposited at different doping concentration (1-5%)

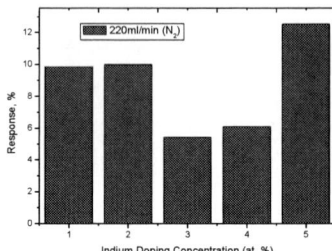

Fig.6 Gas response of IZO thin films with different doping concentration.

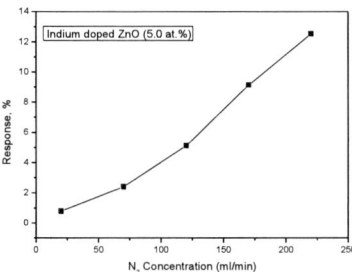

Fig.7 Variation of N_2 response of indium doped ZnO thin film with different concentration of N_2 at room temperature.

IV. CONCLUSIONS

In this study, indium doped ZnO thin films were prepared with different doping concentration. The thin films were prepared by easy and low cost sol-gel spin coating method. The optical and electrical properties of nanostructured indium doped ZnO thin film with different doping concentration was investigated. The optical properties of the thin films exhibit more than 90% of transmittance in the visible and infrared region. The optical transmittance obviously difference in the near ultraviolet (200–380nm) range. The sensitivity of the indium thin films was obtained 5 at.% of indium doped ZnO thin film exhibit much higher increase in the sensor resistance after exposed to N_2 gas. For indium doping at 1,2,3,4 and 5 at.%, the sensitivities of the sensor were 9.8, 10, 5.4, 6.1 and 12.5 respectively.

The study of indium ZnO thin film based gas sensing application indicates that 5 at.% of indium doping is potential material to be used as effective N_2 sensor.

ACKNOWLEDGEMENTS

The author would like to thank Research Management Institute of Universiti Teknologi MARA (Uitm) for providing financial support under grant 100-RMI/SF 16/6/2(15/2013) , UKM and Ministry of Higher Education (MOHE) for their financial support. My thanks also to supervisors, technicians and all that involve in the Nano-ElecTronic Centre, UiTM.

REFERENCES

[1] M. Fleischer, M. Seth, C. D. Kohl, and H. Meixner, "A selective H2 sensor implemented using Ga2O3 thin-films which are covered with a gas-filtering SiO2 layer," *Sensors and Actuators B: Chemical,* vol. 36, pp. 297-302, 1996.

[2] B. Mondal, B. Basumatari, J. Das, C. Roychaudhury, H. Saha, and N. Mukherjee, "ZnO–SnO2 based composite type gas sensor for selective hydrogen sensing," *Sensors and Actuators B: Chemical,* vol. 194, pp. 389-396, 2014.

[3] D. C. Look, B. Claflin, Y. I. Alivov, and S. J. Park, "The future of ZnO light emitters," *Phys. Status Solidi A: Appl. Res.,* vol. 201, pp. pp. 2203–2212., 2004.

[4] P. Feng, Q. Wan, and T. H. Wang, "Contact-controlled sensing properties of flowerlike ZnO nanostructures," *Applied Physics Letters,* vol. 87, pp. 213111-213111-3, 2005.

[5] S. Yi, S. Tian, D. Zeng, K. Xu, S. Zhang, and C. Xie, "An In2O3 nanowire-like network fabricated on coplanar sensor surface by sacrificial CNTs for enhanced gas sensing performance," *Sensors and Actuators B: Chemical,* vol. 185, pp. 345-353, 2013.

[6] M. Vishwas, K. Narasimha Rao, K. V. Arjuna Gowda, and R. P. S. Chakradhar, "Influence of Sn doping on structural, optical and electrical properties of ZnO thin films prepared by cost effective sol–gel process," *Spectrochimica Acta Part A: Molecular and Biomolecular Spectroscopy,* vol. 95, pp. 423-426, 2012.

[7] Y. Huawa, W. Jing, Y. Yangan, W. Xin, G. Bin, L. Hanchen, and D. Yali, "ZnO thin films produced by the RF magnetron sputtering," in *Electronic and Mechanical Engineering and Information Technology (EMEIT), 2011 International Conference on,* 2011, pp. 2486-2489.

[8] B. S. Li, Y. C. Liu, Z. S. Chu, D. Z. Shen, Y. M. Lu, J. Y. Zhang, and X. W. Fan, "High quality ZnO thin films grown by plasma enhanced chemical vapor deposition," *Journal of Applied Physics,* vol. 91, pp. 501-505, 2002.

[9] Y. R. Ryu, S. Zhu, J. D. Budai, H. R. Chandrasekhar, P. F. Miceli, and H. W. White, "Optical and structural properties of ZnO films deposited on GaAs by pulsed laser deposition," *Journal of Applied Physics,* vol. 88, pp. 201-204, 2000.

[10] T. Ohgaki, N. Ohashi, H. Kakemoto, S. Wada, Y. Adachi, H. Haneda, and T. Tsurumi, "Growth condition dependence of morphology and electric properties of ZnO films on sapphire substrates prepared by molecular beam epitaxy," *Journal of Applied Physics,* vol. 93, pp. 1961-1965, 2003.

[11] M. J. Alam and D. C. Cameron, "Preparation and properties of transparent conductive aluminum-doped zinc oxide thin films by sol gel process," *Journal of Vacuum Science & Technology A: Vacuum, Surfaces, and Films,* vol. 19, pp. 1642-1646, 2001.

[12] L. Xu, X. Li, Y. Chen, and F. Xu, "Structural and optical properties of ZnO thin films prepared by sol–gel method with different thickness," *Applied Surface Science,* vol. 257, pp. 4031-4037, 2011.

[13] S. S. Badadhe and I. S. Mulla, "H2S gas sensitive indium-doped ZnO thin films: Preparation and characterization," *Sensors and Actuators B: Chemical,* vol. 143, pp. 164-170, 2009.

[14] M. A. Chougule, S. Sen, and V. B. Patil, "Fabrication of nanostructured ZnO thin film sensor for NO2 monitoring," *Ceramics International,* vol. 38, pp. 2685-2692, 2012.

[15] P. Bhattacharyya, P. K. Basu, B. Mondal, and H. Saha, "A low power MEMS gas sensor based on nanocrystalline ZnO thin films for sensing methane," *Microelectronics Reliability,* vol. 48, pp. 1772-1779, 2008.

Switching Behavior of Lateral-Structured Zinc Oxide-Based Memristive Device

Raudah Abu Bakar, Mohd Nur Nazmi, 'Awatif Harun,
Nur Syahirah Kamarozaman, Nor Azira Akmar Shaari,
Shafaq Mardhiyana Mohamat Kasim
NANO-Electronic Center, Faculty of Electrical Engineering
Universiti Teknologi MARA,
40450 Shah Alam, Selangor, Malaysia
raudah088@salam.uitm.edu.my

Sukreen Hana Herman
CoRe of Frontier Materials & Industry Applications,
Universiti Teknologi MARA
40450 Shah Alam, Selangor Darul Ehsan, Malaysia
hana1617@salam.uitm.edu.my

Abstract— Lateral-structured zinc oxide (ZnO) based memristive device was studied. The effect of oxide and electrode size variations on the switching behavior was investigated. The lateral structure was formed by depositing the metal electrodes at the right and left ends of a glass substrate. The ZnO thin films were deposited right at the center of the metal coated substrate using sol-gel spin coating technique. The oxide and electrode widths were varied. It was observed that using the smallest oxide width in the combination with wide electrode size resulted in better memristive behavior. The values of both on and off resistances (R_{ON} and R_{OFF}) were found to be decreased as the width of the electrode increasing. The hysteresis curve on the other hand became wider with the increases of electrode width. The R_{OFF}/R_{ON} ratio was calculated to be 1.044.

Keywords— *Memristive behaviour, zinc oxide, lateral, sol-gel and spin coating.*

I. Introduction

Memristor (memory resistor) is a two terminal circuit element which was theoretically introduced by Prof. L. Chua in 1971 [1]. This fourth fundamental passive element however, has been only successfully realized in 2008 by HP labs due to the fact that the memristive behavior can only be seen in a micro- or nano-scale device [2]. The signature of a memristive device is the current-voltage (I-V) switching loop shown in Fig. 1. Thus the I-V characteristic is the fundamental characterization used to determine the memristive behavior in this work.

Recently the memristive device has attracted much attention because of its advantages of non-volatile, fast programming, small size and low power consumption [3-4]. Of the available metal oxide materials, titanium dioxide (TiO_2) has been extensively used as the active layer in memristor. This is because of its excellent memristive behavior and compatibility with the standard complementary metal oxide semiconductor (CMOS) technology [5-6]. On the other hand, since ZnO can be made CMOS compatible as well and widely used in various applications such as photovoltaic device, gas sensor, flexible memory application (RRAM) and surface acoustic wave device, several studies have been carried out to study the performance of ZnO based memristive devices [7-9].

In general the configuration of a memristor consisted of metal insulator metal (MIM) layers in which an oxide layer is sandwiched between the top and bottom metal electrodes (Fig. 2(a)). Because of this, the fabrication process of memristor is simpler compared to transistor. However, in the case of exploring the use of memristor as a sensing device, the conventional MIM structure obstruct the fully utilization of the oxide layer. Therefore, in order to make the full used of the layer, a new lateral configuration shown in Fig. 2(b) is proposed in this study where the oxide layer can be exposed to measurand, thus probably allowing a memristive device being applied as a sensing device.

In this study the ZnO based memristive device was deposited using sol-gel spin coated technique onto glass substrate. The oxide and electrode widths were varied to study the effect of the lateral configuration on the memristive behavior of ZnO thin films. The IV characteristic was measured using two point probe measurement system and the resistance ratio (R_{OFF}/R_{ON}) was calculated.

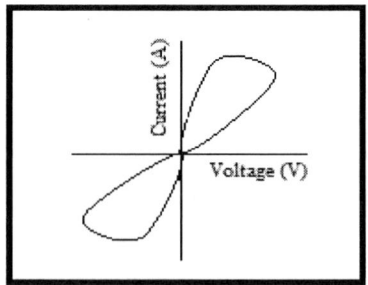

Fig. 1. Current-voltage (I-V) switching loop

II. Methodology

A lateral-structured ZnO memristive device was fabricated by varying the widths of the oxide and electrode regions. The oxide width was varied for 5, 6 and 7 mm while the widths of electrode were varied for 5, 7, 9 and 10 mm respectively. The

978-1-4799-5761-3/14 $31.00 © 2014 IEEE   448

lateral configuration of ZnO memristive thin film is illustrated in Fig. 2b(i) and b(ii). The width and length of the electrodes are represented by x and y respectively while w represents the width of oxide layer. The x and w were varied while y was remained constant (y = 20 mm).

The ZnO thin films were prepared by the sol-gel spin coating technique. For solution preparation, zinc acetate dehydrate ($Zn(CH_3COO)_{22}H_2O$), 2-methoxyethanol ($C_3H_8O_2$) and monoethanolamine (C_2H_7NO, MEA) were used as the precursor, solvent and stabilizer respectively. The 0.4 M solution was produced by first dissolving the zinc acetate dehydrate in 2-methoxyethanol. The ethanolamine was added in order to stabilize the solution. The ZnO solution was then sonicated for 30 minutes at 50°C. After that, the solution was subjected to stir-heated process for 3 hours at 80°C. The stirring was continued for another 24 hour in a room temperature environment. After spin-coating the ZnO solution on the Pt metal electrodes to form the lateral-structure, the samples were dried in a furnace for 10 minutes at 150˚C. Finally, the films were then annealed at 450˚C for 1 hour.

The electrical characterization was carried out by the two point-probe using Keithley 4200 semiconductor characterization system connected to a probe station by sweeping the bias voltage from 0 V to 5 V, 5 V to -5 V, -5 V to 0. The resistance ratio was defined by calculating the resistance from the current-voltage measurement produced before R_{OFF}/R_{ON} ratio was determined.

Fig. 2. Device configuration of (a) MIM conventional structure (b) lateral-structured (i) the upper view and (ii) the device cross section.

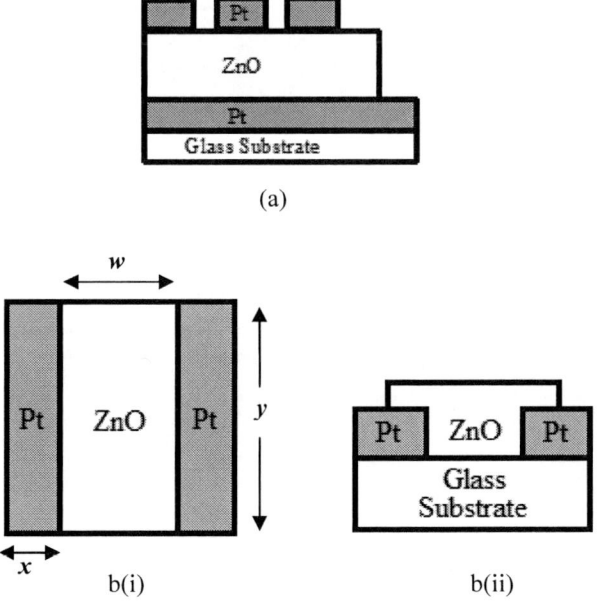

III. Results and Discussion

A. Effect of oxide width

Fig. 3 (a-c) show the IV characteristics of lateral ZnO based memristive device for various oxide layer widths. The electrode width on the other hand was kept constant (x = 10 mm). The results show that memristive behavior can be obtained using lateral MIM configuration as achieved in conventional MIM structure. All the samples with varied oxide layer widths exhibit resistive switching with symmetrical I-V characteristics. The curves obtained were pinched and crossed the origin. As compared to others, a better memristive behavior was obtained when the oxide layer width is 5 mm. This is may be due to the shortest path that can be created within the oxide layer between the two electrodes on the both sides. It is predicted that the path created would allow the oxygen ions to travel between the metal electrodes. Because of good memristive behavior achieved by the smallest oxide width (w = 5 mm), thus this width was used for investigation on the electrode widths.

Fig. 3. I-V characteristics of a lateral ZnO memristor with (a) 5 mm (b) 6 mm and (c) 7 mm oxide layer widths.

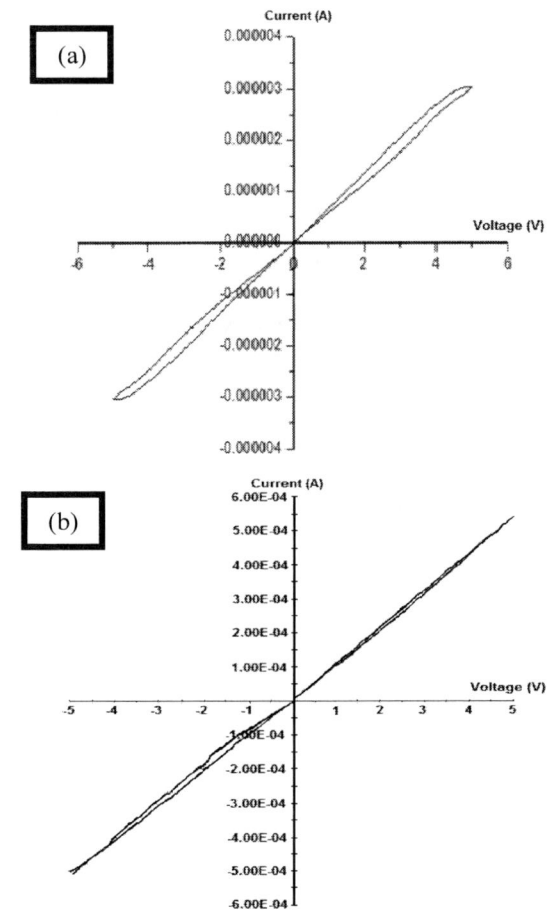

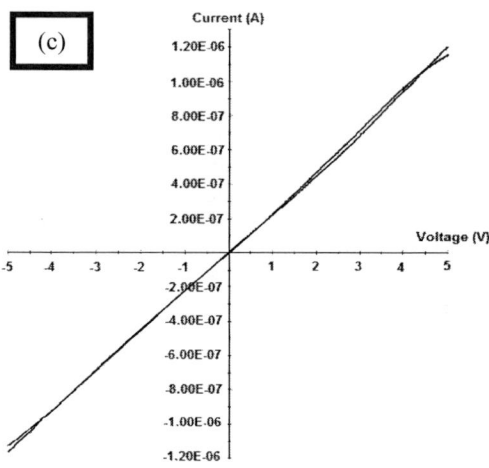

similar effect of varying electrode width on the hysteresis loop was expected to be achieved.

Fig. 4. I-V characteristics of a lateral ZnO memristor with different electrode sizes (a) 5 mm (b) 7 mm (c) 9 mm and (d) 10 mm

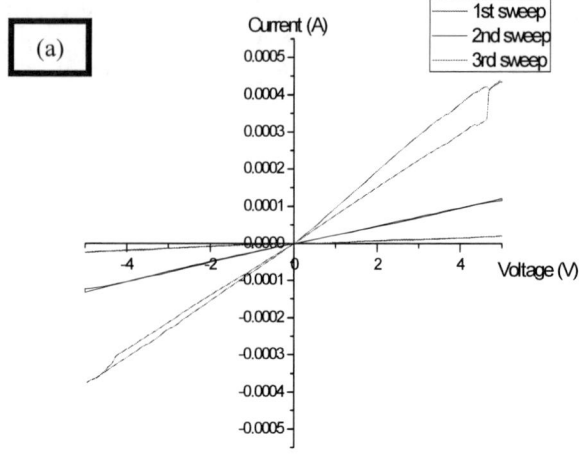

B. Effect of electrode width

The IV characteristics of lateral ZnO based memristive device with various electrode widths are shown in Fig. 4(a-d). The electrode widths were varied from 5 to 10 mm while the oxide width was kept constant at w = 5 mm. The measurement was performed for three times in order to test the repeatability of the fabricated devices. Based on the result obtained, it is believed that the first and second measurements are initially performed to trigger the switching behavior of the device since no electroforming was performed to the virgin sample. A broader hysteresis loop was observed after the third measurement suggesting that the existence of a complete channel that formed a path for the ion flowing within the device. It is also suggested that the broadening of hysteresis loop is attributed to the increases of charge displacement [10]. However, by performing consecutive voltage sweeping across the ZnO film resulted in the variation in resistance values.

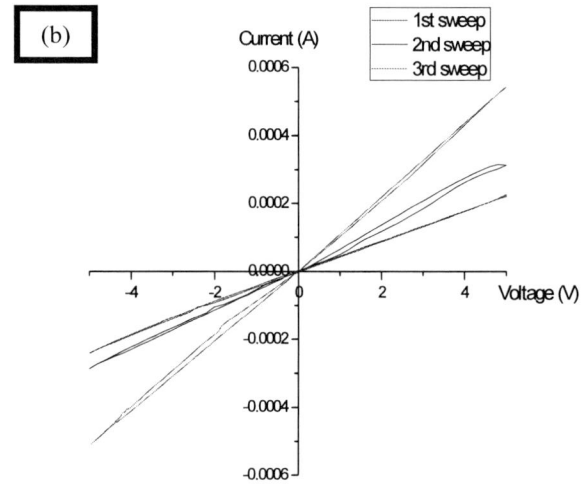

As also can be seen from the graph a symmetrical and broader hysteresis curve can be obtained if wider electrode width is used. The memory-conservation model developed by E. Gale et al. [11] explained how the hysteresis could be affected by the electrode size in the conventional MIM structure. In the model she has correlated the hysteresis and electrode size as,

$$H = m_F F + C_F \qquad \text{(i) [11]}$$

where, H = hysteresis, m_F = -1.68×10^{-8}, F = width of top electrode and C_F = 1.39×10^8. Both m_F and C_F values were found from the experimentally fitted data carried out by [11]. So, based on the equation (i), H is linearly proportional to F, therefore it can be concluded that the hysteresis size of conventional MIM memristor can be changed by controlling the size of the metal electrode. Furthermore the increases in the top electrode size will lead to the increase of the active area that will then affect the performance of a memristive device. Although a lateral configuration was used in study, the

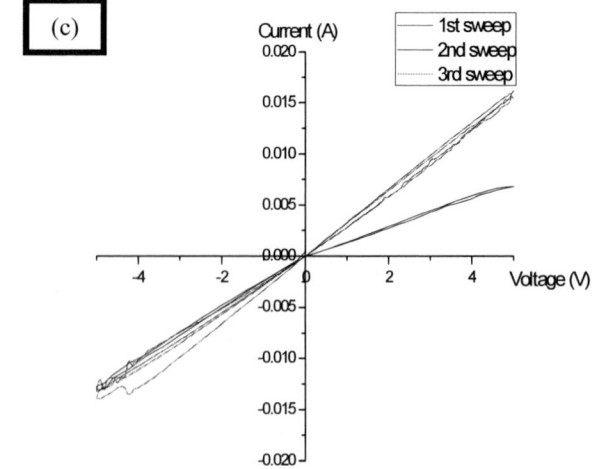

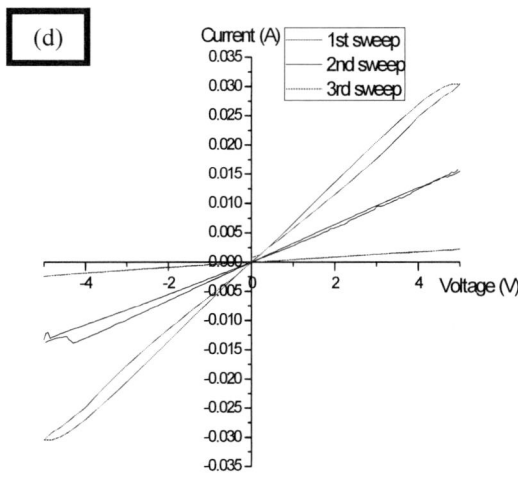

Acknowledgment

This work is partially supported by Ministry of Education Malaysia (MOE) under the Niche Research Grant Scheme (NRGS) (Project code: 600-RMI/NRGS 5/3 (7/2013)).

References

[1] L. Chua, "Memristor- The missing circuit element," IEEE Trans. on Circuit Theory, Vol.18, pp. 507-519, Sept. 1971.

[2] D. B. Strukov, G. S. Snider, D. R. Stewart, and R. S. Williams, "The missing memristor found," Nature, vol. 453, no. 7191, pp. 80–83, May 2008.

[3] R. S. Williams, "How we found the missing memristor," in IEEE Spectrum.

[4] Q. Xia, "Nanoscale Resistive Switches: Devices, Fabrication and Integration," Appl. Phys. A, Vol. 102, pp. 955-965, 2011.

[5] C. Yakopcic, A. Sarangan, J. Gao, T. M. Taha, G. Subramanyam and S. Roger, "TiO₂ Memristor Devices," Proc. IEEE, pp. 101-104, 2011.

[6] N. S. Kamarozaman, Z. Aznilinda, S. H. Herman, R. A. Bakar and M. Rusop, "Effect of annealing duration on the memristive behavior of Pt/TiO₂/ITO memristive device" in 10th IEEE Int. Conf. on Semicond. Electron. (ICSE) 2012, Kuala Lumpur, 2012, pp. 703-706.

[7] N. S. Kamarozaman, Z. Aznilinda, S. H. Herman, R. A. Bakar and M. Rusop, "Memristive behavior of NaOH-immersed titania nanostructures" J. Mech. Eng. & Sciences. (JMES), Vol. 5, pp. 688-695, 2013.

[8] N. Gergel-Hackett, B. Hamadani, B. Dunlap, J. Suehle, C. Richter, C. Hacker and D. Gundlach, "A flexible solution-processed memristor," IEEE Elect. Device Lett., Vol. 30, pp. 706-708, 2009.

[9] Wang, X.L.et al., "Non-volatile, reversible switching of the magnetic moment in Mn-doped ZnO films" J. Appl. Phys., Vol. 113, pp. 17C301 - 17C301-3, 2013.

[10] Zheng Fan et al., "Resistive switching in copper oxide nanowire-based memristor" 12ᵗʰ IEEE Conference on Nanotechnology (IEEE-NANO) 2012, Birmingham, 2012, pp. 1-4.

[11] Chang, Yao-Fenget al., "Study of SiOx-based complementary resistive switching memristor" 2012 70ᵗʰ Annual Device Research Conference (DRC), University Park, TX, 2012, pp. 49-50.

[12] Hu, S.G. et al., "Emulating the paired-pulse facilitation of a biological synapse with a NiOx-based memristor" Appl. Phys. Lett., Vol .102, pp. 183510-183510-4, 2013.

[13] Mamat, M.H. et al., "Electrical characteristics of sol-gel derived aluminum doped zinc oxide thin films at different annealing temperatures" Int. Conf. on Electron. Devices, Syst. and Appl. (ICEDSA), 2010, Kuala Lumpur, 2010, pp. 408 – 411.

[14] A. Shih, W. Zhou, J. Qiu, H-Jen Yang, S. Chen, Z. Mi and I. Shih, "Highly Stable Resistive Switching on Monocrystalline ZnO," Nanotechnology, Vol. 21, pp. 125201, 2010.

[15] Y.-T. Li, S.-B. Long, H.-B. Lv, Q. Liu, Q. Wang, Y. Wang, S. Zhang, W.-T. Lian, S. Liu, and M. Liu, "A low-cost memristor based on titanium oxide," 10ᵗʰ IEEE Int. Conf. on Solid State and Integr. Circuit Technol 2010, Nov. 2010, pp. 1148–1150.

[16] Z. Fang, H. Y. Yu, W. J. Liu, R.Wang, X. A. Tran, B. Gao and J. F. Kang, "Temperature instability of resistive switching of HfOx-based RRAM devices," IEEE Electron Device Letts., Vol. 31, pp. 476-478, 2010.

[17] J. Qiu, A. Shih, W. Zhou, Z. Mi and I. Shih, "Effects of metal contacts and dopants on the performance of ZnO-based memristive devices," J. Appl. Phys., Vol. 110, 2011.

[18] N. S. Kamarozaman, M. A. R. Rashid, M. Z. Musa, S. H. Herman, R. A. Bakar, W. F. H. Abdullah, and M. Rusop, "Effect of film thickness on the memristive behavior of spin coated titanium dioxide thin films," 2013 IEEE Regional Symp. on Micro and Nanoelectronics (RSM), Langkawi, Malaysia, 2013, pp. 155-158.

The ratio between the OFF and ON resistance (R_{OFF}/R_{ON}) was calculated at a particular voltage point that gives the largest hysteresis loop width. The R_{OFF}/R_{ON} ratios for lateral-structured zinc oxide (ZnO) based memristive device is summarized in Table 1. The result shows that the R_{OFF} and R_{ON} values decrease as the electrode width increases except for the electrode width is 7 mm. This result was agreed well with the prediction hypothesized by [11] and [12].

TABLE I. R_{OFF}/R_{ON} RATIO OF A ZNO MEMRISTIVE DEVICE

Electrode size (mm)	R_{OFF} (kΩ)	R_{ON} (kΩ)	R_{OFF}/R_{ON} ratios
5	9.210	9.202	1.001
7	11.360	11.323	1.003
9	0.325	0.320	1.016
10	0.165	0.157	1.044

IV. Conclusion

A lateral-structured ZnO based memristive device has been successfully fabricated using sol-gel spin coating technique. The oxide and electrode widths were varied to study the effect of the oxide and electrode dimensions on the memristive behavior of lateral ZnO based configuration. It was demonstrated that small oxide width in combination with wide electrode width resulted in better memristive behavior. The R_{OFF} and R_{ON} values were decreased as the electrode width increased. Using wider electrode however formed a broader hysteresis loop. The resistance ratio for the widest electrode was calculated to be 1.044.

Annealing Temperature Dependence of Resistive Switching Behavior for Sol-gel Spin Coated Zinc Oxide Thin Films

Raudah Abu Bakar, Ahmad Faiz Mohamad Zohaimi,
Nur Syahirah Kamarozaman, Nor Azira Akmar Shaari,
Shafaq Mardhiyana Mohamat Kasim
NANO-Electronic Center, Faculty of Electrical Engineering,
Universiti Teknologi MARA,
40450 Shah Alam, Selangor, Malaysia
raudah088@salam.uitm.edu.my

Sukreen Hana Herman
CoRe of Frontier Materials & Industry Applications,
Universiti Teknologi MARA
40450 Shah Alam, Selangor Darul Ehsan, Malaysia
hana1617@salam.uitm.edu.my

Abstract—**This work focuses on the resistive switching behavior of sol-gel spin coated zinc oxide (ZnO) thin films on ITO substrate. The deposited ZnO thin films were annealed at various temperatures from 300°C to 500°C in a furnace for 60 minutes in order to study the effect of annealing temperature on the resistive switching behavior of ZnO thin film. The electrical property of the thin film was characterized using 2-point probe current-voltage (I-V) measurement. The surface morphology and film thickness were examined and measured using atomic force microscopy (AFM) and surface profiler respectively. The I-V characteristic showed that the heat treatment on the ZnO thin films at 300 and 400°C resulted in the resistive switching characteristic behavior. Further increasing the temperature up to 500°C on the other hand leads to the formation of asymmetrical hysteresis loop.**

Keywords— Zinc oxide thin film, resistive switching behaviour, sol-gel spin coating, ITO, annealing process.

I. Introduction

Memristor is the fourth fundamental passive circuit element discovered by Prof. Leon Chua in 1971 [1]. It is a two terminal circuit that creates a relationship between charge and magnetic flux [1]. A device is a memristive device when the I-V characteristic exhibits a resistive switching in a shape of pinched hysteresis loop (the loop crosses at origin point). Since Strukov *et al.* succeeded in fabricating a memristor based on titanium dioxide (TiO_2) film [2-3], many studies have been initiated in this field. The term memristor is derived from the word memory resistor which combines the behavior of a memory and a resistor. Hence, as the name implies a memristor has the capability to memorize its current state even after removing the external voltage [2-4].

Recently, many studies have been carried out to study the memristive behavior of TiO_2 thin film [5-8]. A memristor based on TiO_2 film can be fabricated using various deposition techniques including RF magnetron sputtering [5-7], atomic layer deposition (ALD) [2] and sol-gel [8]. Sol-gel is one of the most attractive method compared to other method because

it does not require expensive and complicated equipment and easy to be coated on the desired shape and area.

Other than TiO_2, memristors can also be fabricated based on ZnO thin film [9], copper oxide nanowire (CuONW) [10], SiOx-based [11] and NiOx-based [12]. Of the various available materials for memristor, ZnO has received the considerable attention. ZnO is an n-type semiconductor which possesses a wide direct band gap of 3.37 eV at room temperature. ZnO has free-exciton binding energy of 60 meV [13]. Since this non-toxic and stable thermally and chemically material has reported to exhibit high R_{OFF}/R_{ON} ratios and large degree of controllability [14], thus it has the advantages to be used as an oxide layer for memristor application.

Previous studies have shown that the fabrication technique, process parameters, thin film thickness, electrode size and material as well as oxygen concentration played the important role in determining the resistive switching behavior of an oxide thin film [15]. As reported by other researchers [14-16], the annealing step in the thin film deposition process could result in the formation of oxygen vacancies at the interface between the bottom electrode and the oxide layer that will then affect the resistance properties of the thin film.

In this work the effects of annealing temperature on the resistive switching behavior of ZnO-based thin film will be studied. The thin films were prepared using sol-gel spin coating technique on ITO substrate. The annealing temperature was varied between 300 and 500°C. The I-V characteristic was measured using two point probe measurement system.

II. Methodology

In this work ZnO thin films were deposited onto ITO substrate via sol-gel spin coating technique. Zinc acetate dehydrate ($Zn(CH_3COO)_2 2H_2O$), 2-methoxyethanol ($C_3H_8O_2$) and monoethanolamine (C_2H_7NO, MEA) were used as the precursor, solvent and stabilizer respectively. In order to prepare 0.4 molar ZnO solution, the precursor was first dissolved into the solvent. Then the stabilizer was added to achieve a stable solution. Before subjecting the ZnO solution to stir-heated process, the solution was sonicated for 30 minutes at 50°C. The stir-heated process was then performed at 80°C for 3 hours to increase reaction process between all the materials in the solution. The stirring process was continued for another 24 hours at room temperature to reinforce gel network and yield clear solution.

The ZnO solution was then spin-coated onto the ITO substrate at 3000 rpm for 1 minute. Followed by the drying process in a furnace at 150°C for 10 minutes to evaporate the solvent and eliminate organic component in the film. The deposited ZnO thin films were then subjected to annealing process at various temperatures of 300, 400 and 500°C for 60 minutes.

The electrical characterization was performed using 2-points probe I-V measurement system. To study the ZnO based thin film memristive behavior, the bias voltage was swept from 0 V to 5 V, then 5 V to -5 V and back to 0 V while measuring the current simultaneously. The structural property and film thickness were examined and measured using atomic force microscopy (AFM) and surface profiler respectively.

III. Results and Discussion

A. I-V Characteristic

The I-V characteristics of ZnO-based thin film annealed at 300, 400 and 500°C are shown in Fig. 1(a)-(c) respectively. The results show that the ZnO annealed at 300 and 400°C exhibit memristive behaviors. Both hysteresis loops cross at the origin point. At 3.5 V the R_{OFF}/R_{ON} ratios for the ZnO films were calculated to be 10 and 28 respectively. Annealed the film at 500°C however resulted in asymmetrical I-V characteristic. As shown in Fig. 1 (c), the hysteresis loop was only found when the ZnO thin film was negatively biased. As also can be seen the increases in the annealing temperature leads to the increases of the conduction of current during negative voltage application.

As suggested by the previous studies [5, 14-16], the observed switching phenomenon occurs in memristive-based device is due to the formation of filament that can be formed by the movement of oxygen vacancies within the oxide layer. Basically, the oxygen vacancies can be existed at the ITO/ZnO interface after subjecting to the thermal or high voltage treatment process [17]. Fig. 2(a)-(c) illustrate the movement of oxygen vacancies when the applied voltage is zero, negative and positive biased respectively. Without the application of voltage, the oxygen vacancies are pinned at the ITO/ZnO interface (Fig. 2(a)) causing the current to be zero. The moment the device is negatively biased, the positively charge oxygen ions move towards the top electrode and form the filament within the oxide layer (Fig. 2(b)). Since a path between the electrodes has been created, the current can be easily conducted. The application of positive voltage however repels the oxygen vacancies towards the bottom electrode layer and subsequently leads to the existence of wider insulating region (Fig. 2(b)). This will then increase the resistivity and cause a small current to conduct.

Fig. 1 .The I-V characteristics of ZnO thin films annealed at (a) 300°C (b) 400 °C and (c) 500°C.

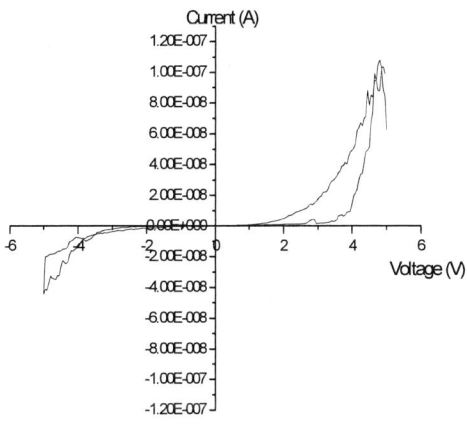

(a)   Annealing temperature = 300°C

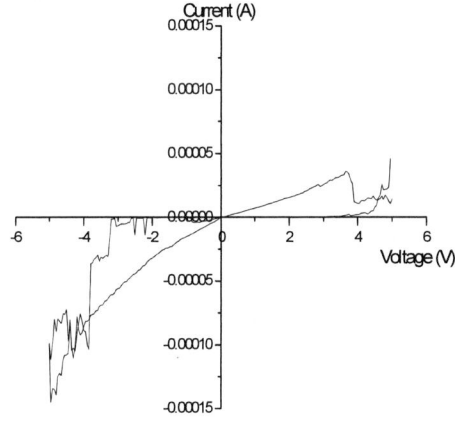

(b)   Annealing temperature = 400°C

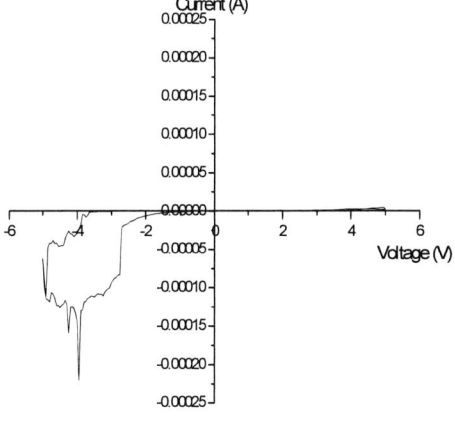

(c)   Annealing temperature = 500°C

Fig. 2. The illustration of the movement of oxygen vacancies with (a) no applied voltage (b) negatively biased and (c) positively biased

Fig. 3. The three dimensional (3-D) and two dimensional (2-D) AFM images for the ZnO thin films annealed at (a) 300°C (b) 400°C and (c) 500°C

(a) No applied voltage

(b) Negative voltage application

(c) Positive voltage application

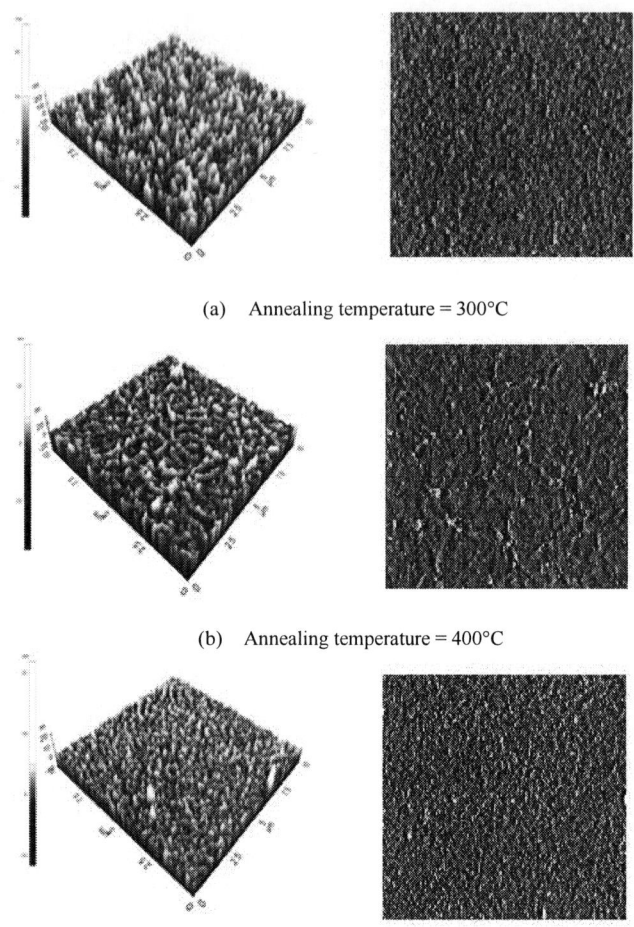

(a)   Annealing temperature = 300°C

(b)   Annealing temperature = 400°C

(c)   Annealing temperature = 500°C

B. Physical Properties

The surface morphologies of ZnO-based thin films annealed at 300, 400 and 500°C examined using AFM is shown in Fig. 2 (a)-(c) respectively. As can be seen the morphology of the ZnO thin film changes as the annealing temperature increases. A smooth film can be obtained if the thin film annealing temperature is high. The ZnO thin film root mean square surface roughness is summarized in Table I. As shown in the Table 1 the film annealed at 300, 400 and 500°C resulted in the thin film roughness of 5.170, 4.571 and 2.501 nm respectively.

The thickness of ZnO thin film is summarized in Table II. The thin film thickness was measured using surface profiler. As can be observed, the thickness for the film annealed at 300, 400 and 500°C was measured to be 106.85, 80.60 and 95.95 nm respectively. The thickness of the film was found to decrease after performing the annealing process at 400°C. Further increasing the annealing temperature to 500°C resulted in thicker film compared to the film annealed at 400°C.

As reported by the previous studies [18], the increase in thin film thickness resulted in a small value of resistance ratios and vice versa. In this study the same behavior has been observed in which thinner thin film thickness leads to the highest resistance ratios. Furthermore as suggested by [2], a good memristive behavior could be obtained if thin oxide layer has been used.

TABLE I.   THE ROUGHNESS OF ZnO-BASED THIN FILM

Temperature (°C)	Roughness (nm)
300	5.170
400	4.571
500	2.501

TABLE II.   THE THICKNESS OF ZnO-BASED THIN FILM

Temperature (°C)	Thickness (nm)
300	106.85
400	80.60
500	95.95

IV. Conclusion

This work is carried out to study the effect of annealing temperature on the resistive switching behavior of ZnO thin film on ITO substrates. The results showed that the film annealed at 300 and 400°C exhibited symmetrical resistive switching behavior. As annealing temperature increased to 500°C, an asymmetrical hysteresis loop was obtained. The ZnO film annealed at 400°C also resulted in the highest R_{OFF}/R_{ON} ratios.

Acknowledgment

This work is partially supported by Ministry of Education Malaysia (MOE) under the Niche Research Grant Scheme (NRGS) (Project code: 600-RMI/NRGS 5/3 (7/2013)) and the supporting grant by UiTM Research Management Institute (RMI) (Project code : 600-RMI/DANA 5/3/PSI (248/2013)).

References

[1] L. Chua, "Memristor- The missing circuit element," IEEE Trans. on Circuit Theory, Vol.18, pp. 507-519, Sept. 1971.

[2] D. B. Strukov, G. S. Snider, D. R. Stewart, and R. S. Williams, "The missing memristor found," Nature, vol. 453, no. 7191, pp. 80–83, May 2008.

[3] R. S. Williams, "How we found the missing memristor," in IEEE Spectrum.

[4] Q. Xia, "Nanoscale Resistive Switches: Devices, Fabrication and Integration," Appl. Phys. A, Vol. 102, pp. 955-965, 2011.

[5] C. Yakopcic, A. Sarangan, J. Gao, T. M. Taha, G. Subramanyam and S. Roger, "TiO$_2$ Memristor Devices," Proc. IEEE, pp. 101-104, 2011.

[6] N. S. Kamarozaman, Z. Aznilinda, S. H. Herman, R. A. Bakar and M. Rusop, "Effect of annealing duration on the memristive behavior of Pt/TiO$_2$/ITO memristive device" in 10th IEEE Int. Conf. on Semicond. Electron. (ICSE) 2012, Kuala Lumpur, 2012, pp. 703-706.

[7] N. S. Kamarozaman, Z. Aznilinda, S. H. Herman, R. A. Bakar and M. Rusop, "Memristive behavior of NaOH-immersed titania nanostructures" J. Mech. Eng. & Sciences. (JMES), Vol. 5, pp. 688-695, 2013.

[8] N. Gergel-Hackett, B. Hamadani, B. Dunlap, J. Suehle, C. Richter, C. Hacker and D. Gundlach, "A flexible solution-processed memristor," IEEE Elect. Device Lett., Vol. 30, pp. 706-708, 2009.

[9] Wang, X.L.et al., "Non-volatile, reversible switching of the magnetic moment in Mn-doped ZnO films" J. Appl. Phys., Vol. 113, pp. 17C301 - 17C301-3, 2013.

[10] Zheng Fan et al., "Resistive switching in copper oxide nanowire-based memristor" 12th IEEE Conference on Nanotechnology (IEEE-NANO) 2012, Birmingham, 2012, pp. 1-4.

[11] Chang, Yao-Fenget al., "Study of SiOx-based complementary resistive switching memristor" 2012 70th Annual Device Research Conference (DRC), University Park, TX, 2012, pp. 49-50.

[12] Hu, S.G. et al., "Emulating the paired-pulse facilitation of a biological synapse with a NiOx-based memristor" Appl. Phys. Lett., Vol .102, pp. 183510-183510-4, 2013.

[13] Mamat, M.H. et al., "Electrical characteristics of sol-gel derived aluminum doped zinc oxide thin films at different annealing temperatures" Int. Conf. on Electron. Devices, Syst. and Appl. (ICEDSA), 2010, Kuala Lumpur, 2010, pp. 408 – 411.

[14] A. Shih, W. Zhou, J. Qiu, H-Jen Yang, S. Chen, Z. Mi and I. Shih, "Highly Stable Resistive Switching on Monocrystalline ZnO," Nanotechnology, Vol. 21, pp. 125201, 2010.

[15] Y.-T. Li, S.-B. Long, H.-B. Lv, Q. Liu, Q. Wang, Y. Wang, S. Zhang, W.-T. Lian, S. Liu, and M. Liu, "A low-cost memristor based on titanium oxide," 10th IEEE Int. Conf. on Solid State and Integr. Circuit Technol 2010, Nov. 2010, pp. 1148–1150.

[16] Z. Fang, H. Y. Yu, W. J. Liu, R.Wang, X. A. Tran, B. Gao and J. F. Kang, "Temperature instability of resistive switching of HfOx-based RRAM devices," IEEE Electron Device Letts., Vol. 31, pp. 476-478, 2010.

[17] J. Qiu, A. Shih, W. Zhou, Z. Mi and I. Shih, "Effects of metal contacts and dopants on the performance of ZnO-based memristive devices," J. Appl. Phys., Vol. 110, 2011.

[18] N. S. Kamarozaman, M. A. R. Rashid, M. Z. Musa, S. H. Herman, R. A. Bakar, W. F. H. Abdullah, and M. Rusop, "Effect of film thickness on the memristive behavior of spin coated titanium dioxide thin films," 2013 IEEE Regional Symp. on Micro and Nanoelectronics (RSM), Langkawi, Malaysia, 2013, pp. 155-158.

Effect of V/III Ratios on Surface Morphology in a GaSb Thin Film Grown on GaAs Substrate by MOCVD

Chih-Wen Hsiao[a], Chun-Yuan Liu[b], Sa-Hoang Huynh[c], Thien-Huu Ha Minh[c], Hung-Wei Yu[c], Hong-Quan Nguyen[c], Yer-Shen Laa[b], Shoou-Jinn Chang[a], IEEE Fellow and Edward Yi Chang[b,c], IEEE Fellow

[a]Institute of Microelectronics and Department of Electrical Engineering, Center for Micro/Nano Science and Technology, Advanced Optoelectronic Technology Center, National Cheng Kung University, No.1 University Road, Tainan 70101 Taiwan
[b]Institute of Photonic System and Institute of Lighting and Energy Photonics, National Chiao Tung University, No.1001 University Road, Hsinchu 30010 Taiwan
[c]Department of Materials Science and Engineering and Department of Electronic Engineering, National Chiao Tung University, No.1001 University Road, Hsinchu 30010 Taiwan

— The epitaxial growth of GaSb thin film with different V/III ratios on high-lattice-mismatched GaAs (001) substrates by metal organic chemical vapor deposition (MOCVD) was investigated. Under optimal V/III ratio of 2.5, we found that there are many periodic 90° interfacial misfit dislocation (IMF) arrays existing at the GaAs/GaSb interface. The surface roughness is about 3.6nm, 2.2nm, 3.8nm, respectively while different V/III ratios (1.25, 2.5, 5) were adopted. These results demonstrated that the hill-and valley structure on the surface of GaSb/GaAs heterostructure can be effectively improved, and formed smooth surface morphology.

—

I. INTRODUCTION

The AlGa$_{1-x}$Sb/GaAs heterostructure has been attracted attention from both academic or industrial world because of its a wide range of electronic band gaps, band-gap offsets and electronic carriers along with extremely high electron mobility [1]. Consequently, it can be used in high electron mobility transistors (HEMTs) and infrared-photodetector (IR-PDs).

However interfacial misfit dislocations in high lattice mismatched GaSb epitaxial layer grown on GaAs substrate containing periodic 90° misfit for strain relaxation, they can effectively terminate the propagation of micro-twin induced structural defects into several monolayers (MLs) and with the low defect material grown on a highly mismatched substrate with a relatively IMF technique [2]. Several approaches have been proposed to prepare IMF growth mode for the growth of GaSb on GaAs depends on the growth condition and may achievable under specific growth parameters can be prepared by molecular beam epitaxy (MBE) [3,4]. So far, there have been a few studies based on IMF growth mode by metalorganic chemical vapor deposition (MOCVD) [5]. Compared with other methods, MOCVD can be used to grow samples in IMF mode, which is crucial when preparing high strain relaxation samples. By taking advantage of its have great potential for mass production compared to MBE system.

II. EXPERIMENTAL PROCEDURE

In this paper, we present the use of an interfacial misfit dislocations or periodic 90° misfit dislocation based growth of GaSb on GaAs with various V/III ratios of 1.25 (Sample A), 2.5 (Sample B) and 5 (Sample C), respectively. All samples in this study were grown by a low-pressure metal organic chemical vapor deposition (MOCVD, AIXTRON 2400). Trimethylgallium (TMGa) was used as group III sources, whereas arsine (AsH$_3$) and trimethylantimony (TMSb) were used as group V sources. The surface morphology and RMS roughness were measured by optical microscope and atomic force microscope (AFM). The structural properties of GaSb on GaAs layers were investigated using X-ray Diffraction (XRD). The interface of GaAs/GaSb was examined by Transmission Electron Microscope (TEM). The detailed description of the growth of GaSb epitaxy on GaAs substrate can be found elsewhere [5]. The GaSb epitaxy was grown on the GaAs heterostructure by IMF growth technique and growth temperature at 550°C.

III. RESULT AND DISCUSSION

A. Surface Morphology

The factor of V/III ratios is one of the main determinants for GaSb epitaxy grown on GaAs substrate. Depending on the V/III ratios, there are several different epitaxial reactions on the surface during growth process. Different V/III ratios used during material growth strongly affect the activation energy of surface and precursors coverage, and thus the results of surface morphology would be distinct.

The GaSb layer were measured by AFM with scan size of $5\mu m$ x $5\ \mu m^2$ to investigate the effect of the V/III ratio. Figure 1 shows the AFM images of the GaSb thin film grown on GaAs (001) substrate. The V/III ratio was varied in a restricted range of 1.25, 2.5 and 5 for samples A, B and C. And the surface roughness is about 3.6nm, 2.2nm, 3.8nm, respectively while different V/III ratios (sample A, B and C) were adopted. We observed some Ga droplets remained on the surface of sample A in Figure 1(a). This phenomenon we found is apparently related to Ga-rich condition. On the other hand, in the growth condition of V/III ratio of 2.5 shows the smooth surface (Figure 1(b)), it means that the growth condition is GaSb free surface condition. In addition, we can observe some hillock in sample C (Figure 1(c)), it may due to the Sb-rich growth condition [6].

And the difference of the island shapes in Ga-rich or Sb-rich growth condition can be attributed to anisotropy of the diffusion length of adatoms or growth speed along different crystallographic directions [7].

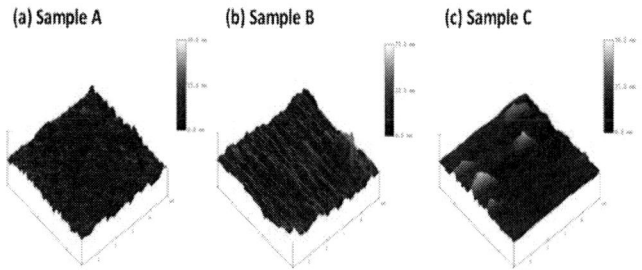

(a) Sample A (b) Sample B (c) Sample C

Figure 1. AFM images of GaSb/GaAs heterostructure grown at V/III ratio of (a) 1.25 (b) 2.5 (c) 5

Figure 2 shows the average height of island and the island density as a function of the V/III ratio. It is reported that a high Sb flux could suppress surface migration of Ga adatoms [8]. Due to the atomic motion of self-atoms depends on the equilibrium concentrations of native defects. Besides the V/III ratio, many other factors can affect the initial growth mode, such as growth rate, growth temperature, lattice mismatch and surface energy. Moreover, the initial growth is not only concerned with the adatom diffusion.

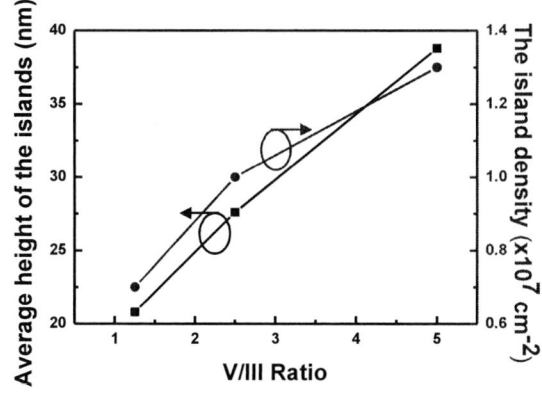

Figure 2. The average height of island and the island density as a function of the V/III ratio

B. Full Width at Half Maximum (FWHM) and Strain Relaxation

The strain relaxation and crystal quality, which are depend on the different growth condition it can be determined by HR-XRD. Figure 3 shows the ω-2θ (004) scan of GaSb/GaAs layer grown with IMF array technique. The GaSb peak is at − 80 arcseconds in Sample B from the GaAs substrate peak indicating almost full relaxation (7.4%). The full width at half maximum (FWHM) of the rocking curve at ~1184 arcseconds when V/III ratio was 2.5 (insert Figure 3). We can found that the carbon concentration increases, because of a increase in the number of Sb vacancies on which C can be incorporated in GaSb/GaAs with decreased the V/III ratio.

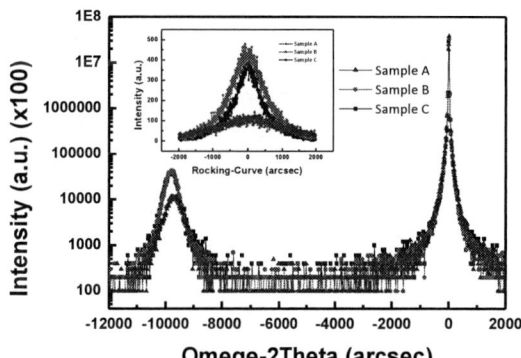

Figure 3 The ω-2θ symmetric (004) HR-XRD spectra of three samples with various V/III ratios

C. Interfacial Misfit Dislocations (IMF) and Defects

Figure 4 shows the TEM images of the GaSb epitaxial (Sample B) grown on the GaAs (001) heterostructure using interfacial misfit dislocations (IMF) growth mode as technique. It can be seen that 90° interfacial misfit dislocation (IMF) arrays were formed in the GaSb layer on the GaAs (001) substrate for thickness is 100 nm, as shown in Figure 4.(a). For the V/III ratio of 2.5 (Sample B), a strain-free GaSb epitaxy with periodic 90° interfacial misfit dislocation (IMF) arrays was obtained (insert Figure 4(a)). The formation of periodic 90° misfit dislocation was result from reaction of 60° dislocations from different glide systems [9].

Also, defects like threading dislocation, stacking fault and microtwins, are observed in the Figure 4.(b), which may be attributed to relaxation of strains on the GaSb/GaAs interface. This is a serious problem in both simulation and HEMTs devices [10], its attributed to both surface state effects as well as substrate deep trap effects. Due to the term gate lag and drain lag are used to describe the slow transient response of the drain current when the gate voltage is changed abruptly [11]. Though surface state hole traps are the major cause of gate lag, simulations have shown that the influence of electron traps cannot be ignored [12]. The electron traps cause a small decrease of drain current immediately after a turn-on transient of the device.

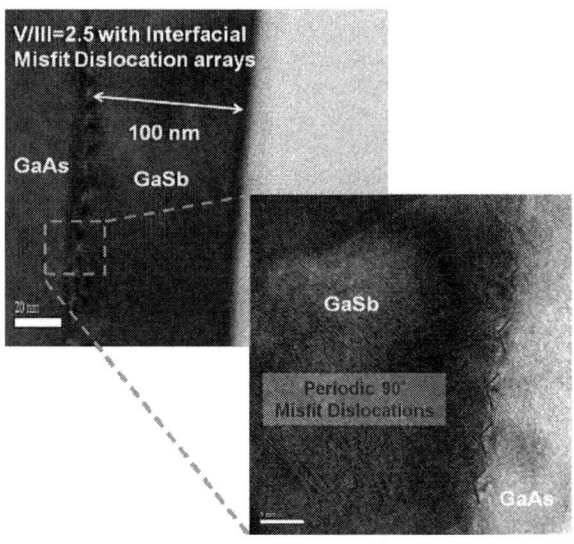

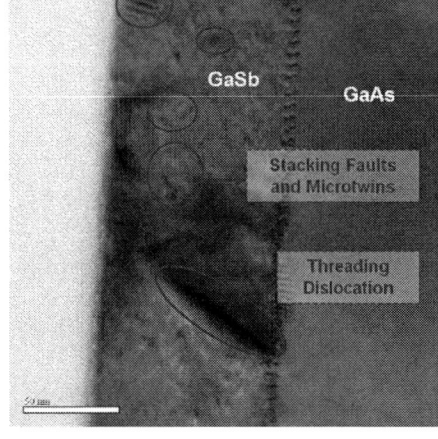

Figure 4. TEM images of the GaSb epitaxy with 2.5 grown on the GaAs substrate (a) with interfacial misfit dislocation arrays (b) with defects

V. CONCLUSIONS

In summary, we demonstrate that appropriate V/III ratio of 2.5 can effectively improve the quality of GaSb epitaxy on GaAs substrate. However, it is not the only key factor controlling GaSb epi-layer growth. Under specific growth condition, its shows in high-relaxation GaSb epi-layer with periodic 90° misfit dislocations can be formed and effectively reduce the difference between surface energy and strain energy during material growth, and promote formation of GaSb epi-layer.

ACKNOWLEDGMENT

This work was sponsored by the NCTU-UCB I-RiCE program, Ministry of Science and Technology, Taiwan, and TSMC, under Grant No. NSC 103-2911-I-009-302.

REFERENCES

[1] S. M. Sze, *Semiconductor Devices*, Wiley, New York (2002).

[2] A. Rocher, "Interfacial Dislocations in the GaSb/GaAs (001) Heterostructure," *Solid State Phenom.*, **19**, 563 (1991).

[3] S. H. Huang et al., "Strain relief by periodic misfit arrays for low defect density GaSb on GaAs," *Appl. Phys. Let.*, **88**, 131911 (2006).

[4] K. Kallipalli et al., "Atomistic modeling of strain distribution in self-assembled interfacial misfit dislocation (IMF) arrays in highly mismatched III-V semiconductor materials," *J. Cryst. Growth*, **303**, 449 (2007).

[5] Wei Zhou et al., "High Hole Mobility of GaSb Relaxed Epilayer Grown on GaAs Substrate by MOCVD through interfacial Misit Dislocations Array," *J. Mater. Sci. Technol.*, **28** (2), 132 (2012).

[6] R.M. Biefeld et al., "The metal-organic chemical vapor deposition and properties of III-V antimony-based semiconductor materials," *Mater. Sci. Eng. R*, **36**, 105 (2002).

[7] B. R. Bennett et al., "Microstructural evolution of GaSb self-assembled islands grown by metalorganic chemical vapor deposition," *Appl. Phys. Let.*, **73**, 1233 (1998).

[8] E. Suzuki et al., "Density Control of GaSb/GaAs Self-assembled Quantum Dots (25nm) Grown by Molecular Beam Epitaxy," *Japan J. Appl. Phys.*, **37**, 203 (1998).

[9] E. Karagan et al., "Formation of misfit dislocations in thin film heterostructures," *J. Appl. Phys.*, **92**, 7122 (2002).

[10] G. Meneghesso et al., "Surface-Related Drain Current Dispersion Effects in AlGaN-GaN HEMTs," *IEEE Trans. Electron Devices*, **51**, 10, 1554, 2004.

[11] J. Kumi et al., "Gate-lag and drain-lag effects in AlGaN/InAlGaN and InAlN/AlGaN HEMTs," *Phys. Stat. Sol. (a)*, **204**, 6, 2011 (2007).

[12] K. Horio et al., "Two-dimensional analysis of surface state effects on turn-on characteristics in GaAs MESFETs," *IEEE Trans. Electron Devices*, **46**, 648 (1999).

Electrical and Optical Properties Characterization of MEH-PPV Thin Film using Sol-Gel Method

H. Hashim, S. S. Shariffudin, A. M. Khairuddin, M. S. P. Sarah and M. Rusop

Faculty of Electrical Engineering
Universiti Teknologi MARA
40450 Shah Alam, Selangor, Malaysia
hashimah655@salam.uitm.edu.my

Abstract— Light-emitting diode (LED) applications consist of various materials. One of the materials used is polymer. In this study, MEH-PPV known as poly[2-methoxy-5-(2'-ethyl-hexyloxy)-1, 4-phenylenevinylene] was used. The objective of this paper is to characterize the electrical and optical properties of MEH-PPV thin film by using sol-gel method. The scope is to study only the MEH-PPV thin film without implementation to any device. The MEH-PPV thin film thicknesses were varied from 10 to 100nm. The experiment was started by stirring the sol-gel solution with toluene. The spin-coating technique was used to deposit the MEH-PPV thin film on a glass substrate. All samples were characterized using Atomic Force Microscopy (AFM), Surface Profiler, Two-point Probe, Raman PL Dispersive and UV-Vis Spectroscopy for the surface morphologies, thin film thickness, electrical and optical properties respectively. From the current-voltage (I-V) measurement, it show that symmetrical line plotted at low-voltage ranges. Moreover, the calculated conductivity was inversely proportional with the thin film thickness. The results from photoluminescence (PL) spectra showed that the intensity reached optimum peak at 38nm thickness and quenched for other samples. At 108nm of thickness, absorption reached the highest peak compared to other samples of different thickness. The film was non-uniformed for the thickness at 134nm, due to aggregation phenomenon.

Keywords— *spectra, Polymer (MEH-PPV), thin film thickness, electrical properties, optical properties*

I. INTRODUCTION

Conducting polymers consist of group of compounds and materials with very specific properties [1]. There are many similarities in electronic field that caught the attention of researchers. For example, MEH-PPV thin film was fabricated for OLEDs and solar cells. According to A. Petrella et al. [2] organic solar cells had an excitation diffusion length that was not ideally shorter than the ability to absorb energy or photon that makes it easier to bond. This theory was applied into this experiment to optimize the characteristics need for optoelectronic devices. Most of researchers concluded that organic semiconductors have been restricted by the fabrication process [3]. The problems could be solved by investigating the electrical and optical properties of MEH-PPV thin film by using sol-gel method. This study was focusing to characterize the properties of the organic polymer material rather than synthesizing the device itself. In this work, the suitable

solution to bond with MEH-PPV thin film was toulene. Toulene is more red-shifted organic solvent that provides more reliable conductivity and optical properties. Ten samples of different thickness were prepared with two different architectures. Five samples were prepared for non-metal contact surface while the other five samples with metal contacts for electrical measurements.

II. EXPERIMENTAL METHOD

The experimental work began with the procedure of fabrication the MEH-PPV thin film. It took about two days to fully-bond with the polymer and synthesized it with spin-coating method. Spin-coating method is the solution for low-cost manufacturing of solar-cell because of organic polymers behavior that can easily be formed [4]. The schematic structure of the sample is shown in Fig. 1 that consists of glass substrate, MEH-PPV thin film and gold (Au) contact. Glass was used as the bulk substrate for the sample. MEH-PPV thin film was synthesized on top of the bulk surface for the optical investigations. After that, Au contact was deposited on top of the MEH-PPV thin film for electrical measurements.

Fig. 2 shows the process flow of the experiment. Glass substrate was prepared by cleaning it with acetone, methanol and de-ionized water by immersed in the ultrasonic cleaner instrument. MEH-PPV powder with the average molecular weight between 40,000 to 70,000%wt was prepared for the sol-gel solution. The MEH-PPV powder with concentration of 100mg was dissolved in 20ml toulene and stirred for 48 hours. The sol-gel solution was stirred using the hot-stirrer instrument and the thin film was deposited by using spin-coating method. The deposition spin speed was set at 2000 rpm with the duration time of 60 seconds. After depositing a layer, the sample was dried at 50°C for 10 minutes. This procedure was repeated until 10 layers of thin film in order to obtain variations of thickness. Sputter beam technique was used for the metal contact layer through a mask layout for the preferred architecture. In this process, Au (60nm thickness) was used as the metal contact for the characterization of electrical properties. The film thicknesses and morphology were measured by using the Surface Profiler (Veeco Dektak 150) and AFM (Park System XE-100) respectively.

978-1-4799-5761-3/14 $31.00 © 2014 IEEE

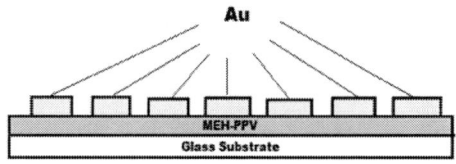

Fig. 1. Schematic of sample structure with Au metal contact.

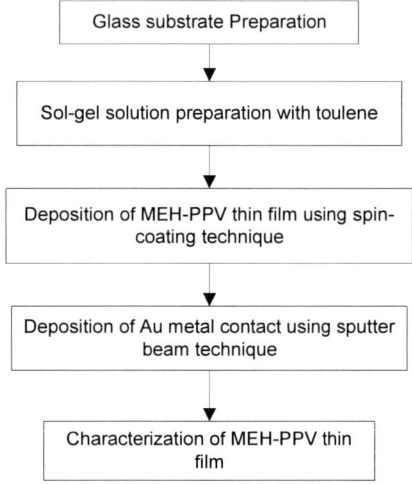

Fig. 2. The flow chart of the experiment.

The electrical measurements were obtained by using the two-point probe solar simulator (Bukoh Keiki CEP 2000) with the range of -5 to 5V while the PL spectra were obtained by using Raman PL Dispersive (Horiba Jobin Yvon) with He:Cd emitting at 325 nm as laser source and the transmittance spectra were measured by using UV-Vis Spectroscopy (JASCO, FLH-740). The absorption waveform was calculated from transmittance spectra for the optical properties.

III. RESULTS AND DISCUSSION

A. Electrical Properties

This investigation was conducted by using five samples with Au contact. Table I shows the tabulated samples with different film thickness. The results show that the film thickness increases with deposition layer. According to N. Juhari [5], spin-coating technique usually forming thickness at sub-micron range. For the I-V measurement, the experiment was conducted by using the two-point probe method in dark condition. According to A. H. Reshak [6], the probe acts as the carrier that transport the amount of voltage to thin film itself. Fig. 3 shows the measurement result from the two-point probe technique. By referring to the graph, symmetrical line spotted at low voltages range. As for the proof of the situation by solid-state theory, localized gap states formed with the occurrence of defects [7].

TABLE I. SURFACE THICKNESS AT DIFFERENT DEPOSITION LAYER

Samples	Deposition Layer	Thickness (nm)
A	1	26.88
B	3	38.57
C	5	62.00
D	7	79.94
E	10	134.00

The plotted graph sampled at voltage range within -5 to 5V to provide much more exact visual pattern. Sample A shows a different pattern than the other samples which is similar to typical diode behavior. Other samples obtained more rigid pattern within the range 5nA and below. Furthermore, it shows similar typical characteristics as the thickness greater than 38nm. Fig. 4 shows the plotted graph of conductivity with the analysis from I-V measurements in Fig. 3. The conductivity was calculated using its relationship with resistivity. The resistivity can be calculated using the Equation 1 below:

$$\rho = R.(\frac{w \times t}{l}) \tag{1}$$

where R is the resistance of the films, w is the width, t is the thickness and l is the length between the metal contacts [8], [9]. The values of w and l are 2.0cm and 0.0455cm respectively. The formation of electrical conductivity in MEH-PPV thin film is due to the formation of delocalized and π-electron systems [10]. The conductivity was calculated using the Equation 2 below:

$$\sigma = \frac{1}{\rho} \tag{2}$$

From the plotted graph, the conductivity reduces as the thin film thicknesses getting thicker. Higher conductivity achieved for sample A with 326×10^{-7} S.m^{-1} while the lowest value obtained at sample C with 5.19×10^{-7} S.m^{-1} as shown in Table II.

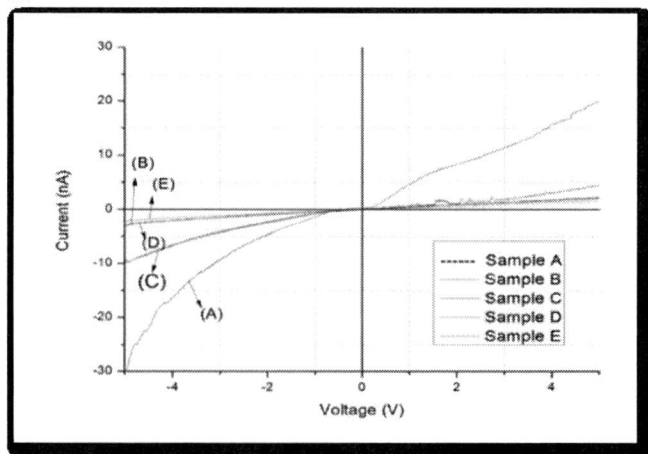

Fig. 3. The plot of I-V measurements of MEH-PPV thin film.

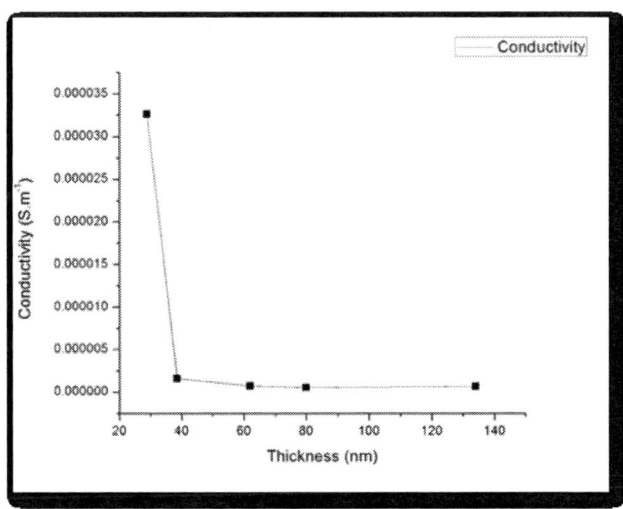

Fig. 4. Conductivity of MEH-PPV thin film with different thickness.

TABLE II. THE SUMMARIZED CONDUCITVITY DATA

Samples	Thickness (nm)	Conductivity X 10^{-7} (S.m^{-1})
A	26.88	326
B	38.57	15.8
C	62.00	7.13
D	79.94	5.19
E	134.00	6.82

B. Optical Properties

Optical investigations were divided into two main categories which are PL spectra and absorption of the thin film (Fig. 5 and 6 respectively). PL investigation conducted for the wavelength between 350 and 800nm. Based on Fig. 5, it shows that the peak intensity obtained from sample B while the sample A shows more flattened pattern. All samples peak intensity exhibits at wavelength ranging from 550 to 630nm. The shoulder peak occurred after the peak intensity because of the aggregation phenomenon. The factors of aggregation formation is affected on both concentration and solvents as well [11].

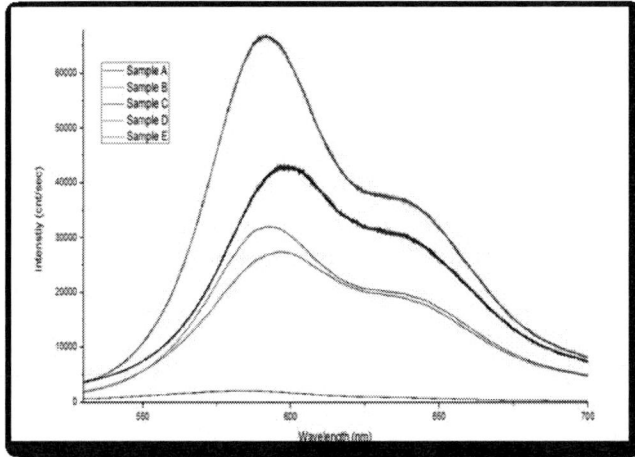

Fig. 5. The plot of PL spectra at different thickness.

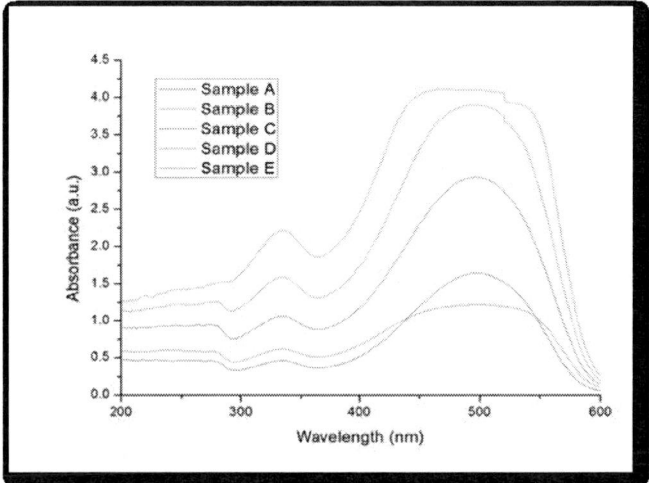

Fig. 6. Absorption of MEH-PPV at different thickness.

TABLE III. THE SUMMARIZED ELECTRICAL AND OPTICAL PROPERTIES AT DIFFERENT THICKNESS

Samples	Thickness (nm)	Conductivity X 10^{-7} (S.m^{-1})	PL peak (cnt/sec)	Absorption peak (a.u.)
A	26.88	326	2120.43	1.64
B	38.57	15.8	67205.71	1.21
C	62.00	7.13	43639.41	2.93
D	79.94	5.19	32121.06	3.90
E	134.00	6.82	27620.70	4.12

Highest peak intensity achieved at sample B with the amount of 67205.71cnt/sec. This occurrence leads to OLED capability to emit light and it posses' great intensity. Absorption is the reverse analysis of PL spectra. If the thin film exhibits high photoluminescence, the absorption of light will indicates low absorbance capabilities. Based on Fig. 6, sample E provides high ability of absorb light rather than emission of photon. It is reasonable for this sample to act as the solar cell device due to great absorption abilities to convert it into electrical energy.

Table III shows the summarized characterization for all samples. Sample D possess the lowest conductivity while sample E has the highest absorption peak. Sample B has the optimum PL peak intensity rather than other samples.

C. Surface Morphology

Surface morphologies of MEH-PPV thin film were characterized by using AFM method. This instrument visualizes the surface images of the thin film itself. Based on Fig. 7, the morphology samples of non-metal contact were captured. As the thickness increases, the morphology visualizes random uniformity. At 134nm, a lot of bumpy spotted rather than slimmer thickness of thin film. White bumpy images on all of the samples occurred because of aggregation using spin-coating technique.

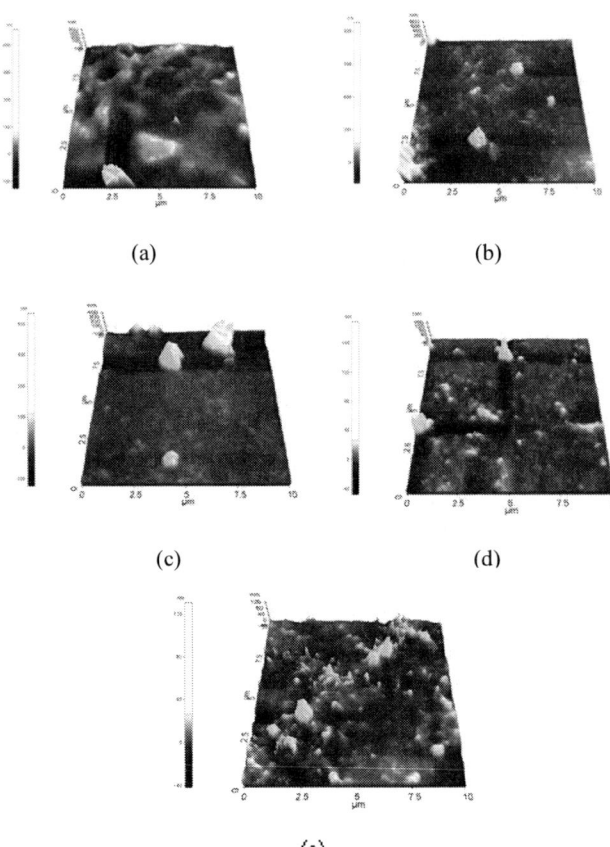

(a) (b)

(c) (d)

(e)

Fig. 7. The AFM surface morphology images at thickness of (a) 28nm (b) 38.57nm (c) 62.00nm (d) 79.94nm and (e) 134.00nm.

TABLE IV. SURFACE ROUGHNESS AT DIFFERENT THICKNESS

Samples	Thickness (nm)	Roughness, Ra (nm)
A	26.88	19.43
B	38.57	1.52
C	62.00	12.12
D	79.94	37.26
E	134.00	3549.90

The analysis of the captured images was tabulated in Table IV. Sample B achieved lesser roughness while sample E indicated large amount of roughness. This is because of the agglomeration and aggregation occurred during the fabrication process.

IV. CONCLUSION

The electrical and optical properties characterization of MEH-PPV thin film was completely conducted. The I-V measurement results show that the thin layer of MEH-PPV film show a typical diode behavior. The calculated conductivity reached the highest level for this layer. For optical properties, PL spectra reached its high intensity at

thickness greater than 38.57nm which is suitable for OLED applications. However, the PL intensity quenched as it increased above 40nm. Moreover, great absorption capability recorded at thicker layer that suitable for the solar-cell applications. Other than that, lack of uniformity in morphology investigation at 134nm was due of aggregation by using the spin-coating technique. For further investigations, this study should proceed to optoelectronic device fabrication process.

ACKNOWLEDGMENT

The author would like to thank the staff of NanoelecTronic Centre (NeT) and Nano-SciTech Centre (NST) of Universiti Teknologi MARA for their support of granting the permission to use all the equipment.

REFERENCES

[1] G. Inzelt, *Conducting polymers: a new era in electrochemistry*. 2012, p. 1.

[2] A. Petrella, M. Tamborra, P. . Cozzoli, M. . Curri, M. Striccoli, P. Cosma, G. . Farinola, F. Babudri, F. Naso, and A. Agostiano, "TiO2 nanocrystals – MEH-PPV composite thin films as photoactive material," *Thin Solid Films*, vol. 451–452, pp. 64–68, Mar. 2004.

[3] M. Road, C. C. B. O. E, and J. H. Burroughes, "Light-Emitting Diodes Based on Conjugated Polymers," vol. 347, no. Nature, pp. 539–541, 1990.

[4] N. Mustapha, K. H. Ibnaouf, Z. Fekkai, a. Hennache, S. Prasad, and a. Alyamani, "Improved efficiency of solar cells based on BEHP-co-MEH-PPV doped with ZnO nanoparticles," *Optik - International Journal for Light and Electron Optics*, vol. 124, no. 22, pp. 5524–5527, Nov. 2013.

[5] N. Juhari, W. H. A. Majid, and Z. Abidin, "Structural and Optical Studies Of MEH-PPV using Two Different Solvents Prepared by Spin-Coating Technique," *Solid State Science and Technology*, vol. 15, no. 1, pp. 141–146, 2007.

[6] A. H. Reshak, M. M. Shahimin, N. Juhari, and S. Suppiah, "Electrical behaviour of MEH-PPV based diode and transistor.," *Progress in biophysics and molecular biology*, vol. 113, no. 2, pp. 289–94, Nov. 2013.

[7] M. S. P. S. and M. R. F.S.S. Zahid, "The effect of TiO2 nanoparticle on the photoconductivity properties of organic MEH-PPV thin films for organic photovoltaic cell applications," *IEEE Sympsomium on Humanities, Science and Engineering Research*, vol. 0, pp. 513–517, 2012.

[8] F. S. S. Zahid, M.S.P.Sarah, and M.Rusop, "Photoconductivity of MEH-PPV: TiO 2 nanocomposite thin film," *International Conference on Electronic Devices, Systems and Applications (ICEDSA)*, pp. 263–266, 2011.

[9] C. Chen, F. Wu, H. Geng, W. Shen, and M. Wang, "Analytical model for the photocurrent-voltage characteristics of bilayer MEH-PPV/TiO2 photovoltaic devices.," *Nanoscale research letters*, vol. 6, no. 1, p. 350, Jan. 2011.

[10] A. H. Reshak, M. M. Shahimin, N. Juhari, and R. Vairavan, "Photovoltaic characteristics of hybrid MEH-PPV-nanoparticles compound," *Current Applied Physics*, vol. 13, no. 9, pp. 1894–1898, Nov. 2013.

[11] T. Q. Nguyen, V. Doan, and B. J. Schwartz, "Conjugated polymer aggregates in solution: Control of interchain interactions," *The Journal of Chemical Physics*, vol. 110, no. 8, p. 4068, 1999.

Dye-Sensitized Solar Cell Using Aligned ZnO Nanorod Grown on SZO Films at Different Solution Molarities

I. Saurdi[1, 3,*], A.K. Shafura[1]
[1]NANO - ElecTronic Centre (NET),
Faculty of Electrical Engineering,
Universiti Teknologi MARA (UiTM),
[3]Faculty of Electrical Engineering,
UiTM Sarawak Kampus Kota Samarahan Jalan Meranek,
Sarawak, Malaysia
Email: [*]saurdy788@gmail.com,

M.H. Mamat[1], and M. Rusop[1, 2, **]
[1]NANO - ElecTronic Centre (NET),
Faculty of Electrical Engineering,
Universiti Teknologi MARA (UiTM),
[2]NANO - SciTech Centre (NST), Institute of science,
Universiti Teknologi MARA (UiTM), Shah Alam, Selangor,
Malaysia
Email: [**]nanouitm@gmail.com

Abstract — In this work, the aligned ZnO Nanorod were grown on ITO-coated glass substrates with Sn-doped ZnO films as a seed layer at different zinc acetate solution molarities of 0.05M, 0.04M, 0.03M and 0.02M by Sol gel immersion ultrasonic assisted. The aligned ZnO nanorods grown at different solution molarities had different of diameter and interspaces between nanorod. The ZnO nanorod grown at 0.03M Zinc acetate exhibits high surface area, whereby bigger interspaces between nanorod and had smaller of nanorod diameter. The nanorod grown at 0.05M had bigger of nanorod diameter and slightly lower of surface area as compared to nanorod grown at 0.03M. Meanwhile, the nanorod grown at 0.02M shows a scattered of nanorod growth with low surface area and lesser density. From the solar simulator measurement the solar energy conversion efficiency (η) of 0.989% under AM 1.5 was obtained for 0.03M aligned ZnO nanorod photoanode DSSC, which higher than other ZnO nanorod photoanode. The improvement which was due to higher surface area of smaller diameter nanorod that had bigger interspaces between nanorod for better dye absorption.

Keywords — Zinc acetate; Structural properties; Optical properties, Photovoltaic properties, Dye-Sensitized Solar Cell

I. INTRODUCTION

Nowadays, there are extensive study on ZnO and their application for electronic devices such as optoelectronic devices, gas sensor and dye-sensitized solar cell. It's due to the uniqueness of ZnO that can be tuned to various forms of nanostructures such as nanorod, nanowire, nanosheet and nanocone that have attracted much attention for various applications [1]. Zinc oxide is a wide-bandgap material, high electron mobility, non-toxic and stable. In future, ZnO can be considered as one of most promising material for device applications. There are many of researchers have reported dye-sensitised solars (DSSCs) cell using variety of ZnO nanostructures [2-4].The 1D-ZnO nanorods are attracted much attention from researcher for the application in dye-sensitized solar cell. This is due to the feasible of the electron transfer in 1-dimensional structure. There are many techniques have been used to prepare the ZnO nanorod such as chemical vapor deposition (CVD) [5], electrode deposition [6] and hydrothermal [7]. Among of these methods, hydrothermal based of simple solution method has been considered by many of researchers due to its low processing condition, homogeneity and toward cost effective of fabrication of thin films with large area.

Meanwhile, the use of metal doped ZnO thin film is one of feasible way to grow variety of ZnO nanostructures by simple solution based method. In this paper, we report the ZnO nanorods grown on Sn-doped ZnO (SZO) thin films by a low cost technique of simple sol-gel immersion method with ultrasonic assisted at different of zinc acetate solution. To our best of knowledge, the ZnO nanorods grown on Sn-doped ZnO thin films at different zinc acetate solution using a simple sol-gel immersion ultrasonic assisted method and applied in DSSCs were rarely reported by others. The characteristics of photovoltaic properties of ZnO nanorod at different zinc acetate solution have been investigated.

II. EXPERIMENTAL DETAILS

The ITO substrates were cleaned with acetone followed by methanol, and deionized water for 10 min respectively in Ultrasonic Cleaner (power Sonic 405) in order to remove all the contamination. The ITO substrates were finally blown with nitrogen gas.

The Sn-doped ZnO thin film were prepared using zinc acetate dihydrate ($Zn (CH_3COO)_2 \cdot 2H_2O$) as a precursor and the 2-methoxyethanol ($C_3H_8O_2$) as a solvent for the solution, while monoethanolamine (MEA) as a stabilizer. Then, Sn doping were achieved by adding 2 at.% of Tin (IV) chloride pentahydrate to the precursor. The solutions were stirred and heat for 3 hours before aged for 24 hours at room ambient. Subsequently, the solutions were spin-coated on glass and ITO substrates. The as-deposited films were heated at 150 °C for 10 minutes. The coating procedures were repeated for five times to increase the film thicknesses. The thin films were finally annealed for 1 hour at 500 °C in ambient air. The zinc acetate solutions were composed of the mixture of zinc acetate dihydrate ($Zn(CH_3COO)_2 \cdot 2H_2O$), hexamethylenetetramine (HMT) and deionized (DI) water. The zinc acetate solutions were prepared at different molarities of 0.05M, 0.04M, 0.03M and 0.02M. Ultrasonic assisted was introduced to the solution for 30 minutes before stirred and aged for 3 hours. The water bath instrument was set at 95 °C. Subsequently, Schott bottles contain of zinc acetate solution and seed layered-coated ITO glass substrates were immersed into the water bath. After one hour, the samples were taken out from the Scoot bottles and dried in air for 15 minutes and then annealed at 500 °C in air. The structural and optical were characterized by field emission scanning electron microscopy (FE-SEM, ZEISS Supra 40VP) and JASCO UV-Vis-NIR spectrophotometer (V670 EX), respectively.

To fabricated the dye-sensitized solar cell devices, the ZnO Nanorods at different solution molarities were immersed in 0.5mM ethanolic solution of (Ru[LL'(NCS)$_2$], L=2,2'-bypyridyl-4,4'-dicarboxylic acid, L'=2,20-bypyridyl-4,4'-ditetrabutylammonium carboxylate) dye (N719) at room temperature for 24h. The sputtered platinized (Pt) on ITO was used as the counter electrode. Meanwhile, the holt-melt spacer was used to bond the ZnO Nanorods electrode and dye with sputtered counter electrode. Sealing was accomplished by pressing the two electrodes at about 100°C a few seconds. The electrolytes were composed of 0.5 M LiI, 0.05 M I$_2$ and 0.5M 4-*tert*-butyl pyridine (TBP) in acetonitrile.

The photovoltaic characteristics of dye sensitized solar cell devices were carried measured using a Solar Simulator (150 W simulator, Bunkoh-Keiki CEP 2000) under Simulated solar light with a Xenon Lamp power supply (AM 1.5, 100mW/cm^2). The Incident photon conversion efficiency (IPCE) measurement was carried out by using a multimeter (Keithley 2400). The black mask with area 0.25cm^2 was used for DSSC measurement.

III. RESULTS AND DISCUSSIONS

A. Surface Morfhology

Fig.1 (a), (b), (c) and (d) show the images of the ZnO Nanorod deposited on SZO films by Sol gel immersion ultrasonic assisted. The FESEM images show that the nanorod growths at different solution molarities have different of diameter and interspaces between nanorod that reflect the surface area of nanorods. The nanorod diameter decrease from 120nm to 70nm as the molarities decreased from 0.05M to 0.02M. At 0.02M the diameter of ZnO nanorod is about 70nm, while the 0.03M is about 80nm. However, at 0.02M the surface area is slightly lower than 0.03M, this is due to the scattered of nanorod growth as can be observed from the FESEM images Fig.1. Meanwhile, at 0.05M and 0.04M the nanorod diameter is about 120nm and 90nm, respectively. The surface area of nanorods grown with these two molarities is also lower then 0.03M of zinc acetate solution whereby the bigger diameters with densely packed structure were observed from FESEM images in Fig.1 as compared to 0.03M ZnO nanorod. The length of nanorod is almost similar for nanorod growth for 1 hour at 0.05 to 0.03M except 0.02M of zinc acetate solution. There nanorod length is about 1.8um, 1.75um, 1.7um and 1.2um for the nanororod growth at 0.05, 0.04, 0.03 and 0.02M, respectively. It can be seen that different of nanorod diameter and length for nanorod growth at different zinc acetate solution molarities. It suggested that the molarities of different zinc aceteate solution might influence the morphology and density of ZnO nanorod grown on SZO films. Furthermore, the aligned ZnO Nanorods grown on ITO substrates are crystallized along the ZnO [0001] direction as shown in Fig. 2. The nanorad is also known as hexagonal prisms that also reported by others [8, 9].

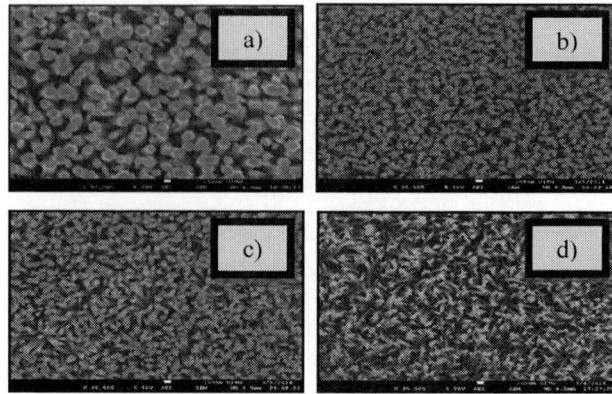

Fig.1 FESEM images of ZnO nanorod at (a) 0.05M (b) 0.02M (c) 0.03M (d) 0.04M by magnification of 30k and 5.0kV

B. Optical Propertie

Fig. 2 shows the transmittance spectra of aligned ZnO nanorod (no N719 dye). It can be seen that the nanorod transparency is about 40-50% between visible wavelengths and the regular wave shape of the transmittance suggests the thickness of ZnO nanorod arrays is uniform. The transparency of the ZnO Nanorod at visible region for 0.05M is about 51%. Meanwhile, the transparency for ZnO nanorod at 0.04M and 0.03M is about 45% and 43%, respectively. The decreasing of transparency phenomenon is might be due to surface roughness and verticality of ZnO nanorod. Therefore, the decreasing of transmittance occurs is due to light scattering effect from the high surface roughness of ZnO Nanorod which is more interspaces between nanorod that might increase the amount of absorption of dye. However, the ZnO nanorod at 0.02M has slightly higher transmittance as compared 0.03M and 0.04M. This might be due to the scattered growth of ZnO nanorod that contributed to less efficient for dye absorption. Furthermore, the amount of dye absorbed by films can affect the photogenaration in DSSCs [10-11].

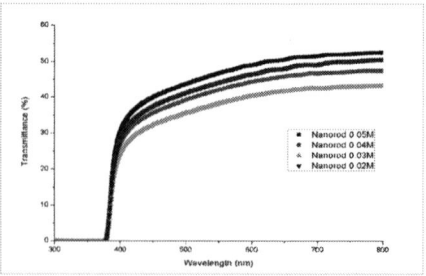

Fig.2 Transmittance spectra of ZnO nanorod

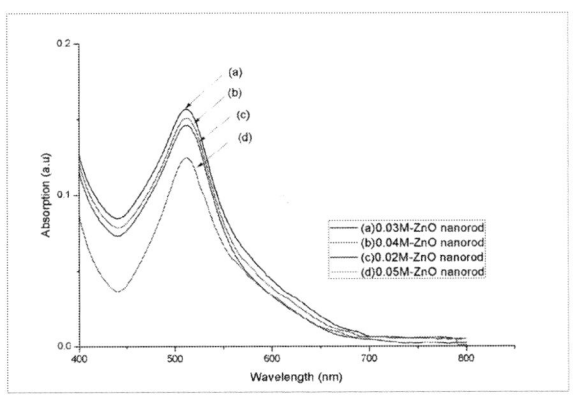

Fig. 3 The UV-Vis absorption spectra of desorption dye from N719-sensitized 0.05M, 0.04M, 0.03M and 0.02M-Nanorod, respectively.

In dye-sensitized solar cell the amount of dye that absorbed by film might affect the overall performance of DSSCS. Lihong Qi. *et.al* [12] reported that the dye adsorption is correlated with the photoelectrochemical properties of DSSCs. Therefore, understanding the absorption properties of dye in DSSCs photoanode is useful. In order to determine the dye-absorption in ZnO nanorod, the loading/unloading measurement has been carried out. Fig. 5 shows the UV-vis absorption spectra of desorption dye from N719-sensitized 0.05, 0.04, 0.03, and 0.02M, respectively. The calculation of molar concentration of dyes unloading is based on Lambert-Beer Law [12]. The molar concentration of dye unloading from 0.05, 0.04, 0.03 and 0.02M-Nanorod is 0.87, 1.04, 1.10 and 1.01 x 10^{-7} mol/cm^2, respectively. The 0.03M-Nanorod has a higher of molar concentration of dye unloading as compared to others. This may be due to large of surface area for 0.03M-Nanorod that absorbed more dyes.

C. Photovoltaic characteritics of ZnO Nanorod

Fig. 3 shows the I-V characteristics for ZnO Nanorod DSSCs at different solution molarities on SZO film. The ZnO nanorod photoanode grown at 0.03M showed higher efficiency as compared to ZnO nanorod photoanode grown at 0.05M, 0.04 and 0.02M. It can be concluded that the ZnO nanorod grown at 0.03M has larger of surface area as compared to others. Furthermore, higher photocurrent is attributed from efficient of photogeneration due to better of dye absorption that react with light illumination for generate current. The larger surface area of ZnO nanorod might enhance the absorption of photon energy due to higher dye absorbed [13-15]. Moreover, decreasing of transmittance ZnO nanorod 0.03M as indicated in Fig.2 is due to high scattering effect from interspaces between nanorod that might enhance the absorption of dye. Meanwhile, the ZnO nanorod grown at 0.02M showed a slightly lower photocurrent as compared to ZnO Nanorod grown at 0.03M and 0.04M attributed of low surface area and also much bigger interspaces between nanorod due to

scattered nanorod growth. The ZnO nanorod 0.05M has a lowest photocurrent and also much higher of transmittance, indicated that low of scattering effect occurs and inefficient of photogeneration from absorbed dye.

TABLE I
PHOTOVOLTAIC PERFORMANCE OF ZnO NANOROD DSSCs

Photoanode	J_{sc}(mA/cm^2)	V_{oc}	Fill factor	Efficiency (percent)
0.05M-Nanorod	2.840	0.552	0.381	0.599
0.04M-Nanorod	3.732	0.554	0.389	0.836
0.03M-Nanorod	4.319	0.555	0.395	0.989
0.02M-Nanorod	3.395	0.553	0.385	0.747

Table I shows the photovoltaic characteristics of fabricated DSSCs using ZnO nanorod at different solution molarities as a photoanode. From these results, the J_{sc}, V_{oc} and energy conversion efficiency η of DSSC are higher for the ZnO nanorod photoanode grown at 0.03M zinc acetate solution. This might be due to a larger surface area of ZnO nanorod thus enriched the absorption of light from absorbed dye and then contributed to higher photocurrent density as well as higher of conversion efficiency. Moreover, the multi-scattering effect in nanorod may enhance the incident light in them. Meanwhile, the J_{sc} and V_{oc} for ZnO nanorod photoanode of 0.05M, 0.04 and 0.02 M DSSC is smaller as compared to ZnO nanorod photoanode 0.03M. Therefore, less surface area for efficient dye absorption is one of factor that contributes to the lower of J_{sc} and V_{oc} as well as the performance of DSSCs. Other than that, the ZnO nanorod photoanode 0.02M has a lower surface area due to scattered ZnO nanorod growth as indicated in Fig.1. Therefore, the improvement of ZnO nanorod quality due to high surface area which is much smaller ZnO Nanorod diameter with efficient dye absorption due to better interspaces between nanorod had contributed to higher photocurrent and efficiency ZnO nanrod photoanode that had been proven in this work. Furthermore, the ZnO nanorod was grown using SZO seed layer with better properties as compared to undoped ZnO.

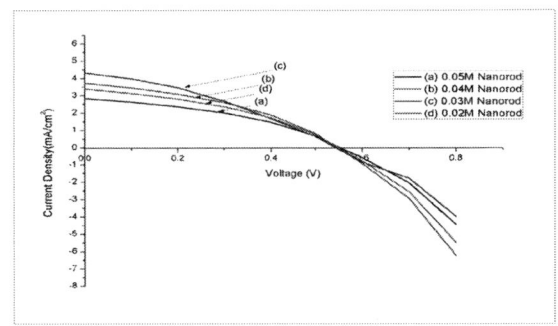

Fig.4 I-V Characteristics of ZnO Nanorod DSSCs

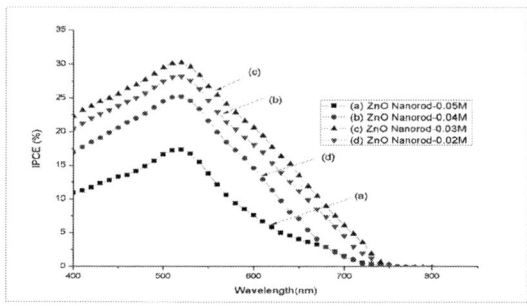

Fig.5 IPCE of ZnO Nanorod DSSC fabricated with different photoelectrodes

Fig.5 depicted the incident photon conversion efficiency (IPCE) of aligned ZnO Nanorod photoelctrodes. From Fig. 5, the value of IPCE of about 520nm is a maximal IPCE that contributed to dye absorption, resulting from the visible t_2-π^* metal-to-ligand charge transfer (MLCT) [16]. It could be clearly seen that the IPCE based 0.04M-ZnO Nanorod, 0.03M-ZnO Nanorod and 0.02M-ZnO Nanorod photoanodes had been improved as compared with DSSC 0.05M-ZnO Nanorod. Moreover, the higher IPCE of 0.03M- Nanorod grown with SZO film at different solution molarities is attributed to the better dye absorption due to high surface area with more pores of film which might increase the incident light intensity in the N719 dye.

IV. CONCLUSIONS

The aligned ZnO nanorods were successfully grown on SZO films. The results revealed that the nanorod grown at 0.03M Zinc acetate solutions exhibits high surface area, whereby bigger interspaces between nanorod and had a smaller of nanorod diameter. The nanorod grown at 0.05M had bigger of nanorod diameter and slightly lower of surface area. Meanwhile, the nanorod grown at 0.02M shows a scattered of nanorod growth with low surface area and lesser density. From the solar simulator measurement the solar energy conversion efficiency (η) of 0.989% under AM 1.5 was obtained with the 0.03M aligned ZnO nanorod photoanode DSSC which correspond to the short-circuit photocurrent density (J_{sc}) and open-circuit voltage (V_{oc}) of 4.32mA/cm2 and 0.555 V, respectively. While 0.747% conversion efficiency (η) obtained from 0.02M ZnO nanorod photoanode. The 0.03M ZnO nanorod photoanode was efficiently in the fabrication process of dye-sensitized solar cells (DSSCs), where the improvement which was due to higher surface area of smaller diameter nanorod that had bigger interspaces between nanorod for better dye absorption. Moreover, the 0.03M ZnO nanorod had better of IPCE and higher of dye desorption that proven the higher of photocurrent density generated as compared to others.

ACKNOWLEDGEMENT

The authors express their gratitude to the Ministry of Higher Education Malaysia for the scholarship, NANO–ElecTronic Centre at Faculty of Electrical Engineering, NANO-SciTech Centre at Institute of Science, and Research Management Institute Universiti Teknologi MARA for the facilities.

REFERENCES

[1] X.D. Wang, J.H. Song and Z. L Wang., "Nanowire and nanobelt arrays of zinc oxide from synthesis to properties and to novel devices," *Journal of Material Chemistry*, Vol. 17, pp. 711-720, 2007.

[2] S.H. Jeong, J.W. Lee, S.B. Lee and J.H. Boo, "Deposition of aluminum-doped zinc oxide films by RF Magnetron sputtering and study of their structural, electrical and optical properties," Thin solid Films, Vol. 435, pp. 78-82, 2003.

[3] D. Raoufi and T. Raoufi, " The effect of heat treatment on the physical properties of sol-gel derived ZnO thin films,'' Applied Surface Science, Vol. 255, pp. 5812-5817, 2009.

[4] D. wei, " Dye-sensitized solar cell review," Int. J. Mol. Sci. 2, Vol.11, pp1103-1113, 2010.

[5] M. H. Lai and W. Lee, "Photovoltaic Performance of New-Structure ZnO-nanorod Dye-Sensitized Solar Cells," *International Journal Electrochemical Science*, Vol. 6, pp. 2122-2130, 2011.

[6] Hee Kwan Lee, Myung Sub Kim and Jae Su Yu, "Effect of AZO seed layer on electrochemical growth and optical properties of ZnO nanorod arrays on ITO glass," *Nanotechnology*, Vol. 22, 445602 (8pp), 2011.

[7] Jinjun Qiu, Weidong Yu, Xiandong Gao, Xiamin Li, Weizhen He, Se-Jeong Park, Hyung-Kook Kim and Youn-Hwee Hwang, "Controlled growth of ZnO nanorrod templates and TiO2 nanotube arrays by using porous TiO2 film as mask," *Journal Sol-Gel Science Technology*, Vol. 47, pp. 187-193, 2008.

[8] Zi Qin, Qingliang Lioa, Yunhua Huang, Lidan Tang, Xiaohui Zhang and Yue Zhang, "Effect of hydrothermal reaction temperature on growth, photoluminescence and photoelectrochemical properties of ZnO nanorod arrays," *Material Chemistry and Physics*, Vol. 123, pp. 811-815, 2010.

[9] Ying Tao, Ming Fu, Ailun Zho, Dawei He and Yongsheng Wang, "The effect of seed layer on morphology of ZnO nanorod arrays grown by hyrothermal Method," *Journal of Alloys and Compounds*, Vol. 489, pp. 99-102, 2010.

[10] Kuan Jen Chen, Fei Yi Huang, Yen Ting Chen, Shaoou Jin Chen and JZhan Shou Hu, "Surface characteristic, optical and electrical on sol-gel synthesis Sn-doped ZnO thin film," Material Transaction, Vol. 51, pp. 1340-1345, 2010.

[11] Sheng Xu, Cong-hua Zhaou, Ying Yang, Hao Hu, Bobby Sebo, Bo-lei Chen, Qi-dong Tai and Xiangzhong Zhao, " Efffects of Ethanol on optimizing porous films of dye-sensitized solar cells," *Energy Fuels*, Vol. 25, pp. 1168-1172, 2011.

[12] Lihong Qi, Hailong Yu, Zhenyu Lei, Qingshan Wang, Qiuyun Quyang, Chunyan Li, Yujin Chen, " Dye-sensitized solar cells bsed on ZnO nanowire array/TiO2 nanoparticle composite photoelectrodes with contollable nanowire aspect ratio," *Applied physics A*, Vol. 111, pp. 279-284, 2013.

[13] Yanfeng Gao, Masayuki Nagai, Tien-Chih Chang and Jing-Jong Shyue, "Solution-derived ZnO nanowires array film as photoelectrode in dye-sensitized solar cells," *Crystal Growth & Design*, Vol. 7, pp. 2467-2471, 2007.

[14] Weiguang Yang, Farong Wan, Siwei Chen, Chunhua Jiang, "Hyrothermal Growth and application of ZnO nanowire films with ZnO and TiO2 buffer layers in dye sensitized solar cells." *Nanoscale Research Latter*, Vol. 4, pp. 1486-1492, 2009.

[15] K. H. Kim, K. Utashiro, Z. Jin, Y. Abe, and M. Kawamura, "Dye-Sensitized Solar Cells with Sol-Gel Solution Processed Ga-Doped ZnO Passivation Layer," *International Journal of Electrochemical Science*, Vol. 8, PP. 5183-5190, 2013.

[16] H.M. Cheng, W.H. Chiu, C.H. Lee, S.Y. Tsai, W.F. Hsieh "Formation of branched ZnO nanowires from solvothermal method and dye-sensitized solar cells application, journal of Physical Chemistry C 112 (2008) 16359-16364.

IEEE-ICSE2014 Proc. 2014, Kuala Lumpur, Malaysia

The Effect of Softbaking Temperature on SU-8 Photoresist Performance

Shazlina Johari[1], Nithiyah Tamilchelvan[1], Mohammad Nuzaihan Md Nor[2], Muhammad Mahyiddin Ramli[1],
Bibi Nadia Taib[1], Mazlee Mazalan[1] and Yufridin Wahab[1]

[1]Advanced Multi-disciplinary MEMS-based Integrated Electronics NCER Centre of Excellence (AMBIENCE)
School of Microelectronic Engineering, Universiti Malaysia Perlis (UniMAP)
[2]Institute of Nano Electronic Engineering (INEE), Universiti Malaysia Perlis (UniMAP)
Perlis, Malaysia
shazlinajohari@unimap.edu.my

Abstract— One of the steps required during the fabrication of SU-8 mold for soft lithography is softbaking, which is conducted after the deposition of the photoresist. The purpose of softbaking is to stabilize the resist film and eliminate any remaining solvent through evaporation. This ensures that the resist surface is non-sticking, hence avoiding debris when transferring the patterns later. In this paper, we investigate the effects of softbaking temperature on the polymerization of SU-8 photoresist. The significance of this work is to optimize the fabrication process involved in producing SU-8 mold structures with thickness of 30 μm. This project involves a series of experiments covering softbaking temperatures ranging from 45° to 115° C. Experiments results show that softbaking temperature of 85°C results in completely stick and crack free structures. By this, a huge improvement obtained if compared to the result of processing at the standard soft bake temperature of 95°C. The soft bake temperature should not be taken lightly while optimizing SU-8 processing because it has a big influence on the material properties and the lithographic performance of the resist.

Keywords— SU-8 photoresist; softbaking temperature;

Introduction

Microfluidic system and devices have diverse and widespread applications particularly in the medical industry. The fabrication process of microfluidic devices involve a multiple-step sequence of photolithographic and chemical processing during which microfluidic structures are gradually created on a certain substrate, followed by PDMS replications using soft-lithography. The microstructure, also known as a master mold, is typically produced using SU-8 negative photoresist. For reliable reproduction of uniform features, this mold should be mechanically strong and thermally stable. Formulated at IBM in the early 1980s, SU-8 is a negative epoxy based photoresist which is near UV sensitive. When SU-8 photoresist is exposed to proton or light, photochemical reaction occurs, where an acid which acts as a catalyst in the exposed regions during post baking exposure is produced.

One of the most important parameter that requires attention during SU-8 structure development is the soft baking

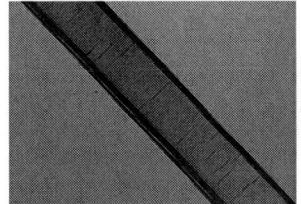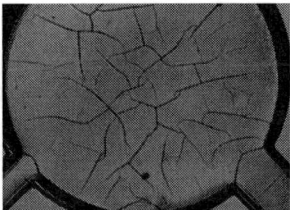

Fig. 1. Example of cracks formed on a microfluidic device inlet from previous work [1]

temperature. Sofbaking can be defined as the process of baking the substrate after the deposition to evaporate the volatile substances or unwanted gases that are present on the wafer, remove solvents from the SU-8 layer and to dry the layer for consequent handling. Insufficient softbake temperature can cause SU-8 sticking on the mask where the substrate will stick on the mask during exposure. This can indirectly decrease the quality of the patterned microstructures produced. Furthermore, the adhesion strength between the SU-8 resist and the substrates is important for the development of microfluidic devices with high quality and functionality. Over baking the substrate may result in poor adhesion strength between the resist and the substrate [2]. This may bring potential hazard and even failure to the subsequent processes. Another major problem in SU-8 processing is the formation of cracks on the structure (Fig. 1). The cracks form due to tensile stress in the SU-8, which builds up during processing, and begin to appear during the development. Cracked structures may lead to leakage in microfluidic systems, and unwanted scattering and increased propagation loss in optical systems [3]. In this paper, preliminary results of our investigation on the influence of soft baking temperature on SU-8 performance are presented. At present, the study is focused on SU-8 2010 thickness of 30 μm deposited on silicon wafers, which forms part of a broad optimization of the fabrication process involved in producing SU-8 microfluidic mold. The design structure used in this work is depicted in Fig. 2, where the device consists of a 'Ý shape' channel with three inlets and one circular channel in the middle.

978-1-4799-5761-3/14 $31.00 © 2014 IEEE   467

Fig 2. The structure of the SU-8 master mold

Materials And Method

Initially, the substrates were cleaned using buffered oxide etch (BOE) in order to provide good adhesion between the substrate material and SU-8 photoresist. Prior to spinning, the wafer was centered onto the chuck and an adequate amount of SU-8 was poured onto the middle of the wafer. This was allowed to settle slightly before spinning. While the adequate amount can be estimated from experience, a simple minimum volume guideline is the product of the desired resist thickness and the surface area of the wafer. The SU-8 2010 resist was then spun for 10 s at 500 rpm (acceleration 100 rpm/s), followed by 30 s at 1000 rpm (acceleration 300 rpm/s). After spinning, the wafer substrate was transferred to a hot plate for the softbaking procedure. A level hotplate with good uniformity and thermal control is required for this procedure. An uneven/tilted hotplate could result in redistributing of SU-8 during softbaking and formation of distorted resist film on the entire substrate. Softbaking using convection oven is also not recommended as it has few disadvantages. According to [4], softbaking using convection oven could form a 'skin' on the resist, which prevent the solvent to evaporate hence resulting in incomplete film drying. In addition, when using an oven, it is hard to guarantee a uniform softbaking temperature due to the big dimension of the oven. In order to investigate the influence of the soft bake temperature on the lithographic performance of SU-8, six coated substrates are soft baked at temperatures range from 45°C to 115° with a step of 10°C for 4 minutes. Once the substrate was allowed to cool down to room temperature, the microfluidic structure/pattern was exposed (without filter) using the mask aligner for 20 seconds. The exposure energy was set to 160 mJ/cm² (365nm wavelength). When the exposure was complete, post exposure bake was performed at 65 °C for 1 min followed by 5 min at 95 °C. Developing the exposed resist consisted of a timed bath in (1-methoxy-2-propyl)-acetate (PGMEA, Sigma-Aldrich) solvent, which removes non-crosslinked photoresist. It will also dissolve the crosslinked areas, albeit very slowly, if left for too long in the developing solution. The substrate was immersed in PGMEA for the development for approximately 5 minutes.

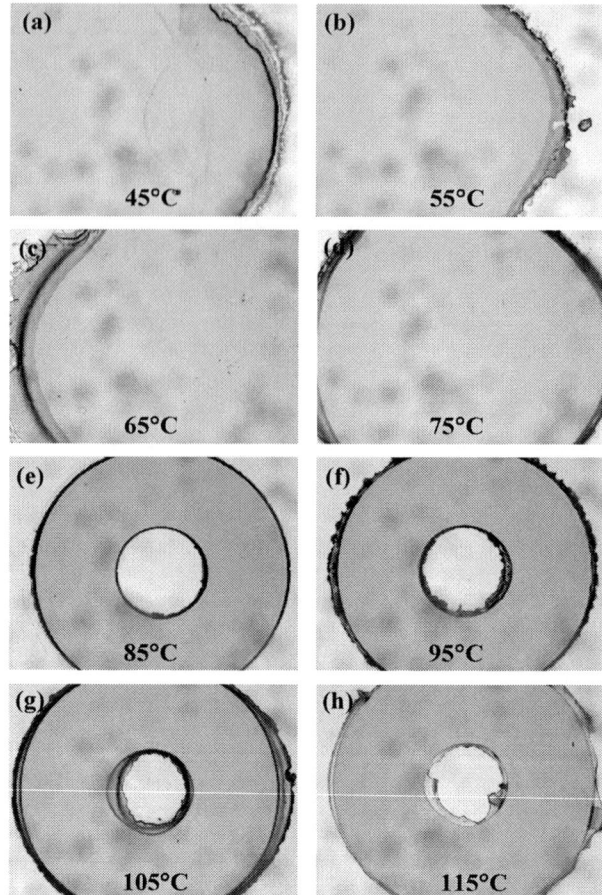

Fig. 3. Optical images of the structure circular chanel softbaked at (a) 45°C, (b) 55°C, (c) 65°C, (d) 75°C, (e) 85°C, (f) 95°C, (g) 105°C and (h) 115°C. For temperature of 45°C to 65°C, full image of the inlet structure could not be captured using microscope magnification of 5x

Results and Discussion

Initially, visual inspection is performed using optical microscope with 5x magnification on the device after going through the development process. Fig. 3 shows the inlet of the device softbaked at six different temperatures. For the softbaking temperature below 85°C, the hole of the middle circular channel is not visible. The edge of the circle is also not properly formed after being developed. We predict that this is due to the low softbaking temperature making the photoresist not fully polymerize. The hole of the middle circular channel is clearly visible when the softbaking temperature was set to 85°C up to 115°C. The edge of the inner circle however is not properly formed for the softbaking temperature of 95 to 115°C. It can be seen that the optimum temperature of 85°C produced the best inlet structure with the middle hole of the circular channel being properly developed with minor resist residue at the edge.

We then investigated the height of the device structure by using a surface profilometer. The deposited SU-8 photoresist is expected to be 30 μm thick based on the step described in the previous section. According to the results shown in Fig. 4, we found that the height of the SU-8 structure varies depending on the softbaking temperature. When baked at lower temperature of 45°C and 55°C, the measured structural height is 27.5 μm and 28.7 μm respectively.

When the temperature is increased to 65°C, the height of the structure also increased by 2 μm. Similar pattern is observed when the softbaking temperature is increased to 75°C. The structural height maintained at 33 μm at the temperature of 85°C. However, it is observed that the height reduce to 27.9 μm, 26.9 μm and 26.6 μm once the temperature is increased from 95°C to 115°C. In average, the structural height varies at least ±8% from its expected height when softbaked from 45°C to 115°C.

We also verify the etch profile of the structure using 3D surface profilometer. Fig. 5 shows the thickness of the structure obtained using 3D Nano Profiler. The red area indicates the maximum thickness of structure while the green area indicates the moderate thickness. Orange area points out the thickness lower than red but higher than green. The blue area shows the minimum thickness obtained using 3D Nano Profiler. As depicted in the figure, for softbaking temperature of 45°C, 55°C and 65°C, it can be seen from the profile that the surface of the measured area is not smooth. This is due to the sticking of the resist after being baked at low softbaking temperature. This verified that insufficient softbake temperature lead to SU-8 sticking, which could make the substrate and mask stick together during exposure. In contrast, at softbaking temperature of 85°C, 95°C and 115°C the surface of the substrate is smooth with no visible sticking as shown in the profile in Fig. 5.

In order to investigate the effect of the softbaking temperature on the SU-8 microstructure pattern, we measured the trench and ridge of the device inlet for each softbaking temperature, as depicted in Fig. 6. The trench is the diameter of the hole of the circular channel while the ridge is the dimension between the outer circle edge and the edge of the inner hole. The actual dimension of the trench and the ridge as designed in the mask is 800 μm and 900 μm respectively. From the plot shown in Fig. 6, it is evident that the softbaking temperature affects the size of the device dimension. At softbaking temperature of 45°C, 55°C, 65°C and 75°C the trench could not be measured because the structure is not properly form due to the low softbaking temperature.

Additionally, lower soft bake temperatures provide higher sensitivity which leads to overexposure causing the trench to degrade. Both trench and ridge are only visible and can be measured when the softbake temperature is increased to 85°C until 115°C. The trench dimension decreases from 900.70 μm, to 890.90 μm, to 833.17 μm and finally to 784.16 μm as the temperature is increased gradually from 85°C to 115°C. However, the ridge dimension fluctuates as the softbaking temperature is varied. At 85°C, the dimension of the ridge is

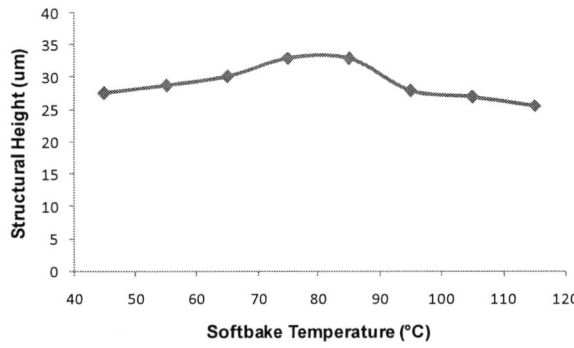

Fig. 4. Structural height versus softbaking temperature

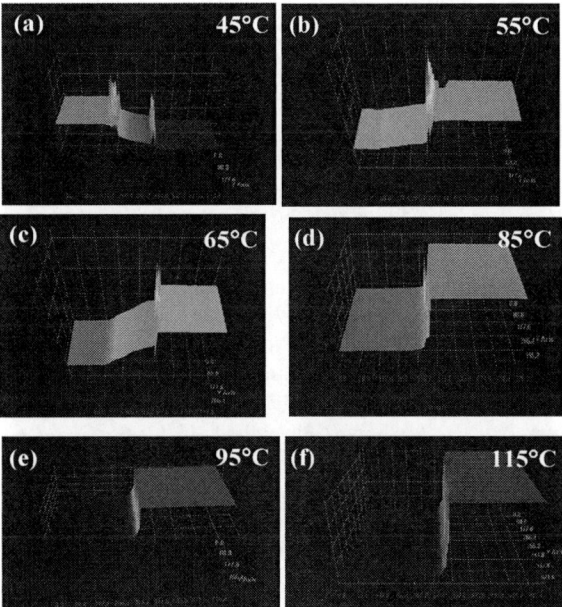

Fig. 5. 3D images of structural height softbaked at (a) 45°C, (b) 55°C, (c) 65°C, (d) 85°C, (e)   95°C, and (f) 115°C obtained using 3D Nano Profiler

close to the expected dimension which is 800.5 μm. The dimension increased to 816.84 μm when the temperature is increased to 95°C. At 105°C, the dimension increased rapidly to 972.58 μm and the dimension reduced to 903.97μm when the temperature is increased to 115°C.

The results obtained from the trench and ridge measurement indicate that the softbaking temperature should not be taken lightly during the fabrication of SU-8 master mold. The inconsistency of the pattern dimension due to the variation of softbaking temperature could affect the final SU-8 master mold in particular when replicated using PDMS. This could lead to inefficiency of the final microfluidic performance when applied in the medical and biological area later.

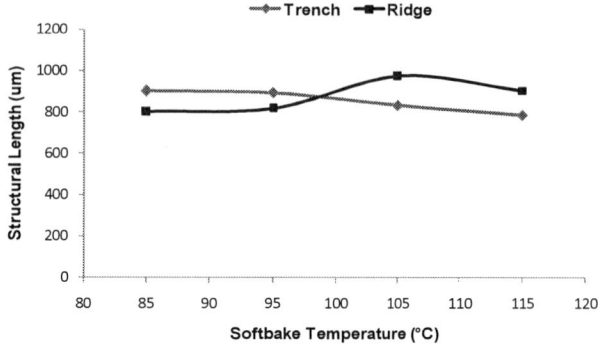

Fig. 6. Trench and ridge dimension versus softbaking temperature

For each substrate, the microstructures were patterned during exposure and the whole structures were inspected optically to observe any sticking, cracking, good structure formation and resist condition. It is well known that softbaking temperature is always induced in the SU-8 layer during the resist baking processes, and it affects the overall pattern quality of the structures. A lower soft baking temperature leaved residual solvent content that let the solvent to evaporate away from the film at additional controlled rate.

Moreover, low softbaking temperatures caused the SU-8 photoresist not to dry completely and some portions on the substrate remain wet which ultimately results in sticking during the exposure process. Sticking pattern can be seen after the development process under visual inspection clearly as shown in Fig. 7 when softbaked at 45°C, 55°C, 65°C and 75°C. Sticking structure can indirectly decrease the quality of the patterned microstructures produced.

At softbaking temperature of 85°C, it can be seen that the microstructure is free from sticking and cracking hence resulting in a good structure formation. Over baking or higher soft bake temperature effect thermal cross linking, formation of crack structure, decreasing in material or substance hardness due to the SU-8 starts losing its composition and reduce the adhesion strength among the SU-8 resist and substrate. Furthermore, adhesion strength occurs between the SU-8 photoresist and substrates are important for the development of devices with high quality and functionality.The higher the baking time, the higher the cross linking rate occur which decrease the sensitivity of the material to the environmental humidity. The microstructures baked at higher softbaking temperature can be seen in Fig. 7 (e) – (h).

Conclusion

Although the recommended softbaking temperature has been provided by the photoresist supplier, it is beneficial to understand the effect of different softbaking temperature on the SU-8 photoresist structure. In summary, we found that the variation in softbaking temperature affects the fabrication

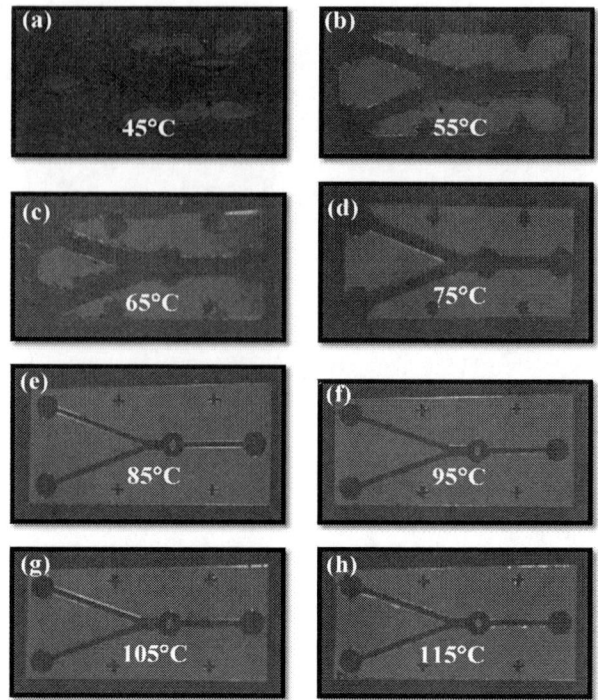

Fig. 7. (a) – (h) Optical images of the SU-8 master mold softbaked at 45°C to 115°C.

process of the SU-8 master mold in terms of the structural height and the dimension of the patterned structure. Results obtained from the experiments verified that softbaking temperature of 85°C results in completely stick free structures and accurate patterned dimension. Future plans include optimizing the softbaking temperature for photoresist with different thickness that suit different microfluidic device application. The findings obtained from this research can give a lot of proficiency, information and method to be used for accurate optimization criteria of lithography process in future work.

Acknowledgment

The authors would like to thank INEE, AMBIENCE, SoME and MoE (FRGS 9003:00405) for their contributions.

References

[1] S. Johari, V. Nock, MM Alkaisi and W. Wang, "On-chip analysis of *C. elegans* muscular forces and locomotion patterns in microstructured environments," Lab Chip, 2013,13, pp. 1699-1707

[2] R. Feng and R.J. Farris, "Influence of processing conditions on the thermal and mechanical properties of SU8 negative photoresist coatings," J. Micromech. Microeng., 2003,13, pp.80-88.

[3] T.A. Anhoj, A.M. Jorgensen, D.A. Zauner and J. Hubner, "The effect of soft bake temperature on the polymerization of SU8 photoresist," J. Micromech. Microeng., 2006,16, pp.1819-1824.

[4] J. Cao, "Experimental study of migration of shape-specific particles and its applications in fluidic self-assembly," ProQuest, 2008.

2.4GHz WLAN RF Energy Harvester for Passive Indoor Sensor Nodes

Fatima Alneyadi, Maitha Alkaabi, Salama Alketbi, Shamsa Hajraf, and Rashad Ramzan
Department of Electrical Engineering,
UAE University, Alain-15551, United Arab Emirates
rashad.ramzan@uaeu.ac.ae

Abstract—this paper presents the design and measurement results of an RF Energy Harvester aimed to power sensor nodes like temperature, humidity, chemical, or radiation in an indoor industrial or residential environment. The harvester operates at 2.42 GHz WiFi-WLAN frequency band. It consists of multiple microstrip patch antennas, power combiner, voltage quadruple Greinacher rectifier circuit, and a super capacitor to store the harvested energy. All elements are designed using low-loss Rogers RO3206 substrate. The impedance matching of the power combiner with a rectifier is a non - trivial issue due to change in diode impedance with the input power. The peak efficiency is measured to be 57.8% at 6 to 8dBm input power. In the presence of realistic −10dBm continuous signal, the system can charge a 33mF super capacitor to 1.6V in 20 minutes. This collected energy is enough to power 10mW sensor node for a period of more than 4 seconds to perform wake up, sense and transmit functions, and put a sensor back to sleep mode.

Keywords — Energy harvesting, WLAN energy harvester, self powered sensors, wireless power, indoor energy harvesting, multiple patch antennas, greinacher rectifier, greinacher quadrupler

I. INTRODUCTION

Modern industrial and residential buildings are installed with many different types of sensors for human safety and comfort. These sensors include temperature, humidity, light, smoke, chemical and radiation sensors. These sensors need two separate types of wiring, one for power delivery and another for signal transmission. Some of these sensors also use the batteries for uninterruptable power supply. These batteries are not being replaced with super capacitors. Super capacitors are similar storage capacity and more environmentally friendly compared to the batteries.

Electromagnetic (EM) power is omnipresent source of energy around us; countless communication devices transmit RF energy from a few MHz to several tens of GHz frequency range. The researchers have extensively measured the available power in the RF spectrum. It is concluded that the ambient RF energy density is so small that it is not possible to power a sensor which continuously dissipates power more than a quarter microwatt [1]. Most of these sensors mentioned above do not work in continuous mode. They read the desired variable after some predetermined interval or work on sense-on-demand principle.

Therefore, energy harvesters which can scavenge and store the energy and then provide this stored energy to power up the

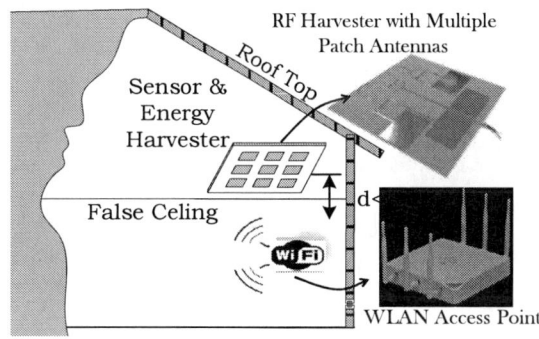

Fig. 1. Concept diagram of RF Energy Harvester

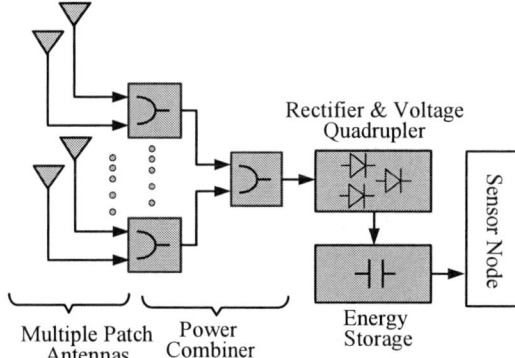

Fig.2 Block diagram of the implemented (gray blocks) energy harvester

sensor node for a fixed duration, are suitable for above mentioned applications.

Our lab measurements show that for indoor environment the most powerful signal available in close physical proximity of the sensor location is 2.42GHz WiFi signal compared to 100-108MHz FM and 900MHz GSM signal.

The modern MIMO type WiFi transceiver transmits the power of the 18-23dBm range from a single antenna [2], if the energy harvester is placed close to a transmitter at the distance less than 50cm as shown in Fig.1, we can receive the signal as powerful as 0dBm. The MIMO system transmits the different phase at different antennas in order to form a beam in a particular direction. The power added without compensating this phase delay using multiple input antennas is not always a linear sum. But for a harvest and store energy system, like ours, the instantiation received power is not as important as the long term average power. This average power over a time is proportional to received energy [3].

In recent literature, a large number of RF energy harvesting schemes are reported. In [4] two different strategies are discusses. In the first case, RF power is added and then rectified, while in the second case first the RF signals are rectified and then added to get the higher DC power. The reported rectifier has a peak efficiency of 68% for the input signal of -5dBm. In [5], an RF harvester design is presented which achieves an efficiency of 68% at 20dBm. Similar designs are also presented in reference [6] and [7]. None of these designs are suitable for indoor RF energy harvesting; the RF signal strength requirements are too high to be met by weak WLAN signals.

In this paper, we present an RF energy harvester to power up a passive sensor node as shown in Fig.1 and Fig.2. The energy harvester is placed close to the WiFi access point in order to receive the large amount of transmitted power. The energy harvester consists of multiple 2.42GHz microstrip patch antennas. The power from each antenna is added using a power combiner and then rectified using Greinacher voltage quadrupler rectifier to a useable DC signal as shown in Fig.2. This DC energy is stored in a super capacitor in order to energize the on-demand self-powered sensor nodes.

The paper is arranged as follows. In Section II, the design and measurement results of microstrip path antenna, power combiner and Greinacher rectifier circuit are discussed. An important issue is the change in input impedance of rectifier with input power. In Section III, the measurements results of a complete RF harvester are presented and last section concludes the paper.

II. ANTENNA, POWER COMBINER AND RECTIFIER

The RF Energy Harvesting Device design consists of three sections; antenna for RF energy collection, power combiner for addition of energy, and rectifier circuit to convert the alternating RF signal to DC power. For a final circuit these elements have to be designed on a single PCB. For ease of testing and debugging we have designed them separately and combined them using the RF cables. The substrate used for all the PCBs is Rogers's low loss material RO3206 with dielectric constant (ε_r) of 6.5 and thickness of 1.6mm.

A. Microstrip Patch Antenna

Among several choices, the rectangular patch antennas are most suitable for a compact, space-saving design and can be easily manufactured on standard PCB. In order to match the impedance of the rectangular patch antenna with a 50-Ω port, an inset line feed was designed based on standard text book design procedure [8][9]. Fig.3 (a) shows the manufactured 48.8×45.1 mm^2 microstrip patch antenna on a standard 1.6mm thick PCB substrate. The measured and simulated S11 plots are shown in Fig.4. The slight shifts in frequency can be

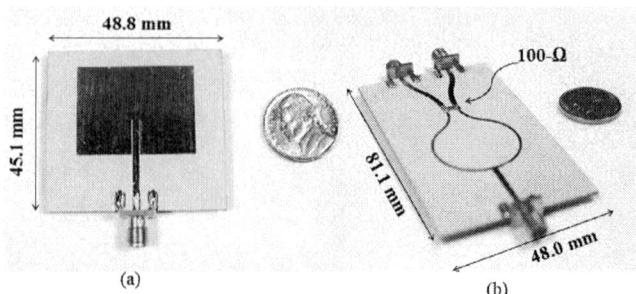

Fig. 3. (a) The photographs of manufactured microstrip antenna and (b) Wilkinson's power combiner on Rogers RO3206 substrate.

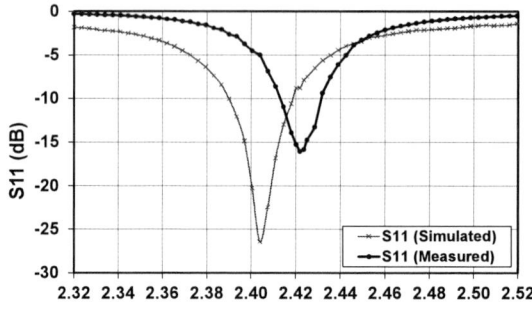

Fig.4. Microstrip patch antenna input reflection coefficient (S11)

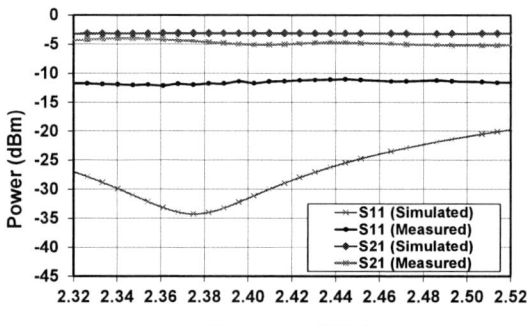

Fig.5. Wilkinson's Power Combiner input reflection coefficient (S11) an coupling coefficient (S21)

attributed to the PCB manufacturing tolerances and slightly misaligned manually soldered connector. Measurement results show the S11 of -16dB at 2.4GHz with 3dB bandwidth of more than 20MHz. These are acceptable values for a WLAN energy harvester front-end working at 2.4GHz.

B. Wilkinson Power Combiner

If a power harvested by a single patch antenna is -10dBm for example, then combing two of them, theoretically, will provide a power of -7dBm and four would provide -4dBm. Out of the many power combiners' types, Wilkinson power combiner was chosen due to its good port isolation, design flexibility, wide-bandwidth, and easy PCB based implementation.

Fig.3 (b) shows the manufactured standard Wilkinson's power combiner with 100Ω resistor which absorbs reflected power which in turn improves the port isolation [10]. The summary of measurement results is plotted in Fig.5. The measured value of S11 is better than -11dB in the 300MHz band from 2.32GHz to 2.52GHz. The coupling coefficient values of S21 (S12) are also plotted. After microstrip width and length optimization, ideal value of -3dB is obtained in simulations as shown in Fig.5. However the measured values deviate by −1.5 to 2dB from the simulated values. It is important to note that S21 value changes with the deviation in the physical soldering location of the 100Ω resistor on the PCB. Since we hand soldered the SMT components, the desired precision is not achieved, which is partially responsible for the discrepancy between the simulation and measurement results. Manufacturing tolerances are yet another factor contributing to this discrepancy. The S31 (& S13) curves are almost same as the S21 (& S12) curves; therefore, only S12 values are plotted.

C. Modified Greinacher Rectifier

A rectifier circuit is a core component which converts the RF signals to a useable DC voltage. Out of many options reported in literature, the Modified Greinacher Rectifier as shown in Fig. 6 was selected for this harvester for our application [4][7]. This circuit has unique ability that it works as voltage quadrupler, using two Greinacher cells of opposite polarities, and as a rectifier at the same time with a minimum number of diodes. The RF signals enter from the RF port, C1 and D1, shift this voltage up to node A then C2 and D2 rectify it to appear across the load. Similarly, C3 and D3 shift the voltage to node B while D4 and C4 rectify it to have a DC voltage across the load [4]. The dual HSMS285B zero bias Schottky diodes were used in this design as rectifiers. They are suitable for 2.4GHz operation. The series and back to back connect diodes are available in a single package, which makes the layout of rectifier simple and efficient due to small package parasitic capacitance. The s-parameter simulation models of these packaged diodes are available from the manufacturer's web page. The major issue in the design is a change in rectifier input impedance. In order to achieve the maximum power transfer, the impedance of the rectifier circuit must be matched with the impedance of input RF port. However, the rectifier circuit consists of nonlinear components and thus the overall impedance varies both with a power level and frequency. In order to match the rectifier impedance to the power combiner's 50-Ω port, the rectifier impedance was estimated under a −20dBm power level of input and then matched using a matching network of short-circuited stubs as shown in Fig.7.

These stubs are placed at both legs and the circuit was optimized for the operating frequency of 2.42GHz and input level of −20dBm. The final artwork work of designed rectifier and its PCB on RO3206 substrate is shown in Fig.7. Multiple samples of rectifiers are tested for input impedance matching with 50Ω with respect to frequency and power level. There is a minor discrepancy between the simulated and measured results (Fig.8 and Fig.9) and these results also changes slightly from sample to sample. The sample to sample variation points

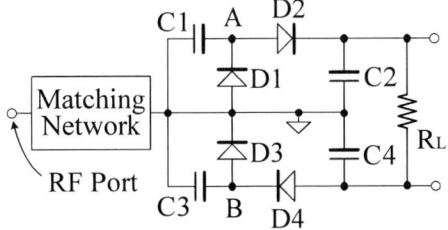

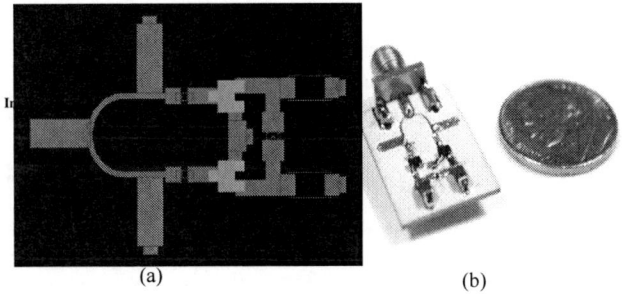

Fig.6 (a) Modified Greinacher Rectifier Circuit as depicted from [6].(b) Rectifier Circuit with a matching network and power combiner.

Fig.7: (a) Modified Greinacher rectifier art work, a snap shot from ADS layout. (b) Photograph of manufactured rectifier PCB

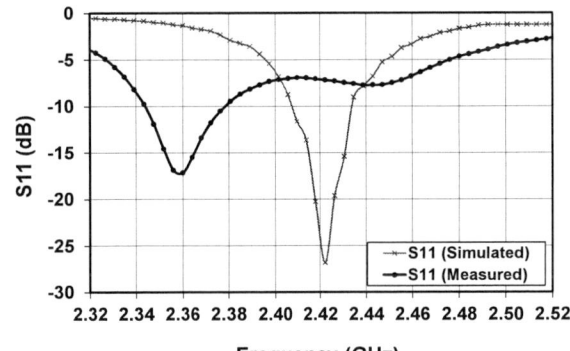

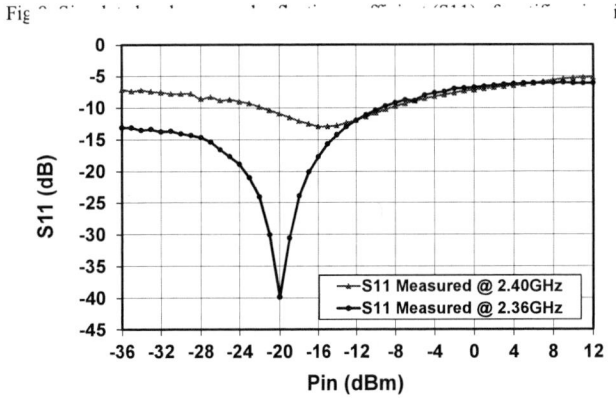

Fig.9. Measured reflection coefficient (S11) of rectifier circuit at frequency of 2.36GHz and 2.40GHz

towards the fact that the major cause for this discrepancy is imprecise manual soldering not the PCB tolerances. As shown in Fig. 8, we get the best matching at 2.36GHz instead of desired simulated and optimized value of 2.42GHz; therefore the complete harvester testing was performed at 2.36GHz. In Fig.9, the rectifier's S11 is plotted with input power from −40dBm to 12dBm. The S11 changes from −25dB to −5dB with change in input power level of −20dBm to 0dBm. There is no simple solution to this mismatch problem which arises due to diode nonlinear characteristics, and inherent in all rectifier circuits; an adaptive input matching network is needed to overcome this problem.

III. COMBINED RF ENERGY HARVESTER

The final Energy Harvester under test consists of two patch antennas connected through a power combiner which is connected to a modified Greinacher rectifier. The rectifier charges a 33mF super capacitor. The complete measurement setup is shown Fig. 10. A 2.4GHz Yagi antenna is used to mimic the WLAN signal, this antenna is placed at the distance 104cm from an RF harvester's dual 2.4GHz patch antenna for initial testing. Fig.11 shows the comparison of power received by the single and dual patch antennas. It is obvious that received signal becomes powerful when two individual antenna signals are added using a power combiner.

Fig.12 shows the measured and simulated efficiency of the RF harvester with respect to incident signal on an antenna port. The peak measured efficiency is 57.8% at 6 to 8dBm input power. At high input power the measured and simulated curves depart due to increased mismatch between combiner and rectifier port. The rectifier input impedance deviated significantly from 50Ω with the increase in input power due to diode non-linear characteristics.

IV. CONCLUSIONS

A Radio Frequency Energy Harvester design was presented in this paper. Using two patch antenna and in presence of a −10dBm signal at antenna port of combiner the system can charge a 33mF super capacitor to 1.6V in 20 minutes. This sum up to the stored energy is around 42mJ, which is enough to power a 10mW sensor node for a period of 4 seconds. Since the signals as powerful as 0dBm are available in close proximity of a WLAN access point, this makes the presented RF harvester feasible for low power wireless sensor application.

REFERENCES

[1] Berini G. A, , "NiPS Summer School, energy Harvesting at micro and nanoscale", Erice (Sicily), Italy, July 23-25, 2012
[2] DS1235007 Data Sheet, "AP350 802.11 a/b/g access point" Aerohive Inc.
[3] D. Cabric, R. W. Brodersen, "Implementation issues in spectrum sensing for cognitive radios," 38th Asilomar Conference on Signals, Systems, and Computers, ,Nov. 2004
[4] Olgun,U, Chi-Chih Chen, Volakis L, "Investigation of Rectenna Array Configurations for Enhanced RF Power Harvesting ,"*IEEE antennas and propagation* letters Vol.10, 2011.
[5] Zbitou, J.; Latrach, M.; Toutain, S. "Hybrid rectenna and monolithic integrated zero-bias microwave rectifier". *IEEE Trans. Microw. Theory. Tech. 54*, 147–152, 2006,

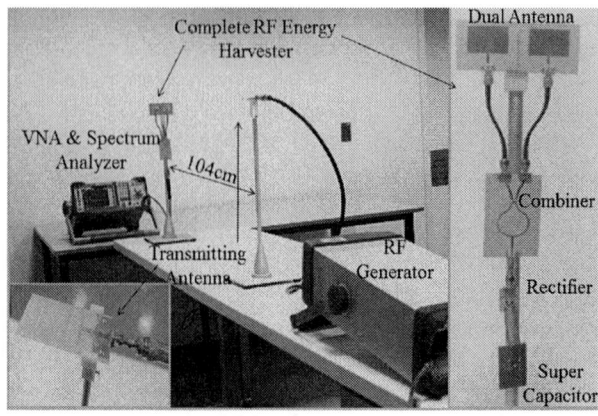

Fig.10. Complete RF energy harvester measurement setup. The inset pictures show the transmitting antenna and RF energy harvester hardware.

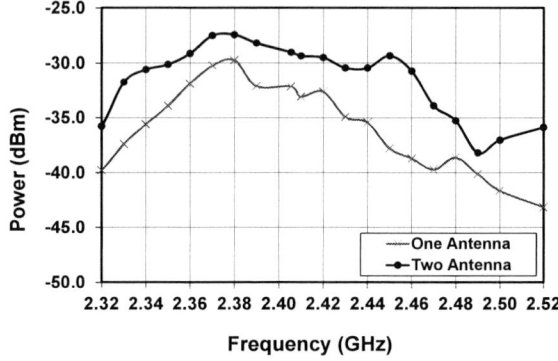

Fig.11: Power received by a single patch antenna and power received by two patch antenna and then added by a combiner.

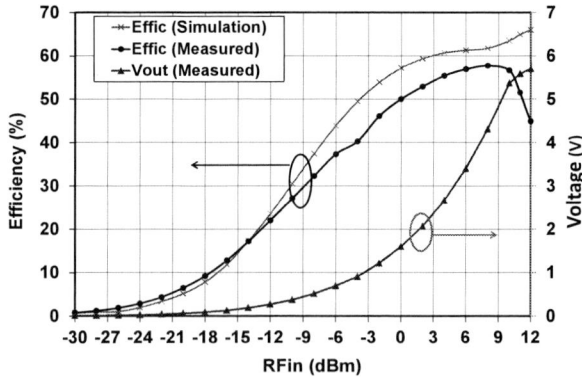

Fig.12. Left Side: Simulated and measured efficiency of the RF harvester; Right Side: The DC voltage level at the output versus RF input power.

[6] Nimo A., Grgić D., Reindl M., "Optimization of Passive Low Power Wireless Electromagnetic Energy Harvesters". Journal of Sensors, 149–153, 2012.
[7] Curty J., "A Model for Power Rectifier Analysis and Design". *IEEE Trans on Circuits and Systems*, vol. *52*, no. 12, Dec 2005.
[8] Balanis, A., "Antenna Theory: Analysis and Design" 3rd ed. New Jersey, John Wiley & Sons, pg:774-780, 2005.
[9] Electromagnetic an d Microwave Engineering, Tools: Microstrip Patch Antenna Calculator, http://www.emtalk.com/mpacalc.php
[10] Bobba,B, "Design and Analysis of an equal split Wilkinson's power Divider" , pg-4, 2010.

IEEE-ICSE2014 Proc. 2014, Kuala Lumpur, Malaysia

Theoretical Study of On-Chip Meander Line Resistor to Improve Q-factor

Wong Goon Weng
Department of Electrical Engineering
Faculty of Engineering
University of Malaya, Malaysia
wonggoonweng@um.edu.my

Norhayati Binti Soin
Department of Electrical Engineering
Faculty of Engineering
University of Malaya, Malaysia
norhayatisoin@um.edu.my

Abstract—In this paper, the theoretical configuration geometry of the layout on-chip meander line resistor was studied and investigated. Various simulation of the geometric design on-chip resistor in a range Giga Hertz frequency are performed. The effect of the quality factor of each design geometry of meander line resistor on high frequency operation was in deep studied and discussed. Besides, parameter extraction geometry of this on-chip meander line resistor was introduced. As a result, the parameter line length (*h*), line segment *(N)* and then following by spacing *(d)* and width (*w*), which are playing an important role on designing the geometry layout to improve the *Q*-factor. Throughout the scaling graphical method, it has been granted out optimize value combination of parameter by improving almost 70% of *Q*-factor and loss of resistance less than 17% of the nominal design. The result of the Design optimization configuration has low *Q*-factor when compared with a nominal Design nominal configuration. This is because of the large value of number segment *(N)* and smaller numbers of line length *(h)*, which has less coupling effect and less resistivity effect. All result base on mathematics computation data was discussed and performed.

Keywords— radio frequency integrated circuits (RFICs); meander line resistor; on chip resistor; radio frequency; geometry; modelling; Q-factor; parameter extraction.

I. INTRODUCTION

Today is rapid growth in RFIC wireless communications, with increasing use of RFIC wireless communication devices, high performance of the quality factor operate at high frequencies should be taken into account. There are numerous papers have been published on new design and fabrication of on-chip resistor for wireless applications [1-7]. Low Q-factors are required to establish a high performance on-chip resistor in new generation of RFIC circuits. Normally the on-chip resistor performance is given in its layout design and its physical characteristics like resistive losses and substrate losses [8-10]. The changing of the characteristic of resistor in the high frequency may be caused by several parameters. The performance of Q-factor can be observed that the behavior of on-chip resistor is operating at high frequency. Resistor values and quality factor of a resistor can be computed based on its equation [11] by using MathCAD. The aims of this paper is to improve the characteristics of on-chip resistor at low Q-factor operate at high frequency by modifying and optimizing the configuration geometry of the layout. The performances in terms of Q-factor, resistance and capacitance Cmf are

compared between the nominal design and the optimization design configuration.

II. DESCRIPTION OF THE ON CHIP RESISTOR

In earlier designs on chip resistors, most resistors are designed in straight line configurations such as in Fig. 1.a. [8] Fig. 1.b shown the meander-line configuration has made it popular and attractive in radio frequency integrated circuits (*RFICs*) due to its allows large area-saving on the chip compact area and can, more efficient in substrate area and provide a possibility for large resistor values by increasing the total length and number segment of the resistor [8, 9, 10].

a.   b.

Fig. 1. Configuration On Chip Resistor. a. straight line; b. meander line

In this paper, the design of the meander-line resistor is investigated [10,11]. It is based on varying geometrical parameters of a number of lengths (*N*), width (*w*), line length (*h*) and spacing (*d*) between the lines to obtain a low *Q*-factor. The structure of the configuration is studied as showed in Fig. 2. Mathematical computations *MathCAD* can make use to calculate and identified through a contour plot of the *Q*-factor and resistance value.

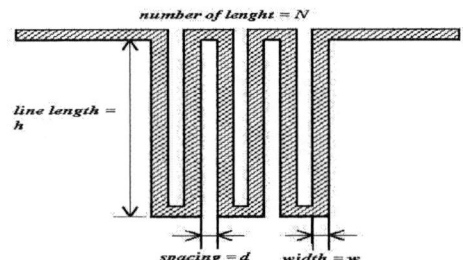

Fig. 2. Conventional Meander-Line Resistor Layout

III. THEORY OF EQUITION

A. Q-factor Effects on Performance On Chip Resistor

The quality factor *Q* describes how well the passive component can perform at a high frequency [9, 13-15]. The

978-1-4799-5761-3/14 $31.00 © 2014 IEEE   475

resistor is frequency dependent during operating at high frequency. This is because parasitic capacitances, inductance, and resistances are associated within the on chip resistor physical layout when operate in high frequency. To obtain Q-factor value, a typical equivalent circuit of pie type on chip resistors are being studied [12]. To improve on chip resistors, the main requirement is obtained low Q-factor. Generally, Q is defined to be.

$$Q = \omega \frac{EnergyStored}{PowerLoss} \tag{1}$$

ω is defined as the angular frequency of the system.

$$Q = \pi \frac{EnergyStored}{EnergyDissipatedPerCycle} \tag{2}$$

However, the conventional meander line resistor layout as showed in Fig. 2 of on chip meander line resistor can represent the equivalent pie type lumped circuit [8, 10], we can calculate the parallel circuit of Q-factor by formula as below

$$Q = \left(\frac{Im[Y_{in}]}{Re[Y_{in}]} \right) = \left(\frac{Im[Y_{11}]}{Re[Y_{11}]} \right) \tag{3}$$

B. Parameter of the lumped equivalent circuit

In the paper by Rizwan Murji, and M. Jamal Deen [11], the N-well meander line resistor is fabricated in a $0.18\mu m$ technologies. The geometric parameter of the meander line resistor are defined by the width (w), line length (h), spacing (d) and additional lengths in the resistor geometry $(a$ and $b)$.

The value of the thickness n-well (t_{nwell}), thickness of the doped epitaxial layer (t_{epi}), substrate thickness (t_{si}), lumped equivalent circuit and formula equation is being used in this paper to perform deep analysis and investigation.

Thus the series resistor is given by [10],

$$R = R_{sh} \frac{hN + d(N+1) + 2(b+a)}{w} \tag{4}$$

And can simplify to

$$N = \frac{\frac{Rw}{Rsh} - (d+h)}{d+h} \tag{5}$$

Where: the length $a \approx 0$ while the length $b = h/2$.

The capacitance, represents the parasitic capacitances is given by [11], $\quad C_{mf} = C_{in} + 2C_{top} \tag{6}$

Which is proportional to $1/N$ [11].

Where: $\quad C_{in} = \varepsilon 0 x \varepsilon 0 \frac{[t_{nwell}]}{dN} \tag{7}$

And $\quad C_{top} = \frac{\varepsilon 0 x \varepsilon 0}{N} In\left|1 + \frac{\omega}{t_{epi} + t_{nwell}}\right| \tag{8}$

And the equivalent circuit for Y-parameters are given by [11]

$$Y_{11} = \frac{I_1}{V_1}\bigg|_{V2=0} = \frac{j\omega C_{ov}}{1 + j\omega R_{epi} C_{ov}} + j\omega C_{mf} + \frac{1}{j\omega L + R} \tag{9}$$

During on chip resistor operate in high frequencies, the value of inductor in the model structure is small and it is dominated by capacitance and resistance. Therefore, equation (9) can be simplified as

$$Y_{11} = \frac{j\omega C_{ov}}{1 + j\omega R_{epi} C_{ov}} + j\omega C_{mf} + \frac{1}{R} \tag{10}$$

IV. RESULTS AND ANALYSIS

The design parameters configuration using *MathCAD* theoretical calculation is being determined through the contour plot of resistor and Q-factor in the range of frequency up to 10G Hz. Detail investigates effects on the width (w), spacing (d), line length (h) & line segments (N), Cmf and Q-factor are being studied. The results of all total 14 designs, parameters configuration have been obtained and tabulated in table 1 to table 5 respectively. Design 1 is defined by the set as the nominal value of Q=1. 481, Cmf=3. 85F, Rsh=953. 125Ω, width $(w$=3um), spacing $(d$=3um), line lengths $(h$=45um) and line segments $(N$=3). All the nominal values are extracted from the paper by Rizwan Murji, and M. Jamal Deen [11]. Design 2 to Design 13 is made with the assumption by varying the parameter width (w), spacing (d), line length (h) & line segments (N).

TABLE1. DESIGN PARAMETER ASSUMPTIONS WIDTH OF CHARATERISTICS BY NOMINAL WITH FIXED SPACING (d=3μm), LINE LENGHTS (h=45μm) AND LINE SEGMENT (N=3).

Design of Experiment	Width (w) um	Capacitor, Cmf (f)	Real Resistor (KΩ)	Q-Factor at cutoff 1Gz	Resistor at cutoff 1Ghz (KΩ)
Design 1	3	3.85	61	1.481	19.11
Design 2	1.5	2.23	122	1.72	30.82
Design 3	6	6.523	30.5	1.253	11.87
Design 4	9	8.68	20.33	1.111	9.103

TABLE2. DESIGN PARAMETER ASSUMPTIONS SPACING OF CHARATERISTICS BY NOMINAL WITH FIXED WIDTH (w=3μm), LINE LENGHTS (h=45μm) AND LINE SEGMENT (N=3).

Design of Experiment	Spacing (d) um	Capacitor, Cmf (f)	Real Resistor (KΩ)	Q-Factor at cutoff 1Ghz	Resistor at cutoff 1Ghz (KΩ)
Design 1	3	3.85	61	1.481	19.11
Design 5	1.5	4.187	59.09	1.56	17.21
Design 6	6	3.681	64.81	1.505	21.17
Design 7	9	3.625	68.62	1.569	19.82

TABLE3. DESIGN PARAMETER ASSUMPTIONS LINR LENGHT OF CHARATERISTICS BY NOMINAL WITH FIXED SPACING (d=3μm), WIDTH (w=3μm) AND LINE SEGMENT (N=3).

Design of Experiment	Line length (h) um	Capacitor, Cmf (f)	Real Resistor (KΩ)	Q-Factor at cutoff 1Ghz	Resistor at cutoff 1Ghz (KΩ))
Design 1	45	3.85	61	1.481	19.11
Design 8	10	3.85	16.52	0.401	14.23
Design 9	90	3.85	118.2	2.869	12.80
Design 10	200	3.85	258	6.263	6.414

TABLE4. DESIGN PARAMETER ASSUMPTIONS LINE SEGMENT OF CHARATERISTICS BY NOMINAL WITH FIXED SPACING (d=3μm), LINE LENGHTS (h=45μm) AND WIDTH (w=3μm).

Design of Experiment	Line segment (N)	Capacitor, Cmf (f)	Real Resistor (KΩ)	Q-Factor at cutoff 1Ghz	Resistor at cutoff 1Ghz (KΩ)
Design 1	3	3.85	61	1.481	19.11
Design 11	1	11.55	30.5	2.216	5.16
Design 12	5	2.31	91.5	1.336	32.86
Design 13	7	1.65	122	1.275	46.46

The computation of resistance values, Cmf and Q-factor for Design 1 to Design 14 are conducted and analyzed by using mathematics computation of Mathcad through the equation (6), equation (11) and equation (12), which operating at cutoff frequency one Giga Hertz.

Table 1 shows the design parameter at a fixed value of space (d=3um), line lengths (h=45um) and line segments (N=3) respectively. Based on observation, when the width (w) is increasing the Q-factor value is decreased, the capacitor Cmf is increased and the resistance loss is improved during operating at high frequency one Giga Hertz. In Design 4, the Q-factor has improved almost 25%, which is from 1.481 to 1.111 and the loss resistance is improved also up to almost 13% from 20.33KΩ to 9.103KΩ, if compare to Design 1. The capacitor Cmf is increased due to high induces coupling effect when the width (w) is increasing. As a summary of the low Q-factor can be obtained & the loss of resistance can be improved during operating at high frequency by increasing the width (w) of geometry layout. However, capacitor Cmf will be increased due to the coupling effect [11].

Table 2 shows design parameters at a fixed value of width (w=3um), line lengths (h=45um) and line segments (N=3). Based on the *MathCAD* theoretical calculation, the capacitor Cmf is reduced while the spacing (d) is increased. This is because the distance between two parallel lines are being separately far off each other due to the coupling less effect. When there are closer parallel lines, the more coupling effect is realized [11]. Based on the table 2 data, there is no specific trend for Q-factor and the loss of resistance. However, we can conclude that there is less coupling effect when the spacing (d) is large.

Table 3 shows design parameters at a fixed value of width (w=3um), spacing (d=3um) and line segments (N=3). During the line length (h) is increasing at 10um to 200um, the Q-factor is increased from 0.401 up to 6.263 and the loss of resistance is reduced from 14.23 6.414. This is because of the resistive effect [11]. However, there is no influence on capacitance, Cmf and maintained value as 3.85fF. Based on equation (6), (7) & (8), only spacing (d) and line segment (N) has effected on the capacitance Cmf. Design 8 has shown the lowest Q-factor at 0.401, which is 72.9% lower than Design 1 and the loss of resistance is minimized about 13.86% from 16.52 kΩ to 14.23 kΩ at 1GHz cutoff frequency, if compare to Design 1, resistance has losses of 68.67% from the real resistor. It can be inferred that, as the line length (h) of the resistor increases, there is a poor performance behavior of resistance during operated at high frequency because of high Q-factor and high losses of resistor. But with a short line length (h), it can enhance the Q-factor and the loss of resistance.

Table 4 shows design parameters at fixed value of width (w=3um), spacing (d=3um), and line lengths (h=45um). Base on equation (6), capacitance is proportional to $1/N$. This shows that the capacitor Cmf value is decreased as a line segment (N) is increased due to less coupling effect [11]. As the line segment (N) increases, the loss of resistance is improved and also the Q-factor shows a significant reduction because of the number of line segments (N) has been increased. In Design 13 with 7 lines segments have the lowest Q-factor at 1.275 and loss of resistance is improved about 6.8% if compare to the Design 1. From the observation, highest number of segments (N) will lead to lowering Q-factor and improve the loss of resistance during operating at high frequency.

Based on a study at table 3 and table 4, by increasing the number of line segments (N) and reducing of the line length (h) the Q-factor has shows significantly improve. Then it is followed by increasing the width (w) and spacing (d) as shown in table 1 and table 2 respectively, but based on theoretical calculation, both parameters have less impact on the performance improvement. Graphical in Fig. 3, Fig. 4 and Fig. 5 have shown that the suitable value combination of each parameter. Four parameter variables; width (w), spacing (d), line length (h), and line segment (N) have been set and scaled to obtain optimized suitable combinations setting value. By using equations (5), optimal values on each parameter setting have been determined. The scaling method in Fig. 3 shows that the line segment (N) is proportional to the width (w) at a fixed the parameter of d=3um and h=45um. Fig 4 shows the relationship between the line segment (N) and spacing (d) with the width (w) and line length (h) fixed to 3um and 45um respectively. It shows that when the distance of space (d) is increasing, while the line segment (N) is decreased. In Fig.5 shows the scaling graphical between number of line segment (h) and the line length (N) fixed the parameters to d=3um and w=3um. The influence between the two variable parameter line segment (h) and line length N is inversely proportional to each other. Fig. 5 shows that the line length (h) is reduced while the number of line segments (N) is increased.

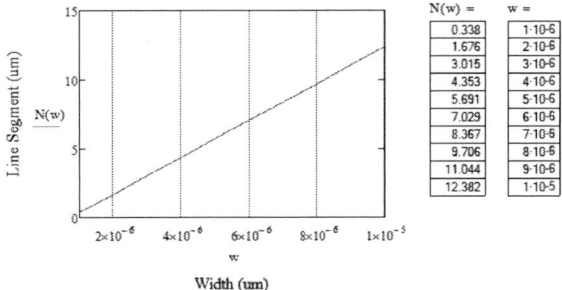

Fig. 3. The suitable value combination of a line segment (N) and width (w) by fixed value spacing (d) = 3um and line length (h) =45um.

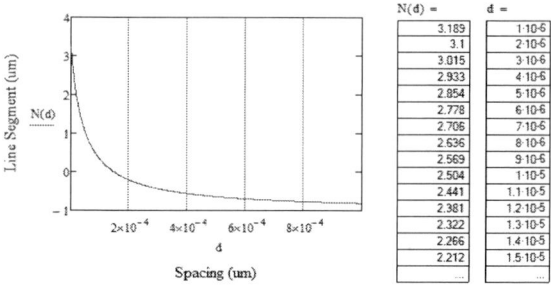

$N(d) =$	$d =$
3.189	1·10-6
3.1	2·10-6
3.015	3·10-6
2.933	4·10-6
2.854	5·10-6
2.778	6·10-6
2.706	7·10-6
2.636	8·10-6
2.569	9·10-6
2.504	1·10-5
2.441	1.1·10-5
2.381	1.2·10-5
2.322	1.3·10-5
2.266	1.4·10-5
2.212	1.5·10-5
	...

Fig. 4. The suitable value combination of a line segment *(N)* and spacing *(d)* by fixed value width *(w)* = 3um and spacing *(h)* = 45um

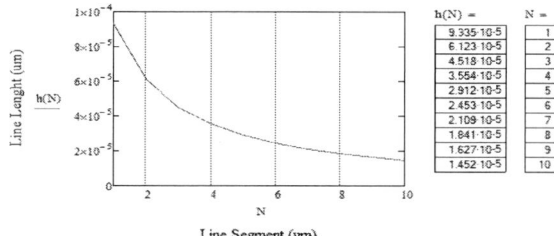

$h(N) =$	$N =$
9.335·10-5	1
6.123·10-5	2
4.518·10-5	3
3.554·10-5	4
2.912·10-5	5
2.453·10-5	6
2.109·10-5	7
1.841·10-5	8
1.627·10-5	9
1.452·10-5	10

Fig. 5. The suitable value combination of *N* and *w* by the fixed value *d* = 3um and *w* = 3um.

Based on scale graphical method study, we can identify the optimal value to create design optimal configuration. Referring to the Fig. 5, by selecting the line segment *(N)* equal to 10um, the line length *(h)* is given by approximated to 14.52um. For the width *(w)* and spacing *(d)*, both parameters remained unchanged as *w*=3um, *d*=3um, due to design the resistor value at around 61K ohm. Table 5 shows the comparison between Design 1 as nominal configurations and the Design 14 as an optimal configuration base on scaling method.

TABLE 5. COMPARISON BETWEEN NOMINAL DESIGN 1 AND OPTIMIZES DESIGN 14 WITH SPACING (*d*=3μm) AND WIDTH (*w*=3μm).

Design of Experiment	Line segment *(N)*	Line length *(h)* um	Capacitor, *Cmf* (f)	Real Resistor *(KΩ)*	Q-Factor at cutoff 1Ghz	Resistor at cutoff 1Ghz *(KΩ)*
Design 1	3	45	3.85	61	1.481	19.11
Design 14	10	14.52	1.16	61.1	0.449	50.9

Design 14 has shown that significant improvement on *Q*-factor and the loss of resistance. *Q*-factor had improved almost 70% from 1.481 to 0.449 and the loss of resistance improves about 17% if compare to the Design 1. Design 14 also shows that capacitor *Cmf* has less coupling effect which is 0.449 while Design 1 is 1.481. Based on this date, Design 14 optimal configuration is much better than Design 1 nominal configuration.

V. CONCLUSION

It can be concluded that the *Q*-factor value can be affected based on the number of line segments *(N)*, width *(w)*, spacing *(d)* and line length *(h)* due to the effect of resistivity resistance and capacitive coupling effects of capacitance. Shorter line length and high number of line segments *(N)* have significantly

improved on the *Q*-factor value, same as well for the loss of resistance. This can be proven through comparison between the Design 14 optimization configuration, with the value of *Q*-factor =0.449, *Cmf*=1.16f and the loss resistance is 10.2K ohm which is customize design based on scale graphical method study and the nominal Design 1 nominal configuration, with the value of *Q*-factor =1.481, *Cmf*=3.85f and the loss resistance is 41.89K ohm. By varying the spacing *(d)* and width *(w)* is much impact on the optimization of the *Q*-factor value, but increasing the spacing *(d)*, the coupling effects are reduced. There are some areas that can be further improved and investigation in order to obtain better performance in on-chip resistor, including substrate material that can reduce resistive loss and different configurations of resistors.

ACKNOWLEDGMENT

The authors want to thanks the government of Malaysia for sponsorship financial of this research project.

REFERENCES

[1] A. Dziedzic, "Electrical and Structural Investigations in Reliability Characterization of Modern Passives and Passive Integrated Components", Vol. 42, 2002, pp. 709-719.

[2] Anvesha. A; Dave. M; Baghini. MS; Sharma. D, "A process and temperature invariant on chip-resistor and its application", Circuit and Systems, IEEE Conference Publications Years 2012, Page (s): 706-709.

[3] Cheng Ying; Li Fule; Zheng Xuqiang; Zhang Chun, "Self-calibrating on-chip termination resistor for high-speed SerDes", Consumer Electronics, Communications and Networks, 2011 International IEEE Conference Publications Page (s): 5207 – 5210.

[4] Bhattar. N; Gupta.A, "On-chip resistor can make a stable current reference", Potentials, IEEE Jornals & Magazines Publication Year 2008, Page(s):31-36.

[5] Xu Hongjie; Zhang Yonghong; Fan Yong, "Ka-Band Wilkinson Power Divider based on Chip Resistor", Microwave and Millimeter Wave Technology, 2007 ICMMT. International Conference on IEEE Publications Years:2007, Page(s):1-4.

[6] Talebbeydokhti.N; Hanumolu. PK;Kurahashi. P; Un-Ku Moon, "Constant transconductance bias circuit with an on-chip resistor", ISCAS 2006 IEEE International Symposium Page(s): 4-2860.

[7] Fan Wu, Morris, J.E., "Integration issue of Crx (SiO) 1-x on-chip resistors", Electronic Components and Technology Conference, 2005. Proceedings. 55th. Page (s): 1659 - 1663 Vol. 2.

[8] M. Engles and R. H. Jansen, "Modeling and Design of Novel Passive MMIC Components with Three and More Conductor Layers", IEEE MTT-S Digest, pp. 1293-6, 1994

[9] Rizwan Murji, and M. Jamal Deen, "A Scalable Mender-Line Resistor Model for Silicon RFIC's" IEEE Transcations on electron devices, Vol.49, No.1, pp. 187-190, January 2002.

[10] C. P. Yue and S. S. Wong, "Physical Modeling of Spiral Inductors on Silicon," IEEE Transactions on Electron Devices, vol. 47, no. 3, pp. 560-568, Mar. 2000.

[11] Rizwan Murji, and M. Jamal Deen, "Accurate Modeling and Parameter Extraction for Meander-Line N-Well Resistors" IEEE Transactions on electron devices, Vol. 52, No7, pp. 1364-1369, July 2005

[12] Charles A. Hasper, & Donald D. Dunlap, Jr, "Passive Electronic Component Handbook", 1997.

[13] Dae-Hyun Elan, Jeom-Young Ahn, "Measurements of High Q On-chip Inductors for Wireless Applications", 2005.

[14] Thomas Bluhm, "S-Parameters- Characteristics of Passive Component", 2001

[15] Heo, Seung Kyun, "The Performance of Passive Lumped Element", 2010.

Temperature Dependence of Ga:ZnO Film Deposited By RF Magnetron Sputtering

Farah Lyana Shain
School of Science and Technology,
University Malaysia Sabah
88400, Kota Kinabalu, Sabah, Malaysia
farahlyanashain@gmail.com

Azmizam Manie @ Mani, Lam Mui Li
Saafie Salleh and Afishah Alias
School of Science and Technology,
University Malaysia Sabah
88400, Kota Kinabalu, Sabah, Malaysia

Abstract— This paper investigate the dependence of substrate temperature onto characteristic of Gallium doped Zinc Oxide (Ga:ZnO). Ga:ZnO films were deposited on a glass substrate by RF Magnetron Sputtering using Ga:ZnO ceramic target with 99.99% purity. Sputtering power, argon flow and target distance were fixed in order to investigate the influence of substrate temperature to the growth characteristic, structural and optical properties of the films. Sputtering was performed with RF power of 100 Watt and the argon flow in was set at 10 sccm. The deposition times were fixed at 40 minute for all films. The result shows growth rate for Ga:ZnO growth at higher temperature are lower than at room temperature. Ga:ZnO thin films on different substrate temperature were successfully deposited onto glass substrate. All films are polycrystalline with (0 0 2) preferential orientation and fully transparent films with high transparency over 80 percent were achieved.

Keywords— *Gallium Zinc Oxide; RF Magnetron Sputtering; Substrate Temperature*

I. Introduction

Indium Tin Oxide (ITO) in semiconductor material is widely used as the fundamental element in Liquid Crystal Displays (LCDs), Plasma Displays Panels, Electronics Paper Displays, Sensors, Photovoltaic Panels [1-3]. However due to the high demand of ITO in line with the development of technology in this century added with the drawbacks of ITO that have been reported such as high cost and shortage, and also ITO have been reported contain hazardous toxics [4-5], it is crucial for the researcher to looks for the suitable substitute for ITO. Doped Zinc Oxide has gain lots of interest for it due to its advantages compared to ITO in term of low cost, and environmental toxic free [6].

This research was fund by project code PHD004-TK-2012.

One of the candidate for doped zinc oxide is Gallium Zinc Oxide (Ga:ZnO). Superiority Ga:ZnO when used as dopant for n-type Zinc Oxide semiconductor is high oxidation resistance during deposition, exhibit better electronics stability in humidity also suitable for stabilization of ZnO lattice system [7-9]. In addition, to improve the electrical conductivity of Doped Zinc Oxide, the Zn^{2+} ions replaced with other ions that have higher valence electrons to act as a shallow donor for example In^{3+}, Al^{3+} and Ga^{3+} and in this report, Ga^{3+} is more favorable among others due to the Zn and Ga similar tetrahedral radii that would contribute to the small lattice distortion and minimal strain when the substitutions in doping process occur [10].

Variety deposition technique have been done for fabrication of Ga:ZnO films such as sol gel method, spray pyrolisis, pulsed laser deposition (PLD) and also magnetron sputter [11-13]. Among of that techniques magnetron sputtering is choose because of ability to produce high quality thin film with a high density and good adhesion interaction between surfaces, and can be obtained at low substrate temperature with good uniformity of the film thickness in a large area [14]. There are five influential deposition parameters in RF magnetron sputtering that is RF power, sputtering pressure, deposition time , deposition temperature and also annealing temperature [15]. In this case substrate temperature is chose as a parameter and fully transparent Ga: ZnO thin films were deposited on glass substrate via RF magnetron sputtering by the functions of substrate temperature. The dependence of growth characteristic, structural, optical and surface roughness on substrate temperature for the films were investigated and the result were reported.

II. Methodology

Ga:ZnO films were deposited onto glass substrate (7.62 cm x 2.54 cm x 0.10 cm) under different substrate temperature using RF magnetron sputter deposition technique.Ga:ZnO ceramic target with 99.99% purity contain 3 % of Gallium from Advanced Technology Material was used as sputter target material. Before the deposition, the substrates were cleaned using an Ultrasonic Branson 3200 Cleaner by immersing the substrate in repeatedly three times in Distilled Water, Ethanol, Acetone and Isopropyl Alcohol respectively for 3 minutes and then rinsed using distilled water before dried out using dry nitrogen gas to eliminated the any unknown layer from the surface of glass substrate. Initially the vacuum chamber of RF magnetron sputter were evacuated to the base pressure of 1.0×10^{-5} to and then the sputtering was performed with RF power of 100 watt and the argon flow was set at 10 sccm. The deposition times were fixed at 40 minute. The thicknesses of Ga:ZnO films was measured using NanoLab 550 Profilometer. The structure of Ga:ZnO was studied using X-Ray Diffraction (Philips Analytical X-Ray).The transmission spectra of Ga:ZnO films were obtain from UV-Vis spectrometer Lambda 25.

III. Result And Discussions

A. Structural Properties

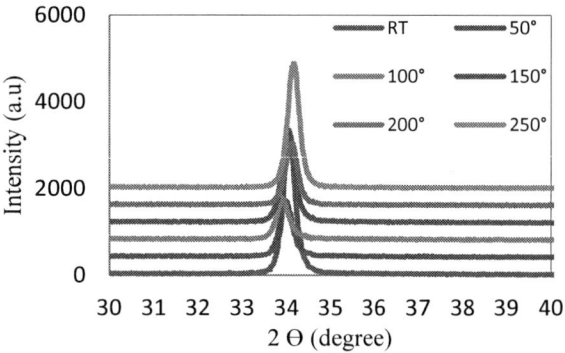

Fig.1. X-ray diffraction pattern for Ga:ZnO films by the function of substrate temperature (°C).

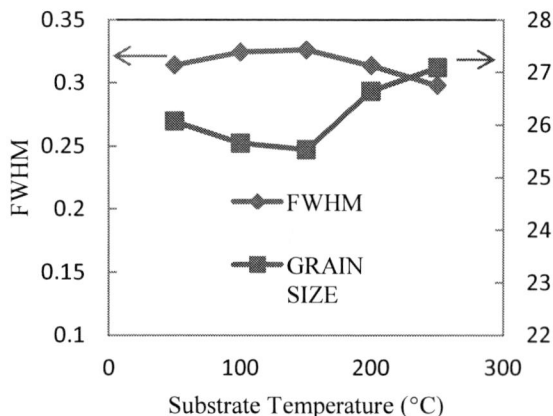

Fig. 2: Variation of crystallite grain size and full width half maximum (FWHM) for Ga:ZnO films by the functions of substrate temperature(°C).

X-Ray Diffraction pattern proved Ga:ZnO films has be successfully deposited on glass substrate (Fig 2) where all the peak observed indicate Ga:ZnO thin film prepared by RF sputtering are polycrystalline with preferential orientation of (0 0 2) similar to the XRD peak pattern for standard ZnO (JCPDS 36-1451) [16].There are no impurity of Ga compound shows in XRD pattern suggesting Ga^{3+} ions filled or substitutes the Zn^{2+} site in ZnO lattice or either Ga^{3+} ions occupy the interstitial sites of ZnO [17]. As the deposition temperature increase from 50° to the 250°, the position of peak of Ga:ZnO peaks shift slightly to the higher angle.

This behavior could be related to the Bragg law where when the Bragg angle θ increase; inter-planar spacing of the films would decrease. Since the result of inter-planar spacing are directly proportional to the lattice constant, therefore as the temperature increase the lattice constant of Ga:ZnO will also decrease. The phenomena could be explain with the smaller atomic radius of Ga^{3+} ions compared with Zn^{2+} ions allowing more substitution process of Ga^{3+} ions into ZnO lattice when the value of heat supply increase [18]. The only small changes of Ga:ZnO peak locations also implied that the tensile strain between film are smaller who will contribute to the smaller lattice mismatch which agreed by Gorrie et al [8]. Fig. 3, shows the fully width half maximum values (FWHM) and crystallite grain size for Ga:ZnO (0 0 2) peaks as a functions of deposition temperature calculated to evaluate the crystal quality of Ga:Zno films. The values of FWHM decrease as the value of crystallite grain size increase where is determine by Scherrer's formula [19] :

$$D = \frac{0.94\lambda}{\beta \cos\theta} \qquad (1)$$

Where X-ray wavelength, λ = 1.54056 Å, β is the FWHM of ZnO diffraction peak, and θ is the Bragg diffraction angle. The result of the crystallite grain size of Ga:ZnO films show that crystal quality of films enhanced due to the temperature changes. The improving crystallite size might happen because of at higher temperature the atoms have sufficient activation energy to fill the vacancies in the crystal lattice and grains with lower surface energy will become larger [20].

B. Growth Characteristic

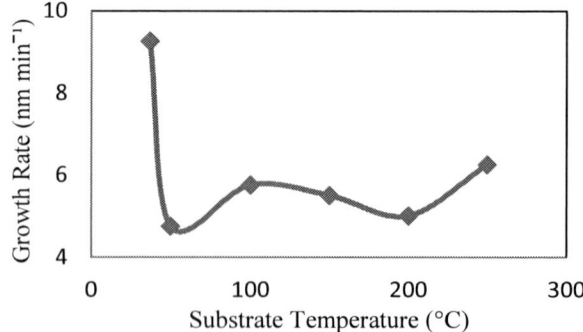

Fig. 3. Ga:ZnO films growth rate as a function of substrate temperature (°C).

Growth rate of Ga:ZnO films are influenced by the functions of substrate temperature as shown in Fig. 3 where the growth rate of films deposited at room temperature are higher than when the substrate heated. The change in growth rate between films deposited in room temperature and at 250 °C can be clearly seen. This might be due to the deposition in higher temperature contributing to the larger crystallite grain size where the increase in activation energy of Ga atoms due to the added thermal energy from as the substrate temperature increase also lead Ga atoms to filled/replace the vacancies in the lattice structure of ZnO contribute to the increased grain size of films which is consistent with FHWM result extract from XRD data in Fig. 2, thus as the deposition temperature increased the crystallite grain size of Ga:ZnO layer on top of substrates becomes densely packed slows the growth rate of Ga:ZnO [9].

C. Optical Properties

Ga:ZnO films fabricated by the RF sputtering method in different substrate temperature show a high transparency in the wavelength ranging from 300 to 900 nm as shown in Fig 4. Ga:ZnO films growth at lower deposition temperature have high transmittance value approximately over 80 %. In addition Ga:ZnO suitable for the devices transparent conductive oxide exposed to high temperature such as solar heater due to the improvement in crystallite grain size obtain from the XRD result. Several interference pattern have been observed suggest that Ga:ZnO films have a good adhesion contact with glass substrate therefore may extend the lifespan of electronics devices. The good optical transmissions properties of Ga:ZnO films in low and high temperature indicated that Ga:ZnO films is an ideal materials for electronics devices applications.

The estimation value for band gap of Ga:ZnO obtain from Tauc's equation for direct optical band gap semiconductor as in equation below [21] :

$$(\alpha h v) = A\left(h v - E_g\right)^{\frac{1}{2}}$$

(2)

Where Eg is the energy band gap, α is the absorption coefficient, hv is the incident photon energy and A the edge width parameter. The value of energy band gap Eg evaluated from the extrapolating linear part of the curve obtain from the graph $(\alpha h v)^2$ versus Photon energy (eV). It is seen that the optical band gap for the Ga:ZnO films becomes broader as the deposition temperature increase. The phenomena are related with the Burstein Moss Effect where the deposition temperature will contributed to the high Fermi energy level in the conduction band of a degenerate semiconductor [22].

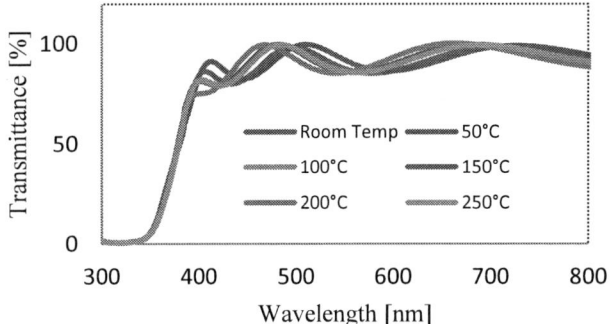

Fig. 4. Transmission spectrums of RF sputter Ga: ZnO in different substrate at different substrate temperature (°C).

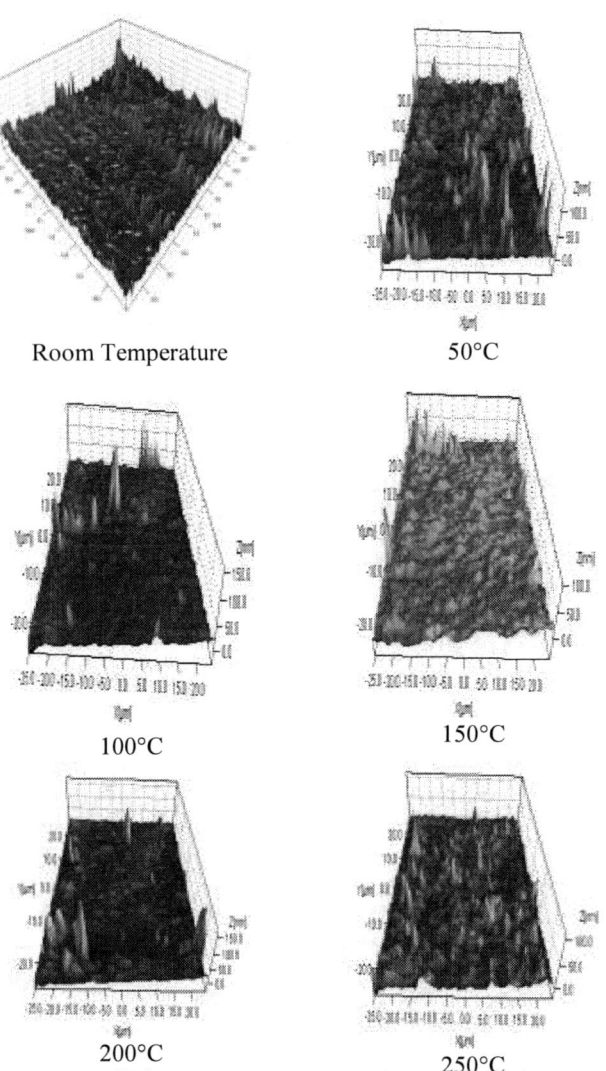

Fig. 5. Comparison of surface morphology obtain from profilometer 3D image for Ga:ZnO films

D. Surface Morphology

The comparison of surface morphology for Ga:ZnO obtain from NanoLab 550 Profilometer 3D image films at different substrate temperature are shown in Fig. 6. For each data, the surface roughness in term of root-mean-square (RMS) is measured. As the deposition temperature increase, the RMS value for the Ga:ZnO films were increases. In room temperature, the RMS value for the film was 1.84 nm whereas the RMS value for Ga : ZnO film deposited at 250 ° was 3.84 nm.The increasing value of RMS were correlated with the enhancement quality of crystallite grain size as the deposition temperature increase where as the crystallite size become larger, the surface roughness also become bigger [22]. This behavior might be influence by the high activation energy of the films atom at higher temperature to filled the proper sites in crystal lattice hence the crystallite grains size with lower surface energy will become larger [20].The result would help searching the better quality of surface morphology will help improved the electrical properties of the films in the future.

IV. Conclusion

The temperature dependence for Ga:ZnO films has been studied. The result shows growth rate for Ga:ZnO were influenced by the substrate temperature. Ga:Zno films on different substrate were fabricated onto glass substrate. Structural analysis revealed films are polycrystalline with (0 0 2) orientation and the crystallite size enhanced on high substrate temperature and optical transmission pattern analysis shows fully transparent films with high transparency achieved.

Acknowledgment

This research was fund by project code PHD004-TK-2012. The author would like to thank University Malaysia Sabah for the hospitality provided.

References

[1] T. Minami and T. Miyata, "Present status and future prospects for development of non- or reduced-indium transparent conducting oxide thin films," Thin Solid Films, vol. 517, no. 4, pp. 1474–1477, Dec. 2008.

[2] M. Neophytou, E. Georgiou, M. M. Fyrillas, and S. a. Choulis, "Two step sintering process and metal grid design optimization for highly efficient ITO free organic photovoltaics," Sol. Energy Mater. Sol. Cells, vol. 122, pp. 1–7, Mar. 2014.

[3] N. G. Patel, P. D. Patel, and V. S. Vaishnav, "Indium tin oxide (ITO) thin film gas sensor for detection of methanol at room temperature," Sensors Actuators B Chem., vol. 96, no. 1–2, pp. 180–189, Nov. 2003.

[4] M. Jiang and X. Liu, "Structural, electrical and optical properties of Al–Ti codoped ZnO (ZATO) thin films prepared by RF magnetron sputtering," Appl. Surf. Sci., vol. 255, no. 5, pp. 3175–3178, Dec. 2008.

[5] "Electronic Waste : Concerns & Hazardous Threats," vol. 4, no. 2, pp. 802–811, 2014.

[6] M. Osada, T. Sakemi, and T. Yamamoto, "The effects of oxygen partial pressure on local structural properties for Ga-doped ZnO

thin films," Thin Solid Films, vol. 494, no. 1–2, pp. 38–41, Jan. 2006.

[7] A. S. Pugalenthi, R. Balasundaraprabhu, V. Gunasekaran, N. Muthukumarasamy, S. Prasanna, and S. Jayakumar, "Effect of thickness on the structural, optical and electrical properties of RF magnetron sputtered GZO thin films," Mater. Sci. Semicond. Process., pp. 1–7, Feb. 2014.

[8] C. W. Gorrie, A. K. Sigdel, J. J. Berry, B. J. Reese, M. F. a. M. van Hest, P. H. Holloway, D. S. Ginley, and J. D. Perkins, "Effect of deposition distance and temperature on electrical, optical and structural properties of radio-frequency magnetron-sputtered gallium-doped zinc oxide," Thin Solid Films, vol. 519, no. 1, pp. 190–196, Oct. 2010.

[9] X. Yu, J. Ma, F. Ji, Y. Wang, C. Cheng, and H. Ma, "Thickness dependence of properties of ZnO:Ga films deposited by rf magnetron sputtering," Appl. Surf. Sci., vol. 245, no. 1–4, pp. 310–315, May 2005.

[10] A. de Souza Gonçalves, S. Antonio Marques de Lima, M. Rosaly Davolos, S. Gutierrez Antônio, and C. de Oliveira Paiva-Santos, "The effects of ZnGa2O4 formation on structural and optical properties of ZnO:Ga powders," J. Solid State Chem., vol. 179, no. 5, pp. 1330–1334, May 2006.

[11] B. Du Ahn, S. Hoon Oh, C. Hee Lee, G. Hee Kim, H. Jae Kim, and S. Yeol Lee, "Influence of thermal annealing ambient on Ga-doped ZnO thin films," J. Cryst. Growth, vol. 309, no. 2, pp. 128–133, Dec. 2007.

[12] V. Fathollahi and M. M. Amini, "Sol – gel preparation of highly oriented gallium-doped zinc oxide thin films," no. September, pp. 235–239, 2001.

[13] F. Wu, L. Fang, K. Zhou, Y. J. Pan, L. P. Peng, Q. L. Huang, X. F. Yang, and C. Y. Kong, "Effect of Thickness on the Properties of Ga-doped Nano-ZnO Thin Films Prepared by RF Magnetron Sputtering," J. Supercond. Nov. Magn., vol. 23, no. 6, pp. 905–908, Jan. 2010.

[14] R. Boukhicha, C. Charpentier, P. Prod'Homme, P. R. i Cabarrocas, J.-F. Lerat, T. Emeraud, and E. Johnson, "Influence of sputtering conditions on the optical and electrical properties of laser-annealed and wet-etched room temperature sputtered ZnO:Al thin films," Thin Solid Films, pp. 8–12, Aug. 2013.

[15] C.-C. Chen, C.-C. Tsao, Y.-C. Lin, and C.-Y. Hsu, "Optimization of the sputtering process parameters of GZO films using the Grey–Taguchi method," Ceram. Int., vol. 36, no. 3, pp. 979–988, Apr. 2010.

[16] N. Manjula, K. Usharani, A. R. Balu, and V. S. Nagarethinam, "Studies on the Physical Properties of three Potentially important TCO Thin Films fabricated by a Simplified Spray Technique under same Deposition conditions," vol. 6, no. 1, pp. 705–718, 2014.

[17] D.-H. Lee, K. Kim, Y. S. Chun, S. Kim, and S. Y. Lee, "Substitution mechanism of Ga for Zn site depending on deposition temperature for transparent conducting oxides," Curr. Appl. Phys., vol. 12, no. 6, pp. 1586–1590, Nov. 2012.

[18] Y.-D. Ko, K.-C. Kim, and Y.-S. Kim, "Effects of substrate temperature on the Ga-doped ZnO films as an anode material of organic light emitting diodes," Superlattices Microstruct., vol. 51, no. 6, pp. 933–941, Jun. 2012.

[19] S. Lee, S.-H. Kim, Y. Kim, A. I. Kingon, D. C. Paine, and K. No, "Structural and electrical properties of transparent conducting Al2O3-doped ZnO thin films using off-axis DC magnetron sputtering," Mater. Lett., vol. 85, pp. 88–90, Oct. 2012.

[20] G. Gonçalves, E. Elangovan, P. Barquinha, L. Pereira, R. Martins, and E. Fortunato, "Influence of post-annealing temperature on the properties exhibited by ITO, IZO and GZO thin films," Thin Solid Films, vol. 515, no. 24, pp. 8562–8566, Oct. 2007.

[21] X.-Y. Li, H.-J. Li, Z.-J. Wang, H. Xia, Z.-Y. Xiong, J.-X. Wang, and B.-C. Yang, "Effect of substrate temperature on the structural and optical properties of ZnO and Al-doped ZnO thin films prepared by dc magnetron sputtering," Opt. Commun., vol. 282, no. 2, pp. 247–252, Jan. 2009.

[22] J. H. Park, "Deposition-Temperature Effects on AZO Thin Films Prepared by RF Magnetron Sputtering and Their Physical Properties," vol. 49, no. December, pp. 584–588, 2006.

Memristor To Control Delay Of Delay Element

Siti Musliha Ajmal Binti Mokhtar
Faculty of Electrical Engineering
Universiti Teknologi MARA (UiTM)
Selangor, Malaysia

Wan Fazlida Hanim Abdullah
Faculty of Electrical Engineering
Universiti Teknologi Mara (UiTM)
Selangor, Malaysia

Abstract— In this work, a SPICE memristor model is combined with CMOS element in order to use the memristor resistance to regulate delay of a current starved inverter. Since memristor resistance changes according to supply voltage, the changes of memristor resistance have been studied and demonstrated using a memristor spice model. A pulse signal is supplied to the memristor to control the resistance of the memristor where the resistance changes according to the pulse width. Due to changes of resistance the current also changes and regulates the delay of delay element. The proposed SPICE model and CMOS combination is important for other memristor application studies.

Keywords— *memristor; memristor spice model; memristor application; memristor resistance*

I. INTRODUCTION

The theory of memristor is originally reported by Leon Chua in 1971 [1-2]. However, memristor remains as theory only until the first physical memristor model is realized by Stan William and his group from HP lab [3-4]. The team has built a nanoscale TiO_2 device that share one common point with memristor, it also exhibits pinched hysteresis loop. William also explains about memristor switching mechanism in the paper. Since the memristor physical model discovery, researchers have been trying to understand the new device and develop its potential applications.

Memristor is used as non-volatile memory device in work [5-6] with high density and speed as fast as DRAM. Memristor switching ability is applied in logic circuit [7] where the resistance is represented as logical states. In nueromorphic system, memristor mimics the role of synapses, where each device may interact with other devices throughout the system [8]. Several programmable analog circuits demonstrating memristor based programming of threshold, gain, and frequency are proposed in work [9]. Another example of memristor-based programmable resistance is reported in work [10].

However, despite such various memristor applications, memristor is not yet available in current market due to the cost and technical difficulties in fabrication nanoscale device. Few papers [11-13] have proposed SPICE macro model to analyze the behavior of memristor. Although these spice models are accurate and the parameters are changeable, they are only intended to exhibit memristor behavior and not applicable for memristor application. In order to study various applications

of memristor, emulators of memristor have to be developed such in work [9, 14-16]. Most memristor emulator circuits are built using CMOS elements and only take the importance of exhibiting pinch off behavior of the memristor. However, since memristor emulators use CMOS elements, they are process dependent and only exhibit memristor pinched off behavior that comparable to memristor found by HP lab. It is hard to adjust the behavior of memristor for any other fabricated memristor behavior.

In this work, a SPICE memristor model is used to exhibit memristor behavior and combined with CMOS element in order to use the behavior for memristor application. Since memristor resistance changes according to supply voltage, the resistance of memristor is used to regulate the delay of a delay element. The idea to control memristor resistance is using pulse signal that increases the resistance according to pulse width. The current from memristor is copied to current mirrors using an opamp and resistor. A capacitor stores charge in order to keep the resistance information when voltage supply is zero. The voltage from capacitor, V_c is converted into control current, I_c. The control current, I_c then regulates the delay of the delay element.

This paper is divided into 5 sessions including the introduction and conclusion. Session 2 briefly explains on the study of memristor variable resistance that is demonstrated using a memristor spice model. Next, session 3 explains on memristor spice model combined with CMOS element to for memristor application. Session 4 describes on memristor to control the delay of delay element that consists of a current starved inverter.

II. MEMRISTOR SPICE MODEL AND VARIABLE RESISTANCE

One of distinguished memristor feature is the ability to change its resistance when voltage/current is supplied to the memristor. In this session, the resistance of memristor is studied using memristor spice model. To demonstrate memristor SPICE model behaviour, a SPICE model of memristor is adopted from [12].

A. Resistance of Memristor Spice Model

First, the memristor spice model is simulated to get the current-voltage (I-V) characteristic of the memristor. Based on relation between the memristor voltages and current from the ohm's law, the memristor resistance is

$$v(t) = R_{MEM}(x)i(t) \qquad (1)$$

The resistance is calculated by dividing voltage over current from equation 1. To demonstrate the memristor behavior and to get the resistance value, memristor spice model from [3] is adopted with spice model parameters as R_{ON}=100, R_{OFF}=16k, R_{INIT}=11k, D=10N, uv=10F and p=10. Fig 1 and Fig 2 shows I-V characteristic of the spice model and the resistance value based on I-V derivation. For easier explanation, the resistance range is only when supply voltage increase from 0 to 1.2 V.

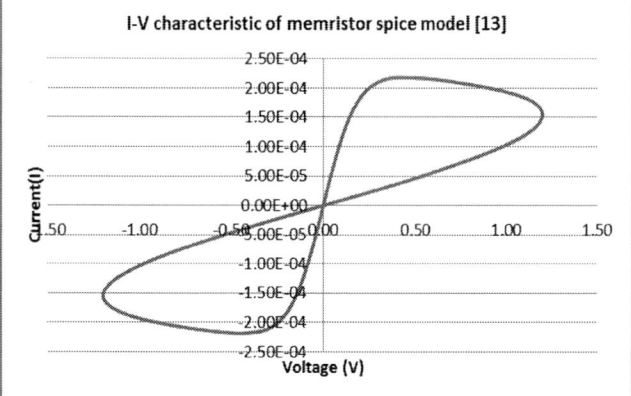

Fig. 1. I-V characteristic of memristor spice model adopted from [13]. Supply voltage is 1 V and frequency is 1 Hz.

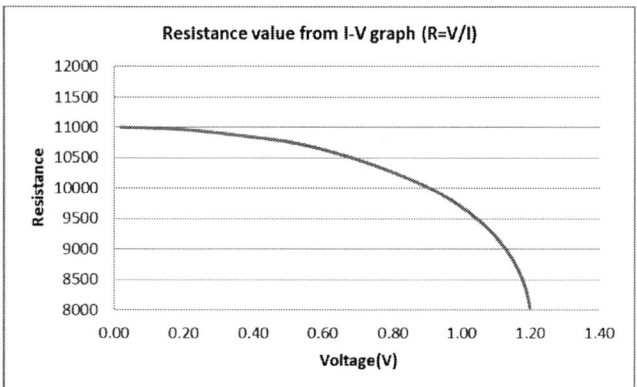

Fig. 2. Resistance of memristor from I-V characteristic data where R= V/I.

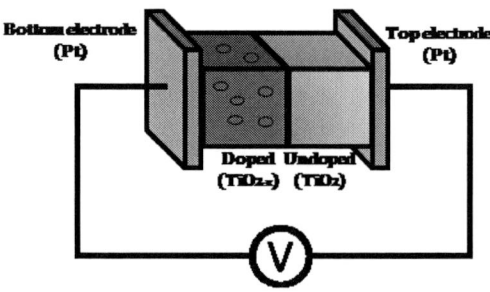

Fig. 3: TiO₂ thin film memristor structure

Fig 3 illustrates the conceptual view of single memristor structure: a thin titanium dioxide (TiO₂) between two metal wires (Pt). The inside TiO₂ thin film has formed into bilayer. The first layer is pure titanium oxide layer. It retains its natural form as an insulator. It is also indicated as undoped region and has high resistance. The second layer is oxygen deficient titanium oxide (TiO₂₋ₓ) layer. It has higher conductivity due to existence of positively charged oxygen vacancies, thus it acts as a conductor. It is also indicated as doped region and has low resistance. When voltage increase from 0V to 1.2V, the oxygen vacancies in the doped TiO₂₋ₓ layer are repelled, moving them towards the undoped TiO₂ layer. As a result, the boundary between the two materials moves, causing an increase in the percentage of the conducting TiO₂₋ₓ layer thus the resistance decreases. This is how a memristor varies its resistance according to the voltage supply.

B. Memristor Variable Resistance by Pulse Signal

As stated before, memristor resistance changes according to supply voltage. Two main factors that control the memristor resistance are the amplitude of the voltage supply and times memristor is supplied with voltage. Fig 2 shows how the resistance changes when amplitude of voltage supply changes. The other factor, time of voltage supplied to memristor is shown in Fig 4. Here, 1.2 V voltage supply is applied to the memristor for 3 sec. Graph shows that resistance decreases over times.

One way to control memristor resistance is proposed in work [10] where pulse signal input is applied to memristor to regulate the memristor resistance. According to [17], memristor resistance change rate depends on the number of pulse, pulse amplitude, and pulse width. High number of pulse, high pulse amplitude value and wide pulse width result in higher resistance. Therefore, to control the memristor resistance, it is achievable by regulating these three factors. A pulse input voltage is applied to the memristor spice model to demonstrate the resistance changes. In Fig 5, pulse input voltage with different widths is applied to the memristor. Smaller pulse width results in smaller resistance change rate and otherwise. Therefore, in this work, pulse input voltage is applied to memristor to control its resistance.

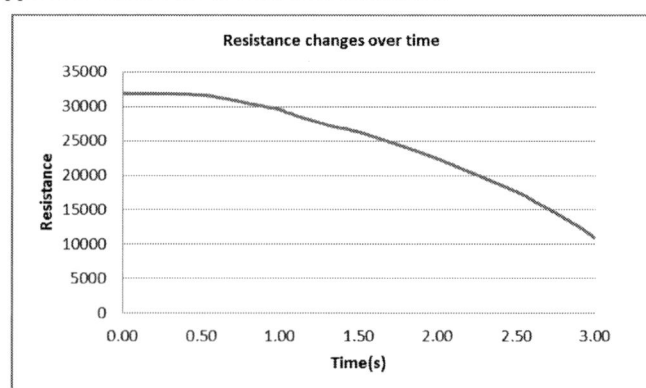

Fig. 4. Resistance changes over time where 1.2V voltage applied to memristor.

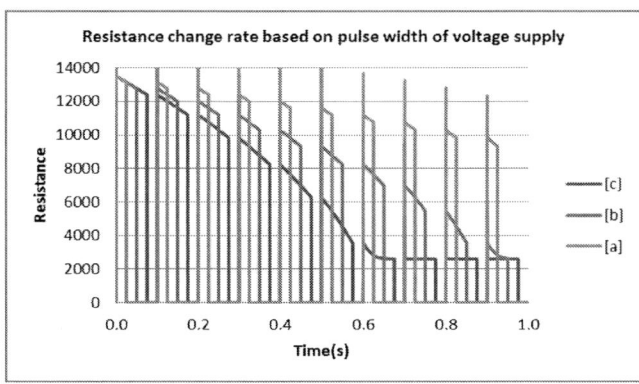

Fig 5. Memristor resistance changes for pulse input voltage. Pulse width: (a) 75ms, (b) 50ms and (c) 25ms

III. MEMRISTOR MODEL COMBINED WITH CMOS ELEMENT

The circuit of memristor model with CMOS is shown in Fig 6. As explain before, memristor resistance change rate increases according to width of pulse signal. Since resistance changes, memristor current i_{MEM} also change. The memristor current and resistor R1 generates opamp input voltage.

$$v_{in}(t) = i_{MEM}(t)R_1 \qquad (2)$$

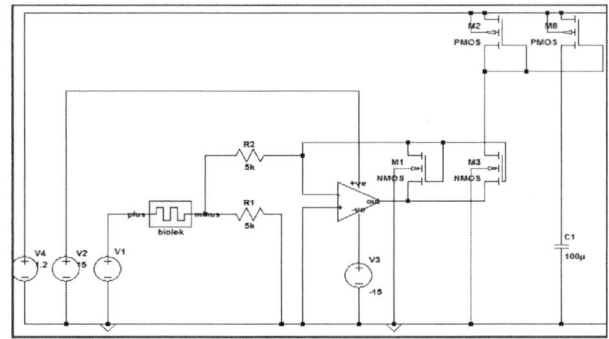

Fig 6. Memristor spice model with CMOS element for memristor application. Current from memristor is copy to the current mirror using opamp.

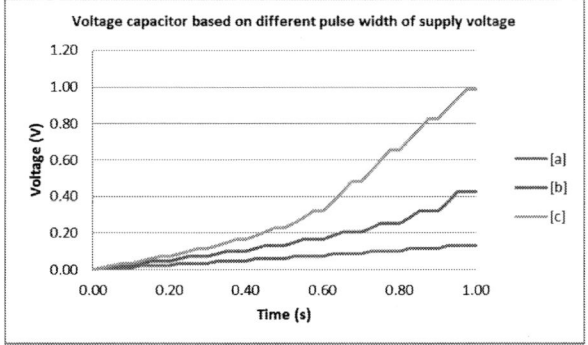

Fig. 7. Voltage capacitor, V_C for three different pulse widths. Pulse width: (a) 75ms, (b) 50ms and (c) 25ms.

Resistor R2 is to convert input voltage into an opamp input current i_{in}. The opamp input current acts as current source and

then copied using current mirror, consists of transistor M1-M3. In order to keep the programmed memristor resistance R_{MEM} before new resistance input is presented, the charge from i_{MEM} is stored by capacitor C1. Fig 7 shows voltage at capacitor C1, Vc. Since memristor model cannot retain its current when input voltage is 0, the resistance information is kept as charge that stored by the capacitor.

IV. MEMRISTOR TO CONTROL DELAY OF DELAY ELEMENT

Using the memristor model combined with opamp, resistor and capacitor, new memristor application as delay element is proposed. The proposed design in shown is Fig 8.

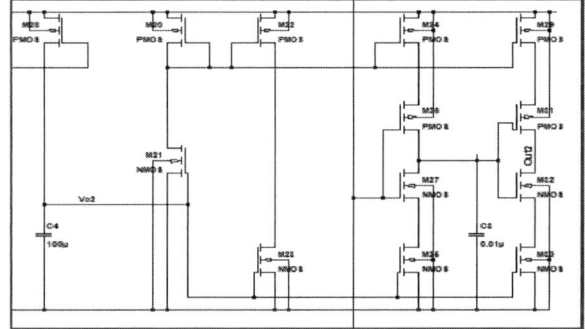

Fig 8. Memristor spice model with CMOS element to control delay of a current starved inverter.

First, voltage from capacitor, Vc is converted into control current, Ic. The control current, Ic then regulates the delay of the delay element. Here, the delay element consists of current starved inverter. The gate voltage of M24 and M26 control the amount of current flow to middle inverter that has parasitic capacitor, C3. The delay is determined by size of capacitor and amount of (dis)charging current, I_{DS} as shown in equation (3).

$$t_{delay} = C\frac{V_{TN}}{I_{DS}} \qquad (3)$$

$$I_{DS} = I_C \qquad (4)$$

As shown in above equation, delay is regulated by controlling I_{DS} since C is constant value when transistor in saturation region. Assuming circuit is in saturation region with no channel length modulation, the I_C is shown below.

$$I_C = \frac{1}{2}\mu_n C_{OX}\frac{W}{L}\left(V_C - V_{TH}\right)^2 \qquad (5)$$

Equation (5) is inserted in equation (3) to roughly estimate the delay.

$$t_{delay} = C\frac{V_{TN}}{\frac{1}{2}\mu_n C_{OX}\frac{W}{L}\left(V_C - V_{TH}\right)^2} \qquad (6)$$

Equation 6 shows how memristor regulates delay of delay element. Result for the delay element with memristor is shown in Fig. 9 (a), (b) and (c). There are three different outputs according to three different pulse width of supply voltage that applied to the memristor. In fig 9 (a) pulse width of supply voltage is 25ms and output signal takes longest time to generate full output signal. In fig 9 (b) pulse width of supply voltage is 50ms and output signal takes shorter time to generate full output signal. In fig 9 (c) pulse width of supply voltage is 75ms and output signal takes shortest time to generate full output signal. The small pulse width of voltage supply results in small resistance change rate and therefore the current produce by memristor is small. Thus, voltage at capacitor, V_c is small. According to equation (6), since Vc is small, the delay is bigger and otherwise.

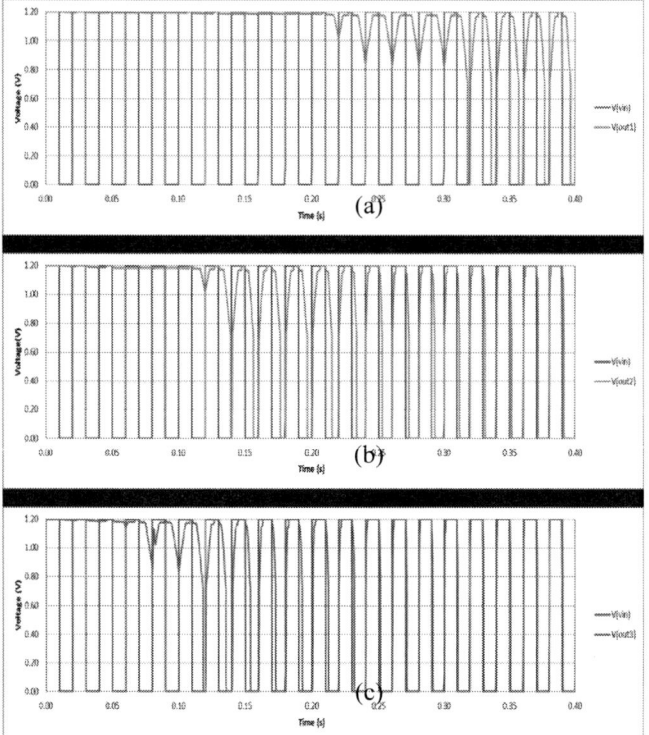

Fig 9. Memristor spice model with CMOS element to control delay of a current starved inverter. Pulse width: (a) 75ms, (b) 50ms and (c) 25ms.

V. CONCLUSION

In this work, resistance changes of memristor has been studied and demonstrated using a SPICE memristor model. One way to control the resistance is using pulse signal for voltage input that applied to the memristor. A small width of pulse input results in small memristor change rate. Next, memristor is combined with opamp, resistor and capacitor to regulate delay of a current starved inverter. The proposed SPICE model and CMOS combination is important for memristor application that works together with CMOS elements.

ACKNOWLEDGMENT *(Heading 5)*

The authors would like to thank the Ministry of Science, Technology and Innovation (MOSTI) and the Research Management Institute of Universiti Teknologi MARA (UiTM) for providing the financial support under the Science Fund grant (100-RMI/SF 16/6/2 (31/2012).

REFERENCES

[1] L.O. Chua, "Memristor-the missing circuit element," IEEE Trans. on Circuit Theory, vol. 18, pp. 507-519, 1971.

[2] L. O. Chua and S. M Kang, "Memristive devices and systems," Proc. Of IEEE, vol. 64, no. 2, pp. 209-223, Feb. 1976.

[3] D.B. Strukov, G.S. Snider, D.R. Stewart and R.S. Williams, "The mising memristor found," Nature (London), vol. 453, pp. 80-83, 2008.

[4] R. S. William, "How we found the missing memristor, " IEEE Spectrum, p. 1-11,2008

[5] Y.C. Yang, F. Pan, Q. Liu, M. Liu, F. Zeng, "Fully roomtemperature-fabricated nonvolatile resistive memory for ultrafast and high-density memory application," Nano lett., vol. 9, pp. 1636-1643, 2009.

[6] Davide Sacchetto, Giovanni De Micheli and Yusuf Leblebici, "Multiterminal Memristive Nanowire Devices for Logic and Memory Applications: A Review," Proceedings of the IEEE, vol. 100, No-6, pp. 2008-2020, 2012

[7] S. Kvatinsky, A. Kolodny, U. C. Weiser and E.G. Friedman, "Memristor-based IMPLY Logic Design Procedure, " IEEE 29th International Conference on Computer Design (ICCD), pp. 142-147, 2011.

[8] H. Wang, H. Li, R.E. Pino, "Memristor-Based Synapse Design and Training Scheme for Neuromorphic Computing Architecture, " The 2012 International Joint Conference on Neural Networks (IJCNN), pp. 1-5, 2012

[9] Y.V. Pershin, M.D. Ventra, "Practical Approach To Programmable Analog Circuit With Memristors," IEEE Transactions on Circuits and Systems, vol. 57, Issue 8, pp. 1857-1864, 2010

[10] S. Shin, K. Kim and Sung-Mo Kang, "Memristor-based Fine Resolution Programmable Resistance and Its Applications", International Conference on Communications, Circuits and Systems, pp. 948–951,2009.

[11] Y.N. Joglekar, S. J. Wolf, "The elusive memristor: properties of basic electrical circuits," Eur. J. Phys. 30, pp. 661-685,2009.

[12] Z. Biolek, D. Biolek, and V. Biolkova, "Spice model of memristor with nonlinear dopant drift," Radioengineering, vol. 18, no. 2, pp. 210–214, 2009.

[13] T. Prodromakis, B. P. Peh, C. Papavassiliou, and S. Member, "A Versatile Memristor Model With Non-linear Dopant Kinetics," vol. 58, no. 9, pp. 1–7,2011.

[14] Jung CM, Jo KH, Min KS: SPICE macromodel and CMOS emulator for memristors. J Nanosci Nanotechnol 2012, 12(2):1487–1491.

[15] Kim H, Sah MP, Yang C, Cho S: Memristor emulator for memristor circuit applications. IEEE Trans Circuits and Syst-I 2012, 59(10):2422–2431.

[16] Choi JM, Shin SH, Cho SI, Min KS: CMOS circuit with small area and low complexity for emulating memristive behavior. In Collaborative Conference on 3D & Materials Research (CC3DMR). Jeju in Korea; 2013.

[17] A.G. Radwan, M.A. Zidan, K.N. Salama,"HP memristor mathematical model for periodic signals and DC," 53rd IEEE International Midwest Symposium on Circuits and Systems (MWSCAS), pp. 861-864, 2010

Design and Fabrication of PCB Based Planar Micro-coil For Magnetic MEMS Actuator

Roer Eka Pawinanto, Jumril Yunas, Muzalifah Mohd. Said, Mimiwaty Mohd. Noor, Burhanuddin Yeop Majlis

Institute of Microengineering and Nanoelectronics (IMEN)
Universiti Kebangsaan Malaysia (UKM)
43600 UKM Bangi, Selangor, Malaysia
Email : jumrilyunas@ukm.my

Abstract— **This paper describes the design and fabrication of planar micro-coil on printed-circuit board (PCB)for magnetic MEMS actuator. A simple and cost effective technique for the fabrication of planar micro-coil is presented. The design and analysis process have been carried out by using COMSOL Multiphysics 4.2. Three different coil dimensions such as width/space= 150μm/100μm; 100μm/100μm and 50μm/100μm, have been analyzed in order to find the optimum coil geometry. The results showed that the coil with dimension of width/space of 50μm/100μm produced the highest magnetic flux density of max 0.0068 T. The optimized design of the planar micro-coil can be used as a reference for the future fabrication of magnetic MEMS actuator.**

Keywords—*Planar micro-coil; magnetic actuator; MEMS*

I. Introduction

Planar micro-coil is one of the most important parts of magnetic MEMS devices that requires thin and compact device structure [1]. As compared to the solenoid coil model, the planar structure can reduce the device volume up to 10 times lower. One of the potential applications of the planar micro-coil is magnetic field generator for actuating purposes or particle capturing in lab on chip systems.

In magnetic actuator, the generated force for mechanical actuation of flexible membrane can be produced through interaction between a current carrying planar micro-coil part on PCB and magnetic field generated by a permanent magnet part. Both part i.e. PCB-based planar micro-coil and a permanent magnet that is attached on a flexible membrane stick together to form an integrated MEMS device. In this case, planar micro-coil has been applied in various magnetic actuator devices for micropump [2, 3], microspeakers [4], flexible beams [5], and relays actuation [6, 7].

Recently, various techniques to realize planar micro-coils have been reported. The surface micromachining has been the common technique in fabricating a planar structure that includes, sputtering of thick copper coil, micromolding using SU-8 and electroplating [8, 9, 10, 11]. Metal deposition technique using sputtering and micromolding are still more expensive compared to planar micro-coil fabricated on electroplated copper material on the PCB.

In other hand, the dimension of planar micro-coil has significantly influenced the magnetic actuation capability. Therefore, this work is also done to find optimum geometry of the coil structure that can generate high magnetic field with maximum magnetic flux and force generation. The results of this study will provide the important parameters that are required for the future fabrication of magnetic MEMS actuator i.e the fabrication of micro-pump integrated in Lab on Chip.

II. Design of Planar Micro-coil

The function of the electromagnetic micro-coil is to generate a magnetic flux when electrical current is supplied to the system. The planar electromagnetic coil produces a magnetic force upon interaction with the magnetic field generated by a permanent magnet. The magnetic force produced can be calculated as equation 1:

$$F_z = B_r \int_z^{z+h_m} S_m \frac{\partial H_z}{\partial z} dz \quad (1)$$

Where H_z is the vertical magnetic field generated by the coil, B_r is the magnets remanence, S_m and $\partial H_z/\partial z$ is the area and the gradient of magnetic field along vertical direction, respectively.

The coil designs are considered as an important step before fabricating the planar micro coil. In this work, basic characteristics of the planar coils such as resistance, inductance and magnetic flux are analyzed by using COMSOL Multiphysics 4.2. The coil resistance is calculated by using equation 2:

$$R = \rho \frac{l}{A} \quad (2)$$

Where, R is the resistance value, ρ is the resistivity of conductor material, l is the length of the coil, and A is the cross section area of the coil. The coil length can be calculated as equation 3:

$$l = 2\pi r \quad (3)$$

Where, l is the length of the coil and r is coil radiant. Meanwhile, the cross-section area of the coil can be calculated as equation 4:

$$A = w.h \quad (4)$$

Where, A is the cross-section area of the planar coil, w is the width of planar coil and h is the thickness of the coil. Based on our design of the planar micro-coil configuration, the total coil resistance as same as like the parallel resistance value of the whole system.

Three different dimension parameters of planar micro-coil have been designed and fabricated in this work as shown in Table 1. The schematic diagram of the micro-coil geometry as shown in Fig. 1. The coil traces are designed with a parallel configuration. The design configuration enables the structure fabrication to be fabricated at only on 1 layer without interconnection.

Table 1. Dimension of micro-coil

Sample	Width (µm)	Space (µm)	Inner diameter (µm)	Outer diameter (µm)
Type 1	150	100	2500	5000
Type 2	100	100	2500	4500
Type 3	50	100	2500	4000

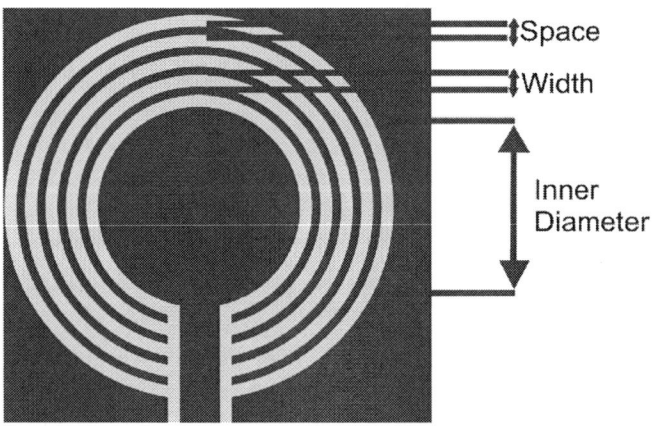

Fig.1. Design of micro-coil: type 1, 2 and 3: inner diameter= 2500 µm, s/w= 100 µm/150 µm; 100 µm/100 µm; 100 µm/50 µm and outer diameter depends on the s/w ratio.

III. Fabrication

Fig. 2 shows the steps of the fabrication process of the PCB based planar micro-coil. The cleaning process starts with the PCB dipped in an acetone and methanol for a 10 minute ultrasonic cleansing. After that, the PCB is rinsed with DI water before being dip into 10% HCl for 10 seconds. Lastly, the PCB is rinsed with DI water again and dried with a nitrogen blow.

The coil pattern is transferred to the PCB by the photolithography process. Firstly, AZ 4620 photoresist is applied on the PCB surface. The photoresist is then exposed to UV light for 80 seconds. The coil pattern is then produced after the developing process by dipping the PCB into 1:2 (AZ400K: water) developer.

For the etching process of the unwanted area of the PCB copper, the PCB is dipped in 70 % FeCl$_3$ solution. The copper etching rate is about ±0.3 µm/min. 4 % HCL has to be added into FeCl$_3$ solution in order to increase the etching rate up to ±3.9 µm/min [12]. The three types of planar micro coil are finally formed after this process.

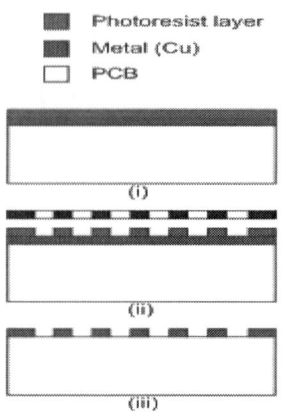

Fig. 2. Coil fabrication process. (i) Photoresist coating. (ii) Photoresist patterning. (iii) Copper etching and photoresist removal

The reactions formula for etching of copper with ferric chloride is $Cu_{(s)}+2Fe^{3+}+3Cl_3^- \rightarrow CuCl_3^- +2Fe^{2+}$. The Fe^{3+} ions are reduced to Fe^{2+}, and remain in the solution, while the copper metal is oxidized to Cu^{2+}. FeCl$_3$ dissociates in water to generate Fe^{3+} and Cl^- ions.

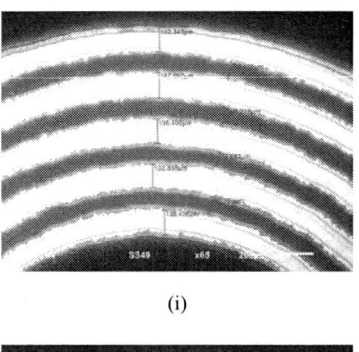

(i)

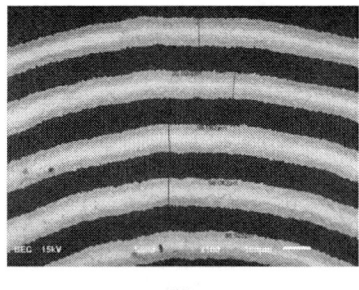

(ii)

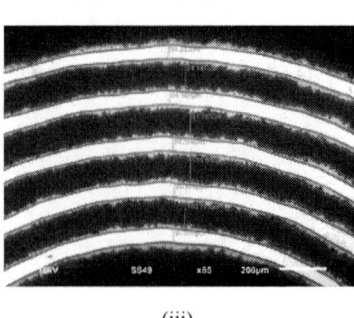

(iii)

Fig. 3. SEM image of planar micro-coil: (i) Type 1, (ii) Type 2, (iii) Type 3

Fig. 3 shows the dimension of the fabricated three types of planar micro-coil. Type 1 was etched for 1 hour 14 minutes, while the micro-coil type 2 and 3 were etched for 1 hour 20 minutes and 1 hour 30 minutes, respectively. The sidewalls of the micro-coil were over etched due to the prolonged immersion, resulting in a grossly wide and irregular pattern of micro-coil.

IV. Result and Discussion

Fig. 4 shows the magnetic flux density of the planar micro-coil type 1 simulated using COMSOL Multiphysics 4.2. It is shown that the generated magnetic flux density of maximum 0.0035 T was observed at inner side of the coil structure. It has been also shown that the magnetic flux density has a maximum value at the edge of the coil due to the interaction of generating magnetic flux density at each turn. Meanwhile, the magnetic flux density generated by a planar micro-coil of type 2 and type 3 are 0.0045 T and 0.0069 T, respectively.

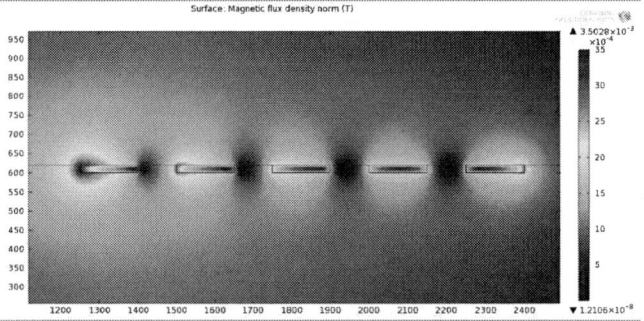

Fig. 4. Simulation result of magnetic flux density micro-coil type 1

To compare the simulation and fabrication results, the fabricated micro-coils have been measured by giving constant DC current of 0.5 A. The inductance was measured by using LCR meter Agilent 4284A and the magnetic flux density was measured by using Tesla Meter 3B scientific U33110. As for the planar micro-coil resistance the measurement was done by using a multi-meter Sanwa CD771.

The measurement results are shown in Table 2. The planar micro-coil type 3 showed the highest inductance value among the others which is 3.6 μH. Meanwhile, the planar micro-coil type 2 and type 1 showed the same inductance value of 3.5 μH each, possibly due to DC-current consumption was supplied to the coil area has the same value.

For the resistance measurements, the planar micro-coil type1 showed a resistance value of 0.2 Ω which is the lowest among the other samples. Based on the equation (2), the resistance value is inversely proportional to the cross sectional area of the coil. However, the resistance value is proportional to the length of the coil. The high resistance value can lead to the overheating effect that can reduce the device efficiency and cause damage to the device.

Table 2. Measurement Result

Sample	Inductance (μH)	Resistance (Ω)	Magnetic flux (T)
Type 1	3.5	0.2	0.00351
Type 2	3.5	0.3	0.004
Type 3	3.6	0.3	0.0068

Fig. 5 shows the comparison results of the simulation and measurement of the generated magnetic field of planar micro-coil. The measured magnetic field graph line has a similar trend to the simulated magnetic field graph line with a deviation value of ±4.96 %. Any difference results between the simulation and measurement are may be due to the measurement losses such as parasitic effect of the substrate and metal, inner resistance of the measuring instrument and wire losses that were not considered in the simulation.

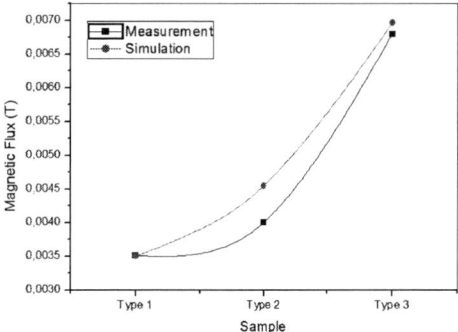

Fig. 5. Comparison of simulation and measurement magnetic flux density result

V. Conclusion

In this paper, the design and fabrication process of PCB based planar micro-coil for magnetic MEMS actuator application has been discussed. The result showed that the planar micro-coils can be easily fabricated by utilizing the electroplated copper layer on the PCB. The process is simpler compared to the deposit copper metal using a sputtering process or planar coil having interconnections.

In this paper, the design and fabrication process of PCB based planar micro-coil for magnetic MEMS actuator application has been discussed. A simple technique of fabrication planar micro-coil using PCB has been presented.

The measurement results showed that the copper micro-coil patterned on the PCB has a good conductivity with low resistance value. The magnetic flux density of 6,8 mT was attained/recorded and the value is significant to generate magnetic force for the electromagnetic actuator on a membrane. The generated magnetic flux density between simulation and measurement results were compared and a small deviation value about 4.96% resulted.

Acknowledgment

We would like to thank Ministry of Science and Technology for the research grant under the project 03-01-02-SF0841 (Development of Intergrated Electromagnetic Micro-pump Based on Embedded Planar Micro-coil for Ultra Low Fluid Injection of Bio-Samples).

References

[1] A. Beyzavi and N-T Nguyen, "Modeling and optimization of planar micro-coils," J. Micromechanics Microengineering, vol 18, no. 9, pp095018, Sep. 2008.

[2] A.T. Al-Halhouli, M.I. Kilani, and S. Büttgenbach, "Development of a novel electromagnetic pump for biomedical applications," Sensors Actuator A Phys, vol 162, no. 62, pp 172-176, Aug. 2010.

[3] H-L. Yin, Y-C. Huang, W. Fang, and J. Hsieh," A novel electromagnetic elastomer membrane actuator with a semi-embedded coil," Sensors and Actuators A, vol. 139, no. 1-2, pp 194–202, Sep. 2007.

[4] J.H. Kwon and S.M. Hwang, "Analysis of acoustic characteristics according to design parameter of diaphragm," Journal of Materials Processing Technology, vol. 187-188, pp 442-446, Jun. 2007.

[5] C.T. Pan, P.J. Cheng, C.K. Yen, and C.C. Hsieh," Application of polyimide to bending-modemicroactuators with Ni/Fe and Fe/Pt magnet," Microelectronics Reliability, vol. 46, no. 8, pp 1369–1381, Aug. 2006.

[6] X. Miao, X. Dai, P. Wang, G. Ding, and X. Zhao,"Design, fabrication and characterization of a bistable electromagnetic microrelay with large displacement," Microelectronics Journal, vol. 42, no. 8, pp 992–998, Aug. 2011.

[7] M. Ruan, J. Shen, and C.B. Wheeler," Latching microelectromagnetic relay," Sensors and Actuators A, vol. 91, no. 3, pp 346–350, Jul. 2001.

[8] T. Kohlmeier, V. Seidemann, S. Buettgenbach, H. H. Gatzen," An investigation on technologies to fabricate microcoils for miniaturized actuator systems," Microsystem Technologies, vol. 10, no. 3, pp. 175–181, Mar. 2004.

[9] M. Woytasik, J. Moulin, E. Martincic, A-Lcoutrot, and E. Dufour-Gergam," Microsyst Technol, vol. 14, no. 7, pp 951–956, Nov. 2007.

[10] J. Yunas, A.A. Hamzah, and B.Y. Majlis," Surface micromachined on-chip transformer fabricated on glass substrate," Microsyst Technol, vol. 15, no. 4, pp 547–552, Oct. 2008.

[11] C-Y. Lee, Z-H. Chen, H-T. Chang, C-Y. Wen, and C-H. Cheng," Design and fabrication of novel micro electromagnetic actuator," Microsyst Technol, vol. 15, no. 8, pp 1171–1177, Nov. 2008.

[12] K.R. Williams, K.Gupta, and M. Wasilik," Etch Rates for Micromachining Processing—Part II," Journal of microelectromechanical systems, vol. 12, Dec. 2003.

TiO₂-based Extended Gate FET pH-sensor: Effect of Annealing Temperature on its Sensitivity, Hysteresis and Stability

*K. A. Yusof, **S. H. Herman, W. F. H. Abdullah

Faculty of Electrical Engineering;
Universiti Teknologi MARA (UiTM);
40450 Shah Alam, Selangor, Malaysia
*khairul.aimi.yusof@gmail.com, **hana1617@salam.uitm.edu.my

Abstract—In this study, titanium dioxide (TiO₂) thin films have been investigated as a sensing membrane of the extended gate field effect transistor (EGFET) for pH detection application. The sol-gel TiO₂ has been prepared and spin coated onto the indium tin oxide (ITO) coated glass as a substrate. Then the TiO₂/ITO test structures thin films were annealed for 15 min at different temperatures; 300 °C and 400 °C under ambient atmosphere. The pH sensing characterizations of TiO₂ thin films were measured by Semiconductor Parametric Device Analyzer in different pH buffer solutions of pH 4, 7, 10 and 12. The sensitivity of TiO₂ thin film annealed at 400 °C exhibited a higher sensitivity that is 51 mV/pH compared to the thin film annealed at 300 °C gave slightly lower sensitivity of 49 mV/pH. The hysteresis and drift effect for TiO₂ thin films also being investigated in this study. TiO₂ thin films annealed at 400 °C obtain better hysteresis and drift value compared to the TiO₂ thin films annealed at 300 °C.

Keywords— extended gate field effect transistor, sol gel, titanium dioxide

I. INTRODUCTION

Since Bergveld has developed ion sensitive field effect transistor (ISFET) in 1970 [1], measurement of pH and hydrogen potential has attract much attention because it can be applied in numerous areas such as in environmental monitoring, aquaculture, agriculture, food inspection technology and also biomedical application [2]. ISFET is a solid-state device which the field-effect-transistor (FET) and sensing membrane was combined together and contact directly with the buffer solution that lead to several problems [3-4]. Thus, extended gate field effect transistor (EGFET) has been proposed by Van der Spiegel et al [5] to overcome ISFET drawbacks. EGFET configuration is the modified ISFET which the FET and sensing membrane was separated into two parts. Only the sensing membrane of EGFET is exposed directly with buffer solution while the FET device is isolated. The EGFET configuration offers some advantages like less sensitive to light, simpler packaging and low cost [6].

Silicon dioxide (SiO₂) was the first material oxide used as a pH sensing membrane [7]. Later, more oxide layer materials been introduced as pH sensing membrane such as aluminum oxide (Al₂O₃), niobium pentoxide (Nb₂O₅), tantalum pentoxide (Ta₂O₅), hafnium oxide (HfO₂), tin oxide (SnO₂) and titanium dioxide (TiO₂) [7-10] because of its good sensing properties [10]. TiO₂ has become an attractive oxide material for biosensor application and electrochemical sensor due to its promising properties like good biocompatibility, chemical stability and high dielectric constant [11].

In this work, the deposited thin films were annealed at two different temperatures to investigate the effect of annealing temperature towards the pH sensitivity, drift and hysteresis characteristics of thin films for pH detection application on EGFET.

II. METHODOLOGY

A. TiO₂ Sol-gel Preparation

The sol-gel method was used in order to prepare the TiO₂ solution. The TiO₂ solution was prepared by mixture of two solutions. For solution A, it was prepared by using three types of chemicals that is absolute ethanol (C_2H_5OH), glacial acetic acid, GAA (CH_3COOH) and titanium (IV) butoxide, TTIB ($Ti(OCH_2CH_2CH_2CH_3)_4$). All of three chemicals were mixed together and stirred at speed of 2000 rpm for one hour.

Next, the solution B was prepared by mixing the absolute ethanol (C_2H_5OH), triton x-100 $C_{14}H_{22}O$ (C_2H_4O) n(n=9-10) and de-ionized water. All of the three chemicals were mixed together and stirred at speed of 2000 rpm for one hour. Then, both solutions A and B were mixed together in one bottle and stirred for another one hour at speed of 2000 rpm at room temperature.

B. TiO₂ Thin Films Deposition

After two hours of mixed and stirred process, the solution was ready to use. The solution was deposited onto the cleaned ITO coated glass by spin coating method. The conductive ITO glass was used in this work both to act as substrate and also the conductive bottom contact to be connected to the gate of the FET. The speed and rotation time was set to 3000 rpm and 60 second accordingly. The deposited thin films then were annealed for 15 min at two different temperatures which are at 300 °C and 400 °C.

C. EGFET Measurement Setup

1) Current-Voltage (I_D-V_G) Measurement Setup

After the deposition process completed, a metal wire was bound to the uncovered ITO area of the sample substrate with silver paste and encapsulated with epoxy resin in order to avoid leakage current during measurement process. The opening sensing window of 1 cm x 1 cm was defined at the completed TiO_2 deposited thin films. This forms the extended gate which was then connected to the gate of commercial MOSFET (NDP 6060L) and was dipped together with reference electrode (Ag/AgCl) into different buffer pH buffer solutions of pH 4, 7, 10 and 12. The current-voltage (I_D-V_G) characteristics of TiO_2 thin films were measured by using Semiconductor Parametric Device Analyzer model Agilent B1500A. EGFET I_D-V_G measurement setup is described elsewhere [12]. All measurements were done in a dark box at room temperature to avoid from light and temperature effects.

2) Drift and Hyteresis Measurement Setup

The drift and hysteresis measurements were conducted by using the constant voltage constant current readout interfacing circuit in different buffer pH solutions. The drift measurement was done by taking the response of output voltage of TiO_2 thin film over sensing time (V_{OUT}-time) in 300 sec for each different buffer solutions of pH 4, 7, 10 and 12. For hysteresis measurement, the TiO_2 thin films were measured at pH loop 7 → 4 → 7 → 10 → 7 at 500 sec loop time. Fig. 1 shows the block diagram of drift and hysteresis measurement setup.

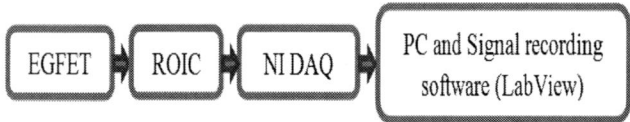

Fig. 1. Block diagram of drift and hysteresis measurement setup

III. RESULT & DISCUSSION

A. Current-Voltage (I_D-V_G) and pH Sensitivity of TiO_2 Thin Films

Fig. 2 (a) and (b) show the transfer characteristic (I_D-V_G) of TiO_2 thin films with different annealing temperature of 300 °C and 400 °C respectively at four different pH buffer solutions. Both thin films show similar behavior of transfer characteristic. From both graphs, it can be seen that the threshold voltage increase as pH value increase due to the decreasing of surface potential. As shown in Fig. 2 (a) and (b), the pH sensitivity of the TiO_2 thin films for EGFET was determined through the shift of the threshold voltage in the various pH buffer solutions.

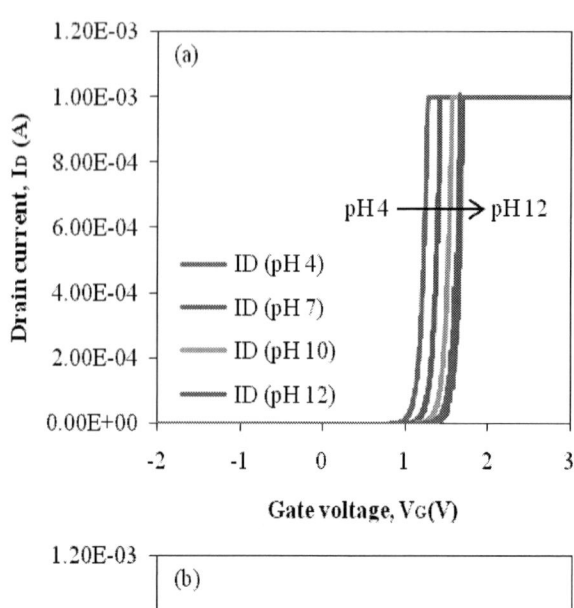

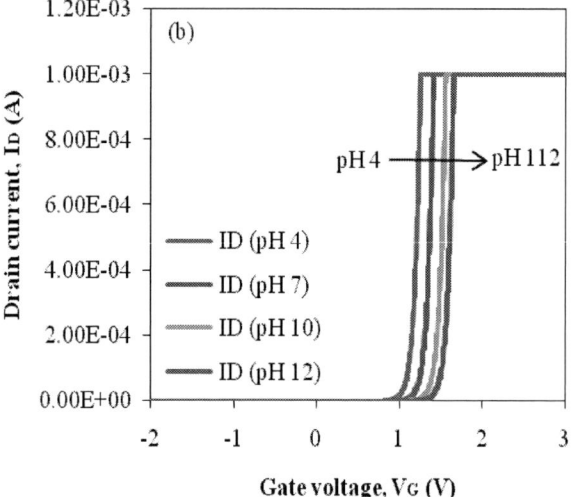

Fig. 2. I_D-V_G curve of TiO_2 thin films with annealing temperature of (a) 300 °C (b) 400 °C toward pH variations

From I_D-V_G curve, the gate voltage was determined at the drain current of 100 μA. The relationship between gate voltage, V_G and pH sensitivity under different annealing temperatures were shown in Fig. 3 (a) and (b). The slope of V_G-pH graph determined the sensitivity of thin films. At annealing temperatures of 300 °C and 400 °C, the pH sensitivities are 49 mV/pH and 51 mV/pH accordingly. Both thin films also exhibited good linearity at 0.999 which is close to 1. According to the obtained sensitivity, annealing temperature gives effect on thin film sensitivity which higher annealing temperature resulting higher sensitivity.

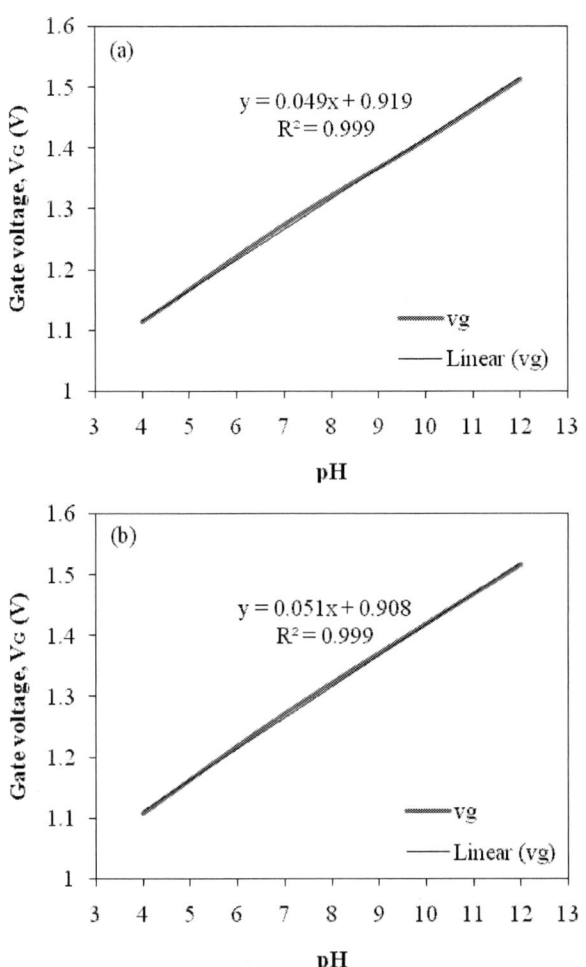

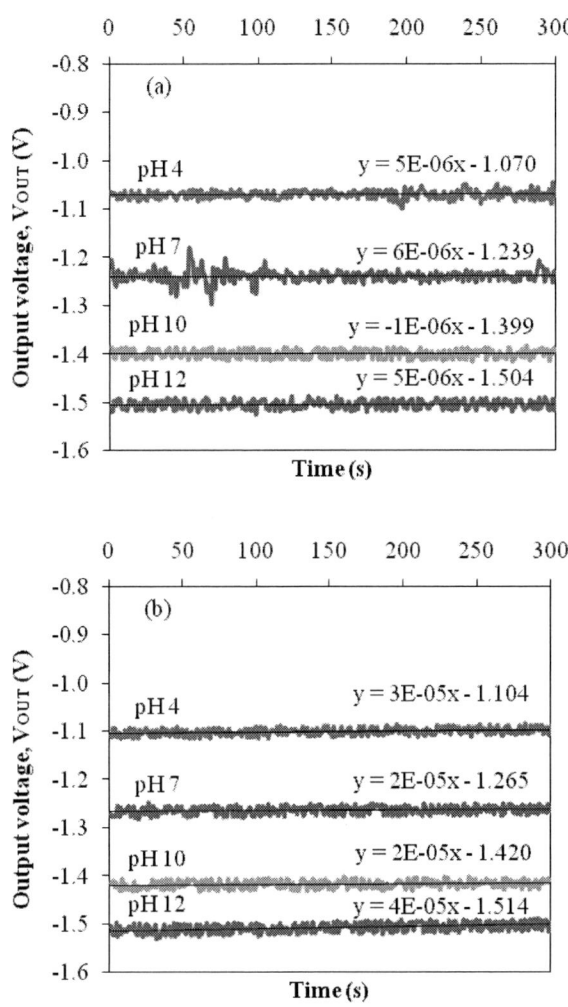

Fig. 3. Sensitivity and linearity of TiO_2 thin films with annealing temperature of (a) 300 °C (b) 400 °C toward pH variations

Fig. 4. Drift characteristics of TiO_2 thin films with annealing temperature of (a) 300 °C (b) 400 °C toward pH variations

B. Drift and Hysteresis Effect of TiO_2 Thin Films

The drift test is used to estimate durability and reliability of electrochemical sensors. Fig. 4 (a) and (b) show the response output voltage of TiO_2 thin films over sensing time (V_{OUT}-time) characteristics for 300 sec duration for each pH buffer solutions in order to observe the drift's characteristic between both samples. The drift rate of the TiO_2 thin films for both different annealing temperatures were measured in different pH buffer solution of pH 4, 7, 10 and 12 at the room temperature. Fig. 4 (a) shows the drift rate of TiO_2 thin film at annealing temperature of 300 °C at pH 4, 7, 10 and 12 are 5 µV/s, 6 µV/s, 1 µV/s and 5 µV/s respectively. Meanwhile, Fig 4 (b) shows the drift rate of TiO_2 thin film at annealing temperature of 300 °C at pH 4, 7, 10 and 12 are 30 µV/s, 20 µV/s, 20 µV/s and 40 µV/s correspondingly. From the results obtained, it show that TiO_2 thin film with annealing temperature of 300 °C has lower drift rate for all pH value compared to the TiO_2 thin film with annealing temperature of 400 °C.

Hysteresis behavior is related to the change in the pH value of the solution and the corresponding change in the output voltage of the sensor. Fig. 5 (a) and (b) show hysteresis characteristics of TiO_2 thin films with different annealing temperature of 300 °C and 400 °C. The thin films is directly been dipped in pH loop of 7 → 4 → 7 → 10 → 7 at 500 sec loop time. The hysteresis voltages were measured in pH buffer solutions with pH level changed every 100 sec in pH loop of 7 → 4 → 7 → 10 → 7. TiO_2 thin film with annealing temperature of 300 °C has hysteresis value of 40 mV while TiO_2 thin film with annealing temperature of 400 °C has hysteresis value of 20 mV. The results indicated that higher annealing temperature caused the hysteresis to decrease. Low hysteresis value indicates better long stability for pH sensing properties.

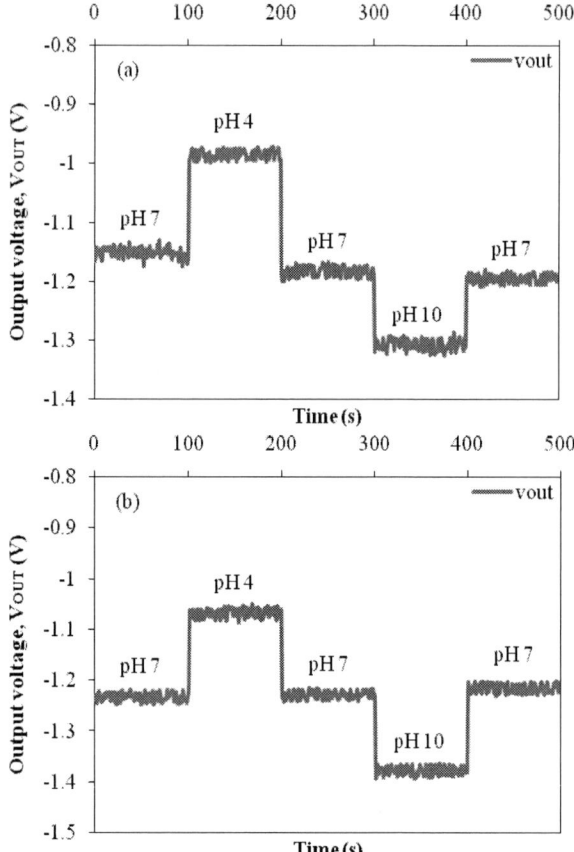

Fig. 5. Hysteresis characteristics of TiO$_2$ thin films with annealing temperature of (a) 300 °C (b) 400 °C

IV. CONCLUSION

Titanium dioxide thin films were prepared using sol-gel spin coating method with different annealing temperatures for the pH sensing membrane of the extended gate field effect transistor. There was a difference in pH sensitivity observed between the TiO$_2$ thin films anneal at 300 °C and 400 °C. TiO$_2$ thin film annealed at 400 °C exhibited slightly higher sensitivity of 51 mV/pH and linearity of 0.999 compared to TiO$_2$ thin film annealed at 300 °C indicated slightly lower sensitivity of 49 mV/pH and linearity of 0.999. The hysteresis value of thin film annealed at 400 °C showed lower value of 20 mV than the thin film annealed at 300 °C. It could be deduced that TiO$_2$ thin film annealed at 400 °C would be an option for pH sensing membrane which has acceptable sensitivity, linearity and long-term stability. Further research is required by varying more deposition and post-deposition treatment parameter in order to enhance the pH sensing properties and better long stability.

ACKNOWLEDGMENT

The authors would like to thank all members of NANO-ElecTronic Center (NET) Universiti Teknologi MARA, UiTM and to NEMS and Photonics Cluster of MIMOS Berhad for all the research facilities. The work is partially supported by Ministry of Education Malaysia under the Principal Investigator Support Initiative scheme (Project Code: 600-RMI/DANA 5/3/PSI (278/2013)).

REFERENCES

[1] P. Bergveld "Development of an ion-sensitive solid state device for neurophysiological measurement," *IEEE Trans. Biomed. Eng., BME-17*, pp. 70–71, 1970.

[2] C. Jimenez-Jorquera, J. Orozco, and A. Baldi, "ISFET based microsensors for environmental monitoring," *Sensors (Basel, Switzerland)*, vol. 10, no. 1, pp. 61–83, Jan. 2010.

[3] S. Casans, Diego Ramírez Muñoz, A.E. Navarro, A. Salazar," ISFET drawbacks minimization using a novel electronic compensation," *Sensors and Actuators B: Chemical*, vol. 99, no.1, pp. 42-49, 15 April 2004.

[4] Wen-Yaw Chung, Yeong-Tsair Lin, Dorota G. Pijanowska, Chung-Huang Yang, Ming-Chia Wang, Alfred Krzyskow, Wladyslaw Torbicz, "New ISFET interface circuit design with temperature compensation," *Microelectronics Journal*, vol. 37, no.10, pp. 1105-1114, October 2006.

[5] J. van der spiegel, I. Lauks, P. Chan, D. Babic, "The extended gate chemically sensitive field effect transistor as multi-species microprobe," *Sensors and Actuators*, vol. 4, pp. 291-298, 1983.

[6] Yun-Shan Chien; Wan-Lin Tsai; I-Che Lee; Jung-Chuan Chou; Huang-Chung Cheng, "A Novel pH Sensor of Extended-Gate Field-Effect Transistors With Laser-Irradiated Carbon-Nanotube Network," *Electron Device Letters, IEEE* , vol.33, no.11, pp.1622-1624, Nov. 2012.

[7] T.N.T. Nguyen, Y.G. Seol, N.-E. Lee, "Organic field-effect transistor with extended indium tin oxide gate structure for selective pH sensing," *Organic Electronics*, vol.12, no.11, pp. 1815-1821, 2011.

[8] Chan-Rok Park, Jin-Ha Hwang, "Effect of double-layered Al2O3/SiO2 dielectric materials on In–Ga–Zn–O(IGZO)-based amorphous transparent thin film transistors," *Ceramics International*, Available online 9 May 2014.

[9] Hung-Hsien Li; Wei-Syuan Dai; Jung-Chuan Chou; Huang-Chung Cheng, "An Extended-Gate Field-Effect Transistor With Low-Temperature Hydrothermally Synthesized SnO$_2$ Nanorods as pH Sensor," *Electron Device Letters, IEEE* , vol.33, no.10, pp.1495,1497, 2012.

[10] Chyuan-Haur Kao, Hsiang Chen, Lien-Tai Kuo, Jer-Chyi Wang, Yun-Ti Chen, Yu-Cheng Chu, Chian-You Chen, Chao-Sung Lai, Shan Wei Chang, Che Wei Chang, "Multi-analyte biosensors on a CF4 plasma treated Nb2O5-based membrane with an extended gate field effect transistor structure," *Sensors and Actuators B: Chemical*, vol. 194, pp. 419-426, 2014.

[11] Kang Li; Mingfang Zhu; Hongwu Zhang; Jie Zhao, "Electrochemical Determination of Phenylephrine Hydrochloride Based on Graphene-TiO2 Modified Glassy Carbon Electrode," *International Journal of Electrochemical Science*, vol. 8, no.3, pp. 4047-4054, 2013.

[12] K.A. Yusof, N.S. Kamarozaman, N.I.M. Noh, S.H. Herman, W.F.H. Abdullah, "Low Temperature Deposited Titania Thin Films for Extended Gate Fild Effect Transistor Sensing Membrane," presented at the 2nd International Conference on Advanced Micro and Nanocomposite for Engineering 2014, Penang, Malaysia, 2014, unpublished.

Modeling and FEM Simulation Using Fluid-Structures Interaction of Flexible Micro-Bridge Bending Within PDMS Micro-Channel

Nadir Belgroune
FUNDAPL MEMS and NEMS Physics Group
Faculty of Sciences University Saad Dahlab Blida 1
Blida, Algeria.
belgroune.nadir@gmail.com

Abdelkader Hassein-Bey
FUNDAPL MEMS and NEMS Physics Group
Faculty of Sciences University Saad Dahlab Blida 1
Blida, Algeria.
ahassei@hotmail.fr

Burhanuddin Yeop Majlis
Institute of Microengineering and Nanoelectronics (IMEN)
Universiti Kebangsaan Malaysia (UKM)
43600 UKM Bangi, Selangor, Malaysia

Mohamed El-Amine Benamar
FUNDAPL
Faculty of Sciences University Saad Dahlab Blida 1
Blida, Algeria.

Abstract— The manipulation of fluid flow in micro-channel with dimension of hundreds of micrometers has emerged as a distinct new field. In this paper, devices modeling and FEM simulation using Comsol Multiphysics of the fluid-structure interaction applied to PDMS micro-bridge deflection study under various fluid flow rates are presented. The introducing of fluid flow into the micro-channel causes the deflection of the micro-bridge. The simulation results predict a sensitive bending for low flow rate with an adapted micro-bridge dimension for each micro-channel depth. Thus demonstrate the feasibility of a new concept and open a new area of research in micro-fluidic devices and BioMEMS sensing to obtain higher sensitivity with low fluid flow rate, low coast and low power consumption.

Keywords—micro-channel; polydimethylsiloxane; PDMS; micro-bridge; fluid flow; Young's modulus; bending; FEM simulation; modeling.

I. Introduction

Micro-fluidic systems have experienced a great development during the last years, leading to more and more complex systems for different application's fields. Biology, chemistry, particle separations, bio-sensing, medical analysis, environmental monitoring, biochemical analysis, and microchemistry have grown as a range of new components and techniques have been developed and implemented for introducing, mixing, pumping, and storing fluids in micro-channels [1].

The study of fluid flow becomes interesting and challenging at such miniature channels as the characteristics of flow vary to a great degree when compared to fluid flow in pipes and tubes. The flow is no longer turbulent and mixing of fluids in such a laminar flow becomes an issue [1].

Recently a very common material used to fabricate micro fluidic systems is the PolyDiMethylSiloxane (PDMS), it

provide several main advantages, low cost, ease of processing techniques, desired mechanical properties, transparent within visible spectrum, bio and chemical-compatible [2]. Hence, PDMS has been extensively used in prototyping but also used as a final material of micro-fluidic systems.

In this context, this paper presents the devices modeling and FEM simulation using Comsol Multiphysics of the fluid-structure interaction and micro-bridge deflection under various fluid flow rates. The introducing of fluid flow into the micro-channel causes the deflection to the micro-bridge. The simulation results for the micro-bridge response demonstrate the feasibility of a new concept and open a new area of research in micro-fluidic devices and Bio-MEMS to obtain higher sensitivity with low coast and low power consumption.

The goal is to show through modeling and simulation the microscopic behavior of the micro-bridge bending under various flow rates within the micro-channel [3,4].

II. Fluid Structure Interaction

Fluid flow through a rectangle micro-channel exhibits a number of characteristic features, the most important of which is laminar flow. The velocity profile of the flow has a parabolic shape with the fastest velocity in the middle of the channel and decreasing velocity towards the channel vertical and horizontal walls as shown in Fig. 1.

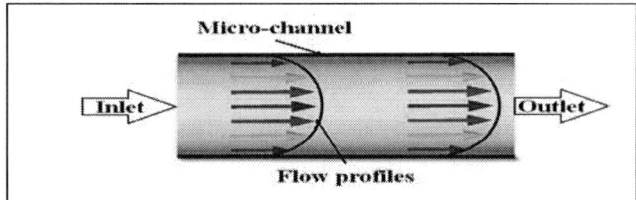

Fig. 1. Laminar flow vertical profile.

The flow is assumed to be laminar Newtonian, viscous and incompressible (material's density is constant or almost constant). The fluid motion is governed by the incompressible Navier-Stokes equations and the continuity equations [5] as follows:

$$\rho(\partial u/\partial t) - \nabla.\left(-pI + \eta\left(\nabla u + (\nabla u)^T\right) + \rho(u\,.\,\nabla)\,.u\right) = F \qquad (1)$$

$$-\nabla.\,u = 0 \qquad (2)$$

Where I is the unit diagonal matrix, $u = (u, v)$ is the velocity field, p is the fluid pressure and F is the volume force affecting the fluid. Since the gravitation and other volume forces affecting on the fluid are neglected, so $F = 0$. For each flow rate, the fluid velocity at the entrance of the micro-channel is provided as the boundary condition for the inlet. The pressure at the outlet is set to atmospheric pressure. At all other boundaries, no-slip condition and defined as $u = 0$.

The structural deformation of the micro-bridge resulting from the moving fluid is given by [6] :

$$F_T = -n.\left(-pI + \eta(\nabla u + (\nabla u)^T)\right) \qquad (3)$$

where n and F_T are the normal vector to the boundary and the fluid loading, respectively. The first term on the right-hand side of equation (3) is the pressure gradient extracted from the fluid simulation results. The second term is the viscous component of the force depending on the velocity and the dynamic viscosity of the fluid.

III. **Design and Modeling of Micro-Fluidic Devices**

The proposed structure consists of thin middle layer PDMS micro-bridge suspended within micro-fluidic device and sandwiched between bottom and top PDMS layers. It is shown in the Fig. 2. This fluid flow enters the micro-fluidic device through the inlet and exit toward the outlet, the fluid flow in the micro-channel, pressure and viscous drag causes bending of the micro-bridge suspended within PDMS micro-channel.

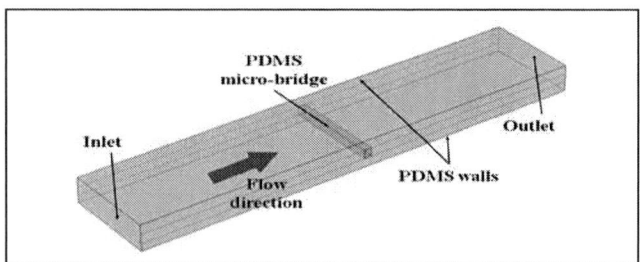

Fig. 2. Schemitic view of micro-bridge embedded in micro-channel.

In order to investigate the interaction between fluids and structures, often named as Fluid-Structure Interaction (FSI) and behavior of the micro-bridge deflection in the micro-channel flow, several simulations were carried. The estimated

bending of the middle of the micro-bridge according to the boundary and fluid loading can be obtained analytically by the following equation [4].

$$\delta = F_T L^3\left(1-v^2\right)/3EI \qquad (4)$$

where E, v and I are the Young's modulus, Poisson's ratio and moment inertia, respectively. F_T modulus is obtained numerically by solving (1) and (3).

The material used is PolyDiMethylSiloxane (PDMS). The Young's modulus value E of PDMS depends on geometric and process of fabrication. This modulus decreases strongly with thickness reduction as shown in Fig.3 [7]. Above this, the increase in the mixing ratio of the curing agent added to the PDMS base, reduce the value of E as shown in Fig.4 [8].

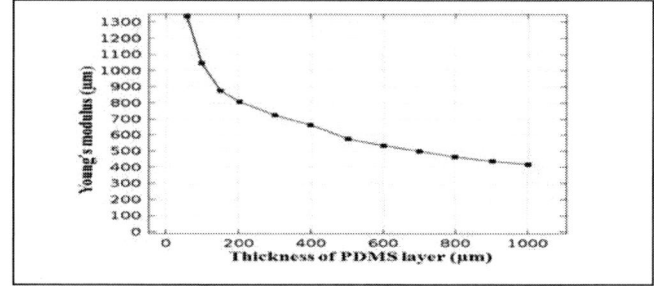

Fig. 3.   Young's modulus as function of the thickness of PDMS layer [7].

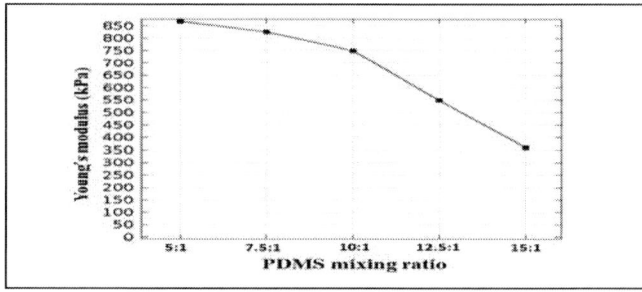

Fig. 4.   Young's modulus as function of Mixing ratio [8].

The main parameters in our model for the micro-bridge (PDMS) and the fluid (water) are provided in Table 1.

TABLE I.   MATERIALS PARAMETERS

Parameters	Materials	
	PDMS (unit)	**Water (unit)**
Young's modulus	300-900 (kPa)	--
Poisson's ratio	0.5	--
Dynamic viscosity	--	1.002×10^{-3} (Pa.s)
Density	970 (kg.m^{-3})	1000(kg.m^{-3})

IV. **FEM Simulation and Results**

This section describes the simulations and results performed to estimate the deflection of micro-bridge, mechanical response and fluid structure interaction using

COMSOL Multiphysics software. Also the analysis of the results obtained from this simulations is discussed.

A. Simulation of PDMS Micro-Bridge Bending

To understand the micro-bridge bending behavior under fluid flow rate, the micro-bridge mechanical response should be studied. The purpose of this analysis is to simulate the mechanical deformation when an external spreading force load is applied on the boundary surface. The PDMS micro-bridge has a typical length of 820 µm, a width of 70 µm and thickness of 80 µm with density of 970 kg.m^{-3}, Poisson ratio of 0.5 and Young's modulus is 803 kPa, fixed constraints at the both end boundaries (left and right side) and free at the other boundaries except the loaded boundary. The Fig.5 shows the structure bending toward the horizontal plan due to the contact pressure along X-axis direction.

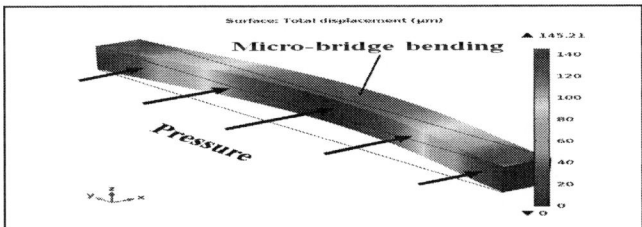

Fig. 5. Simulation of total displacement for applied load pressure.

The numerical results for the micro-bridge response to the various applied pressures are illustrated in the Fig. 6. The pressure range is from 0 to 10 kPa, the maximum deflection of the middle of the micro-bridge rise with the increasing of the applied load pressure. The sensitivity of the deflection is around slope1=0.05 µm/Pa or 50 µm/kPa under 2kPa and close to slope2=0.009 µm/Pa or 9 µm/kPa. The first part of this curve with higher sensitivity is more interesting for flow rate sensing application eventually due to the similar range of contact pressure values.

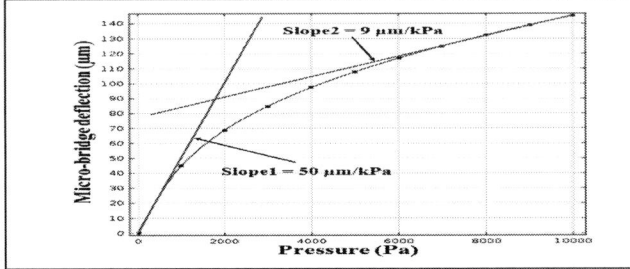

Fig. 6. Simulated deflection as function of various applied pressures.

B. Fluid-Micro-Bridge Interaction Simulation

3D simulation is performed to estimate the deflection of PDMS micro-bridge suspended within rectangular section of PDMS micro-channel under various flow rates using COMSOL including Fluid-Structure Interaction model (FSI). This simulated micro-fluidic device consists of micro-channel with one inlet, one outlet and the micro-bridge. The micro-channel's cross section has 840µm width and 240µm depth and a length of 4000µm. The micro-bridge has 840µm length (Y-axis), 40µm width (X-axis) and 80µm thickness (Z-axis). Water with density of ρ = 1000 kg.m^{-3} and dynamic viscosity

η = 0.001 Pa.s is used as the reference fluid to test the micro-bridge performance.

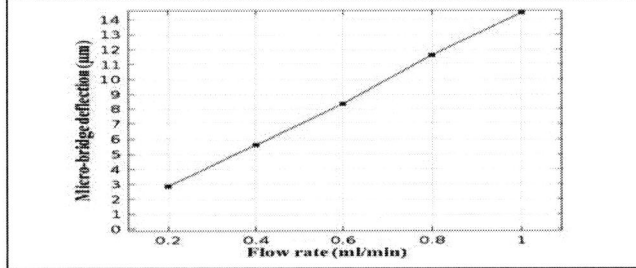

Fig. 7. Simulated deflection as function of various fluid flow rates.

Each micro-fluidic structure was studied for 5 different flow velocity rates. The deflection of the micro-bridge was given for various flow rates between 0.2-1 ml/min using FEM simulation results and is sum up as shown in Fig. 7. We can see a roughly linear behavior with an average slope of 14.5 µm/ml.min^{-1}. The pressure field magnitude in the x-z horizontal plan along the micro-channel at flow rate of 1ml/min is evaluated.

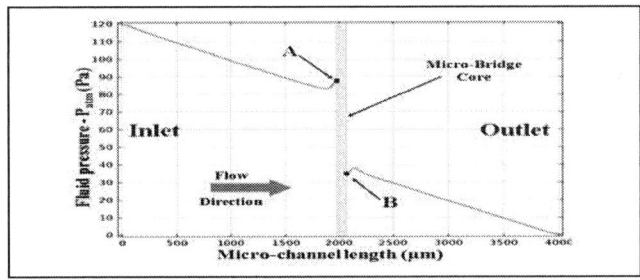

Fig. 8. Middle cut line evaluation (x-z plan) of fluid pressure magnitude along of micro-channel.

The Fig. 8 demonstrates the cut line in the middle along of micro-channel, the fluid pressure is maximum in the inlet and is zero (referenced to atmospheric pressure P_{atm}) in the outlet. Notice that the fluid pressure is the sum of two components : the static pressure when the velocity is zero and the dynamic one dues to kinetic energy of fluid flow. The fluid pressure is highest when the velocity is zero (in the region close to the micro-bridge) then stagnation pressure is at its maximum value point A (stagnation point) as shown in Fig. 8. In the opposite at the trailing edge, we can observe a weak depression at point B as shown in Fig.8.

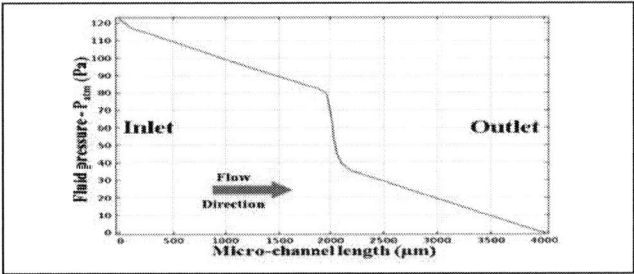

Fig. 9. Above cut line evaluation (x-z plan) of fluid pressure magnitude along of micro-channel.

In Fig. 9, we take a cut line above the middle that doesn't pass through the micro-bridge. In this case, the narrow region close to the micro-bridge the fluid pressure is not zero, and fall down drastically as shown in Fig. 9. This phenomenon can be explained by the fact that the fluid over velocity in the adjacent area to keep the main flow rate constant despite the flow restriction due to the micro-channel.

In the laminar flow regime, the velocity in the walls is zero due to the dominant viscous force and the flow has maximum velocity in the middle of the micro-channel. It is more interesting to study the fluid velocity field in x-z plan along the micro-channel as shown in Fig. 10. In this figure we represent the velocity field with color (rainbow scale). Blue color is for zero velocity and the red color is for the maximum velocity magnitude. Fig. 10 shows that the flow velocity is zero in the region close to the micro-bridge.

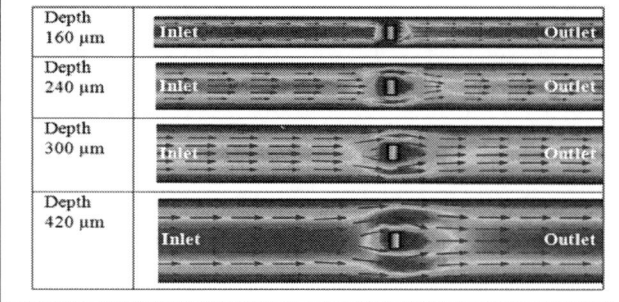

Fig. 10. Velocity magnitude field as function of micro-channel depth.

When the micro-channel depth is equal to 160 μm, we have around the micro-bridge a zero velocity field displayed with a totally blue color that strangles the flow. If the micro-channel depth continues to increase, the zero velocity field vanishes completely. The laminar flow is restored after the strangle caused by the micro-bridge structures. This demonstrates that the micro-bridge don't destroy the laminar aspect of the fluid flow. However, we have a zero velocity field around the micro-bridge that causes the static pressure increase in the leading edge as shown in Fig. 10. This pressure is the source of micro-bridge deflection.

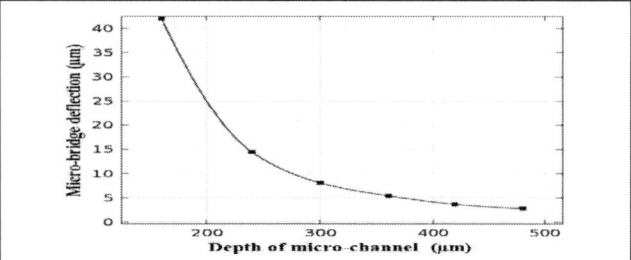

Fig. 11. Micro-bridge deflection as function of micro-channel depth.

Finally, Fig. 11 shows the micro-bridge deflection versus the micro-channel depth. This deflection is maximum around 42 μm with micro-channel depth of 160 μm, and decreases to 3 μm with 500 μm micro-channel depth. Furthermore Fig. 11 shows the micro-bridge bending has more sensitivity when the micro-channel depth is thin. It is clear that we have to take an optimum value to get a good deflection but not a strong strangling in the micro-channel flow.

V. Conclusion

In this paper, a 3D finite element analysis is carried out using Comsol Multiphysics to simulate the fluid-structure interaction and to compute the PDMS micro-bridge deflection under various fluid flow rates. We study the effect of several parameters : micro-channel dimension, flow rate magnitude, Young's modulus of the micro-bridge. The proposed structure consists of PDMS micro-bridge suspended within micro-fluidic device. The fluid flow in the micro-channel causes the bending of the micro-bridge suspended within PDMS micro-channel. The mechanical response of the micro-bridge against flow has been studied by introducing different flow rates into the micro-fluidic device and different micro-channel depth.

The main conclusions obtained are : (i) The sensitivity around 14.5 μm/ml min^{-1} for low flow rate is predicted, despite the importance of dimension of the micro-bridge the laminar aspect of the flow is maintained. (ii) A good equilibrium must be taken between the choice of the micro-channel depth value and the desired range of micro-bridge deflection.

We are currently investigating how to improve the deflection sensitivity for low flow rate applications. We believe this study could be applied to many micro-fluidic devices and BioMEMS sensing.

Acknowledgment

The authors would like to thank the faculty of Sciences and FUNDAPL Laboratory. A great thanks to Asma Hassein-Bey, Ahmed Taharoui and Abderahmane Benhaffaf. We thank the MEMS & Sensors Group at CDTA. Finally, a great thanks for IMEN UKM staff and MEMS and LOC group in IMEN.

References

[1] G. M. Wbitesides, "The origins and the future of microfluidics", Nature, vol. 442, n. 27, pp. 368-373, 2008.

[2] J. C. Löttersy, W. Olthuis, P. H. Veltink, and P. Bergveld, "The mechanical properties of the rubber elastic polymer polydimetylsiloxane for sensor application", J. Micromech. Microeng., vol. 7, pp. 145–147, 1997.

[3] A. S. Nezhad, M. Ghanbari, C. G. Agudelo, M. Packirisamy, R.B. Bhat, and A. Geitmann, "PDMS microcantilever-based flow sensor integration for lab-on-a-chip", IEEE SENSORS , vol. 13, no. 2, pp.601-609, 2013.

[4] M.S. Cheri, H. Latifi, J. Sadeghi, M. S. Moghaddam, H. Shahraki, and H. Hajghassem, "Real-time measurement of flow rate in microfluidic devices using a cantilever-based optofluidic sensor.", Analyst, vol.139, pp. 431-438, 2014.

[5] K. R. Stein and, R. J. Benney, "Fluid-structure interactions of a round parachute: Modeling and simulation techniques", J. Aircraft, vol. 38, no. 5, pp. 800–808, 2001.

[6] S. Basak, and A. Ramana, "Hydrodynamic loading of microcantilevers vibrating in viscous fluids", J. Appl. Phys., vol. 99, no. 11, pp. 1149061–114906-10, 2006.

[7] M. Liu, J. Sun, Y. Sun, C. Bock, and, Q. Chen, "Thickness dependent mechanical properties of polydimethylsiloxane membranes", J. Micromech. Microeng., vol. 19, no. 3, pp. 035028-1–035028-4, 2009.

[8] D. Armani, C. Liu and, N Aluru, "Re-configurable fluid circuits by pdms elastomer Micromachining" Proc. IEEE Int. Conf. on MEMS., 1999.

The effects of Growth Parameters on the Electrical Properties in InAlN/AlN/GaN High-Electron-Mobility Transistors (HEMTs)

Wei-Ching Huang, Yuen-Yee Wong, Kuan-Shin Liu, Chi-Feng Hsieh, Edward Yi Chang
Department of Materials Science and Engineering
National Chiao Tung University
1001 University Rd., Hsinchu, 30010 Taiwan
Email: weiching0928@gmail.com

Abstract—**The electrical properties in the InAlN/AlN/GaN high electron mobility transistor grown by metal-organic vapor deposition (MOCVD) with different growth parameters were investigated in this study. We observed that two-dimensional electron gas (2DEG) channel was influenced by thickness of AlN spacer layer and a stable stage prior to the growth of AlN spacer layer. The TEM images showed the generation of dislocations at interface between AlN/GaN and InAlN/AlN with the thicker AlN spacer layer. These dislocations acted as electron scattering center and degraded the electron mobility in the 2DEG channel. Besides, a too long stable stage also appeared degradation in the electron mobility due to the etching effect of H_2 gas. By optimizing growth parameters, the highest electron mobility of 890 cm^2/V.s**

Keywords—HEMTs;InAlN; GaN; styling; insert (key words)

I. INTRODUCTION

In recent years, the InAlN has been considered to replace the AlGaN as barrier layer in the high electron mobility transistors (HEMTs) due to its superior material properties. Comparing with the conventional AlGaN, the most attractive in the InAlN was the characteristic of lattice match with GaN while In composition at 18%. It also avoided the strain issues such as using higher Al composition in the AlGaN to increase electron concentration in the 2DEG channel. In addition, a stronger spontaneous polarization was built by InAlN and induced more electrons into 2DEG channel even without piezoelectric polarization. However, it was difficult to obtain highly crystallized quality because of a greatly difference in growth parameters between AlN and InN. Moreover, a severe alloy scattering effect was usually observed in the InAlN HEMT because of characteristic of ternary alloy and its growth condition of low temperature. In order to prevent alloy scattering effect efficiently and had better electron mobility, a thin AlN spacer layer was inserted between InAlN and GaN[1].

In this study, in order to understand influence of interface on the 2DEG channel properties, the growth parameters including the thickness of AlN spacer layer and a stable stage prior to the growth of AlN spacer layer in the $In_{0.18}Al_{0.82}N$/GaN HEMTs would be discussed. By hall measurement and transmission electron-beam microscope (TEM), the influences of these growth parameters would be observed.

II. EXPERIMENTAL

The InAlN/AlN/GaN heterostructure were grown on the (0001) sapphire substrate by metal-orangic vapor deposition (MOCVD). The epilayers consisted of 120nm-thick high temperature AlN buffer layer, 3μm-thick undoped GaN, a thin AlN spacer layer and 20nm-thick $In_{0.18}Al_{0.82}N$ barrier layer. The Fig.1 was the evolution of temperature and pressure during the structure growth. During the material growth, samples with different thickness of AlN spacer layer would be prepared by adjusting growth time. In addition, samples with different durations of stable stage prior to the growth of AlN spacer layer were developed as well. The thickness of AlN spacer was examined by the transmission electron microscope (TEM) analysis. The 2DEG properties including electron mobility, electron concentration and sheet resistance were performed with Van De Pauw configuration by Hall measurement system (bio-ray hall measurement system).

III. RESULTS AND DISCUSSION

IEEE-ICSE2014 Proc. 2014, Kuala Lumpur, Malaysia

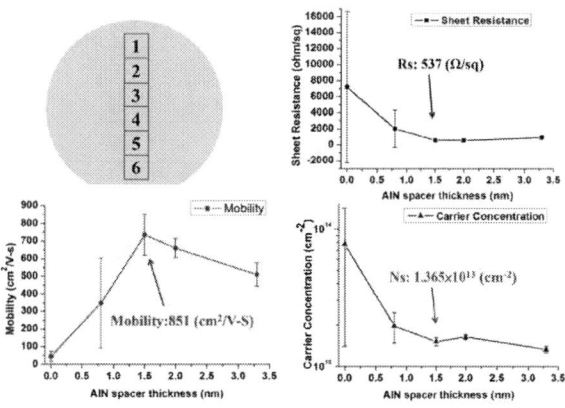

Fig. 1 The results of hall measurement in the samples with different thickness of AlN spacer layer (0nm,0.75nm,1.5nm,2.0nm and 3.25nm)

The Fig.1 shows the results of hall measurement with different AlN spacer thicknesses. The measurement was finished by six measured points on each sample and averaged it. The better properties of 2DEG channel occur at the spacer thickness of 1.5nm. The thinner AlN spacer thickness reveled the degradation in the electron mobility and increase of sheet resistance and electron concentration, while the thicker AlN spacer has a mobility degradation, too, which is less than with thinner spacer thickness.

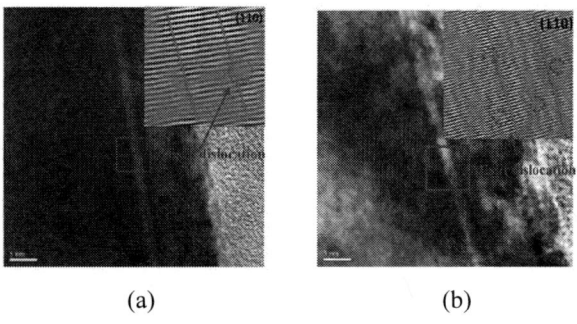

(a)   (b)

Fig, 2 TEM images of InAlN HEMT with different thickness of AlN spacer layer(a) with 1.5-nm AlN spacer layer (b) with 3.3-nm AlN spacer layer (The insets were the reconstructed images after selecting (110) Fourier Filtering)

For further understand, the TEM images was performed to investigate the impacts of spacer thickness on the 2DEG properties. The Fig. 2(a) was the images of InAlN HEMT with 1.5-nm AlN spacer and Fig. 2(b) was the images of InAlN

HEMT with 3.3-nm AlN spacer. Apparently, an obvious difference was observed in the reconstructed images after selecting (110) Fourier Filtering (The insets in both of Fig.2(a) and Fig. 2(b)). As the thickness of AlN spacer was increase to 3.3 nm , the misfit dislocations were generate to relax the mechanical strain as seen in inset of the Fig. 2(b). These

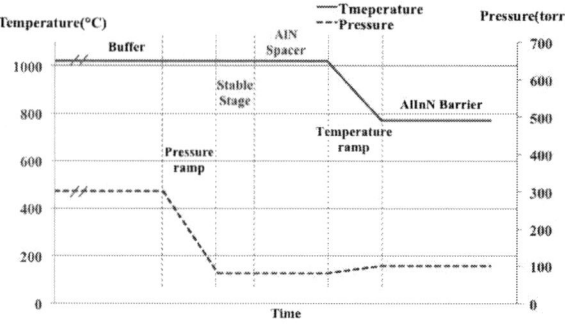

Fig. 3 The evolution of temperature and pressure in the growth of InAlN/AlN/GaN heterosture by MOCVD

misfit dislocations acted as electron scattering center and reduce the electron mobility.

The Fig.3 was the evolution of InAlN/AlN/GaN HEMTs growth. It reveals the relation of temperature& pressure to growth time. A duration of stable stage was defined by the time prior to the AlN spacer growth. In this stable stage, the

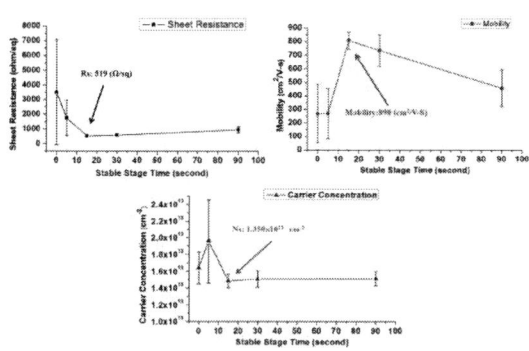

Fig,4 The results of hall measurement in the InAlN/AlN/GaN HEMTs with different duration of stable stage prior to AlN spacer growth.

changed pressure could be stabilize by this stage before.

The influences of stable stage with different duration of time on the 2DEG properties were investigated as seen in the Fig. 4. The results of hall measurement showed a degradation trend in the shorter stable stage and longer stable stage. In the shorter stable stage, the changed pressure might not stable enough to provide a better growth circumstance and created a perfectly interface between AlN spacer and GaN. On the contrary, longer stable stage deteriorated the electron mobility as well due the introduction of H_2 etching effect at high temperature[2].

IV. CONCLUSION

In this study, the effect of growth parameters on the 2DEG properties in InAlN/AlN/GaN HEMT were discussed. A proper thickness for AlN spacer layer was 1.5nm.As the thickness of AlN spacer layer was too thin or too thick, an evident decrease of electron mobility was observed. Thinner AlN spacer layer results in worse electron mobility due to the severe electron scattering effect. The thicker AlN spacer layer induced the generation of misfit dislocations in order to relax mechanical strain. Those misfit dislocations acted as electron scattering center and reduce the electron mobility. Moreover, the duration time of stable stage prior to the growth of AlN spacer was proved a critical influence on the properties of 2DEG channel. The shorter stable stage would not provide enough time to stabilize changed pressure resulting in worse growth environment. Too long of stable stage deteriorated the interface between AlN spacer and GaN because of H_2 etching effect.

ACKNOWLEDGEMENT

This work was supported by Ministry of Science and Technology, Taiwan under Grant Nos. NSC 102-2221-E-009-095-MY2 and NSC 101-2221-E-009 -173 -MY2

REFERENCES

[1] M. Gonschorek, J.-F. Carlin, E. Feltin, M.A. Py, N. Grandjean, High electron mobility lattice-matched AlInN / GaN field-effect transistor heterostructures, Applied Physics Letters, 89 (2006) -.

[2] J. Han, T.-B. Ng, R.M. Biefeld, M.H. Crawford, D.M. Follstaedt, "The effect of H2 on morphology evolution during GaN metalorganic chemical vapor deposition", Applied Physics Letters, 71 (1997) 3114-3116.

High Quality Ge Epitaxial Films Grown on In$_{0.51}$Ga$_{0.49}$P/GaAs and GaAs Substrates by Ultra High Vacuum Chemical Deposition

Yung-Hsuan Su[a], Shih-Hsuan Tang[a], Chi Lang Nguyen[a], Ching-Wen Kuan[a], Hung-Wei Yu[a], IEEE Fellow and Edward Yi Chang[a,b], IEEE Fellow

[a] Department of Materials Science and Engineering and Department of Electronic Engineering
[b] Institute of Photonic System and Institute of Lighting and Energy Photonics
[a,b] National Chiao Tung University
[a,b] No.1001 University Road, Hsinchu 30010 Taiwan

Abstract — The epitaxial growth of high quality Ge thin films on different materials of In$_{0.51}$Ga$_{0.49}$P and GaAs by ultra high vacuum chemical vapor deposition (UHVCVD) system was studied. The crystallinity of high quality Ge layers on In$_{0.51}$Ga$_{0.49}$P and GaAs layers can be proved by X-ray diffraction (XRD) and transmission electron microscopy (TEM). The comparison of surface morphology between Ge grown on In$_{0.51}$Ga$_{0.49}$P and GaAs was also analyzed by atomic force microscopy (AFM). The roughness of Ge on GaAs shows better than that of In$_{0.51}$Ga$_{0.49}$P. Both of these structures were designed for fabricating p-channel metal-oxide-semiconductor field-effect transistor (MOSFET) for the integration of Ge p-channel device with III-V n-channel electronic device.

Keywords—Ge; InGaP; GaAs; Heterostructure; Ultra High Vacuum Chemical Vapor Deposition (UHVCVD)

I. Introduction

In the past several decades, improvement of device performance has gained much attention because of continued reduction of device sizes for silicon devices. In order to fabricate small feature size silicon devices, applying a strain to the channel material has been achieved [1]. In addition to apply a strain to the channel, another way to gain high device performance is to replace Si with some materials with high electron mobility such as III-V materials. However, high-speed devices made from III-V materials are characterized with a high electron mobility but low hole mobility. Therefore, it is still a challenge to find a p-channel material with high hole mobility.

The epitaxial Ge layer on In$_{0.51}$Ga$_{0.49}$P and GaAs substrate as the p-channel material for complementary III-V CMOS logic is investigated in this paper. Device with Ge channel material on Si substrate with higher p-channel hole mobility has been reported before [2]. High quality Ge grown on GaAs used for the integration of Ge p-channel device with GaAs n-channel electronic device has also been studied [3]. The advantages of Ge grown on either In$_{0.51}$Ga$_{0.49}$P or GaAs are: (1)

Ge has a much higher bulk hole mobility (μ_h =1900 cm^2/Vs) as compared to GaAs (μ_h = 400 cm^2/Vs) or other III-V materials, (2) the Ge/GaAs interface has a very small lattice mismatch (~0.08%) while Ge/In$_{0.51}$Ga$_{0.49}$P shows nearly zero lattice mismatch, therefore Ge films can be easily grown on either GaAs or In$_{0.51}$Ga$_{0.49}$P layers with very low or nearly zero threading dislocation density, and (3) the narrower bandgap of Ge (E$_g$ = 0.66eV) lies within a wider band gap of GaAs (1.42eV) and In$_{0.51}$Ga$_{0.49}$P (~1.8eV), resulting in a good confinement of carriers in the Ge layer.

II. Experimental Procedure

In this paper, high quality Ge epitaxial layers were successfully grown on In$_{0.51}$Ga$_{0.49}$P/GaAs (100) substrate with 6°-offcut toward <110> direction and blanket GaAs (100) substrate by ultra high vacuum chemical vapor deposition (UHVCVD) system. The In$_{0.51}$Ga$_{0.49}$P/GaAs (100) wafer was cleaned first by NH$_4$OH+H$_2$O$_2$+DI water (1:1:50) for 10 minutes and HF+HCl+DI water (1:1:30) for 1 minute [4,5] before loading into the load-lock chamber while the GaAs wafer was loaded without any pre-cleaning step. Both wafers were transferred into the growth chamber as soon as the pressure of the load-lock reached 2×10^{-6} Torr. Thermal pretreatments were carried out for two minutes at 500°C for In$_{0.51}$Ga$_{0.49}$P/GaAs and GaAs in order to remove the native oxide on the surface prior to GeH$_4$ flow. The crystallinity of Ge epitaxial layer on In$_{0.51}$Ga$_{0.49}$P and GaAs layers were investigated using X-ray Diffraction (XRD) and transmission electron microscopy (TEM). The surface morphology and RMS roughness were measured by atomic force microscopy (AFM). The interfaces of both Ge/In$_{0.51}$Ga$_{0.49}$P and Ge/GaAs were also examined by cross-sectional images of transmission electron microscope (TEM).

III. Result and Discussion

A. The Purpose of Ge Grown on $In_{0.51}Ga_{0.49}P$ or GaAs

Since continued physical scaling of mainstream silicon CMOS technology has boosted the performance of the devices in last 40 years, increasing carrier mobility is a route to increase the drive current at low bias voltage to reduce the gate delay time for 30 nm and beyond low power logic application. III-V materials in general have significantly higher electron mobility than Si and can potentially play a major role along with Si in future high-speed low-power computing [3]. However, III-V material shows low hole mobility for further CMOS fabrication (shown in Fig. 1). The integration of high hole mobility material such as Ge and III-V becomes an important issue from now on. Fig. 2 shows the band diagram of the narrower bandgap of Ge ($E_g = 0.66eV$) lies within a wider bandgap of $In_{0.51}Ga_{0.49}P$ (~1.8eV), resulting in a better confinement of carriers in the Ge layer than that of Ge/GaAs interface.

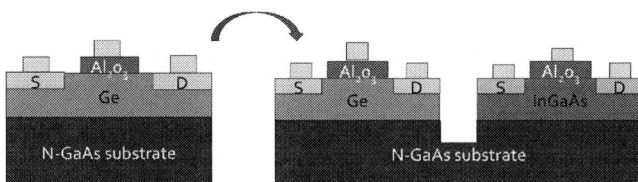

Figure 1. Future CMOS application

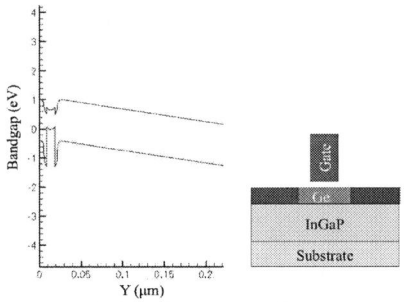

Figure 2. The band diagram of Ge on InGaP

B. Crystallinity of Ge Epitaxial Layer

The Ge films deposited at 500°C on $In_{0.51}Ga_{0.49}P$/GaAs and GaAs substrate were analyzed by the XRD measurement using a Bede D1 XRD system. The scanned result of Ge on $In_{0.51}Ga_{0.49}P$/GaAs is shown in Fig. 2(a) and Ge on GaAs is shown in Fig. 2(b). In Fig. 2(a) the red line shows the result of Ge grown on $In_{0.51}Ga_{0.49}P$/GaAs and the black line shows the data before Ge growth. According to same lattice constants of a= 5.65Å of Ge and $In_{0.51}Ga_{0.49}P$, the peaks of these two materials cannot be distinguished obviously. Good crystal quality still exist after growing at 500°C. The XRD result of Ge grown on GaAs shows in Fig. 2(b), in which Ge and GaAs can obviously be seen because of the 0.08% lattice mismatch. The angle between the peaks of the Ge film and the GaAs substrate is less than 200 arcsec. The appearance of fringes on

both sides of Ge and GaAs peaks implies a parallel and very sharp interface existed in this heterojunction structure [2].

(a)

(b)

Figure 2. The ω-2θ symmetric (004) HR-XRD results of (a) Ge/$In_{0.51}Ga_{0.49}P$/GaAs and (b) Ge on GaAs (100)

Figure 3 are cross-sectional TEM micrographs of 190nm and 100nm Ge thin film deposited on $In_{0.51}Ga_{0.49}P$/GaAs and GaAs substrate at 500°C. No threading dislocations show in Figure 3(a) as a result of nearly same lattice constant of Ge and $In_{0.51}Ga_{0.49}P$. In figure 3(b), a few misfit dislocations are detected at the Ge/GaAs interface, but there is no appearance of any threading dislocation, which is expected. The lattice mismatch between Ge and GaAs is extremely small.

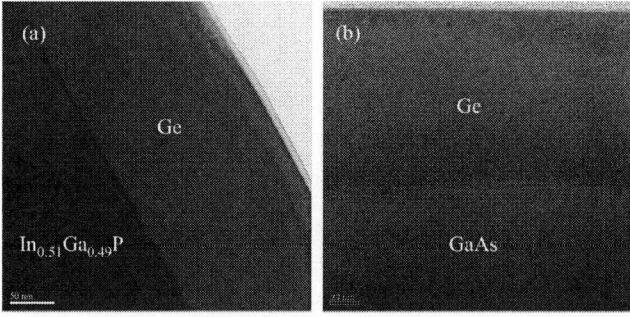

Figure 3. TEM images of the Ge epitaxy grown on the (a) $In_{0.51}Ga_{0.49}P$/GaAs and (b) GaAs substrate

C. The Surface Morphology and Growth Rate of Ge on $In_{0.51}Ga_{0.49}P$ and GaAs

The Ge layers were measured by AFM with scan size of 5μm x 5 μm² to investigate the surface morphology of Ge grown on different substrates. Figure 4 shows the AFM images of the Ge thin films with the thickness of 100 nm grown on $In_{0.51}Ga_{0.49}P$/GaAs and GaAs (100). The surface roughness is about 12nm and 0.132nm, respectively. For Ge frown on GaAs, the roughness is low enough for future device fabrication. The island formation is observed on the Ge film of thickness of 100nm grown on $In_{0.51}Ga_{0.49}P$/GaAs.

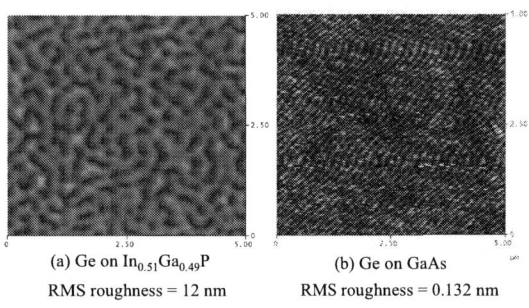

(a) Ge on $In_{0.51}Ga_{0.49}P$
RMS roughness = 12 nm

(b) Ge on GaAs
RMS roughness = 0.132 nm

Figure 4. AFM images of (a) Ge/$In_{0.51}Ga_{0.49}P$ and (b) Ge/GaAs heterostructure

The growth rate of Ge grown on $In_{0.51}Ga_{0.49}P$ and GaAs is shown in Fig. 5 at the growth temperature of 500°C. The growth rates of Ge on GaAs and Ge on $In_{0.51}Ga_{0.49}P$ are 43.7 nm and 26.93 nm, respectively. Mitin et al. has presented that the surface would be rough under low growth rate [6]. We predicted the rough surface of Ge grown on $In_{0.51}Ga_{0.49}P$ because of the lower growth rate than Ge on GaAs.

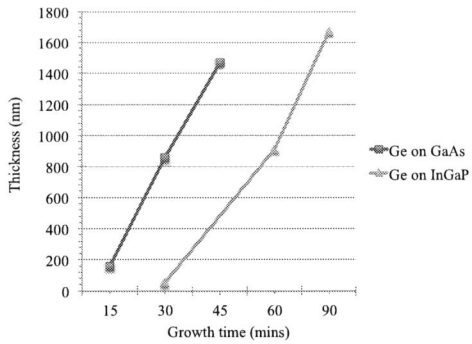

Figure 5. Growth rate of Ge on GaAs and Ge on $In_{0.51}Ga_{0.49}P$

IV. Conclusions

In summary, we demonstrate that Ge can be successfully grown on $In_{0.51}Ga_{0.49}P$ or GaAs with high crystallinity and good interface between these materials that can be used for future integration of Ge p-channel and III-V n-channel

electronic devices on Si template. The smooth surface of Ge film grown on GaAs was detected. However for the surface of Ge/ $In_{0.51}Ga_{0.49}P$, island formation was observed for large critical thickness.

Acknowledgment

This work was sponsored by the NCTU-UCB I-RiCE program, Ministry of Science and Technology, Taiwan, and TSMC, under Grant No. NSC 103-2911-I-009-302.

References

[1] S. E. Thompson et al., " A Logic Nanotechnology Featuring Strained-Silicon," IEEE Electron Device Lett. **25**, 191 (2004).

[2] M. Zhu et al., " Fabrication of p-MOSFETs on Germanium Epitaxially Grown on Gallium Arsenide Substrate by Chemical Vapor Deposition," Electrochem. Soc. **155**, H76 (2008).

[3] S. H. Tang et al., " High quality Ge thin film grown by ultra high vacuum chemical vapor deposition on GaAs substrate," Applied Physics Letters **98**, 161905 (2011).

[4] J. S. Song et al., " Wet chemical cleaning process of GaAs substrate for ready-to-use," Journal of Crystal Growth **264,** 98–103 (2004).

[5] M. Dubey et al., " SINGLE CRYSTAL EPITAXIAL Ge BASED CONTACTS TO GaAs, and InGaP," Mat. Res. Soc. Symp. Proc. Vol. 337 (1994).

[6] V. F. Mitin et al., " Effect of film growth rate and thickness on properties of Ge/GaAs(100) thin films," Thin Solid Films (550) 715–722 (2014).

Preparation and Characterization of MWCNT Dispersed in Various Solutions

[1]M. A. Farehanim, [1]U. Hashim, [1]S. Norhafiezah, [1]M.F. Fatin, [1]RM Ayub

[1]Institute of Nano Electronic Engineering, University Malaysia Perlis (UniMAP), 01000 Kangar, Perlis, Malysia.

E-mail: farehanim88@gmail.com, uda@unimap.edu.my

[2]Norhayati Soin, [2]Fatimah Ibrahim

[2]Centre for Innovation in Medical Engineering, Department of Biomedical Engineering, University of Malaya, 50603 Kuala Lumpur.

Abstract—In modern technology of biomedical applications, the potential of carbon nanotubes based materials has been widely used in recent years. In this paper, the preparation of the multi wall carbon nanotube (MWCNT) with biocompatibility of these composite are investigated, although many aspects have been studied separately by researchers. We have chosen three different solvents; namely chitosan, Sodium Dodecyl Sulfate (SDS), and isopropyl alcohol (IPA) to mix with MWCNT respectively. This functionalized CNT with carboxylic (COOH) groups were prepared in three different liquid forms and further will be dropped on fabricated Interdigitated electrodes (IDEs) as devices. Scanning electron microscopy (SEM) was used to identify the structures effect of synthesized MWCNT in different solvents. The conductivities show the ability of chitosan and SDS to be used as a solvent in order to synthesis MWCNTs and further will be used as a biosensor.

Keywords— MWCNTs_COOH group; various solvents; biosensor

I. INTRODUCTION

The new advanced technology needs new materials with improved lower production cost, duration and higher strength. In order to fulfill these requirements, one of the most commonly used carbon nanotubes performances also have been used in biomedical application in recent years. CNTs can perform in various fields of study, including physic, chemistry and material sciences. Factual performance of CNTs that have a very high tensile strength, thermal stability and electrical conductivity, making them useful for several potential applications [1,2]. However, the multiple features regarding the dispersion, stability and behavior of MWCNTs in biological settings have already been investigated in separate studies, but not all in a concurrent manner or performed such that immediate comparisons in different ambient systems could be made. In this experiment, we used three different solutions; Chitosan, Sodium Dodecyl Sulfate (SDS) and Isopropyl alcohol (IPA). The different solutions used have many different advantages. Chitosan is a naturally polysaccharide bearing amino and hydroxyl group showing wide application potentials in biology, medicine, food and membrane separation due its attractive of low cost and biocompatibility[5]. Moreover, chitosan is good for binding the MWCNTs on wafer surface and working like are glue. An addition Chitosan has a number of commercial and possible biomedical uses (Chitosan). Another solution is sodium dodecyl sulfate (SDS). Functionalized of Sodium Dodecyl Sulfate (SDS) solvent is a comparatively long negative ion surface active agent. It can adsorb on the surface of solution and form diffusion layer, preventing MWCNTs from re-gathering. Thus, MWCNTs have good dispersion property [6]. Isopropyl alcohol (also propan-2-ol, 2-propanol or the abbreviation IPA) is a common name for a chemical compound with the molecular formula C_3H_8O. It is a colorless, flammable chemical compound with a strong odor.

This experiment presents a stability and reproducibility of three different solutions of solvent; isopropanol (IPA), Chitosan and (SDS) with mixed MWCNTs-COOH group. The CNTs without any chemical functionalization, CNTs are hard to be dispersed in any liquid due to the van der Waals force interaction that tend to cause agglomeration of CNTs and prevent them to be used in any application. Thus, bundling and individual application [3]. These samples were analyzed after sonication, 1 hour and 5day. We investigated the stability of MWCNTs dispersions in room temperatures of 37oc with varying solution [4]. Additionally, the combination of CNTs with conducting polymers enables the production of Nano-composites with significant improvement in electrical conductivity and thermal stability [1].

This present study, we inspected the aqueous dispersions of functionalized MWCNTs using three different types of solvent; Chitosan, Sodium Dodecyl sulfate (SDS) and isopropyl alcohol (IPA). The dispersion of MWCNTs was observed under Scanning Electron microscopy (SEM) images result indicated that the procedure was done correctly, as well as MWCNTs structures on sample and electrical conductivity of MWCNTs dispersed in various solutions was studied.

II. EXPERIMENT DETAIL

A. Material

All chemicals used were of analytical reagent grade purchased from Sigma Chemical Co. Malaysia; functionalized MWCNT-COOH (>97% carbon purity), Chitosan (85% purity), Sodium Dodecyl sulfate (SDS) (99% purity) and isopropyl alcohol (IPA).

B. characterization

The dispersion of MWCNTs with varying solvent were characterized using Scanning electron microscope (SEM) image. The experimental setup for ultrasonic process is shown in fig 1. All picture of MWCNTs dispersion in solvent in the bottle was captured using a digital camera for 1 hour and 5 days in fig 2. For more accurate measurement, we used electrical characterization by current-voltage (IV) characterization (KEITHLEY, 6487) to measure the conductivity between different solutions.

III. PREPARETION OPERATION

The process started with the cleaning process where the sample (wafer) is cleaned using RCA/BOE. The RCA/BOE is used to remove the native oxide and organic contaminant from the wafer surface. Later on, the wafer is ready to proceed with the next step.

A. Preparation of Chitosan

2 mg of chitosan was dissolved in 100 mL, 2 wt% acetic acid solution. Then Chitosan solutions were stirred constantly on hot plate to make a homogenous solution at room temperature to ensure a complete reaction. After that, Chitosan solution 5ml was mixed with MWCNTs (1mg) and hot water for several times in an ultrasonic bath to remove the physical adsorbed and make homogenous solution. Then, CHIT-MWCNTs was dropped onto wafers and dried under dryer at 25 sec at room temperature.

B. Preparation of Sodium Dodecyl Sulfate(SDS)

In a typical experiment, we used 33.3mg sodium dodecyl sulfate (SDS), 33.3ml ethanol, 33.3ml DIW. After the solution was already mixed, the SDS solution was stirred until it became homogenous. After that, take the 5ml SDS solution and mixed with MWCNTs (1mg) and were collected by filtration. Then, the solution also makes sonicated to ensure the solution of the SDS - CNT have mixed and became homogenous.

C. Preparation of Isopropyl Alcohol(IPA)

This is a simple process to do, takes 5ml IPA mixed with (1mg) MWCNT and used ultrasonic for sonicated with several times to make sure the solution mixed smooth and homogenous. The experimental setup of ultrasonic process is shown in fig. 1.

IV. RESULT AND DISSCUSION

To facilitate solution in MWCNTs for integration into sample they need to be purified and dispersed into solution. For this dispersion study, it is necessary to conduct an ultrasonic process to ensure that the functionalized MWCNTs were fully dispersed in the solvent. In the other words, the purpose of ultrasonic was to de-bundle the functionalized MWCNTs in the solvent to obtain a transparent solution by utilizing the ultrasonic waves. The functionalization process can be effect of the length and defect density of MWCNTs but the sonication to MWCNTs disperse in any solvent gives the best dispersion. For our knowledge this is the first report of dispersing MWCNTs with various solvent. This important observation is investigate the stability of dispersion and can be used good adhesive layer on the samples.

The result shows that the Chitosan, SDS and IPA solution can be dispersed the functionalization MWCNTs with a most darkness solution after 5 days as compared to other solution in fig 2. We can see at the bottom of bottles for the IPA, some particles of MWCNT has already precipitated out. The fig 3 was observed under SEM image to ensure a stable dispersion.

Fig 1: Experimental setup for ultrasonic process.

Fig 2: preparation of various solvent. (a) 1mg of MWCNT with 10ml Chitosan (left), 1mg MWCNT with 10ml SDS (middle) and 1mg MWCNT with 10ml IPA solution (right), respectively. (b) After 5 days, MWCNT in IPA solution already precipitates out.

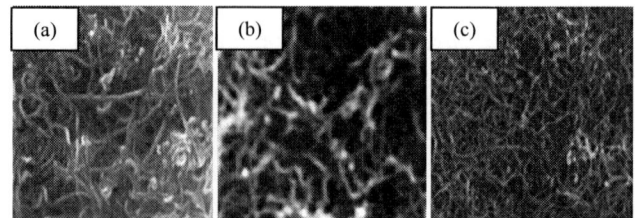

Fig 3: SEM image of three different solvents (a) Chitosan (b) SDS (c) IPA were used to disperse MWCNT.

Initially, characterization of deposition was carried out by scanning electron microscope (SEM). The SEM image widely used to characterize the morphology of dispersion as well as the structure of CNTs. As seen under SEM images shown in fig 3. These images are used primarily to determine structural CNT as received Chitosan, SDS and IPA, which are easy to distinguish is obviously different. In addition, the sonicated have used is too dispersed CNT in bundles/ropes. This bundle problematic for fabricating as they often protrude up through the active layer and cause shorting. In order to improve, dispersion were sonicated for 1 hour prior to deposition. This sonicated process is to break up the CNT in rope/bundles. As seen the MWCNT-Chitosan structure under SEM image shown the CNTs already break up from rope it because the single CNT was appear. In the other word, sonicated time is important to disperse CNTs.

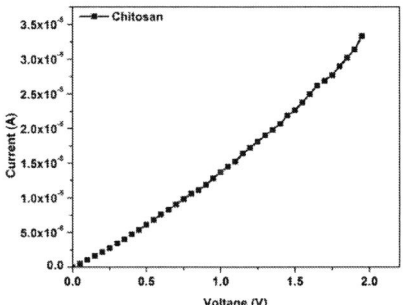

Fig 4: Electrical testing for Chitosan solution.

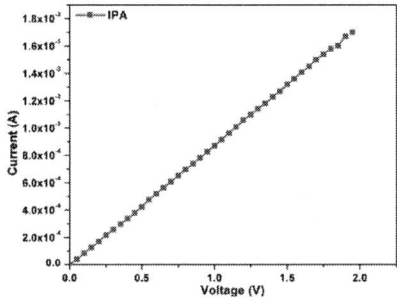

Fig 5: Electrical testing for IPA solution.

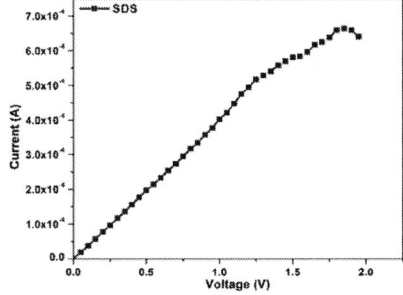

Fig 6: Electrical testing for SDS solution.

As in Fig 4, 5 and 6 shows, the electrical conductivity and resistivity of different solution in MWCNT (Multi Walled Carbon Nano Tube) at a room temperature. In dispersion, the functionalized MWCNT are dispersed in these three types of solvents. The HCI modified SWCNTs contained carboxyl group, COOH and able to react with solvents. For MWCNTs mixed in chitosan (black line), SDS (blue line) and IPA (red line). The result of fig 4 shows, the value of chitosan gradually increased because of the dispersion chitosan solution is homogeneously on the wafer surface. From the chitosan solution observation, the conductivity of chitosan is high and also for the resistivity became lower. Where for the IPA solution result, one can be observed that it has low conductivity. The IPA solution in fig 5 is a chemical component that are not suitable to dilute with CNT because IPA solution is not capable to conduct current flow. Furthermore the IPA solvent evaporate quickly and MWCNTs is easily peeled off when scratched. The graph shows a higher resistivity because of solvent which cannot flow of current. Where the last solution is the SDS solution in fig 6 that get a result identically like chitosan which mean that it also have a lower resistivity and stable to use for surface modification.

The rate of decrease in resistivity becomes mild until the content of MWNTs is greater than 0.75%. This simply means that a very high percentage of electrons are permitted to transfer through the specimen due to the creation of an interconnecting conductive pathway. For SDS-MWNTs/ Chitosan-MWCNT composites, the electrical resistivity can become lower when the amount of the MWNTs becomes greater due to the better dispersion of the nanotubes. However, the SDS solution the nanotubes can decrease the electrical conductivity and increase the resistivity[7].

Although the functionalized nanotubes have a much better dispersion in the composites, the size of the nanotubes has significantly reduced. It appears that the increase of dispersion of the nanotubes in the composites which can lead to increase in the thermal conductivity has offset the effect of shortening of the nanotubes which can reduce the thermal conductivity of the composites[7].

V. CONCLUSION

As conclusion, we have successfully prepared the functionalized MWCNTs dispersed with varying solution; that is chitosan, SDS and IPA. In the other word, we also study the dispersing method using sonication of time. Furthermore, morphological characterization by applying SEM to show the image structure of MWCNTs with a different solution. Finally, the electrical measurement also used to verify the result in this paper and a good solution is found for to get flexible way and can straight proceed our work in order to select solvent apply for our fabricated device.

ACKNOWLEDGMENT

The author would like to thank especially all staff at Nano Technology project of INEE (UniMAP), for their supervision and guidance in regard to processing and fabrication of the devices. And grateful for support financial for this work from University Malaya.

REFERENCES

[1] Q. L. Pham, Y. Haldorai, V. A. N. H. O. A. Nguyen, and D. Tuma, "Facile synthesis of poly (p -phenylenediamine)/ MWCNT nanocomposites and characterization for investigation of structural effects of carbon nanotubes," vol. 34, no. 1, pp. 37–43, 2011.

[2] E. Afsharmanesh, H. Karimi-Maleh, A. Pahlavan, and J. Vahedi, "Electrochemical behavior of morphine at ZnO/CNT nanocomposite room temperature ionic liquid modified carbon paste electrode and its determination in real samples," *J. Mol. Liq.*, vol. 181, pp. 8–13, May 2013.

[3] L. F. Wah, U. Hashim, W. Liu, and T. Adam, "RTICLES The Effect of Chemical Solutions (Isopropyl Alcohol, Dichloromethane, Acetone and Triton X-100) on the Dispersion of Single-Walled Carbon Nanotubes," vol. 9, no. 6, pp. 3411–3416, 2013.

[4] E. Heister, C. Lamprecht, V. Neves, C. Tîlmaciu, L. Datas, E. Flahaut, B. Soula, P. Hinterdorfer, H. M. Coley, S. R. P. Silva, and J. McFadden, "Higher dispersion efficacy of functionalized carbon nanotubes in chemical and biological environments.," *ACS Nano*, vol. 4, no. 5, pp. 2615–26, May 2010.

[5] Y.-L. Liu, W.-H. Chen, and Y.-H. Chang, "Preparation and properties of chitosan/carbon nanotube nanocomposites using poly(styrene sulfonic acid)-modified CNTs," *Carbohydr. Polym.*, vol. 76, no. 2, pp. 232–238, Mar. 2009.

[6] K. Xu, C. Wu, H. Yu, X. Tian, and Z. Dong, "A study on the influence of surface active agent molecules on the AFM scanning and imaging of SWCNTs," *J. Phys. Conf. Ser.*, vol. 188, p. 012013, Sep. 2009.

[7] C. Gau, S.-Y. Chen, H.-L. Tsai, S.-T. Jenq, C.-C. Lee, Y.-D. Chen, and T.-H. Chien, "Synthesis of Functionalized Carbon Nanotubes/Phenolic Nanocomposites and Its Electrical and Thermal Conductivity Measurements," *Jpn. J. Appl. Phys.*, vol. 48, no. 6, p. 06FF10, Jun. 2009.

Design Optimization and Finite Element Analysis of AlN/3C-SiC Piezoelectric Bio-Sensors

Abid IQBAL*, Faisal MOHD-YASIN and Sima DIMITRIJEV
Queensland Micro- and Nanotechnology Centre,
Griffith University, Brisbane,
QLD 4111, Australia

Abstract— In this paper we present the design and simulation of a bio-sensor for pathogens detection based on AlN/3C-SiC/Si piezoelectric cantilever. Cubic silicon carbide (3C-SiC) is chosen as the base layer due to its excellent material properties and chemical inertness over silicon in harsh environmental conditions. Aluminum nitride (AlN) is selected as piezoelectric active layer due to its similar thermal expansion coefficient with silicon carbide to reduce thermal stress. The desired resonant frequency of 157.16 KHz is optimized using Matlab and the finite element analysis is carried out using COMSOL software to verify the shift in the resonant frequency due to the added mass of the bacteria. The surface functionalizations of the SiC as biosensor, as well as the fabrication recipes are also proposed.

Keywords—Microresonator, Piezoelectric, 3C-SiC, Si, Pathogens, Extreme environment.

I. INTRODUCTION

Quartz crystals resonators are being widely used in timing reference due to their high precision and thermal stability. However, the poor integrability of these crystals with CMOS and their large size encouraged various research groups to develop other types of resonators. MEMS resonators have received the most interest as an alternative to quartz crystals due to their smaller size, compatibility with CMOS ICs and low power consumption [1]. Various transduction methods such as piezoelectric, capacitive, piezoresistive and thermal transduction are explored, each possessing relative merits and trade-offs over the other [2-3]. For example, capacitive microresonators have weak electromechanical, while the thermal microresonators require high voltage and power. Similarly the piezoelectric resonators lose out in terms of lower quality factor (Q).

Beside timing reference, these microresonators are used for other applications such as energy harvester and bio sensors. The MEMS-based resonant bio sensors detect the pathogens by measuring the decrease in resonance frequency due to the added mass of the pathogens. This mass-sensing technique is being explored by various research groups. Their smaller sizes and high resonant frequencies provide higher resolution and precision compared to conventional mass sensors [4-10]. Hao et al. employed a length-extensional bulk mode resonator for mass sensing with a mass sensitivity of 215 Hz/pg accompanying with the Quality Factor, Q of 4000 in air [6]. One group has utilized a square-extensional mode resonator as mass sensor with sensitivity of 3.3 Hz/ng and Q exceeding 106 [7].

Heteroepitaxial 3C-SiC and poly-crystalline SiC are normally utilized for micromachining process because the patterning and wet etching of single crystal 4H- and 6H-SiC is much more difficult compared to 3C-SiC thin film. The 3C-SiC thin film is used in various applications due to its better mechanical properties over silicon at high temperature and harsh enviroments. The most common applications are micro jet-engine power generation sources, energy harvester for harsh enviroment and electric micromotors [8]. Also the SiC based harsh sensors are emplyed for the monitoring of combustion engines, to improve the fuel efficiency and pollution reduction. It is also used as gas sensors for monitoring gas emission and fuel leakage detection [9].

In this work, we proposed the cantilever beam resonator sensor based on AlN/3C-SiC for sensing the bio particles e.g. bacteria and viruses etc. The high Young modulus to Poison ratio of 3C-SiC results in high resonant frequency and higher Quality Factor as compared to Silicon on insulator (SOI). The resonant frequency of the proposed resonator based on 3C-SiC is 157 KHz [10]. The 3C-SiC is chosen as structural layer due to its resilient mechanical properties to harsh environment and also the large energy band gap of 2.49eV, which interprets into higher stability of resonance frequency with varying temperature and environmental changes. The finite element analysis is carried out for the design verification and also to measure the frequency shift due to a single bacterium cell. The piezoelectric sensing is proposed to use for the biosensors. Aluminum Nitride is chosen as piezoelectric material. High melting point and chemical stability of AlN has given preference over zinc oxide as the chosen piezoelectric material.

II. FUNDAMENTAL FREQUENCY OF CANTILEVER

The resonance frequency of the cantilever beam depends on material properties and its dimensions. The fundamental resonance frequency of the cantilever beam can be given by eq. 1.

$$f = \frac{1}{2\pi}\sqrt{\frac{k}{m}} \qquad (1)$$

Where "k" is the spring constant of cantilever beam, which depends on the dimension such as length, width and thickness of the beam and "m" is beam mass which also depends on the dimension and density of the material, respectively.

III. MATHEMATICAL MODELLING

The unimorph piezoelectric cantilever beam is proposed to be used as bio sensor as shown in Fig.1. It consists of three layers i.e., the first layer is the substrate which is also employed as bottom electrode; the second layer is the active piezoelectric layer while the third layer is the top electrode. In our case, we proposed 3C-SiC as substrate/bottom electrode, AlN as piezoelectric layer while Mo as top electrode. The cantilever beam is attached to the substrate at one end, while the other end is set as free, at x=L, where "L" and "w" is the length and width of the cantilever beam respectively. The thickness of the substrate is t_s, while the thickness of the piezoelectric is t_p and the thickness of the top electrode layers is denoted as t_e, respectively. The modulus of elasticity for the substrate electrode is E_e, while for the piezoelectric and the top electrodes are represented as E_p and E_t respectively.

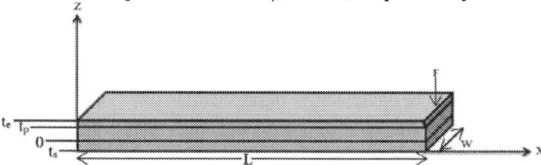

Fig. 1. Unimorph piezoelectric cantilever beam in cartesian system

The output voltage generated from energy harvester can be given by equation 3, which depends on its dimensions, material properties and also on the applied force.

$$\frac{V_{avg}}{F} = \frac{L}{2}g_{31}\frac{E_p}{wD}\left(\frac{t_p^2}{2} - t_n t_p\right) \qquad (3)$$

Where "g_{31}" is the piezoelectric coefficient correlated to piezoelectric strain coefficient, the Young's modulus of the piezoelectric is E_p, the width of the cantilever is "w" and the thickness of the layers is t_p and t_n, respectively.

The maximum displacement at free end can be found as eq. 4. The maximum displacement is proportional to the applied force and the length of the cantilever beams, while inversely proportional to the width and bending modulus respectively.

$$h_{max} = \frac{FL^3}{3wD} \qquad (4)$$

Where D is the bending modulus per unit width and is given as eq. 5:

$$D = \frac{1}{3}\left(E_s t_s^3 + E_p t_p^3 + E_e\left(t_e^3 - t_p^3\right)\right) + t_n(E_s t_s^3 - E_p t_p^2 - E_e\left(t_e^2 - t_p^2\right)) + t_n^2(E_s t_s + E_p t_p + E_e\left(t_e - t_p\right)) \qquad (5)$$

IV. WORKING PRINCIPLE

The working principle of piezoelectric bio sensor is quite simple. The cantilever beam is coated with a sensing layer, which is capable of recognizing specific target molecule or pathogens, i.e. able to recognize target molecules in key-lock processes. The piezoelectric technique is proposed for the sensing. When the cantilever beam resonates, it induces the charges in the piezoelectric layer due to cantilever movement at the resonance frequency.

The resonant frequency of the cantilever beams is inversely proportional to the mass. This resonance frequency shifts towards the lower value with the increase in mass due to pathogen attachment. The mass of the pathogens can also be measured by calculating the frequency shift towards the right.

V. FINITE ELEMENT ANALYSIS

The Finite element Analysis (FEA) is performed to pre-determine and optimize the the resonant frequency of the desired Eigen mode utilizing Comsol, which is shown in Fig 2. The fixed constraint was applied on the pad and the model was meshed using Manhathan bricks with standard physics. The desired first mode is shown in Fig 3. The Eigen frequency is found as 157.16 KHz. The dimension of the proposed cantilever beam resonator is optimized to get the desired resonance frequency of 157 KHz, which can give a high sensitivity of 307 Hz/pg. The thickness of each layer is optimized to 1 micron which gives the desired resonance frequency and maximum open circuit piezoelectric potential. The dimension of the proposed piezoelectric sensor is shown in Fig 3. The design is optimized in terms of thickness of the layers, length and width of the structural and piezoelectric layer and proof mass to obtain the desire resonant frequency.

The FE analysis is performed to verify the design and concept. The polystyrene particle of size 1 micron was placed on the microresonators as shown in Fig 2. The frequency analysis was done by applying boundary load of $1*10^{-5}$ N/m^2 on the resonator to see the effect of particle on the resonance frequency. It can be observed that the resonance frequency decreases by 307 Hz after the particle placement due to increase in mass according to the equation 1. The frequency sweep was done to find the resonance frequency with and without particle. The shift in resonance frequency towards the right due to polystyrene particle can be observed by comparing Fig 4 and Fig 5, which is proportional to the mass of the particle.

IEEE-ICSE2014 Proc. 2014, Kuala Lumpur, Malaysia

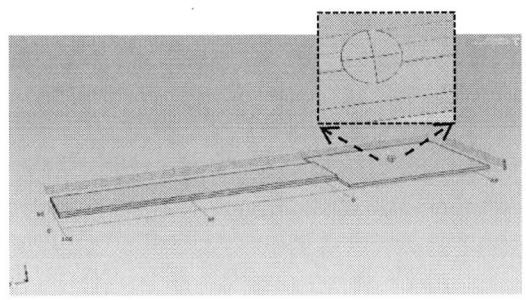

Fig. 2. Schematic of T-shaped cantilever beam resonator with small small particle

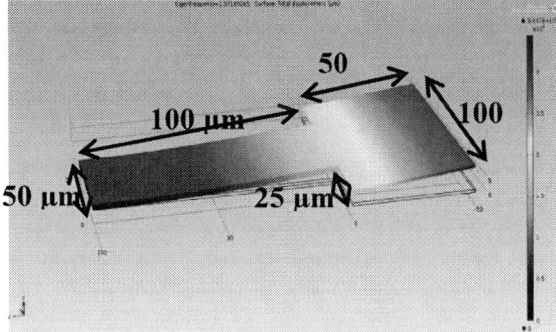

Fig. 3. Optimized dimension of proposed T-shape piezoelectric microresonator

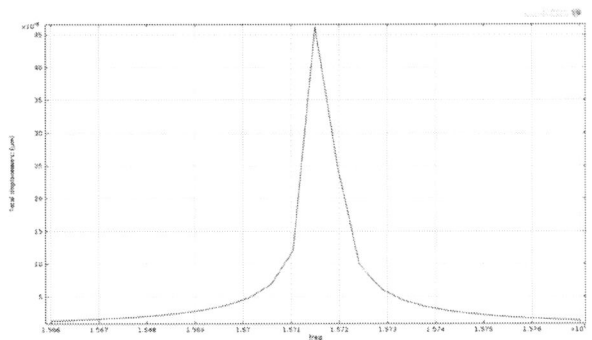

Fig. 4. Frequency sweep of the T-shaper microresonator without the polystyreene

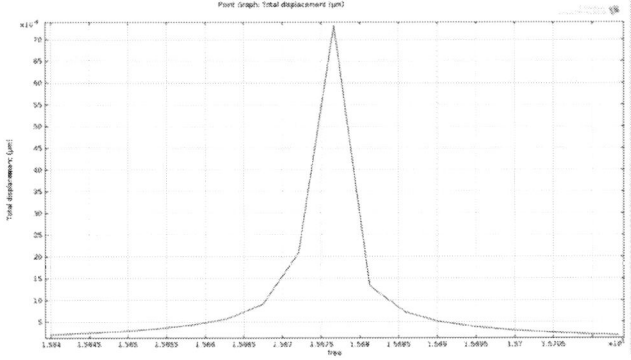

Fig. 5. Frequency sweep of the T-shaper microresonator with the polystyreene

VI. FABRICATION STEPS AND FUNCTIONALIZATION:

The main advantage of 3C-SiC over Si as bio sensor is regarding the immobilization agent. Silicon-based biosensors need an expansive gold layer on top, which is modified for the immobilization for the antibodies. Unlike Si, Silicon carbide can be used for the immobilization of the antibodies on its surface without depositing any gold layers which also reduce the costly and tedious fabrication steps. The immobilization of the antibodies can be easily achieved via aminopropyltriethoxysilane (APTES) linker via covalent conjugation as shown in Figure 6 [7]. The Surface hydroxylation of the SiC (-OH termination) can be done by treating it with a hydrofluoric acid (HF) while the surface functionalization of 3C-SiC with APTES (Sigma- Aldrich) can be achieved by soaking in a 1% APTES/toluene solution for approximately 90 min. The carboxylic group of antibodies can be stimulated through EDC 1-ethyl-3-[3-dimethylaminopropyl] carbodiimide hydrochloride (EDC) and N-hydroxy-sulfo succinimide (Sulfo-NHS).

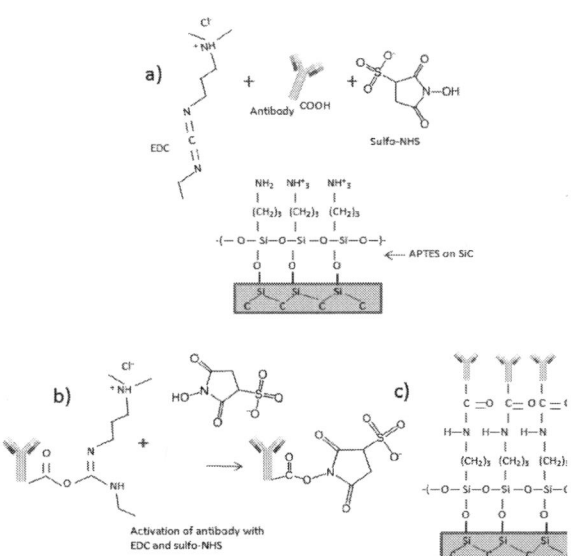

Fig. 6. Process followed for anti-myoglobin immobilization. a) EDC Sulfo NHS solution and anti-myoglobin are deposited on APTES|SiC samples, b) Activation of the antibody carboxylic group with EDC, sulfo- NHS produces a semi-stable amine-

The detail fabrication steps of the proposed design are shown in Fig 7.

978-1-4799-5761-3/14 $31.00 © 2014 IEEE   511

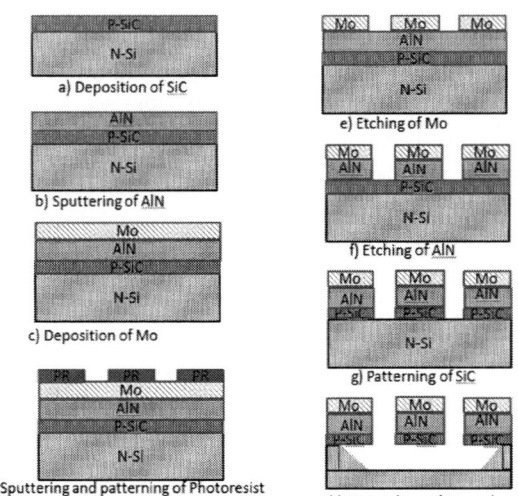

Fig.7. Surface-micromachined 3C-SiC MEMS fabrication process flow. (a) p type doping of Silicon substrate (b) LPCVD 3C-SiC and doping . (c) Sputtering of AlN on SiC. (d) Deposition of Mo (e) Patterning of Photoresist for Pad Metals deposition (f) Etching od Mo (g) Etching of AlN and SiC (h)Wetch etching of Si to release SiC structure

VII. DISCUSSION

The finite element analysis and the mathematical modeling of the bio sensor based on SiC/AlN/Mo cantilever beam are presented. The proposed bio sensor utilizes the electrical actuation while employing the piezoelectric sensing. The SiC is utilized as structural and also as the lower electrode while AlN is used as active piezoelectric material. The similar thermal expansion coefficient of all the layers reduces the residual and thermal stress, and the high curie temperature of AlN enables the operation of designed resonator at harsh environments. As from equation 1, when the proof mass increases due to the pathogens cell attached to it, the resonance frequency decreases which is proportional to the mass. The FEA analysis of the single particle of 1 micron diameter of polystyrene, which is placed on the micro resonator, is performed. The results show the decrease of 307 Hz in resonance frequency due to small particle, which is proportional to the mass of the polystyrene particle. The Von mises stress is also found at the resonant frequency, which is less than the elastic limit of all the structural layers i.e. 3C-SiC, AlN and Mo. Also unlike silicon-based biosensors that need to deposit an expensive gold layer and modify it for the antibodies, the immobilization of the antibodies can be done via aminopropyltriethoxysilane (APTES) linker on 3C-SiC surface which reduces the costly and tedious fabrication steps of gold deposition.

VIII. CONCLUSION:

The mathematical modeling and Finite Element Analysis (FEA) of a bio sensor based on cubic silicon carbide (3C-SiC)

for pathogen detection is presented in this paper. The resonator is proposed to actuate electrically while piezoelectric sensing will be employed to detect the shift in the resonant frequency. The AlN is chosen as piezoelectric material due to its high Curie temperature and similar thermal expansion coefficient with 3C-SiC to reduce the thermal stresses. The design parameters such as length, width and resonance frequency were optimized using Matlab and the design was evaluated using the Comsol and Intellisuite software for finite element analysis. 3C-SiC is chosen due to its excellent material properties and ease in functionalization compared to silicon which reduces the expansive and tedious steps of depositing gold on Si for functionalization. The functionalization of the 3C-SiC is also proposed in this paper. The close agreement between the analytical model and simulation results verify the design and working of the proposed bio sensors.

ACKNOWLEDGMENT

This work is supported by Queensland Micro- and Nanotechnology Centre (QMNC) and Griffith School of Engineering. This work was performed in part at the Queensland node of the Australian National Fabrication Facility, a company established under the National Collaborative Research Infrastructure Strategy to provide nano and microfabrication facilities for Australia's researchers.

REFERENCES

[1] E. Noel du Toit, L.W. Brian and K. Sang-Gook, " Design considerations for MEMS-scale piezoelectric mechanical vibration energy harvesters", Integrated Ferroelectrics, vol. 71, pp.121-160.

[2] S. P. Beeby, M. J. Tudor and N. M. White," Energy harvesting vibration sources for microsystems applications." Meas. Sci.Technol. 2006;17:R175-R195.

[3] S. Roundy, PK Wright, "A Piezoelectric vibration based generator for wireless electronics",Smart Mater. Struct., vol: 13, pp.1131-1142, 2004.

[4] F. Hua-Bin, L Jing-Quan, X Zheng-Yi, Lu D, W .Li and C .Di, " Fabrication and performances of MEMS-based piezoelectric power generator for vibration energy harvesting", Microelectronics Journal., vol:37, pp: 1280-1284, 2006.

[5] Fang Hua-Bin, Liu Jing-Quan, XuZheng-Yi, L. Dong, D. Chen and Cai Bing-Chu, " A MEMS-based piezoelectric powergenerator for low frequency vibration energy harvesting.", Chinese Phys.Lett., vol:23, pp. 732, 2006.

[6] T. H. Lee, S Bhunia and M. Mehregany, "Electromechanical computing at 500C with Silicon Carbide", Science, vol:329, pp:1316-1318, 2010.

[7] A. H. Epstein, "Millimeter-scale, Micro-ElectroMechanical Systems gas turbine engines",ASME journal of Engineering for Gas Turbines and Power, vol:126(2), pp: 204-226, 2006.

[8] G.W. Hunter, P.G. Neudeck and J. Xu, "Developed of SiC-based gas sensors foraerospace applications", Material Research Society Symposium Proceeding, vol.815, 2004.

[9] PM Sarro, "Silicon Carbide as a new MEMS technology", Sensors and Actuators,vol:82, pp:210-218, 2008.

[10] J.M.R. Kudimi, F. Mohd-Yasin and S. Dimitrijev, " SiC-Based Piezoelectric Energy Harvester for Extreme Environment", Procedia Engineering, vol. 47, pp. 1165-1172, 2012.

Synthesizing SnAgCu Nanoparticles by Electrodeposition of Reverse Microemulsion Electrolyte

S.P. Foo, K.S. Siow, A. Jalar
Microelectronics and nanopackaging (MIPAC)
Institute of Microelectronics and Nanopackaging (IMEN)
Universiti Kebangsaan Malaysia, Malaysia
azmn@ukm.edu.my

Abstract— Solder printing faces challenges in print definition and sustainable volume transfer as the microelectronic industry migrates to smaller footprints. Nano-sized alloy particles mitigate these issues by modifying the solder formulation and rheology. We propose an electrodeposition method using reverse microemulsion (without reducing agent) as electrolyte to synthesize Sn-Ag-Cu solder alloy nanoparticle. $SnCl_2.2H_2O$, Ag_2SO_4 and $CuSO_4$ are used as the precursor. N-hexane, TritonX-100 and n-hexanol are used as oil phase, surfactant and co-surfactant, respectively. Three volume ratios of aqueous to surfactant phase, W_o are carried out namely: 0.20, 0.40 and 0.60 but only the ratio of 4.5:1.0 is able to synthesize the nano-sized SAC particles. Scanning electron microscopy shows the spherical SAC particles are well dispersed and their sizes range from 15 to 90 nm. X-ray diffraction spectra show the formation of Sn, Ag_3Sn and Cu_6Sn_5 without any oxide peaks in the synthesized nanoparticles; their absence suggests the effectiveness of the surfactants in protecting the alloy particles from oxidation. This electrodeposition method compares favourably to the microemulsion (with reducing agent) method in producing spherical and well-distributed nanosized solder particles.

Keywords—SnAgCu nanoparticle; SAC; reverse microemulson; electrodeposition

I. INTRODUCTION

Solder paste is typically printed onto solder pad before being reflowed to connect electronic components to boards or silicon dies to the leadframes/substrates. While solder paste may differ in formulations depending on the final applications, particle size is one of the most important parameters to control because of its influence on rheology, printing quality and related down time to clean the stencil [1]. Nano-sized metal particles in the solder paste reduces the melting and reflow temperature to reduce the overall cost of ownership and at the same time, produces sound joint after reflow. High reflow temperature leads to the component damage, delamination, high flux residues, and high maintenance costs for the soldering equipment [2].

The need to lower the reflow temperature coincides with another industry trend to use Pb-free solder. The industry workhorse, Pb-Sn solder, are phased out because of environmental and health hazard. The most common

replacement for Pb-Sn solder is Sn-based alloy with elements of Ag and Cu. The melting temperatures of Sn-Ag-Cu (SAC) alloys are typically higher than Pb-Sn solders by 30°C. Hence, there is a need to reduce the melting temperature of SAC alloy to maintain the thermal budget of the microelectronic package.

In the literature, several "top down" and "bottom-up" techniques have been used to synthesize Sn or Sn based alloy nanoparticles [3]. "Top-down" technique breaks the bulk materials into smaller particle sizes; one such example is ball milling [4]. "Bottom-up" technique implies assembling single atoms and molecules into larger nanostructures; examples are chemical reduction method [5-7] and arc discharged [8-9].

In this paper, we are reporting a "bottom-up" method involving the electrodeposition of "reverse microemulsion" (without reducing agent) as the electrolyte. Similar electrodeposition method has been used to synthesize different metallic elements and alloys [10-12] but none has reported the use of reverse microemulsion (without reducing agent) as electrolyte to produce SAC particles. Reverse microemulsions are "nano-dispersions" of water in oil (W/O) stabilized by surfactant films. The water phase traps the metallic ions that will be reduced to metallic nanoparticle at the cathode during the electrodeposition proces. The size and morphology of synthesized nanoparticles depend on the ratio of water to surfactant. This paper presents the chemistries used in this reverse microemulsion and initial characterization of SAC alloys produced from this technique.

II. METHODOLOGY

A. Materials

$SnCl_2.2H_2O$ (product no: 1001412355), Ag_2SO_4 (product no: 34629), $CuSO_4$ (product no: 35185) were obtained from Sigma-Aldrich; these chemicals were used as precursors for this electrodeposition without further purification. TritonX-100, $C_{14}H_{22}O(C_2H_4O)_n$ (Sigma-Aldrich product no: 1001412355), n-hexanol, $CH_3(CH_2)_5OH$ (Merck product no: S6589793), n-hexane, C_6H_{14} (Merck product no: K4379347424) were used in the microemulsion as surfactant, co-surfactant and oil phase, respectively.

978-1-4799-5761-3/14 $31.00 © 2014 IEEE

B. Experimental method

Similar weight ratio of $SnCl_2.2H_2O$, Ag_2SO_4 and $CuSO_4$ as the alloy composition of Sn3.5Ag1.0Cu was used to prepare the precursor by stirring this mixture in a sealed beaker. After ionic precursor preparation step, reverse microemulsion were prepared with the following ratio of water to surfactant (precursors in the water phase), *Wo* namely 0.20, 0.40 and 0.40. Surfactant, co-surfactant, oil phase and water phases were stirred in a sealed beaker at room temperature. The reverse microemulsion electrolyte was considered suitable for electrodeposition when there were no visible separation layers between the surfactant, oil and water phases. Fig. 1 shows the steps of preparing the electrolyte with relevant mixing ratios and subsequent electrodeposition in the reverse microemulsion electrolyte.

In this study, the Cu wire and Al plate were used as anode and cathode, respectively. Cathode was grinded and polished with Stuers polishing cloth and multicrystal polishing spray 1 μm. Anode and cathode were rinsed with isopropanol and deionized water to clean the surface before immersing into the electrolyte. Potentiostat (model: EKKAJ20D) provided the potential differences of 5V during the electrodeposition process of SAC particles. Electrodeposition was carried out in a sealed beaker at room temperature. Copious amount of isopropanol and deionized water were used to remove residual chemical from these SAC nanoparticles before subsequent analysis were carried out.

C. Characterization

The SAC particles were analyzed with AXS D8 Advance X-ray diffractometer AXS using Cu Kα radiation (λ = 0.15406 nm) with incident angles from 30 to 80° (2Θ). Field emission scanning electron microscope (model: FESEM SUPRA 55VP) was used to observe the size and morphology of the SAC particles at an accelerating voltage of 20 kV under magnification of 100.00K x, 10.00K x and 30.00K x, respectively.

III.   RESULTS AND DISCUSSION

Based on Fig. 2, the XRD spectra of these SAC particles shows the presence of elemental Sn, Cu_6Sn_5 and Ag_3Sn . The formation of Ag_3Sn and Cu_6Sn_5 indicated the successful alloying of Sn-Ag and Sn-Cu during the electrodeposition process [5, 14]. This XRD spectra also did not show any Sn or Ag or Cu oxide peaks; suggesting that the surfactant layer protected the synthesized SAC nanoparticles from oxidation.

Fig. 2. XRD spectra of SAC alloy nanoparticle synthesized by electrodeposition

Fig. 3 (a), (b) and (c) show the SEM results of the SAC nanoparticles produced from electrolyte with the following ratios *Wo* 0.20, 0.40 and 0.60 respectively. Fig. 3 (a) shows the SAC nanoparticles produced from *Wo* ratio of 0.20 compared to the sub-micron sized particles produced from other ratios (Figure 3b and 3c). A larger volume ratio of the water phase to surfactant results in smaller SAC particles [15]. Surfactant and co-surfactant forms the chemical interfacial layer in the reverse microemulsions. This chemical interfacial layer protects the nano-dispersion water phase in the reverse microemulsion. The ratios of 4.5 to 1.0 (Figure 3a) produced nano-sized microemulsions that led to SAC nanoparticles reduced at the cathode during electrodeposition.

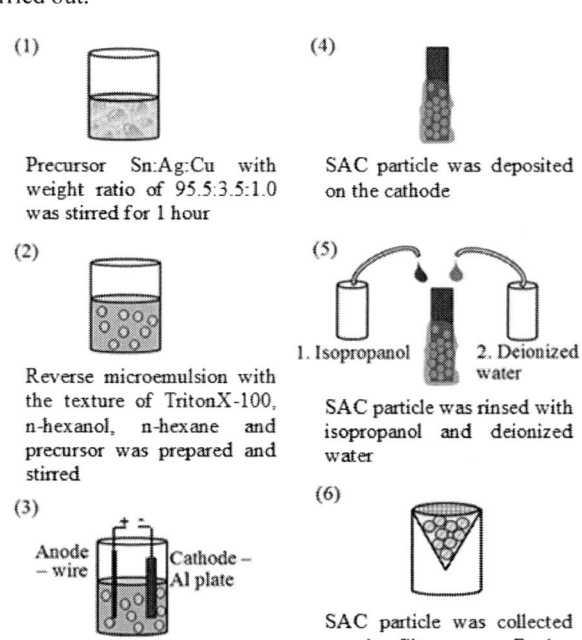

(1) Precursor Sn:Ag:Cu with weight ratio of 95.5:3.5:1.0 was stirred for 1 hour

(2) Reverse microemulsion with the texture of TritonX-100, n-hexanol, n-hexane and precursor was prepared and stirred

(3) Anode – wire / Cathode – Al plate — Electrodeposition with electropotential 5V at room temperature

(4) SAC particle was deposited on the cathode

(5) 1. Isopropanol 2. Deionized water — SAC particle was rinsed with isopropanol and deionized water

(6) SAC particle was collected on the filter paper. Beaker was sealed with filter paper and kept in the auto dry box

Fig. 1. Steps to prepare the reverse microemulsions electrolyte and subsequent electrodeposition to produce the SAC particles

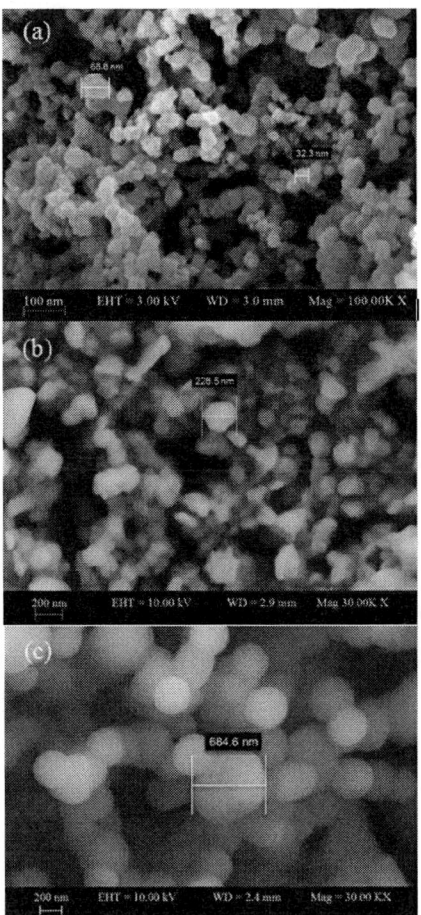

Fig. 3. SEM image of synthesized SAC particles in the microemulson ratio, *Wo* of (a) 0.20 (100K X), (b) 0.40 (10K X), (c) 0.60 (30K X)

Fig. 3 (a), (b) and (c) also show that the particles are slightly spherical in shape and well dispersed. The surfactant shells protected the growing metallic particles within the confine of microemulsion during electrodeposition and reduced the agglomeration of the alloy nanoparticles when the particles were collected with filter papers [5-7].

High surface tension of the chemical interfacial layer affects the materials exchange process during microemulsion stirring process [16]. High rigidity of the interfacial layer affects the deposition of the reduced particles at the cathode surface. The addition of short hydrophobic chain from co-surfactant decreases the surface tension and rigidity of this interfacial layer. Fig. 4 shows the schematic diagram of reverse microemulsion electrolyte. Our surfactant molecules (TritonX-100, $C_{14}H_{22}O(C_2H_4O)n$) possess long hydrophobic chain (i.e. $C_{14}H_{22}O$) which increase the surface rigidity of the microemulsions. Co-surfactant (n-hexanol, $CH_3(CH_2)_5OH$) has a shorter hydrophobic chain $CH_3(CH_2)_5$) which increases the penetration of the oil phase within the surfactant layer to reduce the interfacial rigidity, interfacial tension and increase microemulsion stability [17].

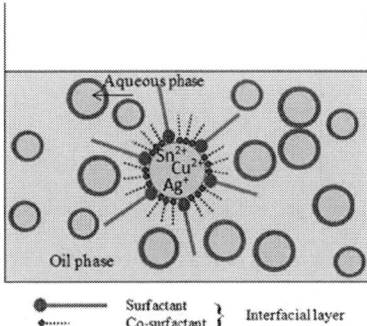

Fig. 4. Schematic diagram of the reverse microemulsion electrolyte

Compared to microemulsion (with reducing agent) method, our electrodeposition method can synthesize the solder nanoparticles without the addition of reducing agent [18-19]. The reducing agent is replaced by a reduction step at the cathode. Electrodeposition of reverse microemulsion electrolyte is a facile method to produce solder nanoparticles of different sizes and morphologies by simply adjusting the *Wo* ratio. The quality of the synthesized SAC particles is comparable with the solder particles synthesized by other method in terms of morphology, size and shape controllable solder nanoparticles.

IV. CONCLUSION

Nano sized solder particles consisting of Sn, Ag and Cu, were successfully synthesized by electrodeposition method with reverse microemulsions as electrolyte. This electrodeposition process produces nanoparticles ranging from 15 to 90 nm when the *Wo* ratio was 0.20. SEM results showed that the particles are well dispersed and slightly spherical in shape. XRD analysis indicated the presence of the components of SAC alloy such as Sn, Ag_3Sn and Cu_6Sn_5 in the synthesized particles. There were no obvious oxidation peaks in the XRD spectra, suggesting the presence and effectiveness of the passivating surfactant layer on the synthesized particles.

ACKNOWLEDGMENT

The authors gratefully acknowledge the financial support from Universiti Kebangsaan Malaysia through research grants AP-2012-016, DPP-2014-041 and GUP-2013-045.

REFERENCES

[1] R. Kay and M. Desmulliez, "A review of stencil printing for microelectronic packaging," Soldering & Surface Mount Technology, vol. 24, iss: 1, pp. 38-50, 2012.

[2] P. Vianco, J. Rejent and R. Grant, "Development of Sn-based, low melting temperature Pb-free solder alloys," Materials Transactions, vol. 45, No.3, pp. 765-775, 2004.

[3] M. Arturo Lopez-Quintela, "Synthesis of nanomaterials in microemulsions: formation mechanisms and growth control," Current

Opinion in Colloid & Interface Science, vol. 8, iss. 2, pp. 137-144, June 2003.

[4] R. Chukka, S. Telu, B. Nrmr, and L. Chen, "A novel method of reducing melting temperatures in SnAg and SnCu solder alloys," Journal of Materials Science: Materials in Electronics, vol. 22, iss. 3, pp. 281-285, March 2011.

[5] L.Y. Hsiao and J.G. Duh, "Synthesis and characterization of lead-free solders with Sn-3.5Ag-xCu (x = 0.2 , 0.5, 1.0) alloy nanoparticles by the chemical reduction method," Journal of The Electrochemical Society, vol. 152, pp. J105-J109, September 1, 2005.

[6] L.Y. Hsiao and J.G. Duh, "Revealing the nucleation and growth mechanism of a novel solder developed from Sn-3.5Ag-0.5Cu nanoparticles by a chemical reduction method," Journal of Electronic Materials, vol. 35, pp. 1755-1760, 2006/09/01 2006.

[7] C.Y. Lin, U.S. Mohanty and J.H. Chou, "Synthesis and characterization of Sn–3.5Ag–XZn alloy nanoparticles by the chemical reduction method," Journal of Alloys and Compounds, vol. 472, pp. 281-285, 2009.

[8] T.T. Bao, Y. Kim, J. Lee and J. G. Lee, "Preparation and Thermal Analysis of Sn-Ag Nano Solders," Materials Transactions, vol. 51, No. 12, pp. 2145-2149, 2010.

[9] X. Xia, C. Zou, G. Y. Gao, J. Liu, and Q. Zhai, "Preparation Techniques and Characterization for Sn-3.0Ag-0.5Cu Nanopowders," in Proceeding of High Density Packaging and Microsystem Integration, 2007, pp. 1-4.

[10] J. Solla-Gullón, A. Rodes, V. Montiel, A. Aldaz, and J. Clavilier, "Electrochemical characterisation of platinum–palladium nanoparticles prepared in a water-in-oil microemulsion," Journal of Electroanalytical Chemistry, vol. 554–555, pp. 273-284, 9/15/ 2003.

[11] X. Zhang, F. Zhang, R. F. Guan, and K. Y. Chan, "Preparation of Pt-Ru-Ni ternary nanoparticles by microemulsion and electrocatalytic activity for methanol oxidation," Materials Research Bulletin, vol. 42, pp. 327-333, 2/15/ 2007.

[12] C. Fu, H. Zhou, W. Peng, J. Chen, and Y. Kuang, "Comparison of electrodeposition of silver in ionic liquid microemulsions," Electrochemistry Communications, vol. 10, iss. 5, pp. 806-809, May 2008.

[13] J. Liang, N. Dariavach and D. Shangguan, "Metallurgy, Processing and Reliability of Lead-Free Solder Joint Interconnections," Micro- and Opto-Electronic Materials and Structures: Physics, Mechanics, Design, Reliability, Packaging, pp A351-A409, 2007.

[14] C. Zou, Y. Gao, B. Yang and Q.Zhai, "Nanoparticles of Sn3.0Ag0.5Cu alloy synthesized at room temperature with large melting temperature depression," J.Mater Sci: Mater Electron , vol. 23, pp. 2-7, 2012.

[15] L. Qi, "Synthesis of inorganic nanostructures in reverse micelles," Encyclopedia of Surface and Colloid Science, vol. 2, pp. 6183-6207, 2006.

[16] M.A. Malik, M.Y. Wani and M.A. Hashim, "Microemulsion method: A novel route to synthesize organic and inorganic nanomaterials: 1st Nano Update," Arabian Journal of Chemistry, vol. 5, iss. 4, pp. 397-417, October 2012.

[17] M.A. Lopex-Quintela, J. Rivas, M.C. Blanco, C. Tojo, "Synthesis of nanoparticles in microemulsions," Nanoscale Materials, pp135-155.

[18] J. Eastoe, M. J. Hollamby, and L. Hudson, "Recent advances in nanoparticle synthesis with reversed micelles," Advances in Colloid and Interface Science, vol. 128–130, pp. 5-15, 12/21/ 2006.

[19] E. Nourafkan and A. Alamdari, "Study of effective parameters in silver nanoparticle synthesis through method of reverse microemulsion," Journal of Industrial and Engineering Chemistry.

Hydrothermal growth of ZnO nanotubes on InGaP/GaAs/Ge Solar Cells

Chen-Chen Chung [a], Kung-Liang Lin [b], Hung-Wei Yu [a], Nguyen-Hong Quan [a],
Chang-Fu Dee [c] and Edward Yi Chang [a]*,IEEE Fellow

[a] Department of Materials Science and Engineering, National Chiao Tung University, Taiwan
[b] Mechanical and Systems Research Labs, Industrial Technology Research Institute (ITRI), Taiwan
[c] Institute of Microengineering and Nanoelectronics (IMEN), Universiti Kebangsaan Malaysia (UKM),
43600, Bangi, Selangor, Malaysia

Corresponding Author: **Edward Yi Chang**
Fax: (886)-3-5751826; Tel: (886)-3-5131536
E-mail: edc@mail.nctu.edu.tw
Department of Materials Science and Engineering National Chiao Tung University
1001 Ta-Hsueh Rd. Hsin-Chu 30050, Taiwan

Abstract — **A new design where ZnO nanotubes were grown on the antireflection (AR) layer coated on triple-junction (T-J) solar cell devices to enhance the light conversion efficiency. Compared to the bare T-J solar cells (without an AR layer), the performance of Si_3N_4 AR coated solar cell showed improvement. The sample with a layer of ZnO nanotubes grown in top of AR layer showed the lowest light reflection compared with the bare and solely AR coated T-J solar cell especially in the spectrum range of 350-500 nm. The use of ZnO nanotubes have increased the conversion efficiency by 4.9% compared with the conventional T-J solar cell. While the Si_3N_4 AR coated sample only increased the conversion efficiency by 3.2%. This result is quite encouraging as further refinement and variation in the experiment procedures could possibly bring more exciting performance in the future.**

Keywords—ZnO nanotube, Antireflection layer, Triple-junctions (T-J) solar cell, Hydrothermal

I. INTRODUCTION

In solar power technology, the III-V solar cell is so far the commercial solar cell with the highest efficiency. It is expected that III-V solar cell will play a major role in the future low-cost and high-efficiency photovoltaic market. Within the various solar power technologies, triple-junction (T-J) GaAs-based solar cells composing of three subcells, which are the top InGaP cell (Eg = 1.9 eV), middle GaAs cell (Eg = 1.42 eV), and bottom Ge cell (Eg = 0.66eV). It has achieved conversion efficiency of over 40% and is already applied extensively for the use in terrestrial and space [1]. For high performance multi-junction solar cell, the antireflection layer plays an important role because it can reduce about 30% light reflection on the top layer. In order to further reduce the reflection, plenty of approaches has been implemented, but most of them need expensive instruments or complicated fabrication processes. Therefore, this has become an attractive issue.

Zinc oxide (ZnO) has attracted worldwide attention for its novel applications on electronic and optical devices. It has wide bandgap (3.37 eV) and exciton binding energy of 60 meV at ambient temperature [2]. It has many different form of nanostructures. The use of nanostructures in solar cell technology may increase the performance. Compared with sole thin film devices, nanostructure based devices give larger response to light ranging from UV to Vis light [3-5], which can increase the top cell light absorption for short wavelengths. ZnO thin film not only acts as transparent conductive oxide (TCO) layer but also as antireflection (AR) layer (refractive index of 2.0).

There are plenty of approaches on fabricating one-dimensional (1D) ZnO nanotubes. The common method is by chemical vapor deposition. Though, it requires high processing temperature, ranging from 400 to 1000°C [6]. However, these high temperatures may degrade the device characteristics. Hydrothermal growth method of ZnO nanotubes therefore becomes one of the most intensively studied method due to its low processing temperature (below 100°C). It will therefore achieve the goal without degrading device characteristics properties [7]. In this paper, a layer of hydrothermal-grown ZnO nanotubes on top of the AR layer coated T-J solar cell was investigated.

II. EXPERIMENTAL

A p-type Ge substrate was used to grow the layer of InGaP / GaAs / Ge epitaxial layers by metal organic chemical vapor deposition system (MOCVD). The sources for MOCVD used for this process were arsine (AsH_3), trimethylgallium (TMGa), trimethylindium (TMIn) and phosphine (PH_3). They provided As, Ga, In and P respectively. Two different types of doping materials (n- and p-type, respectively) used in the growth were silane (SiH₄) and diethylzinc (DEZn). Fig. 1 shows the structure of InGaP / GaAs / Ge after the epitaxial process. P-type Ge substrate was used to grow a T-J solar cell as shown in Fig. 1(a) at a temperature of 650°C under chamber pressure of 50 mbar [8].

The grown epitaxial layers for T-J solar cell was then cleaned by trichloroethylene, then followed by acetone and methanol to remove any possible organic contaminants. Lastly it was rinsed in deionized water and dried using N_2 gas. The cleaning process to ensure a good contact on the following back electrode of Ti (500 Å) / Pt (600 Å) / Au (2500 Å) deposited by e-beam evaporator. It was then annealed at 390°C in H_2 ambient for 3 mins to form an ohmic contact. The top surface contact was achieved by a typical multilayer of Ni/Ge/Au/Ni/Au (60/500/1000/400/2500 Å).

For the second design, a 75 nm silicon nitride anti-reflection film was coated on the top surface by plasma-enhanced chemical vapor deposition (PECVD) to increase the penetration of incident light. The top contact has reduced the illuminated area by 6.22%. The final total area is about 0.125 cm².

In the third design, ZnO nanotubes were grown on top of the antireflection layer using the hydrothermal method. Prior to growth, the sample was placed vertically in a metal container with 60 ml solution. The solution is a mixture of 40 ml zinc nitrate hexahydrate $(Zn(NO_3)_2 \cdot 6H_2O)$, (0.025 mol/l)) and 10 ml hexamethenamine $(C_6H_{12}N_4$, (0.025 mol/l)). The metal container was sealed and heated at 90°C during the growth process.

The grown ZnO nanotubes were characterized by a field-emission scanning electron microscope (Hitachi S-4700I). The reflectivity of the samples were analyzed by ultraviolet-visible (UV-Vis) spectrophotometer using an integrating sphere. For solar cell measurement, the current-voltage (I-V) characteristics of the samples were measured under one sun AM1.5 solar simulator (100 mW/cm²).

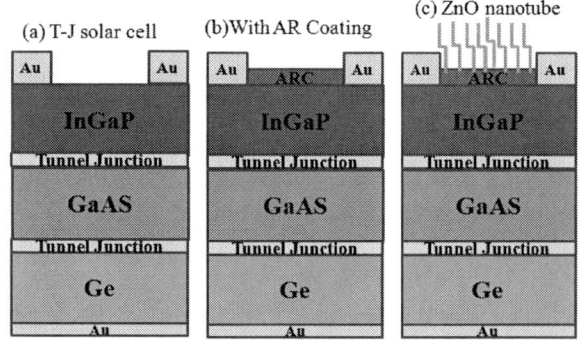

Fig. 1 Schematic diagrams of T-J solar cell and its variants.

III. RESULT AND DISCUSSION

Fig. 2(a) shows the top view of SEM images of the ZnO nanotube structure. The ZnO nanotubes have different diameters ranging from 80 to 100 nm respectively. The

inset shows the cross section of the as-grown ZnO nanotubes. Fig. 2(b) shows the Energy Dispersive Spectrometer (EDS) image of a ZnO nanotubes. It shows clearly the Zn and O elements on the cell.

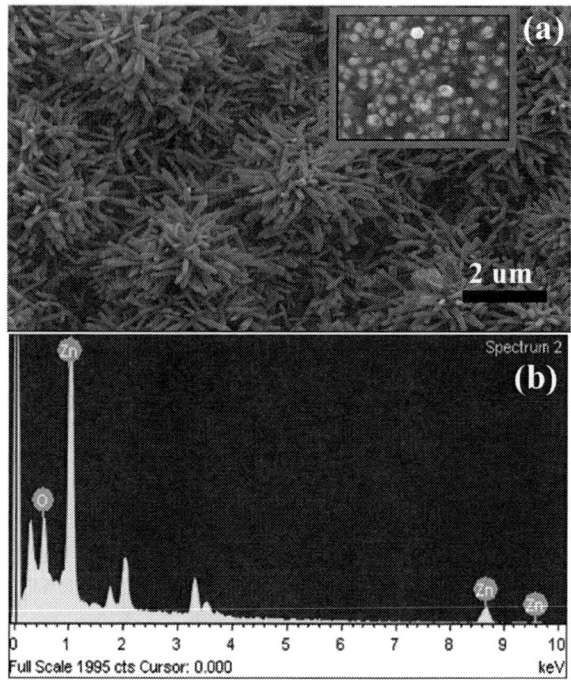

Fig. 2. (a) Plan-view of SEM images of the ZnO nanotubes structure; (b) Energy Dispersive Spectroscopy (EDS) of coated ZnO nanotubes.

The photovoltaic current-voltage (I-V) characteristic was measured in solar simulator under one sun AM1.5 (100 mW/cm²) illumination. The short-circuit current (I_{sc}), conversion efficiency (η), fill factor (FF) and open-circuit voltage (V_{oc}) were determined. The important factor can be explained both in formula (1) [9].

$$V_{OC} = \frac{nkT}{q} \cdot \ln\left(1 + \frac{I_L}{I_0}\right) \qquad (1)$$

where I_L, I_o, n, k and T are the light-generated current, dark saturation current density, diode ideality factor, Boltzmann constant, and absolute temperature.

Fig. 3 and Table 1 show the photovoltaic I-V data of bare T-J solar cells and T-J solar cells with Si_3N_4 and ZnO nanotubes AR coatings. For the device formed by standard process of the bare T-J devices, a short-circuit current (J_{sc}) of 12.48 mA/cm², an open-circuit voltage (V_{oc}) of 2.2 V, and conversion efficiency (η) of 19.3% were obtained. After Si_3N_4 AR coating and growth ZnO nanotubes on Si_3N_4 AR layer, the J_{sc} and solar cells conversion efficiency were increased to 13.2 and

13.9 mA/cm^2, 22.5% and 24.2%, respectively. The performance was improved when the ZnO nanotubes were deposited on the device surface of the T-J solar cells, ZnO nanotubes can improve solar cell efficiency. It increases the top cell light absorption for short wavelengths from 22.5 to 24.2%, corresponded to only 7.5% increment with only the AR coated sample. The short-circuit current (J_{sc}) also was increased from 13.2 to 13.9 mA/cm^2, which is 5.3%. The open-circuit voltage (V_{oc}) and fill-factor (FF) is better than sole AR coated sample. This is an indication for good series resistance of a solar device.

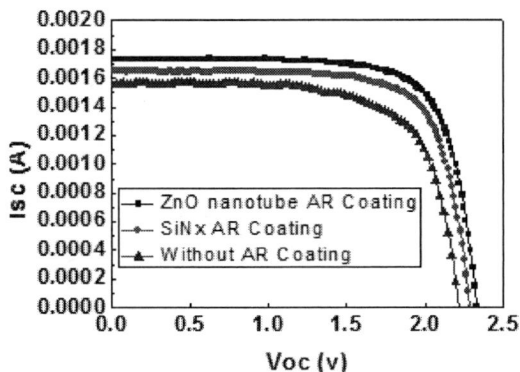

Fig. 3 I-V characteristics of bare T-J, Si$_3$N$_4$ coated and Si3N4 plus ZnO nanotubes coated solar cell.

Table 1. I-V data for T-J, Si$_3$N$_4$ coated and Si$_3$N$_4$ plus ZnO nanotubes coated solar cell.

Sample (solar cell)	V_{oc} (V)	J_{sc} (mA/cm^2)	FF (%)	Efficiency (%)
Bare T-J	2.2	12.48	70	19.3
Si$_3$N$_4$ coated	2.3	13.2	74.5	22.5
Si$_3$N$_4$ plus ZnO nanotubes coated	2.33	13.9	74.8	24.2

Fig. 4 shows the reflectance of a bare T-J, and Si$_3$N$_4$ coated and Si$_3$N$_4$ plus ZnO nanotubes coated solar cells, respectively. Since the ZnO nanotubes can suppress light back scattering especially at short wavelengths, the T-J solar cell with a ZnO nanotubes has the lowest reflectance, especially in the wavelength range of UV to green. The reflectance was ~5.7% for the wavelength region of 300 to1800 nm, which performed better than that of a cell with only Si$_3$N$_4$ AR layer which was ~18.1%. Therefore, the Si$_3$N$_4$ plus ZnO nanotubes coated T-J solar cell exhibited superior antireflective property.

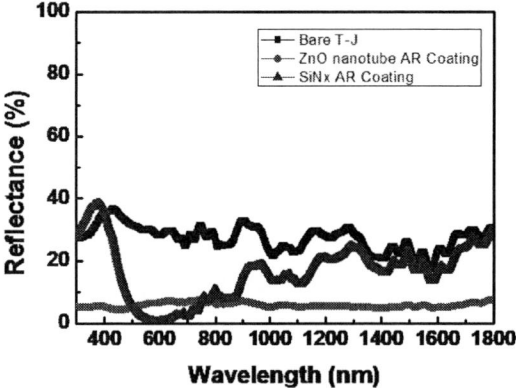

Fig. 4 Graph of reflectance under different wavelengths for bare T-J, Si$_3$N$_4$ coated and Si$_3$N$_4$ plus ZnO nanotubes coated solar cells.

The external quantum efficiency (EQE) of the bare T-J, Si$_3$N$_4$ coated and Si$_3$N$_4$ plus ZnO nanotubes coated solar cells, respectively, are presented in Fig. 5(b). Physically, EQE means the ability to generate electron-hole pairs caused by the incident photon [10]. The cell with ZnO nanotubes coating showed an enhanced EQE in a range from of 350 to 1800 nm. The average EQE enhancements (ΔEQE) of the top and middle cells were 2.5 and 6.6%, respectively due to the low reflection in the wavelengths of 350-500 nm.

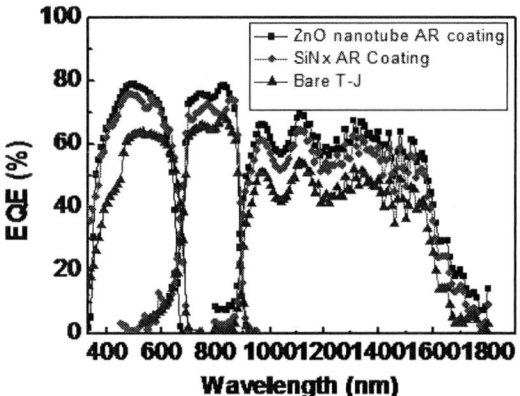

Fig. 5 External quantum efficiency (EQE) of bare T-J solar cell and T-J solar cell with Si$_3$N$_4$ and ZnO nanotube coating, respectively.

IV. CONCLUSIONS

ZnO nanotube has been grown on top of AR layer on a T-J solar cell devices by hydrothermal growth method. Reflectance measurement showed increase in the anti-reflectance properties ranging from visible to ultraviolet region. ZnO nanotubes have the lowest light reflection compared with the bare and solely AR coated T-J solar cell. Solar cells with ZnO nanotubes exhibited a 4.9% increment in conversion efficiency compared with a bare T-J solar cell, whereas T-J solar cells with Si$_2$N$_3$ AR coating had only 3.2%

increment. More experiments should be carried out to fine-tune the parameters in order to achieve higher conversion efficiency so that it could one day be implemented in a commercial solar cell.

ACKNOWLEDGMENT

We would like to give special thanks to the NCTU-UCB I-RiCE program, National Science Council of Taiwan, for sponsorship under Grant No. MOST 102-2911-I-009-302. We also are thankful for the support from the Industrial Technology Research Institute (ITRI).

REFERENCES

[1] A. Luque, A. Marti, and A.J. Nozik: *MRS Bull.* **32**, 236 (2007).

[2] Lyu SC, Zhang Y, RuhH, Lee HJ, Shim HW, Suh EK, Lee CJ. *Chemical Physics Letters* **363**, 134 (2002).

[3] J. Lee, N. Lakshminarayan, S. K. Dhungel, K. Kim, and J.Yi, *Sol. Energy Mater. Sol. Cells* **93**, 256 (2009).

[4] U. Gangopadhyay, S. K. Dhungel, A. K. Mondal, H. Saha, and J. Yi, *Sol. Energy Mater. Sol. Cells* **91**, 1147 (2007).

[5] P. Yu, C. H. Chang, C. H. Chiu, C. S. Yang, J. C. Yu, H. C. Kuo, S. H. Hsu, and Y. C. Chang, Adv. Mater. **21**, 1618 (2009).

[6] Huang MH, Wu YY, Feick H, Tran N, Weber E, Yang PD. *Advanced Materials*; **13**,113 (2001).

[7] Vayssieres L. *Advanced Materials*; **15**, 3870 (2005).

[8] C. Yen, C. T. Lee, *Sol. Energy.* **89**, 17 (2013).

[9] R. R. King, D. C. Law, K. M. Edmondson, C. M. Fetzer, G. S. Kinsey, H. Yoon, R. A. Sherif, and N. H. Karam, *Appl. Phys. Lett.* **90**, 183516 (2007).

[10] S. M. Sze, *Physics of Semiconductor Devices* (Wiley, 2nd Edition, 1981).

Effects of Thermal Cycling on the Mechanical Properties of Gold Wire Bonding

Wan Yusmawati Wan Yusoff, Azman Jalar
Institute of Microengineering and Nanoelectronics
Universiti Kebangsaan Malaysia
Malaysia
yusmawati@upnm.edu.my; azmn@ukm.edu.my

Norinsan Kamil Othman, Irman Abdul Rahman
School of Applied Physics
Faculty of Science and Technology
Universiti Kebangsaan Malaysia, Malaysia
insan@ukm.edu.my; irman@ukm.edu.my

Abstract— The mechanical properties of gold wire bonding are subjected to thermal cycling (TC) test has been investigated. Gold wire bonding was experienced to temperature cycle of (-65) °C to 150 °C for 10, 100, and 1000 cycles. In order to determine the mechanical properties of gold wire, nanoindentation test was performed. A constant load nanoindentation test was carried out at the center of the gold wire to investigate hardness and reduced modulus. The load-depth curve for the thermal cycled gold wire bond displayed apparent discontinuities during loading as compared to the as-received gold wire bond. The hardness value has increased after the gold wire bond subjected to thermal cycle whilst, the hardness value has decreased with the increment of the TC cycle number. For reduced modulus, the values increased with increase of the TC cycle number. The decrease in the hardness value is in line with theoretical grain size coarsening following thermal treatment. These nanoindentation results are important in assessing the strength of gold wire bond after exposure to the thermal cycles.

Keywords— *gold wire; micromechanical properties; nanoindentation; thermal cycle*

I. INTRODUCTION

Today's trends required the size of an electronic device continues to scale down. Quad-Flat No-Lead (QFN) is one of the promising electronic packages which meet the demand of many new mobile devices. In the electronics package industry, gold wire bonding is commonly used as an interconnection material due to its mechanical properties and superior electrical resistivity. The reliability and characterization activities of wire bond are facing more challenges because of the miniaturization of wire and bonding pitch. On the other hand, the conventional tests such as ball shear and wire pull test provides inconsistent and limited information regarding the deformation and dislocation on gold wire bonding [1]. Thus, the nanoindentation test becomes the suitable tool to analyze the deformation and dislocation phenomena on the gold wire bonding. This test provides the micromechanical properties in terms of hardness and elastic modulus values in the nanometer scale [2]. Also known as depth-sensing indentation (DSI), this test method is also used to study both the elastic and plastic properties of material [3,4].

During application and service conditions, an electronic package is subjected to temperature cycling exposure due to cyclic environment such as electronic equipment power switch on – off condition. The failure rate increases with the number of thermal cycles [5]. Previous studies have shown that the mechanical properties of bulk electronic materials such as solders are sensitive to temperature and strain rate [6,7]. Therefore, it is important to understand the ability of the package's material (gold wire) to survive mechanical stresses induced by alternating low and high temperature extreme condition. This research aims is to assess temperature cycling effects on the micromechanical properties of wire bonding, to develop better understanding of the reliability problems faced by the electronic packaging industry.

II. MATERIALS AND METHODS

A. Thermal Cycles

Quad-Flat No-Lead packages bonded with gold wire were used as samples. The thermal cycle test used is based on condition C of the JESD22-A104D [8], achieved using a thermal chamber (TPS, Tenney). The thermal cycles profile used in this investigation was (-65) °C to (+150) °C with upper and lower soak temperature time of 15 minutes. The temperature ramp was 14.3 °C/min and the cycle time lasted for 1 hour. Four categories of samples, based on the differing number of cycles namely as-received, 10, 100 and 1000 cycles of TC were denoted as TC 0, TC 10, TC 100 and TC 1000, respectively were prepared for the present analysis. In this study, the as-received (zero cycle of TC) package was used as control samples.

B. Cross-sectioning

Prior to indentation test, the package was first cross-sectioned perpendicular to the wire bonding interface. The packages were first mounted using cold mounting. Wet grinding was carried out to expose the gold wire from the QFN package using silicon carbide paper of grits 800, 1200, 2400 and 4000. Subsequently, the sectioned area was polished with 1 and 0.25 μm of diamond suspension on silk cloths. The cross-section of ball bond wire bonding is depicted in Fig. 1.

978-1-4799-5761-3/14 $31.00 © 2014 IEEE

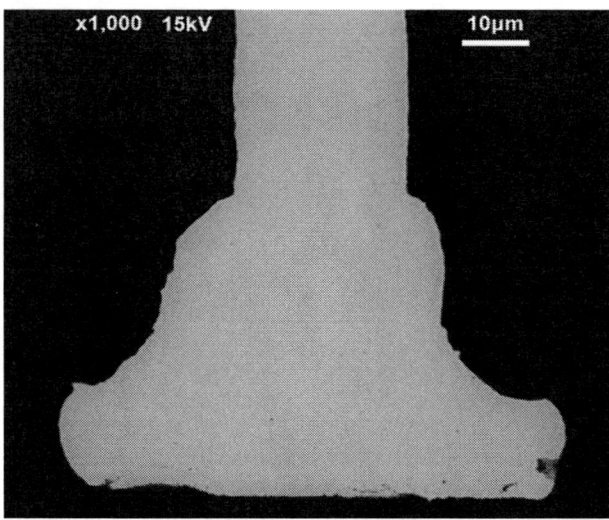

Fig. 1. Cross section of ball bond wire bonding

C. Nanoindentation Test

The indentation test was carried out using nanoindentation Nanotest™ (Micro Materials) with Berkovich diamond tip. The selected location for indentation was at the center of the gold ball bonds, where the microstructure is less influenced by the wire bonding process [9]. Indentations were performed with a constant loading and unloading rate of 0.5 mN/s. In this study, a 10 second hold time at the maximum load (10 mN) and for thermal drift correction, a 60 second hold time at 90 % unload was applied. The hardness and reduced modulus from load-displacement data obtained throughout the nanoindentation test were based on the Oliver and Pharr method [10].

III. RESULTS AND DISCUSSION

During indentation, the load, P and depth, h was recorded concurrently. Fig. 2 shows the P-h curves for the TC 0, TC 10, TC 100 and TC 1000 gold wire bond. Generally, the P-h profile reflects the mechanical behavior fingerprint of a material [10,11]. From this figure, it shows that after being subjected to the thermal cycle with 10 cycles, the gold wire exhibited the occurrence of pop-in events. These pop-in events occurred mainly due to the local discontinuities during loading. This discrepancy may be due to high diffusivity of Au atoms to promote the recrystallization nuclei [12]. The initial pop-in on the P-h profile is reflected by the shift from elastic to plastic behavior [13,14], while the subsequent pop-in corresponds to the nucleation of dislocation which is related to the material's microstructure [15]. Thus, the occurrence of pop-in events are more pronounced for the indentations in the gold wire with 10 cycles of TC as compared to as-received and other number of cycles.

Further analysis, the value of hardness and elastic modulus of gold wire bond can be extracted from the load-depth data using equations (1) and (2) which based on the method suggested by Oliver and Pharr [11]:

$$H = \frac{P_{max}}{A_c} \tag{1}$$

where P_{max} is the maximum load and A_c is the projected contact area of the indentation, which $A_c = 24.5\ h^2$.

$$\frac{1}{E^*} = \frac{(1 - v_{ind}^2)}{E_{ind}} + \frac{(1 - v_{Au}^2)}{E_{Au}} \tag{2}$$

where E is the Young's modulus and v is the Poisson ratio and the subscripts ind and Au are refer to indenter and gold wire bond, respectively. For indenter $E_{ind} = 1000$ GPa and $v_{ind} = 0.07$, and for gold $v_{Au} = 0.42$ [16].

Fig. 3 shows that the variations of hardness value for the as-received and gold wire with thermal cycle. The results show that the hardness of gold wire was inconsistent with the number of cycles. A hardness value of 0.91 GPa was displayed by the as-received sample. This value is still low in comparison to the previous studies done by Shah et al. [16] which found for that 20 mN maximum loads, the hardness value is about 1.35 GPa. This variation might be due to the behavior of small grain size of wire bond. After exposure to thermal cycling, the hardness was increased. However, the hardness was decreased as the cycle number was increased from 10 to 1000 cycles. The decrease in hardness value may be attributed to the grain size of gold experiencing a coarsening behavior of grain structure after being subjected to thermal cycling aging [17,18]. The increasing of gold grain size could be linked with a recrystallization process [19]. The yield strength is reflected to the onset of plasticity and it's closely related to the hardness value obtained using the Tabor equation [20]:

$$H = CY \tag{3}$$

where H is hardness, C is constraint factor and Y is yield strength. The constant factor is represented the hydrostatic component of a material and the value of $C_{metal} = 3$ [18]. Fig. 4 shows the value of yield strength with the number of cycles.

The variation of reduced modulus value (as shown in Fig. 5) shows a different pattern as compared to the hardness. It has been well documented that the reduced modulus is related to the intrinsic properties rather than the changes of microstructural [16]. However, the reduced modulus value for indentations in the gold wire indicates a distinct change. The reduced modulus value of gold wire was increased with increasing the number of thermal cycles as shown in Fig. 5. The reduced modulus suggests that the intrinsic properties such as crystallographic orientation [16] of gold wire bond changes in texture after exposure to thermal cycling. From the results above, it can be deduced that exposure to thermal cycle has a significant effect on the basic micromechanical properties of the gold wire bond. It was found that the hardness and reduced modulus changed after exposure to thermal cycling.

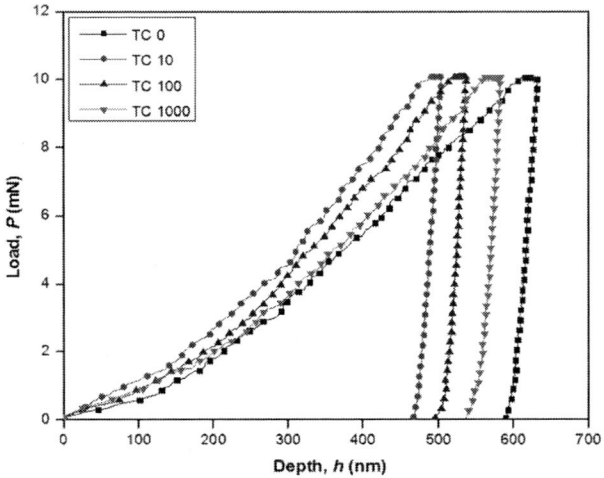

Fig. 2. Load versus depth profile

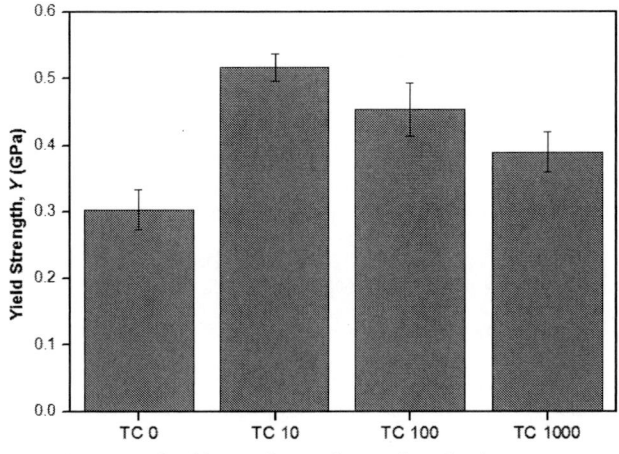

Fig. 4. Variations of yield strength towards a number of cycles

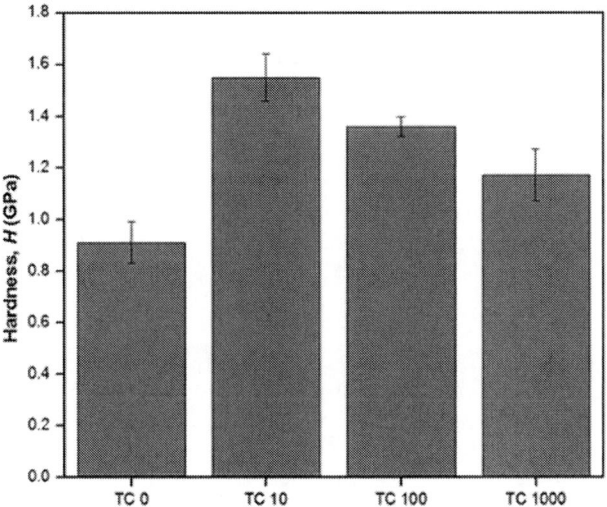

Fig. 3. Variations of hardness towards number of cycles

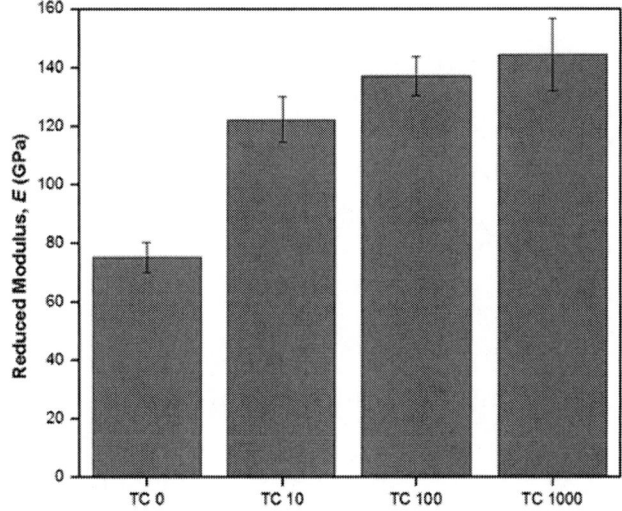

Fig. 5. Variations of reduced modulus towards number of cycles

IV. CONCLUSION

The temperature cycling effects on the micromechanical properties of gold wire were presented in this paper. Nanoindentation test has been proven to provide a more detailed and localized assessment in evaluating the strength of the gold wire bond after being exposed to thermal cycling, TC. It was observed that the *P-h* profiles for gold wire bond with 10 cycles exhibited pop-in events indicting the local discontinuities during loading. The hardness value of gold wire was decreased with increasing of cycles due to coarsening of grain size of gold wire.

ACKNOWLEDGMENT

The authors gratefully acknowledge research funding from OUP-2012-120, DIP-2012-14, DPP-2013-035, DPP-2014-041 and DLP-2013-037. The author Wan Yusmawati Wan Yusoff would also like to thank Universiti Pertahanan Nasional Malaysia and Ministry of Higher Education for her study leave and scholarship.

REFERENCES

[1] C.D. Breach, F.W. Wulff, "A brief review of selected aspects of the materials science of ball bonding. Microelectron. Reliab.," 50, pp. 1-20, January 2010.

[2] E.T. Lilleodden, W.D. Nix, "Microstructural length-scale effects in the nanoindentation behavior of thin gold films," Acta Materialia, 54, pp. 1583-1593, April 2006.

[3] G.M. Pharr, "Measurement of mechanical properties by ultra-low load indentation," J. Mater. Sci. Engin. A, 253, pp. 151-159, 1998.

[4] A.C. Ficher-Cripps, "Introduction Contact Mechanism," New York NY, Springer, 2009.

[5] S. Terashima, Y. Kariya, T. Hosoi and M. Tanaka, "Effect of silver content on thermal fatigue life of Sn-xAg0.5Cu flip-chip interconnects," J. Elect. Materials 32(12), pp. 1527-1533, December 2003.

[6] R. Darveaux, K. Banerji, Constitutive relations for tin-based solder joint, IEEE Trans. Comp. Hybrids Manufact. Technol. 15 (6), pp. 1013–1024, December 1992.

[7] X.Q. Shi, W. Zhou, H.L.J. Pang, Z.P. Wang, Z.P. Wang, Effect of temperaturenand strain rate on mechanical properties," J. Electron. Pack. 121, pp. 179–185, 1999.

[8] JESD22-A103C. 2004. High Temperature Storage Life. JEDEC STANDARD

[9] S. Murali, N. Srikanth and C.J. Vath III, "Effect of wire diameter on the thermosonic bond reliability," Microelectron Reliab 46: pp. 467-75, February-April 2006

[10] W.C. Oliver and G.M. Pharr. "An improved technique for determining hardness and elastic modulus using load and displacement sensing indentation experiments", J. Mater. Res. 7: pp. 1564-1583, 1992.

[11] C.A. Schuh, "Nanoindentation studies of materials", Mater. Today 9: pp. 32-40, May 2006,

[12] T.H. Chuang, C.C. Chang, C.H. Chuang, J.D. Lee and H.H. Tsai, "Formation and growth of intermetallics in an annealing-twinned Ag-8Au-3Pd wire bonding package during reliability tests,", IEEE Trans Comp Packag Manufact Tech 3(1), pp. 3-9, 2013.

[13] A. Gouldstone, H.J. Koh, K.Y. Zeng, A.E. Giannakopoulos and S Suresh, "Discrete and continuous deformation during nanoindentation of thin films," Acta Mater. 48, pp. 2277-2295, 2000.

[14] A. Gouldstone, N. Chollacoop, M. Dao, J. Li, A.M. Minor and Y. Shen, "Indentation across size scales and disciplines: Recent developments in experimentation and modeling," Acta Mater. 55, pp. 4015-4039, 2007

[15] B. Yang and H. Vehoff, "Grain size effect on the mechanical properties of nanonickel examined by nanoindentation, Mat. Sci. Eng. A-Struct. 400-401, pp. 467-470, 2005.

[16] M. Shah, K. Zeng, A.A.O. Tay and S. Suresh, "Mechanical Characterization of the Heat Affected Zone of Gold Wirebonds Using Nanoindentation, J. Electron. Packaging. 126, pp. 87-94, 2002.

[17] H.L.J. Pang, K.H. Tan, X.Q. Shi and Z.P. Wang, "Microstructure and intermetallic growth effects on shear fatigue of solder joints subjected to thermal cycling aging," Mater Sci Eng A 307, pp. 42-50, June 2001.

[18] H.L.J. Pang, T.H. Low, B.S. Xiong, X. Luhua and C.C. Neo. "Thermal cycling aging effects on Sn–Ag–Cu solder joint microstructure, IMC and strength," Thin Solid Films 462-463, pp. 370-375, September 2004.

[19] J.Y. Cho, K. Kim, J. Oh, J. Moon, J. Lee, Y. Cho and A. Rollett, "Recrystallization and grain growth of cold-drawn gold bonding wire," Metall. Mater. Trans. A. 34, pp. 1113-1125, 2003.

[20] D. Tabor, "The physical meaning of indentation and scratch hardness," J. Appl. Phys.7, pp. 159-166, 1959.

Parametric Analysis of Boost Converter for Energy Harvesting using Piezoelectric for Micro Devices

Shafii A. Wahab, M. S. Bhuyan,
Jahariah Sampe
Institute of Microengineering and Nanoelectronics (IMEN)
Universiti Kebangsaan Malaysia
43600 UKM Bangi, Selangor
Malaysia
shafii@ukm.edu.my

Sawal H. Md. Ali
Faculty of Engineering and Built Environment
Universiti Kebangsaan Malaysia
43600 UKM Bangi, Selangor
Malaysia
sawal@eng.ukm.my

Abstract— Lower amount of power delivered from piezoelectric based ambient vibration energy harvester devices is a barrier to adopt the technology for different applications. Energy harvesting circuitry can enhance power output to provide a regulated DC supply to the end application. In this paper, various circuit simulations are carried out to investigate output power enhancement. A parametric analysis of a boost circuit simulation using Cadence OrCAD Capture PSpice software with input less than 1 V is carried out to find the optimum parameters including, the switching frequency rise and fall times, duty cycle, inductance and load capacitance value. Simulation results show that passive component based boost converter can significantly increase the voltage output of an ambient vibration based energy harvester. The output voltage increases linearly with the increase of single supply voltage input range 0.1 V to 0.5 V, to the output voltage range of 7 to 35 V. The optimum parameter found for 10 kΩ load is 100 µH inductor and 1µF load capacitor. A comparison of output performance of the boost circuit with existing literature is presented. The ease of the boost converter circuit will facilitate the development of an efficient piezoelectric energy harvesters for low power applications like automotive, healthcare portable devices, and wireless sensor networks.

Keywords—Energy harvester; boost converter; piezoelectric; parmetric analysis

I. INTRODUCTION

Micro energy harvesting relates to the practice of harvesting small amounts of energy from ambient environmental sources (solar, wind, water, heat, vibration and etc.) in order to power either miniaturized Ultra-Low-Power (ULP) electronic system directly, or to charge a rechargeable battery or capacitor [1]. In general, energy harvesting system consists of three main blocks: the harvesting device (energy transducer), the harvesting circuitry and the application system (load) is shown in Fig. 1.

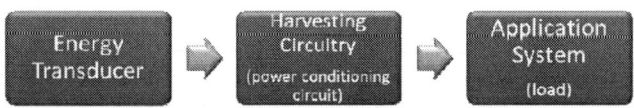

Fig. 1. Basic block diagram of an energy harvesting system.

Many research works for energy harvesting device has been carried out in the last twenty years [2-9], with the prime objective of a built-in energy harvesting system in various electronic applications, for example, Wireless Sensor Network (WSN) nodes. A usual WSN system could be comprised of even hundreds of thousands of nodes that gather and transmit desired data to a neighboring node or to a central collection point. Present design of a WSN nodes usually consist of a sensor and/or actuator, a micro processing unit to manipulate data, and a radio transceiver to receive and send collected data. These devices derive their value from distribution and their ability to be implemented in different locations that are random and difficult to access as being relatively mobile. Such nodes operate in a pulse power fashion with an X % duty cycle, spending (100-X) % of their lifetime in a sleep state where a node power consumption is at a bare minimum and only waking to a peak consumption state to take measurements or transmit and receive data. To power up such remote nodes, an obvious solution to power up is by means wire from a designated power supply source. However, it immediately undermines the value of WSN nodes being random and mobile in distribution and implementation. Another logical alternative could be the use of a battery. However, the choice of a primary battery is also not the optimal one. Because, if a WSN node has an average power consumption of 100 µW that is powered by a 1cm^3 lithium battery containing of 2,880 J energy, then the life expectancy would be less than a year. The cost and time spent replacing the primary batteries in a few devices of a small WSN network every year would be acceptable. However, the task of replacing thousands or hundreds of thousands of batteries in WSN nodes per year are widespread and difficult to access locations, is neither practical nor desirable. Other less common power supplies such as micro-fuel cells , micro-heat engines etc., are being considered for implementation however these technologies are still in development phase and not readily available for practical applications [17]. It occurs that the development of a suitable power supply with sufficient amount of delivered power is the major factor hindering the final development and deployment of WSN node into the industries [10].

978-1-4799-5761-3/14 $31.00 © 2014 IEEE

Energy harvester device optimization is one way in which the power density of a harvesting device can be significantly improved. Another way in which the power output can be enhanced is through the use of the 'harvesting circuitry' that is usually connected to the output of the harvesting device to condition and/or manage the electrical power output. Numerous studies have shown that power densities of energy harvesting devices can be hundreds of µW; however, literature also reveals that the power requirements of many electronic devices are in the mW range. Therefore, a key challenge for the successful deployment of energy harvesting technology remains, in many cases, the provision of adequate power.

II. BOOST CONVERTER CIRCUIT DESIGN AND ANALYSIS

A typical boost converter circuit consists of an inductor, a switch, a diode, and a capacitor as shown in Fig. 2. It can be divided into two modes; Mode 1 begins when the switch SW is turned on, and the input current flows through inductor L and switch SW. During this mode, energy is stored in the inductor. Mode 2 begins when the switch is turned off, and the current that was flowing through the switch SW would now flow through inductor L, diode D, capacitor C, and load resistor R [11]. The inductor current falls until the switch is turned on again in the next cycle. Energy stored in the inductor is then transferred to the load. Therefore, the output voltage is greater than the input voltage and is expressed in equation (1), where V_{out} is the output voltage, k is duty cycle, and V_{in} is input voltage which in this work will be the piezoelectric vibration [12].

$$V_{out} = \frac{1}{1-k} V_{in} \qquad (1)$$

Fig. 2. Schematic diagram of the conventional Boost Converter [4].

The research objectives are to investigate, design and simulate the Boost Converter of Energy Harvesting circuit using piezoelectric vibration for micro devices. The proposed simulation circuit is shown in Fig. 3 for source resistor Rs = 0.1 Ω, duty cycle D = 50 %, input voltage Vs = 0.5 V, load resistor R1 = 10 kΩ and fall time Tf = 10 ns.

The proposed power switch is using MOSFET IRF 540 because it is superior in high frequency applications where switching power losses are important. It also can be categorized as a fast switching device with low rise time, T_r (ns) and fall time, T_f (ns).

The proposed inductor value is chosen to compromise the fact that larger inductance increases the start-up time and small value allows the current to ramp up to higher levels before the switch turns off. The inductor of 180 uH is selected based on the maximum allowed ripple current at minimum duty cycle, at maximum input voltage for optimal operation mode of the converter. The primary criterion for selecting the output filter capacitor is its capacitance and equivalent series resistance, R_1. Since this capacitor affects the circuit efficiency, low capacitor of 1 uF is used for the best performance. The capacitor output will meet an output voltage ripple specifications, as well as its ability to handle the required ripple current stress. The Schottky diode of 1N5822 is used as a DC-DC converter switch because it has small voltage drop as compared to PN diode. The completed model of the Boost converter circuit is simulated using PSpice software.

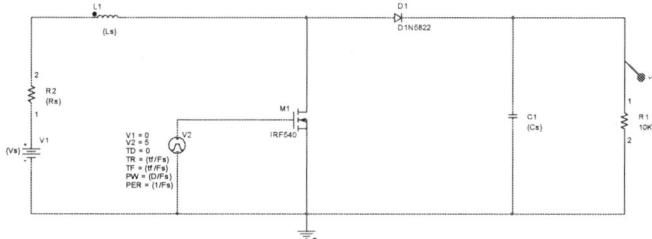

Fig. 3. The proposed Boost Converter Circuit

III. SIMULATION RESULTS ANALYSIS

The designed Boost Converter energy harvester circuit as shown in Fig. 3 has been improved and also validated using PSpice software. The boost converter also compared to previous research parameter which has been done in [13]. The simulation has been performed by varying the circuit parameters such as the input voltage Vs, source resistance Rs and rise time of switching frequency, T_{rise}.

As the main objective of the research is to design the Boost Converter for micro energy harvesting circuit, the input voltage variation has been investigated. Therefore, the input voltage is varied from minimum 0.1 V to 0.5 V and the simulated results is shown in Fig 4. The result showed that the maximum output voltages equal to 7 V and 35 V for 0.1 V and 0.5 V input voltages respectively. The other circuit parameters (Rs = 0.1 Ω, frequency = 1 kHz and duty cycle = 50%) are kept constant during the simulation process. As a result, the design circuit is capable to harvest energy from the vibration as small as 0.1 V. The respective output voltage of 7 V also sufficient to operate the micro integrated circuit device.

It is necessary to include the source resistance, R_s as this parasitic resistance will create uncertainty characteristic in the circuit. The Rs value should be kept smaller for ensuring small power losses in the circuit. Therefore, the source resistance is varied from minimum 0.01 Ω to 10 Ω and the simulated results is shown in Fig 5.

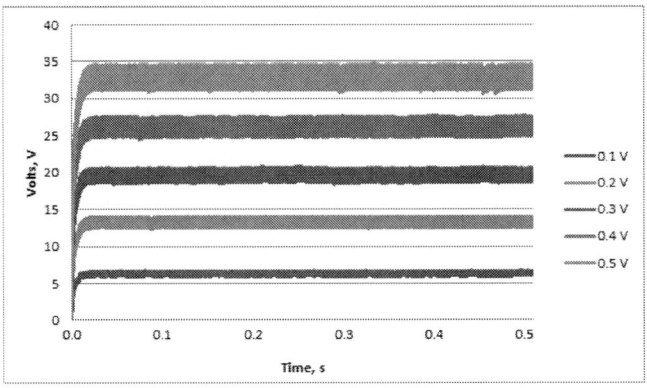

Fig. 4. The output voltages for different input voltages,Vs.

The result showed that the maximum output voltages equal to 1 V and 42 V for 10 Ω and 0.01 Ω source resistance respectively. The other circuit parameters (R_s = 0.1 Ω, frequency = 1 kHz and duty cycle = 50 %) are kept constant during the simulation process. Even though the R_s equal to 0.01 Ω has given the maximum output voltage of 42 V, but this is not a practical value. As a result, the R_s equal to 0.1 Ω which gives maximum output voltage of 35 V is choosing for the designed circuit after compromising the limited factor.

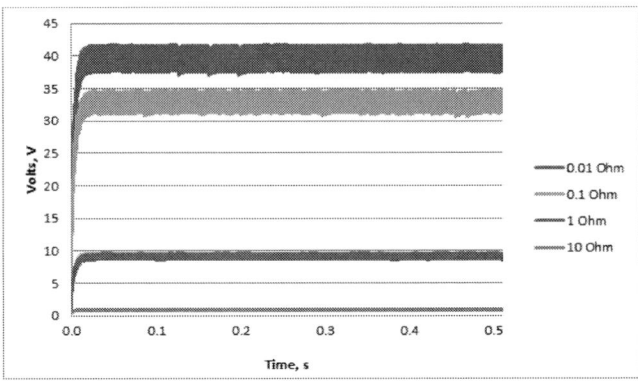

Fig. 5. The output voltages for different source resistance, Rs.

The optimal inductance is determined based on the maximum allowed ripple current at the minimum duty cycle and at the maximum input voltage. Therefore, the simulation is performed by varying the inductance value from the minimum of 10 uH to the maximum of 500 uH. The results as shown in Fig. 6 (a) and (b) show that the maximum output voltage of 35V can be obtained using 100 uH inductor. As a result, the inductor optimal value should be chosen base on the converter operating mode in the required load is equal to 100 uH.

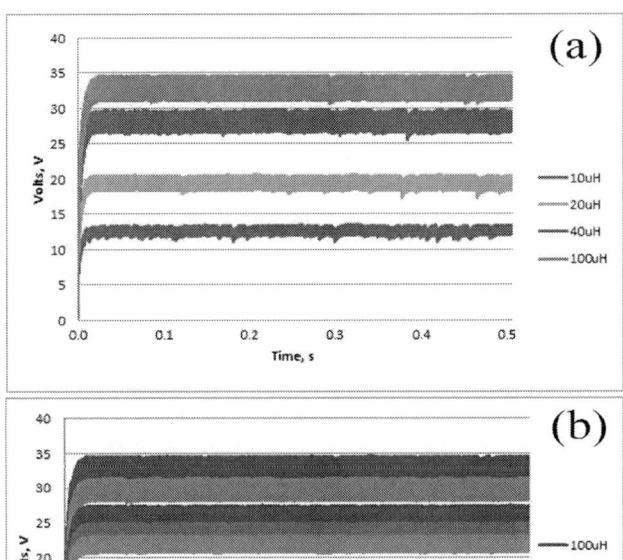

Fig. 6. The output voltages for inductance values, Ls; (a) Less and equal to 100 uH; and (b) More and equal to 100 uH.

TABLE 1. BOOST CONVERTER COMPARISON FROM THE PREVIOUS LITERATURE

DC-DC Boost Characteristic	[14]	[15]	[16]	[13]	This work
Input voltage, Vs	0.12V	0.25-0.4V	1.5V	0.1-0.5V	0.1-0.5V
Output voltage, Vout	1.2V	3.3V	7.5V	4-22V	7-35V
Sampling frequency, Fs	3 MHz	170 kHz	100 kHz	1 k – 10 kHz	1kHz
Inductance, L	4.5 µH & storage capacitor of 300 nF	Not mentioned	470 µH	160 µH	100µH
Load Resistance, R1	10 kΩ	33 kΩ	Not mentione d	10 kΩ	10kΩ
Load Capacitance, C1	30 nF	50 pF	225 mF	10 µF	1µF
Duty Cycle, D	0.75	0.5	0.8	0.5	0.5
Tr and Tf	60 µs	20 ns	Not mentione d	10 µs	10 µs
Method/ topology	Two boost converter	3 boost converter and 1 charge pump circuit	Ideal Switchin g controller	Switching gate boost converter	Switching gate boost converter

IV. CONCLUSION

An ultra-low voltage boost converter circuit for low voltage vibration-based energy harvester is presented. Parametric simulations were carried out to identify the boost circuit's passive components effect to boost output voltage. From the simulation result analysis the optimum value of the inductive resistance, switching rise and fall times, load and inductor were identified. The proposed boost circuit approach can overcome the problem of low input voltage from a low frequency vibration based energy harvester. The optimum boost circuit components for a low voltage vibration-based energy harvester are found L = 100 μH, load = 10 kΩ and switching frequency = 1 kHz with duty cycle D = 0.5 and Rs = 0.1 Ω. The boost circuit is presently under further investigation to integrate it with the power management circuit of a vibration based energy harvester.

REFERENCES

[1] Guyomar D, Badel A, Lefeuvre E, Richard C, "Toward energy harvesting using active materials and conversion improvement by nonlinear processing," IEEE Transactions on Ultrasonics Ferroelectrics and Frequency Control 2005; 52(4) 584-595.

[2] Satyanarayanan, M, "Avoiding dead batteries," *IEEE Pervasive Comput*, 4(1), 2005, pp. 2-3.

[3] Gates, B, The disappearing computer. The world in 2003 (The Economist), 2002. Cited in: Roundy, S, "Energy scavenging for wireless sensor nodes with a focus on vibration-to-electricity conversion," University of California, Berkerley, 2003 (Powerpoint presentation).

[4] Wright, M, "Harvesters gather energy from the ether, power lightweight systems," Electronic Design News, Dec 2006, pp. 57-62.

[5] Brown, C, "Endless energy is Harvesting's Promise," EE Times, Feb 2006, (3 pp.).

[6] Zhang, P.; Sadler, C. M.; Lyon, S. A. and Martonosi, M, "Hardware design experiences in zebranet," In: 2nd ACM conference on embedded networked sensor systems (SenSys '04), Baltimore, MD, USA, Nov 3-4 2004.

[7] Inversen, W, "Scavenging energy for wireless sensor networks," Automation World, March 2005, p. 24.

[8] IET electronic systems & software sector news (19th December 2006) "UK Government puts money into energy scavengers," http://www2.theiet.org/oncomms/sector/electronics/news.cfm/ (accessed 15th August 2007).

[9] Randall, Julian F, Designing Indoor Solar Products: Photovoltaic Technologies for AES. Chichester: John Wiley, 2005.

[10] Elder, M, "Engineering Project Proposal 2011," (University of Queensland), http://ceit.uq.edu.au/system/files/blog/thesis_proposal2.pdf (posted 12th June 2012, accessed 4th November 2013).

[11] Mohd Izam and Jahariah Sampe, "Development of a Microcontroller-Based Boost Converter for Laptop Powering System," International Conference in Engineering Technology, 2011 (ICET 2011), Kuala Lumpur

[12] N. Mohan, T.M. Undeland and W.P. Robbins, Power Electronics. USA: John Wiley & Sons, 2003, pp. 172-178.

[13] Nurul Arfah Che Mustapha, A.H.M. Zahirul Alam, Sheroz Khan, Amelia Wong Azman, "Parametric Analysis for Designing Low Voltage and Low Frequency Energy Harvester Booster," IEEE Regional Symposium on Micro & Nanoelectronics 2013, Langkawi, Malaysia, 2013, pp 122-125

[14] A. Richelli, S. Comensoli, and Z. M. Kovacs-Vajna, "A DC/DC boosting technique and power management for ultralow-voltage energy harvesting applications," IEEE Transactions on Industrial Electronics, vol. 59, no. 6, pp. 2701–2708, Jun. 2012.

[15] A. Bertacchini, S. Scorcioni, M. Cori, L. Larcher, and P. Pavan, "250mv input boost converter for low power applications," IEEE International Symposium on Industrial Electronics, 2010, pp.533–538.

[16] M. Tanaka, P. Hemthavy, and K. Takahashi, "Optimization of control parameters of a boost converter for energy harvesting," Journal of Physics: Conference Series, vol. 379, p. 012022, Aug. 2012.

[17] S. Shukla, Micro Fuel Cells: A Novel Breakthrough in Portable Power, Yale Scientific Magazine, 2013.

Nanostructured Al-doped ZnO-Based Gas Sensor Prepared Using Sol-Gel Spin-Coating Method

[1]A.K. Shafura, [1]I. Saurdi, [1]N.E.A. Azhar, [1]M.H. Mamat, [1]Uzer M and [1,2]M. Rusop
[1]NANO-EleTronic Centre, Faculty of Electrical Engineering,
[2]NANO-SciTech Centre, Institute of Science,
Universiti Teknologi MARA, 40450 Shah Alam, Malaysia
[1]shafura@ymail.com, [1,2]rusop@salam.uitm.edu.my

A. Shuhaimi
Low Dimensional Materials Research Centre, Department of Physics, Faculty of Science, University of Malaya, 50603 Kuala Lumpur, Malaysia

Abstract—**Nanostructured Aluminium (Al) doped zinc oxide (ZnO) was prepared using sol-gel spin-coating method. These films were tested under different exposure of oxygen flow rates at room temperature with bias voltage applied at 5 V. The structural properties were characterized using Atomic Force Microscopy (AFM) and Field Emission Scanning Electron Microscopy (FESEM). The fesem image revealed the surface morphology of nanostructured ZnO. The diameters size of nanostructured Al-doped ZnO thin film was observed in range of 16 – 46 nm. These thin films were tested for oxygen-sensing characteristic by varying the gas flow rates at room temperature. The nanostructured Al-doped ZnO-based gas sensor exhibited good sensitivity at low flow rates of oxygen exposure.**

Keywords—*nanostructured zinc oxide; electrical properties; structural properties; oxygen-sensing characteristic*

I. INTRODUCTION

In recent year, gas sensor gained rapid development in various areas to improve its performance. Gas sensor has been used in many areas such as public health, environmental monitoring, domestic safety and sensors networks. Due to this large application potential the need of low-cost, small, low power consuming and reliable solid state gas sensors, has grown exponentially.

ZnO relative to other materials is a strong candidate for industrial applications because of its low price. Zinc oxide (ZnO) is a unique n-type semiconductor material having direct band gap of 3.37 eV at room temperature[1]. It allows excitonic emission process happens at or even above room temperature due to its larger excitation binding energy of 60 meV at room temperature [2]. ZnO poses the wurzite family of structures which also has numerous types of morphology such as nanocombs, nanowires, nanorods and nanoflakes [3-5]. ZnO thin films was reported can be deposited using magnetron sputtering [6-9], spray pyrolisis[10-12], CVD [13, 14] and sol-gel [15-17]. Y. Hou et al. reported Al-doped ZnO thin film that was deposited using sol-gel process by spin-coating technique. The doped thin film was prepared using zinc acetate-based precursor exhibited better gas sensing properties and higher conductivity. It is might due to Al donors that enhance the electron concentration [17]. H. Minaee et al. reported the impact of different morphology on oxygen sensing properties. The ZnO nanowires resulted the highest gas sensitivity at 50 °C as compared to nanorods and nanoflowers [18].

Also, the concern of electronic device performance limitation using bulk materials has leaded a way to nanotechnology applications. By introducing nanotechnology in material preparation, smaller particles could be synthesized in nanoscale thus can enhance the sensing performance [19]. In this work, nanostructured Al-doped ZnO thin film was prepared using sol-gel spin-coating method which is comparatively low cost and simple technique. We study the effect of oxygen flow rates on the gas sensing properties of the thin film.

II. EXPERIMENTAL PROCEDURE

A. Thin film deposition method

Nanostructured Al-doped ZnO thin films were deposited onto glass substrates using sol-gel spin-coating method. In order to prepare a solution of 0.4 M with 1 at.% doping, zinc acetate dehydrate ($Zn(CH_3COO)_2.2H_2O$), monoethanilamine (MEA, $C_2H_7N_{14}$), aluminium nitrate nanohydrate ($Al(NO_3)_3 \cdot 9H_2O$) and 2-methoxethanol were used as starting material, stabilizer, dopant source and solvent respectively. The solution then was heated at 80 °C and magnetically stirred using hotplate magnetic stirrer for 3 hours to yield a homogeneous and clearly colorless solution. The solution was further stirred for 24 hours for ageing process at room temperature.

The prepared solution was dropped onto glass substrates using spin-coating technique with spin speed of 3000 rpm for 60 s. After that, the substrates were preheated at 150 °C for 10 minutes to evaporate the solvent and eliminate the organic residuals in the film. In this work, the coatings were repeated 5 times to achieve the desired thickness. Finally, the substrates were annealed in air ambient at 500 °C for an hour using a furnace.

B. Fabrication of oxygen sensor

To complete the sensor structure, gold (Au) electrodes were deposited onto the thin films using thermal evaporator with the presence of a mask. The current was set to 75 mA to achieve the desire thickness. In this work, the sensors were tested by varying the gas flow rate that was controlled using mass flow meter controller (MFC). Then the sensors have been characterized using current-voltage (I-V) measurement system (Keithley 2400).

C. Characterization method

There were 2 types of characterization that have been used to study the characteristics of nanostructured Al-doped ZnO thin films. The structural properties were observed using Field Emission Scanning Electron Microscopy (FESEM,JEOL JSM6360LA) and Atomic Force Microscopy (AFM, Park System). Electrical properties were investigated using current-voltage (I-V) measurement system which consists of two probes and power supply system.

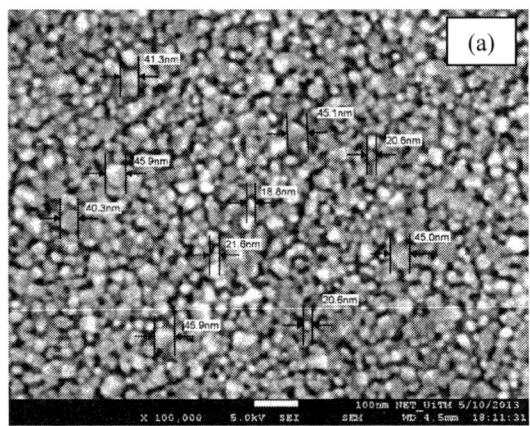

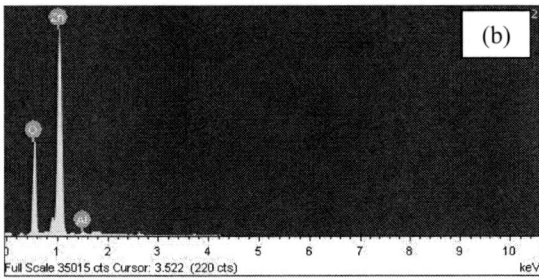

Fig. 1. (a) FESEM image and (b) EDS spectrum of nanostructured Al-doped ZnO thin film annealed at 500 °C.

III. RESULTS AND DISCUSSION

A. Structural properties

Fig. 1(a) shows the FESEM image of the surface morphology at 100000× magnification of the nanostructured Al-doped ZnO thin film. It was observed that the grain size in diameter ranging between 18 – 46 nm. The FESEM image also indicates that the nanoparticles are well-connected to each other since it is very important for electron to move and develop continous transport pathway in the granular film. The optimum annealing temperature (i.e. 500 °C) might helps to reduce the defect and instability between the nanoparticles, as

the atom could move to its favorable position thus increase the electron movement [20]. The porous nature might enhance the sensing properties since the diffusion rate of gas diffused into the pores and the surface reaction become higher hence contributes to faster carrier mobility [21]. Fig. 1(b) shows the EDS spectrum of the thin film which indicates Al doping was successfully performed in this work. According to the EDS spectrum result, the peak of Zn, O and Al were found with atomic ratio of 63.58: 35.91: 0.42. The root mean square roughness of the thin film was approximately 7.157 nm which was measured using AFM.

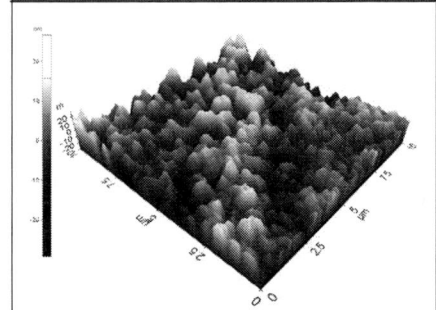

Fig. 2. AFM topography of Al-doped ZnO thin film.

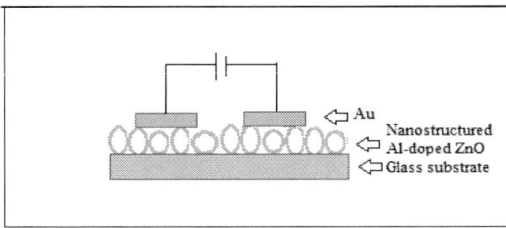

Fig. 3. Schematic cross-sectional illustration of the device configuration of nanostructured Al-doped ZnO.

Y. Hou et al. reported Al-doped ZnO thin film that was deposited using sol-gel process by spin-coating technique. The doped thin film was prepared using zinc acetate-based precursor exhibited better gas sensing properties and higher conductivity. It is might due to Al donors that enhance the electron concentration [17]. H. Minaee et al. reported the impact of different morphology on oxygen sensing properties. The ZnO nanowires resulted the highest gas sensitivity at 50 °C as compared to nanorods and nanoflowers [18].

B. Electrical properties

The Fig. 3 illustrates the cross-sectional view of device configuration of nanostructured Al-doped ZnO thin film. In this work, the sensitivity behavior of the Al-doped ZnO-based gas sensor was investigated by varying the oxygen flow rate ranging between 50 sccm to 150 sccm. The bias voltage that has been applied was set to 5 V. The gas sensitivity (S) is defined as the electric current generated in the gas sensor:

$$S\,(\%) = ((I_{gas} - I_{air})/I_{air}) \times 100 \qquad (1)$$

Fig. 4 shows the current vs time curve of nanostructured Al-doped ZnO when exposed to oxygen at room temperature. It is shown that there is a decreasing trend on the sensitivity behavior of the gas sensor when it was exposed to higher oxygen flow rate which eventually reaching saturation. It is might due to the slow desorption kinetics of the oxygen gas on the surface of diffusion area as the gas flow rate increases.

Fig. 5 shows the real-time sensitivity of nanostructured Al-doped ZnO-based gas sensor in the presence of different oxygen flow rates at room temperature. The sensitivity

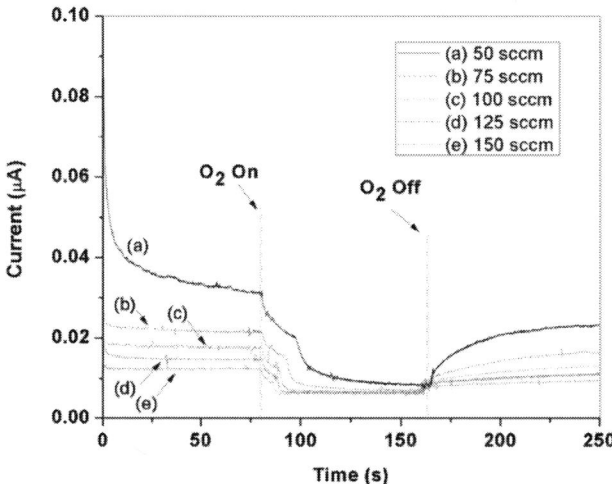

Fig. 4. Current vs time curve of the nanostructured Al-doped ZnO thin film at different oxygen flow rate when 5 V bias voltage was applied.

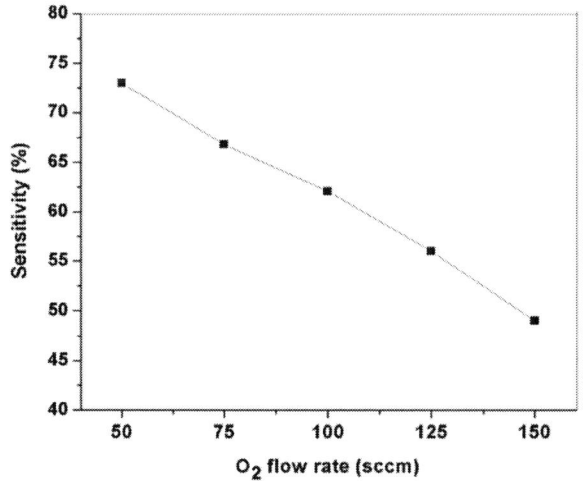

Fig. 5. Real-time sensitivity of nanostructured Al-doped ZnO-based sensor under differenet flow rate of O_2 exposure at room temperature.

reached to 73.01%, 66.79%, 62.09%, 55.99% and 49.01% when exposed to oxygen flow rates of 50, 75, 100, 125 and 150 sccm, respectively, for 80 s. The figure clearly seen that the sensitivity behavior decreases when the gas sensor was exposed to higher oxygen flow rate. It was observed that the

Al-doped ZnO-based gas sensor is more efficient for oxygen-sensing at lower rate of oxygen flow at room temperature.

IV. CONCLUSION

Nanostructured Al-doped ZnO-based gas sensor was successfully prepared on glass substrate using sol-gel spin-coating method. Based on AFM result, the roughness of the Al-doped ZnO thin film was approximately 7.157 nm. The result suggested that the nanostructured Al-doped ZnO thin film exhibited good sensitivity of 73.01% in the presence of 50 sccm of oxygen. The nanostructures might enhance the gas sensing properties due to its high volume-to-surface ratio.

ACKNOWLEDGMENT

The author would like to thank Research Management Institute (RMI), UiTM through the project 600-RMI/DANA 5/3/CIFI (139/2013) and Ministry of High Education (MOHE), Malaysia for the financial support.

REFERENCES

[1] V. A. Coleman and C. Jagadish, "Chapter 1 - Basic Properties and Applications of ZnO," in *Zinc Oxide Bulk, Thin Films and Nanostructures*, J. Chennupati and P. Stephen, Eds., ed Oxford: Elsevier Science Ltd, 2006, pp. 1-20.

[2] C. Klingshirn, "The Luminescence of ZnO under High One- and Two-Quantum Excitation," *physica status solidi (b)*, vol. 71, pp. 547-556, 1975.

[3] M. H. Mamat, Z. Khusaimi, M. Z. Musa, M. F. Malek, and M. Rusop, "Fabrication of ultraviolet photoconductive sensor using a novel aluminium-doped zinc oxide nanorod–nanoflake network thin film prepared via ultrasonic-assisted sol–gel and immersion methods," *Sensors and Actuators A: Physical*, vol. 171, pp. 241-247, 11// 2011.

[4] Y. S. Lim, J. W. Park, S. T. Hong, and J. Kim, "Carbothermal synthesis of ZnO nanocomb structure," *Materials Science and Engineering: B*, vol. 129, pp. 100-103, 4/15/ 2006.

[5] P. S. Venkatesh and K. Jeganathan, "Investigations on the growth and characterization of vertically aligned zinc oxide nanowires by radio frequency magnetronsputtering," *Journal of Solid State Chemistry*, vol. 200, pp. 84-89, 4// 2013.

[6] R. Ondo-Ndong, F. Pascal-Delannoy, A. Boyer, A. Giani, and A. Foucaran, "Structural properties of zinc oxide thin films prepared by r.f. magnetron sputtering," *Materials Science and Engineering: B*, vol. 97, pp. 68-73, 1/15/ 2003.

[7] N. H. Kim and H. W. Kim, "Room temperature growth of zinc oxide films on Si substrates by the RF magnetron sputtering," *Materials Letters*, vol. 58, pp. 938-943, 2// 2004.

[8] Y. Xiaozheng, X. Zheng, and S. Zhigang, "Metal copper films deposited on cenosphere particles by magnetron sputtering method," *Journal of Physics D: Applied Physics*, vol. 40, p. 2894, 2007.

[9] Y. M. Lu, W. S. Hwang, W. Y. Liu, and J. S. Yang, "Effect of RF power on optical and electrical properties of ZnO thin film by magnetron sputtering," *Materials Chemistry and Physics*, vol. 72, pp. 269-272, 2001.

[10] "Physical properties of Ga-doped ZnO thin films by spray pyrolysis," *Journal of Alloys and Compounds*, vol. 506, p. 788, 2010.

[11] N. Jabena Begum, R. Mohan, and K. Ravichandran, "Effect of solvent volume on the physical properties of aluminium doped nanocrystalline zinc oxide thin films deposited using a simplified spray pyrolysis technique," *Superlattices and Microstructures*, vol. 53, pp. 89-98, 1// 2013.

[12] S. M. Rozati, F. Zarenejad, and N. Memarian, "Study on physical properties of indium-doped zinc oxide deposited by spray pyrolysis technique," *Thin Solid Films*, vol. 520, pp. 1259-1262, 12/1/ 2011.

[13] M. Labeau, P. Rey, J. L. Deschanvres, J. C. Joubert, and G. Delabouglise, "Thin films of high-resistivity zinc oxide produced by a modified CVD method," *Thin Solid Films,* vol. 213, pp. 94-98, 5/29/ 1992.

[14] C. L. Wu, L. Chang, H. G. Chen, C. W. Lin, T. F. Chang, Y. C. Chao, *et al.,* "Growth and characterization of chemical-vapor-deposited zinc oxide nanorods," *Thin Solid Films,* vol. 498, pp. 137-141, 3/1/ 2006.

[15] M. H. Mamat, Z. Khusaimi, M. Z. Musa, M. Z. Sahdan, and M. Rusop, "Novel synthesis of aligned Zinc oxide nanorods on a glass substrate by sonicated sol–gel immersion," *Materials Letters,* vol. 64, pp. 1211-1214, 5/31/ 2010.

[16] H. Li, J. Wang, H. Liu, H. Zhang, and X. Li, "Zinc oxide films prepared by sol–gel method," *Journal of Crystal Growth,* vol. 275, pp. e943-e946, 2/15/ 2005.

[17] Y. Hou, A. M. Soleimanpour, and A. H. Jayatissa, "Low resistive aluminum doped nanocrystalline zinc oxide for reducing gas sensor application via sol–gel process," *Sensors and Actuators B: Chemical,* vol. 177, pp. 761-769, 2// 2013.

[18] H. Minaee, S. H. Mousavi, H. Haratizadeh, and P. W. de Oliveira, "Oxygen sensing properties of zinc oxide nanowires, nanorods, and nanoflowers: The effect of morphology and temperature," *Thin Solid Films,* vol. 545, pp. 8-12, 10/31/ 2013.

[19] N. Yamazoe, "Toward innovations of gas sensor technology," *Sensors and Actuators B: Chemical,* vol. 108, pp. 2-14, 7/22/ 2005.

[20] M. Z. Sahdan, M. Mohammad Rusop, M. H. Mamat, M. Awalludin, and Z. Mohamad, "Zinc Oxide Nanorods Characteristics Prepared by Sol-Gel Immersion Method Immersed at Different Times," *Advanced Materials Research,* vol. 667, pp. 375-379, 2013.

[21] H. Shang and G. Cao, "Nanostructured ZnO Gas Sensors," *Environmental Applications of Nanomaterials: Synthesis, Sorbents and Sensors,* p. 315, 2007.

Effects of the fin width variation on the performance of 16 nm FinFETs with round fin corners and tapered fin shape

S.Wan Muhamad Hatta[1], N.Soin[1], S.H.Abdul Rahman[1], Y.Abdul Wahab
[1]Department of Electrical Engineering, Faculty of Engineering, University of Malaya

H.Hussin
[2]Department of Electrical Engineering, Faculty of Engineering, University Technology MARA

Abstract— The rapid scaling of the CMOS technology is causing the evaluation from conventional planar MOSFETs to the FinFET architecture, particularly in the 22 nm and 14 nm technology nodes. FinFETs technologies ensure low power usage and better area utilization, as well as traditional scaling improvements. It was observed that for FinFETs, the smaller the width of the fin, the better the characteristics. It was observed that drain current characteristics of the NFinFET anf PFinFET at both the linear and saturation regime would decrease in magnitude as the width of the fin was decreased. The Ion/Ioff ratio generally decreases as the width of the fin increases. The NFinFET particularly exhibits a significant drop in the Ion/Ioff of to nearly 50% for a change of fin width from 5nm to 15nm.

Keywords—FinFET, TCAD Sentaurus Simulation, MOSFET

I. INTRODUCTION

One approach to sustaining performance gains as CMOS scaling matures in the next decade is a transition from conventional planar MOSFETs to FinFET. A higher performance of FinFET as compared to planar MOSFETS has been successfully demonstrated on a 22nm node FinFET technology [1]. The distinguishing characteristic of the FinFET is that the conducting channel is wrapped by a thin silicon "fin", which forms the body of the device. The thickness of the fin (measured in the direction from source to drain) determines the effective channel length of the device. As we move to FinFET, one of the challenges is the discrete size of the fin. Transistor width (W), which is one of the main variables for modifying transistor sizes, is no longer a continuum. Discrete fin sizing brings a new variable in design, without any easy workarounds, that designers have never had to deal with before [2]. Thus, three dimensional TCAD analysis of such FinFET seem to be crucial in determining the best design, performance, and manufacturability tradeoffs of the next node technologies (14nm and 16nm nodes). The investigation is beneficial in presenting device engineers with comprehensive analysis on the pivotal parameters which affects the performances of leading-edge transistor design

The main objective of this paper is to investigate on the effects of the fin width variation of the FinFET towards the performance of the device. Both the PFinFET and the NFinFET were investigated.

II. DEVICE AND SIMULATION CONDITIONS

A. Device

The device investigated in this work is a 16 nm node FinFET technology whereby the performance of both the PMOS and NMOS FinFETs were studied. The device structure was generated via a 3D process simulation and the process step is illustrated in Fig. 1. The device process features advanced processing steps which includes gate-last high-k metal gate (HKMG) process, *in situ* doped epitaxially grown SiGe source/drain for PMOS and Si:C source/drain for NMOS as well as shallow trench isolation (STI). Fig. 2 depicts the device structure of the FinFET investigated along with the physical parameters of structure while Table 1 gives the nomenclature and values of the parameters used in the simulated device. Similar to other works in TCAD transistor analysis, only half of the structure is simulated to increase simulation efficiency [3-5]. The simulation domain covers from half of the source to half of the drain in the current flow direction, and from middle of the fin to middle of the STI in the direction perpendicular to the current flow.

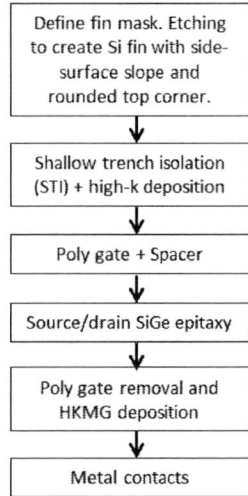

Fig. 1 3D process simulation structure formation flow

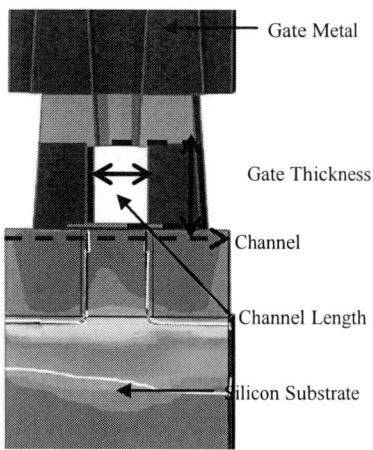

Fig.2 Illustration of the input parameters of the investigated FinFET structure and doping profile

Table 1
Nomenclature and values of the parameters used in the simulated NFinFET and PFinFET [6]

Parameter	Value
Gate length (channel length)	25 nm
Gate Thickness	32 nm
Channel doping	2.0 E+18 cm^{-3}
Channel stop doping	2.0 E+18 cm^{-3}
Fin Height,H (in z-direction)	35 nm
Fin width at the bottom. Wb (y-direction)	15 nm
Fin width at the top, Wt (y-direction)	varied
Vdd	0.8 V
Metal gate workfunction	4.6 eV
Physical thickness of hafnium oxide	1.7 nm
Physical thickness of interlayer oxide	0.6 nm

B. Simulation Details

In this work, the impact of the fin shape and width on the I-V characteristics were investigated. Fig. 3 provides the illustration of the input parameters which define the cross-section of the fin. The fin width was varied from 5nm to 15nm in steps of 5nm, keeping all other device parameters constant. The gate metal work function was selected to be of value 4.6 eV which corresponds to the material properties of Tungsten [7].

The TCAD Synopsys Sentaurus was used in this work to simulate the electrical characteristics of the device. The electrical analysis simulated was in the linear region, which was of low drain bias, and in the saturation regime, which was of high drain bias. The electrical parameters of interest in the

linear and saturation regimes such as the threshold voltage, subthreshold slope, transconductance and the on-state ad off-state currents were extracted. Fig. 4 presents the simulated doping concentration of the p-type FinFET.

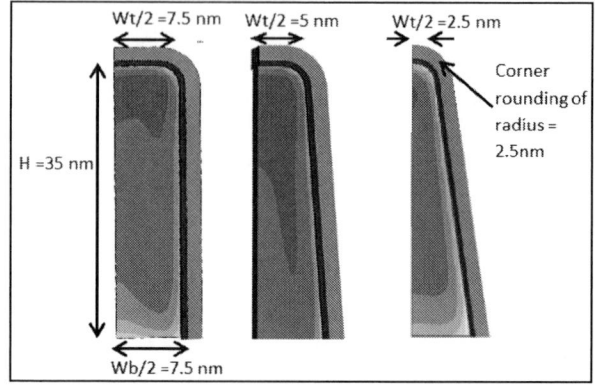

Fig.3 Illustration of the varied fin width at the top, Wt from 5 nm to 15 nm in steps of 5 nm. The other parameters specifically the fin height, H and fin width at the bottom, Wb were maintained constant. Shown here that the only half of the structure was simulated.

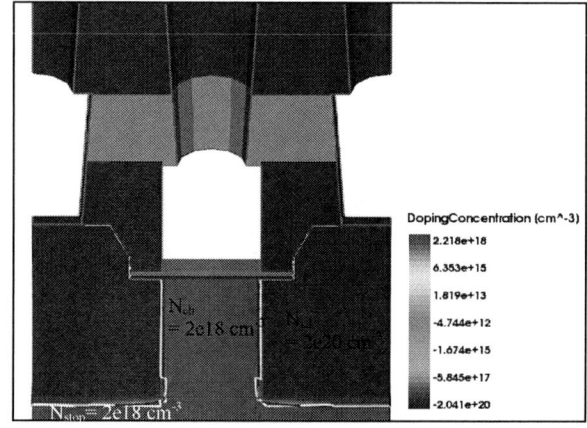

Fig.4 Illustration of the simulated doping concentration of the p-type FinFET. N_{ch} and N_{sd} denote the channel doping concentration and source/drain doping concentration respectively, while N_{stop} denotes the channel stop doping.

In order to simulate the I-V curves of the FinFETs, the TCAD software uses the following set of physical models which takes into account of the evaluation of multiple design trade-offs in analyzing and optimizing the transistor design: the Drift-diffusion (DD) model which takes into account of the high-k degradation mobility and thin-layer mobility, the density-gradient quantum correction with auto-orientation and stress effects models which applies strain engineering to enhance the transistor performance.

III.   SIMULATION RESULTS AND DISCUSSION

This section presents the simulation results of both the p-type and n-type FinFETs. The workfunction of the metal gate is assumed to be 4.6 eV in this work, which is a midgap workfunction

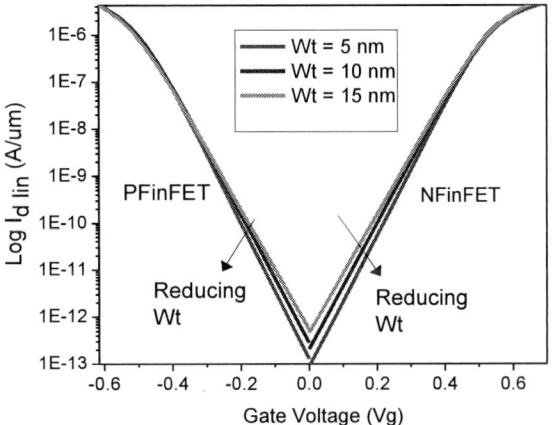

Fig.5.Linear drain current I_{dlin} versus Vg curves for varying fin widths for both PFinFET and NFinFET

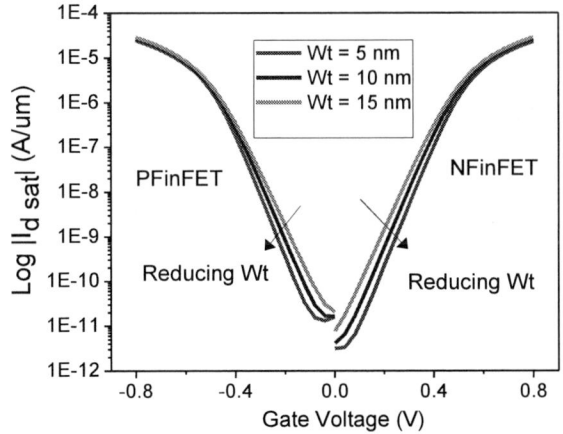

Fig.6 Saturation drain current I_{dsat} versus Vg curves for varying fin widths for both PFinFET and NFinFET

Fig.5 and 6 present the I-V characteristics of the simulated device and the impact of varying fin widths was investigated. It can be observed that as the fin width is reduced, the drain current is reduced particularly at lower magnitude of gate voltage. This is as expected due to the fact that the devices with lower fin widths are exhibiting reduced short channel effects though reducing the fin width increases the source-drain (S/D) resistance [8]. This leads to the reduction of the drain current. Fig. 7 presents the ratio of the I_{on} to I_{off} which indirectly reflects the performance of the device as the fin width is varied. It can be observed, specifically for the n-type FinFET, that as the fin width is increased, the I_{on}/I_{off} reduces. This particular observation is pivotal since it demonstrates how the fin width variation can affect the performance of FinFET in analog applications.

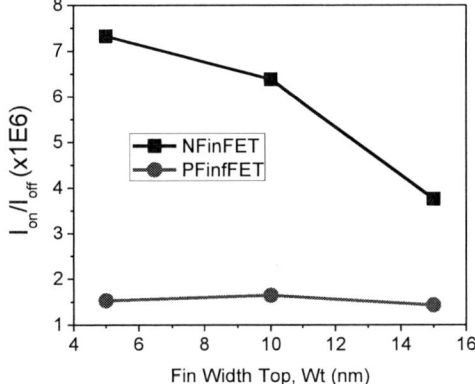

Fig. 7. Variation of I_{on}/I_{off} with respect to fin width

Fig. 8 presents the extracted threshold voltage of simulated devices at the linear region. The Vt_linear denote the extracted threshold voltage defined as Vg at which Id = 100 nA while Vt_gm is the threshold voltage defined as the intersection of the tangent at the maximum gm with the Vg axis [8]. It can be observed that the threshold voltage of the devices generally increases as the fin width increase. Similar variations for threshold voltage is demonstrated in [10] for devices with higher fin width. Fig.9 presents the subthreshold slope as a function of the varying fin width, extracted for both n-type and p-type FinFETs. It is observed that the subthreshold slope of the NFinFETs are larger compared to that of the PFinFET, the same trend as observed in [11].

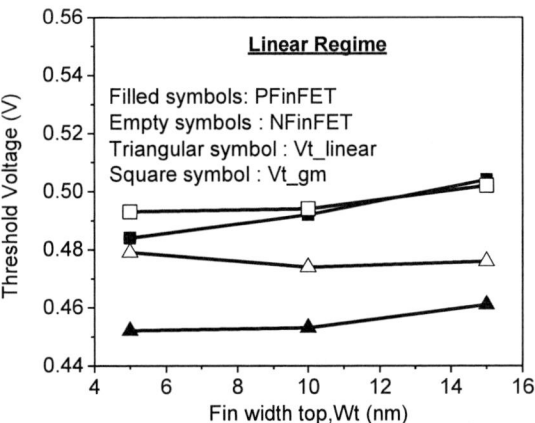

Fig. 8. Fluctuation of the threshold voltage in the linear regime according to the variation of fin width

The drain-induced barrier lowering (DIBL) of the transistors can be extracted from the parameters simulated. DIBL of transistor devices have been studied thoroughly by many authors [8], [12], [13]. It is defined as the threshold voltage difference divided by the drain bias between the linear and saturation regime [14]. From this definition it was observed that the DIBL of the NFinFET is larger compared to that of the PFinFET. This supports the observation of the subthreshold slope for the different types of FinFETs.

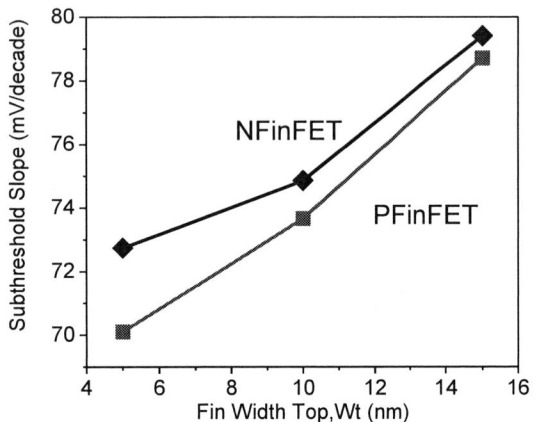

Fig. 9. Subthreshold slope affected from the fin width variation

IV. CONCLUSION

A three-dimensional TCAD analysis on the 16 nm node FinFET technology was carried out in determining the best design of the structure in order to achieve the best performance and manufacturability trade-offs of the next-node technologies. The I-V characteristics of the device were simulated and crictical electrical parameters such as the threshold voltage, subthreshold slope, transconductance and the on-state ad off-state currents are extracted. It was observed that for FinFETs, the smaller the width of the fin, the better the characteristics. It was observed that linear and the saturation drain current characteristics had decrease in magnitude as the width of the fin was decreased. The Ion/Ioff ratio generally decreases as the width of the fin increases. The subthreshold slope would also significantly increase as the width of the fin increases. These observations suggest that the optimization of the FinFET performance can be achieved through the reduction of the width of the fin.

References

[1] V. Subramanian, B. Parvais, J. Borremans, A. Mercha, D. Linten, P. Wambacq, J. Loo, and M. Dehan, "Device and circuit-level analog performance trade-offs : a comparative study of planar bulk FETs

versus FinFETs," in *IEEE IEDM 2005*, 2005, vol. 00, no. c, pp. 898–901.

[2] T. B. Hook, "Fully Depleted Devices for Designers : FDSOI and Finfets," in *Custom Integrated Circuits Conference (CICC)*, 2012, no. Figure 3, pp. 1–7.

[3] Hatta, S. F., Soin, N., Hadi, D. A., & Zhang, J. F. (2010). NBTI degradation effect on advanced-process 45nm high-< i> k</i> PMOSFETs with geometric and process variations. *Microelectronics Reliability*, *50*(9), 1283-1289.

[4] Schwarz, M., Holtij, T., Kloes, A., & Iñíguez, B. (2011). 2D analytical calculation of the electric field in lightly doped Schottky barrier double-gate MOSFETs and estimation of the tunneling/thermionic current. *Solid-State Electronics*, *63*(1), 119-129.

[5] Wang, X., Brown, A. R., Idris, N., Markov, S., Roy, G., & Asenov, A. (2011). Statistical threshold-voltage variability in scaled decananometer bulk HKMG MOSFETs: A full-scale 3-D simulation scaling study. *Electron Devices, IEEE Transactions on*, *58*(8), 2293-

[6] FinFET: Histroy, Fundamentals and Future, http://www.eecs.berkeley.edu/~tking/presentations/KingLiu_2012V LSI-Tshortcourse. July 11, 2012.

[7] Li, Y., & Cheng, H. W. (2012). Random work-function-induced threshold voltage fluctuation in metal-gate MOS devices by Monte Carlo simulation. *Semiconductor Manufacturing, IEEE Transactions on*, *25*(2), 266-271.

[8] J. Joseph and R. Patrikar, "Impact of Fin Width and Graded Channel Doping on the Performance of 22nm SOI FinFET," pp. 153–159, 2013.

[9] C. M. Mezzomo, A. Bajolet, A. Cathignol, G. Ghibaudo, J. Monnet, and P. L. Néel, "Drain Current Variability in 45nm Heavily Pocket-implanted Bulk MOSFET," no. 1, pp. 122–125, 2010.

[10] R.-H. Baek, C. Y. Kang, C.-W. Sohn, D. M. Kim, and P. Kirsch, "Investigation of process-induced performance variability and optimization of the 10nm technology node Si bulk FinFETs," *Solid. State. Electron.*, vol. 96, pp. 27–33, Jun. 2014.

[11] B. Murugan, "Study of Substreshold Behavior of FinFET," University of Nevada, Las Vegas, 1999.

[12] S. Chabukswar, D. Maji, C. R. Manoj, K. G. Anil, V. R. Rao, F. Crupi, P. Magnone, G. Giusi, C. Pace, and N. Collaert, "Implications of fin width scaling on variability and reliability of high-k metal gate FinFETs," *Microelectron. Eng.*, vol. 87, no. 10, pp. 1963–1967, Oct. 2010.

[13] K. Kim, R. Kanj, and R. V. Joshi, "Impact of FinFET technology for power gating in nano-scale design," *Fifteenth Int. Symp. Qual. Electron. Des.*, pp. 543–547, Mar. 2014.

[14] X. Zhou, K. Y. Lim, and D. Lim, "A simple and unambiguous definition of threshold voltage and its implications in deep-submicron MOS device modeling," *IEEE Trans. Electron Devices*, vol. 46, no. 4, pp. 807–809, Apr. 1999.

978-1-4799-5761-3/14 $31.00 © 2014 IEEE

INDEX

'Aqilah binti Abdul Tahrim	44
'Awatif Harun	448
A Makarimi Abdullah	408
A Wesam Al-Mufti	388, 436
A. F. Mahyidin	436
A. Ishak	424
A. Jalar	343, 513
A. K. Shafura	428
A. M. Khairuddin	459
A. R. Mohmad	354
A. R. Nurulfadzilah	92
A. Rahim Ruslinda	377
A. Shuhaimi	529
A. Sudin	104, 384
A. Tijjani	432
A.H. Azman	24, 48, 174, 392, 404
A.K. Shafura	463, 529
A.R Munir	24
A.R Nurulfadzilah	108
A.R Ruslinda	24
Abdelkader Hassein-Bey	495
Abdullah C.W. Noorakma	248
Abid Iqbal	509
Abir Shadman	36, 80
Abu Hanifah Muhamad Ali	267
Adila Syaidatul Azman	138
Afifah Maheran A.H.	178, 232
Afishah Alias	479
AHM Zahirul Alam	197
Ahmad Faiz Mohamad Zohaimi	452
Ahmad Faizal Mohd Zain	32
Ahmad Ismat Abdul Rahim	294
Ahmad Shuhaimi Abu Bakar	1
Aishah Fauthan	5
Akrajas Ali Umar	290
Ali Yeon Md Shakaff	32
Ammar Zakaria	32
Anis Nurashikin Nordin	142, 259
Anthony Holland	366
Arash Dehzangi	170, 408

Arman-Ur-Rashid	150
Asma Fatehi	362
Asrulnizam Abd Manaf	68
AZ MD Rejab	325
Azman Jalar	170, 521
Azmi Ibrahim	294
Azmizam Manie @ Mani	479
Azrif Manut	444
Azrul Azlan Hamzah	52, 72, 119, 193, 236, 240, 362
Azrul Bin Ghazali	60
B. P. Ng	217
B. Y. Majlis	252, 354
B.S Lim	112
B.Y.Lim	412, 420
Bablu Ghosh	154
Bablu Kumar Gosh	205
Badariah Bais	213, 259
Beena Pandey	13
Bibi Nadia Taib	467
Brett C Johnson	366
Bun Seng C.	154
Burhanuddin Bin Yeop Majlis	56
Burhanuddin Y. Majlis	170, 408
Burhanuddin Yeop Majlis	5, 52, 72, 115, 193, 201, 213, 236, 240, 244, 255, 259, 270, 362, 487, 495, 119
C. A. Norhidayah	92, 108
C. J. Hunter	354
C. Senthilpari	332
C.C. Yee	396
C.F. Soon	432
C.H. Voon	104, 384, 412, 420
Chan Yee Kit	9
Chee Lee Cheong	60
Chen-Chen Chung	298, 517
Chi Lang Nguyen	502
Chia-Ao Chang	229
Chien Fat Chau	60
Chi-Feng Hsieh	499
Chih-Jen Hsiao	298, 456
Chin Lin Ng	40
Ching-Hsiang Hsu	298, 336
Ching-Wen Kuan	502

Chong Wee Keat	339
Chun-Kuan Liu	456
Darushini Kunalan	60
Dee Chang Fu	193, 362
Dilla D. Berhanuddin	244
E. Y. Chang	358
Edward Yi Chang	20, 229, 298, 336, 456, 499, 502, 517
Ehsanur Rahman	36, 80
Elena Pirogova	366
Elias B Saion	170, 408
F. Bastiman	354
F. Fotovvatikhah	186
F. H. Wee	384
F. Hamzah	388
F. Lumbantoruan	358
F. Mahmud	274
F.Malek	384
Fahid Algahtani	366
Faisal Mohd-Yasin	52, 509
Farah Lyana Shain	479
Farhad Larki	170, 244, 408
Farseem M. Mohammedy	150
Fatima Alneyadi	471
Fatimah Ibrahim	505
Franky Lumbantoruan	20
Gandi Sugandi	255
Goh, Soon Lock	313
H. C. Wang	358
H. Guliga	440
H. Hashim	158, 459
H. Hussin	209, 533
H. Saim	100
H. W. Yu	358
H.Hisham	104, 384
Halimah Mohamed. K	408
Hanim Abdul Razak	301
Harikrishnan Ramiah	339
Hazura Haroon	301
Hizamel Mohd Hizan	294
Hong-Quan Nguyen	298, 336, 456
Hsun-Jui Chang	336
Hung-Wei Yu	298, 336, 456, 502, 517

I. Abdullah	343
I. Ahmad	178, 232
I. H.H. Affendi	373, 400, 428
I. Saurdi	463, 529
Ibrahim Bin Ahmad	56
Idalailah Dayah	321
Iffa Binti Sharuddin	305
Ili Shairah Abdul Halim	317, 321
Irman Abdul Rahman	521
Ismail Saad	154, 205
J. P. R. David	354
J. S. Ng	354
Jahariah Sampe	525
Jais Lias	1
Jeffrey C McCallum	366
Jer-Shen Ma	336
Jer-Shen Maa	456
Jia Wei Low	32
Jincai Wen	286
Jumril Yunas	115, 213, 270, 487
Jun Liu	17, 286
K Ibrahim	325, 329
K. A. Yusof	491
K. L. Foo	104, 380, 384, 404, 412, 420
K.S. Siow	513
Kalaivani T.	178, 232
Kanak Datta	36, 80
Kenichiroh.Mukai	313
Khairul A.M	154
Khairul Anuar Mohamad	205
Kok, Swee Leong	313
Kuan-Shin Liu	499
Kung- Liang Lin	517
L. Lee	305
Lam Mui Li	479
Lee Lini	332
Lee, Swee Kah	313
Lim Huey Sia	1
Ling Li Ong	9
Lingling Sun	17, 286
Low Jia Wei	1
M. A. Farehanim	377, 388, 505

M. Aminuddin	24
M. F. Fatin	48, 174
M. F. Malek	424
M. F. Omar	388
M. H. Ani	252, 267
M. Hosseinghadiry	186
M. K. Md Arshad	48, 134, 174, 370, 377, 392, 396, 404, 412, 420
M. Khaledian	186
M. M. Ramli	370
M. M. Ramly	225
M. Mazalan	217
M. Mudasir	158
M. N. Hamidon	189
M. Nurfaiz	24, 392
M. Nuzaihan M. N	396, 404
M. R. Zakaria	370, 380, 388, 436
M. Rusop	158, 373, 400, 424, 428, 444, 459, 463, 529
M. S. Alias	416
M. S. Bhuyan	525
M. S. P. Sarah	459
M. Saeidmanesh	186
M. Z. Sahdan	92
M.A. Baqir	282
M.A. Farehanim	48, 174
M.F. Fatin	112, 377, 505
M.F. Mohd Razip Wee	236, 244
M.F. Nurfazliana	100
M.F.M Idros	84
M.F.M. Fathil	24, 112, 134, 392, 396
M.H Wahid	373
M.H. Mamat	463, 529
M.Hannas	444
M.K. Md Arshad	24, 76, 88, 112
M.N. Hamidon	64
M.N.Derman	412, 420
M.S. Nur Humaira	96, 104, 134, 384
M.S.P. Sarah	158, 400
M.Z. Kamarudin	392
M.Z. Sahdan	100, 108, 274, 416, 432
M.Z.M. Kamarudin	24
MA Chik	325, 329
Ma Li Ya	142

Mahmood Goodarz Nasery	408
Maitha Alkaabi	471
Maizan Muhamad	339
Mamat, H.	123, 126, 130
Manizheh Navasery	408
Marianah Masrie	115
Mark Blackford	366
Marlia binti Morsin	290
Mazlee Mazalan	467
Md. Aynal Hossain	150
Md. Shabiul Islam	28, 84, 170, 309
Menon, P.S.	178
Michael Loong Peng Tan	40, 44
Michelle S.M.Lim	28
Mimiwaty Mohd Noor	255, 487
Mohamad Rusop	1
Mohamed El-Amine Benamar	495
Mohammad Nuzaihan Md Nor	467
Mohammad Rashed Iqbal Faruque	182
Mohammad Tariqul Islam	182
MohammadMahdi Ariannejad	408
Mohammadmahdi Vakilian	201
Mohd Ambri Mohamed	244, 252, 267
Mohd Faizal Aziz	72, 193, 255
Mohd H.S Alrashdan	72
Mohd Khairul Ahmad	1, 32
Mohd Nizar Hamidon	5, 170
Mohd Nur Nazmi	448
Mohd Zainizan Sahdan	1, 32
Mohd Zulfadli Mohamed Yusoff	294
Mohd. Zuhir H.	154
Montree Kumngern	146, 166
Muhamad Amri Ismail	162
Muhamad Mat Salleh	193, 290
Muhamad Ramdzan Buyong	236, 240, 259
Muhammad Mahyiddin Ramli	467
Muhammad Zulkhairi Roslan	244
Muzalifah Mohd Said	213, 487
N. A. Yusof	96
N. Aimaier	189
N. Bolong	154
N. E. A. Azhar	428

N. Iqbal	282
N. Nafarizal	92, 100, 416, 432
N. Pan	347
N. Sarip	274
N. Soin	209, 533
N. Sulaiman	189
N.A. Azhar	373
N.A. Yusof	64
N.E.A. Azhar	529
N.H.M. Saleh	412
N.M Salih	432
N.Nafarizal	108
N.Tukemon @ Tukiman	420
Nadir Belgroune	495
Nafarizal Nayan	1, 32
Nasrin Khalilzadeh	408
Nguyen-Hong Quan	517
Niraj Man Shrestra	20
Nithiyah Tamilchelvan	467
Noor Faizah Z. A.	178, 232
Nor Azira Akmar Shaari	448, 452
Nor Shahanim Mohamad Hadis	68
Nor Syazwana Mohd Yusof	317
Noraini Marsi	52
Noraini Othman	76, 88, 112, 134
Norani Bte Atan	56
Norazreen Abd Aziz	236, 240, 259
Norbahiah Misran	182
Norhafizah Burham	119
Norhana Arsad	301
Norhayati Binti Soin	475
Norhayati Soin	142, 248, 339, 505
Norinsan Kamil Othman	521
Norlaili Mohd Noh	339
Norliana Yusof	248
NS Kamarozaman	225
Nur Hidayah Ibrahim	263
Nur Syahirah Kamarozaman	448, 452
Nurul Huda Abdul Rahman	134
P Susthitha Menon	301
P. S. Menon	232
P.K. Choudhury	282

P.R. Apte	178, 232
Qahtan Khalaf Omran	182
Quazi D. M. Khosru	36, 80
R. A. Rahim	370
R. D. Richards	354
R. Haarindra Prasad	380
R. Ismail	343
R. M. Sidek	189
R. Mat Ayub	370, 388, 436
R.M. Ayub	24
R.M. Zawawi	64
Rashad Ramzan	471
Raudah Abu Bakar	448, 452
Razali Ismail	138, 154, 186
Razali Ngah	294
Retnasamy, V.	123, 126, 130
RM Ayub	48, 174, 392, 505
Roer Eka Pawinanto	213, 487
Rosalind Deena Kumari Selvam	332
Rosminazuin Ab Rahim	263
S. Ahmad	96
S. C. Huang	358
S. H. Herman	225, 440, 444, 491
S. Jahariah	28
S. Johari	217
S. N. M. Tawil	92
S. Norhafiezah	48, 174, 377, 392, 505
S. R. Kasjoo	370
S. S. Shariffudin	158, 428, 459
S. Shaari	178
S. Shahabuddin	209
S. T. Ten	96, 104, 384, 412, 420
S.A. Kamaruddin	100, 108, 416
S.F.A. Rahman	64
S.H.Abdul Rahman	533
S.L.Lai	380
S.M.Sultan	221
S.N. Sabki	88
S.N.M Tawil	108, 274
S.P. Foo	513
S.Wan Muhamad Hatta	533
Saadah Abdul Rahman	193

Saafie Salleh	479
Sabar D. Hutagalung	170, 408
SachchidaNand Shukla	13
Sahbudin Shaari	301
Sa-Hoang Huynh	456
Salah Hasan Alkurwy	309
Salama Alketbi	471
Salehuddin, F.	232
Sauli. Z.	123, 126, 130
Sawal Ali	84
Sawal H. Md. Ali	525
Sawal H.M.Ali	28
Sawal Hamid Md Ali	170, 309
Shafaq Mardhiyana Mohamat Kasim	448, 452
Shafii A. Wahab	525
Shahrir R. Kasjoo	436
Shaikat Debnath	278
Shamsa Hajraf	471
Shazlina Johari	467
Shengzhou Zhang	286
Sheroz Khan	197
Shih-Hsuan Tang	502
Shoou-Jinn Chang	456
Sima Dimitrijev	509
Siti Lailatul Mohd Hassan	317, 321
Siti Musliha Ajmal Binti Mokhtar	350, 483
Soheli Farhana	197
SR AB Rahim	325
Sudipta Romen Biswas	36, 80
Suhana Mohd Said	278
Sukreen Hana Herman	68, 448, 452
T. Nazwa	96, 104
Tafadwa Magaya	313
Tai, Min Fee	313
Tamer A. Tabet	205
Taniselass, S.	130
Tanvir Rahman	150
Thien-Huu Ha Minh	456
Tiong Teck Yaw	193
Tran Binh Tinh	20
Tung Tien Luong	20
U. Hashim	24, 48, 64, 76, 88, 96, 104, 112, 134, 174, 325, 329, 370, 377,

	380, 384, 388, 392, 396, 404, 432, 436, 505, 412, 420
Umar Faruk Shuib	205
Umi Milhana Jamain	263
Ummikalsom Abidin	52, 270
Usa Torteanchai	146
Uzer M	529
Vairavan, R.	126, 130
W. C. Huang	358
W. F. H. Abdullah	440, 491
W.H.Khoo	221
W.W. Liu	104, 384
Wan Fazlida Hanim Abdullah	350, 483
Wan Yusmawati Wan Yusoff	521
Wei-Ching Huang	20, 499
Wong Goon Weng	475
Y. Abdul Wahab	209, 533
Y. M. Ang	404
Y. S. Chiu	358
Y. Wahab	217
Y. Y. Wong	358
Y.S. Lee	384
You Ah Heng	9
Young Chul Lee	294
Yuan-Yee Wong	20
Yue-Han Wu	20
Yuen-Yee Wong	499
Yufridin Wahab	467
Yu-Lin Hsiao	229
Yung-Hsuan Su	502
Z. Aznilinda	225
Z. Nurbaya	373, 428
Zaharah Johari	138
Zainab Kazemi	408
Zainab Yunusa	5
Zonghua Zheng	17
Zulkifli Ambak	294